SOLID STATE PHYSICS and ELECTRONICS

A. B. Gupta
Nurul Islam

CELEBRATING 55 YEARS 1960-2015
Since 1960

For students majoring in Physics (CBCS) and also of Engineering and other allied disciplines.

SOLID STATE PHYSICS
and
ELECTRONICS

A. B. Gupta, Ph.D.

Professor and Head (Retired)
Physics Unit and Dean,
Indian Statistical Institute, Kolkata

AND

Nurul Islam, M.Sc. (Cal.), M.Sc. (Mad.)

BOOKS AND ALLIED (P) Ltd.

No. 1-E(1), 'Shubham Plaza' (1st Floor)
83/1 Beliaghata Main Road, Kolkata 700010 (India)
TTel : 8961053844 • 8274085530
e-mail : booksandallied1960@gmail.com • booksmktng@gmail.com

SOLID STATE PHYSICS
and
ELECTRONICS

First Published : September, 2012
Reprinted : 2014
Second Edition (thoroughly revised & updated) **: 2017**
Reprinted : 2019

Publisher : Arunabha Sen
BOOKS & ALLIED (P) LTD.
8/1 Chintamoni Das Lane, Kolkata 700009

Typesetter : BOOKS & ALLIED (P) LTD.
8/1 Chintamoni Das Lane, Kolkata 700009

Printer : EUREKA
91/A, Baithakkhana Road, Kolkata 700009

Project Supervisor : Salil Das

ISBN 81-87134-83-6

Price **: ₹680.00**

In

sacred memory of

Sitangshu Majumdar,

an

unfailing friend

of the

senior author

and a

true teacher

of the

junior one,

a man

of indomitable courage

and a

unique symbol

of

human qualities.

Other Outstanding Books on Higher Physics
*by the same Author**

1. **Thermal Physics**
2. **Modern Optics**
3. **Modern Acoustics**
4. **Classical Mechanics**
5. **Physics of the Atom**
6. **Physics of the Nucleus**
7. **Modern Physics**
8. **Foundation of Quantum Mechanics**
9. **Modern Atomic and Nuclear Physics**
10. **Fundamentals of Mathematical Physics**
11. **Classical Mechanics and Properties of Matter**
12. **Electricity and Magnetism (in Press)**

*From the same House of Publications

Preface to the Second Edition

This book 'SOLID STATE PHYSICS AND ELECTRONICS' was first published in 2012 to cater to the need of undergraduate students majoring in Physics and other allied disciplines of Indian Universities. This was followed by a subsequent reprint of the work in the year 2014. During the period, the book received a wide appreciation of the academic community for simple and lucid but accurate and adequate presentation of the subject matter and other salient features such as coverage of topics, concise explanation of fundamental principles, theoretical derivations, the set of varied solved examples, short and multiple choice review questions.

However, the numerous feedbacks received during the period were not an unalloyed appreciation but also contained useful critical suggestions for improvement of the work, particularly in electronics to make it more comprehensive and useful. In the light of this, we have the pleasure of presenting the second edition of the work in which we hope, with all humility, that we have been able to address almost all the suggestions made by our well-wishers.

And in doing so, major changes have been made throughout the book, especially in electronics. In solid state physics, such changes would be discernible in 'Superconductivity' and 'Magnetic properties of materials', apart from the addition of two new chapters, namely 'Imperfections in Crystals,...' and 'Basics of nanomaterials'. By far most of the changes and new additions are in electronics. These include *inter alia* transformation utilization factor (TUF) for half- and full wave rectifiers and analysis of filter circuits in Chapter 4; transistor manufacturing techniques and transistor as a switch in Chapter 5; simplified hybrid model and expressions for performance of quantities in CE, CB and CC modes and Darlington pair in Chapter 7; transfer characteristics of JFET, pinch-off voltage and CMOS in Chapter 9; different types of regulated power supplies in Chapter 10; monostable and bistable multi-vibrators in Chapter 11; subtractor, logarithmic amplifier, OP AMP as analog computer and solutions of linear algebraic and differential equations in Chapter 12 and details of sequential logic circuits, different types of flip-flops and latches in Chapter 14. The chapter on 'Communication principles' is almost completely rewritten so as to appear as a new chapter. The last chapter 15 is developed out of an expansion of the materials of Appendix in the previous edition.

Naturally, the book has now assumed a goodly girth—an addition of 300 new pages. The number of worked out examples has also been increased from about 200 to 350 and that of the problems from nearly 300 to staggering 500. We have also enormously increased the number of interesting MCQs which would be helpful for facing competitive examinations and serve better as the basis for important self-study. Apart from the major changes, no pains have been spared for academic brush-ups of materials wherever found needed to improve the presentation and conceptual understanding. In the light of all these, it is hoped that this comprehensively revised second edition would make the work more broad-based and useful to develop a realistic impression of the subject by its readers.

In presenting the new edition, we feel indebted to many people to whom we express our most grateful acknowledgement. They are, among many others, (i) the authors of standard works in the field, quoted mostly in the bibliography, from whose work we have drawn freely and heavily; (ii) the students and teachers who kindly sent their valuable feedbacks; (iii) the members of the Staff of the DTP centre of Messers Books and Allied (P) Ltd., particularly Mr. Salil Das, Project Supervisor, who by his magic wand has converted the handwritten obscure manuscript into this fine, final finished form and (iv) Mr. Arunabha Sen, MS (USA), Director, Books and Allied (P) Ltd., who had been silently vocal in every matter, be it academic or organi-zational during the production of the work.

Suggestions and criticisms for further improvement of the work would be most thankfully received.

Independence Day
August 15, 2017

A.B. Gupta
Nurul Islam

Preface to the First Edition

The intention of this title 'SOLID STATE PHYSICS AND ELECTRONICS' is to present the subject matter of solid state physics and of electronics as a text for an introductory one-semester course in each of the two disciplines for the undergraduate students majoring (Hons.) in physical sciences in any Indian University. The basic plan of the book is the following: The *first part* covers the physics of materials in the solid or condensed state—Crytallography, analysis of crystal structure, bonding in solids, the transport phenomena, lattice vibrations (phonons) and thermal properties, band theory of solids, dielectric and magnetic properties of materials and superconductivity. The basic ideas of quantum mechanics have been used whenever and wherever required. The *second part* of the book is devoted to solid-state electronics—Starting from semiconductors, it terminates at digital electronics through the P-N junction, diode circuits, BJT, transistor biasing and stability, transistor hybrid model, different types of amplifiers, FET, MOSFET, oscillators and communication principles. There are some unavoidable areas of overlap that may naturally appear repeatitive but they are also characterised by complementarity.

There are a number of advanced books in the fields and it is needless to mention that we have been much benefitted from some of those. These sources have been listed in the bibliography at the end of the book. Regretably, however, in an introductory work like this, it is neither feasible to include an exhaustive list of major textbooks or references on the subjects nor to mention all the people who have contributed to our understanding of the disciplines concerned. What we at best can and should do is to express and record our indebtedness to one and all of them.

A fundamental necessity and a major oppression the students and the teachers face is that a text book may not be so much subject-friendly as examination-friendly. We consciously endeavoured, during the production of the manuscript, to be balanced on both the counts. We have neither overburdened the work unnecessarily nor indulged in the luxury of deprivation of the essential materials for a sound understanding of the disciplines. In this context, we preferred a simple and lucid style of presentation with a focus and emphasis on the fundamentals. We have tried to provide the readers with an appropriate level of detail in derivation of all equations and results. The algebraic manipulations are also outlined in a manner so as to be followed by an interested student without the risk of getting lost. Textual materials are accompanied by, wherever required, numerous well-drawn illustrations. In electronics, in particular, the operating principles must go hand in hand with circuit configurations as solid-state electronics is essentially circuit-oriented. In this respect, we have been as much liberal as has been possible.

Each chapter of the book contains a large number of solved problems, selected mostly from the papers set in various Indian universities or examinations like GATE, NET etc. to illustrate the theoretical concepts, circuit analysis etc. discussed throughout the book. The numerous short answer and multiple choice type objective questions, all appended at the end of each chapter, would provide the readers a scope for self-evaluation and instil in them a self-confidence.

We sincerely believe that the readers will find in the work a self-contained and self-explanatory book providing a smooth and sound introduction to two vitally important disciplines of physics. We have spared no pains to make the work as much useful to the beginners as practicable and would consider our efforts fully rewarded if the readers enjoy reading it as much as we enjoyed in writing it.

We shall be failing in our duty if we do not record our gratefulness to at least two persons—Mr. Salil Das of the DTP centre of Messrs Books and Allied (P) Ltd. for his untiring efforts in the admirable production of the book and above all to Mr. Arunabha Sen, M.S.(USA), the soul behind all the endeavours of Books and Allied.

Awaiting criticisms and comments—good, bad or caustic—from the academic community of the country. They should be addressed direct to the Publisher.

Jhulanjatra
July 28, 2012

A.B. Gupta
Nurul Islam

Contents

Solid State Physics

Electronics

SOLID STATE PHYSICS

1

Crystal Structure

1.1 Introduction

A ***crystalline solid*** or a ***crystal*** is characterised by a regular or periodic arrangement of constituent particles—atoms, ions or molecules—in three dimensions i.e., it has a reticular structure. Based on the geometric arrangement of the constituents, there exists different types of crystal structures on which many properties of solids depend. Studies and analysis of crystal structures constitute the subject matter of this chapter. We shall proceed from a discussion of various parameters and concepts having bearing on the structures of crystals.

A crystal in which the periodicity of atoms extend throughout the material is called a ***mono-crystal*** or a ***single crystal.*** It is essentially an isolated piece of one crystal of a crystalline solid. Examples are calcite, quartz, diamond, mica etc. Its typical features are *anisotropy* and *sharp melting point.* Single crystals are generally prepared by growing from a saturated solution or pulling from the melt.

A ***polycrystalline solid***, on the other hand, is an aggregate of many small crystallites grown as interlocking mass, oriented at random, by well-defined boundaries. The small crystallites are called ***grains*** and the boundaries the ***grain boundaries.*** Examples are rock, sand, metals etc. A polycrystalline solid is *isotropic*, the distribution of the grains being random.

• ***Amorphous solids*** are non-crystalline and are characterised by random arrangement of atoms or molecules. They show neither the reticular structure of crystals nor the granular structure of polycrystals. Examples are glass, plastic, rubber, resin etc. They show, if at all, *short range* ($\sim 3\,$A) *order* or periodicity and have *no defined* or *sharp melting points.* With rise in temperature, they gradually become soft and behave like viscous liquids.

● The branch of physics in which the properties of solids are discussed is called the ***solid state physics*** or the ***physics of condensed matter.***

1.2 Some terms : Crystal lattice, basis vectors and translation vectors

In a crystal, the arrangement of atoms is periodic and one can readily imagine each atom to be replaced by a geometrical point at the equilibrium position of the atom concerned. The pattern of this imaginary *space points* is called the ***crystal lattice*** or simply ***lattice.*** The points referred to are termed the ***lattice points.*** The arrays of atoms or lattice points in two and three dimensions are respectively known as ***plane lattice*** and ***space lattice.*** In a lattice, any atom has exactly the same environment as every other atom.

● A space lattice is a mathematical concept, representing an infinite number of points in space such that the arrngement of points about a given point is identical to that about any other point.

There are, in fact, *two classes* of lattice—*Bravais* lattice and *non-Bravais* lattice. In a ***Bravais lattice***, all the lattice points are equivalent so that all the atoms in the crystal are of the same type. In a ***non-Bravais lattice***, however, it is not so and some of the lattice points are not equivalent.

Let O be any lattice point in a plane lattice xy (Fig. 1.1). Then, any other lattice point in the lattice can be obtained by repeated translation of vectors $\vec{a}$ $(= \overrightarrow{OD})$ and $\vec{b}$ $(= \overrightarrow{OF})$ along the directions x and y. Such vectors as $\vec{a}$ and $\vec{b}$ are called the ***basis vectors.*** The position of any lattice point with respect to O (as origin) can be expressed as

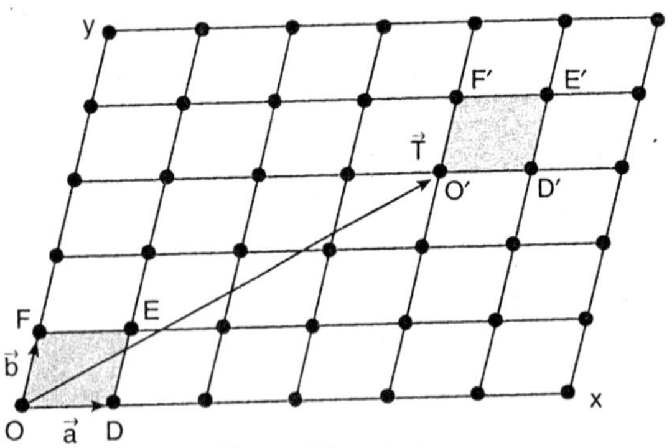

Fig. 1.1 Plane lattice

$$\vec{T} = n_1\vec{a} + n_2\vec{b} \qquad (1.2.1)$$

where n_1 and n_2 are two arbitrary integers. The vector $\vec{T}$ is known as the ***translation vector*** in two dimensions.

In a similar fashion, one can develop only space lattice by employing the basis vectors $\vec{a}, \vec{b}$ and $\vec{c}$ along the directions x, y, z of the crystallographic axis and with respect to any lattice point as origin, the position of any other lattice point may be written as

$$\vec{T} = n_1\vec{a} + n_2\vec{b} + n_3\vec{c} \qquad (1.2.2)$$

where n's are arbitrary integers and $\vec{T}$ the translation vector in three dimensions.

1.3 Crystal structure : Unit cell

A ***crystal structure*** develops only when an atom or a group of atoms or molecules is attached identically to each *lattice point.* The group of these atoms or molecules is called the ***basis***, and it is the repeatition of this basis in space that gives rise to the crystal structure, i.e., basis is the basic building unit for the entire crystal structure.

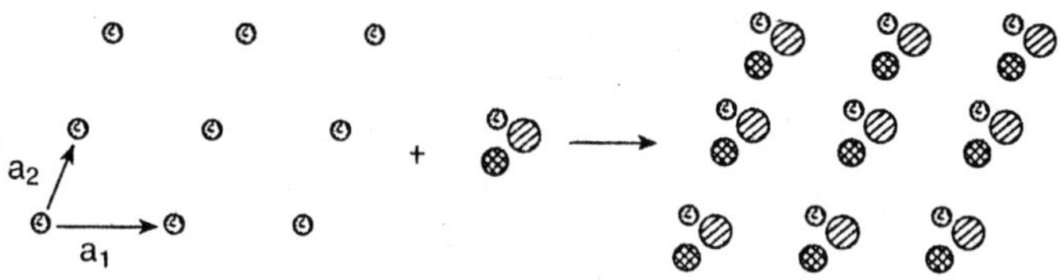

Fig. 1.2 A two-dimensional crystal with a basis of three atoms

$$\therefore \quad \textbf{\textit{Lattice + Basis = Crystal structure}}$$

The above relation is illustrated in Fig. 1.2 for a two-dimensional crystal with a basis of three atoms.

A ***unit cell*** is defined as a parallelopiped formed by the three basis vectors $\vec{a}, \vec{b}$ and $\vec{c}$ along the three crystallographic axes x, y and z respectively. The crystal structure can be generated by the translation of unit cell in three dimensions. To describe a unit cell completely, the three basis vectors $\vec{a}, \vec{b}$ and $\vec{c}$ and the angles α, β and γ between them are to be specified (Fig. 1.3).

The choice of unit cell is not unique.

Quantities specifying the size of a unit cell, e.g., the length a, b, c, are called ***lattice constants*** or ***lattice parameters***.

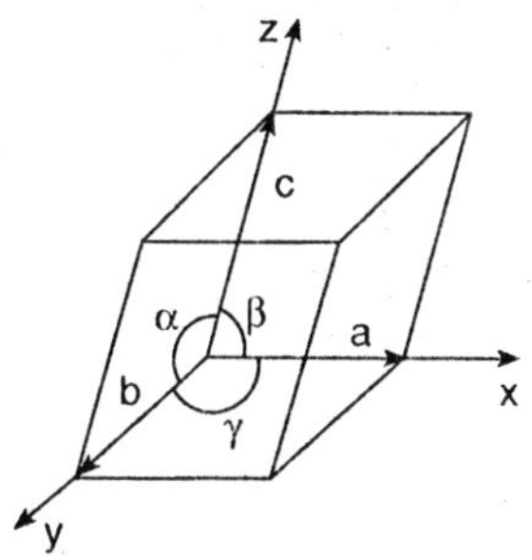

Fig. 1.3a A unit cell with basis vectors and angles between them

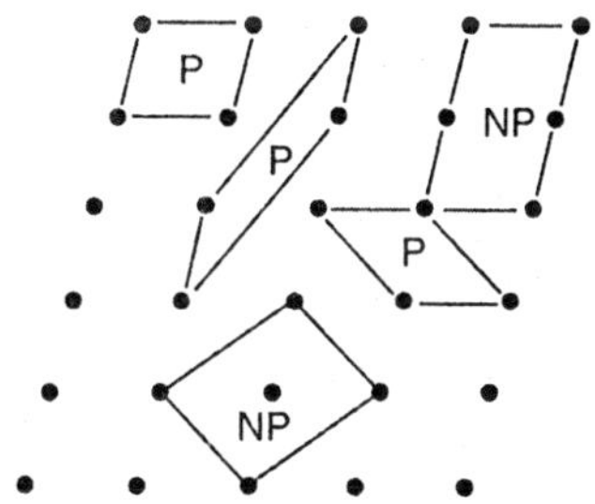

P = Primitimitive cell
NP = Nonprimitive unit cell

Fig. 1.3b Primitive and nonprimitive unit cells

The ***volume*** of the cell is, by definition, $(\vec{a} \times \vec{b}) \cdot \vec{c}$. If this volume contains *only one lattice point*, it is termed ***primitive unit cell.*** Fig. 1.3a illustrates a primitive unit cell. The contribution of each point at the corners is $1/8$ so that the number of lattice points is $8 \times \frac{1}{8} = 1$ in the cell. One can, however, depending on the requirements and the symmetry of lattice, consider larger cell having *more than one* lattice point as a *unit cell.* Such a unit cell is known as a ***non-primitive unit cell.***

• *All primitive cells are unit cells, but all unit cells are not primitive.* The difference is that while a primitive unit cell contains one lattice point, a unit cell may contain more. Fig. 1.3b shows examples of primitive (P) and non-primitive (NP) unit cells.

A different way of choosing a unit cell was proposed by Wigner and Seitz, and this unit cell is called ***Wigner-Seitz unit cell.*** With any lattice point as origin, lines are drawn to all the neighbouring points. Then, lines (for a plane lattice) are drawn or planes (for a space lattice) are constructed normal to and passing through the mid-points of the previous vectors as shown shaded in Fig. 1.4. The *cell thus generated with the smallest area* (in plane lattice) or *smallest volume* (in space lattice) *is the Wigner-Seitz unit cell.*

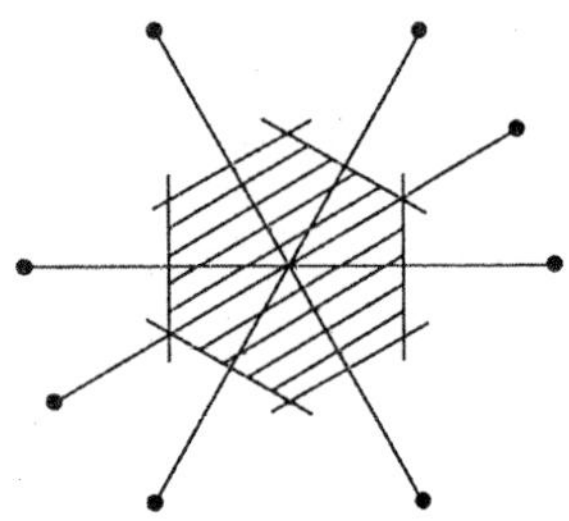

Fig. 1.4 Wigner-Seitz unit cell

1.4 Symmetry operations

The obvious feature of a crystal is its *symmetry* and symmetries form a powerful tool at the hand of crystallographers *to study the internal structure of crystals.*

If by performing an operation in a crystal, the same configuration is obtained so that it is *indistinguishable* from the original one, the *operation* is called a ***symmetry operation.*** A symmetry operation thus leaves the *crystal and its environment invariant.* The locus about which the symmetry operations are made is known as the ***symmetry element.*** We describe hereunder the different types of symmetry operations.

1. Translation—It is defined as the *displacement* of the crystal *parallel to itself* by a translation vector and is the simplest among the types of operations. It is the result of the orderly arrangement of a lattice. A lattice point $\vec{r}$ under translation vector operation $\vec{T}$ gives another lattice point $\vec{r}'$ identical with $\vec{r}$ i.e.,

$$\vec{r}' = \vec{r} + \vec{T}$$

2. Point operations—The point operations are of *three* different kinds and these are: (i) *reflection,* (ii) *inversion* and (iii) *proper rotations.*

(a) Reflection—In reflection operation, the symmetry element is a mirror plane or symmetry plane in the lattice, symbolized by m, that divides it into two identical halves, each being the mirror image of the other. If a structure is left unchanged in reflection in the plane, it is said to have a ***reflection symmetry.***

(b) Inversion—If for every lattice point having a position vector $\vec{r}$, there exists a corresponding lattice point at the position $-\vec{r}$, a crystal structure is said to possess an ***inversion symmetry.*** The origin about which the position vector $\vec{r}$ is selected is known as the ***centre of inversion.*** The operation is symbolised by $\vec{I}$.

• Note the difference between the reflection and the inversion. While the former occurs in a plane through the lattice point, the latter may be considered as equivalent to reflection at a point.

(c) Proper rotations—If a structure, on rotation about an axis, by an angle remains invariant i.e., unchanged in configuration, it is said to possess ***rotational symmetry*** about that axis. The angle of rotation θ must be an integral sub-multiple of 2π i.e., $\theta = 2\pi/n$ where n is an integer and is called the ***throw*** of the axis. The number of successive rotations n required to cause superposition is called the ***fold*** or ***multiplicity*** of an axis.

Because of the recticular structure of the crystal, n can take up values only 1, 2, 3, 4 and 6. So, only one, two, three, four and six-fold rotation axes of symmetry are possible (as we shall prove below), constituting the symmetry elements of the proper rotation operation.

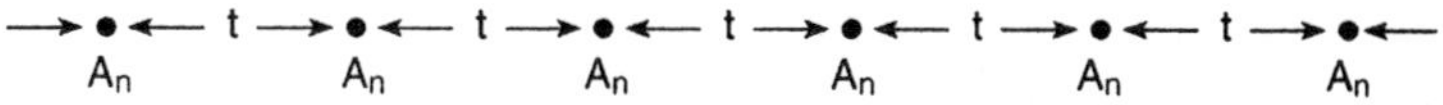

Fig. 1.5 One-dimensional lattice (row lattice)

Proof. Starting from the row lattice shown in Fig. 1.5, we place an n-fold axis, A_n at each lattice point. Since n-rotations, each of θ, cause superposition, the sense of rotation

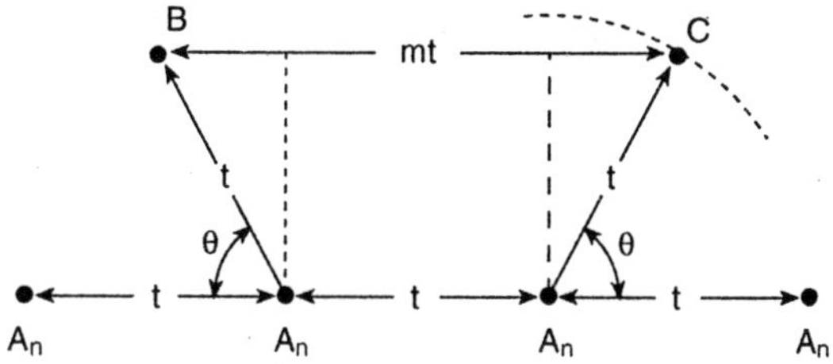

Fig. 1.6 To explore the permissible throws for a rotation axis

is immaterial. Two opposite rotations by θ about two axes (Fig. 1.6) generate two new lattice points B and C which, by construction, must be equidistant from the original row lattice. So, the line BC is parallel to lattice translation t and must be some integral multiple[1] of t. From simple geometry,

$$BC = mt = 2t\cos\theta + t$$

$$\Rightarrow \quad \cos\theta = (m-1)/2$$

where m is an integer: $m = 0, \pm 1, \pm 2, \cdots$, the $\pm$ sign is indicative of measuring translation to the right or to the left.

Since m is an integer, $m-1$ is also an integer, say N.

$\therefore \quad \cos\theta = N/2, \quad N = 0, \pm 1, \pm 2, \cdots$

Since $|\cos\theta| \leq 1$, $N > 2$ is an *impossibility* and the allowed values of N are $0, \pm 1$, and ± 2. The corresponding values of θ and n are as given in Table 1.1.

Table 1.1

N	$\cos\theta$	θ	$n\,(=2\pi/\theta)$
-2	-1	180°	2
-1	$-\frac{1}{2}$	120°	3
0	0	90°	4
$+1$	$+\frac{1}{2}$	60°	6
$+2$	$+1$	360° (or 0°)	1

[1]If it is not, it implies that the line joining B and C is not a translation of the lattice and the array is not periodic.

Since $n \neq 5$, ***a 5-fold rotation is not compatible with lattice translation symmetry.***

• If $n = 1$, the crystal is to be rotated through 360° for congruence. Such an axis is called the ***identity axis.*** Obviously, every crystal has an infinite number of identity axes. For $n = 2$ and hence $\theta = 180°$, the axis is a *diad axis.* Similarly, *triad axis* corresponds to $n = 3$, $\theta = 120°$, etc.

• There may as well be ***hybrid operations*** such as *roto-reflection, roto-inversion, screw translation, glide reflection.* ***Roto-reflection***, as the name suggests, is a combination of an n-fold rotation followed by a reflection in a plane normal to the rotation axis; ***roto-inversion*** similarly is a combination of an n-fold rotation followed by an inversion. In ***screw translation***, the n-fold rotation axis is coupled with translation, parallel to the rotation axis; in ***glide reflection***, a mirror plane is coupled with translation parallel to the reflecting plane.

1.4.1 Point groups and space groups

The symmetry operations about a point or a line are termed *point group symmetry operations.* Reflection, rotation and inversion are the point symmetry operations and they may be combined if compatible. These combinations are known as ***point groups*** and lead to different symmetry points in the crystal. There exists 32 point groups implying 32 allowed combinations.

Again, the point group symmetry operations can be combined with translation symmetry operation if compatible. Such combinations are called ***space groups.*** There are 230 space groups[2]. Once the space group is known, the symmetry of the crystal structure is completely specified.

1.4.2 Seven crystal systems

Based on the lengths of the primitive axes and the angles between them, the Bravais lattice can be grouped into 7 crystal systems and 14 Bravais lattice in three dimensions. They refer to the possible shapes the unit cell can take up and are characterised by a set of symemtry elements associated with each point in the Bravais lattice. The *seven* types of Bravais lattices are

1. **Cubic:** The basis vectors $a = b = c$; $\alpha = \beta = \gamma = 90°$, characterised by four 3-fold axes.
2. **Tetragonal:** $a = b \neq c$ and $\alpha = \beta = \gamma = 90°$, results from orthorhombic symmetry by the addition of a 4-fold rotation axis along one of the original 2-fold rotation axis.
3. **Monoclinic:** $a \neq b \neq c$ and $\alpha = \gamma = 90° \neq \beta$; characterised by one 2-fold axis and a reflection plane normal to this axis.
4. **Triclinic:** $a \neq b \neq c$ and $\alpha \neq \beta \neq \gamma$; *no* rotation or reflection symmetry.

[2]Cubic-36, tetragonal-68, monoclinic-13, triclinic-2, trigonal-25, hexagonal-27, orthorhombic-59.

5. **Trigonal** or **rhombohedral :** $a = b = c$ and $\alpha = \beta = \gamma < 120°, \neq 90°$; characterised by a 3-fold axis of rotation.
6. **Hexagonal :** $a = b \neq c$ and $\alpha = \beta = 90°$, $\gamma = 120°$; has 6-fold, 3-fold and 2-fold rotation axes (along and normal to the hexagonal axis) and also an inversion centre.
7. **Orthorhombic :** $a \neq b \neq c$ and $\alpha = \beta = \gamma = 90°$; characterised by two mutually orthogonal 2-fold axes of rotation and a reflection plane normal to one of the 2-fold axes.

● The quantities that specify the size of a unit cell such as the length a in a cubic crystal, are called *lattice constants* or *lattice parameters*. For a general unit cell, a, b, c and α, β, γ are the *lattice parameters*.

Fig. 1.7 illustrates the seven crystal systems with fourteen Bravais lattices.

1.5 Characteristics of a unit cell

We shall be mainly concerned here with ***cubic crystals***: these are (i) the ***simple cubic*** (*sc*), (ii) the ***body-centred cubic*** (*bcc*) and (iii) the ***face-centred cubic*** (*fcc*), as are illustrated in Fig. 1.7.

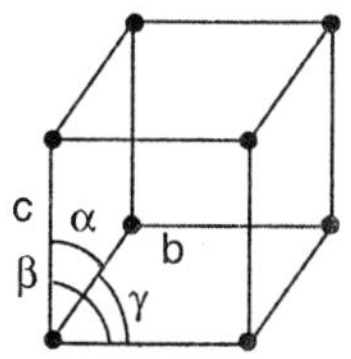

a=b=c, α=β=γ=90°
(i) Simple Cubic (sc)

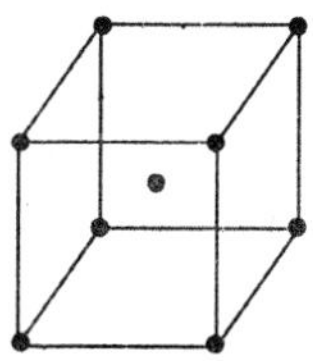

(ii) Body-Centred Cubic (bcc)

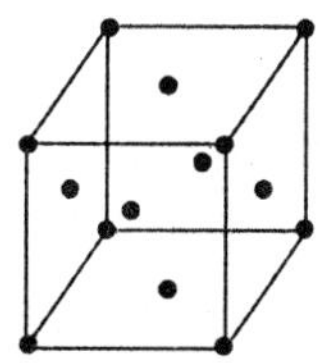

(iii) Face-Centred Cubic (fcc)

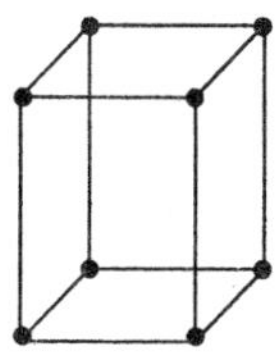

(iv) Simple Tetragonal (st)

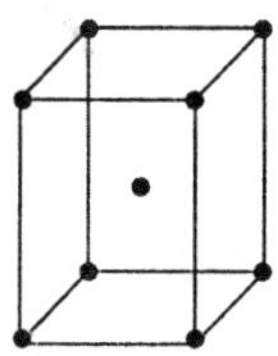

(v) Body-Centred Tetragonal (bct)

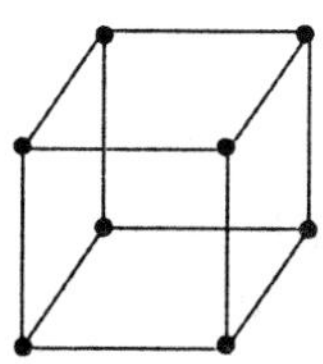

(vi) Simple Orthorhombic

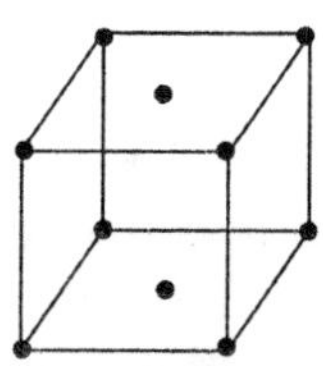

(vii) Base-Centred Orthorhombic

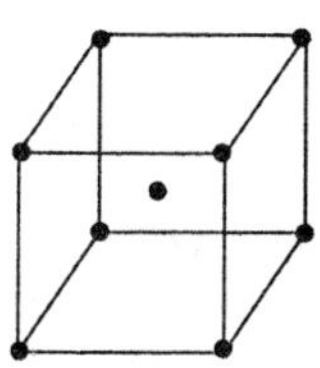

(viii) Body-Centred Orthorhombic

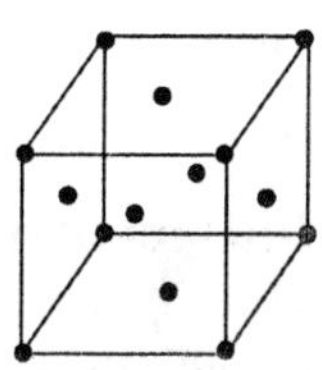

(ix) Face-Centred Orthorhombic

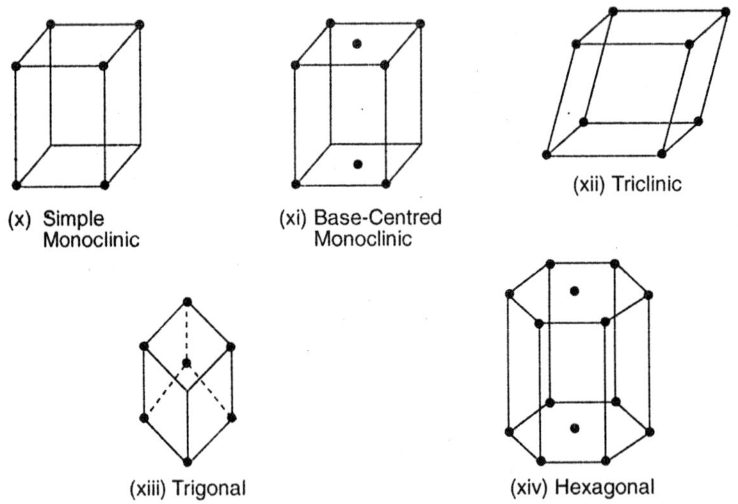

Fig. 1.7 Seven crystal systems : Conventional unit cells of the 14 Bravais space lattices.

The chief ***characteristic features*** of a unit cell are (i) *the volume*, (ii) *the atoms per unit cell*, (iii) *the coordination number*, (iv) *the atomic radius* and (v) *the atomic packing fraction* which we shall now define and discuss.

1. Volume of unit cell—For cubic crystals, we have $a = b = c$.

$$\therefore \text{ Volume of unit cell, } \boldsymbol{V} = a \times b \times c = \boldsymbol{a^3}.$$

2. Atoms per unit cell—In cubic crystals, *every corner atom* in each unit cell is shared equally with 8 units cells and so counts for *one-eighth* of an atom.

An atom lying completely *within a unit cell* however belongs *wholly to that cell.*

An atom *in the face of a unit cell*, on the other hand, belongs to two unit cell and hence is counted as *one-half* for that particular cell.

An atom *on one edge of a unit cell* is shared equally by four unit cells and is thus counted as a *one quarter* of an atom per unit cell.

(a) Simple cubic (*sc*)—The unit cube of this structure has atoms only at the corners of the cube. Each corner atom contributes only 1/8 of its effective part. As there are in all 8 corner atoms, their total contribution is $8 \times \frac{1}{8} = 1$.

So, the number of ***atoms per unit sc-cell*** is $n = \mathbf{1}$.

(b) Face-centred cubic (fcc)—The unit cell is a cube with an atom at each of the eight corners and an atom at each of the six faces. Each face-centred atom is shared by two unit cells and there are six faces of a cube.

$\therefore$ Number of ***atoms per unit cell*** of ***fcc-cube*** is $\frac{8}{8} + \frac{6}{2} = 1 + 3 = \mathbf{4}$.

(c) Body-centred cubic (bcc)—The unit cell is cubic with an atom at each corner and an atom at the centre of the cube. The atom within the body of the unit cell belongs wholly to that unit cell. In addition, there are eight corner atoms.

$\therefore$ Number of ***atoms per unit cell*** of ***bcc-cube*** is $1 + 8 \times \frac{1}{8} = 1 + 1 = \mathbf{2}$.

3. Co-ordination number—The points in a Bravais lattice that are closest to a given point are known as its *neighbours* and the periodic nature of Bravais lattice ensures that each point has the same number of nearest neighbours. This number of nearest neighbours that an atom has, in a given structure, is called the ***coordination number*** and is a characteristic of the lattice, and provides an idea of the closeness of the packing of the atoms.

The distance between the centres of two nearest neighbours is termed ***nearest neighbour distance.***

In a *simple cubic cell,* any corner atom has one close neighbour along $+X$ axis another along $-X$ axis. This is also true for Y and Z-axis as well. Thus each atom has only six neighbours.

$\therefore$ ***Coordination number*** for sc = **6.**

In *body-centred cubic,* the atom inside the cell has all the corner atoms as its close neighbours.

$\therefore$ ***Coordination number*** for bcc = **8.**

In face-centred cubic, each corner atom has four close neighbours in each of the three planes: XY, YZ and ZX (Fig. 1.8)

$\therefore$ ***Coordination number*** for $fcc = 4 \times 3 =$ **12.**

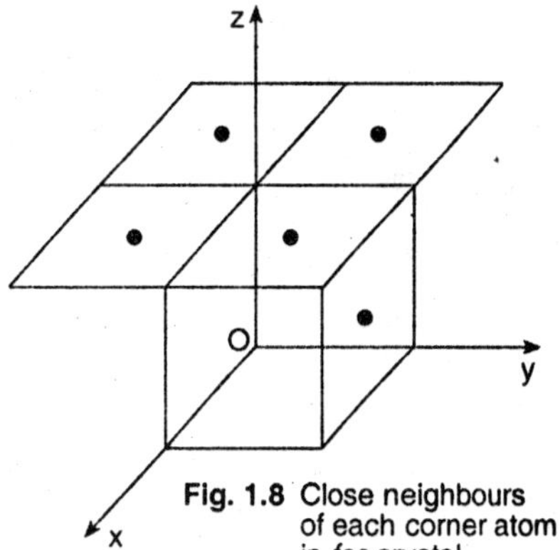

Fig. 1.8 Close neighbours of each corner atom in *fcc* crystal

4. Atomic radius—In a crystal, all the atoms have supposedly the same size and touch each other. The ***atomic radius*** is *defined* as the *distance between the centres of two neighbouring atoms.* And it can be related to lattice parameter a.

In **sc**, if r be the atomic radius and a the lattice parameter, then

$$a = 2r \;\Rightarrow\; \boldsymbol{r = a/2}$$

In **bcc**, the atoms touch each other along the diagonal of the cube (Fig. 1.9).

$$\therefore\; (4r)^2 = (\sqrt{2}a)^2 + a^2 \;\Rightarrow\; \boldsymbol{r = \sqrt{3}a/4}$$

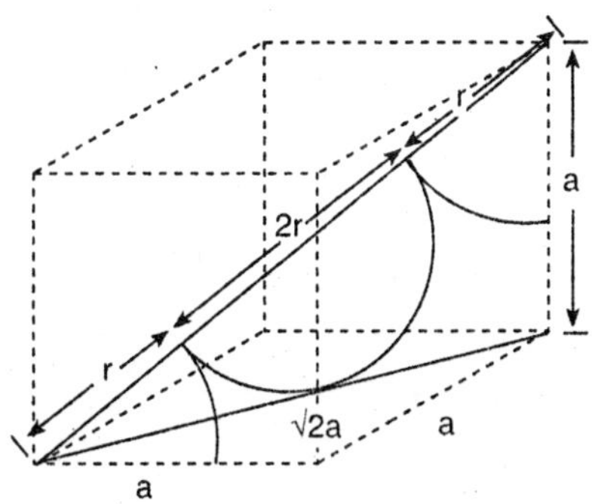

Fig. 1.9 Atomic radius of *bcc* crystal

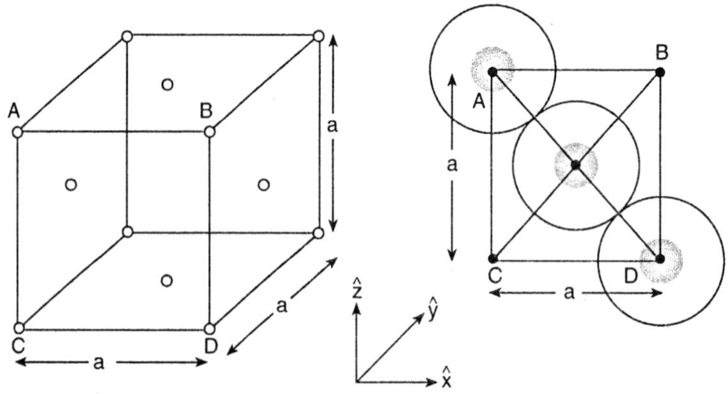

Fig. 1.10 Atomic radius of *fcc* structure

In ***fcc*** (Fig. 1.10), the atoms touch along the diagonal of the faces.

$$\therefore \quad (4r)^2 = 2a^2 \quad \Rightarrow \quad \boldsymbol{r = a/2\sqrt{2}}$$

5. Atomic packing fraction (or **factor**)—This is *defined* as the *ratio of the volume of the atoms occupying the unit cell to that of the unit cell of the crystal structure.* This is also called ***relative packing density.***

In ***sc-structure***, the number of atoms per unit cell = 1; atomic radius, $r = a/2$.

$$\therefore \quad \text{Volume of unit cell occupied by atom} = \frac{4}{3}\pi\left(\frac{a}{2}\right)^3$$

$$\text{Volume of the unit cell} = a^3$$

$$\therefore \quad \textbf{\textit{Packing fraction, f}} = \frac{\frac{4}{3}\pi(a/2)^3}{a^3} = \frac{\pi}{6}$$

$$= \mathbf{0.52 = 52\%}$$

The atomic packing is therefore *loose* in ***sc*** structure.

In ***bcc-lattice***, the number of atoms per unit cell = 2; atomic radius, $r = \sqrt{3}a/4$.

$$\text{Volume of unit cell} = a^3$$

$$\therefore \quad \textbf{\textit{Packing fraction, f}} = \frac{\frac{4}{3}\pi(\sqrt{3}a/4)^3 \times 2}{a^3} = \frac{\sqrt{3}\pi}{8}$$

$$= \mathbf{0.68 = 68\%}$$

In ***fcc-lattice***, the number of atoms per unit cell = 4; atomic radius, $r = a/2\sqrt{2}$.

$$\text{Volume of a unit cell} = a^3$$

$$\therefore \quad \textbf{\textit{Packing fraction, f}} = \frac{\frac{4}{3}\pi(a/2\sqrt{2})^3 \times 4}{a^3} = \frac{\pi}{3\sqrt{2}}$$

$$= \mathbf{0.74 = 74\%}$$

Thus, the ***fcc***-structure has the highest packing fraction among the cubic crystals.

1.6 Miller indices

To express the orientation of a plane in a crystal lattice, with respect to three directions, Miller evolved a scheme according to which a crystal plane is represented by three numbers h, k and l called the ***Miller indices***. They are *the three smallest possible integers, whose ratio is equal to the ratio of the reciprocals of the intercepts on the three rectangular axes X, Y and Z (crystallographic axes) by the plane.* If l_1, l_2 and l_3 be the lengths of the intercepts on the X, Y and Z axes respectively by a crystal plane of Miller indices h, k and l, we have

$$h : k : l = \frac{1}{l_1} : \frac{1}{l_2} : \frac{1}{l_3}.$$

Miller's scheme to obtain the indices runs as follows.

1. Determine the intercepts of the plane on the (crystallographic) axes.

2. Find the reciprocals of these intercepts.

3. Simplify to remove the fractions, if any, so as to reduce the numbers to the smallest three integers h, k and l and enclose them in parentheses without the comas.

Then the *Miller indices* of the plane are given by (hkl).

● In step 1, the intercepts are taken in terms of the length of the fundamental vectors, choosing the origin at one of the lowest points.

● If a plane is parallel to the axis (i.e., no intercept), its intercept with the axis is taken as ∞.

● In step 2, the reciprocals are taken only to avoid the occurrence of ∞ in Miller indices.

Illustrations—Consider the plane ABC (Fig. 1.11) with intercepts 1, 2 and 1 with the crystallographic axes $\vec{a}, \vec{b}$ and $\vec{c}$ respectively. Then the Miller indices of the plane ABC are specified by the following procedure.

(a) intercepts $\cdots$ $1, 2, 1$;

(b) reciprocals $\cdots$ $1, \frac{1}{2}, 1$

(c) simplification $\cdots$ $2, 1, 2$

So, the Miller indices of the plane ABC are (212).

If instead, we consider plane $DEF \parallel$ plane ABC, enclosed by the lattice, its

(a) intercepts $\cdots$ $\frac{1}{2}, 1, \frac{1}{2}$

(b) reciprocals $\cdots$ $2, 1, 2$

(c) simplification $\cdots$ not required.

$\therefore$ Miller indices of the plane DEF are thus (212), the same as those of ABC.

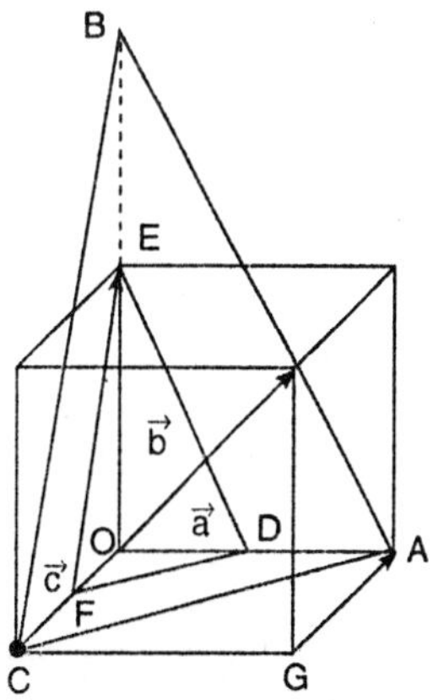

Fig. 1.11 Miller indices of parallel planes and planes passing through the origin

Thus, ***the parallel planes have the same Miller indices.***

• If a plane intercepts on the *negative side* of an axis, the corresponding index is *negative* and a bar is given on the top of that index, e.g., $(h\bar{k}l)$. The intercepts of a plane through the origin cannot as such be determined. How then to fix up its Miller indices? This can be done in *two* ways.

We take another plane parallel to the one through the origin and determine its Miller indices. The given plane has the same Miller indices. Alternatively, we may shift the origin from the plane to some suitable lattice point with no change in orientation of the axes and then find the Miller indices.

Significance of Miller indices—Miller indices *do not* refer to any particular plane but a set of parallel planes with a specific orientation, for the Miller indices refer to the ratio of the reciprocal of intercepts and not to individual intercepts. Parallel planes in a crystal behave identically. So, instead of locating a particular plane, the orientation of an infinite set of parallel equivalent planes is to be specified. The Miller indices provide exactly the same.

In studying X-ray diffraction, it becomes all the more palpable for a diffracted beam is the result of scattering from a large number of equivalent parallel planes, acting collectively to diffract the beam.

• The Miller indices of the six faces of a cube and some other important planes in a cubic crystal are shown in Fig. 1.12.

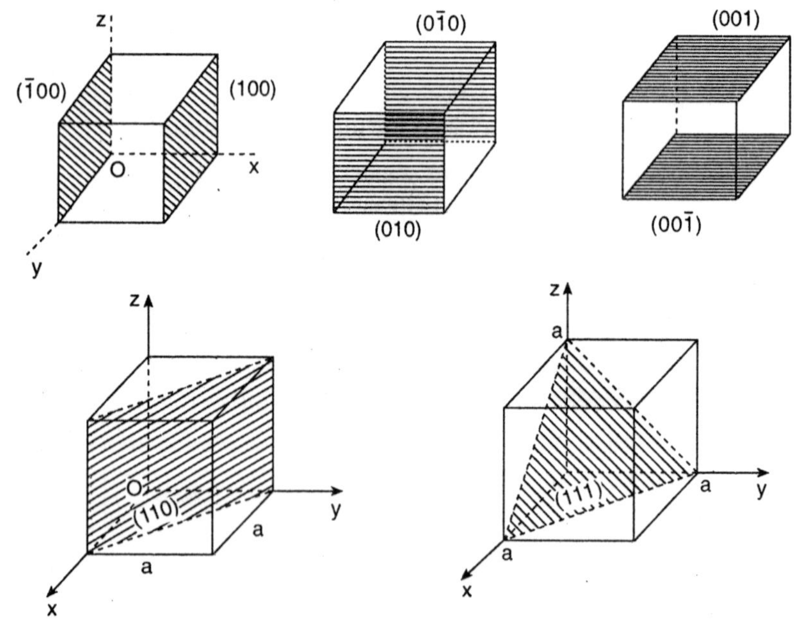

Fig. 1.12 Some important planes in cubic structure

• The *purpose of taking reciprocals* to find Miller indices is to bring all the planes inside a single unit cell so that all the crystal planes may be discussed in terms of planes passing through the single unit cell. And the whole crystal may be generated from a single unit cell that acts as a representative of the whole crystal.

We shall now find an analytic expression for the interplanar spacing.

1.7 Interplanar spacing: Analytic expression

Consider a set of parallel planes with indices (hkl). With origin O on a lattice point of one such plane, draw crystallographic axes Ox, Oy and oz.

The equation of a plane nearest to the origin, but not passing through it, is

$$\frac{x}{(a/h)} + \frac{y}{(b/k)} + \frac{z}{(c/l)} = 1 \qquad (1.7.1)$$

where $(a/h), (b/k)$ and (c/l) are the lengths of the intercepts on $\vec{a}, \vec{b}$ and $\vec{c}$ respectively, since the plane has the Miller indices (h, k, l).

Draw a normal OP from the origin O to this plane, Then the length d of this normal is the spacing between the adjacent parallel planes in the set. If $\cos\alpha, \cos\beta$ and $\cos\gamma$ be the direction cosines of $OP(= d)$, then

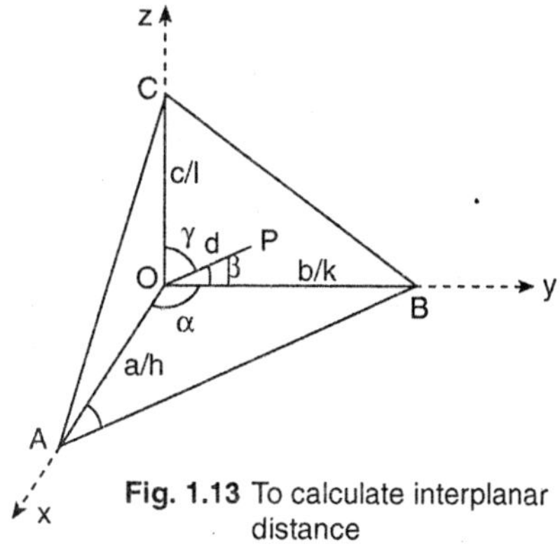

Fig. 1.13 To calculate interplanar distance

$$\cos\alpha = \frac{d}{(a/h)} = \frac{hd}{a}; \quad \cos\beta = \frac{kd}{b}; \quad \cos\gamma = \frac{ld}{c} \qquad (1.7.2)$$

But the direction cosines give: $\cos^2\alpha + \cos^2\beta + \cos^2 y = 1$,

$$\Rightarrow \qquad \left(\frac{h^2}{a^2} + \frac{k^2}{b^2} + \frac{l^2}{c^2}\right) d^2 = 1,$$

$$\Rightarrow \qquad \boxed{\boldsymbol{d = \frac{1}{\sqrt{(h/a)^2 + (k/b)^2 + (l/c)^2}}}} \qquad (1.7.3)$$

This is a general equation and is applicable to an *orthorhombic, cubic*, and a *tetragonal structure*.

- For a ***cubic crystal***, $a = b = c$ and the equation (1.7.3) reduces to

$$\boxed{\boldsymbol{d = \frac{a}{\sqrt{h^2 + k^2 + l^2}}}} \qquad (1.7.4)$$

- For a ***tetragonal crystal***, we have $a = b \neq c$ so that

$$\boldsymbol{d = \frac{1}{\sqrt{\dfrac{h^2 + k^2}{a^2} + \dfrac{l^2}{c^2}}}} \qquad (1.7.5)$$

• Since $(a/h), (b/k)$ and (c/l) are the intercepts on the axis, a zero Miller index implies ∞-intercept. So, $l = 0$ means intercept on c-axis is ∞, i.e., the plane is parallel to c-axis. Similarly, if k and l are both zero, the plane is parallel to both the b and c axes of the crystal.

• Eq. (1.7.3) is valid for ***orthogonal lattices*** only. For *non-orthogonal lattices*, the process to obtain the expression for d is somewhat involved.

1.8 Indices of a direction in a crystal

The concentration of atoms in a crystal is different in different directions and so it is necessary to locate the different directions. But how? This is done by what is called *indices of a direction.*

The ***indices of a direction*** in a crystal is expressed as a set of smallest integers having the same ratios as the components of a vector in the given direction, written as multiplies of base vector $\vec{a}, \vec{b}, \vec{c}$. The direction index of a vector $h\vec{a} + k\vec{b} + l\vec{c}$ is therefore $[hkl]$, where h, k, l are integers with no common factor except unity. The squre brackets enclosing the indices are the *direction indices.* For example, in a cubic crystal, [100] direction is direction of x-axis, $[1\bar{0}0]$ direction is that of $-y$ axis etc.

1.8.1 Direction normal to (hkl)-plane in cubic crystals

A crystal of *cubic symmetry is orthogonal* so that $\alpha = \beta = \gamma = 90°$ and $a = b = c$.

Consider x, y and z-axes along the crystallographic axes (Fig. 1.14) and a plane (hkl). The plane has thus intercepts $\vec{a}/h$, $\vec{b}/k$ and $\vec{c}/l$ on the axes. The equation to the plane is therefore given by

$$f(x, y, z) \equiv \frac{x}{a/h} + \frac{y}{a/k} + \frac{z}{a/l} = 1 \quad (\because\ a = b = c)$$

$$\text{or,}\ f \equiv \frac{h}{a}x + \frac{k}{a}y + \frac{l}{a}z = 1$$

Then, the normal to this plane is along the vector $\vec{\nabla} f$.

$$\text{But,}\quad \vec{\nabla} f = \frac{h}{a}\vec{i}_x + \frac{k}{a}\vec{i}_y + \frac{l}{a}\vec{i}_z$$

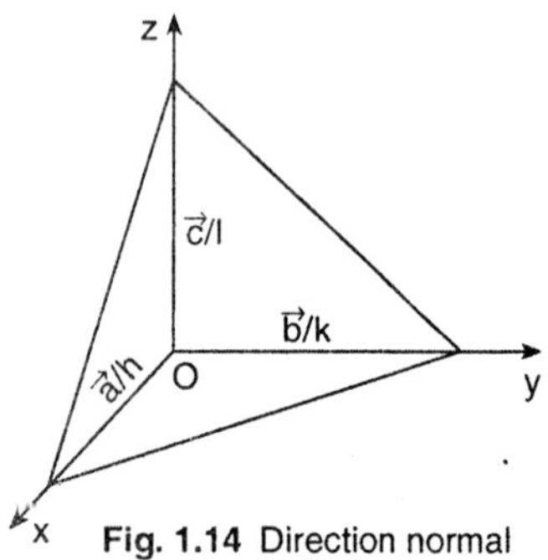

Fig. 1.14 Direction normal to (hkl) plane

It thus appears that the indices of the direction of $\vec{\nabla} f$ are h, k and l, implying that ***the normal to the (hkl) plane is along the direction [hkl].***

• As in crystal planes with different Miller indices, the direction of different direction indices may be crystallographically equivalent, for normals to equivalent planes are equivalent directions. The whole set of crystallographically equivalent directions, of which $[hkl]$ is a member, is written as $\langle hkl \rangle$, that is, by enclosing the direction indices in angle bracket. While in cubic crystals, a direction $[hkl]$ is normal to a plane having Miller indices (hkl), it is *not generally* true for other systems.

1.8.2 Angle between two places of Miller indices $(h_1k_1l_1)$ and $(h_2k_2l_2)$

If θ be the angle between the planes $(h_1k_1l_1)$ and $(h_2k_2l_2)$ of a cubic cyrstal, then

$$\cos\theta = \frac{h_1h_2 + k_1k_2 + l_1l_2}{\sqrt{(h_1^2 + k_1^2 + l_1^2)(h_2^2 + k_2^2 + l_2^2)}}$$

For *proof* of the above relation, see Illustrative Example 12.

1.9 Density of atoms in a crystal plane

A crystal consists of a large number of equidistant parallel planes. If N be the total number of planes separated from one another by an interplanar spacing d and A the area of cross-section of each plane, then the *volume* of this part of crystal is plainly given by NAd.

Now, let δ be the density of atoms in one of these planes. Then, the total number of atoms in this part of the crystal is $NA\delta$.

If V be the volume of the unit cell and n the number of atoms in the unit cell, then

$$NA\delta = \frac{NAd}{V}n \quad \Rightarrow \quad \boxed{\delta = \frac{nd}{V}} \tag{1.9.1}$$

Let us now apply (1.9.1) to find, by way of illustration, the density of atoms in (100), (110) and (111) planes of *sc*-lattice.

Illustration—The interplanar spacing in terms of lattice constant a for the above planes are given by

$$d_{100} = a, \quad d_{110} = a/\sqrt{2}, \quad d_{111} = a/\sqrt{3}$$

Volume of a simple cubic cell, $V = a^3$ and $n = 1$. So, $\delta = d/V = d/a^3$

$$\therefore \quad \delta_{100} = \frac{d_{100}}{a^3} = \frac{a}{a^3} = \frac{1}{a^2}$$

$$\delta_{110} = \frac{d_{110}}{a^3} = \frac{a}{\sqrt{2}a^3} = \frac{1}{\sqrt{2}a^2}$$

$$\delta_{111} = \frac{d_{111}}{a^3} = \frac{a}{\sqrt{3}a^3} = \frac{1}{\sqrt{3}a^2}$$

● The readers should try to find the density of atoms in the above planes for *bcc* and *fcc* structures. The results will be

$$(1/a^2, \sqrt{2}/a^2, 1/\sqrt{3}a^2) \text{ for } bcc \text{ and } (2/a^2, \sqrt{2}/a^2, 4/\sqrt{3}a^2) \text{ for } fcc.$$

Alternatively, the density of atoms in a crystal plane may be defined as the *ratio of the number of atoms in the plane to the area of the plane.*

$$\therefore \quad \text{Density of atoms in a plane} = \frac{\text{Number of atoms in the plane}}{\text{Area of the plane}}$$

Obviously, the same result would be obtained with this definition as well. For instance, let us find δ for the plane (100) of an *sc*-system.

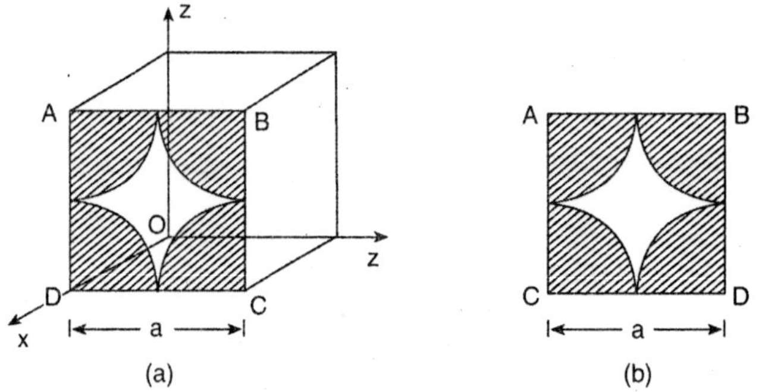

Fig. 1.15 Actual (a) and cross-sectional (b) view of (100) plane of *sc*-structure

Number of atoms in the plane $ABCD(100) = 4 \times \frac{1}{4} = 1$

Area of the plane $ABCD = a \times a = a^2$.

$$\therefore \quad \delta_{100} = \frac{\text{No. of atoms in plane (100)}}{\text{Area of plane (100)}} = \frac{1}{a^2},$$

the same as obtained earlier.

1.9.1 Number of atoms (lattice points) per unit cell in a cubic lattice: Calculation of lattice constant

This is already discussed in Art. 1.5. The number of atoms per unit cell can also be determined if the volume, density and molecular weight of the constituent atom of the cell are known.

Let a (in m) be the lattice parameter of a cubic crystal and n the number of atoms per unit cell. If M be the atomic (or molecular) weight of the material and N_A ($= 6.023 \times 10^{26}$ per k.mole) the Avogadro number, then 1 kg of material has N_A atoms (or molecules).

$\therefore$ Mass of n atoms (or molecules) $= (M/N_A)n$ kg.

$$\therefore \quad \text{Density, } \rho = \frac{(M/N_A)n}{V} \text{ kg/m}^3$$

But the volume of a unit cell of cubic crystal, $V = a^3$ cubic m.

$$\therefore \quad \rho = \frac{Mn}{N_A a^3} \text{ kg/m}^3 \cdot$$

$$\therefore \quad \boxed{n = \frac{a^3 \rho N_A}{M}} \tag{1.9.2}$$

From (1.9.2), we get lattice constant,

$$a = \left(\frac{nM}{\rho N_A}\right)^{\frac{1}{3}} \tag{1.9.3}$$

1.10 Some common and simple crystal structures

We shall discuss hereunder some relatively simple crystal structures in brief. We shall present each atom as a sphere of a definite radius and assume that crystals are formed, in principle, by the most stable packing of the spheres.

1. Simple cubic (*sc*) structure—Fig. 1.7(i) illustrates the simple cubic arrangement of atoms. The unit cube of the structure has atoms only at each corner and they touch each other along the cube edges. The lattice is spanned by three mutually perpendicular primitive vectors of equal length. Each atom has only 6 neighbours and the structure is rather open and *loosely packed.* α-Polonium is the only element that shows this structure.

2. Body-centred cubic (*bcc*) structure—The unit cell is cubic with one atom at each corner and one atom at its centre. Each atom has eight neighbours and the atoms touch along the body diagonals. Number of atoms per unit cell $= (8 \times 1/8) + 1 = 2$ (Fig. 1.7(ii)). Atomic radius $r = \sqrt{3}a/4$.

$$\therefore \text{ Volume of the cube} = a^3 = (4r/\sqrt{3})^3$$

Examples of body-centred cubic crystals are metals like Fe, Mo, Cr, K, Na, V etc.

3. Face-centred cubic (*fcc*) structure—The unit cell here also is cubic with an atom at each of the eight corners and an atom at each of its six faces (Fig. 1.7(iii)).

The number of atoms in unit cell $= 8 \times \frac{1}{8} + 6 \times \frac{1}{2} = 1 + 3 = 4$.

The atoms touch along the diagonal of the faces. Atomic radius $r = a/2\sqrt{2}$.

$$\therefore \text{ Volume of the cube} = a^3 = (4r/\sqrt{2})^3$$

Examples of the face-centred cubic crystals are Ca, Ni, Pt, Pd etc.

4. Sodium chloride structure—The NaCl-structure, illustrated in Fig. 1.16, consists of alternating Na and Cl atoms (ions) at the lattice points of a simple cubic lattice. The atoms of Na lie on the lattice point of an *fcc*-lattice, like the Cl-atoms. In

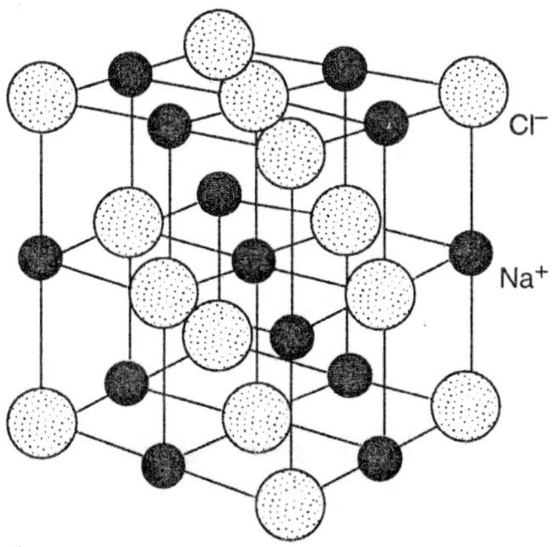

Fig. 1.16 Sodium choride structure ● ≡ Na^+ ○ ≡ Cl^-

a unit cube, there are four units of NaCl and the positions of the atoms are

$$\text{Cl} : (0,0,0);\ (a/2,a/2,0);\ (a/2,0,a/2);\ (0,a/2,a/2)$$

$$\text{Na} : (a/2,a/2,a/2);\ (0,0,a/2);;(0,a/2,0);\ (a/2,0,0)$$

where the lattice constant a is the distance between two successive Cl (or Na) atoms (ions).

5. Diamond structure—It is characterised by a tetrahedral bond arrangement as illustrated in Fig. 1.17. The space lattice is face-centred cubic. In Fig. 1.17b the fractions indicate the height above the base in units of a cube edge. The lattice may be looked upon as formed of two interpenetrating *fcc*-lattices, displaced along the body diagonal by 1/4 of the diagonal length. The points at 0 and 1/2 are on one *fcc* lattice

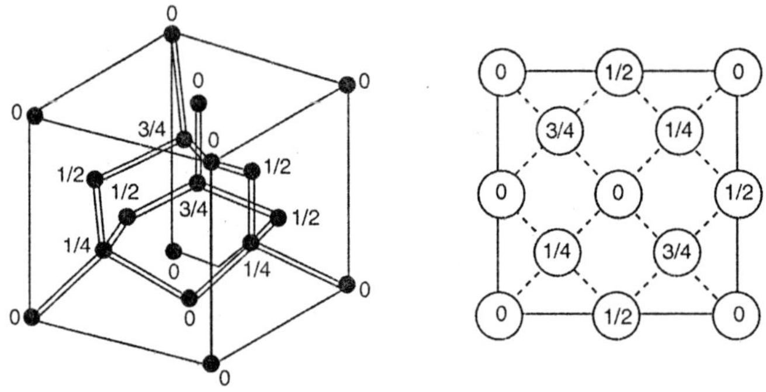

Fig. 1.17 (a) Crystal structure of diamond showing the tetrahedral bond arrangement
(b) Atomic positions in the cubic cell of diamond structure projected on a cube face

and those at 1/4 and 3/4 are on the other. Hence, with a *fcc*-space lattice, the basis consists of two identical atoms at (0 0 0) and (1/4 1/4 1/4). It is not a Bravais lattice as the arrangement of points surrounding a given point is different from that of its nearest neighbours.

There are 4 nearest neighbours of each atom and in a unit cube there are 8 atoms. Its atomic packing factor is only 0.34, less than even half of the close-packed structure (0.74). Examples of diamond structure are C, Si, Ge etc.

For diamond structure, $r = \sqrt{3}a/8$ since the structure consists of two interpenetrating *fcc* lattices, displaced by $\sqrt{3}a/4$ along the body diagonal of length $\sqrt{3}a$. There are 8 close-packed spheres, each of radius r in the cell.

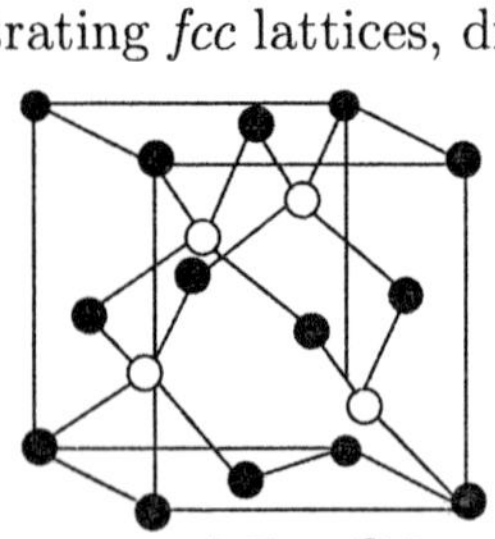

Fig. 1.18 Crystal structure of zinc blende

$$\therefore\ \textbf{\textit{Packing fraction}} = 8 \times \tfrac{4}{3}\pi r^3/a^3$$

$$= \sqrt{3}\pi/16 = \mathbf{0.34}$$

6. Zinc blende (ZnS) structure—Zinc blended structure is identical with the diamond structure except that the two interpenetrating *fcc* lattices are occupied by two different elements, zinc and sulphur (Fig. 1.18). Zn atoms are placed on one *fcc* lattice and S-atoms on the other *fcc* lattice, shifted from the first by 1/4 of the cube diagonal vector. ZnS structure has no centre of inversion symmetry at mid-point of each line joining the nearest neighbour atoms. The diamond structure possesses an inversion symmetry. GaS, CuCl, CdS, GaP etc. all have the ZnS-structure.

7. Hexagonal close packed (*hcp*) structure—We first discuss the ***close packed structure.*** The most efficient way of packing atoms, considered as hard spheres, in a plane is the closest packed arrangement, as shown in Fig. 1.19. The spheres (atoms) may be arranged by placing each sphere in contact with six others.

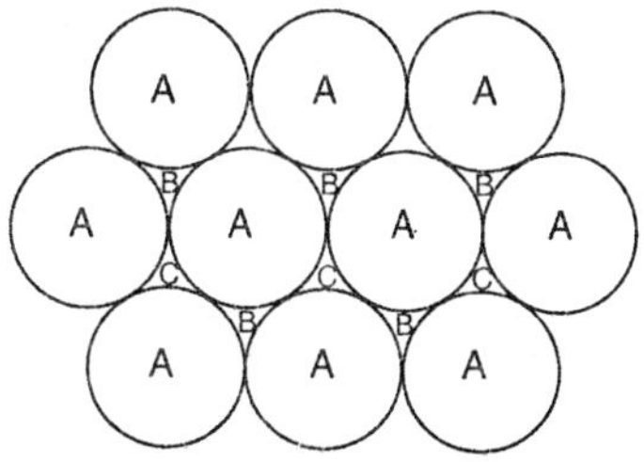

Fig. 1.19 Close packed array of spheres (atoms). A, B and C are different possible positions for the successive layers

The next layer is the second layer that can be made above the basal layer A in space, marked B or C. If B-layer is the second layer, the third layer C is the arrangement of atoms immediately above those of the first layer A, and the fourth layer is layer position B and so on. This gives (Fig. 1.20a) a stacking $AB\,AB\,AB\cdots$ (or equivalently, $AC\,AC\,AC\cdots$).

The ***hexagonal close packed structure*** formed by metals like Mg, Zn, Cd is shown in Fig. 1.20b where the points are representatives of the atom positions, but *not* the lattice points. The unit cell contains *one* atom at each corner, *one* atom each at the centre of the hexagonal faces and *three* more atoms within the body of the cell. Each

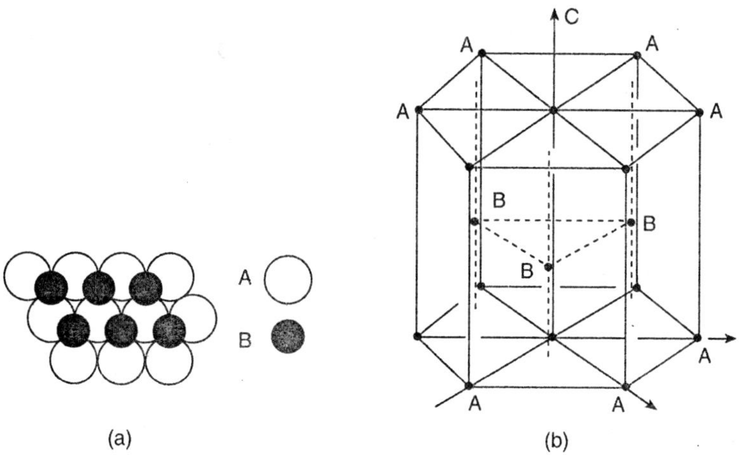

Fig. 1.20 Hexagonal close packing (a) Stacking pattern of layers ABABA... (b) Hexagonal pattern of three dimension arrangement of atom

atom is in contact with (i.e., touches) *three* atoms in the layer below its plane, *six* atoms in its own plane, an *three* atoms in the layer above. The *coordination number* of the structure is, therefore, 12. Again since the atom touch each other along the edge of the hexagon, $a = 2r$ and the height of the unit cell is c.

The top layer has 7 atoms. Of them, each corner atom is shared by 6 surrounding hexagonal cells and the centre atom is shared by 2 surrounding cells. The 3 atoms are in the body of the cell. So, the contribution to the total number of atoms in a unit cell by the top layer $= \left(6 \times \frac{1}{6}\right) + \left(1 \times \frac{1}{2}\right) = \frac{3}{2}$. Similarly, the contribution of the bottom layer$= \frac{3}{2}$. So, the number of atoms per unit cell, $n = \left(\frac{3}{2} + \frac{3}{2} + 3\right) = 6$.

● It can be computed (see Example 21) to show that the c/a-ratio of ***HCP*** structure is $\sqrt{8/3}$ and the packing factor is 74%.

● There is another hexagonal structure—***hexagonal fcc structure***—similar to ***hcp*** structure, but *without* the 3-atoms in the body of the cell (B, B, B). Its packing fraction is low and metals do not crystallize in this structure.

1.11 Illustrated Examples

▸ **Example 1.** *The primitive translation vectors of the hexagonal space lattice are*

$$\vec{a} = (\sqrt{3}a/2)\hat{x} + (a/2)\hat{y}$$
$$\vec{b} = -(\sqrt{3}a/2)\hat{x} + (a/2)\hat{y}$$
$$\vec{c} = c\hat{z}$$

Show that the volume of the primitive cell is $\sqrt{3}a^2c/2$.

Solution: The required volume V is plainly given by

$$\begin{aligned} V &= \vec{a}\cdot(\vec{b} \times \vec{c}) = \vec{a}\cdot\left[(-\sqrt{3}a/2)\hat{x} + (a/2)\hat{y}\right] \times (c\hat{z}) \\ &= \vec{a}\cdot\left[\sqrt{3}ac\hat{y}/2 + ac\hat{x}/2\right] \\ &= \left[(\sqrt{3}a/2)\hat{x} + (a/2)\hat{y}\right]\cdot\left[(ac/2)\hat{x} + (\sqrt{3}ac/2)\hat{y}\right] \\ &= \frac{\sqrt{3}}{4}a^2c + \frac{\sqrt{3}}{4}a^2c = \frac{\sqrt{3}}{2}\boldsymbol{a^2c} \end{aligned}$$

▸ **Example 2.** *Find the Miller indices for planes in each of the following set of the intercepts:* (i) $3a, 3b, 2c$; (ii) $a, 2b, \infty$ *and* (iii) $a, b/2, c$. **(Punjab Univ.)**

Solution: (i) Set of intercepts: $3a, 3b, 2c$

∴ Reciprocal of numbers representing the intercepts $= \frac{1}{3}, \frac{1}{3}, \frac{1}{2}$

Smallest three integers having the same ratio $= 2, 2, 3$

∴ Miller indices are **(223)**

(ii) Set of intercepts: $a, 2b, \infty$

$\therefore$ Reciprocal of numbers representing the intercepts $= 1, \frac{1}{2}, 0$

Smallest three integers of the same ratio $= 2, 1, 0$

$\therefore$ Miller indices are **(210)**

(iii) Set of intercepts: $a, b/2, c$

$\therefore$ Reciprocal of numbers representing the intercepts $= 1, 2, 1$

Smallest three integers of the same ratio $= 1, 2, 1$

$\therefore$ Miller indices are **(121)**

▶ **Example 3.** *Find the Miller indices of a plane having the intercepts* (i) $6a, 2b, 3c$ *and* (ii) $2a, \infty, c/2$ *on the* a, b, c *axes respectively.* **(Burdwan Hons.)**

Solution: (i) Set of intercepts: $6a, 2b, 3c$

Reciprocal of numbers representing the intercepts $= \frac{1}{6}, \frac{1}{2}, \frac{1}{3}$

Smallest three integers of the same ratio $= 1, 3, 2$

$\therefore$ Miller indices are **(132)**

(ii) Set of intercepts: $2a, \infty, c/2$

Reciprocal of numbers representing the intercepts $= \frac{1}{2}, 0, 2.$

Smallest three integers of the same ratio $= 1, 0, 4$

$\therefore$ Miller indices are **(104)**

▶ **Example 4.** *The lattice constant of a cubic lattice is 'a'. Calculate the spacing between* (211), (001), (100) *and* (101) *planes.* **(Punjabi Univ.)**

Solution: The spacing d of the plane (hkl) of cubic lattice of side a is

$$d = \frac{a}{\sqrt{h^2 + k^2 + l^2}}$$

$\therefore$ From (211) plane,

$$d_{211} = \frac{a}{\sqrt{2^2 + 1^2 + 1^2}} = \frac{\boldsymbol{a}}{\sqrt{\mathbf{6}}}$$

Similarly, we obtain

$$d_{001} = \frac{a}{\sqrt{0^2 + 0^2 + 1^2}} = \boldsymbol{a}$$

$$d_{100} = \frac{a}{\sqrt{1^2 + 0^2 + 0^2}} = \boldsymbol{a}$$

$$d_{101} = \frac{a}{\sqrt{1^2 + 0^2 + 1^2}} = \frac{\boldsymbol{a}}{\sqrt{\mathbf{2}}}$$

▶ **Example 5.** *Calculate the separation between lattice planes in a sc, fcc and bcc lattice. For a cubic lattice, calculate the distance of* (123) *and* (234) *planes from a plane passing through the origin.* **(Meerut and Himachal Univ.)**

Solution: (a) A ***simple cubic lattice*** has the planes (100), (110) and (111).

$\therefore$ Separations :

$$d_{100} = \frac{a}{\sqrt{1^2+0^2+0^2}} = a$$

$$d_{110} = \frac{a}{\sqrt{1^2+1^2+0^2}} = \frac{a}{\sqrt{2}}$$

$$d_{111} = \frac{a}{\sqrt{1^2+1^2+1^2}} = \frac{a}{\sqrt{3}}$$

In fcc-lattice, there is one unit situated at each corner and one unit at the centre of each of the six faces. An additional (100) plane develops half-way between the (100) planes. The additional planes pass through the centres of the front and the back face. The distance between (100) planes of an *sc*-lattice is a so that the distance between (100) planes of the *fcc*-lattice is

$$d_{100} = \tfrac{1}{2}a$$

Similarly, an additional set of (110) planes parallel to the first set midway between the planes of *sc*-lattice develops.

$$d_{110} = \frac{1}{2}\frac{a}{\sqrt{2}}$$

The (111) planes of *sc*-lattice passes through the centre of all the faces of *fcc*-lattice and thus no new plane develops in this case

$$\therefore\ d_{111} = \frac{a}{\sqrt{3}}$$

In bcc-lattice, one unit is situated at each corner and one at the center of the cube. So, an additional (100) plane develops half-way between (100) plane of *sc*-lattice.

$$\therefore\ d_{100} = \tfrac{1}{2}a$$

The (110) plane is the same as that in a *sc*-lattice.

$$\therefore\ d_{110} = \frac{a}{\sqrt{2}}$$

The (111) plane of a *bcc*-lattice lies half-way betwen (111) planes of a *sc*-lattice.

$$\therefore\ d_{111} = \frac{1}{2}\frac{a}{\sqrt{3}}$$

(b) From a cubic lattice, the spacing d of the plane (hkl) is given by

$$d = \frac{a}{\sqrt{h^2+k^2+l^2}}$$

$\therefore$ For (123) plane,

$$d_{123} = \frac{a}{\sqrt{1^2+2^2+3^2}} = \frac{a}{\sqrt{14}}$$

For (234) plane,

$$d_{234} = \frac{a}{\sqrt{2^2+3^2+4^2}} = \frac{a}{\sqrt{29}}$$

► **Example 6.** *Draw* (101) *and* (111) *planes in a cubic unit cell. Determine the Miller indices of the directions which are common to both the planes.* **(Lucknow Univ.)**

Solution : For a cubic unit cell, the planes (101) and (111) are shown in Fig. 1.21.

Intercept of plane (101) with the axes are $1, \infty$ and 1.

Intercept of plane (111) with the axes are $1, 1$ and 1.

With O as origin and OA, OB, OC as axes $\vec{a}, \vec{b}, \vec{c}$ respectively, the plane with intercepts $1, \infty, 1$ is the plane $ADGC$ and that with intercept $1, 1, 1$ is the plane ABC. So, the line common to both the planes is AC. It can have two directions AC and CA.

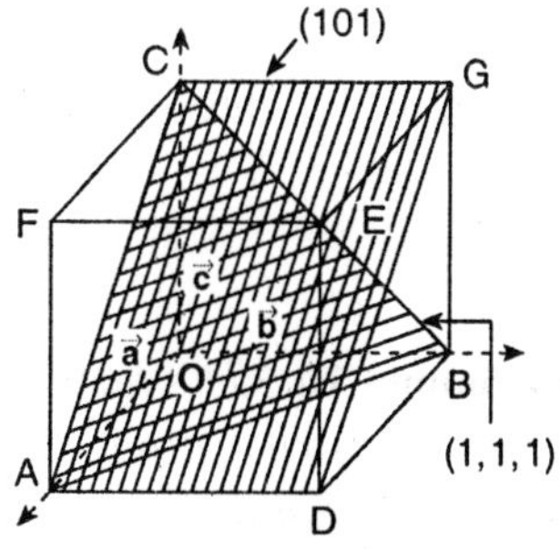

Fig. 1.21

Projections of direction AC on the axes $= -1, 0, 1$.

Projections of direction CA on the axes $= 1, 0, -1$.

Required Miller indices are thus $(\bar{1}\,0\,1)$ and $(1\,0\,\bar{1})$.

► **Example 7.** *Determine the density of diamond crystal from the following data : lattice parameter,* $a = 3.57$ A *and the atomic mass* $A = 12$ u.

Solution : The effective number of atoms in unit (cubic) cell of diamond is

$$n = \frac{1}{8} \times 8 \text{ (corner atoms)} + \frac{1}{2} \times 6 \text{ } (fc\text{-atoms}) + 1 \times 4 (bc\text{-atoms}) = 8$$

$$\therefore \text{ Density, } \rho = \frac{\text{mass of unit cell}}{\text{volume of unit cell}} = \frac{A.n}{N_A a^3}, \quad N_A = \text{Avogadro number}$$

$$= \frac{12 \times 8}{6.023 \times 10^{26} \times (3.577 \times 10^{-8})^3} = \mathbf{3540\ kg/m^3}$$

► **Example 8.** *Cesium metal* (atomic weight = 130) *has a cubic unit cell of side* 6A. *If the density of* Cs *is* $2\,\text{g/cm}^3$, *determine if the unit cell is simple, face-centred or body centred.* **(N. Bengal Hons.)**

Solution : If n be the number of Cs atoms in cubic unit cell of side a, then the number of atoms per unit volume $= n/a^3$. If A be the atomic weight of Cs and N_A, the Avogadro number, then the mass of an atom $= (A/N_A)$ g.

$$\therefore \text{ Density, } \rho = \frac{nA}{N_A a^3} \quad \Rightarrow \quad n = \frac{\rho a^3 N_A}{A}$$

$$\therefore \quad n = \frac{2 \times (6 \times 10^{-8})^3 \times 6.02 \times 10^{23}}{130} = 2$$

The cubic unit cell is therefore ***body-centred.***

► **Example 9.** *If the lattice parameter for a fcc crystal is* 2.14 A, *find the radius of the particle.* **(Nagpur Univ.)**

Solution : For an *fcc*-sturcture, the radius of the atom (particle) is given by

$$r = \frac{a}{2\sqrt{2}}, \quad \text{where } a = \text{lattice parameter.}$$

Again, by the problem, $a = 2.14\,\text{A}$, here. So, the radius of the atom,

$$r = \frac{2.14}{2\sqrt{2}}\,\text{A} = \mathbf{0.76\ A}$$

► **Example 10.** *The lattice constant of an fcc-lattice is* 6.38 A. *Calculate* (i) *the distance between a corner atom at the base and the atom at the centre of the top face, and* (ii) *the maximum distance between two atoms in the cubic cell.*

Solution : From Fig. 1.22, the required distance d $(= OS)$ is given by

$$d^2 = (a/\sqrt{2})^2 + a^2 = 3a^2/2$$

$$\therefore\ d = \sqrt{3/2a} = 1.225 \times 6.38\,\text{A} = \mathbf{7.81\ A}$$

The maximum distance d_m between the two atoms in the cubic cell is plainly the distance between the atoms at the ends of a body diagonal. Therefore,

$$d_m^2 = a^2 + a^2 + a^2 = 3a^2$$

$$\Rightarrow\ d_m = \sqrt{3}a = 1.732 \times 6.38\,\text{A} = \mathbf{11.05\ A}$$

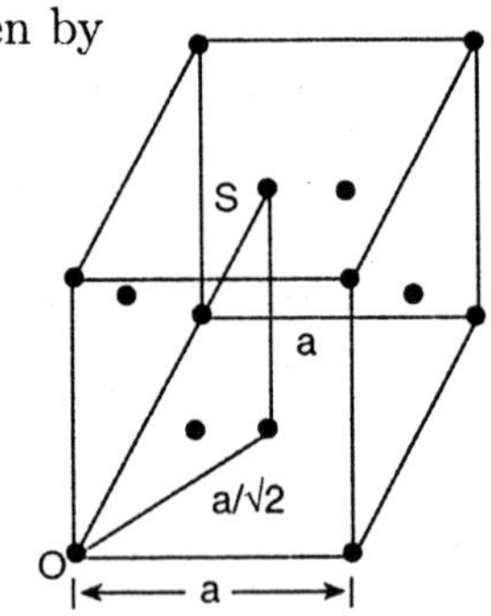

Fig. 1.22

► **Example 11.** NaCl *crystal has a cubic structure. If the molecular weight of* NaCl *is* 58.46 *and its density* 2.17 g/cm^3, *find the distance between the two adjacent atoms of the same kind and the lattice constant in the crystal.* **(Burdwan and Calcutta Hons.)**

Solution : Mass of a NaCl molecule, M = molecular weight in g/Avogadro number

$$= \frac{58.46}{6.02 \times 10^{23}}\,\text{g} = 9.7 \times 10^{-23}\,\text{g}$$

$\therefore$ Number of NaCl molecules per unit volume, $n = \text{density}/M$

$$\therefore\ n = \frac{2.17}{9.7 \times 10^{-23}} = 2.237 \times 10^{22}/\text{cm}^3$$

But NaCl is diatomic. So, the number of atoms per unit volume $= 2 \times 2.237 \times 10^{22} = 4.47 \times 10^{22}/\text{cm}^3$.

Let a be the lattice constant. So, the volume a^3 will contain one atom due to cubic structure implying that the number of atoms per unit volume is $1/a^3$.

$$\therefore \quad \frac{1}{a^3} = 4.47 \times 10^{22} \quad \Rightarrow \quad a^3 = 2.237 \times 10^{-23}\,\text{cm}^3 \quad \Rightarrow \quad a = 2.82\,\text{A}$$

$$\therefore \quad \text{Lattice constant} = \mathbf{2.82\ A}$$

$\therefore$ Distance between two adjacent atoms of the same kind $= 2 \times 2.82 =$ **5.64 A**

► **Example 12.** *If θ be the angle in a cubic crystal between the normals to the plane $(h_1k_1l_1)$ and $(h_2k_2l_2)$, show that*

$$\cos\theta = \frac{h_1h_2 + k_1k_2 + l_1l_2}{\sqrt{(h_1^2 + k_1^2 + l_1^2)(h_2^2 + k_2^2 + l_2^2)}}$$

Solution: In a cubic crystal, the normal to the (hkl) plane is the direction $[hkl]$. So, θ is the angle between the vectors

$$h_1\hat{i}_x + k_1\hat{i}_y + l_1\hat{i}_z = \vec{A}, \text{ say}$$

$$\text{and } h_2\hat{i}_x + k_2\hat{i}_y + l_2\hat{i}_z = \vec{B}, \text{ say.}$$

$\therefore$ From the relation, $\vec{A}\cdot\vec{B} = AB\cos\theta$, we obtain

$$\cos\theta = \frac{\vec{A}\cdot\vec{B}}{AB} = \frac{\vec{A}\cdot\vec{B}}{\sqrt{(\vec{A}\cdot\vec{A})(\vec{B}\cdot\vec{B})}} \tag{1.11.1}$$

Now, $\vec{A}\cdot\vec{A} = (h_1\hat{i}_x + k_1\hat{i}_y + l_1\hat{i}_z)\cdot(h_1\hat{i}_x + k_1\hat{i}_y + l_1\hat{i}_z) = h_1^2 + k_1^2 + l_1^2$

Similarly, $\vec{B}\cdot\vec{B} = h_2^2 + k_2^2 + l_2^2$,

and $\vec{A}\cdot\vec{B} = (h_1\hat{i}_x + k_1\hat{i}_y + l_1\hat{i}_z)\cdot(h_2\hat{i}_x + k_2\hat{i}_y + l_2\hat{i}_z)$

$$= h_1h_2 + k_1k_2 + l_1l_2$$

$$\therefore \quad \boldsymbol{\cos\theta = \frac{h_1h_2 + k_1k_2 + l_1l_2}{\sqrt{(h_1^2 + k_1^2 + l_1^2)(h_2^2 + k_2^2 + l_2^2)}}}, \quad \text{using (1.11.1)}$$

► **Example 13.** *The molecular weight of* KCl *crystal is* 74.45 *and density* 1.98 g/c.c. *Calculate* (i) *the lattice constant,* (ii) *the distance betwen atoms of the same kind and* (iii) *the first nearest heighbour and the second nearest neighbour distance.*

Solution: (i) Mass of KCl molecule $= 74.45 \times \dfrac{1}{6.023 \times 10^{23}}\,\text{g} = 12.37 \times 10^{-23}\,\text{g}$.

$$\therefore \text{ No. of KCl moelcules/c.c.} = \frac{1.98}{12.37 \times 10^{-23}} = 1.6 \times 10^{22}$$

No. of atoms/c.c. $= 2 \times 1.6 \times 10^{22} = 3.2 \times 10^{22}$ ($\because$ KCl is diatomic)

Let a be the distance between adjacent atoms i.e., lattice constant and n be the number of atoms along the edge of a 1-cm cube.

$$\therefore \text{ Volume of unit cube} = (na)^3 = n^3a^3,\ n^3 = \text{No. of atoms/c.c}$$

$$\therefore\ 3.2 \times 10^{22} \times a^3 = 1$$

$$\therefore\ a = \left(\frac{1}{3.2 \times 10^{22}}\right)^{1/3} = 3.15 \times 10^{-8}\,\text{cm.}$$

$$\therefore\ \textit{Lattice constant} = \mathbf{3.15\ A}$$

(ii) Distance between the two atoms of the *same kind* is $2a = 2 \times 3.15 = \mathbf{6.3\ A}$

(iii) The Bravais lattice of KCl crystal is face centred cubic. Therefore,

First nearest neighbour distance $= \sqrt{(a/2)^2 + (a/2)^2} = \dfrac{a}{2} = \dfrac{3.15}{2} = \mathbf{1.575\ A}$

Second nearest neighbour distance $= a = \mathbf{3.15\ A}$.

▶ **Example 14.** *The primitive vector, in an orthorhombic cell, are* 1.27 A, 2.14 A *and* 1.51 A. *Deduce the intercepts of x and y axes if the* (213) *plane cuts an intercept of* 1.51 A *along the z-axis.*

Solution: The reciprocal of the coefficients of unit translation vectors are in the ratio

$$p^{-1} : q^{-1} : r^{-1} = h : k : l = 2 : 1 : 3$$

$$\therefore\ p : q : r = \tfrac{1}{2} : 1 : \tfrac{1}{3}$$

So, the ratio of the actual lengths of the intercepts are given by

$$l_1 : l_2 : l_3 = pa : qb : rc$$

$$= \tfrac{1}{2} \times 1.27 : 1 \times 2.14 : \tfrac{1}{3} \times 1.51$$

Since $l_3 = 1.51$ A (given), we get on multiplying the right hand side by 3,

$$l_1 : l_2 : l_3 = 1.90 : 6.42 : 1.51$$

$$\therefore\ l_1 = \text{intercept of } x\text{-axis} = \mathbf{1.90}$$

$$l_2 = \text{intercept of } y\text{-axis} = \mathbf{6.42}$$

▶ **Example 15.** *Calculate the fraction of volume unoccupied in the unit cell of body-centred cubic lattice.* **(JAM)**

Solution: Volume of the unit cell, $V = a^3$, a being the lattice const. Let r be the radius of the atom.

For *bcc*-structure, $4r = \sqrt{3}a \Rightarrow r = \sqrt{3/4}a$

Number of atoms in *bcc*-unit cell $= 1 + 8 \times \frac{1}{8} = 2$

$$\therefore \text{ Volume occupied by atom} = 2 \times \frac{4}{3}\pi r^3 = \frac{8}{3}\pi \cdot \frac{3\sqrt{3}}{64}a^3 = \frac{\sqrt{3}}{8}\pi a^3.$$

$$\therefore \text{ Fraction of volume unoccupied} = (a^3 - \sqrt{3}\pi a^3/8)a^3 = \mathbf{\frac{8 - \sqrt{3}\pi}{8}}$$

▶ **Example 16.** *In a tetragonal crystal, the lattice parameters* $a = b = 2.42\text{Å}$ *and* $c = 1.74\text{Å}$. *Show that the interplanar spacing between consecutive* (101) *planes is* 1.41Å.

Solution: The interplanar spacing for a tetragonal system, with $a = b$, is given by, using (1.7.5)

$$d_{hkl} = \left(\frac{h^2 + k^2}{a^2} + \frac{l^2}{c^2}\right)^{-1/2}$$

By the problem, $a = 2.42\,\text{A}$, $c = 1.74\text{Å}$, $h = 1, k = 0, l = 1$.

$$\therefore \quad \mathbf{d_{101}} = \left[\frac{1^2 + 0^2}{(2.42)^2} + \frac{1^2}{(1.74)^2}\right]^{-1/2}$$

$$= \mathbf{1.41\text{Å}}, \text{ on simplication.}$$

▶ **Example 17.** *Calculate the lattice constant of rock salt* (NaCl) *from the following data.*

Molecular weight of NaCl = 58.45 kg.mol, *density of* NaCl = 2.17×10^3 kg/m^3, *Avogadro no.* $N_A = 6.023 \times 10^{26}$ per kg.mol. **(Calcutta and Burdwan H.)**

Solution: From eq. (1.9.2), the lattice constant

$$a = \left(\frac{nM}{\rho N_A}\right)^{1/3}, \quad M = \text{mol.wt}, \ \rho = \text{density}, \ N_A = \text{Avogadro number and}$$

n = no. of atoms per unit cell.

Now, NaCl is an *fcc*-crystal and so $n = 4$.

$$\therefore \quad \mathbf{a} = \left(\frac{4 \times 58.45}{2.17 \times 10^3 \times 6.023 \times 10^{26}}\right)^{1/3} \text{m, substituting the data}$$

$$= \mathbf{5.64A}, \text{ on simplification.}$$

▶ **Example 18.** *Calculate the maximum radius of the sphere that can just fit into the void at the body centre of the fcc structure coordinated by facial atoms.*

Solution: Let r be the radius of the atom and R the radius of the sphere that can just fit into the void, as illustrated in Fig. 1.23.

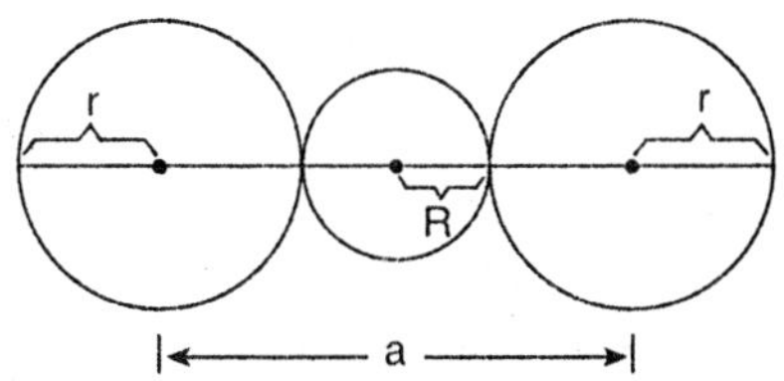

Fig. 1.23

$$\therefore \quad 2r + 2R = a \Rightarrow R = a/2 - r$$

where a is the side of the unit cell.

But, we have, for *fcc*-structure, $a = 4r/\sqrt{2}$.

$$\therefore \quad \boldsymbol{R} = \frac{4r}{2\sqrt{2}} - r = \frac{2r}{\sqrt{2}} - r = r(\sqrt{2}-1) = \mathbf{0.41}\boldsymbol{r}$$

► **Example 19.** *In a tetrahedral lattice* $a = b = 2.5$Å *and* $c = 1.8$Å. *Determine the lattice spacing between the* (111)*-planes.*

Solution: The tetrahedral lattice spacing for a given set of parallel planes (hkl) is

$$d_{hkl} = \left(\frac{h^2}{a^2} + \frac{k^2}{b^2} + \frac{l^2}{c^2}\right)^{-\frac{1}{2}}$$

Here, by the problem, $a = b = 2.5$Å, $c = 1.8$Å, $h = k = l = 1$.

By substituting these given values, we get

$$\boldsymbol{d_{111}} = \left[\frac{1^2}{(2.5)^2} + \frac{1^2}{(2.5)^2} + \frac{1^2}{(1.8)^2}\right]^{-\frac{1}{2}}$$

$$= \left(\frac{2}{6.25} + \frac{1}{3.24}\right)^{-\frac{1}{2}} = \mathbf{1.26\text{Å}}$$

► **Example 20.** *The distance* d_{100} *between* (100)*-plane in bcc-strcuture is* 0.232 nm. *What is the size of the unit cell? What is the radius of the atom?* **(WBUT)**

Solution: For a *bcc*-crystal, the radius of the atom is given by

$$r = \sqrt{3}a/4 \quad (a = \text{lattice const.})$$

For a *bcc*-structure, $\quad d_{100} = a/2 \Rightarrow 0.232\,\text{nm} = a/2$

$$\therefore \quad \text{Size of unit cell} = \boldsymbol{a} = 2 \times 0.232\,\text{nm} = \mathbf{0.464\,nm}$$

Radius of atom $= \boldsymbol{r} = \sqrt{3} \times 0.464/4\,\text{nm} = \mathbf{0.20\,nm}$

► **Example 21.** *Determine* (a) *the* c/a*-ratio and* (b) *the atomic packing factor of an hcp-structure.*

Solution: Let a be the edge, c the height of the unit cell of *hcp* structure and r the radius of an atom. The three body atoms lie in a horizontal plane at $c/2$ from the orthocentres of alternate equilateral triangles at the top or base of the hexagonal cell. They just rest on three atoms at the corners of the triangles.

(a) From ΔABY, $\cos 30° = AY/AB \Rightarrow AY = AB\cos 30° = a\sqrt{3}/2$

And, $AX = \frac{2}{3}AY = \frac{2}{3}\frac{a\sqrt{3}}{2} = a/\sqrt{3}$

Also, from ΔAXZ, $ZX^2 = AZ^2 - AX^2 \Rightarrow (c/2)^2 = a^2 - a^2/3$

So, $c^2/4 = 2a^2/3 \Rightarrow \boldsymbol{c/a = \sqrt{8/3}}$

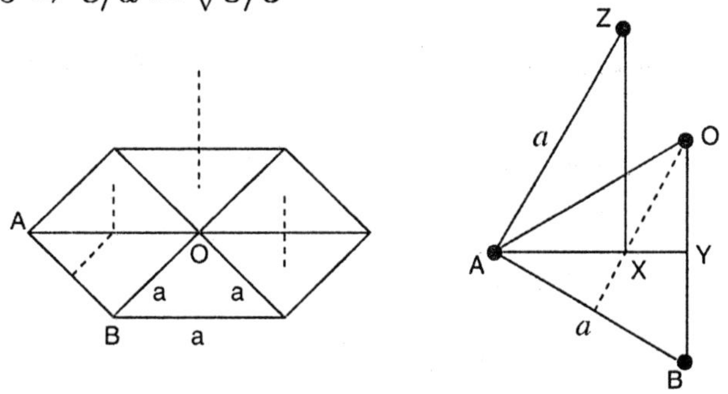

Fig. 1.24 Bottom layer of *hcp*-structure

(b) Area of base of unit cell $= 6 \times$ area of $\Delta AOB = 6 \times \frac{1}{2} BO \times AY$

$$= 3 \times a \times a\sqrt{3}/2 = \tfrac{3}{2}\sqrt{3}a^2$$

Volume of the unit cell, $=$ area of base $\times$ height

$$= \tfrac{3}{2}\sqrt{3}a^2 \cdot c$$

$$\therefore \text{ Atomic packing factor, } APF = \frac{\text{Volume of atoms per unit cell}}{\text{Volme of the unit cell}}$$

$$\therefore \quad \boldsymbol{APF} = \frac{6 \times \frac{4}{3}\pi r^3}{\frac{3\sqrt{3}}{2}a^2 c} = \frac{16\pi^3}{3\sqrt{3}a^2 c} = \frac{16\pi(a/2)^3}{3\sqrt{3}a^2 c} \quad (\because r = a/2)$$

$$= \frac{2\pi a^3}{3\sqrt{3}a^2 c} = \frac{2\pi a}{3\sqrt{3}c} = \frac{2\pi}{3\sqrt{3}}\sqrt{\frac{3}{8}} \quad (\because c/a = \sqrt{8}/3)$$

$$= \frac{\pi}{3\sqrt{2}} = \mathbf{0.74}$$

► **Example 22.** *Diamond crystal structure has the cube edge of* 3.56Å. *Calculate* (a) *the distance between the nearest neighhbours and* (b) *the number of atoms per unit volume.*

Solution: (a) Diamond structure is equivalent to two interpenetrating *fcc* lattices, displaced from each other along the body diagonal by one-fourth of diagonal-length.

$\therefore$ Distance between nearest neightbours, $\boldsymbol{d} = \frac{1}{4} \times$ length of body diagonal.

$$= \frac{\sqrt{3}a}{4} = \frac{\sqrt{3}}{4} \times 3.56 = \mathbf{1.54\ Å}$$

(b) For *fcc*-crystal, number of atoms per unit volume $= 4/a^3$.

For diamond structure, consisting of two interpenetrating *fcc* lattices, therefore, the number of atoms per unit volume,

$$\boldsymbol{n} = 2 \times 4/a^3 = 8/(3.56 \times 10^{-8})^3 \text{ cm}^3$$

$$= 1.77 \times 10^{23} \text{ cm}^{-3} = \mathbf{1.77 \times 10^{29}\ m^{-3}}$$

▶ **Example 23.** *A solid with fcc-structure has a lattice constant* 4.0 Å. *Each lattice point has an atom of mass* 4.0×10^{-26} kg. *Calculate* (i) *the* (*mass*) *density of the solid,* (ii) *the number of conduction electrons in* $1\,\text{m}^3$ *volume of the solid, if each of the solid contributes two electrons to its conduction band.* **(JAM)**

Solution: (i) A unit cell of *fcc* lattice has 4 atoms. So, mass of unit cell

$$m = 4 \times \text{ mass of 1 atom at lattice point}$$
$$= 4 \times 4 \times 10^{-26}$$
$$= 1.6 \times 10^{-25}\,\text{kg}$$

$$\text{Volume of unit cell, } V = a^3 = 64 \times 10^{-30}\,\text{m}^3$$

$\therefore$ Mass density of solid $= m/V = (1.6 \times 10^{-25})/(64 \times 10^{-30})\ \text{kg/m}^3$
$= \mathbf{2.5 \times 10^3\ kg/m^3}$

(ii) No. of atoms per unit volume $= \dfrac{\text{No. of atoms in unit cell}}{\text{Volume of unit cell}}$

$$= \frac{4}{64 \times 10^{-30}} = 6.25 \times 10^{28}\ \text{atoms/m}^3.$$

Each atom contributes 2-electrons to conduction band.

$\therefore$ No. of conduction electrons in $1\,\text{m}^3$ volume, $N = 2 \times 6.25 \times 10^{28} = \mathbf{1.25 \times 10^{29}}$

▶ **Example 24.** *An fcc-lattice is formed by atoms having radius r.* (i) *Obtain the expression, in terms of r, for the areal density of atoms* (atoms/m^2) *for the* (111) *plane.* (ii) *Obtain the relationship between r and the maximum radius R of a sphere that con just occupy the centre of the unit cell* (Fig. 1.25). **(JAM)**

Solution: **(i)** For *fcc*-lattice, $4r = \sqrt{2}a \Rightarrow a = 2\sqrt{2}r$.

Number of corner atoms covered by the plane $= 3 \times \frac{1}{6} = \frac{1}{2}$ atom.

Number of face-centred atoms contained by the plane $= \frac{1}{2} \times 3 = \frac{3}{2}$ atom.

$\therefore$ Total number of atoms contained by the plane $= \frac{1}{2} + \frac{3}{2} = 2$.

$$\text{Areal density (atoms/m}^2) = \frac{\text{Number of atoms}}{\text{Area of the plane}} = \frac{2}{\sqrt{3}a^2/2}$$
$$= \frac{2}{\sqrt{3}(8r^2)/2} = \mathbf{\frac{1}{2\sqrt{3}r^2}}$$

(ii) Let R be the maximum radius of the sphere. From Fig. 1.25, we have

$$OD = \frac{1}{3}AD = \frac{1}{3}\frac{\sqrt{3}}{2}\sqrt{2}a = \frac{1}{\sqrt{6}}a$$

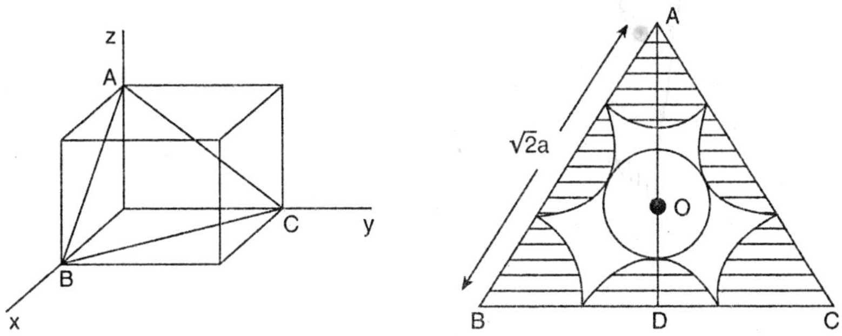

Fig. 1.25 Refers to Example 24

Also, we have
$$OD = R + r$$
$$\therefore \quad \frac{1}{\sqrt{6}}a = R + r \Rightarrow \frac{1}{\sqrt{6}}2\sqrt{2}r = R + r$$
$$\text{or,} \quad \left(\frac{2}{\sqrt{3}} - 1\right) r = R$$
$$\text{or,} \quad \boldsymbol{(2 - \sqrt{3})r = \sqrt{3}R}$$

which is the required relationship between r and R.

Review Questions

1. Define the following terms in relation to crystal structure: lattice, basis, unit cell, primitive unit cell and Wigner-Seitz cell.
2. Define the various symmetry operations in a crystal lattice. Which basic symmetry operation exists for cubic crystals?

   How many types of cubic crystals are known? Name them. **(Calcutta Hons.)**
3. Mention the number of lattice points per unit cell and the coordination number of (a) an *sc*-lattice, (b) an *fcc*-lattice, and (c) a *bcc*-lattice. Mention also some materials having *sc*, *fcc* and *bcc* lattice structure.
4. Define the term 'packing fraction'. What is the packing fraction for (a) a simple cubic lattice, (b) a *bcc*-lattice, (c) an *fcc*-lattice and (d) a diamond lattice.
5. Determine the relationships between the atomic radius r and the lattice parameter a for monatomic *sc*, *bcc* and *fcc* structures.
6. Define Miller indices of a plane in a crystal. What is the utility and the significance of Miller indices? **(Calcutta Hons.)**
7. Explain the concept of Miller indices. How are they calculated? How do they specify the orientation of a plane? **(Guru Nanak and Osmania Univ.)**

8. Draw the planes (100), (010), (001), (110), (101), (011) in a simple cubic cell. **(Meerut and Nagpur Univ.)**

9. Derive the expression for the interplanar spacing between two parallel planes with Miller indices (hkl) and show that for a simple cubic lattice of lattice constant a, it is given by the expression

$$d_{hkl} = \frac{a}{\sqrt{h^2 + k^2 + l^2}}$$ **(Bangalore Univ.)**

10. Show that for a crystal of cubic symmetry the direction $[hkl]$ is perpendicular to the plane (hkl). **(Guru Nanak and Himachal Univ.)**

11. What is a hexagonal close-packed structure? Name some of the materials having this structure.

    Describe the crystal structure of the following two materials: (a) KCl and (b) Diamond.

12. What is the difference between a crystal and an amorphous solid? What is a single crystal? Give example. **(Calcutta Hons.)**

13. Show that an ideal hexagonal close packed (hcp) structure has a c/a ratio of $(8/3)^{1/2} = 1.633$ and the density of atomic packing is equal to that of the face centred cubic structure.

14. Distinguish between the coordination number and the number of atoms per unit cell. Find both these numbers for (i) an *sc*-lattice, (ii) a *bcc*-lattice and (iii) an *fcc*-lattice.

15. Show that the angle θ in a cubic crystal between the planes $(h_1k_1l_1)$ and $(h_2k_2l_2)$ is given by

$$\theta = \cos^{-1}\frac{h_1h_2 + k_1k_2 + l_1l_2}{\sqrt{(h_1^2 + k_1^2 + l_1^2)(h_2^2 + k_2^2 + l_2^2)}}$$

16. Show analytically that only one, two, three, four and six-fold symmetry axes of rotation are possible in a crystal, and a five-fold rotation is not compatible with lattice translation symmetry.

17. Prove that the packing fraction for a simple cubic (sc) structure is 0.52; for body centred cubic (bcc) structure is 0.68 and for a face-centred cubic (fcc) structure is 0.74. **(Calicut Univ.)**

18. Show that for a cubic lattice, the lattice constant a is given by

$$a = \left[\frac{nM_A}{N_A\rho}\right]^{\frac{1}{3}}$$

    where n is the number of atoms per unit cell, M_A is the atomic weight, N_A is the Avogadro number and ρ the density of crystal material.

19. What is coordination number? Work out the coordination number and packing fraction for *bcc* and *fcc* structures.

20. (a) Final an expression for the density of atoms in a crystal plane.

(b) Show that for an orthorhombic crystal with three perpendicular axes of unequal lengths

$$d = \frac{1}{\sqrt{(h/a)^2 + (k+b)^2 + (l+c)^2}}$$

where the symbols have their usual meanings.

Short and Objective Questions

1. What is the difference between a crystalline and an amorphous solid? **(Gauhati and Calcutta Hons.)**

2. Define the single crystal, Bravais lattice and the unit cell.

3. Define crystal lattice, plane lattice, space lattice, basis, unit cell and translational vectors. **(Guru Nanak and Lucknow Univ.)**

4. Define a primitive cell. Distinguish between primitive and non-primitive unit cells. Can a unit cell be primitive? **(Punjabi and Kanpur Univ.)**

5. What do you mean by a symmetry operation? Explain briefly the various symmetry operations. What is a point operation? **(Himachal Univ.)**

6. Mention the number of lattice points per unit cell and the coordination number of (a) an *sc*, (b) an *fcc* and (c) a *bcc* lattice.

7. Name the seven crystal systems. How many types of cubic crystals are known?

8. What do you mean by (a) a simple cubic lattice, (b) a body-centred cubic lattice and (c) a face-centred cubic lattice?

9. Name some of the materials having the *sc*, *fcc* and *bcc* lattice structures.

10. In the following three figures (Fig. 1.26), give the Miller indices of the planes shown.

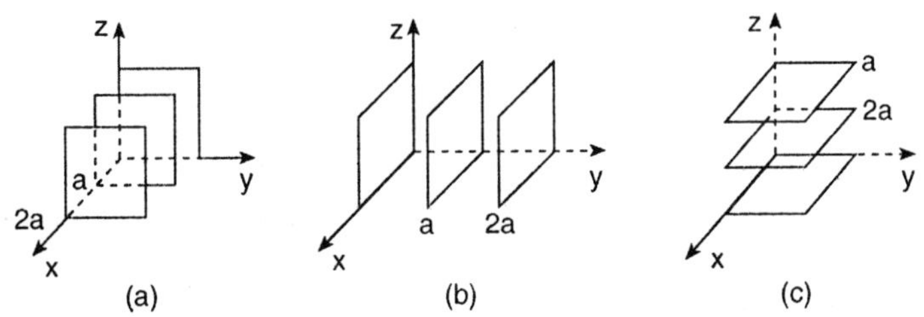

Fig. 1.26

[(100), (010), (001)]

11. What is the Miller indices of the line of intersection of (100) and $(\bar{1}\bar{1}1)$ planes in a cubic system? $[1\bar{1}0]$ or $[\bar{1}10]$

12. The cubic unit cell of a *bcc* lattice contains

(a) one lattice point, (b) 2, (c) 9, (d) none of these.

13. The number of lattice points in the cubic unit cell of an *fcc*-lattice is

(a) 4, (b) 14, (c) 6, (d) 8. **(Nagpur Univ.)**

14. Miller indices refer to

(a) two perpendicular planes in a crystal,
(b) a particular plane in a crystal,
(c) a set of parallel planes in a crystal,
(d) none of the above is true.

15. The primitive unit cell of a crystal lattice contains

(a) 8 lattice points, (b) no lattice point, (c) one lattice point.

16. A rectangular object has 6 faces and 8 corners. The number of edges are

(a) 6, (b) 8, (c) 12, (d) none.

17. In a cubic crystal, the $[hkl]$ direction and the (hkl) plane are

(a) perpendicular, (b) parallel,
(c) oriented at $\angle 45°$, (d) none of these.

18. NaCl crystal has

(a) *fcc*-structure, (b) *sc*-structure,
(c) *bcc*-structure, (d) *hcp*-structure.

(Himachal Univ.)

19. The atomic radius for *fcc*-lattice is

(a) $a/2$, (b) $\sqrt{3}a/4$, (c) $\sqrt{2}a/4$, (d) none of these.

(Himachal Univ.)

20. Packing fraction has the maximum value for the

(a) *fcc*-structure, (b) *sc*-structure, (c) *bcc*-structure. **(Himachal Univ.)**

21. The coordination number of a face-centred cubic structure is

(a) 12, (b) 8, (c) 6, (d) 4.

22. The pure rotational symmetry which is not possible in a crystalline solid is

(a) 5-fold, (b) 2-fold, (c) 4-fold, (d) 3-fold.

(GATE)

23. For a close-packed *bcc* structure of hard spheres, the lattice constant a is related to the radius R of the sphere as

(a) $a = 4R/\sqrt{3}$, (b) $a = 4R\sqrt{3}$, (c) $a = 4R\sqrt{2}$, (d) $a = 2R\sqrt{2}$.

(GATE)

24. If a and r be respectively the lattice constant and radius of an atom in a *bcc*-structure, then

(a) $r = \sqrt{3}a/4$, (b) $r = \sqrt{2}a/4$, (c) $r = a$, (d) None.

(WBUT)

25. The nearest neighbour distance in *bcc*-structure is

(a) $a\sqrt{3}/2$, (b) $a\sqrt{2}/\sqrt{3}$, (c) $2a/\sqrt{3}$, (d) $2a/\sqrt{2}$.

(WBUT)

26. The fraction of volume unoccupied in the unit cell of a *bcc*-lattice is

(a) $(8 - \sqrt{3}\pi)/8$, (b) $\sqrt{3}\pi/8$, (c) $(6 - \sqrt{2}\pi)/6$, (d) $\pi/3\sqrt{2}$.

(JAM)

27. The volume of the primitive unit cell of a *bcc*-structure with lattice constant a is

(a) a^3, (b) $a^{3/2}$, (c) $a^{3/4}$, (d) $a^{3/8}$.

28. The effective number of lattice points in a primitive cell is

(a) 1, (b) 2, (c) 4, (d) 8.

29. A metal of atomic weight W and density ρ has *fcc*-structure. The side a of its conventional cubic cell is given by

(a) $(W/\rho)^{1/3}$, (b) $(2W/\rho)^{1/3}$, (c) $(4W/\rho)^{1/3}$, (d) $(6W/\rho)^{1/3}$.

(JAM)

30. In a single layer of graphite, called *graphene*, the carbon atoms form a hexagonal lattice (Fig. 1.27). How many carbon atoms are there in a unit cell of the lattice?

(a) one, (b) two, (c) three, (d) six.

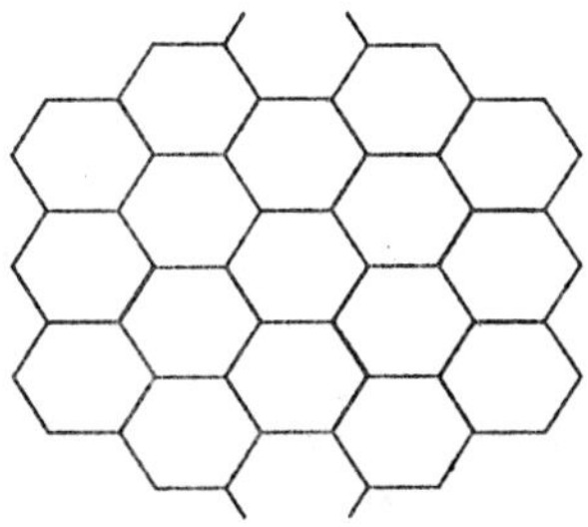

Fig. 1.27

Problems

1. How many atoms are there in *sc*, *bcc* and *fcc* unit cells? Calculate the packing fractions for these structures. [1, 2 and 4; 52%, 68% and 74%]
2. Determine the Miller indices of a plane that makes an intercept of 3 A, 4 A and 5 A on the coordinate axes of an orthorhombic crystal with $a : b : c = 1 : 2 : 5$. [(236)]
3. The diamond crystal structure has the cube edge of 3.56 A. Find the distance between the nearest neighbour and the number of atoms per c.c. [1.54 A, $1.77 \times 10^{23}/\text{cm}^3$]
4. Find the Miller indices of the plane containing the three lattice points $\vec{r}_1 = \vec{a} - \vec{b}$, $\vec{r}_2 = 2\vec{a} + \vec{c}$ and $\vec{r}_3 = 3\vec{b} + \vec{c}$, where $\vec{a}$, $\vec{b}$, $\vec{c}$ are the primitive basis vectors. [$(32\bar{5})$]
5. In NaCl crystal, the radius of Na^+ is 0.98 A and that of Cl^- is 1.82 A. The atomic masses of sodium and chlorine are 22.99 u and 35.45 u. Assuming that Na^+ ions and Cl^- ions touch along the cube edges of unit cell of NaCl, find the packing fraction and density of NaCl. [0.665, 2.21×10^3 kg/m^3]
6. Show that for a simple cubic lattice $d_{100} : d_{110} : d_{111} = \sqrt{6} : \sqrt{3} : \sqrt{2}$ where d_{hkl} is the separation between the adjacent (hkl) parallel planes.
   (Burdwan and Calcutta Hons.)
7. Show that the distances between the adjacent (100), (110) and (111) parallel planes in an *fcc* lattice of cube edge a are $a/2$, $a/2\sqrt{2}$ and $a/\sqrt{3}$.
8. The density of *bcc* iron is 7.9 g/cm^3 and its atomic weight is 56. Calculate the length of the side of the cubic unit cell and the nearest neighbour distance.
   (Calcutta Hons.) [2.87 A, 2.48 A]
9. Polonium has a cubic unit cell of side 3.42Å. If the atomic weight and density of Po are 210 and 8.72 g/cm^3 respectively, show if the unit cell is *sc*, *bcc* or *fcc*. [*sc*]
10. KBr crystal has a cubic structure. Its density is 2.75×10^3 kg/m^3 and its molecular weight is 119.01. Calculate its lattice constant. **(Calcutta Hons.)** [6.6 A]
11. Aluminium (atomic weight = 27) crystallizes in cubic form with $a = 4.05$ A. The density of Al is 2.7 g/cm^3. Find the type of the unit cell. [*fcc*]
12. Calculate the number of atoms per unit area in (100), (110) and (111) planes of a *bcc* crystal with lattice parameter 2.5Å.
    [16×10^{18}, 22.63×10^{18}, 9.24×10^{18} atoms/m^2]
13. The smallest distance between any two atoms in a *fcc* crystal is 0.2 nm. Find the size of the unit cube. Also, calculate the lattice spacing for (100), (110) and (111) planes. **(Calcutta Hons.)** [0.0226 nm^3, 0.1414 nm, 0.1 nm, 0.1633 nm]
14. In a tetragonal lattice $a = b = 25$ A, $c = 1.8$ A. Deduce the lattice spacing between the (111) planes. [1.26 A]

15. If $|\vec{a}| = 4.50\,\text{A}$, $|\vec{b}| = 3.50\,\text{A}$ and $|\vec{c}| = 2.50\,\text{A}$, calculate the lattice spacing for (110) planes in an orthorhombic lattice. [0.36 A]

16. The cube edge of a simple cubic structure is 5 A. Calculate the number of atoms per m^2 in the (110) plane. **(Burdwan Hons.)** [$2.828 \times 10^{18}/\text{m}^2$]

17. Assuming the molecular weight of the KCl crystal as 74.45 and the density $1.98\,\text{g/cm}^3$, determine the following:

(a) the distance between the adjacent atoms (lattice constant).

(b) the distance between the atoms of the same kind.

(c) the spacing between (100) planes, (110) planes and (111) planes.

(d) the distance of the first and the second nearest neighbour.

[3.15 A; 6.30 A; 3.15 A, 2.23 A, 1.82 A; 1.575 A; 3.15 A]

18. The lattice parameters of a crystal are in the ratio $a : b : c = 0.52 : 1.00 : 0.48$. A plane having the Miller indices (121) cuts an intercept of 2.6 A along the x-axis. Compute the lengths of the intercepts along y and z-axes. [2.5 A, 2.4 A]

19. KF has NaCl structure. The atomic mass of K = 39 and that of F = 19. If its density is $2480\,\text{kg/m}^3$, find the distance between K^+ and F^- ions in KF. [2.69 A]

20. Find the angle between [111] and [100] directions in a cubic lattice. [$54°44'$]

21. Calculate the atomic density in (100) plane of copper (*fcc*) with the lattice parameter of 3.61Å. [1.53×10^{19} atoms/m^2]

22. Chromium is a *bcc* Bravais lattice having the edge length of 2.88 A for the cubic cell. Find the number of chromium atoms per unit volume of the crystal.

[$8.37 \times 10^{22}/\text{cm}^3$]

23. Show that, if a is the edge of a unit cell cube of an fcc lattice, the concentration of lattice points on (111) plane is $4/\sqrt{3}a^2$, that on a (110) plane is $\sqrt{2}/a^2$ and that on a (100) plane is $2/a^2$.

24. Write the indices of the different planes belonging to the form (110) in a cubic system where x, y, z directions are four-fold axes. **(Calcutta Hons.)**

[(110), $(\bar{1}10)$, $(\bar{1}\bar{1}0)$, $(1\bar{1}0)$, (011), $(0\bar{1}1)$, $(0\bar{1}\bar{1})$, $(01\bar{1})$, (101), $(\bar{1}01)$, $(\bar{1}0\bar{1})$, $(10\bar{1})$]

25. Metallic copper forms cubic crystals and the lattice is about 0.36 nm. If the specific gravity of coper is 8.96 and its atomic weight is 63.5, then show that the crystal is face-centred cubic type. **(TIFR)**

[**Hints.** Follow the Illustated Ex. 8 to get $n \simeq 4$.]

26. The density of *bcc* iron is $7900\,\text{kg/m}^3$ and its at. wt. is 56. Calculate the side of the unit cell and the interatomic spacing. **(Burdwan Hons.)** [2.86 A, 2.48 A]

27. Show that the angle between the planes $(10\bar{1})$ and (111) in a cubic crystal is $\pi/2$.

28. In a triclinic crystal, a lattice point makes intercepts of lengths $a, 2b$ and $-(3c/2)$. Find the Miller indices of the plane. $[(63\bar{4})]$

29. For a cubic crystal of Cu it is given that $a = 3.60$A, atomic weight of Cu $= 63.6$, density of Cu $= 8960\,\text{kg/cm}^3$ and the Avogadro number $N_A = 6.023 \times 10^{26}$. Show that the lattice of Cu is body centred cubic.

30. Aluminimum has *fcc*-structure. Its density is $2.7 \times 10^3\,\text{kg/m}^3$. Determine the unit cell dimensions and the atomic diameter. [$a = 4.05$ A, 2.86 A]

31. An orthorhombic crystal has primitive translations $a = 1.21$Å, $b = 1.84$Å, and $c = 1.97$Å. If a plane with Miller indices $(23\bar{1})$ cuts an intercept of 1.21Å along the x-axis, find the lengths of intercepts along y and z-axes. [1.23Å, −3.94Å]

32. The distance between consecutive (111) planes in a cubic crystal is 2Å. Find the lattice parameter. [3.46Å]

33. In a tetragonal crystal, the lattice parameters are $a = b = 2.42$Å and $c = 1.74$Å. Determine the interplanar spacing between consecutive (101) planes. [1.426Å]

34. Titanium is known to have *hcp*-structure. Its molecular weight is 47.90 and density $4.51\,\text{g/cm}^3$. If the c/a ratio is 1.58, determine the values of c and a.

35. Fe forms a *bcc* crystal with $a = 2.86$ A. The mass of Fe-atom is 9.27×10^{-26} kg. Determine the mass density of Fe-crystal. [$7.93 \times 10^3\,\text{kg/m}^3$]

[**Hints.** The number atoms per unit cell for *bcc* crystal $= 2$. Volme of the primitive cell $= a^3$. So, the number of atoms per unit volume is

$$n = \frac{2}{a^3} = \frac{2}{(2.86 \times 10^{-10})^3} = 8.55 \times 10^{28}\ \text{atom/m}^3$$

$\therefore$ Mass density of Fe,

$$\rho = n \times \text{mass of Fe-atom} = (8.55 \times 10^{28}) \times (9.27 \times 10^{-26})\ \text{kg/m}^3 = \mathbf{7.93 \times 10^3\ kg/m^3}$$

36. Calculate the areal density ρ of lattice points for (100) planes in a simple cubic lattice having lattice constant of 2.5 A. [$1.6 \times 10^{19}\,\text{m}^{-2}$]

37. A box of volume $1.0\,\text{m}^3$ is filled with balls of diameter 1.0 mm. If the balls are packed in an *sc*-lattice, how many can we get into the box? If the process is repeated for *bcc* and *fcc* lattice, what would be the results?

[1.0×10^9, 1.3×10^9, 1.41×10^9]

38. Show that in simple cubic lattice $d_{100} : d_{110} : d_{111} = \sqrt{6} : \sqrt{3} : \sqrt{2} = 1 : 0.71 : 0.8$ where d_{hkl} is the separation of the adjacent (hkl) parallel planes.

(Calcutta and Vidyasagar Hons.)

39. Consider a *bcc* crystal with lattice constant a. Determine (i) Miller indices for the (100) plane and (ii) the number of atoms per unit area in the (111) plane.

(JAM) [(100), $\sqrt{3}/a^2$]

2

Analysis of Crystals and Reciprocal lattice

2.1 Introduction

Analysis of crystals is essentially the X-ray analysis of their structures using the phenomenon of diffraction of X-rays by them. But why only X-rays and not other radiations like visible, UV or γ-rays? We seek for answer to it.

The interatomic spacing in crystals is of the order of 1 A. To explore the structure of crystals, we require wavelengths of the incident waves comparable with the interatomic spacing of the crystal. X-radiations produced by acceleration of electrons through a potential difference V have a wavelength λ given by, with usual notation,

$$eV = h\nu = hc/\lambda$$

$$\Rightarrow \boxed{\lambda = \frac{\boldsymbol{hc}}{\boldsymbol{eV}}} \tag{2.1.1}$$

For $V = 10\,\text{kV}$, we obtain $\lambda = \dfrac{6.62 \times 10^{-34} \times 3 \times 10^{8}}{1.6 \times 10^{-19} \times 10^{4}}\ \text{m}$

$$= 1.24 \times 10^{-10}\,\text{m} = \mathbf{1.24\ A}$$

This is comparable to the interatomic spacing d of an actual crystal and X-rays can thus be used for crystal-diffraction works.

• Visible or UV-radiations have much larger wavelengths ($\lambda \sim 5000\,\text{A}$) and *cannot* therefore be used for exploring the structures of the crystals. The γ-radiations have however wavelengths somewhat shorter than X-rays and are thus diffracted through small angles that cannot be conveniently determined. Hence γ-rays are also not used to study the crystal structure.

● It is only X-rays that are scattered *elastically* without any change of wavelength by the charged particles of the atoms. Usually, X-radiations obtained by voltages 10–50 kV are regarded as satisfactory for *crystal diffraction*, i.e., wavelengths produced by voltages of this range show observable diffraction effects with reasonable values of the angle of diffraction θ and d. Wavelength of X-rays can also be determined very accurately. X-ray diffraction, in fact, is the *cheapest, most convenient* and *most widely used* method of crystals analysis.

2.2 Bragg's diffraction law

W.L. Bragg in 1913 developed a relation for the condition of diffraction, based on a simple model. He pointed out that the lattice planes in a crystal behave to X-rays as lines in a diffraction grating, and incident monochromatic X-rays are reflected specularly from these planes. Strong diffracted beams are obtained when the reflection from parallel planes interfere constructively. Scattering of X-rays by the atoms is perfectly elastic. So, the wavelength of X-rays does not change on reflection. On this basis, Bragg derived a relation as a condition for reflection. This condition is called ***Bragg's law***..

Derivation—Let a parallel beam of monochromatic X-rays of wavelength λ penetrate a stack of planes of interplanar spacing d in a crystal at a glancing angle of incidence θ (called the *Bragg angle*), such that the rays lie on the plane of the paper.

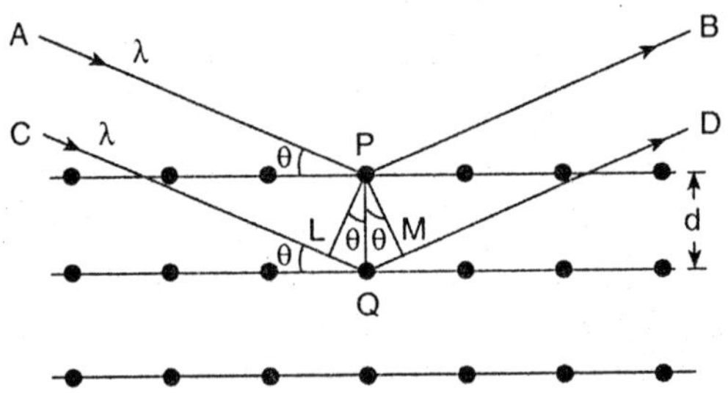

Fig. 2.1 Bragg's diffraction law

Consider two such rays AP and CQ that strike the first two planes and get scattered (reflected) at the same angle θ as rays PB and QD (Fig. 2.1). The diffraction results from the in-phase or constructive interference of these reflected rays.

Draw $PL \perp CQ$ and $PM \perp QD$. The path difference between the two rays is given by

$$\Delta = LQ + QM = d\sin\theta + d\sin\theta = 2d\sin\theta$$

If this path difference is an *integral* multiple of λ, constructive interference will occur. Thus, the intensity will be *maximum* if the following condition

$$\boxed{2d\sin\theta = n\lambda} \quad n = 1, 2, 3\cdots, \tag{2.2.1}$$

gets satisfied. n being the order of diffraction.

This boxed equation (2.2.1) is the well-known ***Bragg's diffraction law.***

For $n = 1$, we get the ***first order*** diffraction, for $n = 2$, the ***second order*** and so on.

Discussion—The law states the condition in which X-ray diffraction in terms of in-phase reflections from successive parallel planes having the spacing d occurs for a monochromatic beam of wavelength λ incident at a glancing angle θ. For angles other than θ, no diffraction of the beam takes place. Atomic planes do not act like mirrors.

For a given d and λ, the higher order reflections occur for larger θ-values. With *increasing* n or θ, the intensity of the reflected beam *decreases* and conversely. Since $\sin\theta$ *cannot* exceed the value unity, i.e., $\sin\theta \leq 1$, the wavelength λ should be $\leq 2d$, for Bragg reflection to occur, i.e., λ must not be greater than twice the *interplanar spacing*. Otherwise, no diffraction will occur.

• Bragg's law reflects the periodic nature of the space lattice. It is not indicative of either the orientation or the basis of the space lattice.

• We have *disregarded the refraction* of X-rays at the surface of the crystal. The *correction term* for such refraction is usually quite small to affect the result and becomes progressively smaller at higher orders. Nevertheless, we derive hereunder the corrected Bragg's equation, taking refraction into account.

2.2.1 Correction in Bragg's law due to refraction

Let a parallel X-ray beam of wavelength λ be incident on the crystal surface (Fig. 2.2) at a glancing angle θ. The glancing angle of refraction θ' is slightly less than θ, for the refractive index μ (for all substances) is slightly less than 1 for X-rays. Let λ' be the wavelength of the refracted ray.

$$\therefore \quad \mu = \frac{\lambda}{\lambda'} = \frac{\cos\theta}{\cos\theta'} \tag{2.2.2}$$

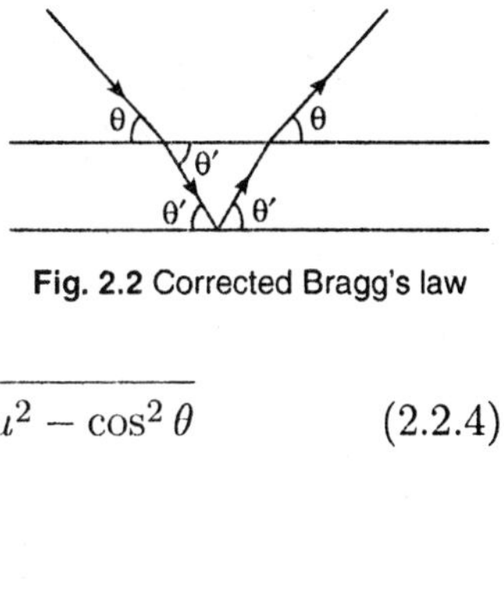

Fig. 2.2 Corrected Bragg's law

Bragg's equation inside the crystal is given by

$$2d\sin\theta' = n\lambda' \tag{2.2.3}$$

But,

$$\sin\theta' = \sqrt{1-\cos^2\theta'} = \sqrt{1-\cos^2\theta/\mu^2} = \frac{1}{\mu}\sqrt{\mu^2-\cos^2\theta} \tag{2.2.4}$$

Using the equations (2.2.2) to (2.2.4), we obtain

$$n\lambda' = \frac{n\lambda}{\mu} = \frac{2d}{\mu}\sqrt{\mu^2-\cos^2\theta}$$

$$\Rightarrow \quad n\lambda = 2d\sqrt{\mu^2-\cos^2\theta} = 2d\sin\theta\sqrt{1-\frac{1-\mu^2}{\sin^2\theta}} \tag{2.2.5}$$

But, $1-\mu^2 = (1+\mu)(1-\mu) = 2(1-\mu)$, since $\mu \simeq 1$. We have, therefore,

$$n\lambda = 2d\sin\theta\sqrt{1-\frac{2(1-\mu)}{\sin^2\theta}} \simeq 2d\sin\theta\left(1-\frac{1-\mu}{\sin^2\theta}\right) \tag{2.2.6}$$

Now, $(1-\mu) \ll 1$, so that $\sin^2\theta$ in the parenthesis can also be taken as $n^2\lambda^2/4d^2$ as it is in the correction term. Finally, therefore,

$$n\lambda = 2d\sin\theta\left[1 - \frac{4d^2(1-\mu)}{n^2\lambda^2}\right] \tag{2.2.7}$$

which is the ***corrected*** or the ***modified Bragg's equation.***

2.3 Laue condition of diffraction : Laue equations

von Laue examined the phenomenon of diffraction in a more general way by assuming that X-rays get scattered from individual atoms or ions in the crystal. The scattered rays subsequently reunite to give the directions of diffraction maxima. The maxima appear, as we shall see, in some specific directions that obey certain conditions. These *conditions* are known as the ***Laue equations*** that confirm Bragg's treatment and, in fact, the Bragg's law can be derived from the Laue's equations.

Laue equations—Let A, B be two identical scattering centres (Fig. 2.3) separated by $\vec{r}$ in a crystal and a parallel beam of X-rays be incident on A, B along the direction

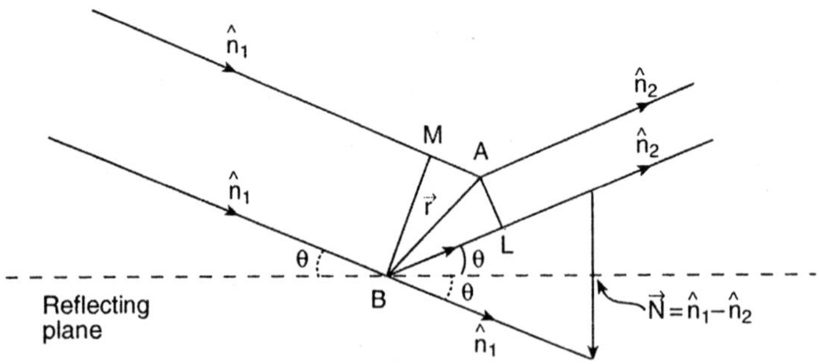

Fig. 2.3 Laue equations

of unit vector $\hat{n}_1$. The scattered beam is also parallel in any arbitrary direction given by unit vector $\hat{n}_2$. Draw BM perpendicular to the incident beam and AL perpendicular to the scattered beam.

$\therefore$ Path difference between the rays scattered from A and B is

$$\Delta = AM - BL = \vec{r}\cdot\hat{n}_1 - \vec{r}\cdot\hat{n}_2 = \vec{r}\cdot(\hat{n}_1 - \hat{n}_2) = \vec{r}\cdot\vec{N} \tag{2.3.1}$$

where $\vec{N} = \hat{n}_1 - \hat{n}_2$ is a vector directed normal to the reflected plane.

If θ be the angle the incident beam makes with the reflecting planes, then the scattered (reflected) beam also makes an angle θ with it. So, the angle between $\hat{n}_1$ and $\hat{n}_2$ is plainly 2θ.

$$\therefore \quad |\vec{N}| = 2\sin\theta \tag{2.3.2}$$

The phase difference between the waves scattered from A and B is

$$\phi = \frac{2\pi}{\lambda}(\vec{r}\cdot\vec{N}) \tag{2.3.3}$$

The radiations from A, B will constructively interfere only if ϕ be an integral multiple of 2π.

$$\therefore \quad \phi = 2\pi n, \quad \text{where } n = \text{an integer.}$$

Due to the periodic arrangement of atoms, other nearest neighbour atoms in the same direction will also scatter radiations exactly in phase with those scattered from A, B. So, $\vec{r}$ may be replaced by $\vec{a}, \vec{b}, \vec{c}$ in (2.3.3) since the nearest neighbour are separated by distanes a, b and c.

$\therefore$ *Diffraction maxima* (i.e., constructive interference) dictate the simultaneous satisfaction of the following three conditions:

$$\left.\begin{aligned} \frac{2\pi}{\lambda}(\vec{a}\cdot\vec{N}) &= 2\pi h' = 2\pi nh \\ \frac{2\pi}{\lambda}(\vec{b}\cdot\vec{N}) &= 2\pi k' = 2\pi nk \\ \frac{2\pi}{\lambda}(\vec{c}\cdot\vec{N}) &= 2\pi l' = 2\pi nl \end{aligned}\right\} \tag{2.3.4}$$

where h', k' and l' are three integers.

Now, the atoms A, B were assumed to be nearest neighbours and so $\vec{a}, \vec{b}$ and $\vec{c}$ are the primitive lattice vectors. Also h', k', l' differ from h, k, l only by a common factor $n \geq 1$. Thus h, k, l can have only unity as a common factor and thus resemble the Miller indices of a plane—the reflecting plane in the sense of Bragg's treatment.

Finally, therefore, we obtain

$$\left.\begin{aligned} \vec{a}\cdot\vec{N} &= aN\cos\alpha = 2a\sin\theta\cos\alpha = h\lambda \\ \vec{b}\cdot\vec{N} &= bN\cos\beta = 2b\sin\theta\cos\beta = k\lambda \\ \vec{c}\cdot\vec{N} &= cN\cos\gamma = 2c\sin\theta\cos\gamma = l\lambda \end{aligned}\right\} \tag{2.3.5}$$

where α, β and γ are the angles between the normal to the reflecting plane and $\vec{a}, \vec{b}$ and $\vec{c}$ respectively.

The set of three equations in (2.3.5) are known as ***Laue equations*** for X-ray diffraction, representing the conditions for the diffraction to occur. The equations serve to determine the unique value of θ, thus defining a scattering direction. All the three equations must be satisfied *simultâneously* to get the most intense beam.

2.4 Bragg's law from Laue equations

From Laue equations, for a fixed θ-value, the direction cosines $\cos\alpha, \cos\beta$ and $\cos\gamma$ of the scattering normal are proportional respectively to

$$\frac{h}{a}, \frac{k}{b} \text{ and } \frac{l}{c}.$$

Also, we learnt, the direction cosines of the normal to any arbitrary plane (hkl) are proportional to

$$\frac{h}{a}, \frac{k}{b} \text{ and } \frac{l}{c}$$

It thus follows that the scattering normal $\vec{N}$ is identical with the normal to the plane (hkl). Hence, the arbitrary plane (hkl) is the *reflecting plane* of Bragg.

If d be the interplanar spacing between two adjacent planes of a set (hkl), then we shall have

$$d = \frac{a}{h}\cos\alpha = \frac{b}{k}\cos\beta = \frac{c}{l}\cos\gamma$$

In combination with equation (2.3.5), therefore, we obtain that

$$\boxed{2d\sin\theta = n\lambda}$$

which is the well-known *Bragg's law* or *Bragg's equation.* Bragg's law is thus derived from Laue equations. This incidentally shows the *equivalence* of Bragg and Laue equations.

2.5 Reciprocal lattice

For a deeper understanding of X-ray diffraction by crystals, the concept of *reciprocal lattice* was first introduced by P. P. Ewald. Bragg analysed X-ray diffraction in terms of orientations or *slopes* of different sets of lattice planes that makes the analysis rather tedious since it becomes difficult to visualise all such planes due to their two-dimensional character. Ewald showed that it would be easier if one considers normal to various sets of planes instead of the planes themselves and that led to the development of the concept of ***reciprocal lattice.***

But ***how to obtain a reciprocal lattice*** to a given space lattice? The following procedure will enlighten:

1. *A normal is drawn to each set of parallel planes of the direct lattice from a common arbitrary origin.*

2. *The length of each normal is made equal to (or 2π times) the reciprocal of interplanar spacing d_{hkl} for the corresponding set of planes.*

3. *The point at the end of each such normal represents the crystal plane.*

The collection of points thus obtained corresponds to different lattice planes and forms an array which is termed the ***reciprocal lattice.***

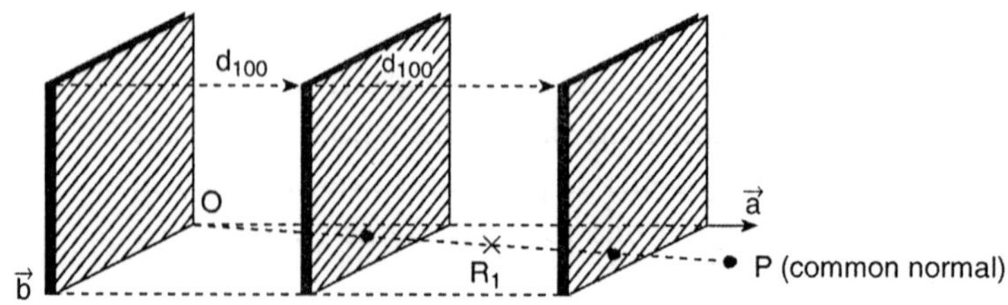

Fig. 2.4 A set of (1 0 0) planes

To explain, consider a set of parallel equidistant planes (100) as shown in Fig. 2.4. Draw a common normal OP to the planes from an origin O, any arbitrary lattice point and locate a point R_1 on OP at a distance inversely proportional to the interplanar spacing d_{100} i.e.,

$$OR_1 = \frac{C}{d_{100}}; \quad C = \text{constant of proportionality.}$$

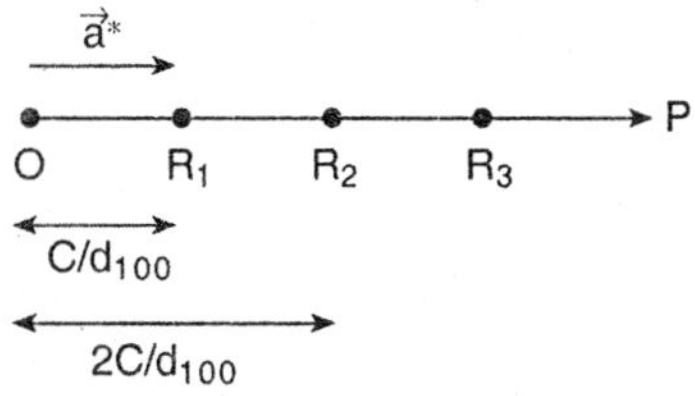

Fig. 2.5 Representation of set of lattice planes by equivalent points R_1, R_2, R_3 ... etc.

The entire set of planes is represented just by the single point R_1. Similarly, one obtains a reciprocal point for the set (200) planes. The (200) planes are parallel to (100) planes, but the spacing is $d_{100}/2$ so that the reciprocal point R_2 is along OP at a distance $C/d_{200} = 2C/d_{100}$ (Fig. 2.5).

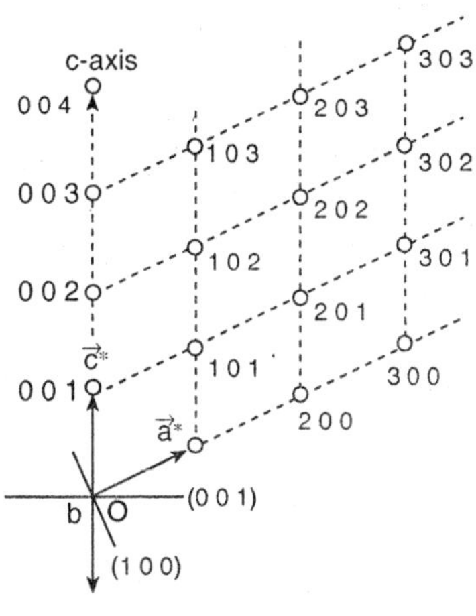

Fig. 2.6 Reciprocal lattice

The unit vectors in reciprocal lattice are denoted by $\vec{a}^*, \vec{b}^*, \vec{c}^*$ and the lattice points are denoted by

$$|\vec{a}^*| = |\overrightarrow{OR_1}| = \frac{C}{d_{100}}. \tag{2.5.1}$$

Thus, we have the a^* axis when we consider all planes $\|$ to the plane of the crystal lattice. Similarly, we shall obtain

$$|\vec{b}^*| = \frac{C}{d_{010}} \text{ (normal to } ac\text{-plane)} \tag{2.5.2}$$

$$|\vec{c}^*| = \frac{C}{d_{001}} \text{ (normal to } ab\text{-plane)} \tag{2.5.3}$$

The reciprocal points R_1, R_2 etc. may be denoted by 100, 200 etc. as they denote the sets of (100), (200) etc. planes of the direct lattice. The array of points $R_1, R_2, \cdots$ defined by unit translations $|\vec{a}^*|, |\vec{b}^*|, |\vec{c}^*|$ forms the reciprocal lattice. Fig. 2.6 gives the reciprocal lattice for a monoclinic crystal.

b is the unique axis and a and c are the unit cell edges of the direct lattice. The array of points such as $100, 101, 102, \cdots, 200, 201, \cdots$, etc. give the reciprocal lattice for the monoclinic crystal. The points are obtained thus: O is a common origin whence normals are errected to each of lattice planes (hkl), etc. and points are placed on each such normals at distances $1/d_{hkl}$.

● Each reciprocal lattice point preserves all the important features of the stack of planes it symbolizes—the direction and the distance of the point from the origin O preserve respectively the orientation of the plane and the interplanar spacing.

The vector distance of reciprocal lattice point $(h00)$ is given by

$$\vec{r}^*_{h00} = h\vec{a}^*, \quad h = 1, 2, 3, \cdots$$

Similarly, we have

$$\vec{r}^*_{0k0} = k\vec{b}^* \quad k = 1, 2, 3, \cdots$$

$$\vec{r}^*_{00l} = l\vec{c}^*, \quad l = 1, 2, 3 \cdots,$$

In general, therefore, we shall have

$$\vec{r}^*_{hkl} = h\vec{a}^* + k\vec{b}^* + l\vec{c}^* \tag{2.5.4}$$

where $\vec{a}^*, \vec{b}^*$ and $\vec{c}^*$ are basic unit translation vectors in the reciprocal lattice.

For $\vec{a}^*$ to be normal to $\vec{b}$ and $\vec{c}$, $\vec{b}^*$ to be normal to $\vec{c}$ and $\vec{a}$, and $\vec{c}^*$ to be normal to $\vec{a}$ and $\vec{b}$, the following relations (*normalization condition*) should hold.

$$\left.\begin{aligned} \vec{a}^*\cdot\vec{a} &= \vec{b}^*\cdot\vec{b} = \vec{c}^*\cdot\vec{c} = 2\pi \\ \text{Also,} \quad \vec{a}^*\cdot\vec{b} &= \vec{b}^*\cdot\vec{c} = \vec{c}^*\cdot\vec{a} = 0 \\ \vec{a}^*\cdot\vec{c} &= \vec{c}^*\cdot\vec{b} = \vec{b}^*\cdot\vec{a} = 0 \end{aligned}\right\} \tag{2.5.5}$$

Expressions of reciprocal unit vectors $\vec{a}^*, \vec{b}^*, \vec{c}^*$ in terms of $\vec{a}, \vec{b}, \vec{c}$ of the direct lattice—Let a unit cell be defined in direct lattice by $\vec{a}, \vec{b}, \vec{c}$ (Fig. 2.7)

$\therefore$ Volume of unit cell, V = (area of base) $\times$ (height) = $A\cdot OP$

But, the area of base, $A = b \times c$ so that $V = \vec{a}\cdot(\vec{b}\times\vec{c})$

Now, height OP = spacing of bc-planes = d_{100}

$$\therefore \quad V = A\cdot d_{100} \Rightarrow \frac{1}{d_{100}} = \frac{A}{V}$$

$$\text{or,} \quad \frac{\hat{n}}{d_{100}} = \frac{b\times c}{a\cdot(b\times c)} \tag{2.5.6}$$

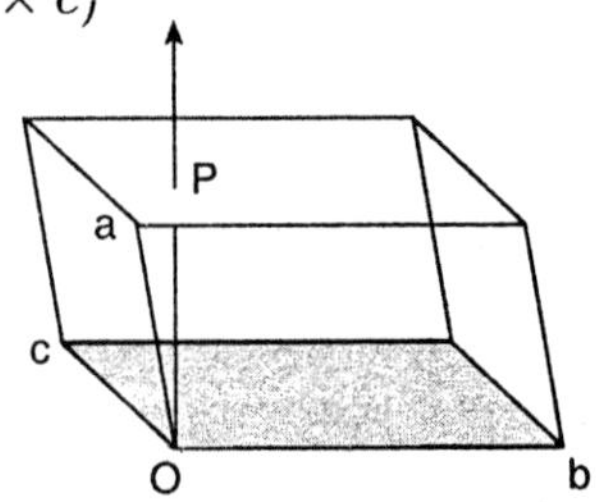

Fig. 2.7 Calculation of volume of unit cell

where $\vec{n}$ = unit vector-normal to the shaded area in Fig. 2.7.

From the definition, $|\vec{a}^*| = |\vec{d}^*_{100}| = \dfrac{C}{d_{100}}$. We therefore have,

$$\therefore \quad \vec{d}^*_{100} = \vec{a}^* = C\frac{\hat{n}}{d_{100}} = C\frac{\vec{b}\times\vec{c}}{\vec{a}\cdot(\vec{b}\times\vec{c})} \tag{2.5.7}$$

Similarly, $$\vec{d}^*_{010} = \vec{b}^* = C\frac{\vec{c}\times\vec{a}}{a\cdot(b\times c)}; \quad \vec{d}^*_{001} = \vec{c}^* = C\frac{\vec{a}\times\vec{b}}{\vec{a}\cdot(\vec{b}\times\vec{c})} \tag{2.5.8}$$

Applying the normalization conditions (2.5.5), we get $C = 2\pi$.

$$\left.\begin{aligned} \therefore\quad \vec{a}^* &= 2\pi\frac{\vec{b}\times\vec{c}}{\vec{a}\cdot(\vec{b}\times\vec{c})} \\ \vec{b}^* &= 2\pi\frac{\vec{c}\times\vec{a}}{\vec{a}\cdot(\vec{b}\times\vec{c})} \\ \vec{c}^* &= 2\pi\frac{\vec{a}\times\vec{b}}{\vec{a}\cdot(\vec{b}\times\vec{c})} \end{aligned}\right\} \tag{2.5.9}$$

which are the required expressions for *reciprocal unit vectors.*

● Note that the denominator in each equation of the set (2.5.9), viz $\vec{a}\cdot(\vec{b}\times\vec{c})$ is the volume of the unit cell in direct lattice.

● Some authors prefer to use capital $\vec{A}, \vec{B}$ and $\vec{C}$, instead of $\vec{a}^*, \vec{b}^*$ and $\vec{c}^*$, for the reciprocal unit vectors.

Properties of reciprocal lattice—The following important properties of reciprocal lattice are worth-remembering.

1. Every vector of the reciprocal lattice is normal to a set of lattice planes of the direct lattice.
2. The magnitude of every reciprocal vector is inversely proportional to the interplanar spacing of the corresponding set of lattice planes.
3. The volume of a unit cell in reciprocal lattice is inversely proportional to the volume of a unit cell of the crystal lattice.
4. The direct lattice is the reciprocal lattice to its own reciprocal lattice.
5. The simple cubic lattice is self-reciprocal, that is, the reciprocal lattice to *sc*-lattice is also simple cubic.
6. The *bcc* and the *fcc*-lattices are reciprocal to each other, that is, the reciprocal lattice to a *bcc* lattice is an *fcc* lattice and conversely.

We shall now prove the above properties one after the other.

Property 1. ***Every reciprocal lattice vector is perpendicular to a set of direct lattice planes.***

Let the reciprocal lattice vector be given by

$$\vec{G} = h'\vec{A} + k'\vec{B} + l'\vec{C}$$

and (hkl) the set of planes of the direct lattice.

$$\therefore\quad h'/h = k'/k = l'/l = n, \text{ say}$$

n being the highest common factor.

Now the intercepts on X-axis, Y-axis and Z-axis are respectively

$$\overrightarrow{OA} = \vec{a}/h, \quad \overrightarrow{OB} = \vec{b}/k \text{ and } OC = \vec{c}/l$$

as shown in Fig. 2.8.

But, $\overrightarrow{AC} = \overrightarrow{OA} - \overrightarrow{OC} = \vec{a}/h - \vec{c}/l$; $\overrightarrow{AB} = \overrightarrow{OA} - \overrightarrow{OB} = \vec{a}/h - \vec{b}/k$ where both $\overrightarrow{AC}$ and $\overrightarrow{AB}$ lie in (hkl) plane.

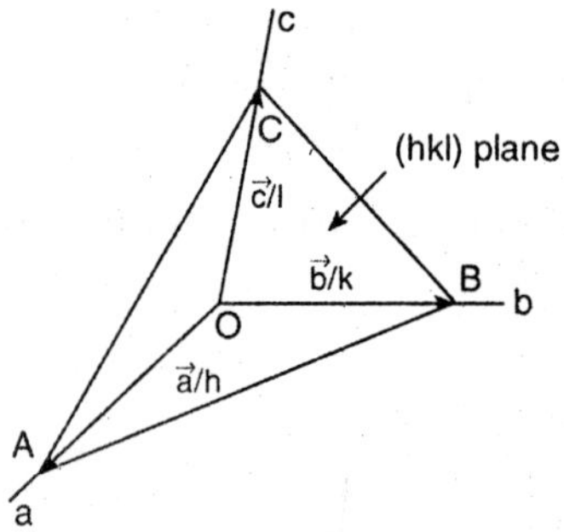

Fig. 2.8 An (hkl) plane in the direct lattice

Now, $\quad \vec{G}\cdot\overrightarrow{AC} = (h'\vec{A} + k'\vec{B} + l'\vec{C})\cdot(\vec{a}/h - \vec{c}/l)$

$$= \frac{h'}{h}\vec{A}\cdot\vec{a} - \frac{l'}{l}\vec{C}\cdot\vec{c} \qquad (\because\ \vec{A}\cdot\vec{c} = \vec{B}\cdot\vec{a} = \text{etc.} = 0)$$

$$= (h'/h - l'/l)\cdot 2\pi \qquad (\because\ \vec{A}\cdot\vec{a} = \vec{C}\cdot\vec{c} = 2\pi)$$

$$= \mathbf{0}. \qquad (\because\ h'/h = l'/l = n)$$

Similarly, we shall get $\vec{G}\cdot\overrightarrow{AB} = \mathbf{0}$.

Vector $\vec{G}$ is thus normal to two linearly independent vectors $\overrightarrow{AC}$ and $\overrightarrow{AB}$ which lie in (hkl) plane This implies that *every reciprocal lattice vector $\vec{G}$ is normal to a set of direct lattice plane (hkl).*

Property 2. ***The magnitude of every reciprocal lattice vector is inversely proportional to the spacing of the corresponding set of lattice planes.***

Let the reciprocal lattice vector be $\vec{G} = h'\vec{A} + k'\vec{B} + l'\vec{C}$ and $\hat{n}$ be the unit vector normal to (hkl) plane of direct lattice.

Since, the reciprocal lattice is normal to direct lattice

$$\hat{n} = \vec{G}/|\vec{G}| = \vec{G}/G \tag{2.5.10}$$

If $\vec{a}, \vec{b}, \vec{c}$ be the primitive translation vectors, then the spacing d between the adjacent (hkl) planes is given by

$$d = \frac{\vec{a}\cdot\hat{n}}{h} = \frac{\vec{a}\cdot\vec{G}}{hG}, \quad \text{using (2.5.10)}$$

$$= \frac{\vec{a}\cdot(h'\vec{A} + k'\vec{B} + l'\vec{C})}{hG}$$

$$= \frac{2\pi}{G}\cdot\frac{h'}{h} \quad (\because\ \vec{a}\cdot\vec{A} = 2\pi;\ \vec{a}\cdot\vec{B} = \vec{a}\cdot\vec{C} = 0)$$

$$= \frac{2\pi n}{G}, \quad n = \text{an integer.} \quad (\because\ h'/h = n) \tag{2.5.11}$$

$$\therefore\ \mathbf{G \propto 1/d} \tag{2.5.12}$$

Hence, the magnitude of every reciprocal lattice vector is inversely proportional to the spacing of the corresponding set of lattice planes.

Property 3. ***The volume of a unit cell in reciprocal lattice is inversely proportional to the volume of a unit cell of the crystal lattice.***

The primitive unit cell is defined in reciprocal lattice by the primitive basis vectors $\vec{a}^*, \vec{b}^*$ and $\vec{c}^*$. Its volume is given by

$$\begin{aligned}
\boldsymbol{V^*} &= \vec{a}^* \cdot \vec{b}^* \times \vec{c}^* \\
&= (2\pi)^3 \frac{\vec{b} \times \vec{c}}{\vec{a} \cdot \vec{b} \times \vec{c}} \cdot \left[\frac{(\vec{c} \times \vec{a}) \times (\vec{a} \times \vec{b})}{(\vec{a} \cdot \vec{b} \times \vec{c})^2} \right] \\
&= \frac{(2\pi)^3}{(\vec{a} \cdot \vec{b} \times \vec{c})^3} (\vec{b} \times \vec{c}) \cdot [(\vec{c} \times \vec{a}) \times (\vec{a} \times \vec{b})] \\
&= \frac{(2\pi)^3}{(\vec{a} \cdot \vec{b} \times c)^3} (\vec{b} \times \vec{c}) \cdot [\vec{a}(\vec{c} \times \vec{a} \cdot \vec{b}) - b(\vec{c} \times \vec{a} \cdot \vec{a})]
\end{aligned}$$

using the formula for the vector triple product.

$$\begin{aligned}
&= \frac{(2\pi)^3}{(\vec{a} \cdot \vec{b} \times \vec{c})^3} (\vec{b} \times \vec{c} \cdot \vec{a})(\vec{c} \times \vec{a} \cdot \vec{b}) \qquad (\because\ \vec{c} \times \vec{a} \cdot \vec{a} = 0) \\
&= \boldsymbol{\frac{(2\pi)^3}{\vec{a} \cdot \vec{b} \times \vec{c}}}
\end{aligned}$$

since, $a \cdot b \times c = c \times a \cdot b = b \times c \cdot a =$ volume of unit cell in direct lattice.

● Since $\vec{a} \cdot \vec{b} \times \vec{c}$ is the volume $\boldsymbol{V}$ of the unit cell of the crystal lattice, ***the volume of a primitive cell in a reciprocal lattice is*** $\boldsymbol{8\pi^3/V}$. Thus, $\boldsymbol{V^* \propto 1/V}$. *Hence.*

Property 4. ***The direct lattice is the reciprocal of its own reciprocal lattice.***

We have: $\quad \vec{a}^* = 2\pi \dfrac{\vec{b} \times \vec{c}}{\vec{a} \cdot \vec{b} \times \vec{c}}$. So, the reciprocal of $\vec{a}^*$ is given by

$$\begin{aligned}
(\boldsymbol{\vec{a}^*})^* &= 2\pi \frac{\vec{b}^* \times \vec{c}^*}{\vec{a}^* \cdot \vec{b}^* \times \vec{c}^*} = 2\pi(\vec{a} \cdot \vec{a}^*) \cdot \frac{\vec{b}^* \times \vec{c}^*}{\vec{a}^* \cdot \vec{b}^* \times \vec{c}^*} \qquad (\because\ \vec{a} \cdot \vec{a}^* = 1) \\
&= 2\pi \vec{a} \frac{\vec{a}^* \cdot \vec{b}^* \times \vec{c}^*}{\vec{a}^* \cdot \vec{b}^* \times \vec{c}^*} = \boldsymbol{2\pi \vec{a}}. \quad \text{Hence.}
\end{aligned}$$

We are now left with ***two more properties*** and they are proved in the following *three articles.*

2.6 Reciprocal lattice to a simple cubic (sc) lattice

The primitive translation vectors of a simple cubic lattice may be represented as

$$\vec{a} = a\hat{i}, \quad \vec{b} = a\hat{j}, \quad \vec{c} = a\hat{k}$$

where a is the length of the cube edge and $\hat{i}, \hat{j}, \hat{k}$ are the orthogonal unit vectors along the edges of the cube.

$\therefore$ Volume V of the simple cubic unit cell is given by

$$V = \vec{a} \cdot \vec{b} \times \vec{c} = a^3(\hat{i} \cdot \hat{j} \times \hat{k}) = a^3$$

Using the above equations, the reciprocal lattice vector to simple cubic lattice are given by

$$\vec{a}^* = 2\pi\frac{\vec{b}\times\vec{c}}{\vec{a}\cdot\vec{b}\times\vec{c}} = 2\pi\frac{a\hat{j}\times a\hat{k}}{a^3} = \frac{2\pi}{a}\hat{i}$$

Similarly,

$$\vec{b}^* = 2\pi\frac{\vec{c}\times\vec{a}}{\vec{a}\cdot\vec{b}\times\vec{c}} = \frac{2\pi}{a}\hat{j}$$

$$\vec{c}^* = 2\pi\frac{\vec{a}\times\vec{b}}{\vec{a}\cdot\vec{b}\times\vec{c}} = \frac{2\pi}{a}\hat{k}$$

All the three reciprocal lattice vectors $\vec{a}^*, \vec{b}^*$ and $\vec{c}^*$ are thus found to be equal in magnitude, implying that ***the reciprocal lattice to simple cubic lattice is also simple cubic*** but with lattice constant $2\pi/a$. ***A simple cubic lattice is thus self-reciprocal in character.***

2.7 Reciprocal lattice to a bcc-lattice

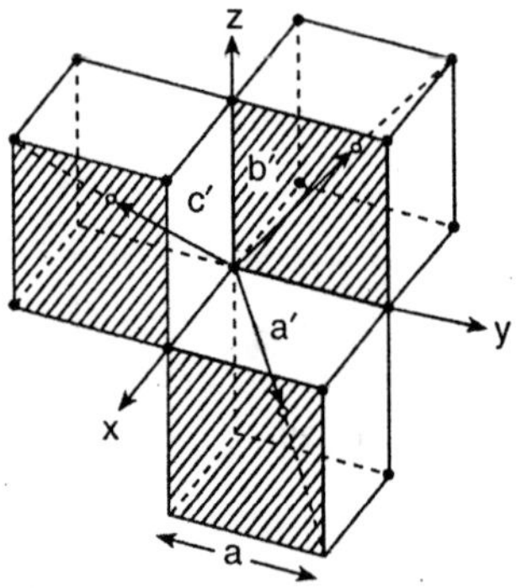

Fig. 2.9 Primitive translation vectors of a *bcc* lattice

The primitive translation vectors of a *bcc* lattice are

$$\vec{a}' = \frac{a}{2}(\hat{i}+\hat{j}-\hat{k})$$

$$\vec{b}' = \frac{a}{2}(-\hat{i}+\hat{j}+\hat{k})$$

$$\vec{c}' = \frac{a}{2}(i-\hat{j}+\hat{k})$$

where a is the length of the cube edge and $\hat{i},\hat{j},\hat{k}$ are the orthogonal unit vectors along the edges (Fig. 2.9).

$\therefore$ Volume V of the primitive cell is given by

$$V = \vec{a}'\cdot\vec{b}'\times\vec{c}'$$

$$= \frac{a}{2}(\hat{i}+\hat{j}-\hat{k})\cdot\left[\frac{a^2}{4}(-\hat{i}+\hat{j}+\hat{k})\times(\hat{i}-\hat{j}+\hat{k})\right]$$

$$= \frac{a}{2}(\hat{i}+\hat{j}-\hat{k})\cdot\frac{a^2}{4}(\hat{i}+\hat{j}) = \frac{a^3}{4}$$

$\therefore$ The reciprocal lattice vectors are given by

$$\vec{a}^* = 2\pi\frac{\vec{b}'\times\vec{c}'}{\vec{a}'\cdot\vec{b}'\times\vec{c}'} = 2\pi\frac{(a^2/4)}{a^3/4}(\hat{i}+\hat{j}) = \frac{2\pi}{a}(\hat{i}+\hat{j})$$

Similarly,

$$\vec{b}^* = 2\pi\frac{\vec{c}'\times\vec{a}'}{\vec{a}'\cdot b'\times\vec{c}'} = \frac{2\pi}{a}(\hat{j}+\hat{k})$$

$$\vec{c}^* = 2\pi\frac{\vec{c}'\times\vec{b}'}{\vec{a}'\cdot\vec{b}'\times\vec{c}'} = \frac{2\pi}{a}(\hat{k}+\hat{i})$$

These, in fact, are the primitive translation vectors of an *fcc* lattice. Thus, ***the reciprocal lattice to a bcc lattice is an fcc-lattice.***

2.8 Reciprocal lattice to an fcc-lattice

The primitive translation vectors of an *fcc* lattice (Fig. 2.10) may be written as

$$a' = \frac{a}{2}(\hat{i}+\hat{j});\ \ b' = \frac{a}{2}(\hat{j}+\hat{k});\ \ c' = \frac{a}{2}(\hat{k}+\hat{i})$$

$\therefore$ Volume V of the primitive cell is given by

$$V = \vec{a}'\cdot\vec{b}'\times\vec{c}' = \frac{a}{2}(\hat{i}+\hat{j})\cdot\frac{a^2}{4}[(\hat{j}+\hat{k})\times(\hat{k}+\hat{i})]$$

$$= \frac{a}{2}(\hat{i}+\hat{j})\cdot\frac{a^2}{4}(\hat{i}+\hat{j}-\hat{k}) = \frac{a^3}{4}$$

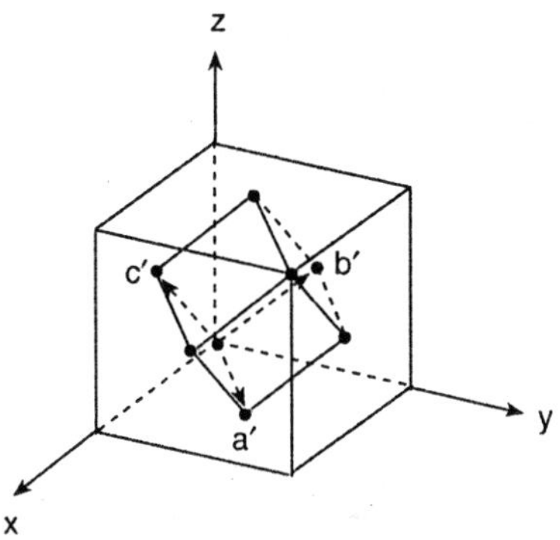

Fig. 2.10 Primitive translation vectors of an *fcc* lattice

$\therefore$ The reciprocal lattice vectors are obtained from (2.5.9) as

$$\vec{a}^* = 2\pi\frac{\vec{b}'\times\vec{c}'}{\vec{a}'\cdot\vec{b}'\times\vec{c}'} = 2\pi\frac{(a^2/4)(\hat{i}+\hat{j}-\hat{k})}{a^3/4}$$

$$= \frac{2\pi}{a}(\hat{i}+\hat{j}-\hat{k})$$

$$\vec{b}^* = 2\pi\frac{\vec{c}'\times\vec{a}'}{\vec{a}'\cdot\vec{b}'\times\vec{c}'} = \frac{2\pi}{a}(-\hat{i}+\hat{j}+\hat{k});$$

$$\vec{c}^* = 2\pi\frac{\vec{a}'\times\vec{b}'}{\vec{a}'\cdot\vec{b}'\times\vec{c}'} = \frac{2\pi}{a}(\hat{i}-\hat{j}+\hat{k})$$

These are the primitive translation vectors of a *bcc*-lattice with $2\pi/a$ as the length of the cube edge. So, ***the reciprocal lattice to an fcc lattice is a bcc lattice.***

2.9 Ewald's constuction : Bragg's law in reciprocal lattice

Ewald's construction is a geometrical construction to obtain a relation between the incident wave vector $\vec{k}$ and the diffracted wave vector $\vec{k}'$ with the help of reciprocal lattice and thus derive the ***Bragg's law*** in *vector form.*

Draw a wave vector $\overrightarrow{OA} = \vec{k} = 2\pi/\lambda$ in the direction of X-ray beam that originates at O (Fig. 2.11) and terminates at a reciprocal lattice point A. With O as centre and k as radius draw a sphere—the so-called ***reflex sphere***—whose section in a plane is a circle. Let the sphere intersect at some point B of the reciprocal lattice whose indices are $(h'k'l')$. So, $\overrightarrow{AB}$ represents the reciprocal vector $\vec{G}$, being normal to some set of direct lattice planes such as ON. Now,

$$AB = |\vec{G}| = \frac{2\pi n}{d},$$

where d is the interplanar spacing of the set and n the highest integral factor common to the numbers $(h'k'l')$.

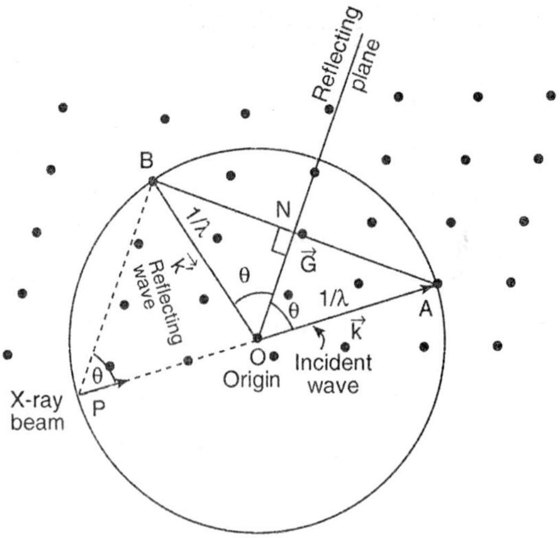

Fig. 2.11 Ewald construction in the reciprocal lattice

Now, $\overrightarrow{OA} = \vec{k}$ = incident wave vector so that $\overrightarrow{OB} = \vec{k}'$ = diffracted wave vector and plainly from figure

$$\vec{k}' = \vec{k} + \vec{G}.$$

The scattering thus changes only the direction of $\vec{k}$ and the scattered wave vector differs from the incident wave vector by the reciprocal lattice vector $\vec{G}$.

For the purpose diffraction, it is required that

$$|k'| = |k|$$

$$|\vec{k} + \vec{G}| = |\vec{k}'|$$

and we thus have

$$\vec{k}'^2 = (\vec{k} + \vec{G})^2 = (\vec{k} + \vec{G})\cdot(\vec{k} + \vec{G})$$

$$= \vec{k}^2 + 2\vec{k}\cdot\vec{G} + \vec{G}\cdot\vec{G}$$

or,

$$\mathbf{2}\vec{\boldsymbol{k}}\cdot\vec{G} + \vec{G}\cdot\vec{G} = 0 \qquad (\because\ \vec{k}'^2 = \vec{k}^2) \tag{2.9.1}$$

or,

$$\boxed{\left(\vec{k} + \frac{\vec{G}}{2}\right)\cdot\vec{G} = 0} \tag{2.9.2}$$

The above expression (2.9.1) or (2.9.2) gives the ***Bragg's law in vector form*** in terms of the reciprocal lattice vector $\vec{G}$.

The concept of the reciprocal lattice and Ewald's geometrical construction help in the determination of the crystal structure through the vector form of Bragg's law. Also, it is observed that Bragg's equation has a simple ***geometrical significance*** in the reciprocal lattice space.

- It is to be remembered, however, that the centre O of the reflex sphere is *not necessarily* a reciprocal lattice point.

● If the sphere does not pass through a reciprocal lattice point, it means that a particular wavelength in question would not be diffracted by the crystal in that orientation.

● It is rather easy to switch over to the ***scalar form*** from the boxed relation (2.9.2). From Fig. 2.11, we have

$$\overrightarrow{ON} = \overrightarrow{OA} + \overrightarrow{AN} = \vec{k} + \vec{G}/2;$$

$$ON \perp AB.$$

Also, $AB = 2AN = 2OA\sin\theta = 2(2\pi/\lambda)\sin\theta$, where $\theta = \angle AON$.

But, $AB = |\vec{G}| = 2\pi n/d$, using (2.5.11). Therefore, we obtain

$$\frac{2\pi n}{d} = 2\frac{2\pi}{\lambda}\sin\theta$$

$$\therefore \quad \boxed{\mathbf{2d\sin\theta = n\lambda}}$$

which is the usual ***scalar form*** of the ***Bragg's law*** of diffraction.

● **Advantages of reciprocal lattice over direct ones**—The reciprocal lattice plays an important role in crystal analysis. The use of reciprocal lattice over the direct space lattice has the following advantages.

1. The orientation of the different sets of crystal planes and their spacings are most conveniently represented by the reciprocal lattice.

2. It simplifies the understanding of the diffraction photographs of the crystals and also gives a rather simple method for obtaining the diffraction condition.

3. The wave mechanical behaviour of electrons in periodic crystal lattices is easily understood in terms of the reciprocal lattice.

2.10 Brillouin zones

Ewald's construction shows that all the $\vec{k}$-values for which the reciprocal lattice points intersect the reflex sphere are Bragg reflected. *A **Brillouin zone** is the locus of those $\vec{k}$-values in the reciprocal lattice which are Bragg reflected.*

We shall consider here a ***simple square lattice*** of side a in order to indicate how the Brillouin zones are in fact constructed. Such a lattice possesses the primitive translation vectors

$$\vec{a} = a\hat{i}; \ \vec{b} = a\hat{j}.$$

$$\therefore \quad \vec{a}^* = (2\pi/a)\hat{i}; \ \vec{b}^* = (2\pi/a)\hat{j}$$

are the corresponding vectors of the reciprocal lattice.

∴ Reciprocal lattice vector, $\vec{G} = (2\pi/a)(h\hat{i} + k\hat{j})$ where h and k are integers.

$$\text{But, } \vec{k} = k_x\hat{i} + k_y\hat{j}$$

Bragg's condition dictates $\qquad 2\vec{k}\cdot\vec{G} + G^2 = 0$

$$\text{or,}\quad \frac{4\pi}{a}[(k_x\hat{i} + k_y\hat{j})\cdot(h\hat{i} + k\hat{j})] + \frac{4\pi^2}{a^2}(h^2 + k^2) = 0$$

$$\text{or,}\quad hk_x + kk_y = (-\pi/a)(h^2 + k^2) \qquad (2.10.1)$$

The $\vec{k}$-values which are Bragg reflected are obtained by having all the possible combinations of h and k.

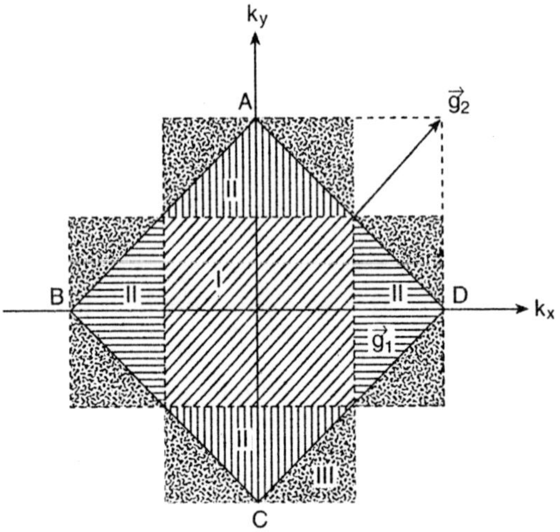

Fig. 2.12 First,second and third Brillouin zones (marked I, II and III, respectively) for a two-dimensional square lattice

For $h = \pm 1$ and $k = 0$, $k_x = \pm\pi/a$, and k_y arbitrary.

For $h = 0$, and $k = \pm 1$, $k_y = \pm\pi/a$ and k_x arbitrary.

By plotting these four lines, $k_x = \pm\pi/a$ and $k_y = \pm\pi/a$, we get a bounded square, (Fig. 2.12) called the ***first Brillouin zone***. In addition, some other sets of lines are also possible to satisfy (2.10.1). For example:

For $h = \pm 1$ and $k = \pm 1$, we readily obtain from equation (2.10.1), the following set of four lines

$$\pm k_x \pm k_y = 2\pi/a$$

which are also plotted in the same Fig. 2.12.

The additional area bounded by the above four lines constitutes the ***second Brillouin zone***.

Similarly, one can as well construct other Brillouin zones.

- The Brillouin zones for a three-dimensional cubical lattice may similarly be constructed using the generalized equation

$$hk_x + kk_y + lk_z = (-\pi/a)(h^2 + k^2 + l^2)$$

where a is the length of the cube edge.

• The boundaries represent the reflecting planes, the first zone for the first order, the second zone for the second order and so on.

• If however a $\vec{k}$-vector does not terminate at a zone boundary! Then what does it mean? Such a $\vec{k}$-vector cannot give the Bragg reflection. The Brillouin zone pattern can therefore be utilized in determining the X-ray diffraction pattern of a crystal and conversely.

• We shall return to Brillouin zones again while discussing 'Band theory of solids' in Chapter 7.

2.11 Atomic scattering factor or form factor

Bragg's law is simply a statement giving the condition for constructive interference of waves scattered from the lattice points. In fact, however, crystals contain atoms arranged in the lattice points and to find the scattering intensity (i.e., the square of scattered wave amplitude), scattering from both the nuclei and the electrons should be considered. But the interaction of X-rays with nuclei is not significant. So, scattering is essentially by the cloud of electrons, which again depends on the number density of electrons and their spatial distribution within each cell.

Computation—The ***atomic scattering factor*** or the ***form factor f*** describes the scattering efficiency of a single atom in relation to that of a single electron. It is defined as

$$f = \frac{\text{radiation amplitude scattered from an atom}}{\text{radiation amplitude scattered from an electron}}$$

To compute f, consider an atom with its electrons arranged in a spherically symmetric configuration around its centre O as origin. Let r be the radius of the atom and dV an elemental volume at $\vec{r}$ so that the charge at $\vec{r}$ is $\rho(\vec{r})dV$ where $\rho(\vec{r})$ is the charge density at $\vec{r}$ (Fig. 2.13).

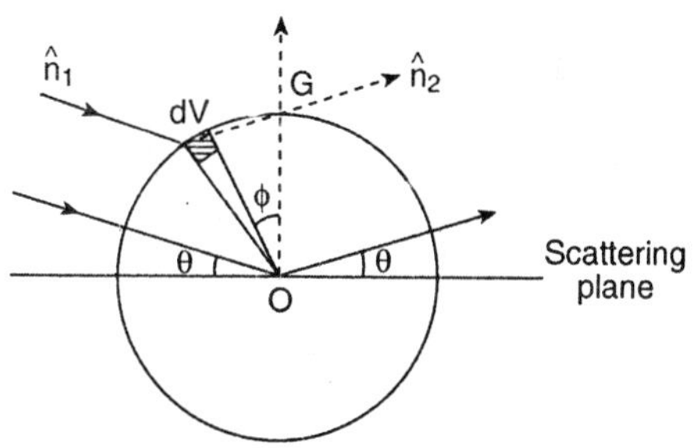

Fig. 2.13 Calculation of atomic scattering factor

Let us first consider the scattering from the charge $\rho(\vec{r})dV$ and an electron at the origin. If $\hat{n}_1$ and $\hat{n}_2$ be the directions of the incident and the scattered beam respectively (Fig. 2.14), the phase difference between the waves

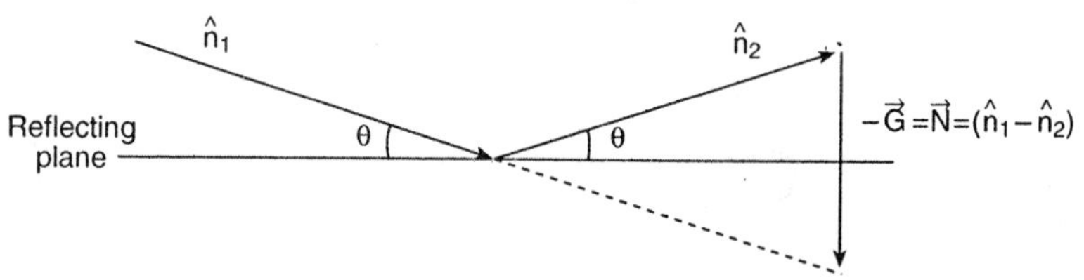

Fig. 2.14 Relationship between the incident beam and the scattered beam

scattered from dV and from the center is given by

$$\phi_r = \frac{2\pi}{\lambda}(\vec{r}\cdot\vec{N}) \tag{2.11.1}$$

where $\vec{N}$ represents the scattering normal.

If the scattering amplitude from the point electron at O along $\hat{n}_2$ be $A\exp[i(kx-\omega t)]$ where x is the distance covered along $\hat{n}_2$ and k the wave number, then the scattering amplitude from the charge $\rho(\vec{r})dV$ will be $Ae^{i(kx-\omega t)+i\phi_r}\rho(\vec{r})dV$.

$$\therefore\ df = \frac{Ae^{i(kx-\omega t)+i\phi_r}}{Ae^{i(kx-\omega t)}}\rho(\vec{r})dV \tag{2.11.2}$$

Thus the ratio of amplitude from the whole atom to that from an electron is

$$f = \int_V \rho(\vec{r})e^{i\phi_r}dV, \quad \text{where } V = \text{volume of the atom.} \tag{2.11.3}$$

From *spherical symmetry* and hence using spherical polar coordinates, we get $dV = 2\pi r^2 \sin\phi d\phi dr$, where $\phi =$ polar angle. Also, we have $|\vec{N}| = 2\sin\theta$. So, from (2.11.1), $\phi_r = (2\pi/\lambda)rN\cos\phi = (4\pi/\lambda)r\sin\theta\cos\phi = \mu r\cos\phi$, where $\mu = (4\pi/\lambda)\sin\theta$.

$$\therefore\ f = \int_{r=0}^{\infty}\int_{\phi=0}^{\pi}\rho(\vec{r})e^{i\mu r\cos\phi}\,2\pi r^2\sin\phi d\phi dr$$

$$\text{Now,}\quad \int_0^{\pi} e^{\mu r\cos\phi}\sin\phi d\phi = 2\left(\frac{\sin\mu r}{\mu r}\right)$$

$$\therefore\ \boxed{f = \int_0^{\infty} 4\pi r^2\rho(\vec{r})\frac{\sin\mu r}{\mu r}dr} \tag{2.11.4}$$

Expression (2.11.4) represents the *general expression* for the ***atomic structure factor***. Further evaluation requires information about the charge distribution. But since for $\theta \to 0, \mu \to 0$ and $(\sin\mu r)/\mu r \to 1$, we obtain from (2.11.4),

$$\begin{aligned} f &= \int_0^{\infty} 4\pi r^2\rho(\vec{r})dr \\ &= \text{Total electronic charge stored inside the atom} \\ &= Z, \text{ the number of atomic electrons (atomic number).} \end{aligned}$$

• The significance of f is that for $\theta \to 0$, it is equal to the total number of electrons Z in the atom. For $\theta > 0$, $f < Z$. A knowledge of f thus gives the charge distribution in the atom, the integrand $\int_0^{\infty} 4\pi r^2\rho(r)dr$ giving the charge of a spherical shell of radius r and $r + dr$.

2.12 Geometrical structure factor

Apart from the atomic scattering factors of various atoms, the intensity of an X-ray beam diffracted from a crystal depends also on the number, type and distribution of atoms within the cell. This is taken into account by the total scattering amplitude or ***structure factor*** $F(h'k'l')$ for the reflection $(h'k'l')$ and is defined as *the ratio of the amplitude of radiation scattered by the entire unit cell to the amplitude of radiation scattered by a single point-electron at the origin for the same wavelength.* It is therefore given by

$$F(h'k'l') = \sum_j f_j e^{i\phi_j} = \sum_j f_j e^{i(2\pi/\lambda)(\vec{r}_j \cdot \vec{N})}$$

where f_j is the atomic scattering factor for the jth atom, ϕ_j the phase difference between the radiation scattered from the jth atom of unit cell and that from the electron at the origin.

If now (u_j, v_j, w_j) be the coordinates of the jth atom, then we have

$$\vec{r}_j = u_j\vec{a} + v_j\vec{b} + w_j\vec{c}$$

$$\therefore \quad \vec{r}_j \cdot \vec{N} = \lambda(\mu_j h' + v_j k' + w_j l') \quad (\because \vec{a} \cdot \vec{N} = h'\lambda \text{ etc.})$$

$$\therefore \quad F(h'k'l') = \sum_j f_j e^{i2\pi(u_j h' + v_j k' + w_j l')} \tag{2.12.1}$$

When the atoms are identical, all the f_i's are equal, say equal to f, then (2.12.1) becomes

$$\therefore \quad F(h'k'l') = fS \tag{2.12.2}$$

$$\text{where} \quad S = \sum_j e^{2\pi i(u_j h' + v_j k' + w_j l')}, \tag{2.12.3}$$

and is called the ***geometrical structure factor*** for it is dependent on the geometrical arrangement of atoms within the unit cell. Eq. (2.12.2) defines the *geometrical structure factor* S as the ratio of total scattering amplitude F to the atomic structure factor f.

● Since the intensity is proportional to the square of the amplitude, the intensity I of the diffracted beam is given by

$$I = |F|^2 = F^* F$$

where F^* is the complex conjugate of F.

Structure factor of cubic cyrstals—We shall now calculate the structure factor of *cubic crystals.*

Simple cubic (sc)—For *sc*-crystals, there is only one atom per unit cell at the origin. So, $u_j = v_j = w_j = 0$ and from (2.12.3), $S = 1$ and the diffraction amplitude $F(h'k'l') = f$ which is independent of (h', k', l').

This ensures that all the diffraction lines dictated by Bragg's law would appear in the diffraction pattern if f is large enough to produce an observable intensity.

Body centred cubic (bcc)—There are two identical atoms per unit cell, one contributed by corner atoms and the other at the centre. Let, for the corner atoms, $u_j = v_j = w_j = 0$, when for the centre atom they would be $u_j = v_j = w_j = \frac{1}{2}$.

$$\therefore \quad F(h'k'l') = f\sum e^{2\pi i/(u_j h' + v_j k' + w_j l')} = f\left(1 + e^{\pi i(h'+k'+l')}\right)$$

For a $(h'k'l')$-*reflection* for which $(h'+k'+l')$ is an *odd* integer, $e^{\pi i(h'+k'+l')} = -1$.

$$\therefore \quad F(h'k'l') = f(1-1) = 0$$

For a reflection for which $(h'+k'+l')$ is an *even* integer

$$F(h'k'l') = f(1+1) = 2f$$

Face-centred cubic (fcc)—There are four identical atoms per unit cell, one contributed by corner atoms and three by face-centred atoms. Their coordinates are:

$$(0\,0\,0),\ \left(0\,\tfrac{1}{2}\,\tfrac{1}{2}\right),\ \left(\tfrac{1}{2}\,0\,\tfrac{1}{2}\right),\ \left(\tfrac{1}{2}\,\tfrac{1}{2}\,0\right)$$

$\therefore$ Diffraction amplitude for $(h'k'l')$-reflection is

$$F(h'k'l') = f\left[1 + e^{\pi i(h'+k')} + e^{\pi i(k'+l')} + e^{\pi i(l+h')}\right]$$

If (h', k', l') are all *odd*, then $F(h'k'l') = f(1+1+1+1) = 4f$ and the intensity is proportional to $(4f)^2 = 16f^2$.

If only one, say h' is *odd*, the other two k', l' are *even*, then $e^{\pi i(h'+k')} = -1$, $e^{\pi i(l'+h')} = -1$ and $e^{\pi i(k'+l')} = +1$.

$$\therefore \quad F(h'k'l') = f(1-1-1+1) = 0 \text{ and so also the } intensity.$$

So in *fcc* crystals, no reflection occurs for *partly even* and *partly odd* integers.

• The space lattice of a crystal structure may be obtained from the position of diffraction lines in the diffraction pattern. But the determination of the basis is complicated in that it requires a knowledge of the intensity of various diffraction lines. In this context, the concept of structure factor has a special importance in structure determination.

2.13 Experimental methods: X-ray diffraction

The X-ray diffraction experiments require the following basic essentials:

(i) the radiation from an X-ray source,

(ii) the sample under investigation, and

(iii) a photographic film (or a detector) to record the diffraction beam.

The radiation may be monochromatic or of variable λ, the sample may be a single crystal or polycrystal (powder) or a solid piece (amorphous) and the detector may be a film or a radiation counter.

We shall describe hereunder the *three* standard experimental methods of X-ray diffraction. These are (i) the ***Laue method***, (ii) the ***rotating crystal method***, and (iii) the Debye-Scherrer ***powder method***.

1. Laue method—In this method, a single crystal is mounted on a *goniometer* that enables the crystal to be rotated through known angles in two perpendicular planes, and is held stationary in a beam of X-rays of continuous wavelength i.e., white radiation. The crystal selects out and diffracts the appropriate wavelengths for which planes exist, of spacing d and glancing angle θ, satisfying the Bragg's equation.

Fig. 2.15 schematically illustrates the experimental set-up which is relatively simple. The white radiation from the X-ray source is well-collimated by a pin hole collimator.

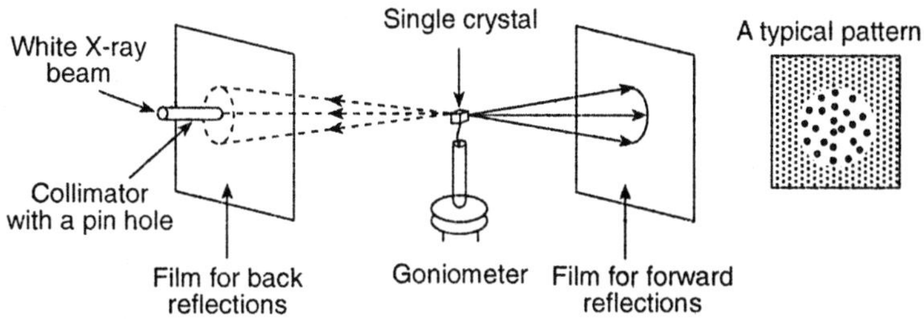

Fig. 2.15 Schematic representation of Laue technique

A flat photographic film is properly placed to receive and record either the transmitted diffracted beam or the reflected diffracted beam.

The resulting Laue pattern consists of a series of spots the positions of which depend on the orientation of the crystal relative to the incident beam. Sharp, well-defined spots provide good evidence of a perfect crystal structure. Diffuse or broken ones indicate lattice distortion. The pattern shows the crystal-symmetry in the orientation used, e.g., a [100] axis parallel to the incident beam, the pattern would show the four-fold symmetry appropriate to the axis.

● When the structure of the crystal is not known, the set of spots appears highly confusing for several λ's of different orders for the same set of planes may be reflected to the same position. Laue method is therefore rarely used for finding the unknown structure. It is now mainly used for determining the routine orientation of known crystals only.

2. Rotating-crystal method—In this method also, a small single crystal (~ 1 mm dimensions) is mounted on a *goniometer* which is rigidly fixed to a *spindle*. The crystal can thus be made to rotate, by coupling the goniometer head with a motor, about a fixed axis in a monochromatic beam of X-radiation. The specimen is usually oriented with one of the crystallographic axes parallel to the axis of rotation. The rotation brings different lattice planes into position for reflection. A photographic film placed cylindrically and coaxially with the rotating spindle records the diffracted images. The arrangement is illustrated in Fig. 2.16.

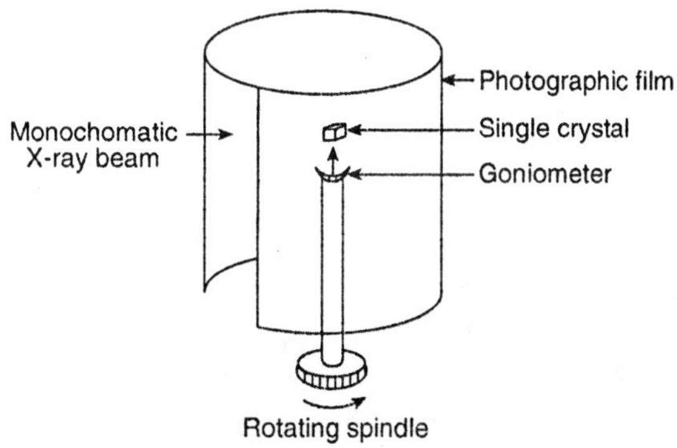

Fig. 2.16 Schematic representation of rotating crystal technique

After unrolling and on developing and flattening out the film, the diffracted images are found to be on a series of lines called *layer lines* (Fig. 2.17). The images on zero layer lines are from planes parallel to the axis of rotation i.e., planes with $l = 0$. So, if the crystal is positioned with c-axis coincident with the axis of rotation, all planes having Miller indices of the type $(hk0)$ will generate the central line. Layers above and below the central line will be produced by planes with indices of the type (hkl) and $(hk\bar{l})$ respectively and so on. Images from planes with same h, k but different l's lie on a series of curves called *row lines* transverse to the layer lines.

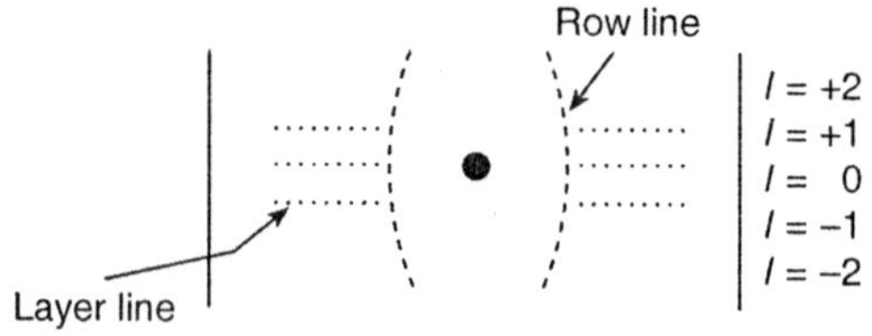

Fig. 2.17 Diffraction images in rotating crystal technique

• Modern diffractometers record the diffracted images by using scintillation counter and proportional counter automatically and collect the data.

As the wavelength λ is known, the spacing d can be evaluated from Bragg's relation : $2d\sin\theta = n\lambda$.

The vertical spacing between the layer lines in the photograph is a measure of the translational vector $\vec{c}$. Similarly, $\vec{a}$ and $\vec{b}$ can also be measured by mounting the crystal along $\vec{a}$ and $\vec{b}$ axes respectively. Thus, the dimensions of the unit cell may be determined by this method.

3. Debye-Scherrer powder method—This method was introduced by Debye and Scherrer in 1916 and is most widely used in crystal structure analysis since it is both quick and convenient for it does not require any single crystal. The arrangement is known as Debye-Scherrer camera.

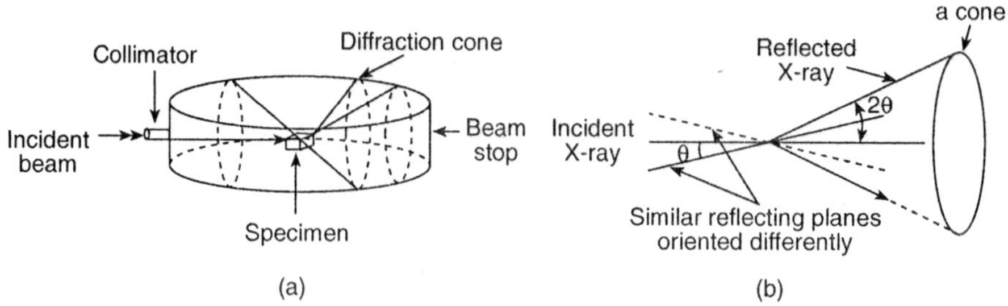

Fig. 2.18 Experimental arrangement in powder method

In this method, as illustrated schematically in Fig. 2.18, a monochromatic parallel X-ray beam is allowed to fall normally on a small specimen, finely powdered and kept in a thin glass capillary. The orientation of the minute crystal fragments is completely random so that a certain number of them will lie, with a given set of planes, making exactly the correct angle with the incident beam so as to satisfy the condition for reflection. There would still however be a number of combinations of d and θ that would satisfy Bragg's law. For each combination of d and θ, one cone of reflection results. The cone is coaxial with the axis of the incident beam and has a semivertical angle 2θ, i.e, twice the Bragg angle. Many such cones of reflection are emitted by the powdered specimen.

The specimen is completely surrounded by a cylindrical photographic film on which portions of each cone are recorded as pair arcs. When the strip of the film is taken out, developed and spread, it looks like the Fig. 2.19a.

The Bragg angle θ of the various reflections can be easily evaluated by measuring the separation of the pair of arcs for, from geometry of Fig. 2.19b, we get

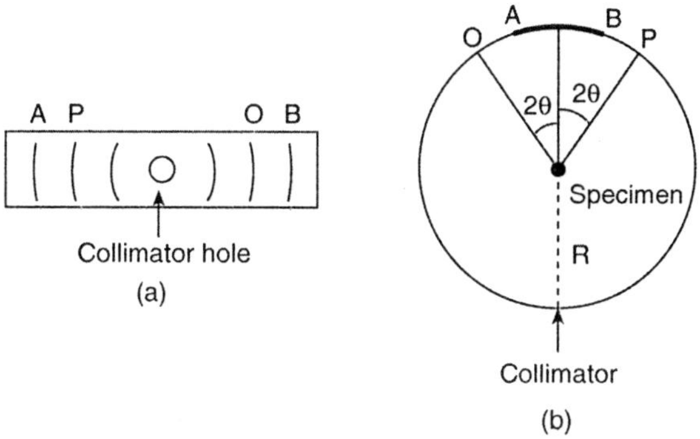

Fig. 2.19 (a) Developed film, (b) diffraction geometry

$$\frac{2\pi - 4\theta}{2\pi} = \frac{OP}{2\pi R}$$

The rereflection can thus be properly indexed and the unit cell parameters may be easily determined.

2.14 Determination of crystal structure of KCl

X-ray spectrometer—Bragg designed an X-ray spectrometer—***Bragg spectrometer***—to study the crystal structures using his law.

It consists of a collimated beam of X-rays, from an X-ray tube, that falls on the crystal C under test. The crystal, in its turn, is mounted on a table T capable of rotation about a vertical axis causing the glancing angle θ to change relative to the incident beam of X-rays. The angle θ is measured with a circular scale and vernier. The reflected X-ray beam finally enters an ionisation chamber and the resulting ionisation current is measured by an electrometer E (Fig. 2.20). The angle of setting of the ionisation chamber can be changed and measured with a circular scale and vernier and for each setting the angle of reflection is made equal to the angle of incidence. The ionisation current is recorded for different θ-values.

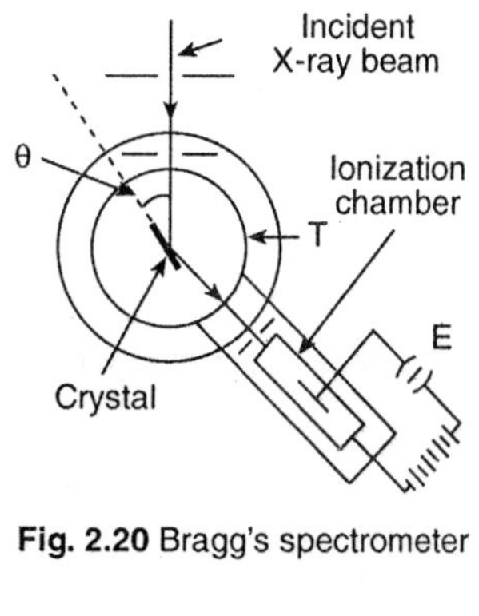

Fig. 2.20 Bragg's spectrometer

Analysis—The crystal structure of KCl (sylvine) can be determined with the Bragg's X-ray spectrometer.

The K_α-line of X-ray from palladium gives peak ionisation current at $\theta = 5°23'$ corresponding to the first order $n = 1$ when the (100) face of KCl is exposed to the X-ray. When however the (110) and the (111) planes are exposed, the values of θ with $n = 1$ are $7°37'$ and $9°25'$ respectively.

If d_{hkl} be the lattice spacing for the (hkl) planes, then, from Bragg's law, we have for KCl crystal

$$\begin{aligned} d_{100} : d_{110} : d_{111} &= (1/\sin 5°23') : (1/\sin 7°37) : (1/\sin 9°25') \\ &= 1 : 0.7076 : 0.5731 \end{aligned} \tag{2.14.1}$$

For a *cubic crystal*, using equation (2.7.4), we obtain

$$\begin{aligned} d_{100} : d_{110} : d_{111} &= 1 : 1/\sqrt{2} : 1/\sqrt{3} \\ &= 1 : 0.7071 : 0.5773 \end{aligned} \tag{2.14.2}$$

Comparing (2.14.1) and (2.14.2), it is concluded that the KCl-crystal is ***cubic.*** Similarly, the other forms of crystal can also be investigated.

● For NaCl, the peak intensity of the reflected beam occurs at $\theta = 6°$, corresponding to the first order (i.e., $n = 1$) for the (100) planes. Conversely, by putting $n = 1$ and $\theta = 6°$ for the (100) planes of NaCl, we can find the wavelength of K_α-line of palladium using the Bragg's diffraction law:

$$\begin{aligned} \lambda &= 2d_{100} \sin\theta \\ &= 2 \times 2.82 \times \sin 6° \text{ A} \\ &= 0.59 \text{ A} \end{aligned}$$

We can thus determine, by using Bragg's law, (i) the *lattice spacing d* of the crystal with a known wavelength and (ii) the *wavelength of an incident radiation* with a known lattice spacing.

2.15 Illustrated Examples

▶ **Example 1.** *If the potential difference across the X-ray tube is* 5000 V *and the current through it is* 2 mA, *calculate the number of electrons striking the target per second and the speed at which they strike it.* **(Punjab Univ.)**

Solution: By the problem, $V = 5000$ V, current $i = 2$ mA $= 2 \times 10^{-3}$ A.

Charge on electron, $e = 1.6 \times 10^{-19}$ C, mass of electron, $m = 9 \times 10^{-31}$ kg.

∴ No. of electrons hitting the target *per sec* is

$$n = \frac{2 \times 10^{-3}}{1.6 \times 10^{-19}} = \mathbf{1.25 \times 10^{16}}$$

Maximum kinetic energy of electron, $\frac{1}{2}mv^2 = Ve$

$\therefore$ Speed of electrons, $$v = \sqrt{\frac{2Ve}{m}} = \sqrt{\frac{2 \times 5000 \times 1.6 \times 10^{-19}}{9 \times 10^{-31}}}$$
$$= \mathbf{4.2 \times 10^{-7}\ m/s}$$

► **Example 2.** *The distance between adjacent atomic planes in* $CaCO_3$ *is* 3Å. *Find the smallest angle Bragg scattering for* 0.3 Å *X-rays.*

Solution: From Bragg's law, $2d\sin\theta = n\lambda$. Angle $\theta = \theta_0$ is smallest when $n = 1$

$$\therefore\ 2d\sin\theta_0 = \lambda \Rightarrow \sin\theta_0 = \frac{0.3}{2 \times 3} = \frac{0.3}{6} = 0.05. \quad \therefore\ \boldsymbol{\theta_0 = 2.9^\circ}$$

► **Example 3.** *The spacing between successive* (100) *planes in* NaCl *is* 2.82 A. *X-rays incident on the surface of the crystal is found to give rise to first order Bragg reflection at glancing angle* 8.8°. *Calculate the wavelength of X-rays.*

(Himachal Univ. and Burdwan Hons.)

Solution: From Bragg's law, with usual symbols, we obtain

$$2d\sin\theta = n\lambda$$

By the problem, here, $d = 2.82\,\text{A} = 2.82 \times 10^{-10}\,\text{m}$; $\theta = 8.8^\circ$, $n = 1$.

$$\therefore\ \sin\theta = \sin 8.8^\circ = 0.153$$
$$\therefore\ 2 \times 2.82 \times 10^{-10} \times 0.153 = \lambda$$
$$\therefore\ \lambda = 0.863 \times 10^{-10}\,\text{m} = \mathbf{0.863\ A}$$

► **Example 4.** *If X-rays of wavelength* 0.5 A *are diffracted at an angle of* 5° *in the first order, what is the spacing between the adjacent planes of the crystal? At what angle will the second maximum occur?* **(Bangalore Univ.)**

Solution: Here $\lambda = 0.5\,\text{A} = 0.5 \times 10^{-10}\,\text{m}$, $\theta = 5^\circ$ and $n = 1$.

$$\therefore\ \sin\theta = \sin 5^\circ = 0.0871$$

From Bragg's law, $2d\sin\theta = n\lambda$, we obtain

$$d = \frac{n\lambda}{2\sin\theta} = \frac{1 \times 0.5 \times 10^{-10}}{2 \times 0.0871} = 2.87 \times 10^{-10}\,\text{m} = \mathbf{2.87\ A}$$

Let θ' be the Bragg angle for the second order maximum.

$$\therefore\ 2d\sin\theta' = 2\lambda$$
$$\Rightarrow \sin\theta' = \frac{\lambda}{d} = \frac{0.5 \times 10^{-10}}{2.87 \times 10^{-10}} = 0.1742$$
$$\therefore\ \theta' = \sin^{-1} 0.1742 = \mathbf{10.03^\circ}$$

► **Example 5.** *The first order* (100) *reflection angle is* 18° *for a cubic crystal using X-rays of wavelength* 1.54 A. *Determine the distance between the* (100) *planes and the* (111) *planes of the crystal.* **(Calcutta Hons.)**

Solution : From Bragg's law, $2d\sin\theta = n\lambda$, where d is the separation between the (100) planes.

Here, $n = 1$, $\lambda = 1.54$ A and $\theta = 18° \Rightarrow \sin\theta = \sin 18° = 0.309$.

$$\therefore \quad d = \frac{\lambda}{2\sin\theta} = \frac{1.54}{2 \times 0.309} = \mathbf{2.49\ A}$$

This separation d between the (100) planes gives the length of the cube edge of the unit cell, i.e., a.

∴ Distance between the (111) planes is given by

$$d_{111} = \frac{a}{\sqrt{h^2+k^2+l^2}} = \frac{2.49}{\sqrt{1^2+1^2+1^2}} = \frac{2.49}{\sqrt{3}} = \mathbf{1.44\ A}$$

► **Example 6.** *The Bragg angle for the first order reflection from* (111) *planes in a crystal is* 60°. *Calculate the interatomic spacing if X-rays of wavelength* 1.8×10^{-10} m *are used.* **(Kanpur Univ.)**

Solution : Here $\theta = 60° \Rightarrow \sin\theta = \sin 60° = 0.866$, $\lambda = 1.8 \times 10^{-10}$ m and $n = 1$.

∴ From Bragg's law, $2d\sin\theta = n\lambda$, we have

$$2d \times 0.866 = 1.8 \times 10^{-10}$$

$$\therefore \quad d = \frac{1.8 \times 10^{-10}}{2 \times 0.866} = 1.039 \times 10^{-10}\,\text{m}$$

$$\therefore \quad d_{111} = 1.039 \times 10^{-10}\,\text{m}$$

If the interatomic spacing for a cubic lattice $d_{100} = a$, then

$$d_{111} = \frac{a}{\sqrt{1^2+1^2+1^2}} = \frac{a}{\sqrt{3}}$$

$$\therefore \quad a = d_{100} = \sqrt{3}d_{111} = \sqrt{3} \times 1.039 \times 10^{-10}\,\text{m} = \mathbf{1.8 \times 10^{-10}\ m}$$

► **Example 7.** *The wavelength of* K_α*-line of silver is* 0.563 A. *The radiation from a silver target is analysed with a Bragg spectrometer using a calcite (a simple cubic of lattice constant* = 3.02945 A). *Determine the angle of reflection for the first order. What is the highest order for which the line may be observed?* **(Calcutta Hons.)**

Solution : Here, $n = 1$, $d = 3.02945$ A and $\lambda = 0.563$ A. From the Bragg's law,

$$2d\sin\theta = n\lambda$$

$$\Rightarrow \quad \sin\theta = \frac{n\lambda}{2d} = \frac{1 \times 0.563}{2 \times 3.02945}$$
$$= 0.0929$$

$\therefore$ Angle of reflection, $\theta = \sin^{-1}(0.0929) = \mathbf{5.33^\circ}$

With increase in n, θ increases and $\theta_{\text{max}} = 90^\circ \Rightarrow \sin\theta = \sin 90^\circ = 1$

$$\therefore \text{ Highest value of } n = \frac{2d}{\lambda} = \frac{2 \times 3.02945\,\text{A}}{0.563\,\text{A}} = 10.8$$

$$\therefore \quad \text{Highest possible order} = \mathbf{10}$$

► **Example 8.** *The first order Bragg reflection in* Po *appear at* $\sin\theta$*-values* 0.225, 0.317 *and* 0.389 *for reflections from planes* (100), (110) *and* (111) *respectively. If the wavelength of X-radiation used be* 1.54 A, *prove that the unit cell is cubic and determine the length of the side of the unit cell.*

Solution: The ratio of the $\sin\theta$-values is

$$0.225 : 0.317 : 0.389 = 1 : 1.41 : 1.73 = 1 : \sqrt{2} : \sqrt{3}$$

Corresponding interplanar spacing ratio is given by

$$d_{100} : d_{110} : d_{111} = 1 : 1/\sqrt{2} : 1/\sqrt{3}$$

Hence the *unit cell* is ***simple cubic*** in nature.

If a be the length of the side of cubic unit cell, then for the reflection from (100) plane, the Bragg's law dictates

$$\boldsymbol{a} = d_{100} = \frac{\lambda}{2\sin\theta} = \frac{1.54}{2 \times 0.225} = \mathbf{3.42\ A}$$

► **Example 9.** *In a powder diffraction experiment with X-rays of* $\lambda = 1.54\,\text{A}$, *the first order Bragg reflections occur from a monatomic cubic crystal for* θ*-values:* $17.4^\circ, 20.3^\circ, 29.3^\circ, 35.1^\circ$ *and* 36.9°. *Determine* (i) *the Miller indices of diffraction planes,* (ii) *the crystal structure and* (iii) *the lattice parameter.*

Solution: First order reflection from (hkl) planes, for the lattice parameter a, satisfies

$$2d\sin\theta = \lambda$$

$$\Rightarrow \quad 2\frac{a}{\sqrt{h^2+k^2+l^2}}\sin\theta = \lambda \qquad (2.15.1)$$

$$\text{or,} \quad \sin^2\theta = \frac{\lambda^2}{4a^2}(h^2+k^2+l^2)$$

$$\Rightarrow \quad \sin^2\theta \propto (h^2+k^2+l^2)$$

Corresponding to the given θ-values, $\sin^2\theta$'s are: 0.09, 0.12, 0.24, 0.33 and 0.36.

$$\therefore \quad \sin^2\theta\text{'s are in the ratio: } 3 : 4 : 8 : 11 : 12$$

The Miller indices (hkl) giving these ratio are:

$$(\mathbf{111}),\ (\mathbf{200}),\ (\mathbf{220}),\ (\mathbf{311})\ \textbf{and}\ (\mathbf{222})$$

The indices are either *all odd* or *all even.* So, it is an ***fcc-structure.***

For (111) reflection, $\theta = 17.4°$. With $\lambda = 1.54$ A, we get from Bragg's law (2.15.1)

$$a = \frac{\lambda}{2\sin\theta}\sqrt{h^2+k^2+l^2} = \frac{1.54\sqrt{3}}{2\sin 17.4°} = \mathbf{4.46\ A}$$

▶ **Example 10.** *Given a square piece of X-ray film* 10 cm ×10 cm, *copper radiation of* $\lambda = 1.52$ A, *and powdered* NaCl (cubic) *with lattice parameter* 4.57 A, *devise an experiment in such a fashion that the rays from the* (111) *planes will produce a circle of dia* 10 cm ($\sin 16.75° = 0.288$, $\tan 33.5° = 0.662$). **(Calcutta Hons.)**

Solution: The appropriate device is the 'powder method' of Debye and Scherrer.

For the first order, the Bragg's law of reflection dictates

$$2d\sin\theta = \lambda$$

By the problem, $d = 4.57/\sqrt{1^2+1^2+1^2} = 4.57/\sqrt{3}$ A and $\lambda = 1.52$ A.

$$\therefore\ \sin\theta = \frac{\lambda}{2d} = \frac{1.52\times\sqrt{3}}{2\times 4.57} = 0.288 = \sin 16.75°\ \text{(given)}$$

$$\therefore\ \theta = 16.75°$$

The film is to be placed normally to the incident X-ray beam at a distance D cm from the sample such that the light cone (diffracted) intersects the X-ray film in a circle of radius 5 cm.

$$\therefore\ \tan 2\theta = \frac{5}{D} \Rightarrow D = \frac{5}{\tan(2\times 16.75)} = \frac{5}{\tan 33.5°} = \frac{5}{0.662} = \mathbf{7.55\ cm}$$

▶ **Example 11.** *The molecular weight of rock salt (a simple cubic) crystal is* 5.85 kg/k.mole *and its density is* 2.16×10^3 kg/m^3. *Calculate the grating space* d_{100} *of rock salt, taking Avogadro number* $= 6.02\times10^{26}$.

Using the value of lattice constant, calculate the wavelength of X-rays in the second order, if the angle of diffraction $\theta = 26°$. **(Calcutta Hons. and Bangalore Univ.)**

Solution: Mass of a NaCl molecule $= \dfrac{58.5}{6.02\times10^{26}}$ kg.

$$\therefore\ \text{Volume of a NaCl molecule} = \frac{\text{mass}}{\text{density}}$$

$$= \frac{58.5}{6.02\times10^{26}\times2.16\times10^3}\ \text{m}^3$$

$$= 4.5\times10^{-29}\ \text{m}^3$$

Now, in a NaCl molecule, there are 2 atoms since NaCl is diatomic.

$$\therefore \quad \text{Volume per atom of NaCl} = \tfrac{1}{2} \times 4.5 \times 10^{-29} = 2.25 \times 10^{-29}\,\text{m}^3$$

If d be the atomic spacing for NaCl crystal, then the volume per atom $= d^3$

$$\therefore \quad d^3 = 2.25 \times 10^{-29} \quad \Rightarrow \quad d = 2.82 \times 10^{-10}\,\text{m} = 2.82\,\text{A}$$

$$\therefore \quad d_{100} = \mathbf{2.82\ A}$$

From Bragg's law, we have $\qquad 2d \sin\theta = n\lambda$

Here, $d = 2.82\,\text{A}$, $n = 2$, $\theta = 26° \Rightarrow \sin\theta = \sin 26° = 0.4383$

$$\therefore \quad 2\lambda = 2 \times 2.82 \times 0.4383 = 2.48\,\text{A}$$

$$\therefore \quad \lambda = \mathbf{1.24\ A}$$

▶ **Example 12.** *A crystal plate is mounted on an X-ray spectrometer. The glancing angles of incidence for three reflections are* $5°58', 12°01'$ *and* $18°12'$. *Show that they are successive orders of reflections from the same crystal plane. Also, evaluate the interplanar spacing, if the wavelength of X-rays used is* 0.586 A.

Solution: The given glancing angles are: $\theta_1 = 5°58' = 5.966°$, $\theta_2 = 12°01' = 12.01°$ and $\theta_3 = 18.11' = 18.18°$. Also, $\lambda = 0.586\,\text{A} = 0.586 \times 10^{-10}\,\text{m}$. To show that the given angles represent the successive, i.e., first, second and third order of reflections from the same crystal plane, we check the ratio

$$\begin{aligned} \sin\theta_1 : \sin\theta_2 : \sin\theta_3 &= \sin 5.966° : \sin 12.01° : \sin 18.18° \\ &= 0.1040 : 0.2080 : 0.3120 \\ &= 1 : 2 : 3 \end{aligned}$$

This implies that θ_1, θ_2 and θ_3 are angles of incidence corresponding to **1st, 2nd** and **3rd order** of reflections respectively.

From Bragg's law, $2d \sin\theta = n\lambda$, we have for

$$\text{1st order,} \quad d = \frac{1\lambda}{2\sin\theta_1} = \frac{0.586 \times 10^{-10}}{2 \times 0.1040} = 2.817 \times 10^{-10}\,\text{m}$$

$$\text{2nd order,} \quad d = \frac{2\lambda}{2\sin\theta_2} = \frac{2 \times 0.586 \times 10^{-10}}{2 \times 0.2080} = 2.817 \times 10^{-10}\,\text{m}$$

$$\text{3rd order,} \quad d = \frac{3\lambda}{2\sin\theta_3} = \frac{3 \times 0.586 \times 10^{-10}}{2 \times 0.3120} = 2.817 \times 10^{-10}\,\text{m}$$

Hence the ***interplanar spacing*** (mean value) $= \mathbf{2.817 \times 10^{-10}\ m}$.

▶ **Example 13.** KCl *crystallizes in simple cubic structure. Find the interatomic spacing d and the glancing angle at which an X-ray of* $\lambda = 1.787$ A *is reflected in the third order. Given: density of* KCl $= 1990\ \text{kg/m}^3$ *and its molecular weight is* 74.6.

Solution: By the problem, $\lambda = 1.787$ A, n = order of reflection = 3, $\rho = 1990$ kg/m^3, mol. wt. = 74.6, structure = *sc*. The number atom per unit cell, $n' = 1$.

If a be the lattice parameter, $a^3 = \dfrac{Mn'}{N\rho} = \dfrac{74.6 \times 1}{6.023 \times 10^{26} \times 1990} = 62.24 \times 10^{-30}$.

$$\therefore \quad a = (62.24 \times 10^{-30})^{1/3} = 3.96 \times 10^{-10} = \mathbf{3.96\ A} = \boldsymbol{d}, \text{ in } sc$$

Using Bragg's law for 3rd order reflection,

$$\theta = \sin^{-1}\left(3\frac{\lambda}{2d}\right) = \sin^{-1}\frac{3 \times 1.787}{2 \times 3.96} = \mathbf{42.6^\circ}$$

► **Example 14.** *Calculate, from the following data, the wavelength of the neutron beam and its speed. Spacing between successive* (100) *planes* = 3.84 A; *grazing angle* = 30°; *order of Bragg reflection* = 1. **(Himachal Univ.)**

Solution: Here, we have, $d = 3.84\,\text{A} = 3.84 \times 10^{-10}$ m; $\theta = 30^\circ$, $n = 1$.

$\therefore$ From Bragg's law, $2d\sin\theta = n\lambda$, we have

$$2 \times 3.84 \times 10^{-10} \times \sin 30^\circ = 1 \times \lambda$$

$$\Rightarrow \quad \lambda = 2 \times 3.84 \times 10^{-10} \times \tfrac{1}{2} = 3.84 \times 10^{-10}\,\text{m} = \mathbf{3.84\ A}$$

From de Broglie's relation, we obtain

$$\lambda = \frac{h}{mv} \quad \Rightarrow \quad v = \frac{h}{m\lambda}$$

where v = speed of neutron, m = mass of neutron, h = Planck's constant.

$$\therefore \quad v = \frac{6.62 \times 10^{-34}}{1.67 \times 10^{-27} \times 3.84 \times 10^{-10}} = \mathbf{1.03 \times 10^3\ m/s}$$

► **Example 15.** *An X-ray beam of wavelength* $\lambda = 0.71$ A *is diffracted by a cubic* KCl-*crystal of density* 1.99×10^3 kg/m^3. *Calculate the interplanar spacing for* (200) *planes and the glancing angle for the second order reflection from these planes. Given: molecular weight of* KCl = 74.6 u *and Avogadro number* = 6.023×10^{26} per k. mole.

Solution: For a cubic crystal, $a^3 = nM/N_A\rho$ where M is the molecular weight, N_A the Avogadro number, a the lattice constant, ρ the density and n the number of molecules in unit cell.

KCl has the same structure as NaCl. So, $n = 4$

$$\therefore \quad a^3 = \frac{4 \times 74.6}{6.023 \times 10^{26} \times 1.99 \times 10^3} = 0.249 \times 10^{-27}\ \text{m}^3$$

$$\therefore \quad a = 6.29 \times 10^{-10}\,\text{m}$$

∴ Interplanar spacing for (200)-plane is

$$\boldsymbol{d_{200}} = \frac{a}{\sqrt{2^2+0^2+0^2}} = \frac{a}{2} = \frac{6.29\times10^{-10}}{2}\,\text{m} = \mathbf{3.145\ A}$$

From Bragg's law, $2d\sin\theta = n\lambda$. For the second order $n = 2$.

$$\therefore\quad 2d\sin\theta = 2\lambda \Rightarrow \sin\theta = \frac{\lambda}{d} = \frac{0.71}{3.145} = 0.4610$$

$$\therefore\quad \text{Glancing angle, } \boldsymbol{\theta} = \mathbf{27.5°}$$

▶ **Example 16.** *A powder camera of radius* 57.3 mm *is used to obtain diffraction pattern of gold* (fcc) *having a lattice parameter of* 4.08 A. *The radiation used is* Mo-K_α *of* $\lambda = 0.71$ A. *Calculate the first two S-values, where S is the distance between the arcs.*

Solution: Here $R = 57.3$ mm, $\lambda = 0.71$ A, $a = 4.08$ A.

The $(h^2+k^2+l^2)$ values for the first two reflections for an fcc-crystal are 3 and 4.

From Bragg's law, for first order, $2d\sin\theta = \lambda$

And for cubic crystal, $d_{hkl} = a/\sqrt{h^2+k^2+l^2}$

$$\therefore\quad \sin^2\theta = \frac{\lambda^2}{4d^2} = \frac{\lambda^2}{4a^2}(h^2+k^2+l^2)$$

For $h^2+k^2+l^2 = 3$, $\quad \sin^2\theta_1 = \dfrac{3\times(0.71)^2}{4\times(4.08)^2} = 0.0227 \Rightarrow \theta_1 = 8.67°$

For $h^2+k^2+l^2 = 4$, $\quad \sin^2\theta_2 = \dfrac{4\times(0.71)^2}{4\times(4.08)^2} = 0.0303 \Rightarrow \theta_2 = 10.02°$

Now, S (in mm) $= 4\theta$ (dgrees)

$$\therefore\quad \boldsymbol{S_1} = 4\times 8.67 = \mathbf{34.68°}$$

$$\boldsymbol{S_2} = 4\times 10.02 = \mathbf{40.08°}$$

▶ **Example 17.** *The primitive translation vectors of a hexagonal space lattice are*

$$\vec{a} = (a/2)\hat{x} + (a/16)\hat{y};\quad \vec{b} = -(a/2)\hat{x} + (a/16)\hat{y};\quad \vec{c} = c\hat{z}.$$

Find (i) *the volume of the primitive cell and* (ii) *the primitive translations of the reciprocal lattice.* **(Himachal Univ.)**

Solution: (i) The volume V of the primitive cell is given by

$$V = \vec{a}\cdot(\vec{b}\times\vec{c})$$

But, $\vec{b}\times\vec{c} = [-(a/2)\hat{x} + (a/16)\hat{y}]\times c\hat{z} = (ac/2)\bar{y} + (ac/16)\hat{x}$ (2.15.2)

$$\therefore\quad \boldsymbol{V} = \vec{a}\cdot(\vec{b}\times\vec{c}) = [(a/2)\hat{x} + (a/16)\hat{y}]\cdot[(ac/2)\hat{y} + (ac/16)\hat{x}]$$

$$= \frac{a^2c}{32} + \frac{a^2c}{32} = \frac{\mathbf{a^2c}}{\mathbf{16}} \qquad (2.15.3)$$

(ii) Let $\vec{A}, \vec{B}, \vec{C}$ be the primitive translations of reciprocal lattice.

$$\therefore \quad \vec{A} = 2\pi \frac{\vec{b} \times \vec{c}}{\vec{a} \cdot (\vec{b} \times \vec{c})}$$

Using (2.15.2) and (2.15.3) above, we get

$$\vec{A} = 2\pi \frac{(ac/2)\hat{y} + (ac/16)\hat{x}}{a^2c/16} = \boldsymbol{(2\pi/a)\hat{x} + (16\pi/a)\hat{y}}$$

$$\text{Again,} \quad \vec{B} = 2\pi \frac{\vec{c} \times \vec{a}}{\vec{b} \cdot (\vec{c} \times a)}$$

Now, $\quad \vec{c} \times \vec{a} = c\hat{z} \times [(a/2)\hat{x} + (a/16)\hat{y}] = (ac/2)\hat{y} - (ac/16)\hat{x}$

$$\therefore \quad \vec{B} = \frac{2\pi(ac/2)\hat{y} - 2\pi(ac/16)\hat{x}}{a^2c/16} = \boldsymbol{-(2\pi/a)\hat{x} + (16\pi/a)\hat{y}}$$

$$\text{Also,} \quad C = 2\pi \frac{\vec{a} \times \vec{b}}{\vec{c} \cdot (\vec{a} \times \vec{b})}$$

Again, $\quad \vec{a} \times \vec{b} = [(a/2)\hat{x} + (a/16)\hat{y}] \times [-(a/2)\hat{x} + (a/16)\hat{y}]$

$$= (a^2/32)\hat{x} + (a^2/32)\hat{z} = (a^2/16)\hat{z}$$

$$\therefore \ \vec{C} = 2\pi(a^2/16)\hat{z}/(a^2c/16) = \boldsymbol{(2\pi/c)\hat{z}}$$

► **Example 18.** *Two dimensional lattice has the basis vector $\vec{a} = 2\hat{x}$; $\vec{b} = \hat{x} + 2\hat{y}$. Determine the reciprocal lattice vectors.* **(Punjab Univ.)**

Solution: The plane of the reciprocal lattice vector $\vec{A}$ and $\vec{B}$ will be the plane of the basis vector $\vec{a}$ and $\vec{b}$.

But, $\quad \vec{A} = \dfrac{\vec{b} \times \vec{c}}{\vec{a} \cdot (\vec{b} \times \vec{c})} \quad$ and $\quad \vec{B} = \dfrac{\vec{c} \times \vec{a}}{\vec{b} \cdot (\vec{c} \times \vec{a})}$.

Now, $\quad \vec{b} \times \vec{c} = (\hat{x} + 2\hat{y}) \times \hat{z} = -\hat{y} + 2\hat{x}$

$$\vec{c} \times \vec{a} = \hat{z} \times 2\hat{x} = 2\hat{y}$$

$$\therefore \quad \vec{a} \cdot (\vec{b} \times \vec{c}) = 2\hat{x} \cdot (-\hat{y} + 2\hat{x}) = 4$$

$$\vec{b} \cdot (\vec{c} \times \vec{a}) = (\hat{x} + 2\hat{y}) \cdot 2\hat{y} = 4$$

$$\therefore \quad \boldsymbol{\vec{A}} = \frac{-\hat{y} + 2\hat{x}}{4} = \boldsymbol{\frac{1}{2}\hat{x} - \frac{1}{4}\hat{y}}; \quad \boldsymbol{\vec{B}} = \frac{2\hat{y}}{4} = \boldsymbol{\frac{1}{2}\hat{y}}$$

$\boldsymbol{\vec{A}}$ and $\boldsymbol{\vec{B}}$ are the required reciprocal vectors in two dimensions.

● Note that by assuming the third dimensional lattice basis vector $\vec{c} \parallel z$-axis, i.e., $\vec{c} = \hat{z}$, we used three-dimensional relation in the given two-dimensional problem.

▶ **Example 19.** *If* $\vec{A} = \dfrac{\vec{b}\times\vec{c}}{\vec{a}\cdot(\vec{b}\times\vec{c})}$; $\vec{B} = \dfrac{\vec{c}\times\vec{a}}{\vec{b}\cdot(\vec{c}\times\vec{a})}$ *and* $\vec{C} = \dfrac{\vec{a}\times\vec{b}}{\vec{c}\cdot(\vec{a}\times\vec{b})}$, *then prove that*

$$\vec{a} = \frac{\vec{B}\times\vec{C}}{\vec{A}\cdot(\vec{B}\times\vec{C})},\quad \vec{b} = \frac{\vec{C}\times\vec{A}}{\vec{B}\cdot(\vec{C}\times\vec{A})}\quad \text{and}\quad \vec{c} = \frac{\vec{A}\times\vec{B}}{\vec{C}\cdot(\vec{A}\times\vec{B})}.$$

(Allahabad Univ.)

Solution: Here, we have $\vec{A} = \dfrac{\vec{b}\times\vec{c}}{\vec{a}\cdot(\vec{b}\times\vec{c})} \Rightarrow \vec{A}\cdot\vec{a} = \dfrac{\vec{a}\cdot(\vec{b}\times\vec{c})}{\vec{a}\cdot(\vec{b}\times\vec{c})} = 1$

$$\therefore\ \vec{a} \text{ will be} = \frac{\vec{B}\times\vec{C}}{\vec{A}\cdot(\vec{B}\times\vec{C})},\quad \text{if } \vec{a}\cdot\vec{A} = 1$$

Now, $\vec{a}\cdot\vec{A} = \vec{A}\cdot\vec{a} = \dfrac{\vec{A}\cdot(\vec{B}\times\vec{C})}{\vec{A}\cdot(\vec{B}\times\vec{C})} = 1 \Rightarrow \boldsymbol{\vec{a} = \dfrac{\vec{B}\times\vec{C}}{\vec{A}\cdot(\vec{B}\times\vec{C})}}$

Similarly, $\boldsymbol{\vec{b} = \dfrac{\vec{C}\times\vec{A}}{\vec{B}\cdot(\vec{C}\times\vec{A})}}$ and $\boldsymbol{\vec{c} = \dfrac{\vec{A}\times\vec{B}}{\vec{C}\cdot(\vec{A}\times\vec{B})}}$

▶ **Example 20.** *The spacing of planes in a crystal is* 1.2 A *and the angle for the first order of Bragg reflection is* 30°. *Determine the energy of X-rays.* **(WBUT)**

Solution: For Bragg reflection, $2d\sin\theta = n\lambda$, with usual notations.

Here, $\theta = 30°$, $d = 1.2$ A $= 1.2\times10^{-8}$ cm and $n = 1$. Therefore,

$$\lambda = \frac{2d\sin\theta}{n} = \frac{2\times1.2\times10^{-8}\times\sin30°}{1} = 1.2\times10^{-8}\text{ cm}$$

$\therefore$ Energy of X-rays, $E = h\nu = hc/\lambda$

$$= \frac{(6.62\times10^{-27})\times(3\times10^{10})}{1.2\times10^{-8}}\text{ erg}$$

$$= \frac{(6.62\times10^{-27})\times(3\times10^{10})}{(1.2\times10^{-8})\times(1.6\times10^{-12})}\text{ eV}$$

$$= 10.34\times10^{3}\text{ eV} = \mathbf{10.34\ eV}$$

▶ **Example 21.** *Prove that in determining the lattice parameter, the greater the diffraction angle, greater is the accuracy.* **(Punjab Univ.)**

Solution: By Bragg's law, $2t\sin\theta = n\lambda$, where t is the interplanar spacing in the crystal.

$$\therefore\ t\sin\theta = n\lambda/2.$$

Differentiating, we get:

$$t\cos\theta\,d\theta + \sin\theta\,dt = 0$$

$$\therefore\ \sin\theta\,dt = t\cos\theta\,d\theta \quad \text{(considering magnitude only)}$$

$$\therefore\ \frac{dt}{t} = \frac{\cos\theta}{\sin\theta}d\theta$$

Now, dt/t is the *error* in determining the lattice parameter t, the distance between two nearest parallel planes.

As $\theta \to 90°$, $\cos\theta \to 0$ and $\sin\theta \to 1$. So the *error in t* approaches zero. Hence, *greater the value of diffraction angle θ*, smaller is the error and *greater the accuracy.*

► **Example 22.** *A set of crystal planes are separated by 1.95 Å. If X-rays of wavelength λ = 1.542 Å is used, find all possible Bragg angles for reflection from these planes.*

Solution: For Bragg reflection, with usual notation, we have

$$2d\sin\theta = n\lambda$$

For $n = 1$, $\theta = \theta_1$, say. We then have, $\sin\theta_1 = \lambda/2d = \dfrac{1.542}{2 \times 1.95} = 0.3953$

$$\therefore\ \theta_1 = 23°$$

For $n = 2$, $\theta = \theta_2$, say. We then have, $\sin\theta_2 = 2(\lambda/2d) = 0.7906$

$$\therefore\ \theta_2 = 52°$$

For $n = 3$, $\theta = \theta_3$, say. We then have, $\sin\theta_3 = 3(\lambda/2d) = 1.1859$ which is impossible, since sine of an angle cannot exceed 1.

∴ Possible Bragg angles are **23°** and **52°**

Review Questions

1. Discuss the Bragg's treatment of X-ray diffraction by crystals. How does Bragg reflection differ from an ordinary reflection?

2. Discuss briefly the method of determination of crystal structure by Bragg's law. **(Bangalore Univ.)**

3. Sate and prove Bragg's law and verify it for NaCl crystal from the following data: Bragg angles are 11.8°, 23.5° and 36°. **(Allahabad Univ.)**

4. Derive Bragg's law of X-ray diffraction $2d\sin\theta = n\lambda$. How is the law modified due to refraction at the crystal surface? **(Calcutta Hons.)**

5. Name the following standard experimental methods of X-ray diffraction and describe in brief (a) the Laue method, (b) rotating crystal method and (c) the powder method.

6. What is reciprocal lattice? Derive the expression for the primitive translation vectors of reciprocal lattice. How is reciprocal lattice constructed? List its important properties. **(Punjabi and Himachal Univ.)**

7. Derive Laue's equation of diffraction of X-rays and obtain Bragg's diffraction condition from them. How does the Laue approach differ from that of Bragg?
(Calcutta Hons. and Himachal Univ.)

8. Show that every reciprocal lattice vector is perpendicular to a direct lattice plane. Calculate the spacing between the adjacent planes of a crystal in terms of reciprocal space lattice.

9. Discuss how the reciprocal lattice concept helps in the Ewald's construction and the determination of the crystal structure. **(Lucknow Univ.)**

10. State the merits of the reciprocal lattice over the direct space lattice in structure analysis of crystals.

11. What is atomic scattering factor? Derive an expression for it. Explain its significance. **(Kurukshetra Univ.)**

12. Define geometrical structure factor. Derive an expression for the scattering amplitude in terms of geometrical structure factor.

13. Calculate the geometrical structure factor for simple cubic, body-centred cubic and face-centred cubic lattices.

14. Show that any reciprocal lattice vector is normal to a lattice plane of the crystal lattice. Also, prove that the spacing d of (hkl) planes of a crystal lattice is $2\pi n/K$, where K is given by
$$K = |h\vec{a}^* + k\vec{b}^* + l\vec{c}^*|$$
and n is an integer. **(Burdwan Hons.)**

15. Show that in vector form, the Bragg's law is given by
$$G^2 + 2k \cdot G = 0$$
where k represents the wave vector and G is the recirpcal lattice vector.

16. Show from Bragg's ideas and experiments that in NaCl structure, Na and Cl atoms are arranged in *fcc* lattices.

17. On the basis of Ewald's construction, find the geometrical significance of the Bragg equation in the reciprocal lattice space. **(Burdwan Hons.)**

18. Obtain the geometrical structure factor for the *fcc* crystal having identical atoms. Determine the possible and missing X-ray reflections for the crystal.

19. If V is the volume of a primitive unit cell in the direct lattice, show that the volume of a primitive unit cell in the reciprocal lattice is $8\pi^3/V$. **(Burdwan Hons.)**

20. Deduce Bragg's law in vector form in terms of the reciprocal lattice. Also, convert the vector form back into the usual scalar form of Bragg's law.

21. Show that the reciprocal lattice of an fcc lattice is a bcc lattice. **(Calcutta Hons.)**

22. Describe briefly the basic principles of X-ray powder diffraction. **(Burdwan Hons.)**

23. Describe how the structural arrangement of KCl and NaCl are ascertained through X-ray diffraction. **(Vidyasagar Hons.)**

24. Draw a sketch, with necessary labelling, of an X-ray spectrometer and discuss its working.

25. What are Brillouin zones? Explain their significance.
Discuss how Brillouin zones of BCC and FCC lattice can be constructed.

26. Write short explanatory notes on:

(a) Ewald's construction and its utility,

(b) Reciprocal lattice and its properties,

(c) Bragg's law and its importance,

(d) Structure factor for the *fcc*-lattice,

(e) Atomic scattering factor.

Short and Objective Questions

1. Why are X-rays used for the crystal structure analysis? **(Himachal Univ.)**

2. (a) State Bragg's law of X-ray diffraction and its importance in crystal structure analysis. **(N. Bengal Hons. and Lucknow Univ.)**

(b) Deduce Laue equation for X-ray diffraction for periodic crystals. **(Vidyasagar Hons.)**

3. Write down Laue's equations and hence obtain Bragg's condition. **(Lucknow Univ.)**

4. Name the standard experimental methods of X-ray diffraction. **(Guru Nanak Univ.)**

5. Mention, in brief, the relative usefulness of Laue, powder and rotating crystal methods in studying the crystal structure. **(Himachal Univ.)**

6. Define atomic scattering factor. Name the factors on which its value depends.

7. What is geometrical structure factor? Calculate its value for simple cubic lattice.

8. Explain the fact that (100) reflection line vanishes for metallic sodium, but not for ceasium chloride although both have *bcc* structure.

9. What is reciprocal lattice? List its important properties. **(Allahahad Univ.)**

10. Show that the reciprocal lattice to the *sc* direct space lattice of lattice constant a is itself an *sc* lattice of constant $2\pi/a$.

11. Show that every reciprocal lattice vector is perpendicular to a direct lattice plane.

12. Prove that the reciprocal lattice to the body-centred cubic lattice is face-centred cubic lattice and conversely. **(Calcutta, Burdwan Hons. and Punjab Univ.)**

13. *bcc* lattice is the reciprocal of

(a) *bcc* lattice itself, (b) *fcc* lattice,
(c) *sc* lattice, (d) none of these. **(Himachal Univ.)**

14. The primitive translation vectors have the dimensions of

(a) length, (b) $(\text{length})^2$, (c) $(\text{length})^{-1}$, (d) $(\text{length})^{-2}$.
(Himachal Univ.)

15. Can the X-radiation having wavelength greater than $2d$ be diffracted from the crystal?

(a) Yes, (b) No, (c) May be in some situations.

16. If the angle between the incident X-ray beam and the diffracted one be 16°, the angle of incidence will be

(a) 32°, (b) 90°, (c) 82°, (d) 24°.

17. For an *fcc* crystal, which one of the following reflections is absent?

(a) (111), (b) (200), (c) (100), (d) None of these.

18. On being subjected to a monochromatic X-ray beam, a crystal gives the first order diffraction at an angle of 15°. If the same beam is used, what would be the angle corresponding to the third order?

(a) 15°, (b) 31°, (c) 51°, (d) none of these.

19. The volume of the primitive cell of the fcc reciprocal lattice is

(a) $4\left(\frac{2\pi}{a}\right)^3$, (b) $4\left(\frac{\pi}{a}\right)^3$, (c) $4\left(\frac{\pi}{2a}\right)^3$, (d) $4\left(\frac{3\pi}{a}\right)^3$. **(GATE)**

20. For a NaCl crystal, the cell-edge $a = 0.563$ nm. The smallest angle at which Bragg reflection can occur corresponds to a set of planes of indices

(a) 100, (b) 110, (c) 111, (d) 200. **(GATE)**

21. Monochromatic X-rays of wavelength 1 Å are incident on a simple cubic crystal. The first order Bragg reflection from (311) plane occurs at an angle of 30° from the plane. The lattice parameter of the crystal in Å is

(a) 1, (b) $\sqrt{3}$, (c) $\sqrt{11/2}$, (d) $\sqrt{11}$. **(JAM)**

22. In an X-ray diffraction experiment on a material with lattice parameter 4.0 Å, the first peak obtained is at $2\theta = 10°$. If the same experiment (with the same wavelength of X-rays) is performed on a second material with the same crystal structure but a lattice parameter of 2.0 Å, the first peak will appear at a value of 2θ most nearly equal to

(a) 5°, (b) 10°, (c) 20°, (d) 40°. **(IISc)**

Problems

1. Calculate the Bragg angles for X-rays having wavelength 1.51 A in different orders 1, 2 and 3 if the inerplanar distance is 2.67 A.
(Allahabad Univ.) $[16.76°, 35.22°, 59.9°]$

2. A certain crystal reflects monochromatic X-rays strongly when Bragg glancing angle (first order) is 15°. What are the glancing angles for the second and the third order spectrum? **(Himachal Univ.)** $[31.17°, 50.93°]$

3. When X-rays from Pd are diffracted from the crystal, the glancing angle for the first order is measured to be 3.66°. Hence calculate the wavelength of the incident X-rays, assuming the reflection to occur from the (100) plane. **(Calcutta Hons.)**

4. From the following data, calculate the wavelength of neutron beam and its speed. Spacing between successive (100) planes = 3.84 A; grazing angle 30°; order of Bragg reflection = 1. **(Himachal Univ.)** [3.84 A, 1.03×10^3 m/s]

5. Copper (*fcc*) has a lattice parameter of 3.61 A. The first order Bragg reflection from (111) planes appears at an angle of 21.7°. Determine the wavelength of X-rays. [1.54 A]

6. Consider X-rays of wavelength 1.54 A incident on a single cubic crystal of lattice constant 4 A. Calculate the angle for the first order ($n = 1$) reflection.
(Calcutta Hons.) [11°]

7. In an X-ray diffraction experiment using radiations of $\lambda = 1.54$ A, the first reflection from an *fcc* crystal occurs when $2\theta = 84°$. Find the indices of this reflection and the corresponding interplanar spacing. Show that only one more reflection is possible. Find the indices of that reflection and the corresponding interplanar spacing. [(111), 1.15 A; (200), 0.996 A]

8. The cubic unit cell of *fcc* copper is 3.6 A. Calculate the longest wavelength of X-rays which will produce diffraction from the most closely packed planes.
(Calcutta Hons.) [4.17 A]

[**Hints.** For *fcc* lattice, the (111) plane is most closely packed]

9. The line A of an X-ray beam gives a first order reflection maximum when the glancing angle is 30° to the smooth face of a crystal. Line B of $\lambda = 0.97$ A gives a third order reflection maximum at an angle of 60° from the same face of the crystal. Find the wavelength of line A. **(Kanpur Univ.)** [1.68 A]

10. The spacing of the planes in a crystal is 1.2 A. The angle for the fist order reflection is 30°. Find the energy in eV of the X-rays used. If the spacing changes by ±0.01 A, determine the energy spread in the diffracted beam.
[10.3 keV, ±85.8 eV]

[**Hints.** Energy spread $\Delta E = -(hc/\lambda^2)\Delta\lambda = -(hc/\lambda)\frac{\Delta\lambda}{\lambda} = -E\cdot\frac{\Delta\lambda}{\lambda}$, since $E = h\nu = hc/\lambda$. $E = hc/\lambda = 10.3 \times 10^3$ eV, putting the values of h, c and λ. Since $\lambda = 2d\sin\theta$, $\Delta\lambda = 2(\Delta d)\sin\theta = 2(\pm 0.01)\sin 30° = \pm 0.01$. So, $\Delta E = -E\Delta\lambda/\lambda = -\frac{10.3\times10^3\times(\pm0.01)}{1.2}$ eV $= \mp 85.8$ eV]

11. On an X-ray powder photography of a cubic substance with copper K_α radiation ($\lambda = 1.542$ A) lines are observed at the following Bragg angles (θ in degrees): 12.3, 14.1, 20.2, 24.0, 25.1, 29.3 and 33.2. Assign the Miller indices to these lines and decide whether the lattice *sc*, *bcc* or *fcc*. Calculate also the unit cell dimension. **(Calcutta Hons.)**

[(111), (200), (220), (311), (222), (400), (331); *fcc*; 6.3 A]

[**Hints.** From Bragg's law, $\sin^2\theta = (n\lambda/2a)^2 = (\lambda/2a)^2(n_1^2+n_1^2+n_3^2)$. So, $(n_1^2+n_1^2+n_3^2)$ is proportional to 3, 4, 8, 11, 12, 16, 19 for the seven lines. Hence, the Miller indices are respectively those given in the result. For the (111) plane, $\sin 12.3° = 3(\lambda/2a)^2$ and using the given λ-value, $a = 6.3$ A]

12. The primitive basis vectors of a lattice are $\vec{a} = \vec{i} + 2\vec{j}$; $\vec{b} = 4\vec{j}$ and $\vec{c} = \vec{k}$. Show that the primitive translation vectors of its reciprocal lattice are

$$\vec{a}^* = 2\pi\vec{i}; \quad \vec{b}^* = \pi(-2\vec{i} + \vec{j})/2; \quad \vec{c}^* = 2\pi\vec{k}.$$

13. An *sc* crystal, illuminated with X-rays of $\lambda = 0.09$ nm, is rotated till the first order Bragg reflection occurs at a minimum glancing angle of 8.8°. Determine the set of planes responsible for this reflection. Find the spacing between these planes. Find also the angle for the first order reflection from (110) crystal planes.

[(100), 0.294 nm, 12.5°]

14. The first order Bragg reflection from the most closely packed planes of a *bcc* crystal of lattice parameter 2.87 A appears at the glancing angle of 22.3°. Calculate the wavelength of X-rays. [1.54 A]

[**Hints.** The planes (110) are most closely packed in *bcc* lattice]

15. Compute the energies in eV corresponding to a wavelength of 1 A for the X-rays, electrons and neutrons. Given $h = 6.6 \times 10^{-27}$ erg.sec, neutron mass $= 1.7 \times 10^{-24}$ g, electron mass $= 9 \times 10^{-28}$ g and 1 eV $= 1.6 \times 10^{-12}$ erg.

[12370 eV, 151.2 eV, 0.08 eV]

16. The primitive translation vectors of hexagonal space lattice may be assumed as

$$\vec{a} = (\sqrt{3}a/2)\hat{i} + (a/2)\hat{j}; \quad \vec{b} = -(\sqrt{3}a/2)\hat{i} + (a/2)\hat{j}; \quad \vec{c} = c\hat{k}.$$

(a) Show that the volume of the primitive cell is $3^{1/2}a^2c$.

(b) Show that the primitive translations of the reciprocal lattice are

$$\vec{a}^* = (2\pi/\sqrt{3}a)\hat{i} + (2\pi/a)\hat{j}; \quad \vec{b}^* = -(2\pi/\sqrt{3}a)\hat{i} + (2\pi/a)\hat{j}; \quad \vec{c}^* = (2\pi/c)\hat{k}.$$

17. Compute the angle between the reciprocal lattice vectors G_{100} and G_{111} corresponding to a simple cubic lattice. [$\cos^{-1}(1/\sqrt{3})$]

18. The second order Bragg reflection from a set of parallel planes in a crystal with monochromatic X-ray appears at a glancing angle of 31.2°. What is the glancing angle for the first order reflection? [15°]

19. An X-ray tube operation at 30 kV emits continuous X-radiations with $\lambda_{\text{min}} =$ 0.414 A. Compute the value of Planck's constant. [6.624×10^{-34} Js]

[**Hints.** $eV_{\text{max}} = h\nu_{\text{max}} = hc/\lambda_{\text{min}}$. So, $h = (eV_{\text{max}} \times \lambda_{\text{min}})/c$. Substitute the given values and those of fundamental constant to get h.]

20. Calculate the Bragg angle at which electrons accelerated from rest through a potential difference of 80 V will be diffracted from (111) planes of a face-centred cubic crystal of lattice parameter 3.5 A. [19°40′]

21. Find the wavelength of short-wave limit of continuous X-ray spectrum if it shifts by 0.5 A when the voltage applied to X-ray tube is doubled.

(Calcutta Hons.) [0.5 A]

[**Hints.** Since $eV = hc/\lambda$, $\lambda_1 = hc/eV_1$ and $\lambda_2 = hc/eV_2 \Rightarrow \lambda_1/\lambda_2 = V_2/V_1 = 2/1$. So, $\lambda_1 = 2\lambda_2$. By the problem, $\lambda_1 - \lambda_2 = 0.5$ A $\Rightarrow 2\lambda_2 - \lambda_2 = 0.5$ A $\Rightarrow \lambda_2 = 0.5$ A and $\lambda_1 = 2\lambda_2 = 1$ A]

22. Starting from $2\vec{K}\cdot\vec{G} + G^2 = 0$, where the symbols carry usual meanings, determine the Bragg's diffraction condition, $2d\sin\theta = n\lambda$.

23. The molecular weight of KBr (a simple cube) is 119 and its density is 2.75 g/cm^3. Taking Avogadro number as 6.022×10^{23} per mole, calculate the grating space of the Laue grating for this crystal. If a narrow beam of K_α radiation ($\lambda = 1.541$ A) from Cu is reflected from (100)-planes of such a crystal, find the glancing angle for first order spectrum. **(Calcutta Hons.)** [4.157 A, 10.69°]

24. Two dimensional lattice has the basis vectors $\vec{a} = 2\hat{x}$, $\vec{b} = \bar{x} + 2\hat{y}$. Show that the reciprocal lattice vectors are $\pi(\hat{x} - \hat{y}/2)$ and $\pi\hat{y}$.

25. Copper (fcc) has a lattice parameter of 3.61 A, The first order Bragg reflection from (111)-planes appears at an angle 21.7°. Determine the wavelength of X-rays.

[1.54 A]

26. The primitive translation vectors of a two dimensional lattice are $\vec{a} = 2\hat{i} + \hat{j}$, $\vec{b} = 2\hat{j}$. Determine the primitive translational vectors of its reciprocal lattice.

[$\vec{a}^* = \pi\hat{i}$, $\vec{b}^* = \frac{\pi}{2}(\hat{i} + 2\hat{j})$]

27. The S values for the first three lines in the powder pattern of a cubic crystal are 34.88, 40.36 and 54.40 mm respectively. Given the wavelength of X-radiation used is 0.71 Å and the camera radius 57.30 mm, determine the crystal structure and the lattice parameter.

28. In an X-ray tube, electrons are accelerated by 844 volt to produce X-rays which are reflected from a crystal. The reflection maximum occurs when the glancing angle is 58°. Find the interplanar spacing in the crystal. [0.248 Å]

29. Consider a monatomic *fcc* solid with lattice constant $\sqrt{3}$ Å. Find (i) the interplanar spacing of a set of parallel (111) planes and (ii) the incident angle θ for which the first order peak would be observed of monochromatic X-rays of $\lambda = 1$ Å incident on these planes. (JAM) [1 Å, $\theta = \pi/6$]

3

Imperfections in Crystals, Liquid Crystals and Quasicrystals

This chapter[1] is intended to describe and discuss some special topics such as defects or imperfections in crystals, liquid crystals and quasicrystals. The discussion would be rather in an introductory level and in no way can be claimed to be extensive.

3.1 Imperfections in a crystal: Introduction and classification

In an earlier chapter, we described the ideal crystals having perfectly periodic arrangement of atoms in the lattice. Based on the ideal crystal structure, many *structure-insensitive* properties like density, specific heat, elastic properties etc. can be explained. But, the ideal concept is true only at absolute zero of temperature when there is no entropy contribution to disorder. At any finite temperature, however, certain structural disorder is bound to be introduced as a consequence of the laws of thermodynamics. This leads to the development of various imperfections or defects in the structure of the crystal. These defects strongly influence *structure-sensitive* properties of solids such as strength, electric conductivity, hysteresis loss of ferromagnets etc.

But what is meant by *imperfections* in a crystal? By ***imperfections** or **defects** in a crystal, we mean any deviation from the ideal crystal structure of perfect periodic arrangement of its constituents.*

[1]May be skipped in the first reading without the loss of continuity.

Crystal imperfections or defects that are normally observed can be broadly classified into the following different types. Under each type, again, there may be a number of sub-classes.

(a) *Point* (or zero-dimensional) *imperfections*,

(b) *Line* (or one-dimensional) *imperfections*,

(c) *Surface* (or two-dimensional) *imperfections*, and

(d) *Volume* (or three-dimensional) *imperfections.*

We shall essentially restrict ourselves to the *first two* kinds of defects : *Point defects* and *line defects. Surface imperfections* refer to regions of distortions that lie about a surface with a thickness of few atomic diameters and include grain boundary, tilt and twist boundary, twin boundary and stacking faults. *Volume imperfections* refer to particle inclusion, large voids or non-crystalline regions having dimensions of at least ten of angstrom.

3.2 Point imperfections

If the deviation from the ideal crystal structure is due to missing atoms (vacancies), displaced atoms or extra atoms, the imperfection is known as a ***point imperfection*** or ***point defect***. Such defects may arise due to imperfect packing of atoms during the process of crystallization or may be due to thermal vibrations of atoms at elevated temperatures when some atoms may jump out of their positions of lowest energy.

Most common among the point defects are (a) *vacancy defects*, (b) *Schottky defect* and (c) *Frenkel defects*. Comparatively less common are (d) *metal excess defects* and (e) the *metal deficiency defects*.

3.2.1 Vacancy defects

A *vacancy defect* is the simplest point defect in a crystal and refers to a missing atom or a vacant atomic site. Such defects are rather common in elemental solids. They may arise from an imperfect packing during crystallization or from thermal vibrations at high temperatures.

Vacancy defects in elemental solids

A *vacancy* is an atom site, normally occupied in a perfect crystal, from which an atom is missing. Formally, such a defect may be developed in taking out an interior atom of its lattice site. This requires some energy and the disorder increases the entropy. In thermal equilibrium, a certain number of vacancies are always present and may be estimated as follows.

Estimation of vacancies—Let E_v be the energy required to take an atom from its lattice site inside the crystal to produce a vacancy and n be the number of isolated vacant sites generated in the solid. So, the associated energy-increase is nE_v.

The total number of ways of picking up n-atoms from the crystal consisting of N atoms ($N \gg n$) is

$$W = \frac{N!}{(N-n)!n!}$$

The corresponding increase in entropy is given by

$$S = k \ln W = k \ln \frac{N!}{(N-n)!n!}$$

This in turn produces a change in free energy

$$F = U - TS = nE_v - kT \ln \frac{N!}{(N-n)!n!}$$
$$= nE_v - kT[N \ln N - (N-n)\ln(N-n) - n \ln n],$$

using Stirling's approximation: $\ln x! \simeq x \ln x - x$ and simplifying.

Free energy F in *thermal equilibrium* at a given temperature T is a constant.

$$\therefore \quad \left(\frac{\partial F}{\partial n}\right)_T = 0 = E_v - kT \ln \frac{N-n}{n}$$

$$\therefore \quad \frac{E_v}{kT} = \ln \frac{N-n}{n} \simeq \ln \frac{N}{n} \quad (\because\ N \gg n)$$

$$\therefore \quad \boxed{n = Ne^{-E_v/kT}}$$

which gives the *equilibrium concentration* of vacancies.

Note and check that the *concentration of vacancies decreases with decrease in temperature and conversely.*

3.2.2 Schottky defects

In this defect, some of the lattice in a crystal remain unoccupied, that is, missing or vacant. They are called *lattice vacancies.* Such defects arise either due to imperfect packing during crystallization or thermal vibrations at high temperatures. Vacancies may be single or two or more. We illustrate two such vacancies in a crystal of NaCl in Fig. 3.1. One of the vacancies is due to missing Na^+ ion and the other due to a missing Cl^- ion. The crystal however remains electrically neutral as the number of missing positive and negative ions is the same.

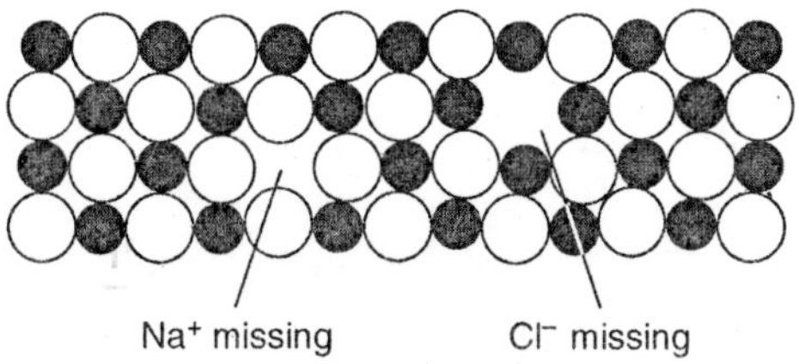

Fig. 3.1 Schottky defects in a crystal of NaCl

Such defects generally appear in ionic crystals in which the two different types of atoms do not differ much in size. NaCl, CsCl are good examples of ionic crystals where Schottky defects are found to occur.

Due to Schottky defects, the crystal may ionically conduct electricity to a small extent. What happens is this: With the application of the electric field, a nearby ion moves from its lattice to occupy a vacancy and create thereby a new vacancy to which another ion moves and the process thus goes on. This results in the migration of the vacancy from one end of the crystal to the other.

Number of Schottky defects in ionic crystals

Let n Schottky defects be produced by the removal of n cations and n anions from a binary ionic crystal containing a total number of N cation-anion pairs ($N \gg n$).

The different ways in which n separated *ion pairs* can be created is

$$W = \left[\frac{N!}{(N-n)!n!}\right]^2 \tag{3.2.1}$$

∴ Increase in entropy in the process is, by Boltzmann equation,

$$S = k \ln W = k \ln \left[\frac{N!}{(N-n)!n!}\right]^2 \tag{3.2.2}$$

This in turn produces a change in free energy

$$F = U - TS = nE_p - kT \ln \left[\frac{N!}{(N-n)!n!}\right]^2 \tag{3.2.3}$$

where E_p is the energy of formation of a pair.

Applying Stirling's approximation to the factorial term in (3.2.3), we have

$$\begin{aligned} \ln \left[\frac{N!}{(N-n)!n!}\right]^2 &= 2\left[\ln N! - \ln(N-n)! - \ln n!\right] \\ &= 2\left[N \ln N - N - (N-n)\ln(N-n) + (N-n) - n \ln n + n\right] \\ &= 2\left[N \ln N - (N-n)\ln(N-n) - n \ln n\right] \end{aligned}$$

∴ Free energy, $F = nE_p - 2kT\left[N \ln N - (N-n)\ln(N-n) - n\ln n\right]$ (3.2.4)

by using (3.2.3).

Differentiating (3.2.4) with respect to n, we get

$$\begin{aligned} \left(\frac{\partial F}{\partial n}\right)_T &= E_p - 2kT[0 + \ln(N-n) + 1 - \ln n - 1] \\ &= E_p - 2kT \ln \frac{N-n}{n} \end{aligned} \tag{3.2.5}$$

At *equilibrium*, the free energy F at a given temperature is constant. This implies $(\partial F/\partial n)_T = 0$. Hence, from (3.2.5), we get

$$E_p - 2kT \ln \frac{N-n}{n} = 0$$

$$\text{or,} \quad \frac{N-n}{n} = \exp(E_p/2kT) \tag{3.2.6}$$

$$\text{or,} \quad \frac{N}{n} = \exp(E_p/2kT) \quad (\because N \gg n)$$

$$\therefore \quad \boxed{n = Ne^{-E_p/2kT}} \tag{3.2.7}$$

which gives the *number of Schottky defects* in binary ionic crystals.

Thus, the number of Schottky defects depends on:

(i) the total number of ion pairs N, i.e., on the mass of the crystal,

(ii) the average energy E_p required to produce a Schottky defect, and

(iii) the temperature T; the fraction of defects increases exponentially with increasing temperature and conversely.

• To illustrate the equilibrium concentrations of vacancies, consider NaCl crystal for which $E_p = 2.02\,\text{eV}$ and at room temperature (300 K).

$$n/N = \exp(-2.02/(2 \times 0.025)) = e^{-40.4}$$
$$= 2.8 \times 10^{-9} \quad (\because kT = 0.025)$$

3.2.3 Frenkel defects

These defects occur when *an ion occupies an interstitial position between the lattice points* i.e., *a position which is not a normal lattice site.* This is illustrated in Fig 3.2 for Ag Br crystal.

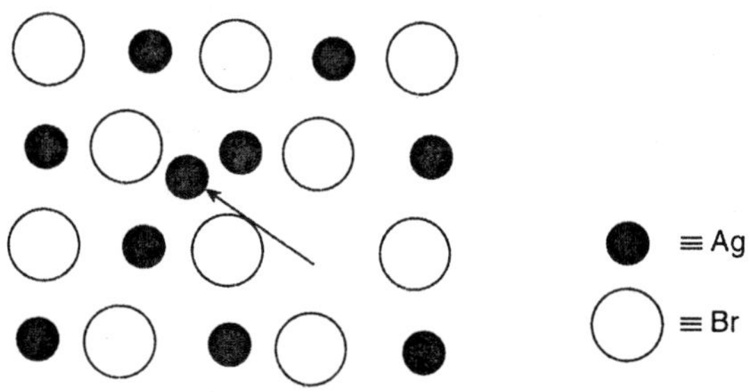

Fig. 3.2 Frenkel defect

Note from the figure that one of the Ag^+ ions occupies a position in the *interstitial position*, instead of its appropriate site in the lattice, and thereby create a vacancy in the lattice. Since the number of positive ions is equal to the number of negative ions, the crystal remains neutral.

Frenkel defects are observed in crystals where the negative ions are much larger in size than the positive ions. As in Schottky defects, the Frenkel defects are also responsible for electric conduction in crystals and also for the phenomenon of *diffusion* in solids.

Examples of crystals showing Frenkel defects are Ag Br, ZnS etc.

● The presence of Ag^+ ions in the interstial space of Ag Br crystal gives rise to the formation of *photographic image* on exposure of Ag Br crystals (i.e., photographic plate) to light.

Number of Frenkel defects

Let an ionic crystal have N number of ions and N_i interstitial spaces in its structure. The number of ways in which n Frenkel defects can be created is

$$W = \frac{N!}{(N-n)!n!} \cdot \frac{N_i!}{(N_i-n)!n!} \tag{3.2.8}$$

∴ Increase in entropy in the process is

$$S = k \ln W = k \ln \left[\frac{N!}{(N-n)!n!} \cdot \frac{N_i!}{(N_i-n)!n!}\right] \tag{3.2.9}$$

This in turn produces a change in free energy

$$F = U - TS = nE_i - kT \ln \left[\frac{N!}{(N-n)!n!} \cdot \frac{N_i!}{(N_i-n)!n!}\right] \tag{3.2.10}$$

Using Starling's approximation to the logarithmic term and simplifying,

$$\ln \left[\frac{N!}{(N-n)!n!} \cdot \frac{N_i}{(N_i-n)!n!}\right] = N \ln N + N_i \ln N_i - (N-n)\ln(N-n) \\ -(N_i-n)\ln(N_i-n) - 2n \ln n$$

Substituting this value in (3.2.10) for free energy and differentiating,

$$\left(\frac{\partial F}{\partial n}\right)_T = E_i - kT \ln \frac{(N-n)(N_i-n)}{n^2} \tag{3.2.11}$$

At *equilibrium*, the free energy F is constant $\Rightarrow (\partial F/\partial n)_T - 0$.

$$\therefore \text{ From (3.2.11), } E_i = kT \ln \frac{(N-n)(N_i-n)}{n^2} \simeq kT \ln \frac{NN_i}{n^2} \quad (\because N \gg n) \tag{3.2.12}$$

$$\therefore E_i = kT\left[\ln(NN_i) - 2\ln n\right]$$

$$\Rightarrow \ln n = \tfrac{1}{2}\ln(NN_i) - E_i/2kT$$

$$\therefore \boxed{\boldsymbol{n = \sqrt{NN_i}e^{-E/2kT}}} \tag{3.2.13}$$

which gives the number of Frenkel defects in a crystal.

3.2.4 Metal excess defects: Colour centres

If a crystal of sodium chloride be heated in sodium vapour, it is found to acquire a *yellow colour* due to the formation of a non-stoichiometric NaCl compound, having a slight *excess* of sodium ions. What happens is this: some sodium metal gets *doped* into sodium chloride crystal, each atom of which becomes ionized into Na^+ and e^- by the crystal energy. This electron occupies a site in the crystal which would otherwise have been filled by Cl^- (Fig.3.3).

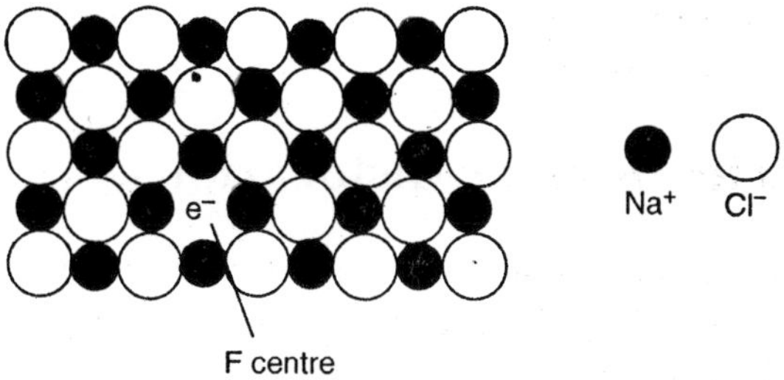

Fig. 3.3 F-centre in a sodium chloride crystal

Plainly, there is an excess of Na^+ ions although the crystal as a whole is neutral, A careful thinking shows that there are, in fact, six Na^+ sites adjacent to the vacant site occupied by e^-. The electron is thus shared by six Na^+ ions implying that the electron under reference is not localized at the vacant Cl^- site. It is similar to de-localized π-electrons in molecules having conjugate double bonds. Light is absorbed when the delocalized electron transits from the ground state to an excited one and colour appears in sodium chloride. This is why the sites occupied by extra electrons are called ***colour centres*** or ***F-centres***.

3.2.5 Metal deficiency defects

In this kind of defects, one of the positive ions is found missing from its lattice site and the extra negative charge is balanced by some metal ion in proximity, acquiring two charges instead of one. Plainly, there is a *deficiency of metal ions* although the crystal, as a whole, is neutral.

Fe^{2+}	O^{2-}	Fe^{2+}	O^{2-}	Fe^{2+}	O^{2-}
O^{2-}	Fe^{2+}	O^{2-}	Fe^{2+}	O^{2-}	Fe^{2+}
Fe^{2+}	O^{2-}	Fe^{2+}	O^{2-}	●	O^{2-}
O^{2-}	Fe^{2+}	O^{2-}	Fe^{2+}	O^{2-}	Fe^{2+}
Fe^{2+}	O^{2-}	Fe^{2+}	O^{2-}	Fe^{2+}	O^{2-}

Fig. 3.4 Metal deficiency defects in a crystal of FeO

Metal deficiency defects generally occur in compounds of transition metals that can show variable valency. Typical examples of crystals showing this type of defects are FeO, NiO and FeS. Fig. 3.4 illustrates such defects in the crystal of FeO.

3.2.6 Compositional defects

These arise from impurity atoms during original crystallization. It is the impurity atoms, considered as defects in an ideal lattice, that are behind the functioning of semiconductor devices. They occur at a lattice point as *substitutional impurity* or as *interstitial impurity*

In substitutional impurity, a foreign atom substitutes for or replaces parent atom in the lattice. For example, zinc is a substitutional atom in copper lattice in brass. An interstitial impurity is a small-sized foreign atom that occupies a space between the regularly positioned parent atoms.

3.3 Line imperfections or dislocations

In addition to point imperfections, another type of imperfections arise when deviation from the perfectly periodic arrangement of atoms occurs *along a line* i.e., along a *certain direction* in a crystal. Such imperfections are called ***line imperfections*** or ***line defects***. Since in this case, the distortion is only along a line, the imperfection may be considered as the boundary between two regions of a surface which are themselves perfect but out of register with each other. The *line of imperfection*, acting as boundary between the slipped and unslipped regions, lies in the slip plane and is called a ***dislocation***. The presence of dislocations and their movement result in easy deformation of a crystal under the influence of shear stress.

- The sliding of one part of the crystal as a unit, relative to another, leads to plastic flow in crystals. The sliding process is termed *slip*. The plane in which slip occurs is known as the *slip plane*.

The *two* common and extreme types of dislocations are (a) the ***edge dislocation*** and (b) the ***screw dislocation***. Any particular dislocation is a mixture of these two extreme ones. These dislocations are the most striking defects responsible for the property called ductility in metals, ceramics and crystalline polymers. We shall describe here the edge dislocation and the screw dislocation with the help of models based on simple cubic lattice.

1. Edge dislocations—An ideal crystal is illustrated in Fig. 3.5. While (a) gives the three-dimensional view, (b) shows the atoms on the front face. An ideal crystal is made-up of vertical planes parallel to one another and to the side faces. The atoms are all in equilibrium positions and all the bond lengths have the equilibrium value.

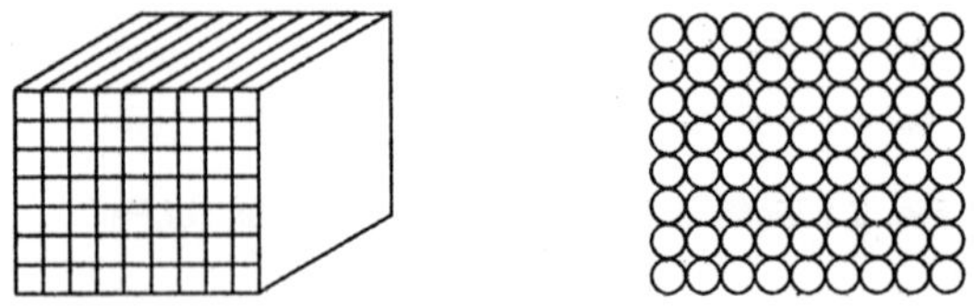

(a) (b)

Fig. 3.5 Ideal crystal with no edge dislocation (a) three-dimensional view, (b) atoms on the front face

If the edge of one of the vertical planes do not extend all the way from the top to the bottom of the crystal, but terminates partway within it (Fig. 3.6a), a dislocation is present. In such an imperfect crystal, the atoms just above the edge

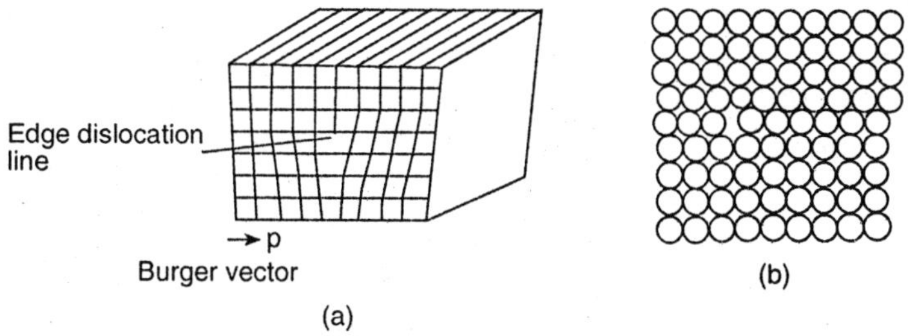

Fig. 3.6 Incomplete plane in a crystal resulting in an edge dislocation (a) termination of a vertical plane partway in the crystal, (b) states of compression and tension of atoms above and below the edge

of the incomplete plane, are squeezed to a state of *compression* (Fig. 3.6b), the bond lengths are compressed to less than the equilibrium value. The atoms just below the edge, however, are pulled apart to a state of *tension* (Fig. 3.6b), the bond lengths being stretched to above normal values. This distorted configuration that extends along the edge into the crystal represents a *line of imperfection*, called ***edge dislocation***.

Due to the impact of the shear, the dislocation moves across the crystal in such a fashion that the top half of the crystal gets displaced by *one lattice distance* relative to the lower half.

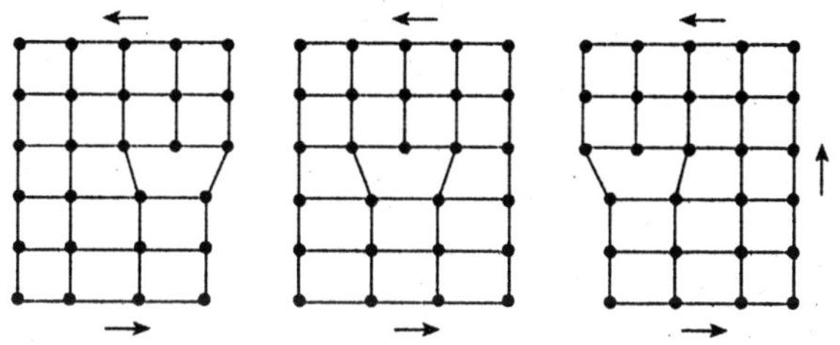

Fig. 3.7 Motion of an edge dislocation

Fig. 3.7 illustrates the motion of an edge dislocation under the influence of shear.

- Since near the edge dislocation, the atoms are pushed together above the edge and pulled apart below the edge, the impurity atoms of dia larger than the parent atoms tend to concentrate below the edge and those of smaller dia concentrate above the edge. Due to binding of impurity at the dislocation, it is more difficult to move a dislocation in an impure material. This explains why *alloys* require greater shear force for permanent deformation compared to pure metals.

- The direction of the line of atoms in the incomplete plane determines the line (or direction) of dislocation. The vector representing the magnitude and direction of the lattice displacement is termed *Burger vector*, which is normal to the dislocation line.

2. Screw dislocation—The formation of a *screw dislocation* is illustrated by the crystal model as shown in Fig. 3.8 where the crystal is cut along $ABEF$ and then its right half is pushed down relative to left half by one atomic spacing. As a result, a step of height equal to the slip vector or Burger vector b is created on the surface of the crystal. It does not extend throughout the surface but is limited to the point F to the edge of the crystal. EF is the boundary of the slipped and unslipped regions—the *dislocation line.* If initially the crystal contained atoms at regular lattice points, the

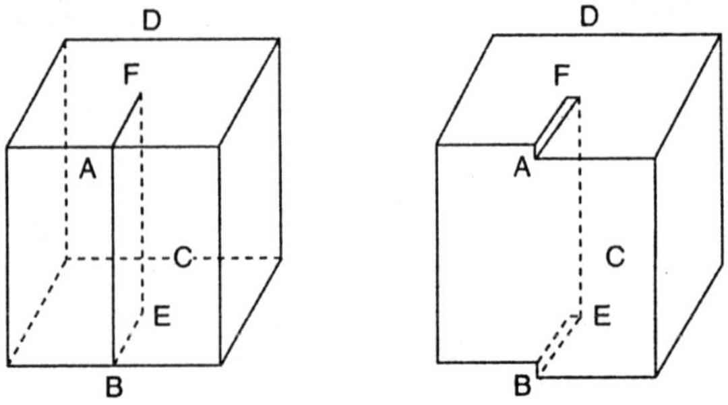

(a) (b)

Fig. 3.8 The formation of screw dislocation

deformation would convert the parallel plane of atoms into a kind of *spiral ramp* and hence such a displacement is termed ***screw dislocation.***

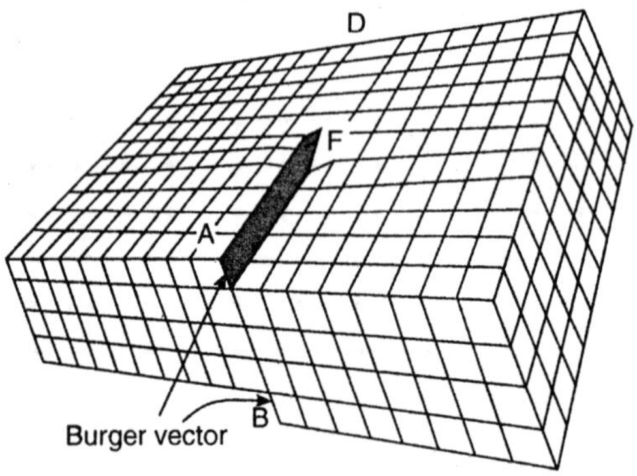

Fig. 3.9 Formation of screw dislocation using simple cubic model

Fig. 3.9 shows the formation of screw dislocation using a simple cubic model where the dislocation line is parallel to the slip vector.

- A type of ***imperfections due to transient atomic displacement*** may also occur in a crystal. Like those of simple harmonic oscillators, the vibrations of a crystal have three normal modes—one *longitudinal* and two *transverse* (mutually perpendicular). When a transition occurs from a higher vibrational to a lower vibrational state, a quantum of thermal energy—***phonon***—is emitted. Absorption of phonons by a crystal may produce atomic displacement, resulting in imperfections. But atomic displacements produced by phonons are *time-dependent* and hence *transient.*

3.4 Liquid crystals

Liquids are spatially *homogeneous* and rotationally *isotropic.* This implies that the average environment of any point in a liquid is identical to that of any other point and is independent of the direction. A crystal however, unlike a liquid, is *not* isotropic and homogenous. It is only invariant under such operations as translations by a lattice spacing that leave the periodic lattice unchanged, rather than with respect to any *arbitrary* displacements and rotations. As the crystal is invariant under fewer operations, it has a lower symmetry compared to a liquid and transition from liquid to crystal breaks the symmetry of the liquid state. Hence a crystal is often called a ***broken symmetry phase.***

Thus, there are two extreme forms of condensed matter: *homogeneous isotropic liquids* and *crystalline solids.* Between these two extremes, there are systems having symmetries intermediate between those of liquids and the crystals. The widely studied systems showing intermediate order are the ***liquid crystals*** that display the properties both of a liquid and of a crystalline medium in a certain range of temperature or concentration of a solvent. These are, in general, formed by stretched molecules, mostly of aromatic compounds. The benene ring gives the rigidity to the molecules.

Classification—Liquid crystals, i.e., substances showing mesomorphic state, can be classified into *three* different phases: *nematic, smectic* and *cholesteric.*

1. Nematic—Nematic means *thread-like.* This phase has the simplest structure with rod-shaped molecules and shows near-normal flow-behaviour of liquids. The flow is Newtonian, but the viscosity is rather low. The molecules are aligned (Fig. 3.10) parallel along a preferred *director axis* ($\vec{L}$) but are spatially distributed at random in different small regions. There is an orientational order and the molecules can move around from one region to another leading to high *optical inhomogeneity* of the crystal. Due to this, the crystal appears *opaque* in both reflected and transmitted lights.

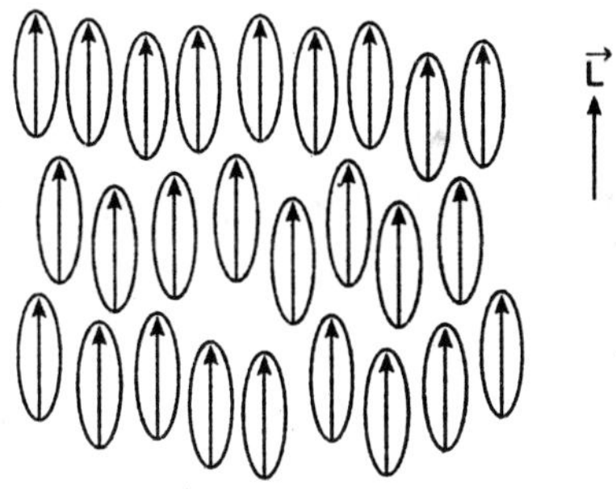

Fig. 3.10 Nematic liquid crystal

An external field E can make the molecules of different regions align in a particular direction and the crystal becomes transparent. Molecules can be aligned *parallel* (planar) and *perpendicular* (hemotropic) to the conducting plates by placing a thin

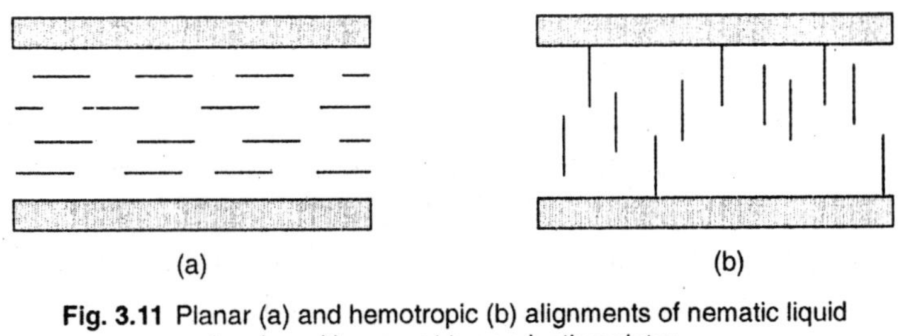

Fig. 3.11 Planar (a) and hemotropic (b) alignments of nematic liquid molecules with respect to conducting plates

layer of nematic liquid between the plate electrodes (Fig. 3.11). The hemotrophic structure has *no effect* on polarisation of transmitted light through the cell at right angles to the liquid crystal layer. The planar structure however has *significant effect* on the polarisation of light. In polarised light, nematic crystals appear *thread-like*, but they are *uniaxial.*

Pure nematic crystals are insulators, but can be made conducting by doping ionic impurities. Examples of nematic crystals are: *p*-Azoxy anisole, *p*-Azoxy phenetole, dibenzyl benzidine etc.

2. Cholesteric—In the *cholesteric phase* also, the molecules are aligned parallel to the *director axis* $\vec{L}$, but the direction of alignment *twist* progressively to have a *helical* structure (Fig. 3.12). The substances thus consist of sheets or layers in each of which the molecules are arranged parallel to each other. The vector $\vec{L}$ rotates by a certain angle in passing from plane to plane and the end of $\vec{L}$ describes a *helix* with a pitch of 0.2–20 μm. The structural periodicity along the helical axis gives rise to ***Bragg reflection*** of λ = pitch/refractive index.

The *optical activity* of cholesteric crystals is *extremely high* and the activity rotates the plane of polarisation through unusually large angles.

If subjected to an external electric field, the *pitch* of the helix *changes.* The field turns the helical axis and the cholesteric crystal gets converted back to nematic one. So, the transmission of light through the liquid crystal between two polarisers will change with change in the field.

Examples of cholesteric crystals are: Cholesteryl benzoate, *l*-cholesteryl esters etc.

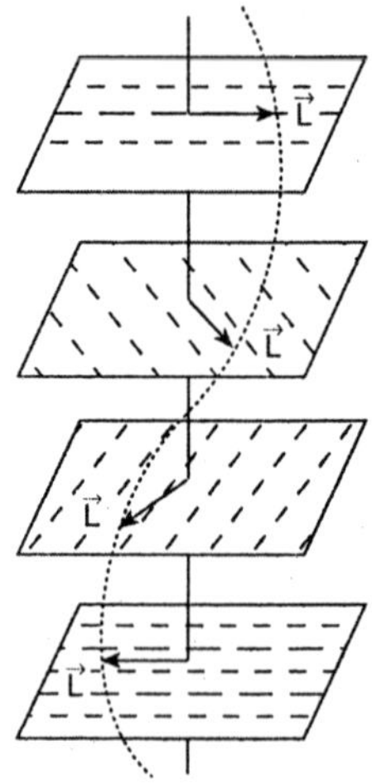

Fig. 3.12 Cholesteric phase of liquid crystal; rotation of $\vec{L}$ from plane to plane

3. Smectic—The *smectic phase* consists of a series of layers wherein the molecules are aligned parallel and normal to the lower plane. The layers *interact* only *weakly* and can *slip* or *glide* past each other or *rotate* relative to each other. This promotes the fluidity of the system, but they do not flow as normal liquids, the flow being non-Newtonian.

The long axis of the molecules are either perpendicular to the planes or rotated through an angle ϕ with normal to the planes (Fig. 3.13). The *smectic axis* is represented by $\vec{L}$ and the shear plane by AA. The existence of layers suggests that the crystal lattice still exists though in somewhat distorted form. The lamellar structure is closest to a solid and it is essentially a *two-dimensional solid.* Smectic crystals also give X-ray diffraction patterns but in one direction only viewed in polarised light. Smectic phases appear to have fan-like appearance. They are always *uniaxial* and *not* affected by electric field.

Examples are: ethyl *p*-azoxy benzoate, ethyl *p*-azoxy cinnamate etc.

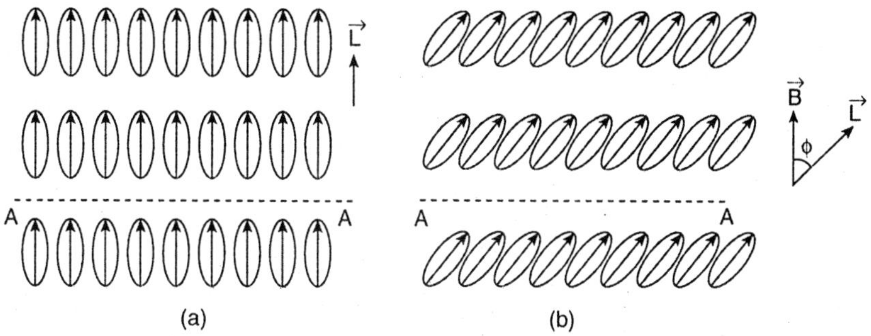

Fig. 3.13 Smectic phase showing gliding of layers with smectic axis $\vec{L}$ and shear plane AA

Applications—Liquid crystals find *applications* in low power visual displays as in electric watches, calculators, digital electronic meters etc.

The *advantage* of liquid crystals is that they can be made of any size or shape, consume less power, cheap and of variety of colours. The *disadvantage* is that they operate in a limited range of temperature and have large response time.

3.5 Quasicrystals

Natural crystals cannot have lattice periodicity other than 1, 2, 3, 4 and 6-fold symmetries. The essential characteristics for crystal formation is the existence of translational periodicity which a solid with a 5-fold symmetry cannot have. So, it was thought to be non-existent.

But in 1984 Schechman *et al* reported that there are physical systems with many properties of crystalline state but without the three-dimensional lattice periodicity. They cannot thus be indexed by Bravais lattice but show, like natural crystals, sharp Bragg peaks in their X-ray diffraction pattern, indicating high degree of three-dimensional regularity. A few such materials are alloys $Al_{0.86}Mn_{0.14}$ and $Cu_{0.20}Al_{0.65}Fe_{0.15}$. They do not have crystal-like lattice periodicity but has some type of quasi-translational periodicity and a long range orientational order but with 5-fold symmetry axes. Such quasi-periodic crystals are known as ***quasicrystals***.

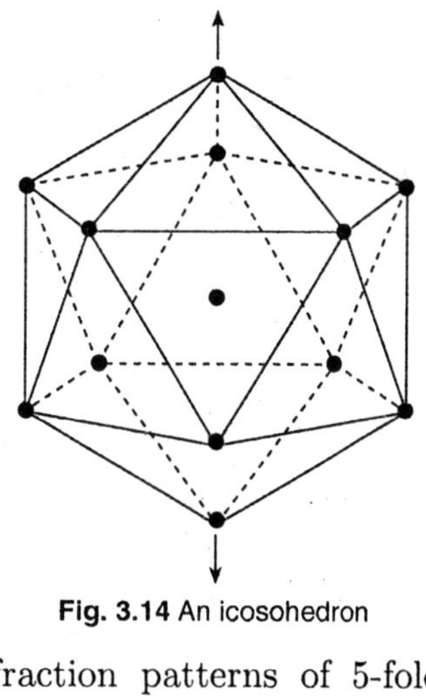

Fig. 3.14 An icosohedron

The full point symmetry group of the observed diffraction patterns of 5-fold symmetry corresponds to a platonic solid with ***icosohedral symmetry***. Fig. 3.14 shows an icosohedron with its 12 vertices, 20 triangular faces and 30 edges, the 5-fold axes passing through the vertices (as one is shown in Fig. 3.14—the arrow), the three-fold axes through the centres of the faces and the two-fold axes through the centres of the edges.

A quasicrystal may be regarded as a three-dimensional generalisation of two dimensional ***Penrose tiling pattern*** (Fig. 3.15) constituting of rhombi of angles 144°

and 108°. Although there is no pure translational symmetry in the lattice pattern, the regular decagon are similarly oriented and lines can be drawn through the corners (atom positions) at 72° (= 360°/5). In three-dimensions, two different rhombohedral cells (compressed cubes) can be packed to give an icosohedral structure.

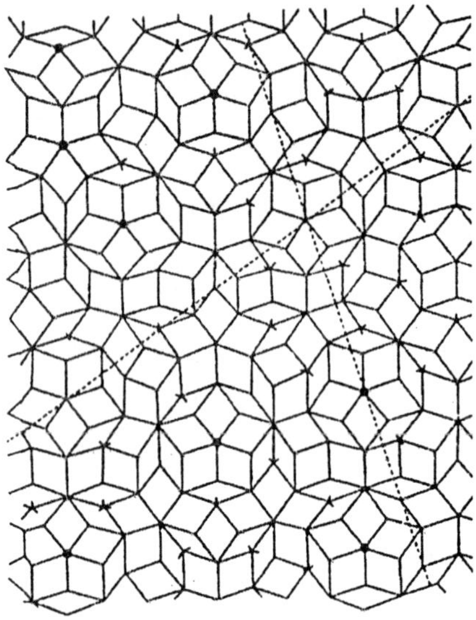

Fig. 3.15 Two-dimensional Penrose tiling pattern

• Quasicrystals require fresh investigations on them. For instance, the electronic band structure of solids rests on Bloch theorem which is based on strict translational periodicity. To get the electronic band structure of quasicrystals, new methods are to be devised that may lead to new electronic properties.

3.6 Illustrated Examples

► **Example 1.** *If* 1 eV *is required to move an atom from the interior of a crystal to its surface, then determine the proportion of vacancies in the crystal at* 300 K *and* 1000 K. *Comment on the result.*

Solution: Here $E = 1\,\text{eV} = 1.6 \times 10^{-19}\text{C}$, $T_1 = 300\,\text{K}$, $T_2 = 1000\,\text{K}$.

Also, k = Boltzmann constant $= 1.38 \times 10^{-23}$.

$\therefore$ Proportion of vacancies is: $\dfrac{n}{N} = \exp(-E/kT)$

$$\therefore \text{ At 1000 K,} \quad \frac{n}{N} = \exp\left(\frac{1.6 \times 10^{-19}}{1.38 \times 10^{-23} \times 1000}\right) - e^{-11.6} = \mathbf{10^{-5}}$$

$$\text{At 300 K,} \quad \frac{n}{N} = \exp\left(\frac{1.6 \times 10^{-19}}{1.38 \times 10^{-23} \times 300}\right) = e^{-38.55} = \mathbf{1.66 \times 10^{-17}}$$

• It is observed that about a *three-fold* increase in temperature increases the vacancies by about 12 *orders of magnitude.*

▶ **Example 2.** *Given that the energy of vacancy formation for metallic silver is* 1.1 eV. *Determine the fraction of lattice sites that are vacant at the melting point of silver* (960°C). *Atomic weight of silver* = 107.9, *density*, $\rho = 10.500\ \text{kg/m}^3$.

Solution: The vacancy fraction is given by $n_v/N = Ne^{-E_v/kT}/N = e^{-E_v/kT}$.

By the problem, $E_v = 1.1\,\text{eV}$, $T = 960 + 273 = 1233\,\text{K}$. Also k = Boltzmann constnat $= 1.38 \times 10^{-23}\,\text{J/K} = 8.63 \times 10^{-5}\,\text{eV/K}$.

∴ Vacancy fraction at 960°C is:

$$e^{-E_v/kT} = \exp\left[\frac{1.1\,\text{eV}}{8.63 \times 10^{-5}\,\text{eV/K} \times 1233\,\text{K}}\right]$$

$$= e^{-10.34} \simeq \mathbf{3 \times 10^{-5}}, \text{ on simplification.}$$

▶ **Example 3.** *If the average energy required to develop a Frenkel defect in an ionic crystal* $A^{2+}B^{2-}$ *is* 1.4 eV, *determine the ratio of Frenkel defects at* 300 K *and* 600 K *in* 1 g *of the crystal.*

Solution: By the problem, $E_f = 1.4\,\text{eV}$, $T_1 = 300\,\text{K}$, $T_2 = 600\,\text{K}$. To find the ratio (n_{300}/n_{600}).

Using the relation, $n = \sqrt{NN_i}e^{-E_f/2kT}$, we get

$$n_{300} = \sqrt{NN_i}\,e^{-1.4/600\,\text{K}}; \quad n_{600} = \sqrt{NN_i}\,e^{-1.4/1200\,\text{K}}$$

$$\therefore \quad \frac{n_{300}}{n_{600}} = e^{-1.4/k(1/600-1/1200)}$$

$$= \mathbf{1.316 \times 10^{-6}}, \text{ on simplification.}$$

▶ **Example 4.** *The concentration of Schottky defects in an fcc crystal is* 1 *in* 10^{10} *at* 300 K. *Estimate the average separation between the defects in terms of lattice spacing at* 300 K.

Solution: By the problem, $n/N = 1/10^{10} = 10^{-10}$.

$$\therefore \quad n = N \times 10^{-10} = 6.023 \times 10^{23} \times 10^{-10} \simeq 10^{14}$$

Let the interatomic spacing $\simeq 1\,\text{A} = 10^{-10}\,\text{m}$. So, volume of unit cube $= 10^{-30}\,\text{m}^3$.

Now, in an *fcc* crystal, the number of atoms per unit cell = 4, that occupies a volume of $10^{-30}\,\text{m}^3$. In one mole, therefore, 10^{23} atoms occupy of volume of

$$10^{-30} \times 10^{23} \simeq 10^{-7}\,\text{m}^3$$

Since 10^{14} defects are available in the same volume, i.e., within $10^{-7}\,\text{m}^3$, the volume per defect is $V = a^3 \simeq 10^{-7}/10^{14} = 10^{-21} \Rightarrow \boldsymbol{a = 10^{-7}\,\textbf{m}}$.

▶ **Example 5.** *A Burger lineage boundary in a simple cubic crystal creates a line of etchpits spaced* 2×10^{-4} cm *apart when the crystal is etched. If the lattice constant of this crystal is* 4×10^{-8} cm, *find the angle of misfit between the sections of the crystal on opposite sides of the lineage boundary,.*

Solution: A line boundary orientation is represented by an array of edge dislocations of spacings D.

$$\therefore \quad D = b/\theta$$

where b = Burger's vector or slip vector and θ is the angle of misfit. The minimum value of $b = 1$ lattice constant.

Here, by the problem, $D = 2 \times 10^{-4}$ cm and $b = 4 \times 10^{-8}$.

$$\therefore \quad \theta = \frac{b}{D} = \frac{4 \times 10^{-8}}{2 \times 10^{-4}} = \mathbf{2 \times 10^{-4}\ rad.}$$

Review, Short and Objective Questions

1. What do you mean by crystal imperfections? Name the different types of crystal imperfections.

2. What are point defects and line defects? Name the different types of each.

3. Explain the Schottky and Frenkel defects. Calculate the equilibrium concentration of Schottky defects in a binary ionic crystal.

4. There are N atoms in a metal crystal and N_i interstitials in its structure. If E_f be the average energy required to create one Frenkel defect and n the equilibrium number of defects, show that

$$n = (NN_i)^{1/2} e^{-E_f/2kT}$$

where k is Boltzmann constant.

5. Explain the terms: edge dislocation, screw dislocation, Schottky defects and Frenkel defects.

6. What are F-centres? Why does their presence impart colour to an otherwise colourless ionic crystal such as NaCl?

7. Explain the metal excess defect and metal deficiency defect. Give examples where such defects are observed.

8. Write short notes on: (i) Liquid crystals, (ii) Quasicrystals.

9. What are the different types of liquid crystals? Name them with examples.

10. Explain the characteristics of nematic, cholesteric and smectic phases of liquid crystals.

11. Which of the following has Frenkel defect?

    (a) NaCl, (b) AgBr, (c) Graphite, (d) Diamond.

12. The flame colours of metal ions are due to

(a) metal excess defect, (b) metal deficiency defect,
(c) Schottky defect, (d) Frenkel defect.

13. In a solid lattice, cation is absent from lattice site and present at an interstitial position. The lattice defect is

(a) Frenkel, (b) Interstitial, (c) Schottky, (d) None.

14. In Ag Br, there can occur

(a) only Schottky defect, (b) only Frenkel defect,
(c) both (a) and (b), (d) none of these.

15. The structure shown beside (Fig. 3.16) represents the

(a) Schottky defect,
(b) Frenkel defect,
(c) metal excess defect,
(d) none of these.

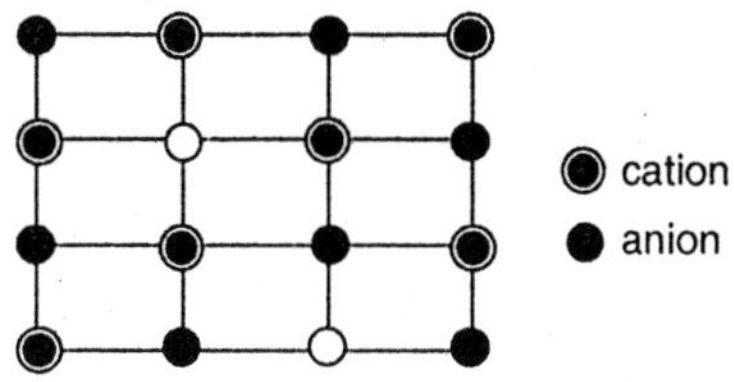

Fig. 3.16

16. Point defects in a crystal involve

(a) only vacancies in lattice sites,
(b) vacancies and impurity atoms,
(c) impurity atoms and interstitials,
(d) vacancies, impurity atoms and interstitials.

17. The equilibrium concentration of Schottky defects in crystals is given by

(a) $n_s = e^{-E_s/kT}$, (b) $n_s = Ne^{-E_s/k}$,
(c) $n_s = Ne^{\pm E_s/kT}$, (d) $n_s = Ne^{-E_s/kT}$.

Problems

1. Determine the fraction of atoms in a given solid with the energy equal to or greater than 1.5 eV at 300 K and at 1500 K. [6.185×10^{-26}, 9.084×10^{-6}]

2. The energy required to remove a pair of Na^+ and Cl^- from NaCl is 2 eV. Determine the approximate number of Schottky imperfections in NaCl crystals at room temperature of 27°C. [0.35×10^6 per cm^3]

3. The energy of vacancy formation for silver is 1.1 eV. Find the number of vacancies per cubic meter at 300°C. Given: atomic weight of silver $A = 107.9$ and density $\rho = 10{,}500$ kg/m^3. [$1.096 \times 10^{19}/m^3$]

4. The density of Schottky defects in a certain sample of NaCl is 5×10^{11} per m^3 at 300 K. If the interionic separation be 2.82 A, what is the average energy required to create one Schottky defect? [1.984 eV]

5. A strip of iron of dimensions $1 \times 2 \times 15$ cm is bent into a radius of curvature of 12 cm. What is the dislocation density if [111] edge dislocations line up with their Burger vector along the strip? [$3.42 \times 10^{10}/m^2$]

6. The fraction of vacancy sites in a certain metal is 1×10^{-10} at a temperature of 500°C. Determine the fraction of vacancy sites at 1000°C. [8.41×10^{-7}]

7. If the energy required to create a vacancy in a solid be 2 eV, show that the relative density of vacancies to atoms will be always less than 10^{-8} per cent unless the melting point is higher than 1000 K.

8. Calculate the atomic percentage of interstitials and vacancies at the melting point in copper. Take the formation energies as 4.5 and 1.5 eV respectively.

4

Bonding in Solids

While atoms combine to form molecules, the molecules in their turn combine to form solids. *The phenomenon or the ability to hold or bind the atoms or ions of the crystal together is known as* ***bonding*** and the atoms are held together by what is called *chemical bonds.* The bound states are *stable*, for such states have the minimum energy. So, in the process of bond formation, the system goes to the *minimum energy state.*

Electrical forces are operative behind this bonding. In any solid, the mutual interatomic forces are basically electrostatic in nature. The different charge distributions in space, particularly of the valence electrons of the atoms, determine the bonding type. The physical properties of solids are, to a great extent, dependent on the electron distribution.

Before we launch upon discussing the various types of bonding, we would consider the general nature of atomic bond between two atoms in a solid.

4.1 General nature of cohesion between two atoms

From the very existence of solids, the following two conclusions are inevitable in regard to the type of forces acting between the atoms.

1. *Attractive forces* that keep the atoms together to form the solid and arise from the interaction of negative electron cloud of one atom with the positive nuclear charge on the other. The magnitude of the force is proportional to some power of the interatomic distance r.

2. *Repulsive forces* are apparent when we find that large external prerssure is required to compress a solid. These forces develop when the interatomic distance decreases to such an extent that their electron clouds start overlapping, violating the exclusion principle of Pauli. The repulsion between positively charged nuclei also contributes. The magnitude of the total repulsive force is also proportional to some power of r.

Let two atoms A and B exert both attractive and repulsive forces on each other. So the net potential energy of B in the field of A may be expressed by a general form

$$U(r) = -\frac{\alpha}{r^m} + \frac{\beta}{r^n} \tag{4.1.1}$$

where m, n, α, β are all positive constants characteristic of the atoms. The α and β are known as *attractive and repulsive constants* respectively.

• The attractive part of potential is conventionally taken as *negative* since atoms themselves do work of attraction. The repulsive part is with *positive* sign for in binding the atoms closer, work is to be done against the repulsive force by an external agency.

The condition in which the particles will form a stable molecule is that $U(r)$ should show a minimum for some values of r, say r_e. If r_e exists, it is called the ***equilibrium spacing*** and $U(r_e)$ the ***binding*** or ***bond energy***. This minimum energy is negative so that the energy required to dissociate the molecules is the positive quantity $-U(r_e)$, the ***dissociation energy***.

The force (cohesive) force between the two atoms is given by

$$F = -\frac{dU}{dr} = \frac{m\alpha}{r^{m+1}} - \frac{n\beta}{r^{n+1}} \tag{4.1.2}$$

At $r = r_e$, the attractive and the repulsive forces balance each other and $F(r_e) = 0$. So, at $r = r_e$, we have

$$\frac{m\alpha}{r_e^{m+1}} = \frac{n\beta}{r_e^{n+1}}$$

$$\Rightarrow \quad \boxed{r_e = \left(\frac{\alpha}{\beta}\frac{m}{n}\right)^{1/(m-n)}} \tag{4.1.3}$$

The corresponding equilibrium potential energy U_e is obtained by substituting this value of r_e in (4.1.1) and simplifying and we get

$$U_e = U_{r=r_e} = -\frac{\alpha}{r_e^m}\left(1 - \frac{m}{n}\right)$$

When the net force is zero, the potential energy is minimum and is negative.

A positive quantity $D = -U_e$ is, as already stated, the ***dissociation energy*** of the molecule.

$$\therefore \quad \boxed{D = \frac{\alpha}{r_e^m}\left(1 - \frac{m}{n}\right)}$$

It is to be noted that the *attractive and repulsive forces are equal at equilibrium separation*, but the *attractive and repulsive energies are not*, for $m \neq n$.

For a minimum of $U(r)$, we must have $d^2U/dr^2 > 0$.

$$\therefore \text{ From (4.1.2), we get} \qquad -\frac{\alpha m(m+1)}{r_e^{m+2}} + \frac{\beta n(n+1)}{r_e^{n+2}} > 0$$

$$\text{or,} \qquad \beta n(n+1) > \alpha m(m+1) r_e^{n-m}$$

Substituting for r_e from (4.1.3) and simplifying, we obtain $n > m$. So, *the formation of a chemical bond requires that the repulsive force must be of short range than the attractive one.*

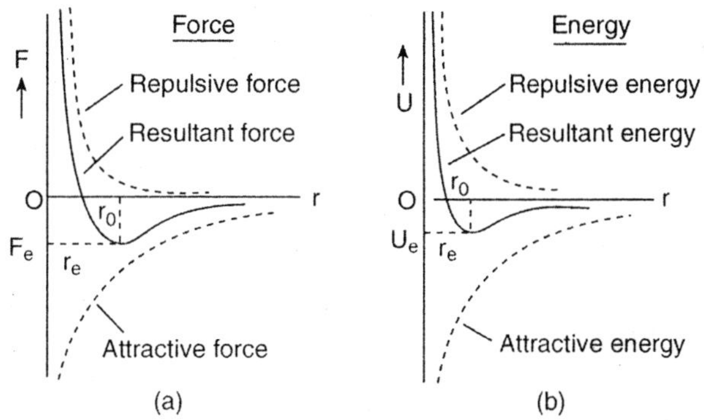

Fig. 4.1 Force (a) and potential energy (b) as a function of distance between two atoms

Thus, a minimum of the energy curve exists if $n > m$, that is, the formation of a stable bond requires that the attractive forces should vary more slowly with r than the repulsive forces. Fig. 4.1a and 4.1b illustrate the variation of the force F and the potential energy U respectively as a function of the separations of the atoms.

● Can you show that the ratio of atttractive to repulsive energy in the stable condition of atom is $-n/m$, i.e., $(U_{\text{att}})_{r_e}/(U_{\text{rep}})_{r_e} = -n/m$?

4.2 Types of bonding

In crystals, every atom is surrounded by a number of other atoms and the simple general expressions for the attractive and the repulsive energies as given in (4.1.1) do not, in fact, hold. The different charge distributions, particularly of the valence electrons of the atoms, the sharing of the electrons among the neighbouring atoms etc. give rise to different types of bonding which may be conveniently classified broadly in the following *five* categories.

1. ***Ionic bonding***, e.g., in the alkali halides
2. ***Covalent bonding***, e.g., in diamond, silicon etc.
3. ***Metallic bonding***, e.g., in Cu, Ag, Fe etc.
4. ***van der Waals bonding***, e.g., in solid argon, O_2 etc.
5. ***Hydrogen bonding***, e.g., in ice, some fluorides.

The inner (core) electrons have compensated spins and they do not therefore participate in the bonding.

We shall now discuss the above bondings in some detail.

● Although divided into five categories, the boundaries between them are not always distinct. There are many crystals that exhibit mixed bonding, e.g., ZnS which is believed to be partly ionic and partly covalent. The bond-based crystal classification is thus somewhat arbitrary.

● Based on the bond strength, the bondings can be grouped again into (i) ***primary bonding*** and (ii) ***secondary bonding***. Ionic, covalent and metallic bondings are all *primary* with bond energies between 0.1–10 eV/bond. Hydrogen bonding and van der Waals bonding are *secondary* having energies between 0.01–0.5 eV/bond, much less than those of the primary. Among the primary bonding again, the ionic and the covalent ones are usually much stronger than the metallic one. van der Waals bond is generally the weakest of all.

4.3 Ionic bonding

This type of bonding can only be formed between an electropositive atom (e.g. Na, K, Ba, etc. of Gr I or II elements) and an electronegative atom (e.g. Cl, Br, etc. of Gr. VI or VII elements). The electropositive atom readily loses electrons and the electronegative one readily takes them up. This actual *electron transfer* enables each atom to attain a *stable outer shell electronic configuration*. The electron transfer leaves the electropositive atom with a net positive charge (positive ion) and makes the electronegative atom acquire a net negative charge (negative ion). The electrostatic

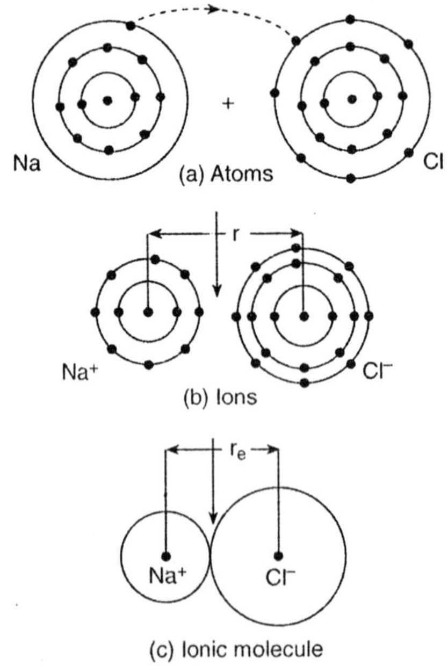

Fig. 4.2 Schematic representation of the formation of an ionic molecule of sodium chloride

Coulomb attraction between the two oppositely charged ions bind them together and the resulting bond is called the ***ionic bond.*** In the process, each atom attains a stable electronic configuration similar to the nearest inert gas atom and the charge distribution of each ion is spherically symmetric. As a result, the bond is *non-directional.*

For instance, take the ionic crystal NaCl. The electronic configurations of Na and Cl before and after the transfer are as under.

$$\left.\begin{aligned} \mathrm{Na}\,(11) &= 1s^2 2s^2 2p^6 3s^1 \\ \mathrm{Cl}\,(17) &= 1s^2 2s^2 2p^6 3s^2 3p^5 \end{aligned}\right\} \quad \cdots \quad \text{Before electron transfer}$$

$$\left.\begin{aligned} \mathrm{Na}^+ &= 1s^2 2s^2 2p^6 \text{ (Ne-structure)} \\ \mathrm{Cl}^- &= 1s^2 2s^2 2p^6 3s^2 3p^6 \text{ (Ar-structure)} \end{aligned}\right\} \quad \cdots \quad \text{After electron transfer}$$

Hence, an ionic bond exists between two atoms when one has as many valence electrons as are required by the other to complete its outermost valence orbit. Fig. 4.2 schematically represents the formation of the ionic molecule of sodium chloride, NaCl.

- In general, the force of attraction is given by

$$F_a = \frac{(+Z_1 e)(-Z_2 e)}{4\pi\varepsilon_0 r^2} \tag{4.3.1}$$

where $Z_1 e, Z_2 e$ are ionic charges, e the electronic charge and r the separation between the two ions.

Apart from attraction, a force of repulsion will also act as the ions are brought nearer, due to Pauli's exclusion principle and the force may be written as

$$F_r = \frac{nB}{r^{n+1}} \tag{4.3.2}$$

$\therefore$ The bond energy,

$$U = \int_{\infty}^{r} \left[\frac{(+Z_1)(-Z_2)e^2}{4\pi\varepsilon_0 r^2} - \frac{nB}{r^{n+1}} \right] dr \tag{4.3.3}$$

- The attractive potential (energy) V is given by

$$V = -\frac{Z_1 Z_2 e^2}{4\pi\varepsilon_0 r} \tag{4.3.4}$$

with the same notations, as have been used before.

If both the atoms are monovalent, $Z_1 = Z_2 = 1$ and $V = -e^2/4\pi\varepsilon_0 r$; if both are divalent, $Z_1 = Z_2 = 2$, and $V = -4e^2/4\pi\varepsilon_0 r$. In general, the potential of any one ion due to all others in the crystal is given by

$$V = -\frac{\alpha Z_1 Z_2 e^2}{4\pi\varepsilon_0 r} \tag{4.3.5}$$

where α is a constant, called ***Madelung constant***, to which we shall return shortly.

4.3.1 Equilibrium or cohesive energy of ionic crystals

Born and Meyer in deriving for the first time an expression for the cohesive energy assumed that the potential energy $V(r)$ of an ionic crystal of ionic charges $+Z_1e$ and $-Z_2e$ respectively is the sum of the Coulomb attractive and repulsive energies that prevents the lattice from breakdown.

$$\therefore\ V(r) = -\frac{\alpha Z_1 Z_2 e^2}{4\pi\varepsilon_0 r} + \frac{B}{r^n} \tag{4.3.6}$$

For example, for NaCl crystal, $Z_1 = Z_2 = 1$ and hence the total energy

$$V(r) = N\left[-\frac{\alpha e^2}{4\pi\varepsilon_0 r} + \frac{B}{r^n}\right] \tag{4.3.7}$$

N being the Avogadro number.

Now, the potential energy is minimum at $r = r_e$, the equilibrium distance.

$$\therefore\ \left(\frac{dV}{dr}\right)_{r=r_e} = N\left[\frac{\alpha e^2}{4\pi\varepsilon_0 r_e^2} - \frac{nB}{r_e^{n+1}}\right] = 0$$

$$\Rightarrow\ \frac{\alpha e^2}{4\pi\varepsilon_0 r_e^2} = \frac{nB}{r_e^{n+1}} \quad \Rightarrow\ B = \frac{\alpha e^2 r_e^{n-1}}{4\pi\varepsilon_0 n}$$

$$\therefore\ \boldsymbol{V(r_e)} = -\frac{N\alpha e^2}{4\pi\varepsilon_0 r_e} + \frac{N\alpha e^2}{4\pi\varepsilon_0 n r_e}$$

$$= -\frac{\boldsymbol{N\alpha e^2}}{\boldsymbol{4\pi\varepsilon_0 r_e}}\left(\boldsymbol{1 - \frac{1}{n}}\right) \tag{4.3.8}$$

which is the *potential energy per ion pair* and is called ***lattice energy*** of an ionic crystal. It is defined as the energy released when the constituent ions are placed in their positions in the crystal lattice.

Apart from the *potential energy*, we are to consider the energy required to transfer an electron from Na-atom to a Cl-atom to form Na^+ and Cl^- ion pairs. This energy for transfer is given by the difference:

ionisation energy of Na^+ – electron affinity of Cl^-.

$$\therefore\ \textbf{Cohesive energy} = -\frac{\boldsymbol{N\alpha e^2}}{\boldsymbol{4\pi\varepsilon_0 r_e}}\left(\boldsymbol{1 - \frac{1}{n}}\right) + \textbf{\textit{ionisation energy}} - \textbf{\textit{electron affinity.}}$$

- On evaluation, the value of n is found to be so large that most of the lattice energy is due to *Cloulomb interaction.*

- *Cohesive energy* is *defined* as the energy released during the formation of a crystal from neutral atoms or, in other words, the energy required per ion to break the crystal into individual atoms. (see Illustrated Example 2)

4.3.2 Lattice energy of ionic crystals: Born's theory

Following Born, let W_{ij} be the interaction energy between the ith and the jth ions and W_i be that on ion i due to all others.

$$\therefore \quad W_i = \sum_j W_{ij} \tag{4.3.9}$$

where the summation extends over all the ions *except* $j = i$.

Now, W_i consists of two parts: (i) a *Coulomb potential* $\pm q^2/4\pi\varepsilon_0 r$ and (ii) a central field *repulsive potential* which, from quantum consideration, is of the form $\lambda e^{-r/\rho}$ where λ a measure of the *strength* and ρ the *range of repulsive interaction.* Plainly, the repulsive interaction reduces to $1/e$ of its value at $r = \rho$.

$$\therefore \quad W_{ij} = \lambda e^{-r_{ij}/\rho} \pm q^2/4\pi\varepsilon_0 r_{ij} \tag{4.3.10}$$

If the crystal contains N molecules (i.e., $2N$ ions), the total lattice energy is

$$W_T = NW_i \tag{4.3.11}$$

Since the interaction of *each pair* is to be considered *once* only, we used N and *not* $2N$ in the above expression.

We put, for convenience, $r_{ij} = p_{ij}R$ where p_{ij} is a dimensionless quantity, being the distance between any two ions in terms of nearest neighbour distance R in the crystal. Restricting to repulsive interaction among the nearest neighbours only, $r_{ij} = R$ or $p_{ij} = 1$.

$$\therefore \quad W_{ij} = \lambda e^{-R/\rho} \pm q^2/4\pi\varepsilon_0 R, \quad \text{using (4.3.10)}$$

$$\therefore \quad W_T = NW_i = N\sum W_{ij}, \quad \text{using (4.3.11) and (4.3.9)}$$

$$= N\left(z\lambda e^{-R/\rho} \pm \sum \frac{1}{p_{ij}} \frac{q^2}{4\pi\varepsilon_0 R}\right) \tag{4.3.12}$$

$$= N\left(z\lambda e^{-R/\rho} - \frac{\alpha q^2}{4\pi\varepsilon_0 R}\right) \tag{4.3.13}$$

where z = number of nearest neighbours of the ith ion and $\alpha = \sum_j \mp 1/p_{ij}$ = *Madelung constant.*

If the reference ion is negative, the positive sign is used for a positive ion and the negative sign for a negative ion for α.

At equilibrium separation $R = r_e$, the crystal energy W_T is a minimum.

$$\therefore \quad \frac{dW_T}{dR} = 0$$

$$\text{or,} \quad N\frac{dW_i}{dR} = -\frac{Nz\lambda}{\rho}e^{-R/\rho} + \frac{N\alpha q^2}{4\pi\varepsilon_0 R^2} = 0$$

$$\text{or,} \quad 4\pi\varepsilon_0 r_e^2 e^{-r_e/\rho} = \frac{\rho\alpha q^2}{z\lambda}, \quad \text{putting } R = r_e. \tag{4.3.14}$$

If the values of ρ and λ are known, the equilibrium separation r_e can be evaluated.

$\therefore$ Total energy of the crystal at equilibrium separation r_e of the ions is

$$\boldsymbol{W_T} = N\left(z\lambda e^{-r_e/\rho} - \alpha\frac{q^2}{4\pi\varepsilon_0 r_e}\right), \quad \text{from (4.3.13)}$$

$$= N\left(\frac{\rho\alpha q^2}{4\pi\varepsilon_0 r_e^2} - \frac{\alpha q^2}{4\pi\varepsilon_0 r_e}\right), \quad \text{using (4.3.14)}$$

$$= -\frac{\boldsymbol{N\alpha q^2}}{\boldsymbol{4\pi\varepsilon_0 r_e}}\left(\boldsymbol{1} - \frac{\boldsymbol{\rho}}{\boldsymbol{r_e}}\right) \tag{4.3.15}$$

$$\therefore \quad W_T \text{ per ion pair} = -\frac{\alpha q^2}{4\pi\varepsilon_0 r_e}(1 - \rho/r_e) \tag{4.3.16}$$

The term $\boldsymbol{-N\alpha q^2/4\pi\varepsilon_0 r_e}$ of (4.3.15) is known as the ***Madelung energy*** or *electrostatic interaction energy*. The term in the bracket, $(1 - \rho/r_e)$, then gives the contribution of the *short range repulsive interaction.*

● For NaCl, $\rho/r_e \sim 1/9$, q = electronic charge e, $\alpha = 1.748$ and $r_e = 2.81$ A. So from (4.3.16), W_T per ion pair $= -7.97$ eV, in close agreement with the experimentally observed value.

4.3.3 Expression for bulk modulus : Range of repulsive interaction of ionic crystals

The bulk modulus $B = -dp/(dV/V)$ where dp is a change in pressure producing a volumetric strain dV/V.

From the first law of thermodynamics,

$$dQ = dU + pdV \tag{4.3.17}$$

For an *adiabatic compression*, $dQ = 0$ so that from (4.3.17)

$$dU = -pdV \;\Rightarrow\; p = -\frac{dU}{dV} \;\Rightarrow\; \frac{dp}{dV} = -\frac{d^2U}{dV^2}$$

$\therefore$ Bulk modulus, $$B = V\frac{d^2U}{dV^2}$$

Now, $$\frac{dU}{dV} = \frac{dU}{dr}\cdot\frac{dr}{dV}$$

$$\therefore \quad \frac{d^2U}{dV^2} = \frac{d^2U}{dr^2}\left(\frac{dr}{dV}\right)^2 + \frac{dU}{dr}\frac{d^2r}{dV^2} \tag{4.3.18}$$

At the *equilibrium distance* $r = r_e$, $dU/dr = 0$.

$$\therefore \quad \frac{d^2U}{dV^2} = \frac{d^2U}{dr^2}\left(\frac{dr}{dV}\right)^2, \quad \text{for } r = r_e.$$

$$\therefore \quad B = V\frac{d^2U}{dr^2}\left(\frac{dr}{dV}\right)^2, \quad \text{at } r = r_e. \tag{4.3.19}$$

NaCl structure For this structure, $r = a/2$ where a is the lattice parameter. The volume of unit cell containing 4 molecules of NaCl is $a^3 = 8r^3$.

$\therefore$ Volume occupied by N-molecules, $V = (N/4)8r^3 = 2Nr^3$

$$\Rightarrow \frac{dV}{dr} = 6Nr^2$$

Substituting this value of dV/dr in (4.3.19), we obtain,

$$B = 2Nr_e^3 \frac{d^2U}{dr^2}\left(\frac{1}{36N^2r_e^4}\right) = \frac{1}{18Nr_e}\left[\frac{d^2U}{dr^2}\right]_{r=r_e} \tag{4.3.20}$$

From equation (4.3.13),

$$\frac{dU_T}{dr} = -\frac{N\lambda z}{\rho}e^{-r/\rho} + \frac{N\alpha q^2}{4\pi\varepsilon_0 r^2}$$

$$\therefore \frac{d^2U_T}{dr^2} = \frac{N\lambda z}{\rho^2}e^{-r/\rho} - \frac{2N\alpha q^2}{4\pi\varepsilon_0 r^3}$$

Substituting this value in (4.3.20), we get

$$B = \frac{1}{18Nr_e}\left[\frac{N\lambda z}{\rho^2}e^{-r_e/\rho} - \frac{2N\alpha q^2}{4\pi\varepsilon_0 r_e^3}\right]$$

But, we have already seen in (4.3.14) that

$$e^{-r_e/\rho} = \frac{\rho\alpha q^2}{4\pi\varepsilon_0 r_e^2 z\lambda}$$

$$\therefore B = \frac{1}{18Nr_e}\left[\frac{N\lambda z}{\rho}\cdot\frac{\rho\alpha q^2}{4\pi\varepsilon_0 r_e^2 z\lambda} - \frac{2N\alpha q^2}{4\pi\varepsilon_0 r_e^3}\right]$$

$$\therefore \text{Bulk modulus, } B = \frac{1}{4\pi\varepsilon_0}\frac{\alpha q^2}{18r_e^4}\left[\frac{r_e}{\rho} - 2\right] \tag{4.3.21}$$

This equation can also be used to find the value of ρ, the ***range of repulsive interaction***, if r_e, α, q and B are known.

4.3.4 Evaluation of Madelung constant for NaCl structure

The Madelung constant α is a function of the crystal structure and may be evaluated from the geometrical arrangement of ions in the crystal. We shall evaluate here the Madelung constant for the NaCl structure.

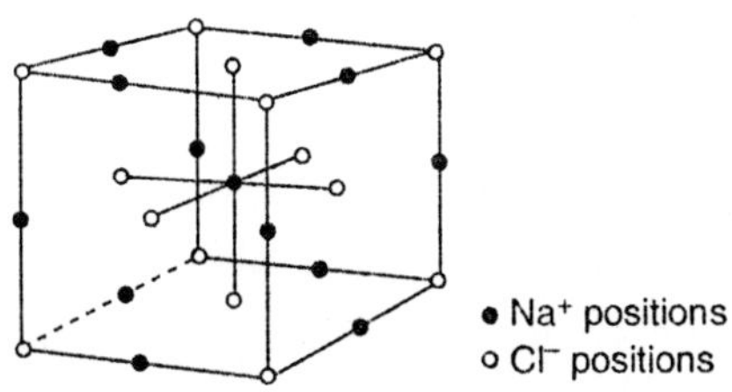

Fig. 4.3 Equilibrium positions of ions in NaCl crystal

Let a central Na^+ ion be the reference ion which is surrounded by $6Cl^-$ ions (Fig. 4.3) as the first nearest neighbours, $12Na^+$ ions as second nearest, $8Cl^-$ ions as third nearest neighbours and so on. The distances between the ions, in terms of equilibrium separation, are r_e, $\sqrt{2}r_e$, $\sqrt{3}r_e$ etc. (Fig. 4.4).

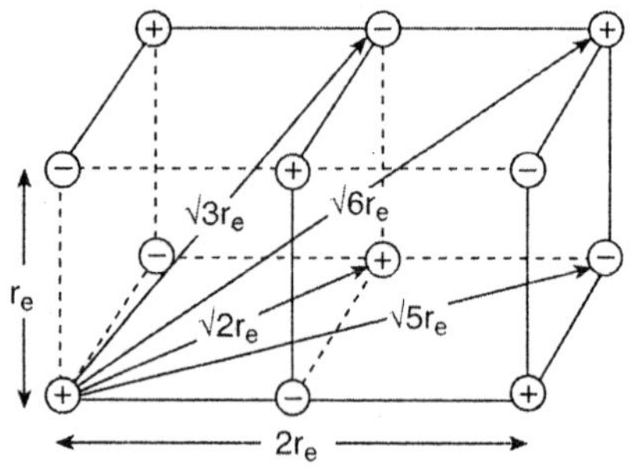

Fig. 4.4 Distances of neighbouring ions in NaCl lattice

$$\therefore \text{ Madelung constant, } \alpha = \frac{6}{1} - \frac{12}{\sqrt{2}} + \frac{8}{\sqrt{3}} - \frac{6}{\sqrt{4}} + \frac{24}{\sqrt{5}} - \frac{24}{\sqrt{6}} + \cdots$$
$$= 1.74756.$$

as the series is convergent. The convergence however is conditional as the summation may stop at any finite point. The value obtained at finite point is characteristic of a finite crystal. The above value of α is true for nearly infinite crystals.

● For the ***Madelung constant for an infinite linear chain of ions*** of alternating unit charge (e.g. 1d NaCl crystal) at a separation r_e, the worked out Example 8 may profitably be seen.

● Madelung constant, α is a dimensionless constant i.e., it is a *pure number* having *no unit.* It has the same value for all crystals of the same structure. For CsCl, $\alpha = 1.763$ and for ZnS, $\alpha = 1.638$.

4.3.5 Characteristics of ionic solids

The ***chief characteristics*** of ionic crystals may be summarised as:

1. The ionic bond is *fairly strong* (binding energy $\sim$ 5 to 10 eV) and so ionic compounds are *rigid* and *crystalline* and generally show *fairly high melting* and *boiling points.*

2. At normal temperatures, they are rather *poor conductors* of electricity as the outer electrons of ions are tightly bound and cannot contribute to transport of charges. At elevated temperatures, the conductivity increases due to increased ionic mobility under the influence of the field. This *ionic conductivity* is many orders of magnitude smaller than the *electronic conductivity* of metals.

3. They are found to be *easily soluble in polar solvents* like water due to high dielectric constant and are ordinarily *diamagnetic* since the electronic configuration is that of closed shells, leaving no unpaired electron spins.

4. Ionic crystals are in general *transparent* to visible light but strongly ***absorb*** *infrared radiation* (absorption peaks) at frequencies at which the ions vibrate. At frequencies higher than these, they are opaque.

5. The ionic bond is *non-directional* due to the spherical charge distribution.

6. If stressed, the ionic crystals prefer to *break* along certain planes rather than to get deformed.

4.4 Covalent bonding

A ***covalend bond*** is formed by an equal sharing (as against transfer) of one or more pair of valence electrons between two neighbouring atoms, each with incomplete outermost shell, so as to acquire a stable closed shell electronic configuration. Unlike ionic bonding, the atoms in the covalent bond are of such electronic configuration that their octets cannot be completed by electron transfer. Thus there is no charge with any atom of the crystal. When two atoms, each with a deficiency of one electron, come so close that their electronic shells start to overlap, their original charge distributions get distorted and each atom transfers its unpaired electron to the common space between them. The common space has now a pair of electrons that belongs to both the atoms and serves to complete the outermost shell of each atom. This is what is meant '*sharing of electrons*'. The spins of the two electrons in a shared pair are *anti-parallel* in order to satisfy the Pauli's exclusion principle.

The simplest covalent bonding occurs in hydrogen molecule where two hydrogen atoms contribute their $1s^1$ electrons to form an electron-pair covalent bond Fig. 4.5(a) and is usually indicated by dot notation as shown below (Fig. 4.5b).

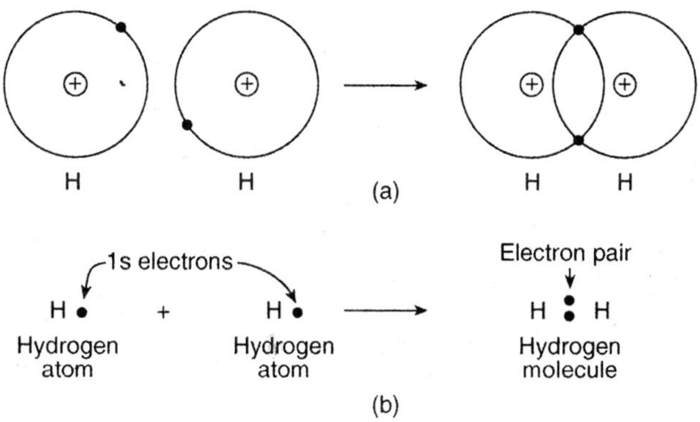

Fig. 4.5 (a) Single covalent bond in hydrogen molecule
(b) Covalent bond shown by electron-dot notation

Here the two involved atoms in bond formation share a single pair of electrons and the resulting bond is called a ***single covalent bond*** (Fig. 4.5a).

Similarly, a ***double*** or a ***triple bond*** results when the two coordinating atoms share *two* or *three* pairs of electrons. An example of a double covalent bond is the formation of an oxygen molecule and a triple covalent bond is formed in nitrogen molecule (Fig. 4.6).

The oxygen atom has an electronic configuration $1s^2 2s^2 2p^4$, i.e., two more electrons are needed to complete the shell. Hence, two O-atoms share two pairs of electrons forming a covalent double bond (Fg. 4.6a). In a nitrogen molecule, a triple bonding is formed by shairing of 3 pairs of electrons (Fig. 4.6b). In general, if in a neutral atom, the outermost orbit contains n electrons, $(8 - n)$ electrons are required for stability. Thus, $(8 - n)$ gives the ***coordination number*** in a covalent bond.

A covalent bond is also formed when two or more dissimilar atoms share one or more pairs of valence electrons. Examples of this type are H_2O, CO_2, HCl, NH_3. The

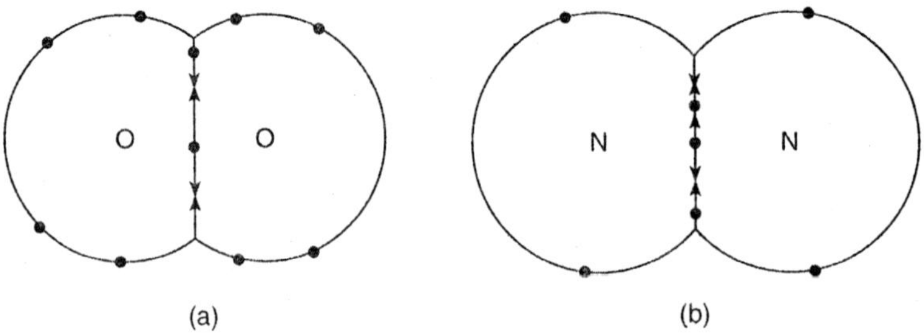

Fig. 4.6 Covalent bonds: (a) double bond (b) triple bond

formation of covalent bonds between an oxygen atom and two hydrogen atoms in H_2O is illustrated below by electron dot notation.

$$\mathrm{H}\bullet + \bullet\ddot{\underset{\bullet\bullet}{\mathrm{O}}}\bullet + \bullet\mathrm{H} \longrightarrow \mathrm{H}:\ddot{\underset{\bullet\bullet}{\mathrm{O}}}:\mathrm{H}$$

The electron-dot notation, although useful, does not give the valence electron density distribution. When the two atoms, say of hydrogen, come close to other to form hydrogen molecule, their electron clouds interact and overlap. The probability of finding the $1s^1$ electrons is maximum between the two nuclei of the hydrogen molecule as illustrated in Fig. 4.7.

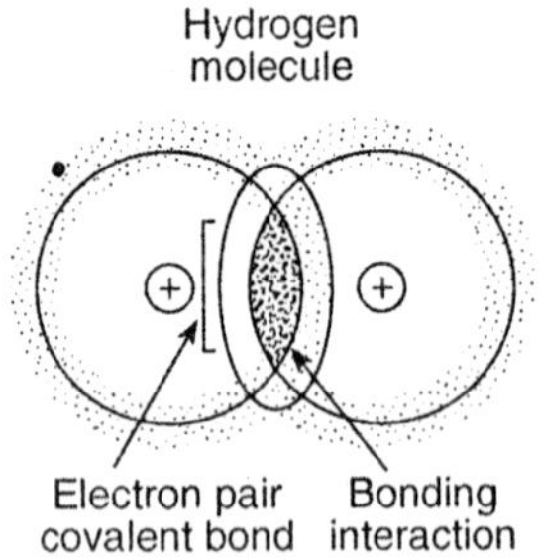

Fig. 4.7 Electron density distribution

Pauli's exclusion principle dictates that the shared electrons must have opposite spins. The interaction of these anti-spin electrons results in the force of attraction between the atoms forming the covalent bond. The electrostatic repulsion between the electrons however keeps the two atoms at equilibrium distance where the attractive and repulsive forces are equal or, in other words, the potential energy of the system is minimum. As a result of mutual sharing of the electrons by the two atoms, the configuration becomes stable and the bond between the two H-atoms is said to be saturated. If a third H-atom is brought near H_2-molecule, the exchange of electron spin of H-atom with the parallel spin of H_2-molecule results in repulsion. And the exchange of electron spin of H-atom with antiparallel spin of H_2-molecule is not allowed by Pauli's principle. And, therefore, an H_3-molecule *cannot* be formed.

A covalent bond is always *directional.* The directional nature of the bond is the outcome of the restricted orbital motion of the electrons. The shapes of atomic orbitals s (i.e., $l = 0$) and p (i.e., $l = 1$) are shown in Fig. 4.8 where the p-orbitals do not have spherical symmetry but show angular-dependent distribution of electron density, and hence the directionality.

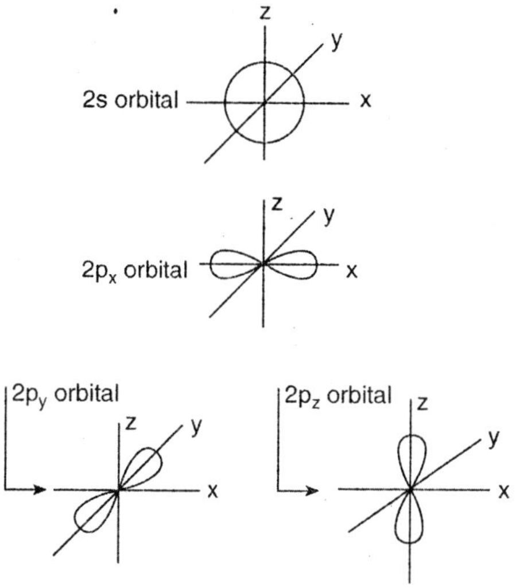

Fig. 4.8 Showing shapes of different atomic orbitals

The covalent bonds are formed not only due to overlap of pure s-orbitals or pure p-orbitals, but also by the overlapping of s and p-orbitals. Such covalent bonding is known as ***hybrid bonding***, as found in carbon . The carbon atom in the ground state has the electronic configuration $1s^2 2s^2 2p^2$ and the electron spin distribution is

$$\begin{array}{ccccc} \downarrow\uparrow & \uparrow\downarrow & \uparrow & \uparrow & \\ \hline 1s & 2s & 2p_x & 2p_y & 2p_z \end{array}$$

The electron spins are thus paired only in $1s$ and $2s$ orbitals and $2p$ electrons are unpaired. These unpaired electrons are available for bond formation so that only two bonds are expected from carbon atom. But when carbon atoms approach each other, one of the electrons from $2s$ orbital is excited to $2p$ state. This modifies the electronic configuration of carbon to $1s^2 2s^1 2p_x^1 2p_y^1 2p_z^1$ and the electron spin distribution is

$$\begin{array}{ccccc} \downarrow\uparrow & \uparrow & \uparrow & \uparrow & \uparrow \\ \hline 1s & 2s & 2p_x & 2p_y & 2p_z \end{array}$$

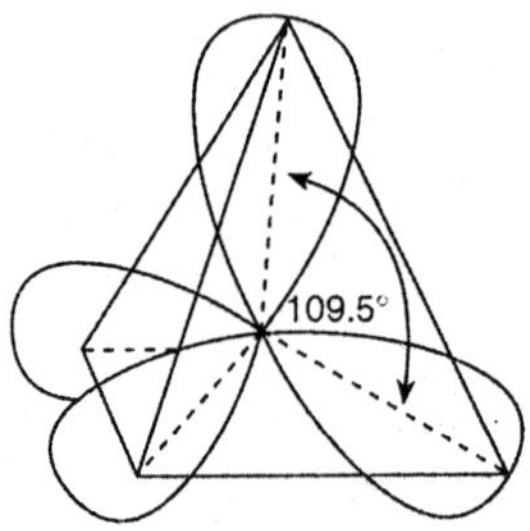

Fig. 4.9 sp^3 orbitals directed symmetrically towards the corners of a tetrahedron

Thus in the excited state, the carbon atom has four unpaired electron spins. The favoured bonding directions of these orbitals are towards the four corners of a regular tetrahedron (Fig. 4.9) with bond angles 109.5°. This arrangement of orbitals is known as ***hybridization*** and the four orbitals the ***sp^3-hybrids*** (Fig. 4.9).

Typical covalent solids are formed by Group IV elements in the periodic table, e.g., Carbon, Silicon, Germanium etc.

• Why two or more unpaired eletrons of the same atom do not form pairs? This is due to Hund's rule that states that spins of electrons in a shell add in such a way as to give maximum angular momentum and magnetic moment. Nitrogen, for instance, has seven eletrons: $1s^2 2s^2 2p^3$. The arrangement of spins is as shown.

$$1s^2\ \boxed{\uparrow}\ \boxed{\downarrow}\quad 2s^2\ \boxed{\uparrow}\ \boxed{\downarrow}\quad 2p^3\ \boxed{\uparrow}\ \boxed{\uparrow}\ \boxed{\uparrow}$$

The p-shell has six states, three with spin 'up' and three 'down'. As there are only three eletrons in p-sub-shell, all will have spins in the same direction, say 'up' which is the maximum angular momentum a p-subshell can have. These three electrons of nitrogen are unpaired and can form three pair bonds with another nitrogen atom in N_2-molecule.

Covalency is a purely quantum mechanical phenomenon having no classical analogy.

4.4.1 Characteristics of covalent bonding

Some of the ***chief characteristics*** of covalent bonding are:

1. The covalent bond is a *strong bond*, the cohesive energy of covalent crystals is of the same order as of ionic crystals $\sim 10\,\text{eV/atom}$. So, the covalent crystals are in general *hard and brittle* and also have *high melting and boiling points.*

2. Covalent substances are *insoluble in polar solvents* like water but are *soluble in non-polar solvents* like benzene, carbon di-sulphide etc. Giant molecules however are not soluble in any solvent because of their size.

3. This bonding is *strongly directional* in character and hence exhibit structural and space isomerism.

4. Covalent crystals are *transparent to long wavelength radiation* ($\lambda \geq$ infrared) but *opaque to shorter wavelength*, the transition being abrupt.

5. Multiple bonds have shorter bond lengths and double bonds are stronger than single bonds, a fact of importance in the formation of polymers and plastics.

6. Covalent crystals are typically *semiconductors* (e.g., Si, Ge) which are electrical insulators at 0 K and the conductivity rises exponentially with increase in temperature as the bonds get successively broken up. The conductivity is sensitive to the presence of tiny amount of impurity atoms.

4.5 Metallic bonding

It is difficult to describe *metallic bonding* in terms of a simple model consisting of pair of atoms, unlike ionic and covalent bondings. The bonding in a metal must be considered in terms of all the atoms of the solid with their valence electrons belonging to the crystal as a whole, and not to any specific atom. The valence electrons form a common pool of electrons enjoying the freedom to move throughout the structure of positive ion cores. They form the free '*electron gas*' or '*electron cloud*' acting as a mobile 'glue' to bind or hold together the ion cores by electrostatic attraction between the positive ions and negative electron cloud. The bond so formed in materials is called ***metallic bonding*** since the materials are invariably metals.

In such bonding, the atoms can share the required number of electrons with their neighbours *via* the common pool to complete their octets. Metallic bonding is thus formed by *partial sharing* of the valence electrons by neighbouring atoms. Unlike covalent bonding, this sharing is *not localized.* The free nature of the electron gas makes the bonding electrons resonate between different atoms and necessarily therefore the sharing *changes with time.*

The metallic bonding may be looked upon in two different ways: (i) as *a limiting case of ionic bonding* and (ii) as *a special case of covalent bonding.*

It is a limiting case of ionic bonding where ions are simply electrons. For instance, while NaCl has equal numbers of Na^+ ion and Cl^- ion, metallic sodium has equal numbers of Na^+ and e^-. But the striking difference is that the mass of an electron is quite small compared to that of Cl^- ion. This results in a large zero-point motion of electron. So, it is not localized to a lattice site and is free to move in an array of positive metal ions. A metallic structure is thus essentially determined by the packing of positive metal ions, electrons merely serving as a negatively charged *glue.* Fig. 4.10 illustrates a metal crystal as an array of positive ion cores embedded in a 'cloud' of free electrons.

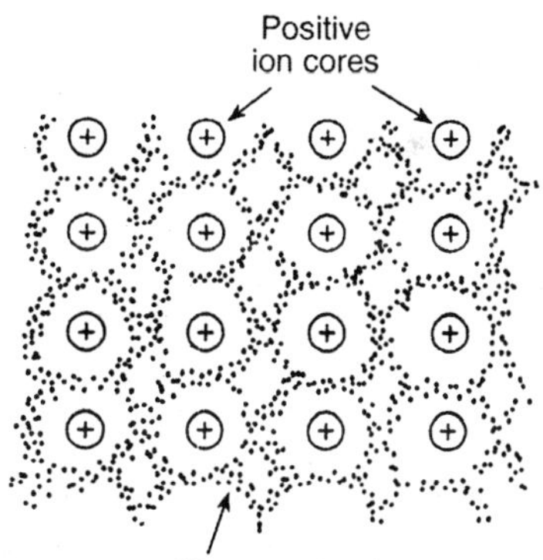

Fig. 4.10 Atomic arrangement in metal crystals

To understand metallic bonding as a *special case of covalent bonding,* take Na-atom with electronic configuration: $1s^2 2s^2 2p^6 3s^1$. So, an unpaired valence electron exists in $3s$-orbital. When two Na-atoms approach, their $3s$-orbitals start overlapping to form an electron-pair bond if the electrons are of opposite spins. If now a third Na-atom approaches such a pair, its valence electron may occupy the $3p$ state without violating the Pauli's exclusion principle. So, the third Na-atom may also form an electron-pair bond with either of the two Na-atoms, but not two electron pair bonds with two atoms simultaneously, for each atom has one unpaired electron only. So, on the average, one Na-atom forms one-half of an electron-pair bond with each of the other two. Na has body-centred cubic structure, each atom forming one-eighth of an electron-pair bond with each of its neighbours. Such bonds are thus *unsaturated covalent bonds* and the valence electrons are free and non-localized. The unsaturated nature of metallic bonding accounts for the alloying properties of metals. For instance, when Cu and Ni are mixed, each atom reacts unspecifically to the other since they are held by the common free electron gas to which both contribute.

The binding energy of metallic bonds usually ranges between 1–3 eV per bond.

4.5.1 Characteristics of metallic bonding

The ***main characteristics*** of the metallic bonds are:

1. The free electrons cause *high electrical and thermal conductivity* of metals.

2. The *unsaturated nature* of metallic bonding accounts for the *alloying properties* of the metals.

3. The metallic bonding is *insensitive* to any particular metal atom and this imparts the ability of metals to be joined by *welding* and *soldering.*

4. The *non-directional nature* of metallic bond influences the *surface* and *mechanical properties* of metals. It is easier in metals, than in most other substances, for atoms to slide over one another without any breakage of bonds. The *ductility* and the *toughness* of metals are due to this property.

5. Metals have high tensile strength due to the stronger attraction between the positive ions and the electrons.

6. The *opacity* and *lustre* of metals are due to the presence of free electrons in bonding. The electrons first absorb e. m. radiation to re-radiate the same, producing lustrous appearance.

7. The metallic bond is weaker than ionic or covalent bond and the melting and boiling points of metallic solids are lower than ionic or covalent solids.

4.6 Hydrogen bonding

Hydrogen molecule forms covalent bonding. But the *atomic hydrogen* is somewhat unique in the following sense.

1. The ion core of an H-atom is a bare proton of radius $\sim 10^{-13}$ cm and smaller than any other ion core by a factor of 10^5.

2. An H-atom has just one electron less to from the stable configuration of the inert gas He.

3. The first ionization potential of atomic hydrogen is unusually high: 13.59 eV.

These *characteristics* impart hydrogen to play a role in crystal structure, unlike any other element. For instance, consider the bonding of an H-atom with a highly electronegative atom like O, F, N, Cl etc. When a pair of electrons is shared between an

H-atom and say an electronegative O-atom, the shared pair is attracted more towards the electronegative atom that gets slightly negatively charged. And as a result, the H-atom gets an equal amount of positive charge. The molecule is thus polarized and acts like a *permanent electric dipole* (Fig. 4.11) with H-atom at the positive end. Many such dipoles get attracted to one another by Coulomb forces. This type of dipole bonding in permanently polarized molecules, each having an H-atom, is termed ***hydrogen bonding***. Thus, the *hydrogen bonding is the interaction between the oppositely charged ends of permanently polarized molecules, each having an* H-*atom.*

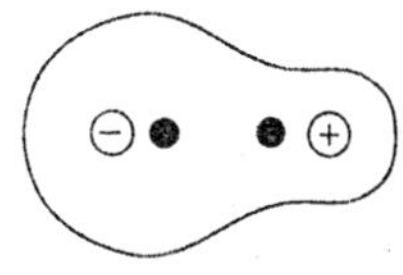

Fig. 4.11 An electric dipole

Hydrogen bonding plays an important role in water molecule, especially in the crystal structure of ice. Water molecules are extremely prone to form hydrogen bonds

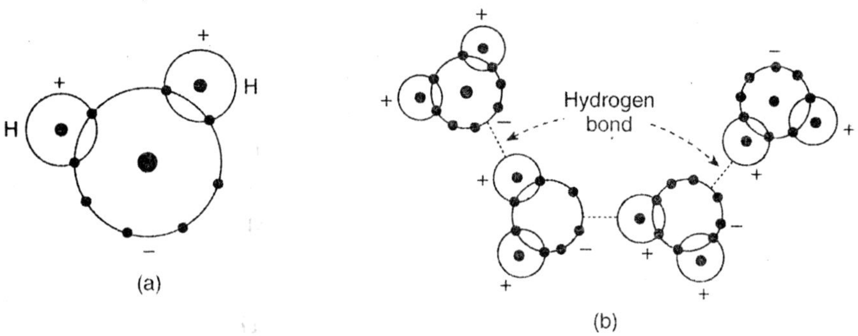

Fig. 4.12 (a) Sharing of electrons between hydrogen and oxygen atoms in a water molecule (b) Scheme of bonding in different molecules

since the electrons in O-atom are *not* distributed *symmetrically*. They have a high probability of being found in certain regions which project outward towards the vertices of a regular tetrahedron (Fig. 4.12).

Each water molecule can form hydrogen bonds with four other H_2O molecules, H-atoms are at two of these vertices and hence show localized positive charges and the other vertices show somewhat more diffuse negative charges. Each water molecule thus can form H-bonds with four other H_2O molecules.

Characteristics—Since the proton in H-atom is not shielded by surrounding electron, such bonds are *relatively strong* and hence the melting and the boiling points of hydrogen-bonded crystals are higher than those of molecular crystals.

Since the hydrogen bond is formed between the oppositely charged ends of two permanent electric dipoles, this is *directional* in nature. Hydrogen bonding is usually formed between the most electronegative atoms like F, O, N because these atoms produce strong electric dipoles.

- This bonding is of particular importance in many biological molecules such as DNA, where it helps to control the possible pairing between the two strands of the molecule and also in some ferroelectric crystals.

4.7 van der Waals bonding

Atoms of rare gases such as He, Ar, Ne etc. are chemically inactive and do not form compounds with other atoms. So, they are called *inert gases.* The inert gases are characterized by closed outer shells (no valence electron) so that there is very little attraction between their atoms. That is why these substances remain in the gaseous state at normal temperature. With the lowering of temperature, the gases condense (liquefy and crystallize) and a gas like argon forms a solid and the question arises what keeps these chemically inactive atoms together.

The answer to this is the *van der Waals forces* which are *weak short range* forces operating between atoms and molecules. They are weaker than atomic forces due to bonding. While the electron distribution around the nucleus is spherically symmetric, an uneven distribution of electron around the nuclei at a certain instant may occur during the molecular motion (Fig. 4.13b). Then the centres of the positive and the negative charges no longer coincide. This displacement of electron cloud gives rise to *fluctuating temporary dipole.* The displacement at an instant in one atom or

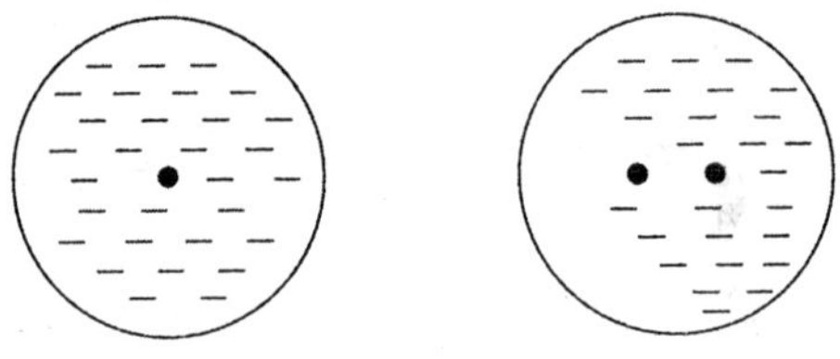

(a) (b)

Fig. 4.13 Distribution of electrons in a noble gas atom (a) idealized symmetrical distribution of an isolated atom (b) real nonsymmetrical distribution around two neighbouring atoms generating fluctuating dipoles

molecule induces another in the neighbour and the dipoles attract each other. Thus the attraction between two atoms become possible. This *dipole-induced* ***dipole interaction*** or ***London interaction***, as it is called, leads to ***van der Waals bonding*** and the crystals formed by such bonding are called ***molecular crystals***. The attractive forces however are *non-directional.*

4.7.1 Binding energy of inert gas crystals: Lenard-Jones potential

Let U_{ij} be the potential energy of interaction between the ith and jth inert gas atoms of the crystal and U_i be that on atom i due to all others.

$$\therefore \quad U_i = \sum_j U_{ij}$$

where the summation extends over all the atoms, *except* $j = i$.

U_i consists of two terms corresponding to two types of interactions: (i) *repulsive interaction* and (ii) *attractive interaction.*

The *repulsive enerегy* is given by the empirical relation

$$U_{\text{rep}} = \frac{B}{r_{ij}} \tag{4.7.1}$$

where r_{ij} is the distance between the ith and jth atom.

The *attractive energy* originates from the *dipole-diple interaction* between the atoms i and j. Let p_i be the instantaneously dipole moment induced on atom i by atom j which produces an electric field $\mathcal{E}$, at the centre of atom j, of magnitude

$$\mathcal{E} = \frac{2p_i}{r_{ij}^3} \quad \text{(in c.g.s unit)} \tag{4.7.2}$$

If α be the electronic polarizability, then the dipole moment induced will be

$$p_j = \alpha\mathcal{E} = \frac{2\alpha p_i}{r_{ij}^3} \tag{4.7.3}$$

The potential energy due to ineraction between p_i and p_j is

$$U_{\text{att}} = -\frac{2p_i p_j}{r_{ij}^3} = -\frac{4\alpha p_i^2}{r_{ij}^6} = -\frac{C}{r_{ij}^6} \tag{4.7.4}$$

where αp_i^2 has the dimensions of (charge)^2 × (length)^5.

U_{att} is known as ***van der Waals*** or ***London interaction***, as stated already.

● Note that $U_{\text{att}} \propto 1/r_{ij}^6$. So, the *van deer Waals interaction is of very short range.*

The total binding energy due to interactions is: $U_{ij} = U_{\text{rep}} + U_{\text{att}}$.

$$\therefore \; U_{ij} = \frac{B}{r_{ij}^{12}} - \frac{C}{r_{ij}^6} \tag{4.7.5}$$

We now introduce two new parameters ε and σ such that $4\varepsilon\sigma^{12} = B$ and $4\varepsilon\sigma^6 = C$. So, from (4.7.5)

$$U_{ij} = \frac{4\varepsilon\sigma^{12}}{r_{ij}^{12}} - \frac{4\varepsilon\sigma^6}{r_{ij}^6} = 4\varepsilon\left[\left(\frac{\sigma}{r_{ij}}\right)^{12} - \left(\frac{\sigma}{r_{ij}}\right)^6\right] \tag{4.7.6}$$

U_{ij} is called the ***Lenard-Jones potential.***

If the crystal is composed of N-atoms, the total binding energy will be

$$U_{\text{tot}} = \tfrac{1}{2}NU_{ij} = \tfrac{1}{2}N(4\varepsilon)\left[\sum_j(\sigma/r_{ij})^{12} - \sum_j(\sigma/r_{ij})^6\right] \tag{4.7.7}$$

where the kinetic energy of the atoms is neglected.

Factor $\frac{1}{2}$ is becaue interaction between atoms i and j is to be counted only once.

Introducing a dimensionless quantity p_{ij} such that $r_{ij} = p_{ij}r$, where r is the interatomic distance, eq. (4.7.7) can be written as

$$U_{\text{tot}} = 2N\varepsilon\left[\sum_j \left(\frac{\sigma}{p_{ij}r}\right)^{12} - \sum_j \left(\frac{\sigma}{p_{ij}r}\right)^{6}\right] \tag{4.7.8}$$

For *fcc*-structure, $\sum\left(\frac{1}{p_{ij}}\right)^{12} = 12.131$ and $\sum\left(\frac{1}{p_{ij}}\right)^{6} = 14.454$

$$\therefore\ U_{\text{tot}} = 2N\varepsilon\left[12.131(\sigma/r)^{12} - 14.454(\sigma/r)^{6}\right] \tag{4.7.9}$$

At *equilibrium interatomic distance*, i.e., at $r = r_e$, the potential energy is *minimum* $\Rightarrow dU_{\text{tot}}/dr = 0$, at $r = r_e$.

$$\therefore\ \left(\frac{dU_{\text{tot}}}{dr}\right)_{r=r_e} = 2N\varepsilon\left[12.131 \times (-12)\frac{\sigma^{12}}{r_e^{1/3}} - 14.454 \times (-6)\frac{\sigma^{6}}{r_e^{7}}\right] = 0$$

$$\Rightarrow r_e/\sigma = 1.09$$

Substituting this value of r_e/σ in (4.7.9), the *cohesive energy* or the ***binding energy*** U_0 of a *molecular crystal* is given by

$$U_{\text{tot}} = U_0 = -2.5 \times 4N\varepsilon = -10N\varepsilon \tag{4.7.10}$$

at zero pressure and zero degree Kelvin.

• A more accurate value for U_0 can be obtained by taking the kinetic energy of the atoms into consideration and applying the quantum mechanical concept.

4.7.2 Characteristics of van der Waals bonding

Some of the ***characteristics*** of van der Waals bonding are:

1. van der Waals attraction is a short range ($\propto r^{-7}$), weak force. The van der Waals bond is therefore *very weak.*

2. The inert gas crystals, due to the characteristic 1 above, have *low melting and boiling points.*

3. The inert gas crystals are found to be *brittle* and lack in strength, their *ionization energy is very high.*

4. The bond is formed due to attraction (electrostatic) between the oscillating or permanent dipoles. The bonds are non-directional and unsaturated, and directional only for dipole bonds.

5. As no valance electron exists, the inert gas crystals are good insulators. *Low density* is the other characteristic of substances bound by this bonding.

6. The inert gas crystals are *transparent* to electromagnetic radiations down to the ultraviolet range.

4.8 A comparison of properties of different bonds

The physical properties of solids are, to a great extent, dependent on the bonding by which the crystal of the solid is formed. A comparative study of the properties of different bonds is embodied in the following table.

	Ionic bond	Covalent bond	Mettalic bond	van der Waals bond
1.	The bonding is due to electro-static force of attraction between the ions.	Bond is formed by sharing of the valence electrons.	Bonding is due to attraction between positive ion cores and valence electrons.	Bonding is formed due to electrostatic attraction between the oscillating and permanent dipoles.
2.	Intermediate density.	Intermediate density.	High density.	Low density.
3.	High melting point.	Intermediate melting point.	Intermediate melting point.	Low melting point.
4.	Low thermal and electrical conductivity.	Low thermal and electrical conductivity.	High thermal and electrical conductivity.	Low thermal and electric conductivity.
5.	Bonds are non-directional.	Bonds are directional.	Bonds are non-directional.	Bonds are directional.

4.9 Illustrated Examples

► **Example 1.** *The energy of two particles in the field of each other, separated by r, is*

$$U = \frac{\alpha}{r} + \frac{\beta}{r^8}$$

where α, β are constants. At what separation will they form a stable compound?

(Himachal Univ.)

Solution: The particles will form a stable compound at a separation $r = r_e$ corresponding to which the energy U is a minimum, i.e.,

$$\left(\frac{dU}{dr}\right)_{r=r_e} = 0$$

Differentiating the given expression for U with respect to r, $\frac{dU}{dr} = -\frac{\alpha}{r^2} - \frac{8\beta}{r^9}$.

$$\therefore \quad \left(\frac{dU}{dr}\right)_{r=r_e} = 0 = -\frac{\alpha}{r_e^2} - \frac{8\beta}{r_e^9} \Rightarrow r_e^7 = 8\frac{\beta}{\alpha}$$

$$\therefore \quad r_e = \left(8\frac{\beta}{\alpha}\right)^{1/7}$$

▶ **Example 2.** *Calculate the cohesive energy per atom of* NaCl *crystal from the following data: Equilibrium distance between ions,* $r_e = 2.81$ Å, *Madelung constant* $\alpha = 1.748$, *ionisation energy of* Na $= 5.1$ eV, *electron affinity for* Cl $= 3.6$ eV *and* $n = 9$.

(Himachal and Punjab Univ.)

Solution: The total potential energy V for ionic bonding that binds NaCl-crystal is

$$V = V_{\text{attraction}} + V_{\text{repulsion}}$$

$$= -\frac{\alpha e^2}{4\pi\varepsilon_0 r}\left(1 - \frac{1}{n}\right), \text{ with usual notation.}$$

$$\therefore \quad V = -\frac{1.748 \times (1.6 \times 10^{-19})^2}{2.81 \times 10^{-10}} \times \left(1 - \frac{1}{9}\right) \times 9 \times 10^9$$

$$[\because \ 1/4\pi\varepsilon_0 = 9 \times 10^9 \text{ Nm/C}^2]$$

$$= -1.27 \times 10^{-18} \text{ J} = -7.97 \text{ eV}$$

$\therefore$ Cohesive energy = P.E. + ionisation energy − electron affinity

$$= (-7.97 + 5.1 - 3.6) \text{ eV} = -6.47 \text{ eV}$$

$\therefore$ Cohesive energy per ion (atom) $= \frac{1}{2} \times (-6.47) = \mathbf{-3.24\,eV}$

▶ **Example 3.** *If the potential energy function is expressed as*

$$U = \frac{\alpha}{r^{12}} - \frac{\beta}{r^6},$$

show that the intermolecular distance for which the potential energy is minimum is $(2\alpha/\beta)^{1/6}$ *and the minimum potential energy* $U_{\min} = -\beta^2/4\alpha$.

(Himachal Univ. and Burdwan Hons.)

Solution: If r_e be the intermolecular distance corresponding to minimum potential energy, we must have $dU/dr = 0$ at $r = r_e$. Imposing this condition, we get

$$\frac{d}{dr}\left(\frac{\alpha}{r^{12}} - \frac{\beta}{r^6}\right) = -12r^{-13}\alpha + 6r^{-7}\beta = 0 \text{ for } r = r_e.$$

$$\therefore \quad -12r_e^{-13}\alpha + 6r_e^{-7}\beta = 0$$

$$\Rightarrow \quad -2\alpha/r_e^6 + \beta = 0 \ \Rightarrow \ \boldsymbol{r_e = (2\alpha/\beta)^{1/6}}$$

Minimum potential energy is obtained by putting $r = r_e = (2\alpha/\beta)^{1/6}$.

$$\therefore \quad U_{\min} = \frac{\alpha}{r_e^{12}} - \frac{\beta}{r_e^6} = \frac{1}{r_e^6}\left(\frac{\alpha}{r_e^6} - \beta\right) = \frac{\beta}{2\alpha}\left(\alpha \times \frac{\beta}{2\alpha} - \beta\right)$$

$$= -\frac{\beta}{2\alpha} \times \frac{\beta}{2} = -\frac{\beta^2}{4\alpha}$$

▶ **Example 4.** *The potential energy of a system of two atoms is given by*

$$U(R) = -\frac{\alpha}{R^4} + \frac{\beta}{R^{12}}$$

Calculate the amount of energy released when the atoms form a stable bond. Determine the bond length.

Solution: The force of attraction between the two atoms is

$$F = -\frac{dU}{dR} = -\frac{4\alpha}{R^5} + \frac{12\beta}{R^{13}}, \text{ using the relation for } U(R)$$

At the equilibrium distance (= *bond length*) $R = R_e$, the net force acting on the atom is zero.

$$\therefore \quad F = -\left(\frac{dU}{dR}\right)_{R=R_e} = 0 \Rightarrow \frac{4\alpha}{R_e^5} = \frac{12\beta}{R_e^{13}}$$

$$\therefore \quad R_e^8 = \frac{12\beta}{4\alpha} = \frac{3\beta}{\alpha}$$

$$\therefore \quad \boldsymbol{R_e = \left(\frac{3\beta}{\alpha}\right)^{1/8}}$$

This is the required ***bond length.***

Substituting this value of R_e in the expression for $U(R)$ and simplifying,

$$U_e = -\frac{\alpha}{R_e^4}\left(1 - \frac{1}{3}\right)$$

$$= -\frac{\alpha}{(3\beta/\alpha)^{1/2}} \times \frac{2}{3} = -\left(\frac{4\alpha^3}{27\beta}\right)^{1/2}$$

which is the ***energy released*** when the atoms form a stable bond.

▶ **Example 5.** *The potential energy of a system of two atoms is given by*

$$U = -A/r^2 + B/r^{10}$$

A stable molecule is formed with release of 8.0 eV *of energy when the interatomic distance is* 2.8 A. *Calculate the values of A and B.*

Solution: The potential energy of the system is

$$U = -A/r^2 + B/r^{10} \tag{4.9.1}$$

∴ Force between atoms at an interatomic distance r is

$$F = -dU/dr = -2A/r^3 + 10B/r^{11} \tag{4.9.2}$$

At equilibrium distance r_e, we have $F = 0$. So, from equation (4.9.2)

$$A = 5B/r_e^8 = 5B/(2.8 \times 10^{-10})^8 = 1.32 \times 10^{77} B \tag{4.9.3}$$

But, the dissociation or bond energy U_e is the energy released during the bond formation. From equation (4.9.1),

$$U_e = -A/r_e^2 + B/r_e^{10} = -(A/r_e^2)[1 - B/Ar_e^8]$$
$$= -(A/r_e^2)(1 - 1/5), \quad \text{using (4.9.3)}.$$

But, $U_e = -8.0\,\text{eV} = -8.0 \times 1.6 \times 10^{-19}$ J, the negative sign shows energy-release.

$$\therefore \quad \boldsymbol{A} = (5/4)(2.8 \times 10^{-10})^2 \times (8.0 \times 1.6 \times 10^{-19})$$
$$= \mathbf{1.256 \times 10^{-37}\,Jm^2}$$
$$\boldsymbol{B} = \mathbf{9.52 \times 10^{-115}\,Jm^{10}}, \quad \text{substituting the value of } A \text{ from (4.9.3)}$$

▸ **Example 6.** *In the above* Example 5, *determine the force necessary to dissociate the molecule into atoms and the interatomic distance at which the dissociation occurs.*

Solution: The magnitude of the force becomes maximum at a certain critical interatomic distance $r_c(> r_e)$. This gives the force necessary to dissociate the molecule. The value of r_c is obtained from the condition

$$\left.\frac{dF}{dr}\right|_{r=r_c} = 0$$
$$\Rightarrow \quad 6A/r_c^4 - 110B/r_c^{12} = 0, \quad \text{using (4.9.2)}$$
$$\Rightarrow \quad \boldsymbol{r_c} = (110B/6A)^{1/8} \tag{4.9.4}$$
$$= \left[\frac{110 \times 9.52 \times 10^{-115}}{6 \times 1.256 \times 10^{-37}}\right]^{1/8} = \mathbf{3.3 \times 10^{-10}}\,\text{m}$$

Substituting $r = r_c$ in (4.9.2), the force needed to dissociate the molecule is

$$F = -(2/r_c^3)(A - 5B/r_c^8)$$
$$= -(2/r_c^3)[A - 5B(6A/110B)], \quad \text{using (4.9.4)}$$
$$= -\left(2/r_c^3)(8A/11\right).$$
$$= \frac{2}{(3.3 \times 10^{-10})^3}\left[\frac{8 \times 1.256 \times 10^{-37}}{11}\right] = \mathbf{-5.08 \times 10^{-9}\,N},$$

the negative sign shows that at $r = r_c$, the force existing is of attractive type.

$\therefore$ Force needed to dissociate the molecule is $\mathbf{5.08 \times 10^{-9}\,N}$.

▸ **Example 7.** *Assuming that the potential energy of two particles in the field of each other is given by*

$$U(r) = -\frac{A}{r} + \frac{B}{r^9}$$

where A, B are constants, show that for stable configuration, the energy of attraction is 9 times the energy of repulsion.

Solution: The potential energy $U(r = r_e)$ is minimum for stable configuration.

$$\therefore \quad \left(\frac{dU}{dr}\right)_{r=r_e} = \frac{A}{r_e^2} - \frac{9B}{r_e^{10}} = 0 \;\Rightarrow\; r_e = \left(\frac{9B}{A}\right)^{1/8}$$

For stable configuration, the attractive and repulsive energies are respectively

$$\phi_{\text{att}} = \frac{A}{r_e} \text{ and } \phi_{\text{rep}} = \frac{B}{r_e^9}.$$

$$\therefore \quad \frac{\phi_{\text{att}}}{\phi_{\text{rep}}} = \frac{A}{B} \times r_e^8 = \frac{A}{B} \times \frac{9B}{A}, \text{ substituting for } r_e.$$

$$\therefore \quad \boldsymbol{\phi_{\text{att}} = 9\phi_{\text{rep}}}$$

▶ **Example 8.** *Calculate the Madelung constant for an infinite linear chain of ions of alternating unit charge (e.g., one dimensional* NaCl *crystal) at a separation R.*

Solution: In the one-dimensional (linear) infinite chain as shown in Fig. 4.14, we take the positive ion as the reference ion. So, we may write, with R as the distance between the adjacent ions,

$$\frac{\alpha}{R} = 2\left[\frac{1}{R} - \frac{1}{2R} + \frac{1}{3R} - \frac{1}{4R} + \cdots\right]$$

The factor 2 is due to the presence of similar ions on both sides of the reference ion.

$$\therefore \text{ Madelung constant, } \alpha = 2\left(\frac{1}{1} - \frac{1}{2} + \frac{1}{3} - \frac{1}{4} + \cdots\right) \tag{4.9.5}$$

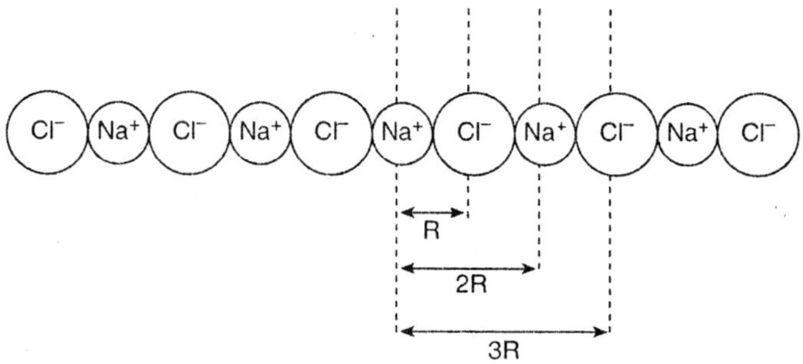

Fig. 4.14 Geometry of a one dimensional NaCl crystal

The expression (4.9.5) is written with a positive sodium ion as the reference, and the factor 2 appears as similar ions are present on both sides of the reference ion.

Using the series, $\ln(1+x) = x - \frac{x^2}{2} + \frac{x^3}{3} - \frac{x^4}{4} + \cdots$, with $x = 1$ and using (4.9.5)

$$\alpha = 2\ln(1+1) = \mathbf{2\ln 2} = 2 \times 0.69 = \mathbf{1.38}$$

▸ **Example 9.** *Given for* NaCl, $r_e = 2.81\,\text{A}$, $\alpha = 1.75$, $\rho = 0.298\,\text{A}$, $e = 4.8 \times 10^{-10}$ esu, *calculate the binding energy per molecule.* **(Punjabi Univ.)**

Solution: The potential or binding energy per ion pair (or molecule) is

$$U_{\text{total}} = -\frac{\alpha q^2}{r_e}\left(1 - \frac{\rho}{r_e}\right)$$

$$= -\frac{1.75 \times (4.8 \times 10^{-10})^2}{2.81 \times 10^{-8}} \times \left(1 - \frac{0.298}{2.81}\right), \quad \text{using the given data.}$$

$$= 12.77 \times 10^{-12} \text{ erg, on simplification.}$$

$$= \frac{12.77 \times 10^{-12}}{1.6 \times 10^{-12}} = \mathbf{7.98\,eV}$$

▸ **Example 10.** *The potential energy of a pair of atoms in a crystal is of the form* $A/r^9 - B/r$, *when the separation is* r. *Assuming the equilibrium separation to be* $2.8\,\text{A}$ *and the dissociation energy* 8×10^{-19} Joule, *calculate* A *and* B. **(Calcutta Hons.)**

Solution: The potential energy is given by

$$U(r) = \frac{A}{r^9} - \frac{B}{r} \tag{4.9.6}$$

At $r = r_e = 2.8\,\text{A}$ and $U(r_e) = -8 \times 10^{-19}\,\text{J} = -\dfrac{8 \times 10^{-19}}{1.6 \times 10^{-19}}\,\text{eV} = -5\,\text{eV}$

But, we have
$$U(r_e) = -5 = \frac{A}{(2.8)^9} - \frac{B}{2.8} \tag{4.9.7}$$

Also, at $r = r_e = 2.8\,\text{A}$, $\dfrac{dU}{dr} = 0$. From (4.9.9), therefore,

$$-\frac{9A}{r_e^{10}} + \frac{B}{r_e^2} = 0 \;\Rightarrow\; B = \frac{9A}{r_e^8} = \frac{9A}{(2.8)^8} \tag{4.9.8}$$

From (4.9.10) and (4.9.11), we obtain,

$$\frac{8A}{(2.8)^9} = 5 \;\Rightarrow\; \boldsymbol{A} = \frac{5 \times (2.8)^9}{8} = \mathbf{6611\,eV\cdot\text{Å}}$$

$$\text{and } \boldsymbol{B} = \frac{9A}{(2.8)^8} = \frac{9 \times 5 \times (2.8)^9}{(2.8)^8 \times 8} = \frac{9 \times 5 \times 2.8}{8} = \mathbf{15.75\,eV\cdot\text{Å}}$$

▸ **Example 11.** *Find the nearest neighbour distance in bcc and fcc structures when the atomic radius is* $10\,\text{A}$.

Solution: The nearest neighbour distance in a *bcc* structure is half the body diagonal.

$$\therefore\; d_{bcc} = \frac{\sqrt{3}a}{2}, \quad a = \text{side of the cube.}$$

In the limiting condition, $\sqrt{3}a = 4R$, where R = the atomic radius.

$$\therefore \quad \frac{\sqrt{3}a}{2} = 2R = 2 \times 10\ \text{A} = \mathbf{20\ A}$$

Similarly, in *fcc* structure, the nearest neighbour distance is half the face diagonal.

$$\therefore \quad d_{fcc} = \frac{\sqrt{2}a}{2} = \frac{a}{\sqrt{2}}$$

In the limiting condition, $\sqrt{2}a = 4R \quad \Rightarrow \quad \sqrt{2}a/2 = 2R = 2 \times 10\ \text{A} = \mathbf{20\ A}$

► **Example 12.** *An overlap interaction between the nearest neighbours is of exponential type given by*

$$U(R) = Be^{-R/\rho}$$

where B and ρ are constants. Show that the equilibrium spacing is given by $R_e = \rho \ln B$.

Solution: We have: $\quad U(R) = Be^{-R/\rho}$

At equilibrium spacing, $\quad R = R_e,\ U(R) = U(R_e) = 0.$

$$\text{Hence,} \quad Be^{-R_e/\rho} = 0$$

$$\therefore \quad \ln B - R_e/\rho = 0$$

$$\therefore \quad \frac{R_e}{\rho} = \ln B \quad \Rightarrow \quad \mathbf{R_e = \rho \ln B}$$

► **Example 13.** *The potential energy of a diatomic molecule in terms of interatomic distance R is given by*

$$U(R) = -\frac{A}{R^2} + \frac{B}{R^{10}}$$

Calculate (i) *the equilibrium separation and* (ii) *the dissociation energy when* $A = 1.44 \times 10^{-39}\ \text{Jm}^2$ *and* $B = 2.19 \times 10^{-115}$.

Solution: (i) For equilibrium separation R_e, we have

$$\left(\frac{dU}{dR}\right)_{R=R_e} = \frac{2A}{R_e^3} - \frac{10B}{R_e^{11}} = 0$$

$$\text{or,} \quad R_e^8 = \frac{5B}{A} = \frac{5 \times 2.19 \times 10^{-115}}{1.44 \times 10^{-39}}$$

$$= 76042 \times 10^{-80}$$

$$\therefore \quad R_e = (76042 \times 10^{-80})^{1/8}$$

$$= \mathbf{4.08 \times 10^{-10}}\ \text{m}$$

$$= \mathbf{4.08\ A}$$

(ii) To obtain the dissociation energy D, we have

$$D = -U_e = \frac{A}{R_e^m}\left(1 - \frac{m}{n}\right)$$

$$= \frac{1.44 \times 10^{-39}}{(4.08 \times 10^{-10})^2}\left(1 - \frac{2}{10}\right) \text{ J} \quad [\because m = 2, n = 10]$$

$$= \frac{4 \times 1.44 \times 10^{-39}}{5 \times (4.08 \times 10^{-10})^2} \cdot \frac{1}{1.6 \times 10^{-19}} \text{ eV}$$

$$= \mathbf{4.33 \times 10^{-2}\ eV}$$

► **Example 14.** *Determine if a gas molecule A^+B^- is stable with respect to ioslated gas atoms A and B. Use the following data: First ionization energy of A* = 502 kJ/mol, *electron affinity of B atom* = −335 kJ/mol, *interionic separation* = 0.3 nm, *permittivity of free space* = 8.85×10^{-12} C/Nm

Solution: Electrostatic attractive energy of A^+B^- molecule is

$$U_{att} = -\frac{e^2}{4\pi\varepsilon_0 r} = -\frac{(1.602 \times 10^{-19})^2}{4\pi \times 8.85 \times 10^{-12} \times 3 \times 10^{-10}}$$

$$= -7.689 \times 10^{-19} \text{ J/ion pair}$$

$$= -7.689 \times 10^{-19} \times 6.02 \times 10^{23} \times 10^{-3} \text{ kJ/mol}$$

$$= -463 \text{ kJ/mol}$$

∴ Net change in energy per mole = 502 − 335 − 463 = **−296 kJ/mol**

Since the change in energy is *negative*, A^+B^- molecule will be ***stable***.

Review Questions

1. Explain the term bonding. Name the different types of bonding, with examples.
2. What do you mean by cohesive energy of a crystal? Discuss the general nature of cohesion between two atoms in a solid.
3. What are ionic crystals and ionic bonding? State the properties of ionic crystals and explain the following:
   (a) ionic crystals are hard and brittle
   (b) ionic solids are poor conductors. **(Bangalore Univ.)**
4. Describe the ionic bonding process between a pair of Na and Cl atoms. Which electrons are involved in the bonding process?
5. If the total energy of an ionic crystal is expressed as

$$U = -Ae^2/r + B/r^n,$$

show that the equilibrium energy U_e at $r = r_e$ is given by

$$U_e = -\frac{Ae^2}{r_e}\left[1 - \frac{1}{n}\right]$$ **(Himachal Univ.)**

6. What is Madelung constant? What is its unit? Evaluate the Madelung constant for a linear ionic crystal. Show that for a one-dimensional array of ions of alternate sign, its value is $\simeq 1.38$. **(Burdwan Hons. and Guru Nanak Univ.)**

7. How does the Madelung constant vary with the number of nearest neighbour? Would you expect this?

8. Consider a line of $2N$ ions of the alternating charge $\pm q$ with a repulsive potential energy A/R^n between the nearest neighbours. Show that at the equilibrium separation, R_e

$$U(R_e) = -\frac{2Nq^2 \ln 2}{R_e}\left(1 - \frac{1}{n}\right)$$

9. Show that the total lattice energy of an ionic crystal assuming a repulsive interaction of the form $\lambda e^{-R/\rho}$ operating between nearest neighbours only, is given by

$$U = -\frac{1}{4\pi\varepsilon_0}\frac{N\alpha e^2}{R_0}\left(1 - \frac{\rho}{R_0}\right)$$

where $2N$ is the number of ions in the crystal, R_0 the equilibrium separation between nearest neighbours and α the Madelung constant. **(Burdwan Hons.)**

10. Explain covalent bonding. What is the cause of directional nature of covalent bond? Give the characteristics of covalent bond. **(Punjab Univ.)**

11. Define cohesive energy for inert gas crystals. State Lenard Jones two body potential and use it to obtain the expression for the cohesive energy of an inert gas crystal. **(Guru Nanak and Himachal Univ.)**

12. What type of interaction is responsible for binding among atoms of the inert gas crystals? Explain the origin of both attractive and repulsive parts of the interaction.

13. What is hydrogen bond? How is it different from a dipole bond? Describe the role of hydrogen bond during the formation of ice.

14. Explain what you understand by metallic bonds. Give the characteristics of metallic bonding.

15. Distinguish between ionic and covalent bonds. What is the nature of bonds in NaCl and diamond?

16. Deduce an expression for the bulk modulus of an ionic crystal assuming a repulsive interaction between nearest neighbours of the form λ/R^n. **(Budwan Hons.)**

Short and Objective Questions

1. Name the different types of bonding, citing examples in each case.
2. Define cohesive energy of a solid. Show that the formation of a chemical bond requires that the repulsive forces be of shorter range than the attractive forces.
3. Briefly discuss the different types of bonding in solids. **(Burdwan Hons.)**
4. Explain the origin of van der Waal's interaction between two atoms of inert gas in the solid state. **(Burdwan Hons.)**
5. Write down the nature of bonding in diamond form of carbon and CO_2 and explain briefly the mechanism of such bonding. **(Calculatta Hons.)**
6. Explain (a) ionic crystals are hard and brittle; (b) ionic solids are poor conductors of electricity.
7. What is Madelung constant? Give its unit. What is its significance? What is Madelung energy?
8. Explain why the inert gas crystals exists only at very low temperature and not at room temperature. **(Himachal Univ.)**
9. All metals are opaque to light of all wavelengths and have high lustre and shear easily. Explain. **(Bangalore Univ.)**
10. Why is NaCl type structure more stable than CsCl type structure although the Madelung constant of CsCl is greater than that of NaCl? **(Lucknow Univ.)**

    [**Hints.** The value of $\alpha = 1.748$ for NaCl-type crystal and $\alpha = 1.763$ for CsCl type. Still the CsCl-type is not more stable because the number of nearest neighbours in CsCl-type structure is more and hence the repulsive energy is more than that in NaCl-type]
11. What are primary and secondary bonds? Give examples for each.
12. Given a solid made up of molecules possessing permanent dipoles and another made up of molecules possessing only induced dipoles. Which will show the most temperature-dependent properties?
13. Describe the covalent-bonding electron arrangement in the following diatomic molecules: (a) fluorine, (b) oxygen, (c) nitrogen.
14. Show that the Madelung constant for an infinite linear chain of ions of alternating unit charge at an equilibrium separation is $2 \ln 2$.
15. What limits the coordination of a cation and an anion in an ionic crystal?
16. Why are closed packed structures experienced in metals and not in ionic crystals?
17. What is van der Waals interaction in crystal binding? Is it present in all types of solids? **(Burdwan Hons.)**

18. At the equilibrium separation of a pair of atoms in a crystal

(a) the potential energy is a maximum,
(b) the potential energy is a minimum,
(c) the interatomic force is a minimum,
(d) none of the above is true.

19. Transfer of electrons takes place in

(a) ionic bonding, (b) covalent bonding,
(c) metallic bonding, (d) none of these. **(Himachal Univ.)**

20. An ionic molecule is a

(a) polar molecule, (b) non-polar molecule,
(c) may be both, (d) none of these.

21. Which of the following bonds are directional?

(a) covalent, (b) ionic, (c) metallic, (d) van der Waals.

22. The bond between the ice-molecules is

(a) covalent bond, (b) hydrogen bond,
(b) metallic bond, (d) none of these.

23. The equilibrium condition of the atoms in a solid is characterized by

(a) maximum kinetic energy, (b) minimum kinetic energy,
(c) minimum potential energy, (d) none of these.

24. Which of the following has hydrogen bonding?

(a) CH_4, (b) CsCl, (c) NaCl, (d) HF.

25. van der Waals' interaction is proportional to

(a) $1/r^2$, (b) $1/r^3$, (c) $1/r^6$, (d) None.

26. The strongest bond is

(a) metallic, (b) ionic, (c) covalent, (d) van der Waals.

Problems

1. The potential energy of a pair of atoms is $U = -A/r^4 + B/r^{12}$ where r is the interatomic distance. Find the value of r where a stable bond is formed. Also, determine the energy released when the atoms form a stable bond.

$[(3B/A)^{1/8}, (4A^3/27B)^{1/2}]$

2. The energy of two particles in the field of each other is given by the expression $U = -\alpha/r^2 + \beta/r^{10}$, where α and β are constants and r, the separation between the particles. Determine the separation between the particles for a stable compound and also determine the cohesive energy of the crystal.

(Guru Nanak Univ.) $[r_e = (5\beta/\alpha)^{1/8}; -\frac{4}{5}(\alpha/r_e^2)]$

3. The interaction energy of a system of two atoms is given by

$$U = -A/r^6 + B/r^{12}$$

The atoms form a stable bond with bond length of 3 Å and bond energy of 1.8 eV. Calculate A and B. $[4.19 \times 10^{-76}\,\text{Jm}^6, 1.53 \times 10^{-133}\,\text{Jm}^{12}]$

4. Given for NaCl crystal $r_e = 2.81$ Å, $\alpha = 1.75$, $n = 9.4$, $e = 4.8 \times 10^{-10}$ esu. Calculate its binding energy/molecule. [8 eV]

5. In Problem 3, compute the forces required to break the molecule and the critical interatomic distance for which it occurs. Also, calculate the force required to reduce the interatomic distance by 5% of the value at equilibrium.

$[2.56 \times 10^{-9}\,\text{N}, 3.33 \times 10^{-10}\,\text{m}, 5.97 \times 10^{-9}\,\text{N}]$

6. The cohesive energy and the nearest neighbour-distance for a LiF molecule are 1.68×10^{-18} J and 2.014 Å respectively. The structure of LiF is the same as that of NaCl. Find the bulk modulus of LiF using the expression for the potential energy $U = -\alpha q^2/R + B/R^n$, the value of α may be taken as 1.75.

$[7.15 \times 10^{10}\,\text{N/m}^2]$

7. In NaCl crystal, the equilibrium distance between the ions is $r_e = 2.31$ Å and $\alpha = 1.748$. Taking $n = 9$, find the potential energy per ion pair and the total cohesive energy per atom in the NaCl crystal. [7.94 eV, −3.22 eV/atom]

8. If an atom A has an ionisation energy 5 eV and another atom B has an electron affinity 4 eV and if they are separated by 5 Å, find the amount of energy required to transfer an electron from A to B. [−1.9 eV]

9. Calculate the compressibility of NaCl if a repulsive potential of the form B/r^9 acts between the nearest neighbours. The distance of the nearest neighbour is $r = 2.81$ Å and the corresponding Madelung constant $\alpha = 1.7476$.

$[3.49 \times 10^{-11}\,\text{m}^2/\text{N}]$

10. Consider a line of $2N$ ions of alternating charge $\pm q$ with a repulsive core potential β/r^n between nearest neighbours. Show that at equilibrium separation

$$U(r_e) = \frac{2Nq^2 \ln 2}{4\pi\varepsilon_0 r_e}\left(1 - \frac{1}{n}\right)$$

(Lucknow Univ.)

11. Calculate the value of n for NaCl crystal having lattice parameter 5.63 Å, binding energy 1.83 k.cal/mol and $\alpha = 1.75$ respectively.

12. The potential energy of a diatomic molecule in terms of interatomic separation r is given by

$$U(r) = -\frac{\alpha}{r^2} + \frac{\beta}{r^{10}}$$

where $\alpha = 1.44 \times 10^{-39}\,\text{Jm}^2$ and $\beta = 2.19 \times 10^{-115}\,\text{Jm}^{10}$. Find the equilibrium spacing r_e and the dissociation energy. **(Calcutta and Burdwan Hons.)**

[4.08×10^{-10} m, 4.33×10^{-2} eV]

13. What is the potential energy of CsCl at equilibrium, if the separation between cesium and chlorine atoms is 3.56 A and n is 11.5? [–6.50 eV]

14. Show that the first five terms in the series for the Madelung constant for NaCl (fcc) are given by

$$\alpha = 6 - 12/\sqrt{2} + 8/\sqrt{3} - 6/2 + 24/\sqrt{5}$$

How can the convergence of the series be improved? **(Burdwan Hons.)**

15. Calculate the binding energy per molecule of NaCl crystal if $r_0 = 2.81$ A, $\alpha = 1.75$, $\rho = 0.298$ A, $e = 4.8 \times 10^{-10}$ esu. **(Himachal Univ.)** [7.98 eV]

16. Find if a gaseous molecule A^+B^- will be stable with respect to the isolated A and B gaseous atom. Given: first ionization energy of A = 502 kJ/mol, electron affinity of B atom = 335 kJ/mol, interionic (A^+B^{-1}) separation = 0.3 nm, permittivity of free-space = 8.85×10^{-12} C/N.m. What is the net change in energy per mole? [stable, –296 kJ/mol]

17. Show that the value of repulsive potential exponent n in the total energy expression of $2N$ ions of NaCl

$$U(r) = -N\left(\frac{\alpha e^2}{r} - \frac{\beta}{r^n}\right) \quad \text{is } 9.4$$

where α = Madelung constant = 1.75, compressibility of NaCl is $K = 3.3 \times 10^{12}\,\text{cm}^2/\text{dyne}$ and the equilibrium nearest neighbour spacing $r = 2.81$ A.

18. Assuming the energy of two particles in the field of each other is $U(r) = -\frac{\alpha}{r} + \frac{\beta}{r^8}$ where α, β are constants and r is the distance between the centres of the particles, show that the two particles form a stable compound for $r = r_0 = (8\beta/\alpha)^{1/7}$ and in the stable configuration, the energy of attraction is 8 times the energy of repulsion.

19. Calculate the nearest neighbour separation r_0 at which the potential energy

$$U(r_0) = -N\left(\frac{\alpha e^2}{r} - \frac{\beta}{r^n}\right)$$

remains unchanged if β/r^n is replaced by $Ce^{-r/\rho}$. [$r_0 = n\rho$]

20. Lattice energy of KCl crystal containing N molecules of KCl is given by

$$U = -N(\alpha q^2/4\pi\varepsilon_r R - \beta/R^n).$$

Determine the repulsive exponent n from the given data:

Nearest neighbour (equilibrium) distance, $r_e = 3.147$ A

Compressibility of KCl, $1/B = 5.747 \times 10^{-11}$ m^2/N,

Madelung constant, $\alpha = 1.748$. [8.63]

21. Calculate the lattice energy per ion of the NaCl-crystal of which the nearest neighbour distance is 2.81 A. Madelung constant for NaCl is 1.748 and the repulsive component is 9. **(Burdwan Hons.)**

22. The repulsive force between Na^+ and Cl^- ions is often written as

$$F = Be^{-r/R}$$

where $B = 4.45 \times 10^{-6}$ N and $R = 0.321$ A.

At equilibrium, this force is equal to the magnitude of the attractive force given by

$$F = \frac{e^2}{4\pi\varepsilon_0 r^2}$$

Find the equilibrium value of r. [2.28 A]

5

Free Electron Theories: Transport Phenomena

The outstanding properties of metals are their high electrical and thermal conductivities. The conductions of heat and electricity are termed ***transport phenomena*** which also include the electric conduction in semiconductors. In the nineteenth century, there was hardly any precise concept of atomic structure. With the discovery of electron in 1897 by Thomson, a number of investigators, particularly Drude and Lorentz, attempted to explain these properties on the basis of free electron model and applied the kinetic theory to the '*free electron gas*'. The success of Drude's model was indeed remarkable, but not without its failures. The failures were however resolved by replacing the classical statistics of Maxwell-Boltzmann by quantum statistics of Fermi-Dirac applied to free electrons moving inside the metal.

We assume that the reader possesses the basic knowledge of ***Fermi-Dirac statistics***. If not, any standard text or Chapter: *Statistical Mechanics* from author's book 'Thermal Physics', may please be consulted for without that the used expressions in the chapter may appear strange.

5.1 Drude's model of free electrons

The starting point of Drude's model is that the valance electrons in a metal are free to move about in the lattice to form an 'electron gas', whereas ions remain intact and play the role of immobile particles. It was suggested by Lorentz that the classical energy distribution of gas particles (MB-distribution) can be applied to this electron gas as well.

When atoms of a metal are brought close to one another to form the metal, the valence electrons get detached and freely move inside the metal. They are termed *free* or *conduction electrons*; the positive metallic ion cores however remain immobile. Unlike the ordinary gas particles, the free electrons move through the background of these immobile ions. The free-electron density may range from $2\times10^{22}\,\mathrm{m}^{-3}$ to upwards, much denser compared to that of an ordinary gas at usual temperatures and pressures.

Basic assumptions—The basic assumptions of Drude's classical free electron model are as follows:

1. The free-electrons are in thermal equilibrium with the environment, colliding constantly with the ion cores elastically, the mutual repulsion between the electrons being ignored.

2. Between two collisions, therefore, the electron moves uniformly in a straight line (free path) in absence of an electric field, the motion being controlled primarily by these collisions.

3. The time taken in a collision is negligible and the velocity of an electron suffers an abrupt change in a collision.

4. If τ be the *mean free time* between two successive collisions of an electron, τ is taken, for simplicity, to be *independent of the position and the velocity of electron.* τ is also called *relaxation time.*

5. The electron-electron collisions are usually unimportant, the metallic ions forming the major source of collisions.

• Note that the *probability of* an electron suffering *a collision* in unit time is $1/\tau$, i.e., the reciprocal of the mean free time.

5.2 Electric conductivity of metals

In the absence of an electric field, the free electrons in a metal move at random due to their thermal energy and the corresponding velocity is called *random velocity.* At each collision, the path of the electron changes abruptly (Fig. 5.1). The number of electrons moving in one direction is equal to that moving oppositely. So, considering

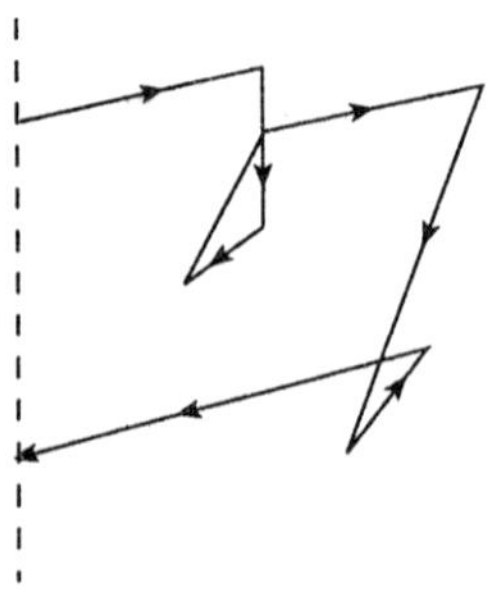

Fig. 5.1 Random electron motion

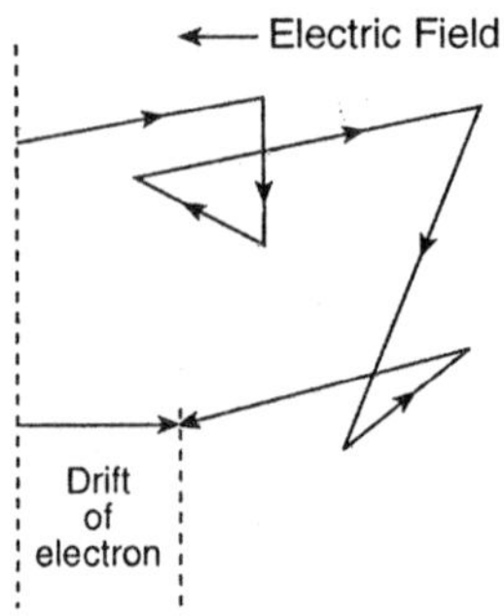

Fig. 5.2 Electron drift in an applied electric field

the electrons in totality, the average velocity is zero. The random thermal velocity of electron is $v_{\rm rms}$, the root mean square velocity.

In presence of an electric field, however, there is in addition to random velocity, an additional net velocity associated with the free electrons and is called the *drift velocity*, v_D (Fig. 5.2). Compared to the random thermal velocity, v_D is very small. The drift gives rise to current flow through the metal.

From kinetic theory, $$\tfrac{1}{2}mv_{\rm rms}^2 = \tfrac{3}{2}kT \Rightarrow v_{\rm rms} = \sqrt{3kT/m} \tag{5.2.1}$$

where k is Boltzmann constant.

If N be the number of ion cores per unit volume, each having an effective scattering radius R, the *mean free path* λ of an electron is obtained purely from geometric consideration, as in kinetic theory, and is given by

$$\lambda = \frac{1}{\pi R^2 N}; \quad \text{also } \tau = \frac{\lambda}{v_{\rm rms}} \tag{5.2.2}$$

Expression for the drift velocity—The drift velocity v_D can be expressed in terms of the applied field E, $v_{\rm rms}$ and the mean free path λ.

When the electric field E is applied, an electron plainly gains an acceleration given by $a = eE/m$. Let an electron just collide with an ion core. This collision would momentarily destroy the tendency of the electron to drift and the electron would move in random direction after the collision. At its next collision, its velocity, on an average, changes by $a\tau$ given by

$$a\tau = a\frac{\lambda}{v_{\rm rms}}.$$

This $a\tau$ is termed the ***drift velocity*** $\boldsymbol{v_D}$ of the electron. So, we have

$$\boldsymbol{v_D} = a\tau = \frac{\boldsymbol{eE}}{\boldsymbol{m}}\tau = \left(\frac{eE}{m}\right)\left(\frac{\lambda}{v_{\rm rms}}\right), \quad \text{using (5.2.2).} \tag{5.2.3}$$

But, the current density, $$J = nev_D = ne^2\tau E/m = \sigma E \tag{5.2.4}$$

where n is the number density of electrons and σ the ***electrical conductivity.***

Electrical resistivity ρ is defined as the inverse of conductivity ($\rho = 1/\sigma$) and is measured in ohm.m.

• Note that (5.2.4) can be written in the *vector form* as $\vec{J} = \sigma\vec{E}$ which, in fact, is the ***vector form*** of the well-known ***Ohm's law.***

$$\therefore \quad \sigma = \frac{J}{E} = \frac{\boldsymbol{ne^2\tau}}{\boldsymbol{m}} = \frac{\boldsymbol{ne^2\lambda}}{\boldsymbol{mv_{\rm rms}}} \tag{5.2.5}$$

using (5.2.4) and (5.2.2).

The ***mobility*** $\boldsymbol{\mu}$ of a free electron (in the metal) is defined as the *steady-state drift velocity* $\langle v_D\rangle$ *per unit electric field.*

$$\therefore \quad \mu = \frac{\langle v_D\rangle}{E} = \frac{\boldsymbol{e\tau}}{\boldsymbol{m}}, \quad \text{using (5.2.3).} \tag{5.2.6}$$

$\therefore$ From (5.2.5), using (5.2.6),
$$\boxed{\sigma = ne\mu} \tag{5.2.7}$$

- From equation (5.2.5), using (5.2.1), we obtain

$$\sigma = \frac{ne^2\lambda}{\sqrt{3mkT}}, \tag{5.2.8}$$

Since the mean free path, λ is essentially independent of temperature, we have from equation (5.2.8)

$$\sigma \propto T^{-\frac{1}{2}}$$

So, *conductivity in metals decreases with rise in temperature as* $T^{-\frac{1}{2}}$.

Experimentally, however, $\sigma \propto T^{-1}$ and the discrepancy between the experimental and the theoretical values may readily be resolved by taking recourse to *FD*-statistics in place of classical *MB* statisics. Then we shall see that all the free electrons of a metal do not participate in the conduction, rather only the electrons near the Fermi level take part in electrical and thermal conduction. The root of the discrepancy lies there and we shall revisit the problem for its complete resolution.

5.3 Thermal conductivity of metals

We shall now develop an expression for the *thermal conductivity* due to the free electrons in a metal.

Let us assume a temperature gradient dT/dx in the x-direction. The thermal conductivity of the electron gas is defined as *the ratio of the energy flux H* (i.e., energy crossing unit area per unit time normal to the direction of flow) *to the temperature gradient* which gives rise to the flow of energy. Based on this definition:

Thermal conductivity, $$K = \frac{-H}{dT/dx}, \tag{5.3.1}$$

the negative sign implies that the direction of the heat flow is opposite to that of the temperature gradient.

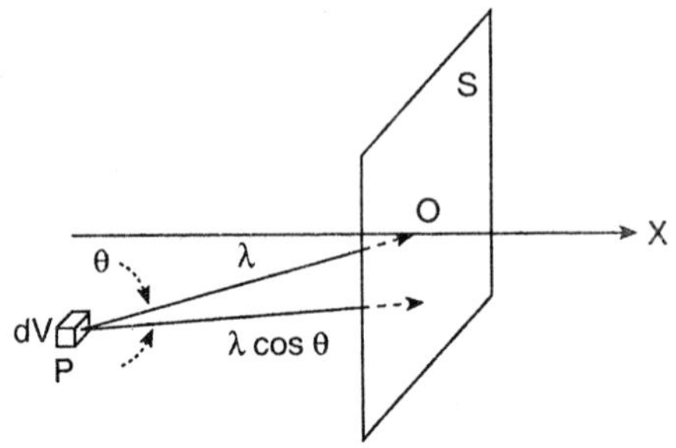

Fig. 5.3 Energy flux crossing a plane

We shall first calculate the energy flux H crossing the plane normal to x-direction. We also assume that, on an average, the component of an electron velocity along x is zero at the scattering centre and it acquires a velocity v_{rms} after it has moved over a distance λ, the mean free path, when it is scattered again by a lattice ion.

The amount of energy flux crossing a plane between the solid angles ω and $\omega + d\omega$ from the scattering centre located in the volume element dV at P is

$$dH = Uf(\omega)d\omega \tag{5.3.2}$$

where U is the energy carried by an electron and $f(\omega)d\omega$ the number of electrons per unit area per unit time in the incremental solid angle $d\omega$.

The velocities of electrons in dV are all taken to be equally distributed with respect to their directions and the velocity vector may then be represented by points on a sphere with centre at P (Fig. 5.3). The number that crosses the plane S along the normal per unit area per unit time is $nv_{\text{rms}}\cos\theta$. Then $f(\omega)d\omega$ is obtained (integrating over ϕ, the azimuthal angle) as

$$f(\omega)d\omega = nv_{\text{rms}}\cos\theta\frac{d\omega}{4\pi} = \tfrac{1}{2}nv_{\text{rms}}\cos\theta\sin\theta\,d\theta \tag{5.3.3}$$

where n is the number of electrons per unit volume.

∴ Energy per electron carried across planes along x is by Taylor series,

$$U = U(0) + \lambda\cos\theta\left(\frac{\partial U}{\partial x}\right)_{x=0} + \cdots \tag{5.3.4}$$

The above series converges rapidly, provided U varies by small amount over distances comparable to λ. Retaining therefore the first two terms only and substituting (5.3.3) and (5.3.4) in (5.3.2), we obtain

$$\begin{aligned} H &= \int dH = \int_0^\pi \left[U(0) + \lambda\cos\theta\left(\frac{\partial U}{\partial x}\right)_{x=0}\right]\left[\frac{1}{2}nv_{\text{rms}}\cos\theta\sin\theta\right]d\theta \\ &= \frac{1}{2}nv_{\text{rms}}\lambda\left(\frac{\partial U}{\partial x}\right)_{x=0}\int_0^\pi \cos^2\theta\sin\theta\,d\theta \\ &= \frac{1}{3}nv_{\text{rms}}\lambda\left(\frac{\partial U}{\partial x}\right)_{x=0} \end{aligned} \tag{5.3.5}$$

$$\left[\because \int_0^\pi \cos^2\theta\sin\theta\,d\theta = \int t^2dt, \text{ putting } \cos\theta = t\right]$$

Now, the variation of electron energy with x is due to a temperature-gradient that causes thermal diffusion of electrons.

So, we have
$$n\left(\frac{\partial U}{\partial x}\right)_{x=0} = n\left(\frac{\partial U}{\partial T}\cdot\frac{\partial T}{\partial x}\right) = C_V\frac{\partial T}{\partial x} \tag{5.3.6}$$

where C_V is the specific heat of the *electron gas*.

$$\therefore \text{ Heat flux, } H = \frac{1}{3}v_{\text{rms}}\lambda C_V\frac{dT}{dx} \tag{5.3.7}$$

$$\therefore \textit{ Thermal conductivity, } \boxed{\mathbf{K = \frac{1}{3}v_{rms}\lambda C_V}} \tag{5.3.8}$$

In case of a classical (ideal) gas, $C_V = \frac{3}{2}nk$, where k is Boltzmann costant.

$$\therefore \; \mathbf{K = \tfrac{1}{2}v_{rms}nk\lambda}$$

Wiedemann-Franz law—In 1853, Wiedemann and Franz discovered that good electrical conductors are also good thermal conductors and at a given temperature, the ratio of the thermal conductivity K to the electrical conductivity σ is the same for al metals. In particular, they observed empirically that

$$\boxed{\frac{K}{\sigma T} = constant} \tag{5.3.9}$$

which is known as the ***Wiedemann-Franz law.***

The remarkable success of Drude's theory is the explanation of the above empirical law of Wiedemann-Franz.

For, using (5.3.8) and (5.2.8), we obtain

$$\frac{K}{\sigma} = \frac{C_V \lambda v_{\text{rms}}}{3}\left(\frac{\sqrt{3mkT}}{ne^2\lambda}\right)$$

$$= \left(\frac{3}{2}\frac{nk}{3}\right)\left(\sqrt{\frac{3kT}{m}}\right)\left(\frac{\sqrt{3mkT}}{ne^2}\right)$$

$$\left[\because C_V = \tfrac{3}{2}nk;\ v_{\text{rms}} = \sqrt{3kT/m}\right]$$

$$= \left(\frac{3}{2}\frac{k^2}{e^2}\right)T = LT, \tag{5.3.10}$$

where $L = \dfrac{3}{2}\dfrac{k^2}{e^2}$ = ***Lorentz number*** = a constant.

$\therefore$ From (5.3.10), we obtain $\dfrac{K}{\sigma T}$ = constant,

which is the ***Wiedemann-Franz law.***

- The Lorentz number, $L = 3k^2/2e^2 = 1.12 \times 10^{-8}\ \text{V}^2\text{K}^{-2}$.

This value of L however *does not agree* with the experimental values and is about one half of the value for most of the metals. To resolve the discrepancy, we are to apply the quantum statistics of Fermi-Dirac in place of classical statistics. We shall revert back to this problem.

5.4 Quantum mechanical model of free electrons

This was first suggested by Sommerfeld for the assembly of free electrons in a metal, where the electrons were assumed to obey Pauli's principle and the *FD*-statistics.

Electrical conductivity—Classically, we obtained (5.2.5),

$$\sigma = \frac{ne^2\tau}{m} = \frac{ne^2\lambda}{mv_{\text{rms}}} \tag{5.4.1}$$

for the electrical conductivity σ.

From quantum consideration, a collision of a free electron can only occur only when an unoccupied state corresponding to the post-collision energy and momentum of the electron exists. The ***Fermi level*** being the highest occupied energy state by an electron at 0 K, the electron at this temperature will occupy states lying in a sphere about the origin of the momentum space. The radius $p_F(0)$ of the sphere, called ***Fermi sphere***, satisfies the relation

$$\frac{p_F^2(0)}{2m} = E_F(0) \tag{5.4.2}$$

where $E_F(0)$ is the ***Fermi energy*** at absolute zero.

The surface of the Fermi sphere which separates the occupied level from the unoccupied ones is known as the ***Fermi surface***. At high temperatures, there is a partial occupation of energy states over a region few times kT about E_F (Fig. 5.4). The quantity p_F is known as the magnitude of ***Fermi momentum***: $p_F = \hbar k_F$ where k_F is *Fermi wave vector*. And the ***Fermi velocity***, $v_F = p_F/m = \hbar k_F/m$.

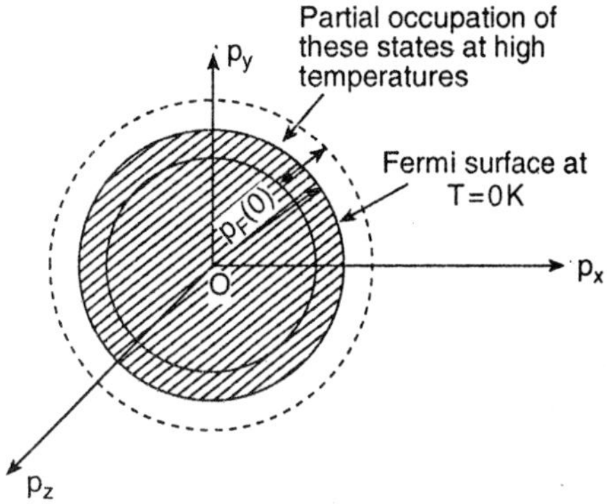

Fig. 5.4 Fermi surface at 0K

Fermi wave vector $k_F = (3\pi^2 n)^{1/3}$, where n is the electron density.

The collisions which the electrons suffer are nearly elastic. So, the change in energy on collision is negligibly small. In momentum space, therefore, an electron on collision is scattered to a state almost equidistant from the origin. Since all states below a few times of kT of E_F are occupied, Pauli's principle will not allow such collisions to occur. Only those electrons having states close to the Fermi level can involve in collisions as they find empty states to receive them after the collision.

When an electric field is applied, the electrons are accelerated during their free motion and the Fermi sphere of occupied states is slightly shifted opposite to the field (Fig. 5.5). All the free electrons thus participate in electric conduction and a mean drift velocity can be defined. However, as stated already, only electrons having states

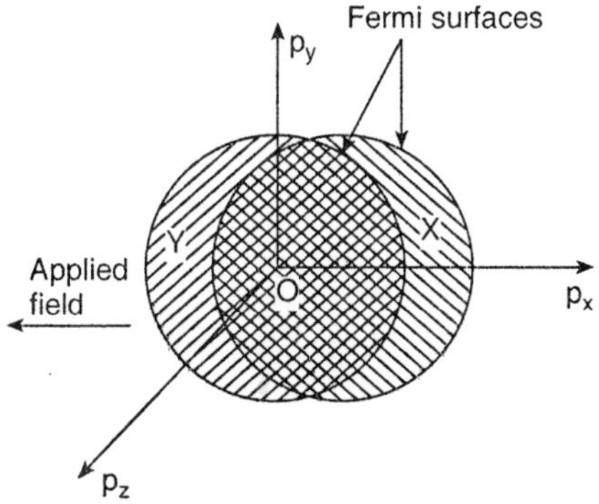

Fig. 5.5 Displacement of Fermi sphere in presence of an electric field

near E_F encounter collisions. Electrons with states in region X can fill up states in regions Y after collision for the states in the two regions have nearly the same energy.

The equilibrium electron distribution tends to get restored by way of such collisions with a characteristic time, known as ***relaxation time*** or ***mean free time***, τ_F of electrons with energy close to E_F. Thus τ_F determines the drift velocity of the whole system of free electrons. So, the quantum theory gives the electrical conductivity

$$\sigma = \frac{ne^2\tau_F}{m} = \frac{ne^2\lambda}{mv_F} = \frac{e^2 \cdot 3\pi^2 n}{3\pi^2 mv_F}\lambda = \frac{e^2}{3\pi^2}\frac{k_F^3\lambda}{mv_F} \tag{5.4.3}$$

where v_F is the ***Fermi velocity*** and $k_F = (3\pi^2 n)^{1/3}$, the radius of Fermi sphere in k-space.

$$\text{or,}\quad \sigma = \frac{e^2 \cdot 4\pi k_F^3\lambda}{12\pi^3\hbar k_F} = \frac{e^2(4\pi k_F^2)\lambda}{12\pi^3\hbar} \qquad (\because\ mv_F = \hbar k_F)$$

$$= \frac{e^2 A_F\lambda}{12\pi^3\hbar^3},$$

where $A_F = 4\pi k_F^2$ = area of the free-electron Fermi sphere of radius k_F.

• Eq. (5.4.3) has been obtained from (5.2.5) by the replacement of τ by τ_F or λ/v_F. The scattering contributing to τ_F is essentially due to lattice vibration that increases with temperature. At normal and high temperatures, it is found that $\tau_F \propto 1/T$ leading to a decrease of electron mobility and electric conductivity σ decreases with increase in temperature.

• At low temperatures, lattice scattering is small. The electrons then get scattered mainly by impurity atom and crystal imperfections.

• The electron concentration in a metal remains unchanged with temperature so that the temperature variation of σ is essentially dictated by τ_F.

Thermal conductivity—The conduction of heat in a metal is mainly due to free electrons. We have just described and discussed that only those electrons having energies near the Fermi level can participate in collisions and their relaxation time τ_F controls the drift velocity of the entire free electron system in the metal.

From the Drude model we had, from (5.3.8), the following expression for the thermal conductivity K.

$$K = \tfrac{1}{3}\lambda v_{\text{rms}} C_V \tag{5.4.4}$$

In quantum theory, this equation can be used with appropriate values of C_V and the mean free path λ.

The *electronic heat capacity* per unit volume C_V is given by (See (5.5.9) later),

$$C_V = \frac{\pi^2}{2} n \frac{k^2 T}{E_F} \tag{5.4.5}$$

But, plainly, the mean free path, $\lambda = \tau_F v_F$. (5.4.6)

Substituting for C_V and λ from (5.4.5) and (5.4.6) respectively in (5.4.4),

$$K = \frac{\pi^2}{6} n \frac{k^2 T}{E_F} \tau_F v_F^2 \quad (\because\ v_{\text{rms}} = v_F,\ \text{here})$$

$$= \frac{\pi^2}{6} n \frac{k^2 T}{\frac{1}{2} m v_F^2} \tau_F v_F^2, \quad \text{using } E_F = \tfrac{1}{2} m v_F^2$$

$$= \frac{\pi^2}{3} n \frac{k^2 T}{m} \tau_F \tag{5.4.7}$$

• Since $\tau_F \propto 1/T$ i.e., $\tau_F T$ = constant, equation (5.4.7) readily explains why at ordinary temperatures, the thermal conductivity of metals is *nearly independent of temperature.*

Ratio of K and σ—From (5.4.7) and (5.4.3), we get

$$\frac{K}{\sigma} = \frac{\pi^2}{3} \left(\frac{k}{e}\right)^2 T = LT$$

which is the ***Wiedeman-Franz law.***

The Lorentz number L is now given by

$$L = \frac{\pi^2}{3} \left(\frac{k}{e}\right)^2 = 2.45 \times 10^{-8}\ \text{V}^2/\text{K}^2,$$

This is in good agreement with experiments and nearly twice the value we had obtained classically.

5.4.1 Electrical conductivity of semiconductors

In a semiconductor, both the free electrons in the conduction band and the holes in the valence band act as carriers of electric current.

If an electric field ξ is applied to a semiconductor, the negatively charged electrons get drifted opposite to the field direction (Fig. 5.6), but the conventional current is in the direction of ξ. The holes however carry positive charge. They are drifted to give a current in the direction of ξ.

If σ_n be the conductivity of a semiconductor due to electrons, then from (5.2.7)

$$\sigma_n = ne\mu_n, \tag{5.4.8}$$

where n = concentration of electrons and μ_n = electron mobility.

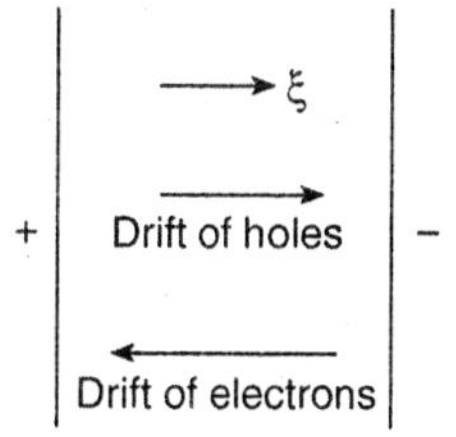

Fig. 5.6 Drift of holes and electrons in an electric field

Similarly, if σ_p be the conductivity of a semiconductor due to holes, then

$$\sigma_p = pe\mu_p \tag{5.4.9}$$

where p = hole concentration and μ_p = hole mobility. Generally, $\mu_n > \mu_p$.

So, the overall conductivity in a semiconductor is

$$\sigma = \sigma_n + \sigma_p = e(n\mu_n + p\mu_p) \tag{5.4.10}$$

since the hole current and electronic current agree in direction.

For an intrinsic semiconductor, $n = p = n_i$, where n_i is the intrinsic carrier concentration.

$\therefore$ Intrinsic conductivity,

$$\begin{aligned}\sigma_i &= e(n_i\mu_n + n_i\mu_p)\\ &= en_i\mu_n(1+r), \quad \text{where } r = \mu_p/\mu_n\end{aligned}$$

Now, the ratio $r < 1$ and hence electrons dominate in the conductivity of intrinsic semiconductor.

For an n-type semiconductor, $n \gg p$ and hence from (5.4.10)

$$\sigma_{n\text{-type}} \simeq ne\mu_n$$

For a p-type semiconductor, $p \gg n$ and if $p\mu_p > n\mu_n$, then from (5.4.10),

$$\sigma_{p\text{-type}} \simeq pe\mu_p$$

- The conductivities σ_n or σ_p is expressed by the relation (5.4.3) :

$$\sigma = \frac{ne^2\tau_F}{m}$$

where n = concentration of electron (or hole), m = the effective mass of electron (or hole). τ_F however is to be replaced by an average value of τ because for a semiconductor the Fermi level is generally inside the band gap and the distribution function is practically Maxwellian. Electrons in all energy states can thus undergo collisions and since τ is a function of energy, an average value of τ is to be used.

5.5 Electronic specific heat : Quantum mechanical

The total number of electron energy states per unit energy range or the *density of states* is given by

$$D(E) = dN/dE \Rightarrow dN = D(E)dE$$

At 0 K, all the energy states up to the Fermi level E_F are full.

$\therefore$ Total (internal) energy at 0 K$= \int_0^{E_F} ED(E)dE$, where E = energy of a state.

At temperature T, the number of filled states between energy E and $E + dE = D(E)f(E)dE$, $f(E)$ being the *Fermi distribution function.*

$\therefore$ Total (internal) energy at T K $= \int_0^{\infty} ED(E)f(E)dE$

Considering unit volume, the net increase in internal energy per unit volume as temperature rises from 0 to T K, is

$$U = \int_0^\infty ED(E)f(E)dE - \int_0^{E_F} ED(E)dE \tag{5.5.1}$$

$\therefore$ Total number of electrons per unit volume, $n = \int_0^\infty D(E)f(E)dE$

$$\therefore \quad E_F n = E_F \int_0^\infty D(E)f(E)dE = \int_0^\infty E_F D(E)f(E)dE \tag{5.5.2}$$

Differentiating (5.5.1) with respect to T, we have

$$\frac{\partial U}{\partial T} = \int_0^\infty ED(E)\frac{\partial f}{\partial T}dE, \tag{5.5.3}$$

the second integral being constant gives zero on differentiation.

Similarly, differentiating equation (5.5.2) with respect to T, we obtain

$$E_F \frac{\partial n}{\partial T} = \int_0^\infty E_F D(E)\frac{\partial f}{\partial T}dE = 0, \quad \text{since } \partial n/\partial T = 0 \tag{5.5.4}$$

Subtracting (5.5.4) from (5.5.3), we obtain

$$\frac{\partial U}{\partial T} = \int_0^\infty ED(E)\frac{\partial f}{\partial T}dE - \int_0^\infty E_F D(E)\frac{\partial f}{\partial T}dE$$

$$= \int_0^\infty (E - E_F)D(E)\frac{\partial f}{\partial T}dE \tag{5.5.5}$$

The derivative $\partial f/\partial T$ has significant value in the region $\pm 2kT$ about E_F and in this region $D(E)$ hardly varies so that we may write $D(E) = D(E_F) =$ constant.

Equation (5.5.5), then, becomes

$$\frac{\partial U}{\partial T} = D(E_F)\int_0^\infty (E - E_F)\frac{\partial f}{\partial T}dE \tag{5.5.6}$$

Now, the Fermi distribution function is: $f(E) = \dfrac{1}{e^{(E-E_F)/kT} + 1}$.

Differentiating the function with respect to T,

$$\frac{\partial f}{\partial T} = \frac{E - E_F}{kT^2}\frac{e^{(E-E_F)/kT}}{\left[e^{(E-E_F)/kT} + 1\right]^2} \tag{5.5.7}$$

Let now $\quad (E - E_F)/kT = x$ so that $dx = dE/kT \Rightarrow dE = kTdx$,

$$\therefore \quad \frac{\partial f}{\partial T} = \frac{x}{T}\frac{e^x}{(e^x + 1)^2}; \qquad \frac{\partial f}{\partial T}dE = kx\frac{e^x}{(e^x + 1)^2}dx$$

Also, $\quad (E - E_F)\dfrac{\partial f}{\partial T}dE = xkT\cdot kx\dfrac{e^x}{(e^x + 1)^2}dx = k^2T\dfrac{x^2e^x}{(e^x + 1)^2}dx$

Substituting in (5.5.6), we have

$$\frac{\partial U}{\partial T} = D(E_F)k^2T\int_{-E_F/kT}^{\infty}\frac{x^2e^x}{(e^x+1)^2}dx$$

The factor e^x in the integral is negligible for $x \leq -E_F/kT$, i.e., at low temperatures. So, the lower limit of integration can be extended to $-\infty$. The integral then takes the standard form

$$\int_{-\infty}^{\infty}\frac{x^2e^x}{(e^x+1)^2}dx = \frac{\pi^2}{3}$$

$$\therefore\quad \frac{\partial U}{\partial T} = C_e\ \text{(electronic sp. heat)} = \frac{\pi^2}{3}D(E_F)k^2T \tag{5.5.8}$$

At Fermi level, $N = \frac{V}{3\pi^2}\left(\frac{2m}{\hbar^2}\right)^{3/2}E_F^{3/2} \Rightarrow n = \frac{N}{V} = \frac{1}{3\pi^2}\left(\frac{2m}{\hbar^2}\right)^{3/2}E_F^{3/2}$

$$\therefore\quad \ln n = \tfrac{3}{2}\ln E_F + \text{constant}$$

Differentiating, $\frac{1}{n}\frac{dn}{dE_F} = \frac{3}{2}\cdot\frac{1}{E_F} \Rightarrow \frac{dn}{dE_F} = \frac{3n}{2E_F} = D(E_F)$

Substituting this value of $D(E_F)$ in (5.5.8), we obtain

$$C_e = \frac{\pi^2}{3}\cdot\frac{3n}{2E_F}\cdot k^2T = \frac{\pi^2}{2}\frac{nk^2T}{E_F} \tag{5.5.9}$$

$$\text{or,}\quad \boxed{C_e = \frac{\pi^2}{2}nk\frac{T}{T_F}} \tag{5.5.10}$$

where $T_F = E_F/k$, the *Fermi temperature.*

Eq. (5.5.10) is the expression for the *specific heat of electrons of metals at low **temperatures***. The linear temperature-dependence and the measured values of electron heat capacity agree well with (5.5.10).

5.6 Hall effect

When a magnetic field is applied perpendicularly to a conductor (a metal or a semiconductor) *carrying current, an additional electrical field develops across the specimen in a direction perpendicular to both the current and the magnetic field.* This phenomenon is known as the ***Hall effect*** and the electric field so developed is called the ***Hall field.***

Explanation—The physical process underlying the Hall effect is illustrated in Fig. 5.7. When a current J_x flows in a wire along $+x$-direction and a magnetic field B_z is applied normal to the wire along $+z$-direction, we shall show the development of an additional electric field perpendicular to both J_x and B_z, i.e., along the y-direction.

With *no applied magnetic field*, the current flows along $+x$ direction implying that the conduction electrons are drifted with a velocity $\vec{v}$ along $-x$ direction. *With the application of magnetic field*, the Lorentz force $\vec{F} = e(\vec{v} \times B)$ causes the electrons to bend downward, resulting in an accumulation of electrons on the lower surface and hence a net negative charge on this surface. Simultaneously, there appears a net positive charge on the upper surface due to the deficiency of electrons there. Plainly, this combination of positive and negative surface charges would create an electric field *downward* along $-y$-direction, the *Hall field* ξ_H.

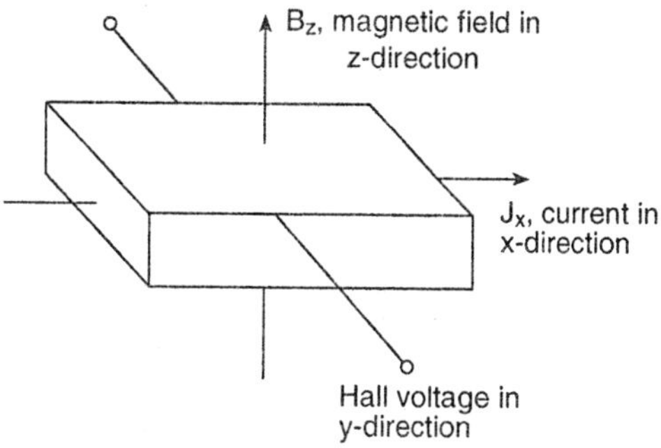

Fig. 5.7 The Hall effect

Evaluation of Hall field—The Lorentz force F, which produces accumulation of electrons, is along $-y$ direction and is given by

$$F = ev_x B$$

Now, the field produced by the surface charges creates a force that opposes this Lorentz force. The charge accumulation continues till the Hall force balances the Lorentz force. Thus, in the steady state, we have

$$F_{\text{Hall}} = F_{\text{Lorentz}}$$

$$\Rightarrow \quad -e\xi_H = -ev_x B$$

$$\text{or,} \quad \boxed{\xi_H = v_x B} \tag{5.6.1}$$

which is the required expression for ***Hall field***.

To express this in terms of measurable quantities, the velocity v_x is expressed in terms of the current density $J_x = n_c(-e)v_x$ where n_c is the concentration of current-carriers.

$$\therefore \quad \xi_H = -\frac{1}{n_c e} J_x B \tag{5.6.2}$$

This shows that *the Hall field is* ***proportional*** *to the current density and also to the magnetic field.*

Hall coefficient or Hall constant—The quantity $\xi_H / J_x B$, giving *the Hall field per unit current density per unit magnetic flux density*, is called the ***Hall constant*** or ***Hall coefficient***. It is usually symbolised by R_H.

$$\therefore \quad \boxed{R_H = \frac{\xi_H}{J_x B} = -\frac{1}{n_c e}} \tag{5.6.3}$$

which is a rather useful relation in practice.

- A rigorous analysis however shows that $R_H = r/n_c e$, where r is a numerical constant close to unity; $r = 1$ for metals, and for semiconductors, $r \simeq 1$.

• The equation shows that the *sign of the Hall coefficient is the same as that of the charge carriers.* If the carriers are holes as in p-type semiconductors, the Hall coefficient would be positive. This is a very valuable information, particularly in semiconductors. R_H is positive for Zn and Cd implying that current in them is carried by holes.

When the carriers are electrons, we have

$$R_H = -1/ne \tag{5.6.4}$$

And when the carriers are holes, we have

$$R_H = 1/pe$$

where n and p are the concentrations of the electrons and the holes respectively.

• Since R_H is inversely proportional to the electron (or hole) concentration, we can determine n (or p) by measuring the Hall constant. In fact, this is the *standard technique* for determining the electron or hole concentration. The technique is all the more valuable as the only other quantity is the electronic charge—a well-known *fundamental constant.*

• Another interesting feature of the transport process in presence of a magnetic field is that the *current* itself, flowing along $+x$ direction, is *not influenced by the field.* It implies that *electrical resistance is independent of the magnetic field.* The result is interesting as it is somewhat unexpected.

Mobility and Hall angle—The *carrier mobility* μ is defined as the *velocity acquired by the current-carrying particles per unit electric field.* So, by definition,

$$\mu = \frac{v_x}{\xi_x} \Rightarrow \boxed{v_x = \mu\xi_x} \tag{5.6.5}$$

If ξ_x be the applied electric field along the x-direction driving the current in the specimen and ξ_T the total electric field in it, then

$$\xi_T = \sqrt{\xi_x^2 + \xi_H^2}$$

ξ_x and ξ_H being normal to each other.

The *angle between ξ_x and ξ_H is known as the* ***Hall angle***, denoted by ϕ_H.

$$\therefore \boxed{\tan\phi_H} = \frac{\xi_H}{\xi_x} = \frac{J_x B}{n_c e \xi_x} \boxed{= \mu B} \tag{5.6.6}$$

$$(\because J_x = n_c e v_x;\ v_x = \mu\xi_x)$$

5.6.1 Hall coefficient when carriers are both electrons and holes

Earlier in Art. 5.6, we assumed either the electrons or the holes as the carriers to derive the Hall coefficient. We shall now derive the appropriate expression for the Hall coefficient when both types of carriers—electrons and holes—are present.

In reference to Fig. 5.7, let an electric field ξ_x be along $+x$ direction along with a magnetic field B_z in the z-direction, normal to the paper. ξ_x would cause the electron drift to the left and the holes to the right. Due to the drift again, the magnetic field would exert Lorentz forces on the carriers resulting in their deflections—the electrons and the holes would be oppositely deflected because of the opposite nature of charges. *Both types of carriers get deflected to the lower surface of the specimen* tending to cancel each other there. But the cancellation in *incomplete* and a net charge accumulates on the lower surface. Since the specimen is neutral, an equal and opposite charge accumulates on the upper surface. These surface charges develop an electric field along the y-direction. This is the ***Hall field***, ξ_H.

Computation of Hall field—The Lorentz force on an electron is given by

$$(F_L)_e = -e(\vec{v}_e \times \vec{B}) = ev_e B_z$$

where v_e is the drift velocity of electrons. Since v_e is negative, $(F_L)_e$ acts downward along the $-y$ direction.

$\therefore$ Lorentz field, $(\xi_L)_e = -v_e B_e$. But $J_e = nev_e \Rightarrow (\xi_L)_e = J_e B_z/ne$, where J_e is the current part carried by electrons.

Similarly, we have, $(\xi_L)_h$ = Lorentez field on the holes acting along the positive x-direction $= -J_h B_z/pe$

$\therefore$ The situation, as it stands, is that the carriers flow along the x-direction, but also feel a number of electric fields along the y-direction. These fields are:

(i) $(\xi_L)_e$, experienced by *electrons*,

(ii) $(\xi_L)_h$, experienced by the *holes*, and

(iii) ξ_H, the Hall field, experienced by *both*.

$\therefore$ Total current density along y-direction is given by

$$J_y = ne\mu_e(\xi_L)_e + pe\mu_h(\xi_L)_h + (ne\mu_e + pe\mu_h)\xi_H$$

But $J_y = 0$, since the carriers are not allowed to flow along the y-direction due to the presence of surfaces of the sample.

$$\therefore \quad (ne\mu_e + pe\mu_h)\xi_H = -ne\mu_e(\xi_L)_e + pe\mu_h(\xi_L)_h$$

$$\therefore \quad \boldsymbol{\xi_H = -\frac{ne\mu_e(\xi_L)_e + pe\mu_h(\xi_L)_h}{ne\mu_e + pe\mu_h}} \tag{5.6.7}$$

which is the required expression for the ***Hall field***.

Hall coefficient—The *Hall coefficient* or the *Hall constant* R_H is defined as the ratio of ξ_H and $J_x B$.

But, we have: $J_x = J_e + J_h$ and $J_e = \dfrac{n\mu_e}{n\mu_e + p\mu_h} J_x$.

Therefore, we get

$$R_H = \frac{p\mu_h^2 - n\mu_e^2}{e(n\mu_e + p\mu_h)^2} \tag{5.6.8}$$

• Note that for n-type semiconductors, $p = 0$ and from (5.6.8), we obtain as a special case,

$$R_e = -\frac{1}{ne}$$

For p-type semiconductors, $n = 0$ and the same relation (5.6.8) gives us

$$R_h = \frac{1}{pe},$$

which is the same expression as was obtained earlier.

• The equation (5.6.8) shows that the *Hall coefficient may either be negative, positive or zero* depending on the values of the carrier concentration and also their mobilities.

• Hall coefficient R_H is negative for metals like Cu, Ag, Au etc implying that current carriers are electrons. But in metals like Zn, Cd, Fe etc. R_H is positive. Why? This is because in metals with nearly filled band, the unfilled states near the top of the band are effective in carrying the current and behave as positive holes.

5.6.2 Experimental determination of Hall coefficient

To determine the Hall coefficient R_H experimentally, the sample of the material is taken in the form of a rectangle of thickness d and width ω. The experimental set up is illustrated in Fig. 5.8. With the help of the battery, a current I is allowed to flow through the sample in the x-direction, as shown. The specimen is properly positioned between the pole pieces of an electromagnet (not shown), such that the flux density B is along the z-direction.

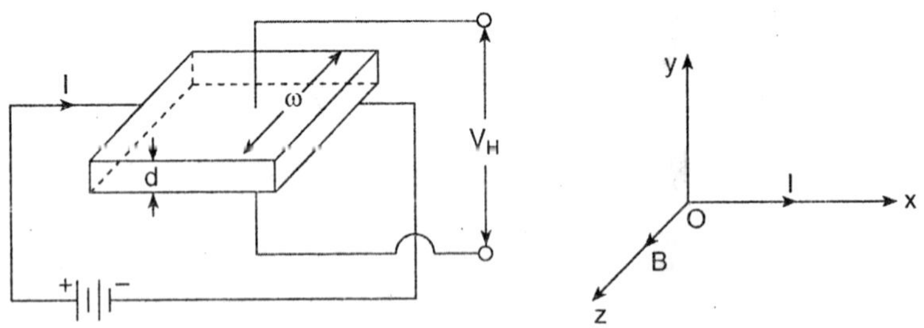

Fig. 5.8 Experimental set-up for measurement of Hall coefficient

The Hall voltage developed in the sample is determined by means of two probes placed at the centres of its top and bottom faces. If ξ_H be the Hall field, we have

$$V_H = \xi_H d = R_H J B d \quad (\because \xi_H = R_H J B) \tag{5.6.9}$$

Now, the cross-sectional area A of the sample under investigation is given by $A = \omega d$. We can therefore write

$$\text{current density, } J = I/A = I/\omega d$$

$$\therefore \quad V_H = R_H(I/\omega d)Bd = R_H IB/\omega, \quad \text{using (5.6.9)}$$

$$\therefore \quad \boxed{R_H = \frac{\omega V_H}{IB}} \tag{5.6.10}$$

If V_H is in volt, ω in metre, I in Ampere and B in Tesla, then R_H, as determined by (5.6.10), will be in **m^3/Coulomb**.

● Polarity of V_H would be opposite for n-type and p-type semiconductors and necessarily therefore R_H also would be of opposite signs for the two types.

5.6.3 Applications of Hall effect

The Hall effect has a number of *important* and *useful applications.* Some of these are enumerated in brief as under.

1. ***Determination of the type of semiconductor***—The sign of the Hall constant R_H determines the type of a semiconductor. It is positive for a p-type semiconductor and is negative for a n-type one.

2. ***Determination of carrier concentration***—From the magnitude of Hall constant R_H, the carrier-concentration i.e., the number of charge carriers per unit volume can be determined, using the formula,

$$n = \frac{1}{|R_H|e} \quad \text{or} \quad p = \frac{1}{R_H e}$$

3. ***Determination of carrier mobility***—If the current is carried by one type of carriers, e.g., electrons, we have the conductivity

$$\sigma = ne\mu_e$$

where μ_e is the electron mobility. But, we have, $R_H = 1/ne$.

$$\therefore \quad \mu_e = \frac{\sigma}{ne} = \sigma|R_H|$$

Thus, by measuring σ and R_H, one can readily obtain the mobility μ_e.

● The mobility determined by R_H is somewhat different from the actual drift mobility and is known as ***Hall mobility.***

4. ***Determination of magnetic flux density***—For a given current through the specimen, the Hall voltage V_H is proportional to the magnetic flux density B. Thus, when R_H and the dimensions of the sample are known, a measurement of I and V_H would give the flux density B.

5. ***Measurement of power in electromagnetic wave***—In an electromagnetic wave in free space, the electric field $\vec{E}$ and the magnetic field $\vec{H}$ are at right angles to each other. The Hall voltage depends on the cross-product of $\vec{E}$ and $\vec{H}$, and can thus be used as a ***microwave power flow detector***. The current in the material is proportional to the electric field so that the resulting Hall voltage $\propto (\vec{E} \times \vec{H})$. So, the Hall voltage of a given crystal can be calibrated for different values of microwave power flow $(\vec{E} \times \vec{H})$. A crystal thus calibrated and supplied with contacts for measuring dc Hall voltage can be inserted in a wave guide. Then, the dc Hall voltage directly gives the microwave power flow.

5.6.4 Quantum Hall effect

A plot of Hall voltage $V_H (= \xi_H d,\ d =$ sample thickness) vs. B the applied magnetic field is a straight line, passing through the origin. In studying the effect in thin semiconductors at very low temperatures, and very large magnetic fields, Klitzing in 1980 noticed that V_H vs. B plot deviates from linearity and shows a number of plateaus pointing to the fact that the Hall voltage is quantized.

If the Hall resistance R_H is defined by $R_H = V_H/I$, it is seen that the Hall resistance is quantized:

$$R_H = R_K/n, \quad n = 1, 2, 3$$

where $R_K =$ a constant $=$ Klitzing constant, related to e and h by

$$R_K = h/e^2 = 25818\,\Omega$$

Tsui and co-workers showed that R_H is quantized in ultrapure semiconductors with $n = 1/2, 1/3, 1/7, 2/3 \ldots$. Quantum Hall effect is used to define the *standard of resistance.*

5.7 Boltzmann transport equation

The distribution function for the velocity components and spatial coordinates of the current carriers in a conductor in a state of steady flow of heat or electric current will be different from that in thermal equilibrium and in absence of flow. The ***Boltzmann transport equation*** (BTE) is concerned with the determination of the steady state distribution function of the system under external fields.

To set up the Boltzmann equation, let a system of particles (e.g., electrons) be in dynamic equilibrium under external forces and the distribution function $f(\vec{k}, \vec{r}, t)$ that gives the probability that the state corresponding to wave vector $\vec{k}$ at the position vector $\vec{r}$ in the crystal be occupied by a current carrier at time t. If there exists a gradient in concentration of the carriers in the crystal, the carriers would diffuse.

In a time-interval Δt, the change in the distribution function is

$$\begin{aligned}\Delta f &= f(\vec{k}, \vec{r}, t + \Delta t) - f(\vec{k}, \vec{r}, t) \\ &= f(\vec{k}, \vec{r} - \Delta\vec{r}, t) - f(\vec{k}, \vec{r}, t)\end{aligned} \tag{5.7.1}$$

since the carrier occupying $\vec{r}$ at $t + \Delta t$ was at $\vec{r} - \Delta\vec{r}$ at t.

Expanding the first term on the right-hand side of (5.7.1) by Taylor's theorem and retaining only the first order term, we have

$$\begin{aligned}\Delta f &= [f(\vec{k},\vec{r},t) - \Delta\vec{r}\cdot\vec{\nabla}_r f] - f(\vec{k},\vec{r},t)\\ &= -\Delta\vec{r}\cdot\vec{\nabla}_r f\end{aligned} \tag{5.7.2}$$

While the carrier velocity is $(1/\hbar)dE/dk$ in one dimension, it is $\vec{v} = (1/\hbar)\vec{\nabla}_k E$ in three dimensions, E being the carrier energy.

Now, $\Delta\vec{r} = \vec{v}\Delta t = (1/\hbar)\vec{\nabla}_k E\cdot\Delta t$. So, from (5.7.2), we obtain

$$\Delta f = -\frac{1}{\hbar}\vec{\nabla}_k E\cdot\vec{\nabla}_r f\Delta t$$

$$\therefore \quad \boxed{\left(\frac{\partial f}{\partial t}\right)_{\text{drift}} = -\frac{1}{\hbar}\vec{\nabla}_k E\cdot\vec{\nabla}_r f}$$

which gives the *contribution* to the time rate of change of f *due to a concentration gradient.*

The distribution function f may also show variation with time caused by fields exerting a force $\vec{F}$ on the carrier. The change in f in time Δt is

$$\begin{aligned}\Delta f &= f(\vec{k},\vec{r},t+\Delta t) - f(\vec{k},\vec{r},t)\\ &= f(\vec{k}-\Delta\vec{k},\vec{r},t) - f(\vec{k},\vec{r},t)\end{aligned} \tag{5.7.3}$$

since the carrier with state $\vec{k}$ at $t+\Delta t$ was at the state $\vec{k}-\Delta\vec{k}$ at t.

Expanding the right-hand side of (5.7.3), as before, by Taylor's theorem and keeping only the first order term, we obtain from equation (5.7.3)

$$\Delta f = -\Delta\vec{k}\cdot\vec{\nabla}_k f = \left(-\frac{1}{\hbar}\vec{F}\Delta t\right)\cdot\vec{\nabla}_k f \tag{5.7.4}$$

since the crystal momentum is $\hbar\vec{k}$ and this implies $\vec{F} = \hbar(\partial\vec{k}/\partial t)$.

$\therefore$ From (5.7.4),

$$\boxed{\left(\frac{\partial f}{\partial t}\right)_{\text{field}} = -\frac{1}{\hbar}\vec{F}\cdot\vec{\nabla}_k f} \tag{5.7.5}$$

which is the *contribution* to the time rate of change of f *due to the force* $\vec{F}$.

The distribution function f may also change from its thermal equilibrium value f_0 due to collisions that tend to restore the equilibrium distribution. So, we may write

$$\boxed{\left(\frac{\partial f}{\partial t}\right)_{\text{collision}} = -\frac{f-f_0}{\tau}} \tag{5.7.6}$$

which is the *contribution* to the time rate of change of f *from collisions.*

∴ The *net* rate of change of f with time is plainly the sum of the three contributions: (i) of drift, (ii) of field and (iii) of collisions.

$$\therefore \quad \frac{\partial f}{\partial t} = -\frac{1}{\hbar}\vec{\nabla}_k E \cdot \nabla_r f - \frac{1}{\hbar}\vec{F}\cdot\vec{\nabla}_k f - \frac{f - f_0}{\tau} \tag{5.7.7}$$

which is known as the ***Boltzmann trnsport equation.***

In *equilibrium* or the *steady state*, f is *independent* of time i.e., $\partial f/\partial t = 0$, and we have from (5.7.7)

$$\frac{1}{\hbar}\vec{\nabla}_k E\cdot\vec{\nabla}_r f + \frac{1}{\hbar}\vec{F}\cdot\vec{\nabla}_k f + \frac{f - f_0}{\tau} = 0$$

● In computing $(\partial f/\partial t)_{\text{drift}}$ and $(\partial f/\partial t)_{\text{field}}$, the time interval dt is assumed to be too small in that the collision-effect on the distribution can be disregarded.

● While the drift variations take place *continuously*, the collisions introduce rather *discontinuous* change in velocity.

● The different transport coefficients such as drift mobility, Hall mobility etc. can be found by solving BTE which, however, is beyond the scope of our discussion.

5.8 Illustrated Examples

► **Example 1.** *Assuming that each atom of copper contributes one free electron, calculate the drift velocity of free electrons in the copper conductor of cross-sectional area* 10^{-4} m^2 *carrying a current of* 200 A. *At. wt. of* Cu = 63.5, *density of* Cu = 8.94×10^3 kg.m^{-3} *and* $e = 1.602 \times 10^{-19}$ C. **(Bangalore Univ.)**

Solution: Volume of 1 kg of atom $= \dfrac{63.5}{8.94 \times 10^3}$ m^{-3}

Number of atoms in 1 kg of atom $= 6.02 \times 10^{26}$ (Avogadro number)

$$\therefore \text{ Number of atoms per m}^3 = \frac{6.02 \times 10^{26} \times 8.94 \times 10^3}{63.5}$$

$$= n_e = \text{No. of electrons per m}^3$$

Area of cross-section, $\alpha = 10^{-4}$ m^2; current through conductor, $I = 200$ A.

∴ Current density, $J_e = I/\alpha = 200/10^{-4} = 2 \times 10^6$ Am^{-2}.

Drift current density due to electrons is given by

$$J_e = n_e e v_D \Rightarrow v_d = J_e/n_e e = I/\alpha n_e e$$

where n_e = number density of electrons, v_d the drift velocity of electrons.

$$\therefore \quad v_D = \frac{200 \times 63.5}{10^{-4} \times 6.02 \times 10^{26} \times 8.94 \times 10^3 \times 1.602 \times 10^{-19}}$$

$$= \mathbf{1.47 \times 10^{-4}\ ms^{-1}}$$

▶ **Example 2.** *There are* 2.5×10^{28} *free electrons per cubic metre of sodium. Calculate the Fermi energy and Fermi velocity.* **(Lucknow Univ.)**

Solution: We have: Fermi energy $E_F = \dfrac{h^2}{8m}\left[\dfrac{3}{\pi}\cdot\dfrac{N}{V}\right]^{2/3}$

Substituting the values of $h = 6.62 \times 10^{-34}$ Js, $m = 9.1 \times 10^{-31}$ kg and $N/V = 2.5 \times 10^{28}/\text{m}^3$, we obtain.

$$\boldsymbol{E_F} = \frac{6.62 \times 6.62 \times 10^{-68}}{8 \times 9.1 \times 10^{31}}\left[\frac{3 \times 2.5 \times 10^{28}}{3.142}\right]^{2/3}$$

$$= 4.97 \times 10^{-19}\,\text{J}, \quad \text{on simplification}$$

$$= \frac{4.97 \times 10^{-19}}{1.6 \times 10^{-19}}\,\text{eV} = \mathbf{3.1\,eV}$$

Again, Fermi velocity, $v_F = \dfrac{\hbar}{m}\left[\dfrac{3\pi^2 N}{V}\right]^{1/3} = \dfrac{h}{2m}\left[\dfrac{3}{\pi}\dfrac{N}{V}\right]^{1/3}$

Substituting the values of h, m etc, as before, we get

$$\boldsymbol{v_F} = \frac{6.62 \times 10^{-34}}{2 \times 9.1 \times 10^{-31}}\left[\frac{3}{3.14} \times 2.5 \times 10^{28}\right]^{1/3}$$

$$= \mathbf{1.048 \times 10^6\,m/s}, \quad \text{on simplification.}$$

▶ **Example 3.** *The number of electrons per unit length of a crystal is* 0.5 *electron per angstrom. Determine the Fermi energy. Given,* $h = 6.62 \times 10^{-34}$ Js, $m = 9 \times 10^{-31}$ kg.

Solution: The Fermi energy E_F is given by

$$E_F = \frac{\hbar^2}{2m}\left(\frac{N}{2}\frac{\pi}{L}\right)^2 = \frac{h^2}{2m}\left(\frac{N}{4L}\right)^2$$

where N is the number of electrons in a length L.

Substituting $N/L = 0.5$ electron/Å$= 0.5/10^{-10}$ electron/m and the given values of h and m in the above relation, we have

$$\boldsymbol{E_F} = \frac{6.62 \times 6.62 \times 10^{-68}}{2 \times 9 \times 10^{-31}} \times \left(\frac{0.5}{4 \times 10^{-10}}\right)^2$$

$$= 3.8 \times 10^{-19}\,\text{J}, \quad \text{on simplification}$$

$$= \frac{3.8 \times 10^{-19}}{1.6 \times 10^{-19}}\,\text{eV} = \mathbf{2.37\,eV}$$

▶ **Example 4.** *Consider silver in the metallic state with one free electron per atom. Calculate the Fermi energy. Given, density of silver* $= 10.5$ g/c.c. *and atomic weight of silver* $= 108$. **(Calcutta, Burdwan, Vidyasagar Hons. and Himachal Univ.)**

Solution: Volume of 1 g-mole of silver, $V = (108/10.50)$ c.c.

Number of atoms per g-atom = Avogadro number $N_A = 6.02 \times 10^{23}$.

$$\therefore \quad N/V = (6.02 \times 10^{23})/(108/10.5)$$

Since there is one free electron per atom, the number of free electrons per unit volume i.e., electron density is given by

$$n = \frac{N}{V} \times 1 = \frac{6.02 \times 10^{23}}{108} \times 10.5 \times 1$$

$$= 5.85 \times 10^{22}/\text{c.c.} = 5.85 \times 10^{28}/\text{m}^3$$

$\therefore$ Fermi energy,
$$E_F = \frac{h^2}{8m}\left(\frac{3}{\pi}n\right)^{2/3}$$

$$= \frac{6.62 \times 6.62 \times 10^{-68}}{8 \times 9.1 \times 10^{-31}}\left[\frac{3 \times 5.85 \times 10^{28}}{3.142}\right]^{2/3}$$

$$= 8.77 \times 10^{-19}\,\text{J}, \quad \text{on simplification}$$

$$= \frac{8.77 \times 10^{-19}}{1.6 \times 10^{-19}}\,\text{eV} = \mathbf{5.48\,eV}$$

▶ **Example 5.** *The atomic radius of sodium is* 1.86 A. *Calculate the Fermi energy of sodium at absolute zero.*

Solution: The Fermi energy at 0 K is given by

$$E_F(0) = \frac{\hbar^2}{2m}\left(\frac{3\pi^2 N}{V}\right)^{2/3}$$

where N = number of free electrons in volume V of the metal.

Let us consider a unit cell of sodium. Given the radius of Na-atom, $r = 1.86\,\text{A} = 1.86 \times 10^{-10}\,\text{m}$.

$\therefore$ Volume of unit cell, $V = a^3 = (4r/\sqrt{3})^3$

$$\therefore \quad V = \left(\frac{4 \times 1.86 \times 10^{-10}}{\sqrt{3}}\right)^3 = 7.93 \times 10^{-29}\,\text{m}^3$$

An atom of sodium has only one free electron. So the number of free electrons in unit cell of sodium is 2. Therefore,

$$\frac{N}{V} = \frac{2}{7.93 \times 10^{-29}} = 2.52 \times 10^{28}\ \text{electrons/m}^3$$

$$\therefore \quad \boldsymbol{E_F(0)} = \frac{(1.05 \times 10^{-34})^2}{2 \times 9.1 \times 10^{-31}} \times (3\pi^2 \times 2.52 \times 10^{28})^{2/3}$$

$$= 4.98 \times 10^{-19}\,\text{J} = \frac{4.98 \times 10^{-19}}{1.6 \times 10^{-19}}\,\text{eV} = \mathbf{3.11\,eV}$$

▶ **Example 6.** *The intrinsic carrier density of* Ge *at* 27°C *is* $2.4 \times 10^{17}\,\text{m}^{-2}$. *Calculate its intrinsic resistivity, if the electron and the hole mobilities are* $0.35\,\text{m}^2\text{V}^{-1}\text{s}^{-1}$ *and* $0.18\,\text{m}^2\text{V}^{-1}\text{s}^{-1}$ *respectively.* **(Bangalore Univ.)**

Solution: The conductivity of an intrinsic semiconductor is

$$\sigma = \sigma_e + \sigma_h = e(n_e\mu_e + n_h\mu_h)$$

But, $n_e = n_h = n_i$, say. So, $\sigma = en_i(\mu_e + \mu_h)$

Given, $n_i = 2.4 \times 10^{17}\,\text{m}^{-2}$, $\mu_e = 0.35\,\text{m}^2\text{V}^{-1}\text{s}^{-1}$, $\mu_h = 0.18\,\text{m}^2\text{V}^{-1}\text{s}^{-1}$.

Also, e = electronic charge $= 1.6 \times 10^{-19}$C.

$$\therefore\ \sigma = 1.6 \times 10^{-19} \times 2.4 \times 10^{17} \times (0.35 + 0.18)$$
$$= 1.6 \times 2.4 \times 0.53 \times 10^{-2}$$

$$\therefore\ \textbf{\textit{Intrinsic resistivity}},\ \mathbf{1/\sigma} = \frac{10^2}{1.6 \times 2.4 \times 0.53}$$
$$= \mathbf{49 \times 10^{-2}\,\Omega m}$$

▶ **Example 7.** *The energy gap in* Ge *is* 0.75 eV. *Compare the conductivities of* Ge *at* 27°C *and* 57°C. *Boltzmann constant* $= 8.61 \times 10^{-5}$ eV. **(Bangalore Univ.)**

Solution: The intrinsic conductivity of Ge is given by

$$\sigma = CT^{3/2}e^{-E_g/2kT}$$

where
$$C = 2e(\mu_e + \mu_h)\cdot (2\pi k/h^2)^{3/2}(m_e^* m_h^*)^{3/4}$$

Let σ_1 and σ_2 be the intrinsic conductivities at temperatures $T_1 = 27 + 273 = 300$ K and $T_2 = 57 + 273 = 330$ K.

$$\therefore\ \frac{\sigma_1}{\sigma_2} = \left(\frac{T_1}{T_2}\right)^{3/2} e^{-E_g/2k(T_1^{-1} - T_2^{-1})}$$

$$\text{or,}\ \ln\frac{\sigma_1}{\sigma_2} = \frac{3}{2}\ln\frac{T_1}{T_2} - \frac{E_g}{2k}\left(\frac{1}{T_1} - \frac{1}{T_2}\right) \tag{5.8.1}$$

We now first compute: $\dfrac{E_g}{2k}\left(\dfrac{1}{T_1} - \dfrac{1}{T_2}\right) = \dfrac{0.75}{2 \times 8.61 \times 10^{-5}}\left[\dfrac{1}{300} - \dfrac{1}{330}\right]$

$= 1.32$, on simplification.

We next compute: $\dfrac{3}{2}\ln\dfrac{T_1}{T_2} = \dfrac{3}{2}\ln\dfrac{300}{330} = \dfrac{3}{2}\ln 0.91 = -0.141.$

$\therefore$ From (5.8.1) $\ln\dfrac{\sigma_1}{\sigma_2} = -0.141 - 1.32 = -1.461$

$$\therefore\ \boldsymbol{\frac{\sigma_1}{\sigma_2} = 0.23}$$

► **Example 8.** *The resistivity of an intrinsic semiconductor is* $4.5\,\Omega$m *at* $20°$C *and* $2.0\,\Omega$m *at* $32°$C. *What is the energy band gap?*

Solution: If σ_1 and σ_2 be the conductivities at temperatures T_1 and T_2 respectively,

$$\frac{\sigma_1}{\sigma_2} = \left(\frac{T_1}{T_2}\right)^{3/2} \exp\left[-\frac{E_g}{2k}\left(\frac{1}{T_1} - \frac{1}{T_2}\right)\right], \quad \text{using (4.8.1)}$$

$$\therefore \quad \ln\left(\frac{4.5}{2.0}\right) = \frac{3}{2}\ln\left(\frac{273+20}{273+32}\right) - \frac{E_g}{2 \times 1.38 \times 10^{-23}}\left(\frac{1}{293} - \frac{1}{305}\right)$$

$$\therefore \quad \frac{E_g}{2 \times 1.38 \times 10^{-23}} \times \frac{305 - 293}{293 \times 305} = \frac{3}{2}\ln\frac{293}{305} - \ln\frac{4.5}{2.0}$$

$\therefore$ Energy band gap, $\quad \boldsymbol{E_g} = 1.54 \times 10^{-19}$ J, on simplification

$$= \frac{1.54 \times 10^{-19}}{1.6 \times 10^{-19}}\,\text{eV} = \mathbf{0.96\,eV}$$

► **Example 9.** *Estimate classically the relaxatinn time* τ *and the mean free path* λ *of copper at* $20°$C *when its resistivity is* $1.69 \times 10^{-8}\,\Omega$m *and the concentration of free electrons in copper,* $n = 8.5 \times 10^{28}/\text{m}^3$. *How far does the estimate agree with the experimentally found value?*

Solution: Mass of electron $m = 9.11 \times 10^{-31}$ kg, $n = 8.5 \times 10^{28}\,\text{m}^{-3}$, charge of electron $e = 1.6 \times 10^{-19}$C, resistivity of copper, $\rho = 1.69 \times 10^{-8}\,\Omega$m, $T = 273 + 20 = 293$.

Now, we have the relaxation time τ given by

$$\tau = \frac{m}{ne^2\rho}$$

$$\therefore \quad \tau = \frac{9.11 \times 10^{-31}}{8.5 \times 10^{28} \times (1.6 \times 10^{-19})^2 \times 1.69 \times 10^{-8}}$$

$$= \mathbf{2.47 \times 10^{-14}\,sec.}$$

But $\tau = \lambda/v_{\text{rms}}$, where $v_{\text{rms}} = \sqrt{3kT/m}$, root mean square velocity.

$$\therefore \quad \boldsymbol{\lambda} = \tau v_{\text{rms}} = 2.47 \times 10^{-14} \times \sqrt{3kT/m}$$

$$= 2.47 \times 10^{-14} \times \left(\frac{3 \times 1.38 \times 10^{-23} \times 293}{9.11 \times 10^{-31}}\right)^{\frac{1}{2}}$$

$$= 2.47 \times 10^{-14} \times 1.154 \times 10^5 = 2.85 \times 10^{-9}\,\text{m}$$

$$= \mathbf{2.85\,nm.}$$

The experimentally found value for λ is about **10 *times*** of this value obtained from classical theory.

▶ **Example 10.** *Show that the wavelength associated with an electron having an energy equal to the Fermi energy is given by* $2(\pi/3n)^{1/3}$.

Solution: The wavelength associated with electron

$$\lambda = h/p = \frac{h}{\hbar k_F} = \frac{2\pi}{k_F}, \text{ with usual notation.}$$

$$= \frac{2\pi}{(3\pi^2 n)^{1/3}} = 2\left(\frac{\pi^3}{3\pi^2 n}\right)^{1/3} = \mathbf{2}\left(\frac{\pi}{3n}\right)^{1/3}$$

▶ **Example 11.** *Derive an expression giving the relation between pressure and volume for a free electron gas at* 0 K. *Hence express the bulk modulus in terms of average kinetic energy.*

Solution: From thermodynamics, we have the expression

$$P = -\frac{\partial E}{\partial V}$$

where E is the internal energy of a system of particles having a volume V at pressure P.

For a free electron gas having N electrons with average kinetic energy $\bar{E}_0$ at 0 K,

$$E = N\bar{E}_0 \Rightarrow P = -N(\partial\bar{E}_0/\partial V)$$

But $\bar{E}_0 = \frac{3}{5}E_F(0) \Rightarrow P = -\frac{3}{5}N\frac{\partial E_F(0)}{\partial V}$ and $E_F(0) = \frac{\hbar^2}{2m}\left(\frac{3\pi^2 N}{V}\right)^{2/3}$

$$\therefore \quad \frac{\partial E_F(0)}{\partial V} = -\frac{2}{3}\frac{\hbar^2}{2m}(3\pi^2 N)^{2/3}\left(\frac{1}{V}\right)^{5/3}$$

$$\therefore \quad \boldsymbol{P} = \frac{2}{5}N\frac{\hbar^2}{2m}(3\pi^2 N)^{2/3}\left(\frac{1}{V}\right)^{5/3} \tag{5.8.2}$$

$$= \frac{\mathbf{2}}{\mathbf{5}}\frac{\boldsymbol{N E_F(0)}}{\boldsymbol{V}} \tag{5.8.3}$$

which is the required pressure-volume relation for a free-electron gas at 0 K.

The bulk modulus, $B = -V\frac{\partial P}{\partial V}$. Differentiating (5.8.2) with respect to V

$$\frac{\partial P}{\partial V} = -\frac{5}{3}\frac{2}{5}\frac{N\hbar^2}{2m}(3\pi^2 N)^{2/3}\left(\frac{1}{V}\right)^{7/3} = -\frac{5}{3}\frac{P}{V} \tag{5.8.4}$$

$\therefore$ Bulk modulus, $\quad \boldsymbol{B} = \frac{5}{3}P = \frac{5}{3}\cdot\frac{2}{5}\frac{NE_F(0)}{V}$

$$= \frac{2}{3}\frac{NE_F(0)}{V} = \frac{\mathbf{10}}{\mathbf{9}}\frac{\boldsymbol{N\bar{E}_0}}{\boldsymbol{V}} = \frac{\mathbf{10}}{\mathbf{9}}\frac{\boldsymbol{E}}{\boldsymbol{V}}$$

using the relation $\bar{E}_0 = \frac{3}{5}E_F(0)$.

▶ **Example 12.** *Calculate the mobility of electrons in silver, assuming that each silver atom provides one conduction electron. Given: density of silver,* $\rho = 10.5 \times 10^3$ kg/m^3, *at wt.,* $M_A = 107.9$ *and conductivity at* $20°\text{C} = 6.8 \times 10^7\,\mho\,\text{m}^{-1}$.

Solution: $(M_A/\rho)\,\text{m}^3$ contains N_A (Avogadro No.) atoms/electron.

$\therefore$ 1 m^3 of silver will contain $n = \rho N_A / M_A$ electrons.

$$\therefore\ n = \frac{\rho N_A}{M_A} = \frac{10.5 \times 10^3 \times 6.02 \times 10^{26}}{107.9}$$
$$= 5.86 \times 10^{28}$$

$\therefore$ ***Mobility of electrons,*** $\mu = \sigma/ne = 6.8 \times 10^7/(5.86 \times 10^{28} \times 1.6 \times 10^{-9})$
$$= \mathbf{0.725 \times 10^{-2}\ m^2V^{-1}s^{-1}}$$

▶ **Example 13.** *An n-type* Ge-*strip* 1 mm *wide and* 1 mm *thick, has a Hall coefficient of* 10^{-2} m^2/C. *For a current of* 1 mA, *the Hall voltage developed inside the strip is* 1 mV. *Determine the strength of the magnetic field.*

Solution: From the relation (5.6.10): $B = \dfrac{V_H \omega}{R_H I}$, with usual symbols.

$$\therefore\ \text{Magnetic field, } \boldsymbol{B} = \frac{1 \times 10^{-3} \times 10^{-3}}{10^{-2} \times 1 \times 10^{-3}} = \mathbf{0.1\ Tesla}$$

▶ **Example 14.** *If* $T \to 0$, *show that* (i) $f(E) = 1$, *for* $E < E_F$ *and* (ii) $f(E) = 0$, *for* $E > E_F$ *where* $f(E)$ *is Fermi distribution function and* E_F *is the Fermi energy.*

Solution: We have $f(E) = \dfrac{1}{e^{(E-E_F)/kT} + 1}$. If $T \to 0$, $f(E) = \dfrac{1}{e^{(E-E_F)/0} + 1}$

(i) For $E < E_F$, $e^{(E-E_F)/0} \to e^{-\infty} = 0$. So, $\boldsymbol{f(E)} = \dfrac{1}{0+1} = \mathbf{1}$

(ii) For $E > E_F$, $e^{(E-E_F)/0} \to e^{\infty}$. So, $\boldsymbol{f(E)} = \dfrac{1}{e^{\infty}+1} = \dfrac{1}{e^{\infty}} = \mathbf{0}$.

▶ **Example 15.** *Prove that the compressibility of the free electron gas is given by* $\beta = 3/2nE_F$.

Solution: From the definition, $\beta = \left(V\dfrac{\partial P}{\partial V}\right)^{-1}$. Now for the free electron gas $E_{\text{Total}} = E_k =$ kinetic energy only $= \frac{3}{5}E_F$.

$$P = \frac{\partial}{\partial V}\left(\frac{3}{5}E_F\right) = \frac{3}{5}\frac{\partial}{\partial V}\left[\frac{\hbar^2}{2m}(3\pi^2 n)^{2/3}\right]$$
$$= \frac{3}{5}\frac{\hbar^2}{2m}(3\pi^2)^{2/3}\cdot\frac{\partial}{\partial V}(V^{-2/3}), \quad \text{since } n = 1/V$$

$$\therefore \quad \frac{\partial P}{\partial V} = \frac{3}{5}\frac{\hbar^2}{2m}(3\pi^2)^{2/3}\left(-\frac{2}{3}V^{-5/3}\right)$$

$$\therefore \quad V\frac{\partial P}{\partial V} = \frac{2}{3}\frac{\hbar^2}{2m}(3\pi^2)^{2/3}V^{-5/3} = \frac{2}{3}\left[\frac{\hbar^2}{2m}(3\pi^2 n)^{2/3}\right]n = \frac{2}{3}nE_F$$

$$\therefore \quad \beta = \left(V\frac{\partial P}{\partial V}\right)^{-1} = \frac{\mathbf{3}}{\mathbf{2}\boldsymbol{n}\boldsymbol{E_F}}$$

Review Questions

1. State the basic postulates or assumptions of the classical Drude theory of metals and deduce expressions for (a) the electrical conductivity and (b) the thermal conductivity.

2. What is Wiedemann-Franz law? On the basis of classical ideas, obtain the Lorentz number for metals.

3. What is meant by 'Fermi surface'? Discuss its significance in the calculations of electrical and thermal conductivities of metal. Establish

$$\frac{K}{\sigma T} = \frac{\pi^2}{3}\left(\frac{k}{e}\right)^2$$

with the usual meanings of the symbols used.

4. Using the free electron theory, find an expression for the electrical conductivity of a metal in terms of Fermi velocity and the mean free path of electrons.

(Calcutta Hons.)

5. How is electrical conductivity of metals described in quantum free electron theory? Using the free electron theory, find an expression for the electrical conductivity of a metal in terms of Fermi velocity of electron and the mean free path of electrons **(Calcutta Hons.)**

6. Describe the basic principle of Hall effect? Show that the Hall coefficient for an intrinsic semiconductor is given by

$$R_H = -\frac{1}{n_e}\left(\frac{\mu_n - \mu_p}{\mu_n + \mu_p}\right)$$

where μ_n, μ_p are the mobilities of electrons and holes and n the intrinsic carrier concentration. **(Calcutta Hons.)**

7. Explain what is Hall effect. How would you experimentally determine the Hall coefficient?

8. Show that the electrical conductivity of a semiconductor is $\sigma = e(n\mu_n + p\mu_p)$, where the symbols have their usual significance.

9. Show that the electrical conductivity of a free electron gas is

$$\sigma = ne^2\tau/m = ne^2\lambda/mv_{\text{rms}} = ne^2\lambda/\sqrt{3mkT}$$

where the symbols have their usual meanings. **(Burdwan Hons.)**

10. Obtain an expression for the electrical conductivity of a metal on the basis of free electron theory. Hence prove the Ohm's law. **(Lucknow and Bangalore Univ.)**

11. Use free electron theory to find an expression for the resistivity of metal in terms of Fermi velocity of free electron and mean free path between the collisions of electrons. **(Calcutta Hons.)**

12. Write the Boltzmann transport equation, and explain the meaning and the origin of the terms in the equation. **(Calcutta Hons.)**

13. (a) Show that the electrical conductivity of a free electron gas is $\sigma = ne^2\tau/m$ where the symbols have their usual meanings.
    (b) Explain why only the electrons near the Fermi-surface suffer collisions to contribute to the electrical conductivity of a metal. **(Burdwan Hons.)**

14. How would you experimentally determine the Hall coefficient? Mention the different uses of Hall effect. Why is the Hall coefficient positive in some metals?

15. Derive, by quantum mechanical calculation, an expression for electronic specific heat in metals. **(Calicut and Lucknow Univ.)**

Short and Objective Questions

1. Give briefly the characteristic properties of a metal.
2. What are the main drawbacks of classical free electron theory.
3. State and explain the Wiedemann-Franz law.
4. Define the terms : mean free time, mean free path, electron mobility and electrical conductivity.
5. Explain drift velocity, relaxation time and collision time of free electrons in metals.
6. What is Fermi energy? Write down its relation with the concentration of electrons in metals.
7. Explain the concept of density of states for a free electron gas. **(Lucknow Univ.)**
8. Show that the average kinetic energy of a free electron at 0 K is $\frac{3}{5}E_F$ where E_F is the Fermi energy and the average speed is $\frac{3}{4}v_F$ where v_F is the velocity at Fermi surface. **(Burdwan Hons.)**

9. Fermi level of an intrinsic semiconductor lies near the middle of the forbidden gap but for an n-type semiconductor it ties near the conduction band. Explain. (Banagalore and Guru Nanak)

10. Does the Fermi energy of a metal depend on temperature and the total number of electrons in the sample? (Punjab Univ.)

11. Show that at absolute zero of temperature, Fermi level lies exactly half-way between the top of the valence band and the bottom of the conduction band. What happens to the Fermi level when the temperature increases? (Guru Nanak Univ.)

12. What is Fermi surface? What is Hall effect? What is quantum Hall effect?

13. Explain Hall effect. Give its theory. Find the Hall coefficient.

14. How would you use Hall effect for the measurement of magnetic field? Why is the Hall coefficient positive in metals? (Calcutta Hons.)

15. Explain why resistivity increases with temperature in metal, while it decreases with temperature in semiconductors. (Calcutta Hons.)

16. Conduction electrons have more mobility than hoes. Explain. (Bangalore Univ.)

17. What is the difference between hole current and electron flow? (Punjab Univ.)

18. Explain why only the electrons near the Fermi surface suffer collisions to contribute to the electrical conductivity of a metal. (Burdwan Univ.)

19. Define Hall coefficient. When is the Hall coefficient zero in a semiconductor?

20. Show that the electrical conductivity of a free electron gas is $\sigma = ne^2\tau/m$, with usual meanings of the symbols.

21. The average kinetic energy of the electron in the ground state in one dimension is given by

    (a) one-third Fermi energy, (b) half of Fermi energy,
    (c) one-fourth Fermi energy, (d) equal to Fermi energy. (Himachal Univ.)

22. If the Fermi energy of silver at 0 K is 5 eV, the mean energy of electron in silver at 0 K is

    (a) 6 eV, (b) 12 eV, (c) 1.5 eV, (d) 3 eV.

23. The Fermi energy of a metal is 1.4 eV. The Fermi temperature of the metal in Kelvin is about

    (a) 1.6×10^3, (b) 1.6×10^4, (c) 1.6×10^5, (d) None of these.

24. The plot of '$\ln \sigma$ against T^{-1}' where σ is the conductivity and T the temperature in degree Kelvin, is a

    (a) circle, (b) parabola, (c) straight line, (d) none.

25. Semiconductors exhibit temperature coefficient

(a) zero, (b) positive, (c) negative, (d) all of these.

26. If the mobility of electrons in a metal increases, the resistivity

(a) decreases, (b) increases, (c) remains unchanged.

27. Resistivity of copper has units

(a) ohm, (b) ohm-metre, (c) ohm^{-1} metre, (d) ohm metre^{-1}.

(Himachal Univ.)

28. In a p-type semiconductor the minority charge carriers are

(a) protons, (b) holes, (c) electrons, (d) neutrons. **(Meerut Univ.)**

29. Fermi level is defined as

(a) Lowest filled level at 0 K, (b) Highest filled level at 0 K,
(c) Lowest filled level at 300 K, (d) Highest filled level at 300 K.

(Purvanchal Univ.)

30. The average drift velocity v_D of electrons in a metal is related to the electric field E and collision time τ as

(a) $\sqrt{eE\tau/m}$, (b) $\sqrt{m/eE\tau}$, (c) $eE\tau/m$, (d) $m/eE\tau$.

31. If K and σ be the thermal and electrical conductivities of a metal at temperature T, then

(a) $KT/\sigma =$ constant, (b) $K\sigma/T =$ constant,
(c) $\sigma/KT =$ constant, (d) $K/\sigma T =$ constant.

Problems

1. The temperature coefficient of resistivity is defined as

$$\alpha = \frac{1}{\rho}\frac{\partial \rho}{\partial T}$$

For a metal, $\rho \propto T$; show that α is inversely proportional to T.

2. Assuming that each silver atom gives one conduction electron, determine the mobility of electrons in silver. Density of silver $= 10.5 \times 10^3$ kg/m^3, atomic weight $= 107.9$ and conductivity $\sigma = 6.8 \times 10^7\,\mho\text{m}^{-1}$. [$0.725 \times 10^{-2}\,\text{m}^2\text{V}^{-1}\text{s}^{-1}$]

3. The density of copper is $8.93\,\text{g/cm}^3$ and its atomic weight is 63. Assuming that each atom contributes one free electron, find the concentration of free electrons in copper. If the resistivity of copper is 1.6×10^{-6} ohm. cm, calculate the Fermi velocity, the mean free time, the mean free path and the radius of Fermi surface.

[$8.53 \times 10^{22}/\text{cm}^3$, $1.568 \times 10^6\,\text{m/s}$, $0.26 \times 10^{-13}\,\text{s}$, $0.41 \times 10^{-7}\,\text{m}$, 14.3×10^{-25} kg.m/s]

4. The resistivity of some material is $1.1 \times 10^{-3}\,\Omega\,\text{m}$ at 20°C. Assuming M-B distribution and a free electron concentration of $10^{22}\,\text{m}^{-3}$, estimate the mean free path λ between the collisions of free electrons in the material at 20°C. Given effective mass of free electrons in the material is 2.733×10^{-31} kg. [1.88×10^{-7} m]

5. Silver is a monovalent metal with atoms arranged in an fcc structure as close packed spheres. If the atomic radius is $r = 1.44$ A, calculate the Fermi energy.

[5.5 eV]

[**Hints.** $4r = \sqrt{2}a$ where a = lattice constant. Volume of cubic unit cell $= V = a^3 = (2\sqrt{2}r)^3$. No. of free electrons in unit cell = 4. So, free electron concentration, $n = 4/V = 1/(4\sqrt{2}r^3)$]

6. The Fermi energy for Li at 0 K is 4.72 eV. Calculate the number of conduction electrons per unit volume in lithium. **(Calcutta Hons.)** [$4.66 \times 10^{23}/\text{cm}^3$]

7. In Example 5 above, calculate the Fermi temperature and Fermi velocity.

[63.85×10^3 K; 1.39×10^6 m/s]

8. Metallic aluminium crystallizes in *fcc*-structure. Each atom contributes a single electron as free and the lattice constant a is 4A. Treating free electrons as Fermi gas, calculate (a) Fermi energy and (b) total kinetic energy of free electron gas per unit volume at 0 K. Given $\hbar = 1.054 \times 10^{-27}$ erg sec, the rest mass of electron $m = 9.11 \times 10^{-28}$ g. **(Punjab Univ.)** [5.74 eV; 21.52×10^{22} eV]

9. At what temperature, we can expect at 10% probability that the electrons in silver have an energy which is 1% above the Fermi energy, $E_F = 5.5$ eV? [290 K]

10. Calculate the concentration of acceptor atoms that have to be added to an intrinsic semiconductor to produce a p-type semiconductor of conductivity $0.05\mho\text{m}^{-1}$. The mobility of holes in p-type semiconductor is $0.4\,\text{m}^2/\text{Vs}$.

(Bangalore Univ.) [$7.8 \times 10^{17}\,\text{m}^{-3}$]

11. Calculate the conductivity of intrinsic Ge at 300 K. Given that at that temperature the intrinsic carrier concentration, electron mobility and hole mobility are respectively $2.4 \times 10^{19}\,\text{m}^{-3}$, $0.38\,\text{m}^2/(\text{Vs})$ and $0.19\,\text{m}^2/(\text{Vs})$. [2.19 $\mho$/m]

12. The ratio of the hole mobility to the electron mobility in a semiconductor is b. If the Hall coefficient is zero, what fraction of the total current is carried by electrons? **(N. Bengal Hons.)** [$b/(1+b)$]

13. The resistivity of aluminum at room temperature is $2.62 \times 10^{-8}\,\Omega$m. Calculate (a) the drift velocity, (b) the mobility and (c) the relaxation time of electrons on the basis of classical free electron model.

[$0.396\,\text{ms}^{-1}$; $3.96 \times 10^{-3}\,\text{m}^2\text{V}^{-1}\text{s}^{-1}$; 2.3×10^{-14}s]

14. At a temperature of 300 K, the intrinsic carrier concentration is $1.6 \times 10^{10}/\text{cm}^3$ for silicon and the electron and hole mobilities are $1500\,\text{cm}^2/(\text{V.s})$ and $500\,\text{cm}^2/(\text{V.s})$ respectively. Find the resistivity of pure silicon at 300 K. Given eletronic charge $= 1.602 \times 10^{-19}$ C. **(Mangalore and Kanpur Univ)** [$2.17 \times 10^5\,\Omega$ cm]

15. The electron and the hole mobilities in a semiconductor are $0.8\,\text{m}^2/(\text{V.s})$ and $0.02\,\text{m}^2/(\text{V.s})$ respectively. The electron concentration in the semiconductor is $2.5 \times 10^{18}/\text{m}^3$, and the Hall coefficient is zero. Find the intrinsic carrier concentration. [$10^{20}/\text{m}^3$]

16. The following data are obtained in an experiment with intrinsic Ge at room temperature (300 K) :

$$n_i = 2.5 \times 10^{13}/\text{cm}^3$$

$$\mu_n = 3450\ \text{cm}^2/(\text{V.s})$$

$$\mu_p = 1600\ \text{cm}^2/(\text{V.s})$$

Compute the conductivity and resistivity of intrinsic Ge. [0.02 ℧/cm, 50 Ω cm]

17. A conductor contains 10^{23} electrons/cm^3. Consider a cube of the conductor of side 0.1 cm carrying a current of 10 A flowing perpendicularly from the face to the opposite one. Calculate the current density and the drift velocity. Compare the drift velocity with the Fermi velocity, assuming that the Fermi energy is 7 eV and electron mass is 9.1×10^{-31} kg. [$10^3\,\text{A/cm}^2$, 6.25×10^{-2} cm/s, 4×10^{-10}]

18. Sodium has desity $0.97\,\text{g/cm}^3$, relative atomic mass 23, and an electrical conductivity 2.1×10^5 ℧/cm. Determine the electron mobility in sodium.

[$5.2 \times 10^{-3}\,\text{m}^2/(\text{V.s})$]

19. Calculate the Fermi energy and Fermi temperature for the valence electrons in sodium metal. Sodium is monovalent and its density is $0.979\,\text{g/cm}^3$, at wt. = 23.

(Vidyasagar Hons.)

20. The donor density of an n-type Ge sample is $10^{21}/\text{m}^3$. If the sample is 3.5 mm thick, find the Hall voltage if the magnetic field $B = 0.5$ Tesla and the current density $J = 500\,\text{A/m}^2$. [5.47 mV (approx)]

21. The density of potassium is $860\,\text{kg/m}^3$. Calculate the Fermi energy and the compressibility of potassium at 0 K. Assume the atomic mass of K = 39 u.

[2.02 eV, $3.51 \times 10^{-10}\,\text{N}^{-1}\text{m}^2$]

22. The resistivity of some material is $1.1 \times 10^{-3}\,\Omega$m at 20°C. Assuming MB distribution and a free electron concentration of $10^{22}/\text{m}^3$, estimate the mean free path $\bar{l}$ between the collisions of free electrons in the material at 20°C. Given: effective mass of free electrons in the material is 2.733×10^{-31} kg.

[**Hints.** $\alpha = ne^2\tau/m^*$ and $\bar{l} = \bar{v}\tau$, where $\bar{v} = \sqrt{8kT/\pi m^*}$] [$1.88 \times 10^{-7}$ m]

23. The resistivity of copper at 20°C is $1.72 \times 10^{-8}\,\Omega$ m. Assuming MB-distribution, estimate the mean free path between the colliusions of free electrons in copper at 20°C. The density of copper is 8.94×10^3 kg/m^3 and its atomic mass is 63.5 u, Boltzmann constant $= 1.38 \times 10^{-23}$ J/K, 1 u $= 1.66 \times 10^{-27}$ kg. **(Calcutta Hons.)**

24. Calculate the electronic specific heat of copper ($E_F = 7.05$ eV) at 600 K.

[2.995 J (k.mol)$^{-1}$ K^{-1}]

25. Show that the wavelength associated with an electron having energy E_F is given by

$$\lambda_F = \left(\frac{8\pi V}{3N}\right)^{1/3}$$

If this wavelength is 4.6Å, find the Fermi temperature. [$T_F = 8.25 \times 10^4$ K]

26. Show that the fraction F of electrons within kT of the Fermi level is given by

$$F = \frac{3}{2}\frac{kT}{E_F}, \quad \text{if} \quad \rho(E) \propto E^{1/2}.$$

27. In Cu, each atom contributes one electron to free-electron gas. Calculate the number of free electrons per unit volume (n). Given: density of Cu is $\rho = 8.94 \times 10^3$ kg/m^3. [8.47×10^{28} m^{-3}]

28. Express electrical conductivity in terms of Fermi energy E_F.

[$\sigma - \frac{e^2\tau}{3\pi^2 m\hbar^3}(2mE_F)^{3/2}$]

[**Hints.** $\sigma = ne^2\tau/m$. $E_F = \frac{\hbar^2}{2m}(3\pi^2 n)^{2/3}$. Apply these two expressions to eliminate n and get the result]

6

Lattice Vibrations and Thermal Properties

6.1 Introduction

Heat energy supplied to a solid increases its internal energy and this increase in internal energy usually appears in the following two forms.

1. The mechanical vibrations of the solid atoms about their mean positions of equilibrium (i.e., Bravais lattice sites) increases.

2. The kinetic energy of the free electrons moving about randomly inside the lattice also increases.

The former is commonly called ***lattice vibrations*** and the associated thermal energy may be visualised as a series of superimposed ***acoustic waves*** or ***lattice waves*** with a frequency spectrum determined by the elastic properties of the crystal. Our aim in this chapter is to study these lattice vibrations and some related topics to understand some thermal properties of solid materials such as specific heat and its variation with temperature—in particular the various proposed models in this context e.g., Einstein's model, Debye's model etc.

An *energy quantum of lattice vibration* is termed a ***phonon***, in analogy with ***photon***, the *quantum of radiation*. Some of the important properties of phonons are:

Characteristics of phonons—1. Like photons, phonons also possess duality, i.e., they behave both as particles and as waves.

2. They are, like photons, indistinguishable particles and obey the Bose-Einstein statistics. Photons are thus *bosons*.

3. The energy of a phonon is $\hbar\omega$ or $h\nu$ where ν is the frequency of lattice vibration and hence the momentum of a phonon is $\hbar k = h/\lambda$.

4. At finite temperatures, the crystals are full of phonons and they are, like photons, also not conserved. An increase in temperature increases the number of phonons (creation) and conversely (destruction). At a given temperature, the total energy of phonons is the thermal energy of the solid.

5. The phonon interaction is essentially a scattering collision process.

6. In an interaction, its energy is always an integral multiple of $\hbar\omega$ and only a single phonon is observed or excited at a time, as confirmed by the neutron scattering experiments.

7. There are *two* kinds of phonons—*acoustical* and *optical*—associated respectively to acoustic and optical modes of vibration. At low temperatures, the acoustic modes far outnumber the optical modes.

● Physically, phonons carry no momentum. But, a phonon of wave vector k interacts with particles like neutrons, electrons etc. and also with fields as if it has a momentum $\hbar k$. Phonon momentum is often called ***crystal momentum.***

Evidences for phonons—While the photoelectric phenomenon provides a direct proof for the existence of photons, such direct evidence for phonons has yet to be observed. But there exists some strong *indirect evidences* for phonons. For instance,

1. At zero temperature, the lattice contribution to heat capacity is practically zero and this is possible only if the lattice waves are quantized.

2. X-radiation and neutrons are inelastically scattered by crystals involving changes in momentum and energy corresponding to the creation or the absorption of one or more phonons. On measuring the recoil of scattered neutron, the properties of individual

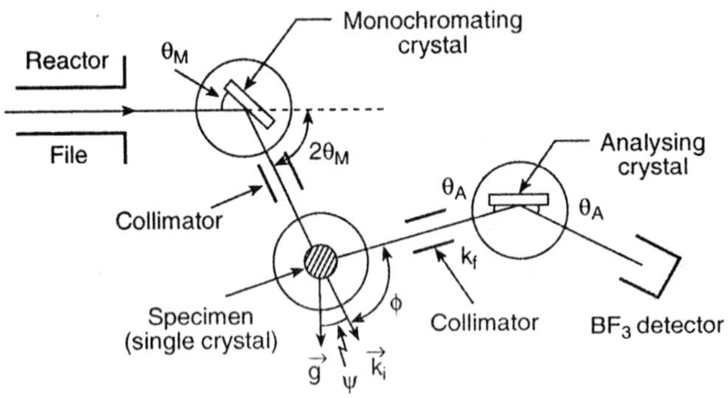

Fig. 6.1 Measurement of phonon spectra using triple-axis neutron spectrometer

phonons may be determined. The dispersion relation (discussed later) between the frequency ω and the wave vector k of phonon can be determined from such experiments that provide the strongest evidence for the existence of phonons.

● At present, inelastic scattering of thermal neutrons by phonons (neutron beam obtained from a reactor and properly thermalized) is the most powerful tool for experimental observation of the lattice waves. A triple axis neutron spectrometer developed by Brockhouse and as illustrated in Fig. 6.1, is used to find the magnitude and direction of the scattered beam, whence the dispersion relation ω vs k for lattice waves can be experimentally determined.

6.2 Vibrations of a monatomic one-dimensional lattice

In order to study the possible modes of vibration of a crystal, we must consider an infinitely long one-dimensional array of identical atoms, each of mass m, connected by ideal springs obeying Hooke's law and having an equilibrium separation a between two consecutive atoms (Fig. 6.2). Only the nearest-neighbour interactions would be taken into account since a solid can be characterized by short range forces which are linear. This is called ***harmonic approximation.***

If x_n is the displacement of the nth atom, then within elastic limit (i.e., force $\propto$ displacement), the force on the nth atom due to $(n-1)$th and $(n+1)$th atoms will be

(i) $\beta(x_n - x_{n-1})$ to the left, from the springs on its left.

(ii) $\beta(x_{n+1} - x_n)$ to the right, from the springs on its right.

where β is the near-neighbour *force constant.*

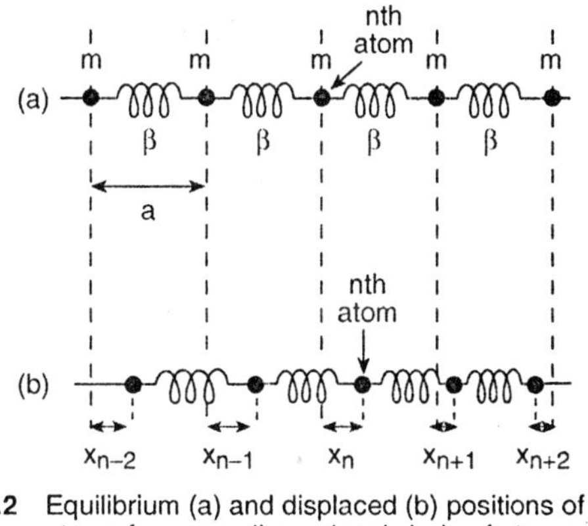

Fig. 6.2 Equilibrium (a) and displaced (b) positions of vibrating atoms for a one dimensional chain of atoms

These forces are oppositely directed so that the net force on the nth atom is

$$F_n = \beta[(x_{n+1} - x_n) - (x_n - x_{n-1})]$$

$\therefore$ Equation of motion of the nth atom is given by

$$m\frac{d^2x_n}{dt^2} = \beta(x_{n+1} + x_{n-1} - 2x_n) \tag{6.2.1}$$

Let the periodic solution of equation (6.2.1) be given by

$$x_n = Ae^{i(kna-\omega t)} \tag{6.2.2}$$

where A is a constant (the amplitude), na the equilibrium position of the nth atom with respect to the origin and k the wave vector.

Eq. (6.2.2) represents a *travelling wave* in which all the atoms oscillate with the same angular frequency ω and having the same amplitude A. Substituting the equation (6.2.2) in (6.2.1),

$$\omega^2 m x_n = -\beta(e^{ika} - 2 + e^{-ika})x_n$$

$$\Rightarrow \quad \omega^2 m = -\beta(e^{ika} - 2 + e^{-ika}) = -2\beta(\cos ka - 1) \tag{6.2.3}$$

since $e^{ika} + e^{-ika} = 2\cos ka$.

$$\therefore \quad \omega^2 m = 4\beta \sin^2 \frac{ka}{2} \tag{6.2.4}$$

$$\therefore \quad \omega = \pm 2\left(\frac{\beta}{m}\right)^{1/2} \left|\sin\frac{ka}{2}\right|$$

$$\text{or,} \quad \boxed{\boldsymbol{\omega = \omega_m \left|\sin\frac{ka}{2}\right|}} \tag{6.2.5}$$

where $\omega_m = 2(\beta/m)^{1/2}$, the *maximum value* (natural cut-off) of angular frequency ω and since ω is assentially positive, irrespective of k being positive or negative.

The maximum frequency ν_{max} through the chain of atoms is:

$$\nu_{\text{max}} = \frac{\omega_m}{2\pi} = \frac{1}{\pi}\sqrt{\frac{\beta}{m}}.$$

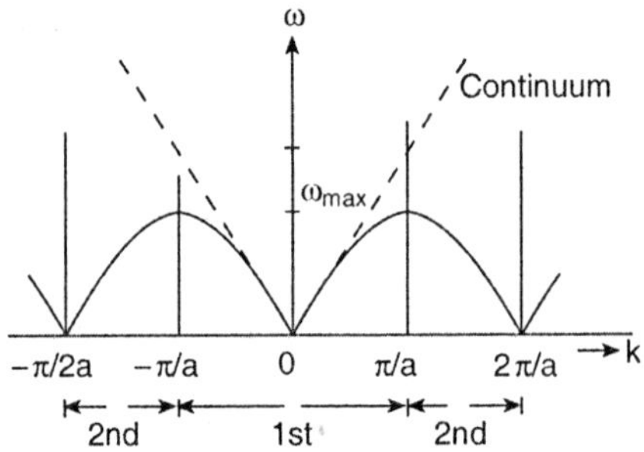

Fig. 6.3 Dispersion curve ω vs. *k* for a one dimensional monatomic lattice with nearest neighbour interactions

This relation between the *frequency* ω and *wave vector* k is known as ***dispersion relation*** and a *plot of* ω vs. k gives the ***dispersion curve*** (Fig. 6.3) which is sinusoidal with a period $2\pi/a$ in k-space. The dispersion relation (6.2.5) has a number of important properties applicable to all the three kinds of lattice—linear, plane and space lattice.

6.2.1 Symmetry in k-space: First Brillouin zone

The dispersion curve (Fig. 6.3) shows two types of symmetry elements—the *translational symmetry* and the *mirror symmetry*—and the frequency ω is a periodic function of k with a period $2\pi/a$. For each ω, below ω_m, there are thus several k-values and the corresponding wavelengths, since $k = 2\pi/\lambda$.

So, for a unique specification, a certain interval in k-space of length equal to the period is to be selected. Usually, *for convenience*, the range

$$-\pi/a < k < \pi/a \tag{6.2.6}$$

is chosen to specify a unique k and hence a unique λ.

The region of k-space defined by (6.2.6) is called the ***first Brillouin zone*** for a one-dimensional periodic lattice. In the region, we obtain

$$k_{\max} = \pm(\pi/a). \tag{6.2.7}$$

The dispersion curve also possesses a mirror symmetry in k-space about the origin $k = 0$ so that we may write

$$\omega(-k) = \omega(k) \tag{6.2.8}$$

where the mode k (> 0) represents a wave that travels through the lattice in the positive x-direction and the mode $-k$ a travelling wave of the same λ but travelling in the opposite direction. As the lattice is symmetric in the two directions, it responds to the two waves referred to, in the same fashion and with identical corresponding frequency as dictated by (6.2.8).

6.2.2 Number of modes in the first zone

The last atom of one-dimensional periodic lattice is joined to the first so as to form the endless chain of atoms, N in number, where $N \gg 1$. The length of the chain is $L = Na$ and the atoms are labelled $1, 2, 3, \cdots, N$ going around the circle. Each atom has the same environment and we may thus apply a *periodic* or *cyclic boundary condition.*

$$x_n(t) = x_{n+N}(t) \tag{6.2.9}$$

the subscripts n and $n + N$ denote the same atom.

Substituting (6.2.2) in equation (6.2.9), we obtain

$$e^{ikna} = e^{ik(n+N)a}$$

$$\Rightarrow \quad e^{ikNa} = 1 \quad \Rightarrow \quad e^{ikL} = 1, \tag{6.2.10}$$

which imposes a restriction on the allowed values of k, namely

$$k = \frac{2\pi}{L}s$$

where s is given by $s = 0, \pm 1, \pm 2, \pm 3, \cdots$ etc.

When these values are plotted along a k-axis, they form a one-dimensional mesh of regularly spaced points, as illustrated in Fig. 6.4, with separation $2\pi/L$ between them. When L is large (as, in practice, for any lattice of macroscopic size), the allowed points are close enough to form a *quasi-continuous* mesh along the k-axis.

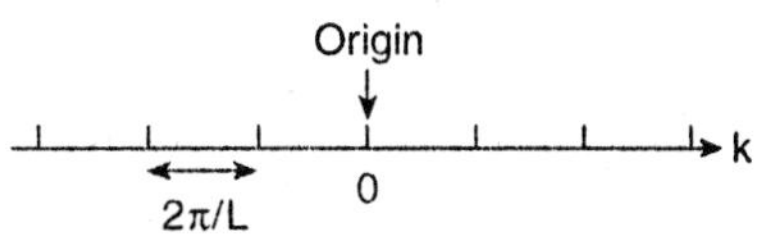

Fig. 6.4 Allowed values of k

Total number of points in the first zone is $\dfrac{2\pi/a}{2\pi/L} = \dfrac{L}{a} = N.$ (6.2.11)

Thus, *there are as many modes of vibration as there are vibrating atoms*, i.e., the number of degrees of freedom in the lattice.

6.2.3 Long wavelength limit

We shall now study the behaviour of the lattice during the propagation of the wave through it. Consider the part of the dispersion curve $0 < k < \pi/a$ in which the frequency of the wave varies in the range $0 < \omega < \omega_m$. This means that a one-dimensional crystal lattice allows the propagation of frequencies between 0 to ω_m only, and strongly attenuates all the other frequencies. It may thus be considered to behave as a ***low pass*** (mechanical) ***filter***.

In the long wavelegth side, i.e., $k \to 0$, the equation (6.2.5) approximates to

$$\omega = \left(\frac{\omega_m a}{2}\right) k \tag{6.2.12}$$

implying that ω is proportional to k. This is rather expected, for in this limit, the lattice behaves as a ***continuum***. Note that the expression in the bracket equals the velocity v_s of sound in the solid medium.

$$\therefore \quad v_s = \frac{\omega_m a}{2} = \sqrt{\frac{Y}{\rho}} \tag{6.2.13}$$

where Y is the Young's modulus and ρ the density of the solid.

Substituting $\omega_m = 2\sqrt{\beta/m}$ and $\rho = m/a^3$ in (6.2.13) above we get

$$\boxed{\boldsymbol{\beta = aY}} \tag{6.2.14}$$

whence the force constant β can be easily estimated.

• The variation of ω with k as dictated by (6.2.12) is *linear* and is shown by the broken lines in Fig. 6.3. The linear relationship ceases when k increases slowly and the dispersion curve starts deviating from the straight path and finally saturates at $k = \pi/a$ with ω_m as the maximum frequency.

6.2.4 Phase and group velocities

We are already familiar with the terms *phase velocity* ($v_p = \omega/k$) and *group velocity* ($v_g = d\omega/dk$). Using the dispersion relation, we have

$$v_p = v\frac{\sin(ka/2)}{ka/2} \tag{6.2.15}$$

$$v_g = v\cos(ka/2) \tag{6.2.16}$$

where $v = \omega_m a/2 = a(\beta/m)^{1/2}$, the ordinary speed of sound in a solid.

• v_p is the velocity of propagation for a pure wave of specified ω and k-values, while v_g is the velocity of a wave group (wave packet) whose average frequency and wave vector are specified by ω and k. Since energy and momentum are transmitted essentially via pulse rather than by waves, group velocity is physically more important.

1. At *low frequencies*, that is, when $k \to 0$, $\dfrac{\sin(ka/2)}{ka/2} \to 1$ and $\cos\left(\dfrac{ka}{2}\right) \to 1$. So from (6.2.15) and (6.2.16), we get

$$v_p = v_g = v = a(\beta/m)^{1/2} = \text{constant}.$$

This result however is expected for waves of *long wavelength* ($k \to 0$) as they are not responsive to the discreteness of the crystal structure; the lattice behaves as a ***continuum*** (homogeneous elastic medium) and *no dispersion* occurs.

2. At *high frequencies*, phase velocity,

$$v_p = \frac{\omega}{k} = 2\sqrt{\frac{\beta}{m}}\left|\frac{\sin(ka/2}{k}\right| = a\sqrt{\frac{\beta}{m}}\left|\frac{\sin(ka/2)}{ka/2}\right|,$$

$$\text{group velocity is}: \; v_g = \frac{d\omega}{dk} = a\sqrt{\frac{\beta}{m}}\cos\frac{ka}{2}.$$

As k increases (Fig. 6.3), the slope $d\omega/dk$ of the dispersion curve, i.e., v_g steadily increases and becomes zero at $k = \pi/a$.

At the zone boundary, $k = \pm\pi/a$, $v_p = 2v/\pi$ from (6.2.15) and $\lambda = 2a$, implying that the nearest-neighbour atoms move *in antiphase*. Also, at $k = \pm\pi/a$, the group velocity $v_g = 0$ and this means that *no energy propagates* and the wave is a ***standing wave***. The situation is equivalent to that of *Bragg reflection* of X-rays from successive lattice planes.

6.3 A diatomic linear chain : Its dynamics

Consider the lattice dynamics of a diatomic chain of masses M and m ($M > m$), placed alternately along the x-axis with an interatomic distance a. The case is a one-dimensional analogue of the three-dimensional NaCl structure.

We number the atoms such that the even numbers correspond to masses M and the odd ones to masses m. The equilibrium and the displaced positions of the system are illustrated in Fig. 6.5. Using the assumptions similar to the monatomic case, the equations of motion—one for the heavier atoms and the other for the lighter ones—are:

$$M\frac{d^2x_{2n}}{dt^2} = \beta\left(x_{2n+1} - 2x_{2n} + x_{2n-1}\right) \tag{6.3.1}$$

$$m\frac{d^2x_{2n-1}}{dt^2} = \beta\left(x_{2n} - 2x_{2n-1} + x_{2n-2}\right) \tag{6.3.2}$$

These equations may be solved, as before, by assuming solutions of the type

$$x_{2n} = A\exp i(2nka - \omega t) \tag{6.3.3}$$

$$\text{and } x_{2n-1} = B \exp i[(2n-1)ka - \omega t] \tag{6.3.4}$$

where k is the wave vector of a particular mode of vibration of frequency ω and A, B are the amplitudes. While the amplitudes are different due to the difference in atomic

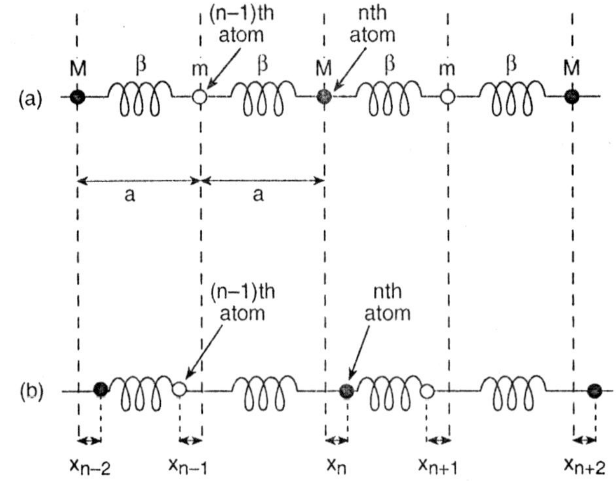

Fig. 6.5 Equilibrium (a) and displaced (b) positions of a vibrating diatomic system

masses, the vibrational frequency is the same as both types of atoms take part in the same wave motion.

Substituting (6.3.3) and (6.3.4) in (6.3.1) and (6.3.2) respectively, we get

$$(2\beta - \omega^2 M)A - (2\beta \cos ka)B = 0 \tag{6.3.5}$$

$$(-2\beta \cos ka)A + (2\beta - \omega^2 m)B = 0, \tag{6.3.6}$$

using the relation: $\quad e^{ika} + e^{-ika} = 2 \cos ka.$

The above two equations form a set of homogeneous simultaneous equations in the unknown A, B and have a non-trivial solution only if the determinant of their coefficients vanishes. This dictates

$$\begin{vmatrix} 2\beta - \omega^2 M & -2\beta \cos ka \\ -2\beta \cos ka & 2\beta - \omega^2 m \end{vmatrix} = 0 \tag{6.3.7}$$

$$\text{or,} \quad (2\beta - \omega^2 m)(2\beta - \omega^2 M) - 4\beta^2 \cos^2 ka = 0$$

$$\text{or,} \quad \omega^4 - \frac{2\beta(M+m)}{Mm}\omega^2 + \frac{4\beta^2 \sin^2 ka}{Mm} = 0 \tag{6.3.8}$$

The two possible solutions of this equation quadratic in ω^2 are

$$\omega^2 = \beta\left(\frac{1}{M} + \frac{1}{m}\right) \pm \beta\sqrt{\left(\frac{1}{M} + \frac{1}{m}\right)^2 - \frac{4\sin^2 ka}{Mm}} \tag{6.3.9}$$

This is the ***dispersion relation*** between the ***angular frequency*** ω ***and the wave vector*** k ***for a linear diatomic lattice.***

• The correspnding relation for a monatomic lattice was

$$\omega^2 = \frac{4\beta}{m}\sin^2\frac{ka}{2},$$

the equation (6.2.4).

From (6.3.9), we have two angular frequencies ω_+ and ω_- for each k. In monatomic lattice, there was one value of ω, since ω is always positive. A plot of ω vs. k is shown in Fig. 6.6 where we observe two clear branches. The upper branch of the curve,

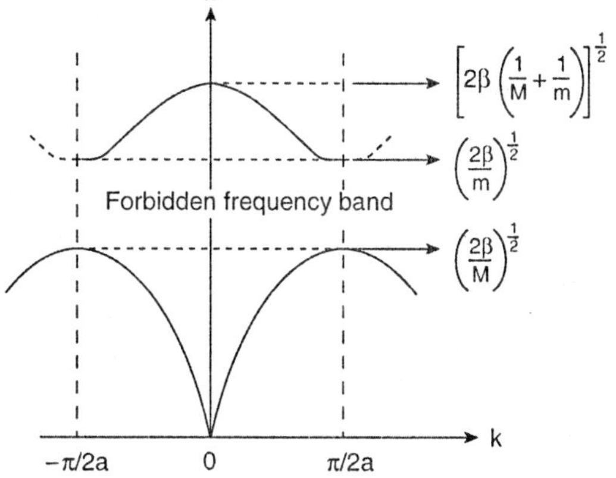

Fig. 6.6 The dispersion curve, ω vs k, for a one dimensional diatomic lattice ($M > m$)

corresponding to ω_+, is called the ***optical branch***, while the lower branch corresponding to ω_- is called the ***acoustic branch***.

Near $k \to 0$, the frequency of the upper branch lies near the *infrared region* of the spectrum, the corresponding wavelength is typically one for infrared absorption due to molecular vibration. Hence it is named ***'optical branch'***.

Near $k \to 0$, the lower branch represents *long wavelength elastic waves* with a frequency independent of the velocity ($\omega \propto k$), and the *longitudinal elastic waves* are identical with sound waves. Hence the name '***acoustic branch***'.

• Note from Fig. 6.6 that the acoustic branch looks qualitatively similar to the dispersion relation for a monatomic lattice while the optical branch shows a different type of wave motion.

There is a band of frequencies between $\sqrt{2\beta/M}$ and $\sqrt{2\beta/m}$ where the wave-like solution does not exist. This is the characteristic feature of elastic waves in diatomic lattice. This shows that it is not possible to excite vibrations of a frequency in this band in the lattice. Hence, it is called ***forbidden frequency band.***

6.3.1 The acoustic branch

The dispersion relation corresponding to the acoustic branch is given by

$$\omega_-^2 = \beta\left(\frac{1}{M}+\frac{1}{m}\right) - \beta\left[\left(\frac{1}{M}+\frac{1}{m}\right)^2 - \frac{4\sin^2 ka}{Mm}\right]^{1/2} \qquad (6.3.10)$$

The acoustic branch starts at $k = 0$ and is defined mathematically as

$$\underset{k \to 0}{\text{Lt}} \, (\omega_-) \to 0 \tag{6.3.11}$$

But from equation (6.3.10), we have

$$\underset{k \to 0}{\text{Lt}} \, \omega_-^2 = \beta\left(\frac{1}{M} + \frac{1}{m}\right) - \beta\left[\left(\frac{1}{M} + \frac{1}{m}\right)^2 - \frac{4k^2a^2}{Mm}\right]^{1/2}$$

$$= \beta\left(\frac{m+M}{Mm}\right)\frac{2k^2a^2Mm}{(M+m)^2}, \quad \text{on simplification.}$$

$$\therefore \underset{k \to 0}{\text{Lt}} \, \omega_- = ka\left(\frac{2\beta}{m+M}\right)^{1/2} \tag{6.3.12}$$

Thus, in a linear diatomic lattice, the acoustic branch is given by (6.3.12).

Phase velocity and group velocity—The phase velocity v_p and the group velocity v_g can be obtained using (6.3.12) and are given respectively by

$$\underset{k \to 0}{\text{Lt}} \, v_p = \underset{k \to 0}{\text{Lt}} \, (\omega_-/k) = a\left(\frac{2\beta}{m+M}\right)^{1/2} = v \tag{6.3.13}$$

and
$$\underset{k \to 0}{\text{Lt}} \, v_g = \underset{k \to 0}{\text{Lt}} \, (d\omega_-/dk) = a\left(\frac{2\beta}{m+M}\right)^{1/2} = v \tag{6.3.14}$$

As k increases, the acoustic branch of the curve first rises *linearly* and then decreases slowly with saturation at $k = \pi/2a$. The maximum allowed frequency for this mode is obtained by substituting $ka = \pm\pi/2$ in (6.3.10) and the result is

$$(\omega_-)_{\text{max}} = (2\beta/M)^{1/2} \tag{6.3.15}$$

which is, interestingly, *independent of the mass of lighter atoms in the chain.*

The *physical distinction* between the acoustical and the optical modes of lattice vibration is highlighted by the ratio of the wave amplitudes A/B and the phases of the solutions of each branch.

From (6.3.5), we have the ratio

$$\frac{A}{B} = \frac{2\beta \cos ka}{2\beta - M\omega_-^2} \tag{6.3.16}$$

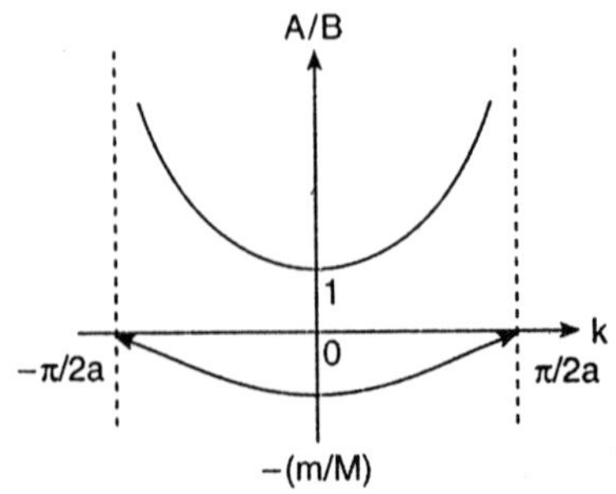

Fig. 6.7 Ratio (A/B) of amplitudes of two branches of dispersion relation for a diatomic linear chain

Using the limit $k \to 0$, the above ratio A/B becomes

$$\boldsymbol{A/B = 1.} \tag{6.3.17}$$

So, at $k = 0$, for the *acoustic* branch, the two atoms in the cell have the same amplitude and are in phase.

From (6.3.6), again

$$\frac{B}{A} = \frac{2\beta \cos ka}{2\beta - m\omega_-^2} \tag{6.3.18}$$

At the first zone boundary, $ka = \pm\pi/2 \Rightarrow B = 0$, indicating that the smaller atoms are stationary in the acoustic branch. So, $A/B \to \infty$.

These two results have been illustrated in Fig. 6.7.

6.3.2 The optical branch

The *optical branch* corresponds to the upper branch of the curve and is given by

$$\omega_+^2 = \beta\left(\frac{1}{M} + \frac{1}{m}\right) + \beta\left[\left(\frac{1}{M} + \frac{1}{m}\right)^2 - \frac{4\sin^2 ka}{Mm}\right]^{1/2} \tag{6.3.19}$$

On simplifying (6.3.19), at $k = \pm\pi/2a$, the minimum allowed frequency corresponding to the optical branch is given by

$$(\omega_+)_{\min} = (2\beta/m)^{1/2} \tag{6.3.20}$$

At $k = 0$, the maximum frequency is

$$(\omega_+)_{\max} = \left[2\beta\left(\frac{1}{M} + \frac{1}{m}\right)\right]^{1/2} \tag{6.3.21}$$

Putting the value of ω from (6.3.20) into (6.3.6), we obtain

$$MA + mB = 0 \;\Rightarrow\; A/B = -m/M \tag{6.3.22}$$

Again, putting $ka = \pm\pi/2$ in equation (6.3.7), we immediately obtain

$$\boldsymbol{A/B} = 0 \;\Rightarrow\; A = 0 \tag{6.3.23}$$

This corresponds to the ***optical branch*** of the dispersion curve.

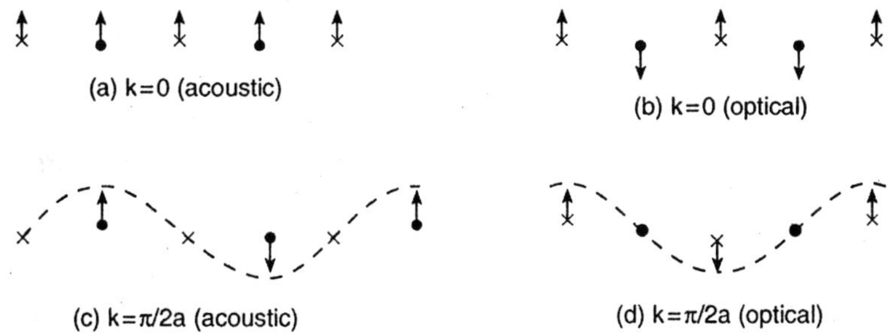

Fig. 6.8 Motion of atoms in acoustic and optical modes for a diatomic linear chain

Summarising, for $k = 0$ in the *acoustic branch*, the two types of atoms vibrate *in phase* (with their centre of mass) as in ordinary elastic wave propagation; in the *optical branch* they vibrate *out of phase* (with the centres of mass fixed). Similarly, at the boundary i.e., for $k = \pi/2a$, only heavy atoms (M) move in the acoustic branch, π out of phase, and in optical branch light atoms (m) move, also π out of phase (Fig. 6.8).

● Acoustic vibrations are exhibited by monatomic crystals. Diatomic crystals show optical vibrations.

6.3.3 Symmetry in k-space : Number of modes in first zone

The diatomic chain, like the monatomic one, also satisfies the symmetry properties in k-space. It has a *reflection symmetry* about $k = 0$ but the periodicity of the dispersion curve has a period $\pi/2a$, unlike monatomic case where it was π/a. So, the range of the first Brillouin zone is $-\pi/2a < k < \pi/2a$. Also, from the periodic boundary conditions, the allowed k-values in the first zone is N, but the total number of modes inside the zone is $2N$, since there are two modes for each k — one *acoustic* and one *optical*. This value is also equal to the number of degrees of freedom in the lattice.

6.3.4 Frequency gap

A critical look at the dispersion curve of Fig. 6.6 shows that for frequencies in the range $0 < \omega < (\omega_+)_{\max}$ at the origin ($k = 0$) and $(\omega_+)_{\min} < \omega < (\omega_-)_{\max}$ at the first zone boundary ($k = \pm\pi/2a$), no solutions exist. The separation between the acoustic and the optical branches is maximum at the origin and minimum at the boundary.

*The gap between the acoustic and the optical branches at the zone boundary is known as the **frequency gap** or **forbidden frequency band**,* as already stated *and this is a characteristic feature of a diatomic lattice.*

● It is possible, however, to get solutions in the region of the gap by introducing what is called an *impurity atom*; for instance, by replacing one of the atoms of mass m by one of mass M. Such solutions go by the name ***gap mode***. Another possibilty to look for the solutions in the gap with real ω will have complex k, implying that the wave is a damped wave in space.

6.4 Vibrational mode as a linear oscillator

It can be easily shown that the energy for a particular mode of vibration for one-dimensional lattice is identical with that of a harmonic oscillator of the same frequency.

The energy of a one-dimensional *harmonic oscillator* with mass M and angular frequency ω is given by

$$E = \frac{p_x^2}{2M} + \tfrac{1}{2}M\omega^2 x^2 \tag{6.4.1}$$

$$= \tfrac{1}{2}M\,(dx/dt)^2 + \tfrac{1}{2}M\omega^2 x^2 \tag{6.4.2}$$

where p_x is its momentum and x is its displacement, counted from the mean position.

The first term in the right-hand side of equation (6.4.2) is the kinetic energy E_{ko} and the second term the potential energy E_{po}. Using $x = A\sin\omega t$, we therefore get

$$\therefore \quad E_{ko} = \tfrac{1}{2}M(dx/dt)^2 = \tfrac{1}{2}MA^2\omega^2\cos^2\omega t \tag{6.4.3}$$

$$E_{po} = \tfrac{1}{2}M\omega^2 x^2 = \tfrac{1}{2}MA\omega^2\sin^2\omega t \tag{6.4.4}$$

Let now a finite ***one-dimensional lattice*** vibrate in a mode corresponding to a standing wave given by

$$x_n(t) = C\sin kna\sin\omega t \tag{6.4.5}$$

So, the K.E. of the atoms in this mode of vibration is given by

$$E_k = \tfrac{1}{2}m\sum_n (dx_n/dt)^2 = \tfrac{1}{2}mC^2\omega^2\cos^2\omega t\sum\sin^2 kna \tag{6.4.6}$$

where m is the mass of each atom.

The net force on the nth atom of the lattice is, as obtained from eqn. (5.2.1), $\beta(x_{n-1} + x_{n+1} - 2x_n)$ so that if E_p be the potential energy of the system in the mode k, we have

$$-\frac{\partial E_p}{\partial x_n} = \beta(x_{n-1} + x_{n+1} - 2x_n)$$

whence $$E_p(x_0, x_1, \cdots, x_n, \cdots) = \tfrac{1}{2}\beta\sum(2x_n^2 - x_n x_{n+1} - x_n x_{n-1}) \tag{6.4.7}$$

Using (6.4.5) in (6.4.7), we obtain

$$\begin{aligned} E_p &= 2C^2\beta\sin^2(ka/2)\sin^2\omega t\sum\sin^2 kna \\ &= \frac{C^2}{2}m\omega^2\sin^2\omega t\sum\sin^2 kna, \quad \text{by (6.2.4).} \end{aligned} \tag{6.4.8}$$

The expression for E_k, given by equation (6.4.6), is identified with E_{ko}, as given by (6.4.3) and E_p, given by equation (6.4.8), is identified with E_{po} as given by (6.4.4), provided, however, we have

$$C^2 m\sum\sin^2 kna = A^2 M \tag{6.4.9}$$

This establishes the *equivalence of the vibrational mode and a harmonic oscillator*.

6.5 Specific heat of solids : Introduction

We know, since our high school days, that the specific heat of a solid is the amount of heat energy required by a mole of the solid to raise its temperature by one degree. When some thermal energy is supplied to a solid, its temperature and internal energy are increased. The increase in internal energy leads to (i) an increase in the vibration of atoms about their mean positions and (ii) an increase in the kinetic energy of free electrons. The specific heat corresponding to (i) is called the *lattice specific heat* and that corresponding to (ii) the *electronic specific heat*. To measure the specific heat, either the volume or the pressure is maintained constant. Usually, it is measured at constant volume and

$$C_V = \left(\frac{dQ}{dT}\right)_V \tag{6.5.1}$$

where Q is the internal energy and T the temperature of the system.

- For *electronic specific heat*, please see Chapter: *Transport phenomena.*

6.5.1 Classical model of specific heat

The kinetic energy of an atom of a monatomic gas at temperature T is, from kinetic theory, $3 \times \frac{1}{2}kT$, where k is the Boltzmann constant and $\frac{1}{2}kT$ is the energy associated with each degree of freedom (dof). If there are N_A (Avogadro number) atoms in a mole of a monatomic solid, the total internal energy is

$$Q = 3 \times 2 \times \tfrac{1}{2}kT \times N_A = 3N_AkT = 3RT, \tag{6.5.2}$$

the factor 2 in (6.5.2) appears since to each translational degree of freedom, there are two associated additional degrees of freedom—the kinetic and the potential.

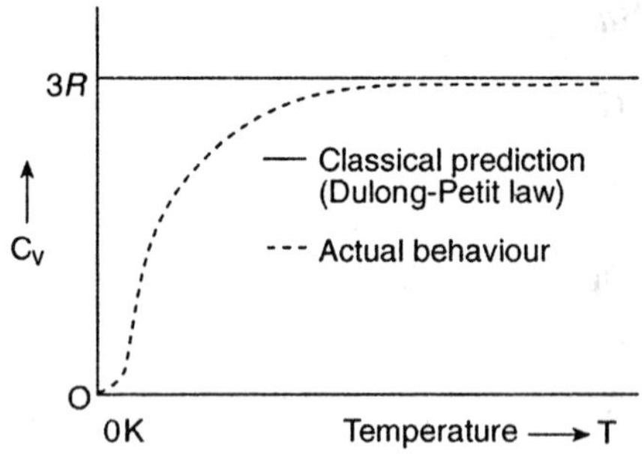

Fig. 6.9 Specific heat of a solid at low temperatures

$$\therefore \text{ Molar specific heat, } C_V = \frac{d}{dT}(3RT) = 3R = 5.76 \text{ cal.mol}^{-1}\text{K}^{-1} \tag{6.5.3}$$

where R ($= N_Ak$) is the universal gas constant.

The above relation is the ***Dulong-Petit's law***, generally correct for elements of atomic weight greater than 40 and at room temperature and above. It however fails under other conditions. For instance,

1. At very low temperatures, the specific heats of all elemental solids decrease and vanish at $T = 0$ (Fig. 6.9). Experimental results show that the specific heat is of the form

$$\boxed{C_V = \alpha T^3 + \gamma T} \tag{6.5.4}$$

where α, γ are constants.

For insulators $\gamma = 0$ and $C_V \propto T^3$. For metals $\gamma \neq 0$ and C_V also contains a linear term in T, being the contribution of the conduction electrons to specific heat.

2. Some low atomic number and high melting point elements (e.g., B, Be, C, Si) show specific heats much lower than that predicted by Dulong's law.

3. In magnetic solids, the specific heat is large near the temperature at which magnetic moments become ordered.

6.6 Einstein's model of specific heat

It was Einstein who first applied the *quantum theory* to account for the variation of specific heat of solids with temperature. The *essential features* of Einstein's model are:

1. A crystalline solid having N atoms, interacting according to Hooke's law, may be regarded as mechanically equivalent to $3N$ one-dimensional *identical* and *independent* harmonic oscillators, each vibrating with the same frequency ν (*monochromatic vibrations*).

2. The oscillators are *quantum oscillators* having *discrete energies* rather than a continuous energy spectrum as in the classical case.

3. The atomic oscillators form an assembly of systems which are *distinguishable* by their locations at distinct and different lattice sites and thus *obey Maxwell-Boltzmann distribution* law of energies.

The quantized energies of the oscillators are given by

$$\varepsilon_n = \left(n + \frac{1}{2}\right) h\nu \simeq nh\nu, \quad n = 0, 1, 2, \cdots \tag{6.6.1}$$

neglecting $\frac{1}{2}h\nu$, the temperature-independent '*zero point energy*' whose inclusion in the energy does not affect the specific heat.

The number of oscillators N_n of each energy state is determined from the Boltzmann function as

$$N_n = N_0 e^{-\varepsilon_n/kT} = N_0 e^{-nh\nu/kT}, \tag{6.6.2}$$

N_0 being the number of oscillators in the zero energy state.

And, the average vibrational energy of an oscillator is given by

$$\bar{\varepsilon} = \frac{\varepsilon}{N} = \frac{\sum_n N_n \varepsilon_n}{\sum_n N_n} = \frac{h\nu}{e^{h\nu/kT} - 1} \tag{6.6.3}$$

For *one mole* of solids, the total energy of the crystal is, taking into account three independent directions,

$$E_{\text{mol}} = 3N_A \bar{\varepsilon} = \frac{3N_A h\nu}{e^{h\nu/kT} - 1}, \quad N = N_A = \text{Avogadro number.}$$

$$\therefore \text{ Molar specific heat, } \boldsymbol{C_V} = \left(\frac{dE}{dT}\right)_V = \boldsymbol{3N_A k \left(\frac{h\nu}{kT}\right)^2 \frac{e^{h\nu/kT}}{(e^{h\nu/kT} - 1)^2}} \tag{6.6.4}$$

Einstein's specific heat curve is fairly close to the experimental curve except at very low temperatures (Fig. 6.10). In the low temperature range, the Einstein's specific heat approaches zero more rapidly than the experimentally observed values. But, at

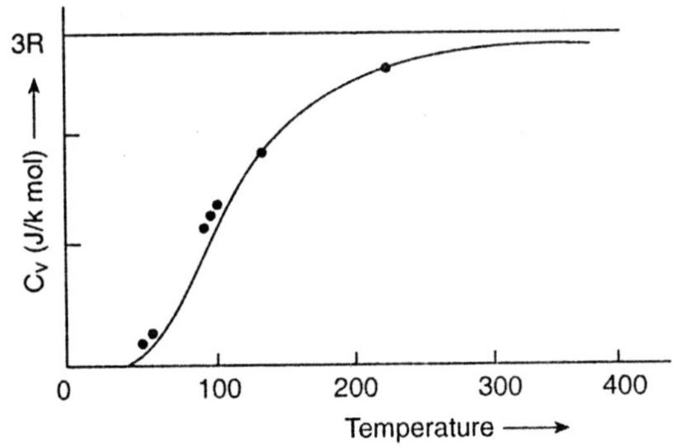

Fig. 6.10 Einstein specific heat curve

high temperatures, it approaches the classical value $3R$. We shall now consider the following *two* limiting cases: (i) ***high temperature range*** and (ii) ***low temperature range*** separately. Before we do so, we give an alternative form to (6.6.4).

Let $\Theta_E = h\nu/k$ where Θ_E, having the unit of temperature, is the *characteristic temperature* known as the ***Einstein temperature*** and the frequency ν, the ***Einstein's frequency***. Introducing Θ_E we get from (6.5.4)

$$C_V = 3N_A k \left(\frac{\Theta_E}{T}\right)^2 \frac{e^{\Theta_E/T}}{(e^{\Theta_E/T}-1)^2} \tag{6.6.5}$$

(i) For high temperatues: For temperatures such that $kT \gg h\nu$ or $T \gg \Theta_E$,

$$e^{h\nu/kT} - 1 = 1 + h\nu/kT + \cdots - 1$$
$$= h\nu/kT$$

From (6.6.3), therefore, $$\bar{\varepsilon} = \frac{h\nu}{h\nu/kT} = kT$$

$$\therefore\; E = 3N_A kT = 3RT$$

$$\therefore\; C_V = \left(\frac{\partial E}{\partial R}\right)_V = 3R, \tag{6.6.6}$$

the same as the classical result—the familiar ***Dulong-Petit's law***.

(ii) For low temperatures: For temperatures such that $h\nu \gg kT$ or $T \ll \Theta_E$, we have $e^{h\nu/kT} \gg 1$.

$$\therefore\; E_{\text{mol}} = 3N_A\bar{\varepsilon} = 3N_A h\nu e^{-h\nu/kT},$$

$$\text{and}\quad C_V = \left(\frac{\partial E}{\partial T}\right)_V = 3N_A h\nu\left[e^{-h\nu/kT}\cdot\frac{h\nu}{kT^2}\right]$$

$$= 3N_A k\left(\frac{h\nu}{kT}\right)^2 e^{-h\nu/kT} \tag{6.6.7}$$

$$= 3N_A k\left(\frac{\Theta_E}{T}\right)^2 e^{-\Theta_E/T}, \tag{6.6.8}$$

in terms of Einstein's temperature Θ_E.

Thus, for $T \ll \Theta_E$, the heat capacity is proportional to $e^{-\Theta_E/T}$ which is more important than the term (Θ_E/T). Thus, with decreasing temperature in the low temperature range, C_V rapidly falls off, almost exponentially. However, experiments show that $C_V \propto T^3$ for most of the solids, i.e., much more slowly.

While the Einstein's model provides a much better explanation for the temperature-variation of specific heat and agrees well with experiments over a wide range of temperaure than the classical theory, it cannot account for, as already stated, the values of specific heat at *very low temperatures*. The discrepancy is due to the oversimplified assumption that the atomic oscillators vibrate *independently* and with the *same frequency*. In fact, the oscillators are *coupled* together and a *number* of vibrational *frequencies*, rather than a single one, is possible. This is accounted for in Debye's model which we shall now describe and discuss.

6.7 Debye's model of lattice heat capacity

In this model, the vibration of the crystal as a whole is considered equivalent to the vibrational motion of a system of coupled harmonic oscillators (as atoms are strongly bound by atomic forces). The coupled harmonic oscillators can generate a wide range of frequencies rather than one. Debye proposed that crystals can propagate elastic waves of low frequencies (sound waves) to high frequencies. There can thus be a number of modes of vibration. The number of vibrational modes per unit frequency range is called the ***density of modes*** $g(\nu)$.

The possible number of modes of elastic standing waves of any kind with frequencies lying between ν and $\nu + d\nu$ is given by

$$g(\nu)d\nu = \left(\frac{4\pi V}{v_s^3}\right)\nu^2 d\nu \tag{6.7.1}$$

where V is the volume of a mole of the solid cube, and v_s is the wave speed.

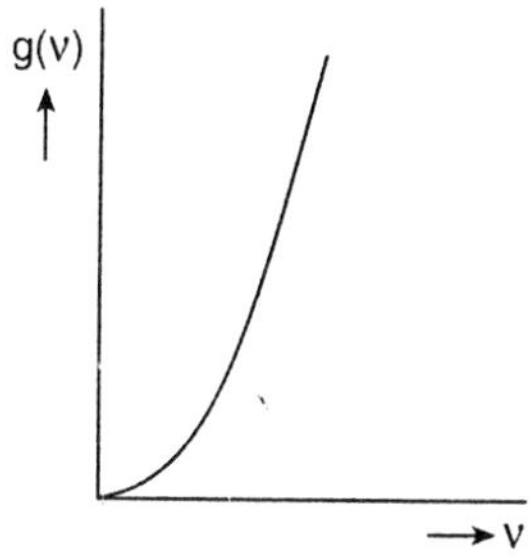

Fig. 6.11 Frequency spectrum

In general, the elastic waves propagating in a solid are of *two* types—*longitudinal* of speed v_l and *transverse* of speed v_t. Also, there are two perpendicular directions of polarisation for a given transverse wave.

∴ The total number of vibrational modes of frequencies lying between ν and $\nu + d\nu$ is given by

$$g(\nu)d\nu = 4\pi V\left(\frac{1}{v_l^3} + \frac{2}{v_t^3}\right)\nu^2 d\nu \tag{6.7.2}$$

• A curve between the density of states $g(\nu)$ and ν (Fig. 6.11) shows that $g(\nu)$ increases as ν^2. The function $g(\nu)$ is also called the ***spectral distribution function.***

Now, the average energy $\bar{\varepsilon}$ of an oscillator of frequency ν at temperature T is

$$\bar{\varepsilon} = \frac{h\nu}{e^{h\nu/kT} - 1} \tag{6.7.3}$$

Associating a harmonic oscillator of the same frequency with each vibrational mode, the vibrational energy of the crystal is

$$E = \int_0^{\nu_D} 4\pi V \left(\frac{1}{v_l^3} + \frac{2}{v_t^3} \right) \frac{h\nu^3}{e^{h\nu/kT} - 1} d\nu \qquad (6.7.4)$$

Here the upper limit of integration is some *maximum frequency*, called the *Debye frequency* ν_D or ***Debye cut-off frequency.*** There cannot thus be an infinite number of standing waves. Debye *assumed* that if there are N number of atoms in a solid, the total number of vibrational modes is limited to $3N$. So, using (6.7.2),

$$\therefore \int_0^{\nu_D} 4\pi V \left(\frac{1}{v_l^3} + \frac{2}{v_t^3} \right) \nu^2 d\nu = 3N$$

$$\text{or,} \quad 4\pi V \left(\frac{1}{\nu_l^3} + \frac{2}{v_t^3} \right) \frac{\nu_D^3}{3} = 3N$$

$$\text{or,} \quad \nu_D^3 = \frac{9N}{4\pi V} \left(\frac{1}{v_l^3} + \frac{2}{v_t^3} \right)^{-1} \qquad (6.7.5)$$

which can be used to determine the Debye frequency, ν_D. It is interesting to note the frequency spectrum according to Debye and Einstein models, as illustrated in Fig. 6.12.

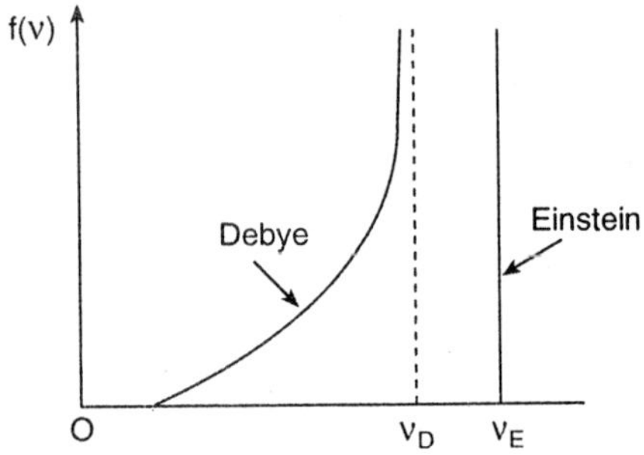

Fig. 6.12 Frequency spectrum according to Debye model and Einstein model

From (6.7.4), using (6.7.5), we obtain

$$E = \frac{9Nh}{v_D^3} \int_0^{\nu_D} \frac{\nu^3 d\nu}{e^{h\nu/kT} - 1}, \quad \text{in terms of } \nu_D. \qquad (6.7.6)$$

Let $h\nu/kT = x$ so that $\nu = xkT/h \Rightarrow d\nu = (kTdx)/h$. Also, let $h\nu_D/kT = x_D$.

$$\therefore \ E = \frac{9Nh}{\nu_D^3} \left(\frac{kT}{h} \right)^4 \int_0^{x_D} \frac{x^3 dx}{e^x - 1} = 9N \left(\frac{kT}{h\nu_D} \right)^3 kT \int_0^{x_D} \frac{x^3 dx}{e^x - 1} \qquad (6.7.7)$$

Let us define, as in Einstein's theory, the ***Debye*** (characteristic) ***temperature***

$$\Theta_D = \frac{h\nu_D}{k} \Rightarrow x_D = \frac{\Theta_D}{T}.$$ So, we have from (6.7.7),

$$E = 9NkT \left(\frac{T}{\Theta_D} \right)^3 \int_0^{\Theta_D/T} \frac{x^3 dx}{e^x - 1} \qquad (6.7.8)$$

which does not depend explicitly on the volume.

$$\therefore \text{ Molar specific heat, } C_V = \left(\frac{\partial E}{\partial T} \right)_V = 9N_A k \left(\frac{T}{\Theta_D} \right)^3 \int_0^{\Theta_D/T} \frac{x^4 e^x}{(e^x - 1)^2} dx$$

$$= 3R \left(\frac{\Theta_D}{T} \right) F_D \qquad (6.7.9)$$

where the function F_D is called the ***Debye function*** and is expressed as

$$F_D = 3\left(\frac{T}{\Theta_D}\right)^4 \int_0^{\Theta_D/T} \frac{x^4 e^x}{(e^x-1)^2}dx \tag{6.7.10}$$

Let us now consider the following two *limiting cases.*

(i) High temperature case: For $T \gg \Theta_D$, $x\ (= \Theta_D/T)$ is small compared to unity for the entire range of integration. So, we have $e^x - 1 \simeq x$. The equation (6.7.8) reduces to

$$E = 9NkT\left(\frac{T}{\Theta_D}\right)^3 \int_0^{\Theta_D/T} x^2 dx = 3NkT$$

$$\therefore\quad C_V = \left(\frac{\partial E}{\partial T}\right)_V = 3Nk = 3R$$

in complete agreement with *Dulong and Petit's law.* It means, in fact, that quantum consideration has hardly any significance at high temperatures for when $T \gg \Theta_D$, every mode of oscillation is completely excited and has an average energy $\bar{\varepsilon} = kT$, the classical value.

(ii) Low temperature case: For $T \ll \Theta_D$, $x_D = \Theta_D/T \to \infty$ and the equation (6.7.8) reduces to the form

$$E = 9NkT\left(\frac{T}{\Theta_D}\right)^3 \int_0^{\infty} \frac{x^3 dx}{e^x - 1}$$

$$= 9NkT\left(\frac{T}{\Theta_D}\right)^3 \times \frac{\pi^4}{15} \quad \left[\because \int_0^{\infty} \frac{x^3 dx}{e^x-1} = \frac{\pi^4}{15}\right]$$

$$= \frac{3}{5}\pi^4 Nk\frac{T^4}{\Theta_D^3} \tag{6.7.11}$$

which shows that the *vibrational energy is proportional to T^4*, analogous to Stefan's law of black body radiation. This implies that *both phonons and photons obey the same statistics* with the difference that while photons obey T^4-law at all temperatures, phonons do so only at low temperatures.

$$\therefore\ \text{Molar specific heat,}\quad C_V = \left(\frac{\partial E}{\partial T}\right)_V = \frac{12}{5}\pi^4 N_A k\left(\frac{T}{\Theta_D}\right)^3$$

$$= \frac{12}{5}\pi^4 R\left(\frac{T}{\Theta_D}\right)^3 \tag{6.7.12}$$

Thus, *at very low temperatures*, $\boldsymbol{C_V \propto T^3}$ and this is the famous expression known as ***Debye's T^3-law*** for specific heat. According to this law, $C_V \to 0$ as $T \to 0$ (the law holds[1] for $T \le \Theta_D/10$). There is an *excellent agreement* between the experimental values of C_V at various temperatures with those calculated from Debye's model (Fig. 6.13) for Al and Cu validating the Debye-approximation at sufficiently low

[1]Can you work it out? See Illustrated Example 19.

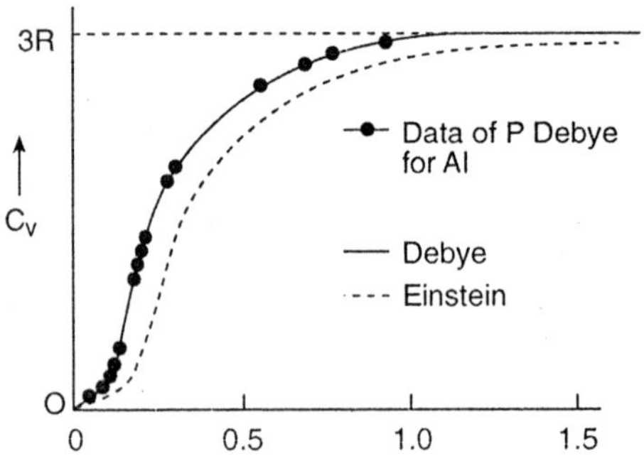

Fig. 6.13 Various specific heat curves of a solid

temperatures. Fig. 6.13 gives a typical Debye plot along with measured data, Einstein's plot and the classical value.

(iii) Intermediate temperature case: Debye's model agrees remarkably well at the two extremes of temperatures, i.e., in the range $T \leq \Theta_D/50$ and $T \geq 0.2\Theta_D$.

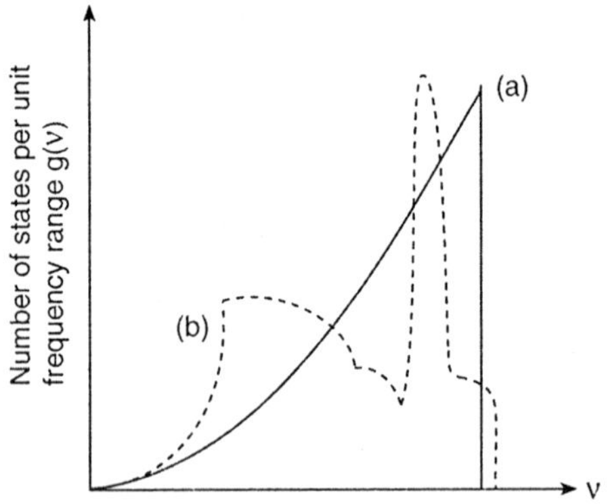

Fig. 6.14 Frequency spectrum (a) according to Debye model, (b) behaviour as expected from a real crystal

But at temperatures intermediate between the above two, where a gradual transition of T^3-law to the classical one occurs, the agreement is *not* satisfactory.

This is because a solid *cannot* be treated as a continuum at frequencies near ν_D as assumed by Debye. In fact, this frequency corresponds to a wavelength of atomic spacing. The frequency spectrum of Debye model is different from that expected from a real crystal, as illustrated in Fig. 6.14. But, interestingly, the area under the two curves are equal. To improve Debye's model, the correct dispersion relation and the corresponding density of states are to be used.

We conclude the section by summarising the different limitations of Debye's model.

Limitations of Debye model—The Debye-model of specific heat of solids however suffers from the following limitations.

1. The Debye's continuum model is *valid for long-wavelengths* (and hence at low frequencies) only.

2. The total number of vibrational modes are assumed to be $3N$, but a solid being an elastic continuum should have *infinite frequencies.*

3. The cut-off frequency is assumed to be the same for both the longitudinal and the transverse waves, which have different velocities and should thus have *different values of cut-off frequency.*

4. According to Debye's theory, Θ_D is independent of temperature but, in fact, it *varies with temperature* by 10% or even more.

5. The Debye model does not take into account the *crystalline nature* of the solid and cannot therefore be applied to crystals which are comprised of *more than one* type of atoms.

6. The model completely *ignores the interaction* among the atoms and the *contribution of electrons* to the specific heat.

6.7.1 Born's modification of Debye's theory

A modification of Debye's theory was suggested by Max Born. In Debye's theory, the cut-off frequency ν_m of vibration of oscillators for both transverse and longitudinal modes is taken to be the same: $\nu_{lm} = \nu_{tm} = \nu_m$ (Fig. 6.15.).

Born asserted that the cut-off for the highest overtone is not limited by an identical ν_m, unlike Debye, but by an identical minimum wavelength λ_{min} for both types of vibration.

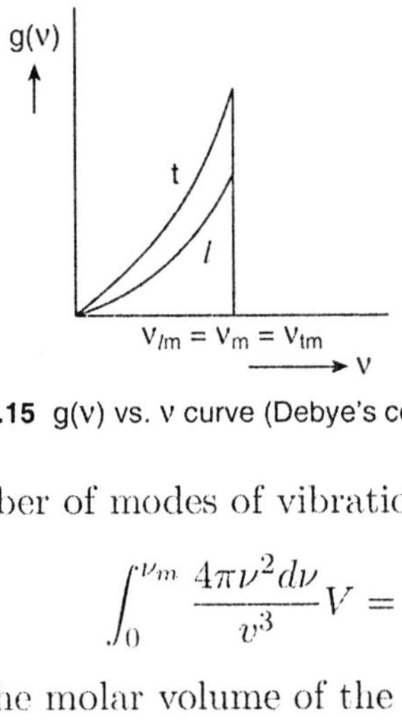

Fig. 6.15 g(ν) vs. ν curve (Debye's concept)

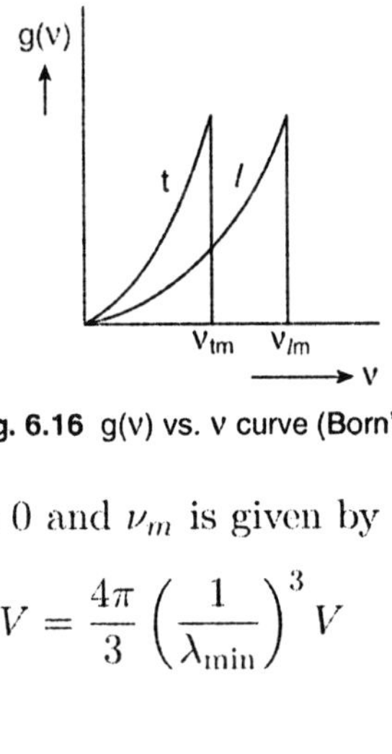

Fig. 6.16 g(ν) vs. ν curve (Born's concept)

The number of modes of vibrations between 0 and ν_m is given by

$$\int_0^{\nu_m} \frac{4\pi\nu^2 d\nu}{v^3} V = \frac{4\pi}{3}\left(\frac{\nu_m}{v}\right)^3 V = \frac{4\pi}{3}\left(\frac{1}{\lambda_{\min}}\right)^3 V$$

where V is the molar volume of the crystal.

Since for a given mode, there are *three* different vibrations—one longitudinal and two transverse.

$$4\pi V\left(\frac{1}{\lambda_{\min}}\right)^3 = 3N_A \Rightarrow \lambda_{\min} = \left(\frac{4\pi V}{3N_A}\right)^{1/3} \tag{6.7.13}$$

Thus, we have two Debye cut-off frequencies, ν_{lm} and ν_{tm} (Fig. 6.16), one for the longitudinal mode and the other for the transverse, given respectively by

$$\nu_{lm} = \frac{v_l}{\lambda_{\min}} = v_l \left(\frac{3N_A}{4\pi V}\right)^{1/3}; \quad \nu_{tm} = \frac{v_t}{\lambda_{\min}} = v_t \left(\frac{3N_A}{4\pi V}\right)^{1/3} \tag{6.7.14}$$

However, the total number of vibrational modes $= 3N_A$, as in Debye's.

$$\therefore \quad 4\pi V \left[\int_0^{\nu_{lm}} \frac{\nu^2}{v_l^3} d\nu + \int_0^{\nu_{tm}} \frac{2}{v_t^3} \nu^2 d\nu\right] = 3N_A \tag{6.7.15}$$

$\therefore$ Total energy per mole of the solid, E is given by

$$E = 4\pi V \left[\int_0^{\nu_{lm}} \frac{\nu^2}{v_l^3} \frac{h\nu}{(e^{h\nu/kT} - 1)} d\nu + \int_0^{\nu_{tm}} \frac{2\nu^2}{v_t^3} \cdot \frac{h\nu}{e^{h\nu/kT} - 1} d\nu\right] \tag{6.7.16}$$

Proceeding as in Debye, we shall finally get

$$C_V = R\left[D(\Theta_{lD}/T) + 2D(\Theta_{tD}/T)\right] \tag{6.7.17}$$

where Θ_{lD}/T and Θ_{tD}/T are the *Debye temperatures* corresponding to the *longitudinal* and the *transverse modes* respectively.

Discussion—Eq. (6.7.17), due to Born, is also *not* free from objections and is *acceptable only for isotropic solids.* Subsequently, further improvements in the theory of specific heat of solids were made by Born and Karman, which was extended later by Blackman. Discussion of the details of these theories is beyond the scope of the book.

6.8 Illustrated Examples

► **Example 1.** *The velocity of sound in a solid is* 3×10^3 m/s *and the interatomic separation is* 3×10^{-10} m. *Calculate the value of the cut-off frequency assuming a linear lattice.*

Solution. By the problem, $v = 3 \times 10^3$ m/s, $a = 3 \times 10^{-10}$ m, $\nu = ?$

The cut-off frequency occurs at $k = \pi/a$, that is, $\lambda \simeq 2a$.

Now, the velocity and the frequency are related as

$$v = \nu\lambda = 2\nu a$$

$$\therefore \quad \nu = \frac{v}{2a} = \frac{3 \times 10^3}{2 \times 3 \times 10^{-10}} = \mathbf{5 \times 10^{12}\ Hz}$$

► **Example 2.** *Show that for a one-dimensional monatomic lattice, the dispersion curve meets the zone boundary in a normal direction.*

Solution. The dispersion curve is given by

$$\omega = 2\left(\frac{\beta}{m}\right)^{\frac{1}{2}} \sin\left(\frac{ka}{2}\right)$$

$$\therefore \quad \frac{d\omega}{dk} = 2\left(\frac{\beta}{m}\right)^{\frac{1}{2}} \cdot \frac{a}{2} \cos\left(\frac{ka}{2}\right)$$

The zone boundary is defined by $k = \pm\pi/a$, for which

$$\cos\left(\frac{ka}{2}\right) = \cos\left(\pm\frac{\pi}{2}\right) = 0$$

Thus, $d\omega/dk = 0$ at zone boundary which implies the ***dispersion curve*** is *normal* to the zone boundary.

▶ **Example 3.** *The unit cell parameter of* NaCl *crystal is* 5.6A *and the Yong's modulus of elasticity along the direction* [100] *is* $5 \times 10^{10}\,\mathrm{N/m^2}$. *Find the wavelength at which an electromagnetic radiation is strongly reflected by the crystal. Atomic weight of* Na $= 23\,\mathrm{u}$ *and of* Cl $= 37\,\mathrm{u}$ *and* $1\,\mathrm{u} = 1.67 \times 10^{-27}\,\mathrm{kg}$.

Solution. By the problem, $2a = 5.6\,\mathrm{A}$, $Y = 5 \times 10^{10}\,\mathrm{N/m^2}$, atomic weight of Na = 23, atomic weight of Cl = 37.

The maximum frequency of radiation (optical range) that gets strongly reflected by the NaCl crystal is given by

$$(\omega_+)_{\max} = \left[2\beta\left(\frac{1}{M} + \frac{1}{m}\right)\right]^{1/2}$$

Assuming that an extension along [100] direction will have hardly any effect on the vertical springs, we may use

$$\beta = aY.$$

$$\therefore\ (\omega_+)_{\max} = \left[2aY\left(\frac{1}{M} + \frac{1}{m}\right)\right]^{1/2}$$

$$= \left[\frac{5.6 \times 10^{-10} \times 5 \times 10^{10}}{1.67 \times 10^{-27}}\left(\frac{1}{37} + \frac{1}{23}\right)\right]^{1/2}$$

$$= 3.44 \times 10^{13}\ \mathrm{rad/sec.}$$

∴ Required wavelength of the radiation for strong reflection is given by

$$\lambda = \frac{c}{\nu} = \frac{2\pi c}{\omega} = \frac{2\pi \times 3 \times 10^8}{3.44 \times 10^{13}} = \mathbf{5.48 \times 10^{-5}}\ \mathrm{m}$$

▶ **Example 4.** *Prove that the gradient of the optical branch of a dispersion curve at the maximum frequency is zero.*

Solution. The dispersion relation corresponding to the optical branch is given by

$$\omega_+^2 = \beta\left(\frac{1}{M} + \frac{1}{m}\right) + \beta\left[\left(\frac{1}{M} + \frac{1}{m}\right)^2 - \frac{4\sin^2 ka}{Mm}\right]^{1/2}$$

At $k = 0$, the freqeuncy is maximum and is, from above,

$$(\omega_+)_{\max} = \left[2\beta\left(\frac{1}{M} + \frac{1}{m}\right)\right]^{1/2} = \text{constant}$$

$$\therefore\ \text{Required gradient} = \frac{d}{dk}(\omega_+)_{\max} = \mathbf{0}$$

▶ **Example 5.** *Prove that in a one-dimensional diatomic lattice, the two kinds of atom oscillate with amplitudes related to each other by*

$$B = A\left(1 - \frac{M\omega^2}{2\beta}\right)\sec ka.$$

Solution. We have, from equation (6.3.5),

$$\frac{B}{A} = \frac{2\beta - M\omega^2}{2\beta\cos ka}$$

$$\Rightarrow \quad \frac{B}{A} = \frac{2\beta - M\omega^2}{2\beta}\sec ka$$

$$= \left(1 - \frac{M\omega^2}{2\beta}\right)\sec ka$$

$$\therefore \quad \mathbf{B = A\left(1 - \frac{M\omega^2}{2\beta}\right)\sec ka}$$

▶ **Example 6.** *In a one-dimensional lattice, $x = m/M \ll 1$. Prove that the squares of the widths of the optical and ocoustic branch are in the ratio $x : 4$.*

Solution. Width of the optical branch is

$$W_{op} = (\omega_+)_{\max} - (\omega_+)_{\min} = \left[2\beta\left(\frac{1}{M} + \frac{1}{m}\right)\right]^{\frac{1}{2}} - \left(\frac{2\beta}{m}\right)^{\frac{1}{2}}$$

$$= \left(\frac{2\beta}{m}\right)^{\frac{1}{2}}\left[1 + \frac{m}{M}\right]^{\frac{1}{2}} - \left(\frac{2\beta}{m}\right)^{\frac{1}{2}}$$

$$= \left(\frac{2\beta}{m}\right)^{\frac{1}{2}}\left(1 + \frac{m}{2M}\right) - \left(\frac{2\beta}{m}\right)^{\frac{1}{2}}, \quad \left[\because \left(1 + \frac{m}{M}\right)^{\frac{1}{2}} = 1 + \frac{m}{2M}, \quad m \ll M\right]$$

$$= \left(\frac{2\beta}{m}\right)^{\frac{1}{2}} \cdot \frac{m}{2M}$$

Width of acoustic branch, $W_{ac} = (\omega_-)_{\max} = \left(\dfrac{2\beta}{M}\right)^{\frac{1}{2}}$.

$$\therefore \quad \frac{W_{op}}{W_{ac}} = \left(\frac{2\beta}{m}\right)^{\frac{1}{2}} \cdot \frac{m}{2M}\left(\frac{M}{2\beta}\right)^{\frac{1}{2}} = \frac{m^{\frac{1}{2}}}{2M^{\frac{1}{2}}} \Rightarrow \frac{W_{op}^2}{W_{ac}^2} = \frac{1}{4}\left(\frac{m}{M}\right) = \frac{x}{4} \quad (\because x = m/M)$$

$$\therefore \quad \mathbf{\frac{W_{op}^2}{W_{ac}^2} = \frac{x}{4}}. \quad \text{Hence.}$$

▶ **Example 7.** *The visible light of wavelength* 5000 A *is scattered from a crystal having refractive index* 1.5. *Find the maximum frequency of the phonon generated. Also, compute the fractional change in frequency of the incident radiation. Given: the velocity of sound in the crystal is* 5000 m/s.

Solution. Freqency ω_p of phonon generated is given by

$$\omega_p = \frac{2v_s\omega n}{c}\sin\frac{\phi}{2}$$

where v_s = velocity of sound in the crystal, ω = frequency of incident radiation, n = refractive index, c = velocity of light in free space and ϕ = angle of scattering of photon from the direction of incidence.

Here, $n = 1.5$, $v_s = 5000$ m/s, $c = 3 \times 10^8$ m/s, $\lambda = 5000$ A $= 5000 \times 10^{-10}$ m.

$\therefore$ Frequency of radiation, $\omega = 2\pi c/\lambda = \dfrac{2\pi \times 3 \times 10^8}{5000 \times 10^{-10}} = 3.77 \times 10^5$ rad/s.

For ω_p to be maximum, $\sin\phi/2 = 1$. Therefore, we have

$$(\omega_p)_{\max} = \frac{2v_s\omega n}{c} = \frac{2 \times 5000 \times 3.77 \times 10^{15} \times 1.5}{3 \times 10^8}$$
$$= \mathbf{1.86 \times 10^{11}\ rad/s}$$

If ω' be the frequency of the scattered radiation, then we have

$$\omega - \omega' = \omega_p$$

$\therefore$ ***Fractional change*** in frequency of incident radiation is

$$\frac{\omega - \omega'}{\omega} = \frac{\omega_p}{\omega} = \frac{1.86 \times 10^{11}}{3.77 \times 10^{15}} = \mathbf{5 \times 10^{-5}}$$

► **Example 8.** *Calculate the Einstein's frequency for C_V for which the Einstein's temperature is* 230 K.

Solution. Einstein's temperature Θ_E is given by

$$\Theta_E = \frac{h\nu}{k} \Rightarrow \nu = \frac{k}{h}\Theta_E = \frac{1.38 \times 10^{-23}}{6.6 \times 10^{-34}} \times 230$$
$$= \mathbf{4.81 \times 10^{12}\ Hz}$$

► **Example 9.** *In aluminium, the transverse and longitudinal waves propagate with velocities* 3111 m/s *and* 6374 m/s *respectively. Assuming that the number of atoms per* m^3 *of aluminium is* 6.02×10^{28}, *find the Debye frequency and Debye characteristic temperature for aluminium.*

Solution. Debye frequency ν_m is given by

$$\nu_m^3 = \frac{9N}{4\pi V\left(\dfrac{1}{c_l^3} + \dfrac{2}{c_t^3}\right)} = \frac{9 \times 6.02 \times 10^{28}}{4 \times 3.14\left[\dfrac{1}{(6374)^3} + \dfrac{2}{(3111)^3}\right]}$$

$$= 614 \times 10^{36} \Rightarrow \nu_m = \mathbf{8.5 \times 10^{12}\ Hz}$$

Debye temperature, Θ_D is given by $\Theta_D = h\nu_m/k$.

$$\therefore \quad \boldsymbol{\Theta_D} = \frac{6.6 \times 10^{-34} \times 8.5 \times 10^{12}}{1.38 \times 10^{-23}} = \mathbf{406.5\,K}$$

► **Example 10.** *For aluminium, the Debye temperature* Θ_D *is* 400 K. *Calculate its specific heat at constant volume at a temperature* 40 K. *Assume* $R = 2$ calorie.

Solution. Since 40 K is quite low a temperature, we may apply Debye's T^3-law.

$$\begin{aligned}\therefore \quad \boldsymbol{C_V} &= 77.9 \times 3R(T/\Theta_D)^3 \\ &= 77.9 \times 3 \times 2 \times (40/400)^3 \\ &= \mathbf{0.467\ cal.}\end{aligned}$$

► **Example 11.** *Estimate the Debye temperature of gold if its atomic weight is* 197, *the density is* 1.9×10^4 kg/m^3 *and the velocity of sound in it is* 2100 m/s.

Solution. At. wt. of Au = 197, density $\rho = 1.9 \times 10^4$ kg/m^3, $v_s = 2100$ m/s, $\theta_D = ?$

$$\therefore \quad \text{Volume, } V = \frac{M}{\rho} = \frac{197}{1.9 \times 10^4} = 103.68 \times 10^{-4}\,\text{m}^3$$

$$\begin{aligned}\text{The Debye temperature, } \Theta_D &= \frac{hv_s}{k}\left(\frac{9N_A}{12\pi V}\right)^{1/3} \\ &= \frac{6.62 \times 10^{-34} \times 2100}{1.38 \times 10^{-23}}\left(\frac{9 \times 6.02 \times 10^{26}}{12\pi \times 103.68 \times 10^{-4}}\right)^{1/3} \\ &= \mathbf{240\,K}\end{aligned}$$

► **Example 12.** *Calculate the value of* C_V *for silver in* calorie per *g*-atom *at* 100 K *from Einstein's equation. Given* $h\nu/k$ *for silver* = 157. **(Calcutta Hons.)**

Solution. From Einstein's relation, the specific heat

$$C_V = 3R \cdot \frac{x^2 e^x}{(e^x - 1)^2}, \quad \text{where } x = h\nu/kT.$$

Here, $T = 100$ K, $h\nu/k = 157$. So, $x = h\nu/kT = 157/100 = 1.57$

$$\begin{aligned}\therefore \quad \boldsymbol{C_V} &= \frac{3 \times 1.99 \times (1.57)^2 \times e^{1.57}}{(e^{1.57} - 1)^2} \\ &= \mathbf{4.89\ cal/g\text{-}atom}\end{aligned}$$

► **Example 13.** *Gold and copper have the same structure. While the velocity of sound in gold is* 2100 m/s, *that in copper is* 3800 m/s. *If the Debye temperature of gold is* 170 K, *find that of copper from the following data:*

Density of Au = 1.93×10^4 kg/m^3, *density of* Cu = 8.96×10^3 kg/m^3, *Atomic weight of* Au = 197 amu, *atomic weight of* Cu = 63.54 amu.

Solution. Debye characteristic frequency ν_m is given by

$$\nu_m^3 = \frac{9N_A}{4\pi V\left(\frac{1}{v_l^3}+\frac{2}{v_t^3}\right)} = \frac{9N_A}{4\pi\frac{M}{\rho}\left(\frac{1}{v_l^3}+\frac{2}{v_t^3}\right)} = v^3\left(\frac{3N_A\rho}{4\pi M}\right)$$

substituting $v_l = v_t = v$ (mean velocity of sound) in the above equation.

Here, V = volume of 1 g-atom, $= M/\rho$, where M = atomic weight and ρ the density, N_A = Avogadro number.

$$\therefore \quad \nu_m = v\left(\frac{3N_A\rho}{4\pi M}\right)^{1/3}$$

Now, Debye temperaturer, $\Theta_D = h\nu_m/k$, h = Planck's constant, k = Boltzmann constant.

$$\therefore \quad \frac{\Theta_D(\text{Cu})}{\Theta_D(\text{Au})} = \frac{\nu_m(\text{Cu})}{\nu_m(\text{Au})} = \frac{v(\text{Cu})}{v(\text{Au})}\left\{\frac{\rho(\text{Cu})\cdot M(\text{Au})}{\rho(\text{Au})\cdot M(\text{Cu})}\right\}^{1/3}$$

$$= \frac{3800}{2100}\left[\frac{8.96\times10^3}{1.93\times10^4}\cdot\frac{197}{63.54}\right]^{1/3} = 2.043$$

$$\therefore \quad \mathbf{\Theta_D(Cu)} = \Theta_D(\text{Au})\times 2.043 = 170\times 2.043 \simeq \mathbf{347\,K}$$

▶ **Example 14.** *The Debye temperature for diamond is 2230 K. Find the highest possible vibrational frequency and the molar heat capacity of diamond at 10 K.*

Solution. By the problem, $\Theta_D = 2230\,\text{K}$, $T = 10\,\text{K}$, $\nu_D =?$. Now, from the relation $h\nu_D = k\Theta_D$, we obtain

$$\nu_D = \frac{k\Theta_D}{h} = \frac{1.38\times10^{-23}\times2230}{6.62\times10^{-34}} = \mathbf{4.64\times10^{13}\,s^{-1}}$$

Also, from the relation, $C_V = \frac{12}{5}\pi^4 R(T/\Theta_D)^3$, we have

$$C_V = \frac{12\times\pi^4\times8.31\times10^3}{5}\left(\frac{10}{2230}\right)^3 = \mathbf{0.1751\,k\,mol^{-1}K^{-1}}$$

▶ **Example 15.** *Show that the zero point energy of a solid according to Debye model of specific heat is given by $\frac{9}{8}R\Theta_D$.* **(Burdwan Hons.)**

Solution. The average energy of an oscillators is given by

$$\bar{\varepsilon} = \frac{1}{2}h\nu + \frac{h\nu}{e^{h\nu/kT}-1}$$

where $\frac{1}{2}h\nu$, the first term, is the zero point energy.

The energy expression, by Debye model, is

$$\bar{E} = \int \bar{E}(\nu)g(\nu)d\nu = \int \tfrac{1}{2}h\nu \cdot 4\pi V\left(\frac{1}{v_l^3}+\frac{2}{v_t^3}\right)\nu^2 d\nu$$

$$= \frac{9N}{2\nu_D^3}\int_0^{\nu_D} h\nu^3 d\nu \quad \left(\because 4\pi V\left(\frac{1}{v_l^3}+\frac{2}{v_t^3}\right) = \frac{9N}{\nu_D^3}\right)$$

$$= \frac{9Nh}{2\nu_D^3}\cdot\frac{\nu_D^4}{4} = \frac{9}{8}Nh\nu_D = \frac{9}{8}Nh\frac{k\Theta_D}{h} \quad (\because \text{Debye temp. } \Theta_D = h\nu_D/k)$$

$$= \frac{9}{8}Nk\Theta_D = \mathbf{\frac{9}{8}R\Theta_D} \quad (\because Nk = R, \text{ the gas constant.})$$

▶ **Example 16.** NaCl *and* KCl *have the same structure. If the lattice heat capacity of* NaCl *at* 5 K *is* 1.6×10^{-2} J mol^{-1} K^{-1}, *estimate the heat capacity of* KCl *at* 5 K *and* 3 K. *Given: Debye temperatures of* NaCl *and* KCl *are* 281 K *and* 230 K *respectively.*

Solution. Here, θ_D (NaCl) = 281 K, θ_D (KCl) = 230 K, C_V (NaCl) at 5 K = 1.6×10^{-2} J mol^{-1} K^{-1}.

Now, Debye specific heat, $C_V = A(T/\theta_D)^3$, where A = constant.

$\therefore$ For NaCl, $1.6 \times 10^{-2} = A(5/281)^3$; for KCl, $C_V = A(5/230)^3$

$$\therefore \frac{1.6 \times 10^{-2}}{C_V(\text{KCl})} = \frac{A(5/281)^3}{A(5/230)^3} = \frac{(230)^3}{(281)^3}$$

$\therefore$ $\mathbf{C_V\ (KCl)_{5\,K}} = 1.6 \times 10^{-2} \times (281/230)^3 = \mathbf{2.91 \times 10^{-2}\ J\ mol^{-1}\ K^{-1}}$

$$\therefore\ 2.91 \times 10^{-2} = A(5/230)^3$$

$$\therefore\ C_V(\text{KCl})_{3\,\text{K}} = A(3/230)^3$$

$$\therefore \frac{C_V(\text{KCl})_{3\,\text{K}}}{2.9 \times 10^{-2}} = \frac{A(3/230)^3}{A(5/230)^5} = \left(\frac{3}{5}\right)^3 = \frac{27}{125}$$

$\therefore$ $\mathbf{C_V\ (KCl)_{3\,K}} = 2.9 \times 10^{-2} \times 27/125 = \mathbf{6.28 \times 10^{-3}\ J\ mol^{-1}\ K^{-1}}$

▶ **Example 17.** *At very low temperatures, the specific heat* C_V *of rock salt varies with temperature* T *according to*

$$C_V = \tfrac{12}{5}\pi^4 Nk(T/\theta_D)^3$$

where θ_D *for rock salt is* 281 K. *How much heat is needed to raise the temperature of* 2 k.mol *of rock salt from* 10 *to* 50 K?

Solution. Here, $\theta_D = 281$ K, $m = 2$ k.mol, $T_1 = 10$ K, $T_2 = 50$ K, $Q = ?$

From Debye's T^3-law, $\qquad C_V = \tfrac{12}{5}\pi^4 Nk(T/\theta_D)^3$

Since the specific heat varies with temperature, we take different ranges of temperatures: 10–20, 20–30, 30–40, 40–50, each at 10 K interval. So, the mean temperatures are 15, 25, 35, 45 K.

$\therefore$ Required amount of heat

$$Q = 2 \times 10^3 \times 10\,(C_V \text{ at } 15\,\text{K} + C_V \text{ at } 25\,\text{K} + C_V \text{ at } 35\,\text{K} + C_V \text{ at } 45\,\text{K})$$

$$= \frac{2 \times 10^4 \times 12\,\pi^4 Nk}{5\theta_D^3}\,(15^3 + 25^3 + 35^3 + 45^3)$$

$$= \frac{2 \times 10^4 \times 12\,\pi^4 R}{5\theta_D^3}\,(3375 + 15625 + 42875 + 91125)$$

$$= \frac{2 \times 10^4 \times 12 \times \pi^4 R}{5 \times 281^3} \times 153{,}000$$

$$= \mathbf{2.68 \times 10^5\,J}, \text{ substituting the values of } \pi \text{ and } R \text{ and simplifying.}$$

▶ **Example 18.** *If Einstein's temperature of a material is* 157 K, *find the value of* C_V *for the material at* 100 K *in* cal/mol/K *using Einstein's formula. Also, calculate Einstein's frequency.* **(North Bengal Hons.)**

Solution. From the relation (6.6.5), the specific heat

$$C_V = 3R\frac{x^2 e^x}{(e^x - 1)^2} \quad \text{where } x = \Theta_E/T \text{ and } R = N_A/k = \text{can const.}$$

By the problem, $\Theta_E = 157\,\text{K}$ and $T = 100\,\text{K}$. So, $x = 157/100 = 1.57$.

$$\therefore\ \boldsymbol{C_V} = \frac{3 \times 2 \times (1.57)^2 e^{1.57}}{(e^{1.57} - 1)^2} \quad (\because\ R \simeq 2\ \text{cal/mol/K})$$

$$= \mathbf{4.89\ cal/mol/K}$$

$$\text{Einstein's frequency, } \boldsymbol{\nu_E} = k\Theta_E/h = \frac{(1.38 \times 10^{-23}) \times 157}{6.6 \times 10^{-34}}$$

$$= \mathbf{3.28 \times 10^{12}\ Hz.}$$

▶ **Example 19.** *Show that Debye's* T^3*-law should hold good for* $T < \theta_D/10$, *where* θ_D *is Debye temperature.*

Solution. While at high temperature, $C_V = 3R$, at low temperature $C_V < 3R$.

$$\therefore\ \frac{C_V}{3R} < 1 \Rightarrow \frac{4\pi^4}{5}\left(\frac{T}{\theta_D}\right)^3, \quad \text{using relation (6.7.12)}$$

$$\text{or, } T^3 < \frac{5\theta_D^3}{4\pi^4} \Rightarrow T < \theta_D/10$$

So, Debye's T^3 law should hold good for $\boldsymbol{T \leq \theta_D/10}$.

▶ **Example 20.** *Potassium has the bcc-structure with lattice constant* 5.3 A. *Find the Debye temperature assuming that the transverse and longitudinal elastic waves in potassium have the same velocity* $v = 1.5$ km/s.

Solution. We have, $g(\nu) = 4\pi V\left(\frac{1}{v_l^3} + \frac{2}{v_t^3}\right)\nu^2 = \frac{9N\nu^2}{\nu_D^3}$, using (6.7.2) and (6.7.5).

Taking $v_l = v_t = v$, $\quad 4\pi V\left(\frac{3}{v^3}\right) = \frac{9N}{\nu_D^3} \Rightarrow \nu_D^3 = \frac{3Nv^3}{4\pi V} = \frac{3v^3}{4\pi(V/N)}$

Now, $V/N = a^3/2$ for *bcc*-stucture. So, $\nu_D^3 = \frac{3v^3}{4\pi(a^3/2)} = \left(\frac{v}{a}\right)^3 \cdot \frac{3}{2\pi}$.

$$\therefore \quad \nu_D = \left(\frac{3}{2\pi}\right)^{1/3} \cdot \frac{v}{a} = 1.02\frac{v}{a}. \tag{6.8.1}$$

Thus, the Debye temperature, $\Theta_D = \frac{h\nu_D}{k} = \frac{h}{k}(1.02)\frac{v}{a}$, using (6.8.1).

$$\therefore \quad \boldsymbol{\Theta_D} = \frac{1.02 \times 6.62 \times 10^{-34}}{1.38 \times 10^{-23}} \times \frac{1500}{5.3 \times 10^{-10}},$$

substituting the given data.

$$= 1.02 \times \frac{990}{7.31}$$

$$= \mathbf{143\,K}$$

▶ **Example 21.** *Calculate the mean velocity of sound in diamond from the following data : Debye temperature of diamond* = 2000 K, *density of diamond* 3500 kg/m^3, *atomic mass of diamond* = 12 amu. **(Burdwan Hons.)**

Solution. Let v_s be the mean velocity of sound, ρ the density and M the atomic mass of diamond, then the Debye temperature

$$\Theta_D = \frac{hv_s}{k}\left(\frac{3N_A\rho}{4\pi M}\right)^{1/3}, \quad \text{where } N_A = \text{Avogadro no.}$$

$$\therefore \quad \boldsymbol{v_s} = \frac{k\Theta_D}{h}\left(\frac{4\pi M}{3N_A\rho}\right)^{1/3}$$

$$= \frac{(1.38 \times 10^{-23}) \times 2000}{6.62 \times 10^{-34}}\left(\frac{4\pi \times 12}{3 \times 6.023 \times 10^{26} \times 3500}\right)^{1/3}$$

$$= \mathbf{1.2 \times 10^4\ m/s}, \text{ on simplification.}$$

Review and Short Questions

1. Derive an expression for the frequency of vibration ω of an infinite chain of identical atoms as a function of the wave vector k.

   Show that the chain behaves as a low pass filter.

   What is the first Brillouin zone? Explain its significance.

2. What are the periodic boundary conditions and the boundary conditions that lead to the formation of standing waves?

3. Find the dispersion relation for a one-dimensional crystal with two types of atoms. Discuss the nature of the optical and the acoustic modes.

4. Derive the vibrational modes of a diatomic linear chain of atoms. What is the difference between the two branches? Why are they so named?

5. Establish the equivalence of a vibrational mode of a linear atomic chain and a harmonic oscillator.

6. Show, by considering a linear chain of atoms, that the number of possible modes of vibrations is equal to the number of vibrating atoms.

7. Discuss, qualitatively, the concept of phonons for elastic vibrations in a solid. Enumerate the properties of phonons. **(Calcutta Hons.)**

8. Obtain the dispersion relation for elastic waves in a linear monatomic chain with nearest neighbour interaction and show that the group velocity vanishes at the zone boundaries. Find the density of vibrational states as a function of the angular frequency ω and sketch the dispersion curve.

9. (a) Name the different branches of the disperson relation curve. What is the difference between the two branches? **(Lucknow Univ.)**

   (b) Obtain the dispersion relation for a one-dimensional diatomic lattice. Hence explain the meaning of optical branch. **(Calcutta Hons.)**

10. Prove that in a one-dimensional diatomic lattice, both acoustic and optical branches in dispesion curve meet the zone boundary normally.

11. Describe the quantum theory of the specific heat of solids emphasising on the shortcomings or inadequacy of classical ideas. **(Calcutta Hons.)**

12. Answer briefly the following questions.

    (a) In what way the Einstein's theory of specific heat superior to classical theory?

    (b) Define the Einstein temperature Θ_E. Does it have the unit of temperature? If so, show it.

    (c) Obtain the values of molar specific heat when $T \gg \Theta_E$ and also when $T \ll \Theta_E$ where Θ_E is Einstein's temperature.

(d) Compare the Einstein's theory of specific heat with that of Debye.

(e) What is Debye temperature?

(f) What are the assumptions and weaknesses of Debye's theory?

13. Show that at very low temperatures, the internal energy of a vibrating lattice following Debye's theory is proportional to T^4. **(Calcutta Hons.)**

14. Explain if the addition of the zero-point energy term influences the Einstein's and the Debye's results of the specific heat of solids.

15. Obtain the vibrational spectrum of a linear diatomic lattice and show that the spectrum consists of two branches. Discuss the main features of these branches.

16. What is phonon? What are its characteristic properties? What information can be obtained by studying the phonon spectrum of solids?

17. Prove that the gradient of optical branch of the dispersion curve at the maximum frequency is zero.

18. Give evidences for the existence of phonons.

19. Briefly answer the following questions.

(a) What are Einstein's temperature and Einstein's frequency?

(b) What are Debye temperature and Debye function?

(c) What do you mean by Debye's T^3 law?

(d) Why does Dulong-Petit's law fail at low temperatures?

20. Consider the expression for internal energy of a lattice in Debye Model

$$U = 9R\frac{T^4}{\theta_D^3}\int_0^{x_D}\frac{x^3\,dx}{e^x - 1}$$

where the symbols have their usual meanings. Find an expression for the specific heat C_V at low temperatures. **(Calcutta Hons.)**

21. A particle moving in one-dimension has energy

$$E = \frac{p^2}{2m} + \lambda q^4$$

where p and q are generalised momentum and coordinate respectively. Show that the heat capacity of a gas comprising of N such particles is

$$C_V = \frac{3Nk_B}{4}$$

where k_B is Boltzmann constant. **(Calcutta Hons.)**

22. A phonon is emitted or absorbed in

(a) elastic scattering of a photon by a crystal,

(b) inelastic scattering of a photon by a crystal,

(c) both type of scattering as indicated in (a) and (b),

(d) none of the above. **(Himachal Univ.)**

23. Quantum of elastic vibration is

(a) photon, (b) phonon, (c) graviton, (d) None. **(Himachal Univ.)**

24. As the temperature approaches zero, the lattice contribution to heat capacity of solids approaches

(a) infinity, (b) zero, (c) any value between $0 \to \infty$. **(Himachal Univ.)**

25. In Einstein's theory of specific heat of solids, the atoms are assumed as

(a) independent oscillators, (b) coupled oscillators, (c) none of these.

(Himachal Univ.)

26. At low temperatures, the specific heat capacity of solids can be expressed in the form

(a) $C_V = aT$, (b) $C_V = aT^3$, (c) $C_V = aT + bT^2$, (d) $C_V = aT + bT^3$.

27. The specific heat at very low temperature ($T \to 0$) is accurately given by

(a) Dulong and Petit's law, (b) Einstein's equation,

(c) Debye's equation, (d) Both Einstein and Debye equations.

28. The common feature in both Einstein's and Debye's theory of specific heat of solids is

(a) the cut-off frequency, (b) monochromatic frequency of vibrations,

(c) assumption of phonons, (d) application of quantum principle.

29. A phonon interacts with

(a) a neutron, (b) another phonon, (c) an electron, (d) all of these.

30. Debye temperature could be derived from

(a) specific heat, (b) elastic data, (c) both of these, (d) none of these.

31. For one-dimension in the Debye model,

(a) $C_V \propto T$, (b) $C_V \propto T^2$, (c) $C_V \propto T^3$, (d) $C_V \propto T^4$.

32. Debye's T^3-law holds for

(a) $T = \theta_D$, (b) $T < \theta_D/50$, (c) $T \leq \theta_D/10$, (d) any temperature.

Problems

1. The atomic frequency is 5.0×10^{14} Hz and the temperature is 5000 K. Calculate the average energy of the oscillator. [2.7×10^{-21} J]

2. The visible light of $\lambda = 5000$ A undergoes scattering from a crystal of refractive index 1.5. Find the maximum frequency of the phonon generated, given the velocity of sound in the crystal as 5000 m/s. [1.86×10^{11} rad/s]

3. Using the relation $\omega = \omega_{\max} \left|\sin \frac{ka}{2}\right|$, determine the phase velocity and the group velocity of the wave motion along a one-dimensional lattice. What does happen to the group velocity when $ka = \pm\pi$? [$v_p = \frac{\nu \sin(ka/2)}{ka/2}$, $v_g = \nu \cos(ka/2)$; $v_g = 0$]

4. Show that for the vibrations of a diatomic molecular chain

$$(\omega_+)^2_{\max} = (\omega_+)^2_{\min} + (\omega_-)^2_{\max},$$

the notations are the same as used in the text.

5. It is given that the vibrations of a monatomic linear chain, for small ka is 5×10^3 m/s. Calculate the group velocity at $k = 2\pi$. [3536 m/s]

6. Prove that the group velocity at the zero boundary of the optical branch and also of the acoustic branch of lattice vibrations of a diatomic linear chain is zero.

7. The Einstein frequency for copper is $\nu_E = 4.8 \times 10^2$ Hz. Calculate its Einstein temperature. [229.6 K]

8. The Debye temperature Θ_D for silver is 210 K.Assuming Debye's T^3-law, find the specific heat of silver at 20 K. Calculate the value of C_V for silver at the same temperature using Einstein's formula assuming Θ_E ($= h\nu/k$) and Θ_D to be equal.

[1.68 J.K^{-1}mol^{-1}; 0.076 J.K^{-1}mol^{-1}]

9. What can we say about the room temperature specific heat if Debye temperature of the solid is 3000 K? **(Himachal Univ.)** [0.46 cal/mol.K]

10. The Debye temperatures of KCl and NaCl having the same structure are 230 K and 281 K respectively. The molar specific heat of KCl is 6.26 mJ/mol/K at 3 K. Find the same for NaCl at 3 K and 5 K. [3.43 and 15.9 mJ/mol/K]

11. Determine the force constant and the frequency for the NaCl crystal of elastic stiffness constant 5×10^{10} N/m^2. Given : lattice constant of NaCl = 5.5 Å, mass of Na-atom = 23 amu, mass of Cl atom = 35.5 amu and 1 amu = 1.66×10^{-27} kg.

[27.5 N/m, 7.75×10^{12} Hz]

12. The specific heat of rocksalt, at very low temperatures, is given according to Debye's T^3-law by

$$C_V = A(T/\Theta_D)^3$$

where $A = 464\,\text{cal.mol}^{-1}\text{K}^{-1}$ and $\Theta_D = 281\,\text{K}$. Find the amount of heat necessary to raise the temperature of 3 moles of rock salt from 10 to 50 K. [97.9 calorie]

[**Hints.** Amount of heat, $Q = (3A/\Theta_D^3)\displaystyle\int_{10}^{50} T^3 dT$]

13. Copper has an atomic weight 63.5, a density of $8.9 \times 10^3\,\text{kg/m}^3$, and $v_l = 4.76 \times 10^3\,\text{m/s}$ and $v_t = 2.32 \times 10^3\,\text{m/s}$. Find the specific heat of copper at 30 K.

[1.328 kJ/k mol-K]

14. In aluminium, $v_l = 6.32 \times 10^3\,\text{m/s}$ and $v_t = 3.1 \times 10^3\,\text{m/s}$. The density of aluminium is $2.7 \times 10^3\,\text{kg/m}^3$ and its atomic weight is 26.97. Estimate the Debye cut-off frequency for aluminium. [8.44×10^{12} Hz]

15. Find the cut-off frequency for a linear monatomic lattice if the velocity of sound and the interatomic spacing in the lattice are 3×10^3 m/s and 3×10^{-10} m respectively. [2×10^{13} Hz]

16. The radii of Na^+ and Cl^- ions are 0.98 and 1.8 A respectively. The Young's modulus of NaCl in [100] direction is $5 \times 10^{10}\,\text{N/M}^2$. Assuming the extension in [100] direction produces negligible contraction in the perpendicular directions, find the wavelength at which the e. m. radiation is strongly reflected by NaCl crystal. Atomic masses of Na and Cl are 23 and 35.5 u respectively. [5.45×10^{-3} m]

17. Estimate the Debye temperature (θ_D) for iron for which the propagation velocities of longitudinal and transverse vibrations are equal to 5850 and 3230 m/s respectively.

Given: $h = 6.62 \times 10^{-34}$ Js, $k = 1.38 \times 10^{-23}$ J/K, atomic weight of Fe = 55.85, density of Fe = $7870\,\text{kg/m}^3$. **(Burdwan Hons.)**

18. The Debye temperature for diamond is 2230 K. Calculate the highest possible vibrational frequency and molar specific heat capacity of diamond at 10 K.

[$4.64 \times 10^{13}\,\text{s}^{-1}$, $0.175\,\text{J.(k mol)}^{-1}\,\text{K}^{-1}$]

19. Visible light of $\lambda = 4000$ A undergoes scattering from a crystal of index 2.42. Determine the maximum frequency of the phonon generated. Also, calculate the fractional change in frequency of the incident light. Velocity of sound in the crystal is 1.2×10^4 m/s. [9.12×10^{11} rad/s; 1.9×10^{-4}]

20. The lattice specific heat of KCl at 5 K is 3.8×0^{-2} J/mol/K. Calculate its specific heat at 2 K. [2.43×10^{-3} J/mol/K]

21. Sound velocities in solids are of the order of $v = 3 \times 10^3$ m/s and interatomic distances are of the order of $a = 3 \times 10^{-10}$ m. Assuming the cut-off frequency to be equal to Debye frequency, calculate Θ_D. [240 K]

[**Hints.** $\nu_D = \nu_{\text{max}} = v/\lambda_{\text{min}} = v/2a$.

Now $\Theta_D = h\nu_D/k = \dfrac{hv}{2ak} = \dfrac{6.62 \times 10^{-34} \times 3000}{1.38 \times 10^{-23} \times 2 \times 3 \times 10^{-10}} = \mathbf{240\ K}$]

22. Prove that a solid comprising of N atoms in a volume V will have a Debye temperature Θ_D given by

$$\Theta_D = \frac{hv_s}{k}\left(\frac{3N}{4\pi V}\right)^{1/3} = \frac{\hbar v_s}{k}\left(\frac{6\pi^2 N}{V}\right)^{1/3},$$

where $v_l = v_t = v_s$.

[**Hints.** $\nu_D^3 = 9N/4\pi V\left(\dfrac{1}{v_l^3} + \dfrac{2}{v_t^3}\right) = 3Nv_s^3/4\pi V$ ($\because$ $v_l = v_t = v_s$). So, $\nu_D = v_s(3N/4\pi V)^{1/3}$.

Hence $\Theta_D = h\nu_D/k = \dfrac{hv_s}{k}\left(\dfrac{3N}{4\pi V}\right)^{1/3} = \dfrac{2\pi\hbar v_s}{k}\left(\dfrac{3N}{4\pi V}\right)^{1/3} = \dfrac{\hbar v_s}{k}\left(\dfrac{8\pi^3 \cdot 3N}{4\pi V}\right)^{1/3}$

$$= \frac{\hbar v_s}{k}\left(\frac{6\pi^2 N}{V}\right)^{1/3}$$]

7

Band Theory of Solids

7.1 Introduction

The free electron theory has been successful in explaining phenomena such as electrical conductivity, thermionic emission, specific heat, paramagnetism etc., but still it was not free from its shortcomings. For instance, it fails to explain why some materials are good electrical conductors while some are electrical insulators and still some others are semiconductors. It could neither account for the variation of resistivity with temperature for insulators nor could explain the properties of semiconductors.

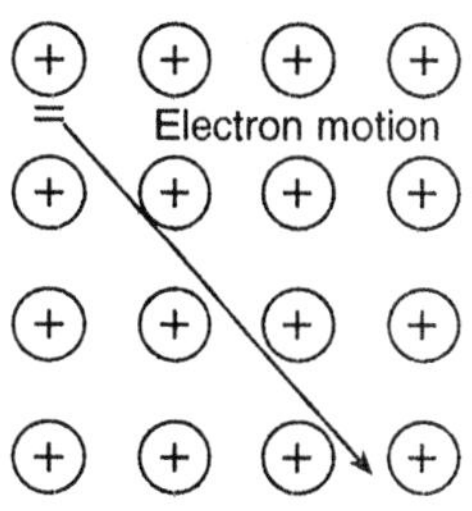

Fig. 7.1 Free electron in a crystal

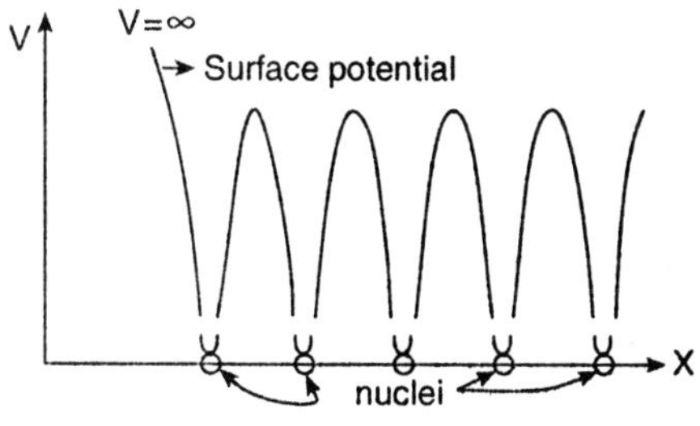

Fig. 7.2 One dimensional periodic potential in a crystal

The failure of the free electron theory lies in its oversimplified assumption that a conduction electron in a metal experiences a *constant* or *zero* potential due to the ion cores and hence freely move about in a crystal (Fig. 7.1). In fact, there is variation of the potential inside a metallic crystal and it appears more realistic to assume the potential to be *periodic* with periodicity of lattice somewhat like the one shown in Fig. 7.2. The potential is minimum at the positive ion sites and maximum between the two ions. Some contributions to potentials may, however, arise because of other electrons present in the crystal. So, the periodic potential may be thought to be due to the fixed nuclei plus some average potential due to all other electrons. The solution of Schrödinger's equation for this potential gives rise to a set of states that may be occupied by electrons according to Pauli principle.

This type of periodic potentials in crystals extends up to infinity in all directions except at the surface of the crystal where the lattice periodicity is interrupted, giving rise to a shape as shown on the edge of Fig. 7.2. This potential irregularity may however be ignored.

The *periodic potential* just referred to forms the very basis of the ***band theory*** of solid. We shall see in what follows that the motion of an electron in a periodic potential leads to the existence of *energy bands* separated by forbidden regions or *band gaps* and the electronic energy functions $E(\vec{k})$ are periodic in the wave vector $\vec{k}$. The introduction of the periodic potential leads naturally to a distinction between conductors, insulators and semiconductors.

7.2 Simple idea on formation of energy bands

Let an isolated H-atom be in the ground state with its electron. Quantum mechanically, the wave function of the electron is given by

$$\psi(r) = (\pi a_0^3)^{-1/2} e^{-r/a_0} \tag{7.2.1}$$

where a_0 is the *Bohr radius.*

Let now two such H-atoms A and B be brought close together. If ψ_1 and ψ_2 be the electron's wave functions for the two atoms, then when they are far apart from each other, they do *not* mutually interact or influence. But on being closer to each other, the wave functions ψ_1 and ψ_2 will overlap. The resultant electronic wave function due to the two nuclei may then be

(i) either, $\psi_1 + \psi_2$, (ii) or, $\psi_1 - \psi_2$.

The above two wave functions are schematically illustrated in Fig. 7.3.

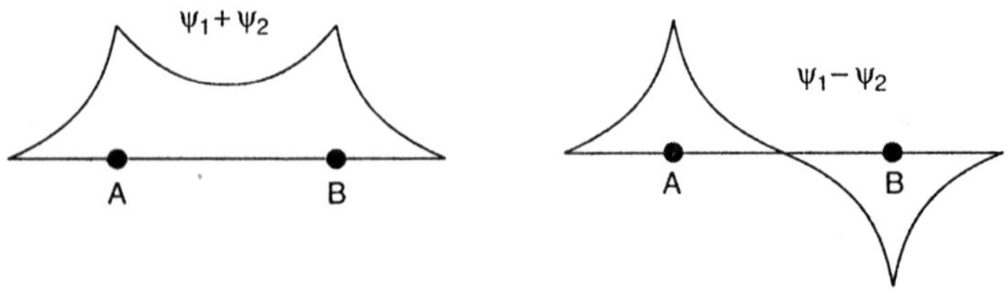

Fig. 7.3 Resultant electronic wave functions for the nuclei at positions marked A and B

For $\psi_1 + \psi_2$, the electron has a finite probability to exist mid-way between the two nuclei. In this region, the binding energy increases, since the binding force on the electron being due to both the nuclei. For $\psi_1 - \psi_2$, the probability density is zero midway between the nuclei, the extra contributory energy vanishes. Consequently, there exists a difference in energy between the states $\psi_1+\psi_2$ and $\psi_1-\psi_2$. So, the energy state splits into two distinct states when the two atoms are brought nearer. If now N atoms are brought close together, each energy state will split into N-energy states. If $N \to$ a large value, the separation between the energy states would be very small so as to develop a *quasi-continuous band*. Thus, each energy level splits into a *band* of energy levels. The formations of band for $1s$ and $2s$ states, with reducing separations between the atoms are illustrated in Fig. 7.4.

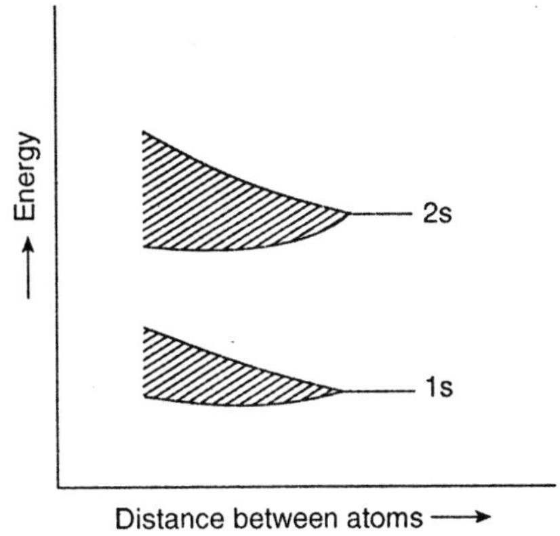

Fig. 7.4 Formation of bands for 1s and 2s states

• The width of a band is a function of (i) the *strength of the interaction* and (ii) the *overlap between the neighbouring atoms.*

7.3 Periodic potential in a crystal

The potential energy (P.E.) of an electron in the vicinity of an atomic nucleus of charge Ze is given by

$$V = -\frac{Ze^2}{4\pi\varepsilon_0 r} \tag{7.3.1}$$

where r is the electron-nucleus separation and ε_0 the permittivity of free space.

The variation of V with distance r for an isolated atom is illustrated in Fig. 7.5. When a number of such nuclei are brought close together to form a crystal, the P.E. of an electron is the sum of the potential energies due to the individual nuclei. The potential energy as a function of distance for an infinite one-dimensional crystal would appear as illustrated in Fig. 7.6 i.e, the variation of potential energy is a periodic function of the distance, with equi-spaced nuclei.

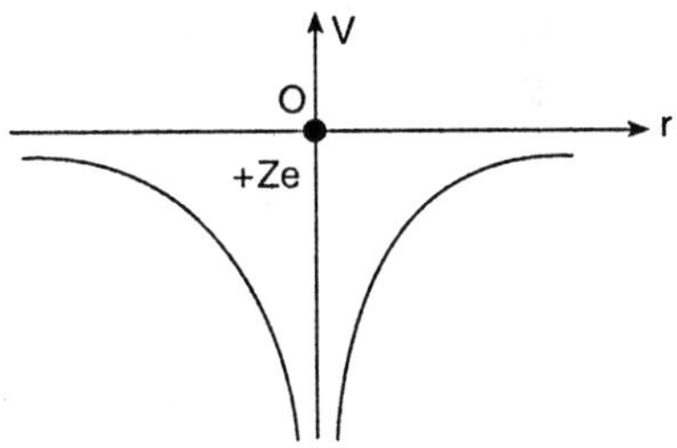

Fig. 7.5 Variation of potential energy with distance for an isolated atom

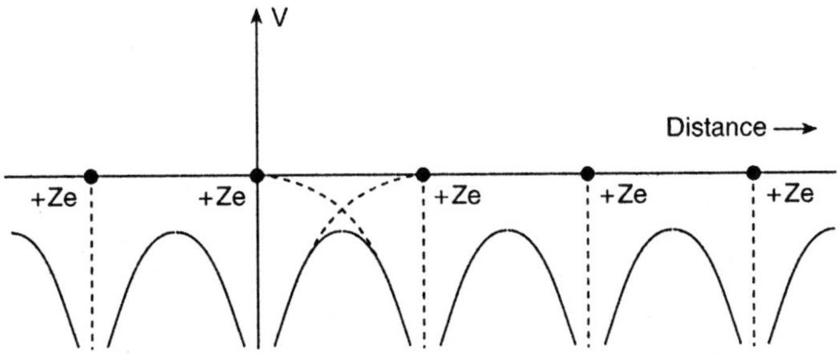

Fig. 7.6 Variation of potential energy with distance for a one-dimensional crystal

7.4 Bloch theorem: Its proof

According to the free electron theory, the electron moves in a *constant potential* V_0. So, the Schrödinger's wave equation for a one-dimensional case is

$$\frac{d^2\psi}{dx^2} + \frac{2m}{\hbar^2}(E - V_0)\psi = 0 \tag{7.4.1}$$

The solution to this equation are the plane waves of the type

$$\psi(x) = e^{\pm ikx} \tag{7.4.2}$$

where $\quad k^2 = \dfrac{2m}{\hbar^2}(E - V_0) \;\Rightarrow\; E - V_0 = \dfrac{\hbar^2 k^2}{2m} = \dfrac{p^2}{2m} = E_{\text{kinetic}}.$

Thus k represents the momentum p of the electron divided by $\hbar$. Each electron is k-specific and has its own energy.

For an electron moving in a one-dimensional *periodic potential* $V(x)$, the Schrödinger wave equation becomes

$$\frac{d^2\psi}{dx^2} + \frac{2m}{\hbar^2}[E - V(x)]\psi = 0 \tag{7.4.3}$$

As the potential is periodic, with a period equal to the lattice constant a,

$$V(x) = V(x + a) \tag{7.4.4}$$

There is an important theorem, called the ***Bloch theorem***, concerned with the solutions of Schrödinger equation (7.4.3), which states that there exists solutions of the form

$$\psi(x) = e^{\pm ikx}u_k(x) \tag{7.4.5}$$

$$\text{where } u_k(x) = u_k(x + a). \tag{7.4.6}$$

Thus the solutions are plane-waves of the type $e^{\pm ikx}$, modulated by a function $u_k(x)$ having the same periodicity as the lattice constant. The wave function of the type (7.4.5) is called the ***Bloch function*** and $\vec{k}$ gives the direction of ***Bloch wave.***

Proof of Bloch theorem—Let $f(x)$ and $g(x)$ be two real and independent solutions of equation (7.4.3). So, the general solution is given by

$$\psi(x) = Af(x) + Bg(x) \tag{7.4.7}$$

where A and B are two arbitrary constants.

Since $V(x) = V(x + a)$, the functions $f(x + a)$ and $g(x + a)$ must also be solutions of (7.4.3) in addition to $f(x)$ and $g(x)$. As (7.4.3), a second order differential equation, can have only two independent solutions, $f(x + a)$ and $g(x + a)$ must be expressible in terms of $f(x)$ and $g(x)$, i.e.,

$$\left.\begin{aligned} f(x + a) &= \alpha_1 f(x) + \alpha_2 g(x) \\ g(x + a) &= \beta_1 f(x) + \beta_2 g(x) \end{aligned}\right\} \tag{7.4.8}$$

where α's and β's are real functions of the energy E.

Another general solution to (7.4.3), as obtained from (7.4.8), is

$$\begin{aligned}\psi(x+a) &= Af(x+a) + Bg(x+a)\\ &= (A\alpha_1 + B\beta_1)f(x) + (A\alpha_2 + B\beta_2)g(x)\end{aligned} \tag{7.4.9}$$

Let us now choose A and B such that we have

$$\left.\begin{aligned}A\alpha_1 + B\beta_1 &= \lambda A\\ \text{and}\quad A\alpha_2 + B\beta_2 &= \lambda B\end{aligned}\right\} \tag{7.4.10}$$

where λ is a constant. Substituting these in (7.4.9) and using (7.4.7)

$$\psi(x+a) = \lambda Af(x) + \lambda Bg(x) = \lambda[Af(x) + Bg(x)] = \lambda\psi(x) \tag{7.4.11}$$

Equations (7.4.10) yield non-zero values for A and B, only if the determinant of their coefficients is zero, i.e.,

$$\begin{vmatrix}\alpha_1 - \lambda & \beta_1\\ \alpha_2 & \beta_2 - \lambda\end{vmatrix} = 0$$

$$\Rightarrow\ \lambda^2 - (\alpha_1 + \beta_2)\lambda + (\alpha_1\beta_2 - \alpha_2\beta_1) = 0 \tag{7.4.12}$$

Now, we shall show that: $\quad \alpha_1\beta_2 - \alpha_2\beta_1 = 1.$

We have
$$\frac{d^2f(x)}{dx^2} + \frac{2m}{\hbar^2}[E - V(x)]f(x) = 0 \tag{7.4.13}$$

Also,
$$\frac{d^2g(x)}{dx^2} + \frac{2m}{\hbar^2}[E - V(x)]g(x) = 0 \tag{7.4.14}$$

since $f(x)$ and $g(x)$ are real and independent solutions of (7.4.3).

Multiplying (7.4.13) by $g(x)$ and (7.4.14) by $f(x)$, and subtracting

$$f(x)\frac{d^2g(x)}{dx^2} - g(x)\frac{d^2f(x)}{dx^2} = 0$$

$$\Rightarrow\ f(x)\frac{dg(x)}{dx} - g(x)\frac{df(x)}{dx} = \text{const.} \tag{7.4.15}$$

$$\text{or,}\qquad W(x) = \text{constant}$$

where $W(x)$ is the left hand side of (7.4.15) and known as *Wronskian* of the solutions.

Further, we shall similarly get, using the relation (7.4.8) in (7.4.15),

$$W(x+a) = (\alpha_1\beta_2 - \alpha_2\beta_1)W(x).$$

$$\therefore\ \alpha_1\beta_2 - \alpha_2\beta_1 = 1 \tag{7.4.16}$$

$\therefore$ From (7.4.12), we obtain $\quad \lambda^2 - (\alpha_1 + \beta_2)\lambda + 1 = 0 \qquad (7.4.17)$

It yields two possible values of λ, say λ_1 and λ_2, and hence two functions $\psi_1(x)$ and $\psi_2(x)$, having the property $\psi(x+a) = \lambda\psi(x)$ as given in (7.4.11). The products of the roots $\lambda_1\lambda_2 = 1$. *Two* situations thus arise.

Case 1. ***For energy ranges such that*** $(\alpha_1 + \beta_2)^2 < 4$.

Eq.(7.4.17) will then have the complex roots and as $\lambda_1\lambda_2 = 1$, they will be complex conjugate:

$$\lambda_1 = e^{ika}, \quad \lambda_2 = e^{-ika}; \quad k = \text{real}. \tag{7.4.18}$$

The corresponding functions $\psi_1(x)$ and $\psi_2(x)$ will have the property

$$\left.\begin{aligned} \psi_1(x+a) &= e^{ika}\psi_1(x) \\ \psi_2(x+a) &= e^{-ika}\psi_2(x) \end{aligned}\right\} \tag{7.4.19}$$

$\therefore$ In general, we shall have, $\psi(x+a) = e^{\pm ika}\psi(x)$.

A function having this property is the *Bloch function* of the type given in (7.4.5) and can be readily checked by replacing x by $x + a$ in (7.4.5). So, the Bloch functions of the type (7.4.5) represent the solutions of Schrödinger equation (7.4.3). *This proves the* ***Bloch theorem.***

Case 2. ***For energy ranges such that*** $(\alpha_1 + \beta_2)^2 > 4$.

The two roots would now be real and may be taken as

$$\lambda_1 = e^{\mu a} \text{ and } \lambda_2 = e^{-\mu a}, \quad \text{where } \mu = \text{real}. \tag{7.4.20}$$

The corresponding solutions to the wave equation: $\psi(x) = e^{\pm\mu x}\, u(x)$, although valid mathematically, are *not acceptable* (forbidden) electron wave functions, as these are *not bounded.*

• The allowed or acceptable roots (7.4.18) correspond to the *allowed energy* regions and the forbidden roots (7.4.20) to *forbidden energy* regions. So, *the energy spectrum of an electron in a periodic potential consists* of *allowed and forbidden energy regions* (or *bands*). This will be further indicated by the Kronig-Penney model that follows.

7.5 Kronig-Penney model

According to this simpler model, the undefined periodic potential-energy curve of Fig. 7.6 is approximated by square wells, as illustrated in Fig. 7.7. Let the period of the periodic potential be $(a + b)$. The potential energy is

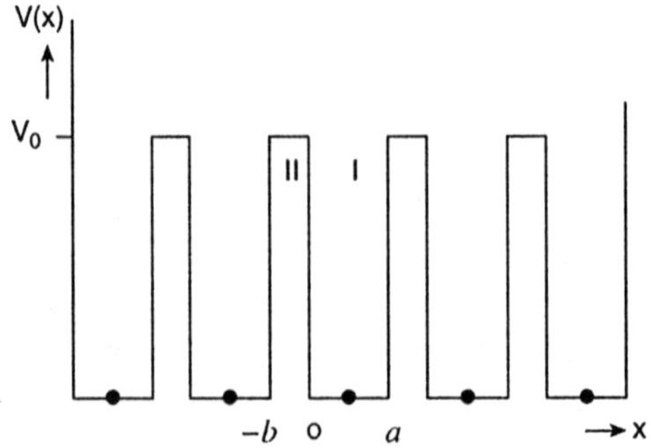

Fig. 7.7 Ideal periodic square well potential suggested by Kronig and Penney

$$V = 0, \qquad 0 < x < a$$
$$\text{and } V = V_0, \quad -b < x < 0$$

The Schrödinger equations can then be written as

$$\frac{d^2\psi}{dx^2} + \frac{2m}{\hbar^2}E\psi = 0, \quad \text{for } 0 < x < a \tag{7.5.1}$$

$$\text{and} \quad \frac{d^2\psi}{dx^2} + \frac{2m}{\hbar^2}(E - V_0)\psi = 0, \quad \text{for } -b < x < 0 \tag{7.5.2}$$

where E is the energy of the electron and $E < V_0$ (assumed).

Putting $\alpha^2 = 2mE/\hbar^2$ and $\beta^2 = 2m(V_0 - E)/\hbar^2$, we get, (7.5.3)

$$\frac{d^2\psi}{dx^2} + \alpha^2\psi = 0, \quad \text{for } 0 < x < a \tag{7.5.4}$$

$$\text{and} \quad \frac{d^2\psi}{dx^2} - \beta^2\psi = 0, \quad \text{for } -b < x < 0 \tag{7.5.5}$$

Since the potential is periodic of period $a + b$, we have

$$V(x) = V(x + a + b) \tag{7.5.6}$$

The wave function must be of the form of Bloch function, and is given by

$$\psi(x) = e^{ikx}u_k(x) \tag{7.5.7}$$

where $u_k(x)$ is a periodic function with the periodicity of the lattice.

$$\therefore \quad u_k(x) = u_k(x + a + b) \tag{7.5.8}$$

From (7.5.4) and (7.5.5.), using (7.5.7), we get

$$\frac{d^2u}{dx^2} + 2ik\frac{du}{dx} + (\alpha^2 - k^2)u = 0, \quad 0 < x < a \tag{7.5.9}$$

$$\text{and} \quad \frac{d^2u}{dx^2} + 2ik\frac{du}{dx} - (\beta^2 + k^2)u = 0, \quad -b < x < 0 \tag{7.5.10}$$

The general solutions of these equations are:

$$u_1 = Ae^{i(\alpha-k)x} + Be^{-i(\alpha+ik)x}, \quad 0 < x < a \tag{7.5.11}$$

$$\text{and} \quad u_2 = Ce^{(\beta-ik)x} + De^{-(\beta+ik)x}, \quad -b < x < 0 \tag{7.5.12}$$

where A, B, C and D are constants that can be determined from the following boundary conditions: based on the fact that ψ and its derivative $d\psi/dx$, u and du/dx must be single-valued and continuous.

$$\therefore \quad u_1(0) = u_2(0); \quad \left(\frac{du_1}{dx}\right)_{x=0} = \left(\frac{du_2}{dx}\right)_{x=0} \tag{7.5.13}$$

$$u_1(a) = u_2(-b); \quad \left(\frac{du_1}{dx}\right)_{x=a} = \left(\frac{du_2}{dx}\right)_{x=-b} \tag{7.5.14}$$

Using these conditions in (7.5.11) and (7.5.12), we obtain,

$$\left.\begin{aligned} A + B &= C + D \\ Ai(\alpha - k) - Bi(\alpha + k) &= C(\beta - ik) - D(\beta + ik) \\ Ae^{i(\alpha-k)a} + Be^{-i(\alpha+k)a} &= Ce^{-(\beta-ik)b} + De^{(\beta+ik)b} \\ Ai(\alpha - k)e^{i(\alpha-k)a} - Bi(\alpha + k)e^{-i(\alpha+k)a} &= C(\beta - ik)e^{-(\beta-ik)b} \\ &\quad -D(\beta + ik)e^{(\beta+ik)b} \end{aligned}\right\} \tag{7.5.15}$$

These equations of set (7.5.15) are *linear* and *homogeneous* and a *non-zero* solution exists only if the determinant of the coefficients of A, B, C and D vanishes. Expanding this determinant and simplifying, we obtain

$$\frac{\beta^2 - \alpha^2}{2\alpha\beta} \sinh \beta b \sin \alpha a + \cosh \beta b \cos \alpha a = \cos k(a + b) \tag{7.5.16}$$

To simplify the equation (7.5.16), Kronig and Penney considered the case when $V_0 \to \infty$ and $b \to 0$ such that the product $V_0 b$ remains finite. This, in fact, is equivalent to a Dirac δ-function type potential energy V_0, separated by a distance a, the potential energy being zero between the δ-function spikes. With this assumption, they obtained

$$\sinh \beta b \to \beta b \quad \text{and} \quad \cosh \beta b \to 1 \quad \text{as} \quad b \to 0.$$

$\therefore$ From (7.5.16),

$$\frac{\beta^2 - \alpha^2}{2\alpha\beta} \beta b \sin \alpha a + \cos \alpha a = \cos ka \tag{7.5.17}$$

Also, from (7.5.3), $\quad \beta^2 - \alpha^2 = \dfrac{8\pi^2 m}{h^2} V_0$, since $V_0 \gg E$.

Substituting this in (7.5.17), we obtain

$$\begin{aligned} \cos ka &= \frac{8\pi^2 m}{2\alpha\beta h^2} V_0 \beta b \sin \alpha a + \cos \alpha a \\ &= \frac{4\pi^2 ma}{h^2} V_0 b \frac{\sin \alpha a}{\alpha a} + \cos \alpha a \\ &= \frac{P \sin \alpha a}{\alpha a} + \cos \alpha a \end{aligned} \tag{7.5.18}$$

$$\text{where } P = \frac{4\pi^2 ma}{h^2} V_0 b \tag{7.5.19}$$

Equation (7.5.18) is represented in Fig. 7.8. While the quantity on the right-hand side is a function of αa, the cosine term on left side can only have values between -1 and $+1$, as shown by the horizontal lines. So, only certain values of α, and hence of E, are permissive. Since $\cos ka$ is an *even* periodic function, it will have the same value whether ka is positive, negative or is increased by integral multiple of 2π.

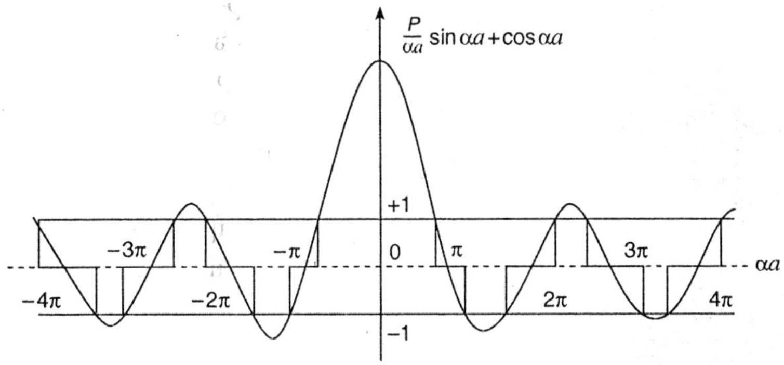

Fig. 7.8 Plot of $\frac{P\sin\alpha a}{\alpha a}+\cos\alpha a$ vs. αa for $P=3\pi/2$

So, the total electron-energy E is an even periodic function of k having a period of $2\pi/a$. Fig. 7.9 is a plot of E vs. k. If k is to be real, $|\cos ka| < 1$ which corresponds to the *allowed energy band.* For values of E for which $|\cos ka| > 1$, only imaginary k-values are possible which correspond to *forbidden bands.* The allowed and the forbidden bands are illustrated in Fig. 7.9.

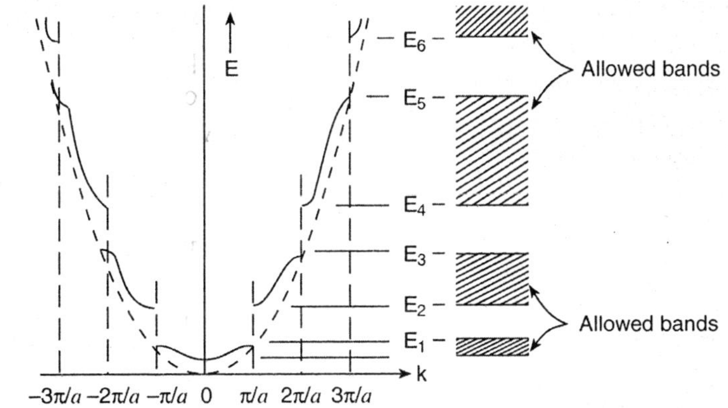

Fig. 7.9 Energy E versus wave vector k for a one dimensional lattice

Analysis of (7.5.18) and conclusions thereof—The following conclusion may be drawn from an analysis of equation (7.5.18).

1. The motion of electrons in a periodic lattice is characterized by alternate bands of *allowed energy* separated by forbidden regions.

2. As αa increases, so does the allowed bandwidth; the bandwidth of a forbidden band, however decreases, since the first term on right side of (7.5.18) decreases with increasing αa, on the average.

3. The quantity P, known to be *a measure of the potential barrier strength*, influences the energy spectrum. Large P implies the potential barrier $V_0 b$ is large (7.5.19), the function on right side of (7.5.18) crosses +1 and −1 region at a steeper angle (Fig. 7.10). So, the allowed bands are *narrower* and the forbidden ones are *wider.*

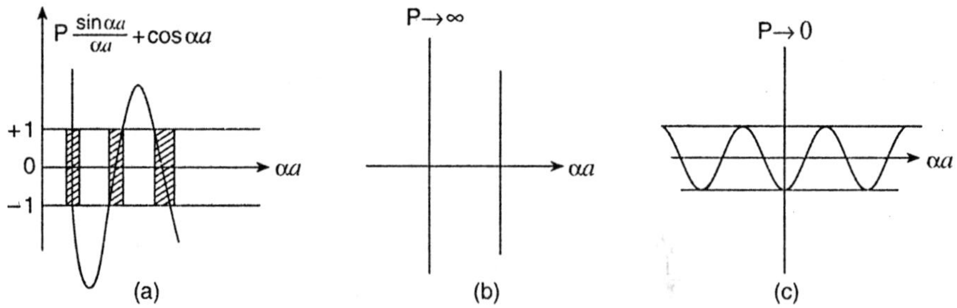

Fig. 7.10 (a) Right side of (7.5.18) for P = 6π as a function of αa (b) P→∞ (c) P→0

If P is very large, i.e., $P \to \infty$, it follows from equation (7.5.18) that to keep its right side within ±1, we shall then have to have

$$\frac{\sin \alpha a}{\alpha a} \to 0 \text{ i.e., } \sin \alpha a \to 0$$

$$\Rightarrow \quad \alpha a = \pm n\pi \Rightarrow \alpha^2 = n^2\pi^2/a^2$$

$\therefore$ From (7.5.4), $$\alpha^2 = \frac{n^2\pi^2}{a^2} = \frac{8\pi^2 mE}{h^2} \quad \Rightarrow \quad E = \frac{n^2h^2}{8ma^2},$$

and we get discrete energy spectrum of a particle in a constant potential box of atomic dimensions , as expected, since for large P, the tunnelling through the barrier is highly improbable. Thus E is independent of k.

4. If $P \to 0$, then from (7.5.18), $\cos \alpha a = \cos ka \Rightarrow \alpha = k$.

$$\therefore \quad \alpha^2 = k^2 = \frac{8\pi^2 mE}{h^2} \quad \Rightarrow E = \frac{h^2}{8\pi^2 m}k^2$$

$$\text{or,} \quad E = \frac{h^2}{8\pi^2 m}\left(\frac{2\pi}{\lambda}\right)^2 = \frac{h^2}{2m}\cdot\frac{1}{\lambda^2} = \frac{h^2}{2m}\cdot\frac{p^2}{h^2} = \frac{p^2}{2m} = \frac{1}{2}mv^2$$

where $\lambda = h/p$ and $p = mv$.

This is equivalent to the case of a *free particle.* Hence, *no energy level exists*, all energies being allowed to electrons.

7.5.1 Reduced zone representation

If we replace k by $k + 2n\pi/a$ where n is an integer, positive or negative, the relevant equations are still satisfied, for the replacement does not change the value of $\cos ka$. So, k is not uniquely determined and it would be useful to translate different segments of E-k curve in Fig. 7.11 parallel to k-axis through distances that are integral multiples of $2\pi/a$ so that k is within the region $-\pi/a \le k \le \pi/a$. The corresponding representation of E-k curve (Fig. 7.11a) is called the ***reduced zone representation.***

• The E-k curve in extended representation (Fig. 7.11b) shows *discontinuities* at k-values given by $k = n\pi/a$, where $n = \pm 1, \pm 2$, etc. and give the boundaries of *Brillouin zones.* But $k = 2\pi/\lambda$. So, the Brillouin zone boundaries in terms of λ is

$$2a = n\lambda,$$

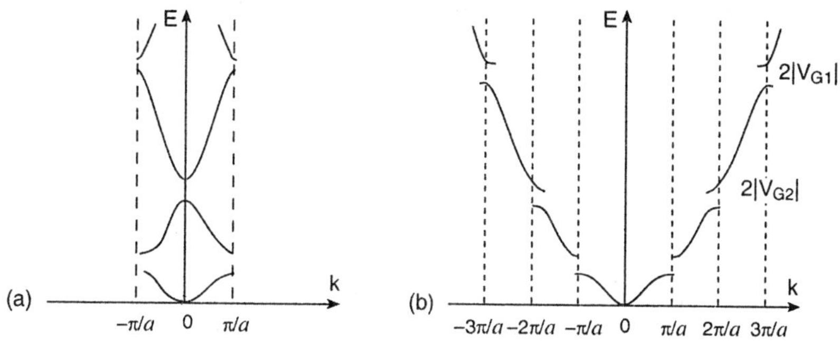

Fig. 7.11 (a) Reduced zone scheme (b) extended zone scheme V_G's

the well-known Bragg's law of reflection for normal incidence. The discontinuities in E-k curve are thus associated with *Bragg reflection* of de Broglie waves representing the electrons for those k-values (Fig. 7.11b). This representation of energy as a function of k is known as ***extended zone representation.***

● Kronig-Penney model deals with an *idealized* periodic potential. Even then, the important features of electrons in crystals, including the three-dimensional ones, are clearly brought out by the model.

7.6 Sketching of Brillouin zones

The Brillouin zone (Bz) is a representation of the allowed values of the wave vector k of electrons in one, two or three dimensions. We discuss below the construction of Bz in one, two or thee dimensional lattices.

Bz in a one-dimensional lattice—We have seen that a direct consequence of periodicity of lattice is that energy spectrum consists of permissive and forbidden regions (Fig. 7.9). The values of k at which discontinuities occur in E is $\cos ka = \pm 1$ or $k = \pm n\pi/a$, where $n = 1, 2, 3, \cdots$.

The region between the values $k = \pi/a$ and $k = -\pi/a$ corresponds to the first Brillouin zone. Next, the region between the first and the second values of k for which discontinues occur is the second Brillouin zone and so on (Fig. 7.12).

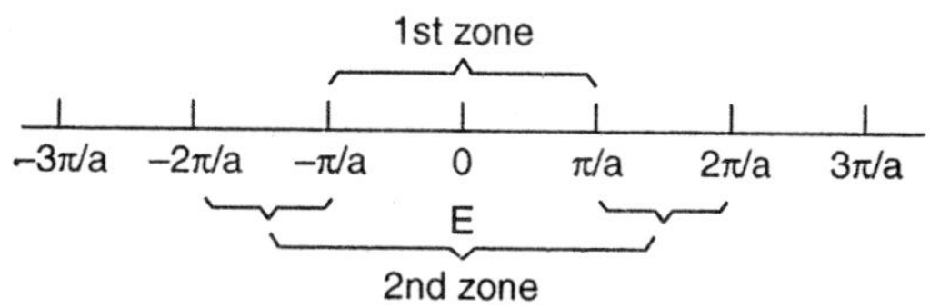

Fig. 7.12 First two Brillouin zones for a one dimensional lattice

● **Brillouin Zones for a two dimensional lattice**—If we follow the procedure of a linear lattice, the *first* zone for a two-dimensional lattice in the XY-plane appears to be the square $ABCD$ (Fig. 7.13), the boundaries of which are given by

$$k_x = \pm\pi/a, \quad k_y = \pm\pi/a$$

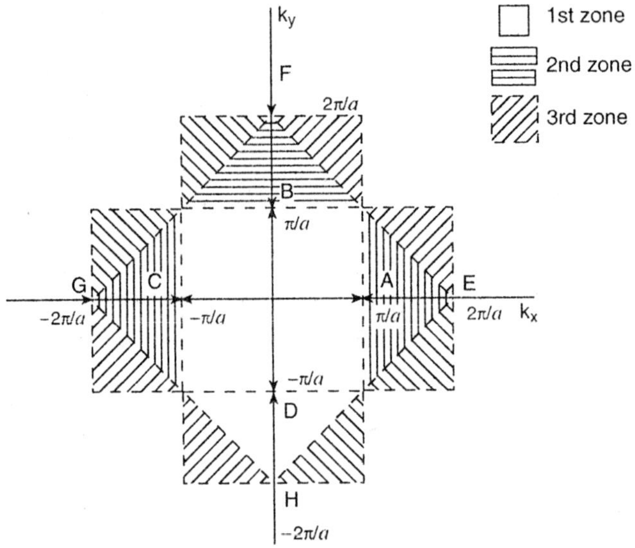

Fig. 7.13 The first three Brillouin zones for a two-dimensional square lattice

The second zone in the XY-plane is the square $EFGH$ whose boundaries are

$$k_x = \pm 2\pi/a, \quad k_y = \pm 2\pi/a$$

● **Three-dimensional lattice**—By the same procedure as adopted in the two-dimensional case, the ***first Brillouin zone*** becomes a cube for a three-dimensional cubic lattice of side $2\pi/a$. It would be better clarified from the treatment in reciprocal lattice in Art. 7.7.

7.7 Brillouin zones in reciprocal lattice

A Brillouin zone is the locus of all those $\vec{k}$-values in the reciprocal lattice which are Bragg reflected.

Let us construct the Brillouin zone for a simple square lattice of side a. The primitive translation vectors of this lattice are

$$\vec{a} = a\hat{i}, \quad \vec{b} = a\hat{j}$$

the corresponding translation vectors of reciprocal lattice are

$$\vec{A} = (2\pi/a)\hat{i}\,; \quad \vec{B} = (2\pi/a)\hat{j}$$

$$\therefore \text{ Reciprocal vector, } \vec{G} = h\vec{A} + k\vec{B} = (2\pi/a)(h\hat{i} + k\hat{j}) \qquad (7.7.1)$$

where h and k are two integers.

The Bragg reflection condition gives:

$$2\vec{k}\cdot\vec{G} + G^2 = 0$$

$$\Rightarrow \quad 2(k_xG_x + k_yG_y) + (G_x^2 + G_y^2) = 0 \quad (\because \; \vec{k} = k_x\hat{i} + k_j\hat{j})$$

$$\Rightarrow\ 2\left(k_x\frac{2\pi}{a}h + k_y\frac{2\pi}{a}k\right) + \left(\frac{2\pi}{a}\right)^2 h^2 + \left(\frac{2\pi}{a}\right)^2 k^2 = 0, \text{ using (7.7.1)}$$

$$\Rightarrow\ \frac{4\pi}{a}(hk_x + kk_y) = -\left(\frac{2\pi}{a}\right)^2(h^2 + k^2)$$

$$\therefore\ \boxed{hk_x + kk_y = -\frac{\pi}{a}(h^2 + k^2)} \tag{7.7.2}$$

Equation (7.7.2) gives a family of straight lines in the k_x and k_y plane; their k_x and k_y intercepts are given by

$$k_x = -\frac{\pi}{a}\frac{(h^2 + k^2)}{h}; \quad k_y = -\frac{\pi}{a}\frac{(h^2 + k^2)}{k}$$

To get the values of $\vec{k}$ that are Bragg reflected, we are to take all the possible combinations of h and k.

For $h = 0$ and $k = \pm 1$, $k_y = \pm\pi/a$, $k_x =$ arbitrary.

For $h = \pm 1$ and $k = 0$, $k_x = \pm\pi/a$, $k_y =$ arbitrary.

The above four lines $k_x = \pm\pi/a$, $k_x = -\pi/a$, $k_y = +\pi/a$ and $k_y = -\pi/a$ are plotted in Fig. 7.14.

With origin at O, all the $\vec{k}$ vectors emanating from it and terminating on these lines will be Bragg reflected. The square, bounded by $k_x = \pm\pi/a$ and $k_y = \pm\pi/a$, is the *first Brillouin zone* (shaded) of a square lattice of side a. It is a square of side $2\pi/a$.

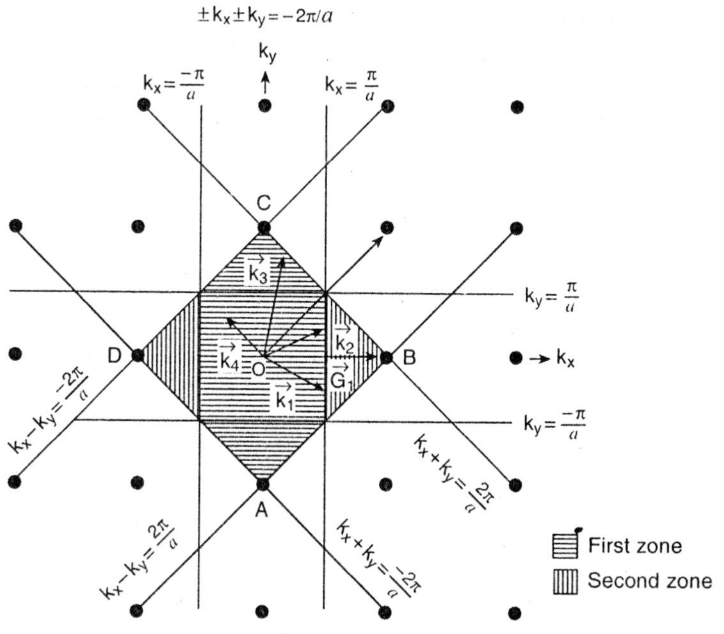

Fig. 7.14 Brillouin zones of a square lattice in its reciprocal lattice. The vectors k_1, k_2, and k_3 are Bragg reflected whereas k_4 is not. The vector k_1 and k_2 have the same reciprocal lattice vector G_1, while G_2 is the reciprocal vectors of k_3

Apart from this set of four lines, other sets of lines may as well satisfy (7.7.2). For instance, for $h = \pm 1$ and $k = \pm 1$, the (7.7.2) provides the following set:

$$\begin{aligned} \pm k_x \pm k_y &= 2\pi/a \\ \text{or,}\quad k_x + k_y &= \pm 2\pi/a \\ k_x - k_y &= \pm 2\pi/a. \end{aligned}$$

The additional area bounded by the above four lines (the square $ABCD$) represents the *second Brillouin zone* and in a like manner one can construct the third, fourth etc. zones. The boundaries of the zones give *the loci of the* $\vec{k}$-values that are *Bragg reflected* and hence may be considered as reflecting planes—the first zone boundary for the first order and the others for other orders.

- A $\vec{k}$-vector not terminating on a zone boundary *cannot be Bragg reflected.*

7.7.1 Brillouin zones in different reciprocal lattices

If we extend the equation (7.7.2) to a three-dimensional case, we obtain

$$hk_x + kk_y + lk_z = -\frac{\pi}{a}(h^2 + k^2 + l^2) \tag{7.7.3}$$

a being the length of the edge of the cube.

We shall now find the Brillouin zones in different reciprocal lattices.

1. Simple cubic lattice—The primitive translation vectors for an *sc*-lattice are:

$$\vec{a} = a\hat{\imath},\quad \vec{b} = a\hat{\jmath},\quad \vec{c} = a\hat{k}$$

The primitive translation vectors of reciprocal lattice are

$$\vec{A} = 2\pi\frac{\vec{b}\times\vec{c}}{\vec{a}\cdot\vec{b}\times\vec{c}} = \frac{2\pi}{a}\hat{\imath},\quad \vec{B} = \frac{2\pi}{a}\vec{\jmath},\quad \vec{c} = \frac{2\pi}{a}\hat{k}$$

The boundaries of the first Brillouin zone are the planes nromal to the six reciprocal lattice vectors $\pm\vec{A}$, $\pm\vec{B}$, $\pm\vec{C}$ at their mid points:

$$\pm\tfrac{1}{2}A = \pm\frac{\pi}{a}\vec{\imath},\quad \pm\tfrac{1}{2}B = \pm\frac{\pi}{a}\vec{\jmath} + \tfrac{1}{2}C = \pm\frac{\pi}{a}\vec{k}$$

The above six planes bound a cube of edge $2\pi/a$ and volume $(2\pi/a)^3$. So, the first Brillouin zone is a ***cube*** (Fig. 7.15a) cutting the axes k_x, k_y and k_z at poitns $\pm\pi/a$.

- The reciprocal lattice vector $\vec{G}$ satisfying $2\vec{K}\cdot\vec{G} + G^2 = 0$ is normal to the zone boundary and all $\vec{K}$'s lying on this boundary have same $\vec{G}$ for reflection. So, one can use only the allowed $\vec{G}$ values and their normal bisectors for the construction of Brillouin zones. The region bounded by normal bisectors to the shortest possible $\vec{G}$ is the *first Brillouin zone.* The region bounded by normal bisectors of next larger $\vec{G}$ corresponds to the *second zone* and so on.

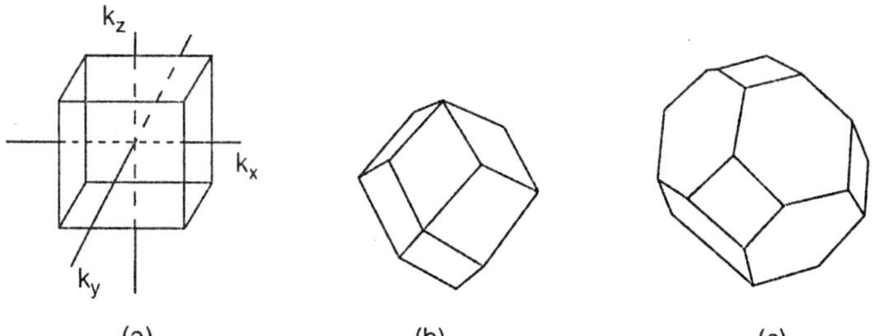

Fig. 7.15 The first Brillouin zone for some simple lattices. (a) simple cubic lattice, (b) face-centred cubic lattice, (c) body-centred cubic lattice

2. Face-centred cubic lattice—The primitive translation vectors of *fcc*-lattice are: $\vec{a} = \frac{a}{2}(\hat{i}+\hat{j})$, $\vec{b} = \frac{a}{2}(\hat{j}+\hat{k})$ and $\vec{c} = \frac{a}{2}(\hat{k}+\hat{i})$.

In reciprocal lattice, the primitive translation vectors are

$$\vec{A} = 2\pi\frac{\vec{b}\times\vec{c}}{\vec{a}\cdot(\vec{b}\times\vec{c})} = \frac{2\pi}{a}(\hat{i}+\hat{j}-\hat{k})$$

$$\vec{B} = 2\pi\frac{\vec{c}\times\vec{a}}{\vec{a}\cdot(\vec{b}\times\vec{c})} = \frac{2\pi}{a}(-\hat{i}+\hat{j}+\hat{k})$$

$$\vec{C} = 2\pi\frac{\vec{a}\times\vec{b}}{\vec{a}\cdot(\vec{b}\times\vec{c})} = \frac{2\pi}{a}(\hat{i}-\hat{j}+\hat{k})$$

$$\therefore\quad \vec{G} = \vec{A}h + \vec{B}k + \vec{C}l$$

$$= \frac{2\pi}{a}[(h-k+l)\hat{i} + (h+k-l)\hat{j} + (-h+k+l)\hat{k}]$$

The shortest non-zero $\vec{G}$'s are the following eight vectors: $(2\pi/a)(\pm\hat{i}\pm\hat{j}\pm\hat{k})$

The boundaries of the *first* Brillouin zone are determined by the normal bisector planes to these eight vectors. The corners of the *octahedron* thus formed are truncated by the planes which are perpendicular bisectors of the six reciprocal lattice vectors:

$$\frac{2\pi}{a}(\pm 2\hat{i});\ \frac{2\pi}{a}(\pm 2\hat{j});\ \frac{2\pi}{a}(\pm 2\hat{k})$$

The shape of the first Brillouin zone is therefore a *truncated* ***octahedron***, as shown in Fig.7.15b. This is one of the primitive cells of a *BCC* lattice since it is reciprocal space and the rreciprocal of the *FCC* lattice is a *BCC* lattice.

(c) Body-centred cubic lattice—The primitive translation vectors of *bcc*–lattice are: $\vec{a} = \frac{a}{2}(\hat{i}+\hat{j}-\hat{k})$, $\vec{b} = \frac{a}{2}(-\hat{i}+\hat{j}+\hat{k})$, $\vec{c} = \frac{a}{2}(\hat{i}+\hat{j}-\hat{k})$.

In reciprocal lattice, the primitive translation vectors are

$$\vec{A} = \frac{2\pi}{a}(\hat{i}+\hat{j});\ \vec{B} = \frac{2\pi}{a}(\hat{j}+\hat{k});\ \vec{C} = \frac{2\pi}{a}(\hat{k}+\hat{l})$$

$$\therefore \quad \vec{G} = h\vec{A} + k\vec{B} + l\vec{C}$$
$$= \frac{2\pi}{a}[(h+l)\hat{\imath} + (h+k)\hat{\jmath} + (k+l)\hat{k}]$$

The shortest non-zero $\vec{G}$'s are the following twelve vectors, where all the choices of signs are independent.

$$\frac{2\pi}{a}(\pm\hat{\imath} \pm \hat{\jmath}); \quad \frac{2\pi}{a}(\pm\hat{\jmath} \pm \hat{k}); \quad \frac{2\pi}{a}(\pm\hat{\imath} \pm \hat{k})$$

The boundaries of the *first Brillouin* as zone are determined by the normal bisector planes to these twelve vectors and the zone is thus a regular twelve faced solid, a ***rhombic dodecahedron***, as shown in Fig. 7.15c.

7.8 Crystal momentum

The free electron wave vector is simply $\vec{k} = \vec{p}/h$, where $\vec{p}$ is the true momentum of the electron. Can we speak of momentum of an electron when it is in motion in the periodic potential of lattice? The answer is 'no'. But why?

Bloch theorem introduces a wave vector $\vec{k}$ that plays the role of $\vec{p}/\hbar$ in free electron theory. But this $\vec{k}$ is not proportional to the momentum of the electron. In other words, $\hbar\vec{k}$ is not true electron momentum. While the Bloch wave functions are the eigenstates of the Hamiltonian, they are not simultaneously the eigenstates of the momentum operator, as shown hereunder.

$$\vec{p}\psi_k(\vec{r}) = \frac{\hbar}{i}\nabla\psi_k(\vec{r}) = \frac{\hbar}{i}\nabla\left[e^{i\vec{k}\cdot\vec{r}}u_k(r)\right]$$
$$= \hbar k\psi_k + e^{i\vec{k}\cdot\vec{r}}\frac{\hbar}{i}\nabla u_k(\vec{r}) \neq \hbar\vec{k}\psi_k(r) \qquad (7.8.1)$$

Neverthelss, $\hbar\vec{k}$ is a natural extension of $\vec{p}$ to the periodic potential of crystal lattice and $\hbar\vec{k}$ for a lattice is known as the ***crystal momentum***, where $\vec{k}$ is the wave vector introduced by Bloch theorem.

- Free-electron momentum is a constant of motion, but $\hbar\vec{k}$ (in presence of lattice potential) is not. However, the crystal momentum $\hbar\vec{k}$ is a *well-defined constant* for a state of given energy as dictated by Bloch theorem. In dealing with electron-lattice interaction, we shall thus use *conservation of crystal momentum* and *not* of the true electron momentum.

7.9 Effective mass of an electron

The energy E of free electron in terms of k is given by

$$E = \frac{\hbar^2k^2}{2m} = \frac{h^2k^2}{8\pi^2m} \qquad (7.9.1)$$

where m is the mass of the electron.

Differentiating (7.9.1) successively twice with respect to k, we get

$$\frac{dE}{dk} = \frac{h^2 k}{4\pi^2 m}; \quad \frac{d^2E}{dk^2} = \frac{h^2}{4\pi^2 m} \tag{7.9.2}$$

$$\therefore \quad m = \frac{h^2}{4\pi^2}\left(\frac{d^2E}{dk^2}\right)^{-1} = \frac{\hbar^2}{d^2E/dk^2} \tag{7.9.3}$$

In free electron model, where E vs. k has a parabolic relationship and d^2E/dk^2 can be taken as constant, since the mass of the electron is considered constant. But, when the electrons move in a periodic potential of the crystal lattice, the parabolic relation between E and k is no longer true. At zone boundaries, the curve breaks up leading to allowed and forbidden bands. Near the forbidden band, the curvature of E-k curve changes and may become negative. For instance, as k approaches the zone boundary, d^2E/dk^2 first reduces to zero and then becomes negative. So, d^2E/dk^2 is not constant and so also the electron mass. In fact, both are functions of k.

Under periodic potential, therefore, the mass of the electron, as given by (7.9.3), is termed the ***effective mass*** m^* of the electron. So,

$$m^* = \frac{h^2}{4\pi^2}\left(\frac{d^2E}{dk^2}\right)^{-1} = \frac{\hbar^2}{d^2E/dk^2} \tag{7.9.4}$$

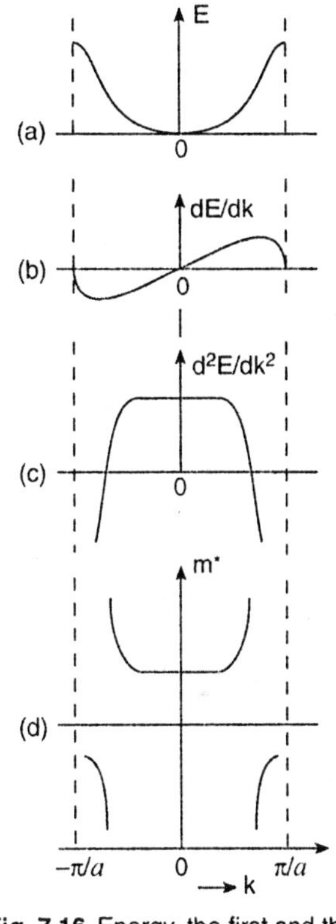

Fig. 7.16 Energy, the first and the second derivartives of energy and the effective mass as a function of k

The result indicates the importance of (E-k) curve for the motion of electrons.

Discussion—The effective mass is totally a ***new concept.*** The response of an electron in the crystal to an applied field is not determined by its actual gravitational mass but by its effective mass, arising out of interaction of electron wave packet with the periodic lattice. If this interaction is large, i.e., if a strong binding force exists between the electron and the lattice, it may be difficult for the electrons to move, implying that the electron has acquired large (even infinite) effective mass.

The effective mass of an electron may be both positive or negative and is illustrated in Fig. 7.16d by plotting a curve between m^* and k in the first Brillouin zone for one-dimensional lattice. The effective mass is positive near the bottom of the band (lower k-values) where the curvature is positive. It is negative close to the zone boundary (near $k \sim \pi/a$) where the curvature is negative. Figure 7.15 also shows the E vs. k, dE/dk vs. k and d^2E/dk^2 vs. k (qualitatively).

• The effective mass m^* equals m only when the energy is *not* near the edge of the band and the E-k curve is parabolic (i.e., under free electron approximation). In most conductors $m^* = m$ as the band is only partially filled. The effective mass m^* however differs from m in semiconductors, insulators and certain conductors like Bi where full or almost full valence bands are involved. It becomes infinite at the inflexion points of the $E(k)$ curve. So, the electron behaves as a positively charged particle in the upper half of the band only.

Degree of freedom—To what extent an electron in the k-state behaves as a free electron can be measured by introducing a factor f_k, called *degree of freedom* and defined as

$$f_k = \frac{m}{m^*} = \frac{m}{\hbar^2}\left(\frac{d^2E}{dk^2}\right) \tag{7.9.5}$$

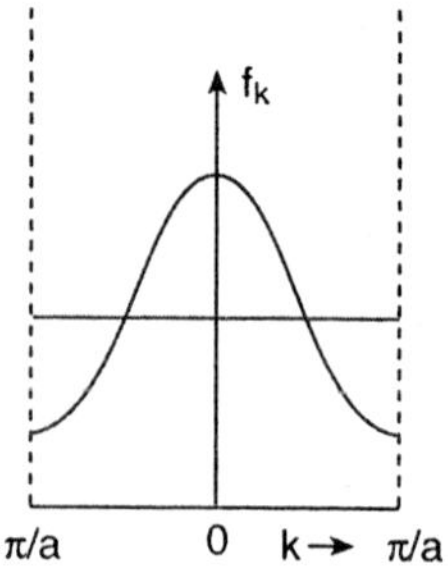

Fig. 7.17 Plot of f_k as a function of k

If $f_k < 1$, i.e., $m^* > m$, the electron behaves as a *heavy* particle and its freedom of motion is *low*. A large value of f_k corresponds to a small value of m^* and the electron acts as a *light particle*, with a greater freedom of motion to contribute to electrical conductivity. If $f_k = 1$, the electron behaves totally as a *free* electron. The variation of f_k with k (Fig. 7.17) shows that f_k is positive in the lower half of the energy band (where m^* is positive) and negative in the upper half (where m^* is negative).

The negative effective mass may well be understood in terms of Bragg reflection when k is close to the value $\pm\pi/a$. Due to the Bragg reflection, a force applied in one direction leads to a gain in momentum in the opposite direction resulting thus in the negative effective mass.

The ***physical significance*** of the concept of effective mass is that it gives a satisfactory description of charge carriers in crystals. In crystals like alkali metals having partially filled energy band, the conduction occurs essentially through electrons. But in crystals for which the energy band is nearly full, the negative charge and negative mass vacancies may be looked upon as positive charge and positive mass particles known as '*holes*' acting as positive charge carries for conduction.

• One may conclude that all the results of free-electron theory are correct provided m in each case is replaced by suitabe m^*.

7.10 Number of possible wave functions in a band

Consider a finite linear crystal of length L, containing N atoms, whose periodicity is a. To compute the number of possible wave functions or k-values per band. For a linear crystal, the periodic boundary conditions can be taken as

$$\psi(x+L) = \psi(x) \tag{7.10.1}$$

The wave functions are the Bloch functions so that we have

$$e^{ik(x+L)}u_k(x+L) = e^{ikx}u_k(x) \tag{7.10.2}$$

$$\Rightarrow\ e^{ik(x+L)} = e^{ikx}\ \Rightarrow\ e^{ikL} = 1 \tag{7.10.3}$$

$$(\because \text{ periodicity of } u_k \text{ gives } u_k(x+L) = u_k(x)) \tag{7.10.4}$$

Thus we have, $kL = 2\pi n \Rightarrow k = 2\pi n/L,$ where $n = \pm 1, \pm 2, \pm 3, \cdots$ (7.10.5)

Now, since $n = kL/2\pi$,

Number of possible wave functions in the range dk is given by

$$dn = \frac{L}{2\pi}dk \tag{7.10.6}$$

Since each state is occupied by two electrons, the 'effective' extent of freedom of electrons populating the band within the limits $-k_0$ to $+k_0$ is

$$\begin{aligned} F_{\text{eff}} &= 2\frac{L}{2\pi}\int_{-k_0}^{k_0} f_k dk \\ &= \frac{2Lm}{\pi\hbar^2}\int_0^{k_0}\left(\frac{d^2E}{dk^2}\right)dk \\ &= \frac{2Lm}{\pi\hbar^2}\left(\frac{dE}{dk}\right)_{k=k_0} \end{aligned} \tag{7.10.7}$$

If the band is filled up to the inflection points of E-k curve, $(dE/dk)_{k=k_0}$ is a maximum. Hence the effective freedom of the electron attains a maximum. If the band is completely filled, $dE/dk = 0$ at the top of the band so that $F_{\text{eff}} = 0$. *The electrons are not thus free in a completely filled band and cannot contribute to electrical conductivity and cannot be a metal.* We can distinguish between metals, insulators and semiconductors on this basis.

- From (7.10.5), we have, $n = \int dn = \frac{L}{2\pi}\int_{-\pi/a}^{\pi/a} dk = \frac{L}{a}.$

$$\therefore\ na = L, \text{ also } Na = L.$$

$$\text{So, } n = N.$$

Thus, the *total number of possible states in an energy band is equal to the number of primitive unit cells in it.* Each state can be occupied by two electrons (Pauli's principle) corresponding to two spin states of an electron, and hence the maximum number of electrons possible in a band is $2N$. This result plays an important role in distinguishing metals, semiconductor and insulators.

7.11 Conductors, semiconductors and insulators

We have seen that the motion of electrons in a periodic potential gives rise to allowed and forbidden bands in solids. Each allowed energy band again has a limited number of energy levels and since each energy level may be occupied, according to Pauli's exclusion principle, by not more than two electrons, each band can accommodate a maximum of $2N$ electrons. In this context, it appears that in solids with a limited number of electrons in atoms, the lower energy bands only will be expectedly filled. The outermost band which remains completely or partially filled is known as the *valence band* in solids. The band above the valence band that remains empty at 0 K is called the *conduction band*. Depending on the nature of occupation of band by electrons, therefore, all solids can be broadly classed into two categories.

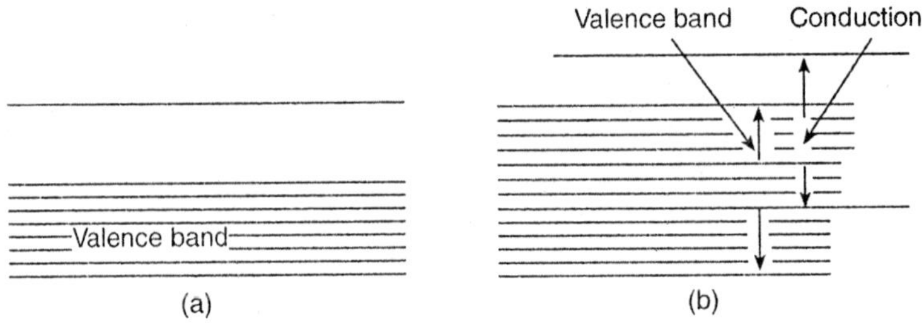

Fig. 7.18 Metals with partially filled or overlapping bands

The first group of solids has a partially filled band immediately above the uppermost filled (valence) band. This may occur in *two* different ways. First, the valence band is only partially filled (Fig. 7.18a) and secondly, a completely filled valence band may overlap the partially filled conduction band (Fig. 7.18b). These solids have metallic character and are ***conductors***.

The second group of solids, on the other hand, has empty bands lying above the completely filled energy bands and can be subdivided into ***insulators*** and ***semiconductors***, based on the width of the forbidden band (Fig. 7.19). The forbidden bands of insulators are comparatively much wider, e.g., the E_g-values of diamond $= 5.4\,\text{eV}$, $Al_2O_3 = 7\,\text{eV}$; for typical insulators, $E_g > 3\,\text{eV}$.

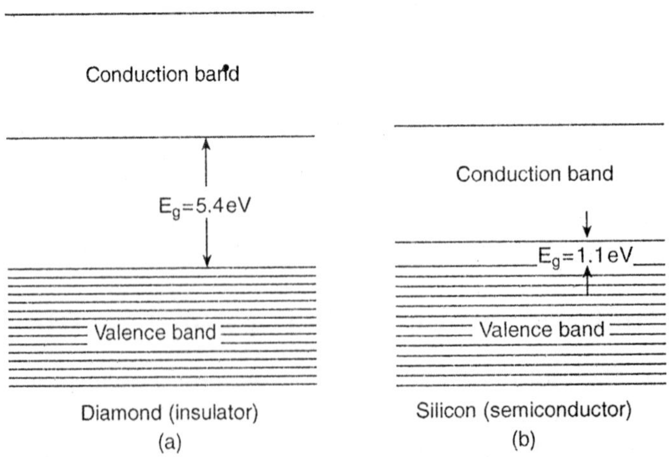

Fig. 7.19 Difference between an insulator and a semiconductor in terms of energy gap

Semiconductor however includes such solids that have relatively much narrower forbidden bands, e.g. for Germanium, $E_g = 0.7\,\text{eV}$; for Silicon, $E_g = 1.1\,\text{eV}$; and for typical semiconductors, $E_g \leq 1\,\text{eV}$.

Insulators thus do not conduct electricity as the number of electrons are just enough to completely fill a number of allowed bands and the energy gap between the topmost filled band (valence band) and the lowermost empty band (conduction band) is so large that at ordinary temperatures, the electrons cannot be thermally excited across the gap from the valence to the conduction band. As the bands are either completely filled or empty, no current can flow.

In semiconductor, the energy gap between the valence and the conduction band is relatively small and an appreciable number of electrons, can be thermally excited from states near the top of the valence band to states near the bottom of the conduction band. Further, semiconductors are classified as *intrinsic* and *extrinsic*. For a complete account of semiconductor read '***semiconductor***' in 'Electronics' part of this book.

• The conductivity of semiconductors increases with temperature owing to increase in the number of conduction-band electrons. The conductivity of metals, however, decreases with increase in temperature because more and more phonons are excited at higher temperatures and they may scatter conduction electrons and reduce their mobility.

7.12 Illustrated Examples

▶ **Example 1.** *Given that the effective mass of the electron is* 0.2 *times its rest mass, calculate the crystal momentum for a free electron of energy* 0.02 eV, *taking the origin at the bottom of the conduction band.*

Solution: With origin at the bottom of the conduction band, the E-k relationship is given by

$$E = \frac{\hbar^2 k^2}{2m^*}$$

Here, $E = 0.02\,\text{eV} = 0.02 \times 1.6 \times 10^{-19}$ joule, $m^* = 0.2m_0 = 0.2 \times 9.11 \times 10^{-31}$ kg.

$$\begin{aligned}\therefore\ \text{Crystal momentum} &= \hbar k = \sqrt{2m^* E}\\ &= (2 \times 0.2 \times 9.11 \times 10^{-31} \times 0.02 \times 1.6 \times 10^{-19})^{1/2}\\ &= \mathbf{3.41 \times 10^{-26}\ kg.m/s}\end{aligned}$$

▶ **Example 2.** *The energy wave vector dispersion relation for a one-dimensional crystal of lattice constant 'a' is given by*

$$E(k) = E_0 - \alpha - 2\beta \cos ka$$

where E_0, α *and* β *are all constants.*

(a) *Find the value of* k *at which the velocity of electron is a maximum. What is the maximum velocity?* (b) *What is the energy at the top and at the bottom of the band?*

Find the difference in energy i.e., energy gap between the top and the bottom of the energy band, and (c) *Obtain the effective mass m^* of the electron at the bottom and the top of the band.* **(Calcutta Hons.)**

Solution: (a) Wave-mechanically, a particle moving with a velocity v is equivalent to a wave packet moving with a group velocity v_g.

$$\therefore \quad v = v_g = \frac{d\omega}{dk} \tag{7.12.1}$$

where ω is the angular frequency of de Broglie waves.

Also, the energy of the particle is

$$E = \hbar\omega \Rightarrow dE/dk = \hbar d\omega \tag{7.12.2}$$

$$\therefore \quad v = \frac{1}{\hbar}\frac{dE}{dk} = \frac{2\beta a}{\hbar}\sin ka$$

$\therefore$ v is maximum when $\sin ka = 1 \Rightarrow ka = n\pi/2 \Rightarrow \boldsymbol{k = n\pi/2a}$. So, $\boldsymbol{v_{\text{max}} = 2\beta a/\hbar}$

(b) At the top of the band, $E(k)$ is a maximum and so $\cos ka$ is a minimum $\Rightarrow \cos ka = -1$.

$$\therefore \boldsymbol{E_{\text{top}} = E_0 - \alpha + 2\beta}$$

At the bottom of the band, $E(k)$ is a minimum and so $\cos ka$ is maximum $\Rightarrow \cos ka = 1$.

$$\therefore \boldsymbol{E_{\text{bottom}} = E_0 - \alpha - 2\beta}$$

$\therefore$ Energy difference or energy gap, $\boldsymbol{E_g} = E_{\text{top}} - E_{\text{bottom}} = 2\beta - (-2\beta) = \boldsymbol{4\beta}$.

(c) $$m^* = \frac{\hbar^2}{d^2E/dk^2} = \frac{\hbar^2}{2\beta a^2 \cos ka}$$

Now, at the bottom of the band, $\cos ka = 1$. So, from above, we get

$$\therefore \boldsymbol{m^* = \hbar^2/2\beta a^2}$$

At the top of the band, $\cos ka = -1$ so that from the same relation

$$\therefore \boldsymbol{m^* = -\hbar^2/2\beta a^2}$$

► **Example 3.** *The energy near the valence band edge of a crystal is given by*

$$E = -Ak^2$$

where $A = 10^{-39}\,\text{Jm}^2$. *An electron with wave vector* $\vec{k} = 10^{10}\hat{k}_x\,\text{m}^{-1}$ *is removed from an orbital in the completely filled valence band. Find the effective mass, velocity, momentum and energy of the hole.*

Solution: Given $E = -Ak^2 \Rightarrow dE/dk = -2Ak \Rightarrow d^2E/dk^2 = -2A$

$\therefore$ Effective mass of electron, $m_e^* = \hbar^2/(d^2E/dk^2) = -\hbar^2/2A$

$$\therefore\ m_e^* = -\frac{(1.05\times10^{-34})^2}{2\times10^{-39}} = -5.5\times10^{-30}\ \text{kg}$$

Since the effective mass of a hole is opposite to that of an electron at the same location in the energy band,

Effective mass of a hole is: $m_h^* = \mathbf{5.5\times10^{-30}\ kg.}$

$$\vec{k}_{\text{hole}} = -\vec{k}_{\text{electron}} = -10^{10}\hat{k}_x\ \text{m}^{-1}$$

$\therefore$ Momentum of the hole, $\hat{p}_h = \hbar\hat{k}_h = -1.05\times10^{-34}\times10^{10}\hat{k}_x$

$$= \mathbf{-1.05\times10^{-24}\hat{k}_x\ Jsm^{-1}}$$

Velocity of the hole, $\vec{v}_h = \dfrac{\vec{p}_h}{m_h^*} = -\dfrac{1.05\times10^{-24}\hat{k}_x}{5.5\times10^{-30}} = \mathbf{-1.9\times10^5\hat{k}_x\ ms^{-1}}$

Energy of the hole, in reference to zero at the top of the valence band, is

$$E_h = -E_e = +Ak^2 = +10^{-39}\times(10^{10})^2 = \mathbf{+10^{-19}\ J}$$

▶ **Example 4.** *Compute the ratio between the kinetic energies of an electron in a two-dimensional square lattice when* (a) $k_x = k_y = \pi/a$ *and when* (b) $k_x = \pi/a$ *and* $k_y = 0$.

Solution: The energy E of an electron inside a Brillouin zone is given by

$$E = \frac{\hbar^2k^2}{2m} = \frac{h^2k^2}{8\pi^2m}$$

$\therefore$ When $k_x = k_y = \pi/a$, $\quad E_1 = \dfrac{h^2}{8\pi^2m}(k_x^2+k_y^2) = \dfrac{h^2}{8\pi^2m}\times\dfrac{2\pi^2}{a^2}$

When $k_x = \pi/a$ and $k_y = 0$, $\quad E_2 = \dfrac{h^2}{8\pi^2m}\times\dfrac{\pi^2}{a^2}$

$\therefore$ The required ratio, $\quad \dfrac{\mathbf{E_1}}{\mathbf{E_2}} = \dfrac{h^2}{8\pi^2m}\cdot\dfrac{2\pi^2}{a^2}\times\dfrac{8\pi^2ma^2}{\pi^2h^2} = \mathbf{2}$

▶ **Example 5.** *Using the Kronig-Penny model, show that for* $P \ll 1$, *the energy of the lowest energy band at* $k = 0$ *is given by*

$$E = \frac{\hbar^2P}{ma^2} = \frac{h^2P}{4\pi^2ma^2}$$

Solution: For $k = 0$, the eq. (7.5.18) reduces to

$$\frac{P}{\alpha a} = \frac{1-\cos\alpha a}{\sin\alpha a} = \frac{2\sin^2(\alpha a/2)}{2\sin(\alpha a/2)\cos(\alpha a/2)} = \tan(\alpha a/2)$$

For $P \ll 1$, we may write, $\alpha a/2 = P/\alpha a \Rightarrow \alpha^2 = 2P/a^2$.

$$\text{But,} \quad \alpha^2 = 2mE/\hbar^2 \Rightarrow 2mE/\hbar^2 = 2P/a^2$$

$$\therefore \quad E = \frac{\hbar^2 P}{ma^2} = \frac{h^2 P}{4\pi^2 ma^2}$$

▶ **Example 6.** *Show that for the free electron model, the effective mass of electron is equal to the free electron mass.*

Solution: For free electron model, the energy of electron in terms of k is given by

$$E = \frac{h^2k^2}{8\pi^2 m} \quad \Rightarrow \quad \frac{d^2E}{dk^2} = \frac{h^2}{4\pi^2 m}, \quad m = \text{mass of free electron}$$

The effective mass expression is given by

$$m^* = \frac{h^2}{4\pi^2}\left(\frac{d^2E}{dk^2}\right)^{-1} = \frac{h^2}{4\pi^2} \times \frac{4\pi^2 m}{h^2}$$

$$= m \text{ (mass of free electron)}$$

▶ **Example 7.** *An insulator has an optical absorption which occurs for all wavelengths shorter than* 1800 A. *Find the width of the forbidden energy band for this insulator.*

(Lucknow Univ.)

Solution: The insulator absorbs all wavelengths shorter than 1800 A. So, the minimum energy needed to move the electron from valence band to conduction band across the energy gap is the energy of light of wavelength 1800 A.

Now, the wavelength $\lambda = 1800\,\text{A} = 1800 \times 10^{-10}$ m.

$\therefore$ Corresponding frequency, $\nu = \dfrac{c}{\lambda} = \dfrac{3 \times 10^8}{1800 \times 10^{-10}} = \dfrac{10^{16}}{6}$

$\therefore$ Energy gap of forbidden band, $E_g = h\nu$.

$$\therefore \quad E_g = \frac{6.62 \times 10^{-34} \times 10^{16}}{6} \text{ Joule}$$

$$= \frac{6.62 \times 10^{-34} \times 10^{16}}{6 \times 1.6 \times 10^{-19}} \text{ eV}$$

$$= \mathbf{6.9\,eV}$$

▶ **Example 8.** *The energy* $E(k)$ *of electrons of wave vector* $\vec{k}$ *in a solid is given by* $E(\vec{k}) = Ak^2 + Bk^2$ *where* A, B *are constants. Show that the effective mass of the electron at* $|\vec{k}| = k_0$ *is* $\hbar^2/(2A + 12Bk_0^2)$. **(GATE)**

Solution: By the problem, $E = Ak^2 + Bk^4$. Differentiating with respect to k,

$$\frac{dE}{dk} = 2Ak + 4Bk^3$$

$$\frac{d^2E}{dk^2} = 2A + 12Bk^2$$

$$\therefore \quad \left.\frac{d^2E}{dk^2}\right|_{k=k_0} = 2A + 12Bk_0^2$$

$$\therefore \text{ Effective mass of electron, } \boldsymbol{m^*} = \frac{\hbar^2}{d^2E/dk^2} = \frac{\boldsymbol{\hbar^2}}{\boldsymbol{2A + 12Bk_0^2}}$$

▶ **Example 9.** *In a crystalline solid, the energy band structure (E-k relation) for an electron of mass m is given by*

$$E = \hbar^2 k(2k - 3)/2m.$$

Find the effective mass of the electron in the crystal. **(JAM)**

Solution: By the problem, we are given that

$$E = \frac{\hbar^2 k(2k-3)}{2m}$$

$$\therefore \quad \frac{dE}{dk} = \frac{\hbar^2(4k-3)}{2m}; \quad \frac{d^2E}{dk^2} = \frac{2\hbar^2}{m}$$

$$\therefore \text{ Effective electron mass,} \quad \boldsymbol{m^*} = \frac{\hbar^2}{d^2E/dk^2} = \hbar^2 \times \frac{m}{2\hbar^2} = \frac{\boldsymbol{m}}{\boldsymbol{2}}$$

▶ **Example 10.** *Using the tight binding model, the energy-wave vector dispersion relation for a one-dimensional crystal of lattice parameter 'a' is given by*

$$E(k) = E_0 - \alpha - 2\beta \cos ka$$

(a) *Find the value of k at which the energy of the electron is maximum.*
(b) *Obtain the value of the effective mass of electron m^* in terms of k.*

(Calcutta Hons.)

Solution: We have, from the given relation,

$$\begin{aligned} E(k) &= E_0 - \alpha - 2\beta \cos ka \\ &= E_0 - \alpha - 2\beta(\cos k_x a + \cos k_y a + \cos k_z a) \end{aligned} \tag{7.12.3}$$

For one-dimensional crystal, $k_y = k_z = 0$ and let $k_x = k$. (7.12.4)

$\therefore$ From (7.12.3), $\quad E = E_0 - \alpha - 2\beta \cos ka$

(a) The energy of electron is maximum when $\cos ka = -1$

$$\therefore \quad E_{\text{max}}^{(k)} = E_0 - \alpha + 2\beta$$

$$\cos ka = -1 = \cos(2n+1)\pi \Rightarrow ka = (2n+1)\pi, \quad n = 0, 1, 2, \cdots$$

$$\therefore \quad \boldsymbol{k = (2n+1)\frac{\pi}{a}}$$

(b) From (7.12.4), $$\frac{d^2E}{dk^2} = 2\alpha^2\beta\cos ka$$

$\therefore$ Effective mass of electron, $$m^* = \frac{\hbar^2}{d^2E/dk^2} = \frac{\hbar^2}{2\alpha^2\beta\cos ka}$$

▶ **Example 11.** *The energy of an electron in an isotropic crystal is given by*

$$E = \alpha_1 k_x^2 + \alpha_2 k_y^2 + \alpha_3 k_z^2.$$

Find the equation of motion of the electron.

Solution: Velocity $v_x = \frac{1}{\hbar}\frac{\partial E}{\partial k_x} = \frac{2\alpha_1 k_x}{\hbar} \equiv \frac{\hbar k_x}{m_1}$ ($x = 1$ for x-axis)

$$\therefore\ m_1 = \frac{\hbar^2}{2\alpha_1} \Rightarrow m_i = \frac{\hbar^2}{2\alpha_i} \quad (i = 1, 2, 3)$$

$\therefore$ Motion along x_i-axis is given by

$$\frac{\partial^2 x_i}{\partial t^2} = \frac{F}{m_i} = \frac{F}{\hbar^2/2\alpha_i}$$

$$\text{or,}\quad \frac{\partial^2 x_i}{\partial t^2} = \frac{2F\alpha_i}{\hbar^2} \quad (i = 1, 2, 3)$$

▶ **Example 12.** *The dispersion relation for a bcc-crystal is given by*

$$E(k) = E_0 - \alpha - 8\beta\cos\tfrac{1}{2}k_x a\cos\tfrac{1}{2}k_y a\cos\tfrac{1}{2}k_z a.$$

Show that the energy bonds are spherical at $k = 0$ and deduce the effective mass of the electron.

Solution: As $k \to 0$, the given dispersion relation may be written as

$$\begin{aligned}
E &= E_0 - \alpha - 8\beta\left(1 - \frac{1}{2}\frac{k_x^2a^2}{4}\right)\left(1 - \frac{1}{2}\frac{k_y^2a^2}{4}\right)\left(1 - \frac{1}{2}\frac{k_z^2a^2}{4}\right) \\
&= E_0 - \alpha - 8\beta\left\{1 - \tfrac{1}{8}(k_x^2 + k_y^2)a^2\right\}\left(1 - \frac{k_z^2a^2}{8}\right) \\
&= E_0 - \alpha - 8\beta\left\{1 - \tfrac{1}{8}(k_x^2 + k_y^2 + k_z^2)a^2\right\} \\
&= E_0 - \alpha - 8\beta + \beta k^2a^2 \qquad (7.12.5)
\end{aligned}$$

So, $E \propto k^2$ and hence the energy bands are ***spherical***.

From (7.12.5), $\frac{\partial E}{\partial k} = 2\beta ka^2$ and $\frac{\partial E}{\partial k^2} = 2\beta a^2$. Therefore:

$$\text{Effective mass of electron, } m^* = \frac{\hbar^2}{\partial^2E/\partial k^2} = \frac{\hbar^2}{2\beta a^2}$$

▶ **Example 13.** *The valence band of a simple cubic metal has the form*

$$E = Ak^2 + B$$
$$\text{where } A = 10^{-38}\,\text{Jm}^2$$

Calculate the value of the ratio m^*/m.

Solution: Since $E = Ak^2 + B$, $\partial E/\partial k = 2Ak$ and $\partial^2 E/\partial k^2 = 2A$.

Now, we have, $m^* = \dfrac{\hbar^2}{\partial^2 E/\partial k^2}$. Therefore,

$$\frac{m^*}{m} = \frac{\hbar^2}{m(\partial^2 E/\partial k^2)} = \frac{(1.05\times 10^{-34})^2}{9.1\times 10^{-31}\times 2A}$$

$$= \frac{(1.05\times 10^{-34})^2}{9.1\times 10^{-31}\times 2\times 10^{-38}} = \frac{1.10}{18.2}\times 10$$

$$\therefore\ \boldsymbol{\frac{m^*}{m}} = \frac{11.0}{18.2} = \mathbf{0.604}$$

▶ **Example 14.** *Find the expression for electron densities at which the free electron spherical Fermi surface first touches the zone bounbdary in a bcc-metal.*

Solution: Let k_m be the minimum distance in $\vec{k}$-space from the origin to the boundary of the first Brillouin zone.

For *bcc*-metal, here $G = \dfrac{2\pi}{a}(1,0,0) = \dfrac{2\pi\sqrt{2}}{a}$. So, $k_m = G/2 = \pi\sqrt{2}/a$.

If the Fermi surface touches the zone boundary, $k_F = k_m$.

$$\therefore\ (3\pi^2 n)^{1/3} = \pi\sqrt{2}/a \quad (\because\ k_F^3 = 3\pi^2 n)$$

$$\therefore\ \boldsymbol{n} = \frac{1}{3\pi^2}\cdot\frac{2\sqrt{2}\pi^3}{a^3} = \boldsymbol{\frac{2\sqrt{2}\pi}{3a^3}}$$

▶ **Example 15.** *In a simple cubic quasi-free electron metal, the spherical free surface just touches the first Brilloiun zone. Calculate the number of conduction electrons per atom.*

Solution: Here $G = \dfrac{2\pi}{a}(1,0,0) = \dfrac{2\pi}{a}$. So, $k_m = G/2 = \pi/a$. By the condition of the problem, we have

$$k_F = k_m \Rightarrow (3\pi^2 n)^{1/3} = \frac{\pi}{a} \Rightarrow n = \frac{\pi^3}{a^3}\cdot\frac{1}{3\pi^2} = \frac{\pi}{3}\frac{1}{a^3}$$

$\therefore$ Number of electrons/atom $= n(a^3) = \pi/3 = 3.14/3 = \mathbf{1.047}$.

Review Questions

1. What are the shortcomings of the free electron theory? What is the main cause of its failure?
2. How does the potential energy of an electron vary in an infinite one-dimensional crystal? How is this potential represented in Kronig-Penney model?
3. Describe the basic principles for the band structure of electronic states in a crystal. **(Calcutta Hons.)**
4. Define the term 'effective mass' of an electron. Is it different from the free electron mass? What is 'crystal momentum'? **(Calcutta Hons.)**
5. In the Kronig-Penney model of electrons in a linear lattice, if the height of the periodic potential increases, how will the widths of the allowed energy bands change? **(Purvanchal Univ.)**
6. Discuss Kronig-Penney model. Using the model, show that the energy spectrum of electron consists of a number of allowed energy bands separated by forbidden regions. **(Patna, Himachal and Lucknow Univ.)**
7. Show graphically the variation of the energy with the wave number k for the first allowed band in a solid. Hence explain the variation in sign of the effective mass of an electron in the band. Explain the concept of holes. **(Calcutta Hons.)**
8. What are the main conclusions of the Kronig-Penney model treatment? Draw schematically the energy-wave vector diagram for a crystal. What do you mean by reduced zone representation?
9. Discuss Kronig-Penney model for the energy band structure of solids. Show that each band can accommodate $2N$ electrons, where N is the total number of atoms in the crystal.
10. State and prove Bloch theorem in periodic potential in a crystal. **(Himachal Univ.)**
11. What do you mean by Brillouin zones? Discuss the formation of Brillouin zones for a (i) linear lattice, (ii) two-dimensional lattice. **(Lucknow and Punjab Univ.)**
12. What is Bloch theorem? Starting from the relation

$$P\frac{\sin \alpha a}{\alpha a} + \cos \alpha a = \cos ka$$

discuss the motion of electrons in one-dimension and the variation of energy, velocity and effective mass as a function of wave vector. **(Vidyasagar Hons.)**

13. Deduce expressions for the concentration of electrons in the conduction band and the concentration of holes in the valence band of an intrinsic semiconductor at a temperature T. Show that the product is independent of the Fermi level. What is the implication of this result (Assume the density of states of electrons is $g(E) = 1/2\pi^2\,(2m/\hbar^2)^{3/2}\,E^{1/2}$). **(Burdwan Hons.)**

14. How does the band theory lead to the classification of solids into metals, semiconductors and insulators?

15. For the first allowed energy band in a crystalline solid in the reduced zone, plot qualitatively the energy E, the velocity v, the effective mass m^* and the extent of freedom f_k of an electron against the wave number k. Hence explain the change in sign of m^*. **(Calcutta Hons.)**

16. Show that the effective freedom of electrons in an energy band in a solid is a maximum when the band is filled up to the inflexion point of E-k curve and zero when completely filled.

17. Distinguish between metals, insulators and semiconductors. Why does a semiconductor act as insulator at 0 K? **(North Bengal and Burdwan Hons.)**

18. What are valence and conduction bands? What do you mean by forbidden band? Explain why a solid whose energy bands are full cannot be a metal. **(Meerut and Punjab Univ.)**

19. Show that the effective mass of an electron in a crystal is given by

$$m^* = \hbar^2/(d^2E/dk^2)$$

where the symbols have their usual meanings. **(Calcutta Hons.)**

20. According to a simple model, the energy of a free electron in a metal is given by the following function of wave number

$$E = A - B\cos ka$$

where A, B are constants and a is the distance between adjacent atoms. Draw the E-k curve. What value of effective mass can you deduce from the above formula? With a spacing $a = 2.0$Å, what is the range of the electron wavelengths in the first Brillouin zone? **(Calcutta Hons.)**

Short and Objective Questions

1. Explain how atomic energy levels split into bands when a number of atoms are brought close together to form a crystal. **(Purvanchal and Bangalore Univ.)**

2. Discuss the periodic character of potential in a crystal. What is the basic assumption in Kronig-Penney model? **(Himachal and Punjab Univ.)**

3. Kronig-Penney model gives a simplified solution of the form

$$\cos ka = P\frac{\sin \alpha a}{\alpha a} + \cos \alpha a \equiv \phi(\alpha, a)$$

Plot $\phi(\alpha, a)$ as a function of ka with $P = 5$. Hence explain 'forbidden' and 'allowed' states. **(Burdwan Hons.)**

4. What is the effect of changing the binding energy of an electron on the energy bands. **(Punjabi Univ.)**

5. Show that the number of possible wave functions in any energy band is equal to the number of unit cells. **(Guru Nanak Univ.)**

6. Show that a band can accommodate $2N$ electrons where N is the number of atoms in the crystal.

7. Show how the energy, velocity and effective mass vary as function of wave vector. **(Lucknow Univ.)**

8. What is the origin of energy gap? Give the approximate values of the energy gap of two semiconductors. **(Punjabi Univ.)**

9. Explain why a solid whose energy bands are full cannot be a metal. Differentiate between a good conductor and a good insulator taking the resistivity range.

10. The conductivity of a metal decreases with rise in temperature whereas that of a semiconductor increases with increase in temperature. Explain.

11. Why does a semiconductor act as an insulator at 0 K? Why are valence electrons called conduction electrons?

12. Fill in the blanks:

    (a) The energy gap in pure Germanium is ____ eV.
    (b) ____ is the quantum of lattice vibrational energy.
    (c) The motion of electron in the periodic crystal lattice gives rise to ____.

    (Meerut Univ.)

13. Decide True or False:

    (a) The effective mass of an electron may be negative.
    (b) Pure semiconductor is insulator at 0 K. **(Meerut Univ.)**

14. A pure semiconductor behaves as an insulator at

    (a) 273 K, (b) −273°C, (c) 373 K, (d) none of these.

15. The energy region separating the highest occupied band and the empty band is called a

    (a) conduction band, (b) valence band,
    (c) forbidden band, (d) none of these.

16. In Kronig-Penney model of electrons in a linear lattice, if the strength of the periodic potential increases, the width of the allowed energy band

    (a) increases, (b) decreases,
    (c) remains constant, (d) none of these. **(Purvanchal Univ.)**

17. The potential of an electron in a one-dimensional arrangement of atoms is identical to that used in the K-P model. If $V_0 ab \ll h^2/4\pi^2 m$, the energy band gap at $k = \pi/a$ is

(a) $\dfrac{2V_0 b}{a}$, (b) $\dfrac{2V_0 a}{b}$, (c) $\dfrac{V_0 b}{2a}$, (d) $\dfrac{V_0 a}{2b}$.

18. The Kronig-Penney potential with $P \ll 1$, the energy of the lowest band at $k = 0$ is

(a) $\dfrac{4\pi^2 ma}{h^2 P}$, (b) $\dfrac{h^2 P}{4\pi^2 ma}$, (c) $\dfrac{4\pi^2 mP}{h^2 a}$, (d) $\dfrac{h^2 a}{4\pi^2 mP}$.

19. In a $2d$-square lattice, the ratio between the kinetic energies of an electron $E_{\text{diag}}/E_{\text{midface}}$ is

(a) π, (b) $1/\pi$, (c) $\frac{1}{2}$, (d) 2.

20. The velocity of electron from (E-k) curve is:

(a) $v = \dfrac{1}{k}\left(\dfrac{dE}{dk}\right)$, (b) $v = \hbar\left(\dfrac{dR}{dk}\right)$,

(c) $v = \dfrac{1}{\hbar}\left(\dfrac{d^2E}{dk^2}\right)$, (d) $v = \hbar\Big/\left(\dfrac{dE}{dk}\right)$.

21. The kinetic energy of a free electron at a corner of the first Brillouin zone of a two-dimensional square lattice is larger than that of an electron at the mid-point of a side of the zone by a factor b. The value of b is

(a) $\sqrt{2}$, (b) 2, (c) 4, (d) 8. **(GATE)**

Problems

1. A one-dimenional lattice of spacing a has a potential distribution of the type as in Kronig-Penney model. If the potential is $-V$ at each lattice point and abruptly changes to zero at a distance of $0.1a$ on either side of the lattice point, find the width of the first energy gap in the electron energy spectrum. [0.37 V]

2. The potential of an electron in a one-dimensional lattice is of the same type as in Kronig-Penney model. Assuming $V_0 ab \ll \hbar^2/m$, prove that the energy band gap at $k = \pi/a$ is $2V_0 b/a$.

3. The energy versus wave vector relationship for a conduction electron in a semiconductor is given by $E = 5\hbar^2 k^2/m_0$. Determine the effective mass of the electron. [$0.1\, m_0$]

4. Assuming copper to be highly pure, find the resistivity of copper at 700 K. Given: resistivity of copper at 300 K $= 1.7 \times 10^{-8}\,\Omega\,$m. [$552 \times 10^{-8}\,\Omega$m]

5. Show that for a simple square lattice, the kinetic energy of a free electron at a corner of the first Brillouin zone is higher than that of electron at mid-point of a side face of the zone by a factor 2, that is, $E_{\text{diagonal}}/E_{\text{mid face}} = 2$.

6. Prove that the total number of states in an energy band of a finite crystal is equal to the number of atoms in the unit cell.

7. The energy of a free elecgtron in a crystal is given by

$$E = A - B\cos ka,$$

where A and B are constants, k is the wave number and a is the distance between adjacent atoms. Draw the E-k curve, and find the effective mass. If $a = 2\text{Å}$, what is the range of the de Broglie wavelengths in the first Brillouin zone?

(Calcutta Hons.) [$\hbar^2/Ba^2 \cos ka$, -4Å to $+4\text{Å}$]

8. Estimate the band gap of an insulated if it absorbs all e. m. radiations of wavelength below 3000Å. [4.14 eV]

9. The energy wave-vector dispersion relation for a one-dimensional crystal of lattice constant 'a' is given by the relation

$$E(k) = E_0 - \alpha - 2\beta\cos ka,$$

where E_0, α, β are constants. Obtain the effective mass of the electron at the bottom and at the top of the band. (Calcutta Hons.)

[**Hints.** Effective mass of electron, $m^* = \hbar^2/(d^2E/dk^2) = \hbar^2/2\beta a^2 \cos ka$. At bottom of band, $\cos ka = 1$. So, $\boldsymbol{m^* - \hbar^2/2\beta a^2}$. At the top of band, $\cos ka = -1$. So, $\boldsymbol{m^* = -\hbar^2/2\beta a^2}$]

10. Obtain expression for the electron densities at which the free electron spherical Fermi surface first touches the zone boundary in a *fcc*-metal. [$n = \sqrt{3}\pi/a^3$]

11. Calculate the ratio k_F/k_m for metals of valency 1 for *fcc*-lattices. [0.901]

12. Show that the number of states in the first Brillouin zone of a simple cubic lattice is equal to the number of lattice sites.

13. Estimate the band gap of an insulator if it absorbs electro-magnetic radiation of the wavelengths below 3000 A. [4.14 eV]

14. Calculate the Fermi radius of two-dimensional square lattice of parameer a.

[$k_F = 0.399\,(2\pi/a)$]

[**Hints.** The reciprocal lattice of a square lattice is a square lattice of dimension $2\pi/a$. So, the area of the first $BZ = 4\pi^2/a^2$. Assuming the metal to be monovalent, the area occupied by electrons will be half the area of BZ (which can be occupied by 2 electrons allowing two possible spin states for each). In two-dimensions, Fermi sphere is a Fermi circle of radius k_F.

$$\therefore\ \pi k_F^2 = \frac{1}{2}\left(\frac{4\pi^2}{a^2}\right) \rightarrow \boldsymbol{k_F} = \frac{1}{\sqrt{2\pi}}\cdot\frac{2\pi}{a} = \boldsymbol{0.399(2\pi/a)}]$$

15. The allowed values of k in a one-dimensional lattice are given by $k = n(2\pi/L)$. Show that the density of the electron states in the lattice of unit length is

$$g(E) = \frac{(1/2\pi)}{(dE/dk)}.$$

16. The dispersion relation for a one-dimensional crystal is

$$E(k) = E_0 - \alpha - 2\beta \cos ka$$

Find (a) the value of k at whch the velocity of an electron is maximum, (b) an expression for m^* as a function of k, and (c) the width of the energy band.

[(a) $k = n\pi/2a$, (b) $\hbar^2/(2\beta a^2 \cos ka)$, (c) 4β]

17. Find the volume V_k of the first Brillouin zone in sodium ($a = 4.3$Å) [6.213 Å]

[**Hints.** Volume V_k of the first Brillouin zone is the volume of a primitive unit cell of the reciprocal lattice. If V be the volume of a primitive unit cell in direct lattice, then

$$V_k = (2\pi)^3/V$$

Sodium is of *fcc* lattice. So, $V = a^3/2$.

$$\therefore\ V_k = \frac{(2\pi)^3}{a^3/2} = \frac{8 \times 30.9}{79.5/2} = \mathbf{6.213\ \mathring{A}^3}, \quad \text{on simplification.}]$$

18. Energy $E(k)$ of electrons of wave vector $\vec{k}$ in a solid is given by

$$E(k) = Ak^2 + Bk^4$$

where $A =$ const. > 0 and $B =$ const. > 0. Find the effective mass of the electron at $|\vec{k}| = k_0$. **(Calcutta Hons.)**

8

Dielectric Properties of Materials

8.1 Introduction

Dielectrics are basically *insulating materials* in which the electrons are very tightly bound and belong to particular atoms and also the forbidden gap in their energy band is quite large. The electrons cannot thus be made free by dislodging them at ordinary temperatures or with ordinary fields to make them wander about other atoms in the material. However, they can be slightly displaced from their positions within the molecules under the application of an external electric field and their cumulative effect accounts for the characteristic behaviour of the dielectrics. When a sufficiently strong external electric field is applied to the dielectric, the centres of the positive charges are slightly displaced in the direction of the field and those of the negative charges opposite to it. This produces *local electric dipole* throughout the dielectric, a *dipole* being an entity of equal and opposite charges separated by a *small* distance. The dielectric is then said to be *polarized*. A dielectric material has interesting electrical properties since an electric field can polarize them. The internal distortion in the lattice can be studied in detail if one has the knowledge of the dielectric constant of the material.

• In the absence of an external field, the centres of mass of the positive and the negative charges in a molecule of a dielectric may or may not coincide. In case the centers coincide, it is said to be a ***non-polar molecule***. If, on the otherhand, they do not, regardless of the external field, the molecule is called a ***polar molecule***, behaving like rigid dipoles.

• Dielectric materials may be *solids* such as glass, porcelain, many solid crystals etc., *liquids* like pure water, methyl alcohol etc., and *gases* such as hydrogen, nitrogen, ammonia etc.

A dielectric is said to be *homogeneous* and *isotropic*, if all its properties are the same at any point and in every direction within it.

8.2 Some conceptual and terminological recapitulation

When a dielectric is placed in an external electric field $\vec{E}_0$, the positive and negative charges are slightly displaced from their equilibrium positions throughout the dielectric. It results in the formation of a large number of dipoles, each with some dipole moment in the field-direction. The dielectric is said to be *polarized* with a ***polarization*** $\vec{P}$, defined as the *dipole moment per unit volume* of the material.

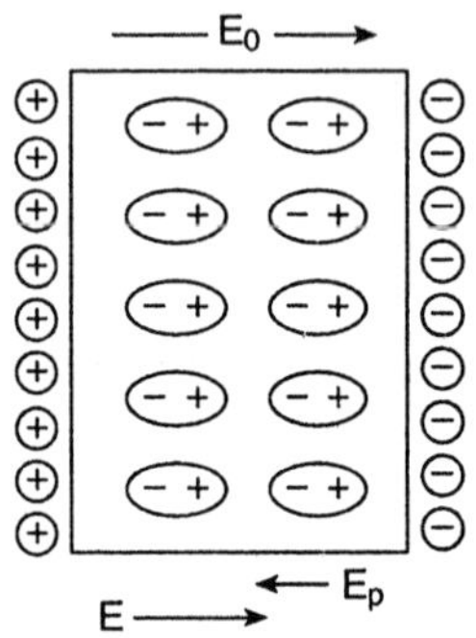

Fig. 8.1 Illustration of dielectric polarization

The effect of polarization is to reduce the magnitude of E_0 so that the resultant field $E < E_0$. Vectorially,

$$\vec{E} = \vec{E}_0 - \vec{E}_p \tag{8.2.1}$$

where the field $\vec{E}_p$ is the *polarization field* that opposes E_0 in the material. For ordinary field, the polarization $\vec{P}$ is proportional to the macrofield E. We write

$$\vec{P} = \varepsilon_0 \chi_e \vec{E} \tag{8.2.2}$$

where ε_0 is the *permittivity* of free space and χ_e the electric *susceptibility*.

• Except for a constant factor, *susceptibility is a measure of the polarization per unit resultant field.*

The *electric displacement vector* $\vec{D}$ for an *isotropic* dielectric medium may be *defined* as

$$\vec{D} = \varepsilon_0 \varepsilon_r \vec{E} = \varepsilon_0 \vec{E} + \vec{P} \quad \text{or,} \quad \vec{P} = \varepsilon_0(\varepsilon_r - 1)\vec{E} \tag{8.2.3}$$

where ε_r is called the *relative permittivity* or *dielectric constant* of the material, a dimensionless scalar quantity for an isotropic medium. Equation (8.2.3) is a fundamental relation between the three vectors $\vec{D}, \vec{E}$ and $\vec{P}$

Eq. (8.2.3) may be recast, using (8.2.2), as

$$\boxed{\varepsilon_r = \frac{\varepsilon_0 \vec{E} + \vec{P}}{\varepsilon_0 \vec{E}} = 1 + \chi_e} \tag{8.2.4}$$

• Like susceptibility, the dielectric constant ε_r is also a measure of the degree of polarization of the material. Larger the polarization per unit resultant field, higher would be the value of the dielectric constant.

• If ε_r depends on the direction of measurement (non-isotropic material), $\vec{D}$ and $\vec{E}$ may not be parallel and the dielectric constant is a *tensor* quantity. We shall however be concerned here with *isotropic materials* only.

The dielectric constant ε_r, of a material can be experimentally measured by using it as a dielectric between the plates of a parallel-plate capacitor. If A be the area of the plate and d the plate-separation, the capacitance C is given by

$$C = \varepsilon_0 \varepsilon_r (A/d) \tag{8.2.5}$$

If C_0 be the capacitance of the same capacitor when the space between the plates is *evacuated*, we get

$$C_{\text{vac}} = C_0 = \varepsilon_0 (A/d) \tag{8.2.6}$$

From (8.2.5) and (8.2.6), $$\varepsilon_r = (C/C_0) \tag{8.2.7}$$

So, the dielectric constant ε_r can be obtained readily by measuring C and C_0 experimentally.

8.3 Dipole moment, polarization and polarizability

Two equal and opposite charges, $+q$ and $-q$, separated by a small distance r, as shown in Fig. 8.2, constitute an ***electric dipole***. The ***moment of the dipole*** is defined is

$$\boxed{\vec{p} = q\vec{r}}$$

and is *directed from the negative to the positive charge*. The dipole moment is thus equal to one of the charges multiplied by the distance between them.

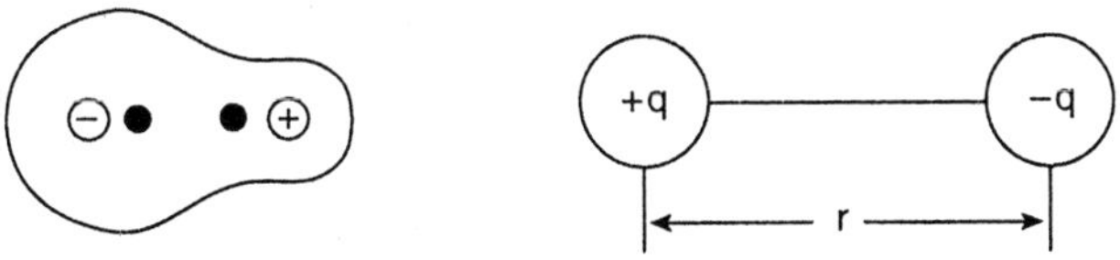

Fig. 8.2 An electric dipole

The total dipole moment of a neutral system of point charges is a vector quantity $\vec{p}$ defined by

$$\vec{p} = \sum_i q_i \vec{r}_i \tag{8.3.1}$$

where $\vec{r}_i$ is the position-vector of the point charge q_i.

Let $-q$ and $+q$ be two point charges separated by a distance r and with position vectors $\vec{r}_1$ and $\vec{r}_2$ respectively, with respect to the origin O.

∴ Dipole moment of the system (Fig. 8.3) is

$$\vec{p} = -q\vec{r}_1 + q\vec{r}_2 = q(\vec{r}_2 - \vec{r}_1) = q\vec{r} \tag{8.3.2}$$

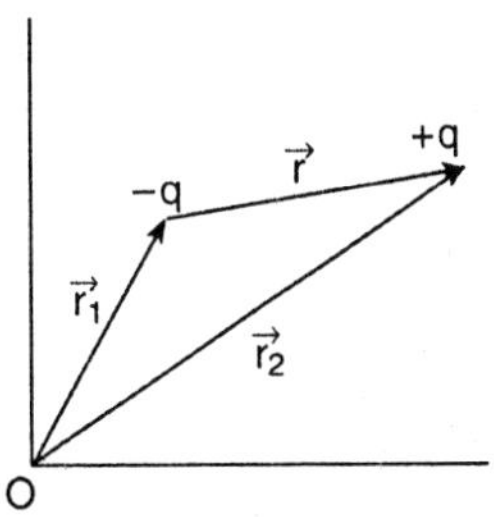

Fig. 8.3 Electric dipole consisting of two points charges $-q$ and $+q$

where $\vec{r}$ is the vector distance from $-q$ to $+q$.

∴ Magnitude of the dipole moment $= qr$.

- The distance r of separation between the two charges is rather small compared to the distance of the observer.

The ***polarization*** $\vec{P}$ *is defined as the dipole moment per unit volume.* If the number of molecules per unit volume is N and if each has a dipole moment $\vec{p}$,

$$\vec{P} = N\vec{p}.$$

- The mechanism of polarization is different for polar and non-polar dielectric materials.

Since $\vec{P} = N\vec{p}$, we have $\vec{p} = \vec{P}/N = \varepsilon_0\chi_e\vec{E}/N = \alpha\vec{E}$, where

$$\alpha = \varepsilon_0\chi_e/N \tag{8.3.3}$$

is called the ***polarizability***.

$$\therefore \quad \boxed{\vec{P} = N\alpha\vec{E}} \tag{8.3.4}$$

using equation (8.2.2).

8.3.1 Field due to an electric dipole

Let a dipole of moment $\vec{p}$ lie at the origin and point along z-direction (Fig 8.4). The potential V at (r, θ) from the centre of the dipole is then given by

$$V(r,\theta) = \frac{\hat{r}\cdot\vec{p}}{4\pi\varepsilon_0 r^2} = \frac{p\cos\theta}{4\pi\varepsilon_0 r^2} \tag{8.3.5}$$

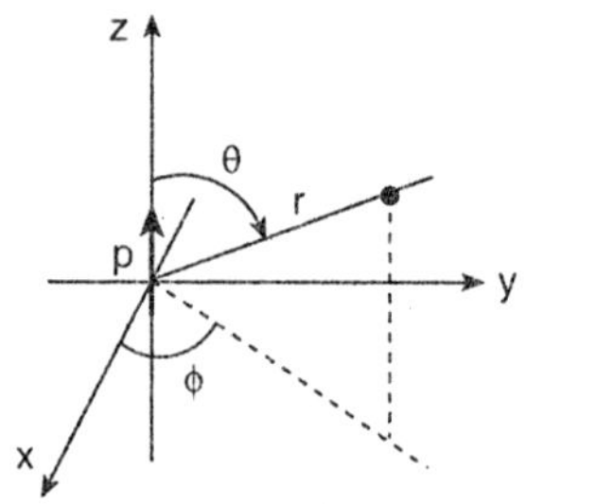

Fig. 8.4 Electric field due to a dipole

In spherical polar coordinates (r, θ, ϕ), the components of the field $\vec{E}$ are:

$$E_r = -\frac{\partial V}{\partial r} = \frac{2p\cos\theta}{4\pi\varepsilon_0 r^3}$$

$$E_\theta = -\frac{1}{r}\frac{\partial V}{\partial\theta} = \frac{p\sin\theta}{4\pi\varepsilon_0 r^3}$$

$$E_\phi = -\frac{1}{r\sin\theta}\frac{\partial V}{\partial\phi} = 0$$

$$\therefore \quad \vec{E}(r,\theta) = \frac{p}{4\pi\varepsilon_0 r^3}(2\cos\theta\hat{r} + \sin\theta\hat{\theta}) = A(2\cos\theta\hat{r} + \sin\theta\hat{\theta}) \tag{8.3.6}$$

where $A = p/4\pi\varepsilon_0 r^3$.

Substituting, $y = r\sin\theta$ and $z = r\cos\theta$, the radius vector $\vec{r}$, lying in the yz-plane, is given by

$$\vec{r} = \hat{j}y + \hat{k}z = \hat{j}r\sin\theta + \hat{k}r\cos\theta$$

$$\therefore \quad \hat{r} = \vec{r}/r = \hat{j}\sin\theta + \hat{k}\cos\theta$$

Similarly, $$\hat{\theta} = \hat{j}\cos\theta - \hat{k}\sin\theta$$

$\therefore$ From (8.3.6), substituting for $\hat{r}$ and $\hat{\theta}$, we get

$$\begin{aligned}\vec{E}(r,\theta) &= A\left[2\cos\theta(\hat{j}\sin\theta + \hat{k}\cos\theta) + \sin\theta(\hat{j}\cos\theta - \hat{k}\sin\theta)\right] \\ &= A\left[\hat{j}3\sin\theta\cos\theta + \hat{k}(2\cos^2\theta - \sin^2\theta)\right]\end{aligned}$$

$$= A\left[\hat{j}3\sin\theta\cos\theta + \hat{k}(3\cos^2\theta - 1)\right]$$

$$= \frac{p}{4\pi\varepsilon_0 r^3}\left[3\cos\theta(\hat{j}\sin\theta + \hat{k}\cos\theta) - \hat{k}\right], \quad \text{substuting for } A$$

$$= \frac{1}{4\pi\varepsilon_0 r^5}\left[3pr^2\cos\theta\hat{r} - \hat{k}r^2 p\right]$$

$$= \frac{1}{4\pi\varepsilon_0 r^5}\left[3(\vec{p}\cdot\vec{r})\vec{r} - r^2\vec{p}\right] \tag{8.3.7}$$

which is the required field due to an electric dipole of moment $\vec{p}$.

8.3.2 Mechanisms of polarization

Both in molecules and atoms, an external electric field displaces the electrons relative to the corresponding nuclei (Fig. 8.5), leading to what is called *electronic polarization.* Apart from the electronic displacement, the external field may as well cause a displacement of atoms or ions within the molecule. For instance, in an HCl molecule, the field may to some extent change the interionic distance giving rise to a change in dipole moment. This is *atomic* or *ionic polarization.* Over and above these two, there may also occur what is termed *dipolar* or *orientational polarization* in substances composed of polar molecules possessing a permanent electric dipole moment. The

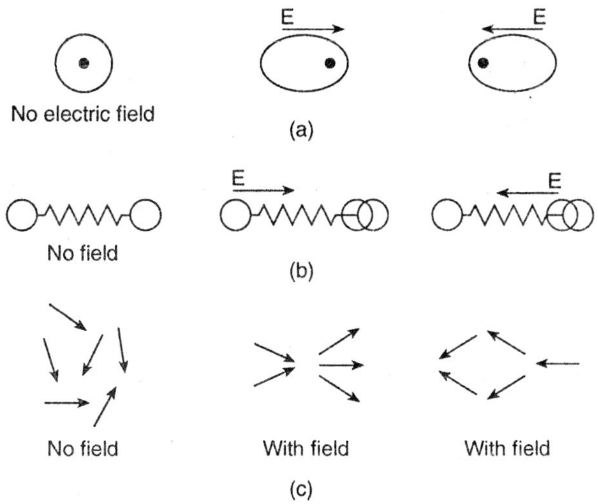

Fig. 8.5 Mechanisms of polarization: (a) electronic, (b) ionic, (c) dipolar or orientational

external electric field tends to rotate and orient the polar molecules in the field direction and this tendency of alignment is opposed by thermal agitation of the molecules. Higher the temperature, greater is the thermal agitation towards randomness and smaller is the degree of orientation. Dipolar polarization is thus *strongly temperature-dependent* and occurs mainly in fluids where the molecules are fairly free to rotate.

In the context of the above discussions, the total polarization

$$P = P_e + P_i + P_d$$

where the subscripts (e, i, d) refer to electronic, ionic and dipolar polarizations.

Also, from equation (8.3.3)

$$\left.\begin{aligned} P_e &= \alpha_e NE \\ P_i &= \alpha_i NE \\ P_d &= \alpha_d NE \end{aligned}\right\}$$

$$\therefore \quad \boldsymbol{\alpha = \alpha_e + \alpha_i + \alpha_d}$$

where $\boldsymbol{\alpha}$ gives the ***total polarizability*** and α_e the *electronic*, α_i the *ionic* and α_d the *dipolar* or *orientational* polarizabilities.

We shall describe and discuss, in what follows, these different types of polarization in somewhat greater details.

8.4 Electronic polarization

We shall discuss the matter first on the basis of *classical atom model*, followed by the *quantum mechanical* one. Further, we shall assume that there is no interaction between the adjacent dipoles which only implies that our treatment will be valid only for *gases* where the interatomic distances are too large.

Classical model—Let the atomic nucleus consist of total positive charge Ze where Z is the atomic number and e the electronic charge. For electrical neutrality of the atom, the total electronic charge $-Ze$ outside the nucleus is uniformly distributed inside a sphere of radius r, the atomic radius. The volume density of charge within the sphere is, plainly, given by

$$\rho = -\frac{Ze}{(4/3)\pi r^3} \tag{8.4.1}$$

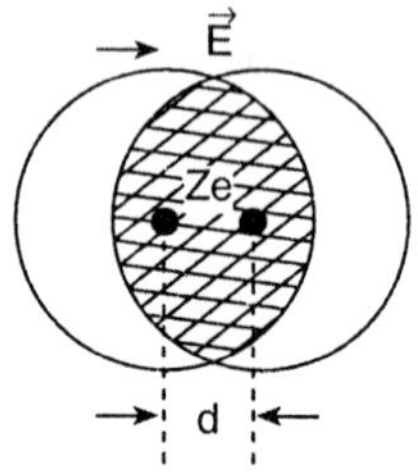

Fig. 8.6 Slight shifting of electron-cloud relative to nucleus

Let now the atom be placed in an electric field $\vec{E}$ (Fig. 8.6). The positively charged nucleus would experience a force in the direction of $\vec{E}$, and so would also the negatively charged electron-cloud, but in the opposite direction. This would result in a relative shift between the electron-cloud and the nucleus. But the Coulomb attraction between them would tend to restore the electron-cloud in its original position with nucleus at the centre. Equilibrium would be established if the Coulomb force and that due to the applied field get balanced with the nucleus shifted by a small distance d relative to the electron-cloud.

Force on the electron-cloud due to the field is $-Ze\vec{E}$. But, by Gauss theorem, only the portion of electron-cloud not surrounding the nucleus experiences the Coulomb force. So, the Coulomb force acts only on the electron-cloud contained in a sphere of radius d.

Charge within the sphere of radius r is given by

$$q = \rho\left(\tfrac{4}{3}\pi d^3\right) = -Ze(d/r)^3, \quad \text{by using (8.3.1)}, \tag{8.4.2}$$

$$\therefore \text{ Coulomb force is: } \; -Ze(d/r)^3 \frac{Ze}{4\pi\varepsilon_0 d^2}\vec{a} \tag{8.4.3}$$

where $\vec{a}$ is the unit vector opposite to the field $\vec{E}$.

In equilibrium, we get $$ZeE = Ze\left(\frac{d}{r}\right)^3 \frac{Ze}{4\pi\varepsilon_0 d^2}$$

$$\Rightarrow \boxed{d = \frac{4\pi\varepsilon_0 r^3}{Ze}E} \tag{8.4.4}$$

From the boxed equation (8.4.4), we observe that the shift d of electron cloud is proportional to the field E. This shift produces a dipole of moment p given by

$$p = Zed = 4\pi\varepsilon_0 r^3 E = \alpha_e E, \quad \text{using (8.4.4)}. \tag{8.4.5}$$

where $\boldsymbol{\alpha_e = 4\pi\varepsilon_0 r^3}$ is the ***electronic polarizability.***

If N be the number of atoms per unit volume, the *induced* dipole moment per unit volume or ***electronic polarization*** is given by

$$\boldsymbol{P_e} = Np = N\alpha_e E = \boldsymbol{4\pi\varepsilon_0 N r^3 E} \tag{8.4.6}$$

• Inert gases such as He, A, Ne show only electronic polarization P_e in presence of a field and P_e is *dependent on atomic structure but is independent of temperature.*

Quantum model—For simplicity, we shall consider the H-atom in the ground state. The volume charge density $\rho(r)$ of an electron, distant r from the nucleus, is given quantum mechanically by

$$\rho(r) = -\frac{e}{\pi a_0^3}e^{-2r/a_0} \tag{8.4.7}$$

where a_0 = the first Bohr radius.

In so far as the perturbing effect of the field is concerned, the charges are confined within very small r, $r \ll a_0$, and in the situation the exponential term in (8.4.7) = 1.

$$\therefore \; \rho(r) = -\frac{e}{\pi a_0^3} \tag{8.4.8}$$

The field displaces the electron cloud by d. Assuming the charge distribution remains unchanged, even in presence of E, the charge within a sphere of radius d is given by

$$\int_0^d \rho(r)4\pi r^2 dr = -\frac{4e}{3a_0^3}d^3 \tag{8.4.9}$$

Coulomb attraction between the nucleus and the electron-cloud is thus

$$\left(\frac{4e}{3a_0^3}d^3\right)\frac{e}{4\pi\varepsilon_0 d^2} \quad (\because \; Z = 1)$$

In equilibrium, we get $$eE = \frac{4e}{3a_0^3}d^3\frac{e}{4\pi\varepsilon_0 d^2}$$

or, $$\boxed{d = \frac{3\pi\varepsilon_0 a_0^3}{e}E} \tag{8.4.10}$$

• Comparing the expressions (8.4.4) and (8.4.10), it is observed that the factor 4 in the *classical expression* for d has been replaced by 3 in *quantum mechanical one* with no essential change in the order of magnitude of electronic polarization. So, quantum mechanically, $\boldsymbol{\alpha_e = 3\pi\varepsilon_0 a_0^3}$.

• Note that α_e has the dimensions of volume and for $r = 10^{-10}$ m, $\alpha_e \sim 10^{-40}$ m^3, and is independent of temperature at moderate temperatures.

8.5 Ionic polarization

In an ionic compound such an NaCl and other alkali halides, the application of an external electric field displaces the positive and negative ions and thereby causes ***ionic polarization*** of the medium.

As a typical ionic compound we take NaCl and assume that the intermolecular separation is quite large to disregard dipole-dipole interaction and we also assume that the field on the ions is the same as the applied one.

Let d_0 be the separation between Na^+ and Cl^- ion in NaCl molecule (Fig. 8.7), the bond between Na^+ and Cl^- being ionic in nature. The electric field $\vec{E}$ exerts on the Na^+ ion a force in the the direction of $\vec{E}$ and also a force on Cl^- ion but in the opposite direction. As a result, the two ions are displaced oppositely and this interionic separation *induces* a dipole moment p given by

$$p = \alpha_i E \tag{8.5.1}$$

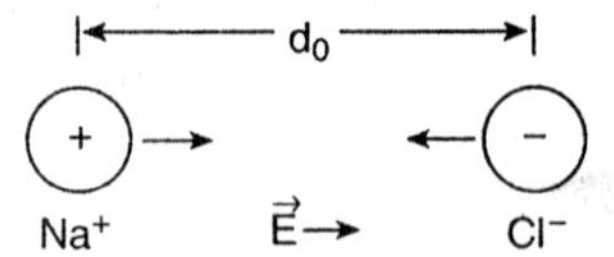

Fig. 8.7 Ionic polarization in NaCl

where $\boldsymbol{\alpha_i}$ is the ***ionic polarizability***.

If Δx be the change in the interionic separation due to the electric field E, we have

$$\beta\Delta x = eE,$$

where $\beta =$ the force constant and e the electronic charge.

$$\therefore\quad \alpha_i = p/E = e\Delta x/E = e^2/\beta$$

$$\therefore\quad \textit{Ionic polarization,}\quad \boxed{P_i = N\alpha_i E} \tag{8.5.2}$$

where N is the number of molecules per unit volume.

• Ionic polarizability is important if the field is steady or slowly time-varying. For a *rapidly varying field*, however, the ionic movements cannot follow the field variations. So, the *ionic polarization* in such a field is *negligible*.

• Note that, similar to the electronic polarizability, the ionic polarizability is also independent of temperature at moderate temperatures.

● It is to be noted that the separation d_0 between Na^+ and Cl^- in a NaCl molecule imparts a *permanent dipole moment* even in absence of the field.

8.6 Dipolar (or Orientational) polarizability

The ***dipolar polarizability***, also called ***orientational polarizability***, is significant only in materials having *permanent dipole moment.* Unlike in a liquid or in a gas, such dipoles are not free to rotate. Only an applied electric field can help to orient them along the field direction and this orientation is opposed by thermal agitation tending to randomize the dipoles. Necessarily, therefore, the calculation of dipolar polarizability is to be made under thermal equilibrium.

Consider a dipole of moment $\vec{p}$ of a dipolar system placed in a uniform electric field $\vec{E}$ (Fig. 8.8) inclined at θ with the direction of $\vec{p}$. The force, due to the field, acting on the positive and negative charges $\pm q$ of the dipole is equal but oppositely directed, forming a couple. So, the dipole experiences a torque τ given by

$$\vec{\tau} = \vec{p} \times \vec{E} \tag{8.6.1}$$

tending to rotate the dipole in the direction of the applied electric field.

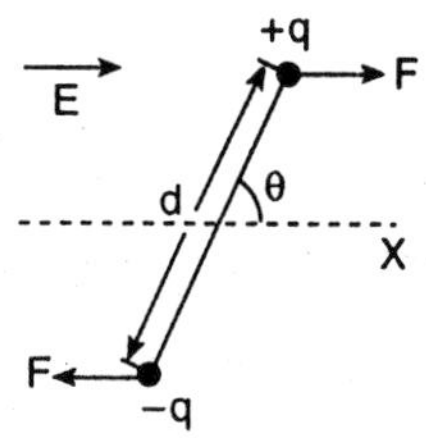

Fig. 8.8 Torque on a dipole due to an external field $\vec{E}$

∴ Magnitude of $\tau = pE\sin\theta$. The direction of τ is such as to tend the dipole turn in the field direction to reduce θ.

The potential energy U of the dipole in the field is given by

$$U = -\vec{p}\cdot\vec{E} = -pE\cos\theta \tag{8.6.2}$$

and $U_{\min} = -pE$, where $\theta = 0$, i.e., when the dipole is aligned parallel to the field $\vec{E}$, and $U_{\max} = pE$, when $\theta = 180°$, i.e., when the dipole is antiparallel to $\vec{E}$.

The potential energy is thus a function of the angle θ. It is to be remembered however that corresponding to the minimum potential energy, $\theta = 0$ does not mean that all the dipoles point to the field direction. This is because of the thermal agitation. A dipole of moment p making θ with the field direction contributes to the polarization a component $p_x = p\cos\theta$, the x-component.

By *MB*-statistics, the probability of finding a dipole within a solid angle $d\Omega$ (i.e., angle between θ and $\theta + d\theta$) is proportional to the Boltzmann factor

$$f(\theta) = e^{-U/kT} = e^{+pE\cos\theta/kT} \tag{8.6.3}$$

∴ Average moment contribution *per dipole* lying within the solid angle $d\Omega$ is

$$\bar{p} = \frac{\displaystyle\int_0^{\pi} p_x f(\theta)d\Omega}{\displaystyle\int_0^{\pi} f(\theta)d\Omega}$$

where $p_x = p\cos\theta$ and $d\Omega = 2\pi\sin\theta\, d\theta$.

$$\therefore \quad \bar{p} = \frac{\int_0^{\pi} p\cos\theta \cdot 2\pi \sin\theta \, e^{pE\cos\theta/kT} d\theta}{\int_0^{\pi} 2\pi \sin\theta \, e^{pE\cos\theta/kT} d\theta}$$

$$= \frac{p\int_{-1}^{+1} e^{ta} t dt}{\int_{-1}^{+1} e^{ta} dt}, \quad \text{putting } a = pE/kT \text{ and } t = \cos\theta.$$

$$= p\frac{d}{da} \log \int_{-1}^{+1} e^{ta} dt$$

$$= p\left[\frac{d}{da}\log(e^a - e^{-a}) - \frac{d}{da}\log a\right]$$

$$= p\left[\coth(a) - \frac{1}{a}\right] \equiv pL(a) \tag{8.6.4}$$

where $L(a) = \coth a - \dfrac{1}{a}$ is the well-known ***Langevin function.***

Fig. 8.9 shows a plot of Langevin function $L(a)$ as a function of a. Note that when a is very large, i.e., *at very high field strength* or *at low temperatures*, $L(a)$ approaches the saturation value unity, implying maximum alignment of dipoles along the field direction, when $\bar{p}$ approaches p.

For small values of a (i.e., $a \ll 1$, or $pE \ll kT$), we obtain, on expansion,

$$\coth a = \frac{1}{a} + \frac{a}{3} + \frac{a^3}{45} + \cdots$$

$$\Rightarrow \quad L(a) \simeq a/3 \tag{8.6.5}$$

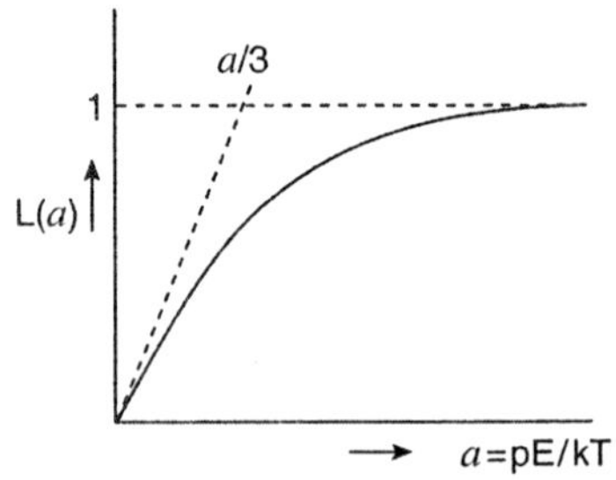

Fig. 8.9 Langevin function L(a) as function of a

$\therefore$ From (8.6.4), we have $\quad \bar{p} = p\dfrac{a}{3} = p\dfrac{pE}{3kT} \quad (\because\ a = pE/kT)$

$$\therefore \quad \boxed{\bar{p} = \frac{p^2 E}{3kT}} \tag{8.6.6}$$

$\therefore$ ***Dipolar polarizability*** or *orientational polarizability* per dipole is given by

$$\alpha_d = \frac{\bar{p}}{E} = \frac{p^2}{3kT} \tag{8.6.7}$$

Thus, the *dipolar polarizability is inversely proportional to the absolute temperature* of the dielectric.

If α_0 be the sum of electronic and ionic polarizabilities i.e., $\alpha_0 = \alpha_e + \alpha_i$, then the total polarizability

$$\boxed{\alpha = \alpha_0 + \frac{p^2}{3kT}} \qquad (8.6.8)$$

This is known as ***Langevin-Debye equation.***

In interpreting molecular structure, the Langevin-Debye equation is of paramount importance to the chemists. Fig. 8.10 shows the plot of molar polarizability as a function of $1/T$ for CH_4, CH_3Cl etc. The plot enables to evaluate p and by determining p, it is possible to know the molecular structure.

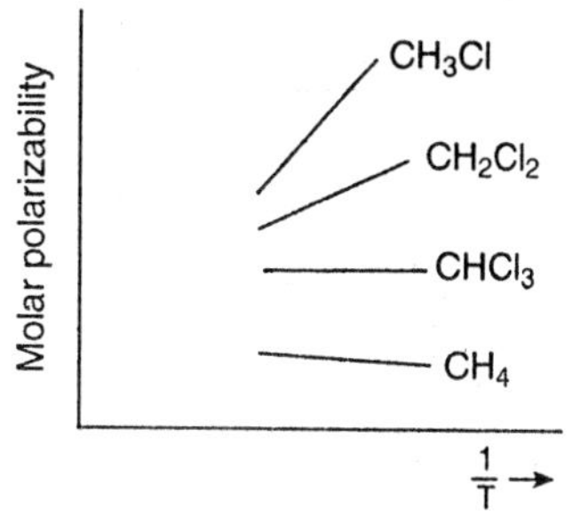

Fig. 8.10 Plot of molar palarizability against 1/T

8.7 Internal field in solids

The intermolecular distance in solids and liquids, unlike gases, is small. A dipole, therefore, is affected not only by the externally applied field, but also by fields due to the dipoles of other particles. So, the effective field $\vec{E}_i$ is different from the externally applied field $\vec{E}$. The field $\vec{E}_i$ is called the ***internal*** or the ***local field*** (also called the ***Lorentz field***).

So, replacing the field $\vec{E}$ by the effective field $\vec{E}_i$ we may write,

$$\vec{P} = N(\alpha_e + \alpha_i + \alpha_d)\vec{E}_i \qquad (8.7.1)$$

$$= N\left(\alpha_e + \alpha_i + \frac{p^2}{3kT}\right)\vec{E}_i \qquad (8.7.2)$$

It can be shown that the ***internal field $\vec{E}_i$*** is in general given by

$$\vec{E}_i = \vec{E} + \frac{\beta}{\varepsilon_0}\vec{P} \qquad (8.7.3)$$

where β is termed ***internal field constant.*** And for a *cubic crystal,* $\beta = 1/3$, as observed by Lorentz.

$$\therefore \quad \boxed{\vec{E}_i = \vec{E} + \frac{\vec{P}}{3\varepsilon_0}} \qquad (8.7.4)$$

This important relation is called the ***Lorentz relation*** and holds for crystals having *cubic symmetry.*

Proof of equation (8.7.4) : Let a dielectric slab be placed between the plates of a capacitor. A spherical cavity is inside the dielectric around an atom A. If the dia of the cavity is greater than the atomic radius, the *internal field* $\vec{E}_i$ at the site of A is:

$$\vec{E}_i = \vec{E}_1 + \vec{E}_2 + \vec{E}_3 + \vec{E}_4$$

The field $\vec{E}_1$ is due to the charge density on the capacitor plates and is given by

$$\vec{E}_1 = \vec{D}/\varepsilon_0 = (\vec{P} + \varepsilon_0 E)/\varepsilon_0 = \vec{E} + \vec{P}/\varepsilon_0 \qquad (8.7.5)$$

The field $\vec{E}_2$ is due to the induced charge density on the two sides of the dielectric

$$\therefore \quad \vec{E}_2 = -\vec{P}/\varepsilon_0 \tag{8.7.6}$$

The field E_3 is the field at A due to other atoms in the cavity, and because of symmetry of cubic structure, $E_3 = 0$.

Finally, we are to find $\vec{E}_4$, the field intensity at A due to polarisation charges on the surface of the cavity.

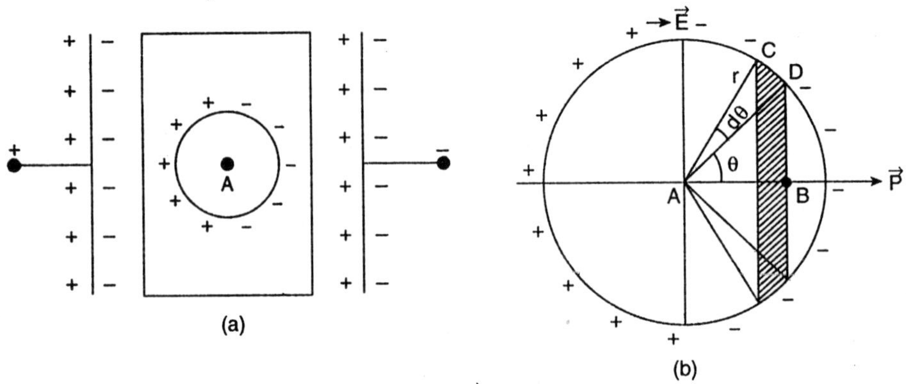

Fig. 8.11 (a) Calculation of $\vec{E}_i$ for a cubic structure
(b) Enlarged cavity around A

Fig. 8.11 illustrates an enlarged view of the cavity inside the dielectric around A. Let dS be an elemental surface area of the sphere of radius r, lying between θ and $\theta + d\theta$ where θ is the angle made by AD with the direction of the field $\vec{E}$.

$$\therefore \quad dS = 2\pi \cdot BD \cdot DR = 2\pi(r\sin\theta)\cdot(rd\theta)$$
$$= 2\pi r^2 \sin\theta\, d\theta$$

Now, the charge on dS is $dq = \sigma dS$, where σ, the surface density of polarisation charge, is $P\cos\theta$.

$$\therefore \quad dq = P\cos\theta \cdot dS = P(2\pi r^2 \sin\theta\cos\theta\, d\theta)$$

The field at A due to dq is along the direction of $\vec{E}$ (i.e., $\theta = 0$)

$$\therefore \quad dE_4 = \frac{dq\cos\theta}{4\pi\varepsilon_0 r^2} = \frac{P\cos\theta \times 2\pi r^2 \sin\theta\cos\theta\, d\theta}{4\pi\varepsilon_0 r^2}$$
$$= \frac{P}{2\varepsilon_0}\sin\theta\cos^2\theta\, d\theta$$

$$\therefore \quad \vec{E}_4 = \frac{\vec{P}}{2\varepsilon_0}\int_0^{\pi}\cos^2\theta\sin\theta\, d\theta = \frac{\vec{P}}{2\varepsilon_0}\int_0^{\pi}\cos^2\theta\, d(-\cos\theta)$$
$$= \frac{\vec{P}}{2\varepsilon_0}\left[\frac{\cos^3\theta}{3}\right]_0^{\pi} = -\frac{\vec{P}}{6\varepsilon_0}(-1-1) = \frac{\vec{P}}{3\varepsilon_0}$$

$$\therefore \text{ Internal field, } \vec{E}_i = (\vec{E} + \vec{P}/\varepsilon_0) + (-\vec{P}/\varepsilon_0) + 0 + \vec{P}/3\varepsilon_0$$

$$= \vec{E} + \frac{3\vec{P}}{\varepsilon_0}, \text{ the } \textit{Lorentz relation.}$$

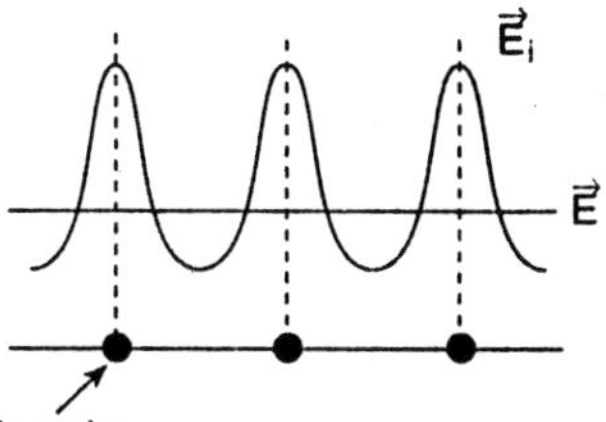

Fig. 8.12 Difference between $\vec{E}$ and $\vec{E}_i$

• The difference between $\vec{E}$, the Maxwell field and the local field $\vec{E}_i$ is that $\vec{E}$ is macroscopic in nature and is an average field and, in this case, constant throughout the medium. $\vec{E}_i$ is a microscopic field and is periodic in nature (Fig. 8.12).

8.8 Clausius-Mossotti relation

We consider dielectrics of cubic structure like diamond, carbon, silicon, germanium for which polarizability is solely due to electronic polarization. All the atoms here are alike without ions or permanent dipoles. So, we have $\alpha_i = \alpha_0 = 0$.

$$\therefore \quad \vec{P} = \vec{P}_e = N\alpha_e\vec{E}_i \tag{8.8.1}$$

where α_e is electronic polarizability.

Applying the Lorentz relation, as the crystal structure has cubic symmetry,

$$\vec{P} = N\alpha_e(\vec{E} + \vec{P}/3\varepsilon_0) \tag{8.8.2}$$

$$\Rightarrow \quad \vec{P} = \frac{N\alpha_e\vec{E}}{1 - N\alpha_e/3\varepsilon_0} \tag{8.8.3}$$

$$\text{But,} \quad \vec{P} = \varepsilon_0(\varepsilon_r - 1)\vec{E}, \text{ from (8.2.3).} \tag{8.8.4}$$

$$\therefore \text{ From (8.7.3) and (8.7.4),} \quad \varepsilon_0(\varepsilon_r - 1) = \frac{N\alpha_e}{1 - N\alpha_e/3\varepsilon_0} \tag{8.8.5}$$

$$\therefore \quad \boxed{\frac{\varepsilon_r - 1}{\varepsilon_r + 2} = \frac{N\alpha_e}{3\varepsilon_0}} \tag{8.8.6}$$

which is the celebrated ***Clausius-Mossotti relation.***

This relates the dielectric constant to the atomic polarizability, provided cubic symmetry holds good. It helps to determine the dielectric constant ε_r, from a knowledge of α_e and conversely the polarizability from the dielectric constant for elemental dielectrics. Again, since the polarizabilities are additive, the relation can also be used to predict dielectric constants of new materials if the individual polarizabilities of atoms are known. The dielectric constants we are considering remain constant even for frequencies in the visible region.

If M be the molecular weight of the substance, ρ its density and N_A the Avogadro number, then $N = (N_A\rho)/M$. So, the equation (8.8.6) gives

$$\frac{M}{\rho}\left(\frac{\varepsilon_r - 1}{\varepsilon_r + 2}\right) = \frac{N_A\alpha_e}{3\varepsilon_0}, \tag{8.8.7}$$

as ***an alternative form*** of Clausius-Mossotti equation.

• If α_i, the ionic polarizability is also taken under consideration, the Clausius-Mossotti relation takes up the form, corresponding to (8.8.6),

$$\frac{\varepsilon_r - 1}{\varepsilon_r + 2} = \frac{N(\alpha_e + \alpha_i)}{3\varepsilon_0} \tag{8.8.8}$$

8.8.1 Frequency-dependence of total polarizability

The electronic polarizability, ionic polarizatbility and the orientational or dipolar polarizability are the *three* factors contributing to the net polarizability of a dielectric.

As already mentioned, the electronic contribution is due to the displacement of electron shell relative to nucleus. The ionic contribution arises from displacement of a charged ion with respect to other ions, and finally the dipolar polarizability is due to molecules with a permanent electric dipole moment, capable of changing orientation in an applied electric field.

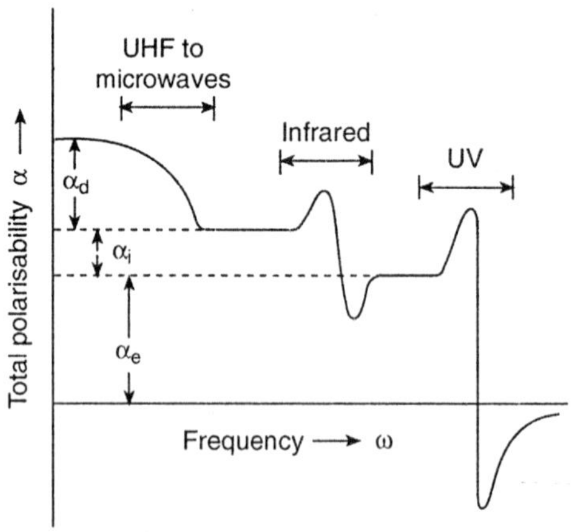

Fig. 8.13 Frequency-dependence of various contributions to total polarizability

The total polarizability of a dielectric is thus given by

$$\alpha = \alpha_e + \alpha_i + \alpha_d \tag{8.8.9}$$

The response of each type of polarization to the frequency of the field is different. While electronic polarization builds up rather quickly within 10^{-14} to 10^{-15} sec., ionic polarization may take up a period $\sim 10^{-11}$ sec. So, while both electronic and ionic polarizability respond to microwave to infrared frequencies, it is the electronic polarizability that responds to optical and UV-regions. Figure 8.13 illustrates the frequency-dependence of the various contributions to the total polarizability.

At optical frequencies, $n^2 = \varepsilon_r$ where n is the refractive index for a non-magnetic material. So, another ***alternative form*** of Clausius-Mossotti relation is

$$\boxed{\frac{n^2-1}{n^2+2} = \frac{N\alpha_e}{3\varepsilon_0}} \tag{8.8.10}$$

8.9 Ferroelectricity: Permanent polarization

Ferroelectricity is the phenomenon that refers to *spontaneous* or *permanent polarization* of a class of materials, i.e., polarization of the material in the absence of electric field ($\vec{E} = 0$), analogous to that of ferromagnetism. The crystals that exhibit ferroelectricity are known as ***ferroelectric crystals***. Examples are barium titanate, $BaTiO_3$ (Fig. 8.14), Rochelle salt ($NaKC_4H_4O_6, 4H_2O$), di-hydrogen phosphate (KH_2PO_4) of alkali metals etc.

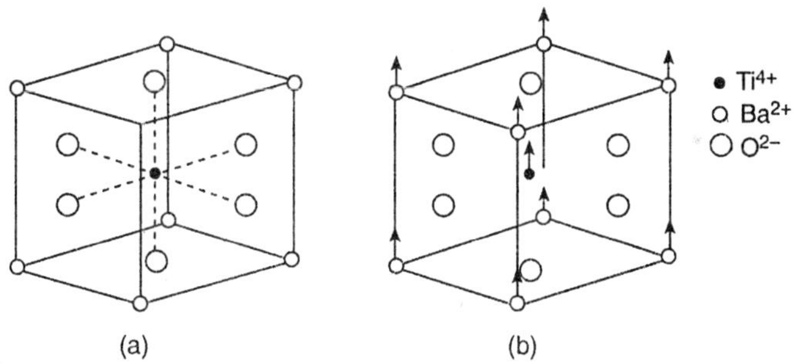

Fig. 8.14 (a) Ideal unit cell structure of $BaTiO_3$, (b) Below Curie temperature (structure slightly deformed and gives rise to ferroelectricity)

The unit cell of $BaTiO_3$ is cubic with Ba^{2+} at the corners, O^{2+} at the face centres and Ti^{4+} at the body centre. So, each Ti^{4+} in the *octahedral* configuration is surrounded by six O^{2+}. At $T > T_c$, the centres of positive and negaive charges exactly coincide and hence no dipole moment. If $T < T_c$, Ti^{4+} and Ba^{2+} slightly move upward whereas O^{2-} slightly downward. The crystal becomes *tetrahedral* with a relative shift in the centres of positive and negative charges. This produces *spontaneous polarization.*

Characteristic features—In such crystals, the centres of positive and negative charges do not coincide even in the absence of the field, thus developing a *non-zero dipole moment.*

The variation of polarization with electric field is *not linear* for ferroelectric crystals but shows, when subjected to an alternating electric field, a ***hysteresis loop***. If a virgin

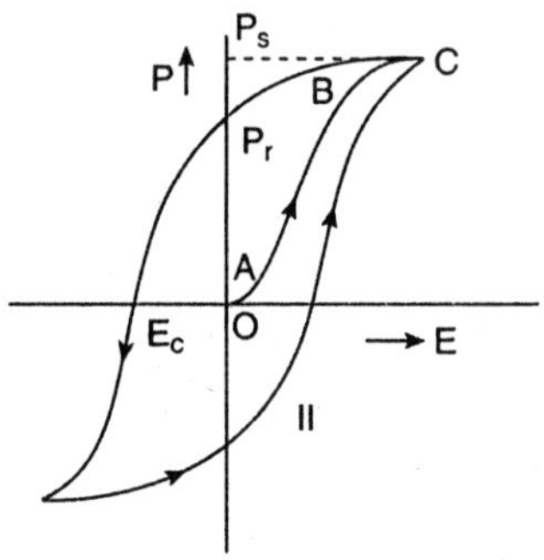

Fig. 8.15 Hysteresis loop for a ferroelectric material

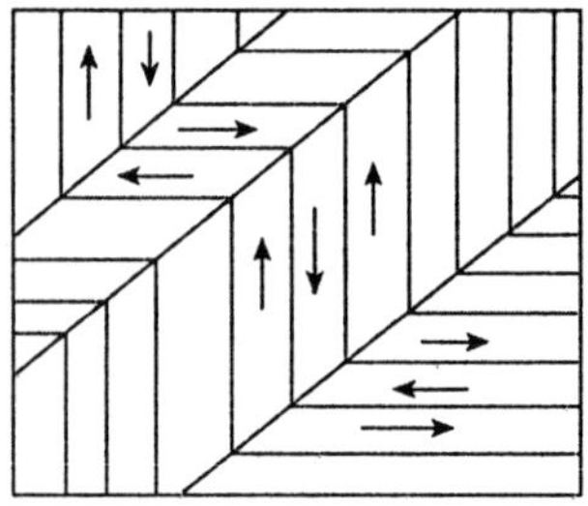

Fig. 8.16 Ferroelectric domains

specimen is under the action of a gradually increasing electric field, the polarization increases along $OABC$ (Fig. 8.15). When the field is decreased, it is found that at $E = 0$, the polarization P does *not* vanish. An electric field in the opposite direction only can make the polarization zero. The amount of polarization OP_r retained by the specimen even at $E = 0$, is called the *residual* or *remanent polarization*, and the oppositely directed field OE_c needed to make OP_r vanish is known as *coercive field.*

The zero polarization of a virgin macro-specimen can be explained from ***domain concept.*** The *ferroelectric domains* are spontaneously polarized but with directions changing from domain to domain (Fig. 8.16). In a virgin macro-specimen, a large number of domains exist and the randomness of polarization of the domains leads to zero resultant polarization. In an electric field, the domains having polarization along the field direction grow while those directed oppositely shrink. The net polarization thus increases with the field (the curve OB in Fig. 8.15).

Ferroelectricity i.e., spontaneous polarizaion usually disappears above a certain critical temperature T_c, called the ***transition temperature*** or ***ferroelectric Curie temperature.*** Above this temperature, the polarization varies with electric field linearly, as illustrated in Fig. 8.17 and the crystal is said to be in *paraelectric*

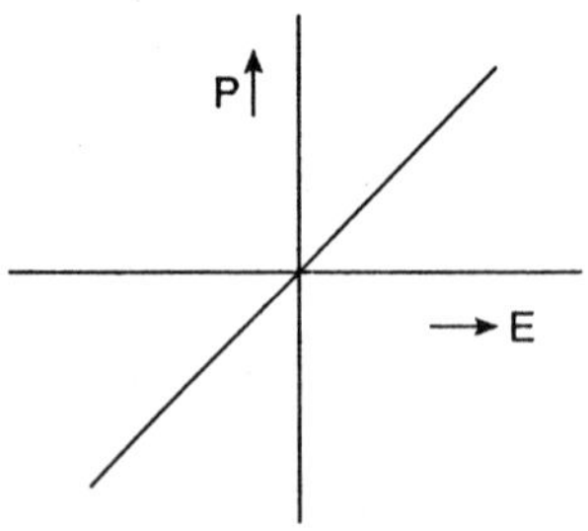

Fig. 8.17 P-E relationship above T_c of ferroelectrics

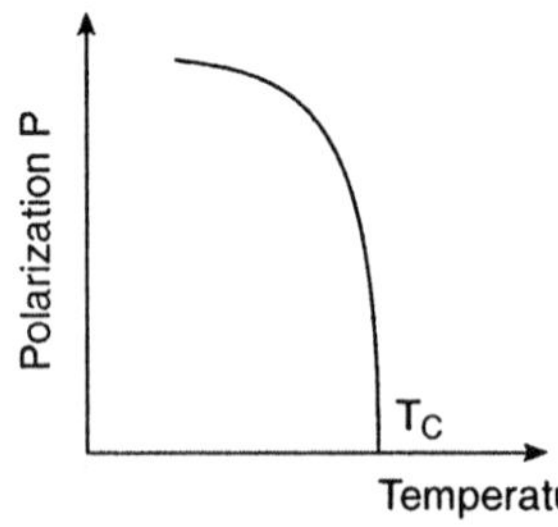

Fig. 8.18 Spontaneous polarization vs. temperature of ferroelectrics

state, analogous to paramagnetism, as indicated by a rapid decrease in the dielectric constant with rise in temperature. Fig. 8.18 illustrates the temperature-dependence of polarization of ferroelectric crystals.

The occurrence of ferroelectricity may be understood in terms of *polarization catastrophe.*

Polarization catastrophe—From Clausius and Mossotti equation, the dielectric constant ε_r is given by

$$\frac{\varepsilon_r - 1}{\varepsilon_r + 2} = \frac{1}{3\varepsilon_0}\sum_i N_i\alpha_i \tag{8.9.1}$$

where α_i is the electronic plus ionic polarizatbility of an ion of type i and N_i the number of ions per unit volume.

The above equation gives

$$\varepsilon_r = \frac{1 + \dfrac{2}{3\varepsilon_0}\sum_i N_i\alpha_i}{1 - \dfrac{1}{3\varepsilon_0}\sum_i N_i\alpha_i} \tag{8.9.2}$$

From (8.9.2), it appears that the dielectric constant ε_r becomes infinite when $\sum N_i\alpha_i = 3\varepsilon_0$ which permits a a finite polarization even for *no applied field* ($E = 0$). This is called the ***polarization catastrophe*** or ***Clausius-Mossotti catastrophe*** and the phenomenon is termed ***spontaneous*** or ***permanent polarization.***

The condition for a permanent polarization is thus $\sum_i N_i\alpha_i = 3\varepsilon_0$. (8.9.3)

In practice, however, it never becomes inifinity, although very high in many cases. The value of ε_r is sensitive to small departures of $\sum N_i\alpha_i$ from the critical value $3\varepsilon_0$. So, let us take

$$\frac{1}{3\varepsilon_0}\sum N_i\alpha_i = 1 - \delta, \quad \text{where } \delta \ll 1 \tag{8.9.4}$$

$\therefore$ From equation (8.9.2),

$$\varepsilon_r = \frac{3 - 2\delta}{\delta} \simeq \frac{3}{\delta} \tag{8.9.5}$$

If near T_c, the critical temperature, the deviation δ in (8.9.5) varies with temperature *linearly* according to

$$\delta \simeq \frac{3}{C}(T - T_c), \quad C = \text{Curie constant},$$

then (8.9.5) assumes the following form:

$$\boldsymbol{\varepsilon_r \simeq \frac{C}{T - T_c}} \tag{8.9.6}$$

This shows that the dielectric constant follows an increase according to the ***Curie-Weiss law.*** This type of variation is experimentally observed in *paraelectric state* of the crystal. At $T = T_c$, the dielectric constant becomes infinite. The temperature-dependence of ε_r in a ferroelectric crystal is illustrated in Fig. 8.19.

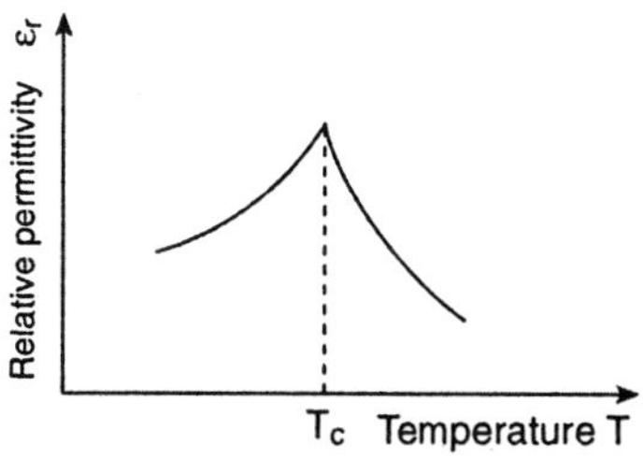

Fig. 8.19 Relative permittivity vs. temperature. ε_r is maximum at T_c

Applications—Ferroelectrics possess high value of dielectric constant and this feature makes them commercially viable in the manufacture of *small size capacitor* having high value of capacitance. They are also in use as *memory devices* in computers. The high value of electromechanical coupling constant of ferroelectrics make them highly useful for piezoelectric transducers. Ferroelectrics are also used as *electrooptic materials* whose refractive indices are regulated by an electric field (e.g., $LiNbO_3$).

• A somewhat opposite phenomenon has also been observed in some substances such as hydrogen halides and ammonium salts at low temperatures where the neighbouring lines of ions are displaced in opposite senses. Their spontaneous polarization thus becomes oppositely directed. The crystal cannot thus be spontaneously polarized on the whole. This phenomenon is called ***antiferroelectricity*** and is accompanied by changes in the values of the dielectric constant. Antiferroelectrics do not exhibit dielectric hysteresis, but may undergo a transition to ferroelectric state under the influence of a strong electric field or on cooling to low enough temperatures.

8.10 Illustrated Examples

► **Example 1.** *The dipole moment of an* HCl *molecule is* 3.3×10^{-30} C-m. *Calculate the amount of charge transferred from the hydrogen to chlorine atom when the separation between hydrogen and chlorine nuclei is* 1.28 A.

Solution. By the problem, $p = 3.3 \times 10^{-30}$ C-m, $d = 1.28\,\text{A} = 1.28 \times 10^{-1}$ m, $q = ?$.

If the amont of charge-transfer be q, the dipole moment is

$$p = q \times d \;\Rightarrow\; q = p/d$$

$$\therefore\; q = \frac{3.3 \times 10^{-30}}{1.28 \times 10^{-10}} = \mathbf{2.58 \times 10^{-20} C}$$

► **Example 2.** *The molecular dipoles in a* 10^{-3} m *radius water drop are pointed in the same direction. Compute the polarization. The dipole moment of the water molecule is* 6×10^{-30} C-m.

Solution. Here $p = 6 \times 10^{-30}$ C-m, $r = 10^{-3}$ m, $p = ?$ Also M = molecular weight = 18 for water, density of water $\rho = 10^3\ \text{kg/m}^3$. So, molar volume of water $V = \frac{M}{\rho} = \frac{18}{10^3}\ \text{m}^3$. Volume of water in the drop $= \frac{4}{3}\pi(10^{-3})^3$.

$$\therefore\; 18/10^3\ \text{m}^3 \text{ water contains } 6.02 \times 10^{26} \text{ molecules.}$$

$$\Rightarrow\; \frac{4\pi}{3}(10^{-3})^3\ \text{m}^3 \text{ will contain } N = \frac{6.02 \times 10^{26} \times 10^3 \times 4\pi \times 10^{-9}}{18 \times 3}$$

$$= 1.40 \times 10^{20}/\text{m}^3 \text{ molecules}$$

$$\therefore\; \text{Polarization, } \vec{P} = N\vec{p} = 1.40 \times 10^{20} \times 6 \times 10^{-30} = \mathbf{8.4 \times 10^{-10}\ C/m^2}$$

► **Example 3.** *For a certain gas molecule, the permanent dipole moment is* 1.35 Debye unit. *Calculate the orientational polarizability at room temperature* (27°C).

Solution. Given, $p = 1.35D = 1.35 \times 3.33 \times 10^{-30}$ C-m, $T = (273 + 27) = 300$ K

$\therefore$ Orientational polarizability is given by

$$\alpha_0 = \frac{p^2}{3kT} = \frac{(1.35 \times 3.33)^2 \times 10^{-60}}{3 \times 1.38 \times 10^{-23} \times 300}$$

$$= \mathbf{1.63 \times 10^{-39}\,Fm^2}$$

► **Example 4.** *Find the total polarizability of carbon di-oxide if its susceptibility is* 0.985×10^{-3}. *The density of carbon di-oxide is* $1.977\,\text{kg/m}^3$.

Solution. Given, $\rho(CO_2) = 1.977\,\text{kg/m}^3$; $\chi = 0.985 \times 10^{-3}$, $\alpha = ?$

The molecular weight of $CO_2 = 44$. So, the number of molecules per unit volume is given by

$$N = \frac{6.02 \times 10^{26} \times 1.977}{44} = 2.70 \times 10^{25}\,\text{m}^{-3}$$

Now, using the relation (8.3.3) : $\alpha = \varepsilon_0 \chi / N$, we have

$$\alpha = \frac{8.85 \times 10^{-12} \times 0.985 \times 10^{-3}}{2.70 \times 10^{25}}$$

$$= \mathbf{3.23 \times 10^{-40}\,Fm^2}$$

► **Example 5.** *Silicon has the dielectric constant* 12 *and the edge-length of the conventional cubic cell of silicon lattice is* 5.43 A. *What is the electronic polarizability of* Si-*atoms?* **(North Bengal Hons.)**

Solution. From Clausius-Mossotti relation, the electronic polarizability α_e is given by

$$\frac{\varepsilon_r - 1}{\varepsilon_r + 2} = \frac{N\alpha_e}{3\varepsilon_0}$$

Given, here, $\varepsilon_r = 12$. Also, we have $\varepsilon_0 = 8.854 \times 10^{-12}\,\text{F/m}$.

Now, silicon has the diamond structure. So, in its conventional cubic cell of volume $(5.43 \times 10^{-10})^3\,\text{m}^3$, there are 8 atoms. So, the number of atoms per unit volume N is given by

$$N = \frac{8}{(5.43 \times 10^{-10})^3} = 4.997 \times 10^{28}\,\text{m}^{-3}.$$

From Clausius-Mossotti relation, therefore,

$$\frac{12 - 1}{12 + 2} = \frac{4.997 \times 10^{28} \times \alpha_e}{3 \times 8.854 \times 10^{-12}}$$

$$\therefore \quad \alpha_e = \frac{11 \times 8.854 \times 10^{-12} \times 3}{14 \times 4.997 \times 10^{28}} = \mathbf{4.17 \times 10^{-40}\,Fm^2}$$

► **Example 6.** *An oxygen atom, on being polarized, develops a dipole moment of* 0.5×10^{-22} C-m. *Find the polarizability of oxygen atom, if the distance of the centre of negative charge-cloud from the nucleus is* 4×10^{-17} m.

Solution. Given, $p = 0.5 \times 10^{-22}$ C-m, $d = 4 \times 10^{-17}$ m, $\alpha = ?$

In equilibrium, both the Coulomb interaction and the Lorentz force will be equal.

$$\therefore \quad 8eE = \frac{(8e) \times (8e)}{4\pi\varepsilon_0 d^2}$$

$$\therefore \quad E = \frac{8e}{4\pi\varepsilon_0 d^2} = \frac{8 \times 1.6 \times 10^{-19}}{4\pi \times 8.85 \times 10^{-12} \times (4 \times 10^{-17})^2}$$

$$= 2.6 \times 10^{24}\ \mathrm{V/m}$$

But, the dipole moment, $p = \alpha E \Rightarrow \alpha = p/E$. Therefore, we have

$$\alpha = \frac{0.5 \times 10^{-22}}{2.6 \times 10^{24}} = \mathbf{1.9 \times 10^{-47}\ Fm^2}$$

▸ **Example 7.** *Determine the percentage contribution of ionic polarizability in* NaCl *crystal having an index of refraction* 1.5 *and static dielectric constant* 5.6.

(Himachal Univ.)

Solution. Given: $n = 1.5$, $\varepsilon_r = 5.6$. Now, from Clausius-Mossotti relation (8.8.8),

$$\frac{\varepsilon_r - 1}{\varepsilon_r + 2} = \frac{N(\alpha_e + \alpha_i)}{3\varepsilon_0} \qquad (8.10.1)$$

with the usual meanings of the symbols used.

At *optical frequencies*, the ionic polarizability α_i is negligible but the dielectric constant $\varepsilon_r = n^2$ and we obtain, from above eq. (8.10.1)

$$\frac{n^2 - 1}{n^2 + 2} = \frac{N\alpha_e}{3\varepsilon_0} \quad (\because\ \varepsilon_r = n^2) \qquad (8.10.2)$$

Dividing (8.10.2) by (8.10.1), we obtain

$$\frac{n^2 - 1}{n^2 + 2} \cdot \frac{\varepsilon_r + 2}{\varepsilon_r - 1} = \frac{\alpha_e}{\alpha_e + \alpha_i} \qquad (8.10.3)$$

The r.h.s. of (8.10.3) gives the fraction of the electronic polarizability.

$\therefore$ Percentage contribution of ionic polarizability is

$$\left(\frac{\alpha_i}{\alpha_e + \alpha_i}\right) \times 100 = \left(1 - \frac{\alpha_e}{\alpha_e + \alpha_i}\right) \times 100$$

$$= \left\{1 - \frac{(n^2 - 1)(\varepsilon_r + 2)}{(n^2 + 2)(\varepsilon_r - 1)}\right\} \times 100$$

$$= \left[1 - \left\{\frac{(1.5)^2 - 1}{(1.5)^2 + 2}\right\}\left\{\frac{5.6 + 2}{5.6 - 1}\right\}\right] \times 100$$

$$= (1 - 0.486) \times 100$$

$$= \mathbf{51.4\%}$$

► **Example 8.** *If the polarizability of oxygen atoms in air molecules is* 9.7×10^{-41} C-m^2/V, *find the average distance of the centre of negative charge-cloud from the nucleus.*

Solution. Given, $\alpha = 9.7 \times 10^{-41}$ C-m^2/V, $d = ?$

In equilibrium both the Coulomb interaction and the Lorentz force will be equal.

$$\therefore \quad 8eE = \frac{8e \times 8e}{4\pi\varepsilon_0 d^2} \Rightarrow E = \frac{8e}{4\pi\varepsilon_0 d^2} \tag{8.10.4}$$

But, the dipole moment $p = \alpha E$ and also $p = qd$

$$\therefore \quad qd = \alpha E \Rightarrow d = \frac{\alpha E}{q} = \frac{\alpha \cdot 8e}{4\pi\varepsilon_0 d^2 \times 8e}, \quad \text{using (8.10.4)} \tag{8.10.5}$$

$$\therefore \quad d^3 = \frac{\alpha}{4\pi\varepsilon_0} = \frac{9.7 \times 10^{-41}}{4\pi \times 8.85 \times 10^{-12}} = 0.8718 \times 10^{-30}$$

$$\therefore \quad \mathbf{d = 9.55 \times 10^{-11}\ m}$$

► **Example 9.** *In* NaCl *vapour, there are* 1.6×10^{20} *molecules/m*^{3}. *If the vapour is subjected to a field of* 5×10^6 V/m, *calculate the orientational polarization at room temperature* (27°C). *Assume that* NaCl *molecule consists of* Na^+ *and* Cl^- *ions distant* 2.5 A *apart.*

Solution. Given, $N = 1.6 \times 10^{20}/\text{m}^3$, $E = 5 \times 10^6$ V/m, $T = 27 + 273 = 300$ K, $r = 2.5\,\text{A} = 2.5 \times 10^{-10}$ m, $P = ?$

The dipole moment of NaCl, $p = qr = 1.6 \times 10^{-19} \times 2.5 \times 10^{-10} = 4 \times 10^{-29}$ C-m.

Also, the dipolar polarizability per dipole is

$$\alpha_d = \frac{p^2}{3kT} = \frac{(4 \times 10^{-29})^2}{3 \times 1.38 \times 10^{-23} \times 300} = 1.29 \times 10^{-37}\ \text{Fm}^2$$

$\therefore$ Orientational polarization,

$$\boldsymbol{P} = N\alpha_d E = 1.6 \times 10^{20} \times 1.29 \times 10^{-37} \times 5 \times 10^6$$

$$= \mathbf{10.32 \times 10^{-11}\ C/m^2}$$

► **Example 10.** *The relative permittivity of Argon at* 0°C *and in one atmosphere is* 1.000435. *Find the electronic polarizability of the argon atom.*

Solution. Given, $\varepsilon_r = 1.000435$ at NTP, $\alpha_e = ?$

The number of argon atoms at NTP is given by

$$N = \frac{6.023 \times 10^{26}}{22.4} = 2.69 \times 10^{25}$$

From the relation $\varepsilon_0(\varepsilon_r - 1) = N\alpha_e$, we obtain

$$\text{Electronic polarizability, } \alpha_e = \frac{\varepsilon_0(\varepsilon_r - 1)}{N}$$

$$\alpha_e = \frac{8.854 \times 10^{-12} \times 0.000435}{2.69 \times 10^{25}} = \mathbf{1.43 \times 10^{-40}\,Fm^2}$$

► **Example 11.** *The atomic polarizability of Neon is* $4.3 \times 10^{-41}\,\mathrm{Fm}^2$. *If a neon atom is placed in an electric rield of* $5 \times 10^4\,\mathrm{V/m}$, *find its dipole moment and the displacement of the centroids of positive and negative charges in it.*

Solution. Here, $\alpha = 4.3 \times 10^{-41}\,\mathrm{Fm}^2$, $E = 5 \times 10^4$ V/m, atomic weight of Ne = 10, $p = ?$ and $d = ?$

Dipole moment of an atom is

$$p = \alpha E = 4.3 \times 10^{-41} \times 5 \times 10^4 = \mathbf{2.15 \times 10^{-36}\,C.m}$$

Since, at. wt. of Ne = 10, the charge

$$q = 10 \times 1.6 \times 10^{-19} = 1.6 \times 10^{-18}\,\mathrm{C}$$

∴ Displacement of centroids of (+) and (−) charges is given by

$$\boldsymbol{d} = p/q = 2.15 \times 10^{-36}\,\mathrm{C.m}/1.6 \times 10^{-18}\,\mathrm{C}$$
$$= \mathbf{1.34 \times 10^{-18}\ m}$$

► **Example 12.** *The polarizability of* NH_3*-molecule is experimentally found to be* $2.42 \times 10^{-39}\,\mathrm{Fm}^2$ *at* 309 K *and* $1.74 \times 10^{-39}\,\mathrm{Fm}^2$ *at* 448 K *respectively. Find the polarizability due to permanent dipole moment and due to deformation of molecules for each temperature.*

Solution. By the problem, $\alpha(309) = 2.42 \times 10^{-39}\,\mathrm{Fm}^2$, $\alpha(448) = 1.74 \times 10^{-39}\,\mathrm{Fm}^2$, $\alpha_i = ?$, $\alpha_d = ?$

Now, $\alpha = \alpha_i + \alpha_d = \alpha_i + \dfrac{p^2}{3kT} = \alpha_i + \beta/T$, where $\beta = p^2/3k$.

∴ At 309 K, $\quad 2.42 \times 10^{-39} = \alpha_i + \beta/309 \quad$ (8.10.6)

At 448 K, $\quad 1.74 \times 10^{-39} = \alpha_i + \beta/448 \quad$ (8.10.7)

Solving (8.10.6) and (8.10.7), $\quad \beta = 677 \times 10^{-39} \quad$ (8.10.8)

Substituting this β-value in (8.10.6), and simplifying, we have

$$\boldsymbol{\alpha_i} = \mathbf{0.23 \times 10^{-39}\,Fm^2}$$

But dipolar or orientational polarizability

$$\alpha_d = \beta/T = 677 \times 10^{-39}/T, \quad \text{using (8.10.8)}$$

$$\therefore \quad \alpha_d\,(309) = \frac{677 \times 10^{-39}}{309} = \mathbf{2.19 \times 10^{-39}\,Fm^2}$$

$$\alpha_d\,(448) = \frac{677 \times 10^{-39}}{448} = \mathbf{1.51 \times 10^{-39}\,Fm^2}$$

Review, Short and Objective Questions

1. What are the different contributions to the total polarization of a dielectric material? **(Burdwan Hons.)**
2. Define polarization and polarizability. Find the relation among the following quantities: electric displacement, field strength and polarization vector.
3. Explain electronic polarization and orientational polarization in a dielectric. **(Calcutta Hons.)**
4. What are various polarizabilities? Discuss the classical theory of electronic polarizability and obtain the correspnding dispersion relation.
5. What are polar molecules and orientational polarization? Neglecting dipole-dipole interaction, obtain an expression for the orientational polarization. Discuss what happens at high and low temperatures.
6. Deduce an expression for the electronic polarizability of an atom on the basis of classical theory. **(Burdwan Hons.)**
7. What is Langevin-Debye equation? How could the equation be used to obtain information regarding the molecular structure?
8. The field seen by dipoles in a solid differ from the applied field. Why? Derive Clausius-Mossotti relationship and name the solids for which it holds.
9. What is ferroelectricity? Show that the dielectric constant of a ferroelectric should follow the Curie-Weiss law above the Curie-temperature.
10. Explain polar and non-polar molecules. Give examples.
11. Name any two ferroelectric materials and their applications. What is the transition temperature of such crystals? Discuss the origin of ferroelectricity.
12. Explain the terms: depolarization field and Lorentz field.
13. Explain: electronic polarization and electronic polarizability. Does the electronic polarization vary with temperature? Is the ionic polarizability temperature-dependent?
14. Do the alkali halides show orientational polarization? Why does the static dielectric constant in such solids differ from the hf-dielectric constant?

15. Distinguish between electronic, ionic and orientational polarizability.

16. Prove that the dipolar polarizability, $\alpha_{\text{dip}} = p^2/3kT$, with usual meanings of the symbols.

17. Explain polarizability of atoms. Derive Clausius-Mossotti equation between polarizability and dielectric constant of solids. **(Punjabi and Patna Univ.)**

18. What is polarization catastrophe? What is the condition for it? Explain.

19. Show that polarization per unit volume $\vec{P} = N\alpha e_0 \vec{E}$ where N is the number of atoms per unit volume, α the atomic polarizability, ε_0 the permittivity of free space and $\vec{E}$ the applied electric field.

20. The dielectric constant of a material at optical frequencies is mainly due to

(a) ionic polarizability, (b) electronic polarizability,
(c) dipolar polarizability, (d) ionic and dipolar polarizability. **(GATE)**

21. Two dielectric materials A and B exhibit both ionic and orientational polarizabilities. The variation of their suspectibilities χ $(= \varepsilon_r - 1)$ with temperature T

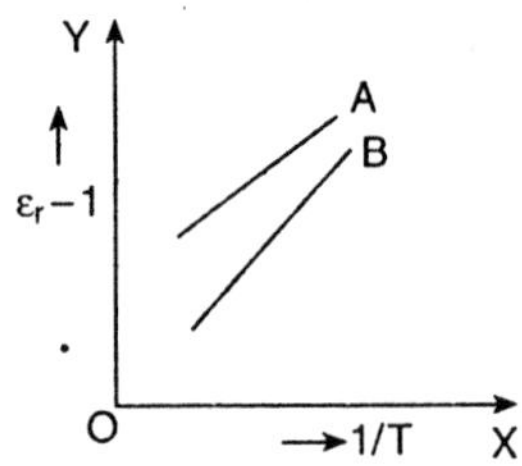

Fig. 8.20

is as in Fig. 8.20 where ε_r is the relative dielectric constant. It can be inferred from the figure that

(a) A is more polar and has a smaller value of ionic polarizability than B,
(b) A is more polar and has a higher value of ionic polarizability than B,
(c) B is more polar and has a higher value of ionic polarizability than A,
(d) B is more polar and has a smaller value of ionic polarizability then A.

(GATE)

Problems

1. The molecular dipoles in a water drop of 10^{-3} m radius are pointed in the same direction. If the dipole moment of the water molecule is 6×10^{-30}C-m., calculate the polarization. $[8.4 \times 10^{-10}\,\text{C/m}^2]$

2. The radius of an argon gas atom is 10^{-10} m. Calculate the electronic polarizability of the atom. $[1.11 \times 10^{-40}\,\text{Fm}^2]$

3. The relative permittivity of germanium is 16. The edge length of a conventional cubic cell for Ge lattice is 5.65×10^{-10} m. Calculate the electronic polarizability of Ge-atoms. [4.96×10^{-40} Fm^2]

4. The optical index of refraction and the dielectric constant for water are 1.33 and 8.1 respectively. Calculate the percentage of ionic polarizability. [71%]

5. For argon gas, $N = 10^{19}\,\text{cm}^{-3}$, $Z = 18$ and $r = 10^{-8}$ cm. Calculate the electronic polarization when the applied electric field is 10 kV/cm. [1.1×10^{-9} C/m^2]

6. Find the total polarizability of CO_2, if its susceptibility is 0.985×10^{-3}. Density of carbon di-oxide is 1.977 kg/m^3. [3.23×10^{-40} Fm^2]

7. The permanent dipole moment for a certain gas molecule is 1.35 D. If there are 10^{27} gas molecules per metre^3, calculate the orientational polarization at room temperature for an applied electric field of 10 kV/cm. [1.63×10^{-6} C/m^2]

8. The polarizability of NH_3 molecule in gaseous state is 2.42×10^{-39} Fm^2 at 309 K and 2×10^{-39} Fm^2 at 448 K respectively. Calculate for each temperature the polarizability due to permanent dipole moment and due to deformation of molecules. [0.23×10^{-39} Fm^2, 2.19×10^{-39} Fm^2; 1.51×10^{-39} Fm^2]

9. The dielectric constant and the density of sulphur at 27°C are 3.75 and 2050 kg/m^3. If its atomic weight is 32, calculate the electronic polarizability of sulphur. Given that the internal field constant $\beta = 1/3$. [3.18×10^{-40} Fm^2]

[**Hints.** Number of sulphur atoms per m^3, $N = (6.02 \times 10^{26} \times 2050)/32 = 385.66 \times 10^{26}$.

Internal field, $E_i = E + \beta P/\varepsilon_0 = E + P/3\varepsilon_0$ $(\because \beta = 1/3)$. From Clausius-Mossotti

$$\frac{\varepsilon_r - 1}{\varepsilon_r + 2} = \frac{N\alpha_e}{3\varepsilon_0} \Rightarrow \alpha_e = \frac{\varepsilon_r - 1}{\varepsilon_r + 2} \times \frac{3\varepsilon_0}{N} = \frac{3.75 - 1}{3.75 + 2} \times \frac{3 \times 8.55 \times 10^{-12}}{385.66 \times 10^{26}}$$

$= \mathbf{3.18 \times 10^{-40}\ Fm^2}$, on simplification.]

10. If there are 10^{27} molecules/m^3 in HCl vapour, calculate the orientational polarization at room temperature when the vapour is subjected to an applied electric field of 10^6 V/m. The dipole moment of HCl-molecule is 3.46×10^{-30} C-m. Show that at this temperature and for such a high field, the value of a where $a = pE/kT$ is much smaller than unity. [9.64×10^{10} C/m^2; $\sim 10^{-3}$]

11. The dielectric constant of He at 0°C and one atmosphere is 1.000074. Calculate the dipole moment induced in each He-atom when the gas is subjected to an electric field of 3×10^4 V/m. [7.3×10^{-37} C-m]

12. Assuming the polarizability of Krypton atom to be 2.18×10^{-40} Fm^2, calculate the dielectric constant at 0°C and at 1 atomospheric pressure. [1.00066]

[**Hints.** Use the relation $\varepsilon_r = 1 + \chi_e = 1 + N\alpha/\varepsilon_0$]

13. If the density of polarization of diamond is 1.32×10^{-10} C/m^2, calculate the shift of the centre of negative cloud of 6 electrons on each atom from the nucleus. The density of carbon is 3500 kg/m^3 and the atomic weight is 12. [7.8×10^{-17} m]

14. There are 1.6×10^{20} molecles/m^3 in NaCl vapour. Find the orientational polarization at room temperature (27°C) if the vapour is subjected to an electric field of 5×10^6 V/m. The NaCl molecule consists of Na^+ and Cl^- separated by 2.5 A. [10.32×10^{-11} C/m^2]

15. Find the total polarization of He-gas when placed in a field of 6×10^5 V/m. The atomic polarizability of He is 0.18×10^{-40} Fm^2 and the concentration of He-atom is 2.6×10^{25}/m^3. Also, evaluate the separation of positive and negative charges in each atom. [2.81×10^{-10} C/m^2; 6.75×10^{-17}m]

16. Calculate the dipole moment of an HCl molecule if a complete transfer of electron from hydrogen to chlorine atom occurs when the separation between their nuclei is 1.28 A. [2.05×10^{-29}C-m]

17. Two parallel plates have equal and opposite charges and they are separated by a dielectric of 5 mm thick whose dielectric constant is 3. If the electric intensity in the dielectric is 10^6 V/m, determine the free charge per unit area on the plate and the polarization P in the dielectric. [26.56×10^{-6} C/m^2; 17.7×10^{-6} C/m^2]

18. The capacitance of a parallel-plate capacitor increases from the value 0.085μF to 0.195μF when a polythene sheet is placed between the plates. Calculate the dielectric constant of polythene. [2.29]

19. The static dielectric constant of water is 8.125 and its refractive index is 1.33. Calculate the percentage contribution of ionic polarizability.

(Burdwan Hons.) [71%]

20. The relative permittivity and the square of refractive index of a dielectric are 4.94 and 2.69 respectively. Find the ratio between the electronic and ionic polarizabilities of the material. [1.738]

21. Helium gas contains 2.7×10^{25} atoms/m^3 and the dielectric constant of He is 1.0000684 at NTP. Calculate the electronic polarizability of He-atoms.

[2.24×10^{-41} Fm^2]

22. The relative permittivity of Ge is 16. The edge length of the coventional cubic cell for Ge-lattice is 5.65×10^{-10} m. Calculate the electronic polarizability of Ge-atoms. [4.96×10^{-40} Fm^2]

23. Calculate the orientational polarizaton at room temperature in HCl vapour containing 10^{27} mlecules/m^3, if subjected to an electric field of 10^6 V/m. The dipole moment of HCl molecule is 3.46×10^{-36} C-m. [9.64×10^{-7} C/m^2]

24. The relative permittivity of argon at 0°C and one atmosphere is 1.000435. Determine the polarizability of the argon atom. [1.43×10^{-40} Fm^2]

9

Superconductivity

9.1 Introduction : Historical backdrop

Kamerling Onnes took a giant step forward in low-temperature physics in 1908 when he could successfully produce liquid helium. Helium boils at 4.2 K. He used liquid helium for a variety of studies of metal at low temperatures and observed that certain metals and alloys exhibit almost *zero electric resistivity* or almost *infinite conductivity* when cooled to the boiling point of helium. This phenomenon of complete disappearance of electrical resistance of materials below a certain temperature is known as ***superconductivity*** and the material in this state is called a ***superconductor***.

The discovery of superconductivity was made by Kamerling Onnes in 1911. While conducting measurements of electrical conductivity of metals close to 4.2 K, he found that the resistance of mercury decreases continuously from its melting point (233 K) to 4.2 K and then, within some hundredths of a degree, abruptly drops for all practical purposes to zero (Fig. 9.1). He concluded that mercury had passed into a new state which he called a *superconducting state*. Similar results were also shown by various other metals such as Pb, Sn etc.

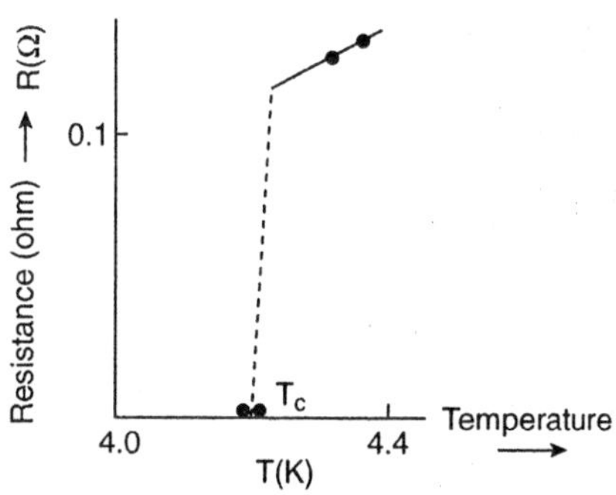

Fig. 9.1 Resistance versus absolute temperature curve for mercury by K. Onnes leading to the discovery of superconductivity

The temperature at which transition from *normal to superconducting* state occurs (in the absence of a magnetic field) is known as the ***transition temperature*** or ***critical temperature*** T_c. Different materials have different transition temperatures. The material is in the *normal state* above T_c, while below T_c it is in the *superconducting state.* For most of the metallic superconductors, the transition temperature lies at about 1–10 Kelvin. While T_c is rather sharp for chemically pure and structurally perfect specimens, it ranges over 0.1 degree or so, for impure and imperfect ones.

In the beginning, superconductivity was only a curiosity among the scientists, but subsequently widespread interest was aroused since materials with zero resistance (hence negligible Joule heat loss) could be exploited to fabricate powerful but economic devices that would consume little electrical energy. But *maintenance of helium temperature* for long is both *difficult and expensive.* Efforts were therefore undertaken to develop a superconductor with as high a critical temperature T_c as possible.

Success to thus effect was slow but steady. With the discovery, in the 1980s, of new ceramic superconductors based on copper oxide the ***high*** T_c ***ceramics***—with transition temperatures about 125 K or above, gave a drammatic turn and a new age of high T_c superconducting devices with commercial applications was opened up. The advantage of high T_c ceramics is that they require *liquid* N_2 *as coolant* which is relatively economical. It also became imperative to investigate the fundamental physics of superconductivity for, unlike conventional superconductors, the electron-lattice interaction is not the origin of superconductivity in these materials.

Superconducting state is a distinct phase of matter with characteristic electrical, magnetic, thermodynamic and other physical properties. The most striking features of bulk superconductors are the *zero electric resistance* and *perfect diamagnetism.*

● Since there is no dc resistance for a superconductor below T_c, a large current should flow through it (Ohm's law). Hence, if a current is set up in a superconductor below T_c, the current will persist without any voltage. Such a persistent current is called ***supercurrent.***

9.2 Response of magnetic field : Meissner effect

In 1933, after extensive studies on the magnetic properties of superconductors, Meissner and Ochsenfeld found that when certain superconductors are *cooled below their critical*

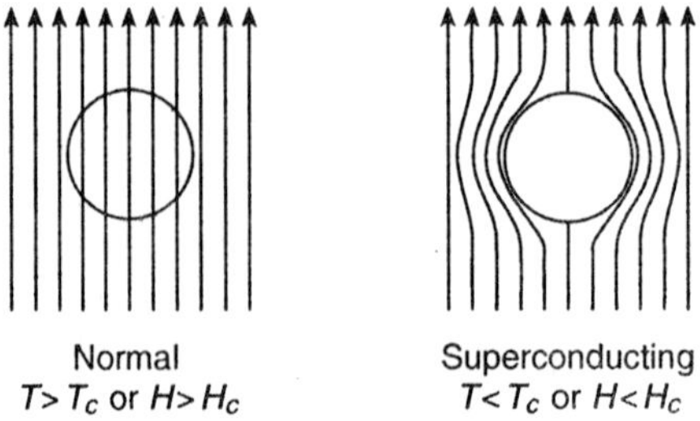

Fig. 9.2 The Meissner effect

temperatures in an external magnetic field, the magnetic flux (i.e., lines of force) initially held *is expelled from the interior of the specimen* and thus behaves as a ***perfect diamagnet.*** This phenomenon of *flux exclusion* from the interior of the specimen below T_c by superconductors is called the ***Meissner effect*** or ***Meissner-Ochsenfeld effect*** shown in Fig. 9.2. The magnetic field and the superconductivity are thus found to be *mutually exclusive.*

They further observed that such flux exclusion also occurs if the superconductor is *first cooled below* T_c and then *placed in a magnetic field.* This points to the fact that *perfect diamagnetism of a superconductor is independent of its past history.* It thus follows that the ***Meissner phenomenon is a reversible one.***

Now, since inside the superconductor, we have, with usual symbols,

$$\vec{B} = 0$$

$$\therefore \quad \mu_0(\vec{H} + \vec{M}) = 0$$

$$\Rightarrow \quad \vec{H} = -\vec{M}, \quad (\because \mu_0 \neq 0) \tag{9.2.1}$$

$\therefore$ Susceptibility, $\chi = \vec{M}/\vec{H} = -1$, and so it corresponds to ***perfect diamagnetism.***

● Interestingly enough, *perfect diamagnetism of a superconductor cannot be explained simply by its zero resistivity* for such a perfect conductor would behave differently under different situations.

For a *perfect conductor*, we have the resistivity

$$\rho = 0, \quad \Rightarrow \quad \vec{E}(= \rho\vec{J}) = 0,$$

using Ohm's law, i.e., *no electric field can exist inside a perfect conductor.*

$$\therefore \text{ From Maxwell's equation, } \vec{\nabla} \times \vec{E} = -\partial\vec{B}/\partial t = 0 \Rightarrow \vec{B} = \text{constant.} \tag{9.2.2}$$

So, *the magnetic flux density through a perfect conductor becomes constant.* This implies that a perfect conductor when cooled in the magnetic field until zero resistance is obtained, the magnetic field in the material gets frozen and cannot change subsequently irrespective of the applied field. *This is in contradiction with* ***Meissner effect.***

It therefore follows that *the behaviour of a superconductor is not the same as that of a perfect conductor.* The ***superconducting state of the material is a characteristic thermodynamic phase of the material in which it cannot sustain steady electric and magnetic fields.***

To *define the superconducting state* of a material, therefore, we require the following *two mutually independent properties.* These are:

1. ***Zero resistivity***, that is, $\rho = 0$.

2. ***Perfect diamagnetism***, that is, $\vec{B} = 0$, from Meissner effect.

● Because of complete expulsion of magnetic fields from the interior of a superconductor, superconducting materials are used to shield unwanted electromagnetic fields.

9.2.1 Critical field and critical temperature

Kamerling Onnes discovered (1913) that a superconductor gets back its normal state below the critical temperature, if placed in a sufficiently strong magnetic field. *The minimum value of the magnetic field at which the superconductivity just gets lost is*

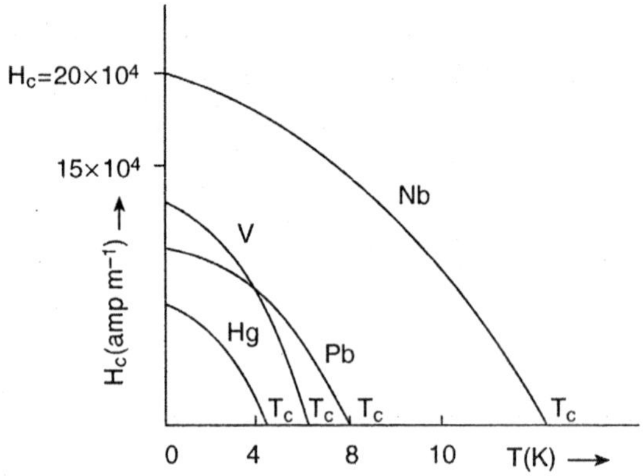

Fig. 9.3 Critical magnetic field as function of temperature for different superconducting elements

known as the ***threshold*** or ***critical magnetic field*** $\vec{B}_c$ or $\vec{H}_c$. It is not however a fixed constant for a material, but is a *function of temperature*, i.e., *changes with change in temperature*. Thus, the superconducting state is stable in some definite ranges of the magnetic field and the temperature.

The functional relationship of the critical field and the temperature for different superconducting elements is shown in Fig. 9.3. Such *a plot of* $\boldsymbol{B_c}$ or $\boldsymbol{H_c}$ vs. ***temperature*** $\boldsymbol{T}$ is termed ***magnetic phase diagram*** which appears to be *almost parabolic* and is represented approximately by the expression

$$\boldsymbol{B_c(T) = B_c(0)\left[1 - \left(\frac{T}{T_c}\right)^2\right]} \tag{9.2.3}$$

where $B_c(0)$ is the critical field value at 0 K, i.e., the *maximum* magnetic field required to destroy the superconductivity in a material at $T = 0\,\text{K}$. The relation (9.2.3) clearly points to the fact that at the critical temperature, i.e., at $T = T_c$,

$$B_c(T_c) = 0, \text{ as stated earlier.}$$

$$\text{and } B_c(T) = B_c(0), \text{ at } T = 0\,\text{K}$$

- $B_c(0)$ and T_c are both constants, characteristics of the material being studied.
- If the field exceeds $B_c(0)$, the metal will not go superconducting at any temperature, not even at $T = 0$. In general, higher the T_c, higher will be $B_c(0)$.
- $B_c(0)$ for pure metals is quite low: $B_c(0) = 10^{-2}$ Tesla around $T_c = 1\,\text{K}$ and 10^{-1} Tesla for higher value of T_c.
- The region below the parabolic curve corresponds to the *superconducting phase* and that outside the curve is the *normally conducting phase*, the curve defining the boundary.

• The critical field that destroys supercondutivity need not necessarily be an external field. It could be the field due to an electric current through the superconducting specimen itself. This restricts the current that can flow in a superconductor and is the main hurdle in developing high field superconducting magnet.

• Magnetic fields are generated both inside and outside the wires carrying current. So, if a superconducting wire is to carry current without any resistance, there must be a limit to the allowed current. The limiting value of the allowed current is called the ***critical current.***

9.3 Types of superconductors

Based on their magnetic behaviour in an external magnetic field, that is, how strictly they follow the Meissner effect, superconductors are classified into *two* groups : ***Type I*** or ***soft*** and ***Type II*** or ***hard*** *superconductors.*

Type I—Superconductors that exhibit complete Meissner effect (i.e., perfect diamagnetic property) are known as *Type I superconductors.* Examples are : Al, Zn, Hg, Sn etc. in pure state. Type I superconductors have the following *features.*

1. ***Perfect diamagnetism below a certain critical magnetic field B_c or H_c.***

2. ***Just beyond the critical magnetic field B_c or H_c, the field completely penetrates the material which abruptly reverts to the normal state.*** The transition from the superconducting to the normal state is sharp.

In most cases, the value of B_c of this type is small ~ 0.1 Tesla. As the loss of superconductivity is at rather lower field strengths, they are also called ***soft superconductors*** and due to this, they have limited technical applications.

Type II—These are superconductors that *do not strictly follow Meissner effect.* Examples are: alloys of transition metals like Ta, V, Nb, etc. Their *characteristic features* are :

1. They possess ***two critical magnetic fields B_{c_1} (or H_{c_1}), and B_{c_2} (or H_{c_2})***, known respectively as the ***lower*** and the ***upper critical fields.***

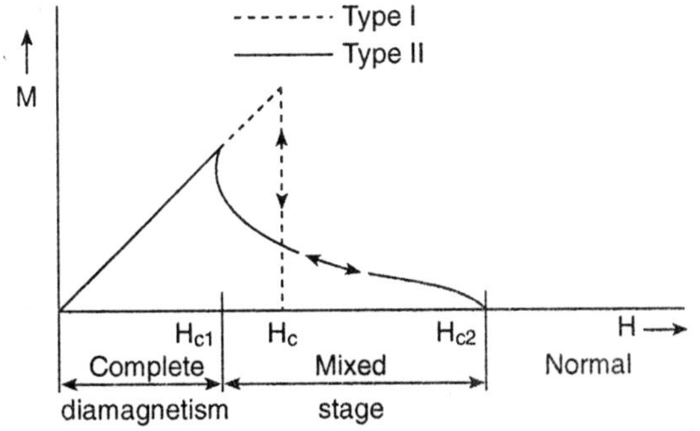

Fig. 9.4 Magnetization curves for Type I and Type II superconductors

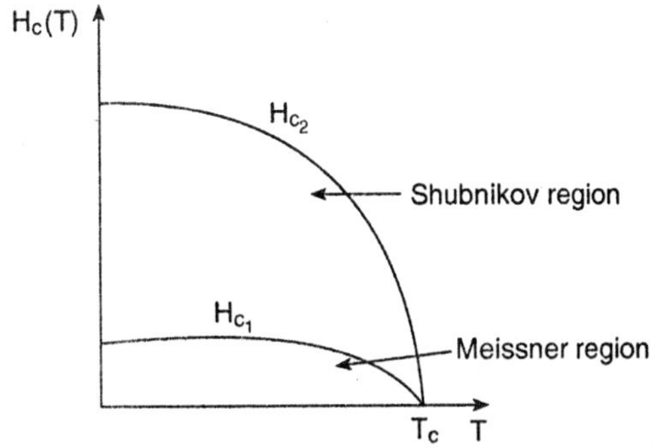

Fig. 9.5 Variation of H_{c_1} and H_{c_2} with temperature

2. For applied fields less than the lower critical field H_{c_1}, the material shows ***perfect diamagnetism*** and there is ***no flux penetration***, thus behaving as Type I.

3. Beyond H_{c_1}, the flux ***begins to penetrate the specimen*** and the penetration increases until H_{c_2} is reached. Above H_{c_2}, the ***flux-penetration is complete*** and the specimen becomes a ***normal conductor***.

Between the fields H_{c_1} and H_{c_2}, the specimen is in a ***vortex*** or ***mixed state***, a mixture of normal and superconducting states.

The typical magnetic behaviour—the *magnetization curves*—of Type I and Type II superconductors is illustrated in Fig. 9.4.

Both H_{c_1} and H_{c_2} vary with temperature qualitatively (Fig. 9.5). Since all of the flux gets expelled below H_{c_1}, the region in the diagram is known as ***Meissner region***. Shubnikov first observed field penetration (for $H > H_{c_1}$) without superconductivity being lost. So, the region is called ***Shubnikov region***.

• The upper critical field may sometimes be very high, hundreds of Tesla or more. Most of the superconducting alloys come under Type II. Type II superconductors are also called ***hard superconductors*** as they require larger fields for bringing back to the normal state. Newly developed copper oxide superconductors having $B_{c_2} \sim 150$ Tesla are of Type II.

• Large *magnetic hysteresis curves* can be induced in Type II superconductors by proper mechanical treatment. They can thus be utilized for large scale production of superconducting wires used to develop high magnetic fields ~ 10 Tesla.

9.4 Thermodynamics of superconducting transition

Experimentally, the transition between the normal and the superconducting state is *thermodynamically reversible* and we may, therefore, apply thermodynamics to this phase transition to get expressions for the difference in *entropy* and *specific heat* between normal and superconducting states near the absolute zero. We shall however restrict ourselves to type I semiconductors only, for which $B = 0$ inside the sample and study the transition under *isothermal* and *isobaric* conditions when the two phases are in equilibrium.

The ***Gibbs free energy*** per *unit volume* in a magnetic field is given by

$$G = U - TS - HM \tag{9.4.1}$$

where U, T, S, H and M are respectively the internal energy, temperature, entropy, magnetic field and magnetization; pV-term is neglected for a magnetic solid.

From the second law,
$$dU = TdS + HdM \tag{9.4.2}$$

where HdM is the work done on superconductor per unit volume.

Differentiating (9.4.1)
$$\begin{aligned} dG &= dU - TdS - SdT - HdM - MdH \\ &= -SdT - MdH, \quad \text{using (9.4.2)} \end{aligned} \tag{9.4.3}$$

Under *isothermal* condition, $dT = 0 \Rightarrow dG = -MdH$ (9.4.4)

Integrating (9.4.4) for the *superconducting state*, we obtain

$$\int_0^H dG = -\int_0^H MdH \Rightarrow G_S(H) - G_S(0) = -\int_0^H MdH \quad (9.4.5)$$

If however the specimen is in the *normal state*, it is a paramagnet, as most superconductors are in normal state implying $M \to 0$ or $\chi \to 0$.

$\therefore$ From (9.4.5), $G_n(H) - G_n(0) = 0 \Rightarrow G_n(H) = G_n(0)$ (9.4.6)

This shows that *in normal state, Gibbs function is invariant under a magnetic field.*

From H_c vs. T-curve, it is seen that the normal and superconducting states are at equilibrium at the same temperature and field. So, the free energies of the two phases at the boundary must be equal.

$$\therefore \quad G_n(T, H_c) = G_s(T, H_c) \quad (9.4.7)$$

Combining (9.4.5) and (9.4.7), we obtain

$$G_n(T, H_c) = G_s(T, H_c) = G_s(0) - \int_0^H MdH \quad (9.4.8)$$

From (9.4.6) and (9.4.8), $G_n(0) = G_s(0) - \int_0^H MdH$

$$\therefore \quad \boxed{\Delta G = G_n(0) - G_s(0) = -\int_0^H MdH} \quad (9.4.9)$$

From Meissner effect, $M = -H \ (B = 0)$,

$$\therefore \quad \Delta G = -\int_0^H MdH = \int_0^{H_C} HdH = \frac{1}{2}H_c^2$$

$$\text{or,} \quad \boxed{\boldsymbol{\Delta G = \frac{1}{2}H_c^2}} \quad (9.4.10)$$

Entropy difference between normal and superconducting states—The entropy of a solid is, from (9.4.3), given by

$$S = -\left(\frac{\partial G}{\partial T}\right)_H \quad (9.4.11)$$

$$\therefore \quad S_n = -\left(\frac{\partial G_n}{\partial T}\right)_H \quad \text{and} \quad S_s = -\left(\frac{\partial G_s}{\partial T}\right)_H$$

$$\therefore \quad S_n - S_s = -\frac{\partial G_n(0)}{\partial T} + \frac{\partial G_s(0)}{\partial T} = -\frac{1}{2}\frac{\Delta G}{\partial T} = -H_c\frac{dH_c}{dT}, \quad (9.4.12)$$

using (9.4.10)

Now, dH_c/dT is always *negative*. So, $S_n > S_s$. This means that the *superconducting state* (having less entropy) *is a more orderly state than the normal one.*

Also, at $T = T_c$, $H_c = 0$. This shows that $S_n - S_s = 0$. So, the entropies of the two phases are equal at the critical temperature. Further, by the third law of

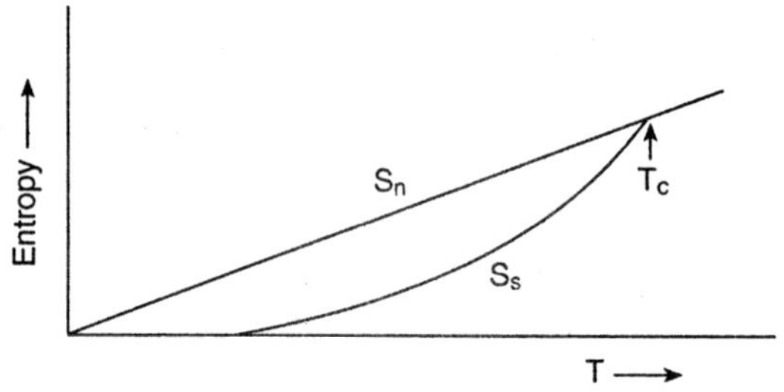

Fig. 9.6 Entropy S of aluminium in normal (S_n) and superconducting (S_s) states as function of temperature

thermodynamics, $dH_c/dT \to 0$ at $T = 0\,\text{K}$ and hence $S_n - S_s = 0$. Thus, between the temperatures $T = 0\,\text{K}$ and $T = T_c$, we find that $S_n - S_s$ is positive, i.e., $S_n > S_s$. This are illustrated in Fig. 9.6.

Latent heat—The latent heat Q of the system is related to entropy-difference as:

$$Q = T(S_n - S_s) = -TH_c\frac{dH_c}{dT}, \tag{9.4.13}$$

using the equation (9.4.12).

But at $T = T_c$, $H_c = 0$ and so $S_n - S_s = 0$. So, from (9.4.13), $Q = 0$. Thus, when the transition occurs at T_c ($H_c = 0$), there is an absence of latent heat. If, however, $H_c \neq 0$, i.e., in the presence of a magnetic field, the transition occurs at some lower temperature $T < T_c$. Between the temperatures $0\,\text{K}$ and T_c, the entropy $S_n > S_s$ and the equation (9.4.13) provides the amount of latent heat to be supplied for the transition to occur at a constant temperature.

Specific heat difference—Differentiating (9.4.12) with respect to T and then multiplying throughout by T, we obtain

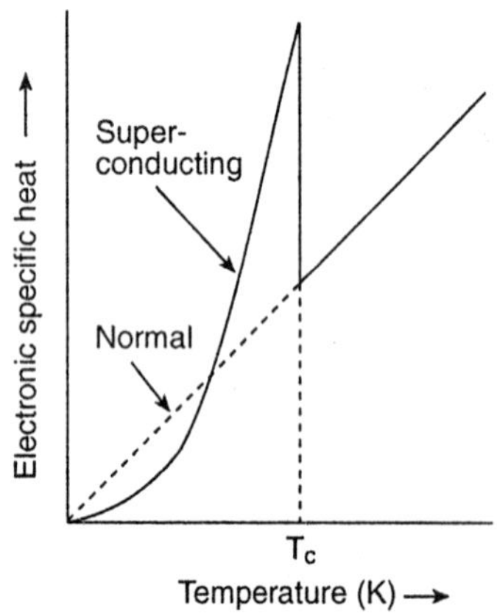

Fig. 9.7 Electronic specific heat vs. temperature for normal and superconductingl states of a sample.

$$T\frac{dS_n}{dT} - T\frac{dS_s}{dT} = -T\frac{d}{dT}\left(H_c\frac{dH_c}{dT}\right)$$

$$\Rightarrow \quad C_n - C_s = -T\left(\frac{dH_c}{dT}\right)^2 - TH_c\frac{d^2H_c}{dT^2} \tag{9.4.14}$$

At $T = T_c$, $H_c = 0$, and the second term on the right hand side of the equation of (9.4.14) vanishes.

$$\therefore \quad \boxed{(C_s - C_n)_{T_c} = T_c\left(\frac{dH_c}{dT}\right)^2_{T=T_c}} \tag{9.4.15}$$

Equation (9.4.15) implies the increase of specific heat as the specimen is cooled below the critical temperature. Equation (9.4.15) is known as ***Rutgers***

equation which shows that if the metal is cooled or warmed in absence of magnetic field, there will be a *discontinuity* of specific heat through transition temperature. This has been verified experimentally. The equation is used to determine $(dH_c/dT)_{T=T_c}$ from the measurements of C_s and C_n calorimetrically. The specific heat vs. temperature (K) curve is illustrated in Fig. 9.7 where the data for normal curve is obtained by applying a field $> H_c$.

9.4.1 Density of states and energy gap

The way the electronic specific heat in the superconducting state indicates the existence of an energy gap E_g in the excitation spectrum of electrons in conduction band. Fig. 9.8 shows this by comparing the difference between the occupancy of states in a normal conductor and a superconductor. While Fig. 9.8a gives density of states in the absence of superconductivity, Fig. 9.8b shows the superconducting ground state for zero temperature, indicating a zero density of states for energies within $\pm\Delta(0)$ or $2\Delta(0)$. This is the energy gap, E_g. So, $E_g = 2\Delta(0)$. At $T = 0$, no electrons are excited

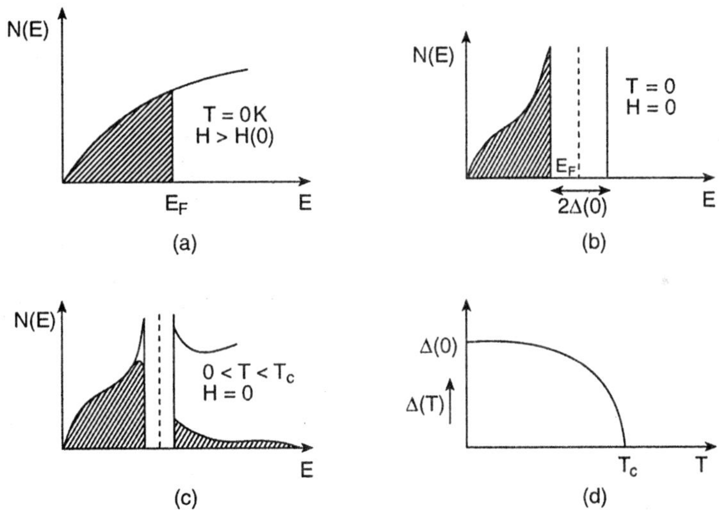

Fig. 9.8 Density and occupancy of states: (a) in normal conductor, (b) in superconducting state at T=0 (c) superconducting state at $T<T_c$, (d) variation of energy gap as a function of temperature

to higher states. At $T < T_c$, the superconducting energy gap is smaller than $\Delta(0)$ (Fig. 9.8c). A fraction of electron number is now in states above $E_F(0) + \Delta(T)$, with some unoccupied states below $E_F(0) - \Delta(T)$.

• The electronic specific heat data can be fitted to a function of the form $Ae^{-2\beta\Delta}$ where A and Δ are parameters, dependent both on the material and the temperature. At low temperatures, both A and Δ become independent of temperature and the specific heat in the superconducting state becomes an exponential function of $1/kT$, indicating an energy gap of $\sim 2\Delta$ in the electron energy spectra (Fig. 9.8).

• The energy gap is situated at Fermi level and is *characteristic* of superconducting state. It is different from the energy gap of insulators. In insulators, the gap is associated with lattice, while here it is centred about Fermi level. The gap $2\Delta \sim kT_c$ prevents the superelectrons from being excited below T_c. Only above T_c, they behave like normal electrons and may be readily excited. The gap is small compared to that of insulators or semiconductors. The variation in the energy gap Δ with temperature T has been illustrated in Fig. 9.8d.

- The existence of an energy gap has been explained in the BCS-theory.

9.5 Isotope effect

A relation between the *transition temperature* T_c and the *isotopic mass* M of the superconductor was observed in 1950 by Frölich. Experimentally, it is found that the transition temperature of many superconductors varies with the average isotopic mass and is given symbolically by

$$T_c \propto M^{-\frac{1}{2}} \quad \Rightarrow \quad \boxed{T_c M^{\frac{1}{2}} = constant} \tag{9.5.1}$$

where M is the mass of the superconducting isotope.

The relation shows that ***larger the isotopic mass, lower is the transition temperature and conversely.*** For instance, when the isotopic mass of mercury is changed from 199.5u to 203.4u, the transition temperature changes from 4.185 K to 4.146 K. *This phenomenon of decrease of critical temperature with increasing atomic mass is known as the* ***isotope effect*** of superconductors.

Again, at low temperatures, the mean square amplitude of the atomic or lattice vibrations is proportional to $M^{-\frac{1}{2}}$, and the Debye temperature Θ_D of the phonon spectrum is related to M as

$$\Theta_D \propto M^{-\frac{1}{2}} \quad \Rightarrow \quad \Theta_D M^{\frac{1}{2}} = \text{constant.} \tag{9.5.2}$$

Combining the relations (9.5.1) and (9.5.2), we obtain

$$T_c/\Theta_D = \text{constant} \tag{9.5.3}$$

$$\therefore \quad \boxed{T_c \propto \Theta_D \propto M^{-\frac{1}{2}}} \tag{9.5.4}$$

The *transition temperature* of superconducting materials is thus *closely related to* the corresponding *Debye temperature.*

The relations (9.5.3) and (9.5.4) also strongly suggest that the ***lattice vibrations might have an important role in causing superconductivity***, in contrast to the classical model of conduction in which zero resistance is due to zero interaction between the electron and the lattice. The electron-phonon interactions have a significant involvement in superconductivity. This was a fundamental discovery. There is no other reason for superconducting transition temperature to depend on the number of neutrons in the nucleus.. From this consideration, Frölich showed that ***two electrons in a metal can effectively attract each other, being mediated by lattice vibrations.*** Such electron-pairs are called the ***Cooper pairs*** (Read also Art. 9.12).

This culminated finally in 1957 in the development of a complete microscopic theory of superconductivity by Bardeen, Cooper and Schrieffer, the so-called ***BCS theory.*** It was based on the formation of Cooper pairs and their coherent superposition into a single quantum state.

● Recent experiments suggest that the above variation of T_c with M can be expressed in the following general form:

$$T_c \propto M^{-\alpha} \quad \Rightarrow \quad M^{\alpha} T_c = \text{constant} \tag{9.5.5}$$

where α varies from material to material; typically α is about 0.5 and is termed ***isotope effect coefficient*** defined as

$$\alpha = -\frac{\partial \ln T_c}{\partial \ln M} \tag{9.5.6}$$

obtained by taking the logarithm of equation (9.5.5).

● More recent experiments have shown a wider variety of power laws, varying from an almost complete absence of isotope effect in Osmium to a dependence of the approximate form $T_c \propto M^2$ in α-uranium. Nothing is sacred about a $\alpha = \frac{1}{2}$ in equation (9.5.1). The absence of isotope effect in Ru and Zr has been accounted for in terms of the electron band structure of the metals.

9.6 London equations

A number of theories has been proposed to explain superconductivity with varying degree of success. These include the phenomenological or *macroscopic theory* by F. London and H. London, the *semi-phenomenological theory* of Ginzburg and Landau and the *microscopic theory* of Bardeen, Cooper and Schrieffer (BCS). We shall hereunder describe the macroscopic theory of superconductivity by London brothers.

A superconducting specimen expels magnetic flux completely from its interior (Meissner effect). But the same is not true in the case of thin films. It is not possible, on the basis of Maxwell's equations alone, to deduce Meissner effect and field penetration in superconductors. London brothers, therefore, first investigated the magnetic aspects quantitatively and showed that two additional equations to the existing four of Maxwell, are needed to explain them completely.

London equations—London assumed that a superconductor has two types of electrons—normal and superconducting. When the temperature is below T_c, almost all the electrons in a superconductor are *superconducting electrons* or *superelectrons* that suffer no scattering by the lattice, and hence there is no damping force. They behave as free electrons and their equation of motion in an applied field $\vec{E}$ is given by

$$m\frac{d\vec{v}}{dt} = -e\vec{E} \tag{9.6.1}$$

where m is the electronic mass and $\vec{v}$ the drift velocity of superelectrons at time t, e being the electronic charge.

$$\therefore \text{ Supercurrent density, } \vec{J}_s = -n_s e\vec{v} \tag{9.6.2}$$

where n_s is the number density of the superconducting electrons.

Differentiating (9.6.2) with respect to t and using (9.6.1),

$$\boxed{\frac{\partial \vec{J}_s}{\partial t}} = -n_s e\frac{d\vec{v}}{dt} = \boxed{\frac{n_s e^2 \vec{E}}{m}} \tag{9.6.3}$$

Equation (9.6.3) is called the ***first London equation.***

The striking aspect of the equation is the *absence of the resistance.* It shows that the presence of an electric field causes the current to change with time. So, steady state currents may be set up in a superconductor *without an electric field* since for $E = 0$, J is finite and constant and vice versa.

• The corresponding expression for normal current density, J_n is $\vec{J}_n = \sigma\vec{E}$ indicating that *no current* is possible *without an electric field*, the behaviour in the normal state of the materials.

Taking the curl of (9.6.3) and using Maxwell's equation, $\vec{\nabla} \times \vec{E} = -\partial\vec{B}/\partial t$,

$$\vec{\nabla} \times \frac{\partial \vec{J}_s}{\partial t} = \frac{n_s e^2}{m}(\vec{\nabla} \times \vec{E}) = -\frac{n_s e^2}{m}\frac{\partial \vec{B}}{\partial t}$$

$$\Rightarrow \quad \frac{\partial}{\partial t}\left[\vec{\nabla} \times \vec{J}_s + \frac{n_s e^2}{m}\vec{B}\right] = 0$$

On integration with respect to t and taking the constant of integration zero (to be consistent with Meissner effect), we have

$$\boxed{\vec{\nabla} \times \vec{J}_s = -\frac{n_s e^2}{m}\vec{B} = \frac{\mu_0 n_s e^2}{m}\vec{H}} \tag{9.6.4}$$

replacing $\vec{B}$ by $\mu_0\vec{H}$ since the material has a linear magnetic character.

Equation (9.6.4) is known as the ***second London equation*** and gives results in agreement with experiments.

9.7 London penetration depth

By introducing the two new equations, London brothers could successfully explain Meissner effect and zero resistivity. According to London equations, the flux does not suddenly drop to zero at the surface of Type I superconductors but decreases exponentially. We shall now see the actual variation of flux with distance.

Using Maxwell's field equation, we have

$$\vec{\nabla} \times \vec{B} = \mu_0 \vec{J}_s. \tag{9.7.1}$$

Taking the curl of the equation (9.7.1), we obtain

$$\vec{\nabla} \times \vec{\nabla} \times \vec{B} = \mu_0 \vec{\nabla} \times \vec{J_s} \tag{9.7.2}$$

$$\Rightarrow \text{ grad (div } \vec{B}) - \nabla^2 \vec{B} = \mu_0 \vec{\nabla} \times \vec{J_s}$$

$$\Rightarrow -\nabla^2 \vec{B} = \mu_0 \vec{\nabla} \times \vec{J_s} = -\mu_0 \frac{n_s e^2}{m} \vec{B}, \tag{9.7.3}$$

using the second London equation (9.6.4).

$$\text{or,} \quad \boxed{\nabla^2 \vec{B} = \frac{1}{\lambda^2} \vec{B}} \tag{9.7.4}$$

$$\text{where } \lambda = \left(\frac{m}{\mu_0 n_s e^2} \right)^{1/2} \tag{9.7.5}$$

The quantity $\boldsymbol{\lambda}$ has the *dimensions* of *distance* and has been termed the ***London penetration depth.***

For a one-dimensional semi-infinite superconductor with surface in yz-plane, the equation (9.7.4) is the form

$$\frac{\partial^2 B_z}{\partial x^2} = \frac{1}{\lambda^2} B_z,$$

having the solution

$$B_z(x) = B_z(0) e^{-x/\lambda} \tag{9.7.6}$$

where $B_z(0)$ represent the magnetic field at the surface.

The equation clearly shows that $\vec{B}$ does not penetrate very deeply into the superconductors, but only a small distance λ within the surface (implying the *Meissner effect*). Such a penetration has indeed been verified experimentally and is illustrated in Fig. 9.9 that shows an *exponential decrease in the magnetic flux density inside the superconductor.* We shall now give a formal definition of the penetration depth.

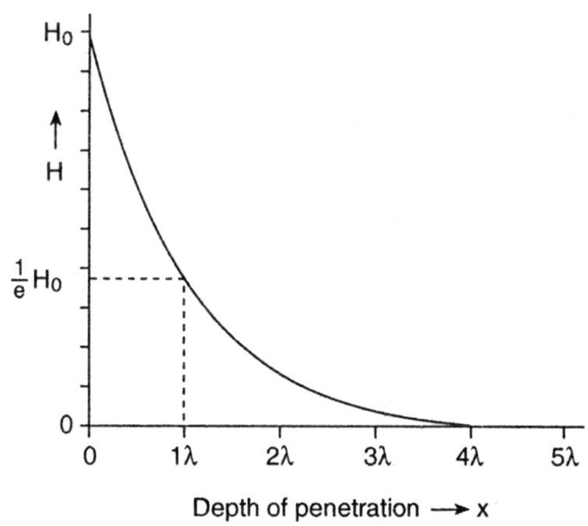

Fig. 9.9 Exponential decrease of magnetic field inside a superconductor

Penetration depth: Definition—*The* ***penetration depth λ*** *is defined as the depth, from the surface of a superconductor at which the magnetic flux density gets reduced to* $1/e$ *of its surface value.*

For superconductors below the critical temperature, the value of $\lambda \sim 10^3$ to 10^4 A, depending on the material.

Temperature dependence of λ—The penetration depth depends also on the temperature and a typical variation of the depth with temperature is illustrated in Fig. 9.10. The penetration depth λ becomes infinite as the temperature approaches the transition temperature T_c.

The temperature variation of λ is given by the empirical relation

$$\lambda(T) = \lambda(0)\left[1 - T^4/T_c^4\right]^{-1/2} \tag{9.7.7}$$

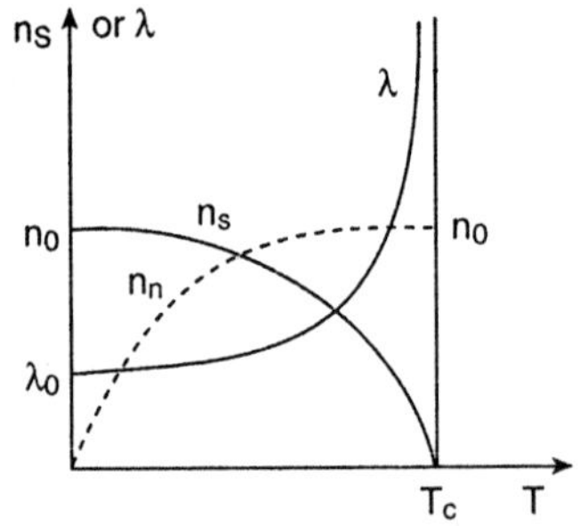

Fig. 9.10 The penetration depth as a function of temperature

where $\lambda(0) = \left(\frac{m}{\mu_0 n_s e^2}\right)^{\frac{1}{2}} = \lambda$ at $T = 0\,\text{K}$.

With tin, placed in a weak magnetic field, the penetration depth changes only slightly with rise in temperature till the critical temperature T_c is reached when the penetration depth increases sharply and abruptly. At critical temperature T_c and above, the *the penetration is complete.*

● The penetration depth and the number density of superelectrons n_s are inversely proportional to each other.

From (9.7.7.), $\left[\frac{\lambda(T)}{\lambda(0)}\right]^2 = \left[1 - \left(\frac{T}{T_c}\right)^4\right]^{-1}$ which implies that

$$n_s = n_0\left[1 - (T/T_c)^4\right]$$

The density superconducting electrons increases from zero at T_c to n_0 at absolute zero as shown in Fig. 9.10.

n_s/n_0 is called the ***order parameter*** as it characterizes the degree of order in the superconducting state.

9.8 Pippard coherence length

We have already learnt that near the surface of a superconductor, there are two types of electrons—*normal electrons* and *super-electrons.* The transition from the normal to the superconducting state or *vice versa* is not abrupt. Pippard in 1953 conceived the idea of a ***transition length*** associated with order parameter n_s/n_0. It corresponds to

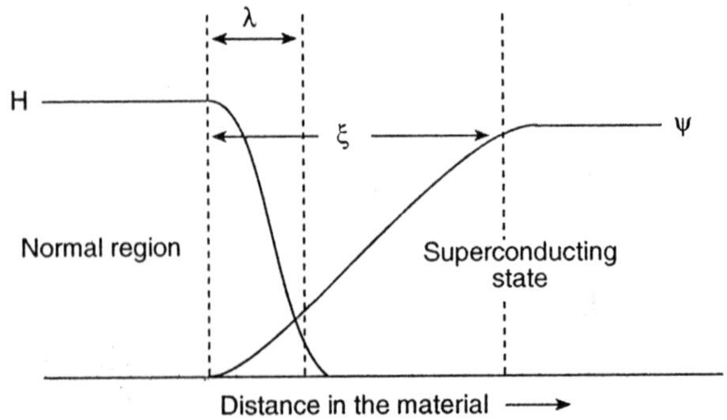

Fig. 9.11 Coherence length ξ : Variation of magnetic field H and the order parameter ψ through an interface between normal and superconducting region in type I superconductor

the distance in which n_s/n_0 changes from its maximum value in bulk superconducting region to zero in the normal region (Fig. 9.11). This distance is termed ***coherence length*** ξ that depends on the purity of the specimen.

Computation—The *intrinsic coherence legnth* ξ_0 for a *pure* superconductor may be computed as follows: The electron states responsible for superconductivity lie within kT_c of Fermi surface. Their life time τ is, from uncertainty principle (energy × time $= \hbar$), given by

$$(kT_c)\tau \simeq \hbar \tag{9.8.1}$$

If v_F be the electron velocity at Fermi surface, the wave function must extend over ξ_0, given by

$$\xi_0 = v_F\tau = \frac{\hbar v_F}{kT_c}$$

$$= \frac{\hbar v_F}{2\Delta} = \frac{\hbar v_F}{E_g}, \quad \text{(using (9.8.1)}$$

where E_g is the energy gap.

A more refined form of ξ_0 is: $\quad \xi_0 = \dfrac{2\hbar v_F}{\pi\Delta} \simeq 10^{-6}$ m.

The coherence length ξ, like the penetration depth λ, is a function of temperature and varies in a similar fashion.

- The ratio of λ to ξ is known as the ***Ginsburg-London parameter*** k, i.e., $k = \lambda/\xi$. This parameter k distinguishes between type I and type II superconductors. $k < 1/\sqrt{2}$ for type I superconductors and $k > 1/\sqrt{2}$ for type II superconductors.

- The coherence length ξ corresponds to spatial extension of a cooper pair i.e., the distance the cooper pairs maintain in the material.

9.9 Flux quantization: Fluxoid

Consider a superconducting ring (hole) and a closed loop (indicated by circle **c**) inside it, as illustrated in Fig. 9.12.

From Maxwell's electromagnetic equation, we have

$$\vec{\nabla} \times \vec{E} = -\frac{\partial \vec{B}}{\partial t} \tag{9.9.1}$$

Integrating (9.9.1) over the surface S bounded by the closed loop, and using Stokes' theorem, we obtain

$$\oint \vec{E}\cdot d\vec{l} = -\int_s (\partial \vec{B}/\partial t)\cdot d\vec{S} \tag{9.9.2}$$

Using the first London equation and equation (9.7.5), we have from above,

$$\mu_0\lambda^2 \oint \frac{\partial \vec{J}_s}{\partial t}\cdot d\vec{l} = -\int_s \frac{\partial \vec{B}}{\partial t}\cdot d\vec{S}$$

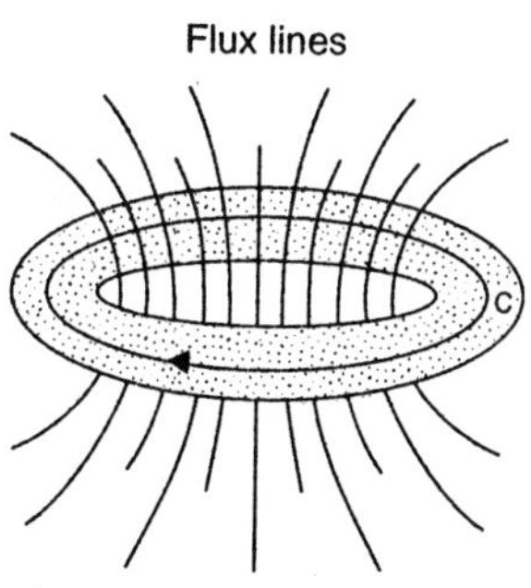

Fig. 9.12 A ring shaped superconductor showing path of integration C through its interior

$$\Rightarrow \quad \frac{\partial}{\partial t}\left(\mu_0\lambda^2 \oint \vec{J}_s \cdot d\vec{l} + \int_s \vec{B}\cdot d\vec{S}\right) = 0$$

$$\text{or,} \quad \frac{\partial \phi_s}{\partial t} = 0, \quad \text{where } \phi_s = \mu_0\lambda^2 \oint \vec{J}_s \cdot d\vec{l} + \int_s \vec{B}\cdot d\vec{S} \tag{9.9.3}$$

$$\therefore \quad \phi_s = \text{constant}$$

If the loop is well within the superconductor, i.e., far removed from the surface, then both $\vec{B}$ and $\vec{J}_s$ are zero at all points on it. For such loops, the first term on the right hand side of (9.9.3) vanishes. So, the total magnetic flux through the loop, given by the second term, is constant. But since very little flux passes through the superconducting material deep inside the surface, the flux is essentially confined to the ring i.e., the hole.

This magnetic flux ϕ_s is quantized and is given by

$$\phi_s = s\left(\frac{h}{2e}\right) = s\phi_0, \quad \text{where } s = \text{an integer}$$

and $\phi_0 = h/2e$ = a fundamental constant $= 2.0678 \times 10^{-15}$ Weber. This ϕ_0 is called the *quantum of flux* or *fluxon* or ***fluxoid*** i.e., the elementary quantity of magnetic flux. The number 2 in the denominator is a consequence of the double charge of a cooper pair.

The flux quantization has been confirmed experimentally.

- The superconducting ring can thus *entrap* flux. Let the magnetic field be switched '*on*' when the superconducting ring is in the normal state and now the temperature is lowered below the transition temperature. So the flux is excluded from the ring but passes through the hole. If the external magnetic field is switched '*off*', the flux through the hole remains unchanged and is trapped. But how is the flux through the hole maintained? The answer is that it is maintained by the field due to the supercurrent around the ring.
- A rather important application of fluxoid is in the *memory element* in computers and in SQUID.

9.10 Josephson junction: Josephson effect

A *thin layer* of insulator sandwiched between two superconductors i.e., a superconductor-insulator-superconductor layer, constitutes a *weak link* between the two and is called a ***Josephson junction*** (Fig. 9.13). In 1962, Brian Josephson suggested that the Cooper

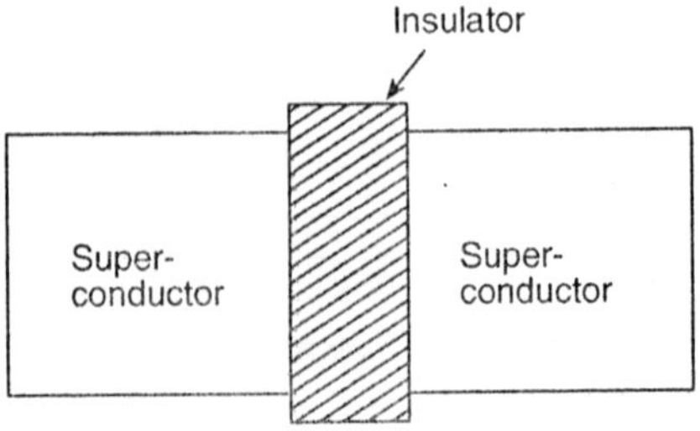

Fig. 9.13 Josephson junction for two superconductors separated by a thin insulator

pairs can tunnel through a sufficiently thin ($\sim$ 1 nm thickness) insulator even when there is no external potential difference between the superconductors; they cannot however get through when the insulator is thick. This is called ***Josephson effect.***

• A phase change arises in the tunnelling of Cooper pairs betwen the two superconductors.

dc Josephson effect—Quantum mechanically, it can be shown that the tunnelling current density is given by the relation

$$J_s = J_0 \sin\delta \tag{9.10.1}$$

where J_0 is the maximum supercurrent density through the insulator with zero voltage across it and δ the change in phase of the order parameter of Cooper pairs across the junction. For wide insulators, J_0 is quite small.

In presence of a dc voltage V across the junction, the current density becomes

$$J_s = J_0 \sin\left(\delta + \frac{q}{\hbar}Vt\right) \tag{9.10.2}$$

where q is the charge of Cooper pair, i.e., $q = 2e$, where e is the electronic charge.

Eq. (9.10.1) and (9.10.2) give the ***dc Josephson effect.*** The current-voltage characteristic is illustrated in Fig. 9.14. At voltages above V_c, the junction has a finite resistance.

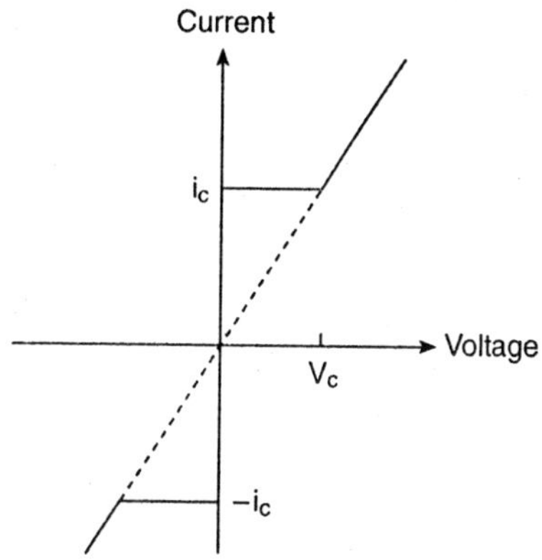

Fig. 9.14 Current-voltage characteristic of a Josephson junction

For *ordinary voltages*, the quantity $qV/\hbar$ gives a very high frequency and consequently the sine function in (9.10.2) oscillates rapidly to provide a zero average. Interestingly, therefore, we obtain a *current for no applied voltage*, and *no current for an applied voltage V*.

ac Josephson effect—Apart from the dc Josephson effect, there is also an ***ac Josephson effect*** which occurs when a high frequency ac voltage, superposed on a dc voltage, is applied across the junction. The applied voltage is therefore

$$V(t) = V_0 + V_a \cos\omega t \tag{9.10.3}$$

where V_0 is the dc voltage and V_a ($\ll V_0$) is the amplitude of ac voltage of angular frequency ω. In such a case, the tunnel current density

$$J_s = J_0 \sin\left(\delta + \frac{q}{\hbar}V_0 t\right) + J_0\left(\frac{q}{\hbar}\right)\left(\frac{V_a}{\omega}\right)\sin\omega t \cos\left(\delta + \frac{q}{\hbar}V_0 t\right) \tag{9.10.4}$$

Now, for ordinary voltages, the first term on rhs of (9.10.4) averages to zero, as in dc. The second term also gives a zero average except however for a special case. To seek for the case, we first write (9.10.4) ignoring the first term and get

$$J_s = J_0\left(\frac{q}{\hbar}\right)\left(\frac{V_a}{\omega}\right)\left[\sin\omega t \cos\delta \cos\left(\frac{q}{\hbar}V_0 t\right) - \sin\omega t \sin\delta \sin\left(\frac{q}{\hbar}V_0 t\right)\right] \tag{9.10.5}$$

The second term on rhs of (9.10.5) would not give a zero average if $\omega = qV_0/\hbar$, since the time functions in the second term give $\sin^2 \omega t$. So, *a current is observed across the junction with frequency* $\boldsymbol{\omega = qV_0/\hbar = 2eV_0\hbar}$. This has been experimentally verified and is analogous to *resonance.*

The tunnelling current I_s depends sensitively on the magnetic field B applied at the junction and is given by

$$I_s = I_0 \frac{\sin(\pi\Phi/\Phi_0)}{\pi\Phi/\Phi_0} \tag{9.10.6}$$

where I_0 depends on the temperature and the structure of the junction, Φ is the magnetic flux at the junction and Φ_0 the flux quantum. The tunneling current I_s thus varies with magnetic field. The variation of I with B is according to (9.10.6) and is illustrated in Fig. 9.15.

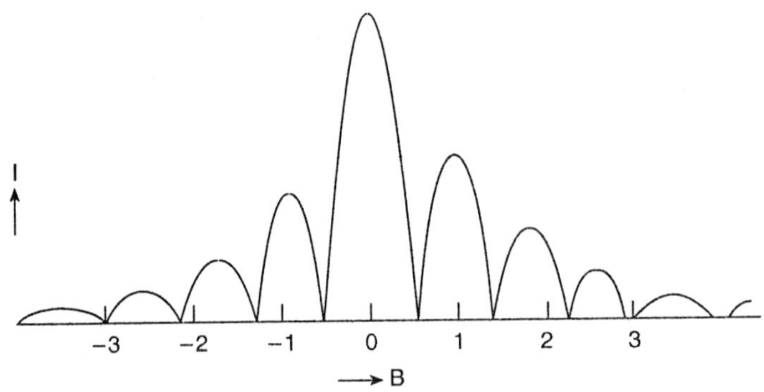

Fig. 9.15 Tunnelling current I versus magnetic field B for a Josephson junction

• In the *ac Josephson effect, the current oscillates with a frequency* $\omega = qV_0/\hbar = 2eV_0/\hbar$ ($\because$ $q = 2e$), *and can thus be utilized to measure precisely the fundamental constant* $e/\hbar$. It also gives a technique for *measuring voltages* very accurately.

• The relation $\omega = 2eV_0/\hbar$ says that a photon of every $\hbar\omega = 2eV$ is emitted or absorbed when an electron pair crosses the junction.

Applications—Josephson effect has a number of applications. Some of them are given as under:

1. This is used to generate *microwave of frequency*, $\omega = 2eV_0/\hbar$.
2. It is usd for switching of signals from one circuit to another. The switching time is $\sim 10^{-12}$s which is very useful in *high speed computers.*
3. A.C. Josephson effect is utilized to *define standard volt.*
4. *Very low temperature can be measured,* using a.c. Josephson effect. The principle is based on frequency variation of emitted radiation with temperature of the source.

9.11 SQUID

SQUID is the acronym of *S*uperconducting *Q*uantum *I*nterference *D*evice consisting of a superconducting ring with either one or two Josephson junctions. Such a device can be used as a sensitive magnetic fluxmeter.

There are two types of SQUIDs—The first type is ***dc SQUID*** (Fig. 9.16a) which consists of two Josephson junctions in a superconducting ring, operated by direct current. The critical current of the SQUID oscillates as a function of Q, the externally applied flux, with a period ϕ_0, one flux quantum. If however the device is biased

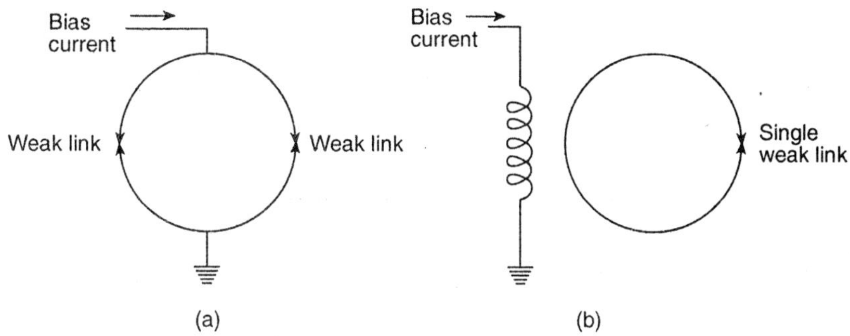

Fig. 9.16 Schematic representation of (a) dc SQUID and (b) rf SQUID

with constant current, the voltage across the SQUID is also periodic in ϕ. The device therefore, is in essence a *flux-to-voltage transducer*.

The second type of device is an ***rf SQUID*** which consists of only one Josephson junction in the superconduting ring, coupled to a tank circuit and is operated by *rf* current. The *rf* voltage across the tank circuit is also periodic in the applied flux ϕ. The *rf SQUID* is illustrated in Fig. 9.16b.

9.11.1 dc SQUID

The Josephson effect arises due to the phase coherence between two superconductors, the dc device would expectedly show interference behaviour similar to optical interference in so far as the critical current is concerned.

Case 1. *Junctions identical, there is no applied magnetic field* (Fig.9.17).

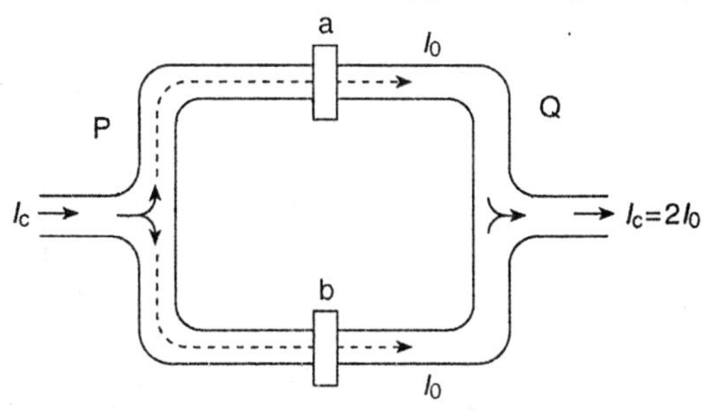

Fig. 9.17 Superconducting ring with two weak links. There is no applied field

There are two possible paths for the (bias) current I_c to get equally divided so that the phase difference across the junctions is the same. Thus, no voltage appears between P and Q, and plainly

$$I_c = 2I_0 \tag{9.11.1}$$

Case 2. *Junctions identical; a magnetic field is applied externally* (Fig. 9.18).

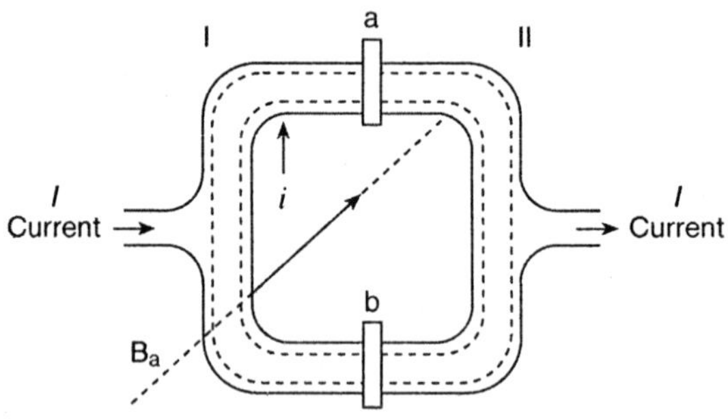

Fig. 9.18 Superconductors I and II with two weak links (a) and (b). A magnetic field B_a is applied

The magnetic field B_a causes a change in the situation in *two* ways. A circulating current i passes along the inner surface of the loop and prevents any flux to pass through the loop and we would get

$$Li = \phi_a = B_a \times \text{area of the loop} \tag{9.11.2}$$

if there were no weak links in the circuit: L is the self-inductance of the loop.

For a single loop of mm-dimensions, the self-inductance L is very small and cancellation of flux is not possible in the circuit with weak links. This is because i will quickly exceed the critical currents I_0 of individual junctions when $LI_0 \ll \phi_0$, the flux quantum or $L \ll \phi_0/I_0$. So, the flux through the loop (i.e., Li) is exactly equal to the applied flux ϕ_a.

9.11.2 rf SQUID

The ***rf SQUID*** consists of a single Josephson link (Fig. 9.19) and the Josephson element is shorted at dc by superconducting loop. The *rf* SQUID is monitored usually at a radio frequency of about 20–30 megahertz by coupling an *rf* current from a coil. The coil in its turn forms a part of a *tank circuit*, resonant at a given frequency supplied by a constant *rf*-drive current I_1.

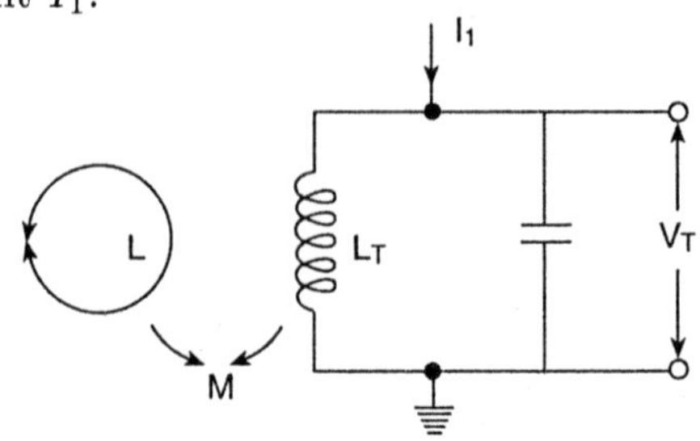

Fig. 9.19 rf SQUID

The amount of loss in *rf* from the resonant circuit due to its coupling to the loop is a periodic function of the *dc flux* enclosed by the loop, the period being ϕ_0. The flux-modulated *rf* loss modulates the Q-value of the tank circuit and thus the magnitude of the *rf*-voltage, V_T across it which is detected to give the flux-dependent read-out from the device.

● With SQUID sensor, one can detect extremely small magnetic signals, even that of the human brain and heart. SQUID magnetometers are used to detect *paramagnetic response* in the liver that gives the amount of ion in it accurately.

9.12 BCS Theory: Qualitative treatment

The fundamental microscropic theory of superconductivity was developed by Bardeen, Cooper and Schrieffer in 1957 and is appropriately called the ***BCS theory***. It describes the coupling of two electrons with opposite spins and momenta by means of a phononon. All observable facts of superconductivity such as zero resistivity, Meissner effect, isotope effect etc. are successfully explained with its help. But it rests on advance quantum mechanics and involved mathematics. We shall present here only a *qualitative account* of the theory, emphasising on its important aspects.

The BCS theory is characterized by the following *essential features.*

1. The electron-lattice interaction leads to an electron-electron attraction, sufficiently strong to overcome their mutual repulsion, and the electrons form bound pairs in the superconducting state. The electrons thus paired propagate throughout the lattice and constitute a single system so that their motion would be correlated.

2. The electron-electron interaction is mediated by the exchange of *virtual phonons* between the two electrons and so their propagation is without resistance. The interaction described by BCS theory is thus also called ***electron-phonon interaction.***

3. The pairing of electrons is such that an electron with *spin-up* and momentum $\vec{k}$ is paired with one of *spin-down* and momentum $-\vec{k}$ and that the angular momentum of the pair is zero. Such a pair of electrons is called a ***Cooper pair***. The formation of Cooper pairs gives rise to the ordering suggested by specific heat data.

Electron-lattice interaction—We shall now explain how the electron-lattice interaction leads to the Cooper pair formation. Consider a metal with conduction electrons inside the Fermi sphere and two electrons marked 1 and 2, lie *just* inside the Fermi surface. If they form, at all, a pair, it leads to BE-condensation of all electrons in the vicinity into a 'sea' of Cooper pairs.

But how is the electron-electron interaction brought about, for a direct Coulomb interaction between them is always a repulsion? Since electron 1 is near the Fermi surface, its speed is high enough, the Fermi velocity. And as it passes through the packing of positive ions it is attracted by the neighbouring positive ions (forming a positive ion-core : Fig. 9.20). But because of the heavy masses, the ions respond rather slowly. And by the time the ions completely respond to electron 1, it has left its initial position, at least partially, stimulating an over-compensation. The attraction between the electron and the ion-core generates an instantaneous local deformation of the lattice.

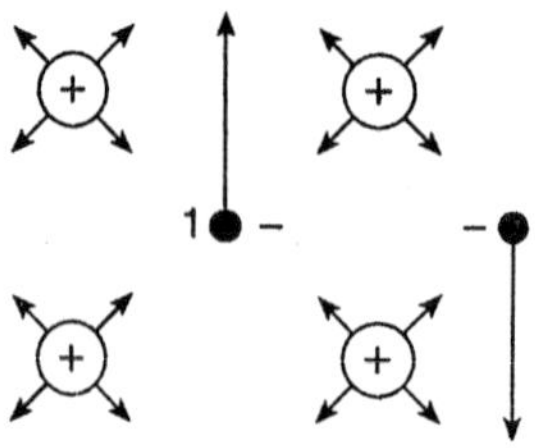

Fig. 9.20 Electron-electron interaction

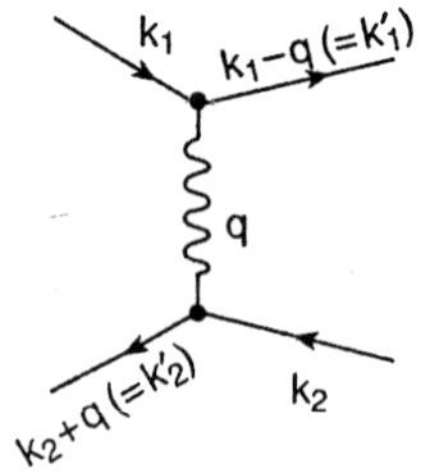

Fig. 9.21 Phonon-mechanism of pairing

If now the electron 2 passes by the side of the assembly of electron 1 and the ion core, it does not 'see' the bare electron 1 alone but a deformed lattice and is attracted to the assembly. It thus interacts with electron 1 *via* lattice deformation. In the language of field theory, the above interaction creates a ***virtual phonon***. Each electron is thus surrounded by a '*phonon cloud*' and the electrons interact by an exchange of a virtual phonon (Fig. 9.21).

The phonon of momentum q emitted by 1 with momentum k_1 is absorbed by 2 with momentum k_2 which gets scattered with momentum $k_2 + q$. The interaction is strongest when 1 and 2 have opposite momenta and spin, i.e.,

$$\vec{k}_1 = \vec{k}\uparrow; \;\; \vec{k}_2 = -k\downarrow$$

In terms of wave vector $\vec{k}$, the interaction process is

$$k_1 - q = k_1'$$

$$\text{and } \; k_2 + q = k_2'$$

$$\therefore \;\; k_1 + k_2 = k_1' + k_2'$$

So, the net k-values of the pair is *conserved* in the process and that leaves the lattice *invariant*, the momentum gets transferred between the electrons.

The mediating phonons have a small energy range

$$\hbar\omega_D = k\Theta_D$$

where ω_D is the Debye resonance frequency, Θ_D the Debye temperature and k the Boltzmann constant.

The pair formation is limited to such k-values that satisfy the momentum conservation. The energy of the pairs is lower than the total energy of two free electrons. The electrons in a Cooper pair are thus *quasi-bound*. As their net energy is positive w.r.t the bottom of the conduction, they are not exactly bound. A '*gap*' is thus formed near the Fermi surface given approximately by

$$\Delta \simeq 1.4kT_c$$

This energy gap determines the physical properties of the BCS superconductor.

Below the critical temperature T_c, the energy gap rises sharply according to the relation given by

$$\Delta(T) \simeq 3.52kT_c\,(1 - T/T_c)^{1/2}$$

For T satisfying the condition $0 < T < T_c$, the pairs get continually scattered by lattice vibrations. Since the pair has high correlation in their motion, the net momentum is always conserved. So, when an electric field is applied, there is no net transfer of momentum from the moving pairs to the lattice and the periodicity of lattice also exerts no influence on the pairs.

The superelectrons, i.e., the pairs give lossless current and the superconducting state offers no electric resistance. The normal electrons although present cannot carry current as they are shorted out by superelectrons.

• Cooper pair (superelectron pair), the quasi-particle of BCS-theory, behaves as a *boson*, since the net *spin* is *zero*, for one electron is with 'spin up' and the other electron is with 'spin down' and the Cooper pairs can be in the same state. The Cooper pair can thus be regarded as a single particle of zero momentum, zero spin and charge $-2e$.

• The superconducting *ground state* is described by saying that each electron is paired and all the pairs formed may occupy the lowest energy state (BE-condensation) characterized by *zero crystal momentum*. They all have a fixed phase relation that may lead to formation of macroscopic quantum states.

• Superconductivity is due to formation of Cooper pairs. The binding energy of a Cooper pair is $\sim 10^{-3}$ eV and corresponds to a temperature ~ 10 K. The pair formation is complete at 0 K and is broken completely at T_c, the critical temperature. So, at higher temperatures, the Cooper pairs do not exist. *Superconductivity* is thus *a low temperature phenomenon.*

• *Good conductors* in normal state are relatively *poor superconductors.* But why? The electron-phonon interaction determines the resistivity of a metal in the normal state. And a strong interaction corresponds to high resistivity. This explains why good conductors in the normal state make relatively poor superconductors.

• The paired electrons, i.e., *the Cooper pair behave in a material as a single unit up to a specific length called the* ***correlation*** or ***coherence length*** ξ. This is the distance within which the gap parameter does not change in a varying magnetic field. Typically, the ξ-value is about 100 μm.

9.12.1 Predictions of BCS theory

The BCS theory successfully explains the *isotope effect* (Eq. 9.5.5) and the variation of critical field with temperature (Eq. 9.2.3). BCS theory predicts the existence of an ***energy gap*** between the ground state (superconducting) and the first excited state. The gap represents the energy needed to break the Copper pair. Larger the gaps, more stable are the superconductors. By BCS theory, the energy gap at $T = 0$ K is

$$E_g(0) = 2\Delta(0) \simeq 3.53\,kT_c \tag{9.12.1}$$

and has been verified by studying the absorption of *e. m. radiation* by superconductors.

A transition from superconducting to normal state occurs on applying a magnetic field $H > H_c$. In presence of the magnetic field, the electron pairs move faster due to increase in K.E and produce a current for expulsion of magnetic flux from the superconductor. At high K.E. of electrons, the phonon scattering occurs with exchange of energy and cooper pair splits into normal electrons. The superconductor transits to the normal conductor.

Another prediction of the theory is the *quantization of magnetic flux* ϕ_s through a superconducting ring. Accordingly,

$$\phi_s = s\phi_0 = s\left(\frac{h}{2e}\right) \tag{9.12.2}$$

where s is an integer and $\phi_0 = h/2e$, the ***fluxoid*** or ***quantum of flux***. Its value is $2.608 \times 10^{-15}\,\mathrm{Tm}^2$. The flux quantization has also been experimentally confirmed.

Another important parameter connected with superconductivity and that follows from BCS theory is the ***coherence length*** ξ *defined* as the distance over which two electrons combine to form a Cooper pair, that is, it gives the smallest dimension over which supercoductivity can be developed or destroyed.

9.13 High T_c superconductivity (HTSC)

A breakthrough in superconductivity occurred after about seven decades of its discovery when in 1986 Bednorz and Muller discovered metallic, oxygen deficient copper oxide compounds of the Ba–La–Cu–O system with transition temperature of about 35 K. Later Wu and Chu discovered Y–Ba–Cu–O system with $T_c = 90\,\mathrm{K}$. Other families of high T_c cuprate oxide materials were discovered in subsequent years and T_c has now been pushed to as high as 133 K. The dream of the scientists is to discover superconductors having a T_c of about 300 K (room temperature) to eliminate the use of cryogenic fluids. Due to lowering of cooling costs, various technological applications of ***high*** T_c ***superconductors*** (HTS) are being potentially considered.

The scope of the book does not permit to do justice to the revolutionary development of HTS and their properties. We shall just mention them and state some salient features of these materials.

Properties—Some of the important properties of high temperature superconductors are mentioned below.

1. They have *layered crystal structure* of one or more CuO_2 layers wherein each Cu-atom is strongly bonded covalently to four O-atoms in a near-square plane, the Cu–O bond length $\simeq 1.9$ A. There is a charge transfer to and from Cu_2O layers, which is induced by doping near metal-insulator phase boundary existing in all oxide HTSCs.

2. In almost all oxide HTSCs, the *Hall coefficient is positive* (i.e., carriers are holes). Exception however is $(Ln_{1-x}Mx)_2\ CuO_{4-y}$ where the carrier is electron.

3. If the rare earth (RE) ions replace Y in $YBa_2Cu_3O_7$, the T_c remains essentially unchanged and so do most of the transport properties. But the specific heat data of

(RE) $Ba_2Cu_3O_7$ compounds which are within 2% of $YBa_2Cu_3O_7$ for $T > T_c$, show wide differences at low temperatures.

4. They are all *type*-II *superconductors* with highly anisotropic lower and upper critical fields. The two fundamental superconducting parameters, namely coherence length and penetration depth, and also the normal state transport properties are also highly anisotropic.

5. In high-temperature cuprate superconductors, the isotope effect is absent for all practical purposes.

6. The value of $2\Delta(0)/kT_c$ is smaller for tunneling of pairs normal to *ab*-plane ($\sim$ 3.8–4), but its value parallel to *ab*-plane is $\sim 5.8 - 6.0$.

7. The low-temperature specific heat of HTSCs in the superconducting state is given by

$$C = \alpha T^{-2} + \beta T + \gamma T^3,$$

the first term arises from magnetic impurities, the second due to free electrons in a metal and the third is due to Debye's lattice contribution. The absence of the exponential dependence of electronic specific heat is a departure of HTSCs from the conventional superconductors.

The phononic mechanism of superconductivity (the BCS theory) *cannot* theoretically explain the superconducting mechanism in cuprate high T_c superconductors. Till to-day, many mechanisms have been suggested but none has been proved completely satisfactory. This perhaps is the most challenging problem in condensed matter physics.

9.14 Fullerenes

Fullerenes are the novel superconductors added, rather recently in 1991, to the list of high temperature superconductors (HTS). The prominent members of fullerenes are C_{60}-based molecular crystals. The structure of a single C_{60}-molecule (Fig. 11.1) consists of 60 carbon atoms, all the valance electrons of each atom being bonded and each carbon atom takes part in one double bond and in two single bonds. It is a cluster of C-atoms in the shape of a truncated icosahedron with 20 hexagonal and 12 pentagonal faces as in graphite. Each hexagonal benzene ring shares sides with 3 hexagonal and 3 pentagonal rings forming a closed structure resembling a football.

Some fullerenes ($K_3\,C_{60}$, $RB_3\,C_{60}$, $RB_{2.7}\,Tl_{0.2}\,C_{60}$) shows superconductivity with T_c ranging from 19 to 48 degree Kelvin. Similar to ceramic HTS, fullerenes also show the gradual onset of diamagnetism when cooled in zero magnetic field to temperatures $T < T_c$.

9.15 Uses and potentialities of superconductors

Ever since the discovery of superconductivity, many possible applications of the phenomenon have been explored. Some of the areas, including the potential ones, are in production of superconducting magnets, in ***electrical power industry***, in medicine and transport, in supercomputers and high energy physics etc.

The current, set up in a closed loop of superconductor, persists almost for ever without decay for there is no i^2R loss or dissipation of energy since $R = 0$. It is possible to carry even 1000 times larger current densities, compared to normal ones, without loss by superconductors. This led to the development and use of compact, light-weight and *high intensity* ***superconducting magnets*** in various fields. For instance, they are employed in NMR-spectrometer and also in NMR-imaging used in medical diagnosis. In this, intense magnetic fields simulate the resonance of H-atoms in human body to give a two-dimensional superior quality picture of human organs. In high energy physics, particles can be accelerated, deflected and focussed by the fields of supermagnets.

The **SQUIDs** are the basis of the most sensitive magnetometers of to-day, capable of detecting magnetic fields less than $10^{-4}\,\text{Am}^{-1}$. SQUIDs made from high-T_c superconductors will extend its use to military and medical fields. The use of **HTSC** for generation, transmission, storage and transformation of power in electrical industry is being seriously assessed the world over. In very large scale integration (VLSI) technology, the HTSC interconnections in high-speed computers and telecommunication would be advantageously used in foreseeable future. The possibility of applications of HTSC in magnetically elevated trains is being pursued to make it cost-effective. With modified ceramics, it may be possible to have superconductors at room temperature to open up new fields of applications, e.g., medical diagnosis, under-sea communications, submarine detection, geophysical prospecting and running non-polluting vehicles like electric trains, cars and trucks at a cheaper cost.

In ***other applications***, Josephson junction may find application in integrated circuits and is used to measure and maintain voltage standards. Meissner effect is used to levitate the transport system. A ***mag***netic ***lev***itation (***maglev***) train in Japan has reached a speed of 550 km/h.

All the possibilities are being hotly pursued, particularly from the consideration of cost-effectiveness but till now they appear to be somewhat remote.

9.16 Illustrated Examples

► **Example 1.** *The critical temperature of lead, in the superconducting state, is* 6.2 K *at zero magnetic field. It has a critical field of* 0.064 Tesla *at* 0 K. *Calculate the critical field at* 4 K.

Solution. By the problem, $H_c(0) = 0.064\,\text{Tesla},\ T = 4\,\text{K},\ T_c = 6.2\,\text{K}.$

From the relation:

$$H_c(T) = H_c(0)\left[1 - (T/T_c)^2\right],$$

$$\boldsymbol{H_c(4K)} = 0.064\,[1 - (4/6.2)^2]$$

$$= \mathbf{0.037\ Tesla}$$

► **Example 2.** *The transition temperature of mercury with an average atomic mass of* 200.59 u *is* 4.153 K. *Determine the transition temperature of one of its isotope* ${}^{204}_{80}\text{Hg}$.

(Punjab Univ.)

Solution. The transition or critical temperature T_c is related to isotopic mass M as

$$T_c \propto \frac{1}{\sqrt{M}} \quad \Rightarrow \quad \frac{(T_c)_2}{(T_c)_1} = \sqrt{\frac{M_1}{M_2}} \quad \Rightarrow \quad (T_c)_2 = (T_c)_1\sqrt{M_1/M_2}$$

Now, by the problem, $(T_c)_1 = 4.153\,\text{K}$, $M_1 = 200.59\,\text{u}$, $M_2 = 204\,\text{u}$.

$$\therefore \quad \boldsymbol{(T_c)_2} = 4.153\sqrt{200.59/204} = \mathbf{4.118\,K}$$

► **Example 3.** *For niobium, the critical field at* $0\,\text{K}$ *is* $1.5 \times 10^5\ \text{Am}^{-1}$. *Calculate the maximum current that a niobium wire of dia* $0.1\,\text{mm}$ *at* $0\,\text{K}$ *can carry.*

Solution. From Ampere's circuital law, the field H at a distance r from the axis of the wire is given by

$$H = I/2\pi r$$

where I is the current carried in the wire.

Here, $r \geq R$, the radius of the wire. H is maximum at $r = R$, and when the current $I = I_{\text{max}}$, $H(R) = H_c$, the critical field.

$$\therefore \quad \boldsymbol{I_{\text{max}}} = 2\pi R H_c = 3.14 \times 0.1 \times 10^{-3} \times 1.5 \times 10^5 = \mathbf{47.1\,A}$$

► **Example 4.** *Assuming that each atom gives one conduction electron, estimate the London penetration depth for tin from the following data: density,* $\rho = 7.3\,\text{g/c.c.}$, $T_c = 3.7\,\text{K}$, *at. wt.* $M_A = 118.7$ *and the effective mass,* $m^* = 1.9\,m_e$, m_e *being the mass of a free electron.*

Solution. Plainly, $M_A/\rho\ \text{m}^3$ contains 6.02×10^{26} electrons (Avogadro no. N_A).

$$\therefore \quad 1\,\text{m}^3 \text{ will contain } \rho N_A/M_A \text{ number of electrons.}$$

$\therefore$ Number density of superelectrons, n_s is given by

$$n_s = \frac{7.3 \times 10^3 \times 6.02 \times 10^{26}}{118.7} = 3.7 \times 10^{28}$$

Effective mass, $m^* = 1.9 m_e = 1.9 \times 9.1 \times 10^{-31}\,\text{kg} = 17.3 \times 10^{-31}\,\text{kg}.$

$$\therefore \text{ Penetration depth, } \lambda(0) = \left(\frac{m^*}{\mu_0 n_s e^2}\right)^{1/2} \quad \text{at } 0\,\text{K}$$

$$= \left(\frac{17.3 \times 10^{-31}}{4\pi \times 10^{-7} \times 3.7 \times 10^{28} \times (1.6 \times 10^{-19})^2}\right)^{1/2}$$

$$(\because \mu_0 = 4\pi \times 10^{-7}\,\text{H/m})$$

$$= 0.0015 \times 10^{-7}\,\text{m}$$

$$= 1.5 \times 10^{-10}\,\text{m} = \mathbf{1.5\,A}$$

► **Example 5.** *The penetration depth of mercury at* 3.5 K *is* 750 Å. *Calculate its penetration depth at* 0 K. *Also calculate the superconducting electron density, given the density of mercury* = 13.55×10^3 kg/m^3, mol. weight = 200.6 *and* $T_c = 4.12$ K.

Solution. Here, $T = 3.5$ K; $\lambda(3.5) = 750$ Å and $T_c = 4.12$ K, $\lambda(0) = ?$

From the relation $\lambda(T) = \lambda(0)[1-(T/T_c)^4]^{-1/2}$, we have

$$\boldsymbol{\lambda}(\mathbf{0}) = \lambda(T)\left[1-(T/T_c)^4\right]^{\frac{1}{2}} = 750\left[1-\left(\frac{3.5}{4.12}\right)^4\right]^{\frac{1}{2}} = \mathbf{520\ A}.$$

Now, $n_0 = N_A\rho/M$ where M = mol. weight, N_A = Avogadro no., ρ = density.

$$\therefore\ n_0 = \frac{6.02 \times 10^{26} \times 13.55 \times 10^3}{200.6} = 4.06 \times 10^{28}/\text{m}^3$$

$$\therefore \text{ Superconducting electron density, } \boldsymbol{n_s} = n_0\left[1-(T/T_c)^4\right]$$

$$= 4.06 \times 10^{28}\left[1-\frac{3.5}{4.12}\right]^4$$

$$= \mathbf{1.95 \times 10^{28}/m^3}$$

► **Example 6.** *If the penetration depths for lead are* 396 A *and* 1730 A *at* 3 K *and* 7.1 K *respectively, compute the critical temperature of lead.*

Solution. By the problem, we have $\lambda(3\,\text{K}) = 396$ A, $\lambda(7.1\,\text{K}) = 1730$ A, $T_c = ?$

Penetration depth is related to critical temperature as

$$\lambda(T) = \lambda(0)\left[1-(T/T_c))^4\right]^{-\frac{1}{2}}$$

$$\therefore\ \lambda(3) = \lambda(0)\left[1-(3/T_c)^4\right]^{-\frac{1}{2}}$$

$$\text{and}\ \ \lambda(7.1) = \lambda(0)\left[1-(7.1/T_c)^4\right]^{-\frac{1}{2}}$$

On removal of the power $\left(-\frac{1}{2}\right)$ and taking the ratioos λ's, we obtain

$$\left[\frac{\lambda(7.1)}{\lambda(3)}\right]^2 = \frac{T_c^4-(3)^4}{T_c^4-(7.1)^4} \Rightarrow \left(\frac{1730}{396}\right)^2 = \frac{T_c^4-81}{T_c^4-2541}$$

$$\text{or,}\ 19 = \frac{T_c^4-81}{T_c^4-2541} \Rightarrow T_c^4 = \frac{19 \times 2541 - 81}{18}$$

$$\therefore\ \boldsymbol{T_c} = \mathbf{7.193\,K}$$

► **Example 7.** *Estimate the London penetration depth for metallic tin at* 2.3 K *from the following data.* $T_c = 3.7$ K, *density of tin* = 7.3×10^3 kg/m^3, $M = 118.7$ *and* $m^* = 1.9m$.

Solution. London penetration depth, $\lambda(0) = (m^*/\mu_0 n e^2)^{1/2}$, with usual symbols.

Now, from given data, $\dfrac{118.7}{7.3 \times 10^3}$ m^3 of tin contains $2 \times 6.02 \times 10^{26}$ electrons.

$\therefore$ 1 m^3 contains $(7.3 \times 10^3 \times 12.04 \times 10^{26})/118.7$ electrons $= 7.4 \times 10^{28}$ electrons $= n$.

$$\therefore \quad \lambda(0) = \left[\frac{1.9 \times (9.1 \times 10^{-31})}{(4\pi \times 10^{-7}) \times (7.4 \times 10^{28}) \times (1.6 \times 10^{-19})^2}\right]^{\frac{1}{2}} = 27\,\text{nm}, \quad \text{on simplification.}$$

But, $\lambda(T) = \lambda_0 \left[1 - (T/T_c)^4\right]^{-\frac{1}{2}}$. So, $\boldsymbol{\lambda(2.3)} = \dfrac{27}{[1-(2.3/3.7)^4]^{1/2}} = \mathbf{29.1\ nm}$, on simplification.

► **Example 8.** *The critical fields at* 6 K *and* 8 K *for a superconducting alloy are* 7.616 *and* 4.284 mA/m *respectively. Determine the critical temperature and the critical field at* 0 K. **(Punjabi Univ.)**

Solution. Here, $H_{c_1} = 7.616$ mA/m, and $H_{c_2} = 4.284$ mA/m. So, from the relation

$$H_c = H_0 \left[1 - (T/T_c)^2\right],$$

$$7.616 = H_0 \left[1 - (6/T_c)^2\right] \tag{9.16.1}$$

$$\text{and } 4.284 = H_0 \left[1 - (8/T_c)^2\right] \tag{9.16.2}$$

$$\therefore \quad \frac{T_c^2 - 36}{T_c^2 - 64} = \frac{7.616}{4.284} = 1.78$$

Solving for T_c, we obtain $\quad \boldsymbol{T_c = 10\,\text{K}}$

Again $H_c(6) = H_0\left[1 - (6/10)^2\right] \Rightarrow 7.616 = H_0 \times (64/100)$

$$\therefore \quad \boldsymbol{H_0} = \frac{100 \times 7.616}{64} = \mathbf{11.9\ mA/m}$$

► **Example 9.** *The transition temperature of mercury with an average atomic mass of* 200.59 u *is* 4.153 K. *Determine the transition temperature of one of its isoptopes* $_{80}\text{Hg}^{204}$. **(Punjab Univ.)**

Solution. Transition temperature T_c of a superconductor is related to its atomic mass M as $T_c \propto M^{-1/2}$. Here, $M_1 = 200.59$ u and $M_2 = 204$ u and $T_{c_1} = 4.153$ K, $T_{c_2} = ?$

$$\therefore \quad \frac{T_{c2}}{T_{c_1}} = \sqrt{\frac{M_1}{M_2}} = \sqrt{\frac{200.59}{204}}$$

$$\therefore \quad \boldsymbol{T_{c2}} = \sqrt{\frac{200.59}{204}} \times 4.153 = \mathbf{4.118\ K}.$$

► **Example 10.** *Estimate the intrinsic coherence length, if the size of the average energy gap is* $3.4 \times 10^{-4}\,\text{eV}$ *and the Fermi velocity* $v_F = 2.02 \times 10^6\,\text{m/s}$.

Solution. Intrinsic coherence length, $\xi_0 = \hbar v_F / E_g$, with usual meaning of symbols.

Here, $E_g = 3.4 \times 10^{-4}\,\text{eV} = 3.4 \times 10^{-4} \times 1.6 \times 10^{-19}\,\text{J}$, $v_F = 2.02 \times 10^6\,\text{m/s}$.

$$\therefore \text{ Coherence length, } \boldsymbol{\xi_0} = \frac{h}{2\pi} \cdot \frac{v_F}{E_g} = \frac{6.63 \times 10^{-34} \times 2.02 \times 10^6}{2 \times 3.14 \times 3.4 \times 10^{-4} \times 1.6 \times 10^{-19}}$$

$$= \mathbf{3.9 \times 10^{-6}\,m}$$

► **Example 11.** *Derive the Meissner effect from the second London equation, using the Maxwell's relation* $\vec{\nabla} \times \vec{B} = \mu_0 \vec{J}_s$.

Solution. On taking the curl of Maxwell's equation, $\vec{\nabla} \times \vec{B} = \mu_0 \vec{J}_s$,

$$\vec{\nabla} \times \vec{\nabla} \times \vec{B} = \mu_0 \vec{\nabla} \times \vec{J}_s$$

$$\Rightarrow \text{grad}\,(\text{div}\,\vec{B}) - \nabla^2 \vec{B} = \mu_0 \vec{\nabla} \times \vec{J}_s$$

$$\Rightarrow -\nabla^2 B = \mu_0 \vec{\nabla} \times \vec{J}_s \qquad (9.16.3)$$

Comparing (9.15.3) with the second London equation, we have

$$\nabla^2 \vec{B} = \mu_0 \frac{n_s e^2}{m} \vec{B} = \frac{1}{\lambda^2} \vec{B} \qquad (9.16.4)$$

where $\lambda = (m^* / \mu_0 n_s e^2)^{1/2}$ = London *penetration depth.*

For a superconductor to the right of the plane $x = 0$, the equation (7.15.4) has the solution given by

$$B(x) = B_0 e^{-x/\lambda}$$

which shows that $\vec{B}$ does not penetrate deeply into superconductor but only a very small distance ($\sim$ 1000Å), depending on the material. This implies the ***Meissner effect.***

► **Example 12.** *The critical field and the critical temperature of lead are* 6.5×10^4 A/m *and* 7.18 K *respectively. Find the critical field and critical current density for* 1 mm *dia of lead-wire at* 4.2 K.

Solution. We have: $H_c(0) = 6.5 \times 10^4$ A/m, $T_c = 7.18$ K, $D = 2r = 1\,\text{mm} = 10^{-3}\,\text{m}$, $T = 4.2$ K, $J_c = ?$

Let us first find the critical field at 4.2 K. From the relation

$$H_c(T) = H_c(0)[1 - (T/T_c)^2]$$

$$\text{or, } \boldsymbol{H_c}(4.2) = 6.5 \times 10^4[1 - (4.2/7.18)^2]$$

$$= \mathbf{4.276 \times 10^4\ A/m}$$

Now, the critical current, $I_c = 2\pi r H_c \Rightarrow J_c = I_c/\pi r^2 = 2\pi r H_c/\pi r^2 = 2H_c/r$.

$$\therefore \text{ Critical current density, } \boldsymbol{J_c} = \frac{2 \times 4.276 \times 10^4}{0.5 \times 10^{-3}} = \mathbf{1.71 \times 10^8\ A/m}$$

▶ **Example 13.** *The critical field and the critical temperature of lead are* 6.5×10^4 A/m *and* 7.18 *respectively. To what temperature must it be cooled to become a superconductor in a magnetic field of* 2×10^4 A/m*?*

Solution. Given, $H_c(0) = 6.5 \times 10^4$ A/m, $T_c = 7.18$ K, $H_c(T) = 2 \times 10^4$ A/m, T =?.

We have: $H_c(T) = H_c(0)[1 - (T/T_c)^2]$

$$\Rightarrow 2 \times 10^4 = 6.5 \times 10^4[1 - (T/7.18)^2]$$

$$\Rightarrow 1 - \left(\frac{T}{7.18}\right)^2 = \frac{2 \times 10^4}{6.5 \times 10^4} \Rightarrow T = 5.97\,\text{K}, \quad \text{on simplification}$$

So, lead must be cooled below **5.97 K** to make it a superconductor.

▶ **Example 14.** *How much current can a* Pb-*wire,* 1.0 mm *in dia, carry in its superconducting state at* 4.2 K*? Given,* $B_c(0) = 0.0803$ Tesla *and* $T_c = 7.2$ K *for lead.*

Solution. From the relation

$$B_c(T) = B_c(0)[1 - (T/T_c)^2]$$

$$B_c(4.2) = 0.0803\,[1 - (4.2/7.2)^2] = 0.0548 \text{ Tesla,} \quad \text{on simplification}$$

Now, the current $$I_c = \frac{2\pi r B_c}{\mu_0} = \frac{\pi(2r)B_c}{\mu_0} = \frac{\pi \times 10^{-3} \times 0.0548}{4\pi \times 10^{-7}}.$$

$$= \mathbf{137\ A}, \quad \text{on simplification.}$$

▶ **Example 15.** *Find the wavelength of the photon that would break a Cooper pair in a superconductor whose critical temperature is* 1.2 K.

Solution. By BCS theory, we have, the energy gap

$$\begin{aligned} E_g &= 2\Delta_0 = 3.53\,kT_c \\ &= 3.53 \times 8.614 \times 10^{-5} \times 1.2\,\text{eV} \\ &= 3.65 \times 10^{-4}\,\text{eV} \\ &= 3.65 \times 10^{-4} \times 1.6 \times 10^{-19}\,\text{J} \\ &= 5.84 \times 10^{-23}\,\text{J} \end{aligned}$$

Wavelength corresponding to this energy is given by

$$\lambda = \frac{hc}{E_g} = \frac{6.626 \times 10^{-34} \times 3 \times 10^8}{5.84 \times 10^{-23}} = \mathbf{3.4 \times 10^{-3}\ m}$$

▶ **Example 16.** *Across a Josephson junction, a dc voltage of* 6 μV *is applied. Evaluate the frequency of the alternating current produced.*

Solution. By ac Josephson effect, the angular frequency of ac is

$$\omega = \frac{qV}{\hbar} = \frac{2eV}{\hbar} \quad (\because q = 2e)$$

$$\Rightarrow \quad 2\pi\nu = \frac{2eV}{\hbar}, \quad \text{where } \nu = \text{frequency of ac}$$

$$\therefore \quad \nu = \frac{2eV}{h} = \frac{2 \times 1.6 \times 10^{-19} \times 6 \times 10^{-6}}{6.626 \times 10^{-34}}$$

$$= \mathbf{2.89 \times 10^9\ Hz}$$

► **Example 17.** *Calculate the energy gap for superconducting tin and the minimum frequency of light that can be absorbed, given that the critical temperature of tin is* 3.72 K.

Solution. From the relation, $2\Delta = 3.53\, kT_c$, we get the energy gap

$$\mathbf{2\Delta} = 3.53 \times 1.38 \times 10^{-23}\,\text{JK}^{-1} \times 3.72\,\text{K}$$

$$= 1.81 \times 10^{-22}\,\text{J} = \mathbf{1.13 \times 10^{-3}\ eV}$$

Light is absorbed when the frequency ν satisfies the relation $2\Delta = h\nu$.

$$\therefore \quad \text{Fequency } \boldsymbol{\nu} = \frac{2\Delta}{h} = \frac{1.81 \times 10^{-22}\,\text{J}}{6.63 \times 10^{-34}\,\text{Js}} = \mathbf{2.72 \times 10^{11}\ per\ sec.}$$

- This frequency corresponds to microwave region of electromagnetic spectrum.

► **Example 18.** *For* Hg. *of atomic mass* 200.59 u, *the transition temperature is* 4.153 K *and that for the isotope of mass* 204 u *is* 4.118 K. *Determine the value of the isotope effect coefficeint* α.

Solution. Isotope effect is:

$$T_c M^{\alpha} = \text{constant} = C.$$

$$\therefore \quad \ln T_c + \alpha \ln M = \ln C$$

Differentiating, we get

$$\frac{dT_c}{T_c} + \alpha \frac{dM}{M} = 0$$

$$\therefore \quad \alpha = -\frac{M}{dM} \cdot \frac{dT_c}{T_c}$$

Now, from given data,

$$dM = 204 - 200.59 = 3.41\,\text{u}$$

$$dT_c = 4.118 - 4.153 = -0.035\,\text{K}$$

$$\therefore \quad \boldsymbol{\alpha} = -\frac{200.59}{3.41} \times \frac{-0.035}{4.153} = \mathbf{0.496}$$

Review Questions

1. What is superconductivity? Give the main characteristics of a superconductor. **(Bangalore Univ.)**
2. A superconductor is a perfect conductor. At the same time, it is a perfect diamagnet. Explain.

3. Explain and distinguish between Type I and Type II superconductors. Give examples of each type. **(Punjabi Univ.)**

4. What is the effect of an external magnetic field on the superconducting state of a material? What do you mean by flux exclusion and what is Meissner effect? **(Punjab Univ.)**

5. Discuss the thermodynamic properties of the superconducting state. **(Guru Nanak Univ.)**

6. Give a brief outline of Bardeen, Cooper and Schrieffer (BCS) theory of superconductivity. **(Himachal and Punjab Univ.)**

7. How does the specific heat of a superconductor vary with temperature? What is observed at the superconducting-normal transition temperature? How is this behaviour related to entropy?

8. Describe the Josephon effect underlying a SQUID. Discuss the applications of SQUIDs.

9. How do the electric, magnetic, thermodynamic and optical properties of superconductors differ from those of normal conductors?

10. Write short explanatory notes on:

(a) Meissner effect and Type I and Type II superconductors,

(b) BCS theory of superconductivity,

(c) Josephson effect and the SQUID,

(d) Superconductor and its characteristic features.

Short and Objective Questions

1. What are superconductors? What is the difference between a conductor cooled to 0 K and a superconductor?

2. Show that the materials get cooled when its superconductivity is destroyed by a magnetic field.

3. What are superconductors? Name a few high-T_c superconductors and their properties.

4. Explain the terms: critical field and transition temperature for a superconductor. How are the two related?

5. 'Generally, good conductors are poor superconductors'. Explain why.

6. Derive the London equations and explain the significance of 'penetration depth'.

7. What are Cooper pairs? Which statistics do they obey? Explain the following : superconducting ground state, superconducting energy gap and coherence length.

8. Experiments show that even though a superconductor exhibits zero static resistance, its a.c. resistance is finite, although very small. Explain how this is possible.

9. It was shown by an expression that the critical field $H_c(0)$ is essentially proportional to the critical temperature T_c. Yet the electron concentration n also appears in the expression and this concentration differs from one superconductor to another. Why does the linear relationship still hold true?

10. Prove that the Meissner effect and the disappearance of resistivity in a superconductor are mutually consistent.

11. Describe the major uses and potentialities of superconductors. Why is superconductivity a low temperature phenomenon? **(Guru Nanak Univ.)**

12. Mention the important changes in property that occur in materials when they change from normal to superconducting state.

13. Explain the effect of isotopes on superconductors. What do you mean by coherent length?

14. Give the significance of London equations. How are the Cooper pairs formed?

15. What are Josephson junction and effect ? Mention some of its applications.

16. Distinguish between dc and ac Josephson effects. Give the principle of SQUID and expand the acronym.

17. The transition temperature of mercury is

    (a) 1 K, (b) 1.14 K, (c) 4.12 K, (d) 9.22 K.

18. The transition temperature T_c and the critical field H_c for a superconductor is related as

    (a) $H_c = H_0(T_c - 1)$, (b) $H_c = H_0(T_c + 1)$,
    (c) $T_c = T_0(1 - H_0^2/H_c^2)$, (d) $H_c = H_0(1 - T^2/T_c^2)$.

19. The width of the energy gap of a superconductor is maximum at

    (a) 0 K, (b) room temperature, (c) transition temperature, (d) none.

20. In the superconducting state,

    (a) the entropy increases, the thermal conductivity decreases,
    (b) both the entropy and the thermal conductivity decrease,
    (c) both the entropy and the thermal conductivity increase,
    (d) the entropy decreases, but the thermal conductivity increases.

21. On being subjected to critical field, a superconductor changes to

    (a) a normal state, (b) a critical conductor,

(c) a superconductor independent of temperature,
(d) remains unaffected.

22. The coherence length of paired electrons is

(a) 0.25 nm, (b) 250 nm, (c) 0.01 nm, (d) 0.001 nm.

23. The specific heat of a superconductor shows an abrupt change at the temperature $T = T_c$, jumping to a large value for

(a) $T < T_c$, (b) $T > T_c$, (c) $T = T_c$, (d) $T \geq T_c$.

24. Superconductors are generally

(a) monovalent metals, (b) thin films of barium titanate
(c) amorphous thin films of Bi and Be, (d) ferro- and antiferromagnetic metals.

25. The resistivity of a superconductor as $T \to 0$ becomes

(a) ∞, (b) 0, (c) any value from $0 \to \infty$, (d) none.

26. A Cooper pair behaves as a

(a) fermion, (b) boson, (c) particle of spin $\frac{1}{2}$, (d) none of these.

27. For a conventional superconductor, which of the following statements is *not* true?

(a) Specific heat is discontinuous at T_c,
(b) Resistivity falls sharply at T_c,
(c) It is diamagnetic below T_c,
(d) It is paramagnetic below T_c. **(GATE)**

28. The critical magnetic field for a solid in superconducting state

(a) does not depend on temperature,
(b) does not depend on transition temperature,
(c) increases if the temperature decreases,
(d) increases if the temperature increases. **(GATE)**

29. A superconducting ring is cooled in presence of a magnetic field below its critical temperature T_c. Total magnetic flux through the ring is

(a) zero, (b) $nh/2e$ (c) $nh/4\pi e$, (d) ne^2/hc. **(GATE)**

30. An external magnetic field of magnitude H is applied to a type I superconductor at a temperature below the transition point. Which of the following is *not* true for $H < H_c$, the critical field.

(a) The sample is diamagnetic,
(b) The magnetisation varies linearly with H,

(c) The sample exhibits mixed states of magnetisation near H_c,
(d) Lines of magnetic induction are pushed out of sample. **(GATE)**

31. A solid superconductor is placed in an external magnetic field and then cooled below its T_c. The superconductor

(a) retains its magnetic flux as surface current supports it,
(b) expels out its magnetic flux as it behaves as a paramagnet,
(c) expels out its magnetic flux as it behaves as an antiferromagnet,
(d) expels out its magnetic flux as the surface current induces a field opposite to the applied magnetic field. **(GATE)**

32. Mark, from among those given below, the material which is *not* a superconductor of Type I.

(a) Aluminimum, (b) Niobium,
(c) Lead, (d) Mercury.

33. Mark, from among those given, the material which is a Type II semiconductor.

(a) Cadmium, (b) Vanadium,
(c) Tin, (d) Aluminium.

34. A superconducting state is associated with

(a) only a change in entropy,
(b) only a change in electronic specific heat,
(c) a change in both entropy and electronic specific heat,
(d) none of the above.

35. Cooper pairs are formed at temperatures

(a) $T > T_c$, (b) $T < T_c$, (c) $T = T_c$, (d) None.

36. A solid sample has the property that, when cooled below a certain temperature, it expels any small applied magnetic field from within the material. Which of the following best describes this sample in the cooled state?

(a) Paramagnet, (b) Diamagnet, (c) Ferromagnet, (d) Antiferromagnet. **(TIFR)**

Problems

1. The critical temperature of a superconductor at zero magnetic field is T_c. Determine the temperature at which the critical field becomes half of its value at 0 K. $[0.707T_c]$

2. Determine the frequency of the electromagnetic waves radiated by a Josephson junction across which a dc voltage of 0.5 mV is applied. [2.41×10^{11} Hz]

3. The London penetration depths for lead at 3 K and 7.1 K are respectively 39.6 nm nd 173 nm. Calculate its transition temperature and also the penetration depth at 0 K. [7.19 K, 39 nm]

4. The critical fields at 6 K and 8 K for a NbTi alloy are 7.616 and 4.284 Tesla respectively. Determine the transition temperature and also the critical field value at 0 K. [10 K, 11.9 Tesla]

5. The magnetic field intensity in Sn is zero at 3.69 K and $3 \times 10^5/4\pi$ at 0 K. Calculate the temperature of the superconductor state if the field intensity measured is $2 \times 10^5/4\pi$ Am^{-1} [2.13 K]

6. The critical temperature for mercury with isotopic mass 199.5 is 4.185 K. Determine its critical temperature when its isotopic mass changes to 203.4. [4.139 K]

7. Calculate the penetration depth for lead at 5.2 K if the London penetration depth at 0 K is 37 nm. The critical temperature of lead is 7.19 K. [43.39 nm]

8. The critical field of niobium at 0 K is 1.55×10^5 A/m and the critical temperature in absence of a magnetic field is 9.25 K. What is the critical field at 4 K? Calculate also the maximum supercurrent that a niobium wire of radius 0.1 mm at 4 K can be able to carry. [1.26×10^5 A/m; 7.92 A]

9. Calculate the critical current for a wire of lead having a dia of 1 mm at 4.2 K. Given, the critical temperature for lead is 7.18 K and $H_c(0) = 6.5 \times 10^4$ Am^{-1}. [134.46 A]

10. A superconducting lead has a critical temperature of 7.26 K at zero magnetic field and a critical field of 8×10^5 A/m at 0 K. Find the critical field at 5 K. [4.2×10^5 A/m]

11. The critical field for lead for at 0 K is 6.39×10^4 Am^{-1} and the critical temperature for zero magnetic field is 7.18 K. Calculate the critical field for lead at 4 K. [4.41×10^4 Am^{-1}]

12. A superconductor has critical fields 1.4×10^5 and 4.2×10^5 A/m at 14 K and 13 K respectively. Calculate its transition temperature and the critical field for 4.2 K. [14.47 K; 20.05×10^5 A/m]

13. Estimate the London penetration depth for tin from the following set of data.

    Critical temperature ... ... 3.7 K
    Density ... ... 7.3×10^3 kg/m^3
    Atomic weight 118.7
    Effective mass (m^*) ... ... $1.9\, m_e$

    Assume that each atom provides one conduction electron. [0.0015 nm]

14. For lead, $T_c = 7.2\,\text{K}$. Calculate the ratio of penetration depth at 3 K to that 0 K. [1.105]

15. A superconducting tin has a critical temperature of 3.7 K at zero magnetic field and a critical field of 0.0306 T at 0 K. Find the critical field at 2 K. [0.0216 T]

16. Calculate the frequency of the alternating current produced when a dc voltage of $5\,\mu\text{V}$ is applied across a Josephson junction. [2.41×10^9 Hz]

17. The dia of a lead wire at 4.2 K is 1 mm. If the critical temperature for lead is 7.18 K and $H_c(0) = 6.5 \times 10^4$ A/m, find the critical current. [134.5 A]

18. The density and the atomic number of niobium are $8.57 \times 10^3\,\text{kg/m}^3$ and 93 respectively. It has one conduction electron per atom. Calculate the London penetration depth of niobium. [225.76 A]

19. The critical fields of a specimen are 1.4×0^5 and 4.2×10^5 A/m at 14 K and 13 K respectively. Calculate the transition temperature and critical fields at 0 K and 4.2 K. [14.5 K, 20.7×10^5 A/m, 18.9×10^5 A/m]

20. Calculate the London penetration depth at 0 K for Pb whose density is $11.3 \times 10^3\,\text{kg/m}^3$ and atomic weight 207.19. Given $T_c = 7.22\,\text{K}$, determine the increase in penetration depth at 3.61 K from its value at 0 K. [1 nm]

21. The penetration depth λ of Hg at 3.5 K is about 750Å. Estimate the values of (a) λ and (b) n_s, the superconducting electron density, as T tends to 0 K.

22. Estimate the angle θ for Aluminium at which $B_c - T$ graph meets the T-axis. Given: for Aluminium, $B_0 = 0.99$ Tesla and $T_c = 1.17\,\text{K}$. [$\theta = 149°$]

[**Hints.** We have the relation: $B_c = B_0(1 - T^2/T_c^2)$ and $\tan\theta = (dB_c/dT)_{T=T_c}$ since $B_c - T$ graph meets the T-axis at $T = T_c$. Now, therefore,

$$\tan\theta = (-2B_0T/T_c^2)_{T=T_c} = -2B_0/T_c = -2 \times 0.99/1.17 = -1.692 \Rightarrow \boldsymbol{\theta = 149°}]$$

23. Calculate the superconducting electron density of Hg at 3.5 K. [$1.95 \times 10^{28}\,\text{m}^{-3}$]

24. For a supercondctor, $T_c = 3\,\text{K}$ and $n_s = 10^{28}/\text{m}^3$. Find the penetration depth at 0 K and 3 K. [530Å, 533Å]

25. Find the angle θ which the $\Delta - T$ curve makes with T-axis. Use the formula: $\Delta(T) = 1.76\Delta(0)[1 - T/T_c]^{1/2}$ for the variation of energy gap with temperature. [$\theta = 90°$]

[**Hints.** Given: $\Delta T = 1.76\,\Delta(0)[1 - T/T_c]^{1/2}$. So, we obtain from $\Delta - T$ curve, $\tan\theta = d\Delta/dT = 1.74\Delta(0)\cdot\frac{1}{2}(1 - T/T_c)^{-\frac{1}{2}}(-1/T_c)$. The curve meets T-axis at $T = T_c$. Thus, $\tan\theta = 1/0 = \infty$, Therefore, $\boldsymbol{\theta = 90°}$]

10

Magnetic Properties of Materials

10.1 Introduction

Perhaps, magnetism is that aspect of the solid state which has been known to the mankind since more than 3000 years. The ancient Greek used *lodestone* (magnetite) to attract iron objects. Youngsters are fascinated with magnets, largely because they act at a distance. One can move an iron nail with a nearby magnet even when a piece of wood is in-between. Similarly, a neurosurgeon can guide a pellet through the brain tissue to inoperable tumors of a patient, pull a catheter in position and implant electrodes doing little harm to the brain tissue. The use of magnets is ever increasing.

A vast wealth of experimental data accumulated before any theory of magnetism could be developed. The reason however is not far to seek. A quantitative description of magnetism requires the knowledge of quantum mechanics and the electromagnetic theory of multi-atom system which, in general, is rather complicated. In this chapter, we shall essentially present the observed magnetic properties of materials and the theories to explain them.

10.1.1 Some magnetic terms

To characterise the magnetization of substances, the important quantity is the ***intensity of magnetization*** or simply ***magnetization*** $\vec{M}$ which is the vector sum of the *magnetic moments of the atoms* (or molecules) *per unit volume* of the material.

$$\therefore \quad \vec{M} = \underset{V\to 0}{\text{Lt}} \left[\frac{1}{V} \sum_{i=1}^{N} \vec{P}_i \right]$$

where N is the number of atoms (or molecules) in a volume V of the material, $\vec{P}_i$ the magnetic moment of the ith atom (or molecule). Its unit is A/m.

For a large class of *isotropic* magnetic materials, $\vec{M}$ is proportional to the applied ***magnetic field intensity***, $\vec{H}$ which implies

$$\vec{M} \propto \vec{H} \Rightarrow \vec{M} = \chi \vec{H} \tag{10.1.1}$$

where the constant χ is called the ***magnetic susceptibility*** of the material and is a scalar quantity.

$$\therefore \quad \boxed{\chi = \vec{M}/\vec{H}} \tag{10.1.2}$$

The ***magnetic flux density***, also called the ***magnetic induction*** $\vec{B}$, developed inside the material due to the applied magnetic field $\vec{H}$, is defined as

$$\vec{B} = \mu_0(\vec{H} + \vec{M}) \Rightarrow \mu_0 = \vec{B}/(\vec{H} + \vec{M}) \tag{10.1.3}$$

where μ_0 is the *permeability* of free space. Its value is $4\pi \times 10^{-7}$ Hm^{-1} (Henry/meter). $\vec{B}$ is measured in Wb m^{-2} (Weber/metre^2) or T (Tesla).

$\therefore$ From equation (10.1.3), $\quad \vec{B} = \mu_0(\vec{H} + \chi\vec{H})$, using (10.1.1)

$$= \mu_0(1 + \chi)\vec{H} = \mu\vec{H} \tag{10.1.4}$$

where $\quad \mu = \mu_0(1 + \chi)$.

The μ is called the *absolute permeability* of the medium. The *relative permeability* μ_r of the medium is defined by

$$\mu_r = \mu/\mu_0 = 1 + \chi \tag{10.1.5}$$

i.e., μ_r is the *ratio of the absolute permeability of the medium to the permeability of free space.*

$$\therefore \quad \boxed{\chi = \mu_r - 1} \Rightarrow \boxed{\mu_r = 1 + \vec{M}/\vec{H}} \tag{10.1.6}$$

Also, from (10.1.4), and (10.1.5)

$$\vec{B} = \mu_0(1 + \chi)\vec{H} = \mu_0\mu_r\vec{H} = \mu\vec{H} \quad (\because \mu = \mu_0\mu_r) \tag{10.1.7}$$

• For free space, $\boldsymbol{\vec{B} = \mu_0\vec{H}}$. Equation (10.1.6) connects the permeability and the susceptibility of the medium.

• A magnetic field can be described by either of the vectors $\vec{B}$, the *magnetic induction* or the vector $\vec{H}$, the *magnetic field intensity.*

10.2 Broad classes of magnetic materials

Magnetic materials are usually classified into different classes on the basis of their χ-values and the way they vary with magnetic field strength and temperature. These are: (i) *diamagnetics*, (ii) *paramagnetics*, (iii) *ferromagnetics*, (iv) *anti-ferromagnetics* and (v) *ferrimagnetics*. We shall, in this section, briefly discuss in a general way these classes of magnetic materials.

Diamagnetism—For diamagnetic substances, the mgnetization M is proportional to the applied field intensity H, but the susceptibility $\chi < 0$ i.e., *negative* and both χ and μ_r are *independent of temperature.* The atoms and molecules of such materials have zero magnetic moment in the absence of an external magnetic field. With the

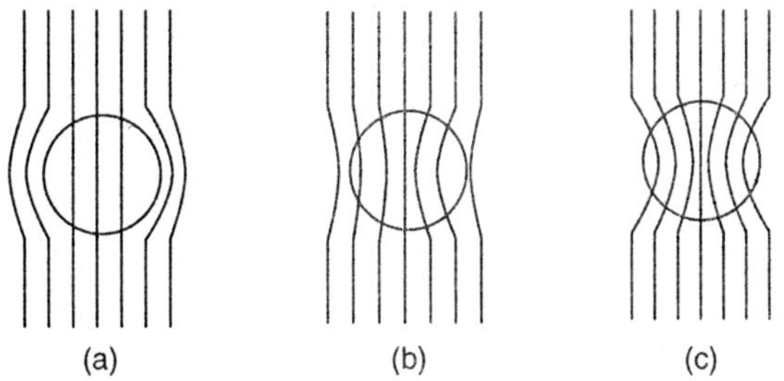

Fig. 10.1 Behaviour of magnetic lines of force in applied magnetic field in (a) diamagnet (b) paramagnet and (c) ferromagnet

application of an external field, a net dipole moment opposing the field is *induced* in the atoms or molecules. The motion of orbital electrons is modified in such a way that a weak magnetic moment opposing the field is induced. The diamagnetic property arises due to this induced magnetic moment.

A diamagnetic solid tends to repel the magnetic lines of force due to an external field (Fig. 10.1a). A superconductor that repels all the lines of force is a perfect diamagnet.

Examples of diamagnetic materials are Cu, Au, Ge, Si, Bi, NaCl, Al_2O_3 etc. Inert gases are also diamagnetic for their atoms have completely filled electronic shells, and hence there is no permanent magnetic dipole moment.

Paramagnetism—Paramagnetic substances get attracted towards the region of stronger magnetic field. Their magnetic susceptibility, although small, is positive i.e., greater than zero, i.e., $\chi > 0$ and *depends on the temperature* according to Curie's law $\chi = C/T$ where C is the *Curie constant.* Their permeabilities are slightly more than 1, between 1 and 1.001 and are *independent of field strength.* They do not show *hysteresis.*

In paramagnetic substances, the atoms, ions or molecules possess a permanent dipole magnetic moment, randomly oriented, in the absence of external magnetic field, due to thermal agitation. When a field is applied, the atomic dipoles tend to orient themselves parallel to the field and make a positive contribution to the χ-value. At low temperature, the thermal agitation is less, larger portion of the dipoles are aligned parallel to the field and the magnetization is high for a given field. So, for ordinary fields and temperatures, $\chi \propto T$ but for large fields at low temperatures, magnatization $\vec{M}$ is not proportional to $\vec{H}$ but tends to a constant value indicating saturation in the alignment of dipoles. Because of susceptibility, a paramagnet weakly attracts the lines of force (Fig. 10.1b).

Examples of paramagnetic materials are Al, O_2, Pt, Fe_2O_3, $FeSO_4$, Cr_2O_3, $NiSO_4$, $MnSO_4$ etc., the rare earth elements, alkali and alkaline earth metals, etc.

• An atom with *permanent dipole moment* will also have the diamagnetic effect. However, in such cases, the permanent magnetic moment will tend to rotate in the direction of the applied field yielding paramagnetic property and would mask the diagmagnetic action.

Ferromagnetism—Ferromagnetic materials are *very strongly magnetic.* For instance, for a given field, the magnetization in ferromagnetics is very high, about 10^6 times more than that of paramagnetics. The χ-values are necessarily also very high and the ferromegnetics thus strongly attract the lines of force (Fig. 10.1c). As temperature increases, the value of χ-decreases and above a certain temperature T_c, known as the *Curie temperature* i.e., at $T > T_c$, ferromagnetics are changed into paramagnetics. One of the characteristics of ferromagnetics is the phenomenon of ***hysteresis.***

Familiar ***examples*** of ferromagnetics are Fe, Co, Ni, and their alloys and some rare earth elements Gd and Dy.

• Ferromagnetics possess *spontaneous magnetization*, i.e., magnetization even in the absence of an external magnetic field below the Curie temperature.

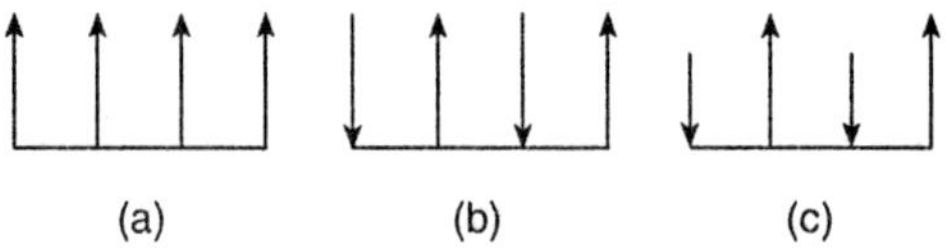

Fig. 10.2 Arrangements of magnetic dipoles in (a) Ferromagnet (b) Antiferromagnet and (c) Ferrimagnet

In fully magnetized state, a ferromagnet has all its dipoles aligned exactly in the same direction (Fig. 10.2a). An ***anti-ferromagnetic*** material has dipoles with equal moments but the alternate dipoles point in opposite directions (Fig. 10.2b). Consequently, the moments balance each other leading to a *zero* net magnetization. In ***ferrimagnetic*** materials, the neighbouring dipoles also point in opposite directions but they are *not* equal (Fig. 10.2c). Hence they cannot completely nullify each other and has a net finite magnetization.

10.3 Atomic theory of magnetism

Let an electron of mass m and charge $-e$ move in a circular loop of radius r with a linear velocity v (Fig. 10.3).

Now, a moving charge constitutes an electric current

$$I = \frac{dq}{dt}, \quad \text{where } dq \text{ is the charge.}$$

$$\therefore \quad Ids = dq\frac{ds}{dt} = -ev$$

where ds = distance covered by electron, $dq = -e$ and $v = ds/dt$ = linear velocity.

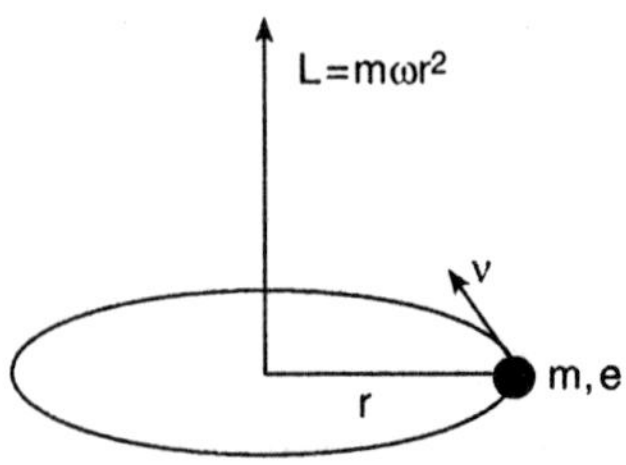

Fig. 10.3 Angular momentum of an orbiting electron

For a *circular orbit*, $ds = 2\pi r$ and $v = r\omega$, where $\omega =$ angular velocity.

$$\therefore \quad I \cdot 2\pi r = -e\omega r \Rightarrow I = -e\omega/2\pi \tag{10.3.1}$$

From e.m. theory, the magnetic field due to the current I flowing through a stationary loop of sectional area A at right angles to the plane of the loop is the same as the field developed by a magnetic dipole measured at distances $\gg$ loop radius.

$\therefore$ Magnitude of the magnetic moment of such a dipole

$$\mu_m = IA = -\frac{e\omega}{2\pi} \times \pi r^2 \quad (\because A = \pi r^2)$$

$$= -\frac{e\omega r^2}{2} = -\frac{em\omega r^2}{2m} = -\left(\frac{e}{2m}\right) L \tag{10.3.2}$$

where $L = mr^2\omega$ is the *orbital angular momentum* of electron and is directed normal to the orbital plane. The negative sign implies that the magnetic moment μ_m is *antiparallel* to the angular momentum L as shown in Fig. 10.4.

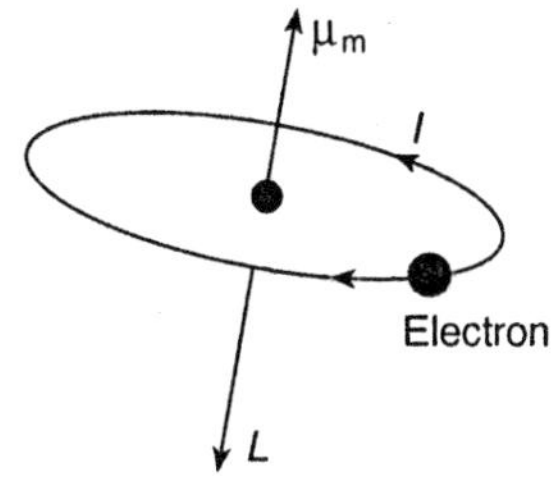

Fig. 10.4 Magmetic dipole moment and angular momentum associated with a current loop

The ratio of the magnetic dipole moment and the angular momentum of orbiting electron is called the ***gyromagnetic ratio.***

$$\therefore \quad \text{Gyromagnetic ratio, } \frac{\mu_m}{L} = -\frac{e}{2m} \tag{10.3.3}$$

The gyromagnetic ratio is thus *independent of angular velocity* of the electron and the *radius* of the orbit.

Equation (10.3.2) above is valid only for the *orbital motion* of the electron and not for *spin* of the electron and of the nucleus.

● To understand the origin of magnetic moments of an atom completely, we must take into account the *spin motions of the electron and the nucleus.* For this, we shall now briefly review the quantum numbers needed to specify the state of a system in quantum mechanics.

10.3.1 Quantum numbers

These are integral or half-integral numbers specifying the state of an atomic system and run as follows.

1. Principal quantum number, *n*—This number n specifies the energy level of the electron in an atom and can take up only integral values: $n = 1, 2, 3, \cdots$, the levels being termed $K, L, M, \cdots$, respectively. Each n value represents a *shell* and determines the size of the electron orbit. The number of electrons in a given shell n is $N = n^2$.

2. Orbital quantum number, *l*—This number l determines the electronic state of an atom related to the shape of the electron orbital. It can also take up only integral values and for a given n, the l-values are

$$l = 0, 1, 2, \cdots, (n-1),$$

the corresponding electrons are called $s, p, d, \cdots$ etc. electrons respectively. Each l-value represents a *sub-shell.* The number of electrons in each sub-shell is $(2l + 1)$.

The *total angular momentum* associated with a given l-value is

$$L_l = \frac{h}{2\pi}\sqrt{l(l+1)} = \hbar\sqrt{l(l+1)} \tag{10.3.4}$$

3. Magnetic quantum number, m_l—An atom when placed in an external magnetic field, the angular momentum vector $\vec{L}_l$ (and hence the magnetic moment) precesses about the field direction and can have only specific spatial orientation. The orientation of L_l in an external field of induction B is characterized by its projection $L_{l,B}$ (Fig. 10.5) along the direction of B:

$$L_{l,B} = L_l \cos\alpha \tag{10.3.5}$$

$\vec{L}_l$ can have such orientations that the projection $L_{l,B}$ is an integral multiplie of $\hbar$ $(= h/2\pi)$. So, we get

$$L_{l,B} \equiv m_l\hbar = m_l(h/2\pi) \tag{10.3.6}$$

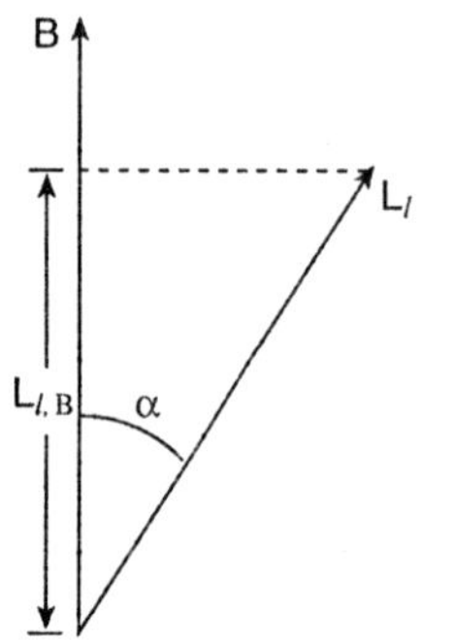

Fig. 10.5 Projection of angular momentum along field

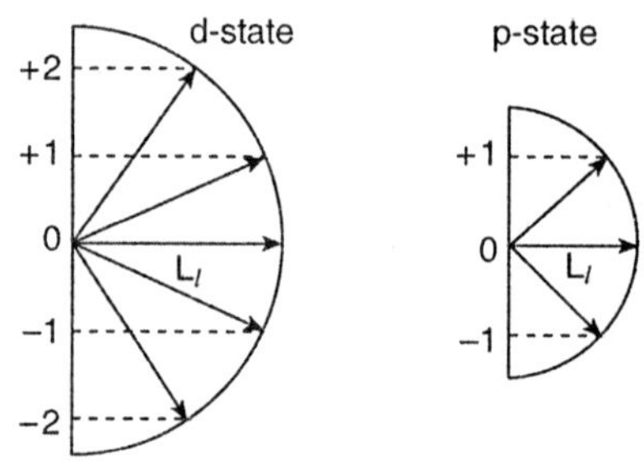

Fig. 10.6 Possible orientations of angular momentum defined by quantum number l in an external magnetic field

The integer m_l that determines the possible values of $L_{l,B}$ is called the *magnetic quantum number* having values

$$\begin{aligned} m_l &= -l, -(l-1), \cdots, 0, \cdots (l-1), l \\ &= 0, \pm 1, \pm 2, \cdots, \pm l \end{aligned} \tag{10.3.7}$$

l being the usual orbital quantum number.

$\vec{L}_l$ vector can have, from above, $(2l+1)$-values of possible orientations.

For an electron in p and d-state, i.e., $l = 1$ and 2, the possible orientations of $\vec{L}_l$ are shown in Fig. 10.6.

4. Magnetic spin quantum number, m_s—In addition to the oribital motion, electrons have also spin motion. It is a fundamental property of an electron and has its own angular momentum. In a field $\vec{B}$, the *spin angular momentum* S is so oriented that its projections along the field $\vec{B}$ can take up only two values:

$$S = m_s\hbar = m_s(h/2\pi) \tag{10.3.8}$$

S points either *upward* (spin *up*) or *downward* (spin *down*) and the corresponding spin quantum number $m_s = +\frac{1}{2}$ or $m_s = -\frac{1}{2}$ respectively. Spin quantization is shown in Fig. 10.7.

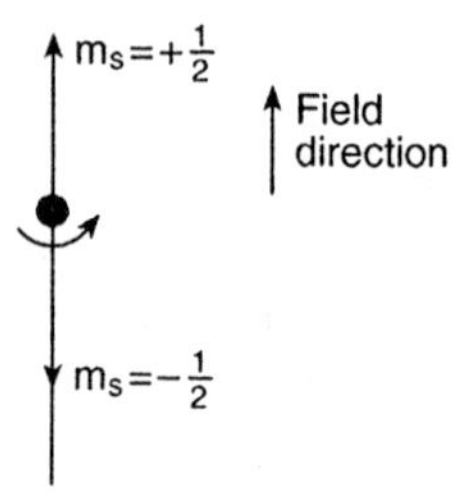

Fig. 10.7 Spin quantization

10.4 Origin of permanent magnetic moments

There are *three* different sources that can provide permanent magnetic moment in an atom. These are as follows:

(a) orbital magnetic moment of electrons due to orbital motion,

(b) spin magnetic moment of electrons due to spin motion, and

(c) spin magnetic moment of nucleus due to its spin motion.

We shall now describe and discuss each of the above sources one after the other.

1. Orbital magnetic moment of electrons—From the atomic theory of magnetism, we just had the following expression for the magnetic moment.

$$\mu = \mu_l = -\left(\frac{e}{2m}\right) L_l \tag{10.4.1}$$

Quantum mechanically, however,

$$\mu = -\left(\frac{e}{2m}\right) L_{l,B} = -\left(\frac{e}{2m}\right) m_l\hbar = -\left(\frac{eh}{4\pi m}\right) m_l = -m_l\mu_B \tag{10.4.2}$$

where $\boldsymbol{\mu_B = e\hbar/2m = eh/4\pi m}$ $= 9.27 \times 10^{-24}\,\text{Am}^2$ and is called the ***Bohr magneton***, the accepted unit of magnetic moment of atomic systems.

• In multi-electron complex atoms, the total orbital magnetic moment is the algebraic sum of the magnetic moments of individual electrons. The *moment of a completely filled shell is zero*, so that an atom having only *partially filled* shells will give a non-zero orbital magnetic moment.

2. Spin magnetic moment of electrons—The spin magnetic moment may be similarly determined by substitution of magnetic spin angular momentum $S = m_s\hbar$ into (10.4.2) so that

$$\mu_s = -\left(\frac{e}{2m}\right) S = \pm\tfrac{1}{2}\left(\frac{e}{m}\right)\hbar = \tfrac{1}{2}\mu_B \quad \left(\because\ m_s = \pm\tfrac{1}{2}\right) \tag{10.4.3}$$

i.e., *spin magnetic moment is half of a Bohr magneton.* This is true only approximately for (10.4.2) is not for electron spin.

The magnetic moment component μ_{sz} due to electron spin along the field direction z is, in fact, given generally by the expression

$$\mu_{sz} = g\left(\frac{e}{2m}\right) m_s\hbar = g\left(\frac{eh}{4\pi m}\right) m_s \tag{10.4.4}$$

where g is called the ***Lande splitting factor*** or ***Lande g-factor*** and determines the amount by which the original level splits up in a magnetic field. $g = 2.0023$ for electron spin. So, the magnetic moment component $\mu_{sz} \simeq \mu_B$ due to electron spin along the field direction z.

3. Spin magnetic moment of nucleus—The nucleus, like an electron, has intrinsic spin, called the *nuclear spin* and consequently an associated magnetic moment.

But the nuclear mass is more than 10^3 times greater than that of an electron so that the nuclear magnetic moment is approximately three order smaller in magnitude compared to that associated with the electron.

While the electronic magnetic moment is measured in the unit of Bohr magneton, the nuclear magnetic moment is expressed and measured in unit of ***nuclear magneton*** which is defined as

$$\mu_N = \frac{eh}{4\pi m_p} = 5.05 \times 10^{-27} \text{ Am}^2$$

where m_p represents the mass of a proton.

● Sometimes $\vec{L}$ and $\vec{S}$ may be combined vectorially (LS-couplng) to get the total angular momentum $\vec{J}$ of the whole electron system of the atom. For such atoms, g is given by ***Lande formula***:

$$g = 1 + \frac{J(J+1) + S(S+1) - L(L+1)}{2J(J+1)} \tag{10.4.5}$$

$\therefore$ *Total magnetic moment* of atomic system is: $\mu = gJ\mu_B$ (10.4.6)

Spectroscopic data provide the values of g and J. And μ, in principle, can be determined.

● For an electron with only spin angular momentum ($L = 0$) in a magnetic field $\vec{B}$, the energy difference between the levels (Fig. 10.8) is given by $g\mu_B B$. So, the split of energy levels is determined by g and hence the name *spliting factor*.

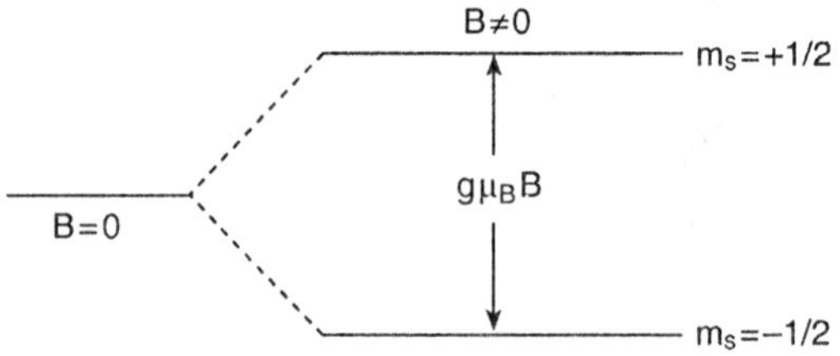

Fig. 10.8 Splitting of energy level of the spin 1/2 electron for B≠0

Hund's rules

The filled electron shells do not contribute to the *magnetic moment* of the atom. It must therefore arise from the *partially-filled shell*. This follows from Pauli's principle that no two electrons in an atom can have the identical set of quantum numbers n, l, m_l and m_s. For the *ground state* of such atoms, ***Hund's rules*** for incomplete shells state

(1) Electron spins add to give the maximum possible S-value.

(2) Orbital moments combine to give a maximum L-value, consistent with (1) above.

(3) For an incomplete shell, the the J-value is given by:

$J = L - S$ ⋯ for shell less than half-occupied,

$J = L + S$ ⋯ for shell more than half-occupied.

Illustrative example—The following example will illustrate the rule. Consider the electron configuration of Cr^{3+}. The electron configuration of Cr-atom ($Z = 24$) is:

$$1s^2\,2s^2\,2p^6\,3s^2\,3p^6\,3d^5\,4s^1$$

When 3 electrons are removed from Cr-atom, it becomes Cr^{3+} of electronic configuration

$$1s^2\,2s^2\,2p^6\,3s^2\,3p^6\,3d^3.$$

Thus, in Cr^{3+}, all the shells *except* $3d$ are completely filled. So, only the $3d$ shell with 3 electrons will contribute to the permanent dipole moment.

For a d-shell, $l = 2$ and the possible m_l-values are $2, 1, 0, -1, -2$ each of which can accommodate 2 electrons with $m_s = \pm\frac{1}{2}$. So, in $3d$ shell, the maximum number of electrons $= 5 \times 2 = 10$. Hence $3d$ shell of Cr^{3+} is less than half occupied.

By Hund's rule (1), $S = 3 \times \frac{1}{2} = \frac{3}{2}$, since 3 electrons are in d-shell.

The possible m_l-values are $2, 1, 0, -1, -2$. If 3 electrons each with $s = \frac{1}{2}$ be placed in the first three of these, then

$$L = 2 + 1 + 0 = 3$$

which is the maximum possible L-value, consistent with (1).

So, by (3) of Hund's rule, $J = L - S = 3 - \frac{3}{2} = \frac{3}{2}$.

$$\therefore\quad g = 1 + \frac{\frac{3}{2}\times\frac{5}{2} + \frac{3}{2}\times\frac{5}{2} - 3\times 4}{3\times\frac{5}{2}} = 0.4$$

The magnetic moment μ_m is then obtained by the relation

$$\mu_m = g(e/m)\hbar/2$$

A. Classical Theory of Magnetism

10.5 Langevin's theory of diamagnetism

Let an electron rotate around the nucleus in a circular orbit (Fig. 10.9) of radius r_0 and ω_0 be its angular velocity. Then, we have

$$\frac{mv^2}{r_0} = \frac{Ze^2}{4\pi\varepsilon_0 r_0^2} \Rightarrow mr\omega_0^2 = \frac{Ze^2}{4\pi\varepsilon_0 r_0^2} \qquad (10.5.1)$$

where m = mass of electron, e = electronic charge, Ze = charge of the nucleus, ε_0 = permittivity of free space, v = linear velocity of electron.

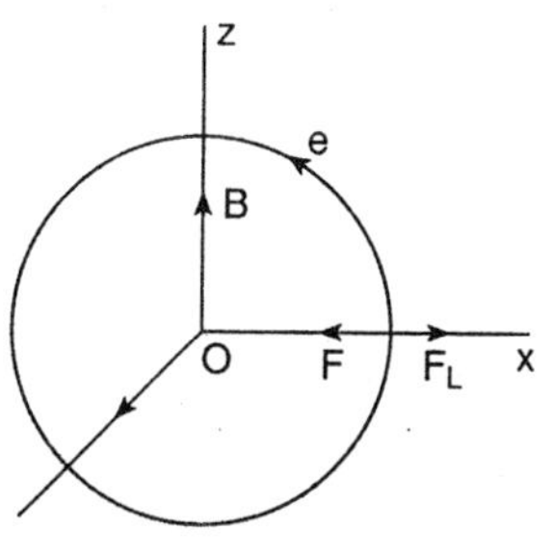

Fig. 10.9 Motion of an electron in xy-plane

The magnetic moment of the electron

$$\vec{\mu} = \text{current} \times \text{loop area} = e\frac{\omega_0}{2\pi} \times \pi r_0^2 = \tfrac{1}{2}er_0^2\omega_0 \quad (10.5.2)$$

Let a magnetic field of induction $\vec{B}$ be applied normal to the plane (xy-plane) of the orbit. An additional force, the ***Lorentz force*** F_L, will now act on the electron away from the centre. Let the force merely accelerate or decelerate the electron with no change in the radius of the orbit. The condition of stable motion of electron is now given by

$$mr_0\omega^2 = \frac{Ze^2}{4\pi\varepsilon_0 r_0^2} \pm er_0\omega B \tag{10.5.3}$$

ω being the new angular velocity due to the change in the net force, and the Lorentz force being $e(\vec{v} \times \vec{B}) = er_0\omega B$.

$$\therefore \ \omega^2 = \omega_0^2 \pm \frac{eB}{m}\omega, \quad \text{using (10.5.1)} \tag{10.5.4}$$

$$\Rightarrow \ \omega = \omega_0\left(1 \pm \frac{eB}{m\omega_0}\cdot\frac{\omega}{\omega_0}\right)^{1/2}$$

$$= \omega_0 \pm \frac{eB}{2m}\cdot\frac{\omega}{\omega_0} \quad (\because \ eB/m \ll \omega_0)$$

$$= \omega_0 \pm \omega_L\frac{\omega}{\omega_0} \tag{10.5.5}$$

where $\boldsymbol{\omega_L = eB/2m}$ is called ***Larmor precessional frequency.***

On solving (10.5.5), we obtain

$$\omega = \frac{\omega_0}{1 \pm \omega_L/\omega_0} \simeq \omega_0(1 \pm \omega_L/\omega_0)$$

$$= \omega_0 \pm \omega_L \tag{10.5.6}$$

The frequency-change ω_L produces the induced magnetic moment (Eq. (10.5.2))

$$\vec{\mu}_{\text{ind}} = -\frac{1}{2}er_0^2\vec{\omega}_L = -\frac{e^2}{4m}r_0^2\vec{B} \tag{10.5.7}$$

Summing over all the electrons in the atom, the induced moment per atom,

$$\vec{\mu}_{\text{ind}} = -\frac{\vec{B}e^2}{4m}\sum r_0^2 \tag{10.5.8}$$

If N be the *number of atoms per unit volume*, the magnetization M is given by

$$M = -\frac{NBe^2}{4m}\sum r_0^2 \tag{10.5.9}$$

But since all the electron orbits are not normally oriented with respect to B, r_0^2 in (10.5.9) is to be replaced by the mean square of the projection of orbital radii for various electrons in a plane perpendicular to B i.e., by $\overline{r_0^2}$. If the coordinates of a point of orbit of radius r_0 be x, y, z, then $\overline{r_0^2} = \overline{x^2} + \overline{y^2} + \overline{z^2}$. But the distribution of orbiting electron is spherically symmetric $\Rightarrow$ $\overline{x^2} = \overline{y^2} = \overline{z^2}$. If the magnetic field is along z-axis, $\overline{r_0^2} = \overline{x^2} + \overline{y^2} = \overline{2x^2} = \frac{2}{3}r_0^2$.

$$\therefore \ M = -\frac{NBe^2}{6m}\sum r_0^2 \tag{10.5.10}$$

The summation extends over all the Z orbital electrons in the atoms of atomic no. Z. The diamagnetic susceptibility of the material, χ_{dia}, is therefore given by

$$\chi_{\text{dia}} = \frac{M}{H} = -\frac{NBe^2}{6mH}\sum r_0^2 = -\frac{\mu_0 Ne^2}{6m}\sum r_0^2 \tag{10.5.11}$$

$$= -\frac{\mu_0 e^2}{6m} NZ\,\langle r_0^2\rangle, \text{ since } B = \mu_0 H \tag{10.5.12}$$

where $\langle r_0^2\rangle$ is the mean square distance of an arbitrary electron from the nucleus.

If instead of per unit volume, magnetizationn per mole is considered, we would get the molar susceptibility by dividing (10.5.12) by N_A/N, where N_A = Avogadro no.

$$\chi_{\text{dia}}^{\text{molar}} = -\frac{\mu_0 e^2}{6m} N_A Z\,\langle r_0^2\rangle \tag{10.5.13}$$

Eq. (10.5.12) is known as the classical ***Langevin's equation.***

The equation shows that (i) χ_{dia} increases with Z and the number of atoms N, that is, with the size of the atom. But it is *independent of the field strength and temperature* in agreement with Curie's results. (ii) The equation also shows that the outer electrons make the maximum contribution to χ_{dia} since it is proportional to $\bar{r}^2$. (iii) The diamagnetic susceptibility is always negative. A magnetic moment is induced on each atom in a direction opposite to external field and the substance acquires a weak magnetism opposite to field. All these characteristics are observed experimentally.

- Inert gases are diamagnetic since their atoms have completely filled electron shells and hence no permanent magnetic dipole moment.

- Diamagnetism is common to all elements. However, it is usually too weak to be detected in para and ferromagnetic cases.

10.6 Langevin's theory of paramagnetism

Langevin's classical theory of paramagnetism is based on the following *assumptions*: (i) each atom has a permanent magnetic moment, (ii) the separation between the dipoles are such that their mutual interactions are negligible and (iii) in absence of the field, the dipoles can have all possible orientations.

When a field $\vec{B}$ is applied, each dipole experiences a torque $\vec{\mu} \times \vec{B}$, where μ is the dipole moment. While the torque tends to bring an order by rotating the dipoles in the field direction, the thermal agitation tends to randomise the positions of dipoles.

The *potential energy* of a dipole in a field $\vec{B}$ is $-\vec{\mu}\cdot\vec{B} = -\mu B\cos\theta$, where θ is the angle between $\vec{\mu}$ and $\vec{B}$. By MB-statistics, the probability of finding the dipole at θ is $p(\theta)$ that varies as $e^{\mu B\cos\theta/kT}$, where k is the Boltzmann constant and T the absolute temperature.

If $n(\theta)d\theta$ be the number of dipole moments in a direction between θ and $\theta + d\theta$,

$$n(\theta)d\theta \propto p(\theta)d\Omega$$

where $d\Omega$ is the solid angle between the two hollow cones with θ and $\theta + d\theta$ as the semivertical angles.

$$\therefore \quad d\Omega = 2\pi \sin\theta \, d\theta$$

$$\therefore \quad n(\theta)d\theta = 2\pi A \sin\theta e^{\mu B\cos\theta/kT} d\theta$$

$$= C' e^{a\cos\theta} \sin\theta \, d\theta \tag{10.6.1}$$

where A = proportionality constant, $C' = 2\pi A$ and $a = \mu B/kT$.

If N be the number of dipole moments *per unit volume*, then we have

$$N = \int_0^\pi n(\theta)d\theta = C' \int_0^\pi e^{a\cos\theta} \sin\theta \, d\theta$$

$$= C' \int_{+1}^{-1} -e^{ax} dx = \frac{C'}{a}(e^a - e^{-a}),$$

where $\cos\theta = x \Rightarrow \sin\theta \, d\theta = -dx$. Therefore, we have

$$C' = \frac{Na}{e^a - e^{-a}} \tag{10.6.2}$$

The component of each dipole moment parallel to B is $\mu\cos\theta$, the components perpendicular to B will cancel out.

$\therefore$ Total magnetic moment of N dipoles, i.e., the magnetization M of the material is given by

$$M = \int_0^\pi \mu\cos\theta \, n(\theta)d\theta$$

$$= \int_0^\pi \mu\cos\theta \, C' e^{a\cos\theta} \sin\theta \, d\theta$$

$$= -\mu C' \int_{+1}^{-1} x e^{ax} dx, \quad \text{substituting } \cos\theta = x.$$

$$= -\mu C' \left[e^a \left(\frac{1}{a} - \frac{1}{a^2} \right) + e^{-a} \left(\frac{1}{a} + \frac{1}{a^2} \right) \right]$$

$$= \mu N \left(\coth a - \frac{1}{a} \right), \quad \text{using (10.6.2)}$$

$$= \mu N L(a) \tag{10.6.3}$$

where $L(a) = \coth a - \dfrac{1}{a}$ and is called the ***Langevin function.***

Fig. 10.10 illustrates the variation of $L(a)$ with a. The maximum value of M is given by μN when all the dipoles are aligned parallel to $\vec{B}$ and corresponds to the *saturation magnetization*, M_s. So, $M_s = \mu N$.

$$\therefore \quad M = M_s L(a) \tag{10.6.4}$$

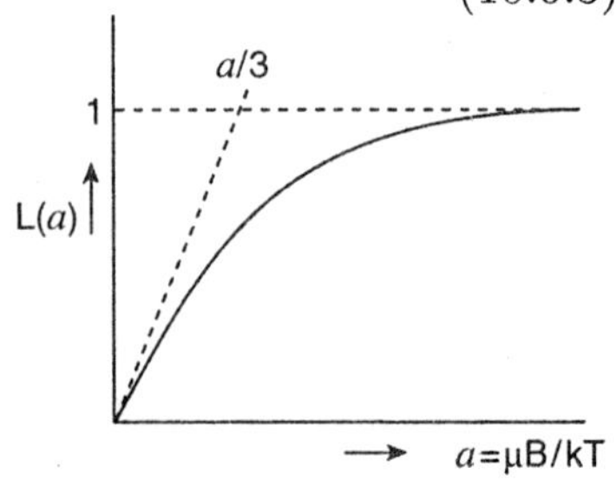

Fig. 10.10 Variation of Langevin function $L(a)$ with a

We shall now discuss the following *two extreme cases* in this context.

Case 1. $a = \mu B/kT \gg 1$. This makes $\coth a \to 1$ and $L(a) = \left(\coth a - \frac{1}{a}\right) \to 1$.

So, from (10.6.4), $M \to M_s$ and this shows that saturation occurs when the temperature T is low, applied magnetic field B is high, and/or the dipole moment μ is large.

Case 2. $a = \mu B/kT \ll 1$ or $\mu B \ll kT$. In this case, we have

$$L(a) = \coth a - \frac{1}{a} = \frac{1 + (a^2/2!) + (a^4/4!) + \cdots}{a + (a^3/3!) + (a^5/5!) + \cdots} - \frac{1}{a}$$

$$\simeq \left(1 + \frac{a^2}{2!}\right)\frac{1}{a}\left(1 - \frac{a^2}{3!}\right) - \frac{1}{a}$$

$$\simeq \frac{1}{a}\left(1 + \frac{a^2}{2} - \frac{a^2}{6}\right) - \frac{1}{a} = \frac{a}{3}$$

Thus the variation of $L(a)$ with a, near the origin, is linear having a slope = 1/3. So, in this region, the magnetization M is given by

$$M = \mu N\frac{a}{3} = \frac{\mu^2 NB}{3kT} = \frac{\mu^2 N\mu_0 H}{3kT} \tag{10.6.5}$$

$$\therefore \text{ Susceptibility, } \chi_{\text{para}} = \frac{M}{H} = \frac{\mu^2\mu_0 N}{3kT} = \frac{C}{T} \tag{10.6.6}$$

where $C = N\mu^2\mu_0/3k$ is a constant and is called the ***Curie constant.***

Eq. (10.6.6) is known as ***Curie's law*** which shows that *paramagnetic susceptibility is inversely proportional to the absolute temperature.* The susceptibility vs. temperature curve is illustrated in Fig. 10.11. If N in (10.6.6) is replaced by Avogadro number, N_A, we would get what is called the ***molar susceptibility*** given by

$$\chi_{\text{para}}^{\text{molar}} = \frac{\mu^2\mu_0 N_A}{3kT}$$

Note that the Curie's law of paramagnetism holds good for $\mu B \ll kT$.

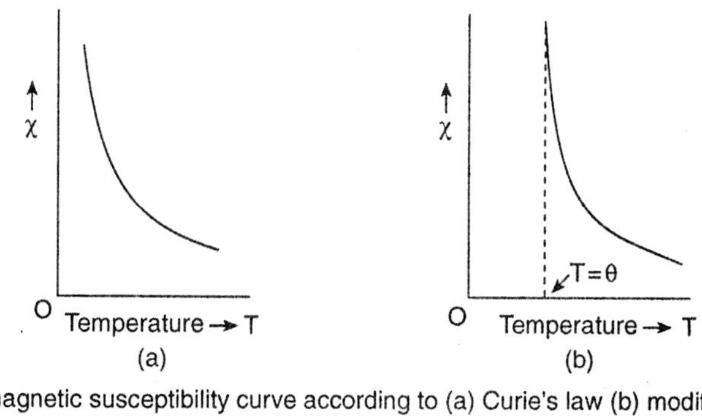

Fig. 10.11 Paramagnetic susceptibility curve according to (a) Curie's law (b) modified Curie law (10.6.7)

• Langevin's theory of paramagnetism applies strictly to gases where the mutual interactions of the molecules are negligible. In liquids and solids, however, the interactions may be large and many obey the *modified Curie law* or ***Curie-Weiss law***, having the following form.

$$\chi = \frac{C}{T-\theta} \tag{10.6.7}$$

where θ is called ***paramagnetic Curie temperature*** or simply ***Curie temperature*** which is a characteristic of the material concerned. Eq. (10.6.7) holds only at $T > |\theta|$ and is of the same form as (10.6.6) only with a shift in the origin of the temperature (Fig. 10.11b) from 0 to θ.

• The magnitude of χ_{para} is $\sim 10^{-3}$ while that of $\chi_{\text{dia}} \sim -10^{-5}$. So, the paramagnetic susceptibility is numerically two orders of magnitude greater than the diamagnetic susceptibility.

• The electrons in a metal are considered free, like the molecules in a gas. They tend to pair, but few unpaired electrons may always be there to give rise to a weak temperature-dependent paramagnetism known as ***Pauli paramagnetism*** to which we shall soon return.

10.7 Ferromagnetism: Spontaneous magnetization

As already stated, ferromagnetic materials are *strongly magnetic*, possessing *permanent magnetic moments* and show ***spontaneous magnetization***, i.e., magnetization even in the absence of external magnetic field, below a certain critical temperature—the *Curie temperature.* The existence of this spontaneous magnetization is characteristic of ***ferromagnetism***. Above the critical temperature, the specimen behaves like a paramagnet. Experimentally, it is also found that magnetization M increases when an external field is applied. Moreover, the relation between the magnetization M and the applied field H is *non-linear* and shows the ***hysteresis effect.***

The atoms or molecules of ferromagnetic substances have a net intrinsic magnetic dipole moment due mainly to spin of the electrons. The interaction between the neighbouring atomic dipoles is very strong and it occurs even with zero external field. It is known as ***spin exchange interaction*** and is quantum mechanical in origin. It is found that the energy of two neighbouring atoms due to this interaction is least when their magnetic moments are parallel and therefore the neighbouring magnetic moments are constrained to take up parallel orientations.

This effect of the exchange interaction (also called ***Weiss field***) to align the neighbouring moments parallel spreads over a small but finite volume of the bulk called a ***domain.*** Ferromagnetics are made up of a large number of *domains.* The size of a domain is about 10^{-6} m or large, each containing about 10^9–10^{15} atoms, and the material in the domain is magnetized to its full value, all the dipoles in the domain being aligned in the same direction, for a given material and temperature

(Fig. 10.12). The direction of magnetization varies from domain to domain so that the net magnetization of the sample as a whole is zero and the sample is said to be in a *demagnetized state* where its free energy is minimum.

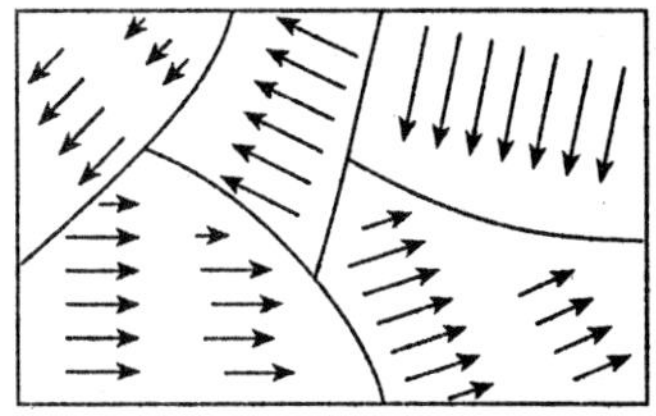

Fig. 10.12 Ferromagnetic domains

When an external field is applied to a ferromagnetic specimen, those domains that are aligned with the field increase in size at the cost of those domains having unfavourable alignment with respect to the field. On increasing the field above a certain value, the whole sample is covered by a single domain magnetized in the direction of the applied field and this corresponds to the *saturation magnetization* of the sample.

• The direct experimental evidence for the physical existence of the domains was first provided in 1931 by F. Bitter.

10.7.1 Weiss classical theory of ferromagnetism

Weiss in 1907 modified Langevin theory to explain ferromagnetism by postulating that the *effective magnetic field* H_{eff} consists of the externally applied field H which orient the molecules and the internal or molecular Weiss exchange field H_{ex} due to interaction between neighbouring atomic dipoles. Weiss assumed H_{ex} to be proportional to the magnetization $\vec{M}$ within a domain i.e., $H_{\text{ex}} = \beta M$ where the proportionality constant β is known as the ***Weiss factor.***

$$\therefore \quad H_{\text{eff}} = H + H_{\text{ex}} = H + \beta M \tag{10.7.1}$$

According to Weiss, the externally applied field H must therefore be replaced by H_{eff} in the expression for M for a paramagnetic material at high T-values. From equation (10.6.5), therefore,

$$\text{Magnetization, } M = \frac{\mu^2 N\mu_0}{3kT}(H + \beta M) = \frac{\mu^2\mu_0 NH}{3k\left(T - \dfrac{\mu_0\beta\mu^2 N}{3k}\right)} \tag{10.7.2}$$

$$\therefore \text{ Susceptibility, } \quad \chi = \frac{M}{H} = \frac{\mu^2\mu_0 N}{3k\left(T - \dfrac{\mu_0\beta\mu^2 N}{3k}\right)} = \frac{C}{T-\theta} \tag{10.7.3}$$

where C is the *Curie constant*, defined already, and $\theta = \mu_0\beta\mu^2 N/3k$ $(= \beta C)$, known as the ***Curie temperature***, a constant for given conditions. Equation (10.7.3) is the celebrated ***Curie-Weiss law*** of ferromagnetism.

• Since $\theta = \mu_0\beta\mu^2 N/3k \Rightarrow \beta = 3k\theta/\mu_0\mu^2 N$.

$$\text{Let us substitute } a = \frac{\mu_0\mu H_{\text{eff}}}{kT} \Rightarrow \beta M = \frac{akT}{\mu\mu_0} - H, \quad \text{using (10.7.1).} \tag{10.7.4}$$

$$\text{When } H = 0 \text{ or, } H_{\text{ex}} \gg H, \text{ we have from above, } \quad M = \frac{akT}{\mu\beta\mu_0} \tag{10.7.5}$$

At the *saturation magnetization*, $M_s = N\mu$, where N is the number of atoms (dipole) *per unit volume* and μ the dipole moment *per atom*, and we have

$$\frac{M}{M_s} = \frac{akT}{\mu_0 \beta N \mu^2} \tag{10.7.6}$$

Since again, $\theta = \dfrac{\mu_0 \beta \mu^2 N}{3k}$, we obtain from equation (10.7.6),

$$\frac{M}{M_s} = \frac{akT}{3k\theta} = \left(\frac{T}{\theta}\right)\frac{a}{3} \tag{10.7.7}$$

This shows that the plot of M/M_s against a is a *straight line* with a slope $T/3\theta$ for ferromagnetics, i.e., $M \neq 0$ even when $H = 0$ or nearly zero.

Langevin's equation, however, gave us (vide (10.6.4))

$$\frac{M}{M_s} = L(a) = \coth a - \frac{1}{a} \tag{10.7.8}$$

Thus, as the pre-requisite for spontaneous magnetization, both the equations (10.7.7) and (10.7.8) should be simultaneously satisfied. The simultaneous solutions of (10.7.7) and (10.7.8) are obtained graphically by plotting M/M_s vs. a separately (Fig. 10.13). Both the plots pass through the origin O which is a trivial solution. The

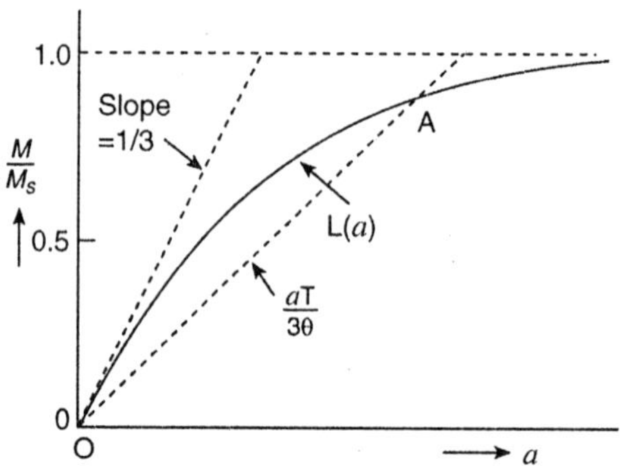

Fig. 10.13 Plots to determine the solution for spontaneous magnetization

other intersection of the curves is the point A, if the slope of the line $M/M_s = aT/3\theta$ is less than that of $L(a)$ at the origin. The slope of $L(a)$ at the origin is already known to be 1/3.

So, the intersection A occurs under the condition given by

$$\frac{T}{3\theta} < \frac{1}{3} \Rightarrow T < \theta \tag{10.7.9}$$

Thus, the ferromagnetic feature of *spontaneous magnetization occurs* only *if the temperature is below the Curie temperature. Above the Curie temperature, the spontaneous magnetization is destroyed* by large thermal agitation in ferromagnetic materials and the paramagnetic feature is exhibited.

The ***Curie point*** or ***Curie temperature θ*** thus *represents the temperature at which the transition from ferromagnetism to paramagnetism occurs in a material with the rise in temperature.*

● Since $\theta = \mu_0\mu^2\beta N/3k$, the Curie point increases with the increasing values of both β and μ. Curie temperature for Fe is 1043 K, that of Ni is 631 K.

Determination of Weiss constant and Curie temperature—We have: Curie temperature, $\theta = \dfrac{N\mu^2\mu_0}{3k}\beta = C\beta$, where $C = N\mu^2\mu_0/3k$.

$$\therefore \quad \chi = \frac{C}{T-\theta} = \frac{\theta/\beta}{T-\theta}$$

$$\therefore \quad \frac{1}{\chi} = \left(\frac{\beta}{\theta}\right)T - \beta \tag{10.7.10}$$

If, therefore, $1/\chi$ is plotted against T at high temperatures ($T \geq \theta$), we shall obtain a *straight line* having a slope β/θ and an intercept $-\beta$ on $1/\chi$-axis (Fig. 10.14). So, with the help of this method, both β, the Weiss constant and θ, the Curie temperature may be evaluated by measuring the susceptibility χ experimentally at different temperatures.

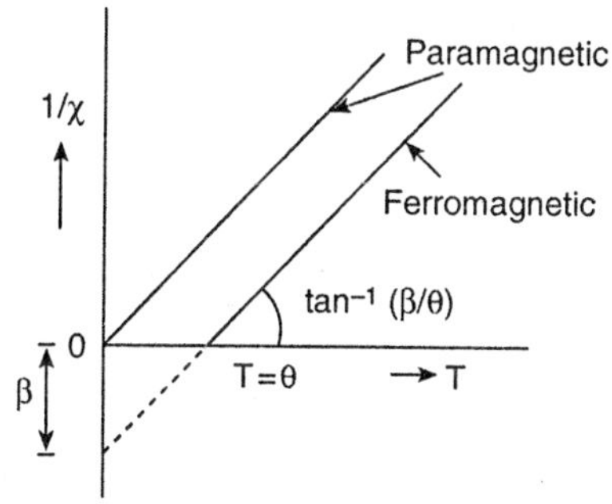

Fig. 10.14 1/χ versus *T* plot for para and ferromagnetic materials for *T*>θ

Variation of M/M_s with T/θ—We have, from the equations (10.7.8) and (10.7.7) respectively

$$\frac{M}{M_s} = \coth a - \frac{1}{a}$$

$$\text{and} \quad \frac{M}{M_s} = \left(\frac{T}{\theta}\right)\frac{a}{3}$$

On elimination of a from the above two equations, we shall obtain

$$\frac{M}{M_s} = f\left(\frac{T}{\theta}\right), \quad \text{i.e., some function of } (T/\theta).$$

But how to obtain the function f? The function f can be theoretically obtained by drawing a number of straight lines from the origin with T as a parameter in Fig. 10.13

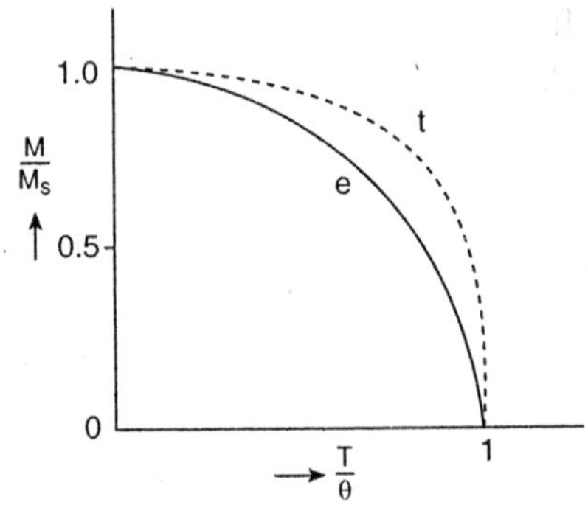

Fig. 10.15 Experimental (e) and theoretical (t) plots of M/Ms versus T/θ

and finding the corresponding point of intersection like A. The theoretically obtained functional relation is the dotted curve t in Fig. 10.15. The solid curve e represents the experimental one. The difference between the two can be reduced by using the quantum mechanical theory to be taken up later.

● The internal field in the saturated state is βM_s and its typical values are of the order of 10^9 A/m.

B. Quantum Theory of Magnetism

10.8 Quantum theory of diamagnetism

The effect of a magnetic field $\vec{B}$ on the Hamiltonian $\mathcal{H}$ of an electron is to add a small perturbation $\mathcal{H}'$, given from quantum mechanics, by

$$\mathcal{H}' = \frac{ie\hbar}{m}\left(\vec{\nabla}\cdot\vec{A} + \vec{A}\cdot\vec{\nabla}\right) + \frac{e^2}{2m}A^2 \tag{10.8.1}$$

where A is magnetic vector potential given by $\vec{B} = \vec{\nabla} \times \vec{A}$.

If the field $\vec{B}$ is uniform, $\vec{A} = \frac{1}{2}\vec{B} \times \vec{r}$, which implies that

$$\vec{A} = \frac{1}{2}\begin{vmatrix} \vec{i} & \vec{j} & \vec{k} \\ B_x & B_y & B_z \\ x & y & z \end{vmatrix} \tag{10.8.2}$$

If $\vec{B}$ is along the z-direction, then $B_x = B_y = 0$, $B_z = B$.

$$\therefore \quad A_x = -\tfrac{1}{2}By, \quad A_y = \tfrac{1}{2}Bx, \quad A_z = 0 \tag{10.8.3}$$

Hence, we have

$$\vec{\nabla}\cdot\vec{A} = \frac{\partial A_x}{\partial x} + \frac{\partial A_y}{\partial y} + \frac{\partial A_z}{\partial z} = 0 \tag{10.8.4}$$

$$\text{and } \vec{A}\cdot\vec{\nabla} = A_x\frac{\partial}{\partial x} + A_y\frac{\partial}{\partial y} + A_z\frac{\partial}{\partial z}$$

$$= \tfrac{1}{2}B\left(x\frac{\partial}{\partial y} - y\frac{\partial}{\partial x}\right), \quad \text{using (10.8.3)} \tag{10.8.5}$$

$$\therefore \text{ Using (10.8.3)}, \; A^2 = A_x^2 + A_y^2 + A_z^2 = \tfrac{1}{4}B^2y^2 + \tfrac{1}{4}B^2x^2 = \tfrac{1}{4}B^2(x^2+y^2) \tag{10.8.6}$$

$$\therefore \; \mathcal{H}' = \frac{ie\hbar}{m}\cdot\frac{B}{2}\left(x\frac{\partial}{\partial y} - y\frac{\partial}{\partial x}\right) + \frac{e^2}{2m}\cdot\tfrac{1}{4}B^2(x^2+y^2) \tag{10.8.7}$$

using (10.8.1) and (10.8.6).

Note that the first term on the right hand side of (10.8.7) is proportional to the orbital angular momentum component L_z. In mono-nuclear systems, this term gives rise to paramagnetism. The second term gives, for a spherically symmetric system, a contribution

$$E' = \frac{e^2B^2}{8m}(x^2+y^2) = \frac{e^2B^2}{12m}\overline{r^2}, \quad \left(\because\; x^2+y^2 = \tfrac{2}{3}\overline{r^2}\right)$$

by first order perturbation energy.

The associated magnetic moment is diamagnetic.

$$\mu_m = -\frac{\partial E'}{\partial B} = \frac{e^2B}{6m}\overline{r^2}$$

$\therefore$ Atomic succeptibility, χ_{at} is given by

$$\chi_{\text{at}} = \frac{\mu_m}{\mathcal{H}} = -\frac{\mu_0 e^2}{6m}\overline{r^2}, \quad \text{as } B = \mu_0\mathcal{H}$$

If N be the number of *atoms per unit volume* and n the *nunber of electrons* in each atom, then

$$\boldsymbol{\chi_{\text{at}} = -\frac{Nn\mu_0 e^2}{6m}\overline{r^2}}$$

where $\overline{r^2}$ is the mean square distance of an arbitrary electron from the nucleus.

The above relation is in agreement with the classical result.

10.9 Quantum theory of paramagnetism

Langevin in his classical theory assumed that all orientations of magnetic dipoles relative to the field direction are possible. Quantum theory, however, permits only certain discrete angles of orientation of atomic dipoles with respect to the field direction and are given by

$$\cos\theta = M_J/J$$

where J is the resultant angular momentum quantum number ($= L + S$; spin-orbit coupling) and M_J the magnetic quantum number having $2J+1$ possible values from $-J$ to $+J$, in steps of unity.

Consider a paramagnetic substance having N number of atoms *per unit volume* and let J be the total angular momentum quantum number associated with each atom. The component of atomic dipole moment for a given M_J in the direction of field $\vec{B}$ is $M_J g\mu_B$ where g is Lande splitting factor and μ_B the Bohr magneton. The potential energy of magnetic dipole is $W = -M_J g\mu_B B$. As M_J depends on θ, so does W.

The number of atomic dipoles with quantum number M_J is proportional to Boltzmann factor $e^{-W/kT}$, where k is Boltzmann constant and T the temperature.

$$\therefore \text{ Magnetization, } M = N\left[\frac{\sum\limits_{-J}^{+J} M_J g\mu_B B e^{M_J g\mu_B B/kT}}{\sum\limits_{-J}^{+J} e^{M_J g\mu_B B/kT}}\right] \tag{10.9.1}$$

The term in [ ] in (10.9.1) is the *statistical average* of atomic magnetic moment component along $\vec{B}$.

Evaluation of M—In evaluating M, only the components of dipole moments along $\vec{B}$ are to be considered, since the perpendicular components cancel out due to symmetry.

Putting $\dfrac{g\mu_B B}{kT} = z$ in (10.9.1), we obtain

$$M = Ng\mu_B\left[\frac{\sum\limits_{-J}^{+J} M_J e^{M_J z}}{\sum\limits_{-J}^{+J} e^{M_J Z}}\right] \tag{10.9.2}$$

$$= Ng\mu_B\frac{d}{dz}\left(\ln\sum_{-J}^{+J} e^{M_J z}\right), \quad \text{since } \frac{d}{dz}\left(e^{M_J z}\right) = M_J e^{M_J z}$$

$$= Ng\mu_B\frac{d}{dz}\left[\ln\left\{e^{Jz} + e^{(J-1)z} + \cdots + e^{-(J-1)z} + e^{-Jz}\right\}\right]$$

$$= Ng\mu_B\frac{d}{dz}\left[\ln\frac{e^{Jz}\left\{1 - e^{-(2J+1)z}\right\}}{1 - e^{-z}}\right]$$

$$= Ng\mu_B\frac{d}{dz}\left[\ln\left\{\frac{e^{(J+\frac{1}{2})z} - e^{-(J+\frac{1}{2})z}}{e^{z/2} - e^{-z/2}}\right\}\right]$$

$$= Ng\mu_B\frac{d}{dz}\left[\ln\left\{\frac{\sin\left(J+\frac{1}{2}\right)z}{\sinh(z/2)}\right\}\right]$$

$$= Ng\mu_B J\left[\frac{2J+1}{2J}\coth\left(\frac{2J+1}{2J}\right)a - \frac{1}{2J}\coth\left(\frac{a}{2J}\right)\right] \tag{10.9.3}$$

where $a = Jz = Jg\mu_B B/kT$. The quantity in [ ] on right-hand-side of (10.9.3) is called *Brillouin function* $B_J(a)$.

$$\therefore \quad M = Ng\mu_B J B_J(a) \tag{10.9.4}$$

The maximum value of M_J is J. So, the saturation magnetization $M_s = Ng\mu_B J$, corresponding to the alignment of dipole moments along the field lines.

$$\therefore \quad M = M_s B_J(a) \tag{10.9.5}$$

We shall now consider the following *two* cases that may arise.

Case 1. Field high, temperature low—At high magnetic fields and low temperatures, $a \gg 1$ and $B_J(a) \to 1$.

When $J \to \infty$, that is, all orientations are possible, then

$$(2J+1)/2J \to 1, \quad \coth(a/2J) \to 2J/a$$

So, $B_J(a)$ approaches $\coth a - 1/a = L(a)$, the Langevin function.

Experimental values of saturation magnetization are in good agreement with equation (10.9.5) and are more accurately described by Brillouin function than by Langevin function. Hence quantum theory is favoured in explaining paramagnetism.

Case 2. Field low, temperature high—At low magnetic fields and high temperatures, $a \ll 1$ and

$$\coth\left(\frac{a}{2J}\right) = \frac{2J}{a}\left(1 + \frac{1}{3}\cdot\frac{a^2}{4J^2}\right),$$

$$\coth\left(\frac{2J+1}{2J}\right)a = \frac{2J}{2J+1}\cdot\frac{1}{a}\left[1 + \frac{a^2}{3}\left(\frac{2J+1}{2J}\right)^2\right].$$

$$\therefore \quad B_J(a) = \frac{1}{a} + \frac{a}{3}\left(\frac{2J+1}{2J}\right)^2 - \frac{1}{a}\left(1 + \frac{a^2}{12J^2}\right) = \frac{a}{3}\left(\frac{J+1}{J}\right)$$

So, from (10.9.5),

$$\begin{aligned} M &= Ng\mu_B(J+1)a/3 \\ &= \frac{Ng^2\mu_B^2 B}{3kT} J(J+1), \quad \text{using } a = Jg\mu_B B/kT. \end{aligned} \tag{10.9.6}$$

Hence, the paramagnetic successptibility is given by

$$\chi_{\text{para}} = \frac{M}{H} = \frac{\boldsymbol{Ng^2\mu_B^2\mu_0 J(J+1)}}{\boldsymbol{3kT}} \quad (\because B = \mu_0 H) \tag{10.9.7}$$

Molar susceptibility is obtained by replacing N by N_A, the Avogadro number.

Since the total magnetic moment μ_J associated with J is given by

$$\mu_J^2 = g^2 J(J+1)\mu_B^2,$$

we have, from (10.9.7),

$$\chi = \frac{\boldsymbol{N\mu_0\mu_J^2}}{\boldsymbol{3kT}} = \frac{\boldsymbol{C}}{\boldsymbol{T}}, \quad \text{where } C = N\mu_0\mu_J^2/3k$$

which is identical with the classical expression.

The ratio μ_J/μ_B gives the ***effective number*** of Bohr magneton, p_{eff}. Therefore, $p_{\text{eff}} = g\sqrt{J(J+1)}$.

$$\therefore \quad \chi_{\text{para}} = \frac{N\mu_0 p_{\text{eff}}^2 \mu_B^2}{3kT} \tag{10.9.8}$$

J can be determined from the Hund's rule and g from Lande formula. Hence, p_{eff} may be evaluated which agrees well with values from experimental measurements of χ. The quantum theory of paramagnetism is thus well supported.

10.10 Pauli paramagnetism: Paramagnetism of free electrons

According to Langevin's theory, paramagnetic susceptibility is inversely proportional to temperature. But some metals exhibit paramagnetism *independent of temperature.* Pauli in 1927 showed that it is the *paramagnetism of free electrons* constituting the '*electron gas*'. Electrons can orient themselves only in *two* directions—either along or against the magnetic field.

To understand this Pauli paramegnetism, we recall density of states vs. energy curve at absolute zero. It may be split into *two* parts: (i) one with spin pointing along $+z$ direction and (ii) the other with spin oppositely directed (Fig. 10.16). With no external field ($B = 0$), the number of electrons with spins parallel to $+z$ axis equals the number of electrons with opposite spin (Fig. 10.16a). And the net magnetic moment of electron gas is zero. With an external field B along $+z$, the energy of spins aligned parallel to B is lowered by μ_B and that of spins opposite to B is equally increased (Fig. 10.16b). The Fermi level of two spin distributions thus shifts relative to each

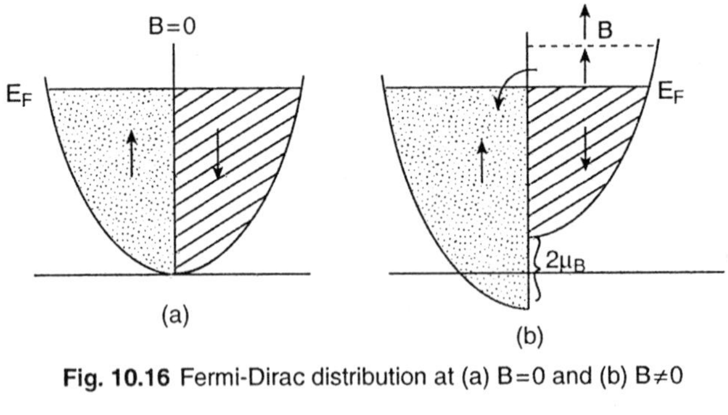

Fig. 10.16 Fermi-Dirac distribution at (a) B=0 and (b) B≠0

other and an energetically unstable situation arises. To restore stability, the electrons near the Fermi level with antiparallel spins flip into region of parallel spins till the two Fermi levels become equal again (Fig. 10.16b).

Now, the number of electrons effectively changing their direction = the density of states at the energy level in one half of spin distribution × the change in energy.

$$\therefore \quad N_{\text{eff}} = \tfrac{1}{2} g(E_F) \times \mu_B B \tag{10.10.1}$$

On application of the field, the number of electrons with spins parallel to the field outweighs those with opposite spin by N_{eff}, developing a *net magnetization*. Since each flip increases magnetization from $-\mu_B$ to $+\mu_B$, that is, by $2\mu_B$, the net magnetization,

$$M \simeq N_{\text{eff}} \times 2\mu_B = g(E_F)\mu_B^2 B \tag{10.10.2}$$

$\therefore$ Pauli spin susceptibility, $\chi_P = M/H \simeq \mu_0\mu_B^2 g(E_F) \quad (\because\ B = \mu_0 H)$ (10.10.3)

χ_P is thus practically *temperature-independent*; *practically* because temperature has a very marginal effect on F.D. distribution of electrons.

Now, from the theory of free electrons in crystals, we have

$$g(E_F) = \frac{3N}{2E_F}$$

$$\therefore\ \chi_P = \frac{3\mu_0 N\mu_B^2}{2E_F} = \frac{3\mu_0 N\mu_B^2}{2kT_F} = \frac{3\mu_0 N\mu_B^2}{2kT_F} \tag{10.10.4}$$

where $E_F = kT_F$, T_F being the Fermi temperature.

- In terms of classical susceptibility χ $(= \mu_0 N\mu_B^2/kT)$ for $J = \frac{1}{2}$, equation (10.10.4) may be rewritten as

$$\chi_P = \frac{3}{2}\chi\frac{T}{T_F} \tag{10.10.5}$$

T_F is normally very high and hence $\chi_P < \chi$, by about two orders of magnitude, in agreement with the experimental data. In the transition metals, however, χ_P is exceptionally high, for $g(E_F)$ is quite large by virtue of narrow and high $3d$ band.

10.11 Temperature dependence of spontaneous magnetization (Ferromagnetics)

The magnetic moment due to the orbital motion of free electrons in a metal is, as just discussed, negligible compared to the spin moments. It is thus possible to study the magnetic behaviour of ferromagnetic materials below the Curie temperature by assuming only the spin system.

With no external field, i.e., with $H = 0$ and from (10.7.1), we have

$$H_{\text{ex}} = \beta M \tag{10.11.1}$$

where $M = N\bar{\mu}$ is the spontaneous magnetization at temperature, say $T < \theta$, the Curie temperature.

From the quantum theory of paramagnetism, the magnetization M for spin-half particles $(S = J = 1/2)$ can be shown to be given by

$$M = Ng\mu_B \tanh(x), \quad \text{where } x = g\mu_B B/kT \tag{10.11.2}$$

For ferromagnetics, the field B in (10.11.2) is to be replaced by $\mu_0 H_{\text{ex}}$.

$$\therefore\quad x = \mu_0 g\mu_B\beta M/kT, \quad \text{using (10.11.1)}$$

$$\Rightarrow\quad M = (kT/\mu_0 g\mu_B\beta)x \tag{10.11.3}$$

Equations (10.11.2) and (10.11.3) can be solved simultaneously by plotting them on M vs. x graph (Fig. 10.17). The magnetization M at a given temperature is given by the point of intersection of the two curves. From (10.11.3), we see that on increasing the temperature T, the slope of the straight line also increases. At temperatures $T < \theta$, the two curves intersect at a point that gives a non-vanishing spontaneous magnetization, which attains its maximum value at temperature $T = 0\,\text{K}$ and is given by

$$M_s(0) = Ng\mu_B \tag{10.11.4}$$

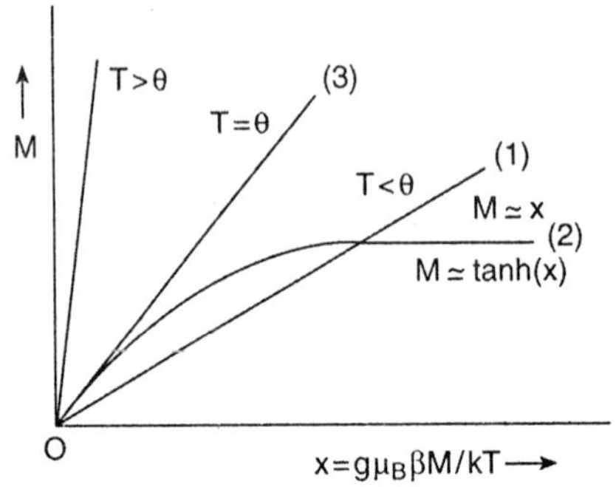

Fig. 10.17 M versus x and M versus tanh(x) curves. Point of intersection represents spontaneous magnetization, θ is paramagnetic Curie temperature

At $T = \theta$, the straight line (3) is tangential to the hyperbolic curve (2) at the origin where the spontaneous magnetization just reduces to zero. And it remains zero at all temperatures $T > \theta$. Also, at temperature $T = \theta$ for $x \ll 1$, the hyperbolic function approximates to

$$\tanh(x) \simeq x \tag{10.11.5}$$

Equating the equations (10.10.2) and (10.10.3), we readily obtain

$$\beta = \frac{k\theta}{\mu_0 N (g\mu_B)^2} \tag{10.11.6}$$

To express the temperature-dependence of spontaneous magnetization conveniently, we rewrite (10.11.2) in the following form, using (10.11.4) and (10.11.6)

$$\frac{M_s(T)}{M_s(0)} = \tanh\left(\frac{T}{\theta}\right) \tag{10.11.7}$$

This describes the relation between normalized magnetization $M_s(T)/M_s(0)$ and normalized temperature T/θ and is known as the *thermomagnetic equation of state* for ferromagnetic phase of materials. This is illustrated in Fig. 10.18 for Ni and Gd

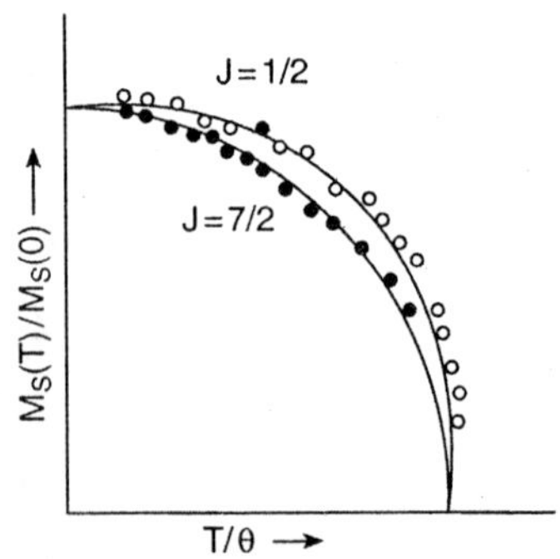

Fig. 10.18 A curve between normalized magnetization Ms(T)/Ms(0) and normalized temperature T/θ for Ni and Gd with J=1/2 and 7/2

with $J = 1/2$ and $J = 7/2$. The curve is in a sense *universal*, for the reduced parameters are common to all the ferromagnetic materials for the same value of J.

At 0 K, the M-value is maximum and the normalised magnetization $M_s(T)/M_s(0)$ is unity. With increasing temperature, the thermal energy randomises the magnetic system and hence the spontaneous magnetization ceases to exist above the Curie temperature.

10.12 Weiss molecular field : Physical origin

While the molecular field of Weiss could successfully explain many features observed in ferromagnetism, the nature of the field remained a mystery for long. It was first supposed to be the dipole-dipole interaction of nearest neighbour. Such an interaction would give a field $\sim \mu_B/a^3 \simeq 0.1$ Tesla, a being the interatomic distance in a ferromagnetic crystal. But at Curie temperature $T_c = 1000$ K, the exchange field is approximately $H_{\text{ex}} \simeq \beta M_s(0) \simeq 10^3$ Tesla.

This discrepancy points to the fact that instead of the simple model of a dipole-dipole interaction, what is required is a much stronger interaction to account for the ferromagnetism.

The first attempt to this effect was made by Heisenberg in 1928. He pointed out that the large Weiss field can be explained in terms of the so-called ***exchange interaction*** between electron spins. For ions i, j of spins S_i, S_j, the (exchange) energy of interaction can be written as follows.

$$E_{\text{ex}}^{ij} = -2JS_i \cdot S_j \tag{10.12.1}$$

Thus the total exchange energy of spin i with its nearest neighbours $j_1, j_2, \cdots$ etc. is given by

$$E_{\text{ex}}^{ij} = -2J\sum_{j=1}^{n} S_i \cdot S_j \tag{10.12.2}$$

where J, termed as ***exchange integral***, measures the strength of interaction.

Generally, the *exchange integral* is negative so that a non-ferromagnetic state is favoured . For ferromagnetics, however, it must be positive. The dot product, in the above equations, has been used to include the fact that the exchange energy is governed by the relative orientations of spins. In ferromagnetism, when spins are parallel, $S_i \cdot S_j = S^2$ and E_{ex} is minimum; when the spins are antiparallel $S_i \cdot S_j = -S^2$ and E_{ex} is maximum.

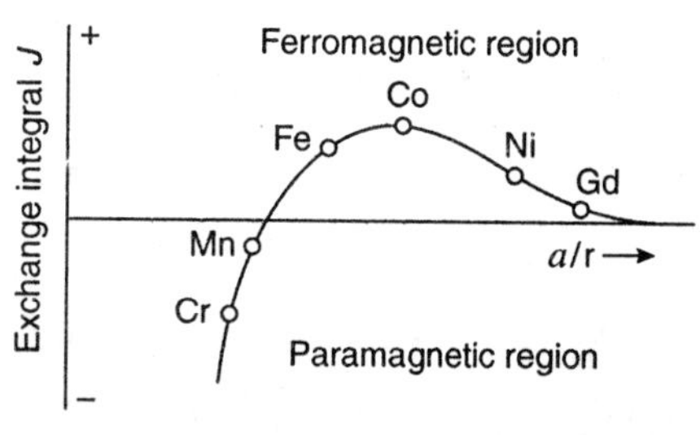

Fig. 10.19 Variation of exchange integral J with a/r where a is interatomic distance and r_0 the atomic radius

The exchange integral can be evaluated experimentally as a function of the ratio a/r, where a is the interatomic spacing and r the radius of the unfilled shell. The curve, in Fig. 10.19

illustrates this for transition metals. The ratio a/r for the transition elements of iron group is

Elements	→	Fe	Co	Ni	Cr	Mn	Gd
a/r ratio	→	3.26	3.64	3.94	2.60	2.94	3.10
Ferromagnetic ?	→	Yes	Yes	Yes	No	No	Yes

Slater (1930) suggested the criterion that for ferromagnetism the ratio $a/r \geq 3$, but not very large, and accordingly, from above, it is seen that Fe, Co, Ni and Gd satisfy

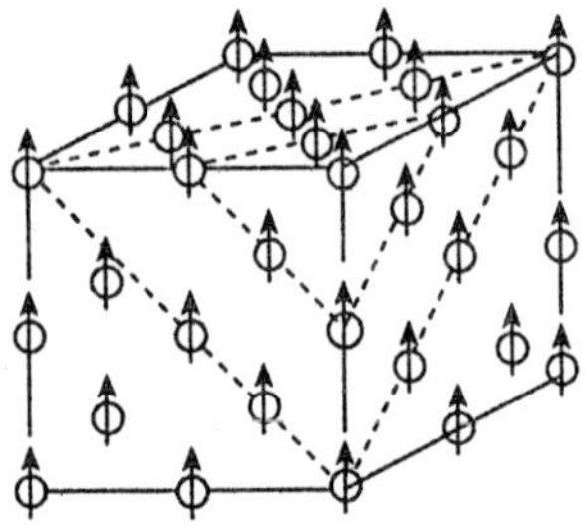

Fig. 10.20 Spontaneous magnetization of a ferromagnetic substance. Exchange forces cause parallel orientation of spins of electrons of partially filled inner shells

Slater's criterion and they are ferromagnetic. But Cr and Mn are *not* as they do not satisfy the criterion. They can however be made ferromagnetic by alloying them with appropriate non-ferromagnetic elements to make compounds whose a/r ratio is greater than 3, e.g., MnSb, CuMn etc.

● From above, it follows that *the necessary and sufficient conditions* for ferromagnetism are given by: (i) the existence of *incompletely filled atomic shells* and (ii) the *positive exchange integral*, causing parallel orientation of spins, as has been illustrated in Fig. 10.20.

● Slater's criterion underscores the importance of 3*d* shell in the origin of ferromagnetism. The fact that the radius of 3*d* shell is small plays a crucial role. The same applies to 4*f* shell in rare-earth ferromagnets. In Fe, Co, Ni, the inner 3*d* sub-shells are incompletely filled, whereas in rare earth elements like Gd, the inner 4*f* shells are incompletely filled.

10.13 Ferromagnetic domains revisited

We have just tried to understand why Fe, Co, Ni are ferromagnetic and shall now discuss why a fresh specimen of those materials is non-magnetised and they exhibit *hysteresis loop* when a magnetic field is applied to them. Domain hypothesis of Weiss may throw light on to these aspects.

According to Weiss, ferromagnetics contain small domains, each magnetized to saturation, but the net maroscopic magnetization may vary from zero to saturation. A fully demagnetized specimen, represented schematically by four equal domains (Fig. 10.21a) has zero resultant magnetic moment. If now a field is applied in some arbitrary direction

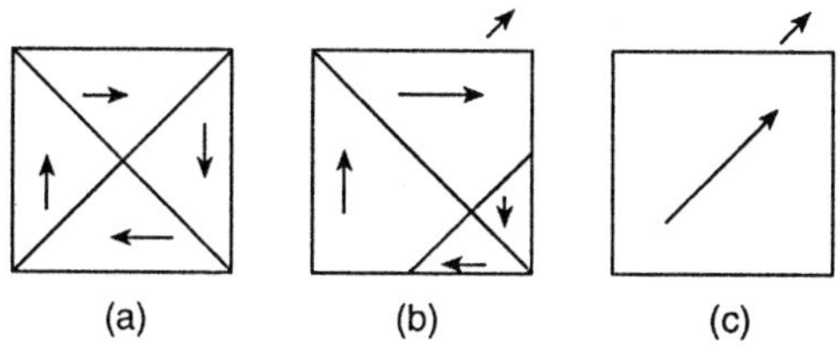

Fig. 10.21 Domain structure (a) corresponds to non-magnetized state, (b) magnetization is due to wall motion (c) magnetization is due to domain rotation

(Fig. 10.21b) and gradually increased, domains having magnetic moments along the field direction (i.e., the so-called favourably oriented domains) will grow at the cost of others (i.e., unfavourably oriented domains), due to the motion of domain walls. With further increase in field, a sudden rotation of unfavourably oriented domains occurs and the magnetization thereby gets saturated. The whole specimen may become one single domain with its moment directed parallel to the field (Fig. 10.21c). Any further increase in field causes no further change in magnetization.

At low temperatures, the dependence of magnetization M on field strength H is shown by the usual ***hysteresis curve*** (Fig. 10.22). This is associated with the motion of domain walls and the domain rotation and has therefore *three* distinct regions.

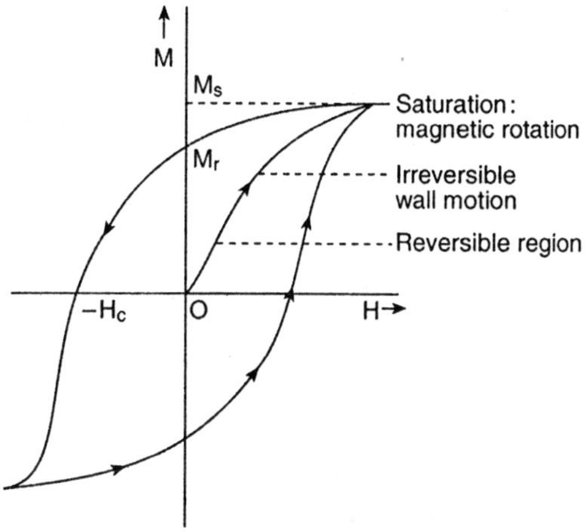

Fig. 10.22 Hysteresis (M–H) loop of a ferromagnetic material

With small applied magnetic field, the induced magnetization varies *linearly* and *reversibly* with the field. The region in which the magnetization is reversible is called the ***reversible region***.

When the field is gradually increased, the magnetization shows a *non-linear* and *irreversible* variation with the field. This is the region of ***irreversible wall motion***.

If the field is further increased gradually, the magnetization also increases gradually and eventually it attains a saturation value M_s. This is the region where the ***rotation of domains*** occur. A further increase in field makes no change in magnetization.

Let us now follow what happens when the field is gradually decreased. When the *field is gradually decreased*, the magnetization also decreases, but interestingly and importantly this decrease follows a different path. Corresponding to even a zero field, the specimen retains a certain amount of magnetization, called the *residual* or *remanent magnetization* M_r. A finite opposing field is required to bring the magnetization back to zero and this is called the *coercive field*, $-H_c$.

A further increase in the reverse field leads to a complete reversal of M_s. A second reversal of the field finally completes the *hysteresis curve*.

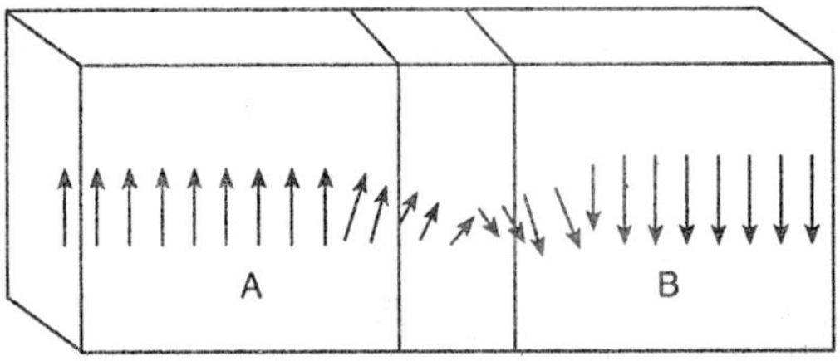

Fig. 10.23 Schematic representation of a Bloch wall separating two domains

- In a crystal specimen, the domains are separated from one another by boundaries (Fig. 10.23) which are the regions in which the moments (spins) change their spatial directions. This transition layer between the two domains is known as the *domain wall* or a *Bloch wall* (Fig. 10.23). The transition from a domain with '*spin up*' to one with '*spin down*' has been illustrated in Fig. 10.23.

10.14 Origin of the domains

One can understand the origin of domains from the general thermodynamic principle that the free energy, (the symbols having their usual meanings)

$$F = E - TS \tag{10.14.1}$$

of a system (a solid, here) is *minimum*.

Again, it has been seen that well below T_c, $M \to M_s(0)$ in a ferromagnet and the degree of order in such a magnetic system is very high. So, the entropy term in (10.14.1) may be neglected. Thus, free energy here means *minimum total internal energy E* and this is enough to understand the existence of domains.

Landau and Lifshitz showed that the domain structure is a natural consequence of the various energies constituting the total energy of a ferromagnet. The energies are: (i) the *exchange energy*, (ii) the *magnetic field energy* and (iii) the *anisotropy energy* which we shall now discuss one after the other.

1. Exchange energy—It is already seen that the exchange energy E_{ex} is minimum when spins are parallel. This energy gives rise to spontaneous magnetization. It establishes the existence of a single domain in a ferromagnetic material (Fig. 10.24a) and favours an indefinitely large domain as it is minimum when all spins are parallel.

2. Magnetic field energy—The free magnetic poles at the ends of a specimen with a single domain (Fig. 10.24a) will have the highest magnetic energy. This energy can be reduced to about half by dividing the crystal into two domains magnetized in opposite directions (Fig. 10.24b). Dividing the crystal into N domains (Fig. 10.24c), the magnetic energy can be reduced by $1/N$. Since some energy is required to create a domain bounding wall, the division process can be continued till the energy needed to create an additional Bloch wall is greater than the reduction in energy. Taking this aspect into account, the domain arrangements in Fig. 10.24d and 10.24e will have zero

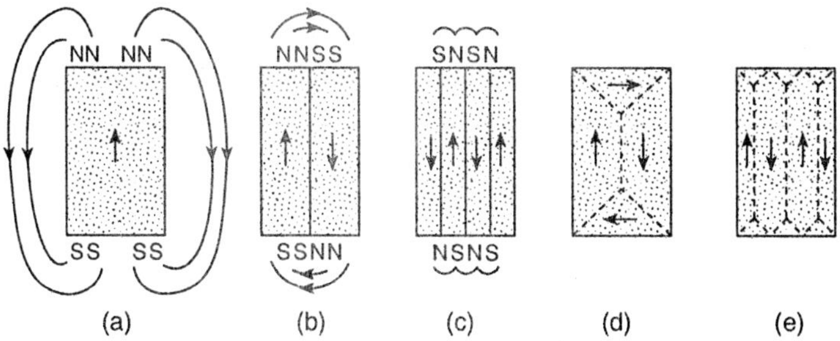

Fig. 10.24 The origin of domains

magnetic field energy. The arrangement consists of triangular prism domains near the end faces. This type of domains is called ***closure domains***, the boundary walls of which make equal angles (45°) with the direction of magnetization each due to closure domains themselves and vertical domains. So, the normal component of magnetization across the boundary wall is continuous. There are no free poles near the end faces of the crystal and hence there is no magnetic field energy.

3. Anisotropy energy—In studying single crystals of ferromagnets, it is found that the field required to magnetize them is direction-dependent. Fig. 10.25 illustrates the M–H curve of a single crystal of iron at 18°C, with field directed along the different crystallographic directions. Plainly, a very small field along $\langle 100\rangle$ is enough for a large (saturation) magnetization, but in order to produce the same magnetization along the direction $\langle 111\rangle$, a large field is required. This difference in the magnetic energy to produce saturation in an 'easy' and 'hard' directions is called the ***anisotropy energy***, E_{an}. For a cubic crystal, this anisotropic energy is given by

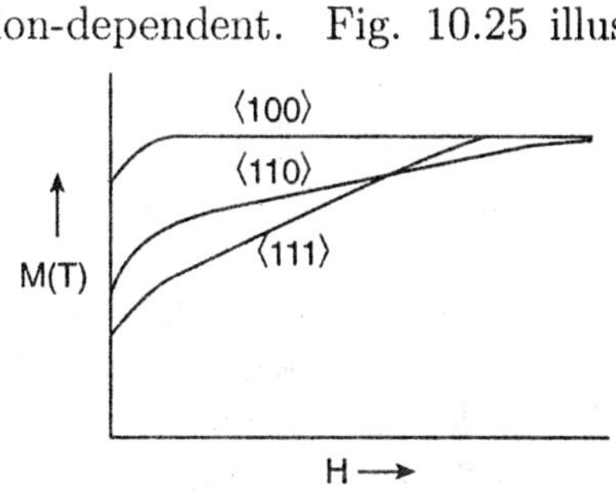

Fig. 10.25 Magnetization curves for single crystal of iron at 18°C

$$E_{\text{an}} = K_1(\alpha^2\beta^2 + \beta^2\gamma^2 + \gamma^2\alpha^2) + K_2\alpha^2\beta^2\gamma^2 \tag{10.14.2}$$

where the K's are called *anisotropy constants* and α, β and γ are the *direction cosines* of the magnetization vector relative to the axes of the cube.

- These three sources are the chief determinants of the domain characteristics. While a balance struck between the exchange energy and the magnetic field energy controls the domain size, the domain thickness however is determined by the balance struck between the exchange energy and the anisotropy energy.

10.15 Antiferromagnetism

In ferromagnetism, the exchange energy H_{ex} is minimum when the neighbouring spins are parallel and the exchange integral is positive. But, interestingly, there are many solids where H_{ex} is found to be minimum when the neighouring spins are antiparallel and the exchange integral is negative favouring a non-ferromagnetic state. If the neighbouring spins are balanced, the material is said to be *antiferromagnetic* and the phenomenon ***antiferromagnetism***. If unbalanced, the material is ***ferrimagnetic*** to which we shall refer later.

Antiferromagnetism was first found in MnO in 1938. Others are Mn and Cr metals, many of rare earths and some compounds of transition elements. Many newer antiferromagnetic materials have been discovered by neutron diffraction technique. The first theoretical investigation of such systems was made by Neel and Bitter.

The susceptibility χ vs. temperature T curve (Fig. 10.26) of antiferromagnetic material shows the distinguishing features of para-, ferro- and antiferromagnetics. The ferromagnetic susceptibility increases enormously at the transition temperature (Curie temperature). In antiferromagnetics, however, the susceptibility at first increases with temperature and reaches a maximum at a certain temperature known as ***Neel temperature***, T_N. Above the Neel temperature T_N, the susceptibility-temperature relation, i.e., (χ - T) relation runs as

$$\chi = \frac{C}{T+\theta} \tag{10.15.1}$$

where C is the Curie constant and θ the paramagnetic Curie temperature.

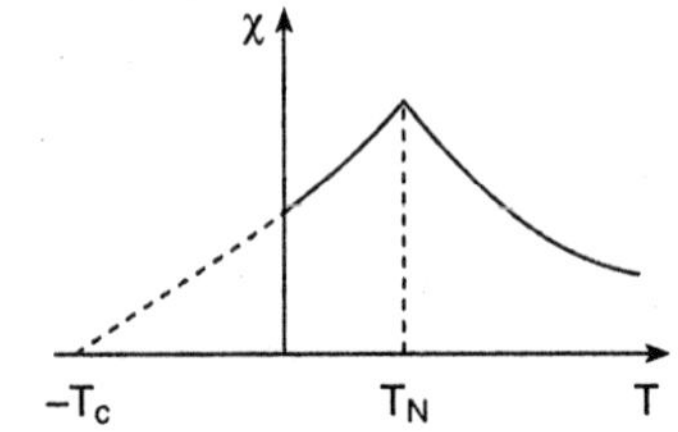

Fig. 10.26 Antiferromgnetic susceptibility: temperature variation

Qualitative explanations—Qualitatively, the characteristics of antiferromagnetics may be explained using *two sublattice model.* Take a lattice system with magnetic atoms arranged in a *bcc*-structure. One may look at all the body-centred ions as having 'spin up' and all the corner atoms as having 'spin down' (Fig. 10.27). This is a simple magnetic structure with two interpenetrating simple cubic sublattices and the spin moments of atoms in one sublattice array are antiparallel with respect to the spin moments of atoms in the other sublattice (Fig. 10.28) array.

At 0 K, the spin moments of both the arrays are ordered (Fig. 10.28) and are mutually compensated with *net magnetization* zero. With increasing temperature,

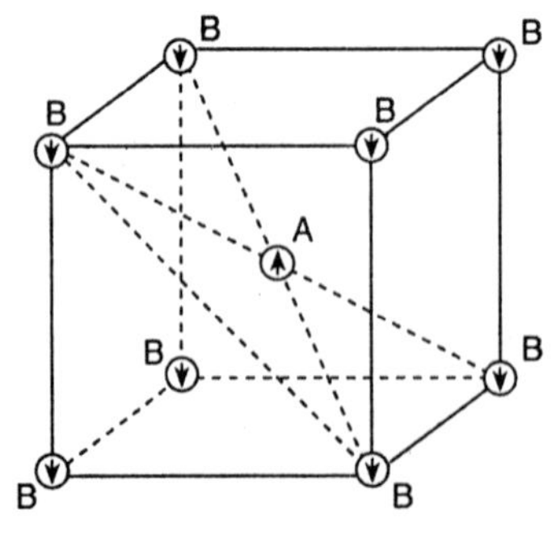

Fig. 10.27 Representation of two sublattices A and B, showing antiparallel spins

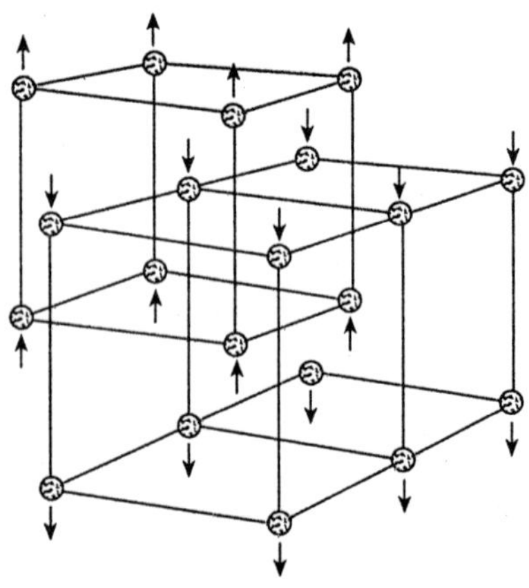

Fig. 10.28 Two interpenetrating simple cubic sublattices with antiparallel spins

the antiparallel spin arrangement starts getting disturbed and the magnetization rises so as to become ***maximum at the Neel point above which the material becomes paramagnetic.***

The simplest antiferromagnetic mateials are crystalline ionic compounds of transition elements, where the metallic ions are so arranged that they can be divided into two equivalent and interpenetrating sublattices A and B where B's are A's nearest neighbours and conversely. Introducing the internal field H_A and H_B for A ions and B ions respectively and since magnetization M_A of A ions is opposite to the internal field of B ions,

$$\begin{aligned} H_A &= H - \beta M_B \\ H_B &= H - \beta M_A \end{aligned} \tag{10.15.2}$$

β being the internal field constant, a measure of the strength of exchange interaction.

So, proceeding as in ferromagnetics, we have

$$\left.\begin{aligned} M_A &= Ng\mu_B \tanh\left(\frac{\mu_0 g\mu_B}{kT}(H - \beta M_B)\right) \\ M_B &= Ng\mu_B \tanh\left(\frac{\mu_0 g\mu_B}{kT}(H - \beta M_A)\right) \end{aligned}\right\} \tag{10.15.3}$$

Making, as before, the approximation $\tanh(x) \simeq x$ at high temperatures, we get

$$\left.\begin{aligned} M_A &= \frac{N\mu_0(g\mu_B)^2}{kT}(H - \beta M_B) \\ M_B &= \frac{N\mu_0(g\mu_B)^2}{kT}(H - \beta M_A) \end{aligned}\right\} \tag{10.15.4}$$

$\therefore$ Total magnetization, $M = M_A + M_B = \dfrac{N\mu_0(g\mu_B)^2}{kT}(2H - \beta M)$ (10.15.5)

If M and H are assumed parallel, the above equation becomes a scalar equation. So, by solving for M/H, we obtain

$$\chi = \frac{M}{H} = \frac{2C}{T + C\beta} = \frac{2C}{T + \theta} \tag{10.15.6}$$

where $C = N(g\mu_B)^2\mu_0/k$ and $\theta = C\beta$.

Comparing equation (10.15.6) with the one of ferromagnetic case, we see that instead of $T - \theta$, we have here $T + \theta$ and Curie constant is *twice* that in ferromagnetics.

The Neel temperature T_N can be determined using equation (10.15.3) since it corresponds to the temperature for which M_A or M_B has a finite value in absence of the external magnetic field.

$\therefore$ From equation (10.15.3), for $H = 0$, we have

$$\left.\begin{aligned} M_A + \frac{C\beta}{T}M_B &= 0 \\ \text{and} \quad \frac{C\beta}{T}M_A + M_B &= 0 \end{aligned}\right\} \tag{10.15.7}$$

For $T > T_N$, these equations have the trivial solution $M_A = M_B = 0$, i.e., there is no spontaneous magnetization of the sublattices above the Neel temperature. If however spontaneous magnetization of sublattices is supposedly set at $T = T_N$, then (10.15.7) must have non-trivial solution for M_A and M_B at that temperature. This implies that the determinant of the coefficients of M_A and M_B vanishes.

$$\therefore \quad \frac{C\beta}{T_N} = 1 \Rightarrow T_N = C\beta = \theta \tag{10.15.8}$$

The *Neel temperature is the same as the paramagnetic Curie temperature.* Experimentally, however, $T_N < \theta$, generally. This is due to the fact that *two sublattices model* is too simplified and the real situation is far more complicated.

Antiferromagnetic susceptibility—The susceptibility depends on the direction of the external magnetic field. We shall consider here *two* directions of interest: (i) field direction *normal* to spin axis and (ii) field direction *parallel* to spin axis.

Case 1. H is $\perp$ to both M_A and M_B (spin axis). Here both M_A and M_B tend to align along the field direction resulting in their rotation through a very small angle 2ϕ with each other (Fig. 10.29a). Then, the molecular field acting on the sublattice B, parallel to H, will be $-\beta M_A \cdot 2\phi$. At equilibrium, this must be equal and opposite to the field H so that

$$2\beta M_A \phi = H, \quad \beta = \text{positive Weiss factor.} \tag{10.15.9}$$

Since $|M_A| = |M_B|$, the total magnetization along the field is given by

$$M = (M_A + M_B)\phi = 2M_A\phi \tag{10.15.10}$$

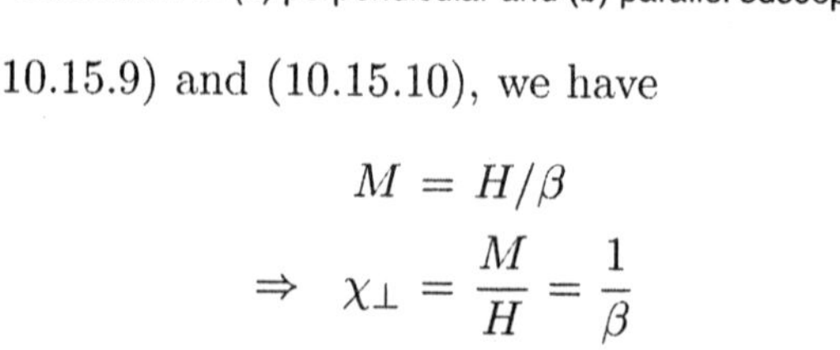

Fig. 10.29 Calculation of (a) perpendicular and (b) parallel susceptibilities at 0 K

Using equations (10.15.9) and (10.15.10), we have

$$M = H/\beta$$

$$\Rightarrow \quad \chi_\perp = \frac{M}{H} = \frac{1}{\beta} \tag{10.15.11}$$

which is *temperature-independent*. It can also be shown that $\chi_\perp$ is equal to the susceptibility at the Neel temperature.

Case 2. H is $\parallel$ to M_A or M_B (spin axis). The calculation in this case is rather complicated involving the Brillouin functions. We shall however try to understand it qualitatively.

When the field is parallel to one sublattice magnetization and antiparallel to the other sublattice magnetization (Fig. 10.29b), the applied magnetic field produces no resultant moment on them. So, there is no net magnetization at 0 K so that

$$\chi_{\parallel} = 0$$

Now, as the temperature is gradually raised, the alignment of spin gets slightly upset and the field produces rotation of spins and thereby a small susceptilibity develops. $\chi_{\parallel}$ increases to $\chi(T_N)$ as temperature rises to T_N, the Neel temperature. The χ - T curve of an antiferromagnetic crystal is illustrated in Fig. 10.30. Below T_N, χ of a material in polycrystalline form is obtained by taking the average of χ's over all the directions. At the *absolute zero*, this is given by

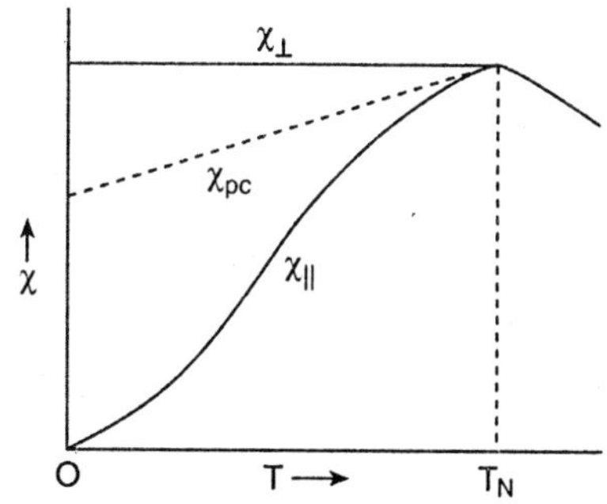

Fig. 10.30 Magnetic susceptibility of antiferromagnet as a function of temperature. T_N is Neel temperature

$$\chi_{pc} = \tfrac{1}{3}\chi_{\parallel}(0) + \tfrac{2}{3}\chi_{\perp}(0) = \tfrac{2}{3}\chi_{\perp}(T_N) \quad (10.15.12)$$

At the Neel temperature T_N, the χ_{pc} is given by

$$\chi_{pc} = \tfrac{2}{3}\chi_{\perp}(0) = \tfrac{2}{3}\chi_{\perp}(T_N) = \tfrac{2}{3}\chi(T_N) \quad (10.15.13)$$

10.16 Ferrimagnetism and ferrites

We have seen that in antiferromagnetic materials, the opposing antiparallel magnetic moments of sets of atoms in each of the two sublattices are exactly equal and totally compensated and hence the net magnetization is zero. However, there are materials where such compensation of antiparallel magnetic moments is not complete for different reasons such as the difference in number or the nature of atoms constituting the two sub-lattices. This *special case of uncompensated form of antiferromagnetism is called **ferrimagnetism***, as proposed by Neel and the corresponding materials are known as ***ferrimagnetic materials*** or ***ferrites***. They show spontaneous magnetization below the ferrimagnetic ***Neel temperature*** T_{FN}, above which the net magnetic moment disappears as thermal energy randomizes the individual magnetic moments and the material becomes *paramagnetic*.

The external behaviour of a ferrimagnet is *similar* to that of metallic ferromagnet. For instance, both show hysteresis, spontaneous magnetization and related phenomena. But the coercive force and saturation magnetization are smaller than those of ferromagnets. However, their internal structure and spontaneous magnetization, particularly the temperature-dependence of the latter are *quite different*. The magnetization of a ferrimagnet does not necessarily decrease monotonically with rise in temperature, but may pass through zero even before reaching the Neel point.

Ferrimagnetic susceptibility—We shell now find an expression for *ferrimagnetic susceptibility*.

Proceeding as in the case of antiferromagnet, we have

$$M_A T = C_A(H - \beta M_B) \tag{10.16.1}$$

$$M_B T = C_B(H - \beta M_A) \tag{10.16.2}$$

where H is the applied magnetic field, β the Weiss factor, C's the Curie constants and M's the magnetizations of the two sub-lattices A and B.

For $H = 0$, $\quad M_A T = -\beta C_A M_B;\ M_B T = -\beta C_B M_A$

$$\therefore\ T = T_F = \beta\sqrt{C_A C_B} \tag{10.16.3}$$

where T_F is called the *ferrimagnetic Curie temperature.*

From (10.16.1), $$M_A T + \beta C_A M_B = C_A H \tag{10.16.4}$$

From (10.16.2), $$M_B T + \beta C_B M_A = C_B H \tag{10.16.5}$$

Multiplying (10.16.4) by T and (10.16.5) by βC_A, we obtain

$$M_A T^2 + \beta T C_A M = C_A T H \tag{10.16.6}$$

$$\beta T C_A M_B + \beta^2 C_A C_B M_A = \beta C_A C_B H \tag{10.16.7}$$

Subtracting (10.16.7) from (10.16.6), we have

$$(T^2 - \beta^2 C_A C_B) M_A = C_A T(1 - \beta C_B) H$$

$$\therefore\ M_A = \frac{C_A T(1 - \beta C_B)H}{T^2 - T_F^2}, \quad \text{using (10.16.3)} \tag{10.16.8}$$

Similarly, we shall get $$M_B = \frac{C_B T(1 - \beta C_A)H}{T^2 - T_F^2} \tag{10.16.9}$$

$$\therefore\ M_A + M_B = \frac{[T(C_A + C_B) - 2\beta C_A C_B]H}{T^2 - T_F^2}$$

$\therefore$ Susceptibility, $$\chi = \frac{M_A + M_B}{H} = \frac{\boldsymbol{T(C_A + C_B) - 2\beta C_A C_B}}{\boldsymbol{T^2 - T_F^2}}$$

For $T \to T_F$, $\chi \to \infty$. This implies that *for $T < T_{FN}$, there is spontaneous magnetization.*

10.16.1 Ferrites and their applications

We first give some examples of *ferrites.* The iron ore magnetite (Fe_3O_4) is the oldest ferrite known to us bearing the formula $Fe^{2+}Fe_2^{3+}O_4^{2-}$. The general chemical formula for ferrites is $X^{2+}Fe_2^{3+}O_4^{2-}$ where X stands for a divalent metal like Fe, Mn, Ni, Co, Mg, Cd or Zn. A complicated class of ferrites, called ***mixed ferrites***, is formed when X^{2+} is replaced by mixture of ions. The most important example of *mixed ferrites* are MnZn-ferrites (*Ferroxcube* IV).

Applications—Ferrites are important for their technical applications.

1. Ferrites are insulators and possess *dc resistivity* about 10 order higher than that of ferromagnetic iron. So, the power losses due to *eddy currents* in ferrites are much less and hence they are suitable for *hf*-applications such as in *hf-transforms* and *hf-generators.*

2. Ferrites are also used as *core computer memories* due to their *low coercive force and short switching time.*

3. Ferrites are used as *microwave devices.* They have high microwave dielectric constant of the order of 10–12 (low dielectric loss) and also high magnetic permeability (500–1000) for mixed ferrites. Mixed ferrites have great technical importance.

4. Gyromagnetic properties of ferrites are utilized in developing *circulators* and *isolator* in radar communication.

● Ferrites are hard, brittle and difficult to machine.

10.17 A comparison of basic magnetic materials

It would be worthwhile to compare the basic differences of diamagnetic, paramagnetic and ferromagnetic materials. We do this in the following tabular form.

Diamagnetism	Paramagnetism	Ferromagnetism
1. Diamagnetics are usually repelled by magnets.	**1.** Paramagnetics are weakly attracted by magnets.	**1.** Ferromagnetics are strongly attracted by magnets.
2. No magnetic moment in absence of the applied magnetic field.	**2.** A net magnetic moment in absence of the applied magnetic field.	**2.** A high magnetic moment even in absence of the magnetic field.
3. Induced magnetic moment is directed opposite to that of applied field.	**3.** Induced magnetic moment is in the direction of the applied field.	**3.** Magnetic moment aligns itself along the direction of the field.
4. Sysceptibility has a small negative value and is independent of temperature.	**4.** Susceptibility is positive but small and varies inversely as temperature.	**4.** Susceptibility has a very high positive value and decreases with temperature.
5. Relative permeability is slightly less than unity.	**5.** Relative permeability is slightly greater than 1.	**5.** Relative permeability has a very large value.
6. Temperature-changes do not influence the magnetic property.	**6.** At very high temperatures, a paramagnet behaves as a diamagnet.	**6.** Above the Curie point, ferromagnets transit to paramagnets.

(*Cont. to next page*)

7. Diamagnetics do not show hysteresis and hence there no is remanent magnetization or any coercive field.	**7.** Paramagnetics also show no hysteresis, no remanent magnetization and no coercive field.	**7.** Hysteresis is characteristic of ferromagnetics and show remanent magnetization and coercive field.
8. Since the electronic shells are completely filled, the atoms or the molecules of diamagnets have no permanent magnetic dipole moment.	**8.** Some partially filled electronic shells give rise to permanent dipole moments for atoms but there is no spontaneous magnetization.	**8.** Some partially filled electronic shells give rise to permanent dipole moments for atoms. Strong interacton of neighbours causes $\uparrow\uparrow$ orientation of electron spins and spontaneous magnetization.

10.18 Illustrated Examples

► **Example 1.** *Find the magnetizing force and the relative permeability of a magnetic material having a magnetization* 3380 A/m *and flux density* $0.0048\,\text{Wb/m}^2$.

Solution. We have the relation $B = \mu_0(H + M) \Rightarrow H = (B/\mu_0) - M$. It is given that $B = 0.0048\,\text{Wb/m}^2$, $M = 3380\,\text{A/m}$, and $\mu_0 = 4\pi \times 10^{-7}$ H/m.

$$\therefore\ \boldsymbol{H} = \frac{0.0048}{4\pi \times 10^{-7}} - 3380 = 3820 - 3380 = \mathbf{440\ A/m}$$

We have, again, $\quad M = H(\mu_r - 1) \Rightarrow \mu_r = (M/H) + 1$

$$\therefore\ \boldsymbol{\mu_r} = \frac{3380}{440} + 1 = 7.68 + 1 = \mathbf{8.68}$$

► **Example 2.** *Estimate the order of diamagnetic susceptibility of* Cu *from the following data: radius of* Cu*-atom* = 1 A, *lattice parameter* = 3.608 A. *Assume that only one electron per atom makes the contribution.*

Solution. Given, $r_0 = 1\,\text{A} = 10^{-10}\,\text{m}$, $a = 3.608 \times 10^{-10}\,\text{m}$, $\chi_{\text{dia}} = ?$

Now copper is a face-centred cubic crystal. It has therefore 4 atoms per unit cell.

$$\therefore \text{ No. of electrons per unit volume, } N = \frac{n}{a^3} = \frac{4}{(3.608 \times 10^{-10})^3} = 8.5 \times 10^{28}/\text{m}^3.$$

$$\therefore\ \boldsymbol{\chi_{\text{dia}}} = -\frac{\mu_0 e^2 N Z \langle r_0^2 \rangle}{6m} = -\frac{4\pi \times 10^{-7} \times (1.6 \times 10^{-19})^2 \times 8.5 \times 10^{28} \times 10^{-20}}{6 \times 9.1 \times 10^{-31}}$$

$$= \mathbf{-5 \times 10^{-6}}$$

► **Example 3.** *A paramagnetic salt has* $10^{28}\,\text{ions/m}^3$ *with magnetic moment equal to* 1 Bohr magneton. *Calculate the paramagnetic susceptibility and magnetization in an external field of* 10^6 A/m *at a temperature of* 300 K.

Solution. The relation between the susceptibility χ and the other involved quantities, with usual meanings of the symbols, is given by the relation

$$\chi = \frac{\mu_0 \mu^2 N}{3kT}$$

It is given: $N = 10^{28}\,\text{ions/m}^3$, $\mu = \mu_B = 9.27 \times 10^{-24}\,\text{A/m}^2$, $H = 10^6\,\text{A/m}$ and $T = 300\,\text{K}$.

$$\therefore\ \chi = \frac{4\pi \times 10^{-7} \times 10^{28} \times (9.27 \times 10^{-24})^2}{3 \times 1.38 \times 10^{-23} \times 300}$$

$$= \mathbf{0.87 \times 10^{-4}}$$

$$\therefore\ \text{Magnetization, } \boldsymbol{M} = \chi H = 0.87 \times 10^{-4} \times 10^6 = \mathbf{87\,A/m}$$

▶ **Example 4.** *A paramagnetic material has* 10^{28} atoms/m^3. *Its susceptibility at* 350 K *is* 2.8×10^{-4}. *Calculate the susceptibility at* 300 K. **(JNTU)**

Solution. Paramagnetic susceptibility, $\chi = C/T$ where C = Curie constant, T = temperature.

∴ If χ_1 and χ_2 be the susceptibilities at temperature T_1 and T_2 respectively, then

$$\chi_1/\chi_2 = T_2/T_1 \Rightarrow \chi_2 = \chi_1 T_1/T_2.$$

Here $\chi_1 = 2.8 \times 10^{-4}$, $T_1 = 350\,\text{K}$ and $T_2 = 300\,\text{K}$, $\chi_2 = ?$

$$\therefore\ \chi_2 = \frac{(2.8 \times 10^{-4}) \times 350}{300} = \mathbf{3.267 \times 10^{-4}}$$

▶ **Example 5.** *Assuming the magnetic moment* μ_m *to be* 5 *times the Bohr magneton* μ_B, *compute approximately how large should be the magnetic induction for the orientation energy to be comparable to thermal energy at room temperature* (27°C).

Solution. By the problem, $\mu_m = 5\mu_B$, $T = 27 + 273 = 300\,\text{K}$.

If B be the magnetic induction, the orientation energy is $\mu_B B$. The thermal energy $= kT$ where k is Boltzmann constant.

By the condition, $\mu_m B \simeq kT \Rightarrow B = kT/\mu_m$.

$$\therefore\ \boldsymbol{B} = \frac{kT}{\mu_m} = \frac{kT}{5\mu_B} = \frac{1.38 \times 10^{-23} \times 300}{5 \times 9.27 \times 10^{-24}}$$

$$= 89.32\,\text{Wb/m}^2 \simeq \mathbf{90\,Wb/m^2}$$

▶ **Example 6.** *The Curie temperature of iron is* 1043 K *and each iron atom has a magnetic moment* $2\mu_B$, *where* μ_B = *Bohr magneton. If the saturation magnetization of iron is* 1.75×10^6 A/m, *calculate the Weiss constant and the Curie constant.*

Solution. Weiss constant, $\beta = \dfrac{3k\theta}{\mu_0 N \mu^2} = \dfrac{3k\theta}{\mu_0 M_s \mu}$ $\quad (\because\ M_s = N\mu)$

Here $\theta = 1043\,\text{K}$, $\mu_0 = 4\pi \times 10^{-7}\,\text{H/m}$, $k = 1.38 \times 10^{-23}\,\text{J/K}$, $\mu = 2\mu_B = 2 \times 9.27 \times 10^{-24}\,\text{Am}^2$, $M_s = 1.75 \times 10^6\,\text{A/m}$.

Substituting the given and the known values of the quantities involved, we have

$$\beta = \frac{3 \times 1.38 \times 10^{-23} \times 1043}{4\pi \times 10^{-7} \times 1.75 \times 10^6 \times 2 \times 9.27 \times 10^{-24}} = \mathbf{1059}$$

Curie constant, $C = \dfrac{\mu_0 N \mu^2}{3k} = \dfrac{\mu_0 M_s \mu}{3k}$. Substituting the values, we obtain

$$C = \frac{4\pi \times 10^{-7} \times 1.75 \times 10^6 \times 2 \times 9.27 \times 10^{-24}}{3 \times 1.38 \times 10^{-23}}$$
$$= \mathbf{9848\,K}$$

► **Example 7.** *Given that the diamagnetic susceptibility of* He *is* -1.9×10^{-6} per mole, *estimate the radius of the* He-*atom.* **(Burdwan Hons.)**

Solution. Given for He, $\chi_{\text{dia}}^{\text{molar}} = -1.9 \times 10^{-6}$.

Now, we have, $\chi_{\text{dia}}^{\text{molar}} = \dfrac{Ze^2 N_A \mu_0}{6m} \langle r_0^2 \rangle$, with usual symbols.

For He, we have, $Z = 2$. We have, therefore,

$$\langle r_0^2 \rangle = \frac{\chi \times 6m}{Ze^2 N_A \mu_0} = -\frac{-1.9 \times 10^{-6} \times 6 \times 9.1 \times 10^{-31}}{2 \times (1.6 \times 10^{-19})^2 \times 6.03 \times 10^{23} \times 4\pi \times 10^{-7}}$$
$$= 0.268 \times 10^{-21}\,\text{m}^2$$
$$\therefore \langle \mathbf{r_0} \rangle = 1.64 \times 10^{-9}\,\text{cm} = \mathbf{0.0164\,nm}$$

► **Example 8.** *The wavefunction of the hydrogen atom in its ground state is given by*

$$\psi(r) = \frac{1}{(\pi a_0^3)^{1/2}} e^{-r/a_0}$$

Show that $\langle r^2 \rangle = 3a_0^2$ *and calculate the molar diamagnetic susceptibility of atomic hydrogen at STP where* a_0 = *atomic radius* = 0.46 A. **(Burdwan Hons.)**

Solution. The ground state wavefunction of H-atom is

$$\psi(r_0) = \frac{1}{(\pi a_0^3)^{1/2}} e^{-r_0/a_0}$$

The mean distance of electronic charge distribution from the nucleus is given by

$$\langle r_0^2 \rangle = \int \psi^* r_0^2 \psi d\tau = \int_0^\infty \psi^* r_0^2 \psi \cdot 4\pi r_0^2 dr$$
$$= 4\pi \int_0^\infty \psi^* r_0^2 \psi r_0^2 dr = \frac{4\pi}{\pi a_0^3} \int_0^\infty r_0^4 e^{-2r_0/a_0} dr$$

$$= \frac{4}{a_0^3}\left(\frac{a_0^4}{16}\right)\left(-\frac{a_0}{2}\right)\int_0^\infty t^4 e^{-t}dt, \text{ putting } r_0 = -\tfrac{1}{2}a_0 t,$$

$$= \frac{4}{a_0^3}\frac{a_0^4}{16}\left(-\frac{a_0}{2}\right)\Gamma(4) = \frac{4}{a_0^3}\frac{a_0^4}{16}\left(-\frac{a_0}{2}\right) \times 24$$

$$\therefore\ \langle \boldsymbol{r_0^2} \rangle = \boldsymbol{3a_0^2} \text{ (in magnitude)}$$

$$\therefore\ \chi_{\text{dia}} = -\frac{\mu_0 e^2}{6m}NZ\langle r_0^2\rangle = -\frac{\mu_0 e^2}{2m}Na_0^2 \quad (\because\ Z = 1 \text{ for H-atom})$$

Now, $N = (6.023 \times 10^{26})/(2.24 \times 10^{-2}) = 2.69 \times 10^{28}\,\text{m}^{-3}$, $a_0 = 0.46 \times 10^{-10}\,\text{m}$, $e = 1.6 \times 10^{-19}\text{C}$, $m = 9.1 \times 10^{-31}\,\text{kg}$.

$$\therefore\ \boldsymbol{\chi_{\text{dia}}} = -\frac{2.69 \times 10^{28} \times (4\pi \times 10^{-7}) \times (1.6 \times 10^{-19})^2 \times (0.46 \times 10^{-10})^2}{2 \times 9.1 \times 10^{-31}}$$

$$= \boldsymbol{-1.01 \times 10^{-6}} \textbf{ (SI units)}$$

▶ **Example 9.** *A ferromagnetic material with $J = 3/2$ and $g = 2$ has a Curie temperature of 125 K. Calculate the intrinsic flux density near 0 K. Also, calculate the ratio of magnetization at 300 K in presence of an external field of 1 mT to the spontaneous magnetization at 0 K.* **(Burdwan Hons.)**

Solution. The intrinsic flux density near 0 K is:

$$\boldsymbol{B_{\text{ex}}} = \frac{3k\theta}{g\mu_B(J+1)} = \frac{3 \times 1.38 \times 10^{-23} \times 125}{2 \times 9.27 \times 10^{-24} \times 5/2}, \text{ using the given data.}$$

$$= \mathbf{111.6\,T}$$

Also, the magnetization for $T > \theta$ is given by

$$M = \frac{Ng^2\mu_B^2 J(J+1)B}{3k(T-\theta)}; \quad M_s(0) = NgJ\mu_B$$

$$\therefore\ \frac{\boldsymbol{M}}{\boldsymbol{M_s(0)}} = \frac{g\mu_B B(J+1)}{3k(T-\theta)} = \frac{2 \times 9.27 \times 10^{-24} \times 10^{-3} \times (5/2)}{3 \times 1.38 \times 10^{-23} \times (300-125)} = \mathbf{6.4 \times 10^{-6}}$$

▶ **Example 10.** *The energy of a ferromagnet as a function of magnetization M is*

$$F(M) = F_0 + 2(T - T_c)M^2 + M^4, \quad F_0 > 0$$

Determine the number of minima in the function $F(M)$ for $T > T_c$. **(GATE)**

Solution. We have $F(M) = F_0 + 2(T - T_c)M^2 + M^4,\ F_0 > 0$

For *minimum* of $F(M)$, $\quad \dfrac{d}{dM}F(M) = 0$

Now, $$\frac{d}{dM}F(M) = 0 + 4(T - T_c)M + 4M^3 = 0, \quad \text{at } minimum.$$

$$\Rightarrow \quad M^2 + T - T_c = 0 \quad (\because \; T > T_c)$$

So, the real root is $M = 0$. Again, we have

$$\left.\frac{d^2}{dM^2}F(M)\right|_{M=0} = 4(T - T_c) > 0 \quad (\because \; T > T_c)$$

and is thus *positive*, implying that at $M = 0$, $F(M)$ is *minimum.*

$\therefore$ ***Number of minima*** in function $F(M)$ is **1**.

► **Example 11.** *Calculate the Pauli's paramagnetic susceptibility from the following data : a paramagnetic substance contains* 6.02×10^{28} atoms/m^3 *and its Fermi energy is* 11.63 eV.

Solution. Here, $N = 6.02 \times 10^{28}$ atoms/m^3, $E_F = 11.63$ eV $= 11.63 \times 1.6 \times 10^{-19}$ J. Also, $\mu_B = 9.27 \times 10^{-24}$ J/T. To find χ_P.

Pauli's paramagnetic susceptibility

$$\chi_P = \frac{3\mu_0 N \mu_B^2}{2E_F} = \frac{3 \times (4\pi \times 10^{-7}) \times (6.02 \times 10^{28}) \times (9.27 \times 10^{-24})^2}{2 \times 11.63 \times 1.6 \times 10^{-19}}$$

$$= \mathbf{5.24 \times 10^{-5}}, \text{ on simplification.}$$

► **Example 12.** *A system of electron spins, placed in a magnetic field of* 5 Tesla *at temperature T, shows that the number of spins parallel to the magnetic field is twice the number of antiparallel spins. Calculate the value of the temperature.*

Solution. Here, $B = 5$ T, temperature $= T$ K, no. of the parallel spins $= N_p = 2N_{ap}$, where N_{ap} = no. of antiparallel spins.

Using MB-statistics, $$\frac{N_p}{N} = \frac{e^{\mu_B B/kT}}{e^{\mu_B B/kT} + e^{-\mu_B B/kT}}; \quad \frac{N_{ap}}{N} = \frac{e^{-\mu_B B/kT}}{e^{\mu_B B/kT} + e^{-\mu_B B/kT}}$$

where μ_B = Bohr magneton, k = Boltzmann constant.

$$\therefore \quad \frac{N_p}{N_{ap}} = 2 = \frac{e^{\mu_B B/kT}}{e^{-\mu_B B/kT}} = e^{2\mu_B B/kT}$$

$$\therefore \quad \ln 2 = 2\mu_B B/kT$$

$$\therefore \quad \boldsymbol{T} = \frac{2\mu_B B}{k \ln 2} = \frac{2 \times 9.27 \times 10^{-24} \times 5}{1.38 \times 10^{-23} \times \ln 2} = \mathbf{9.69 \ K}$$

► **Example 13.** *An atom with* $L = 2$ *and* 0 *spin angular momentum is placed in a uniform magnetic field of induction* 2 Wb/m^2. *Determine the rate of procession in* Hz *of the resultant magnetic moment.*

Solution. Rate of precession = Larmor frequency, given by

$$\omega_L = g\,\frac{eB}{2m},$$

where g = Lande g-factor $= 1 + \dfrac{J(J+1) + S(S+1) - L(L+1)}{2J(J+1)}$

Here, $L = 2$, $S = 0 \Rightarrow J = L + S = 2$

$$\therefore\ g = 1 + \frac{2 \times 3 + 0 - 2 \times 3}{2 \times 2 \times 3} = 1$$

$$\therefore\ \omega_L = \frac{1 \times (1.6 \times 10^{-19}) \times 2}{2 \times 9.1 \times 10^{-31}} = 1.76 \times 10^{11}\ \text{rad/s}$$

$\therefore$ Linear frequency, $\quad \nu_L = \dfrac{\omega_L}{2\pi} = \dfrac{1.76 \times 10^{11}}{2 \times 3.14} = \mathbf{2.8 \times 10^{10}\ Hz}$

► **Example 14.** *The Curie temperature of iron is* 1043 K. *Estimate the magnitude of Weiss internal field, if for* Fe-*atom,* $\mu = 2\mu_B$.

Solution. The internal field at the saturation state is

$$\boldsymbol{H}_{\text{ex}} = \beta M_s = \frac{3k\theta}{\mu_0\mu} = \frac{3k\theta}{\mu_0 \times 2\mu_B} = \frac{3 \times 1.38 \times 10^{-23} \times 1043}{4\pi \times 10^{-7} \times 2 \times 9.27 \times 10^{-24}}$$

$$= \mathbf{1.86 \times 10^9\ A/m}, \text{ on simplification}$$

► **Example 15.** Dy^{3+} *has outer electronic configuration of* $4f^9 6s^0$. *Calculate the magnetic susceptibility for a salt containing* 1 kg mole *of* Dy^{3+} *ions at* 300 K.

Solution. Number of unpairs electrons in $Dy^{3+} = 5$.

From Hund's rule, we have $\quad S = 5 \times \left(\frac{1}{2}\right) = 5/2$

$$L = 3 + 2 + 1 + 0 - 1 = 5$$

$$J = L + S = (5/2) + 5 = 15/2$$

$$\therefore\ g = 1 + \frac{J(J+1) + S(S+1) - L(L+1)}{2J(J+1)}$$

$$= 1 + \frac{\frac{15}{2} \times \left(\frac{15}{2} + 1\right) + \frac{5}{2}\left(\frac{5}{2} + 1\right) - 5(5+1)}{2 \times \frac{15}{2} \times \left(\frac{15}{2} + 1\right)} = 1.33$$

$\therefore\ \chi = \dfrac{\mu_0 N \mu_B^2}{3kT} g^2 J(J+1)$. For 1 kg mole salt, $N = N_A$ = Avogradro no.

$$\therefore\ \chi = \frac{4\pi \times 10^{-7} \times 6.023 \times 10^{26} \times (9.27 \times 10^{-24})^2 \times (1.33)^2}{3 \times 1.38 \times 10^{23} \times 300} \times \frac{15}{2}\left(\frac{1}{2} + 1\right)$$

$$= \mathbf{5.9 \times 10^{-4}}$$

► **Example 16.** *The saturation magnetization of iron is* 1.75×10^6 A/m. *Assuming that iron has a bcc-structure, with* $a = 2.925$ Å, *determine the average number of Bohr magnetons contributing to the saturation magnetization per atom.*

Solution. In *bcc*-structure no. of atoms per unit cell = 2.

$$\therefore \text{ No. of atoms per unit volume, } N = \frac{2}{a^3} = \frac{2}{(2.925)^3 \times 10^{-30}} = \frac{2 \times 10^{30}}{25}$$

$$= 8 \times 10^{28}\,\text{m}^{-3}$$

Saturation magnetisation, $M_s = N\mu \Rightarrow \mu = M_s/N$

$$\therefore \textit{ Number of Bohr magnetons} = \frac{\mu}{\mu_B} = \frac{M_s}{N\mu_B} = \frac{1.75 \times 10^6}{8 \times 10^{28} \times 9.27 \times 10^{-24}} = \mathbf{2.36}$$

Review and Short Questions

1. Distinguish between diamagnetism, paramagnetism and ferromagnetism.
2. Derive an expression for diamagnetic susceptibility on the basis of classical theory.
3. Deduce, using classical laws, the expression for paramagnetic susceptibilty of free electrons. Discuss its shortcomings.
4. Deduce an expression for molar diamagnetic susceptibility of atoms. Does the diamagnetic susceptibility vary with temperature? Give reasons for your answer. **(Burdwan Hons.)**
5. Give an account of the Weiss theory of ferromagnetism. Explain the hysteresis effect and the Curie point on its basis. **(Burdwan Hons.)**
6. Explain clearly the basic difference between paramagnetism and ferromagnetism. Give two examples of each of paramagnetic and ferromagnetic materials.
7. Draw a typical M vs. H curve for a ferromagnetic substance and explain the various stage of magnetization on the basis of the domain theory.
8. Discuss the physical origin of Weiss molecular field.
9. Distinguish between ferromagnetism and antiferromagnetism. Explain theoretically the observed variation of the susceptibility of antiferromagnets with temperature.
10. Distinguish between ferro-, ferri- and anti-ferromagnetism. On the basis of two sublattice model, give the χ value of an antiferromagnetic material, above and below the Neel temperature.
11. How does Heisenberg's exchange interaction explain the phenomenon of ferromagnetism? Relate the exchange integral of Heisenberg to the Weiss ferromagnetic Curie temperature.

12. What is ferrimagnetism? How does the molecular field theory account for the susceptibility of ferrimagnets? Explain ferrites.

13. What are ferromagnetic Curie temperature and Neel temperature? Point out the essential difference between the two.

14. Give the quantum theory of paramagnetism and explain how it removes the shortcomings of Langevin's theory. Under what condition does it reduce to the classical expression of Langevin?

15. Name an element that exhibits paramagnetism. Explain why the inert gases do not show paramagnetism. **(Calcutta Hons.)**

16. Derive Curie's law of paramagnetism from Langevin's theory. **(Calcutta Hons.)**

17. What is Pauli paramagnetism? Name at least two paramagnetic elements. **(Calcutta Hons.)**

18. Show that diamagnetic susceptibility increases with number of atoms per unit volume but is independent of temperature. **(Budwan Hons.)**

19. Discuss briefly the Weiss theory of ferromagnetism. Why is energy loss observed in case of magnetic hysteresis? **(Burdwan Hons.)**

20. Draw the B–H curve for a ferromagnetic material and identify the retentivity and the coercive field on the curve.

    Explain ferromagnetism and antiferromagnetism. **(JNTU)**

21. Explain the phenomenon of antiferromagnetism. What are ferrites? Mention a few uses of ferrites. **(North Bengal Hons.)**

22. The paramagnetic susceptibility of a solid varies with absolute temperature T as

    (a) T, (b) T^2, (c) T^0, (d) $T^{1/2}$.

23. Ferromagnetism in solids originates from

    (a) Orbital motions of atomic electrons, (c) Both (a) and (b),
    (b) Spinning motion of atomic electrons, (d) None of (a) and (b).

24. Magnetic susceptibility χ of a magnetic material is given by

    (a) $\chi = (\mu_r - 1)$, (b) $\chi = M/H$, (c) $\chi = (\mu - \mu_0)/\mu_0$, (d) all of these.

25. For copper the resultant spin in Bohr magneton is

    (a) 0, (b) 1, (c) 2, (d) 3.

26. At Curie temperature, the spontanous magnetization for ferromagnetics is

    (a) infinity, (b) 1, (c) zero, (d) none of these.

27. The temperature (T)-dependence of magnetic susceptibility χ of a ferromagnetic substance with Curie-temperature T_c is given by

(a) $C/(T - T_c)$ for $T < T_c$, (b) $C/(T - T_c)$ for $T > T_c$,
(c) $C/(T + T_c)$ for $T > T_c$, (d) $C/(T - T_c)$ for all temperatures.

(GATE)

28. The plot of inverse magnetic susceptibility $1/\chi$ vs. temperature T of an antiferromagnetic material

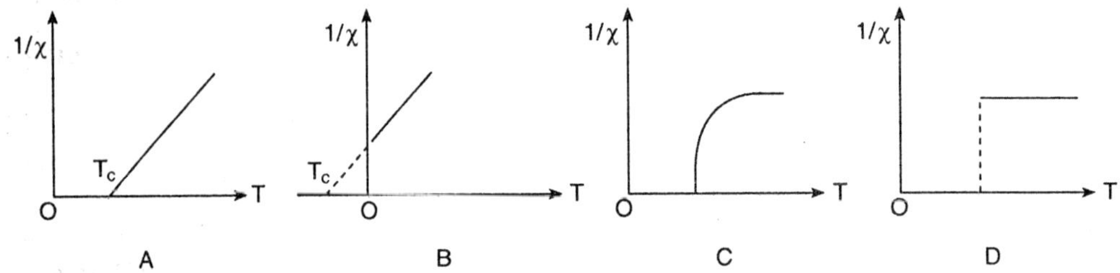

Fig. 10.31

corresponds to (Fig. 10.31)

(a) A, (b) B, (c) C, (d) D. **(GATE)**

29. The susceptibility of a diamagnetic material is

(a) positive and proportional to temperature,
(b) negative and independent of temperature,
(c) negative and inversely proportional to temperature,
(d) positive and inversely proportional to temperature. **(JAM)**

Problems

1. A magnetic material has a magnetization of 3300 A/m and a flux density of 0.0044 Tesla. Compute the values of the magnetic field and the relative permeability of the material. [203 A/m, 17.26]

2. If the magnetization and the flux density of a magnetic material are 3200 A/m and 0.005 Wb/m^2, calculate the relative permeability of the material. [5.1]

3. The magnetic field strength in a specimen is 10^6 Amp/m. Calculate the flux density and the magnetization if the magnetic susceptibility of the specimen is -0.5×10^{-5}. [−5 A/m, 1.257 T]

4. The magnetic susceptibility of Cu is -0.5×10^{-5}. Find the magnetic moment per unit volume when it is placed in an external field at 10^4 A/m. [0.05 A/m]

5. A paramagnetic substance has 10^{28} atoms/m^3. The magnetic moment of each atom is 1.8×10^{-23} Am^2. Calculate the paramagnetic susceptibility at 300 K.
(Burdwan Hons.) [3.28×10^{-4}]

6. The saturation magnetization of iron is 1.75×10^6 A/m. Assuming that iron has a *bcc*-structure with edge length of 2.87 A, calculate the average number of Bohr magnetons contributing to saturation magnetism per atom. [2.23]

7. The paramagnetic susceptibility of Fe_2O_3 at room temperature is 1.4×10^{-3}. If the dipole moment of each atom is of the order of Bohr magneton, find the number of atoms per unit volume of the material. [1.68×10^{29}]

8. The magnetic field strength in a piece of Fe_2O_3 at 27°C is 1.2×10^6 A/m. If the magnetic susceptibility of Fe_2O_3 at 27°C is 1.4×10^{-3}, calculate the flux density and magnetization in Fe_2O_3. **(N. Bengal Hons.)** [1.51 T, 1.68×10^3 A/m]

9. Approximately how large must be the magnetic induction B for the orientation energy to be comparable to the thermal energy at 27°C? Assume $\mu = 5\mu_B$.
[$\simeq 10^2$ W/m^2]

10. Consider a helium atom in its ground state (1s). The mean radius in Langevin formula may be approximated by the Bohr radius, 0.0528 nm. Density of helium is 0.178 kg/m^3. Find the diamagnetic susceptibility of a He-atom.
(Burdwan Hons.) [-0.88×10^{-9}]

11. A system of electron spins is placed in a magnetic field of 2 Wb/m^2 at a temperature T. The number of spins parallel to the magnetic field is twice as large as the number of antiparallel spins. Determine the value of T. [3.8 K]

12. The magnetic field strength in silicon is 1000 A/m. If the magnetic field susceptibility is -0.3×10^{-5}, calculate the magnetization and flux density in silicon. [-3×10^{-3} A/m, 1.257×10^{-3} T]

13. A paramagnetic system of electron spin magnetic dipole moment is placed in an external magnetic field of 10^5 A/m. What would be the average magnetic moment per dipole at a temperature of 300 K?

14. A ferromagnetic material with $J = 3/2$ and $g = 2$ has a transition temperature T_c or $\theta = 200$ K. Estimate the internal field near 0 K. **(Burdwan Hons.)**

15. What would be the dipole moment of a bar of the material referred to Q. 5 above if the dimensions of the bar be 0.1 m in length and 1 sq.cm in cross-section, when placed in a field of 8×10^4 A/m. **(Burdwan Hons.)** [2.62 A–m^2]

16. Iron is a ferromagnet with Curie temperature of 1043 K and an effective moment of $2.2\mu_B$ per atom. Compute the internal field for iron. [2100 T]

17. Gadolinium is ferromagnetic below 16°C with an average magnetic moment per atom 7.1 Bohr magneton. Calculate (i) magnetic moment per g, (ii) saturation magnetization. Atomic weight of Gd = 157.26, its density = $7.8 \times 10^3\,\text{kg/m}^3$.

[1966.85 Am^2, 2.47 Wb/m^2]

18. The paramagnetism in copper sulphate arises mainly from the copper ions with $s = \frac{1}{2}$. Show that the magnetization is given by

$$M = N\mu_B \tanh(\mu_B B/kT)$$

with usual meaning of the symbols used.

19. Cu^{2+} has 9 electrons in $3d$ shell. What magnetic field must be applied to a salt containing Cu^{2+} ions at 1 K so that 90% of the ions are in the ground state?

[3 T]

20. Calculate the change in magnetic moment of a circulating electron in an applied field of 2 T acting perpendicular to the plane of the orbit.

(JNTU) [$3.94 \times 10^{-29}\text{Am}^2$]

21. A paramagnetic system of electronic spin dipole moment is placed in an external field of 10^5 A/m. What is the average magnetic moment per dipole at 300 K? Also, determine the fractional number of spins parallel to the external field.

[$2.6 \times 10^{-27}\,\text{Am}^2, 0.5$]

22. Prove that a superconductor is a perfect diamagnet.

[**Hints.** A superconductor excludes magnetic flux completely from its bulk. So, $B = 0 \Rightarrow \mu_0(H + M) = 0 \Rightarrow H + M = 0 \Rightarrow M = -H$. So, $\chi = M/H = -1$. So, a superconductor is a perfect diamagnet.]

23. Calculate the Lande g-factor and the effective number of Bohr magneton of a paramagnetic gas with $L = 0$ and $S = \frac{1}{2}$. **(Burdwan Hons.)**

24. Sm^{3+} has five electrons in the f-shell ($l = 3$). What are the values of L, S and J? Hence calculate $\mu_{\text{effective}}$. [$L = 5$, $S = 5/2$, $J = 5/2$, $\mu_{\text{eff}} = 2.32\mu_B$]

25. Calculate the magnetic susceptibility at 300 K for a salt containing one g-mole of Cr^{2+} ions, Cr^{2+} has four electrons in the $3d$ shell. [1.25×10^{-7}]

26. A paramagnetic gas with $L = 0$ and $S = \frac{1}{2}$ contains 10^{28} atoms per m^3. Calculate the populations of the levels and the magnetization at 300 K.

[$N_{1/2} = 4.97 \times 10^{27}\,\text{m}^{-3}$, $N_{-1/2} = 5.03 \times 10^{27}\,\text{m}^{-3}$, $M = 519.1\,\text{Am}^{-1}$]

11

Basics of Nanomaterials

11.1 Introduction

Nano is a Greek word meaning *dwarf*. In measurement, it is used as a prefix for 10^{-9} e.g., 1 nanometer (= 1 nm) = 10^{-9} m (the billionth of a meter). Materials having at least one of the dimensions less than 100 nm are called *nanomaterials*. In other words, nanomaterials are materials with a *characteristic length* within about 100 nm. To get an idea of the smallness of nanoscale, we provide below some examples of common small structures with their appropriate sizes.

Material	Size (nm)	Material	Size (nm)
Micromolecule	1–80	Red blood cell	10^4
Colloid silica	5–80	Microparticle	10^3–10^4
Virus	10–100	Bacteria	10^2–10^4
Cooking smoke	10–10^3	Pollen grain	10^4–10^5
Macromolecule	10^2–10^3	Sand	$\sim 10^5$
Clay	10^2–10^3	Human hair (dia)	10^4–10^5

The smallest structure we can see with naked eye is around 10^4 nm across.

While the properties of bulk-structured materials are retained up to the micrometer range, the nanostructured materials develop remarkably new properties. Just for an instance, a nanoscale wire or circuit component does not necessarily obey the venerable Ohm's law. The current through a wire depends on the flow of electrons through it and if the wire is just one atom wide, the electrons can traverse it only one-by-one.

Thus, while at *macroscale*, the properties are *not size-dependent, the coupling of size of materials with their physical and chemical properties is the key to all nanoscience.*

This change in properties is linked to the nature of interactions that make up the material. In bulk-structured materials, the interactions are averaged out of existence. In nanomaterials, however, the interactions are of atomic scale and the nanomaterial is quite different from the bulk-material in colour, melting point, chemical properties etc. Gold is yellow in bulk but depending on the size of the nanoparticles, gold may be red, orange, purple or greenish. This is true for all materials such as iron, lead, plastics etc. since a particle scatters only that wavelength of light that compares with its size.

Nanotechnology is a field of science and technology where devices of size ranging form 1–100 nm are fabricated. Naturally, it is highly *interdisciplinary*. Different fields like applied physics, device physics, material science, electrical and mechanical engineering, molecular chemistry have their respective contributions.

The present chapter is intended to summarise the fundamentals and technical approaches in processing and behaviour of nanomaterials to provide a brief but systematic information in the challenging field of maomaterials. In fact, we shall give here an introduction on the subject for students of physical science and technology, especially of material science, and for those who are just entering the field.

11.2 Origin of nanotechnology

Nanotechnology is undoubtedly a new word and a new area of research, but is not entirely a new field. Nanomaterials are known to be used for centuries. The Chinese used gold nanoparticles as an inorganic dye to infuse red colour to ceramic porcelains, more than thousand years ago. Roman glass artifacts contained metallic nanoparticles that provided beautiful colours to glass. Nanoparticles were also used in mediaval times for decorating the cathedral windows.

Richard Feynmann, the famous physicist, in a lecture in Caltech in 1959 described a process by which the ability to manipulate individual atoms or molecules might be developed using one set of precise tools to build and operate another proportionally smaller set and so on, till the required scale is achieved. He thus heralded the new *concept of* ***nanotechnology***.

The field of nanoscience emerged out of basic studies of stable atomic clusters, only with a specific number of atoms, called *magic number*, and having the corresponding crystalline structure of minimum energy. These clusters show physico-chemical properties depending on their crystallite size, compared to the corresponding bulk one.

The number of free electrons in the outermost orbit of an element decides its chemical behaviour. So, by controlling the number of atoms taking part in forming nanostructure, their chemical behaviour could be varied widely, keeping the basic chemical constutients unchanged. Nanotechnology is thus essentially *molecular manipulation.*

Although the study on nanometerials can be traced back to centuries, the current fever of nanotechnology is, at least partly, driven by the ever shrinking of devices in the semiconductor industry. The continued decrease in device dimension has followed the well-known law of Gordon Moore (***Moore's law***), co-founder of Intel, predicted in 1965. It runs as: *The amount of space required to instal a transistor on a ship shrinks roughly by half every* 18 *months.*

11.3 Nanoparticles

A ***nanopaticle*** is a group of atoms or molecules, 10^6 or less in number, clustered together to a radius $\sim$ 100 nm or less. Nanoparticles show unique properties, different from those of the bulk, when their size is made smaller than certain *critical lengths.* For instance, the metal oxide manoparticles of different sizes in stained glass windows impart different beautiful colours to them, as stated already.

The understanding of the structure and properties of nanoparticles is essential for interpreting the characteristic phenomena shown by them.

11.3.1 Metal nanoclusters

To produce metal manoclusters, the metal surface is melted and the atoms evaporated with a high power laser beam. A burst of He-gas carries the evaporated atoms through a narrow opening to an evacuated chamber where the vapour of atoms expands and cools to form clusters. The atoms combine in different numbers to produce stable clusters of different sizes. The different numbers are termed *magic numbers.*

The reduction in size of a metal particle with bulk properties brings about, below a certain level, a dramatic change in the *density of states* in the *conduction band.* At a size when a particle has a few hundred atoms or less, the *continuous* density of states changes to a set of *discrete* energy levels. If the spacing between the levels

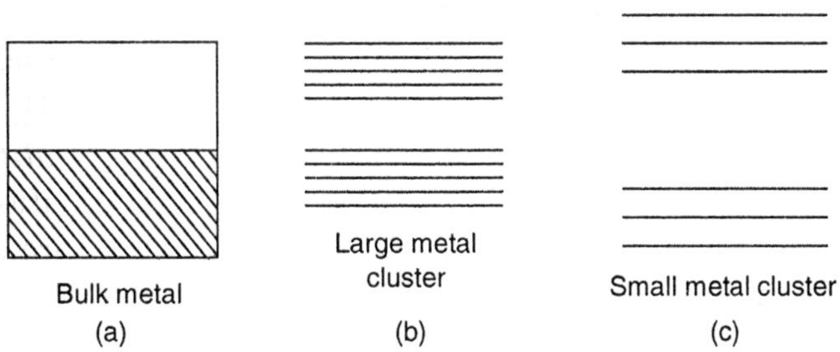

Fig. 11.1 Change in valence band energy levels with size in a metal sample (a) Bulk sample shows no energy gap, (b) Large cluster of 100 atoms shows distinct gap, (c) For a small cluster of only three atoms, the gap grows further

exceeds the thermal energy kT, a gap starts showing up. The bulk sample does not show the gap (Fig. 11.1a). In a large cluster, containing 100 atoms, the gap becomes distinct (Fig. 11.1b) and it grows still further for a small cluster (Fig. 11.1c), of only say 3 atoms.

Except minor changes in the lattice parameters, there is hardly any difference in the crystal structure of a large nanoparticle and that of its bulk sample. If however the diameter of the particles is reduced to less than 5 nm, a different crystal structure may develop. For example, gold nanoparticles of size 3 to 5 nm have an icosahedral structure and not the bulk *fcc*-structure.

Also, the magnetic properties of nanoparticles are remarkable. A net magnetic moment in clusters of non-magnetic atoms is perhaps unique. As the number of atoms in a cluster becomes less than 20, the magnetic moment of these clusters increases significantly.

11.3.2 Semiconductor nanoparticles

Nanoparticles of semiconducting elements like Si, Ge, etc. or compounds like GaAs, CdSe etc. constitute what may be termed ***semiconductor nanoparticles***, which may not behave as semiconductors. These nanoparticles may be synthesized by following the same method as used in metal nanoparticles.

These particles show interesting optical properties. By adjusting the particle size, the colour of the luminescent light of certain semiconductors (e.g., CdS) may be tuned over the whole useful range of visible light. The excitation of excitons modifies the absorption spectra of these particles. A *blue-shift* in the spectrum occurs when the particle-radius exceeds the exciton size.

The nanoparticles of Ge and Si get fragmented on irradiation with a Q-switched Nd-YAG laser. A Si_{12}-particle fissions into two Si_6-particles. The products depend on the cluster-size, and the wavelength and intensity of the laser beam. A cluster with more than 30 atoms tends to fission *explosively*—the *Coulombic explosion.* When the Coulomb repulsion between atoms in a cluster grows to more than the binding energy between the atoms, the cluster explodes and atoms fly apart with great speed.

11.3.3 Other nanoparticles

Nanoparticles may also be had of rare gases and molecules such as carbon dioxide. Examples of such large clusters are $(Kr_{73})^{2+}$, $(Xe_{206})^{4+}$ and $(CO_2)^{4+}_{216}$.

Nanoparticles also show some unusual properties like ***superfluidity***, as observed in He-clusters with 64 and 128 atoms. In the state of BE-condensation, (when all the bosons occupy the lowest energy level), liquid He^4 at a temperature of 2.2 K becomes a *superfluid* and loses completely its viscous property.

11.4 Nanostructure

11.4.1 Carbon clusters : Fullerenes

Fullerenes constitute a family of carbon allotropes. In structure, fullerenes are similar to graphite, being composed of a sheet of linked hexagonal rings. But, they also contain pentagonal rings that prevent the sheet from being planar.

Buckminster fullerene, C-60, named after Buckminster Fuller, was first discovered in 1990 and it looked in shape similar to that of a dome. Other members of the fullerene family discovered after C-60, are C-70, C-76, C-84. We describe below different types of fullerenes. Of these C-60 is most stable.

Buckyballs—When in *spherical* configuration, C-60 is sometimes called ***buckyballs*** because of its shape. It is the smallest fullerene molecule where no two pentagons share an edge and is the commonest fullerene found in natural soot.

C-60, based on the analysis of mass spectrum, has shown to have the structure of a soccer ball made of 20 hexagons and 12 pentagons, with a carbon atom at the vertices of each polygon (Fig. 11.2) and a bond along each polygon edge. A C-60 molecule has a diameter $\sim$ 1 nm with about 0.7 nm internuclear distance. It has two bond lengths. The 6 : 6 ring bonds between two hexagons may be taken as 'double bonds' and are shorter than 6 : 5 bonds between a hexagon and a pentagon. The crystals have *fcc*-structure and retain many of the molecular properties in contrast to the behaviour of other crystalline forms of carbon—diamond and graphite.

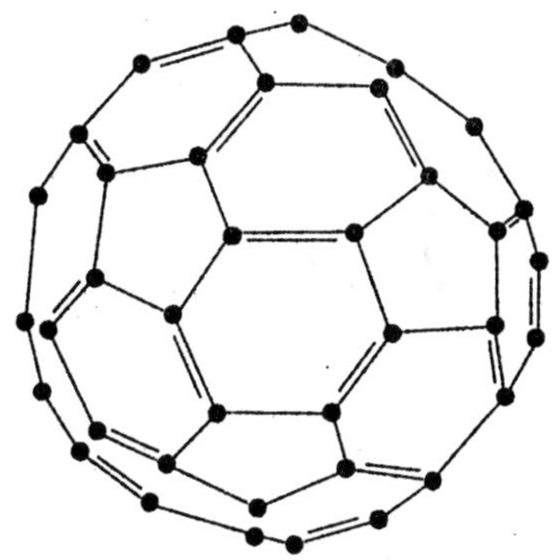

Fig. 11.2 Structure of fullerene C60 moelcule

Solid C-60 (also called '***fullerite***') are highly resistant to photofragmentation and resilient to collisions against surfaces. Fullerites are mechanically light, weak and soft compared to other forms of carbon. Fullerite is a semiconductor with large bandgap and is electrically poor conductor. But, it turns to a *superconductor* if doped with alkali metals.

• A new type of buckyballs with *boron* atoms in place of carbon has been discovered. This ***boron buckyballs***, B-60, is expected to be more stable compared to C-60.

• ***Other common buckyballs*** are fullerenes with 70, 72, 76, 84 and even up to 100 carbon atoms, i.e., C-70, C-72, C-76, C-84, etc.

• Fullerenes have not as yet been applied on a large scale. Their potential lies in their use as building blocks for synthesizing organic chemicals and solid materials. Projected applications include photochromic goggles, xerographic materials, sensors, lubricants, ion rocket propulsion etc.

11.4.2 Carbon nanotubes

Carbon nanotubes or CNTs (Fig. 11.3) are the allotropes of carbon with a *cylindrical* nanostructure of length to dia ratio more than 10^6 and were discovered in 1991 as a minor byeproduct in fullerene synthesis. CNTs have same novel properties useful in many applications in nanotechnology, electronics, optics and many other fields of material science.

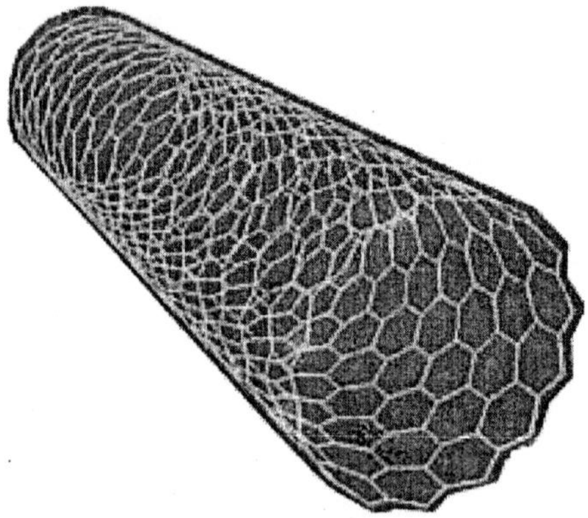

Fig. 11.3 A typical carbon nanotube

Generally, CNTs are formed as *multiple-walled nanotubes* (MWNT). *Single-walled nanotubes* (SWNT) are usually fabricated by mixing traces of catalysts (Co, Ni, Fe) with graphite.

CNTs exhibit extraordinary strength, unique electric properties and are efficient thermal conductors. While buckyballs are spherical, a nanotube is cylindrical in shape with at least one end capped typically with a hemisphere of buckyball structure. They have strength of diamonds and is being considered to weave them with cloths for *bullet proof* clothing.

CNTs may be *synthesized* in sizable quantities by arc discharge, laser ablation, chemical vapour deposition (CVD) etc. Most of the processes occur either in vacuum or process gases.

• New properties can be added to the properties of CNT by grafting chemical functions on their surface. The covalent functionalization in SWNT breaks some carbon-carbon double bonds and leave holes in the structure leading to changes in both the mechanical and the electrical properties.

• Quantum wires (see later) may be made out of metallic carbon nanotubes. Such wires have high electric conductivity, light weight, small dia, low chemical reactivity and high tensile strength.

11.4.3 Carbon nanobuds

Carbon nanobuds, discovered in 2007, form a material that combines the two allotropes of carbon—carbon *nanotubes* and *fullerenes*. The fullerenes are covalently bonded to the outer sidewalls of the underlying nanotube.

Nanobuds therefore show properties of both the carbon nanotubes and the fullerenes. The electrical conductivity and the mechanical properties of nanobuds are similar to those of the corresponding nanotubes . For field emission of electrons, the randomly oriented nanobuds show extremely low work function ($\sim 0.65\ \mathrm{Vm^{-1}}$). Chemical reactivity, good dispersion and variable band gap electronic structure of nanobuds suggest its wide applicability in industry.

11.4.4 Graphenes

Graphene is a sheet of carbon atoms, just one atom thick and is considered as a monolayer of carbon atoms, packed tightly into a two-dimensional honeycomb lattice (Fig. 11.4). It was discovered purely accidently by two Russian scientists Geim and Novoselov who, by stacking cellotape to a piece of graphite, could successfully peel and separate out a one-atom thin flake of the element.

Graphene possesses revolutionary properties. It conducts electricity 100 times faster than silicon and is 200 times stronger than steel and has equally remarkable optical and thermal properties. It is being projected as the real-life miracle material. Graphene shows much higher thermal conductivity compared to CNT, diamond and graphite. At

Fig. 11.4 A Graphene with its honeycomb crystal lattice

room temperature, its thermal conductivity ranges over (4.84 to 5.30) $\times 10^3$ $\mathrm{Wm^{-1}K^{-1}}$. As already stated, graphene is extremely strong and rigid but it is so light as to weigh $\sim 0.77\,\mathrm{mg/m^2}$ only.

For its remarkable electric properties, graphene would have the majority of applications in electronics, particularly to see how it can replace silicon. For its mechanical properties, we may soon see graphene contributing to lighter and stronger cars and aircrafts, ecologically sustainable and cheaper to run. The breaking strength of a graphene sheet is more than 100 times of the strength of a hypothetical steel film of the same thickness. Graphene is also an excellent candidate for conducting electrodes needed in applications, such as touchscreens, liquid crystal displays and the like. *Graphene quantum dots* are ready for their use in integrated circuits applied in various fields.

11.5 Nanostructured crystals

If the symmetry axes are randomly oriented, their spatial positions in a solid do not show any symmetry and the solid is structurally disordered. The nanoparticles, when order themselves in an infinite regular arrangement, give rise to crystals. Examples of natural nanocrystals are those of boron and carbon which have already been discussed. A special class of these crystals—*nanostructred zeolites*—is being treated below.

Nanostructured zeolites—Mineral *faujasite* (Na$_2$, Ca) (Al$_2$ Si$_4$) O$_2$, 8H$_2$O of cubic structure is a typical representative of zeolites. They are *porous* and the pores are distributed in a *regular array.* Small clusters can therefore be accommodated in these pores where the van der Waals force between zeolite and the clusters stabilizes the latter. The material to be accommodated in the pores is added to the melt of zeolite. Zeolites are often called *molecular sieves* as their structure can sift materials. The structure (Fig. 11.5) of zeolite resembles nanoscopic chambers interconnected by nanoscopic pores. It has been possible to grow CNTs of the smallest diameter (~ 0.4 nm) inside the zeolite cavities.

Zeolite structures are being used for directed catalysis to produce petroleum more efficiently and the desired products may be selected out from the distribution of components of petroleum. In the catalysis, the particles of a catalyst are found

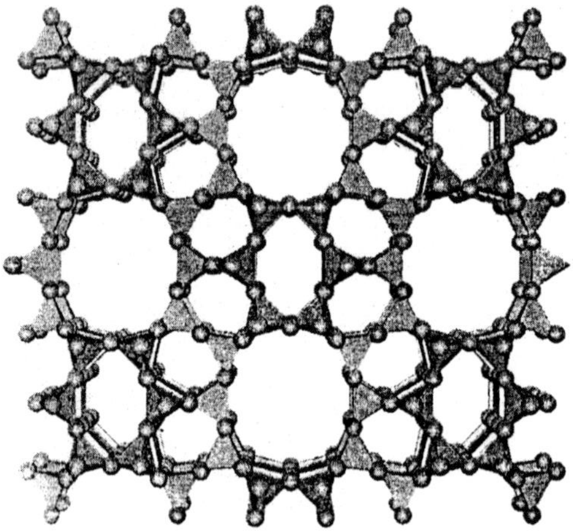

Fig. 11.5 A complex zeolite structure showing holes representing various chambers

within the chambers of zeolites. The locally controlled reactivity by catalyst particles, the physical constraints of the chamber and the pore-size decide the formation of hydrocarbons of specific shape and composition. Some of the important features of this process of catalysis are low cost, less wastage and efficient use of feedstocks.

Zeolites are being used for long as *water softeners* to reduce domestic water hardness by utilizing its property of facilitating exchange of Ca-ions by Na-ions. In fact, one of the first broad-scale, profitable applications of nanotechnology is zeolites.

11.6 Quantum nanostructures

There are three limiting classes of nanostructures, based on if only one or only two or all the three dimensions fall in nanometer scale. If *only one dimension* is reduced to nanometer range, two others remaining large, the material is called a *quantum well.* If however *only two dimensions* are in nanometer range and the other one large, it is known as a *quantum wire.* When *all the three dimensions* are in nanorange, it is termed

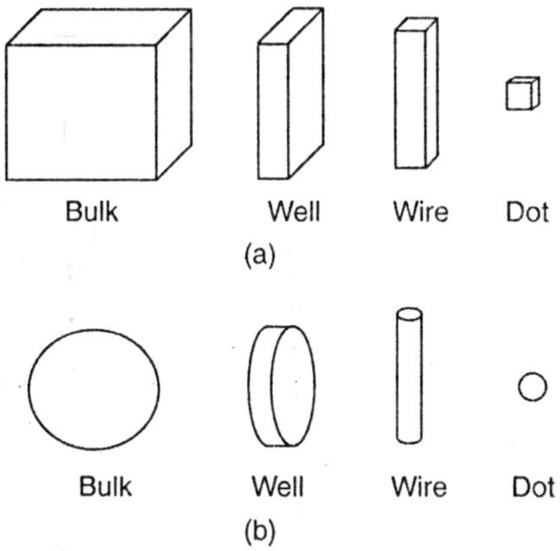

Fig. 11.6 Formation of quantum structure by progressive reduction of size : (a) rectangular and (b) curvilinear

a *quantum dot.* The rectangular and curvilinear quantum nanostructures are shown in Fig. 11.6. The word 'quantum' signifies that such structured materials exhibit new physical properties or phenomena which they do not show while in bulk-structure.

Quantum well—A ***quantum well*** is a potential well that confines particles, originally free to move in three dimensions, to two and forces them to occupy a planar region. When the thickness of the well compares to the de Broglie wavelength of the carriers, the effects of quantum confinement occur and as a result the carriers can have only discrete energy values.

A quantum well confines the motion of electrons or holes in one dimension, allowing free motion in two other dimensions.

In semiconductors, the quantum wells are ***fabricated*** by having a material like GaAs, sandwiched between two layers of a material with a wider bandgap like AlAs. Such structures can be grown by molecular beam epitaxy or chemical vapour deposition, with control of the layer thickness down to monolayers.

Quantum wire—A ***quantum wire*** (Fig. 11.7) is an electrically conducting wire in which the transport properties are influenced by quantum effects. The conduction electrons are confined in the transverse direction of the wire and their transverse energy is quantized in a series of discrete values E_0 (ground state energy), E_1, $E_2, \cdots, E_n$ in ascending order. The quantization invalidates the classical formula $R = \rho l/A$, for computation of resistivity for quantum wires. It is to be computed by using transverse

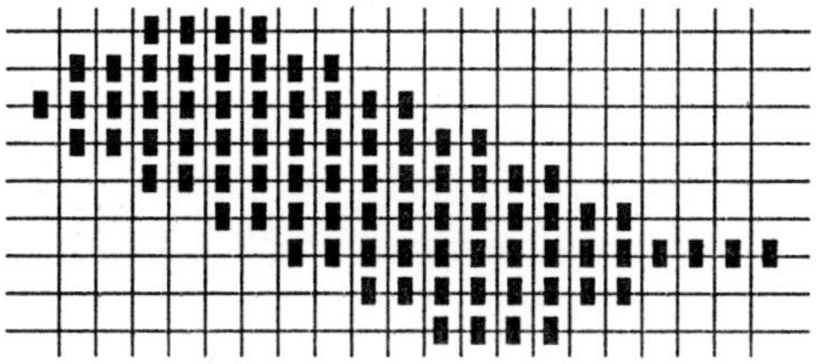

Fig. 11.7 A typical quantum wire

energies of confined electrons, when the resistance of quantum wire is also found to be *quantized.*

A quantum wire confines the motion of electrons (or holes) in two dimensions, allowing free motion in the third. Semiconductor nanowires are mostly cylindrical with diameter $\sim$ 20 to 100 nm.

Quantum dot—A ***quantum dot*** (Fig. 11.8) is made out of semiconductor nanostructure which confines the motion of a small number (1 to 100) of conduction band electrons, valence band holes or excitons in all the three dimensions. Electrostatic potentials or the presence of an interface between different semiconductors or a semiconductor surface or a combination of these may accomplish the confinement.

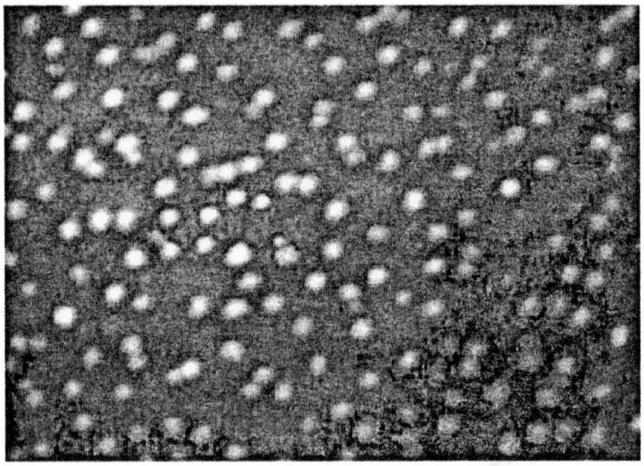

Fig. 11.8 A quantum dot

A quantum dot has a discrete set of quantized energy with wave functions spatially localized within it but extending over many periods of crystal lattice. Quantum dots can be compared to atoms. Both have discrete energy spectrum and bind a small number of electrons. But the dots do not necessarily show spherical symmetry unlike atoms. The confined electrons also do not move in free space but in the host crystal whose band structure plays an important role for the properties of quantum dots. While the typical energy scales are $\sim$ 10 eV in atoms, it is only 1 meV in quantum dots. Nearly spherically symmetric dots or flat dots with near cylindrical symmetry show shell-filling somewhat like Hund's rules in atoms. Quantum dots are thus often called *artificial atoms.* Unlike in atoms, the energy spectrum of a dot can be engineered by controlling its shape and size of confinement potential. Quantum dots of the same material but of different sizes can emit light of different colours.

The quantum wires and quantum dots are ***synthesized*** by attrition of a thin film (quantum well) of the material supported by its substrate, following the procedure of *electron beam lithography.* First, the quantum well is coated with a radiation-sensitive resist (e.g., polymethyl methacrylate). Then, the central surface region of the resist is irradiated by an electron beam through a template. Irradiation transforms the resist material chemically into a form soluble in a developer by which the irradiated part of the resist is removed. Next, an etching mask is planted into the groove formed due to removal of the resist. The residual part of resist is next taken out, the region of quantum well uncovered by etching mask is etched away chemically. The structure left is a *quantum nanostructure.* Depending on the shape and size of the groove, it may be a quantum wire or a quantum dot.

Applications—There is a large number of applications of quantum nanostructures in electronics, quantum electronics and photonics. Quantum wells are in wide use in diode lasers, specifically blue lasers. Quantum well infrared photodetectors are used for infrared imaging. Quantum dot crystals are used in biological research as markers for particular kind of cells. Discrete energy levels of quantum wells are used for fabricating modern infrared detectors. Quantum well and quantum wire-based lasers are the other important applications. Quantum dots in LED ("QD-LED") is a very potential field. Organic dyes using quantum dots is superior to traditional organic dye. Quantum dot technology may be used in future faster computers—*quantum computers.*

11.7 Nanofabrication: Physical techniques

Essentially, there are *two* different approches for the fabrication or synthesis of nanomaterials. These are called *top-down* and *bottom up* techniques. In the former, nanomaterials are constructed from larger entities by gradual reduction without atomic level control. In the bottom-up approach, the materials and devices are built from molecular components. And the components assemble themselves chemically by way of molecular recognition into a specific pattern or structure.

11.7.1 Top-down fabrication

In this method, a piece of the base material is eroded gradually until the desired shape is obtained. The process resembles that of sculping from a block of stone where one starts at the *top* of the block piece and works all the way *down* removing materials not needed. Hence the name *top-down.* Techniques for top-down nanofabrication vary but can be split into mechanical and chemical. Attrition or milling describes best the top-down approach to make nanoparticles.

Nanolithography is the technique of most top-down fabrications. Here, the required material is protected by means of a mask and the exposed material is gradually *etched* away. *Etching* of the base material again can be made either *chemically* using acids or *mechanically* by using X-rays or electron beams. The method of etching depends on the resolution required for the features in the final product. Computer chips are manufactured by this technique.

Classification—The top-down fabrication methods may be generally classified into (a) *bulk* or *film machining,* (b) *surface machining* and (c) *mould machining.*

In ***film machining***, a trench is created or deposited on the subtrate surface of the material, by standard photolithography along with wet/dry film etching of the substrate. This is followed by closure of the structured substrate by another wafer. A trench of 500 nm depth with an accuracy of few per cent and etch-roughness less than 0.5 nm can be readily created by film machining.

In ***surface mechining***, the surface of the deposited surface is etched out to produce the nanostructure by the method of sacrificial layer etching.

Mould machining is commonly done by soft lithography. The advantage of the method is that the nanostructures can be made on any surface—plane or curved. For developing of complex nanoelectronic devices, however, the method is unsuitable.

11.7.2 Bottom-up fabrication

In the bottom-up fabrication, the desired nanostructure is built up by placing atoms/molecules *one at a time*—a process that resembles building of a brick house by placing bricks after bricks, but only one at a time. As the process is time-consuming, the *self-assembly* technique is utilized where atoms arrange themselves as required.

Here, smaller components are arranged to build more complex assemblies. By controlled chemical reactions, the atoms and the molecules are carefully deposited. The process is thus relatively much cheaper. Structures of 2 to 10 nm can be easily developed and nanostructures can be made by *self-assembly* of a few hundred molecules. Material scientists often take the advantage of biologically and nature-inspired self-assembly to build the nanostructures.

- The colloidal dispersion is the simplest example of the bottom-up method.

11.8 Areas of nanotechnology

Nanotechnology is a multidisciplinary field and consists essentially of *three* broad and overlapping areas. These are: *nanomaterials* (nanoparticles), *nanobiotechnology* and *nanoelectronics.*

11.8.1 Nanomaterials

Both the statistico-mechanical and quantum mechanical effects become pronounced with the decrease in size of a system. Materials, with their reduction to the nanoscale size, start showing properties very different compared to what they show in bulk structure or macroscale. For instance, opaque substances turn transparent (Cu), stable materials turn combustible (Al), inert materials become catalysts (Pt), solids become liquids at room temperature (Au), insulators turn into conductors (Si) and many more. Becuase of these unique quantum and surface phenomena at nanoscale, the study of nanotechnology is highly fascinating.

Nanomaterials are catagorized into the following:

1. Carbon nanomaterials—These nanoparticles are entirely carbon-based and are in the form of a hollow, an ellipsoid or a tube. Examples are spherical fullerenes or buckyballs, cylindrical fullerenes or nanotubes, as discussed already.

2. Metal-based materials—Examples are *quantum dots*, nanogold, nanosilver, titanium oxide. Quantum dots are closely packed semiconductor crystals that have hundreds or thousands of atoms and can be manipulated to change their physical, particularly the optical properties. The small size implies that more than two-third of the atoms of the material are at surface sites where chemical changes allow emission of multiple colours from single dots.

3. Dendrimers—They are nanosized polymers, their surfaces being the numerous chain ends that can be tailored for specific chemical functions. Three-dimensional dendrimers have interior cavities into which other molecules e.g., drugs can be put and made them useful for drug delivery.

4. Composites—Composites are materials that combine nanoparticles with other nanoparticles or bulk-type materials. Nanosized clays are already being combined to products ranging from auto-parts to packaging materials to increase their mechanical, thermal barrier and flame-retardant properties.

• These various types of nanomaterials with their unique properties are highly desirable for their applications in sectors such as commerce, medicine, environment and military.

11.8.2 Nanobiotechnology

Nanobiotechnology is the promising field of applications of nanoscale engineering to manipulate living systems directly to build biologically inspired materials and devices at molecular level. Life is governed by molecular behaviour at nanoscale. Proteins, nucleic acids, lipids, carbohydrates are the molecular building blocks of life and show unique properties by their nanoscale size, folding and patterns. With nanofabricated surfaces and devices, the process of genome sequencing and detection of genes' expression can be made remarkably more efficient. As a matter of fact, if the genetic make up of an individual can be characterized, the specificity of diagnosis and therapeutics would bring forth revolution.

Basic studies on cell biology and pathology may also be benefitted by nanotechnology that can provide new formulations and routes for drug delivery. The biological nanoworld may be probed with new analytic tools to study the mechano-chemical properties of cells, cell division and locomotion, and of single molecules.

Applications of nanotechnology in medicine and biology have already become a fast growing field. The attractive application in nanomedicine is the creation of nanodevices of improved therapy and diagnosis. Such nanorobots serve as vehicles for drug delivery, as detectors or as guardians against early disease and perhaps repair of metabolic or genetic defects. The challenge is however new instruments to analyse tissues down to molecular level, sensors smaller than a cell to look at ongoing functions, machines that would circulate within human body pursuing pathogens and neutralizing toxins.

The road map of next several years include: advanced drug-delivery systems, implantable devices to automatically administer drugs, sense drug levels, medical diagnostic tools like cancer-tagger mechanisms and 'Lab-on-a chip' diagnostics for the physicians.

11.8.3 Nanoelectronics

We have already stated that the current fever of nanotechnology is driven by the ever shrinking of devices in semiconductor industry. The continued improvement in miniaturization, speed, power consumption in information processing devices, logic devices for processing, storage devices for memory, transmission devices for communication etc. are some of the features in the road map of the industry. This may lead to shrinkage to a few tens of a nanometer. But with this size, quantum effects are bound to appear and the effective functioning of transistor in high density ICs may be prevented.

In this context, alternative means are being investigated and they include:

(i) quantum effect of solid state nanoelectronics, (ii) CNT based nanoelectronics and (iii) molecular nanoelectronics.

Quantum dots, nanowires, single electron transistor (SET) etc. fall under (i) above. Carbon nanotubes under (ii) show extraordinary electronic properties, apart from stiffness and strength and thermal stability. Their strength is higher than any other material, thermally stable up to 2800°C in vacuum. In current carrying capacity, they are better than copper wires and have thermal conductivity twice than that of diamond. Nanocomputers, using CNTs, have already been demonstrated. They are much smaller in size and more powerful. Molecular nanoelectronics referred to in (iii) above is based on molecules of nanometric size where covalently bonded individual molecules play the role of wires and switching devices.

Nanoelectronics, in the foreseeable days to come, would revolutionize computers, telephones, automobiles, household appliances etc. The research and development to this aspect is steadily in progress.

11.9 Micro-electro-mechanical system

Micro-electro-mechanical system or the ***MEMS*** is the technology of the very small and merges at nanoscale into *nano-electro-mechanical system* or *NEMS*. The MEMS devices range from a micrometer to a millimeter in size where classical physics may not always hold good. Microelectronic ICs are as if the *brains* of a system and MEMS augments this decision-making capacity to enable the microsystem to sense and control the environment. Sensors collect information from the environment through measurements of mechanical, thermal, optical, magnetic, chemical and biological phenomena. Electronics then appear in the scene to process those informations collected by sensors and through some decision-making capacity direct actuators to respond by moving, positioning, etc. and thereby control the very environment for some desired outcome.

MEMS devices are manufactured by batch fabrication technique as in ICs. So, with a silicon chip, an unprecedented level as regards function, reliability and sophistication can be achieved at a low cost. MEMS technology can be implemented by a number of different materials such as silicon, polymers and metals.

Silicon—It is the material used in most ICs in consumer electronics. Silicon is highly attractive for many MEMS applications due to its ready availability, cheapness, high quality, electronic functionality and economies of scale. A single Si-crystal is almost perfect Hookean and hence does not show any hysteresis and associated energy dissipation. Si-based MEMS devices are made by deposition of material layers photolithographically, followed by etching to the desired shapes.

Polymers—MEMS devices may also be made from polymers by processes like injection moulding, embossing or stereo-lithography. Crystalline silicon is still a complex and relatively expensive material. But polymers has the advantage of being produced in huge volumes with a variety of characteristics at a relatively low cost.

Metals—MEMS devices may also be produced using metals like silver, gold, aluminium, nickel, platinum, chromium and tungsten. They can be deposited through electroplating, evaporation and sputtering processes. Although metals do not have the advantages of silicon in terms of mechanical properties, they show, within their limitations, high degrees of reliability.

Applications—While there are many applications of MEMS and nanotechnology, few of current interest are in the fields of biotechnology and communications.

In ***biotechnology***, Polymerase Chain Reaction (PCR) microsystems for DNA amplification and identification, micromachined Scanning Tunneling Microscopes (STMs), biochips for detecting hazardous chemical and biological agents are worth mentioning.

In ***communications***, the advent of RF-MEMS technology will highly benefit *hf*-circuits. If made by using MEMS and nanotechnology, electrical components such as inductors and tunable capacitors will be significantly improved and in turn improve the performance of communication circuits. Also, the total circuit area, power consumption and cost will be greatly reduced.

MEMS and nanotechnology are already being used in situations ranging from monitoring in-dwelling blood pressure to active suspension systems for automobiles.

Notwithstanding the fact that they have made possible electrically driven motors of size smaller than the dia of human hair, they are not primarily about size. MEMS and nano together form an altogether *new manufacturing technology* with diverse useful applications. The new technology unifies electro-mechanical elements with electronics and is far more pervasive than even IC microchips.

11.10 Nanotechnology in Nature: Limit to smallness

Perhaps, Nature is the greatest nanotechnologist being capable of constructing objects atom by atom, molecule by molecule, with self-assembly to the desired pattern. DNA is the perfect self-assembling machine in which a single molecule can replicate and produce incredibly complex organism. It is also a very compact, long term information storage-device.

Numerous examples of sophisticated nanomachines are available in cell biology. Molecular motors make up our muscles which convert chemical energy into mechanical one highly efficiently. The molecular structures of ribosomes can construct protein molecules, amino acid by amino acid, with ultimate precision, according to the dictation of DNA.

Over the years, Nature has produced, through evolution, many natural systems like birds, animals and plants in which its intelligence lies hidden. In nature, a single cell supplied with food and oxygen leads to the most complex system—the human being. Nature uses during this evolutionary process certain principles which we call natural laws and the process is nature's technology. Nanotechnology thus already exists in nature and evolution produced astonishingly efficient nanomachines.

In this context, the question arises as to what sets the limit to smallness in nanotechnology. We seek an answer to it from nature. As already indicated, nature provides examples of nanoscale motors and devices. The contraction of muscle occurs through the concerted effort of many myosin molecules. Other types of motors rotate flagella to let the bacteria move through liquid media. These motors are in extremely primitive life forms. They set the limits to smallness of machines and devices, the nanotechnology may produce. The practical limits, in fact, are many orders of magnitude higher, except some atomic-sized assembles developed by using STM.

• Recently, a nanoturbine wheel (radial flow) has been made from silicon (Fig. 11.9) using deep reactive ion etching at MIT, USA. It is believed that following this approach,

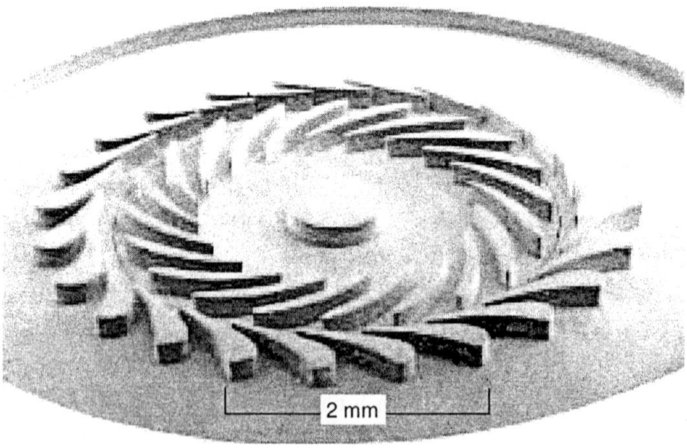

Fig. 11.9 Radial flow nanoturbine made from silicon

it is possible to construct a hydrocarbon burning gas turbine-electric generator with less than 1 cm^3 volume. It would generate power up to 50 W at an energy density 10 to 30 times that of materials used in advanced batteries.

11.11 Present and future of nanotechology

Nanotechnology is hardly three decades old. Nanobased consumer goods and appliances have just started reaching our home. Coating and laminates deserve special mention. Coated bathrooms and kitchen tiles to which dirt and grit particles cannot stick have been made. They contain suspensions of antimicrobial nanoparticles and inhibit growth of rots and fungi and thus maintain a high level of sanitation. Coating windows in cars is protecting us from the harmful UV-radiation—the source of skin cancer.

Applications of nanotechnology in industry are widespread. Mention may be made of the replacement of catalysts with zeolites (molecular sieves) which is highly useful in extracting gasoline from the crude. Computers, communication, consumer electronics and medicine are the various major fields of nanotechnology, CRTs are being replaced by compact and flat LCDs causing less strain on eyes and having higher energy efficiency. Shrinking of screen in laptops will soon be used. New digital displays, giving a paper-like look—*digital paper*—will bring revolution in display technology by creating a paperless world.

Nanotechnology has also entered the field of sports equipments and our fashion arena. Carbon fiber and graphite composites are used in light-weight motor-bikes. Better footballs and hockey-pads have been made. Larger bouncing balls and stronger rackets are already in tennis courts. Nanotubes would be applied in foreseeable future in the manufacture of sports goods.

In the fashion arena, a new generation of clothing has been produced using molecular-scale composite materials. They are resistant to stains, strong and durable and sharper and smarter in look. Biological agents, as in kitchen tiles, are being mixed in manufacturing some fabrics, useful in hospitals to reduce the risk of infection.

Nanotechnology has come in a big way to stay in the field of medicine and serve our health. This has already been discussed under '***Nanobiotechnology***' in Art. 11.8.2. We take here the risk of being repeatitive. One of the attractive applications in medicine is the creation of nanodevices for improved therapy and diagnostics. Such nanodevices or nanorobots serve as vehicles for delivery of drugs, detectors or guardians against early disease and repair of metabolic or genetic defects. The major challenge is 'miniaturization': new instruments for tissue-analysis, down to the molecular level, sensors smaller than a cell to look into ongoing functions, and small machines that would literally circulate within the body pursuing pathogens and neutralizing toxins. To add further, we mention that many tests on health have now become easy and speedy and can be conveniently conducted at home. It is believed that tests for AIDS and anthrax could be made simpler for self-application.

The advent of silicon electronics, with associated miniaturization, is the outstanding technical advancement during the second-half of the last century. To-day, one can place more than 100 million transistors in less than 1 cm^2 of silicon surface area. Unbelievable! This has been made possible only due to the availability of nanosized transistors.

In short, nanotechnology and its impact on our everyday life clearly demonstrates that the age of nano is in the offing and the future of nanotechnology is indeed rosy.

● Apart from what has been stated above, the subject matter is strewn throughout the chapter and may please be noted.

11.12 Conclusions

We are finally at our journey's end and cry halt by making a number of remarks.

First, there may be some problems associated with nanotechnology. With the decrease in particle size, the processes of *thermal instability* and *phase transitions* occur. This may result in non-durability of nanosystems. The cosmic radiation background may knock out atoms from nanostructures leading to *degradation* of their properties and worsening of nanodevice operations. Also, the *thermal noise* and *vibrations* would significantly influence the characteristics of nanodevices. Even a negligible concentration of *inherent impurities* and *irremovable contamination* may destroy the assembling processes and as such super-high-quality reagents and clean-room processes are warranted.

Secondly, the use of new materials is fraught with the danger of damaging our health. Small quartz particles, asbestos fibres, smoke and soot are the known potential health hazards, from allergy to cancer. Similarly, the products with nanostructures may also pose problems or risk during their fabrication and subsequent use. Attention must therefore be paid to the likely health-hazards of nanomaterials applied either on skin or inhaled.

Thirdly, nanotechnology has the undeniable potential to change the world. It is creating both new profiteers and losers and thereby posing socio-ethical threats. The applications of nanoelectronics would also require proper regulations over the protection of fundamental rights of the people against the new type of criminality (cyber crime) in the use of internet by way of hacking personal information.

Fourthly, the field of nanotechnology is replete with exaggerated claims and is being hyped. A case in hand is the concept of *molecular assembler* which was later declared flawed. Caution must thus be observed in making judgement and claims, particularly in the light of huge amount of money and energy that are being made available to the fascinating field of nanotechnology. The 'shrinking' in nanotechnology is not to come *free*. Moore's second law states: *The cost of building a chip fabricating plant doubles every* 36 *months.* The soring manufacturing cost is posing a big challenge.

Finally, nanotechnology is only helping to improve the existing products rather than creating completely new ones. We emphasize that nanotechnology is not intended to replace the existence of our micro-technologies. It has come to stay in close relation with them and to complement them in the deeper and advanced studies and control of the nanoworld.

The *new about nanoscience* is the combination of our *ability to see* and *manipulate matter* on nanoscale and our understanding of *atomic scale interactions.*

Review and Short Questions

1. What is meant by nanomaterials? Describe, in brief, the various types of nanoparticles.
2. Give an account of the structure and properties of different nanoparticles.
3. What is a quantum dot? What are its properties? How can it be synthesized? Describe its possible applications.
4. Describe the structure and properties of carbon clusters.
5. What are fullerenes? Why are they so called? Describe the various types of fullerenes.
6. Describe, in detail, the structure and properties of fullerite. Mention its present applications and its future potentials.
7. Describe briefly how carbon nanotubes are formed. Discuss their structure, characteristics and applications.
8. Describe the structure and the method of procuring graphene.
9. Discuss the properties and applications of graphene.
10. What is meant by quantum nanostructures? Briefly describe their fabrication.
11. What are the salient features of quantum nanostructures and their potential applications?
12. Describe the properties and applications of zeolites.
13. What is nanotechnology? Give a brief account of its origin.
14. Give a critical review of the present state and the future of nanotechnology.
15. There are two different approaches for nanofabrication. These are

    (a) Top-down and bottom-down, (b) Bottom-up and top-up,
    (c) Top-down and bottom-up, (d) None of these.

16. Quantum nanostructures refer to

    (a) Quantum dots, (b) Quantum wires,
    (c) Quantum wells, (d) All of the three.

17. Write in full the following acronyms.

    (a) MEMS, (b) NEMS, (c) CNT, (d) SET.

18. Write the following materials in the ascending order of their size.

    (a) Virus, (b) Micromolecule, (c) Red blood cell, (d) Human hair (dia)

ELECTRONICS

1

Semiconductors

1.1 Introduction

Based on electrical conductivity, solids are usually classified as conductors, insulators and semiconductors. The applications of semiconductors brought about such a revolution in the field of electronics that a world without it cannot be conceived at all.

Semiconductors have an electrical conductivity ($\sim 10^5 - 10^{-4}$ S/m), intermediate between those of conductors ($\sim 10^8$ S/m) and insulators ($\sim 10^{-11}$ S/m). The conductance varies depending on the current and voltage applied to control electrodes or on the intensity of irradiation by infrared, visible, ultraviolet or X-radiation.

The specific properties of a semiconductor depend on the impurities or dopants added to it. The variation of electrical properties by changes in temperature, optical excitation and impurity contents makes semiconductors natural choices for electronic devices. The following properties of semiconductor are important for understanding the functions and fabrication of electronic devices.

(a) Semiconductors, unlike metals, have a negative temperature coefficient of resistivity i.e., the resistivity decreases with increase in temperature.

(b) All semiconductors behave like a perfect insulator at absolute zero.

(c) They have a considerably lower number-density of charge carriers than metals.

(d) Both electrons and holes are the charge carriers, unlike in metals.

(e) The forbidden energy gap of a semiconductor lies in the range 0.2 eV to 2.5 eV which is much smaller than that of insulators like diamond (~ 6 eV).

(f) They may have a positive or a negative thermoelectric power w.r.t metals.

Germanium (Ge) and Silicon (Si) are commonly used elemental semiconductors which belong to the group IV of the periodic table. Others include Boron, Aluminium, Gallium, Indium (group III); Phosphorus, Arsenic, Antimony (group V); Selenium, Tellurium, Sulphur (group VI). While Germanium was the key materials used in the early days of semiconductor, Si is now the best-known of elemental semiconductor forming the basis of most integrated circuits (ICs), rectifiers, transistors etc. for its thermal stability, cheapness and availability in plenty. Common semiconductor compounds include Gallium Arsenide (GaAs), Indium Antimonide and oxides of most metals. The compound semiconductors are mainly used in high speed devices related to emission or absorption of light. GaAs is widely used in low noise, high gain and weak signal amplifying devices.

1.2 Metals, insulators and semiconductors

Metals are generally excellent conductors of electricity. Among them, silver has the highest electrical conductivity followed by copper. The higher conductivity of metals can be understood from their electronic configuration. For example, copper of $Z = 29$ has the electron configuration $1s^2 2s^2 2p^6 3s^2 3p^6 3d^{10} 4s^1$. When an atom has eight electrons in its outermost shell it becomes most stable. The outermost shell containing electrons is known as valence orbit which is of special interest because the electrical and chemical properties of an element are determined by this orbit. If we draw the simplified shell structure of Cu atom (Fig. 1.1), it is seen that the valence orbit contains one electron known as *valence electron* of charge $-e$ and the core has a net charge of $+e$ since the core has 29 protons and 28 electrons. Since the valence electron is in a large orbit around a core of charge of $+e$ it feels a negligible amount of force of attraction to a particular nucleus. Hence it is known as *free electron*, meaning not bound to any particular atom. These electrons, also called *conduction electrons*, take part in electrical conduction even under the influence of very small electric field. The motion of the conduction electrons in the sample is just like the motion of free molecules of a gas in a container. Hence they constitute the so-called *electron gas*. The attraction between the free electrons and the fixed nuclei is responsible for the interatomic bonds. The monovalent metals have 10^{29} atoms/m^3. At room temperature, almost all the valence electrons are free to take part in electrical conduction. As a consequence, metals have resistivities in the range of 10^{-8} ohm-m to 10^{-6} ohm-m.

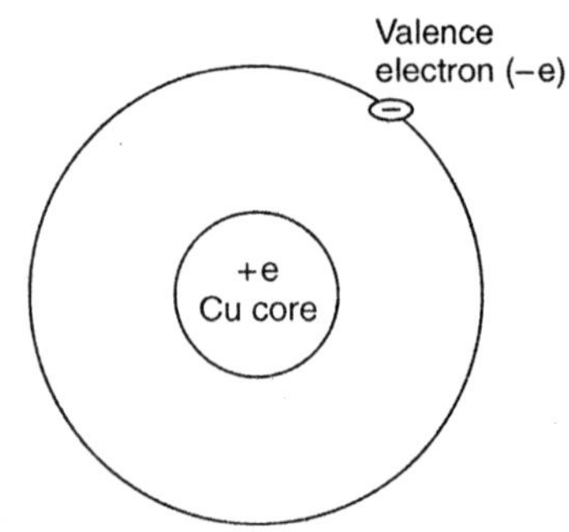

Fig. 1.1 Simplified shell structure of Cu atom

Ge and Si, are available in the form of crystalline solids having diamond lattice structure i.e., each atom of Ge or Si is surrounded by four nearest neighbours lying at the corner of a tetrahedron. (Fig. 1.2(a)). On sharing of valence electrons with opposite spin, Ge or Si atom forms four covalent bonds which are of great significance in semiconductor action. At 0 K the electrons are firmly attached to the covalent bond and are not available for conduction and hence a semiconductor behaves as a perfect insulator at 0 K. When the temperature is increased, the atoms vibrate about their lattice position by absorbing thermal energy. Few of the electrons might get sufficient thermal energy to break their covalent bonds to take part in electric conduction. The minimum energy that is required to break the covalent bonds is 1.1 eV for Si and 0.72 eV for Ge. At room temperature ($\sim$ 300 K), the number of broken bonds per cubic meter is $\sim 10^{19}$ although there are 10^{29} atoms per cubic meter. So, there is one broken bond in 10^{10} atoms i.e., the rupture of bonds is rather rare. The free electrons and the broken bonds enable the material to behave like a semiconductor with resistivity in the range 0.4 ohm-m to 2500 ohm-m. With increase in temperature, the number of broken bond increases rapidly. So does the number of charge carriers and the resistance of the semiconductor decreases. Hence semiconductors have a *negative temperature coefficient* of resistance.

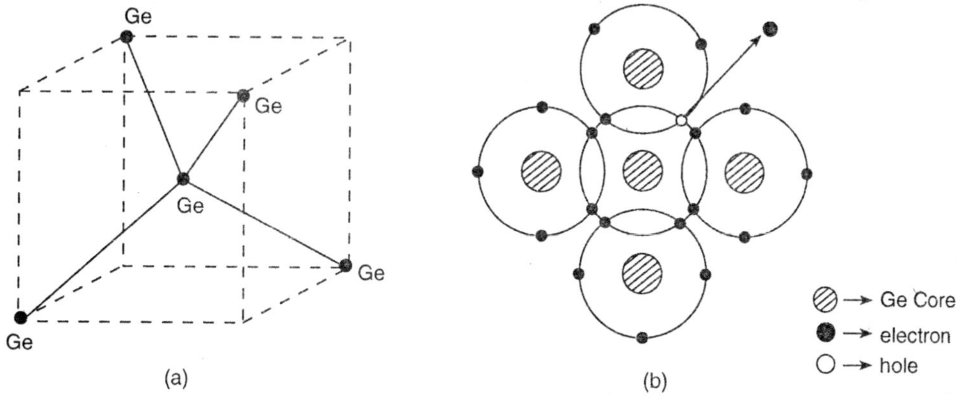

Fig. 1.2 (a) Tetrahedral lattice of Ge (b) Generation of electron-hole pair

When a covalent bond is broken by thermal agitation, an 'electron-vacancy' is created in the bond. The vacant site behaves like a positive charge of $+e$ and is called a *hole*. It also takes part in electric conduction. When an external electric field is applied, an electron having the same energy as the hole may occupy the hole, leaving a new hole behind. Thus the hole may be assumed to move in the direction of electric field i.e., opposite to the direction of motion of electrons. Electrons and holes both contribute to the current in the same direction. So, in electric conduction, a hole behaves like a charge carrier having a charge equal and opposite to that of an electron.

The materials in which the valence electrons are all rigidly held in the interatomic covalent bond even at higher temperature are known as *insulators*. The probability of breaking the covalent bond to make an electron free is practically zero in insulators. The insulators like plastic, polystyrene, quartz etc. have resistivities in the range 10^7 to 10^{18} ohm-m at 300 K.

1.3 Intrinsic semicondutors

The semiconductors, just discussed, are intrinsic semiconductors. A pure semiconductor crystal with no impurities or lattice defects is called an *intrinsic semiconductor*. As we have just seen, it has no free charge carrier at 0 K, but at higher temperatures some covalent bonds are broken and *electron-hole pairs* are produced simultaneously. These electron-hole pairs take part in electrical conduction and are the only charge carriers in intrinsic semiconductors.

The process of generation of electron-hole pairs is illustrated qualitatively in Fig. 1.2(b). If one of the valence electrons is stripped off the covalent bond, it becomes free to move in the lattice as conduction electron and the broken bond creates a deficiency in electron which is called the *hole*. The holes so generated are in a sense mobile (Fig. 1.3). Let a hole be created at A. An electron at B having a comparable energy can move from B to fill the hole at A leaving a hole at B. Another electron may similarly move from a bond at C to the hole at B and a hole appears at C and the process may go on, so that the hole has moved from A to C stepwise.

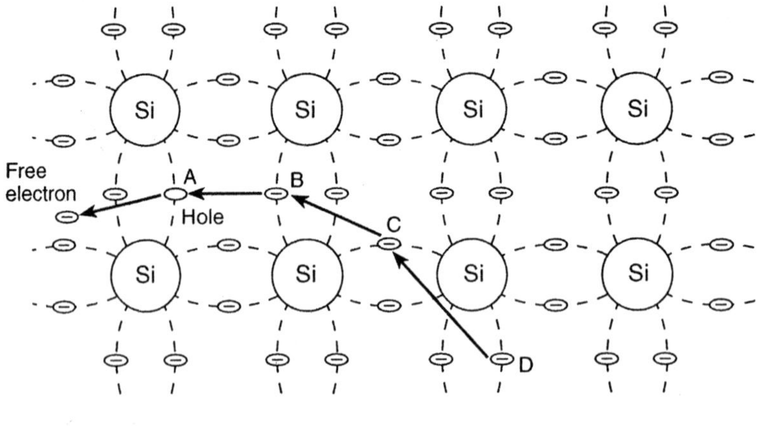

Fig. 1.3 Movement of hole

In the above process, it appears that holes are transported through the lattice. Actually both the electrons and the holes simultaneouly take part in electric conduction. When no electric field is applied, the electrons and the holes are in random thermal motion in the semiconductor. But a directed component is added to the charge in motion if an electric field is applied. When thermally generated electrons take part in electric conduction, it is known as *intrinsic conduction*.

The simultaneous generation of a free electron and a hole by thermal excitement on breakage of covalent bond is called *electron-hole pair* (EHP) *generation*. The rate $g(t)$ of continuous generation of electron-hole pair depends on the temperature and the energy gap (E_g) of the material. There occurs also another process known as *recombination* which is just opposite to the EHP generation. When free electrons move randomly in the semiconductor crystal, some electrons may collide with some broken bonds and may combine with the holes resulting into bound electrons of the covalent bonds. The recombination depends on the number of encounters of the holes and the electrons. The rate of recombination $r(t)$ is thus proportional to the densities of the electrons and the holes and also to the temperature. In an equilibrium condition, both the rate of production of electron-hole pair and the rate of recombination are the same and, therefore, we shall have

$$g(t) = r(t) = kn_i p_i$$

where n_i and p_i are the electron and the hole densities in intrinsic condition. In an intrinsic semiconductor both n_i and p_i are equal.

1.4 Extrinsic semiconductors : Doping

At ordinary temperature, the number of free electrons and holes in an intrinsic semiconductor are too small to produce any cognisable current. Apart from thermal generation, it is also possible to create charge carriers non-thermally in semiconductors by introducing a small percentage of substitutional impurities into the crystal in a controlled way to change significantly the characteristics of intrinsic semiconductors. This process of addition of the impurities is termed ***doping*** which is the most common technique for varying the conductivity of semiconductors. By doping, a crystal may have a predominance of either electrons or holes. Thus there are two types of doped semiconductors, the *n-type* (majority carriers: electrons) and the *p-type* (majority carriers: holes). When an intrinsic semiconductor is doped, it is said to be an *extrinsic semiconductor.* Thus the doped semiconductors are extrinsic semiconductors.

Usually, a very small proportion of impurities (only one impurity-atom per 10^6 to 10^{10} semiconductor atoms) are added to the highly purified semiconductors. The impurities are mainly trivalent (e.g., Ga, B, In from Gr. III) or pentavalent (e.g., As, P, Sb from Gr. V). The scarcity of impurity atoms exerts no influence on the crystal structure. So, an impurity atom must enter the crystal by substituting for one of the Ge or Si atoms.

To prepare this type of semiconductors, the pure Si or Ge crystal is first melted into liquid. To increase the number of free electrons, pentavalent impurities are added to the molten mass. If trivalent impurities are added, then the number of holes increases. The amount of impurity added controls the level of concentration of the charge carriers, known as the *doping level.*

1.4.1 n-type semiconductors

Let a small percentage of pentavalent impurity like As (or Sb or P) be added to pure Si at the time of growth of the crystal. The As atoms diffuse uniformly throughout the molten Si and displace some of the atoms in the lattice i.e., most of the atoms in the crystal are of Si but occasionally we may find an As atom surrounded by four

Si atoms as shown in Fig. 1.4. Four of the valence electrons of pentavalent As atom form covalent bonds with four adjacent Si atoms and the fifth one remains loosely bound to the As atom. The minimum energy required to detach this electron from the impurity atom is ~ 0.01 eV for Ge and 0.05 eV for Si. This is known as the *ionisation energy* of the impurity atom. Since the fifth electron of the impurity atom moves in a dielectric medium (with $k = 11$ for Si) instead of vacuum ($k = 1$) as in hydrogen, the ionisation energy is very low compared to that of hydrogen. Moreover, the electron mass is replaced by its *effective mass.* So almost all the impurity atoms become ionized by absorbing a small amount of thermal energy. This enables the fifth electron to overcome its Coulomb binding to the impurity atom. Thus each impurity atom donates one electron to lattice and the free electron participates in current conduction as conduction electron. The impurity atom however remains fixed in the lattice as a positive ion. Due to thermal breakage of a few covalent bonds, equal number of electrons and holes are also generated. The presence of a large number of free electron increases the rate of recombination. As a result, hole concentration decreases below the intrinsic level and the electron concentration is much higher. So while electrons are the majority carriers, holes are the minority carriers and the resulting semiconductor is called an *n-type* or *donor type* semiconductors.

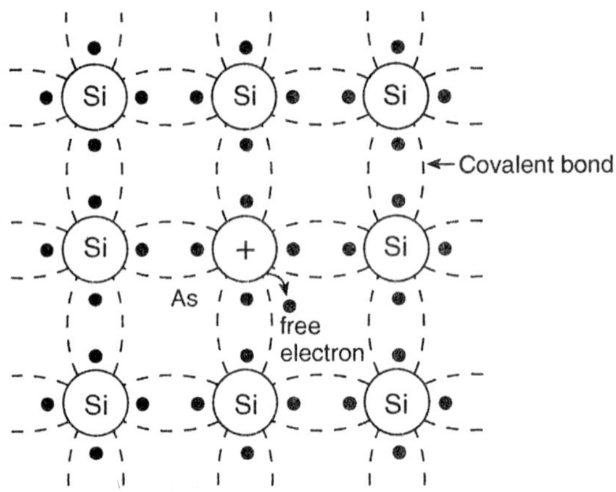

Fig. 1.4 Si crystal doped with As as impurity

1.4.2 *p*-type semiconductors

Let a trivalent impurity like Al (or B, Ga, In) be added in small proportion to pure Si at the time of crystal growth. Al atoms displace some of the Si atoms and each Al atom is surrounded by four Si atoms as shown in Fig. 1.5. Since Al is trivalent, the three valence electrons form covalent bonds with three adjacent Si atoms and there is a lack of one electron to form the fourth covalent bond with the fourth adjacent atom. This means that the impurity atom (i.e., Al atom) is still capable of accepting one electron or in other words, a hole appears in each valence orbit of the Al atom. Moreover, equal number of holes and electrons are produced due to thermal breakage of a few covalent bonds. The presence of large number of holes increases the rate of recombination. As a consequence of this, the electron concentration decreases below the intrinsic level. Thus in this type of semiconductor, the concentration of holes far exceeds that of the electrons. So, the holes are regarded as the majority carriers and the electrons the minority carriers. The resulting semiconductor is known as a *p-type* or an *acceptor type* semiconductor.

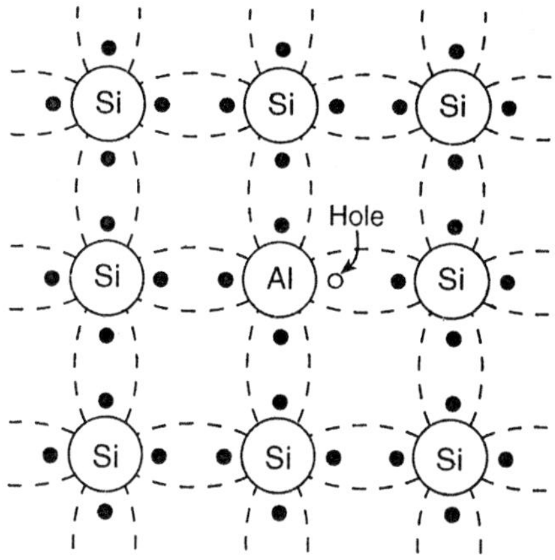

Fig. 1.5 Crystal structure of Si doped with Al-impurity

1.5 Energy bands in solids

To understand the electrical properties of semiconductors, it is important to know the energy bands in solids especially in semiconductors. X-ray crystallography and other studies regarding the structure of solids reveal that most of the metals and semiconductors are crystalline. A regular and repetitive arrangement of atoms and molecules (or ions) in space constitute a *crystal. Lattice points* are the points which represent the position of the atoms in the crystal. The fixed distance between the atoms in a crystal is called the *lattice constant* of the crystal. In a gas, the molecules hardly exert any force on one another as they are sufficiently far apart. The electronic energy levels for a single free atom differ from those for an atom bound in a crystal.

If isolated atoms are brought together to form a crystalline solid, different interactions occur between the adjacent atoms including the ionic bonding, covalent bonding etc. An equilibrium is established between the forces of attraction and repulsion at the proper interatomic spacing. As a result, important changes occur in the electronic energy level configuration that results in varied electrical properties of the solids.

To study the behaviour of an electron in a crystal, consider an atom of atomic number Z. When isolated, its electrostatic potential at a distance r from the nucleus is given (in S.I. units) by

$$V(r) = \frac{Ze}{4\pi\varepsilon_0 r} \tag{1.5.1}$$

where Ze is the charge of the nucleus, ε_0 the free space permittivity.

The potential energy of the electron at the point is given by

$$E(r) = -e \cdot V(r) = -\frac{Ze^2}{4\pi\varepsilon_0 r} \tag{1.5.2}$$

From equations (1.5.1) and (1.5.2), it is seen that the potential energy $E(r)$ is negative although $V(r)$ is positive. Fig. 1.6 and Fig. 1.7 illustrate the variation of $V(r)$ and $E(r)$ respectively with r. These two curves show that both $V(r)$ and $E(r)$ approach zero as r tends to infinity.

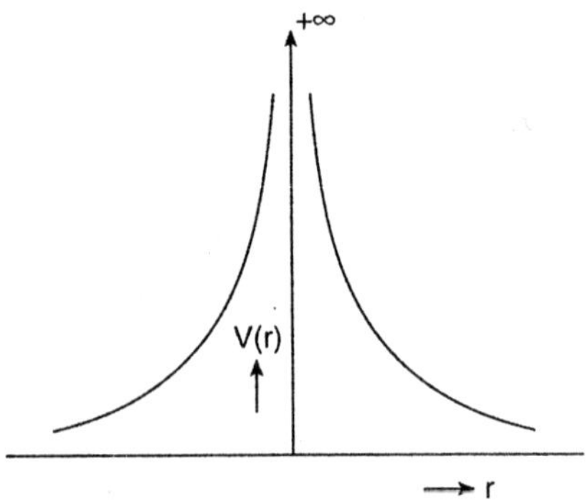

Fig. 1.6 Variation of V(r) with distance from the nucleus

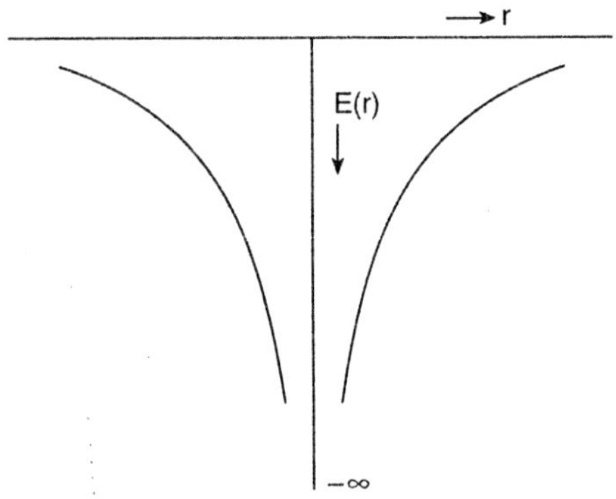

Fig. 1.7 Variation of P.E. [E(r)] with distance from the nucleus

When two atoms are placed close to each other, the net potential energy of electron is smaller than that of an electron of an isolated atom since the net potential energy of an electron is the sum of the potential energies due to the individual nuclei. Fig. 1.8 shows the variation of potential energy with distance between two identical nuclei.

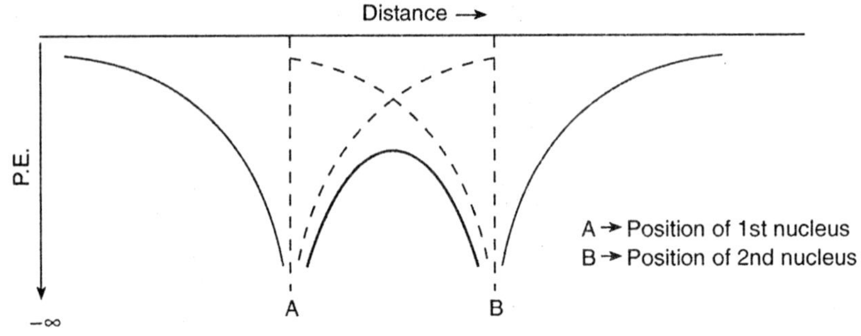

Fig. 1.8 Variation of P.E. of two identical nuclei

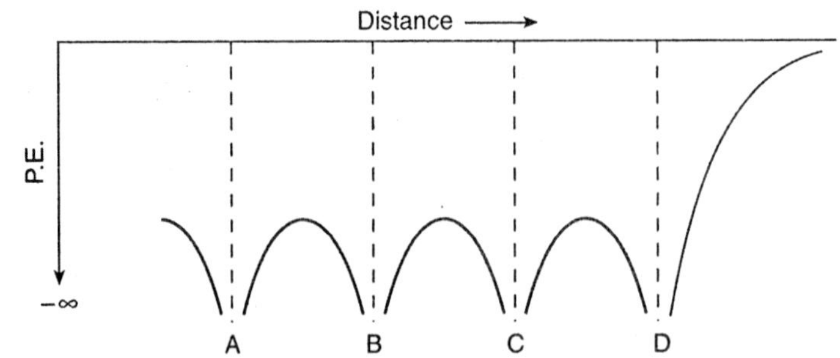

Fig. 1.9 Potential energy (P.E.) of an electron in a crystal: A, B, C, D correspond to positions of nuclei

When the atoms are placed along a line through a row of equispaced points in a crystal, the potential energy consists of a series of humps as shown diagrammatically in Fig. 1.9, there being no atom outside the boundary. The potential energy increases and approaches zero at infinity.

The total energy of an electron in an orbit of radius r and having principal quantum number n is given by

$$E = -\frac{mZ^2e^4}{8\varepsilon_0^2 n^2 h^2} \tag{1.5.3}$$

and shows that an electron in an isolated atom can have only certain discrete energies.

Splitting of energy levels into bands—In Fig. 1.10, the horizontal lines correspond to the discrete energy levels in an isolated atom. When two monovalent identical atoms are slowly brought close to each other, the electron-electron interaction and electron-ion interaction occur due to overlapping of the electronic shells of the two atoms. Such interactions cause the splitting of each individual energy levels into two.

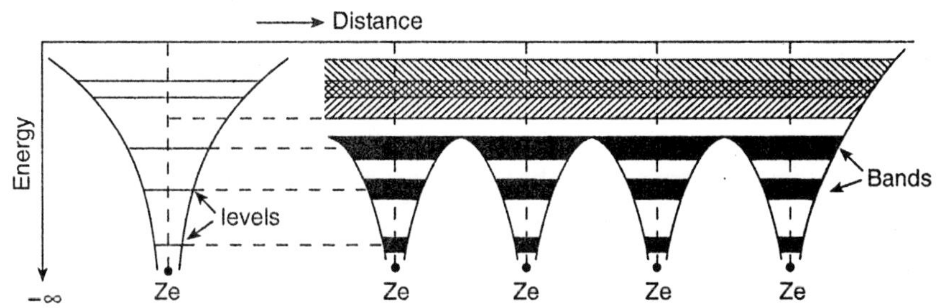

(a) Energy levels of an isolated atom

(b) Energy bands in a crystal

Fig. 1.10 Splitting of energy levels into energy bands

Pauli's exclusion principle states that *no two electrons in an electronic system can occupy the same quantum state.* In this case, the interacting electrons form a single system and according to Pauli's principle an electron may occupy any one of the two levels. In a crystal where a large number of atoms are closely packed, each atom exerts an electric influence on its adjacent atoms. So, the crystal is formed in a single electronic system under interatomic coupling, obeying Pauli's exclusion principle. When a crystal is formed each energy level of the individual atom splits into as many energy levels as the number of atoms present in the crystal so that there is no violation of Pauli's principle. In crystalline solids where a large number of atoms are closely packed, the split up energy levels are so close that they may essentially be considered as continuous. This large number of discrete and closely spaced energy levels form what is called a *band*, known as ***energy band***. Fig. 1.10(b) represents the energy band schematically by the shaded region. The band width depends on the nature of the crystal and the energy level of the isolated atom. In lower energy levels, the effect of interactions among the atoms is not much great and hence the lower energy levels form narrow bands. But the higher energy levels produce wide bands because of the greater influence of the interatomic interactions. The width of the band is large for a crystal with a small interatomic spacing and is independent of the number of atoms in the crystal.

The lower energy bands are generally completely filled by the electrons due to their tendency to occupy the lowest available energy states. When the partly filled energy levels or the partly filled and completely empty energy levels overlap, we get partially filled energy band. So, the higher energy bands may be totally empty or partly filled, the number of electrons in the band being determined by Pauli's principle. The uppermost band that is totally filled up with maximum allowable number of electrons is called ***valence band.*** The band just above the valence band is the ***conduction band*** because the energy states in the valence band being completely occupied, conduction is possible only if electrons are available either in the empty or the partly filled energy band just above the valence band.

1.6 Solids in the light of band theory

Every solid is characterized by its own structure of energy band. The variation in band structure is the important factor responsible for a wide range of electrical properties observed in various materials. Crucial to the conduction process is whether or not there are electrons in the conduction band. So, crystals can be classified into metals, insulators and semiconductors on the basis of the band structure.

A. Metals—A ***metal*** is a crystalline solid whose conduction band is partly filled or whose valence band and conduction band overlap in energy. In terms of band theory, metals are unique as *good conductors* of electricity as their valence electrons are free. Due to the overlap of valence band and conduction band, at least a fraction of the valence electrons can move through the material. The band structure of metal is shown in Fig. 1.11(a).

B. Insulators—The forbidden energy gap E_g is one of the key factors that determine the electrical properties of solids. In some solids, the value of E_g is very large ($\sim$ 3 eV). So, at ordinary temperature an electric field cannot impart enough energy to the electrons in the valence band to transit to the conduction band. The conductivity of this type of materials is very low and the conduction is practically

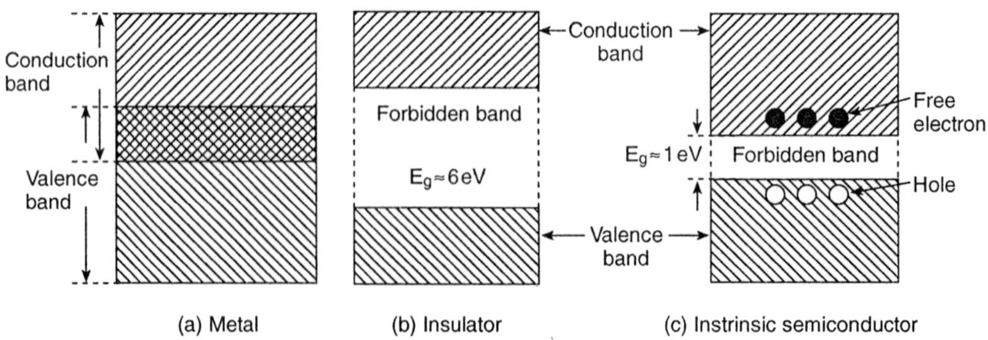

Fig. 1.11 Energy bands in solids

zero. Such solids are known as ***insulators***. The band structure of insulators is shown schematically in Fig. 1.11(b).

C. Intrinsic semiconductor—In some materials, the value of the forbidden energy gap is relatively small ($\sim 1\,\text{eV}$) compared to that of insulators. This type of materials is known as ***intrinsic semiconductors***. Ge and Si are the most typical intrinsic semiconductors having forbidden energy gap 0.785 eV and 1.21 eV respectively at 0 K. In spite of these low values of E_g, Ge and Si behave as insulators at very low temperature, because at low temperature a valence electron cannot normally acquire even this amount of energy from an applied field or by thermal excitation. Hence the valence band remains full and conduction band empty. However, when the temperature is increased, some of the valence electrons move into the conduction band by absorbing thermal energy greater than E_g. So, the conductivity increases with temperature. The electrons in conduction band can move freely under the influence of even a small applied field. These electrons behave as ***free electrons***. When an electron transits from valence band to conduction band, a state in the valence band remains empty. Such an empty state is called a ***hole***. In Fig. 1.11(c), the schematic band structure of an intrinsic semiconductor is shown. The black dots in conduction band and the small circle in the valence band represent the electrons and the holes respectively. The holes can contribute to the current flow. A hole may be assumed to be a quasiparticle with a charge $+e$.

D. Extrinsic semiconductors

(i) n-type semiconductors—When donor impurities i.e., pentavalent impurities are added to an intrinsic semiconductor, allowable energy levels are introduced in a very small distance below the conduction band as shown in Fig. 1.12(a). This type of energy level is known as *donor energy level*. Since the added impurity atoms are far apart in the crystal structure and have small interaction, the new allowable energy levels are essentially a discrete level. In the case of Si, the donor energy level is only 0.05 eV below the conduction band. So, at room temperature, the fifth electron of the donor atom acquires sufficient thermal energy to jump from valence band to conduction band.

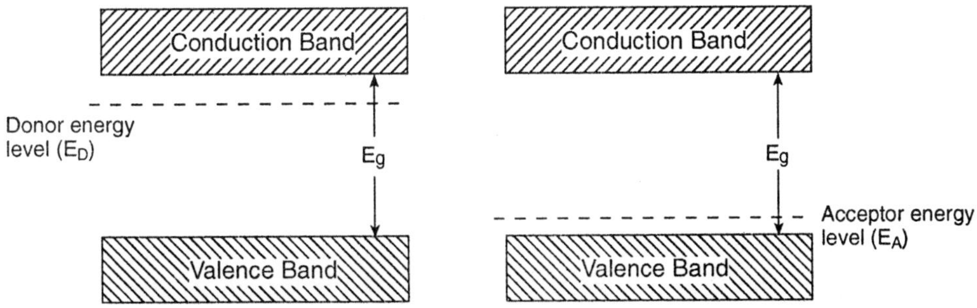

(a) n-type semiconductor (b) p-type semiconductor

Fig. 1.12 Energy band diagram of extrinsic semiconductor

(ii) *p*-type semiconductors—When trivalent or *p*-type or acceptor type impurities are doped in the intrinsic semiconductor, allowable discrete energy levels are produced just above the valence band (Fig. 1.12(b)). This type energy level is known as *acceptor energy level.* Since the added impurity atoms are very small in amount and are far apart, the distance between the acceptor energy level and the valence band is very small. So to occupy the acceptor energy level, an electron requires a very small amount of energy for leaving the valence band. Hence a large number of carriers is generated in the semiconductor as the electrons occupy the acceptor energy level, leaving holes in the valence band.

1.7 Direct and indirect semiconductors

When isolated atoms are brought together to form a solid, the different atomic interactions are responsible for the existence of energy bands. The band structure can also be explained in a different way by quantitative calculations. In this calculation, a single electron is assumed to travel through a perfectly periodic lattice. The corresponding wave function is assumed to be in the form of a plane wave moving along the x-direction with the ***propagation constant k*** (also called the ***wave vector***). The space-dependent wave function for the electron is given by

$$\psi_k(x) = U(k_x, x)e^{jk_x x}$$

where $U(k_x, x)$ regulates the wave function $\psi_k(x)$, depending on the periodicity of the crystal lattice.

To get the full relationship between $\boldsymbol{E}$ and $\boldsymbol{k}$, the $\boldsymbol{E}$-$\boldsymbol{k}$ diagram should be plotted for various crystal directions because of the variation in periodicity of lattice in different directions. The $\boldsymbol{E}$-$\boldsymbol{k}$ diagram is a complex surface if visualized in three dimensions. On the basis of the energy band structure, the semiconductors can be classified into *two* different classes. These are:

(i) Direct band gap semiconductor and

(ii) Indirect band gap semiconductor.

We shall discuss these two types of semiconductor briefly in the following two consecutive sections.

1.7.1 Direct band gap semiconductor

The semiconductor for which the minimum of conduction band and the maximum of valence band are at the same value of wave vector $\boldsymbol{k}$, is known as ***direct band gap semiconductor*** (Fig. 1.13(a)). GaAs, InP are some typical examples of this type of semiconductors. So, in this case, the transition of an electron from the minimum of conduction band to the maximum of valence band takes place without a change in $\boldsymbol{k}$.

Thus, the momentum is automatically conserved and the corresponding transition is a direct recombination of electron and hole. The excess energy is emitted in the form of photon with a minimum frequency (ν) given by

$$E_g = h\nu \quad \text{or} \quad \nu = \frac{E_g}{h}$$

where h is the Planck's constant.

LED and Laser diodes are constructed by this type of semiconductors.

1.7.2 Indirect band gap semiconductor

The semiconductor for which the minimum of conduction band and the maximum of valence band do not occur at the same point in k-space is called ***indirect band gap semiconductor*** (Fig. 1.13(b)). So, the momentum changes when transition from the minimum of conduction band to the maximum of valence band takes place. So, direct recombination of electron and hole never occurs in this type of semiconductors. Here recombination takes place *via* trap which contributes defect states in the bandgap, the energy being liberated as heat in the crystal. This type of semiconductor is used for the construction of diodes and transistors.

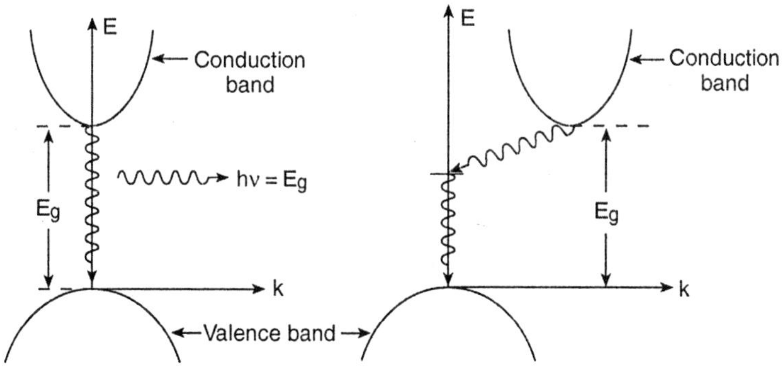

Fig. 1.13 (a) E–k diagram of a direct gap semiconductor
(b) E–k diagram of an indirect gap semiconductor

1.8 Effective mass

When the electrons are inside a crystal, they are not completely free as they interact with the periodic potential of the crystal. According to wave-mechanical analysis, a bound electron or a valence electron cannot be treated as a classical particle. So, to avoid the quantum nature of the problem, we must have the effective mass approximation for using Newtonian mechanics to study the effect of the external electrical field applied to semiconductor crystal.

The ***effective mass*** of *a carrier is the modified mass* (usually different from the eletronic rest mass in vacuum) *when the externally applied fields are much weaker* *than the internal periodic fields of the crystal structure and that the carriers respond (assumed) only to the external field.*

The momentum of an electron is given by: $p = mv = \hbar k$

So, the energy E is given by: $E = \frac{1}{2}mv^2 = \frac{p^2}{2m} = \frac{\hbar^2 k^2}{2m}$

with usual meaning of the symbols used.

Therefore, E-k diagram shows the parabolic nature of the curve (Fig. 1.14).

$$\text{Now,} \quad \frac{d^2E}{dk^2} = \frac{\hbar^2}{m} \quad \text{or} \quad m = m_e^* = \frac{\hbar^2}{d^2E/dk^2} \tag{1.8.1}$$

which shows that the electronic mass is inversely proportional to the curvature (second derivative) of E-k relationship.

For the conduction band, the energy bands are close to parabolic at their minima and for valence band this occurs at the maxima. So, the equation (1.8.1) shows that the electron's effective mass is much smaller at strong curvature (in conduction band) than at weaker curvature. The value of d^2E/dk^2 is positive at the minimum of conduction band and it is negative at the maximum of valence band. Therefore, the electrons have positive effective mass near the bottom of the conduction band and negative effective mass near the top of of the valence band. So, *the motion of an electron with negative charge and negative mass in an electric field is identical with the motion of a hole with positive charge and positive mass.* The charge transport in a semiconductor can be fully accounted for by considering only the hole motion.

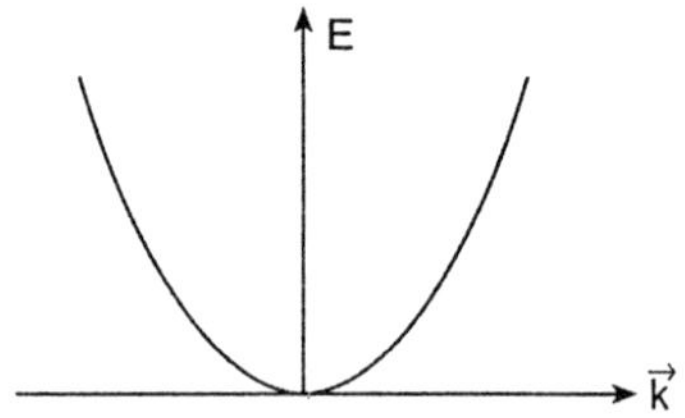

Fig. 1.14 Electron energy vs. wave vector graph

1.9 Carrier concentration in intrinsic semiconductors

In semiconductors, the electrical conduction is due to the movement of electrons in the conduction band and holes in the valence band when an electric field is applied. So, to calculate the conductivity and study the electrical properties of semiconductor, we are to know the concentration of electrons in the conduction band and the concentration of holes in the valence band.

Let us consider an intrinsic semiconductor at temperature T K in thermal equilibrium. To calculate the electron concentration in conduction band, the following assumptions are made.

(i) The energy is calculated w.r.t the top of the valence band.

(ii) The influence of the lattice is avoided by considering the electrons and holes as imaginary classical particles with effective masses m_e^* and m_h^* respectively.

(iii) The carrier density and the temperature are so small that the electron gas in the conduction band and the holes in the valence band are non-degenerate.

1.9.1 Concentration of electrons in the conduction band

The electron density states in the conduction band is given by

$$N(E) = \frac{1}{2\pi^2}\left(\frac{2m_e^*}{\hbar^2}\right)^{3/2}(E - E_c)^{1/2} \tag{1.9.1}$$

where E_c is the energy of the bottom of the conduction band.

The probability of occupancy of a state with energy E is given by the Fermi function

$$f(E) = \frac{1}{1 + e^{(E-E_F)/kT}} \tag{1.9.2}$$

where E_F is the well-known Fermi energy.

The number dn of electrons per unit volume with energy lying between E and $E + dE$ is given by

$$dn = f(E)N(E)dE$$

The total electron concentration n is the integral of the above expression over the entire conduction band i.e., the concentration of electrons in the conduction band is given by

$$\begin{aligned} n &= \int_{E_c}^{\infty} f(E)N(E)dE \\ &= \int_{E_c}^{\infty} \frac{1}{2\pi^2}\left(\frac{2m_e^*}{\hbar^2}\right)^{3/2}(E - E_c)^{1/2} f(E)dE \end{aligned} \tag{1.9.3}$$

At low temperature and for $E \geq E_c$, $E - E_F \gg kT$, so that the equation (1.9.2) may be approximated to

$$f(E) = \left[1 + e^{(E-E_F)/kT}\right]^{-1} = e^{-(E-E_F)/kT}$$

$$\begin{aligned} \therefore\ n &= \frac{1}{2\pi^2}\left(\frac{2m_e^*}{\hbar^2}\right)^{3/2}\int_{E_c}^{\infty}(E - E_c)^{1/2}e^{-(E-E_F)/kT}dE \\ &= \frac{1}{2\pi^2}\left(\frac{2m_e^*}{\hbar^2}\right)^{3/2} e^{-\frac{(E_c-E_F)}{kT}}\int_0^{\infty}(kT)^{3/2}x^{1/2}e^{-x}dx, \quad \text{putting } \tfrac{E-E_c}{kT} = x \\ &= \frac{1}{2\pi^2}\left(\frac{2m_e^*}{\hbar^2}\right)^{3/2} e^{-\frac{(E_c-E_F)}{kT}}(kT)^{3/2}\,\Gamma\left(\frac{3}{2}\right) \\ &= \frac{1}{2\pi^2}\left(\frac{2m_e^*}{\hbar^2}\right)^{3/2} e^{-\frac{(E_c-E_F)}{kT}}(kT)^{3/2} \times \frac{\sqrt{\pi}}{2} \\ &= \frac{1}{2\pi^2}\left(\frac{2m_e^* kT}{\hbar^2}\right)^{3/2} e^{-\frac{(E_c-E_F)}{kT}} = 2\left(\frac{m_e^* kT}{2\pi\hbar^2}\right)^{3/2} e^{-(E_c-E_F)/kT} \\ &= N_c e^{-(E_c-E_F)/kT} \end{aligned} \tag{1.9.4}$$

where $N_c = 2\left(\frac{m_e^* kT}{2\pi\hbar^2}\right)^{3/2}$.

It has the dimensions of concentration and is known as the ***effective density of states*** at the conduction band edge E_c.

Eqn. (1.9.4) represents the expression for the concentration or the density of electrons in the conduction band of an intrinsic semiconductor.

1.9.2 Concentration of holes in the valence band

To calculate the concentration of holes in the valence band, we note that the density of hole-states in the valence band is given by

$$N(E) = \frac{1}{2\pi^2}\left(\frac{2m_h^*}{\hbar^2}\right)^{3/2}(E_v - E)^{1/2} \text{ per unit volume}$$

where m_h^* is the *effective mass* of the hole and E_v is the energy of the top of the valence band.

Since an empty state in the valence band represents a hole, the probability of finding a hole with energy E is $1 - f(E)$, where $f(E)$ is the Fermi function.

Now, therefore, we have

$$\begin{aligned} 1 - f(E) &= 1 - \frac{1}{1 + e^{(E-E_F)/kT}} \\ &= 1 - \left[1 + e^{(E-E_F)/kT}\right]^{-1} \\ &= 1 - \left[1 - e^{(E-E_F)/kT}\right] \\ &= e^{(E-E_F)/kT} \end{aligned}$$

Hence the hole concentration p in the valence band is given by

$$\begin{aligned} p &= \int_{\infty}^{0} \frac{1}{2\pi^2}\left(\frac{2m_h^*}{\hbar^2}\right)^{3/2}(E_v - E)^{1/2} e^{\frac{E-E_F}{kT}}\, dE \\ &= \frac{1}{2\pi^2}\left(\frac{2m_h^*}{\hbar^2}\right)^{3/2} e^{(E_v-E_F)/kT}\int_0^{\infty}(kT)^{3/2}x^{1/2}e^{-x}dx \quad \text{where } \frac{E_v-E}{kT} = x \\ &= \frac{1}{2\pi^2}\left(\frac{2m_h^*}{\hbar^2}\right)^{3/2} e^{\frac{E_v-E_F}{kT}}(kT)^{3/2}\,\Gamma\left(\frac{3}{2}\right) \\ &= \frac{1}{2\pi^2}\left(\frac{2m_h^*}{\hbar^2}\right)^{3/2} e^{\frac{E_v-E_F}{kT}}\cdot(kT)^{3/2}.\frac{1}{2}\sqrt{\pi} \\ &= 2\left(\frac{m_h^*kT}{2\pi\hbar^2}\right)^{3/2} e^{\frac{E_v-E_F}{kT}} \\ &= N_v e^{-(E_F-E_v)/kT} \end{aligned} \tag{1.9.5}$$

where $N_v = 2(m_h^* kT/2\pi\hbar^2)^{3/2}$.

This is the expression for ***hole concentration*** in the valence band of an intrinsic semiconductor.

In equation (1.9.5), N_v has the dimensions of concentration and it is known as the ***effective density of states*** in the valence band edge.

• The two expressions for the concentration—one for electrons and the other for holes—as given by equations (1.9.4) and (1.9.5) respectively should profitably be observed with a comparative look.

1.10 Calculation of Fermi level in intrinsic semiconductors

We are now in a position to calculate the Fermi level in intrinsic semiconductors.

The concentration of electrons and holes represented by (1.9.4) and (1.9.5) respectively, are valid for both the intrinsic and extrinsic semiconductors in thermal equilibrium condition. Since the carriers are created in pairs in intrinsic semiconductors, the concentration of electron and hole concentration are equal

$$\text{i.e., } n = p = n_i \text{ (say).}$$

So, from (1.9.4) and (1.9.5), we get

$$N_c e^{-(E_c - E_F)/kT} = N_v e^{-(E_F - E_v)/kT}$$

$$\text{or,} \quad \ln \frac{N_v}{N_c} = \frac{1}{kT}(-E_c + 2E_F - E_v)$$

$$\text{or,} \quad \frac{kT}{2} \ln \frac{N_v}{N_c} = E_F - \frac{E_c + E_v}{2}$$

$$\text{or,} \quad E_F = \frac{E_c + E_v}{2} + \frac{kT}{2} \ln \frac{N_v}{N_c}$$

$$= \frac{E_c + E_v}{2} + \frac{3kT}{4} \ln \left(\frac{m_h^*}{m_e^*}\right) \tag{1.10.1}$$

on substituting the values of N_v and N_c in terms of the effective masses of the electrons and the holes.

If it is assumed that $m_h^* = m_e^*$, then we have from equation (1.10.2) above

$$\boxed{\boldsymbol{E_F = \tfrac{1}{2}(E_c + E_v)}} \tag{1.10.2}$$

i.e., *the Fermi level of an intrinsic semiconductor lies exactly at the middle of the forbidden energy gap* $\boldsymbol{E_g}$ as is illustrated in Fig. 1.15.

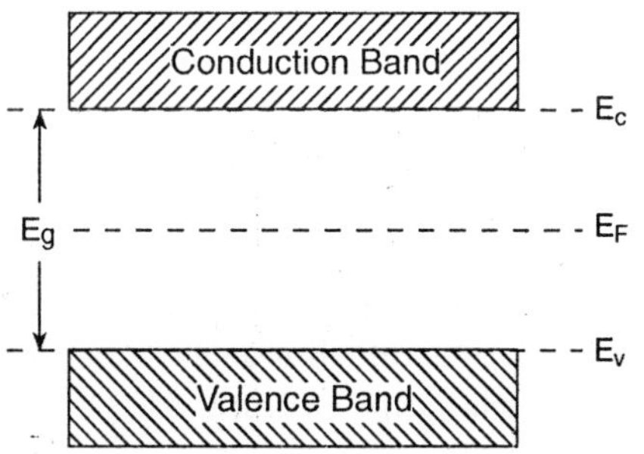

Fig. 1.15 Fermi level E_F in intrinsic semiconductor

In general, however, the Fermi level is raised slightly with temperature, because $m_h^* > m_e^*$.

1.11 Temperature dependence of carrier concentration in intrinsic semiconductors

For intrinsic semiconductor, $n = p = n_i$ and $E_F = \frac{1}{2}(E_c + E_v)$. So, from the equations (1.9.4) and (1.9.5), we obtain

$$n_i^2 = N_c N_v e^{-(E_c - E_v)/kT}$$

$$\therefore \quad n_i = \sqrt{N_c N_v} e^{-E_g/2kT}$$

where $E_c - E_v = E_g$ represents the band gap.

$$\therefore \quad n_i = 2\left(\frac{kT}{2\pi\hbar^2}\right)^{3/2} (m_e^* m_h^*)^{3/4} e^{-E_g/2kT} \tag{1.11.1}$$

Experimental results show that the band gap for most of the semiconductors decreases linearly with temperature so that we may write

$$E_g = E_{g0} - \alpha T$$

where E_{g0} is the band gap at 0 K and and α represents a constant, characteristic of the specific semiconductor.

Therefore (1.11.1) can be rewritten in the following form.

$$\begin{aligned} n_i &= 2\left(\frac{kT}{2\pi\hbar^2}\right)^{3/2} (m_e^* m_h^*)^{3/4} e^{-(E_{g0} - \alpha T)/2kT} \\ &= AT^{3/2} e^{-E_{g0}/2kT} \end{aligned} \tag{1.11.2}$$

where $A = 2\left(\frac{k}{2\pi\hbar^2}\right)^{3/2} (m_e^* m_h^*)^{3/4} e^{\alpha/2k}$.

Eqn. (1.11.1) or (1.11.2) points to the fact that the carrier concentration in intrinsic semiconductor is strongly temperature-dependent.

1.12 Law of mass action

Eqn. (1.9.4) and (1.9.5) can be applied for both intrinsic and extrinsic semiconductors in thermal equilibrium. On multiplication of (1.9.4) and (1.9.5), we get

$$\begin{aligned} np &= 4\left(\frac{kT}{2\pi\hbar^2}\right)^3 (m_e^* m_h^*)^{3/2} e^{-E_g/kT} \\ &= n_i^2, \text{ using equation (1.11.1).} \end{aligned} \tag{1.12.1}$$

The above relation is rather important and is called the ***law of mass action***. It shows that the product of the electron and the hole concentrations *depends on temperature* T and the *forbidden energy gap* E_g. The product however is *independent of* the Fermi level E_F.

1.13 Carrier concentration in extrinsic semiconductors

Let us now calculate the carrier concentration in extrinsic semiconductor.

Extrinsic semiconductors are prepared by doping n-type or p-type impurities in the pure or intrinsic semiconductors. When n-type impurities (pentavalent) are added, the number of free electrons increases over the intrinsic electron concentration but with the excess electrons the increased rate of recombination decreases the number of holes. In a similar way, the addition of p-type impurities (trivalent) increases the number of holes and decreases the number of electrons. The donor atoms are positive ions because they have given up electrons and the acceptor atoms are negative ions as they have accepted electrons. Since the semiconductor as a whole is electrically neutral, we may write the condition of the charge neutrality by assuming that all the donor and acceptor impurity atoms are ionized, as

$$p + N_D = n + N_A \tag{1.13.1}$$

where N_D and N_A are the concentration of donor and acceptor impurity atoms respectively. Law of mass action gives

$$np = n_i^2$$

$$\therefore \quad n = \frac{n_i^2}{p} \quad \text{and} \quad p = \frac{n_i^2}{n} \tag{1.13.2}$$

For an n-type semiconductor $n \gg p$ and only the donor impurities are added, so that $N_A = 0$ and the donor density is made larger than the density of holes, that is $N_D \gg p$.

So, from (1.13.1), we get
$$n \approx N_D \tag{1.13.3}$$

From (1.13.2), we have
$$p = \frac{n_i^2}{N_D} \tag{1.13.4}$$

Similarly for p-type semiconductor $p \gg n$ and $N_A \gg n$, with $N_D = 0$ and we can write from (1.13.1) and (1.13.2),

$$p \approx N_A \tag{1.13.5}$$

$$\text{and } n = \frac{n_i^2}{N_A} \tag{1.13.6}$$

The above equation (1.13.6) signifies that the increase in majority carrier density raises the probability of recombination in such a fashion that the minority carrier density decreases, maintaining the product of the electron and the hole density as n_i^2, a constant.

1.14 Fermi level in extrinsic semiconductors

The electrical properties of semiconductors depend on the concentrations of electrons and holes, and the Fermi level indicates the probability of occupancy of a level. In an intrinsic semiconductor, as we have already observed, the Fermi level lies exactly at the middle of the energy gap between the valence band and the conduction band. The position indicates equal concentration of electrons in conduction band and that of holes in valence band. In an extrinsic semiconductor, the concentration of the electrons and the holes are not the same. So, the position of Fermi level may be shifted towards the conduction band or the valence band.

1.14.1 Fermi level in n-type semiconductors

In an n-type semiconductor, all the donor impurity atoms are assumed to be ionized. So, many of the energy states near the bottom of the conduction band are occupied by the donor electrons and a few holes exist in the valence band. So, electrons from valence band cannot cross the energy gap easily by thermal agitation. Since E_F is the probability of occupancy of the allowed energy states, E_F must move closer to the conduction band (Fig. 1.16(a)).

For n-type semiconductor, $n \gg p$ and $n \approx N_D$. Here $N_A = 0$. So, from the equation (1.9.4) we obtain

$$N_D = N_c e^{-(E_c - E_F)/kT}$$

$$\frac{N_c}{N_D} = e^{(E_c - E_F)/kT}$$

$$\therefore \quad \ln \frac{N_c}{N_D} = \frac{E_c - E_F}{kT}$$

$$\boxed{\boldsymbol{E_F = E_c - kT \ln \frac{N_c}{N_D}}} \tag{1.14.1}$$

The above relation clearly exhibits the following worth noting features.

1. The Fermi level lies below the bottom of the conduction band.
2. With the increase in doping concentration N_D, the Fermi level moves closer to the conduction band.
3. As the temperature increases, the Fermi level shifts away from E_c.

1.14.2 Fermi level in p-type semiconductors

In a p-type semiconductor, all the acceptor atoms are assumed to be ionized by accepting the electrons from the valence band so that there are many holes in the valence band. Since $p \gg n$ in a p-type semiconductor, the E_F must move closer to valence band (Fig. 1.16(b)).

For a p-type semiconductor $N_D = 0$ and $p \gg n$ so that $p \approx N_A$ and (1.9.5) gives

$$N_A = N_v e^{-(E_F - E_v)/kT}$$

$$\text{or,} \quad \frac{N_v}{N_A} = e^{(E_F - E_v)/kT}$$

Taking the logarithm from both sides, we get

$$\ln \frac{N_v}{N_A} = (E_F - E_v)/kT$$

$$\therefore \quad \boxed{E_F = E_v + kT \ln \frac{N_v}{N_A}} \qquad (1.14.2)$$

The above relation furnishes us the following important facts.

1. The Fermi level E_F lies above the top of the valence band.
2. The increase in N_A causes the E_F to move closer to the valence band.
3. As the temperature increases, the E_F shifts away from E_v.

If the temperature of the n-type or the p-type semiconductors is increased, the concentrations of the electrons and the holes become equal at high temperatures. So, it can be said that the extrinsic semiconductor behaves like an intrinsic semiconductor at high temperature and the Fermi level gets shifted towards the centre of the forbidden energy gap.

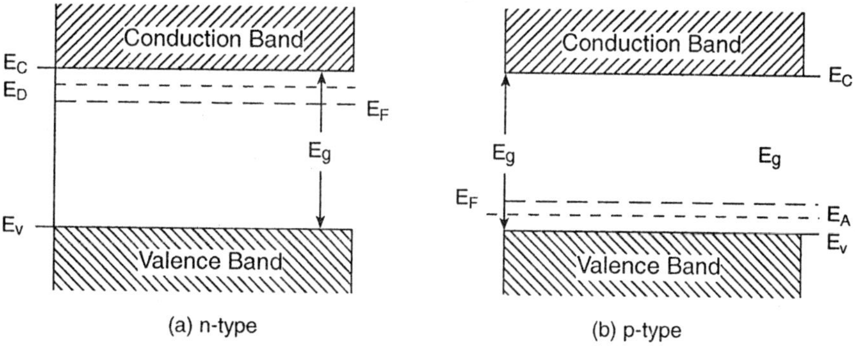

Fig. 1.16 Fermi level in extrinsic semiconductor

1.15 Electrical conduction in semiconductors

In metals, the electric current is carried by only the drift motion of the free electrons under the influence of an applied electric field. But in semiconductors, the current is carried by two types of carriers, namely the electrons and the holes, by two different methods—the *drift* and the *diffusion* of the charge carriers. In both the methods, electrons and holes participate in the conduction process.

1.15.1 Drift current

We shall first discuss the drift current, followed by a discussion of the diffusion current and finally the total current.

In semiconductors, the charge carriers i.e., the free electrons and the holes, move randomly in all possible directions due purely to thermal motion in the absence of any applied electric field (Fig. 1.17). The random motion is caused by scattering due to the vibration of the lattice atoms (i.e., phonons) and the Coulomb field of the ionized impurity atoms. Other imperfections such as dislocations may also contribute to scattering but to a lesser degree. If we consider a certain area inside the semiconductor, it is found that the number of carriers passing the area in any direction is exactly equal to the number of carriers passing the same area in the opposite direction resulting in zero average current. When an external electric field is applied, the random motion of the carriers gets a component along the direction of the field. This directed motion is called *drift motion.* In steady state condition, the energy and momentum acquired by carriers by the field is equally used up through scattering in the crystal. As a result, a steady value of drift velocity and hence a steady electric current is attained.

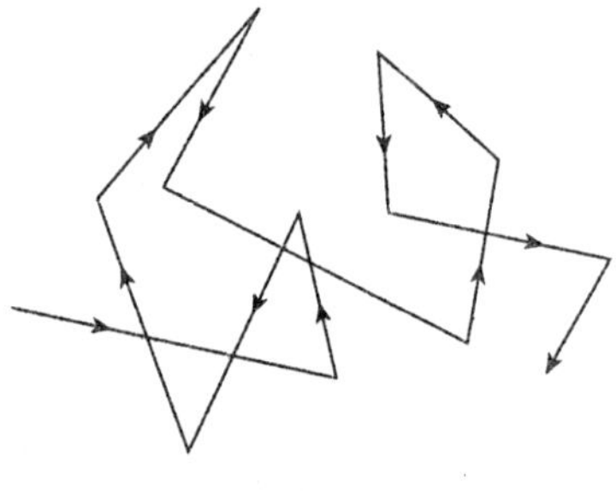

Fig. 1.17 Random thermal motion

Conductivity—To find an expression for the conductivity of semiconductor, consider a cylindrical type semiconductor of uniform cross-sectional area A through which current carriers move with the drift velocity v under the influence of an electric field E (Fig. 1.18). Since the the drift velocity is v, the carriers travel a distance v in one second and hence all the carriers within the cylinder of length v pass through the cross-section A in one second. If n be the number of carriers per unit volume then the amount of charge passing the cross-section in one second i.e., the electric current flowing through the semiconductor is

$$I = nevA \tag{1.15.1}$$

The current density i.e., the current through unit cross-sectional area is

$$J = \frac{I}{A} = nev \tag{1.15.2}$$

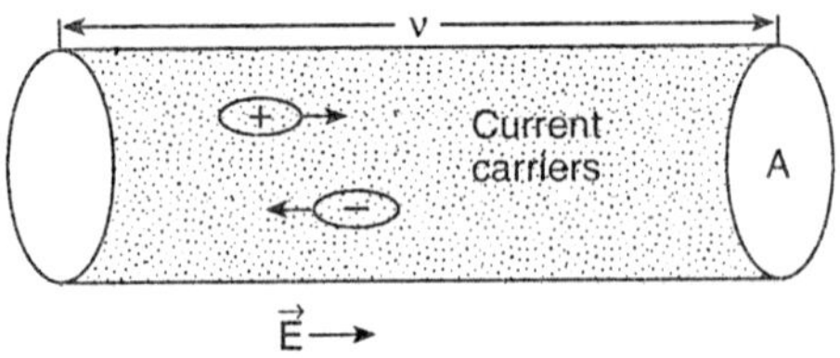

Fig. 1.18 Drift of current carriers

When low electric field is applied, the average drift velocity is proportional to the electric field E i.e.,

$$v \propto E \Rightarrow v = \mu E$$

$$\text{or,} \quad \boxed{\mu = \frac{v}{E}} \tag{1.15.3}$$

where the constant μ is called the ***mobility*** of the carriers, defined as the *average drift velocity per unit electric field.* The mobility μ is inversely proportional to the effective mass of a particle. In general, $m_h^* > m_e^*$ so that $\mu_n > \mu_p$.

From (1.15.2) and (1.15.3) we get, $J = ne\mu E$ (1.15.4)

From (1.15.4) we can write, the drift current density due to electrons as

$$J_e = en\mu_n E \tag{1.15.5}$$

and that due to holes is: $$J_h = ep\mu_p E \tag{1.15.6}$$

Since the electrons and the holes both contribute to current in the same direction, the total drift current density in semiconductor is given by

$$J = J_e + J_h = e(n\mu_n + p\mu_p)E \tag{1.15.7}$$

The electrical *conductivity* σ is defined as the current density per unit electric field so that we have

$$\sigma = J/E \Rightarrow J = \sigma E \tag{1.15.8}$$

Comparing (1.15.4) and (1.15.8), $$\boxed{\sigma = ne\mu} \tag{1.15.9}$$

If the current carriers are electrons and holes, then the relevant conductivities are given respectively by

$$\sigma_n = ne\mu_n \tag{1.15.10}$$

$$\text{and} \quad \sigma_p = pe\mu_p \tag{1.15.11}$$

where n and p are the concentrations of electrons and holes respectively.

Since electrons and holes are oppositely charged and their drift velocities are in opposite direction, they *contribute* to the current in the same direction. So, the overall conductivity of the semiconductor is

$$\sigma = \sigma_n + \sigma_p = e(n\mu_n + p\mu_p) \tag{1.15.12}$$

For intrinsic semiconductor $n = p = n_i$ and the conductivity is given by

$$\boxed{\sigma_i = en_i(\mu_n + \mu_p)} \tag{1.15.13}$$

Now, using (1.11.1) we get,

$$\begin{aligned}\sigma &= 2e\left(\frac{kT}{2\pi\hbar^2}\right)^{3/2}(m_e^* m_h^*)^{3/4}(\mu_n + \mu_p)e^{-E_g/2kT} \\ &= Ae^{-E_g/2kT}, \text{ where} \\ A &= 2e\left(\frac{kT}{2\pi\hbar^2}\right)^{3/2}(m_e^* m_h^*)^{3/4}(\mu_n + \mu_p)\end{aligned} \tag{1.15.14}$$

The variation of A with temperature is relatively slow compared to that of the exponential term for moderate temperatures and hence it may be considered as constant. Taking logarithm on both sides of (1.15.14) we get,

$$\ln \sigma = \ln A - \frac{E_g}{2kT} \tag{1.15.15}$$

The plot of, $\ln \sigma$ vs. $1/T$ is a straight line as shown in Fig. 1.19. From the slope of this *straight line*, one can find the forbidden energy gap E_g.

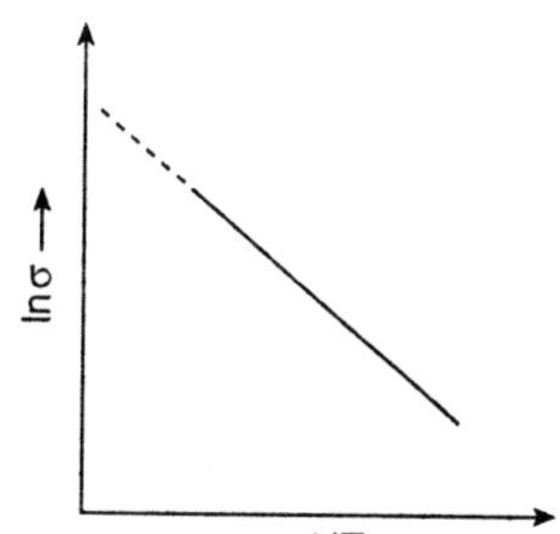

Fig. 1.19 Plot of ln σ vs. 1/T

1.15.2 Diffusion current

The motion of particles from a region of higher concentration to a region of lower one to destroy the non-uniformity of the particle concentration is called ***diffusion***. In semiconductor, it is possible to have a non-uniform concentration of charge carriers (i.e., electrons and holes) by applying different methods. So, by the process of diffusion, charge carriers will move from higher concentration region to lower one constituting a current known as ***diffusion current***.

Let, in a semiconductor, the concentration of the holes p decrease with increase in x. So, there is a concentration gradient of holes dp/dx which is negative in sign (Fig. 1.20).

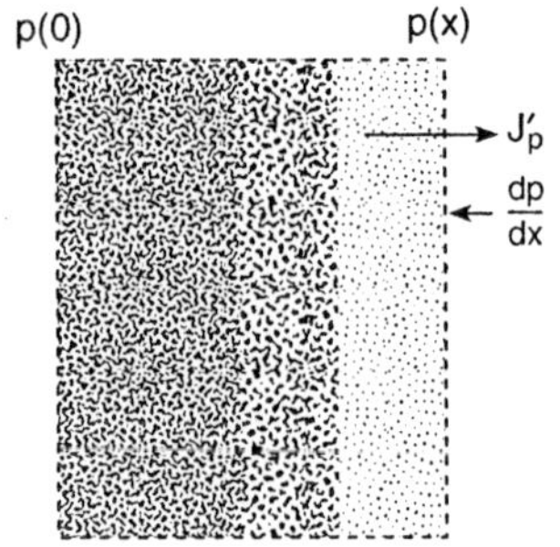

Fig. 1.20 Hole diffusion

The diffusion hole-current density is given by

$$J_p' = -eD_p\frac{dp}{dx} \tag{1.15.16}$$

where D_p is called the ***diffusion constant*** or ***diffusivity*** of the holes.

Since dp/dx is negative, the relation (1.15.16) ensures that J_p' is positive along the positive direction.

Similarly, the diffusion electron-current density is given by

$$J_n' = eD_n\frac{dn}{dx} \tag{1.15.17}$$

where D_n and dn/dx are the *diffusivity* and *concentration gradient* respectively of electrons.

1.15.3 Total current

When an external electric field is applied across a semiconductor in which a concentration gradient is also present, both the drift and diffusion currents exist. The total hole and the electron currents are then plainly the sum of the drift current and the diffusion current. The total hole current and the electron current densities are given respectively by the following equations

$$J_h = J_p + J_p' = e\mu_p pE - eD_p(dp/dx) \tag{1.15.18}$$

$$\text{and} \quad J_e = J_n + J_n' = e\mu_n nE + eD_n(dn/dx) \tag{1.15.19}$$

with the usual meanings of the symbols.

When no external field is applied, the resultant diffusion current density is

$$J_D = J_n' + J_p' = e\left(D_n\frac{dn}{dx} - D_p\frac{dp}{dx}\right) \tag{1.15.20}$$

The ***total current density J*** in a semiconductor is thus given by

$$J = J_h + J_e = e(n\mu_n + p\mu_p)E + e\left(D_n\frac{dn}{dx} - D_p\frac{dp}{dx}\right) \tag{1.15.21}$$

1.15.4 Einstein's relationship between diffusivity and mobility

The mobility and diffusivity are controlled by the collisions experienced by the charge carriers in the crystal. They are thus statistical thermodynamic phenomena and would somehow be interrelated.

To find the relationship between diffusivity D and mobility μ, let us consider a n-type semiconductor bar with non-uniform charge distribution along its length placed along the x-axis. The flow of electrons produces uncompensated space charge which results in an internal electric field E. This field causes an ohmic current which flows against the diffusion current. In equilibrium, the two currents are equal. Therefore,

$$en\mu_n E = eD_n\frac{dn}{dx} \tag{1.15.22}$$

Assuming the electrons as particles of classical electron gas, the pressure may be expressed as

$$p = nkT$$

where n is the number of electrons per unit volume, k the Boltzmann constant.

So, the force per unit volume is

$$F = \frac{dp}{dx} = kT\frac{dn}{dx} \tag{1.15.23}$$

The force on the electrons contained in unit volume due to electric field E is

$$F = en\cdot E = \frac{eD_n}{\mu_n}\frac{dn}{dx} \tag{1.15.24}$$

using equation (1.15.22).

Comparing (1.15.23) and (1.15.24), we have

$$e\frac{D_n}{\mu_n} = kT$$

$$\text{or,}\quad \frac{D_n}{\mu_n} = \frac{kT}{e} \tag{1.15.25}$$

Similarly, for holes we may write,

$$\frac{D_p}{\mu_p} = \frac{kT}{e} \tag{1.15.26}$$

From (1.15.25) and (1.15.26) we get,

$$\frac{D_p}{\mu_p} = \frac{D_n}{\mu_n} = \frac{kT}{e} \tag{1.15.27}$$

The above equation is known as ***Einstein's relationship*** *between diffusivity and mobility.*

The factor kT/e is known as the *volt equivalent* of temperature V_T and its value is 26 mV at room temperature (300 K). So, equation (1.15.27) may be written as

$$\frac{D_p}{\mu_p} = \frac{D_n}{\mu_n} = 0.026\,\text{volt} \tag{1.15.28}$$

• Einstein relation is very useful in determining diffusivity when the measured value of mobility is known. The electron mobility in Ge at room temperature is $\mu_n = 0.38\ \text{m}^2/\text{V.s}$ and the diffusivity $D_n = 0.026\mu_n\,\text{V} = 0.026 \times 0.38\ \text{m}^2/\text{s}$.

1.15.5 Generation and recombination : Lifetime of minority carriers

In an intrinsic semiconductor, the electron concentration n is equal to the hole concentration p. Due to thermal agitation there occurs a continuous generation of new electron-hole pairs (*EHPs*). On the other hand, there exists an opposite process known as *recombination* in which a free electron jumps to an empty covalent bond resulting in the disappearance of an electron-hole pair and releasing some amount of energy equal to E_g. At equilibrium, the rate of generation g_i of *EHPs* is equal to the rate of recombination r_i of *EHPs*, When an external agent causing the generation of excess carrier is stopped to function, the excess concentration quickly reduces to equilibrium value. The average time of existence of a carrier (hole or electron) before recombination is called *mean lifetime* τ of that carrier.

When the minority carriers in an extrinsic semiconductor increases over the thermal equilibrium value and the source of external energy ceases to act, the number of excess carriers decays with time due to recombination.

Let n_0 and p_0 be the thermal equilibrium concentrations of electrons and holes respectively in an extrinsic semiconductor. If Δn and Δp be the changes of electron concentration and hole concentration from the thermal equilibrium values, then the rate of recombination is given by

$$r' = k(n_0 + \Delta n)(p_0 + \Delta o) \tag{1.15.29}$$

where k is the proportionality constant.

In thermal equilibrium, the rate of generation g_0 is equal to the rate of recombination r_0. So, we have

$$g_0 = r_0 = kn_0p_0 \tag{1.15.30}$$

since $\Delta n = \Delta p = 0$.

Therefore, the rate of recombination r for the excess carriers is given by

$$\begin{aligned} r &= r' - r_0 \\ &= k\left[(n_0 + \Delta n)(p_0 + \Delta p) - n_0p_0\right] \\ &= k(n_0\Delta p + p_0\Delta n) \end{aligned} \tag{1.15.31}$$

where $\Delta p\Delta n$ is neglected for its small value.

Now, for n-type semiconductor, $n_0\Delta p \gg p_0\Delta n$ because $n_0 \gg p_0$ and $\Delta n \approx \Delta p$. So, the equation (1.15.31) becomes

$$r = kn_0\Delta p \tag{1.15.32}$$

Similarly, for p-type material we may write

$$r = kp_0\Delta n \tag{1.15.33}$$

Both the equations (1.15.32) and (1.15.33) indicate that the rate of recombination r is proportional to the change in concentration of the minority carriers i.e., Δn for p-type and Δp for n-type semiconductor.

Again, for an n-type semiconductor, we may write,

$$r = \frac{d\Delta p}{dt} = \frac{\Delta p}{\tau_p} \tag{1.15.34}$$

where τ_p is the average lifetime of holes.

Comparing (1.15.32) and (1.15.34) we obtain,

$$\tau_p = \frac{1}{kn_0} \tag{1.15.35}$$

Now, from equation (1.15.34), we have

$$\frac{d\Delta p}{\Delta p} = \frac{1}{\tau_p}dt$$

Integrating,
$$\Delta p(t) = \Delta p(0)e^{-t/\tau_p} \tag{1.15.36}$$

where $\Delta p(0)$ corresponds to the change in hole concentration at time $t = 0$.

Equation (1.15.36) indicates that Δp decays exponentially with time and τ_p is the time corresponding to *time constant.*

Similarly, one can show that the excess electron concentration in a p-type semiconductor varies with time as

$$\Delta n(t) = \Delta n(0)e^{-t/\tau_n} \tag{1.15.37}$$

where $\tau_n = 1/kp_0$ is the lifetime of the excess electron or the time constant.

1.16 Hall effect

When a magnetic field is applied perpendicular to a current-carrying specimen (metal or semiconductor), an electric field is developed across the specimen in the direction perpendicular to both the current and the magnetic field. This phenomenon is called the ***Hall effect*** after its discoverer. The field so developed is known as the ***Hall field*** or the ***Hall voltage***.

Explanation—When a charged particle moves in a magnetic field, it experiences a *Lorentz force* : $\boldsymbol{F} = q(\boldsymbol{v} \times \boldsymbol{B})$. The moving charge-carriers, constituting the current inside the specimen, are forced to one side by the Lorentz force when they pass through the magnetic field and a potential called *Hall voltage* is developed in the specimen due to the shifting of carriers. The action of the Hall voltage is to oppose the deflection of the charge due to the magnetic field. So, an equilibrium is established when the force by magnetic field is balanced by the force applied by the Hall field.

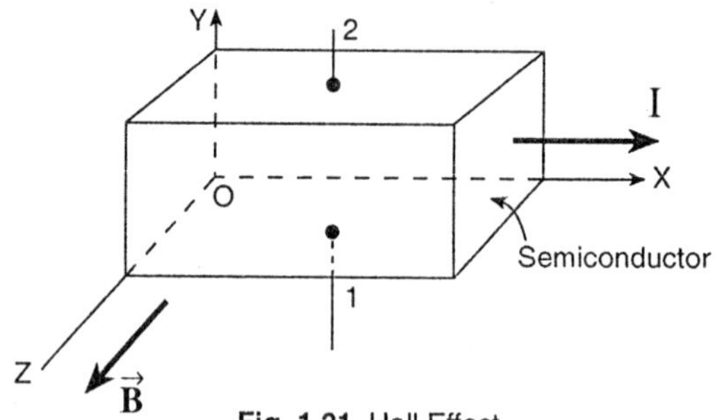

Fig. 1.21 Hall Effect

Expression for Hall voltage—To obtain an expression for the Hall voltage let us consider a p-type semiconductor slab (Fig. 1.21) subject to an external electric field along the x-axis. Let the current density along x-axis be J_x and the magnetic field applied along the z-axis be $\vec{B} = B\hat{k}$. If $\vec{v}_x = v_x\hat{i}$ be the drift velocity of the charge-carrier, then the average force on a carrier in the magnetic field is given by

$$\begin{aligned}\vec{F} &= e(\vec{v}_x \times \vec{B}) \\ &= e(v_x\hat{i} \times B\vec{k}) \\ &= ev_xB\hat{i} \times \vec{k}\end{aligned}$$

$$\therefore \quad \boxed{\vec{F} = -ev_xB\hat{j}} \qquad (1.16.1)$$

The above boxed equation indicates that the holes are deflected towards the negative y-axis under the action of the force. As a result, the side 1 and side 2 of the slab becomes +ve and −ve respectively, giving rise to an electric potential, called the *Hall voltage*. Let the Hall field along the +ve y-axis be denoted by E_H. So, a hole will experience a force eE_H along +ve y-axis. This force tries to prevent further shift of holes by opposing the magnetic force. Therefore, equilibrium is established when the magnetic force and the force due to Hall field balance each other.

$$\therefore \quad eE_H = ev_xB$$

$$\text{or,} \quad \boxed{E_H = v_xB} \qquad (1.16.2)$$

which gives the Hall field.

If p be the concentration of the holes, then the current density along x-axis is

$$J_x = epv_x \qquad (1.16.3)$$

From (1.16.2) and (1.16.3), we obtain

$$E_H = \frac{J_x B}{ep} = R_H J_x B \tag{1.16.4}$$

where
$$\boxed{R_H = \frac{1}{pe}} \tag{1.16.5}$$

This R_H is called the ***Hall coefficient*** for the substance of the slab.

The expression (1.16.4) shows that the ***Hall field is proportional to the current as well as to the magnetic field.***

From (1.16.5), ***Hall coefficient*** may be defined as *the Hall field developed when unit current density flows through the specimen with unit applied magnetic field.* R_H is +ve when the current carriers are holes.

● For a n-type semiconductor, the Hall field is opposite to that of the holes and the Hall coefficient is therefore given by

$$\boxed{R_H = -\frac{1}{ne}} \tag{1.16.6}$$

which is plainly negative.

● Note that the *sign of the Hall coefficient is the same as that of the charge carriers.*

● Since the Hall coefficient R_H is inversely proportional to the electron (or hole) concentration n (or p), we can determine n (or p) by measuring Hall coefficient. In fact, that is the *standard technique* for determining the electron concentration. The technique is all the more valuable since the only other quantity is the electronic charge, a *fundamental constant of known value.*

● Another interesting feature of the transport process in presence of a magnetic field is that the *current* flowing along the $+x$ direction is *not influenced by the field.* This implies that the *electric resistance is independent of the magnetic field.* This is somewhat unexpected.

Relation between mobility and Hall coefficient—The mobility of the charge carriers is given by

$$\mu = \frac{v_x}{E_x} \Rightarrow v_x = \mu E_x \tag{1.16.7}$$

where E_x is the external applied field along the x-axis.

From (1.16.2), using (1.16.7), we get

$$E_H = \mu E_x B \tag{1.16.8}$$

Comparing the equations (1.6.4) and (1.16.8), we have

$$R_H J_x B = \mu E_x B$$

$$\text{or,} \qquad R_H = \mu \frac{E_x}{J_x}$$

$$\text{or,}\quad R_H = \frac{\mu}{\sigma}, \quad \text{since } J_x = \sigma E_x$$

where σ is the electrical conductivity of the material.

Thus, finally, we have the following relation between the mobility and the Hall coefficient.

$$\boxed{\mu = \sigma R_H} \tag{1.16.9}$$

1.16.1 Experimental determination of Hall coefficient

The Hall coefficient of a metal or semiconductor may be experimentally determined by the following procedure.

A thin metallic or semiconductor slab of several mm in width and several cm in length is placed in such a way that a current flows along the length of the specimen, assumed to be x-direction. A magnetic field $\vec{B}$ is applied along the z-axis as shown

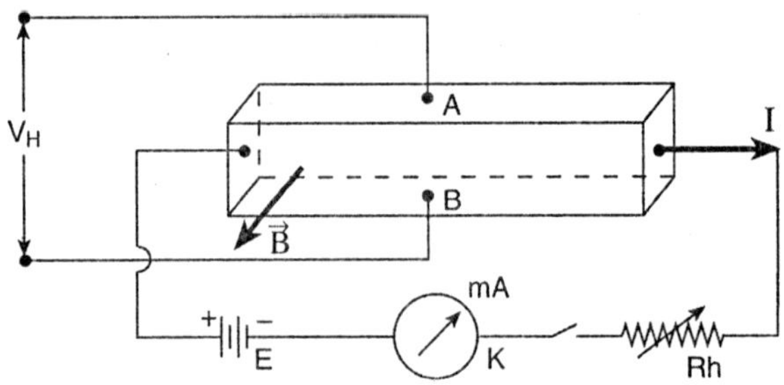

Fig. 1.22 Experimental arrangement for Hall effect

in Fig. 1.22. The current through the specimen can be adjusted by the rheostat Rh. A sensitive calibrated potentiometer is connected through the two port leads placed between the points A and B for measurement of the Hall voltage V_H. By measuring V_H, the Hall coefficient can be calculated as discussed below :

The Hall field is given by

$$E_H = \frac{J_x B}{pe} \tag{1.16.10}$$

If b and d be respectively the breadth and thickness of the specimen, then the current density is given by

$$J_x = \frac{I}{bd} \tag{1.16.11}$$

where I is the current flowing through the specimen.

From (1.16.10) and (1.16.11), we obtain

$$E_H = \frac{IB}{pebd} \tag{1.16.12}$$

Again, the Hall voltage $V_H = E_H d = \dfrac{IB}{peb} = R_H \dfrac{IB}{b}$, using (1.16.5) and (1.16.12).

$\therefore$ Hall coefficient is:

$$\boxed{R_H = \frac{V_H b}{IB}} \tag{1.16.13}$$

So, by measuring the Hall voltage V_H with the help of a calibrated potentiometer and knowing the values of I and B, we can readily evaluate the value of the Hall coefficient R_H.

1.16.2 Applications of Hall effect

The useful applications of Hall effect are, in fact, many. We enumerate some of them in what follows.

1. *Determination of the nature of the semiconductor*—If the current carriers are negatively charged electrons, the measured value of R_H is negative which indicates that the semiconductor is an n-type one. If, on the other hand, R_H is positive the semiconductor is of p-type having holes as majority carriers.

2. *Determination of carrier concentration*—The concentration of the holes and the electrons can be obtained from either (1.16.5) or (1.16.6), provided we measure the Hall coefficient R_H experimentally.

3. *Determination of carrier mobility*—If ρ be the charge density of a semiconductor, then the mobility μ and the conductivity σ are related by the equation

$$\sigma = \rho\mu.$$

The relation between Hall coefficient R_H and mobility μ is given by

$$\mu = \sigma R_H$$

So, by measuring σ and R_H, the mobility μ can be readily determined.

• The mobility determined by R_H is somewhat different from the actual drift mobility and is known as ***Hall mobility***.

4. *Determination of magnetic field*—From (1.16.13), we have the following expression for the magnetic field B.

$$B = \frac{V_H b}{I R_H},$$

the symbols having their usual significance. So, for a specimen of known dimensions, if the Hall voltage and the Hall coefficient are known, the magnetic field can be determined by using the above equation.

5. *Power in e.m wave*—In an e.m. wave in free space, the electric field $\vec{E}$ and the magnetic field $\vec{H}$ are at right angles to each other. Hall voltage depends on the cross-product of $\vec{E}$ and $\vec{H}$, and can thus be used as a *microwave power flow detector*. Current in the material is proportional to the electric field so that the Hall voltage is $\alpha\vec{E} \times \vec{H}$. So the Hall voltage of a given crystal can be calibrated for different values of microwave power flow $(\vec{E} \times \vec{H})$. A crystal thus calibrated and supplied with contacts for measuring the dc Hall voltage can be inserted in a waveguide. Then the dc voltage directly gives the microwave power flow.

1.17 Thermistors, sensistors and photoconductors

The conductivity of metal decreases with increase in temperature because the enhanced thermal vibrations of electrons reduce the mobility. So, the temperature coefficient is positive for a metal. But, the electrical conductivity of a semiconductor changes significantly with temperature. This is due to the increase in the electron and hole concentrations with increasing temperature. This property is utilized with convenience in a device called ***thermistor***.

A *thermistor* is a semiconductor device which is thermally sensitive and has usually a negative temperature coefficient of resistivity. The insertion of a thermistor in an electrical circuit can compensate for temperature variations in a range of 100°C. Thermistors are extensively used in control systems operated by temperature changes, in accurate measurement of temperatures, in measurement of micro-wave power, in thermometry and also as a thermal relay.

A heavily doped semiconductor behaves as a metal. It exhibits positive temperature coefficient of resistivity because of the decrease of the carrier mobility with increasing temperature. Such a device is called a ***sensistor***.

When an electromagnetic wave is incident on semicounductor, the electron-hole pairs in excess of those generated thermally are produced. So, the conductivity of the material increases due to increased number of current carriers and the conductivity increases almost linearly with the intensity of incident radiation. Hence, such a device is called ***photoresistor*** or ***photoconductor***. It is used in measurement of illumination, as in lightmeter, to record the modulating light intensity on a sound track and relay in a control circuit.

1.18 Illustrated Examples

▶ **Example 1.** *The conductivity of* Ge *is* $0.7\,\Omega^{-1}\,\text{m}^{-1}$ *at* $0°$ C *and* $2\Omega^{-1}\,\text{m}^{-1}$ *at* 20°C. *What is the average band gap of* Ge*? Given Boltzmann constant* $k = 1.38\times10^{-23}\,\text{JK}^{-1}$.

(Burdwan Hons.)

Solution: The conductivity of a semiconductor is given by

$$\sigma = Ae^{-\frac{E_g}{2kT}}$$

$$\therefore\ \ln\sigma = \ln A - \frac{E_g}{2kT}$$

Let σ_1 and σ_2 be the conductivities at T_1K and T_2K respectively.

$$\therefore\ \ln\sigma_1 = \ln A - \frac{E_g}{2kT_1} \tag{1.18.1}$$

$$\text{and}\ \ln\sigma_2 = \ln A - \frac{E_g}{2kT_2} \tag{1.18.2}$$

Subtracting (1.18.1) from (1.18.2), we obtain

$$\ln \sigma_2 - \ln \sigma_1 = \frac{E_g}{2k}\left(\frac{1}{T_1} - \frac{1}{T_2}\right)$$

$$\begin{aligned} \text{or,} \quad \boldsymbol{E_g} &= 2k\frac{(\ln \sigma_2 - \ln \sigma_1)}{1/T_1 - 1/T_2} \\ &= 2 \times 1.38 \times 10^{-23} \times \frac{\ln 2 - \ln 0.7}{(1/273) - (1/293)}\,\text{J} \\ &= \mathbf{0.72 \ eV} \end{aligned}$$

▶ **Example 2.** *Find* (i) *the conductivity and* (ii) *the resistance of a bar of pure silicon of length* 1 cm *and cross-sectional area* $1\,\text{mm}^2$ *at* 300 K. *Given* $\mu_n = 0.13\,\text{m}^2/\text{Vs}$, $\mu_p = 0.05\,\text{m}^2/\text{Vs}$, $n_i = 1.5 \times 10^{16}\,\text{m}^{-3}$ *and* $e = 1.6 \times 10^{-19}\,\text{C}$. **(Uttaranchal Univ.)**

Solution: (i) The conductivity of silicon is given by

$$\begin{aligned} \boldsymbol{\sigma_i} &= n_i e(\mu_n + \mu_p) \\ &= 1.5 \times 10^{16} \times 1.6 \times 10^{-19}(0.13 + 0.05)\,\text{Sm}^{-1} \\ &= \mathbf{4.32 \times 10^{-4}\,Sm^{-1}} \end{aligned}$$

(ii) The resistance of the pure silicon bar is given by

$$\begin{aligned} \boldsymbol{R} &= \frac{1}{\sigma_i}\cdot\frac{\ell}{A} \\ &= \frac{0.01}{4.32 \times 10^{-4} \times 10^{-6}}\,\Omega \\ &= \mathbf{23.15\,M\Omega} \end{aligned}$$

▶ **Example 3.** *An insulator has an optical absorption which occurs for all wavelengths shorter than* 1800 Å. *Find the width of the forbidden energy band for this insulator.* **(Lucknow Univ.)**

Solution: As the insulator has an optical absorption for all wavelengths shorter than 1800 Å, the forbidden energy band for the insulator is plainly the energy of a photon of wavelength 1800 Å.

∴ The width of the forbidden energy band is given by

$$\begin{aligned} \boldsymbol{E_g} &= h\nu = \frac{hc}{\lambda}, \quad \text{with usual symbols.} \\ &= \frac{6.626 \times 10^{-34} \times 3 \times 10^8}{1800 \times 10^{-10}}\,\text{J} \\ &= \mathbf{6.9\,eV}, \quad \text{on simplification.} \end{aligned}$$

► **Example 4.** *Find the conductivity of a p-type* Ge *crystal which is doped with acceptor atoms of concentration* 2×10^{17} atoms/cm^3 *and all the acceptor atoms are active. Given* $\mu_h = 1900\ \text{cm}^2/\text{Vs}$ *and* $e = 1.6 \times 10^{-19}$ C. **(Rajasthan Univ.)**

Solution: In a p-type semiconductor, holes are the majority carriers. So the hole concentration n_h is much greater than the electron concentration n_e and $n_h \approx N_a = 2 \times 10^{17}/\text{cm}^3 = 2 \times 10^{23}/\text{m}^3$ where N_a represents the concentration of the acceptor atoms.

∴ The required conductivity of the given crystal is

$$\begin{aligned} \sigma &= n_h \mu_h e \\ &= 2 \times 10^{23} \times 0.19 \times 1.6 \times 10^{-19}\, \Omega^{-1}\text{m}^{-1} \\ &= \mathbf{6100\, \Omega^{-1}\, m^{-1}} \end{aligned}$$

► **Example 5.** *The band gap of a specimen of gallium arsenide phosphide is* 1.98 eV. *Determine the wavelength of the electromagnetic radiation that is emitted upon direct recombination of electrons and holes in this sample. What is the colour of the emitted radiation?* **(Calcutta Hons.)**

Solution: The relation between the band gap energy and the wavelength of the radiation is given by

$$E_g = h\nu = \frac{hc}{\lambda}$$

$$\text{or,} \quad \lambda = \frac{hc}{E_g}$$

where

$$\begin{aligned} c &= \text{velocity of light in vacuum} = 3 \times 10^8\ \text{ms}^{-1} \\ h &= \text{Planck's constant} = 6.6 \times 10^{-34}\ \text{Js} \\ E_g &= \text{band gap energy.} \end{aligned}$$

$$\therefore\ \boldsymbol{\lambda} = \frac{6.6 \times 10^{-34} \times 3 \times 10^8}{1.98 \times 1.6 \times 10^{-19}}\ \text{m} = \mathbf{6250\ \text{Å}}$$

This wavelength belongs to the reddish region of the spectrum of visible light. So the colour of the emitted radiation is ***reddish.***

► **Example 6.** *Assuming that each atom of copper contributes one free electron, calculate the drift velocity of free electrons in the copper conductor of cross-sectional area* $10^{-4}\ \text{m}^2$ *carrying a current of* 200 A (*Atomic weight of copper* = 63.5, *Density of copper* $= 8.94 \times 10^3\ \text{kg m}^{-3}$, $e = 1.602 \times 10^{-19}$ C). **(Bangalore Univ.)**

Solution: The drift current density J_e due to electrons is given by

$$J_e = n_e e v_d \tag{1.18.3}$$

where n_e = concentration of electrons, e = electronic charge and v_d = drift velocity of electrons.

Now, atomic weight of Cu = 63.5. So, 1 mol. of Cu = 63.5 g = 63.5×10^{-3} kg. Density of Cu = 8.94×10^3 kg m^{-3}. No. of atoms per mole = 6.022×10^{23}.

∴ No. of atoms per m^3 is

$$n_e = \frac{6.022 \times 10^{23} \times 8.94 \times 10^3}{63.5 \times 10^{-3}} = \frac{6.022 \times 8.94}{63.5} \times 10^{29}$$

Area of cross-section $\alpha = 10^{-4}\,\text{m}^2$.

∴ Current density $J_e = \dfrac{I}{\alpha} = \dfrac{200}{10^{-4}}\ \text{Am}^{-2} = 2 \times 10^6\ \text{Am}^{-2}$

∴ From (1.18.3), we have the drift velocity v_d given by

$$v_d = \frac{J_e}{n_e e} = \frac{2 \times 10^6 \times 63.5}{6.022 \times 8.94 \times 10^{29} \times 1.6 \times 10^{-19}}\ \text{ms}^{-1}$$

$$= \mathbf{1.47 \times 10^{-4}\ ms^{-1}}$$

▶ **Example 7.** *Calculate the concentration of acceptor atoms that have to be added to an intrinsic semiconductor to produce a p-type semiconductor of conductivity* $0.05\,\Omega^{-1}\,\text{m}^{-1}$. *The mobility of holes in p-type semiconductor is* $0.4\,\text{m}^2\text{V}^{-1}\text{s}^{-1}$.

(Bangalore Univ.)

Solution: In p-type semiconductors, the concentration of the acceptor atoms N_A is equal to the concentration of holes p in valence band i.e., $N_A = p$.

If μ_p and σ_p are the hole mobility and the hole conductivity respectively, then we shall have

$$\sigma_p = pe\mu_p = N_A e\mu_p$$

with the usual meanings of the symbols.

The concentration of the acceptor atoms is thus given by

$$N_A = \frac{\sigma_p}{e\mu_p} = \frac{0.05}{1.6 \times 10^{-19} \times 0.4}\ \text{m}^{-3}$$

$$= \mathbf{7.8 \times 10^{17}\ m^{-3}}$$

▶ **Example 8.** *The energy gap in* Ge *is* 0.75 eV. *Compare the intrinsic conductivity of* Ge *at* 27°C *and* 57°C. *Boltzmann constant* = 8.61×10^{-5} eV. **(Bangalore Univ.)**

Solution: The intrinsic conductivity σ may be given by

$$\sigma = CT^{3/2}e^{-E_g/2kT}$$

$$\text{where}\quad C = 2e(\mu_e + \mu_h)\left(\frac{2\pi k}{h^2}\right)^{3/2}(m_e^* m_h^*)^{3/4}$$

For the two temperatures, T_1 and T_2, therefore

$$\frac{\sigma_1}{\sigma_2} = \left(\frac{T_1}{T_2}\right)^{3/2} e^{-(E_g/2k)\left(\frac{1}{T_1}-\frac{1}{T_2}\right)}$$

$$\therefore \quad \ln\frac{\sigma_1}{\sigma_2} = \frac{3}{2}\ln\frac{T_1}{T_2} - \frac{E_g}{2k}\left(\frac{1}{T_1} - \frac{1}{T_2}\right)$$

Substituting $E_g = 0.75$, $k = 8.61 \times 10^{-5}$, $T_1 = 273 + 27 = 300\,\text{K}$, $T_2 = (273 + 57)\,\text{K} = 330\,\text{K}$, in the above expression, we get,

$$\ln\frac{\sigma_1}{\sigma_2} = \frac{3}{2}\ln\frac{300}{330} - \frac{0.75}{2 \times 8.61 \times 10^{-5}}\left(\frac{1}{300} - \frac{1}{330}\right)$$

$$= -1.461, \quad \text{on simplification,}$$

$$\therefore \quad \boldsymbol{\sigma_1/\sigma_2 = 0.23}$$

▸ **Example 9.** *The Hall cofficient of a body-centred cubic metal having electrons as charge carriers is* $-2.36 \times 10^{-10}\,\text{m}^3/\text{C}$. *To measure the Hall voltage of the metal using a thin rectangular foil of thickness* 1 mm, *a magnetic field of strength* 1.5 T *was applied perpendicular to the foil. If a current of* 10 A *flows along the length of the foil, determine the magnitude of the Hall voltage developed across the width of the foil.* **(GATE)**

Solution: We have the relation: $R_H = \dfrac{V_H d}{IB}$

$$\text{or,} \quad V_H = R_H\frac{IB}{d}$$

Here, R_H = Hall coefficient $= -2.36 \times 10^{-10}\,\text{m}^3/\text{C}$,

I = current = 10 A, B = magnetic field = 1.5 T

and d = thickness = 1 mm = 10^{-3} m.

∴ The Hall voltage developed across the width of the foil is, with the usual symbols, given by the relation

$$\boldsymbol{V_H} = R_H\frac{IB}{d} = 2.36 \times 10^{-10} \times \frac{10 \times 1.5}{10^{-3}}\ \text{volt}$$

$$= 3.54 \times 10^{-6}\,\text{V}$$

$$= \mathbf{3.54\,\mu V}$$

▸ **Example 10.** *At room temperature, the diffusion and drift constants for holes in a p-type semiconductor were measured to be* $D_p = 10\,\text{cm}^2/\text{s}$ *and* $\mu_p = 1200\ \text{cm}^2/\text{V.s}$ *respectively. If the diffusion constant of electrons in an n-type semiconductor at the same temperature is* $D_n = 20\,\text{cm}^2/\text{s}$, *find the drift constant for electrons.*

Solution: Given, diffusion constant for holes, $D_p = 10\ \text{cm}^2/\text{s}$,

drift constant for holes, $\mu_p = 1200\ \text{cm}^2/\text{V.s}$,

and diffusion constant for electrons, $D_n = 20\ \text{cm}^2/\text{s}$.

From Einstein's relation one obtains

$$\frac{D_p}{\mu_p} = \frac{D_n}{\mu_n} = \frac{kT}{e}$$

$$\therefore \quad \mu_n = \frac{D_n}{D_p}\mu_p = \frac{20}{10} \times 1200 \text{ cm}^2/\text{V.s}$$

$$= 2400 \text{ cm}^2/\text{V.s}$$

$\therefore$ The drift constant for electrons, $\mu_n = \mathbf{2400\ cm^2/V.s.}$

▶ **Example 11.** *Silicon is doped with boron to a concentration of* 4×10^{17} atoms/cm^3. *Assume the intrinsic concentration of silicon to be* $1.5\times 10^{10}/\text{cm}^3$ *and the value of* kT/e *to be* 25 mV *at* 300 K. *Find the position of Fermi level with respect to the intrinsic level.*

Solution: When Silicon is doped with boron, p-type semiconductor is produced. For a p-type semiconductor $N_D \approx 0$ and $p \gg n$ so that $p \approx N_A$. So, the concentration of holes $p = N_V e^{-(E_F - E_V)/kT}$ reduces to

$$N_A = N_V e^{-(E_F - E_V)/kT}$$

$$\text{or,} \quad E_F - E_V = kT \ln \frac{N_V}{N_A}.$$

$$\text{or,} \quad E_V - E_F = kT \ln \frac{N_A}{N_V}$$

In the unit of electron-volt, the above equation can be written as

$$E_V - E_F = \frac{kT}{e} \ln \frac{N_A}{N_V} \text{ electron-volt}$$

$$\therefore \quad E_V - E_F = 25 \text{ mV} \times \ln\left(\frac{4 \times 10^{17}}{1.5 \times 10^{10}}\right) \text{eV}$$

$$= 0.4274 \text{ eV}.$$

So, the Fermi level E_F *goes down* by **0.427 eV** with respect to E_V, the intrinsic level.

▶ **Example 12.** *The band gap in germanium is* $\Delta E = 0.68\,\text{eV}$. *Assuming that the number of hole-electron pairs is proportional to* $e^{-\Delta E/2kT}$, *find the percentage increase in the number of charge carriers in pure germanium as the temperature is increased from* 300 K *to* 320 K.

Solution: According to the problem, the number of hole-electron pairs is proportional to $e^{-\Delta E/2kT}$.

Let N_1 and N_2 be the number of charge carriers at temperature T_1 and T_2 respectively. Then we have,

$$\frac{N_2}{N_1} = \frac{e^{-\Delta E/2kT_2}}{e^{-\Delta E/2kT_1}} = e^{\frac{\Delta E}{2k}\left(\frac{1}{T_1} - \frac{1}{T_2}\right)}$$

∴ The percentage increase in the number of chagre carriers as the temperature is raised from T_1 to T_2, is given by

$$f = \frac{N_2 - N_1}{N_1} \times 100$$

$$= \left(\frac{N_2}{N_1} - 1\right) \times 100$$

$$\left[e^{\frac{\Delta E}{2k}\left(\frac{1}{T_1} - \frac{1}{T_2}\right)} - 1\right] \times 100$$

Given, $\Delta E = 0.68$ eV, $T_1 = 300$ K and $T_2 = 320$ K.

$$\therefore \quad f = \left[e^{\frac{0.68 \times 1.6 \times 10^{-19}}{2 \times 1.38 \times 10^{-23}}\left(\frac{1}{300} - \frac{1}{320}\right)} - 1\right] \times 100\%$$

$$= \left(e^{0.82} - 1\right) \times 100\% = 127\%$$

∴ The required ***perecentage increase*** in the numbe of charge carrier is **127%**.

▶ **Example 13.** *A silicon sample is doped with* 10^{15} donors/cm^3. *Calculate the excess electron and hole densities required to increase the sample conductivity by* 15%. *What carrier generation rate is required to maintain this excess concentration? Assume* $\mu_n = 0.3\mu_p$, $\tau_p = 10^{-6}$ s *and* $T = 300$ K.

Solution: The conductivity of the sample is given by

$$\sigma = e(n\mu_n + p\mu_p)$$

$$= e\left[n\mu_n + \frac{n_i^2}{n}\mu_p\right], \quad \left(\because p = n_i^2/n\right)$$

Differentiating the above equation with respect to n, we have

$$\Delta\sigma = e\left[\mu_n \cdot \Delta n - \frac{n_i^2}{n^2}\mu_p \cdot \Delta n\right]$$

$$= e\mu_n \cdot \Delta n\left[1 - \frac{\mu_p}{\mu_n} \cdot \left(\frac{n_i}{n}\right)^2\right]$$

$$= e\mu_n \cdot \Delta n\left[1 - 0.3\left(\frac{n_i}{n}\right)^2\right] \qquad (1.18.4)$$

Again, we have

$$\sigma = en\mu_n\left[1 + \frac{n_i^2}{n^2} \cdot \frac{\mu_p}{\mu_n}\right]$$

$$= en\mu_n\left[1 + 0.3\left(\frac{n_i}{n}\right)^2\right] \qquad (1.18.5)$$

Dividing (1.18.4) by (1.18.5) we obtain

$$\frac{\Delta\sigma}{\sigma} = \frac{\Delta n}{n} \cdot \frac{\left[1 - 0.3\left(\frac{n_i}{n}\right)^2\right]}{\left[1 + 0.3\left(\frac{n_i}{n}\right)^2\right]} \tag{1.18.6}$$

But in Si, n_i ($\sim 10^{10}$ cm^{-3}) is $\ll n$ ($\sim 10^{15}$/cm^3). So, (1.18.6) reduces to

$$\frac{\Delta\sigma}{\sigma} = \frac{\Delta n}{n} = 15\% = \frac{15}{100}$$

$$\therefore \quad \Delta n = \frac{15}{100} n = 0.15 \times 10^{15} \text{ cm}^{-3} = 1.5 \times \times 10^{14} \text{ cm}^{-3}$$

So, the excess electron and hole densities required to be produced are given by

$$\boldsymbol{\Delta n = \Delta p = 1.5 \times 10^{14}} \textbf{ cm}^{-3}$$

The carrier generation rate required to maintain the excess concentration is

$$G = \left(\frac{\Delta p}{\tau_p}\right) = \frac{1.5 \times 10^{14}}{10^{-6}} \text{ cm}^{-3}\text{s}^{-1}$$

$$= \mathbf{1.5 \times 10^{20}} \textbf{ cm}^{-3}\textbf{ s}^{-1}$$

▶ **Example 14.** *Show that for a semiconductor specimen, the ratio of the maximum to intrinsic resistivities is given by*

$$\frac{\rho_{\max}}{\rho_i} = \left[\frac{(\mu_n + \mu_p)}{2(\mu_n\mu_p)^{1/2}}\right]$$

where the symbols have their usual meanings. Assuming $\mu_n = 3\mu_p$*, solve for the doping levels which results in* $\rho_{\max}$.

Solution: The resistivity of any doping level of a semiconductor sample is given by

$$\frac{1}{\rho} = \sigma = e(n\mu_n + p\mu_p)$$

$$= e\left(n\mu_n + \mu_p \cdot \frac{n_i^2}{n}\right), \quad \left[\because \; np = n_i^2\right]$$

To obtain $\rho_{\max}$, we differentiate the above equation with respect to n and equate $d\sigma/dn$ to zero. So,

$$\left(\frac{d\sigma}{dn}\right) = e\left[\mu_n - \mu_p \cdot \frac{n_i^2}{n^2}\right] = 0$$

$$\text{or,} \quad \mu_n = \frac{n_i^2}{n^2}\mu_p \tag{1.18.7}$$

$$\text{Again,} \quad \frac{d^2\sigma}{dn^2} = 2\mu_p \cdot \frac{n_i^2}{n^3} \tag{1.18.8}$$

From (1.18.7) and (1.18.8) we get

$$\frac{d^2\sigma}{dn^2} = 2\mu_n \cdot n_i \left(\frac{\mu_n}{\mu_p}\right)^{1/2} \tag{1.18.9}$$

which is always positive. This implies that for μ_n given by (1.18.7), σ would be a minimum i.e,, ρ would be maximum.

Now from (1.18.7) $$n = n_i \left(\frac{\mu_p}{\mu_n}\right)^{1/2}$$

and $$p = n_i \left(\frac{\mu_n}{\mu_p}\right)^{1/2}, \quad \text{since } np = n_i^2. \tag{1.18.10}$$

$$\therefore \quad \frac{1}{\rho_{\max}} = e(n\mu_n + p\mu_o) = 2en_i\sqrt{\mu_n\mu_p} \tag{1.18.11}$$

The intrinsic resistivity of a semiconductor sample is given by

$$1/\rho_i = \sigma_i = en_i(\mu_n + \mu_p) \tag{1.18.12}$$

Dividing (1.18.12) by (1.18.11) we have

$$\frac{\rho_{\max}}{\rho_i} = \left[\frac{(\mu_n + \mu_p)}{2(\mu_n\mu_p)^{1/2}}\right]$$

When $\mu_n = 3\mu_p$, the equations (1.18.7) and (1.18.10) give $n = n_i/\sqrt{3}$ and $p = \sqrt{3}n_i$.

Review Questions

1. What is a semiconductor? What is an intrinsic semiconductor? What is meant by a hole? Show that in an intrinsic semiconductor both the electrons and the holes contribute towards electrical conductivity. **(Guru Nanak Univ.)**

2. What do you understand by the term 'doping' of a semiconductor material? Why is it essential? Explain the different types of conduction in semiconductors. **(Bangalore Univ.)**

3. What are extrinsic semiconductors? Describe n-type and p-type semiconductors, clearly explaining the term 'donor' and 'acceptor'. Draw the energy band diagram for the n-type and p-type semiconductors. **(Osmania Univ.)**

4. Distinguish between metals, insulators and semioconductors on the basis of band theory of solids. **(Calcutta and Burdwan Hons.)**

5. (a) Derive expression for the density of electrons in the conduction band and density of holes in the valence band of an intrinsic semiconductor. Show that their product is independent of Fermi level.

(b) On the basis of deductions made in part (a), find the location of the Fermi level in an intrinsic semiconductor. Draw neat diagrams showing the location of the Fermi level in n-type and p-type semiconductors. **(Burdwan and Calcutta Hons.)**

6. Write and explain how resistivity of a conductor vary with temperature. **(Calcutta and Burdwan Hons.)**

7. What do you mean by the effective mass of a current carrier in a semiconductor? Show graphically the variation of energy with the wave number k for the first allowed band in a solid. Hence explain the variation in sign of the effective mass for an electron in the band. What is a hole? **(Calcutta Hons.)**

8. (a) What are direct band gap and indirect band gap semiconductor? Give examples.

(b) With suitable diagrams, explain the difference between direct and indirect band gap semiconductors. Mention their uses. **(Burdwan Hons.)**

9. What is meant by Hall effect? Discuss the experiment with suitable diagram for finding the Hall coefficient and the Hall voltage. Mention two applications of the Hall effect. **(Burdwan Hons.)**

10. Define Hall constant. Find an expression for it in the case of a solid with only electrons as carriers. **(Calcutta Hons.)**

11. Show that for an intrinsic semiconductor, the Fermi level lies at the middle of the band gap.

12. What is mobility? Establish a relation between conductivity and mobility in an extrinsic semiconductor containing both electrons and holes.

13. What is Einstein's relation connecting the diffusion and the mobility? What does it signify?

14. Write short notes on

(a) Thermistors,

(b) Photoconductors,

(c) Sensistors.

Short and Objective Questions

1. What is a semiconductor? What are the commonly used semiconductors?

2. Differentiate a good conductor and a good insultator taking resistivity range.

3. Explain the effect of temperature on semiconductor conductivity.

4. What do you understand by intrinsic and extrinsic semiconductor?

5. Why valence electrons are called conduction electrons?

6. What do you understand by the term 'doping' of a semiconductor material?
7. What is 'Donor' and 'Acceptor' and what happens when such impurities are added to semiconductor materials?
8. What is meant by a 'hole'? **(Lucknow Hons.)**
9. What is the nature of charge on n-type semiconductor?
10. What is Fermi level? What happens to it as a result of doping?
11. What happens to the Fermi level when the temperature increases.
12. Define drift velocity and mobility of charge carriers. **(Purvanchal Hons.)**
13. Conduction electrons have more mobility than the holes. Explain.
14. What do you mean by effective mass of an electron?
15. What is meant by direct gap semiconductor? **(Burdwan Hons.)**
16. What is Hall effect? **(Burdwan and Calcutta Hons.)**
17. A pure semiconductor behaves as an insulator at

    (a) 273 K, (b) −273 °C, (c) 373 K, (d) none.
18. The energy region separating the highest occupied band and the empty band is called a

    (a) conduction band, (b) valence band, (c) forbidden band. **(Himachal Hons.)**
19. The separation between valence band and conduction band is measured in

    (a) metres, (b) electron volt, (c) metre^{-1}, (d) none.
20. In a p-type semiconductor, the minority charge carriers are

    (a) protons, (b) holes, (c) electrons, (d) neutrons. **(Meerut Univ.)**
21. Resistivity of copper has the unit

    (a) ohm, (b) ohm metre, (c) ohm^{-1} metre, (d) ohm metre^{-1}. **(Himachal Univ.)**
22. A n-type germanium is obtained on doping intrinsic germanium by

    (a) Phosphorus, (b) Aluminium, (c) Boron, (d) Gold. **(Purvanchal Univ.)**
23. An n-type semiconductor has an electron concentration of $3 \times 10^{20}\,\text{m}^{-3}$. If the electron drift velocity is $100\,\text{ms}^{-1}$ in an electric field of $200\,\text{Vm}^{-1}$, the conductivity in $\Omega^{-1}\,\text{m}^{-1}$ of this material is

    (a) 24, (b) 36, (c) 48, (d) 96. **(GATE)**

24. The effective mass of an electron in a semiconductor

(a) can never be positive, (b) can never be negative,
(c) can be positive or negative, (d) depends on its spin. **(GATE)**

25. The effective mass of an electron in a semiconductor can be

(a) negative near the bottom of the band,
(b) a scalar quantity with a small magnitude,
(c) zero at the centre of the band,
(d) negative near the top of the band. **(GATE)**

26. A sample of Silicon of thickness $200\,\mu$m is doped with 10^{23} phosphorous atoms per m^3. If the sample is kept in a magnetic field of $0.2\,Wb/m^2$ and a current 1 mA is passed through the sample, the Hall voltage produced is

(a) $62.5\,\mu$V, (b) $-6.25\,\mu$V, (c) $+6.25\,\mu$V, (d) $-62.5\,\mu$V. **(GATE)**

27. A piece of semiconductor material is introduced into a circuit. If the temperature of the material is raised, the circuit current will

(a) increase, (b) remain the same, (c) decrease, (d) cease to flow. **(GATE)**

28. On lightly doping an intrinsic semiconductor with donor impurities, the electron carrier concentration in the conduction band is found to increase from n to nf. If μ_n and μ_p are the mobilities of electrons and holes respectively, the conductivity of the doped semiconductor will be

(a) $ne(f\mu_n+\mu_p/f)$, (b) $ne(f\mu_n+\mu_p)$, (c) $nef\mu_n$, (d) $nef(\mu_n+\mu_p)$. **(GATE)**

29. The band gaps at 300 K of diamond cubic crystals of C, Si, Ge and Sn are 5.5, 1.2, 0.67 and 0.1 eV respectively. Identify the crystal which will have the highest electrical conductivity at 300 K.

(a) Ge, (b) Sn, (c) C, (d) Si. **(GATE)**

30. The electron and hole concentrations in an intrinsic semiconductor are n_i per cm^3 at 300 K. Now, if acceptor impurities are introduced with a concentration of N_A per cm^3 (where $N_A \gg n_i$), the electron concentration per cm^3 at 300K will be:

(a) n_i, (b) $n_i + N_A$, (c) $N_A - n_i$, (d) n_i^2/N_A. **(GATE)**

31. Which of the following is true?

(a) A silicon wafer heavily doped with boron is a p^+ substrate.
(b) A silicon wafer lightly doped with boron is a p^+ substrate.
(c) A silicon wafer heavily doped with arsenic is a p^+ substrate.
(d) A silicon wafer lightly doped with arsenic is a p^+ substrate. **(GATE)**

32. In an n-type silicon crystal at room temperature, which of the following can have a concentration of 4×10^{19} cm^{-3}?

(a) Silicon atoms, (b) Holes,
(c) Dopant atoms, (d) Valence electrons. **(GATE)**

33. The ratio of the mobility to the diffusion coefficient in a semiconductor has the unit

(a) V^{-1}, (b) cm.V^{-1}, (c) V.cm^{-1}, (d) V.s. **(GATE)**

34. Drift current in semiconductors depends upon

(a) only on the electric field,
(b) only on the carrier concentration gradient,
(c) both the electric field and the carrier concentration,
(d) both the electric field and the carrier concentration gradient. **(GATE)**

35. At room temperature which one of the following band gap energy Eg corresponds to that of a semiconductor?

(a) $E_g = 0$, (b) $E_g = 1.2\,\text{eV}$, (c) $E_g = 5\,\text{eV}$, (d) $E_g = 9\,\text{eV}$. **(DRDO)**

36. The diffusion constant and mobility for electron in a semiconductor material at a given temperature are 20 cm^2/s and 1600 cm^2/V.s respectively. The thermal voltage V_T for a diode made of this material and at the same temperature is

(a) 125 mV, (b) 32 mV, (c) 12.5 mV, (d) 3.2 mV. **(DRDO)**

37. The conductivity of the intrinsic germanium at 300 K is ________. When n_i at 300 K is 2.5×10^{13}/cm^3 and μ_n and μ_p in germanium are 3800 and 1800 cm^2/V.s respectively

(a) 0.224 S/cm, (b) 0.0224 S/cm,
(c) 2.24 S/cm, (d) 0.00224 S/cm. **(ISRO)**

38. The bonding forces in compound semiconductors such as GaAs arise from

(a) ionic bonding, (b) metallic bonding,
(c) covalent bonding, (d) combination of ionic and covalent bonding.

39. The unit of Hall-coefficient is

(a) Vm^3 A^{-1} Wb^{-1}, (b) Vm^{2}A Wb^{-1}
(c) Vm^3 A Wb^{-1}, (d) Vm^2 A^{-2} Wb. **(Chhatrapati Univ.)**

40. Consider a doped semiconductor having the electron and hole mobilities μ_n and μ_p respectively. Its intrinsic carrier density is n_i. The hole concentration p for which the conductivity is minimum at a given temperature is

(a) $n_i\sqrt{\dfrac{\mu_n}{\mu_p}}$, (b) $n_h\sqrt{\dfrac{\mu_p}{\mu_n}}$, (c) $n_i\sqrt{\dfrac{\mu_p}{\mu_n}}$, (d) $n_h\sqrt{\dfrac{\mu_n}{\mu_p}}$.

(IIT JAM)

41. The silicon sample with unit cross-sectional area shown below is in thermal equilibrium. The following information is given:

$T = 300\,\text{K}$, electron charge $= 1.6 \times 10^{-19}\,\text{C}$, thermal voltage $= 26$ mV.

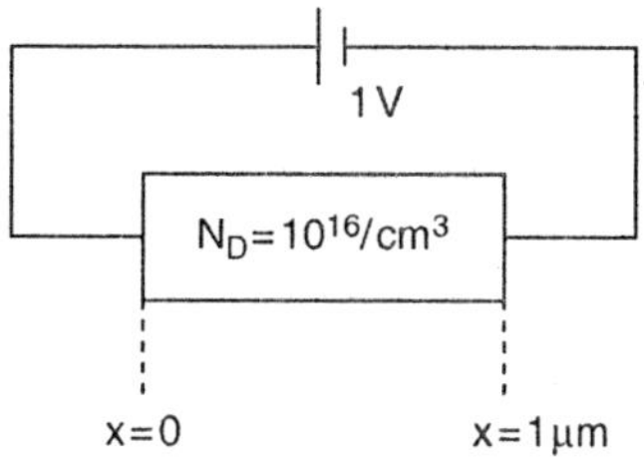

Fig. 1.23

(i) The magnitude of the electric field at $x = 0.5\,\mu$m is

(a) 1 kV/cm, (b) 5 kV/cm, (c) 10 kV/cm, (d) 26 kV/cm.

(ii) The magnitude of the electron drift current density at $x = 0.5\,\mu$m is

(a) 2.16×10^4 A/cm^2, (b) 1.08×10^4 A/cm^2,
(c) 4.32×10^3 A/cm^2, (d) 6.48×10^2 A/cm^2. **(GATE)**

Problems

1. The band gap of diamond and silicon are 5.4 and 1.1 eV respectively. Estimate the temperature at which diamond has the same conductivity as silicon at 27 °C.

(Punjabi Univ.) [1200 °C]

2. The intrinsic carrier density of Ge at 27 °C is $2.4 \times 10^{17}\,\text{m}^{-2}$. Calculate its intrinsic resistivity, if the electron and hole mobilities are $0.35\,\text{m}^2\text{V}^{-1}\text{s}^{-1}$ and $0.18\,\text{m}^2\text{V}^{-1}\text{s}^{-1}$. **(Bangalore Univ.)** [$4.9 \times 10^{-3}\,\Omega$m]

3. Calculate the conductivity of germanium. Given, mobilities of electrons and holes in a sample of germanium at room temperature are $0.54\,\text{m}^2/\text{Vs}$ and $0.18\,\text{m}^2/\text{Vs}$ respectively. Assume that electron and hole densities are each equal to

$3.6 \times 10^{19}\,\text{m}^{-3}$. If a potential difference of 2 volt is applied across the germanium plate of thickness 0.2 mm and area $1\,\text{cm}^2$, calculate the current produced in the plate. **(Bangalore Univ.)** [$4.15\Omega^{-1}\text{m}^{-1}, 4.17\,\text{A}$]

4. A sample of Ge is doped to the extent of 10^{14} donor atoms/cm^3 and 5×10^{13} acceptor atoms/cm^3. At 300 K, the resistivity of intrinsic Ge is $60\,\Omega$ cm. If the applied electric field is $2\,\text{V cm}^{-1}$, find the total conduction current density. Assume $\mu_p/\mu_n = 1/2$ and $n_i = 2.5 \times 10^{13}/\text{cm}^3$ at 300 K.
(Uttaranchal Univ.) [$0.058\,\text{A/cm}^2$]

5. A silicon wafer is doped with phosphorus of concentration 10^{13} atoms/cm^3. If all the donor atoms are active, what is the resistivity at room temperature? The electron mobility is $1350\,\text{cm}^2/\text{Vs}$. The electronic charge is 1.6×10^{-19}C.
(Rajasthan Univ.) [$4.63\,\Omega\text{m}$]

6. Find the resistance of an intrinsic germanium rod 1 cm × 1 mm × 1 mm at 300 K. For Ge, $n_i = 2.5 \times 10^{13}/\text{cm}^3$, $\mu_e = 3900\,\text{cm}^2/\text{Vs}$, $\mu_h = 1900\,\text{cm}^2/\text{Vs}$ at 300 K.
(Rajasthan Univ.) [$4.3 \times 10^3\,\Omega$]

7. For an intrinsic semiconductor, show that the Hall coefficient may be given by

$$R_H = -\frac{1}{n_i e}\left(\frac{\mu_n - \mu_p}{\mu_n + \mu_p}\right)$$

where μ_n andd μ_p are the mobilities of the electrons and the holes and n_i is the intrinsic carrier concentration. **(Calcutta Hons.)**

8. Show that, at a given temperature, the conductivity of a semiconductor will be minimum if $n = n_i\sqrt{\mu_p/\mu_n}$ or $p = n_i\sqrt{\mu_n/\mu_p}$, where the symbols have their usual meanings. Find the corresponding conductivity. [$2en_i\sqrt{\mu_n\mu_p}$]

9. A doped germanium crystal of length 2 cm, breadth 1 cm and width 1 cm, carries a current of 1 mA along its length parallel to $+x$ axis. A magnetic field of 0.5 T is applied along $+z$ axis. Hall voltage of 6 mV is measured with negative polarity at $y = 0$ plane. Find the sign and concentration of the majority charge carriers. Given : $e = 1.6 \times 0^{-19}$C. **(GATE)** [negative, $5.2 \times 10^{19}/\text{m}^3$]

10. There is a sample of n-type semiconductor having Hall coefficient $-0.0125\,\text{m}^3/\text{C}$. An electric field of 100 V/m is applied to it. Assuming the mobility $\mu = 0.36\,\text{m}^2\text{V}^{-1}\text{s}^{-1}$, find the current density. [2880 Am^{-2}]

11. In an intrinsic semiconductor, the effective mass of the electron is $0.09\,m_e$ and that of the hole is $0.4\,m_e$, where m_e is the electron's rest mass. Calculate the intrinsic carrier concentration at 300 K, when the energy gap $E_g = 0.7\,\text{eV}$.[$2.3 \times 10^{18}/\text{m}^3$]

12. Calculate the position of Fermi level at $T = 300\,\text{K}$ for Ge crystal containing 5×10^{22} atoms/m^3. [$E_c - E_F = 0.16\,\text{eV}$]

13. Find the conductivity (i) when 1 donor impurity is added in 10^8 silicon atoms and (ii) when 1 acceptor impurity is added in 10^8 silicon atoms. Given that Si has 5×10^{20} atoms per m^3 and the hole concentration is 1.5×10^{16} per m^3.

[$10.4\,\Omega^{-1}\text{m}^{-1}$, $4\,\Omega^{-1}\text{m}^{-1}$]

14. Find the length of a round copper wire of diameters 1 mm which will have a resistance of 1 kΩ and conductivity 5.8×10^7 S/m. If the wire is subjected to an electric field of 10 mV/m, what is the current density?

(Pune Univ.) [45.5 km, 5.8×10^5 A/m^2]

15. A copper wire of 2 mm diameter with conductivity of 5.8×10^7 S/m and electron mobility of $0.0032\,\text{m}^2/\text{V.s}$ is subjected to an electric field of 20 mV/m. Find (a) the charge density of free electrons, (b) the current density, (c) the current flowing in the wire, (d) the electron drift velocity.

[$1.132 \times 10^{29}/\text{m}^3$, 1.6×10^6 A/m^2, 3.67 A, 64 m/s]

16. A conduction wire has a resistivity of $1.54 \times 10^{-8}\,\Omega$m at room temperature. There are 5.8×10^{28} conduction electrons per m^3. Calculate the mobility and relaxation time of electrons. **(UPSE Engg. Services)** [0.699 m^3/V.S, 3.976 ps]

17. Calculate the intrinsic conductivity of silicon at the room temperature if $n_i = 1.41 \times 10^{16}/\text{m}^3$, $\mu_n = 0.145\,\text{m}^2/\text{V.s}$, $\mu_p = 0.05\,\text{m}^2/\text{V.s}$ and $q = 1.6 \times 10^{-19}$C. What are the individual contributions made by electrons and holes?

(Nagpur Univ.)

[$0.337 \times 10^{-3}(\Omega\text{m})^{-1}$, $\sigma_n = 0.325 \times 10^{-3}(\Omega\text{m})^{-1}$, $\sigma_p = 0.112 \times 10^{-3}(\Omega\text{m})^{-1}$]

18. Determine the concentration of the free electrons and holes in a sample of Ge at 300 K which has concentration of donor atoms equal to 2×10^{14} atoms/cm^3 and a concentration of acceptor ions equal to 3×10^{14} atoms/cm^3.

(Nagpur Univ.) [$n = 1.76 \times 10^{24}\,\text{cm}^{-3}$, $p = 2.65 \times 10^{24}\,\text{cm}^{-3}$]

19. A cylindrical piece of silicon having a diameter of 1 mm is doped with $10^{20}\,\text{m}^{-3}$ atoms of phosphorus which are fully ionized. What length of this silicon would be required to give a resistance of 1 kΩ if electron mobility in silicon is $0.1\,\text{m}^2/\text{V.s}$.

(Pune Univ.) [1.26 mm]

20. Resistivity of a sample semiconductor is $9\,\Omega$m. Its holes have mobility of $0.03\,\text{m}^2/\text{V.s}$. Calculate the value of Hall-coefifcient of the sample. State the assumptions made. [$2.7 \times 10^{-4}\,\text{m}^3\text{C}$]

21. Find the concentration of holes and electrons in a p-type germanium at 300 K, if the conductivity is 100 per ohm per cm. Also find these values for N-type silicon, if the conductivity is 0.1 per ohm per cm. Given, for germanium, $n_i = 2.5 \times 10^{13}/\text{cm}^3$, $\mu_n = 3800\,\text{cm}^2/\text{V.s}$, $\mu_p = 1800\,\text{cm}^2/\text{V.s}$ and for silicon $n_i = 1.5 \times 10^{10}/\text{cm}^3$, $\mu_n = 1300\,\text{cm}^2/\text{V.s}$ and $\mu_p = 500\,\text{cm}^2/\text{V.s}$.

(Bangalore Univ.)

[$3.47 \times 10^{17}/\text{cm}^3$, $1.8 \times 10^9/\text{cm}^3$, $4.7 \times 10^5/\text{cm}^3$, $4.8 \times 10^{14}/\text{cm}^3$]

22. Find the diffusion coefficients of holes and electrons for germanium at 300 K. The carrier mobilities in $cm^2/V.s$ at 300 K for electrons and holes are respectively 3600 and 1700.

Density of carriers is 2.5×10^{23}, Boltzmann constant $k = 1.38 \times 10^{23}$.

(UPSE Engg. Services)

$[D_p = 4.36 \times 10^{-3}\, m^2/s,\ D_n = 9.2 \times 10^{-3}\, m^2/s]$

23. The Hall coefficient of a specimen of doped silicon is found to be $3.66 \times 10^{-4}\, m^3C$, the resistivity of the specimen is $8.93 \times 10^{-3}\Omega m$. Find the mobility and density of charge carriers assuming single carrier conduction.

$[0.041\, m^2/V.s,\ 1.7 \times 10^{22}/m^3]$

24. At a given temperature, a semiconductor with intrinsic carrier concentration $n_i = 10^{16}/m^3$ is doped with a donor dopant of concentration $N_D = 10^{26}/m^3$. If the temperature remains the same, calculate the concentration of hole in the doped semiconductor. $[10^6/m^3]$

25. Resistivity of an n-type Si material at a particular temperature is $0.625\,\Omega.cm$. The electron and the hole mobilities in that material at the same temperature are $\mu_n = 1600\, cm^2/V.s$ and $\mu_p = 500\, cm^2/V.s$ respectively. Calculate the donor dopant concentration N_D. $[6.25 \times 10^{15}/cm^3]$

26. The Fermi level of an n-type semiconductor at a temperature of 300 K is at a position 0.3 eV below the conduction band. (a) Find the position of Fermi level at a temperature of 320 K, (b) What would be the position of Fermi level if the doping concentration is doubled, keeping the temperature fixed at 300 K?

[0.32 eV below C.B, 0.282 eV below C.B]

2

Metal-Semiconductor Junction

2.1 Introduction

When a metal is brought into close contact with a n-type or a p-type semiconductor, a junction between the metal and the semiconductor is thereby formed. This is known as the ***metal-semiconductor junction*** or ***metal-semiconductor contact***. It is quite attractive for its simplicity in fabrication and its importance lies in studying the fundamental parameters of many semiconductor devices. It is very useful for high speed rectification. Sometimes, non-rectifying or ohmic contacts to semiconductors are also formed by metal-semiconductor junction. In this chapter, we shall devote ourselves to describe and discuss the properties of the metal-semiconductor junction, along with rectifying and ohmic contacts.

2.2 Energy-band diagrams of metal-semiconductor junction

When a metal and a semiconductor are isolated and the system is not in thermal equilibrium, the energy-band diagrams of the metal and the semiconductor show (Fig. 2.1(a)) that the Fermi levels do not coincide. To remove an electron at the Fermi level to vacuum outside the metal or the semiconductor, a *minimum* amount of energy is to be supplied to the electron. This energy is known as the ***work function*** of the material. So, the *work function is the energy difference between the Fermi level and the vacuum level.*

If e be the electronic charge and ϕ_m and ϕ_s are the volt-equivalent of the work function of the metal and the semiconductor respectively, then the energies corresponding to work functions of the metal and the semiconductor are given by $e\phi_m$ and $e\phi_s$. Typical values of ϕ_m for very clean surfaces are 4.3 V for Al and 4.8 V for Au. For semiconductors, the $e\phi_s$ may be written as $e\phi_s = e(\chi + V_n)$ where $e\chi$ is the energy difference between the conduction band edge E_C and the vacuum level, known as ***electron affinity*** and $eV_n = E_C - E_F$. The energy difference $e\phi_m - e\phi_s$ is called the ***contact potential.***

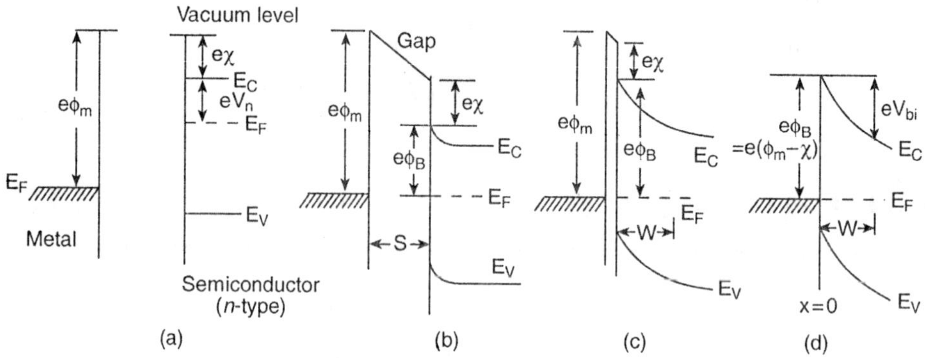

Fig. 2.1 Energy-band diagrams for contact between a metal and an *n*-type semiconductor in absence of surface states

In Fig. 2.1(a), $\phi_m > \phi_s$ and the semiconductor Fermi level is initially higher than that of the metal before the contact is made. If we assume the ideal situation of *absence of surface states* that can trap charges, then the electrons begin to flow from the semiconductor to the metal when they are brought in contact. The transfer of charges occur until the Fermi levels E_F in the two materials coincide at thermal equilibrium (Fig. 2.1(b)) i.e., to align the two Fermi levels, the electron energies are lowered in the semiconductor w.r.t. the metal. So, on decreasing the separation between the metal and the semiconductor surfaces, electrons are accumulated on the metal surface and an equal amount of positive charge due to the uncompensated donor ions is built up in the semiconductor. As a result, a *depletion region* of width W is formed in the n-type semiconductor near the junction and a potential barrier is established at the interface (Fig. 2.1(c)).

Fig. 2.1(d) shows the limiting situation when the separation between the metal and the semiconductor surface becomes comparable to the interatomic separation. The potential barrier in the semiconductor side, measured from the conduction band edge is V_{bi}, In this case, the equilibrium contact potential is $V_0 = \phi_m - \phi_s$ which prevents further electron diffusion from semiconductor conduction band into the metal. The potential barrier height is given by

$$e\phi_B = e(\phi_m - \chi) \tag{2.2.1}$$

So, the ***barrier height*** *in metal-semiconductor junction is the difference between the work function of the metal and the electron affinity of the semiconductor.* This barrier height is also called the ***Schottky barrier height.***

The electric field that is developed at the interface of the metal-semiconductor junction is called ***built-in field*** which opposes the flow of electron from the semiconductor to the metal in equilibrium.

For the contact of metal and a p-type semiconductor (Fig. 2.2), the barrier height is given by

$$e\phi'_B = E_g - e(\phi_m - \chi) \tag{2.2.2}$$

where E_g is the forbidden energy gap.

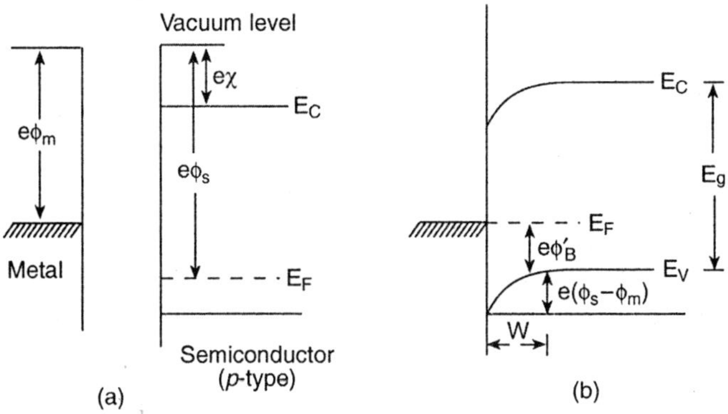

Fig. 2.2 Contact of metal and p-type semiconductor (a) before, (b) after

Adding Eq. (2.2.1) and Eq. (2.2.2), we get

$$E_g = e(\phi_B + \phi'_B) \tag{2.2.3}$$

So, for a given pair of metal and semiconductor, the *band-gap energy* is equal to the sum of the barrier heights for n-type and p-type semiconductors.

In Fig. 2.1(d) is shown the energy band diagram of ideal metal-semiconductor junction at thermal equilibrium in absence of surface states with no applied electric field. But, in actual case, the surface states exist due to imperfections at the interface between the metal and the semiconductor. The surface states act as donors or acceptors and influence the behaviour of the metal-semiconductor junction. The barrier height is determined by the simultaneous influence of the work function of the metal and the surface states. It is mainly controlled by the semiconductor surface if the density of surface states is very high.

Biasing—On application of the *external potential*, the barrier height may be increased or decreased. When the metal is connected with a positive potential with respect to the semiconductor, the electrons are attracted towards the metal resulting in lowering of the barrier height. Such a bias condition is called ***forward bias***. If V_F be the forward bias voltage, then the Fermi level is raised by eV_F and the barrier height in the semiconductor becomes $e(V_{bi} - V_F)$. In idealised case, the quantity ϕ_B remains constant. As the barrier height is reduced by eV_F, the electrons (for n-type semiconductor) can easily flow from semiconductor to metal resulting in a forward current flow from metal to semiconductor. The energy band diagram for the forward bias condition is shown in Fig. 2.3(a).

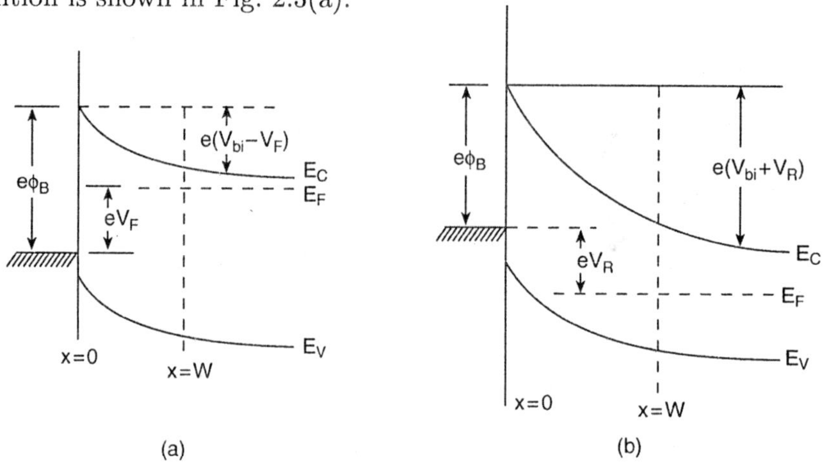

Fig. 2.3 Metal-semiconductor (*n*-type) contact under (a) forward bias and (b) reverse bias

When one applies to the semiconductor a positive voltage V_R with respect to the metal, the Fermi level is lowered by eV_R and the barrier height becomes $e(V_{bi} + V_R)$ where ϕ_B remains fixed. In this case, the electrons are repelled by the metal and the barrier height becomes higher. This bias condition is called ***reverse bias***. The energy-band diagram for reverse-bias condition is shown in Fig. 2.3(b).

For the contact of metal and a p-type semiconductor, the energy-band diagrams under forward bias and reverse bias are shown in Fig. 2.4(a) and 2.4(b) respectively.

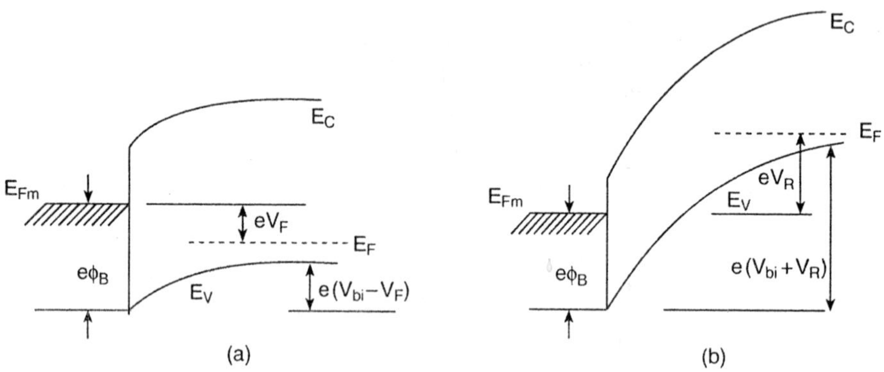

Fig. 2.4 Energy band diagrams of metal and *p*-type semiconductors under (a) forward bias and (b) reverse bias

2.3 Characteristics of depletion region

When a metal and a n-type semiconductor are brought into close contact to form a metal-semiconductor junction, the Fermi levels of the two materials coincide at thermal equilibrium by removal of electrons from the semiconductor side to the metal at the interface. As a result, a barrier potential which opposes further movement of electrons from semiconductor to metal is established. So, the semiconductor side near the interface is depleted of the mobile charge carriers (i.e., electrons) and the ionized donor atoms create a positive space charge in this region. This region is known as the ***depletion region*** or the ***space-charge region.***

If $\rho(x)$ be the volume density of space charge and ε the permittivity of the semiconductor, the differential form of the Gauss's law can be written as

$$\frac{d\mathcal{E}}{dx} = \frac{\rho(x)}{\varepsilon} \tag{2.3.1}$$

where $\mathcal{E}$ is the electric field in the region.

If we assume that (i) doping is uniform with donor atom concentration N_D and (ii) the boundary condition that electric field $\mathcal{E} = 0$ at space charge edge $x = \omega$ in the semiconductor, then by integrating (2.3.1), we get

$$\int_0^{\mathcal{E}} d\mathcal{E} = \int_\omega^x \frac{\rho(x)}{\varepsilon} dx$$

$$= \frac{eN_D}{\varepsilon} \int_\omega^x dx, \quad \text{since } \rho = eN_D$$

$$\therefore \; \mathcal{E} = -\frac{eN_D}{\varepsilon}(\omega - x) \tag{2.3.2}$$

Eqn. (2.3.2) shows that the electric field $\mathcal{E}$ is directed from semiconductor to metal and its value is maximum at $x = 0$ i.e., at the interface and decreases linearly with increase in x. The maximum barrier field is given by

$$\boxed{|\mathcal{E}| = \frac{eN_D\omega}{\varepsilon}} \tag{2.3.3}$$

Since the electric field is zero inside a metal, a negative space charge must exist in the metal at the metal-semiconductor junction.

The electric field at any point is the negative potential gradient at that point. So,

$$-\frac{dV}{dx} = \mathcal{E} = -\frac{eN_D}{\varepsilon}(\omega - x)$$

$$\text{or,} \quad \frac{dV}{dx} = \frac{eN_D}{\varepsilon}(\omega - x) \tag{2.3.4}$$

Integrating (2.3.4) and imposing the boundary condition $V = 0$ at $x = 0$,

$$\int_0^V dV = \frac{eN_D}{\varepsilon}\int_0^x (\omega - x)dx$$

$$\text{or,} \quad V = \frac{eN_D}{\varepsilon}\left(\omega x - \frac{x^2}{2}\right) \tag{2.3.5}$$

If we substitute $x = \omega$ in (2.3.5), we get the expression for the barrier potential in the semiconductor side.

$$\therefore \quad V_B = \frac{eN_D}{\varepsilon}\left[\omega^2 - \frac{\omega^2}{2}\right]$$

$$\text{or,} \quad \boxed{V_B = \frac{eN_D\omega^2}{2\varepsilon}} \tag{2.3.6}$$

$$\text{or,} \quad \boxed{\omega = \sqrt{\frac{2\varepsilon V_B}{eN_D}}} \tag{2.3.7}$$

Eqn. (2.3.6) is the expression for the barrier potential V_B and (2.3.7) is that for the width ω of the depletion layer.

When no bias voltage is applied $V_B = V_{bi}$. For a forward bias voltage V_F and a reverse bias voltage V_R, we substitute $(V_{bi} - V_F)$ and $(V_{bi} + V_R)$ respectively for V_B in (2.3.7). So, it is clear that the thickness of *the depletion layer decreases in forward bias and increases in reverse bias over the unbiased value.*

The space charge per unit area of the semiconductor surface for an applied reverse bias V_R is

$$Q = eN_D\omega = eN_D\sqrt{\frac{2\varepsilon(V_{bi} + V_R)}{eN_D}} = \sqrt{2e\varepsilon N_D(V_{bi} + V_R)} \tag{2.3.8}$$

$\therefore$ The junction capacitance per unit area is

$$C = \frac{\partial Q}{\partial V_R} = \sqrt{\frac{e\varepsilon N_D}{2(V_{bi} + V_R)}} = \frac{\varepsilon}{\omega}. \tag{2.3.9}$$

So, the junction capacitance is given by

$$C_j = \frac{\varepsilon A}{\omega} = A\sqrt{\frac{e\varepsilon N_D}{2(V_{bi} + V_R)}} \tag{2.3.10}$$

Now, from (2.3.9), we have

$$\frac{1}{C^2} = \frac{2(V_{bi} + V_R)}{\varepsilon e N_D} \tag{2.3.11}$$

Clearly, the plot of $1/C^2$ versus V_R will be a straight line of slope $2/\varepsilon e N_D$ which gives the value of N_D. The intercept $2V_{bi}/\varepsilon e N_D$ on the V_R axis yields the value of V_{bi}.

2.4 Schottky effect

When a negative charge (say, an electron) is brought near the metal surface, a positive image charge is induced on the surface. The force of attraction between the electron and the induced image charge reduces the effective work function to some extent when it is combined with an applied electric field. This lowering of the potential barrier induced by the image force is called the ***Schottky effect.*** It is relevant for the discussion of potential barriers arising in metal-semiconductor junction.

When an electron of charge $-e$ is placed in the semiconductor at a distance x from the metal-semiconductor junction, a positive charge of $+e$ is induced on the metal surface. The force of attraction between the electron and the image charge inside the metal at a distance $-x$ is given by

$$F = -\frac{1}{4\pi\varepsilon}\cdot\frac{e^2}{(2x)^2} = -\frac{e^2}{16\pi\varepsilon x^2} \tag{2.4.1}$$

The work done in bringing the electron from infinity to a point at a distance x from the metal surface is the potential energy of the electron and is given by

$$E_{P1}(x) = -\int_{\infty}^{x} F\,dx = -\frac{e^2}{16\pi\varepsilon x} \tag{2.4.2}$$

The potential energy is negative for finite values of x and $E_{P1}(x) \rightarrow 0$ when $x \rightarrow \infty$. The variation of potential energy with x is shown in Fig. 2.5.

When an external electric field $\mathcal{E}$ is applied, the electron gets an additional potential energy given by

$$E_{P2}(x) = -e\mathcal{E}x \tag{2.4.3}$$

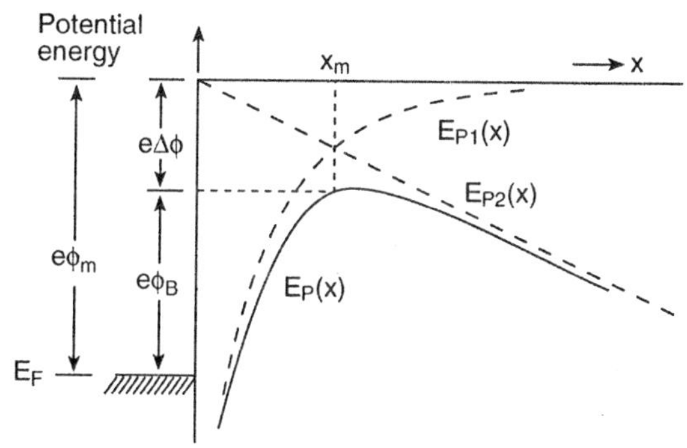

Fig. 2.5 Variation of potential energy with x due to lowering of potential barrier caused by the combined effect of image force and electric field

∴ Total potential energy is given by

$$E_P(x) = E_{P1} + E_{P2} = -\left(\frac{e^2}{16\pi\varepsilon x} + e\mathcal{E}x\right) \tag{2.4.4}$$

In Fig. 2.4, the variations of $E_{P1}(x)$ and $E_{P2}(x)$ with x are shown by the broken curve, and the solid curve shows the variation of $E_P(x)$ which is the sum of the ordinates for the two broken curves. It is clear from the figure that the barrier potential is lowered from $e\phi_m$ to $e\phi_B$ due to Schottky effect.

Differentiating (2.4.4), we get

$$\frac{dE_P(x)}{dx} = \frac{e^2}{16\pi\varepsilon x^2} - e\mathcal{E} \tag{2.4.5}$$

Putting $dE_P(x)/dx = 0$, we can obtain the distance $x = x_m$ at which the Schottky barrier lowering $\Delta\phi$ occurs.

$$\therefore \quad \frac{e^2}{16\pi\varepsilon x_m^2} = e\mathcal{E}$$

$$\text{or,} \quad x_m = \sqrt{\frac{e}{16\pi\varepsilon\mathcal{E}}} \tag{2.4.6}$$

From Fig. 2.4, plainly

$$\Delta\phi = \frac{1}{e}[e\phi_m - e\phi_B] = \left|\frac{1}{e}E_P(x_m)\right|$$

$$= \frac{e}{16\pi\varepsilon x_m} + \mathcal{E}x_m, \text{ using (2.4.4)}$$

$$= 2\mathcal{E}x_m, \text{ using (2.4.6).} \tag{2.4.7}$$

For semiconducting material, $\Delta\phi$ is very small and becomes smaller for lower applied fields. However, a small change in barrier height can influence significantly the current in metal-semiconductor junction.

The energy-band diagram incorporating the Schottky effect for a metal on n-type semiconductor under different biasing conditions are shown in Fig.2.6.

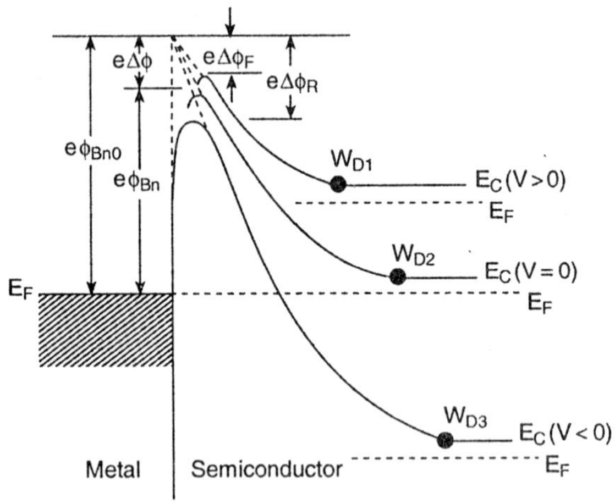

Fig. 2.6 Energy band diagram related to Schottky effect for a metal on n-type semiconductor contact under different biasing conditions

For forward bias ($V > 0$), the field and the image force are smaller and the barrier height $e(\phi_{Bn0} - \Delta\phi_F)$ is slightly larger than the barrier height at zero bias of $e\phi_{Bn} = e(\phi_{Bn0} - \Delta\phi)$. For reverse bias ($V_R > 0$), the barrier height $e(\phi_{Bn0} - \Delta\phi_R)$ is slightly smaller. Practically, the barrier height is bias-dependent.

2.5 Current transport in metal-semiconductor junction

The transport of current in metal-semiconductor contacts is mainly due to the majority carriers of the system. When a metal is in contact with an n-type semiconductor, the electrons with sufficient thermal energy over the potential barrier are transported by the process of thermoionic emission. So, to describe the current transport in metal-semiconductor junction, the following assumptions are to be made.

(i) The thermal energy (kT) is much smaller than the barrier height ($e\phi_B$) so that Maxwell-Boltzmann distribution holds at thermal equilibrium.

(ii) Thermal equilibrium exists at the interface and is not perturbed during the current flow.

The current density due to the flow of electrons from the semiconductor to the metal is given by

$$J_1 = AT^2 e^{-\frac{e(V_{bi}-V)+E_c-E_F}{kT}} \tag{2.5.1}$$

$$\text{where } A = \frac{4\pi e m_e^* k^2}{h^3}. \tag{2.5.2}$$

A is called the ***effective Richardson constant*** for thermoionic emission and m_e^* is the ***effective mass of the electron.***

From Fig. 2.1(d), the barrier height can be written as

$$e\phi_B = e(V_{bi} + E_c - E_F) \tag{2.5.3}$$

$$\therefore \quad J_1 = AT^2 e^{-\frac{e\phi_B}{kT}} \cdot e^{\frac{eV}{kT}}, \quad \text{using (2.5.1).} \tag{2.5.4}$$

If we put $V = 0$ in (2.5.4), we obtain the current density J_2 due to electrons moving from the metal to the semiconductor since the barrier height is not affected by this movement of electrons.

$$\therefore \quad J_2 = -AT^2 e^{-\frac{e\phi_B}{kT}}, \tag{2.5.5}$$

the negative sign in (2.5.5) indicates that the conventional current is in the direction opposite to the motion of the electrons.

∴ The total current density is given by

$$\begin{aligned} J &= J_1 + J_2 \\ &= AT^2 e^{-\frac{e\phi_B}{kT}} \left[e^{\frac{eV}{kT}} - 1\right] \end{aligned}$$

$$\therefore \quad \boxed{J = J_s \left[e^{\frac{eV}{kT}} - 1\right]} \tag{2.5.6}$$

$$\text{where} \quad J_s = AT^2 e^{-\frac{e\phi_B}{kT}} \tag{2.5.7}$$

- When the bias $V \gg kT/e$, then $\exp(eV/kT) \gg 1$ and we have from (2.5.6),

$$J \simeq J_s e^{eV/kT}$$

When a reverse bias is applied, we substitute $-V$ for V in (2.5.6). For large value of reverse bias voltage V, $e^{-\frac{eV}{kT}}$ can be neglected compared to unity and (2.5.6) gives $J = -J_s$. Clearly, J_s is *independent of the bias voltage* V. J_s is known as the ***reverse saturation current density***. So, the current density at reverse bias remains constant at the saturation current density.

• Eq. (2.5.6) is known as the ***rectifier equation***.

2.6 Ohmic contact

The metal-semiconductor junction can be used both as rectifying contact and non-rectifying ohmic contact. The rectifying contact gives a special type of diode known a *Schottky diode* which will be discussed in Chapter 3. Ohmic contacts are mainly used for connection of a semiconductor device to outside circuit. The ohmic contact in metal-semiconductor junction is the low resistance contact between the metal and the semiconductor. This allows electrical conduction in either direction when the voltage across the junction is very small and the current is a linear function of the applied voltage.

Ohmic contacts can be classified in two categories : (i) *non-rectifying barrier* and (ii) the *tunneling barrier.*

2.6.1 Non-rectifying barrier

Non-rectifying barrier is a special type of ohmic contact. The energy band diagrams for a metal and an n-type semiconductor where $\phi_m > \chi + V_n$ is shown in Fig. 2.1. In the case of metal-semiconductor contacts where $\phi_m < \chi + V_n$, the energy band diagrams are shown in Fig. 2.7. Fig. 2.7(a) shows the energy band diagram before the contact is made and Fig. 2.7(b) shows the barrier after close contact at thermal equilibrium.

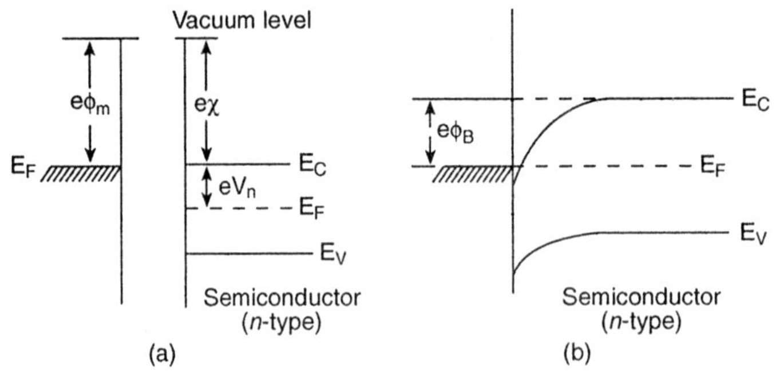

Fig. 2.7 Energy-band diagram for the metal-semiconductor contact (a) before contact (b) after contact in equilibrium when $\phi_m < \chi + V_n$

To attain the thermal equilibrium, electrons move from the metal to the lower energy states of the semiconductor. As a result, electrons accumulate on the semiconductor surface. When a positive potential is applied to the metal w.r.t the semiconductor, electrons flow from semiconductor to metal without experiencing any

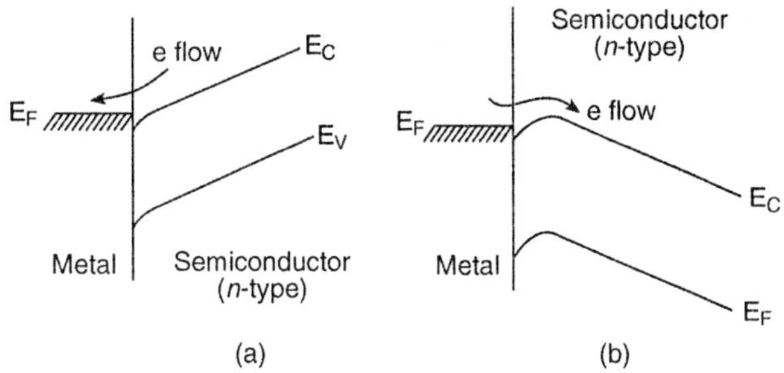

Fig. 2.8 Energy-band diagram for ohmic contact

barrier. When the connection is reversed i.e., the semiconductor is connected to the positive voltage, the electrons move from metal to semiconductor by overcoming a very small potential barrier $e\phi'_B$ for heavily doped semiconductor. So, electrons can flow easily at the metal-semiconductor junction in either direction. The junction is thus ohmic contact. Fig 2.8(a) and Fig. 2.8(b) respectively show the energy band diagram when (i) a positive voltage is applied to the metal w.r.t the semiconductor and (ii) a positive voltage is applied to the semiconductor w.r.t the metal.

2.6.2 Tunneling barrier

The width of the depletion region depends on the concentration of the donor atoms. In a rectifying metal-semiconductor contact, the width is inversely proportional to the square root of the donor concentration. Since the width of the depletion region decreases with increase in donor concentration, it enhances the probability of quantum mechanical tunneling through the barrier. For a metal and a heavily doped n-type semiconductor, the width of the depletion region may be reduced to the order of some nanometer and the tunneling through the narrow barrier takes place. Fig. 2.9 shows the energy-band diagram of the junction between a metal and a heavily doped n-type semiconductor.

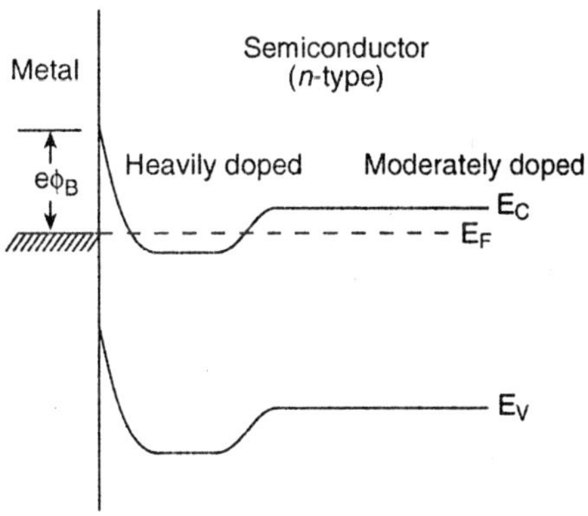

Fig. 2.9 Energy-band diagram for electron tunneling through the barrier of metal-semiconductor (heavily doped) contact

2.6.3 Specific contact resistance

The quality of an ohmic contact is determined by specific contact resistance R_c. The reciprocal of the derivative of current density J with respect to the voltage V under zero bias condition is called the ***specific contact resistance.*** Mathematically,

$$R_c = \left(\frac{\partial J}{\partial V}\right)^{-1}\Bigg|_{V=0} \tag{2.6.1}$$

The unit of R_c is ohm-cm^2. *Smaller the value of* R_c *better is the ohmic contact.*

Differentiating (2.5.6), we get the expression for R_c for contacts with low to moderate doping concentrations and is given by

$$R_c = \frac{k}{eAT} e^{\frac{e\phi_B}{kT}} \tag{2.6.2}$$

The above equation (2.6.2) shows that the specific contact resistance R_c decreases rapidly with decreasing barrier height.

2.7 Illustrated Examples

▶ **Example 1.** *A restifying contact (assumed ideal) is fabricated by depositing gold on n-type silicon having a donor density of* $10^{21}\,\mathrm{m}^{-3}$. *At a temperature of* 300 K, *calculate the barrier height from the semiconductor side, the width of the space region and the maximum electric field at the junction under equilibrium conditions. Work function of* Au = 5.1 eV, *electron affinity for* Si = 4.01 eV, *dielectric constant of* Si = 11.7, *energy difference between conduction band edge of* Si *and Fermi level* = 0.26 eV.

Solution: The barrier height, as seen from the metal side, is

$$\begin{aligned} e\phi_B &= e(\phi_m - \chi)\,\mathrm{eV} \\ &= 5.1 - 4.01, \quad (\because\ e\phi_m = 5.1\,\mathrm{eV} \text{ and } e\chi = 4.01\,\mathrm{eV}) \\ &= 1.09\,\mathrm{eV} \end{aligned}$$

∴ The barrier height, as seen from the semiconductor side, is

$$\begin{aligned} e\boldsymbol{V_{bi}} &= e\phi_B - (E_c - E_F) \\ &= 1.09 - 0.26\,\mathrm{eV} \quad (\because\ E_c - E_F = 0.26\,\mathrm{eV}) \\ &= \mathbf{0.83}\,\mathrm{eV} \end{aligned}$$

The width of the space charge region is given by

$$\omega = \sqrt{\frac{2\varepsilon V_{bi}}{eN_D}} = \sqrt{\frac{2k\varepsilon_0 V_B}{eN_D}},$$

$(\because\ \varepsilon = k\varepsilon_0 \text{ and } V_B = V_{bi} \text{ for no applied voltage})$

$$= \left(\frac{2 \times 11.7 \times 8.854 \times 10^{-12} \times 0.83}{1.6 \times 10^{-19} \times 10^{21}}\right)^{1/2} \mathrm{m} = \mathbf{1.04\,\mu m}$$

The maximum electric field at the junction under equilibrium condition is

$$\begin{aligned} |\mathcal{E}_m| &= \frac{eN_D\omega}{\varepsilon} = \frac{1.6 \times 10^{-19} \times 10^{21} \times 1.04 \times 10^{-6}}{11.7 \times 8.854 \times 10^{-12}}\,\mathrm{V/m} \\ &= \mathbf{1.6 \times 10^6\,V/m} \end{aligned}$$

▶ **Example 2.** *A Schottky contact between gold and n-type silicon with a donor concentration of* $10^{15}\,\mathrm{cm}^{-3}$ *has a cross-sectional area of* $3\times10^{-4}\,\mathrm{cm}^2$. *The barrier height from the semiconductor side is* $0.83\,\mathrm{eV}$. *Find the junction capacitance for an applied reverse bias of* $4\,\mathrm{V}$. *For silicon* $\varepsilon_s/\varepsilon_0 = 11.7$.

Solution: The space charge per unit area of the semiconductor surface for an applied reverse bias V_R is given by

$$Q = eN_D\omega = eN_D\sqrt{\frac{2\varepsilon(V_{bi}+V_R)}{eN_D}} = \sqrt{2e\varepsilon N_D(V_{bi}+V_R)}$$

∴ The junction capacitance per unit area is

$$C = \frac{\partial Q}{\partial V_R} = \sqrt{\frac{e\varepsilon N_D}{2(V_{bi}+V_R)}} = \frac{\varepsilon}{\omega}$$

The junction capacitance,

$$\begin{aligned} C_j &= \frac{\varepsilon A}{\omega} = A\sqrt{\frac{e\varepsilon N_D}{2(V_{bi}+V_R)}} \\ &= 3\times10^{-8}\left[\frac{1.6\times10^{-19}\times11.7\times8.854\times10^{-12}\times10^{21}}{2\times(0.83+4)}\right]^{1/2}\mathrm{F} \\ &= 1.243\times10^{-12}\,\mathrm{F} = \mathbf{1.24\,pF} \end{aligned}$$

▶ **Example 3.** *For a Schottky diode,* $\phi_{Bn} = 0.668\,\mathrm{V}$ *and* $J_s = 6\times10^{-5}\,\mathrm{A\,cm^{-2}}$ *at* $300\,\mathrm{K}$. *Determine the effective Richardson constant.*

Solution: The reverse saturation current density is given by

$$J_s = AT^2e^{-\frac{e\phi_B}{kT}}, \quad \text{with usual symbols,}$$

A being the effective Richardson constant.

$$\begin{aligned} \therefore\ \boldsymbol{A} &= \frac{J_s}{T^2}e^{\frac{e\phi_B}{kT}} \\ &= \frac{6\times10^{-15}}{300^2}\exp\left(\frac{1.6\times10^{-19}\times0.668}{1.38\times10^{-23}\times300}\right) \\ &= \mathbf{1.086\times10^2\,A\,cm^{-2}\,K^{-2}}, \text{ on simplification.} \end{aligned}$$

▶ **Example 4.** *Determine the electric field intensity that is to be applied normal to a metal surface so that its work function gets reduced to half the original value. Assume* $\phi_m = 5.0\,\mathrm{eV}$.

Solution : The external electric field reduces the barrier height due to Schottky effect. The Schottky lowering is given by

$$\Delta\phi = \left(\frac{e\varepsilon}{4\pi\varepsilon_0}\right)^{1/2} \tag{2.7.1}$$

To find ε, such that

$$\Delta\phi = \phi_m/2. \tag{2.7.2}$$

$$\therefore \quad \left(\frac{e\varepsilon}{4\pi\varepsilon_0}\right)^{1/2} = \frac{\phi_m}{2} \Rightarrow \varepsilon = \frac{\pi\varepsilon_0\phi_m^2}{e} \tag{2.7.3}$$

Substituting the appropriate values in (2.7.3), we have

$$\begin{aligned}\varepsilon &= \frac{3.14 \times 8.854 \times 10^{-12} \times (5)^2}{1.6 \times 10^{-19}} \quad (\because \phi_m = 5\,\text{eV}) \\ &= \mathbf{4.34 \times 10^9\ V/m}\end{aligned}$$

► **Example 5.** *What is the conductivity of a piece of* Ge ($n_i = 2.5 \times 10^{13}$ cm^{-3}) *doped with* 5×10^{13} cm^{-3} *donors and* 2.5×10^{13} cm^{-3} *acceptors ?* ($D_n = 100$ cm^2/S, $D_p = 50$ cm^2/S). *If the electron affinity of* Ge = 4.0 eV, *and we put down a metal electrode with work function* = 4.5 eV, *what is the work function difference? Do you expect this to be a Schottky barrier or an ohmic contact?*

Solution : From the relation, $n + N_A = p + N_D = \dfrac{n_i^2}{n} + N_D$, we have

$$n + 2.5 \times 10^{13} = \frac{(2.5 \times 10^{13})^2}{n} + 5 \times 10^{13}$$

Solving for n we obtain,

$$n = 4.04 \times 10^{13}\ \text{cm}^{-3} \quad \text{and} \quad p = \frac{n_i^2}{n} = 1.54 \times 10^{13}\ \text{cm}^{-3}$$

$\therefore$ The conductivity $\boldsymbol{\sigma} = e(n\mu_n + p\mu_p) = \dfrac{e}{kT/e}(nD_n + pD_p)$

$$\begin{aligned}&= \frac{1.6 \times 10^{-19}}{0.0259}\left(4.04 \times 10^{13} \times 100 + 1.54 \times 10^{13} \times 50\right) \\ &= \mathbf{0.0297\ (\Omega\,cm)^{-1}}, \text{ on simplification.}\end{aligned}$$

The work function difference,

$$\begin{aligned}\boldsymbol{\phi_F} = E_F - E_i &= kT \ln\frac{n}{n_i} = 0.0259 \ln\left(\frac{4.04 \times 10^{13}}{2.5 \times 10^{13}}\right) \\ &= \mathbf{0.0124\ eV}, \text{ on simplification.}\end{aligned}$$

So, for the n-type semiconductor, the Fermi level is above the intrinsic Si Fermi level by an amount $\phi_F = 0.0124$ eV.

Now, we have
$$\phi_{ms} = \phi_m - \left(\chi + \frac{E_g}{2} - \phi_F\right)$$
$$= 4.5 - \left(4.0 + \frac{0.67}{2} - 0.0124\right) \text{ eV}$$
$$= 4.338 \text{ eV}$$

∴ Electrons move from Ge to the metal. So, we loose majority carriers (i.e., electrons) in the semiconductor making the contact ***Schottky barrier.***

▶ **Example 6.** *From comparison of the de Broglie wavelength of electron with the depletion width of a contact metal-n* Si*, estimate the electron concentration at which Schottky diode loses its rectifying characteristics. For the estimate, assume that the height of the potential barrier at the contact is half the value of the band gap at room temperaeture* ($E_g = 1.12$ eV)*,* $m_e^* = m_0$*,* $T = 300$ K *and* $\varepsilon_{\text{Si}} = 11.9$.

Solution: When de Broglie wavelength λ of an electron becomes comparable with the depletion width ω of the junction, a Schottky diode loses its rectifying properties.

Now, de Broglie wavelength of an electron $\lambda = 2\pi\hbar/\sqrt{2m_0E}$ and depletion width of the junction $\omega = \sqrt{\varepsilon_{\text{Si}}E_g/4\pi e^2 n}$.

Accoding to the given problem we assume that the mean energy of electron is $E = 3kT/2$ and the potential barrier at the contact is $\phi_B = E_g/2e$.

Now, from the condition of $\lambda \ll \omega$, we obtain
$$n \ll \frac{3\varepsilon_{\text{Si}} m_o kTE_g}{16\pi^3 e^2\hbar^2}$$

On substitution of the numerical values in the above expression one can find that for proper functioning of the Schottky diode, the electron concentration n must be less than 2×0^{19} cm^{-3}.

Review Questions

1. What do you mean by metal-semiconductor contact? Mention some of its applications.
2. For the metal-semiconductor contact, explain the following terms:

   (a) Contact potential (b) Depletion region
   (c) Potential barrier (d) Barrier height
   (e) Built-in-field (f) Image-force lowering of Schottky effect
   (g) Effective Richardson constant (h) Ohmic contact
   (i) Non-rectifying barrier (j) Specific contact resistance

3. Define the work function of a metal or a semiconductor and explain its significance.

4. Draw the energy-band diagrams of electrons for a contact between a metal and an n-type semiconductor in absence of surface states, labelling the work functions of the metal and the semiconductor and the barrier height.

5. What do you mean by electron affinity? How is it related to the work function of a semiconductor?

6. Do the electronic states on the interface between a metal and a semiconductor influence the barrier height? What happens to the Fermi level when a metal and a semiconductor are brought in contact?

7. When is the contact between a metal and an n-type semiconductor said to be (a) forward biased and (b) reverse biased? Draw the energy band diagrams under these biased conditions.

8. Draw the energy-band diagrams of electrons when a metal and a p-type semiconductor are (a) well separated and (b) in close contact.

9. For the junction between a metal and an n-type semiconductor, find the expressions for

   (a) maximum barrier field (b) barrier voltage

   (c) width of the depletion layer.

10. What do you mean by Schottky effect? Derive an expression for the Schottky barrier lowering when a metal-semiconductor junction is reverse biased by an electric field. Show graphically the variation of potential energy with x due to lowering of the potential barrier caused by the combined effect of image force and electric field.

11. (a) Write down the expression relating the current density J with the applied voltage V in a metal-semiconductor junction.

    (b) What is reverse saturation current density? Does it depend on applied bias voltage? Is it a function of temperature?

12. (a) What is Ohmic contact? Explain the different types of Ohmic contact.

    (b) Draw the energy-band diagram for (i) Ohmic contact and (ii) the junction between a metal and a heavily doped n-type semiconductor.

13. (a) What is specific contact resistance? What is its significance?

    (b) How can you construct a good Ohmic contact?

Objective Questions

1. When a metal-semiconductor junction is in thermal equilibrium,

(a) the Fermi levels of the two materials coincide,
(b) the Fermi level in the metal is higher than that in the semiconductor,
(c) the Fermi level in the semiconductor is higher than that in the metal.

2. The depletion region of a metal-semiconductor junction

(a) is practically confined to the metal,
(b) is practically confined to the semiconductor,
(c) extends equally in the metal and the semiconductor.

3. For a given pair of metal and semiconductor, the sum of the barrier heights for n-type and p-type seconductors is

(a) equal to E_g, the band-gap energy, (b) less than E_g,
(c) greater than E_g, (d) zero.

4. When a reverse bias is applied to a metal-semiconductor junction, the barrier height

(a) does not change, (b) increases, (c) decreases.

5. When doping is increased, the width of the space-charge region in a metal-semiconductor contact

(a) decreases, (b) increases, (c) remains the same.

6. A good ohmic contact requires

(a) a low barrier height, (b) a wide band-gap semiconductor,
(c) low damping, (d) none of these.

7. The unit of specific contact resistance is

(a) Ohm-cm^2, (b) Ohm-cm, (c) Ohm, (d) None of these.

Problems

1. In a metal and an n-type semiconductor contact, a reverse bias field of 7×10^6 V/m is applied. Calculate the barrier lowering and the position of the maximum barrier height. Assume, the dielectric constant of the semiconductor = 13.1.

[0.0277 V, 1.98 nm]

2. In an n-type silicon, the Fermi level lies within the forbidden gap 0.21 eV below the conduction band edge for a donor concentration of $10^{16}\,\text{cm}^{-3}$. For its contact with tungsten, calculate the barrier height as seen from the metal and the semiconductor. Also, find the depeletion region width and the maximum electric field for the junction with no applied bias voltage. Given, work function of tungsten = 4.55 eV, electron affinity for silicon = 4.01 eV, dielectric constant of silicon = 11.7. [0.54 eV, 0.33 eV, 2.067×10^{-7} m, 3.2×10^{6} V/m]

3. A Schotky diode has the reverse saturation current density of $0.61\,\text{Am}^{-2}$ at 300 K. Calculate the current for an applied forward bias of 0.32 V, if the cross-sectional area is $4 \times 10^{-4}\,\text{cm}^2$. [5.72 mA]

4. A Schottky diode is formed between tungsten and an n-type silicon with a cross-sectional area of $4 \times 10^{-8}\,\text{m}^2$. Calculate the forward bias voltage required for a current of 2 mA and 10 mA at 300 K. Given, the effective Richardson constant = $114\,\text{A cm}^{-2}\,\text{K}^{-2}$ and barrier height = 0.67 eV. [0.53 V, 0.57 V]

5. In a metal and n-type semiconductor, the reverse saturation current density at 300 K is $5\mu\text{A/m}^2$. Find the voltage to be applied across the junction to obtain a forward current density of $50\mu\text{A/m}^2$. [0.238 V]

6. The plot $1/C^2$ vs. V_R of a Schottky contact between tungsten and an n-type silicon has a slope of $4.4 \times 10^{15}\,\text{cm}^4\,\text{F}^{-2}\text{V}^{-1}$ and an intercept of 0.4 V on the V_R axis. If the dielectric constant of Si = 11.7, calculate the barrier height V_{bi} and the donor concentration N_D in silicon. [0.4 V, $2.74 \times 10^{21} \times \text{m}^3$]

7. A sample of n-type ($N_d = 10^{20}/\text{m}^3$) Si is brought into intimate contact with a metal of $\phi_m = 5\,\text{eV}$. Calculate the following quantities.

   (a) the contact potential,
   (b) the barrier height from semiconductor end,
   (c) the barrier height from the matal end,
   (d) the depletion layer width. (Given: $\chi = 4.05$ eV, $\phi_s = 4.06$ eV, $k = 12$)

   [0.94 eV, 0.94 eV, 0.95 eV, 3.53 μm]

   [**Hints.** (a) Contact pontential, $V_c = \phi_m - \phi_s = (5 - 0.46)$ eV = **0.94 eV**.
   (b) $\phi_{bi} = \phi_{bn} - V_n = (\phi_m - \chi) - (\phi_s - \chi) = \phi_m - \phi_s = (5 - 0.46)$ eV = **0.94 eV**
   (c) $\phi_{bn} = \phi_m - \chi = (5 - 4.05)$ eV = **0.95 eV**
   (d) $W = \left(\dfrac{2k\varepsilon_0\phi_{bi}}{eN_d}\right)^{1/2} = \left(\dfrac{2 \times 12 \times 8.854 \times 10^{-12} \times 0.94}{1.6 \times 10^{-19} \times 10^{20}}\right)^{1/2} = \mathbf{3.53\,\mu m}$]

8. Find the height of the potential barrier for a Au-n Ge Schottky contact at room temperature ($T = 293$ K) if $\rho = 1\Omega$ cm, $\phi_{\text{Au}} = 5.1$ eV, and $\chi_{\text{Ge}} = 4.0$ eV. Electron mobility in Ge is $3900\,\text{cm}^2\,\text{V}^{-1}\,\text{S}^{-1}$, density of the states in the conduction band is $N_C = 1.98 \times 10^{15} \times T^{3/2}\,\text{cm}^{-3}$. [0.88 eV]

9. What is the electric field E for a Schottky contact of Au-n Si at $V = -5\,\text{V}$ at the distance of $1.2\,\mu\text{m}$ from the interface at room temperature if $\rho = 10\,\Omega\,\text{cm}$, $\mu_n = 1400\ \text{cm}^2\,\text{V}^{-1}\,\text{S}^{-1}$, $N_c = 6.2 \times 10^{15} \times T^{3/2}\ \text{cm}^{-3}$. [$2 \times 10^4\ \text{V}\,\text{cm}^{-1}$]

10. For a Schottky contact Au-GaAs, calculate the maximum electric field within the space charge region at $V = 0$, $+0.3\,\text{V}$ and $-100\,\text{V}$. Given, $N_d = 10^6\ \text{cm}^{-3}$, $\chi_{\text{GaAs}} = 4.07\ \text{eV}$, $\varepsilon_{\text{GaAs}} = 12.9$, work function of Au is $5.1\,\text{eV}$, $T = 300\ \text{K}$, density of states in the conduction band is $N_c = 8.63 \times 10^{13} \times T^{3/2}\ \text{cm}^{-3}$.

[5.1×10^4 V/cm, 4.2×10^4 V/cm and 5.1×10^5 V/cm]

11. Calculate the depletion width for a Pt-n Si Schottky contact ($T = 300$ K) at $V = 0$, $+0.4$ V and $-2\,\text{V}$. Concentration of doping impurity in Si equals $4 \times 10^{16}\ \text{cm}^{-6}$. Work function of Pt is 5.65 eV, electron affinity of Si is 4.05 eV, $\varepsilon_{\text{Si}} = 11.9$, density of the states in the conduction band is $N_c = 6.2 \times 10^{15} \times T^{3/2}\ \text{cm}^{-3}$.

[$0.22\,\mu\text{m}$, $0.19\,\mu\text{m}$ and $0.34\,\mu\text{m}$]

12. Find current densities J at room temperature for a Schottky diode Pt-n GaAs at $V = +0.5\,\text{V}$ and $-5\,\text{V}$ if $\rho = 50\,\Omega$ cm, $\mu_n = 8800\ \text{cm}^{-1}\text{V}^{-1}\text{S}^{-1}$, $m_n/m_0 = 0.063$, work function of Pt is 5.65 eV, $\chi_{\text{GaAs}} = 4.07$ eV, $N_c = 8.63 \times 10^{13} \times T^{3/2}\ \text{cm}^{-3}$. Apply thermoionic-emission theory. [$1.5 \times 10^{-13}\ \text{A/cm}^2$, $3 \times 10^{-22}\ \text{A/cm}^2$]

[**Hints.** We have:

$$n = \frac{1}{\rho e \mu_n} = 1.4 \times 10^{13}\ \text{cm}^{-3}.$$

$$e\phi_d = \phi_{\text{Pt}} - \chi_{\text{GaAs}} - kT \ln \frac{N_c}{n} = 1.32\ \text{eV}$$

$$v_T\ \text{(average thermal vel.)} = \sqrt{(8kT/\pi m_n)} = 4.6 \times 10^7\ \text{cm/s}$$

$$\therefore\ J_s = \tfrac{1}{4} e n\, v_T e^{-\frac{e\phi_d}{kT}} = \mathbf{3 \times 10^{-22}\ A\,cm^{-2}}$$

From $J = J_s\left(e^{\frac{eV}{kT}} - 1\right)$, $J = \mathbf{1.5 \times 10^{-13}\ A/cm^2}$ for 0.5 V and $\boldsymbol{J = J_s}$ for -5 V.]

3

The P-N Junction

3.1 Introduction

When a p-type and an n-type semiconductor are brought in close contact, the boundary of the interface where the two types meet is called the ***p-n junction***.

A useful p-n junction cannot be constructed merely by welding a piece of p-type material against another piece of n-type semiconductor as this will introduce a grain boundary i.e., a discontinuity across the junction. There are two principal techniques of fabricating a p-n junction: (i) *grown junction* and (ii) *fused junction*, into the details of which we shall not enter. The p-n junctions are created by doping one side of a single crystal semiconductor with acceptors and the other side with donors. A p-n junction is thus a two-terminal device and is known as a ***junction diode***. Most of the semiconductor devices contain one or more p-n junctions because of its effective role as a control element. So, the p-n junctions are elementary 'building blocks' of almost all the semiconductor devices like transistors, LEDs, ICs, solar cells etc.

3.2 Unbiased p-n junction

In a p-type semiconductor, *holes* are the majority carriers and in an n-type semiconductor *electrons* are the majority carriers. When a p-n junction is formed (Fig 3.1(a)), electrons from the n-region diffuse into the p-region and holes diffuse from p-region into n-region due to the difference in carrier concentration. During the process of diffusion, the electrons recombine with the holes in the p-side after crossing the junction, leaving behind the positively charged immobile donor ions in n-region. The hole in their turn cross the junction to recombine with the electrons in the n-region, leaving behind the negatively charged immobile acceptor ions in the p-region.

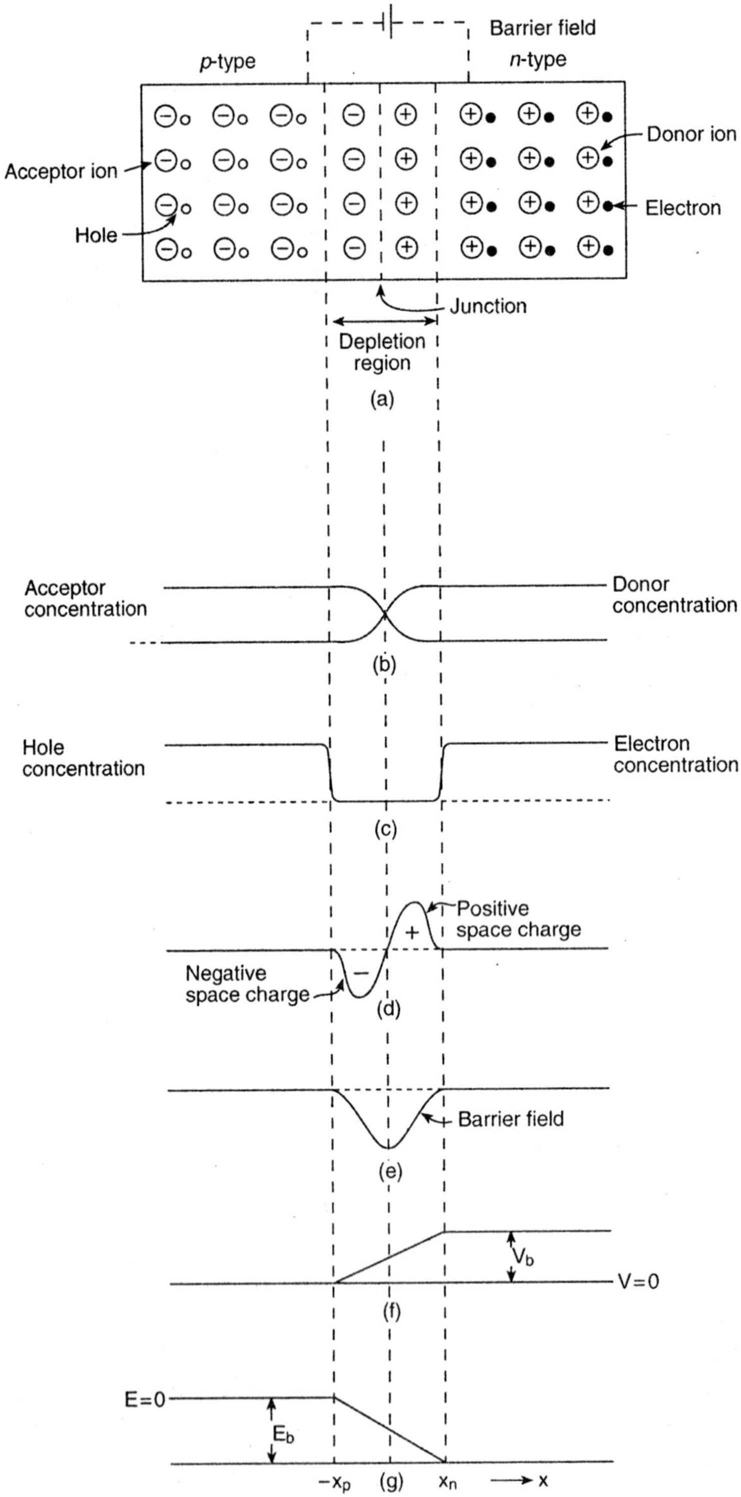

Fig. 3.1 Schematic diagram of a *p-n* junction [(a)] and variation of different parameters [(b) to (g)]

As a consequence of the diffusion and recombination, non-neutralized ions are left in the neighbourhood of the junction. Since the region near the junction becomes deprived of the mobile charge carriers (electrons or holes) and contains only the immobile ions, it is called the ***depletion region*** or *the space charge region*. It is also called *transition region*. The thickness of depletion region is $\sim 10^{-6}$ m. The non-neutralized positive and negative charges produce an electric field across the junction directed from n-side to the p-side. This field is called *potential barrier* or ***barrier field*** which opposes further diffusion of electrons and holes through the junction. In equilibrium, the internal barrier field is just sufficient to prevent diffusion of majority charge carriers. However, it is to be noted that the field helps the minority carriers to move across the junction and a very small amount of current flows. The potential barrier V_b acts as a *fictitious* battery connected respectively with its positive terminal to the n-side and negative terminal to the p-side.

The variation of the acceptor concentration and the donor concentration in a p-n junction is shown in Fig. 3.1(b). Fig. 3.1(c) shows the variation of the mobile charge concentration. It is zero in the depletion region. To the left of this region, the concentration of holes is equal to the concentration of acceptors and to the right of this region, the concentration of electrons equals the donor concentration. The unneutralized ions in the neighbourhood of the junction are referred to as *uncovered charges*. The uncovered charge density (ρ) is zero at the junction, positive to its right and negative to its left. The general shape of the charge distribution is as shown in Fig. 3.1(d).

The Poisson equation (in SI) is given by

$$\frac{d^2V}{dx^2} = -\frac{\rho}{\varepsilon} \tag{3.2.1}$$

where V is the electrostatic potential at x and ε the permittivity of semiconductor.

Integrating (3.2.1), we have the barrier field as

$$E = -\frac{dV}{dx} = -\int_{x_0}^{x} \frac{\rho}{\varepsilon} dx \tag{3.2.2}$$

where E is zero at $x = x_0$.

Fig. 3.1(e) shows the variation of the barrier field across the p-n junction.

From equation (3.2.2), we have

$$V(x) = -\int_{x_0}^{x} E\,dx \tag{3.2.3}$$

The variation of $V(x)$ is shown in Fig. 3.1(f). It indicates that a potential barrier V_b which prevents further diffusion of carriers across the junction is established. To transfer an electron from the n-side to the p-side of the junction an energy $E_b = eV_b$ is to be spent. So, E_b is the ***barrier energy***. An equal amount of energy is required to transfer a hole from p-side to the n-side. It is to be noted that the magnitude of E_b depends on the semiconductor, the level of doping and the width of the depletion region. The plot of the potential energy of an electron at the p-n junction is shown in Fig. 3.1(g).

The *circuit symbol* of a p-n junction diode is shown in Fig. 3.2 where the arrowhead is in the direction of current flow when the p-side and the n-side are connected respectively to the positive and negative pole of a battery.

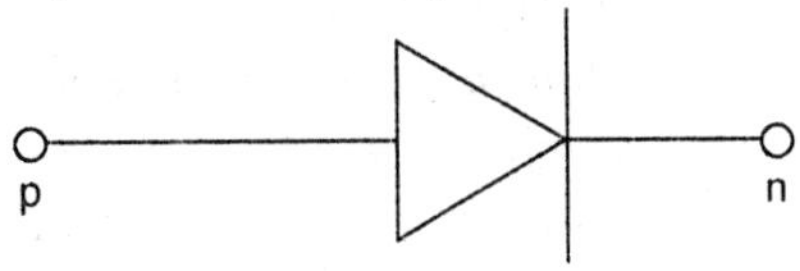

Fig. 3.2 Circuit symbol of a p-n junction diode

- It is worthnoting that the depletion region is a region over which the bulk properties of the p or the n-region are *not* applicable.

3.2.1 Potential barrier height of an unbiased p-n junction

To find the height of potential barrier across an unbiased p-n junction diode, we assume that (i) the p-side is doped uniformly with acceptor concentration N_A and the n-side with uniform donor concentration N_D, (ii) there is an abrupt change from p-type to n-type impurity i.e., the junction is abrupt.

Because of the variation of space charge density, a potential barrier (V_b) develops across the p-n junction. When the p-n junction is open circuited, the net hole current or net electron current must be zero. Considering the hole current, we have $J_h = 0$ i.e.,

$$e\mu_p pE - eD_p\frac{dp}{dx} = 0$$

$$\text{or,} \quad \frac{dp}{p} = \frac{\mu_p E}{D_p}dx \tag{3.2.4}$$

where E is the built-in electric field, μ_p the mobility and D_p the diffusivity of holes.

Einstein relation gives,

$$\frac{D_p}{\mu_p} = \frac{kT}{e} \tag{3.2.5}$$

Using (3.2.5) in (3.2.4), we get

$$\frac{dp}{p} = \frac{eE}{kT}dx \tag{3.2.6}$$

Since $E = -\dfrac{dV}{dx}$, the equation (3.2.6) reduces to

$$\frac{dp}{p} = -\frac{e}{kT}dV \tag{3.2.7}$$

Let $-x_p$ and x_n correspond respectively to left edge of p-region and right edge of n-region of the depletion region of the p-n junction. Integrating (3.2.7) within the entire depletion region, we have

$$\int_{p_p}^{p_n} \frac{dp}{p} = -\frac{e}{kT} \int_{-V_p}^{V_n} dV$$

$$\text{or,} \quad \ln \frac{p_n}{p_p} = -\frac{e}{kT}(V_n + V_p) = -\frac{e}{kT} V_b \tag{3.2.8}$$

where $V_b = V_n + V_p$ = potential barrier and p_n and p_p are respectively the thermal equilibrium concentration of holes in the n-region and p-region.

So, from (3.2.8), we have

$$p_n = p_p e^{-\frac{eV_b}{kT}} \tag{3.2.9}$$

For moderate temperature, the equilibrium concentration of holes in p-side is $p = N_A$ and that of electron in n-region is $n_n = N_D$. So, the law of mass-action gives

$$p_n = \frac{n_i^2}{n_n} = \frac{n_i^2}{N_D}$$

Therefore, the equation (3.2.8) gives

$$V_b = \frac{kT}{e} \ln \frac{p_p}{p_n}$$

$$\text{or,} \quad \boxed{V_b = \frac{kT}{e} \ln \frac{N_A N_D}{n_i^2}} \tag{3.2.10}$$

This is the expression for the *height of the potential barrier.*

The height of the potential barrier depends on the concentrations of donor and acceptor impurity, the temperature and the nature of the semiconductor. At room temperature (300 K), $kT/e \approx 0.026$ eV; for Ge with $n_i = 2.5 \times 10^{19}\,\text{m}^{-3}$, $N_A = N_D = 10^{22}\,\text{m}^{-3}$, the value of V_b is approximately 0.3 V.

● The potential barrier V_b appearing across a p-n junction is a *built-in* potential barrier which is necessary to maintain equilibrium at the junction. So, it does not imply an external potential. When a voltmeter is placed across a p-n junction to measure V_b, new contact potentials are developed at each probe, just cancelling V_b and no net current can result from it. So, V_b cannot be measured directly by a voltmeter.

3.2.2 Width of the depletion region

To find the expression for width of the depletion region, we consider Poisson's equation in one dimension for the depletion region [Fig. 3.1(a)].

$$\frac{d^2V}{dx^2} = \frac{eN_A}{\varepsilon} \quad \text{for } x < 0 \tag{3.2.11}$$

$$\text{and} \quad \frac{d^2V}{dx^2} = -e\frac{N_D}{\varepsilon} \quad \text{for } x > 0 \tag{3.2.12}$$

where the symbols have their usual significance.

Integrating (3.2.11) and (3.2.12), we have

$$\frac{dV}{dx} = \frac{eN_A}{\varepsilon}x + C_1 \tag{3.2.13}$$

$$\frac{dV}{dx} = -\frac{eN_D}{\varepsilon}x + C_2 \tag{3.2.14}$$

Since $\dfrac{dV}{dx} = 0$ for $x = -x_p$ and $x = x_n$ we get from (3.2.13) and (3.2.14),

$$C_1 = \frac{eN_A x_p}{\varepsilon} \quad \text{and} \quad C_2 = \frac{eN_D}{\varepsilon}x_n.$$

So, the equations (3.2.13) and (3.2.14) give,

$$\frac{dV}{dx} = \frac{eN_A}{\varepsilon}(x_p + x) \tag{3.2.15}$$

and
$$\frac{dV}{dx} = \frac{eN_D}{\varepsilon}(x_n - x) \tag{3.2.16}$$

Now, integrating (3.2.15) and (3.2.16), we have

$$V = \frac{eN_A}{\varepsilon}\left[x_p x + \frac{x^2}{2}\right] + C_3 \tag{3.2.17}$$

and
$$V = \frac{eN_D}{\varepsilon}\left[x_n x - \frac{x^2}{2}\right] + C_4 \tag{3.2.18}$$

Let us now apply the following boundary conditions:

$$V = -V_p \text{ at } x = -x_p \quad \text{and} \quad V = V_n \text{ at } x = x_n.$$

Then, $C_3 = -V_p + \dfrac{1}{2}\dfrac{eN_A x_p^2}{\varepsilon}$ and $C_4 = V_n - \dfrac{1}{2}\dfrac{eN_D x_n^2}{\varepsilon}$

Substituting the value of C_3, we have from (3.2.17)

$$\begin{aligned} V &= \frac{eN_A}{\varepsilon}\left[x_p x + \frac{x^2}{2}\right] - V_p + \frac{1}{2}\frac{eN_A x_p^2}{\varepsilon} \\ &= \frac{eN_A}{2\varepsilon}\left[2x_p x + x^2 + x_p^2\right] - V_p \\ &= -V_p + \frac{eN_A}{2\varepsilon}(x_p + x)^2 \end{aligned} \tag{3.2.19}$$

Similarly, on substitution of the value of C_4 in equation (3.2.18), we get

$$V = V_n - \frac{eN_D}{2\varepsilon}(x_n - x)^2 \tag{3.2.20}$$

Now, at $x = 0$, the solutions of both the Poisson equations should give the same value of dV/dx and V.

So, equating $\dfrac{dV}{dx}$ at $x = 0$, we have

$$\frac{eN_A}{2\varepsilon}\cdot 2(x_p) = \frac{eN_D}{2\varepsilon}\cdot 2x_n$$

$$\therefore\ N_A x_p = N_D x_n \tag{3.2.21}$$

$$\text{or,}\quad \frac{x_n}{x_p} = \frac{N_A}{N_D}$$

$$\text{or,}\quad \frac{x_n}{x_n + x_p} = \frac{N_A}{N_A + N_D} \tag{3.2.22}$$

$$\text{Also,}\quad \frac{x_p}{x_p + x_n} = \frac{N_D}{N_A + N_D} \tag{3.2.23}$$

Equating the values of V at $x = 0$, we get

$$-V_p + \frac{eN_A}{2\varepsilon}x_p^2 = V_n - \frac{eN_D}{2\varepsilon}x_n^2$$

$$\text{or,}\quad V_p + V_n = \frac{e}{2\varepsilon}\left[N_A x_p^2 + N_D x_n^2\right]$$

$$\text{or,}\quad V_b = \frac{e}{2\varepsilon}\frac{N_A N_D}{N_A + N_D}(x_p + x_n)^2,\ \text{ using (3.2.22).}$$

$$\therefore\ x_p + x_n = \sqrt{\frac{2\varepsilon V_b(N_A + N_D)}{e\cdot N_A N_D}}$$

Width of the depletion layer is thus

$$W = x_p + x_n = \sqrt{\frac{2\varepsilon V_b}{e}\frac{(N_A + N_D)}{N_A N_D}} \tag{3.2.24}$$

Since $V_b = \dfrac{kT}{e}\ln\dfrac{N_A N_D}{n_i^2}$, we get from equation (3.2.24),

$$W = \left[\frac{2\varepsilon kT}{e^2}\left(\frac{1}{N_A} + \frac{1}{N_D}\right)\ln\frac{N_A N_D}{n_i^2}\right]^{1/2} \tag{3.2.25}$$

Thus the width of the depletion region depends on the acceptor concentrations and donor concentrations on both sides of the p-n junction. Clearly, the width of the depletion region decreases with increase in doping concentrations.

3.2.3 Energy band diagram of an unbiased *p-n* junction

In p-type semiconductors, the Fermi level E_F is closer to the valence band edge E_{vp} and in n-type semiconductors, it is closer to the conduction band edge E_{cn}. When a p-n junction is formed, the majority carriers cross the junction creating space charges which raise the position of all the energy levels, including E_F, on p-side and lowers it on the n-side so that equilibrium is established on equating E_F in both sides such that no current flows through the junction. The energy band diagram of an unbiased p-n junction is shown in Fig. 3.3.

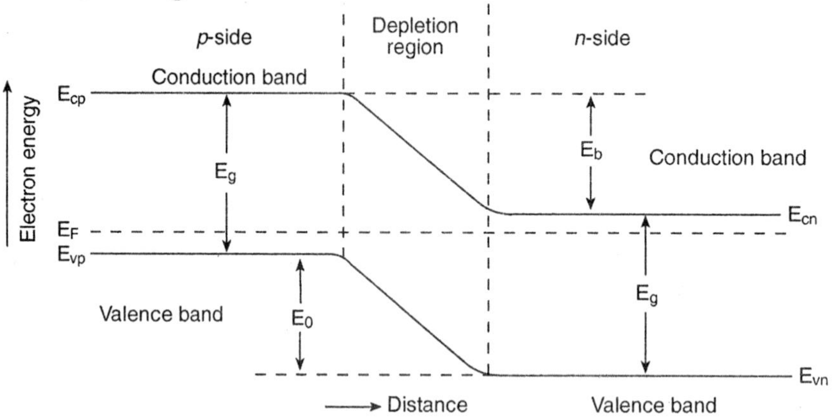

Fig. 3.3 Energy-band diagram of an unbiased p-n junction

The energy barrier E_b for electrons in crossing the junction from the n-side to the p-side is given by

$$\boxed{E_b = E_{cp} - E_{cn}} \tag{3.2.26}$$

Now, we have
$$E_{cn} = E_F + kT \ln \frac{N_c}{N_D} \tag{3.2.27}$$

and
$$E_{vp} = E_F - kT \ln \frac{N_v}{N_A}, \tag{3.2.28}$$

using equations (1.14.1) and (1.14.2).

We know that $E_g = E_{cp} - E_{vp}$. Therefore,

$$E_{cp} = E_g + E_{vp}$$
$$= E_g + E_F - kT \ln \frac{N_v}{N_A} \tag{3.2.29}$$

Substituting (3.2.27) and (3.2.29) in (3.2.26), we get

$$E_b = E_g - kT \ln \frac{N_v}{N_A} - kT \ln \frac{N_c}{N_D} \tag{3.2.30}$$

Thus the band gap E_g is: $\quad E_g = kT \ln \dfrac{N_c N_v}{n_i^2}$

$$\therefore \; E_b = kT \left(\ln \frac{N_c N_v}{n_i^2} - \ln \frac{N_v}{N_A} - \ln \frac{N_c}{N_D} \right), \text{ using (3.2.30)}$$
$$= kT \ln \frac{N_A N_D}{n_i^2} \tag{3.2.31}$$

$\therefore$ Potential barrier, $$V_b = \frac{E_b}{e} = \frac{kT}{e}\ln\frac{N_A N_D}{n_i^2} \tag{3.2.32}$$

Clearly, the equations (3.2.32) and (3.2.10) are identical.

3.3 Biasing of a *p-n* junction

When a *p-n* junction is in an unbiased condition, the intrinsic potential barrier V_b tends to oppose the movement of the majority carriers through the junction. When a battery of voltage V is connected in such a way that the *p*-side is connected to the positive pole and *n*-side to the negative one, the intrinsic potential barrier V_b is reduced to $V_b - V$. The junction is then said to be *forward biased* (Fig. 3.4). If the forward bias is greater than the potential barrier, the majority carriers from both sides cross the junction and constitute a large current in the same direction i.e., from *p*-side to *n*-side. This current is known as *forward current* which increases exponentially with the forward bias.

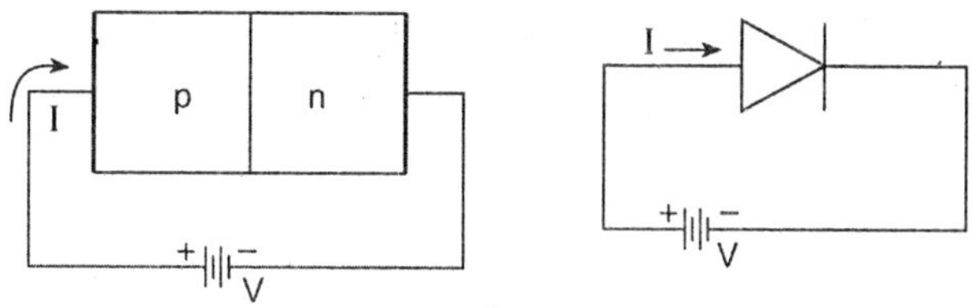

Fig. 3.4 Forward biased p-n junction

When however a battery is connected to a junction diode in such a way that the *p*-side is connected to the negative pole and *n*-side to the positive one, then the intrinsic potential barrier is increased to $V_b + V$. The junction is then called *reverse biased.*

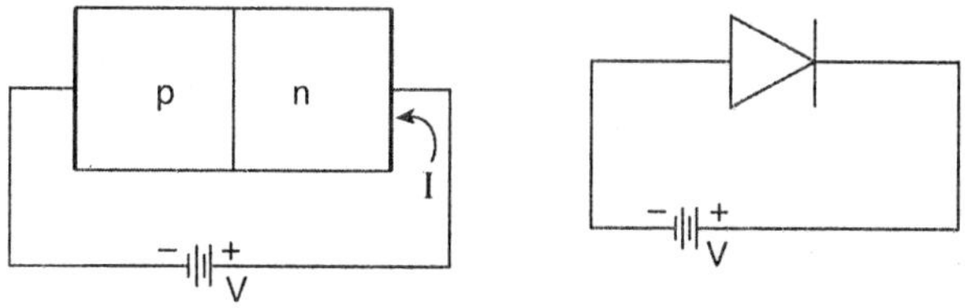

Fig. 3.5 Reverse biased p-n junction

Fig. 3.5 shows the reverse biased circuit representation of a *p-n* junction diode. The increased potential barrier height reduces the current flow due to the majority carriers to a negligible value. But the flow of the minority carriers in the opposite direction is *not* affected by the barrier height. So, a small curent in the reverse direction is obtained. This current is known as *reverse saturation current* which increases with the increase in temperature.

3.3.1 Energy band diagram of a biased *p-n* junction

In an unbiased *p-n* junction under thermal equilibrium, the energy bands are separated by a barrier energy E_b as shown in Fig. 3.3. When the junction is forward biased by a voltage V, the energy barrier is reduced to a value $E_b - eV$ and the Fermi levels E_F on either side are shifted by an energy eV. This causes the majority carriers on either side to cross the reduced barrier and diffuse into other side so that a large component of diffusion current-flow in the forward direction is obtained. The minority carriers constitute a very small current due to their random thermal motion. The energy band diagram of a forward biased p-n junction is shown in Fig. 3.6(a).

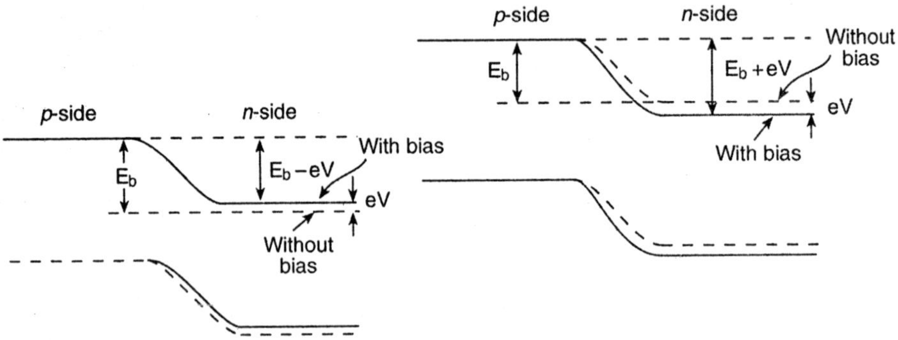

(a) Forward biased (b) Reverse biased

Fig. 3.6 Energy diagram of a biased p-n junction

When however a p-n junction is reverse biased by a voltage V, the energy barrier is raised to $E_b + eV$ which opposes the majority carriers on either side to cross the junction. So, only a very small number of majority carriers may acquire enough energy to cross the barrier and a negligibly small diffusion current is obtained. But the motion of the minority carriers is *not* influenced by the increased barrier. So, they give a very small reverse current which is *independent* of the reverse bias. The energy band diagram of a reverse biased p-n junction is shown in Fig. 3.6(b).

3.4 Diode equation

To derive the diode equation relating current-voltage relationship in a p-n junction diode, the following assumptions are made.

(i) The level of injection of minority carriers is very low.

(ii) There is no generation and recombination within the depletion region.

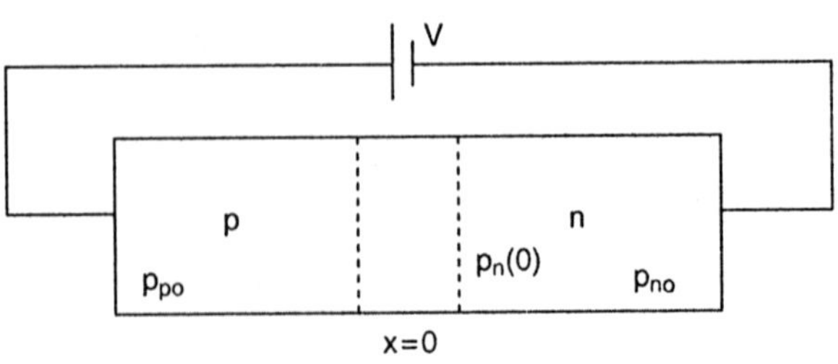

Fig. 3.7 Hole concentration of forward biased p-n junction

(iii) The voltage applied to diode is completely dropped across depletion region.

(iv) The widths of p and n-regions are larger than the diffusion lengths.

Let an external voltage V be applied to forward bias the diode. This reduces the height of the potential barrier to $V_b - V$. So, the concentration of holes (p_n) on the n-side edge ($x = 0$) of the depletion region, according to Maxwell-Boltzmann statistics, is given by

$$p_n(0) = p_{no} e^{\frac{eV}{kT}} \tag{3.4.1}$$

where p_{no} = thermal equilibrium concentration of holes in n-region.

The excess concentration of hole over the equilibrium concentration at the edge of the depletion region (i.e., $x = 0$) due to carrier injection is

$$\begin{aligned} p'_{ne} &= p_n(0) - p_{no} \\ &= p_{no}\left[e^{\frac{eV}{kT}} - 1\right] \end{aligned} \tag{3.4.2}$$

The injected holes in the n-side travel through the n-side and recombine with electrons there.

The excess concentration of holes at a distance x in this region is

$$p'_n(x) = p'_{ne} e^{-\frac{x}{L_p}} \tag{3.4.3}$$

where L_p is the diffusion length of holes.

∴ The net hole concentration at any point x is

$$\begin{aligned} p_n(x) &= p_{no} + p'_n(x) \\ &= p_{no} + p_{no}\left(e^{\frac{eV}{kT}} - 1\right) e^{-\frac{x}{L_p}} \end{aligned}$$

So, the diffusion current density at the junction ($x = 0$) is

$$J_{pn}(0) = -e\cdot D_p \frac{dp_n(x)}{dx}\bigg]_{x=0}$$

$$\text{or,}\quad J_{pn}(0) = \frac{eD_p p_{no}}{L_p}\left(e^{\frac{eV}{kT}} - 1\right) \tag{3.4.4}$$

where D_p = diffusivity of holes.

Similarly, the electron current density injected from the n-side to the p-side of the diode can be obtained. Hence, the electron current density at the junction ($x \geq 0$) is

$$J_{np}(0) = \frac{eD_n n_{po}}{L_n}\left(e^{\frac{eV}{kT}} - 1\right) \tag{3.4.5}$$

where D_n = diffusivity of electrons, n_{po} = equilibrium concentration of electrons in the p-region, L_n = diffusion length of electron.

Since low level injection minority drift currents can be neglected compared to diffusion current, we have the total forward current density

$$\begin{aligned} J &= J_{pn}(0) + J_{np}(0) \\ &= J_0\left(e^{\frac{eV}{kT}} - 1\right) \end{aligned} \tag{3.4.6}$$

where $$J_0 = e\left(\frac{D_p p_{no}}{L_p} + \frac{D_n n_{po}}{L_n}\right)$$

If α be the cross-sectional area of the junction, then the *diode current* is

$$I = J\alpha = I_s\left(e^{\frac{eV}{kT}} - 1\right)$$

or, $$\boxed{I = I_s\left(e^{\frac{eV}{kT}} - 1\right)} \tag{3.4.7}$$

where $$I_s = \alpha J_0 = e\alpha\left(\frac{D_p p_{no}}{L_p} + \frac{D_n n_{po}}{L_n}\right) \tag{3.4.8}$$

Equation (3.4.7) is known as ***rectifier equation*** or ***Shockley equation*** or simply ***diode equation***.

3.5 Current-voltage characteristics of a *p-n* junction

The total current I flowing through a *p-n* junction under the application of a voltage V across it, is given by the following equation known as ***diode equation*** :

$$I = I_s\left(e^{\frac{eV}{\eta kT}} - 1\right) \tag{3.5.1}$$

where I_s is the reverse saturation current, e is the electronic charge, k is the Boltzmann constant, T the absolute temperature of the junction and η is a dimensionless constant known as *emission coefficient* or *ideality factor* having a value between 1 and 2, depending on the semiconductor material. For Ge, $\eta \approx 1$ and for Si, $\eta \approx 2$. The plot of current *versus* voltage across the junction is called the *I-V characteristic* of a *p-n* junction diode.

3.5.1 Forward biased characteristics

To understand the nature of the forward bias characteristics, we use positive value of V in (3.5.1). When the forward voltage V is much greater than kT/e, equation (3.5.1) is approximated as

$$I = I_s e^{\frac{eV}{\eta kT}} \tag{3.5.2}$$

which shows that the forward current increases exponentially with the bias voltage.

• The quantity (kT/e) is called the ***thermal voltage*** (V_T).

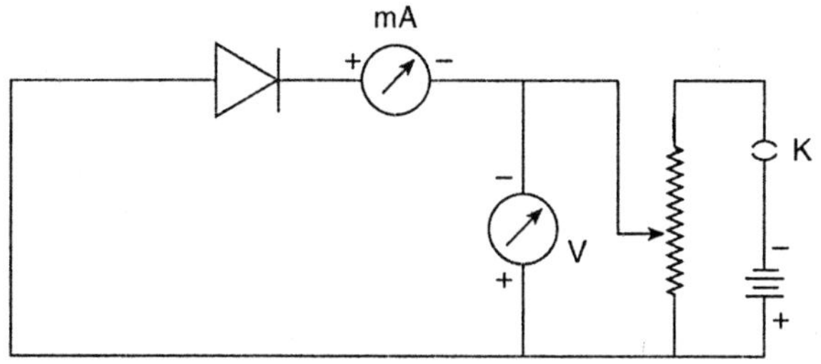

Fig. 3.8 Circuit diagram for studying forward biased characteristics

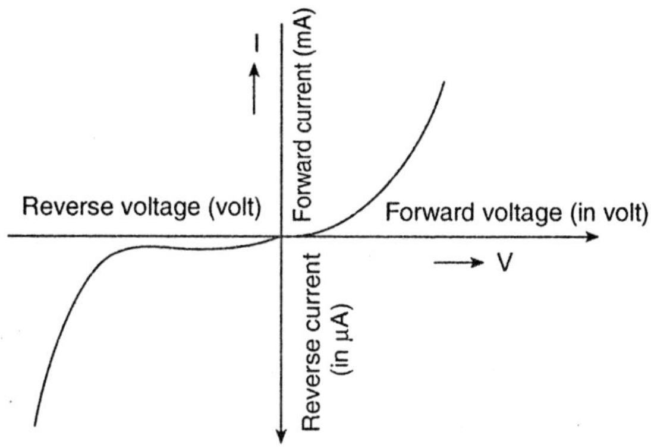

Fig. 3.9 I-V characteristics of a p-n junction diode

The circuit diagram for plotting the graph between forward bias voltage and forward current is shown in Fig. 3.8. The milliammeter and the voltmeter are used to measure respectively the forward current and the p.d. across the diode. The forward bias voltage is increased in steps and corresponding current is recorded for drawing the I-V characteristic. The curve in the first quardrant of Fig. 3.9 is the forward characteristic. It shows that when the external voltage exceeds the potential barrier, the forward current starts to flow and increases exponentially with forward bias.

3.5.2 Reverse biased characteristics

For a reverse biased *p-n* junction, V is negative. When its magnitude is much greater than kT/e, then we obtain from equation (3.5.1),

$$I = -I_s \tag{3.5.3}$$

So, the current becomes the reverse saturation current which is *independent* of the reverse voltage but *dependent* on temperature. I_s is a few mA for Ge-diode and a few nA for Si-diode.

The circuit diagram for plotting the graph between the reverse bias and the reverse current is shown in Fig. 3.10. The voltmeter V and the microammeter μA are used to measure in steps the reverse voltage and the corresponding current respectively for

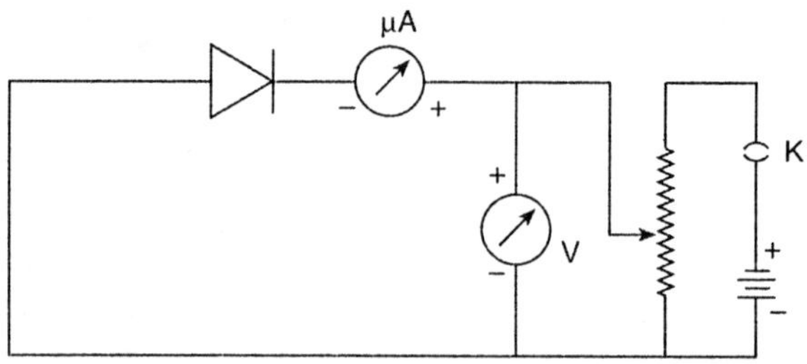

Fig. 3.10 Circuit diagram for studying reverse biased characteristics

drawing the reverse biased characteristic. The curve in the third quadrant of Fig. 3.9 is the *reverse biased characteristic.* It shows that the reverse current is very small and remains almost constant over a long range. On increasing the reverse bias, a junction diode may *breakdown.* The following are the two mechanisms that might cause this breakdown.

(A) Avalanche breakdown—The electric field across the junction increases with increasing reverse bias. At a certain value of reverse bias voltage, the thermally generated charge carriers get sufficient amount of energy from the field to disrupt a covalent bond by collision with a crystal ion and produces an electron-hole pair. These carriers may further produce new electron-hole pairs. This cumulative process develops an avalanche of carriers in a very short time-interval. As a result, the reverse current increases abruptly to a large value. This mechanism is known as ***avalanche multiplication*** and the diode is then said to work in the avalanche breakdown region.

(B) Zener breakdown—When the reverse bias voltage across a *p-n* junction is so high that the electric field imparts sufficient amount force to the bound electron to tear it out from the covalent bond, the Zener breakdown takes place. So, Zener breakdown is the *process of direct rupture of the covalent bond and thus produces a large number of electron-hole pairs and increases the reverse current abruptly.* It may however be noted that the Zener breakdown does not involve any collision of carriers with the crystal ions.

3.6 Diode resistance

The *I-V* characteristic of a *p-n* junction shows that the junction does not obey Ohm's law. So, the resistance offered by a junction diode is not a constant but varies with the applied voltage.

The ***static*** or the ***dc resistance*** r_{dc} of a diode is defined as the ratio of the voltage V across the junction to the current I flowing through the junction.

$$\therefore \quad \boxed{r_{dc} = \frac{V}{I}} \tag{3.6.1}$$

The ***dynamic*** or ***ac resistance*** r_{ac} of a diode is defined as the ratio of the small change in voltage to the small change in diode current. Thus

$$\boxed{r_{ac} = \frac{\Delta V}{\Delta I}} \tag{3.6.2}$$

The inverse of the slope of the I-V characteristic of a p-n junction diode may also define r_{ac} as

$$\boxed{r_{ac} = \frac{dV}{dI}} \tag{3.6.3}$$

Plainly, therefore, the dynamic resistance r_{ac} of a diode is determined by the operating voltage.

3.7 Junction capacitance of a p-n junction diode

A junction diode is composed of two pieces of semiconducting materials of resonable conductivity, separated by an insulating depletion layer. The depletion region of a p-n junction consists of immobile negatively charged acceptor ions in the p-side and immobile positively charged donor ions in the n-side of the junction. When the p-n junction is reverse biased, the width of the depletion region as well as the space charge region increases. Thus the voltage-dependent positive and negative charges make the p-n junction behave as a capacitor. This capacitance is referred to as the *transition capacitance* or *space-charge capacitance* or *depletion-region capacitance.*

If a change in voltage dV results in a change in charge by dQ, the transition capacitance is given by

$$C_T = \left|\frac{dQ}{dV}\right| \tag{3.7.1}$$

For an abrupt or step-graded junction (with junction area A), the transition capacitance is given by

$$C_T = \varepsilon A\sqrt{\frac{eN_D}{2\varepsilon(V_b + |V_a|)}} \tag{3.7.2}$$

where V_a is the applied reverse bias.

When a p-n junction is forward biased, the variation of forward bias voltage causes changes in concentration of injected charge carriers near the junction. This injected charge on either side of the junction gives rise to a capacitance, called the *diffusion capacitance* or *storage capacitance*, C_D. The diffusion capacitance is much larger than the transition capacitance and may be defined as the rate of change of injected charge with voltage.

The diffusion capacitance of a p-n junction diode is given by

$$C_D = g_p\frac{L_p^2}{D_p} + g_n\frac{L_n^2}{D_n} \tag{3.7.3}$$

where g_p and g_n are the conductance due to holes and electrons respectively.

- The transition capacitance decreases as the reverse voltage increases. This phenomenon is utilized in *voltage variable capacitance* or ***varactor diode***.

3.8 Zener diode

When a conventional p-n junction diode is reverse biased below its reverse breakdown voltage, the reverse current is not significant. But when the reverse bias exceeds the reverse breakdown voltage, a huge current flows through the diode due to avalanche breakdown. This current is limited by appropriate circuitry to avoid any permanent damage to the diode.

A ***Zener diode*** is a specially designed p-n junction diode which operates in the reverse breakdown region with a greatly reduced breakdown voltage, the so-called ***Zener voltage***. A reverse biased Zener diode exhibits a controlled breakdown and allows current to maintain the Zener voltage across it. The operation of a Zener diode depends on the level of doping of the p-n junction. For heavily doped Zener diode, the breakdown occurs at voltages below 6 V. For lightly doped diodes, the breakdown occurs above 6 V and the avalanche effect is prominent. Ordinarily, the breakdown diodes are referred to as ***Zener diodes***. The circuit symbol representing a Zener diode is illustrated in Fig. 3.11 below.

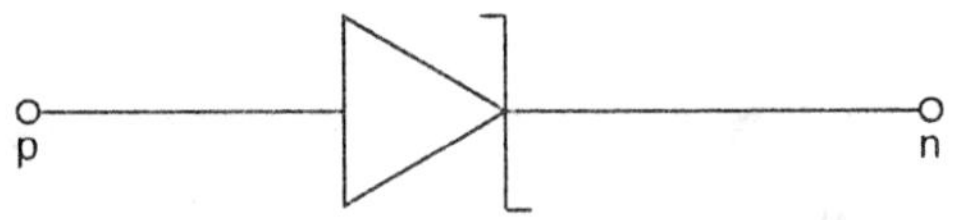

Fig. 3.11 Circuit symbol of a Zener diode

3.9 Characteristics of a Zener diode

A Zener diode exhibits almost the same properties as that of a p-n junction diode. The I-V characteristic of a Zener diode is shown in Fig. 3.12.

The forward biased characteristic is similar to that of an ordinary p-n junction diode. Under reverse bias, the I-V characteristic is similar to an ordinary p-n junction diode up to the breakdown voltage. In Zener diodes, the doping is controlled in such a way that the breakdown voltage is sharp and distinct. Beyond the breakdown voltage, the reverse current increases sharply to high value which indicates that the voltage across the diode is almost independent of the diode current in this region. This breakdown voltage is known as ***Zener voltage V_z*** and the corresponding current is ***Zener current I_z***. Clearly, V_z remains constant even when I_z increases rapidly. This makes a Zener diode extremely useful as a *voltage regulator device* or reference diode.

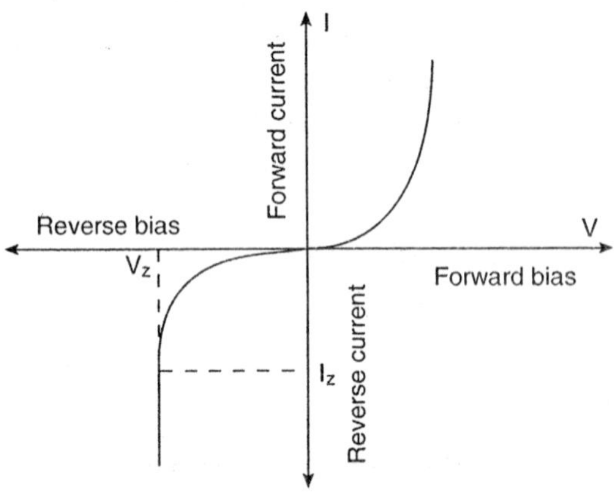

Fig. 3.12 *I-V* characteristics of a Zener diode

3.10 Zener diode as a voltage regulator

The working of a Zener diode is based on the fact that an extremely small voltage change in the reverse breakdown region produces a very large change in Zener current i.e., the Zener voltage V_z remains constant irrespective of the change in Zener current I_z. Moreover, Zener diode is prepared as highly doped p-type and n-type semiconductor and the power rating is so high that it is not damaged by the large reverse current. So, Zener diodes are used as *voltage regulator.* The circuit diagram of a simple Zener voltage regulator is shown in Fig. 3.13.

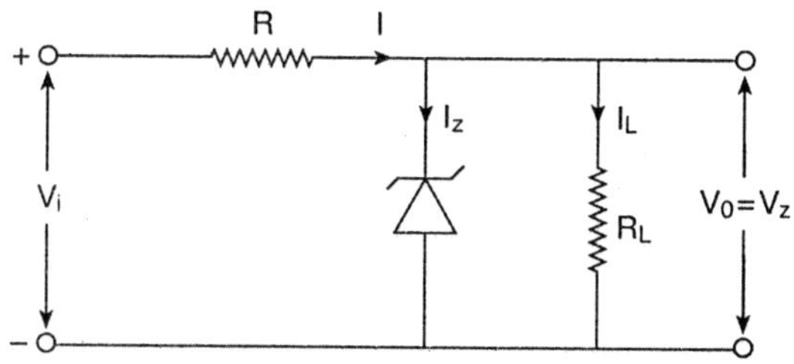

Fig. 3.13 Simple Zener voltage regulator

The unregulated dc supply voltage V_i ($> V_z$) and the series resistance R are so chosen that the diode operates in the reverse breakdown region. The Zener diode keeps a constant output voltage $V_0 = V_z$ across the load resistance R_L, although the input voltage V_i or the load resistance R_L, may vary over a wide range.

Let I and I_L be the supply current and the load current respectively. From Fig. 3.13, we get, by applying the KCL and the KVL, we get

$$I = I_z + I_L \tag{3.10.1}$$

$$V_z = V_i - IR \tag{3.10.2}$$

$$V_z = V_0 = I_L R_L \tag{3.10.3}$$

The output voltage V_0 may be varied (i) by varying the load resistance R_L and (ii) by variation of the input voltage V_i.

(A) Load regulation—In this case, the supply voltage V_i is kept *constant* and the load current I_L is changed by changing R_L. For an ideal case, V_z is *constant.* So, $\Delta V_z = 0$ and $\Delta V_i = 0$.

∴ Eq. (3.10.2) gives, $\Delta I = 0$

i.e., $\Delta I_z + \Delta I_L = 0$, using (3.10.1)

$$\text{or, } \Delta I_z = -\Delta I_L \tag{3.10.4}$$

Thus, if the load resistance is increased, keeping the supply voltage constant, the load current decreases with an equal increase in Zener current so that the output remains constant at V_z.

(B) Line regulation—When the load resistance R_L is kept fixed and the supply voltage V_i is varied, we get from (3.10.2)

$$\Delta V_i = \Delta I \cdot R \tag{3.10.5}$$

since $\Delta V_z = \Delta V_0 = 0$ and I_L is constant. From (3.10.1), therefore,

$$\Delta I = \Delta I_z \tag{3.10.6}$$

Thus, when the input voltage V_i is varied, keeping the load resistance R_L fixed, the supply current and Zener current both change equally so that I_L and V_0 remain constant. The change in V_i appears across the series limiting resistance R.

3.11 Special purpose diodes

The most common application of the *p-n* junction diodes is the *rectification* which will be dealt with in the next chapter (Chapter 4). Rectification converts ac voltage to dc voltage. But various specific types of diodes are also manufactured to serve the specific needs in the fast developing world. Several optoelectronic devices are fitted in our domestic applications like clocks, calculators, remotes of TV's etc. where special purpose diodes are used. A Zener diode is a special kind of diode used for voltage regulation. Common special purpose diodes also include (a) Light emitting diode (LED), (b) Photo diode, (c) Solar cell, (d) Schottky diode, (e) Tunnel diode etc. Some of them are described in the sections that follow.

3.11.1 Light Emitting Diode (LED)

Light emitting diode (abbreviated as LED) is a special type semiconducting *p-n* junction diode which emits light under forward biased condition.

Construction—The construction of a typical LED can be followed from Fig. 3.14(a). First, one *n*-type layer of gallium arsenide (GaAs) or gallium arsenide phosphide (GaAsP) or some other suitable semiconducting material is grown on a substrate. Then, a *p*-type layer of very small thickness is deposited onto it by diffusion so that

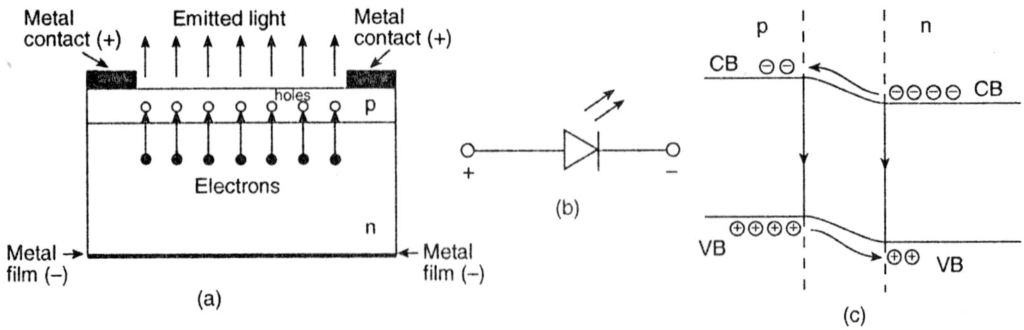

Fig. 3.14 (a) Construction of LED (b) Symbol of LED (c) Electron-hole recombination and emission of light

light energy may emit from the depletion region after travelling a very short distance. The metal anode connection is made at the p-layer so as to allow more surface area for light emission and the cathode connection is made at the bottom of the substrate coated with metal film. Most of the LEDs contain a domed lens that gathers and intensifies light. Fig. 3.14(b) shows the symbol of a LED.

Principle of working—The principle underlying the operation of a LED is the emission of radiation due to recombination of holes and electrons. When a p-n junction is forward biased, the barrier height decreases and electrons from the n-side and the holes from the p-side move towards the narrow depleted region. This results in the electron-hole recombination. Since the electrons make transitions from the conduction band to the valence band during this process, energy is released in the form of a photon of frequency ν given by

$$E_g = h\nu$$

where E_g is band gap energy.

The corresponding wavelength λ is given by

$$\lambda = \frac{hc}{E_g} \quad (\because \nu = c/\lambda)$$

The downward transition of electrons from the conduction band to the valence band and subsequent emission of photons due to recombination is shown in Fig. 3.14(c). The intensity of emitted light depends on the number of photons which in turn depends on the number of recombination that occurs.

For GaAsP, the band gap energy $E_g = 1.9\,\text{eV}$. So we have

$$\lambda = \frac{hc}{E_g} = \frac{6.62 \times 10^{-34} \times 3 \times 10^8}{1.9 \times 1.6 \times 10^{-19}}\,\text{m} = 6533\,\text{Å (reddish)}$$

Uses and advantages of LED—The applications of LED are rather too many. Some of the uses are as under:

1. LEDs are mainly used in signal lamps and display devices, in digital watches, in TV remotes, in calculators and multimeters, in telephone switchboards, etc.
2. The infrared LEDs are used in optical fibre communication and alarm systems where invisible signals are essential.
3. LED can also be used as a laser, known as Injection Junction Laser.

Advantages—LEDs have the following advantages over the traditional lamps with filaments.

1. LEDs consume relatively much smaller power for their operation.
2. They have also quite long life span ($\sim$ 100,000 hours)
3. They are very fast in action, small in size and cheap in price.
4. They are available to emit light of all colours.

3.11.2 Solar cell

An unbiased *p-n* junction when designed in such a way that it can *directly* convert a high percentage of the solar radiation falling on its large surface area (i.e., absorbed optical energy) into useful electrical energy is known as a ***solar cell.*** In fact, a solar cell is a special type of photovoltaic cell.

The first solar cells were developed in USA in 1954 and were made of silicon. Silicon cells are the only ones, even to-day, that enjoy the commercial status.

Construction—A solar cell is basically an unbiased *p-n* junction diode made of silicon. Fig. 3.15(a) shows schematically the constructional details of a typical solar cell

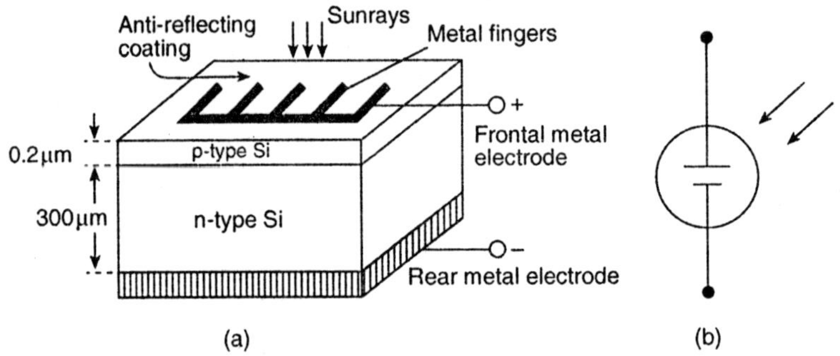

Fig. 3.15 (a) Construction of solar cell (b) Symbol of solar cell

and Fig. 3.15(b) gives its symbolic representation. The upper electrode of the cell is in the form of a ring or a grid with fingers so that sunlight may easily pass through it. To prevent the loss of light by reflection, the upper surface is coated by an anti-reflecting layer. The *p*-layer of the junction diode is much thinner compared to the *n*-layer for an easy diffusion of majority carriers before recombination. The lower electrode (metal) completely covers the surface. Both the *p* and the *n*-regions are heavily doped to obtain a large amount of photovoltage.

Principle of working—The schematic diagram of a *p-n* junction solar cell with a load resistance R_L is shown in Fig. 3.16(a). Since the band gap for silicon is 1.1. eV, the

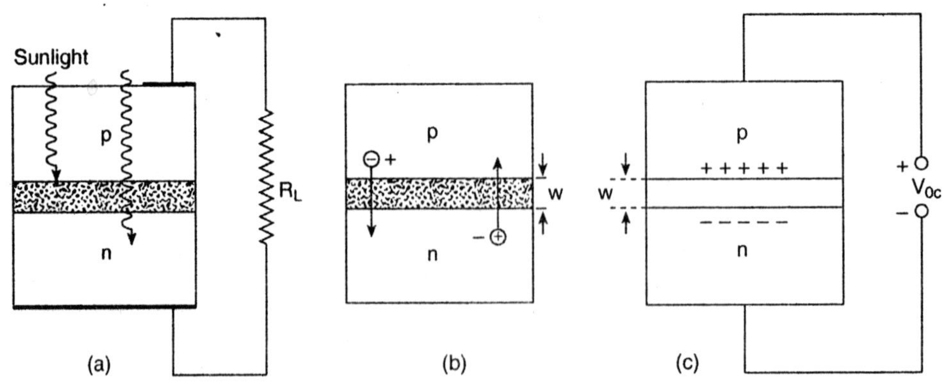

Fig. 3.16 (a) Incident photons (b) absorption of photons and creation of e-h pairs (c) deposition of charges

photons of sunlight incident on the diode, if absorbed, can produce electron-hole pairs in both the p and n-side of the junction. The electrons and holes produced near the depletion layer reach finally the depletion region W by diffusion and are separated by strong barrier field that exists there. The electrons in the p-side and the holes in the n-side are swept away in the depletion region (Fig. 3.16(b)) and are deposited at the edge of the depletion layer producing an open circuit voltage V_{oc}(Fig. 3.16(c)). When a load is connected, a current flows through the load so long the diode is exposed to sunlight. The open circuit voltage is quite low ($\sim 0.5\,\text{V}$) and the maximum value of current i.e., the short circuit current is $\sim 10^{-3}\,\text{A}$. So, for practical applications, a large number of solar cells are to be combined in series and parallel to produce appreciable amount of voltage or current. Such a combination of solar cells is known as a ***solar panel.***

Uses of solar cells—Solar cells have already a number of applications and they are on the increase. Some are enumerated below.

1. Solar cells are extensively used to recharge the storage batteries in satellites and communication equipments.
2. They are used to power watches and calculators.
3. They are also capable of detecting very low level of light intensity.
4. They are being increasingly used to provide domestic and commercial electricity in an eco-friendly way, particularly in areas where the sunshine is plentiful.

3.11.3 Photodiode

A reverse biased p-n junction diode, designed to respond to optical signals by absorption of photons, is known as a ***photodiode.***

Construction—The construction of a typical photodiode is shown schematically in Fig. 3.17(a). The reverse biased p-n junction is so designed that light falling on it may easily reach the depleted region. The p-n junction is embedded in a plastic capsule where light is allowed to fall upon one surface across the junction and all the other sides

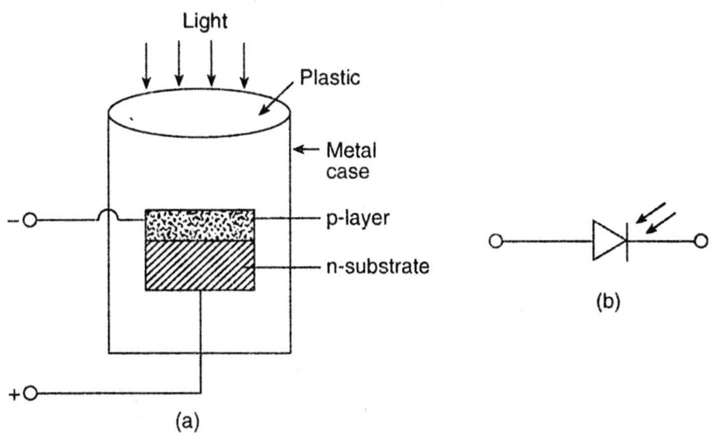

Fig. 3.17 (a) Construction of a photodiode (b) Symbol of a photodiode

are painted black or enclosed in a metal case. For the construction of a photodiode, GaAs is preferred to Si. This is because Si is an indirect-gap semiconductor whereas GaAs is a direct-gap semiconductor. GaAs photodiodes have therefore a greater speed of operation and a higher sensitivity to the intensity of incident radiation. The circuit symbol of a photodiode is shown in Fig. 3.17(b).

Principle of working—When the reverse biased p-n junction is kept under dark condition, a reverse saturation current or leakage current,mainly due to the flow of

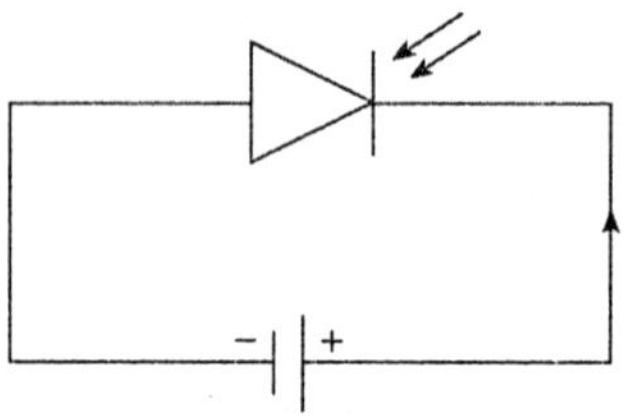

(a)

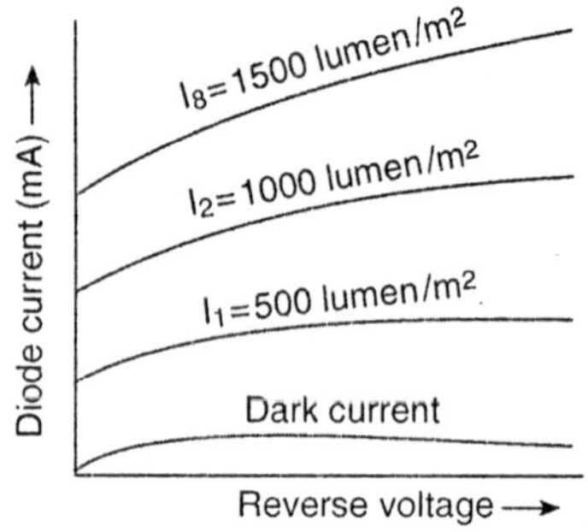

(b)

Fig. 3.18 (a) Circuit diagram of a photodiode (b) *I-V* characteristics for different light intensities

thermally generated minority carriers, is obtained. This current is called the ***dark current***. When light is allowed to fall on the reverse biased p-n junction, additional electron-hole pairs are produced in the p and the n-regions. As a result, a significant change in minority carrier concentration and a negligible change in majority carrier concentration take place. The generated minority carriers increase the leakage current that is found to vary almost linearly with the intensity of light (Fig. 3.18(b)). The total reverse current is given by

$$I = I_s + I_d = I_s + I_0\left(1 - e^{\frac{eV}{\eta kT}}\right)$$

where I_s is the short-circuit current proportional to the light intensity and I_d is the dark current.

Uses of photodiode—Photodiode is used for the detection of visible and invisible optical signals, in fast reading of tape and film sound tracks, and also in optical communication, optical switching etc.

3.11.4 Schottky diode

A rectifying metal-semiconductor contact which has the current-voltage characteristic similar to a p-n junction diode is known as ***Schottky diode***.

Construction—A metal-semiconductor contact may be rectifying or non-rectifying (i.e., ohmic), depending on the nature of the metal and the concentration of majority carriers in the semiconductor. Aluminium is an acceptor impurity in silicon. When it is attached to a highly n-type (i.e., n^+) Silicon as a lead, the n-region near the surface of contact behaves as a non-rectifying contact. If however the concentration is not so high, a rectifying contact is obtained. Fig. 3.19(a) shows schematically the construction of a Schottky diode. The contact A is rectifying and behaves as the anode whereas the contact B is non-rectifying. The circuit symbol of a Schottky diode is shown below in Fig. 3.19(b).

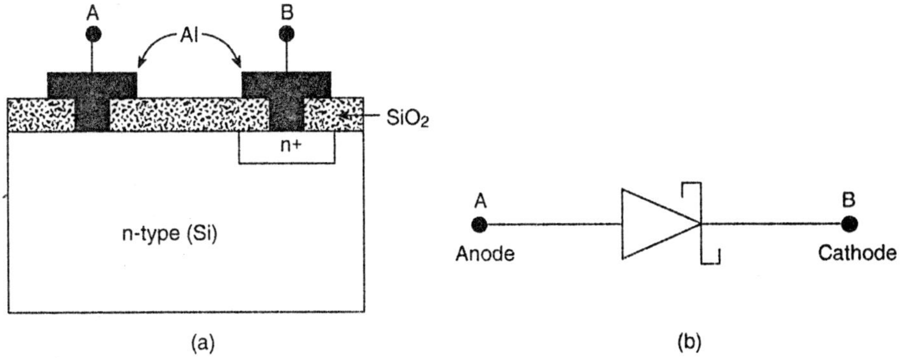

Fig. 3.19 (a) Construction of a Schottky diode (b) symbol of a Schottky diode

The *working principle* of a Schottky diode is similar to a p-n junction diode. The current transport in this diode is determined by the majority carrier electrons. It is thus a *majority-carrier device.*

Uses—Schottky diode is used in faster-switching and high-frequency applications and also in integrated circuits (ICs).

Advantages—Although the I-V characteristics of a Schottky diode and a p-n junction diode are rather similar, the former possesses the following advantages over the latter.

1. The fabrication of a Schottky diode is comparatively much easier than that of a p-n junction diode.
2. Schottky diode is a majority-carrier device and the minority carrier storage does not occur; the storage time is thus greatly reduced. In a p-n junction diode, the minority carriers have a significant role in charge-storage and the storage time is much higher than that of a Schottky diode. Hence, Schottky diodes can be used in faster-switching and higher frequency application than the p-n junction diodes. Switching time for a typical Schottky diode is $\sim 10^{-12}$s while that for a p-n junction diode is $\sim 10^{-9}$ s.

3.11.5 Tunnel diode

When the concentration of the impurity atoms is extremely high ($\sim 10^{24}\,\text{m}^{-3}$) in both p and n regions of a p-n junction diode, the width of the depletion region becomes very small ($\sim 10^{-8}$ m). When the diode is forward biased, the typical I-V characteristic

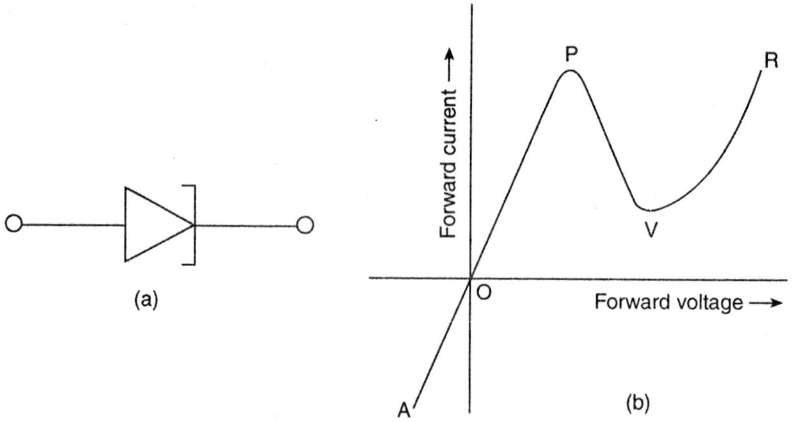

Fig. 3.20 (a) Symbol of tunnel diode; (b) *I-V* characteristics of a tunnel diode

shows a negative slope that can be explained by *quantum mechanical tunneling.* This type of diode is called a ***tunnel diode***. The symbol for a tunnel diode is shown in Fig. 3.20(a).

***I-V* characteristics**—The I-V characteristics of a tunnel diode is shown in Fig. 3.20(b). When the tunnel diode is reverse biased, the reverse tunnel current flows. This current increases with the increase of reverse bias. This is shown by the portion AO of the graph. When the forward bias across the junction is increased from zero, the forward tunnel current increases rapidly and attains the *peak value* (the point P). With further increase in the forward voltage, the tunnel current decreases and reaches a minimum value at V, known as the *valley point.* The region between P and V is called the *negative resistance region* as the reciprocal of the slope of the curve at any point in this region gives the negative differential resistance. When the forward bias is further increased beyond the valley point, the tunnel current increases again exponentially with the applied voltage as shown by the portion VR of the graph.

Explanation—The mechanism of a tunnel diode can be explained on the basis of energy band diagrams as shown in Fig. 3.21. At the absolute zero, the Fermi levels E_F lie within the bands of the semiconductor for high doping in the p-n junction. When no bias voltage is applied, the Fermi levels across the junction coincide and remain at

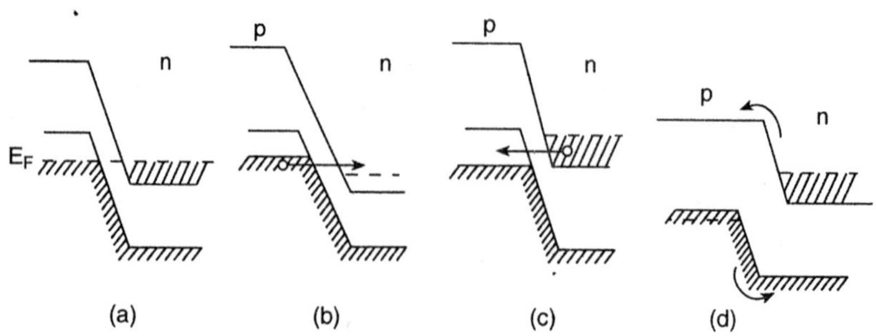

Fig. 3.21 Mechanism of a tunnel diode

thermal equilibrium (Fig. 3.21(a)) and no current flows. When a reverse bias is applied, the E_F in the p-side moves up relative to that of the n-side (Fig. 3.21(b)). The electrons in valence band of p-side find empty allowed energy states in the conduction band of the n-side across the junction. Hence the electrons can tunnel from the valence band of the p-side to the conduction band of n-side producing the reverse tunnel current. With increasing reverse bias, the reverse tunnel current increases and this explains the portion AO of the graph.

When the junction is forward biased, the Fermi level E_F in the n-side goes up relative to that in the p-side (Fig. 3.21(c)). The electrons in the conduction band of the n-side find the available energy states in the valence band of the p-side across the junction. So, the electrons from n-side can tunnel through to the p-side resulting in a forward tunnel current which increases with forward bias, till eventually the maximum value is attained when the conduction band edge of the n-side coincides with the Fermi level in the p-side. This furnishes the explanation for the portion OP of the I-V characteristic of a tunel diode.

With further increase in the forward bias voltage, the number of available unoccupied states in the p-side decreases. This in turn decreases the electron transition from the conduction band of n-side to the valence band of p-side. Hence the tunnel current decreases and becomes minimum when all the electrons in the n-side face no available unoccupied states in the valence band of the p-side. This accounts for the portion PV of the graph.

When the forward bias is increased further, the normal p-n junction current is obtained (Fig. 3.21(d)) as shown by VR of the characteristic curve.

Uses—As the tunnelling process is very fast, tunnel diodes are used in the construction of amplifiers, oscillators or switching devices operated in the frequency range $\sim$ 10 GHz.

3.12 Illustrated Examples

▶ **Example 1.** *A* Si *p-n junction diode carries a current of* 60 mA *at a forward bias of* 0.9 V. *Determine the* d.c *and* a.c *resistance of the diode at* 27° C. **(Punjabi Univ.)**

Solution: Given, $I = 60\,\text{mA} = 60 \times 10^{-3}\,\text{A}$ and $V = 0.9\,\text{Volt}$, $T = 27 + 273 = 300\,\text{K}$.

$$\therefore \text{ The d.c resistance is } \boldsymbol{r_{\text{dc}}} = \frac{V}{I} = \frac{0.9}{60 \times 10^{-3}}\Omega = \mathbf{15\Omega}$$

When the p-n junction is forward biased and applied voltage is larger than a few kT/e, then the forward current is given by

$$I = I_s e^{\frac{eV}{\eta kT}}$$

$$\Rightarrow \quad r_{\text{ac}} = \frac{dV}{dI} = \frac{\eta kT}{eI}$$

$$\therefore \text{ a.c resistance, } r_{ac} = \frac{2 \times 1.38 \times 10^{-23} \times 300}{1.6 \times 10^{-19} \times 60 \times 10^{-3}} \text{ ohm}$$
$$= \mathbf{0.87\Omega}$$

▶ **Example 2.** *The saturation current density of a p-n junction germanium diode is* 200 mA/m² *at* 300 K. *Find the voltage that would have to be applied to cause a forward current density of* 10^5 A/m² *to flow.*

Solution: For a *p-n* junction diode, the current density can be writen as

$$J = J_s\left(e^{\frac{eV}{kT}} - 1\right)$$

If the applied voltage is large compared to kT/e, the above equation becomes

$$J = J_s e^{\frac{eV}{kT}}$$

$$\therefore \ \ln J = \ln J_s + \frac{eV}{kT}$$

$$\therefore \ V = \frac{kT}{e}(\ln J - \ln J_s), \text{ from above}$$

$$= \frac{1.38 \times 10^{-23} \times 300}{1.6 \times 10^{-19}}(\ln 10^5 - \ln 0.2)$$

$$= \mathbf{0.34 \ volt}$$

▶ **Example 3.** *For the circuit of a Zener diode, series resistance* $R_s = 30\,\Omega$, *Zener voltage* $Z_z = 18$ V *and load resistance* $R_L = 200\,\Omega$. *If the input voltage* V_i *can vary from* 20 V *to* 30 V *find* (i) *the minimum and maximum current in the diode and* (ii) *the minimum and maximum power dissipated in the diode and (iii) the minimum and maximum power dissipated in the series resistance.*

Solution: (i) Given $R_s = 20\,\Omega$, $R_L = 200\Omega$, $V_z = 18$ V.

The output voltage $V_0 = V_z = 18$ V as V_i varies from 20 V to 30 V.

$$\text{The load current } I_L = \frac{V_L}{R_L} = \frac{V_z}{R_L} = \frac{18}{200} A = 0.09 \text{ A}$$

$$\text{Now, } V_0 = V_i - IR_s \text{ gives } \quad I = \frac{V_i - V_0}{R_s}$$

$$\therefore \quad \text{For } V_i = 20 \text{ V, we get } I = \frac{20 - 18}{20} = 0.1 \text{ A}$$

$$\text{and} \quad \text{for } V_i = 30 \text{ V}, \quad I = \frac{30 - 18}{20} = 0.6 \text{ A}$$

So, for $V_i = 20$ V, $I_z = I - I_L = (0.1 - 0.09)$ A $= 0.01$ A

and for $V_i = 30$ V, $I_z = 0.6 - 0.09 = 0.51$ A

∴ The ***minimum and maximum currents*** in the diode are **0.01 A** and **0.51 A**.

(ii) When $V_i = 20\,V$, the power dissipated in the diode is

$$V_z I_z = 18 \times 0.01\,W = 0.18\,W$$

For $V_i = 30\,V$, the required power is $V_z I_z = 18 \times 0.51\,W = 9.18\,W$

∴ The ***minmum and maximum power*** dissipated in the diode are **0.18 W** and **9.18 W** respectively.

(iii) When $V_i = 20\,V$, $I = 0.1\,A$. So, power dissipated in resistance $R_s(20\Omega) = I^2R = (0.1)^2 \times 20 = 0.2\,W$. When $V_i = 30\,V$, $I = 0.6\,A$. So, power dissipated in $R_s = (0.6)^2 \times 20 = 7.2\,W$. So, ***minimum and maximum power*** dissipated in series resistance R_s are **0.2 W** and **7.2 W** respectively.

▶ **Example 4.** *In the following circuit, calculate* (i) *the output voltage,* (ii) *the voltage drop across the series resistance* R_s *and* (iii) *the current through the Zener diode.*

(Gharwal Univ.)

Solution: Here, input voltage $V_i = 120\,V$ and the output voltage $\mathbf{V_0 = V_z = 50\,V}$.

∴ The voltage drop across the series resistance R_s is

$$\mathbf{V_s = (120 - 50)\,V = 70\,V}$$

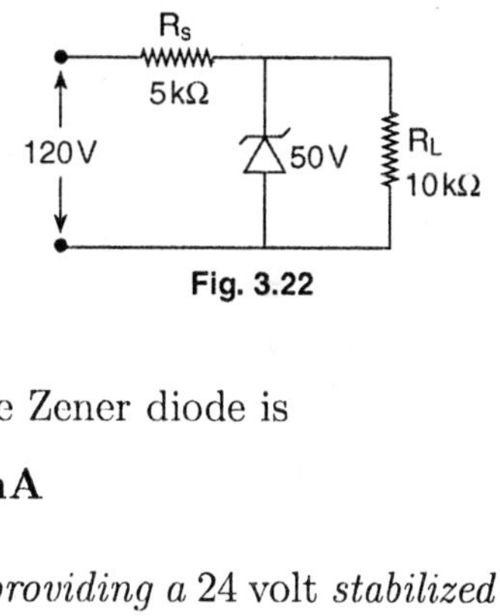

Fig. 3.22

The current through R_s is: $I = \dfrac{70}{5 \times 10^3} = 14\,mA$

Current through R_L is: $I_L = \dfrac{50}{10 \times 10^3} = 5\,mA$

Now, $I = I_z + I_L$. So, the current through the Zener diode is

$$\mathbf{I_z = I - I_L = (14 - 5)\,mA = 9\,mA}$$

▶ **Example 5.** *A* 24 V, 600 mW *Zener diode is to be used for providing a* 24 volt *stabilized supply to a variable load. If the input voltage is* 32 volt, *calculate* (i) *the value of the resistance to be connected in series and* (ii) *the current in the diode when* $R_L = 1.2\,k\Omega$.

(Bangalore Univ.)

Solution: (i) Here, input voltage $V_i = 32\,V$

output voltage $V_0 = V_z = 24\,V$

Power of the diode is 600 mW ⇒ $V_z I_z = 600\,mW$

$$\therefore\ I_z = \frac{600 \times 10^{-3}}{24}\,A = 25\,mA$$

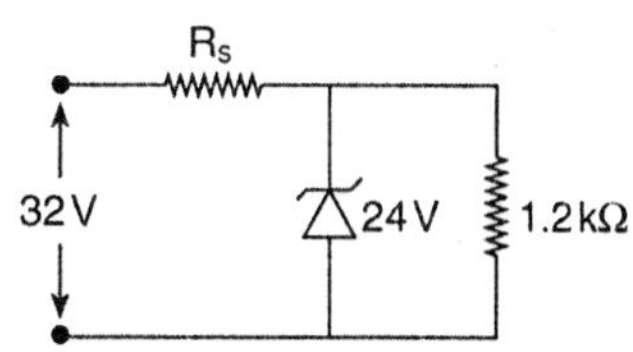

Fig. 3.23

So, the maximum allowable current through the diode is 25 mA.

∴ The value of the resistance to be connected in series is given by

$$R_s = \frac{V_i - V_0}{I_z} = \frac{32 - 24}{25 \times 10^{-3}} = \mathbf{320\,\Omega}$$

(ii) When $R_L = 1.2\,\text{k}\Omega$, the current through the load is plainly

$$I_L = \frac{V_0}{R_L} = \frac{24}{1.2 \times 10^3} = 20\,\text{mA}$$

∴ The ***current*** through the diode is $(25 - 20)\,\text{mA} = \mathbf{5\,mA}$

▶ **Example 6.** *What are the currents that flow through* $1\,\text{k}\Omega$, $10\,\text{k}\Omega$ *resistor and the Zener (assumed ideal) of breakdown voltage* 6 V *in the circuit of* Fig. 3.24? *How will the answer change if* $10\,\text{k}\Omega$ *resistor is replaced by* 500Ω. *What is the voltage across Zener now? Assume the current through Zener is zero when the voltage across it is less than breakdown voltage.* **(Calcutta Hons.)**

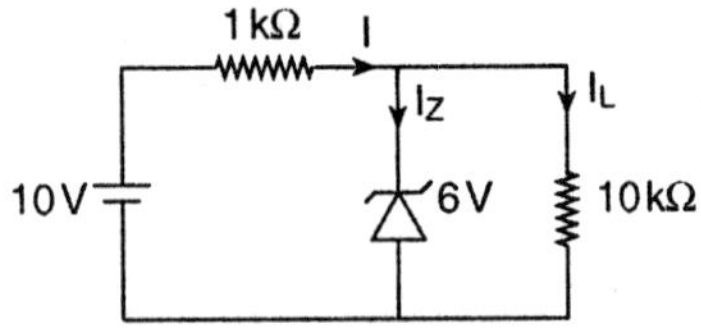

Fig. 3.24

Solution: Assuming the Zener to be in breakdown region, $I_L = 6\,\text{V}/10\,\text{k}\Omega = \mathbf{0.6\,mA}$. Potential difference across $1\,\text{k}\Omega = 10\,\text{V} - 6\text{V} = 4\,\text{V}$. So, $I = 4V/1\,\text{k}\Omega = \mathbf{4\,mA}$. Hence $I_z = I - I_L = 4\,\text{mA} - 0.6\,\text{mA} = 3.4\,\text{mA}$.

With 500Ω, $I_L = 6\,\text{V}/500\Omega = 12\,\text{mA}$. So, the required p.d across $1\,\text{k}\Omega = 12\,\text{mA} \times 1\,\text{k}\Omega = 12\,\text{V}$ and this is not possible as Zener will not be in breakdown region. Then, $I_L = 10\,\text{V}/(1\,\text{k}\Omega + 500\Omega) = \mathbf{6.67\,mA}$ and the voltage across Zener $= 6.67\,\text{mA} \times 500\Omega = 3.3\,\text{V}$, less than the breakdown voltage.

▶ **Example 7.** *The maximum power dissipation of a Zener diode having* 9.1 V *breakdown voltage is* 364 mW. *What is the maximum current possible?* **(Pubjab Univ.)**

Solution: Maximum voltage, $V_m = 9.1$ V (given)

Maximum power, $W_m = 364$ mW (given).

∴ Maximum current, $I_m = \frac{W_m}{V_m} = \frac{364 \times 10^{-3}}{9.1} = 40 \times 10^{-3}\,\text{A} = \mathbf{40\ mA}$

▶ **Example 8.** *A simple Zener diode regulator uses a Zener diode having Zener voltage* 6 V. *The supply voltage is* 15 V *and the series resistance is* $100\,\Omega$. *The load resistance is* $80\,\Omega$. *Calculate the Zener current, if the supply voltage varies between* 15 V *and* 20 V. *Find the maximum value of Zener current.*

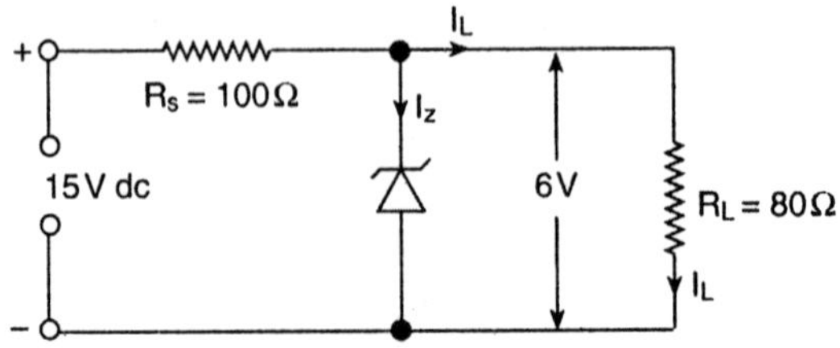

Fig. 3.25

Solution: The load current $I_L = \dfrac{6\,\text{V}}{80\Omega} = 75$ mA

$$\text{Supply current, } I_s = \frac{V_s - V_z}{R_s} = \frac{15-6}{100} = \frac{9}{100} = 90 \text{ mA}$$

$$\therefore \quad I_z = I_s - I_L = 90 - 75 = \mathbf{15\ mA}$$

The Zener current is maximum when the supply voltage is maximum and load current is minimum or I_L is zero, given output open circuit.

I_s at maximum supply voltage is $I_s = \dfrac{20-6}{100} = \dfrac{14}{100} = 140$ mA when $I_L = 0$, $\boldsymbol{I_z = I_s =}$ **140 mA**

$\therefore$ Maximum value of Zener current = **140 mA.**

▶ **Example 9.** *In a Zener regulator, the Zener voltage* V_z *is* 9 V. *The supply voltage varies between* 15 V *and* 24 V. *The load resistance varies between* 900Ω *to infinity. Find the maximum value of series resistance.*

Solution: Maximum value of load current $= \dfrac{V_z}{900\,\Omega} = \dfrac{9\,\text{V}}{900\,\Omega} = 10$ mA

Maximum series resistance permitted with minimum supply voltage and maximum load current is

$$(R_s)_{\text{max}} = \frac{(15-9)\,\text{V}}{(I_L)_{\text{max}}} = \frac{6\,\text{V}}{10\,\text{mA}} = \mathbf{600\,\Omega}$$

If the series resistance is more than 600 Ω, then at supply voltage 15 V and load current 10 mA, the voltage supply across Zener will be less than 9 V and the Zener will not be in breakdown region. Thus, the regulation action will stop as the output voltage drops below 9 V.

▶ **Example 10.** *The barrier capacitance of an abrupt p-n junction diode is* 4 pF *at* 4V. *Find the change in capacitance when the reverse biasing changes to* 4.5 V.

Solution: The barrier capacitance of an abrupt *p-n* junction diode is given by

$$C_T = \varepsilon A \sqrt{\frac{eN_D}{2\varepsilon(V_b + V_a)}}$$

where $V_b + V_a$ = intrinsic potential + applied reverse bias
= effective barrier voltage
= V_T (say)

$$\text{So,} \quad C_T \propto \frac{1}{\sqrt{V_T}}$$

$$\therefore \quad \frac{C_T}{C_T'} = \sqrt{\frac{V_T'}{V_T}}$$

Assuming, $C_T = 4\,\text{pF}$, $V_T = 4\,\text{V}$ and $V'_T = 4.5\,\text{V}$, we have from above

$$\frac{4}{C'_T} = \sqrt{\frac{4.5}{4}} \Rightarrow C'_T = 3.77\,\text{pF}$$

$\therefore$ The change in capacitance is $(4 - 3.77)\,\text{pF} = \mathbf{0.23\,pF}$

▶ **Example 11.** *Calculate the barrier capacitance of a* Ge *p-n junction whose area is* 1 mm *by* 1 mm *and whose depletion region is* 2×10^{-3} mm *thick. Assume the dielectric constant of* Ge *as* 16.

Solution: Given area of the *p-n* junction, $A = 1\,\text{mm} \times 1\,\text{mm} = 10^{-6}\,\text{m}^2$.

Thickness of the depletion region, $W = 2 \times 0^{-3}\,\text{mm} = 2 \times 10^{-6}\,\text{m}$, and dielectric constant of Ge, $k = 16$.

$\therefore$ The barrier capacitance is

$$C = \frac{\varepsilon_0 k A}{W} = \frac{8.85 \times 10^{-12} \times 16 \times 10^{-6}}{2 \times 10^{-6}}\,\text{F}$$
$$= 70.8 \times 10^{-12}\,\text{F}$$
$$= \mathbf{70.8\,pF}$$

▶ **Example 12.** *The current-voltage characteristic of a p-n junction diode is given by the relation*

$$I = I_0(e^{eV/\eta kT} - 1)$$

The diode current is 0.5 mA *at* $V = 340\,\text{mV}$ *and* 15 mA *at* $V = 440\,\text{mV}$. *Determine the value of* η. *Assume* $kT/e = 25\,\text{mV}$.

Solution: Given: Initial current $I_1 = 0.5\,\text{mA}$ and initial voltage $V_1 = 340\,\text{mV}$

Final current $I_2 = 15\,\text{mA}$ and final votlage $V_2 = 440\,\text{mV}$.

$kT/e = 25\,\text{mV}$

Now, using the diode curent relation $I = I_0(e^{eV/\eta kT} - 1) \approx I_0 e^{eV/\eta k.T}$ for forward bias in the two cases, we have,

$$0.5 \times 10^{-3} = I_0 e^{340/25\eta} \tag{3.12.1}$$

$$\text{and } 15 \times 10^{-3} = I_0 e^{440/25\eta} \tag{3.12.2}$$

Dividing equation (3.12.2) by (3.12.1), we get

$$30 = e^{4/\eta}$$

Taking the natural logarithm on both sides, we have

$$\frac{4}{\eta} = \ln 30$$

$$\text{or, } \eta = \frac{4}{\ln 30} = \mathbf{1.18}$$

Review and Short Questions

1. (a) What is a *p-n* junction diode? Explain how a contact potential is developed across a *p-n* junction. Explain the term potential barrier.
   (b) Draw the energy band diagrams for an unbiased, a forward and a reverse biased *p-n* junction. **(Calcutta Hons.)**
   (c) For an unbiased *p-n* junction, sketch the variation of the space charge, electric field, electrostatic potential and electron energy as a function of distance across the junction. **(Calcutta Hons.)**
2. (a) How is depletion region formed in a *p-n* junction? Explain the variation of its width with biasing.
   (b) The barrier potential across a *p-n* junction diode cannot be measured by placing a voltmeter across the diode terminals—Explain. **(Calcutta Hons.)**
3. Briefly explain the working of a *p-n* junction diode when
   (a) it is on no external bias
   (b) it is forward biased
   (c) it is reverse biased

   Sketch the relevant characteristics. **(Bangalore Univ.)**
4. (a) What do you mean by the transition capacitance of a *p-n* junction diode? Does it depend on the depletion width?
   (b) What is diffusion capacitance?
5. Making use of the diode current equation, explain the V-I characteristics of the diode under forward bias and reverse bias conditions.
6. Define the static (D.C) and the dynamic (A.C) resistance of a diode. Obtain expressions for them. **(Himachal Univ.)**
7. What is the difference between avalanche breakdown and Zener breakdown of a *p-n* junction? **(Lucknow Univ.)**
8. Derive an exprerssion for the width of depletion region across a *p-n* junction in terms of impurity concentrations.
9. Establish the reactifier or Shockley equation relating current and voltage in a *p-n* junction.
10. What is the difference between an ordinary *p-n* diode and a Zener diode?
11. What is Zener voltage? Explain with a circuit diagram the use of a Zener diode as a reference diode in voltage stabilization. **(Calcutta Hons.)**
12. What is LED? Explain the principle of operation of a LED. Why is silicon not preferred as a LED material?

13. Explain the construction and working of a tunnel diode. Draw a typical ampere-voltage characteristics of a tunnel diode. What happens to tunneling at (a) zero bias, (b) reverse bias, (c) small forward bias, (d) large forward bias? Discuss the region of negative resistance? **(Bangalore and Himachal Univ.)**

14. What are the advantages of a LED? **(Calcutta Hons.)**

15. Give the structural diagram of a Schottky diode formed with aluminium and silicon. Also draw the cricuit symbol of the diode. Why is the device called a majority carrier device?

16. State the advantages of the Schottky diode. Mention some of its applications.

17. What is a solar cell? Explain the construction and working of a solar cell? Mention some of its applications.

18. What is a photodiode? Draw its circuit symbol. Explain briefly the construction and working principle of a photodiode.

19. Draw the volt-ampere characteristics of a photodiode. Mention some of the applications of a photodiode.

20. Write short explanatory notes on:

(a) Light emitting diode, (b) Photodiode,
(c) Solar cell, (d) Schottky diode,
(e) Tunnel diode, (e) Zener diode as voltage regulator.

Objective Questions

1. The avalanche effect is observed in a diode when

(a) forward voltage exceeds the breakdown voltage,
(b) the heavily doped diode is forward biased,
(c) reverse voltage exceeds the breakdown voltage,
(d) majority carriers have enough energy to dislodge valence electrons. **(GATE)**

2. A Zener diode is used as a voltage regulator across an unregulated d.c source. Overloading will damage

(a) dc source only, (b) Zener diode only,
(c) both (a) and (b), (d) neither (a) nor (b). **(GATE)**

3. The storage or diffusion capacitance in a *p-n* diode arises due to

(a) forward biasing and contribution from majority carriers,

(b) forward biasing and contribution from minority carriers,
(c) reverse biasing and contribution from majority carriers,
(d) reverse biasing and contribution from minority carriers. **(GATE)**

4. A Zener diode can be used as
(a) rectifier, (b) transmitter, (c) voltage regulator, (d) oscillator.
(Himachal Univ.)

5. The Zener voltage in a Zener diode is
(a) any forward biasing voltage,
(b) any reverse biasing voltage,
(c) breakdown voltage in forward biasing region,
(d) breakdown voltage in reverse biasing region. **(Meerut Univ.)**

6. Depletion region is a zone which contains
(a) holes only, (b) electrons only,
(c) both electrons and holes, (d) neither electrons nor holes.
(Purvanchal Univ.)

7. The reverse saturation current in a *p-n* junction is due to the flow of
(a) majority carriers, (b) minority carriers,
(c) both majority and minority carriers, (c) impurity ions.

8. Depletion region is formed in a *p-n* junction due to
(a) diffusion, (b) recombination,
(c) ionization, (d) all of these.

9. The temperature coefficient of an avalanche breakdown voltage of a *p-n* junction diode is
(a) zero, (b) negative, (c) positive, (d) undefined.

10. The tunnel diode has a
(a) positive resistance, (b) negative resistance,
(c) negative resistance at low temperature,
(d) positive as well as negative resistance. **(Himachal Univ.)**

11. In a tunnel diode, a decrease in current causes
(a) votlage constancy, (b) decrease in voltage,
(c) increase in voltage, (d) none of these. **(Purvanchal Univ.)**

12. In a tunnel diode, there is
(a) a decrease of current with increase of voltage,
(b) an increase of current with increase of voltage,

(c) no change in current with increase of voltage,
(d) nothing can be said. **(Himachal Univ.)**

13. A photodiode is normally

(a) forward biased, (b) reverse biased,
(c) neither (a) and (b), (d) emitting light.

14. A *p-n* junction that radiates energy as light, instead of heat, is called

(a) LED, (b) photodiode, (c) Zener diode, (d) tunnel diode.

15. A semiconductor device which operates with a forward biased metal-semiconductor junction is called a

(a) Tunnel diode, (b) Schottky diode,
(c) LED, (d) Solar cell.

16. A *P-N* junction diode used in remote controlled devices is called

(a) photo-diode,, (b) Zener diode,
(c) LED, (d) Schottky diode.

17. When operating as a voltage regulator, the breakdown in a Zener diode occurs due to the

(a) tunneling effect, (b) avalanche breakdown,
(c) impact ionization, (d) excess heating of the junction.

(DRDO)

18. A common LED is made up of

(a) intrinsic semiconductor, (b) direct semiconductor,
(c) degenerate semiconductor, (d) indirect semiconduct. **(DRDO)**

19. The correct match between the following two columns is

P. Tunnel diode, (i) Microwave amplification,
Q. Zener diode, (ii) Voltage regulation,
R. PIN diode, (iii) Photo detection,
S. Schottky diode, (iv) High speed switching.

(a) P–(iii), Q–(iv), R–(ii), S–(i), (b) P–(i), Q–(iii), R–(ii), S–(iv),
(c) P–(iv), Q–(ii), R–(i), S–(iii), (d) P–(i), Q–(ii), R–(iii), S–(iv).

(DRDO)

20. In a forward biased *p-n* junction diode, the sequence of events that best describes the mechanism of current flow is

(a) injection, and subsequence diffusion and recombination of minority carriers,
(b) injection, and subsequent drift and generation of minority carriers,

(c) extraction, and subsequent diffusion and generation of minority carriers,

(d) extraction, and subsequent drift and recombination of minority carriers.

(GATE)

21. The I–V characteristics of devices can be divided into quadrants as shown below.

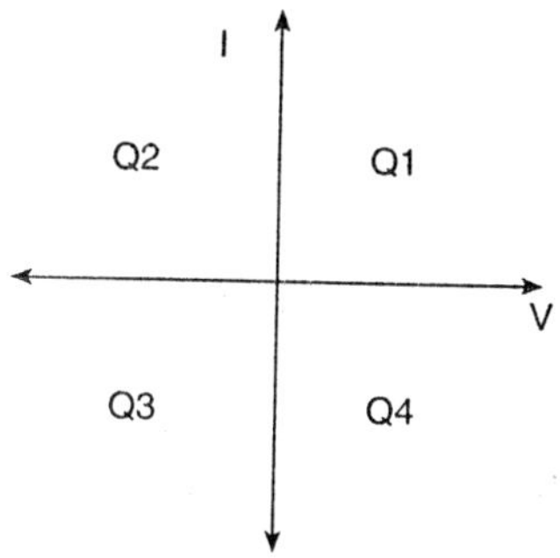

Fig. 3.26

Photodiodes and solar cells are normally operated in quadrants

(a) Q3 and Q4, respectively, (b) Q1 and Q2, respectively,
(c) Q1 and Q3, respectively, (d) Q2 and Q3, respectively. **(DRDO)**

22. A silicon p-n junction is forward biased with a constant current at room temperature. When the temperature is increased by 10°C, the forward bias voltage across the p-n junction

(a) increases by 60 mV, (b) decreases by 60 mV,
(c) increases by 25 mV, (d) decreases by 25 mV. **(GATE)**

23. Which of the following is NOT associated with a p-n junction?

(a) Junction capacitance, (b) Charge storage capacitance,
(c) Depletion capacitance, (d) Channel length modulation.

(GATE)

24. In a p^+n junction diode under reverse bias, the magnitude of electric field is maximum at

(a) the edge of the depletion region on the p-side,

(b) the edge of the depletion region on the n-side,

(c) the p^+n junction,

(d) the centre of the depletion region on the n-side. **(GATE)**

25. ________ current is the leakage current that flows through a photodiode with no input used in as light detectors

(a) Leakage, (b) Dark,
(c) Saturation, (d) Detection. **(ISRO)**

26. The values of voltage (V_D) across a tunnel-diode corresponding to peak and valley currents are V_p and V_v respectively. The range of tunnel-diode voltage V_D for which the slope of its I-V_D charateristics is negative would be

(a) $V_D < 0$, (b) $0 < V_D < V_p$
(c) $V_P \leq V_D < V_v$, (d) $V_D \geq V_v$

Fill in the gaps of the following.

27. The Zener diode is a diode which has been optimized for operation in the ______ region.

28. In a voltage regulator circuit, the Zener diode is in ______ bias mode.

29. The barrier field in a *p-n* junction diode is directed from ______ to ______.

30. The voltage-ampere characteristic of a *p-n* junction diode ______ Ohm's law. **(GATE)**

Problems

1. Calculate the static and dynamic resistances of a *p-n* junction germanium diode if the room temperature is 27°C and reverse saturation current $I_s = 1\mu$A when a forward bias of 0.2 V is applied. **(Himachal Univ.)** [88Ω, 11.4Ω]

2. Calculate the value of the applied forward voltage for a *p-n* junction diode if $I_s = 50\,\mu$A, $I = 2$A and $e/kT = 40$. **(Guru Nanak Univ.)** [0.265 V]

3. Current through a *p-n* junction silicon diode is 45 mA at a forward bias of 0.5 volt. Calculate r_{dc} and r_{ac} at 27°C. Given $\eta = 2$ for Si. **(Himachal Univ.)** [11.1Ω, 1.15Ω]

4. A silicon diode (knee voltage 0.78 V) has a forward current of 100 mA and a reverse current of 1 mA at 50 V. Find the value of bulk resistance R and the reverse resistance. **(Punjab Univ.)** [492.2Ω, 5×10^4Ω]

5. A 10 V Zener diode is used to regulate the voltage across a variable load resistor. The input voltage varies between 13 V and 18 V and load current varies between 10 mA and 90 mA. The minimum Zener curent is 15 mA. Calculate the value of series resistance R_s. [28.57Ω]

6. A 12 V Zener diode is connected in series with a resistance of 150 Ω and a load resistance of 1 kΩ is connected across the Zener diode. The minimum Zener current is nearly zero and maximum Zener current should not exceed 20 mA. Calculate the operating range of the input voltage.

(Calcutta Hons.) [13.8–16.8 V]

7. In a Zener diode regulator the series resistance is 25 Ω, Zener voltage 20 V and load resistance 200 Ω. The input voltage varies from 25 V to 30 V. Calculate (a) the minimum and the maximum current in Zener diode and (b) the minimum and the maximum power dissipated in the diode.

(Bangalore Univ.) [0.1A, 0.3A; 2W, 6W]

8. A Zener diode of of breakdown voltage 4.7 V is connected in series with a resistance 100 Ω and a power supply of 10 Volt. The Zener is connected in such a way that it is in reverse bias. Calculate the current through the Zener.

(Calcutta Hons.)

9. Suppose, in Fig. 3.12, the Zener diode is of type 5 V, 1 W. The input V_i is an ideal voltage source capable of supplying up to 15 V. If the maximum load current is 75 mA, calculate the values of R and R_L which will keep the voltage across R_L fixed. With these values of R and R_L, find the supply voltage below which the circuit fails to regulate the output voltage. Assume minimum Zener current for regulation to be 10 mA. **(Calcutta Hons.)** [50Ω, 66.6Ω, 9.25V]

10. Design a 10 V regulated power supply of 10 mA current rating using a pair of Zener diodes and an unregulated supply 12 to 15 V. Assume $I_{z,\min} = 0.2$ mA.

(Calcutta Hons.)

[**Hints.** Use Fig. 3.13 and two Zener diodes in series. So, $V_{Z1} + V_{Z2} = 10$ V. Thus $R = (12\,\text{V} - 10\,\text{V})/(10\,\text{mA} + 0.2\,\text{mA}) = 196\Omega$; $I_{ZM} = (15\,\text{V} - 10\,\text{V})/196\Omega = 25.5$ mA etc.]

11. A light emitting diode is made up of GaAsP having a band gap of 1.9 eV. Determine the wavelength of the radiation emitted. **(Punjabi Univ.)** [6533 A]

12. The wavelength of light emitted by a certain LED is 600 nm. Find the energy gap in eV. **(Madurai Univ.)** [2.07 eV]

13. A Zener diode rated 30 V is connected across a supply of 50 V d.c having resistance of 2 kΩ. A load R_L is connected across the diode. Find the value of current through the Zener diode if (i) R_L is 30 kΩ and (ii) R_L is 4 kΩ.

[9 mA, 2.5 mA]

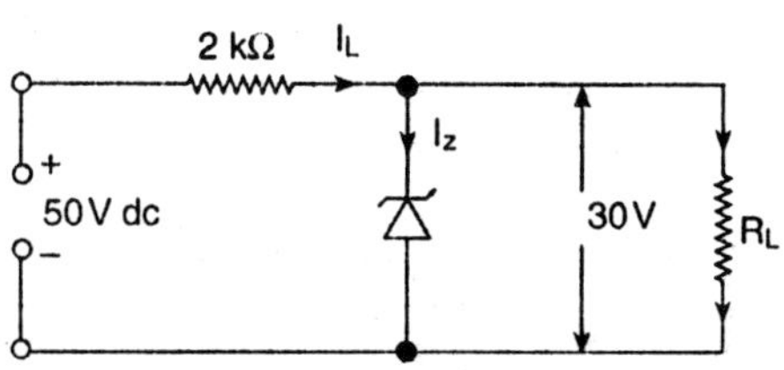

Fig. 3.27

14. What value of series resistor is required to limit the current through a LED to 20 mA, with forward voltage drop of 1.6, when connected to a 10 V supply?

[420 Ω]

15. Show that the junction resistance for a silicon or germanium diode can be given by the relation

$$r_j = \frac{26}{I_F}$$

where I_F is the diode current expressed in milliampere.

16. An ideal germanium diode at a temperature of 125°C has a reverse saturation current of 30μA. At a temperature of 125°C, find the dynamic resistance for a 0.2 V bias in (a) the forward direction and (b) the reverse direction.

[3.36Ω, 0.38 MΩ]

17. The reverse saturation curent in a Ge diode increases from 10 μA at 300 K to 105 muA at 330 K. Find the band gap of the semiconductor. **(GATE)** [0.67 eV]

18. The *I-V* characteristic of the diode in Fig. 3.28 is given by

$$I = \begin{cases} \dfrac{V - 0.7}{500} \text{ A}, & V > 0.7 \text{ volt} \\ 0 \text{ A}, & V < 0.7 \text{ volt} \end{cases}$$

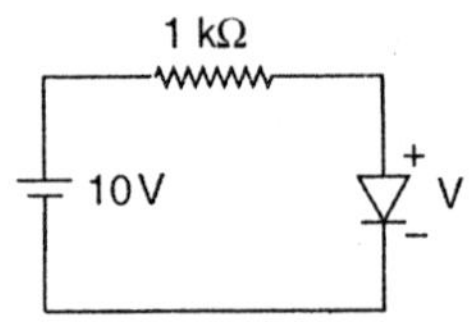

Fig. 3.28

Find the current in the circuit. **(GATE)** [6.2 mA]

4

Diode Circuits: Rectifiers and Filters

The components used to construct a circuit are called ***circuit elements***. Resistor, capacitor, inductor etc. are typical circuit elements. A *p-n junction diode* is also considered as a circuit element with some special properties. Based on the *I-V* characteristics, the circuit elements are classified into *two* categories : (i) ***linear circuit elements*** and (ii) ***non-linear circuit elements***. For the linear circuit element, as the very name suggests, the voltage-current (i.e., volt-ampere) characteristic is a straight line, but that of a non-linear circuit element is not. The relationship between the r.m.s voltage V and the r.m.s current I for a resistor R, a capacitor C and an inductor L are given respectively by $V = RI$, $V = (1/\omega C)I$ and $V = \omega LI$ where ω is the frequency of the supply current. So, if R, C and L are independent of voltage or current, the *I-V* characteristics are straight lines indicating that, under the given condition, the resistor, the capacitor and the inductor are linear circuit elements. However, a resistor used in large current and an iron-core inductor behave as non-linear circuit elements.

4.1 Diode as a circuit element

A ***p-n junction diode*** is a ***non-linear*** circuit element since its *V-I* characteristic is not a straight line. Again, the *p-n* junction capacitance depends on the applied voltage. So, a *p-n* junction capacitance is also a non-linear capacitance.

If the *V-I* relationship of a circuit element remains unchanged when the polarities of the voltage or the direction of current is reversed, the circuit element is called ***bilateral***. For instance, a resistor is a bilateral element. When the *V-I* relation changes with the direction of the current, the circuit element is called ***unilateral***. *A p-n junction diode is a unilateral element* since it behaves completely differently under forward biased and reverse biased conditions.

Finally, therefore, it follows that a *p-n* junction diode is a *non-linear* and *unilateral* circuit element.

4.2 Load-line concept

Since a *p-n* junction diode is a non-linear and unilateral circuit element, it does not obey the Ohm's law. For this reason, various techniques are used to analyse the circuits involving a diode. The basic diode circuit is illustrated in Fig. 4.1 which consist of a diode D in series with a load resistance R_L and an input voltage V_i.

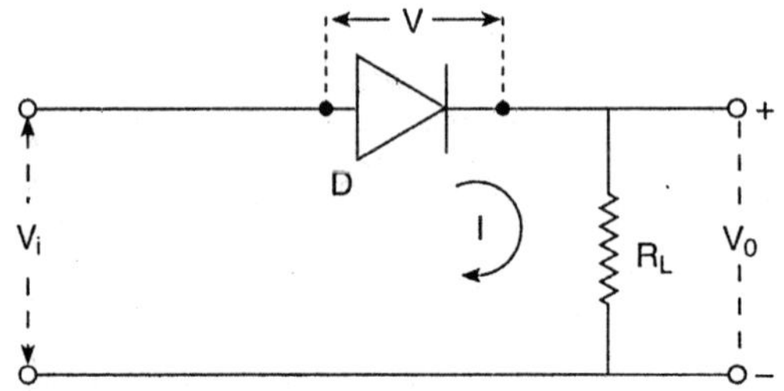

Fig. 4.1 The basic diode circuit

Applying Kirchhoff's voltage law (KVL), we get from Fig. 4.1

$$V = V_i - IR_L \tag{4.2.1}$$

where V and I are respectively the instantaneous voltage and instantaneous diode current and V_i, the instantaneous input voltage applied.

To find the instantaneous value of I and V for a given value of R_L and V_i, the equation (4.2.1) is not sufficient. So we use (i) non-linear mathematical technique and (ii) graphical method or dynamic characteristic for the purpose.

4.2.1 Non-linear mathematical technique

Eq. (4.2.1) is an equation of a straight line. A second equation that shows the relation between V and I is given by

$$I = I_s\left(e^{\frac{eV}{\eta kT}} - 1\right) \tag{4.2.2}$$

Since this relation (4.2.2) is non-linear, the solution can be obtained by iterative numerical analysis technique.

4.2.2 Graphical method

The slope of the straight line of (4.2.1) is determined by the value of R_L. This straight line that intersects the coordinate axes at $(V_i, 0)$ and $(0, V_i/R_L)$ is known as ***load line*** (Fig. 4.2). The point of intersection of load line and the static curve gives the instantaneous value of V and I. The point $Q(I, V)$ is known as *d.c operating point, quiescent point* or simply ***Q-point*** (Fig. 4.2).

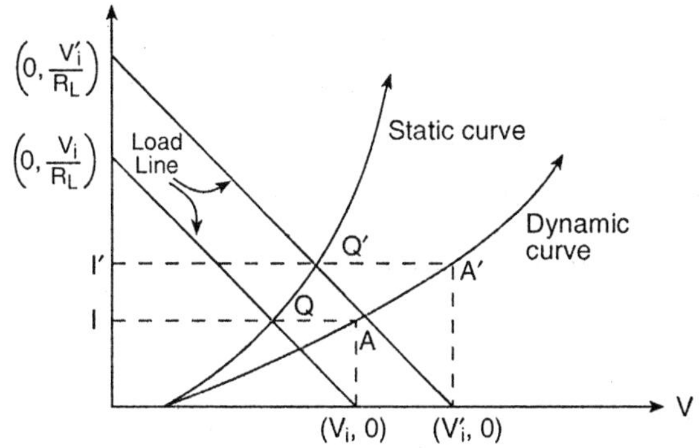

Fig. 4.2 Load line and dynamic characteristic

If the source voltage V_i is varied, the different parallel load lines and hence different Q points are obtained. Q' is such a Q-point for input voltage V_i' and diode current I'. The curve obtained by plotting the variation of diode current I with input voltage V_i for a given value of load resistance R_L is called ***dynamic characteristic curve.*** The curve passing through A, A' is the dynamic characteristic curve of the diode for the given load resistance. Clearly, the dynamic characteristic curve is flatter than the static characteristic curve. For a given input voltage, the diode current can be directly read from the dynamic characteristic. The curve which shows the relation between output voltage V_0 and the input voltage V_i is known as ***transfer*** or ***transmission characteristic.*** Since $V_0 = IR_L$, the transfer characteristic curve has the same shape as the dynamic characteristic.

4.3 Rectification

The process of *converting alternating current (or voltage) into direct current (or voltage) is called* ***rectification.*** The device which is used for the rectification is called the ***rectifier.*** Electrical power is generally produced in the form of a.c for certain advantages such as easy transmission, distribution etc. But most of the electronic apparatus require d.c power supply. *Rectifiers are thus essential for the d.c power supplies* where a.c is converted into d.c.

A *p-n* junction diode offers a *very low resistance* to current *when* it is *forward biased* and a *high resistance when* it is *reverse biased.* This property of diode is used for rectifcation. Depending on the duration of conduction, the rectifiers are mainly classified into two categories : (i) ***Half-wave rectifier*** and (ii) ***Full-wave rectifier***, which we shall now discuss.

4.3.1 Half-wave rectifier

A ***half-wave rectifier*** is an electrical circuit where the output current flows only for the positive half cycles of the input voltage i.e., when the *diode rectifies half of the a.c* *wave*, it is called *half-wave rectifier*. The basic circuit diagram of a half-wave rectifier is illustrated in Fig. 4.3(a).

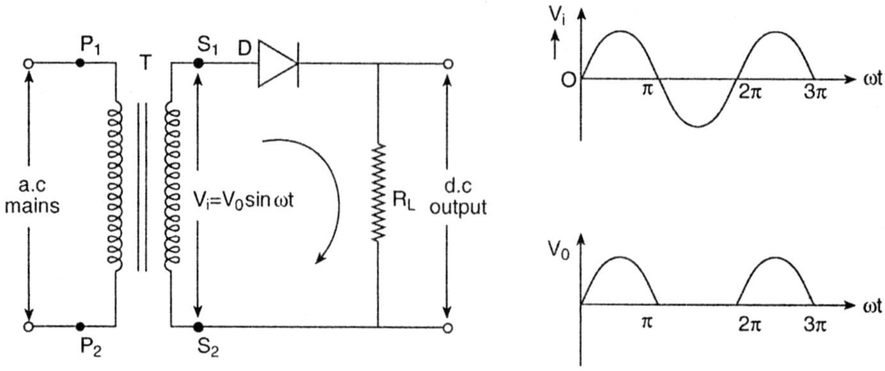

(a) Half-wave rectifier

(b) Wave forms of the input and output voltages

Fig. 4.3

An alternating input voltage if fed to the primary P_1P_2 of a transformer T, which may be a step up or a step down transformer, an alternating voltage is induced at the secondary S_1S_2 due to mutual induction. The *p-n* junction diode D is connected across the secondary in series with a load resistance R_L. When the D is forward biased i.e., S_1 is positive and S_2 is negative, the current flows through the load resistance R_L in the direction of arrows (Fig. 4.3(a)) and an output voltage is obtained across the load. During the next half cycle, S_1 becomes negative and S_2 positive so that the diode is reverse biased. In this case, no output current and hence no output voltage, is obtained. Fig. 4.3(b) shows the a.c input voltage and the unidirectional pulsating output voltage.

4.3.2 Full-wave rectifier

A ***full-wave rectifier*** is an electrical circuit consisting of two identical *p-n* junction diodes such that one conducts during one half of the input a.c signal and the other conducts during the next half of the input so that a unidirectional current flows through the load resistance during the full cycles of the input voltage. So, *a full-wave rectifier rectifies the whole of the a.c wave.* There are two types of full wave rectifiers : (i) ***Centre-tapped*** and (ii) ***Bridge rectifier***.

(i) Centre-tapped full-wave rectifier—Fig. 4.4(a) shows a full-wave rectifier circuit which consists of two identical diodes D_1 and D_2 connected to a *centre-tapped* transformer T in such a way that the induced voltage in each half of the secondary is equal in magnitude and opposite in phase.

The a.c input signal is fed to the primary P_1P_2 of the transformer and the induced voltage is produced in its secondary S_1S_2. The output voltage is measured across the load resistance R_L.

Let at any instant of the input half cycle, S_1 has a positive voltage and S_2 has a negative one with respect to the centre C. In this condition, the diode D_1 is forward

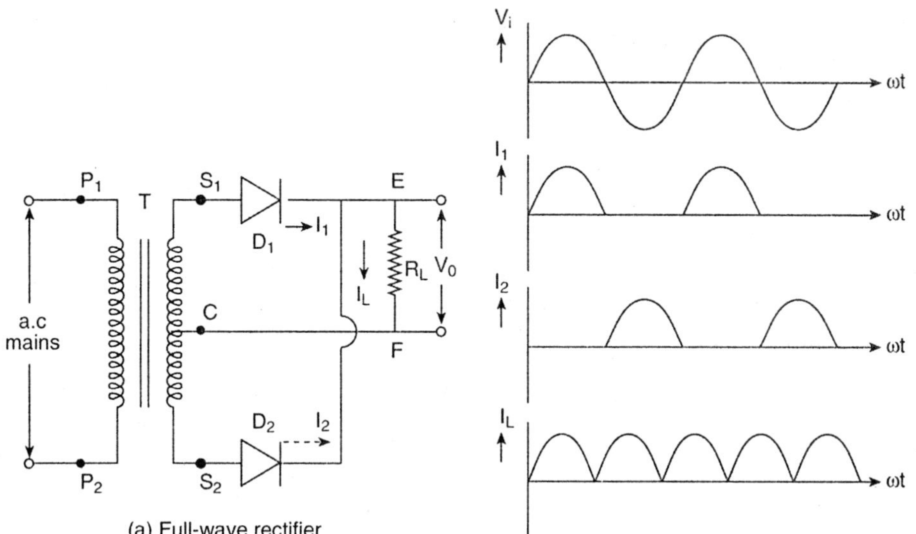

(a) Full-wave rectifier

(b) Wave forms of the input and output signals

Fig. 4.4

biased and D_2 is reverse biased. So, only the diode D_1 conducts current I_L through the load resistance R_L from E to F producing an output voltage $V_0 = V_L = I_L R_L$. During the next half cycle, D_2 conducts and D_1 remains off. In this case, the current-flow is also in the same direction i.e., from E to F. Since the load current I_L is the sum of the currents through D_1 and D_2, an unidirectional current through the load resistance R_L is obtained by utilizing the full signal and the above circuit behaves as a full-wave rectifier. The waveforms for input and output signals are shown in Fig. 4.4(b).

(ii) Bridge rectifier—The full-wave rectifier circuit, consisting of four diodes arranged in the form of a Wheatstone Bridge network, is known as ***bridge rectifier***. This rectifier is widely used in power circuits and rectifier type a.c meters. The circuit diagram of the bridge rectifier is shown in Fig. 4.5.

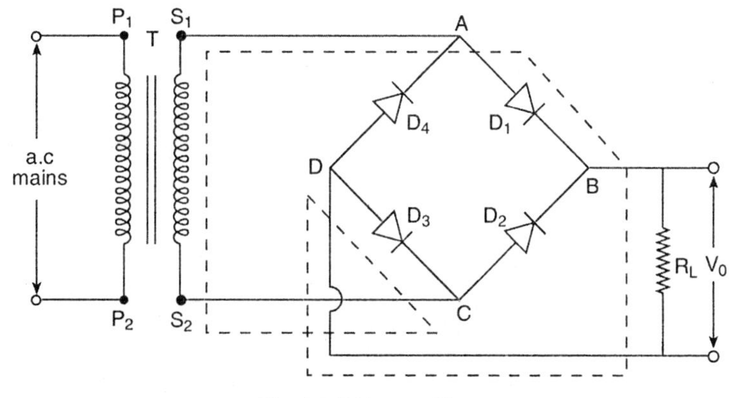

Fig. 4.5 Bridge rectifier

Let at any instant, S_1 is positive and S_2 is negative. For this polarity of the voltage of transformer secondary, the diodes D_1 and D_3 conduct and D_2 and D_4 remain off. As a result, a current flows in the direction as shown by the dashed lines in Fig. 4.5. In the next half cycle, S_1 becomes negative and S_2 positive and hence the diodes D_2 and D_4 conduct and D_1 and D_3 remain off. So, current flows in the direction of $CBR_LDAS_1S_2C$ i.e., the direction of current in both the halves is the same. Thus, a d.c output is obtained and the circuit behaves as a full-wave rectifier.

4.4 Mathematical analysis

It has been found that there are some *factors* which determine the performance of a rectifier. We shall now go into it. We are interested in finding the expressions for those factors separately for half-wave and full-wave rectifiers by way of mathematical analysis in what follows.

4.4.1 Analysis of half-wave rectifier

The instantaneous transformer secondary voltage can be represented by

$$V_i = V_0 \sin \omega t \tag{4.4.1}$$

where the peak value V_0 of the voltage is usually very large compared to the cut-off voltage of the diode.

The instantaneous output current through the diode and the load resistance R_L is given by

$$I = \begin{cases} I_0 \sin \omega t & \text{when } 0 \le \omega t \le \pi \\ 0 & \text{when } \pi \le \omega t \le 2\pi \end{cases} \tag{4.4.2}$$

where I_0 represents the peak value of the load current and is given by

$$\boxed{I_0 = \frac{V_0}{R_L + R_f}} \tag{4.4.3}$$

R_f being the dynamic forward resistance of the diode.

(A) Average or DC value of output current—The output of the half-wave rectifier is of half-wave form but unidirectional and fluctuating.

The average or the d.c value of the output current I_{av} over a complete cycle is plainly given by

$$\begin{aligned} I_{\text{av}} = I_{\text{dc}} &= \frac{1}{2\pi} \int_0^{2\pi} I d(\omega t) \\ &= \frac{1}{2\pi} \left[\int_0^{\pi} I_0 \sin \omega t d(\omega t) + \int_{\pi}^{2\pi} 0 \cdot d(\omega t) \right] \\ &= \frac{I_0}{2\pi} [-\cos \omega t]_0^{\pi} = \frac{I_0}{2\pi} 2 \end{aligned}$$

$$\therefore \quad \boxed{I_{dc} = \frac{I_0}{\pi}} \qquad (4.4.4)$$

So, the d.c output voltage across the load resistance R_L is given by

$$\boxed{V_{dc}} = I_{dc} \cdot R_L = \boxed{\frac{I_0 R_L}{\pi}} \qquad (4.4.5)$$

(B) DC output power—The d.c power across the load is

$$\boxed{P_{dc}} = I_{dc}^2 \cdot R_L = \boxed{\frac{I_0^2}{\pi^2} R_L} \qquad (4.4.6)$$

(C) RMS or effective output current—The effective or rms value of the load current is given by

$$I_{rms} = \sqrt{\frac{1}{2\pi} \int_0^{2\pi} I^2 d(\omega t)}$$

$$= \sqrt{\frac{1}{2\pi} \int_0^{\pi} I_0^2 \sin^2 \omega t d(\omega t)}$$

$$= \sqrt{\frac{1}{2\pi} \cdot I_0^2 \cdot \frac{1}{2}\pi}$$

$$\therefore \quad \boxed{I_{rms} = \frac{I_0}{2}} \qquad (4.4.7)$$

(D) AC input power—The input power to the circuit from the a.c source is

$$P_{in} = I_{rms}^2 (R_L + R_f)$$

$$\text{or,} \quad \boxed{P_{in} = \frac{I_0^2}{4}(R_L + R_f)} \qquad (4.4.8)$$

(E) Rectifier efficiency—The ***efficiency of rectification*** is *defined* as the *ratio of the d.c output power to the a.c input power supplied to the rectifier circuit.* So, the rectifier efficiency measures the fraction of the input power which is converted into dc. It is denoted by η and is expressed in percentage.

$$\therefore \quad \eta = \frac{P_{dc}}{P_{in}} \times 100\%$$

$$= \left(\frac{I_{dc}}{I_{rms}}\right)^2 \cdot \frac{R_L}{R_L + R_f} \times 100\%$$

$$= \frac{4}{\pi^2} \cdot \frac{1}{1 + R_f/R_L} \times 100\%$$

$$\text{or,} \quad \boxed{\eta = \frac{\mathbf{40.6}}{\mathbf{1 + R_f/R_L}}\%} \tag{4.4.9}$$

For maximum efficiency $R_f \ll R_L$ and theoretically the *maximum efficiency* is given by $\eta_{\max} = 40.6\%$.

(F) Ripple factor—The output of a rectifier, though unidirectional, is pulsating. The output current (or voltage) thus contains d.c and a large number of fluctuating harmonic components known as ***ripples***. ***Ripple factor***, denoted by γ, is *a measure of the fluctuating component or simply the lack of smoothness of the output waveform.*

The ***ripple factor*** is *defined* as

$$\gamma = \frac{\text{r.m.s value of the a.c component of output current (or voltage)}}{\text{d.c value of the output current (or voltage)}}$$

$$= \frac{V'_{\text{rms}}}{V_{\text{dc}}} = \frac{I'_{\text{rms}}}{I_{\text{dc}}}$$

where V'_{rms} and I'_{rms} are the rms values of the a.c component of the output voltage and output current respectively.

The total load current $I = I_{\text{dc}} + I_{\text{ac}} \Rightarrow I_{\text{ac}} = I - I_{\text{dc}}$ where I_{ac} is the a.c component of the load current I.

$$\therefore \quad I'_{\text{rms}} = \sqrt{\frac{1}{2\pi}\int_0^{2\pi} I_{\text{ac}}^2\, d(\omega t)}$$

$$= \sqrt{\frac{1}{2\pi}\int_0^{2\pi} (I - I_{\text{dc}})^2\, d(\omega t)}$$

$$= \sqrt{I_{\text{rms}}^2 - I_{\text{dc}}^2}$$

$$\therefore \quad \textbf{\textit{Ripple factor,}}\ \gamma = \frac{I'_{\text{rms}}}{I_{\text{dc}}} = \sqrt{\left(\frac{\mathbf{I_{rms}}}{\mathbf{I_{dc}}}\right)^{\mathbf{2}} - \mathbf{1}} \tag{4.4.10}$$

Using equations (4.4.4) and (4.4.7) we obtain,

$$\gamma = \sqrt{\frac{\pi^2}{4} - 1} = 1.21$$

Since the rms value of the a.c component exceeds the d.c component in a half-wave rectifier, it is a *poor device* for converting a.c into d.c.

(G) Peak inverse voltage—The maximum reverse voltage which a diode can withstand without being damaged during the non-conducting period of operation is called the ***peak inverse voltage*** (PIV). For a half-wave rectifier, it is equal to the peak value of the transformer secondary i.e.,

$$PIV = V_0$$

(H) Load regulation—The variation of d.c output with d.c load current for a given input is known as ***load regulation*** of the rectifier. In a half-wave rectifier,

$$I_{dc} = \frac{I_0}{\pi} = \frac{V_0}{\pi(R_f + R_L)}$$

$$\text{or,} \quad \frac{V_0}{\pi} = I_{dc} \cdot R_f + V_{dc}$$

$$\text{or,} \quad V_{dc} = \frac{V_0}{\pi} - I_{dc} \cdot R_f \tag{4.4.11}$$

The plot of V_{dc} against I_{dc} (Fig. 4.6) shows that V_{dc} *decreases linearly* with *increasing* I_{dc}.

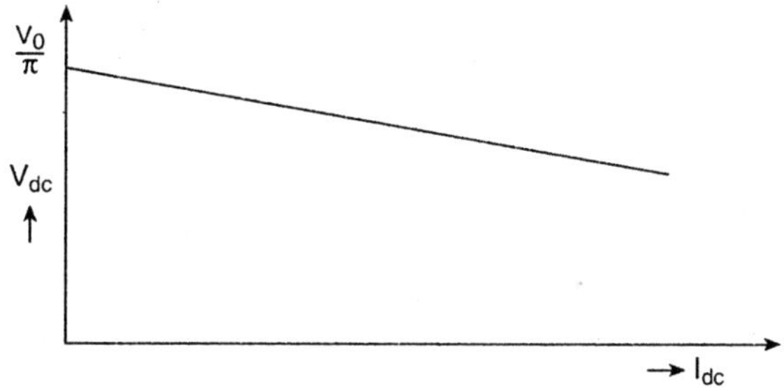

Fig. 4.6 Load regulation

Fig. 4.6 also shows that the regulation of a half-wave rectifier is *poor.* The degree of stabilization is expressed in percentage regulaton S_L which is given by

$$S_L = \frac{V_{NL} - V_L}{V_L} \times 100\% \tag{4.4.12}$$

where V_{NL} is the *no load voltage* and V_L is the *output voltage at a particular load current.* Smaller the value of S_L, better is the regulation.

(I) Transformer utilization factor (TUF)—The amount of dc power obtained from the maximum possible power available at the transformer's secondary is determined by ***transformer utilization factor*** (TUF). This is defined by

$$\text{TUF} = \frac{\text{dc power supplied to the load}}{\text{ac power rating of the transformer}} = \frac{P_{dc}}{P_{dc}\,(\text{rated})}$$

For a half-wave rectifier, the rated secondary voltage is $V_0/\sqrt{2}$ and the rms current through the secondary winding is $I_0/2$. [vide Eq. (4.4.7)]. So, for a half-wave rectifier

$$\text{TUF} = \frac{I_{dc}^2 R_L}{I_{rms} \cdot V_{rms}} = \frac{(I_0/\pi)^2 R_L}{\frac{I_0}{2} \cdot \frac{V_0}{\sqrt{2}}} = \frac{2\sqrt{2}}{\pi^2} \cdot \frac{1}{1 + (R_f/R_L)}$$

substituting $V_0 = I_0(R_f + R_L)$.

If $R_f \ll R_L$ we have $\qquad \text{TUF} = \dfrac{2\sqrt{2}}{\pi^2} = 0.287.$

This implies that the dc power that could be delivered to a resistive load in a half-wave rectifier is much smaller than the ac transformer rating.

4.4.2 Analysis of a full-wave rectifier

In a full-wave rectifier circuit, both the diodes are identical having the same value of dynamic forward resistance R_f.

The current pulses in the two diodes may be represented by

$$\left.\begin{aligned} I_1 &= I_0 \sin \omega t \\ I_2 &= 0 \end{aligned}\right\} \quad \text{for } 0 \leq \omega t \leq \pi$$

$$\text{and} \quad \left.\begin{aligned} I_1 &= 0 \\ I_2 &= -I_0 \sin \omega t \end{aligned}\right\} \quad \text{for } \pi \leq \omega t \leq 2\pi$$

where I_1, I_2 are the currents through D_1 and D_2 respectively and $I_0 = V_0/(R_L + R_f) =$ the peak value of load current.

(A) Average or D.C output current—The average current or *d.c output current* is given by

$$\begin{aligned} I_{\text{dc}} &= \frac{1}{2\pi} \int_0^{2\pi} I d(\omega t) \\ &= \frac{1}{2\pi} \left[\int_0^{\pi} I_1 d(\omega t) + \int_{\pi}^{2\pi} I_2 d(\omega t) \right] \\ &= \frac{1}{2\pi} \left[\int_0^{\pi} I_0 \sin \omega t \, d(\omega t) - \int_{\pi}^{2\pi} I_0 \sin \omega t \, d(\omega t) \right] \\ &= \frac{I_0}{2\pi} \left[-\cos \omega t \right]_0^{\pi} + \frac{I_0}{2\pi} \left[\cos \omega t \right]_{\pi}^{2\pi} \\ &= \frac{I_0}{2\pi} [2 + 2] \end{aligned}$$

$$\therefore \quad \boxed{I_{\text{dc}} = 2\frac{I_0}{\pi}} \tag{4.4.13}$$

The d.c output voltage across the load is therefore given by

$$V_{\text{dc}} = I_{\text{dc}} \cdot R_L = 2\frac{I_0}{\pi} \cdot R_L \tag{4.4.14}$$

(B) D.C output power—The d.c *output power* is plainly given by

$$P_{\text{dc}} = I_{\text{dc}}^2 \cdot R_L = 4 \cdot \frac{I_0^2}{\pi^2} \cdot R_L \tag{4.4.15}$$

(C) RMS or effective output current—The effective output current or the *rms value* of the load current is given by

$$I_{\rm rms} = \sqrt{\frac{1}{2\pi}\int_0^{2\pi} I^2 d(\omega t)}$$

$$= \left[\frac{1}{2\pi}\int_0^{\pi} I_1^2 d(\omega t) + \frac{1}{2\pi}\int_{\pi}^{2\pi} I_2^2 d(\omega t)\right]^{\frac{1}{2}}$$

$$= \left[\frac{1}{2\pi}\int_0^{\pi} I_0^2 \sin^2 \omega t\, d(\omega t) + \frac{1}{2\pi}\int_{\pi}^{2\pi} I_0^2 \sin^2 \omega t\, d(\omega t)\right]^{\frac{1}{2}}$$

$$\therefore \quad \boxed{I_{\rm rms} = \frac{I_0}{\sqrt{2}}} \tag{4.4.16}$$

(D) A.C input power—The *a.c input power* to the rectifier is given by

$$P_{\rm in} = I_{\rm rms}^2 (R_f + R_L) = \frac{I_0^2}{2}(R_f + R_L) \tag{4.4.17}$$

● The students should preferably memorise the important relations deduced in this section as they are too *often required.*

(E) Rectifier efficiency—The rectifier efficiency of a full-wave rectifier is

$$\eta = \frac{P_{\rm dc}}{P_{\rm in}} \times 100\%$$

$$= \frac{I_{\rm dc}^2 \cdot R_L}{I_{\rm rms}^2 (R_f + R_L)} \times 100\%$$

$$= \frac{8}{\pi^2} \cdot \frac{1}{1 + \frac{R_f}{R_L}} \times 100\%$$

$$= \frac{81.2}{1 + R_f/R_L}\% \tag{4.4.18}$$

Note that the maximum value of η is 81.2% which is *twice* the rectifying efficiency of a half-wave rectifier.

(F) Ripple factor—The *ripple factor* γ for a full-wave rectifier is

$$\gamma = \sqrt{\left(\frac{I_{\rm rms}}{I_{\rm dc}}\right)^2 - 1} = \sqrt{\frac{\pi^2}{8} - 1} = 0.48$$

So, the ripple factor of a full-wave rectifier is much smaller compared to that (1.21) of a half-wave rectifier. Hence a full-wave rectifier is far *more useful* in converting a.c into d.c for the purpose of practical applications.

(G) Peak inverse voltage (PIV)—When the diode D_1 is 'on', the diode D_2 remains 'off' i.e., non-conducting (Fig. 4.4(a)). In this case, the peak inverse voltage across D_2 is given by

$$\begin{aligned} PIV &= p.d \text{ across } R_L \\ &= V_0 + V_0 = 2V_0 \end{aligned}$$

So, the peak inverse voltage across each diode is *twice* the maximum transformer voltage, measured with respect to the centre-tap to either end.

(H) Load regulation—We have, the expression for I_{dc} as:

$$I_{dc} = \frac{2I_0}{\pi} = \frac{2V_0}{\pi(R_f + R_L)}$$

$$\text{or,} \quad I_{dc} \cdot R_L = \frac{2V_0}{\pi} - I_{dc} \cdot R_f$$

$$\text{or,} \quad V_{dc} = \frac{2V_0}{\pi} - I_{dc} \cdot R_f \tag{4.4.19}$$

From the above equation it is seen that when I_{dc} increases from 0 to higher values, V_{dc} decreases linearly from $2V_0/\pi$ to lower values. So, the regulation is *poor* in a full-wave rectifier.

(I) Transformer unitilization factor (TUF)—A full-wave rectifier consists of two half-wave rectifiers feeding the same load resistance one at a time. When any one of the diodes conducts, only half or the centre-tapped transformer secondary remains active. So, in a full-wave rectifier, the average TUF can be calculated by considering the primary and the secondary windings separately. The TUF of the primary winding is, therefore, double the TUF of a half-wave rectifier i.e. $2 \times 0.287 = 0.574$. The rated secondary voltage from the centre tap to each end and the r.m.s current through each half of the secondary winding are given by $V_0/\sqrt{2}$ and $I_0/\sqrt{2}$ respectively. So, the TUF of the secondary winding is given by

$$\begin{aligned} \frac{P_{dc}}{P_{ac}\text{ (rated)}} &= \frac{I_{dc}^2 R_L}{\frac{V_0}{\sqrt{2}} \cdot \frac{I_0}{\sqrt{2}}} \\ &= \frac{(2I_0/\pi)^2 R_L}{I_0^2 (R_f + R_L)/2} = 0.812 \end{aligned}$$

using (4.4.13) and (4.4.3) and assuming $R_L \gg R_f$.

So, the average TUF is $(0.574 + 0.812)/2 = 0.693$ which is greater than 0.287 for a half-wave rectifier.

For a bridge rectifier, no centre-tapped transformer is used and a purely alternating current flows through the transformer secondary. So, the TUF is 0.812 for a bridge rectifier. This explains why a bridge rectifier is used in situations where large dc powers are required.

4.5 Comparison of half-wave and full-wave rectifiers

At this stage, we believe, a comparison between a half-wave and a full-wave rectifier would be worth-studying. The important points are:

1. In a half-wave rectifier only one diode is used and the load current is obtained during the positive half-cycles of the input signal. In a full-wave rectifier, two diodes are used in order to obtain the load current during the whole cycle of the input voltage.
2. A full-wave rectifier circuit requires a centre-tapped transformer, whereas in a half-wave rectifier no such transformer is needed.
3. The d.c load current and rectifier efficiency of a half-wave rectifier is half of those of the full-wave rectifier. The ripple factor of half-wave rectifier is also large. Therefore, the performance of a full-wave rectifier is much better than the performance a half-wave rectifier.
4. For a half-wave rectifier, the frequency of the load current and the frequency of the input supply are equal, whereas the frequency of the load current is twice the frequency of the input supply for a full-wave rectifier.

4.6 Comparison of centre-tapped and bridge rectifiers

The centre-tapped full-wave rectifier and the bridge rectifier have the same values of d.c output current I_{dc}, effective output current I_{rms}, ripple factor γ and maximum efficiency. In spite of these similarities, the bridge rectifier has the following advantages over the centre-tapped full-wave rectifier.

Advantages—The advantages of a bridge rectifier are:

1. The bridge rectifier circuit needs no centre-tapped transformer.
2. While the centre-tapped full-wave rectifier uses the two halves of the transformer secondary alternately for diode conduction, the bridge rectifier utilizes both the halves continuously. So, for the same output voltage, a much smaller transformer is required for a bridge rectifier.
3. In the context of transformer utilization being better in bridge rectifier compared to the centre-tapped full-wave rectifier, the former can be used where larger d.c power is required.
4. The peak inverse voltage (PIV) of each diode in bridge rectifier is half that of the centre-tapped rectifier for the same output voltage. So, a bridge rectifier is suitable for high voltage applications.

Disadvantages—The bridge rectifier however suffers from the following disadvantages as well.

1. The centre-tapped full-wave rectifier uses two diodes whereas the bridge rectifier uses four diodes i.e., two extra diodes are required in bridge rectifier. This causes

an additional voltage drop and power dissipation in the internal resistances of the diodes and reduces the output voltage. This cannot be avoided when the rectifier is used in low voltage d.c power supplies.

2. The regulation of the bridge rectifier is rather relatively poorer.

4.7 Filters of rectifiers

Almost all the electronic devices require a steady d.c supply. But the output of a rectifier, as is just observed, is a *unidirectional* but *pulsating* voltage containing a.c components known as *ripples*, superposed on to the d.c component. So, the ripples must be eliminated or reduced to a minimum to make the output of the rectifier a practically useful d.c source. Fortunately, this can be done by using an electrical circuit known as the ***filter circuit***.

A *filter circuit* may be *defined* as an electrical circuit consisting of circuit elements which have frequency-dependent impedances to remove or minimize the ripples of the output voltage (or current) of a rectifier to obtain a steady d.c supply.

There are several types of filters made of simple capacitor or an inductor or a suitable combination thereof. Some of these filters are presented below.

4.7.1 Simple shunt capacitor filter

The ***shunt capacitor filter*** is the simplest type of filter which consists of a capacitor C shunting the load resistance R_L (Fig. 4.7).

The value of the capacitor C is so chosen that its reactance $1/\omega C$ at the frequency of the input voltage is very small compared to the value of the load resistance R_L. As a result, the ripple components of the output voltage are mostly by-passed through the shunt capacitor.

The unidirectional pulsating output voltage of a full-wave rectifier in absence of the capacitor filter is shown by the dashed curve and the output voltage wave-form in presence of the filter is shown by the solid curve in Fig. 4.7(b). It is seen that at the end of the first quarter cycle (point A), the capacitor is charged to the peak voltage of the rectifier output V_0. During the next half cycle, the rectifier output drops to zero and

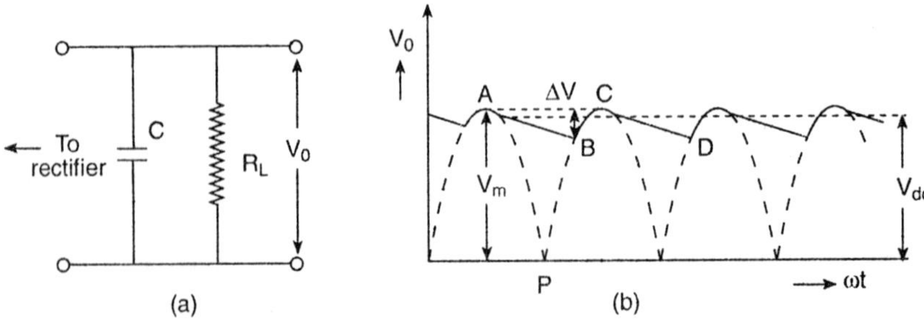

Fig. 4.7 (a) Shunt capacitor filter (b) Output voltage waveform

rises again to peak value (dashed curve $APBC$) when no filter is applied. But in this case, the capacitor discharges slowly through the load which is represented by AB of the solid curve and at B it begins to be recharged to the peak value at C. This process is repeated and the corresponding output voltage is shown by the solid curve. If the *time constant* R_LC of the filter circuit is very large compared to the time period of the alternating voltage, the capacitor discharges very slowly and the average d.c voltage across the load remains nearly constant. The ripple factor γ for a full-wave rectifier with shunt capacitor filter is given by

$$\boxed{\gamma = \frac{1}{4\sqrt{3}fR_LC}}$$

where f is the frequency of the input signal.

Approximate analysis

Since $R_L \gg \dfrac{1}{\omega C}$ i.e., the time constant R_LC of the filter circuit is very large compared to the time period T of the alternating voltage, the discharge interval of the capacitor may be approximated to $T/2$. During this time interval, the charge lost by the capacitor is $I_{dc} \times T/2$ where I_{dc} is the average load current. So, the drop of capacitor voltage in this time interval is $\Delta V = \dfrac{I_{dc} \cdot T}{2C} = \dfrac{I_{dc}}{2fC}$ where f is the frequency of the input supply voltage.

The average or the dc value for the voltage across the capacitor is

$$V_{dc} = V_m - \frac{\Delta V}{2} = V_m - \frac{I_{dc}}{4fC} \tag{4.7.1}$$

The above equation determines the regulation characteristics of the circuit. Clearly, V_{dc} decreses linearly with the increase in I_{dc} and hence the regulation is *poor.*

The fluctuation ΔV of output voltage is a *measure of the ripple voltage* which can be reduced by increasing C and f.

The r.m.s value of the ripple voltage is

$$V'_{rms} = \left[\frac{2}{T}\int_0^{T/2}\left(\frac{\Delta V}{2} - \frac{\Delta V}{T/2}t\right)^2 dt\right]^{1/2}$$

which on evaluation gives

$$V'_{rms} = \frac{\Delta V}{2\sqrt{3}} = \frac{I_{dc}}{4\sqrt{3}fC} \tag{4.7.2}$$

Since $V_{dc} = I_{dc} \cdot R_L$, the ripple factor γ for a full-wave rectifier with shunt capacitor is given by

$$\gamma = \frac{V'_{rms}}{V_{dc}} = \frac{1}{4\sqrt{3}fR_LC} \tag{4.7.3}$$

● The regulation of a full-wave rectifier can be improved by using larger C. To ensure better regulation and low ripple, electrolytic capacitors are used.

- The ripple factor for a half-wave rectifier with shunt capacitor filter is given by

$$\gamma = \frac{1}{2\sqrt{3} f R_L C}$$

- The following features of a C-filter should be noted.

(i) the poor regulation,

(ii) low ripple at low currents,

(iii) output voltage is high even at small load current, and

(iv) undue peaked current passes through the diode.

4.7.2 Series inductor filter

The ***series inductor filter*** consists of an inductor L in series with the load resistance R_L, as illustrated in Fig. 4.8.

The reactance of the inductance ωL at the frequency of the input voltage is very high compared to the value of R_L. So, when the output of a rectifier is fed to the series inductor filter, it offers a high reactance to the ripples and a small resistance to the d.c components. The inductor filter thus reduces the fluctuations of the load voltage. The

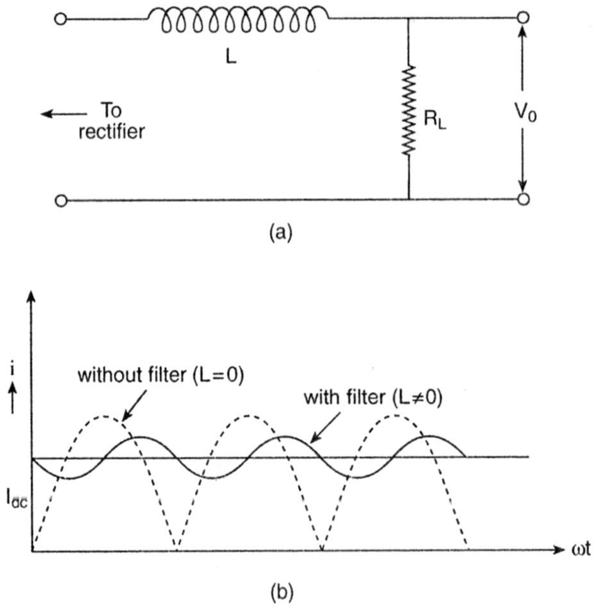

Fig. 4.8 (a) Series inductor filter, (b) Variation of load current with time

ripple factor γ in this filter for a full-wave rectifier is given by

$$\boxed{\gamma = 0.236 \left(\frac{R_L + R}{\omega L} \right)}$$

where R is the resistance of the inductor of inductance L.

Since $\gamma \propto 1/\omega L$, the ripple decreases as $L\omega$ increases. So, a high value of reactance reduces the ripples.

Approximate analysis

For a full-wave rectifier with a series inductor filter, the current in the inductor never falls to zero because of the overlapping of conduction periods of the two diodes. The variation of load current with time in presence of filtering (i.e., $L \neq 0$) and in absence of filtering ($L = 0$) are shown in Fig. 4.8(b). The dashed curve shows the waveform of load current without filter.

To calculate the ripple factor and to study the regulation we have to consider that the rectifier output is composed of harmonic frequencies. A Fourier analysis of the laod current waveform gives

$$i = \frac{2I_0}{\pi} - \frac{4I_0}{3\pi}\cos 2\omega t - \frac{4I_0}{15\pi}\cos 4\omega t - \cdots \tag{4.7.4}$$

Since the inductive reactance $X_L = \omega L$, the third and higher terms on r.h.s of (4.7.4) can be neglected. They are highly blocked by inductance. So, (4.7.4) reduces to

$$i \approx \frac{2I_0}{\pi} - \frac{4I_0}{3\pi}\cos 2\omega t \tag{4.7.5}$$

For simplicity, we consider that the diode resistance, the resistance of the inductor and transformer windings are very small compared to the load resistance R_L. So the steady component (i.e., dc) of the load current corresponding to the first term of r.h.s of (4.7.5) will be

$$I_{\text{dc}} = \frac{2V_0}{\pi R_L} \tag{4.7.6}$$

where V_0 is the no-load peak voltage across one half of the transformer secondary.

The fluctuating component of the current corresponding to the second term of r.h.s of (4.7.5) is

$$I' = \frac{4V_0}{3\pi\sqrt{R^2 + 4\omega^2 L^2}}\cos(2\omega t - \phi) \tag{4.7.7}$$

where ϕ is the phase angle by which the current lags the voltage and is given by

$$\tan\phi = \frac{2\omega L}{R_L} \tag{4.7.8}$$

The term $\sqrt{R_L^2 + 4\omega^2 L^2}$ is the *impedance* offered by L and R_L in series for the angular frequency 2ω.

Therefore, the *load current* with filter is given by

$$i = \frac{2V_0}{\pi R_L} - \frac{4V_0}{3\pi\sqrt{R_L^2 + 4\omega^2 L^2}}\cos(2\omega t - \phi) \tag{4.7.9}$$

The *ripple factor* is $$\gamma = \frac{I'_{\text{rms}}}{I_{\text{dc}}} = \frac{4V_0}{\sqrt{2}\cdot 3\pi\sqrt{R_L^2 + 4\omega^2 L^2}}\cdot\frac{\pi R_L}{2V_0}$$

$$= \frac{\sqrt{2}}{3}\cdot\frac{1}{\sqrt{1+\dfrac{4\omega^2 L^2}{R_L^2}}} \tag{4.7.10}$$

If $2\omega L \gg R_L$, then $$\gamma = \frac{R_L}{3\sqrt{2}\omega L} = 0.236\frac{R_L}{\omega L} \tag{4.7.11}$$

When the resistance of the inductor R is considered, the expression for the ripple factor is given by

$$\gamma = 0.236\left(\frac{R_L + R}{\omega L}\right) \tag{4.7.12}$$

Eqn. (4.7.11) or (4.7.12) indicates that the ripple factor decreases with increasing L and decreasing R_L. Obviously, the inductor filter is suitable for large load current i.e., for small R_L.

Since load voltage $v_L = iR_L$, so v_L has the same waveform as the load current.

∴ The dc component of the output voltage is given by

$$V_0 = I_{\text{dc}}R_L = \frac{2V_0}{\pi} \tag{4.7.13}$$

This relation holds for an ideal inductor with no resistance. But from practical point of view, the effect of the resistance of inductor is to be considered. In that case

$$I_{\text{dc}} = \frac{2V_0}{\pi(R_L + R)}, \quad \text{where } R = \text{resistance of inductor.}$$

$$\text{or,} \quad I_{\text{dc}}(R_L + R) = \frac{2V_0}{\pi}$$

$$\therefore \quad V_{\text{dc}} = I_{\text{dc}}R_L = \frac{2V_0}{\pi} - I_{\text{dc}}R \tag{4.7.14}$$

The above equation shows that the voltage regulation is poor for a series inductor filter having a large resistance.

• Following *important features* of a L-filter may be noted: (i) Regulation is better except for very low current compared to C-filter. (ii) No undue peaked current passes through the diode. (iii) Output voltage is comparatively low. (iv) Low ripple at high load current. (v) The circuit is relatively bulky and costlier due to inductor.

4.7.3 The L-section or inductor input filter

The ***L-section filter*** or ***inductor input filter*** consists of an inductor L, connected in series with the rectifier output and a capacitor C in parallel to the load resistance R_L (Fig. 4.9). It is easy to observe that actually, an L-section filter is a *combination of an inductor filter and a capacitor filter.*

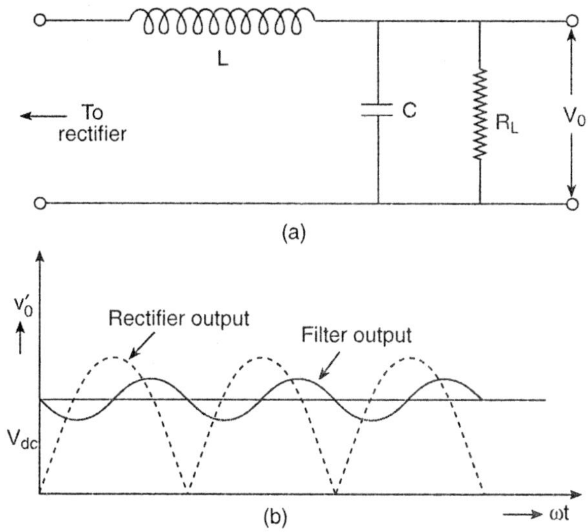

Fig. 4.9 (a) L-section filter, (b) Output voltage wave-form

The series inductor L allows the d.c component to pass through it but offers a very high impedance ωL to the a.c components i.e., ripples. The pulsations that exist after passing through L are largely by-passed through the capacitor C which offers a very low impedance to them but a great resistance to the d.c. Thus, a quite steady d.c output current (or voltage) is obtained.

The ripple factor γ for the full-wave rectifier with L-section filter is given by

$$\boxed{\gamma = \frac{0.47}{4\omega^2 LC - 1}}$$

It is clear from the above expression that *the ripple factor is independent of the load resistance.*

Approximate analysis

In the previous two articles (4.7.1 and 4.7.2), we have seen that for a capacitor filter the ripple factor decreases with increase in R_L and for an inductor filter, the ripple factor increases as R_L increases. So, it is expected that a combination of these two filters should make the ripple factor independent of R_L. A combination of series inductor and a shunt capacitor is found to act more effectively as a filter than with either L or C alone. The variation of load voltage with and without filter is shown in Fig. 4.9(b). The dashed curve corresponds to the waveform of load voltage without filter.

To calculate the ripple factor it is assumed that the load current is continuous and the input voltage to the filter is composed of the Fourier components as

$$v_0' = \frac{2V_0}{\pi} - \frac{4V_0}{3\pi}\cos 2\omega t - \frac{4V_0}{15\pi}\cos 4\omega t - \cdots \tag{4.7.15}$$

For simplicity, we assume that the diode and inductor resistances are negligible compared to R_L and the ripple is predominantly due to second harmonics. Then the dc value of the load voltage is $V_{dc} = 2V_0/\pi$ and the magnitude of ac current through C is given by

$$i = \frac{4V_0}{3\pi} \frac{\cos 2\omega t}{\left(2\omega L - \frac{1}{2\omega C}\right)} \tag{4.7.16}$$

The ac voltage across C is
$$v_{ac} = \frac{4V_0}{3\pi} \frac{\cos 2\omega t}{\left(2\omega L - \frac{1}{2\omega C}\right)} \times \frac{1}{2\omega C}$$

$$= \frac{4V_0 \cos 2\omega t}{3\pi\,(4\omega^2 LC - 1)} \tag{4.7.17}$$

So, the rms value of v_{ac} is
$$v_{rms} = \frac{4V_0}{3\pi\sqrt{2}(4\omega^2 LC - 1)} \tag{4.7.18}$$

Hence, the ripple factor is
$$\gamma = \frac{v_{rms}}{V_{dc}} = \frac{4V_0}{3\pi\sqrt{2}(4\omega^2 LC - 1)} \times \frac{\pi}{2V_0}$$

$$= \frac{\sqrt{2}}{3(4\omega^2 LC - 1)} = \frac{0.47}{4\omega^2 LC - 1} \tag{4.7.19}$$

So, it is seen that the ripple factor is independent of R_L.

• This filter is called L-section filter, since the combination of L and C looks like an inverted L.

4.7.4 The π-section filter or capacitor input filter

A ***π-section filter*** or a ***capacitor input filter*** consists of a capacitor filter followed by a L-section filter (Fig. 4.10a).

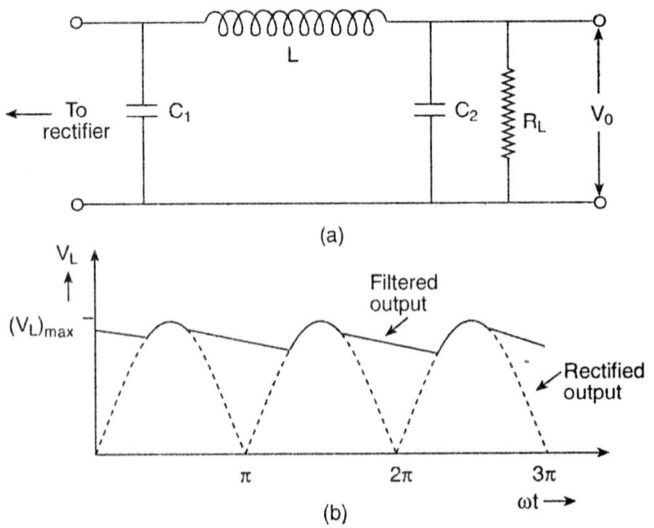

Fig. 4.10 (a) π-section filter, (b) variation of load votlage

The capacitor C_1 is charged to the peak value by the output voltage of the rectifier but it discharges very slowly through the L-section filter and the load. The pulsations are greatly opposed by the inductance L and are largely by-passed through the capacitor C_2. Thus the ripples are largely reduced and a very steady d.c output is obtained.

The ripple factor γ for full-wave rectifier with this kind of filter is given by

$$\boxed{\gamma = \frac{1}{32\sqrt{2}\pi^3 f^3 C_1 C_2 L R_L}}$$

where f is the frequency of the input signal.

Approximate analysis

The variation of load voltage with and without filter is shown in Fig. 4.10(b). The dashed curve corresponds to the waveform of load voltage without filter. Here, the capacitor C_1 is charged periodically to almost the peak value of the rectifier output and discharges through the L-section filter L-C_2 and the load R_L. Thus the filtering occurs in two stages—first through C_1 and then through L-C_2.

To calculate the ripple factor, it is assumed that the load current is continuous. Fig. 4.10(b) shows that the voltage v_0' across C_1 is approximately a triangular wave. So, it can be represented in Fourier components as

$$v_0' = V_{\text{dc}} - \frac{V_r}{\pi}\left(\sin 2\omega t - \tfrac{1}{2}\sin 4\omega t + \cdots\right) \tag{4.7.20}$$

where V_r = the peak-to-peak ripple voltage and $V_{\text{dc}} = V_0 - \frac{V_r}{2}$.

The ripple voltage in case of a full-wave rectifier with shunt capacitor in given by

$$V_r = \frac{I_{\text{dc}} \cdot T/2}{C_1} = \frac{I_{\text{dc}}}{2fC_1} \tag{4.7.21}$$

since the capacitor discharges nearly for a period of $T/2$.

Considering only the second harmonic ripple, the rms value of the ripple voltage is

$$(v_{\text{ac}})_{\text{rms}} = \frac{V_r}{\sqrt{2}\pi} = \frac{I_{\text{dc}}}{2\sqrt{2}\pi f C_1}, \quad \text{using (4.7.21)}. \tag{4.7.22}$$

Usually, the impedance of the parallel combination of R_L and C_2 is much less than that of L. So, the rms value of the current through L corresponding to the second harmonic is

$$i_{\text{rms}} \approx \frac{(v_{\text{ac}})_{\text{rms}}}{2\omega L} = \frac{I_{\text{dc}}}{4\sqrt{2}\pi\omega f C_1 L}, \quad \text{using (4.7.22)}. \tag{4.7.23}$$

Since R_L is much greater than the reactance of C_2, it may be assumed that i_{rms} passes through C_2 only. So, the ripple voltage across R_L or C_2 is

$$V'_{\text{rms}} = i_{\text{rms}} \cdot \frac{1}{2\omega C_2} = \frac{I_{\text{dc}}}{8\sqrt{2}\pi\omega^2 f C_1 C_2 L} \tag{4.7.24}$$

$\therefore$ The ripple factor is $$\gamma = \frac{V'_{\text{rms}}}{V_{\text{dc}}} = \frac{I_{\text{dc}}}{8\sqrt{2}\pi\omega^2 f C_1 C_2 L} \cdot \frac{1}{I_{\text{dc}} \cdot R_L}$$

$$= \frac{1}{4\sqrt{2}\omega^3 C_1 C_2 L R_L}, \quad (\because \omega = 2\pi f) \tag{4.7.25}$$

$$= \frac{1}{32\sqrt{2}\pi^3 f^3 C_1 C_2 L R_L} \tag{4.7.26}$$

• A π-filter has the shape of the Greak letter π and hence it is named π-filter. As $\gamma \propto 1/R_L$, the ripple increases as R_L decreases. So, a *high load resistance helps in reducing ripples.*

4.8 Illustrated Examples

▶ **Example 1.** *A crystal diode having internal resistance $R_f = 20\Omega$ is used for half wave rectification. If the applied voltage is $E = 50 \sin 100\pi t$ and load resistance of 500Ω, find* (i) $I_0, I_{dc}, I_{\text{rms}}$, (ii) *a.c. input power, d.c output power and* (iii) *efficiency.*

(Himachal Univ.)

Solution: Here, the dynamic forward resistance, $R_f = 20\Omega$.

Applied input a.c voltage $E = 50 \sin 100\pi t$

Load resistance, $R_L = 500\Omega$

$\therefore$ The peak value of the load current

$$I_0 = \frac{E_0}{R_f + R_L} = \frac{50}{20 + 500}\text{ A} = 0.096\text{ A} = \mathbf{96\text{ mA}}$$

The average or d.c value of output current is

$$I_{\text{dc}} = \frac{I_0}{\pi} = \frac{96}{\pi}\text{ mA} = \mathbf{30.5\text{ mA}}$$

The r.m.s or effective output current

$$I_{\text{rms}} = \frac{I_0}{2} = \frac{96}{2}\text{ mA} = \mathbf{48\text{ mA}}$$

The a.c input power is $$P_{\text{ac}} = I_{\text{rms}}^2(R_f + R_L)$$

$$= \left(\frac{48}{1000}\right)^2 (20 + 500)\text{ W} = \mathbf{1.198\text{ W}}$$

The d.c output power, $$P_{\text{dc}} = I_{\text{dc}}^2 \cdot R_L = \left(\frac{30.5}{1000}\right)^2 \times 500\text{ W} = \mathbf{0.465\text{ W}}$$

The rectifier efficiency, $$\eta = \frac{P_{\text{dc}}}{P_{\text{ac}}} = \frac{0.465}{1.198} = 0.388 = \mathbf{38.8\%}.$$

▶ **Example 2.** *A half-wave rectifier uses a transformer of turns ratio* 8 : 1. *If the primary voltage is* 230 V (rms), *find* (a) *d.c. output voltage,* (b) *peak inverse voltage.*

Solution: Given $V_{rms} = 230\,\text{V} \Rightarrow$ Maximum primary voltage, $V_p = V_{rms}\sqrt{2} = 230\sqrt{2} = 324.3\,\text{V}$

Primary to secondary turn ratio, $\dfrac{N_1}{N_2} = \dfrac{8}{1}.$

$\therefore$ Maximum secondary voltage, $V_s = V_p \times \dfrac{N_2}{N_1} = 324.3 \times \frac{1}{8} = 40.54\,\text{V}$

(a) d.c. output voltage, $\mathbf{V_{dc}} = I_{dc} \times R = \dfrac{I_0}{\pi} \times R = \dfrac{V_0}{\pi} = \dfrac{40.54}{3.14} = \mathbf{12.9\ V}$

(b) Since the diode is reverse biased, it conducts no current during the negative half-cycle of the a.c. supply. So, the maximum transformer-secondary voltage appears across the diode.

$$\therefore \text{ Peak inverse voltage } (\boldsymbol{PIV}) = V_s = \mathbf{40.54\ V}$$

▶ **Example 3.** *A centre-tapped full wave rectifier has a voltage of* $25 \sin 314t$ *across each half of the secondary winding. A load of* $1\,\text{k}\Omega$ *is connected in the circuit. Determine* (a) *peak current,* (b) *d.c value of current,* (*neglect resistance of diode*), (c) *r.m.s current,* (d) *ripple factor,* (e) *rectification efficiency and* (f) *PIV of diode.*

Solution: Voltage across each half of the secondary is given by

$$V = 25 \sin 314t \;\Rightarrow\; V_0 = 25\,\text{V}$$

(a) peak current, $\boldsymbol{I_0} = \dfrac{V_0}{R} = 25/1000 = \mathbf{25\,mA}$ $\quad (\because\ R = 1\,\text{k}\Omega = 1000\Omega)$

(b) d.c. current, $\boldsymbol{I_{dc}} = \dfrac{2I_0}{\pi} = \dfrac{2 \times 25}{\pi} = \mathbf{15.9\,mA}$

(c) rms current, $\boldsymbol{I_{rms}} = \dfrac{I_0}{\sqrt{2}} = \dfrac{25}{\sqrt{2}} = \mathbf{17.67\,mA}$

(d) ripple factor, $\gamma = \sqrt{\left(\dfrac{I_{rms}}{I_{dc}}\right)^2 - 1} = \sqrt{\left(\dfrac{17.67}{15.9}\right)^2 - 1} = \mathbf{0.482}$

(e) rectification efficiency, $\eta = \dfrac{P_{dc}}{P_{ac}} = \dfrac{I_{dc}^2 R_L}{I_{rms}^2 R_L} = \left(\dfrac{15.9}{17.67}\right)^2 = \mathbf{81.2\%}$

(f) *PIV* of diode is $\quad \boldsymbol{PIV} = 2V_0 = 2 \times 25\,\text{V} = \mathbf{50\ V}$

▶ **Example 4.** *The load resistance applied to a full wave rectifier is* $6\text{k}\Omega$ *and each diode has a resistance of* $1.2\text{k}\Omega$. *If the d.c of peak value of* 300 volt *is applied to each diode of the rectifier, calculate* (i) *the peak value of current,* (ii) *d.c and rms value of current,* (iii) *efficiency of rectifier and* (iv) *the ripple-factor.* (Guru Nanak Univ.)

Solution: Here, the dynamic forward resistance of each diode of the rectifier is

$$R_f = 1.2\text{k}\Omega = 1200\Omega$$

$$\text{Load resistance, } R_L = 6\text{k}\Omega = 6000\Omega$$

$$\text{The applied d.c peak value, } V_0 = 300\,\text{V}$$

(i) The peak value of current, $I_0 = \dfrac{V_0}{R_f + R_L} = \dfrac{300}{1200 + 6000}\,\text{A} = \mathbf{41.7\,mA}$

(ii) The d.c value of current, $I_{\mathbf{dc}} = \dfrac{2I_0}{\pi} = \dfrac{2 \times 41.7}{\pi}\,\text{mA} = \mathbf{26.5\,mA}$

The r.m.s value of current, $I_{\mathbf{rms}} = \dfrac{I_0}{\sqrt{2}} = \dfrac{41.7}{\sqrt{2}}\,\text{mA} = \mathbf{29.4\,mA}$

(iii) Efficiency of rectifier, $\eta = \dfrac{81.2}{1 + R_f/R_L}\% = \dfrac{81.2}{1 + 1200/6000}\% = \mathbf{67.66\%}$

(iv) Ripple factor, $\gamma = \sqrt{\left(\dfrac{I_{\text{rms}}}{I_{\text{dc}}}\right)^2 - 1} = \sqrt{\left(\dfrac{29.4}{26.5}\right)^2 - 1} = \mathbf{0.482}$

▶ **Example 5.** *A full-wave rectifier uses two diodes, the internal resistance of each diode is* 25 ohm. *The transformer secondary voltage (r.m.s.) is* $50\sqrt{2}$ volt *and load resistance is* 1475 ohm. *Find* (i) *the mean load current and* (ii) *the root mean square value of the load current.* **(Punjab Univ.)**

Solution: The forward resistance of each diode, $R_f = 25\Omega$

Load resistance (r.m.s), $R_L = 1475$ ohm.

The transformer secondary voltage (r.m.s), $V_{\text{rms}} = 50\sqrt{2}\,\text{V}$,

∴ The peak value of the transformer secondary voltage,

$$V_0 = V_{\text{rms}} \times \sqrt{2}\,\text{V} = 100\,\text{V}$$

∴ The maximum load current, $I_0 = \dfrac{V_0}{R_f + R_L} = \dfrac{100}{25 + 1475}\,\text{A} = 66.7\,\text{mA}$

The mean load current, $I_{\mathbf{dc}} = \dfrac{2I_0}{\pi} = \dfrac{2 \times 66.7}{\pi}\,\text{mA} = \mathbf{42.4\,mA}$

The r.m.s. value of load current, $I_{\mathbf{rms}} = \dfrac{I_0}{\sqrt{2}} = \dfrac{66.7}{\sqrt{2}}\,\text{mA} = \mathbf{47.3\,mA}$

▶ **Example 6.** *The four diodes used in a bridge rectifier circuit have forward resistance which may be considered constant at* 1 ohm *and infinite reverse resistance. The alternating supply voltage* (r.m.s) *is* 240 V *and load resistance is* 480 Ω. *Calculate* (i) *the mean load current,* (ii) *r.m.s. current through each diode and power dissipated in each.* **(Gauhati Univ.)**

Solution: The input a.c supply voltage = 240 V (r.m.s value).

$\therefore$ The peak value, $V_0 = 240\sqrt{2}\,\text{V}$. Given, $R_f = 1\Omega$, $R_L = 480\Omega$.

In a bridge rectifier, the two diodes in series conduct at a time. So, the total resistance $= 2R_f + R_L$.

$\therefore$ The maximum load current, $I_0 = \dfrac{V_0}{2R_f + R_L} = \dfrac{240\sqrt{2}}{2 \times 1 + 480}\,\text{A} = 0.7\,\text{A}$

The mean load current, $\boldsymbol{I_{\text{dc}}} = \dfrac{2I_0}{\pi} = \dfrac{2 \times 0.7}{\pi}\,\text{A} = \mathbf{0.45\ A.}$

The r.m.s current through each diode is given by

$$\boldsymbol{I_{\text{rms}}} = \frac{I_0}{\sqrt{2}} = \frac{0.7}{\sqrt{2}} = \mathbf{0.5\ A}$$

The ***power dissipated*** in each diode $= I_{\text{rms}}^2 \times R_f = (0.5)^2 \times 1\,\text{W} = \mathbf{0.25\ W}$

▶ **Example 7.** *A power supply provides* 100 mA *at* 20 V *d.c. It uses a capacitative filtering and is driven from a* 50 Hz *source. Calculate the ripple factor for full-wave rectification.* $C = 1000\mu\text{F}$. **(Punjab Univ.)**

Solution: Given, $I_{\text{dc}} = 100\,\text{mA}$, $V_{\text{dc}} = 20\,\text{V}$, $C = 1000\,\mu\text{F}$, $f = 50\text{Hz}$.

$\therefore$ The load resistance, $R_L = \dfrac{V_{\text{dc}}}{I_{\text{dc}}} = \dfrac{20}{100 \times 10^{-3}}\,\text{ohm} = 200\,\text{ohm}$

$$\text{The ripple factor, } \gamma = \frac{1}{4\sqrt{3}fR_LC} = \frac{1}{4\sqrt{3} \times 50 \times 200 \times 1000 \times 10^{-6}}$$
$$= 0.0144 = \mathbf{1.44\%}$$

▶ **Example 8.** *Design an LC filter for a full-wave rectifier circuit supplied with a.c at* 50 Hz. *It has to have a* 3000Ω *load at* 300 V *d.c and a ripple factor of* 0.005. *Use the critical value for the choke.*

Solution: For a choke input (LC) filter, the ripple factor can be given by

$$\gamma = \frac{1}{6\sqrt{2}\omega^2 LC} \quad \Rightarrow \quad LC = \frac{1}{6\sqrt{2}\omega^2\gamma}$$

Here $\gamma = 0.005$, $\omega = 2\pi f = 2\pi \times 50$, $R_L = 3000\Omega$.

$$\therefore \quad LC = \frac{1}{6\sqrt{2} \times (2\pi \times 50)^2 \times 0.005} = 2.4 \times 10^{-4}$$

The critical value of the inductance can be shown as $L_C = R_L/3\omega$.

$$L_C = \frac{1}{3} \times \frac{R_L}{\omega} = \frac{3000}{3 \times 2\pi \times 50} = \mathbf{3.18\,H}$$

$$\therefore \quad C = \frac{2.4 \times 10^{-4}}{L} = \frac{2.4 \times 10^{-4}}{3.18}\,\text{F} = \mathbf{75\,\mu F}$$

▶ **Example 9.** *If the output voltage of a centre-tap full wave rectifier and a bridge type full-wave rectifier is* 100 V, *determine the peak inverse voltage in both cases. Comment on the results obtained.*

Solution: The two rectifier circuits are shown in Fig. 4.11 and Fig. 4.12 respectively.

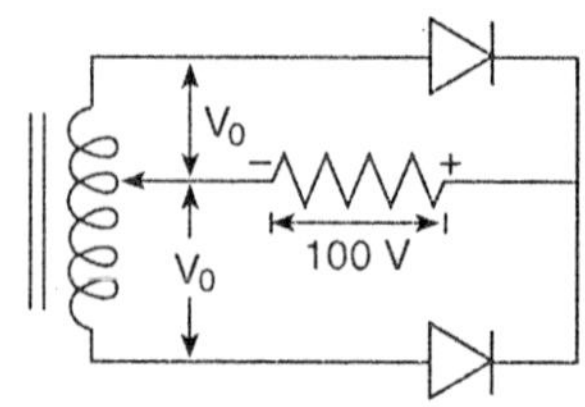

Fig. 4.11 Centre tap circuit

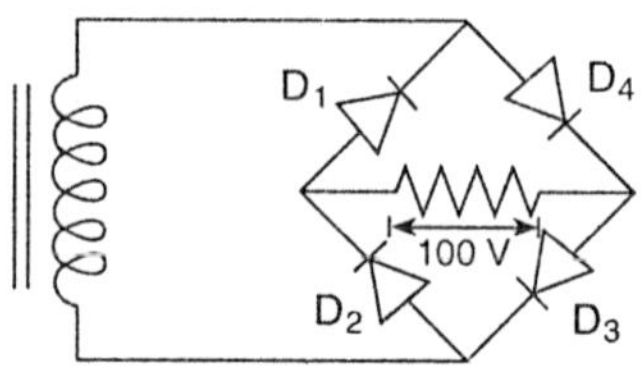

Fig. 4.12 Bridge circuit

(i) Maximum voltage between centre-tap and each side of secondary is

$$V_0 = \frac{\pi V_{dc}}{2} = \frac{\pi \times 100}{2} = 157.08$$

Peak inverse voltage across each diode,

$$\boldsymbol{PIV} = 2V_0 = 2 \times 157.08 = \mathbf{314.16\ V}$$

(ii) Maximum voltage across the secondary,

$$V_0 = \frac{\pi V_{dc}}{2} = \frac{\pi \times 100}{2} = 157.08\,\text{V}$$

Peak inverse voltage across each diode is

$$\boldsymbol{PIV} = V_0 = \mathbf{157.08\ V}$$

• The results show that for the same dc output voltage, PIV in bridge circuit is half that of centre-tap circuit. So, diode of higher PIV is needed for centre-tap full-wave rectifier. This is a distinct *advantage of bridge circuit.*

▶ **Example 10.** *Find the maximum d.c. voltage that can be obtained from the full-wave rectifier circuit, as shown in* Fig. 4.13.

Solution: Here, $V_0 = 220\,\text{V}$, $\nu = 50\,\text{Hz}$, $R_L = 1\,\text{k}\Omega$, $N_1 = 1000$, $N_2 = 100$.

From the relation $\dfrac{V_2}{V_1} = \dfrac{N_2}{N_1}$, the r.m.s secondary voltage is

$$V_2 = V_1 \frac{N_2}{N_1} = \frac{220 \times 100}{1000} = 22\,\text{V}$$

∴ Maximum secondary voltage, $V_{rms} = 22\sqrt{2} = 30.8\,V$

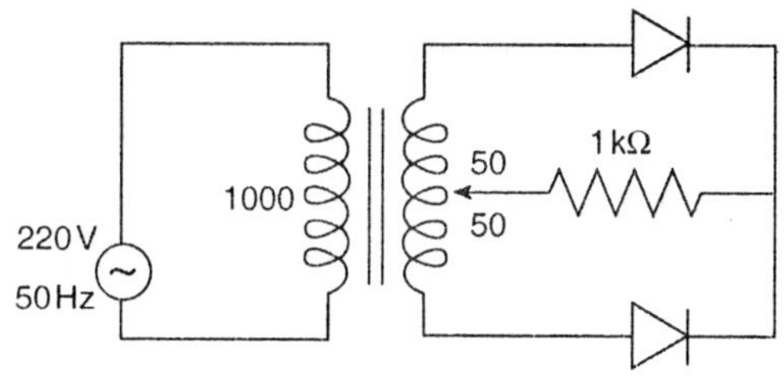

Fig. 4.13

Maximum voltage across half-secondary winding is

$$V_0 = 30.8/2 = 15.4\,V$$

$$\text{Average current, } I_{dc} = \frac{2V_0}{\pi R_L} = \frac{2 \times 15.4}{3.14 \times 1000} = 9.89 \times 10^{-3}\,A$$

∴ Maximum d.c output voltage, $\mathbf{V_{dc}} = I_{dc} \times R_L = 9.89 \times 10^{-3} \times 1 \times 10^3 = \mathbf{9.89\,V}$

Review and Short Questions

1. Explain the behaviour of a diode as a circuit element.
2. What are the different types of circuit elements? Explain with suitable examples.
3. What do you mean by load line and Q-point of a diode circuit? Explain with a suitable diagram.
4. What is a rectifier? Show that a diode is capable of rectification? Draw the circuit for a half-wave rectifier and explain its working. **(Bangalore Univ.)**
5. Draw the circuit diagram of a full-wave rectifier using diodes. Explain the principle of operation of the circuit. **(Burdwan Hons.)**
6. (a) Explain why the peak inverse voltage of a semiconductor diode is an important parameter.
   (b) Draw the circuit diagram of a full-wave rectifier using a centre-tapped transformer and explain its working. Sketch the input and output waveforms. Explain why a bridge rectifier is preferred to a centre-tap rectifier.
   (c) Explain how a capacitor filter smoothens the output of a full-wave rectifier. **(Calcutta Hons.)**
7. (a) What is filter? Explain physically the action of C-filter and L-filter.
   (b) Draw the circuit diagram of a full-wave bridge rectifier with a π-filter. **(Calcutta Hons.)**

8. Define average value, r.m.s. value, efficiency, ripple factor, peak inverse voltage and voltage regulations of a rectifier. Calculate their values for a half-wave rectifier. Show that the ripple factor for a half-wave rectifier is 1.21.

(Himachal and Guru Nanak Univ.)

9. Define and calculate average value, r.m.s. value, efficiency, ripple factor, peak inverse voltage and voltage regulation of a full-wave rectifier. Show that the ripple-factor for a full-wave rectifier is 0.48. **(Allahabad Univ.)**

10. Draw the circuit diagram of a full-wave bridge rectifier using semiconductor diodes. Explain its working principle. **(Lucknow Univ.)**

11. Compare the relative advantages and disadvantages of a centre-tapped and a bridge rectifier. **(Punjabi Univ.)**

12. Explain how the dc output voltage of a full-wave rectifier is improved when a capacitor filter is used. Draw the waveforms of the load voltage and the diode current.

13. Draw the circuit diagram of a full-wave rectifier feeding an inductor filter and a load resistance. Explain the waveform of the load current and give its approximate expression. Calculate the ripple factor.

14. Calculate the ripple factor for an L-type LC filter connected to the output of a full-wave rectifier.

15. Give one example of a linear circuit element.

16. Give one example of a bilateral circuit element.

17. Give one example of a unilateral circuit element.

18. Draw a full-wave rectifier circuit with (i) L-section filter, (ii) π-section filter. Explain the operation of the circuits. Plot the output waveforms. Find approximate expressions for ripple factors.

Objective Questions

1. How many diodes are used in a bridge rectifier?

   (a) 1, (b) 2, (c) 3, (d) 4. **(Meerut Univ.)**

2. Ripple factor for a half-wave rectifier is

   (a) 1.21, (b) 0.81, (c) 0.48, (d) 0.23. **(Meerut Univ.)**

3. In ideal regulated power supply, the regulation is

   (a) positive, (b) negative, (c) 0, (d) ∞. **(Himachal Univ.)**

4. For the rectifier circuit shown in Fig. 4.14, the sinusoidal voltage (V_1 or V_2) at the output of the transformer has a maximum value of 10 V. The load resistance R_L is 1 kΩ. If I_{av} is the average current through the resistor R_L the circuit corresponds to a

   (a) full-wave rectifier with $I_{av} = 20/\pi$ mA,

   (b) half-wave rectifier with $I_{av} = 20/\pi$ mA,

   (c) half-wave rectifier with $I_{av} = 10/\pi$ mA,

   (d) full-wave rectifier with $I_{av} = 10/\pi$ mA. **(GATE)**

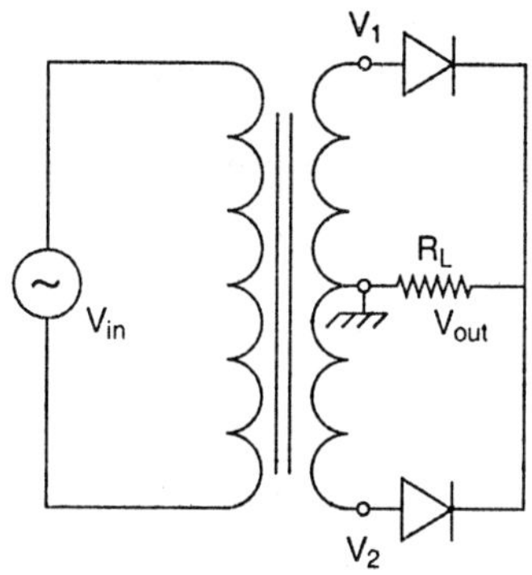

Fig. 4.14

5. If V_0 is the peak value of the input signal of a half-wave rectifier, then the peak inverse voltage with a shunt capacitor filter is equal to

   (a) V_0/π, (b) V_0, (c) $V_0/\sqrt{2}$, (d) $2V_0$.

6. Simple capacitor filters are good for power supply with

   (a) high current, (b) low current, (c) high voltage, (d) low voltage.

7. In a rectifier, no-load condition means

   (a) load resistance $R_L = 0$, (b) output voltage $= 0$,

   (c) load resistance $R_L = \infty$, (d) none of these.

8. If the frequency of a.c input of a bridge rectifier is 50 Hz, then the output lowest ripple component is of frequency

   (a) 100 Hz, (b) 50 Hz, (c) 25 Hz, (d) 200 Hz.

9. The capacitor used in a rectifier circuit is usually

   (a) paper capacitor, (b) mica capacitor,

   (c) electrolytic capacitor, (d) ceramic disc,

10. The curve obtained by plotting the variation of diode current I with input voltage V_i for a given value of load resistance R_L is called

    (a) Static characteristic curve, (b) Load line,

    (c) Dynamic characteristic curve, (c) Input characteristic curve.

11. If a.c input to a half-wave rectifier has an r.m.s value of $400/\sqrt{2}$, then diode PIV rating is

    (a) 400 V, (b) $400/\sqrt{2}$ V, (c) $400\sqrt{2}$, (d) None.

12. The function of a rectifier is to

(a) step up the voltage, (b) convert d.c to a.c,
(c) step down the voltage, (d) convert a.c to d.c

13. The basic function of a filter is to

(a) minimize variations in a.c input signal,
(b) remove ripples from the rectified output,
(c) stabilize the d.c output voltage, (d) none of the above.

14. The efficiency of rectification is given by

(a) $P_{\text{ac}}/P_{\text{dc}}$, (b) $P_{\text{dc}}/P_{\text{ac}}$, (c) $I_{\text{dc}}/I_{\text{rms}}$, (d) $V_{\text{dc}}/V_{\text{rms}}$.

15. In a LC-filter, the ripple factor

(a) increases with the load current, (b) remains unaltered with load current,
(c) has the lowest possible value, (d) increases with load resistance.

16. In a half-wave rectifier, if a resistance equal to the load resistance is connected in parallel with the diode, then

(a) the circuit will stop recifying, (b) output voltage would be halved,
(c) output voltage would be doubled,
(d) output voltage would remain unaltered.

17. As compared to a full-wave rectifier using 2 diodes, the four diode bridge rectifier has the dominant advantage of

(a) Higher current carrying, (b) Lower peak inverse requirement,
(c) Lower ripple factor, (d) Higher efficiency. **(ISRO)**

18. The volt-ampere characteristic of a linear circuit element is a ______.

19. The input to a matched filter is given by

$$S(t) = \begin{cases} 10\sin(2\pi \times 10^6 t), & 0 < t < 10^{-4}\,\text{s} \\ 0, & \text{otherwise.} \end{cases}$$

The peak amplitude of the filter output is

(a) 10 V, (b) 5 V, (c) 10 mV, (d) 5 mV. **(GATE)**

20. When the V-I relation changes with the direction of current, the circuit element is called ______.

Problems

1. A half-wave rectifier is used to supply 25 V d.c to a resistance load of 500 ohm. The diode has forward resistance of 25 ohm. Find the maximum value of a.c voltage required at the input. **(Bangalore Univ.)** [82.48 V]

2. A half-wave rectifier is connected to a 10 : 1 transformer operating at the 250 V, 50 Hz power line. It is used to drive a 20 ohm load ($R_L = 20$ ohm). (a) Sketch the circuit, (b) Sketch the output waveform, (c) Find V_{dc} and I_{dc} across the load and (d) Diode peak inverse voltage. **(Punjab Univ.)** [112.6 V, 5.63 A, 353.6 V]

3. A load of 2.5kΩ is connected across the output of a full-wave rectifier and each diode has a resistance of 500Ω. If $E_0 = 220$ V, find (a) maximum current through the load, (b) average value of output current, (c) rectification efficiency and (d) ripple factor **(Himachal Univ.)** [73.3 mA, 46.6 mA, 66%, 0.482]

4. In a centre-tap full wave rectifier, the load resistance $R_L = 1$kΩ. The voltage across half the secondary winding is $220 \sin 314t$. Find (a) peak value of current, (b) average value of current, (c) ripple factor and (d) rectification efficiency. Forward resistance of each diode is 100Ω. [200 mA, 127.3 mA, 0.482, 74.8%]

5. A power supply is delivering 100 W to a load of 10kΩ. Find the A.C. voltage present across the load if ripple factor is 0.1%. [1 V]

6. In a full-wave rectifier using a bridge circuit, the supply voltage to the transformer is 200 V. If the ratio of primary to secondary turns is 4 : 1, $R_L = 105\Omega$ and input frequency 50 Hz, find (a) d.c output voltage, (b) peak inverse voltage and (c) output frequency. [44.94 V, 70.7 V, 100 Hz]

7. A power supply provides 200 mA at 50 V dc. It uses a capacitative filtering and is driven from a 50 Hz source. Calculate the ripple factor for full wave rectification. $C = 1500\mu$F. **(Punjab Univ.)** [0.77%]

8. A power supply has a voltage regulation of 2%. If no load voltage is 40 V, what is the full load voltage? **(Punjab Univ.)** [39.2 V]

9. A d.c power supply uses full-wave rectification from a 250 V, 50 Hz source. The following are the applicable specifications—Ripple factor not to exceed 0.04, Maximum d.c current $I_{dc} = 0.5$ A, d.c voltage $V_{dc} = 30$ V. The circuit uses a centre-tapped transformer and a single capacitor filter. (a) Draw the circuit, (b) Calculate the values of the load resistance and capacitor and (c) Find the required transformer ratio. [60Ω, 1200 μF, 4 : 15]

10. The output voltage across the load resistance of an inductor filter connected to a bridge rectifier shows a d.c value of 20 V and a peak-to-peak ripple voltage of 1 V. Calculate the ripple factor and the percentage ripple. [0.0177, 1.77%]

11. Prove that the maximum d.c output power occurs in a half-wave rectifier when the load resistance R_L equals the diode resistance R_f.

12. In the circuit of Fig. 4.15, $V_R = 10\,\text{V}$ and v_i is a sinusoidal voltage of peak value 50 V. Taking $R = 10\,\text{k}\Omega$, and the diode as an ideal one, find the maximum diode current. **(Burdwan Hons.)** [4 mA]

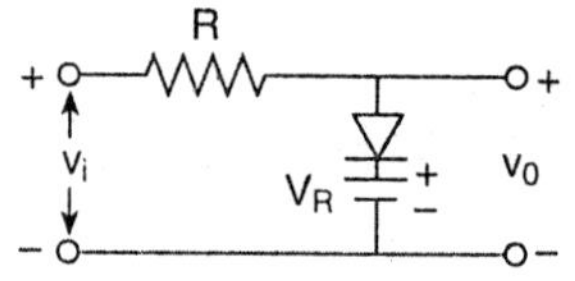

Fig. 4.15

13. A full-wave rectifier uses two diodes, the internal resistance of each is 20Ω. The transformer r.m.s secondary voltage from the centre tap to each end of the secondary is 50 V and load resistance $R_L = 980\Omega$.

Find (i) average value of current, (ii) average dc voltage across R_L, (iii) the rms current I_{rms}, (iv) rms voltage across R_L, (v) maximum rectification efficiency, (vi) ripple factor and (vii) peak inverse voltage.

[45 mA, 44.1 V, 50 mA, 49 V, 79.6%, 0.482, 141.4 V]

14. In a full-wave rectifier with LC-filter, the applied votlage per side of the transformer is 300 V at 50 Hz frequency. Find the value of E_{dc} and ripple factor. Given $L = 10\,\text{H}$, $C = 4\,\mu\text{F}$. **(Punjab Univ.)** [269.95 V, 0.0298]

[**Hints.** Here $L = 10\,\text{H}$, $C = 4\,\mu\text{F} = 4 \times 10^{-6}\,\text{F}$, $E_{rms} = 300\,\text{V}$ and frequency $f = 50\,\text{Hz}$.

$$\therefore \quad \omega = 2\pi f = 2\pi \times 50 = 100\pi. \quad E_0 = \sqrt{2}E_{rms} = \sqrt{2} \times 300\,\text{V}$$

$$\therefore \quad E_{dc} = \frac{2E_0}{\pi} = \frac{2\sqrt{2} \times 300}{\pi} = \mathbf{269.95\,V}$$

$$\text{Ripple factor, } \gamma = \frac{\sqrt{2}}{3} \cdot \frac{1}{4\omega^2 LC} = \frac{\sqrt{2}}{3} \cdot \frac{1}{4 \times (100\pi)^2 \times 10 \times 4 \times 10^{-6}} = \mathbf{0.0298}]$$

15. A full-wave rectifier circuit uses two diodes, each having a forward resistance R_f of 6Ω and infinite reverse resistance. The citcuit is fed from a 200–0–200 V_{rms} secondary winding of a transformer. If the average current in the resistance load be 5*A*, calculate (a) the value of load resistance, (b) *PIV* and (c) the rectification efficiency, [30Ω, 565.68V, 67.54%]

16. A 60 Hz transfomer having 50 V (rms) on each side of centre trap supplies a full-wave recifier circuit having load resistance 150Ω and a shunt capacitor filter of 1000 μF. Calculate the ripple factor. [1.9]

17. A single phase half-wave rectifier supplies power to a 1 kΩ load. If input supply voltage is 200 V (rms) and the forward resistance of diode is negligible, determine (a) d.c current in load, I_{dc}, (b) d.c voltage across load, V_{dc}, (c) ripple factor, γ and (d) peak inverse voltage, *PIV*. [90 mA, 90 V, 1.21, 282 V]

18. A full-wave rectifier is to suppy 30 V d.c across a load of 500Ω. Assuming the forward resistance of the diode to be negligible, calculate the a.c voltage (rms) across the transformer secondary. [33.3 V]

5

Bipolar Junction Transistor

5.1 Introduction

When two *p-n* junctions or point junctions are joined *back to back*, the current through a forward biased junction can control the current through the other reverse biased junction controlling the flow of majority charge carriers. Bardeen and Brattain at Bell Telephone Laboratories, U.S., performed successfully such an experiment in 1947 by using point junctions on semiconductor. The three electrode control device so obtained was named ***transistor*** which is a combination of two words: *transfer* and *resistor*.

The modern version of a transistor was proposed by Shockley in 1951 to overcome problems like poor reliability and reproducibility of point contacts. Since transistors are operated with both holes and electrons, they are known as ***bipolar junction transistor*** (BJT). It is the *solid state version* of a vacuum triode, the device of interest and development during the first half of the last century. It was initially known as *semiconductor triode* or *crystal triode*. Junction transistor is one of the most important elements in modern electronics and provides a number of *advantages* over its predecessor, the vacuum triode. These are:

It is *smaller* in *size* and *light weight*, can *withstand rough handling*, *cheap* and *easily available*, *no filament* and *heating power* are needed, has *longer life* and *greater efficiency* and *saves* a lot of *electrical power*.

Transistors however are not free from disadvantages. The *disadvantages* of a transistor include its *extreme sensitivity to variations in temperature* and the *high noise level*.

5.2 Classification of bipolar junction transistors

A bipolar junction transistor may be considered as an upgradation of junction diode and consists of two opposed *p-n* junctions very close together in a common wafer of semiconducting material (Si or Ge) where the current through one junction controls the current through the other. So, a bipolar junction transistor is a *three* layer sandwich of *p*-type and *n*-type materials. Depending on the positions of the *p*-type and *n*-type layers, the junction transistors can be classified into *two* categories : (i) ***p-n-p transistor*** and (ii) ***n-p-n transistor***.

When a thin layer of *n*-type semiconductor is sandwiched between two much thicker *p*-type semiconductors, the transistor is called *pnp transistor*. A *npn* transistor however is composed of two *n*-type semiconductors separated by a thin layer of *p*-type semiconductor. In practice, the three layers of *p* and *n* in a transistor are grown out of the same crystal by doping corresponding impurities. The *grown junction technique, alloy junction technique, diffusion technique and epitaxial technique* are the four basic techniques for the manufacture of transistors.

Fig. 5.1 shows the schematic diagram and circuit symbols of *p-n-p* and *n-p-n* transistors. A transistor has *three* sections designated as ***emitter, base*** and ***collector***. The middle section of the transistor made of very thin ($\sim 10^{-6}$ m) and lightly doped semiconductors compared to the outer layers is called *base B*. The name 'base' is related to the fact that it was the wafer on which the point electrodes of the original transistor made contact. The two outer sections of the transistor, doped with the same type of impurities and relatively thicker compared to the base, are *emitter E* and the *collector C*. In spite of their similar doping, the conductivities of emitter and collector are made different by different amount of impurities. The emitter layer is heavily doped compared to the collector. There are two junctions present at the interfaces between

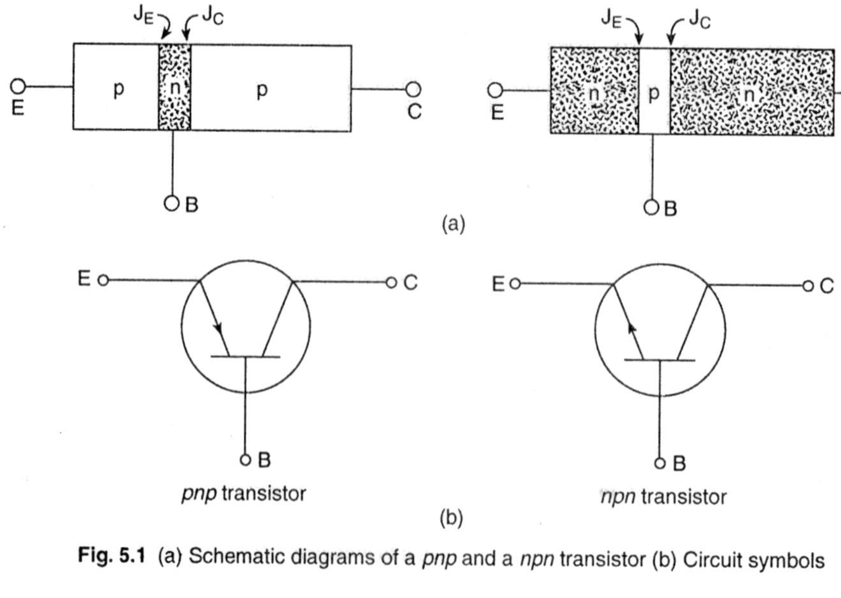

Fig. 5.1 (a) Schematic diagrams of a *pnp* and a *npn* transistor (b) Circuit symbols

the n and p regions : (i) *emitter-base junction* usually known as *emitter junction* J_E and (ii) *collector-base junction* or *collector junction* J_C. For normal operation of a transistor, J_E is forward biased and J_C is reverse biased. So the emitter supplies a large number of majority carriers (holes for *pnp* and electrons for *npn*) to its junction with the base and the charge carriers are finally collected by the collector. Since the collector junction deals with more power than the emitter junctions, the area of collector-base junction is made larger than that of the emitter-base junction in a commercial transistor. So, a normal transistor operation cannot be possible if emitter and collector are interchanged. The arrow head in the circuit symbol of a transistor (Fig. 5.1(b)) shows the direction of current when the emitter-base junction is forward biased.

The structure of a transistor is enclosed in a small sealed metallic or plastic container to protect it from moisture. Three metal leads E, B and C come out of it for connection to the emitter, base and collector respectively. For easier identification, the leads are not equally spaced—E and B are closer and C is little more separated (Fig. 5.2) with a red mark on the container near the collector lead.

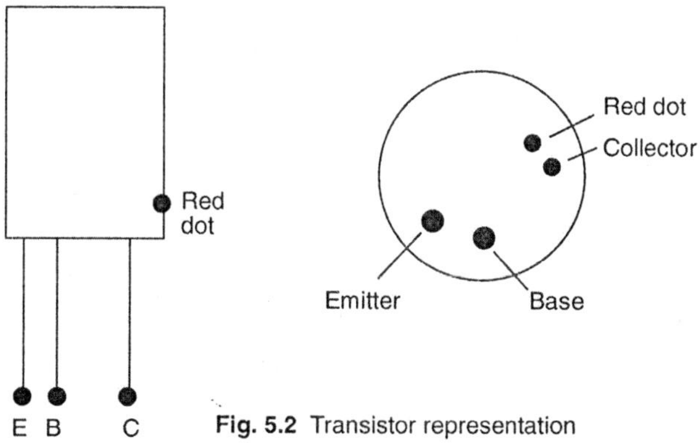

Fig. 5.2 Transistor representation (Longitudinal and transverse view)

Since the emitter-base junction is forward biased and the collector-base junction is reverse biased, the emitter junction J_E has a low dynamic resistance and the collector junction J_C has a high dynamic resistance. By the transistor action, almost the same amount of current passes through the low resistance input to the high resistance output circuits. So, *a transistor transfers a signal from a low resistance to a high resistance.* Hence the name *transistor* which is in fact a combination of two words, namely '*transfer*' and '*resistor*'.

5.3 Transistor manufacturing techniques

There are many ways to construct junction transistors. Among them the following *four* basic techniques are generally used for manufacture of transistors.

(a) Grown-junction technique, (b) Alloy or fused technique,

(c) Diffusion technique, (d) Epitaxial technique.

5.3.1 Grown-junction technique

This technique consists of *three* main steps: (i) *grown* the bulk crystal, (ii) slicing into individual bars and (iii) attaching leads to each region.

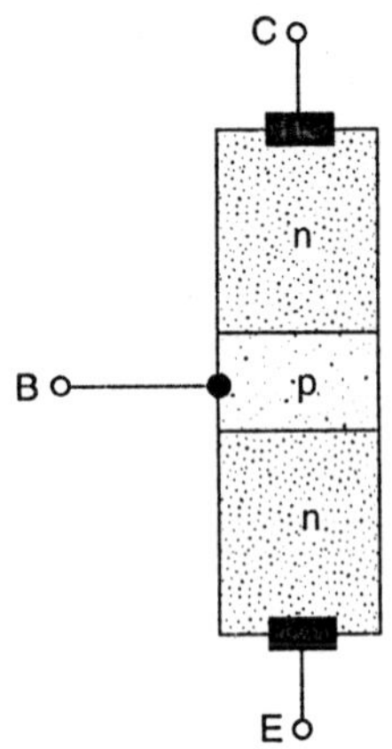

Fig. 5.3 Grown-junction transistor

The grown *npn*-transistor is produced by drawing of a large crystal from a purified semiconductor melt (Ge or Si) weakly doped with *n*-impurity. During the growth process, a *p*-type impurity is added in the melt to produce a *p*-type base. Afer a short interval of time the melt is once again doped with *n*-type impurity to form a *n*-region that acts as emitter. In a similar manner a *pnp* transistor can also be grown.

After the growth of the bulk crystal, it is sliced into small rectangular bars. Then chemical etchning and clearing are done and the connection leads are soldered to each region (Fig. 5.3). The assembly is finally mounted in its housing to complete the construction.

5.3.2 Alloy or fused technique

In this technique, two small dots of same type impurities are placed on opposite sides of a thin semiconductor wafer, doped with opposite impurity. The wafer is taken to serve as the base of the transistor. The system is then heated to a temperature above the melting point of impurity material but below that of wafer so that the impurity elements dissolve and penetrate a little in the wafer and *p-n* junctions are developed on either sides of the base. On cooling, a junction transistor is thus formed and connection leads are joined to the structure to each region.

Fig 5.4 shows a fused-alloy type *pnp*-junction transistor which is formed by using the pellets (dots) of *p*-type material (e.g., Indium) and the wafer of *n*-type semiconductor. An *npn*-transistor can be developed in a similar technique.

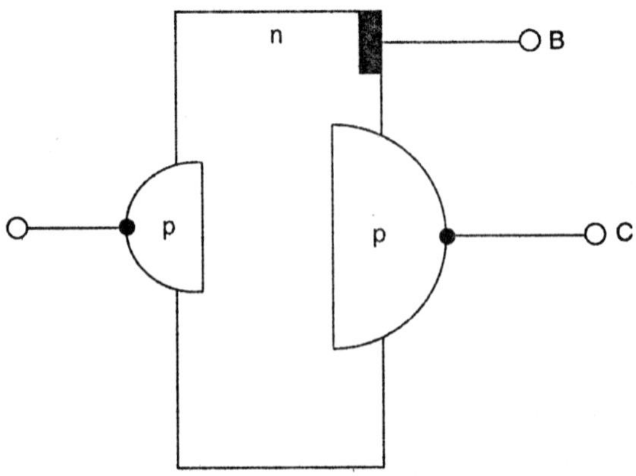

Fig. 5.4 Alloy junction transistor

In fused-alloy type transistors the collector region is made larger than the emitter region so that the collector collects the whole emitter current and hence prevents the current from diffusion into the base region.

5.3.3 Diffusion technique

This technique is applied to form *planar transistors* on a nearly flat semiconductor wafer by gaseous diffusion of impurities. For the construction of an *npn* planar transistor, an *n*-type Si wafer is taken to serve as the collector. The wafer is heated in a furnace containing *p*-type impurities in a gaseous form. The impurities diffuse into the *n*-type wafer surface through a mask and form the *p*-type base of the transistor. It is then covered by a different type mask with aperture and heated in an atmosphere of gaseous *n*-type impurities so that *n*-type atoms may diffuse through the aperture and produce an *n*-layer over the base.

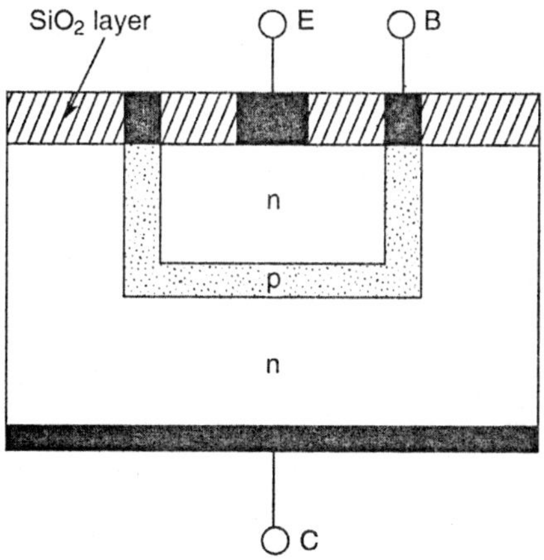

Fig. 5.5 Planar diffusion transistor

This *n*-layer acts as the emitter of the transistor. Finally, a thin layer of SiO_2 is grown over the entire surface and aluminium contacts for the emitter and the base are made by photoetching method as shown in Fig. 5.5.

5.3.4 Epitaxial technique

In this technique, a very thin single-crystal layer of Si (or Ge) is grown on a heavily doped substrate of the same material. This layer serves as a collector on which the base and the emitter regions are diffused. Fig. 5.6 shows an *npn* epitaxial type junction transistor.

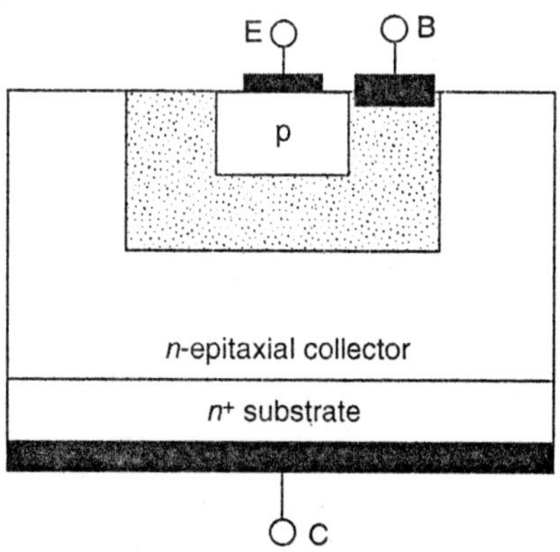

Fig. 5.6 Epitaxial transistor

• The word *epitaxi* has been derived from the Greek words *epi* which means *upon* and *taxis* which means *arrangement*.

• The first transistor invented in 1948 was a point contact type. In that transistor, two tungsten wires were pressed against a semiconductor wafer. Such transistors are now practically obsolete because of their poor relability and reproducibility.

5.4 Working of a transistor and current components

In an unbiased condition, all the currents through a transistor are zero due to the development of an intrinsic potential barrier across the *p-n* junctions (J_E and J_C). Fig. 5.7(a) illustrates the potential distribution in an open-circuit *pnp*-transistor with completely symmetrical junctions, the intrinsic potential barrier being V_b at the junctions.

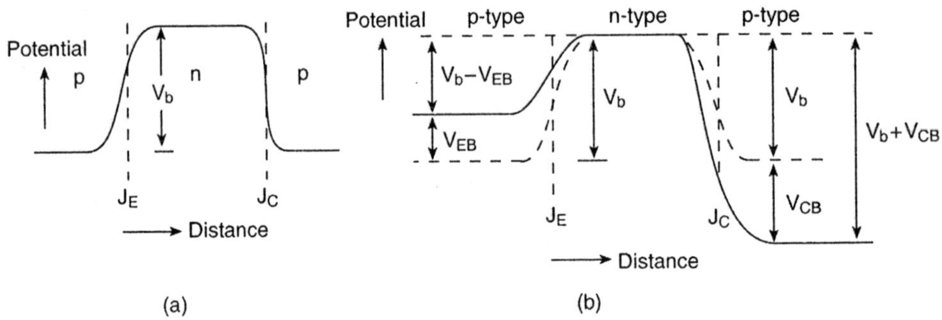

Fig. 5.7 Potential variation in a *pnp* transistor (a) unbiased (b) J_E forward biased and J_C reverse biased

To undertstand the working of a transistor, let us consider a *pnp* transistor with the emitter-base junction J_E forward biased by V_{EB} and the collector-base junction J_C reverse biased by V_{CB}. While the forward biasing voltage V_{EB} reduces the potential at the emitter-base junction to $V_b - |V_{EB}|$, the potential barrier across the collector-base junction is increased to $V_b + |V_{CB}|$ by the reverse biasing voltage V_{CB}.

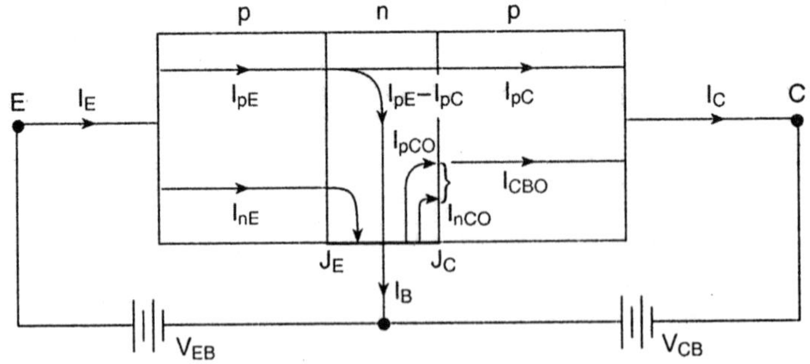

Fig. 5.8 Current components in a *pnp* transistor

The various current components through J_E and J_C are shown in the diagram of Fig. 5.8. The corresponding equivalent circuit diagram with a *pnp*-transistor is illustrated in Fig. 5.9.

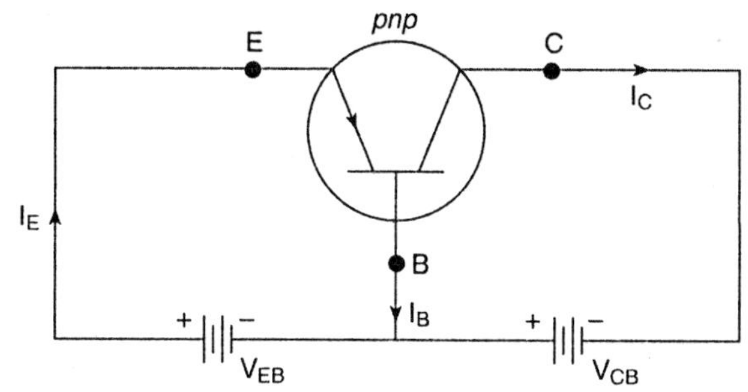

Fig. 5.9 Transistor equivalent circuit of a *pnp* transistor

The lowering of potential barrier of emitter-base junction helps the holes (the majority carriers) in the emitter and the electrons in the base to cross the emitter-base junction. These two flows constitute the emitter current given by

$$I_E = I_{pE} + I_{nE} \tag{5.4.1}$$

where I_{pE} and I_{nE} are the current components due (i) to the holes injected from the emitter to base and (ii) to electrons injected from base to emitter through J_E. Since the doping of emitter is much higher than that of the base in a commercial transistor, $I_{pE} \gg I_{nE}$ and hence the emitter current consists almost of I_{pE} only. The injected holes from the emitter to the base diffuse, through the base, towards the collector-base junction. Since the base region is very thin and slightly doped, only a few holes ($\sim$ 1%) are lost due to the recombination with the electrons (majority carriers in base) during diffusion. As a result, almost all ($\sim$ 99%) of the holes cross the collector-base junction to be collected by the collector. So, the collector current I_C has a current component I_{pC} due to the holes, crossing from base to collector.

Plainly, I_{pC} is slightly smaller than I_{pE}. The difference $(I_{pE} - I_{pC})$ constitutes a part of the base current which arises due to the flow of electron from the battery to the base to maintain the neutrality in the base region during recombination of holes with electrons. Since the collector-base junction is reverse biased by the voltage V_{CB}, a small collector current is obtained due to the flow of minority carriers from both sides of the collector-base junction, even with emitter-base junction in open circuit. So, under open-circuited emitter-base junction and reverse-biased collector-base junction, the collector current is the *reverse saturation current* I_{CBO} of the reverse biased diode with junction J_C. This current is also known as ***leakage current***. It has two components : (i) I_{nCO} due to the minority carriers (electrons) travelling from p-region (collector) to n-region (base) across J_C and (ii) I_{pCO} due to holes (minority carriers) moving from n-side (base) to the p-side (collector) across J_C. The resultant of these two components is

$$I_{CBO} = I_{nCO} + I_{pCO} \tag{5.4.2}$$

I_{CBO} is highly *temperature-sensitive* and is important for an understanding of the transistor characteristics.

Now, we can write the resultant collector current I_C as

$$I_C = I_{pC} + I_{CBO} \tag{5.4.3}$$

and the relation among the terminal currents is given by

$$I_E = I_B + I_C \tag{5.4.4}$$

• The working of a *npn*-transistor is the same as discussed above but only the roles of the electrons and the holes are interchanged. The polarities of the batteries and hence the directions of the currents are also to be reversed in the circuit. This is illustrated in Fig. 5.10(a) and Fig. 5.10(b) represents the corresponding equivalent circuit for an *npn* transistor.

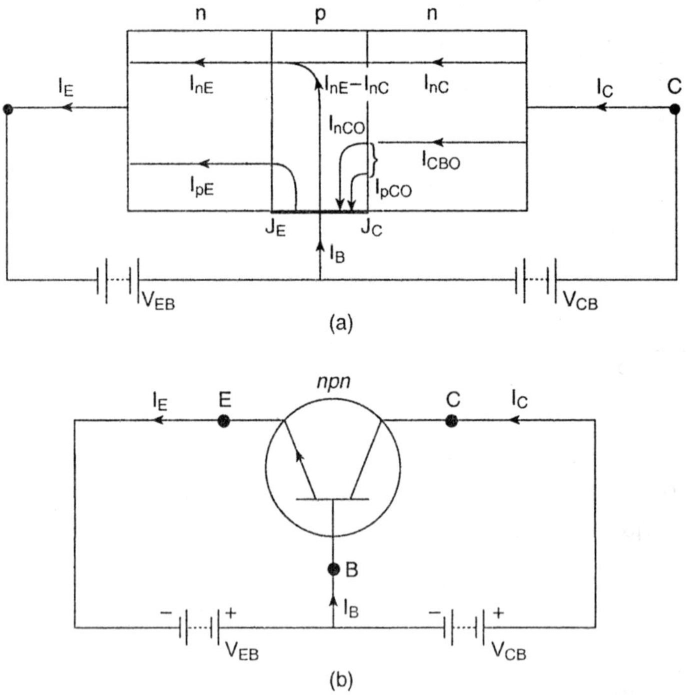

Fig. 5.10 (a) Current components in *npn* transistor
(b) Equivalent circuit diagram of *npn* transistor

For an *npn* transistor the majority electrons from the emitter are injected into the base through J_E. These electrons mainly constitute the emitter current. After diffusion through the base, most of the electrons cross the J_C constituting the collector current I_{nC}. A few of the electrons in the base region are lost due to recombination producing the base current. There is also the leakage current I_{CBO} due to the minority carriers. So, we have the following relations in the case of an *npn* transistor :

$$I_C = I_{nC} + I_{CBO} \tag{5.4.5}$$

$$\text{and} \quad I_E = I_B + I_C \tag{5.4.6}$$

• When the base of a transistor is left unconnected, no current flows through the transistor in any direction. For a *pnp* transistor, if the current tends to flow from E to C, the junction J_C is reverse biased. On the other hand, when the current tends to flow from C to E, the junction J_E becomes reverse biased.

5.5 Modes of transistor connection

There are *three* terminals in a transistor—*emitter, base* and *collector.* So, it can be connected to a circuit if one of the three terminals is made common to both input and output. Thus, three modes of connection or configurations can be applied to operate a transistor in a two-port network depending on the common terminal between the input and the output circuits. These are:

(i) the common base mode,

(ii) the common emitter mode, and

(iii) the common collector mode.

(i) Common base (CB) or Grounded base mode—In this configuration, the base terminal is common to both the input and the output circuits. This mode of operation for a *pnp* and *npn* transistor is shown in Fig. 5.11.

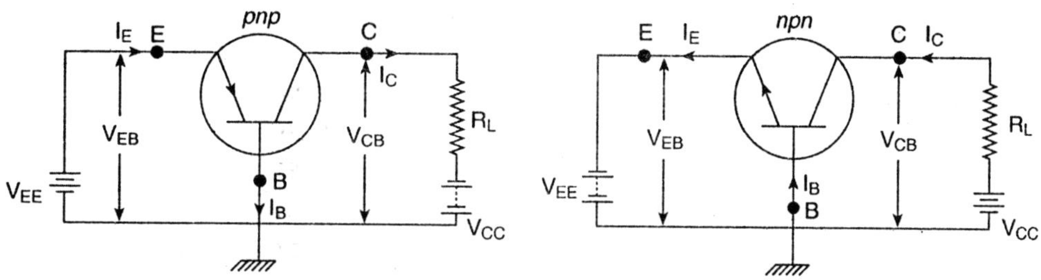

Fig. 5.11 Common base mode for (a) *pnp* transistor, (b) *npn* transistor

(ii) Common emitter (CE) or Grounded emitter mode—In this configuration, emitter terminal is common to the input and output circuits. This mode of operation is illustrated in Fig. 5.12.

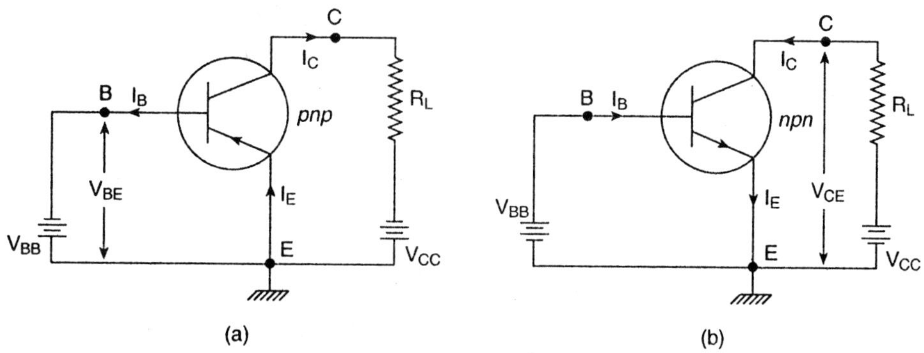

Fig. 5.12 Common emitter mode for (a) *pnp* transistor (b) *npn* transistor

(iii) Common collector (CC) or Grounded collector mode—In this configuration, the collector terminal is common to both the input and output circuit. Fig. 5.13 illustrates this particular mode of operation.

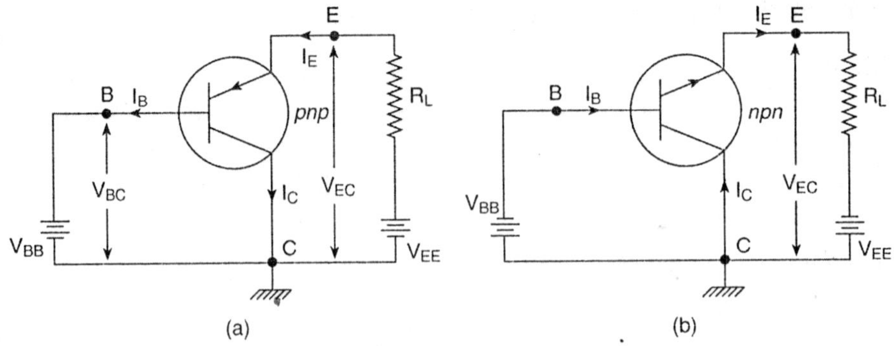

Fig. 5.13 Common collector mode for (a) *pnp* transistor (b) *npn* transistor

Every configuration has some specific *advantages* and *disadvantages*. But, the emitter junction is always forward biased and the collector junction reverse biased without considering the mode of connection.

5.6 Alpha (α) and beta (β) of a transistor

The total collector current of a transistor consists of *two* parts:

(a) the leakage current I_{CBO} due to the motion of the minority carriers across the collector junction when it is reverse biased and the emitter junction is open circuited, and

(b) a fraction of the emitter current which reaches the collector junction.

$$\therefore \quad \text{Collector current, } I_C = I_{CBO} + \alpha I_E \tag{5.6.1}$$

where the fraction α is known as ***transistor alpha***. The α is also called *large signal current gain* of a common-base transistor. Typical values of α lies in the range 0.9 to 0.995. In d.c mode of operation, α is defined as

$$\alpha_{dc} = \frac{I_C}{I_E} \tag{5.6.2}$$

Here, α_{dc} is the ***dc current gain*** of the common-base transistor.

Since $I_{CBO} \ll I_C$, we have, from the relation (5.6.1),

$$I_C \approx \alpha I_E$$

$$\therefore \quad \alpha \approx \alpha_{dc} \tag{5.6.3}$$

When a transistor is operated with a.c in common-base configuration, the parameter α' or α_{ac} is introduced. This α' or α_{ac} in known as the *small-signal short-circuit current transfer ratio* or ***current gain*** of common-base transistor. It is defined as *the ratio of the change in collector current to the change in emitter current at constant collector to base voltage.* Mathematically,

$$\alpha_{ac} = \left.\frac{\Delta I_C}{\Delta I_E}\right|_{V_{CB}=\text{const.}} \tag{5.6.4}$$

For a good transistor $\alpha \approx \alpha_{dc} \approx \alpha_{ac} \approx 1$. The parameter α is *not a constant*, but depends on the emitter current I_E, the collector to the base voltage V_{CB} and also on the temperature.

The current equation of a transistor gives

$$I_E = I_B + I_C \tag{5.6.5}$$

Using the equation (5.6.5) in (5.6.1) we get,

$$\begin{aligned} I_C &= \alpha(I_B + I_C) + I_{CBO} \\ &= \frac{\alpha}{1-\alpha} I_B + \frac{1}{1-\alpha} I_{CBO} \end{aligned} \tag{5.6.6}$$

This is the expression for the *collector current* in terms of the parameter α.

Now, if we put $\beta = \alpha/(1-\alpha)$ in (5.6.6), it becomes

$$\boldsymbol{I_C = \beta I_B + (1+\beta) I_{CBO}} \tag{5.6.7}$$

This equation shows the *output current* I_C of a transistor in CE mode where the parameter β determines the maximum current gain. The parameter $\boldsymbol{\beta}$ is known as ***transistor beta***.

In d.c mode of operation of a common-emitter transistor, β is called *forward current transfer ratio* or the d.c current gain and it is denoted by β_{dc}.

The d.c current gain β_{dc} of a common-emitter transistor is defined as the ratio of the collector current I_C to the base current I_B i.e.,

$$\beta_{dc} = I_C / I_B \tag{5.6.8}$$

Since $I_{CBO} \ll I_B$, we get from equation (5.6.7),

$$\begin{aligned} I_C &\approx \beta I_B \\ \therefore \quad \beta &\approx \beta_{dc} \end{aligned} \tag{5.6.9}$$

Typical values of β_{dc} lies in the range 50 to 400. The parameter β depends on the *temperature* and the *collector current* I_C.

When a transistor in CE mode is operated with a.c, the parameter β_{ac}, known as *small signal short-circuit current gain*, is introduced. It is defined as the ratio of the change in I_C to the change in I_B at a constant collector to emitter voltage V_{CE}. Mathematically,

$$\boldsymbol{\beta_{ac} = \left.\frac{\Delta I_C}{\Delta I_B}\right|_{V_{CE}=\text{const.}}} \tag{5.6.10}$$

5.6.1 Relation between α and β

From the definition of α_{dc} and β_{dc}, we have

$$\alpha_{dc} = \frac{I_C}{I_E} \quad \text{and} \quad \beta_{dc} = \frac{I_C}{I_B}$$

The current equation for a transistor gives,

$$I_E = I_B + I_C \quad \text{or,} \quad \frac{I_E}{I_C} = \frac{I_B}{I_C} + 1$$

$$\text{or,} \quad \frac{1}{\alpha_{dc}} = \frac{1}{\beta_{dc}} + 1$$

$$\text{or,} \quad \beta_{dc} = \frac{\alpha_{dc}}{1 - \alpha_{dc}} \tag{5.6.11}$$

Since $\alpha_{dc} \approx \alpha$ and $\beta_{dc} \approx \beta$, we get from the above equation (5.6.11)

$$\boxed{\beta = \frac{\alpha}{1 - \alpha}} \tag{5.6.12}$$

This is the relation between the parameters α and β and is extremely useful in many computations.

- Note that in order to obtain (5.6.7), the relation (5.6.12) was already used.

5.7 Static characteristics of a transistor

A bipolar junction transistor is a non-linear device with two *p-n* junctions J_E and J_C which can be biased by different methods. So, a transistor can be operated in different zones depending on the biasing of J_E and J_C, and exact analytical relation between the current and voltage of a transistor cannot be obtained. For this, we have to depend on the experimentally drawn current-voltage characteristic curves, known as ***static characteristics*** of a transistor. The graphical form of relationship among the different currents and voltages are represented by these characteristics. The characteristics are of *two* types :

(i) Input characteristics—When the input current is plotted against the input voltage with the output voltage as parameter, the curves so obtained are known as *input characteristics.*

(ii) Output characteristics—The plot of output current versus output voltage with the input current as parameter is called *output characteristics.*

The input characteristics and the output characteristics for a transistor in CE and CB mode are very important for practical applications. We are therefore interested in the characteristics of a transistor in CB and CE configurations only.

5.7.1 Common base (CB) characteristics

Fig. 5.14 shows the circuit arrangement for studying the static characteristics of a *npn* transistor in CB mode. For a *pnp* transistor, the polarities of the batteries and the meters are to be reversed.

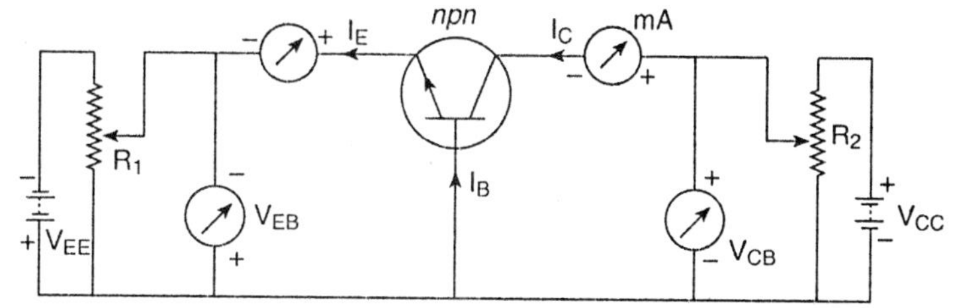

Fig. 5.14 Circuit diagram for characteristics of a *npn* transistor in CB mode

(A) Input characteristics—For the CB configuration of a transistor, the input current and input voltage are the emitter current I_E and the emitter-base voltage V_{EB}. The output voltage is the collector-base voltage V_{CB}. So the input characteristic of CB configuration is the graph between the I_E and V_{EB} with V_{CB} as a parameter. Fig. 5.15 shows the input characteristic curves of a typical *npn* transistor. For normal operation, the emitter-base junction is forward biased. So, the variation of I_E with V_{EB} is similar to the forward characteristic curve of a *p-n* junction diode with a cut-in voltage V_γ ($\simeq$ 0.1 V for Germanium and $\sim$ 0.5 V for Silicon).

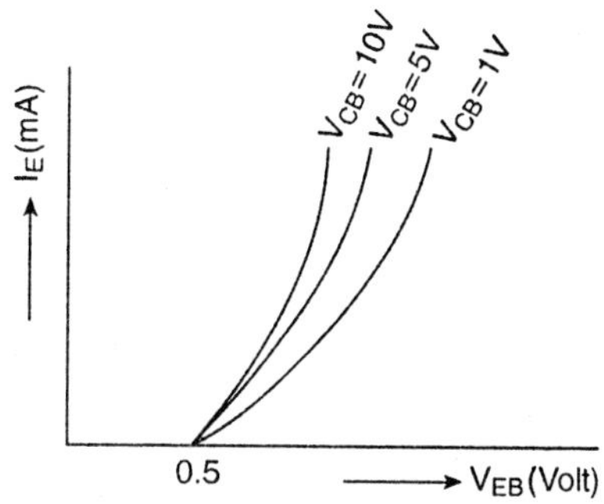

Fig. 5.15 Input characteristics of *npn* Si-transistor in CB mode

An increase in the magnitude of the collector-base voltage causes the width of the depletion region of the collector-base junction to increase. It reduces the effective base width and enhances the gradient of minority carrier resulting in an increase in collector current. The change of the effective base width by the collector-base voltage is called the *early effect* or *base width modulation.*

(B) Output characteristics—For common-base configuration, the collector current I_C and the collector-base voltage V_{CB} are the output current and output voltage

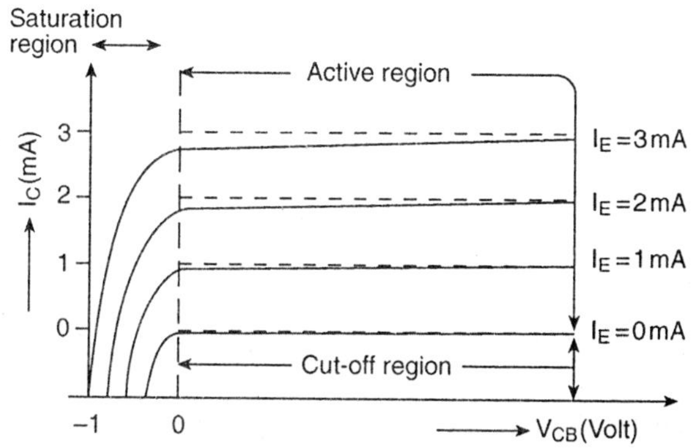

Fig. 5.16 Output characteristics of a *npn* transistor in CB mode

respectively. The input current is the emitter current I_E. So, the graph between the collector current I_C versus the collector-base voltage V_{CB} with emitter current I_E as parameter is the CB output characteristic. Fig. 5.16 shows a typical set of output characteristics of a *npn* transistor in CB mode.

The output characteristics can be divided into *three* distinct regions—*active region, saturation region* and *cut-off region.*

(i) Active region—The normal operating region of a transistor is the active region in which the emitter junction is forward biased and the collector junction is reverse biased. The collector current I_C in CB mode is given by

$$I_C = \alpha I_E + I_{CBO}$$

When the emitter current is zero (at the lower end of the active region), I_C is simply I_{CBO}, the reverse saturation current ($\sim 1\mu$A for Ge and 1 nA for Si). As I_E increases, the collector current becomes $I_C \approx \alpha I_E$ since $I_{CBO} \ll I_E$. As $\alpha \approx 1$, I_C is slightly smaller than I_E. So in the active region, I_C is nearly independent of V_{CB} and depends on I_E. So, the output characteristics are almost parallel lines equally spaced for equal increment in I_E in the active region.

(ii) Saturation region—When both the emitter junction and the collector junction are forward biased, the transistor is operated in the saturation region. This region is located to the left of $V_{CB} = 0$ and above $I_E = 0$ in the output characteristics.

(iii) Cut-off region—The region to the right of $V_{CB} = 0$ and below $I_E = 0$ is the cut-off region. In this region, both the emitter junction and the collector junction are reverse biased.

5.7.2 Common emitter (CE) characteristics

The circuit diagram for studying the static characteristics of a *npn* transistor in CE mode is shown in Fig. 5.17. For a *pnp* transistor, the polarities of the batteries and the meters are simply to be reversed.

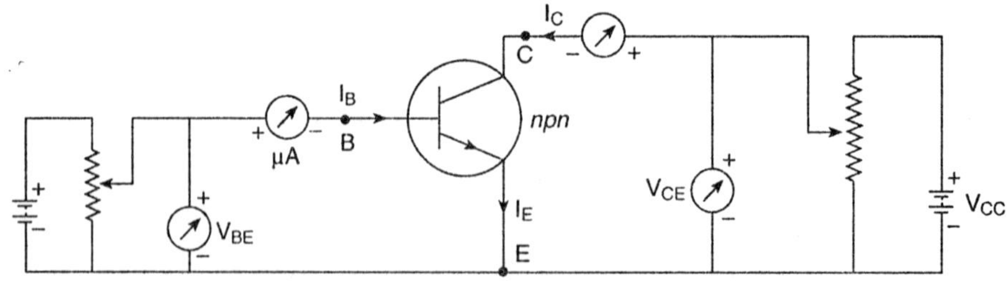

Fig. 5.17 Circuit diagram for CE characteristics of an *npn* transistor

(A) Input characteristics—For CE configuration, base current I_B and emitter-base voltage V_{BE} are the input current and the input voltage respectively and collector-emitter voltage V_{CE} is the output voltage. So, the plot of I_B against V_{BE}

with V_{CE} as a parameter represents the CE input characteristics. Fig. 5.18 shows a set of typical CE input characteristics of a *npn* Si-transistor. The input characteristic curves are similar to that of a forward biased *p-n* junction diode.

With the increase in magnitude of V_{CE}, the effective base width and hence the recombination base-current is reduced due to early effect. So, for a constant value of V_{BE}, the base current I_B decreases with increase in the magnitude of V_{CE}.

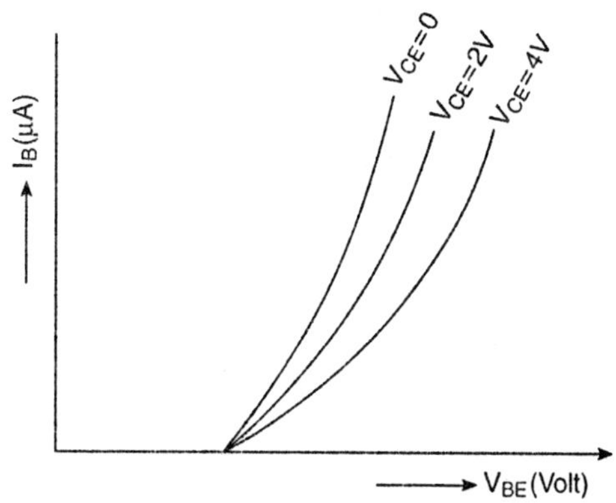

Fig. 5.18 CE input characteristics of an *npn* transistor

(B) Output characteristics—When the output current I_C is plotted against the output voltage V_{CE} taking the input current I_B as parameter, the graph so obtained is called the *output characteristics* for the CE mode. A set of typical CE output characteristics of a *npn* transistor are shown in Fig. 5.19. The output characteristic curves can be divided into *three* regions—*the active region, the cut-off region* and *the saturation region.*

(i) Active region—In this region, the emitter junction is forward biased and the collector junction is reverse biased. In Fig. 5.19, the active region is above $I_B = 0$ and

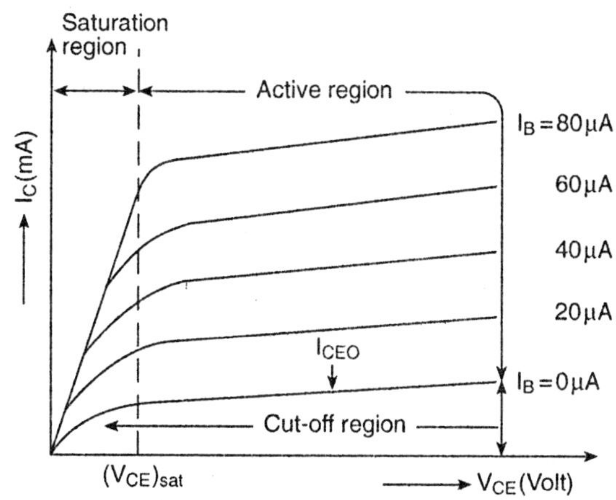

Fig. 5.19 CE output characteristics of a *npn* transistor

to the right of $V_{CE} = (V_{CE})_{\text{sat}}$. In this region, the collector current I_C increases with the increase in V_{CE} according to the relation $I_C = \beta I_B + (1+\beta)I_{CBO}$. So, the curves are almost equispaced straight lines having significant upward slopes.

(ii) Cut-off region : In CE configuration of a transistor, the collector-current $I_C = (1+\beta)I_{CBO} = I_{CEO}$ with $I_B = 0$. So, the transistor is *not* cut-off with $I_B = 0$. Since I_{CEO} is the collector to emitter current due to minority carriers with base open, the emitter junction must be slightly reverse biased in order to cut-off the transistor when $I_B = 0$. For Germanium-transistor, the reverse-bias voltage ~ 0.1 volt and that for Silicon-transistor 0 volt. The cut-off region is shown below the curve $I_B = 0$ in Fig. 5.19.

(iii) Saturation region---The region where both the emitter junction and the collector junction are forward biased by at least cut-in voltage is called the ***saturation region***. The saturation region is very close to the zero voltage axis where all the curves coincide.

5.8 Common collector (CC) connection

When the collector terminal of a transistor is common to both the input and the output circuit, the mode of circuit connection is called the *common collector configuration* or the CC *mode*.

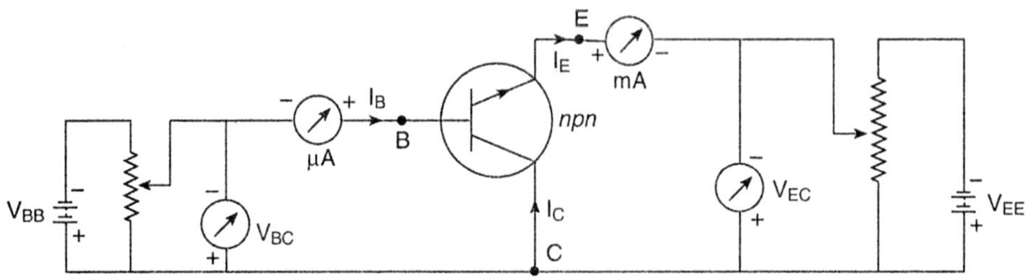

Fig. 5.20 Common collector *npn* transistor circuit

Fig. 5.20 shows the circuit arrangement of a *npn* transistor in CC mode.

Note that the input is applied between the base and the collector while the output is obtained between the emitter and the collector.

Since CC circuit has very high input resistance ($\sim$ 750 kΩ) compared to the output resistance ($\sim$ 25 Ω), the voltage gain is less than 1 and hence this circuit is not suitable for the purpose of amplification.

5.9 Transistor as an amplifier

The volt-ampere characteristics of a transistor indicate that a transistor is capable of producing amplification when it is operated in the active region i.e., when the input circuit is forward biased and the output circuit remains reverse biased. In order to do so, a forward biasing d.c voltage is applied in the input circuit with the a.c signal such that it always maintains the input circuit forward biased, regardless of the polarity of the signal.

Depending on the mode of the circuit connection, a transistor can be used as (i) common-base amplifier and (ii) common-emitter amplifier.

We shall now describe and discuss these *two* types of amplifiers separately in sub-sections **5.9.1** and **5.9.2** that follow.

5.9.1 Transistor as common-base amplifier

The circuit diagram for using a *npn* transistor as a common-base amplifier is illustrated in Fig. 5.21 below.

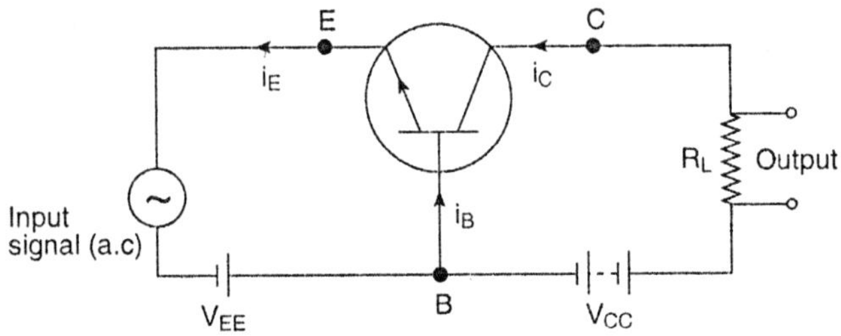

Fig. 5.21 *npn* transistor as CB amplifier

When no input signal (a.c signal) is applied to the input circuit, the p.d. across R_L is $I_C R_L$. When the a.c signal is fed to the input circuit, the emitter becomes less negative w.r.t. the base during the positive half of the input signal i.e., the forward bias is reduced. So the emitter current reduces. It causes nearly an equal amount of decrease in collector current. So, the collector becomes more positive w.r.t the base. During the negative half of the input signal, the emitter current increases due to the increase in forward bias which results in the increase in collector current. As the input circuit has a low resistance and the output circuit has a high resistance, a small change in input current (or voltage) causes an appreciable change in emitter current which in turn causes equal change in the collector current that produces a large voltage across R_L. A transistor can thus act as an *amplifier*.

Expression for power gain—Let ΔV_i be a small change in input voltage which causes a change ΔI_E in the emitter current and consequently an almost equal change ΔI_C in the collector current.

∴ The change in output voltage across R_L is given by

$$\Delta V_0 = \Delta I_C R_L \tag{5.9.1}$$

If r_e be the dynamic resistance of the emitter junction, then the change in the input voltage can be written as

$$\Delta V_i = \Delta I_E \cdot r_e \tag{5.9.2}$$

Dividing (5.9.1) by (5.9.2), we get the ***voltage gain*** as

$$A_V = \frac{\Delta V_0}{\Delta V_i} = \left(\frac{\Delta I_C}{\Delta I_E}\right)\frac{R_L}{r_e}$$

$$= \alpha_{ac}\frac{R_L}{r_e} \tag{5.9.3}$$

where $\alpha_{ac} = \Delta I_C/\Delta I_E$ is the ***a.c current gain.***

As $\alpha_{ac} \approx 1$ and $r_e \ll R_L$, so $A_V \gg 1$ i.e., the amplification is due to the transfer of an almost equal amount of current flow from low resistance input circuit to high resistance output circuit.

The ***power gain*** is: $$A_P = \frac{P_{\text{out}}}{P_{\text{in}}} = \frac{(\Delta I_C)^2 R_L}{(\Delta I_E)^2 r_e} = \alpha_{ac}^2 \cdot \frac{R_L}{r_e} \qquad (5.9.4)$$

A_P may also be written as: $$A_P = \frac{\Delta I_C \cdot (\Delta I_C \cdot R_L)}{\Delta I_E \cdot (\Delta I_E \cdot r_e)}$$

$$\therefore \quad \boxed{A_P = A_I A_V} \qquad (5.9.5)$$

where A_I is the *current gain.*

So, by making $I_C \approx I_E$ i.e., by reducing the width and doping of the base region, one can produce a useful *power amplifying device* by a transistor.

5.9.2 Transistor as common-emitter amplifier

The circuit diagram for using a *npn* transistor as a common-emitter amplifier is shown below in Fig. 5.22.

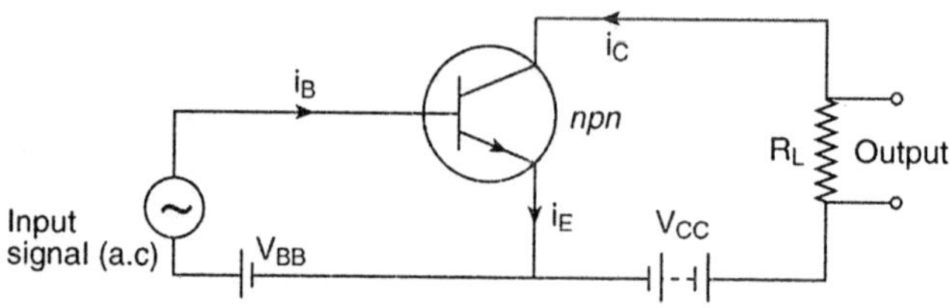

Fig. 5.22 *npn* transistor as CE amplifier

When no input signal (a.c signal) is applied to the input circuit, the p.d. across R_L is $I_C R_L$. When the a.c signal is fed to the input circuit, the emitter and the collector currents increase during the positive half of the input signal due to the increase in forward bias. During the negative half cycle of the input signal, the forward bias across the emitter base junction is decreased and collector current decreases. Hence the output signal becomes more positive and an amplified output is obtained across the load.

Let ΔV_i be a small change in input voltage which causes a change ΔI_B of the base current and consequently a change ΔI_C in collector current.

$\therefore$ The change in output voltage across R_L is given by

$$\Delta V_0 = \Delta I_C \cdot R_L \qquad (5.9.6)$$

If r_e be the dynamic resistance of the emitter-base junction, then the change in input voltage can be written as

$$\Delta V_i = \Delta I_B \cdot r_e \qquad (5.9.7)$$

Dividing (5.9.7) by (5.9.6), we get the voltage gain

$$A_V = \frac{\Delta V_0}{\Delta V_i} = \frac{\Delta I_C}{\Delta I_B} \cdot \frac{R_L}{r_e}$$

$$= \beta_{ac} \frac{R_L}{r_e} \qquad (5.9.8)$$

where $\beta_{ac} = \Delta I_C / \Delta I_B$ is the ***a.c current gain.***

As $\beta_{ac} \gg 1$, $R_L \gg r_e$, so $A_V \gg 1$. The amplification is thus due to the transference of signal from low resistance input circuit to the high resistance output circuit.

For a *pnp* transistor, the polarities of the batteries and the meters are to be simply reversed but the amplifying action can be explained in the same manner as in the case of an *npn* transistor.

5.10 Transistor as a switch

To analyse the behaviour of a transistor in base-biased *CE* configuration, we consider the circuit diagram as shown in Fig. 5.23.

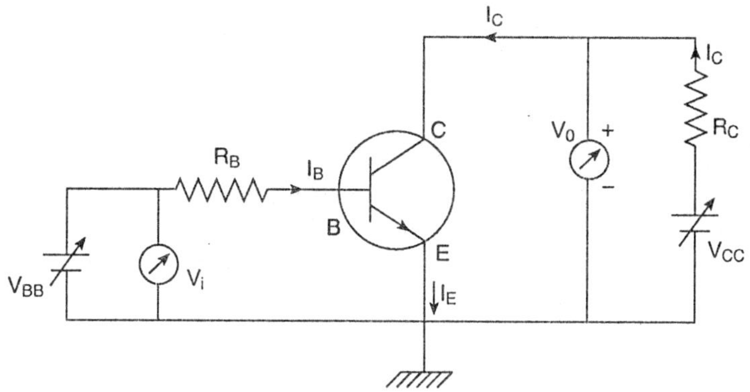

Fig. 5.23 Base-biased *npn* transistor

Applying KVL to the input and the output sides of the circuit, we obtain

$$V_{BB} = V_i = I_B R_B + V_{BE} \tag{5.10.1}$$

$$\text{and } V_{CE} = V_0 = V_{CC} - I_C R_C \tag{5.10.2}$$

where V_i and V_0 are the dc input and dc output voltage respectively.

For a Si-transistor, the cut-in voltage is 0.6 V. So, when the input voltage V_i is increased from zero onwards, the transistor remains in cut-off state as long as the input is less than 0.6 V. In this condition, $I_C = 0$ and equation (5.10.2) reduces to $V_0 = V_{CC}$. When the input voltage V_i exceeds 0.6 V, the transistor goes into the active state and some current I_C flows in the output circuit. With increasing V_i, the output V_0 decreases, following equation (5.10.2), almost *linearly* till its value becomes less than 1.0 V. Beyond this, the transistor goes to saturation region and V_0 changes *non-linearly* towards zero. The variation of V_0 with V_i is shown in Fig. 5.24.

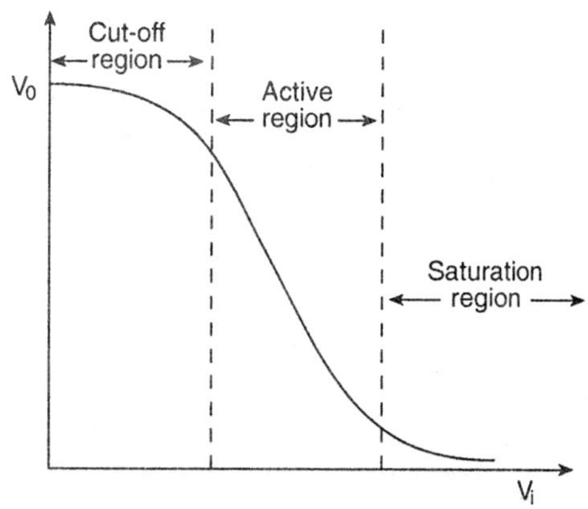

Fig. 5.24 Transfer characteristic of a base-biased transistor

The curve in the figure represents the transfer characteristic of the base-biased transistor. The non-linearity indicates that the transitions from cut-off to active region and from active to saturation region are *not* sharply defined.

From the above discussions, it is seen that a low input voltage makes the transistor non-conducting i.e., *switched off* and a high input drives it to become conducting i.e., *switched on.* Thus *a base-biased transistor in CE mode can be used as a switch.*

5.11 Illustrated Examples

▶ **Example 1.** *When the emitter current of a transistor changes by* 1 mA, *its collector current changes by* 0.995 mA. *Calculate* α *and* β. **(Osmania Univ.)**

Solution: Given, the change in collector current, $\Delta I_C = 0.995\,\text{mA}$,

and the change in emitter current $\Delta I_E = 1\,\text{mA}$.

So, the change in base current is $\Delta I_B = \Delta I_E - \Delta I_C = 1 - 0.995\,\text{mA} = 0.005\,\text{mA}$

$$\therefore \text{ By definition,} \qquad \alpha = \frac{\Delta I_C}{\Delta I_E} = \frac{0.995}{1} = \mathbf{0.995}$$

$$\text{and} \qquad \beta = \frac{\Delta I_C}{\Delta I_B} = \frac{0.995}{0.005} = \mathbf{199}$$

▶ **Example 2.** *Calculate the* α*-parameter of a transistor if* $\beta = 99$. **(Guru Nanak Univ.)**

Solution: The relation between the parameters α and β of a transistor is

$$\beta = \frac{\alpha}{1-\alpha}$$

$$\Rightarrow \qquad 99 = \frac{\alpha}{1-\alpha}$$

$$\text{or,} \quad 100\alpha = 99$$

$$\text{or,} \qquad \alpha = \mathbf{0.99}$$

▶ **Example 3.** *In a common base circuit,* $\alpha = 0.96$. *If the base current is* $90\,\mu\text{A}$, *what are the collector current and the emitter current?* **(Punjab Univ.)**

Solution: Given, the α-parameter of the transistor $= 0.96$ and the base current, $I_B = 90\mu\text{A}$. To find the collector current and the emitter current.

$$\text{Now,} \quad \beta = \frac{\alpha}{1-\alpha} = \frac{0.96}{1-0.96} = 24$$

$$\Rightarrow \quad \frac{I_C}{I_B} = 24 \quad (\because \beta = I_C/I_B)$$

$$\text{or,} \quad I_C = 24 I_B = 24 \times 90\,\mu\text{A} = \mathbf{2.16\,mA}$$

$$\text{Since} \quad \alpha = \frac{I_C}{I_E} \quad \Rightarrow \quad I_E = \frac{I_C}{\alpha} = \frac{2.16}{0.96}\,\text{mA} = \mathbf{2.25\,mA}$$

► **Example 4.** *β of a transistor is 99. Calculate the collector current when the emitter current is* 5 mA. **(Meerut Univ.)**

Solution: Here, by the problem β of the transistor is 99 and the emitter current $I_E = 5\,\text{mA}$. To find the collector current.

$$\text{Now,}\quad \beta = \frac{\alpha}{1-\alpha}$$

$$\therefore\quad 99 = \frac{\alpha}{1-\alpha} \Rightarrow 100\alpha = 99 \text{ or, } \alpha = 0.99$$

$$\text{By definition,}\quad \text{we have}\quad \alpha = I_C/I_E$$

$$\text{or, } 0.99 = I_C/5 \text{ or, } \boldsymbol{I_C} = 5 \times 0.99 = \mathbf{4.95}$$

$$\therefore \text{ The collector current is } \mathbf{4.95\,mA}$$

► **Example 5.** *Calculate the emitter current I_E for a transistor for which $\beta = 50$ and the base current $I_B = 30\,\text{mA}$.* **(Nagpur Univ.)**

Solution: Here, by the problem, $\beta = 50$ and the base current $I_B = 30\,\text{mA}$. To find the emitter current.

$$\text{We have, the current gain,}\qquad \beta = \frac{I_C}{I_B}$$

$$\text{or, } I_C = \beta I_B = 50 \times 30\,\text{mA} = 1500\,\text{mA}$$

$$\therefore \text{ The emitter current } \boldsymbol{I_E} = I_B + I_C = (30 + 1500)\,\text{mA} = 1530\,\text{mA} = \mathbf{1.53\,A}$$

► **Example 6.** *In a common base connection $\alpha = 0.95$. The voltage drop across $1\,\text{k}\Omega$ resistance connected in the collector circuit is $2\,\text{V}$. Find the base current.*

Solution: Here, by the problem, $\alpha = 0.95$ and voltage drop across $1\,\text{k}\Omega$ in the collector circuit is 2 V. To find the base current, I_B.

$$\therefore \text{ Collector current,}\quad I_C = \frac{2}{1000}\,\text{A} = 2\,\text{mA}$$

$$\text{and}\quad \beta = \frac{\alpha}{1-\alpha} = \frac{0.95}{1-0.95} = 19$$

$$\text{Now,}\quad \beta = \frac{I_C}{I_B} \quad \text{or, } I_B = \frac{I_C}{\beta} = \frac{2}{19}\,\text{mA} = 0.105\,\text{mA}$$

$$\therefore \text{ The base current, } \boldsymbol{I_B} = \mathbf{0.105\,mA}$$

► **Example 7.** *In the CB configuration, the current amplification factor is 0.95. If the emitter current is 2 mA, determine the value of the base current.* **(Bangalore Univ.)**

Solution: Here, current amplification factor $\alpha = 0.95$ and emitter current $I_E = 2\,\text{mA}$. To find the base current I_B.

$$\text{Now,} \quad \alpha = \frac{I_C}{I_E}$$

$$\text{or,} \; I_C = \alpha I_E = 0.95 \times 2\,\text{mA} = 1.9\,\text{mA}$$

$$\therefore \text{ The base current, } \boldsymbol{I_B} = I_E - I_C = (2 - 1.9)\,\text{mA}$$
$$= 0.1\,\text{mA} = \mathbf{100\mu A}$$

► **Example 8.** *The current gain* α *of a n-p-n transistor is* 0.98. *It is connected in common base mode and gives a reverse saturation current* $I_{CO} = 12\,\mu\text{A}$. *Find the base and collector current for an emitter current of* 2 mA.

Solution: Given, current gain $\alpha = 0.98$. The reverse saturation current $I_{CO} = 12\mu\text{A} = 12 \times 10^{-6}\,\text{A}$ and the emitter current $I_E = 2\,\text{mA} = 2 \times 10^{-3}\,\text{A}$.

The collector current of a *n-p-n* transistor in *CB* mode is given by

$$I_C = \alpha I_E + I_{CO}$$

$$\therefore \; \boldsymbol{I_C} = 0.98 \times 2 \times 10^{-3} + 12 \times 10^{-6}\,\text{A}$$
$$= (1.96 + 0.012) \times 10^{-3}\,\text{A}$$
$$= \mathbf{1.972\,mA}$$

$$\therefore \text{ Base current, } \boldsymbol{I_B} = I_E - I_C = (2 - 1.972)\,\text{mA} = 0.028\,\text{mA} = \mathbf{28\mu A}$$

► **Example 9.** *Calculate the common base current gain* α *and the base current for a npn-transistor in CB-configuration whose emitter current is* 4 mA, *collector current is* 3.95 mA *and reverse saturation current is* 10 μA.

Solution: In *CB*-configuration, the collector current I_C in terms of emitter current I_E and reverse saturation current I_{CBO} is:

$$I_C = \alpha I_E + I_{\text{CBO}} \tag{5.11.1}$$

By the problem, $I_E = 4\,\text{mA}$, $I_C = 3.95\,\text{mA}$, $I_{\text{CBO}} = 10\,\mu\text{A} = 0.01\,\text{mA}$.

$$\therefore \text{ From (5.10.1)} \qquad \alpha = \frac{I_C - I_{\text{CBO}}}{I_E} = \frac{3.95 - 0.01}{4} = \mathbf{0.985}$$

The base current I_B is related to emitter current I_E and collector current I_C as

$$I_E = I_B + I_C \Rightarrow I_B = I_E - I_C = 4 - 3.95 = \mathbf{0.05\ mA}$$

► **Example 10.** *In a CE-configuration, the collector supply voltage is* 15 V. *When a resistor* $R_C = 3\,\text{k}\Omega$ *is connected in the collector circuit, the voltage drop across it is* 0.5 V. *For* $\alpha = 0.99$, *determine* (i) *collector-emitter voltage and* (ii) *the base current* I_B.

Solution: (i) Collector-emitter voltage, $V_{CE} = V_{CC} -$ drop in R_C.

By the problem, $V_{CC} = 15\,\text{V}$ and drop in $R_C = 0.5\,\text{V}$.

$$\therefore \quad V_{CE} = 15\,\text{V} - 0.5\,\text{V} = \mathbf{14.5\,V}$$

(ii) Collector current, $I_C = \dfrac{\text{drop across } R_C}{\text{resistance of } R_C} = \dfrac{0.5\,\text{V}}{3\,\text{k}\Omega} = 0.16\,\text{mA}$

Also, $$\beta = \frac{\alpha}{1-\alpha} = \frac{0.99}{1-0.99} = 99$$

$\therefore$ Base current, $$I_B = \frac{I_C}{\beta} = \frac{0.16}{99} = \mathbf{0.0016\ mA}$$

▶ **Example 11.** *A transistor has currrent gain of* 0.99 *in CB-configuration. Calculate the current gain in CE-configuration. Another transistor has current gain of* 50 *in CE-configuration. Calculate the current gain in CB-configuration.*

Solution: For the first transistor, the current gain (CB) $\alpha = 0.99$.

$$\text{So,}\ \beta = \frac{\alpha}{1-\alpha} = \frac{0.99}{1-0.99} = \mathbf{99}$$

For the second transistor, the current gain (CE) $\beta = 50$.

$$\text{So,}\ \alpha = \frac{\beta}{1+\beta} = \frac{50}{1+50} = \frac{50}{51} = \mathbf{0.988}$$

▶ **Example 12.** *A* Ge *transistor with* $\alpha = 0.98$ *gives reverse saturation current* $I_{CO} = 12\mu\text{A}$ *when used in common base circuit. When the transistor is connected in common emitter mode with base current of* $0.2\,\text{mA}$, *calculate the collector current.* **(Punjab Univ.)**

Solution: Given, the common base current gain $\alpha = 0.98$, $I_{CO} = 12\mu\text{A} = 12 \times 10^{-6}\text{A}$.

$\therefore$ The common emitter current gain $\beta = \dfrac{\alpha}{1-\alpha} = \dfrac{0.98}{1-0.98} = 49$

The collector current in CE mode is given by

$$I_C = \beta I_B + (1+\beta)I_{CO}$$

Substituting $I_{CO} = 12 \times 10^{-6}\,\text{A}$, $I_B = 0.2\,\text{mA} = 0.2 \times 10^{-3}\,\text{A}$ and $\beta = 49$ in the above equation, we at once get

$$\begin{aligned} I_C &= 49 \times 0.2 \times 10^{-3} + (1+49) \times 12 \times 10^{-6}\,\text{A} \\ &= (9.8 + 0.6) \times 10^{-3}\,\text{A} = 10.4\,\text{mA} \end{aligned}$$

$\therefore$ The collector current $=$ **10.4 mA**

▶ **Example 13.** *A* BJT *with* α *of* 0.98 *is operated as an amplifier in a* CB *mode with* R_L *load of* $5\,\text{k}\Omega$ *and dynamic resistance of emitter-base junction* 70Ω. *Find the voltage gain and power gain of the circuit.* **(Guru Nanak Univ.)**

Solution: Here the dynamic resistance of emitter-base junction $r_e = 70\Omega$.

Load resistance, $R_L = 5\,\text{k}\Omega = 5 \times 10^3\Omega = 5000\Omega$; α = current gain = 0.98.

$$\therefore \text{ Voltage gain, } A_V = \alpha\frac{R_L}{r_e} = 0.98 \times \frac{5000}{70} = \mathbf{70}$$

$$\text{Power gain, } \boldsymbol{A_P} = A_I \times A_V = \text{current gain} \times \text{voltage gain}$$
$$= 0.98 \times 70 = \mathbf{68.6}$$

► **Example 14.** *In a common emitter amplifier, when the signal changes by* 0.01 volt *the base current changes by* 10μA *and the collector current by* 2 mA. *If the load resistance in the collector circuit is* $10\,\text{k}\Omega$, *calculate the input resistance, current gain, voltage gain and the power gain.* **(Lucknow Univ.)**

Solution: The change in input voltage $\Delta V_i = \Delta V_{BE} = 0.01\,\text{V}$

Change in base current $\Delta I_B = 10\mu\text{A} = 10^{-5}\,\text{A}$

Now, $\Delta V_i = \Delta I_B r_e$ where r_e is the input dynamic resistance.

$$\therefore \quad \boldsymbol{r_e} = \frac{\Delta V_i}{\Delta I_B} = \frac{0.01\,\text{V}}{10^{-5}\,\text{A}} = 10^3\Omega = \mathbf{1000\Omega}$$

Change in collector current, $\Delta I_C = 2\,\text{mA} = 2 \times 10^{-3}\,\text{A}$

$$\therefore \quad \text{Current gain, } \boldsymbol{\beta} = \frac{\Delta I_C}{\Delta I_B} = \frac{2 \times 10^{-3}\,\text{A}}{10^{-5}\,\text{A}} = \mathbf{200}$$

$$\begin{aligned} \text{Voltage gain,} \quad \boldsymbol{A_V} &= \beta\frac{R_L}{r_e} \\ &= 200 \times \frac{10 \times 10^3}{1000}, \quad (\because R_L = 10\,\text{k}\Omega) \\ &= \mathbf{2000} \end{aligned}$$

$$\begin{aligned} \text{Power gain,} \quad \boldsymbol{A_P} &= \text{current gain} \times \text{voltage gain} \\ &= 200 \times 2000 = \mathbf{4 \times 10^5} \end{aligned}$$

► **Example 15.** *In a silicon transistor, the base current is changed by* 20μA. *This results in a change of* 0.02 V *in base to emitter voltage and a change of* 2 mA *in collector current. Find the input resistance, a.c current gain and transconductance.*

Solution: Here, change in input voltage $\Delta V_i = \Delta V_{BE} = 0.02\,\text{V}$,

change in base current $\Delta I_B = 20\mu\text{A} = 20 \times 10^{-6}\,\text{A}$,

change in collector current $\Delta I_C = 2\,\text{mA} = 2 \times 10^{-3}\,\text{A}$.

$$\therefore \quad \text{Input resistance, } \boldsymbol{r_e} = \frac{\Delta V_i}{\Delta I_B} = \frac{0.02}{20 \times 10^{-6}}\Omega = \mathbf{1000\Omega}$$

$$\text{a.c current gain, } \boldsymbol{\beta_{ac}} = \frac{\Delta I_C}{\Delta I_B} = \frac{2 \times 10^{-3}}{20 \times 10^{-6}} = \mathbf{100\Omega},$$

$$\text{transconductance, } \boldsymbol{g_m} = \frac{\Delta I_C}{\Delta V_{BE}} = \frac{2 \times 10^{-3}}{0.02} = \mathbf{0.1℧}$$

▶ **Example 16.** *A n-p-n transistor is connected in* CE *configuration in which collector supply is* 8 V *and voltage drop across resistance* R_L *connected in the collector circuit is* 0.5 V. *The value of* $R_L = 800\Omega$. *If* $\alpha = 0.96$, *determine* (i) *the collector-emitter voltage, and* (ii) *the base current.* **(Calcutta Hons.)**

Solution: The circuit diagram of the *n-p-n* transistor connected in CE configuration is illustrated in Fig. 5.25.

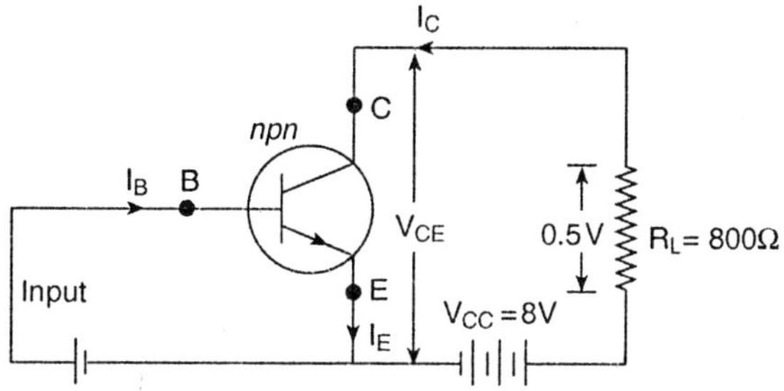

Fig. 5.25

Collector supply voltage, $V_{CC} = 8\,\text{V}$ and voltage drop across $R_L = 0.5 = V_L$ (say)

$\therefore$ The collector-emitter voltage, $\boldsymbol{V_{CE}} = V_{CC} - V_L = (8 - 0.5)\,\text{V} = \mathbf{7.5\,V}$

$$\text{The collector current, } I_C = \frac{V_L}{R_L} = \frac{0.5}{800}\,\text{A} = \frac{5}{8}\,\text{mA} \quad (\because R_L = 800\Omega).$$

$$\text{Now,} \quad \beta = \frac{\alpha}{1-\alpha} = \frac{0.96}{1-0.96} = 24 \quad (\because \alpha = 0.96)$$

$$\Rightarrow \quad \frac{I_C}{I_B} = 24 \Rightarrow \boldsymbol{I_B} = \frac{I_C}{24} = \frac{5}{8 \times 24}\,\text{mA} = \mathbf{0.026\,mA}$$

▶ **Example 17.** *For the circuit shown in* Fig. 5.26, *find the currents* I_B *and* I_C. *Given* $\beta = 100$, $V_{BE} = 0.7\,\text{V}$ *and* $I_{CEO} = 200\mu\text{A}$. **(Bangalore Univ.)**

Solution: Since the base terminal is positive w.r.t. the emitter terminal, we may assume that the transistor is operating in the active region.

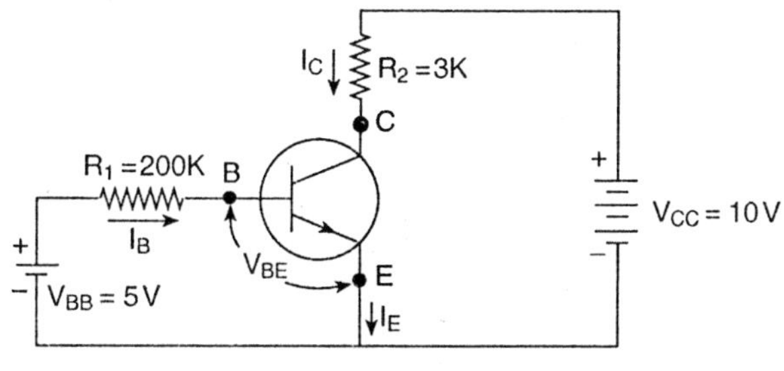

Fig. 5.26

Application of the KVL to the base-emitter circuit results in

$$I_B R_1 + V_{BE} = V_{BB}$$

$$\text{or,} \quad I_B = \frac{V_{BB} - V_{BE}}{R_1} = \frac{5 - 0.7}{200 \times 10^3}$$

$$= 21.5 \times 10^{-6}\ \text{A}$$

$$= \mathbf{21.5\ \mu A}$$

The collector current is plainly given by

$$I_C = \beta I_B + I_{CEO} = (100 \times 21.5)\mu\text{A} + 200\mu\text{A}$$

$$= 2350\mu\text{A} = \mathbf{2.35\ mA}$$

▶ **Example 18.** *Suppose the transistor of* Fig. 5.27 *has* $\beta = 100$. *Find* (i) *whether the transistor is operating in the active, saturation or cut-off region,* (ii) *the value of output voltage* V_0. *Assume* $|V_{BE}| = 0.7\,\text{V}$ *and* $|V_{CE,\text{sat}}| = 0.2\,\text{V}$. **(Calcutta Hons.)**

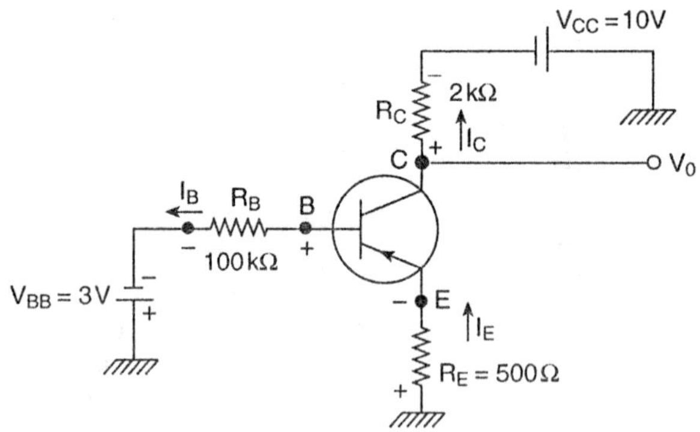

Fig. 5.27

Solution: Let us assume that the transistor is operating in the active region.

Now the current gain, $$\beta = \frac{I_C}{I_B} \Rightarrow 1 + \beta = \frac{I_B + I_C}{I_B}$$

$$\therefore \quad I_E = I_B + I_C = (1 + \beta) I_B$$

Applying the KVL to the input circuit, we obtain

$$I_E R_E - V_{BE} + I_B R_B = V_{BB}$$

$$\therefore \quad I_B = \frac{V_{BB} + V_{BE}}{(1 + \beta) R_E + R_B} = \frac{3 - 0.7}{(1 + 100) \times 500 + 100 \times 10^3}\ \text{A}$$

$$= 15.3 \times 10^{-6}\ \text{A} = 15.3\ \mu\text{A}$$

Applying the KVL to the collector circuit, we get

$$I_E R_E - V_{CE} + I_C R_C = V_{CC}$$

$$\text{or,}\quad V_{CE} = -V_{CC} + I_E R_E + I_C R_C$$
$$= -V_{CC} + I_C(R_E + R_C), \quad [\because I_C \approx I_E]$$
$$= -10V + 1.53 \times 10^{-3}(500 + 2000)\text{ V},$$
$$= -6.175\text{ V}$$
$$[\because I_C = \beta I_B = 100 \times 15.3 \times 10^{-6}\text{ A} = 1.53 \times 10^{-3}\text{ A}]$$

The negative sign in V_{CE} indicates that the collector junction of the given *p-n-p* transistor is reverse biased.

Again $|V_{CE}| > |V_{CE,\text{sat}}|$. So the ***transistor is operating in the active region.***

From the output circuit, we get

$$-V_0 + I_C R_C - V_{CC} = 0$$
$$\therefore\quad \mathbf{V_0} = -V_{CC} + I_C R_C$$
$$= (-10 + 1.53 \times 10^{-3} \times 2000)\text{ V}$$
$$= \mathbf{-6.94\ V}$$

▶ **Example 19.** *A transistor is operating in the* CE *mode* (Fig. 5.28). *Calculate* V_{CE} *if* $\beta = 125$, *assuming that* $V_{BE} = 0.6$ V. **(Calcutta Hons.)**

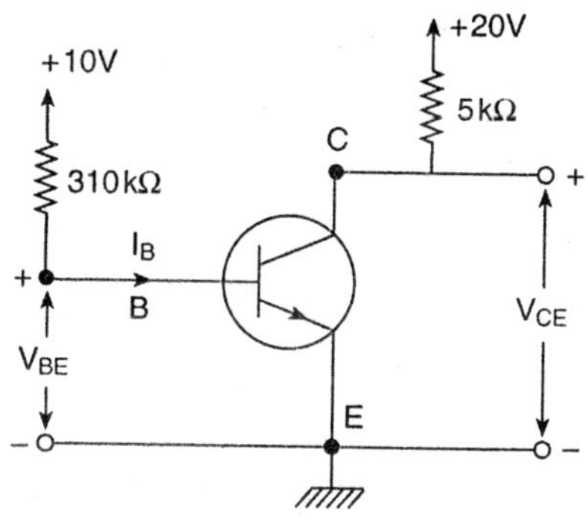

Fig. 5.28

Solution: Applying the KVL to the base-emitter circuit, we get

$$I_B \times 310 \times 10^3 + V_{BE} = 10$$
$$\text{or,}\quad I_B = \frac{10 - V_{BE}}{310 \times 10^3} = \frac{10 - 0.6}{310 \times 10^3}\text{ A} = 30.3\mu\text{A}$$

Again, $\beta = I_C/I_B \Rightarrow I_C = \beta I_B = 125 \times 30.3\mu\text{A} = 3.79\text{ mA}$

$\therefore$ From the output circuit, $\quad V_{CE} + I_C \times 5 \times 10^3 = 20$

$$\text{or,}\quad \mathbf{V_{CE}} = 20 - I_C \times 5 \times 10^3$$
$$= 20 - 3.79 \times 10^{-3} \times 5 \times 10^3$$
$$= \mathbf{1.05\ V}$$

▶ **Example 20.** *Refer to the circuit of* Fig. 5.29. *At saturation,* V_{BE} *and* V_{CE} *are* $V_{BE,\text{sat}} = 0.85\,\text{V}$ *and* $V_{CE,\text{sat}} = 0.22\,\text{V}$. *If* $\beta = 110$, *is the transistor operating in the saturation region?* **(Calcutta Hons.)**

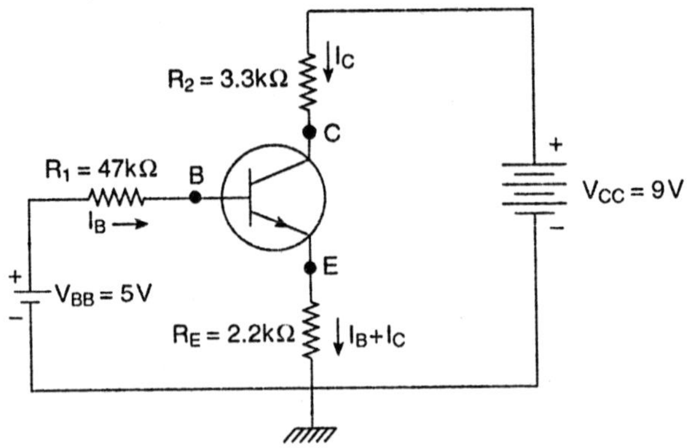

Fig. 5.29

Solution: Let the transistor operate in the saturation region.

Applying KVL to the base circuit and the collector circuit respectively, we get

$$47 \times 10^3 I_B + V_{BE} + 2.2 \times 10^{-3}(I_B + I_C) = 5$$

$$\text{or,}\quad 49.2 \times 10^3 I_B + 2.2 \times 10^{-3} I_C = 4.15 \tag{5.11.2}$$

$$\text{and}\quad 3.3 \times 10^3 I_C + V_{CE} + 2.2 \times 10^3(I_B + I_C) = 9$$

$$\text{or,}\quad 2.2 \times 10^3 I_B + 5.5 \times 10^3 I_C = 8.78 \tag{5.11.3}$$

Solving (5.10.3) and (5.10.4), we get

$$I_B = 13.2 \times 10^{-6}\,\text{A} \quad \text{and} \quad I_C = 1.591 \times 10^{-3}\,\text{A}$$

The minimum base current for saturation is thus

$$I_{B,\text{min}} = \frac{I_C}{\beta} = \frac{1.591}{110}\,\text{mA} = 0.0145\,\text{mA}$$

Since $I_B < I_{B,\text{min}}$, the transistor is ***not operating in the saturation region.***

Review Questions

1. (a) What is a transistor? What is the significance of arrow in a transistor symbol? Why a transistor is so called? Does it transfer anything?
   (b) Explain why the base region of a transistor is made thin as compared to the emitter and collector.
   (c) Draw the circuit diagram of *p-n-p* and *n-p-n* transistors with proper biasing voltages and directions of current. Distinguish between the working of a *p-n-p* and an *n-p-n* type BJT. **(Meerut Univ.)**

2. What are the three transistor configurations? Draw the circuit diagram in each case for *p-n-p* and *n-p-n* transistors. **(Punjab Univ.)**
3. Sketch and explain the input and output characteristic of a common base transistor circuit. How will you study them experimentally? **(Calcutta Hons.)**
4. Sketch the circuit diagram and the input and output characteristics of a common emitter transistor circuit. Explain the conditions to use the transistor in the active, cut-off and saturation regions of output characteristics. **(Calcutta H.)**
5. (a) How does a transistor function as an amplifier?

   (b) Define α and β for a transistor. Deduce a relation between them.

   (Calcuttta Hons.)
6. Derive the accurate expressions for collector currents in CB and CE mode of a bipolar junction transistor and show that

   (a) $I_C = \beta I_B + I_{CEO}$

   (b) $I_C = \dfrac{\alpha}{1-\alpha} I_B + \dfrac{1}{1-\alpha} I_{CBO}$ **(Calcutta Hons.)**
7. Define voltage gain and power gain of a transistor amplifier. Calculate their values for a common base amplifier in terms of current gain. **(Gharwal Univ.)**
8. Obtain expressions for current gain, voltage gain, power gain, input and output resistance for the CE configuration of a transistor. **(Guru Nanak Univ.)**
9. Show the potential variation in a *pnp*-transistor when (a) it is unbiased and (b) when the emitter junction is forward biased and the collector junction is reverse biased.
10. With a block diagram for showing the current components, explain the working of a *pnp*-transistor. Draw the relevant transistor equivalent circuit.
11. Explain the current components in an *npn*-transsistor when its emitter junction is forward biased and the collector junction is reverse biased. Draw the relevant block diagram.
12. In which direction would the current flow in a *pnp* transistor if the base of the transistor is left unconnected?
13. "A base-biased transistor in CE mode can be used as a switch."—Establish the statement with appropriate circuit diagram.

Objective Questions

1. A transistor is also called a

   (a) vacuum triode, (b) vacuum diode,
   (c) semiconductor triode, (d) semiconductor diode. **(Meerut Univ.)**

2. In transistor circuit, the emitter current is 1.0 mA and the collector current is 0.9 mA. Then the base current will be

(a) 0.1 mA, (b) 0.9 mA, (c) 1.0 mA, (d) 1.9 mA.

(Meerut Univ.)

3. In a common emitter transistor amplifier, the phase difference between input and output signal is

(a) 0°, (b) 180°, (c) 270°, (d) 360°. **(Meerut Univ.)**

4. Which of the following characteristics *does not* belong to a common collector transistor amplifier?

(a) Low voltage gain (≈ 1), (b) High current gain,
(c) High input impedance, (d) High output impedance. **(GATE)**

5. A transistor is said to be in saturation region when

(a) both the collector and the emitter junctions are forward biased,
(b) both the collector and the emitter junctions are reverse biased,
(c) the collector junction is forward biased but the emitter junction is reverse biased,
(d) the collector junction is reverse biased but the emitter junction is forward biased. **(GATE)**

6. In the circuit on the right (Fig. 5.30), $R_B = 1\,\text{k}\Omega$ and $R_C = 100\Omega$. If β of the transistor is 100, the current through R_C will be

(a) ≈ 0.43 A,
(b) ≈ 50 mA,
(c) zero,
(d) oscillating between 0 and 50 mA.

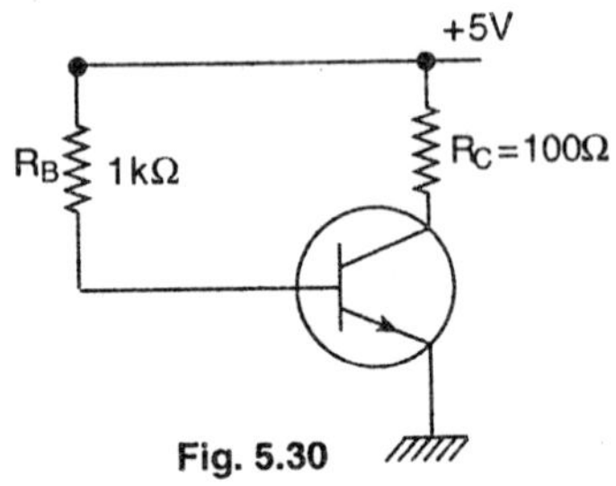

Fig. 5.30

7. In an *n-p-n* transistor, the leakage current consists of electrons moving from

(a) the base to the emitter, (b) the collector to the emitter,
(c) the collector to the base, (d) the base to the collector. **(GATE)**

8. A bipolar junction transistor with one junction forward biased and either the collector or the emitter open, operates in the

(a) cut-off region, (b) saturation region,
(c) pinch-off region, (d) active region. **(GATE)**

9. A bipolar junction transistor used in common collector configuration has

(a) high input impedance and high output impedance,
(b) high input impedance and low output impedance,
(c) low input impedance and high output impedance,
(d) low input impedance and low output impedance. **(GATE)**

10. For normal operation of a transistor

(a) emitter junction is forward biased and collector junction is revese biased,
(b) emitter junction is reverse biased and collector junction is forward biased,
(c) both the emitter and the collector junctions are forward biased,
(d) both the emitter and the collector junctions are reverse biased.

11. The following relationships between α and β are correct, except

(a) $\beta = \frac{\alpha}{1-\alpha}$, (b) $\alpha = \frac{\beta}{1-\beta}$, (c) $\alpha = \frac{\beta}{1+\beta}$, (d) $1 - \alpha = \frac{1}{1+\beta}$.

12. The value of total collector current in a CB-circuit is

(a) $I_C = \alpha I_E$, (b) $I_C = \alpha I_E + I_{\text{CBO}}$, (c) $I_C = \beta I_E$, (d) $I_C = \alpha I_E - I_{\text{CBO}}$.

13. Early effect in BJT refers to

(a) Avalanche breakdown, (b) Thermal breakdown,
(c) Base narrowing, (d) Zener breakdown.

14. If the common base DC current gain of a BJT is 0.98, its common emitter DC current gain is

(a) 51, (b) 49, (c) 1, (d) 0.02. **(DRDO)**

15. In a BJT CASCADE pair, a

(a) common emitter follows a common base,
(b) common base follows a common collector,
(c) common collector follows a common base,
(d) common base follows a common emitter. **(DRDO)**

16. The early effect in a BJT is modelled by the small signal parameter

(a) r_0, (b) r_π, (c) g_m, (d) β. **(DRDO)**

17. When a BJT is in saturation

(a) $V_{CE} = V_{CC}$, (b) $V_{CE} \approx 0\,\text{V}$, (c) $V_{CE} = \frac{1}{2}V_{CC}$, (d) $V_{CC} = \frac{1}{2}V_{CE}$.

18. The band structure of a BJT is illustrated below in Fig. 5.31.

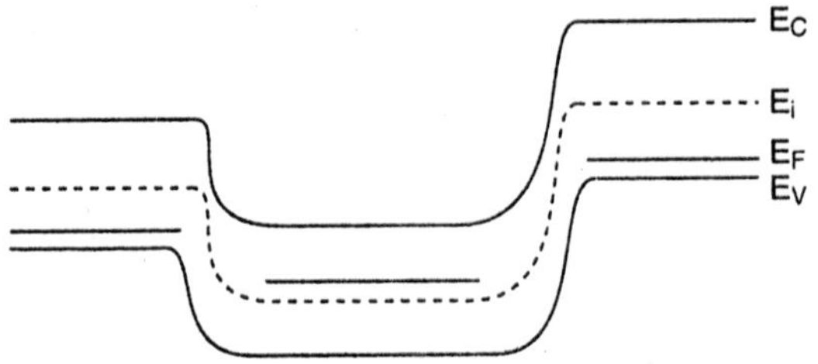

Fig. 5.31

It represents

(a) A *pnp* BJT biased in active mode,
(b) An *npn* BJT biased in sturation mode,
(c) An *npn* BJT biased in reversc-active mode,
(d) A *pnp* BJT biased in cut-off mode. **(DRDO)**

19. When a BJT operates in cut-off region

(a) $V_{CE} \approx 0$, (b) $V_{CE} = V_{CC}$, (c) $V_{CC} = \frac{1}{2}V_{CE}$, (d) $V_{CC} = 2V_{CE}$.

20. When operated between cut-off and saturation a transistor behaves like a

(a) switch, (b) linear amplifier,
(c) variable resistor, (d) variable capacitor.

Problems

1. In a common emitter *n-p-n* circuit, when the collector current is 3.6 mA, the base current is 30μA. The collector current at a fixed collector emitter voltage of 10 volt increases to 4.7 mA when the base current increases to 40μA. Calculate common emitter d.c and a.c current gain and common base a.c current gain. **(Kurukshetra Univ.)** [120, 110, 0.99]

2. The current gain of a transistor in common emitter (CE) configuration is 49. What will be the current gain of the same transistor in common base (CB) configuration? **(Benaras Univ.)** [0.98]

3. If the current gain of a common base amplifier is 0.995, find the current gain of common emitter amplifier. **(Guru Nanak Univ.)** [199]

4. A given transistor has $\alpha = 0.95$. What would be the change in collector current corresponding to a change of 0.4 mA in the base current in a common emitter arrangement? **(Purvanchal Univ.)** [7.6 mA]

5. A transistor having $\beta = 74$ is connected in common-base configuration. Calculate the maximum collector current for an emitter current of 5 mA.
(Himachal Univ.) [4.93 mA]

6. The value of the transistor parameter β is 40. If the base current is 28μA, calculate the value of emitter current. **(Meerut Univ.)** [1.148 mA]

7. In a transistor, $\beta = 45$ and the voltage across 5 kΩ resistance which is connected in the collector circuit is 5 V. Find the base current. [0.022 mA]

8. A certain transistor has $\alpha = 0.98$ and a collector leakage current I_{CO} of 1μA. Calculate the collector current and the base current when $I_E = 1$ mA.
[0.981 mA; 19μA]

9. In a common base transistor amplifier, the input resistance is 200Ω and the load resistance is 500 kΩ. Find the voltage gain. Given $\alpha = 0.85$.
(Gauhati Univ.) [2125]

10. The load resistance in the output of a common base transistor is 1000 times the input resistance and current amplification is 0.96. Calculate the voltage gain of the transistor. [960]

11. A particular transistor has got a value of $\beta = 50$. The voltage across a 5 kΩ resistor in the collector circuit is 5 V. Calculate the base current of the transistor.
(Bangalore Univ.) [20μA]

12. A transistor has $\alpha = 0.99$. What will be the base current if the emitter current is 8 mA? [80μA]

13. For a transistor, the base current is 100μA and the collector current is 2.9 mA. Find α and β. **(Bangalore Univ.)** [0.97, 29]

14. A *p-n-p* transistor in the common base configuration offers dynamic input resitance of 50Ω. If the current gain magnitude of the amplifier is 0.96, find the voltage gain when the load resistance in the collector circuit is 5 kΩ. What is the power gain? [96, 92]

15. (a) Calculate the voltage gain of a single stage transistor amplifier circuit if the collector load $R_L = 2$ kΩ, input resistance $R_i = 1$ kΩ and current gain is 50.
(Punjab Univ.) [100]

(b) A change of 100 mV in base emitter voltage changes the base current by 50 mA. Find the input resistance of the transistor. [2Ω]

16. The value of α is 0.99 for the *p-n-p* transistor. If the input resistance $r_i = 1000\Omega$ and the load resistance $R_L = 2.5$ kΩ, calculate the value of current gain and voltage gain for the transistor in CE configuration. **(Benaras Univ.)** [99, 247.5]

17. For a transistor at quiescent operating point, $I_C = 5.202\,\text{mA}$, $I_{CO} = 2\mu\text{A}$ and $I_B = 50\mu\text{A}$. Determine α, β and I_E. [0.99, 100, 5.252 mA]

18. A transistor in CE mode is connected with a resistance $5\,\text{k}\Omega$ and a power supply of 5 V in the collector circuit. If $\alpha = 0.998$ and the voltage drop across the $5\,\text{k}\Omega$ resistor is 5 V, find the base current. **(Calcutta Hons.)** [2μA]

19. A transistor operating in the CE mode draws a constant base current I_B of 30μA. The collector current I_C is found to change from 3.5 mA to 3.7 mA when the collector-emitter voltage V_{CE} changes from 7.5 V to 12.5 V. Calculate the output resistance and β at $V_{CE} = 12.5$ V. What is the value of α?

(Calcutta Hons.) [$25\,\text{k}\Omega$, 123.3, 0.992]

20. A silicon *n-p-n* transistor having $\beta = 100$ and $I_{CO} = 22\,\text{nA}$ is operated in the CE configuration (Fig. 5.32). Assuming $V_{BE} = 0.7$ V, determine the transistor

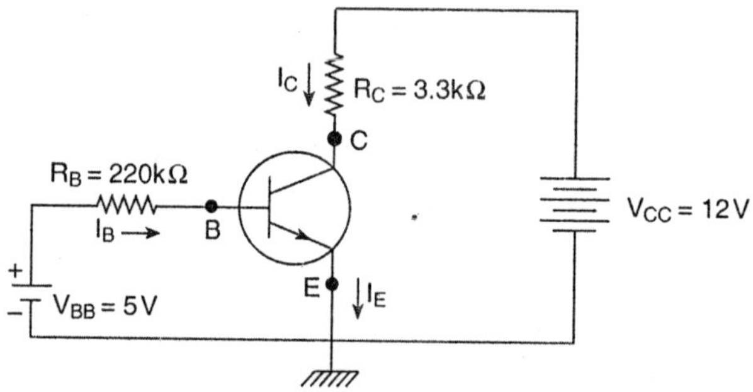

Fig. 5.32

currents and the region of operation of the transistor. What happens if the resistance R_C is indefinitely increased?

(Calcutta Hons.) [$I_B = 19.5\mu\text{A}$, $I_C = 1.95\,\text{mA}$, $I_E = -1.97\,\text{mA}$]

21. Consider an *npn*-transistor in the CE configuration with negligible I_{CO}. Given $\beta = 50$, $V_{BB} = 4$ V, $R_B = 100\,\text{k}\Omega$, $R_C = 4\,\text{k}\Omega$ and $V_{CC} = 10$ V. Draw the circuit diagram. Find I_B and I_C, assuming $V_{BE} = 0.7$ V in the active region. Find R_C for which the transistor will no longer be in active region.

(Calcutta Hons.) [33μA, 1.65 mA, $5.64\,\text{k}\Omega$]

22. A transistor in the CE configuration is connected with a power supply. The voltage drop across a $5\,\text{k}\Omega$ resistance connected in the collector circuit is 5 V. Find the base current. Given $\alpha = 0.998$. **(Calcutta Hons.)** [$2.004\,\mu$A]

23. Calculate β of a transistor for which $\alpha = 0.95$. If α changes by 0.1%, what is the percent change in β? [19, 2%]

24. A transistor is connected in CE configuration in which the collector supply is 8 V and the voltage drop across a resistance 800Ω connected in the collector circuit is 0.5 V. If $\alpha = 0.96$, find (a) the collector-emitter voltage and (b) the base current.

(Calcutta Hons.) [7.5 V, 26μA]

25. A variation of $5\,\mu A$ in the base current produces a change of $1.2\,mA$ in the collector current. The collector-to-emitter voltage remains fixed during the variation. Calculate the current amplification factor β. [$\beta = 240$]

26. The collector current, emitter current and the reverse saturation current flowing through a transistor in CB-configuration are $3\,mA$, $3.05\,mA$ and $52\,\mu A$. If the transistor is now connected in CE-configuraetion with a base current of $20\,\mu A$, calculate the collector current. [$3.53\,mA$]

27. A transistor is connected in CE-configuration. The collector supply voltage V_{CC} is $10\,V$, load resistance $R_L = 800\Omega$. Voltage drop across the load resistance is $0.8\,V$, current gain $\alpha = 0.96$. Determine V_{CE} and base current. [$9.2\,V$, $41.67\,\mu A$]

6

Transistor Biasing and Stability

6.1 Introduction

When any transistor operates, being confined to the active region of its output characteristics, it acts most *linearly* in that region. The main function of a transistor is to *faithfully amplify a weak signal without any change in its general shape.* To confirm this, the following conditions are to be fulfilled.

(i) Appropriate zero signal collector current.

(ii) Proper base-emitter voltage (V_{BE}) at any instant, and

(iii) Proper collector-emitter voltage (V_{CE}) at any instant.

But how to ensure the fulfilment? The above conditions are fulfilled only when the *base-emitter* junction is *forward biased* and the *collector-base* junction is *reverse biased* i.e., when the transistor works in the *active region.* Suitable d.c voltages and currents must therefore be provided to the transistor by external d.c sources in order to establish the proper d.c operating point in the region. The *operating point* represents a point on the characteristic curve indicating the *zero signal d.c output current* and the *d.c output voltage.* On application of the input signal, the total instantaneous current and voltage vary about the d.c values and a magnified *replica of the input signal* is obtained at the output of the amplifier. Usually, the output power (or voltage or current) is greater than the input a.c power (or voltage or current), the additional energy being supplied by the applied d.c sources.

The performance of the transistor amplifier circuit can be analysed in *two* different ways depending on the amplitude and the frequency of the input signal.

1. Equivalent circuit method of analysis—When the amplitude of the input signal is small and the frequency is low, the non-linear transistor amplifier can be replaced by a linear equivalent circuit and it can be analysed by applying the appropriate laws of electricity.

2. Graphical method of analysis—When the amplitude of the signal is large, the operation of the transistor is not confined within the active region but extends to the saturation or the cut-off region. So, the analytical method based on the equivalent linear circuit model fails and the performance of the amplifier is analysed graphically by using static characteristic curves, the load line and the Q-point which will be described and discussed in the section that follows.

6.2 Graphical analysis of CE configuration

For graphical method of analysis, consider a transistor amplifier operating in CE configuration as shown in Fig. 6.1. V_{BB} and V_{CC} are the supply voltages to the base and to the collector respectively relative to the emitter.

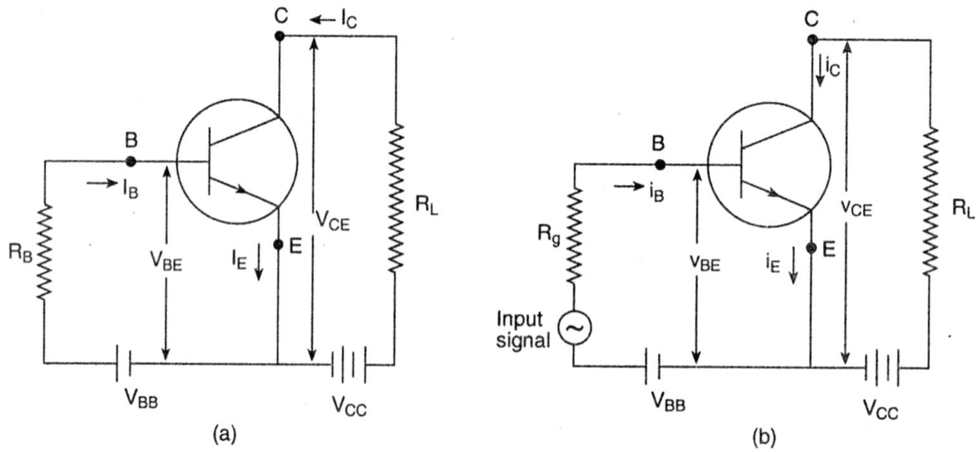

Fig. 6.1 CE amplifier circuit (a) d.c operation (b) with a.c input signal

When an a.c input signal is applied, the currents and voltages in the circuit consist of both d.c and a.c components. According to the notations used in transistor operation, the total instantaneous values of the quantities are represented by small letter with capital subscripts, only a.c values by small letters with small subscript and only d.c values by capital letter with capital subscript. So, the total instantaneous output current i.e., collector current and the total instantaneous output voltage (collector to emitter voltage) can be denoted by i_C and v_{CE}. For the graphical analysis of CE amplifier, we first consider the d.c or no signal operation.

Applying the KVL to the output circuit of Fig. 6.1(a), we get

$$V_{CC} = V_{CE} + I_C R_L$$

$$\text{or,}\quad I_C = -\frac{1}{R_L}V_{CE} + \frac{V_{CC}}{R_L} \tag{6.2.1}$$

where I_C is the collector current through the load resistance R_L and V_{CE} is the collector-emitter voltage.

Eqn. (6.2.1) represents a straight line obtained by plotting I_C against V_{CE} (Fig 6.2).

This straight line is called the ***d.c load line*** because its slope is determined by the value of the load resistance R_L used in the circuit. Thus the d.c load line is *defined as the line on the output characteristics of a transistor that gives the values of I_C and V_{CE} corresponding to zero signal or d.c condition.*

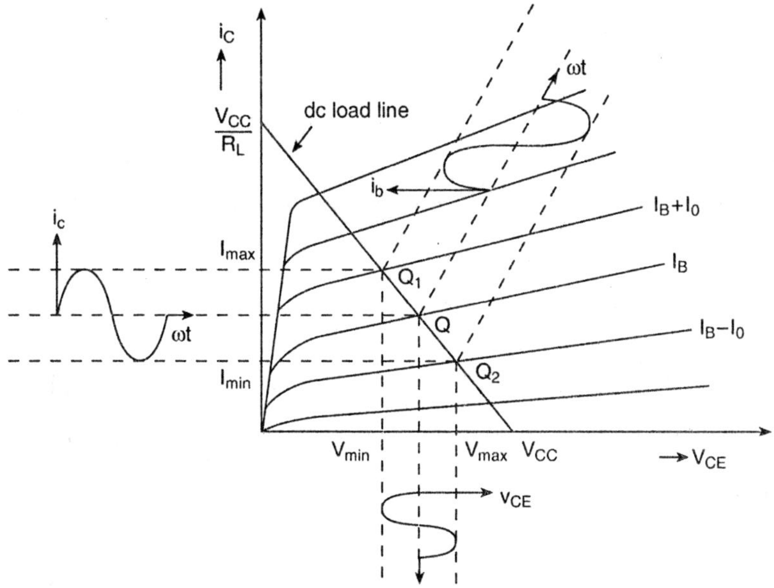

Fig. 6.2 Input-output signal variations in CE amplifier

From (6.2.1), $V_{CE} = V_{CC}$ when $I_C = 0$ and $I_C = V_{CC}/R_L$ when $V_{CE} = 0$.

$\therefore$ The d.c load line has an *intercept* V_{CC}/R_L on the I_C axis and an *intercept* V_{CC} on the V_{CE} axis i.e., the load line can be drawn by simply joining the points $(V_{CC}, 0)$ and $(0, V_{CC}/R_L)$ on the I_C-V_{CE} plane.

The point of intersection of the d.c load line with the static output characteristic curve for a base current I_B gives the ***d.c operating point*** or ***quiescent operating point*** or simply ***the Q-point.*** The base current I_B can be obtained from the input circuit of Fig. 6.1(a), by applying the KVL, as

$$I_B = \frac{V_{BB} - V_{BE}}{R_B} \tag{6.2.2}$$

When no input signal is applied, the transistor and the d.c operating point is said to be silent or quiet. So, the d.c operating point is also known as *Quiescent* (silent) *point* or *Q-point.*

To ensure the *undistorted amplification*, the Q-point is desirably located at the middle of the active region by adjusting the values of the circuit parameters. When an input signal ($I = I_0 \sin \omega t$) is applied, the base current becomes $i_B = I_B + I_0 \sin \omega t$ which varies about I_B between $I_B - I_0$ and $I_B + I_0$ with the frequency of the input signal. As a result, the operating point moves to and fro along the load line between Q_1 and Q_2 as shown in Fig. 6.2.

The corresponding variations in collector current and collector-emitter voltage are also shown in Fig. 6.2 whence it is clear that a sinusoidal variation of base current produces a sinusoidal output waveform of current i_C and voltage v_{CE} such that the input and the output currents are in phase, but the two voltages are in opposition.

6.3 Transistor biasing

A transistor is basically used as an amplifier that amplifies the weak signal given to the base to the amplified output obtained in the collector circuit without change in signal shape. This is known as *faithful amplification.* The conditions of faithful amplification are mentioned in Art. 6.1. All those conditions are satisfied when the d.c operating point or Q-point is properly established. *The establishment of the quiescent point at a suitable location of the active region of output characteristics of a transistor for faithful amplification of the signal is known as* ***transistor biasing.***

For the choice of the Q-point, the following factors on which Q depends are to be particularly noted.

(i) the amplitude of the signal,

(ii) available supply voltages,

(iii) value of the load resistance, and

(iv) allowable distortion of the output signal.

The operating point selected must importantly be consistent with the minimum distortion, the minimum variation of the transistor parameter and the minimum power dissipation.

6.4 Bias stability

On completion of the selection of the Q-point at a desired location, if the transistor is replaced by another of the same type and having the same designation, the operating point changes due to the inherent variations of transistor parameters that result in unfaithful amplification.

Again, a transistor being a semiconductor device and sensitive to the variation in temperature, the operating point is also *temperature-dependent.* The I_{CBO}, β and V_{BE} are functions of temperature and have great influence on I_C. So, the collector current changes rapidly when (i) the temperature changes and (ii) a transistor is replaced by another one of same designation. Hence the change in temperature and the replacement of transistor can alter the operating point (i.e., zero signal I_C and V_{CE}). But for a faithful amplification, the operating point must remain fixed. This requires to make the operating point independent of these variations by proper biasing. This is known as *stabilization.* So, *the process of making the operating point independent of temperature change or variations in transistor parameter is called* ***stabilization*** i.e., the biasing stability means the ability of a biasing arrangement to counteract any shifting of the Q-point.

The reverse saturation current I_{CBO} doubles for every 10°C increase in temperature, the V_{BE} decreases at the rate of 2.5 mV/°C and the value of β rises with temperature. The variation of V_{CE} however is very small when the transistor is operated in the active region. The biasing should thus be so arranged that it can reduce the effect of variations of transistor parameters. The *three* stability factors S, S_V and S_β are used to measure the variation of collector current I_C with respect to I_{CBO}, V_{BE} and β respectively. The various stability factors are defined and discussed below.

The rate of change of the collector current (I_C) w.r.t. the reverse saturation current I_{CBO}, with β and V_{CE} remaining constant, is denoted by the ***stability factor S***. Symbolically, we may write

$$\boxed{S = \frac{\partial I_C}{\partial I_{CBO}}} \tag{6.4.1}$$

For better stability, S is expected to be smaller and the lowest possible value for S is unity when $I_C = I_{CBO}$.

The *stability factor* S_V represents the rate of change of I_C with respect to the base-emitter voltage V_{BE}, the β and I_{CBO} remaining constant. So,

$$\boxed{S_V = \frac{\partial I_C}{\partial V_{BE}}} \tag{6.4.2}$$

The *stability factor* S_β is the rate of change of I_C with respect to β, when V_{BE} and I_{CBO} are kept constant. Thus, we have

$$\boxed{S_\beta = \frac{\partial I_C}{\partial \beta}} \tag{6.4.3}$$

6.5 Thermal runaway

The collector current for a transistor in CE configuration is given by

$$I_C = \beta I_B + (1+\beta)I_{CBO} \tag{6.5.1}$$

When biasing arrangement fails to maintain thermal stabilization, an increase in temperature causes an increase in I_{CBO} and hence also in the collector current I_C. This heats up the collector junction which in turn increses I_{CBO}. From (6.5.1), it is clear that if I_{CBO} increases, then I_C increases further and results in more heating effect at the collector junction and an increase in I_{CBO}. This is a *cumulative process* in nature causing the junction temperature to exceed its rated value and the damage of the transistor within a few second. This phenomenon is known as ***thermal runaway***. It is therefore mandatory to stabilize the operating point in order to prevent the thermal runaway.

6.6 Methods of transistor biasing

A good biasing arrangement of a transistor expects stabilization of the Q-point so that it can be used as an amplifier in a changing temperature environment. Moreover, it is desirable that the transistor circuit should have a single source of supply voltage for simplicity and economy. Since the CE-configuration is the most useful amplifier, the most commonly used methods of transistor biasing from one source of supply V_{CC} for CE amplifier are discussed in the following sections.

6.6.1 Fixed bias or base bias

The fixed bias or base bias circuit arrangement for an *n-p-n* transistor in CE mode is illustrated in Fig. 6.3.

The single d.c voltage source V_{CC} is used to make the emitter base junction forward biased and the collector-base junction reverse biased so that the required quiescent base current I_B and collector current I_C are obtained. R_B is known as *base bias resistance* and R_L, the *load resistance*. The capacitors C_1 and C_2 are called *coupling capacitors*. Their reactances are so small at the signal frequency compared to other resistances in the circuit that they offer practically short-circuited path for a.c and open-circuited path for d.c. As a result, the d.c biasing of this circuit does not change when its output is connected to the input of the other amplifier.

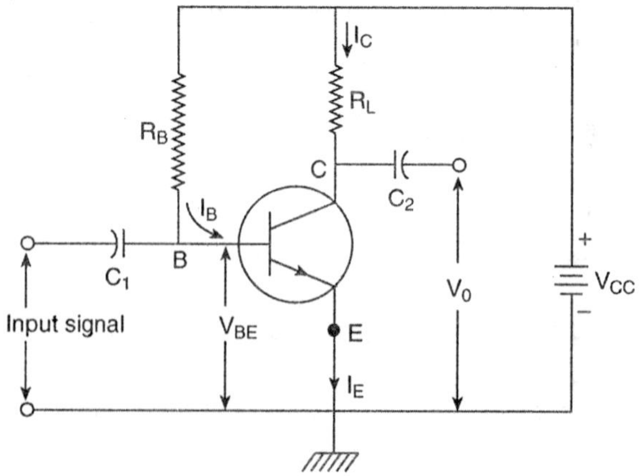

Fig. 6.3 Fixed bias or base bias in CE mode

Analytical determination of Q-point and stability factors

Applying the KVL to the base-emitter loop, we obtain

$$V_{CC} = I_B R_B + V_{BE}$$

$$\text{or,}\quad I_B = \frac{V_{CC} - V_{BE}}{R_B} \tag{6.6.1}$$

As $V_{CC} \gg V_{BE}$, the above equation (6.6.1) reduces to the form

$$I_B \approx \frac{V_{CC}}{R_B} \tag{6.6.2}$$

So, the change of V_{BE} practically does not have any effect on I_B. Since the base current I_B is fixed by the constant supply voltage V_{CC} and the base resistance R_B, the circuit is known as a *fixed bias circuit.* As the Q-point is selected in the active region, we have the collector current

$$\begin{aligned} I_C &= \beta I_B + (1+\beta) I_{CBO} \\ &= \beta \frac{V_{CC} - V_{BE}}{R_B} + (1+\beta) I_{CBO} \end{aligned} \tag{6.6.3}$$

The collector-emitter voltage is given by

$$V_{CE} = V_{CC} - I_C R_L \tag{6.6.4}$$

The various stability factors for this case are given as under.

$$S = \frac{\partial I_C}{\partial I_{CBO}} = 1 + \beta \tag{6.6.5}$$

$$S_V = \frac{\partial I_C}{\partial V_{BE}} = -\frac{\beta}{R_B} \tag{6.6.6}$$

$$\text{and}\quad S_\beta = \frac{\partial I_C}{\partial \beta} = \frac{V_{CC} - V_{BE}}{R_B} + I_{CBO}$$

$$\approx \frac{V_{CC}}{R_B} + I_{CBO} \tag{6.6.7}$$

From (6.6.6) and (6.6.7), it is clear that stabilization due to S_V and S_β can be much improved by using a high value of R_B. But S being controlled by β, the fixed bias circuit does not provide stable operating point and the transistor may be damaged by thermal runaway. Since S and S_V depend on β, the replacement of a transistor by another of the same type may change the Q-point. The Q-point is determined by equations (6.6.1), (6.6.3) and (6.6.4).

Advantages and disadvantages : The fixed bias or base bias circuit arrangement has both its advantages and disadvantages. The *advantages* include

1. Biasing circuit does not impose any load on the source.

2. Since only one resistor is required to bias the base, the circuit is rather very simple to construct.
3. The biasing conditions can be easily and readily set-up.

The arrangement suffers mainly from *two disadvantages.* These are:

1. Thermal runaway is highly probable.
2. Stabilization is highly unstable.

6.6.2 Collector to base bias

The circuit arrangement for the collector to base bias for an *n-p-n* transistor in CE mode is shown below in Fig. 6.4. This is an attempt to improve the stability of the fixed bias circuit.

Here, the resistance R_B is connected between the base and the collector terminals of the transistor and a *negative voltage shunt-feedback* is supplied from the collector to the base. If the collector current I_C increases due to increase in temperature or replacement of transistor by another of the same type, the potential drop across R_L increases and V_{CE} decreases. This in turn diminishes the base current and hence also the collector current I_C (Eqn. 6.6.3). Thus any change in I_C due to change in I_{CBO} is counterbalanced by a change in I_B and the Q-point is automatically *stabilized.*

We shall now determine the Q-point analytically and the stability factors.

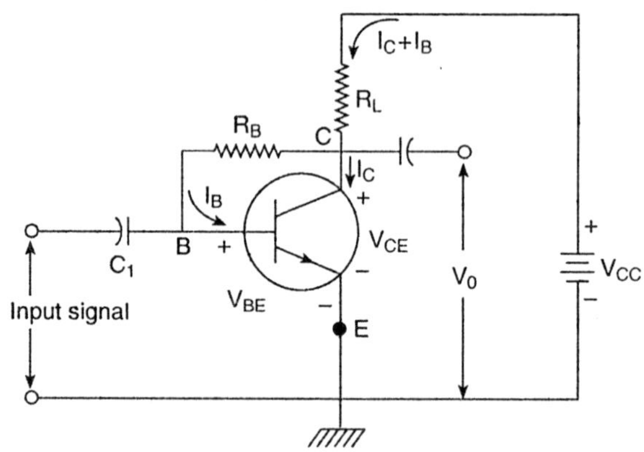

Fig. 6.4 Collector to base bias in CE mode

Analytical determination of Q-point and stability factors

The potential difference across the collector-emitter junction is given by

$$V_{CE} = I_B R_B + V_{BE} \tag{6.6.8}$$

Appling the KVL to the collector-emitter loop, we obtain

$$V_{CC} = (I_C + I_B)R_L + V_{CE} \tag{6.6.9}$$

From (6.6.8) and (6.6.9), we have

$$V_{CC} = (I_C + I_B)R_L + I_B R_B + V_{BE} \tag{6.6.10}$$

$$\text{or,} \quad I_B = \frac{V_{CC} - V_{BE} - I_C R_L}{R_B + R_L} \tag{6.6.11}$$

The collector current is given by

$$I_C = \beta I_B + (1+\beta) I_{CBO}$$

Substituting I_B from (6.6.11) in the above equation, we get,

$$I_C\left(1 + \frac{\beta R_L}{R_B + R_L}\right) = \frac{\beta(V_{CC} - V_{BE})}{R_B + R_L} + (1+\beta) I_{CBO} \tag{6.6.12}$$

The various *stability factors* in this case are obtained from (6.6.12) as

$$S = \frac{\partial I_C}{\partial I_{CBO}} = \frac{1+\beta}{1 + \frac{\beta R_L}{R_L + R_B}} \tag{6.6.13}$$

$$S_V = \frac{\partial I_C}{\partial V_{BE}} = -\frac{\beta}{R_B + (1+\beta)R_L} \tag{6.6.14}$$

$$\text{and} \quad S_\beta = \frac{\partial I_C}{\partial \beta} = \frac{V_{CC} - V_{BE} - I_C R_L + (R_B + R_L) I_{CBO}}{R_B + (1+\beta) R_L}$$

Since $V_{CC} \gg V_{BE}$, the above equation becomes

$$S_\beta \approx \frac{V_{CC} - I_C R_L + (R_B + R_L) I_{CBO}}{R_B + (1+\beta) R_L} \tag{6.6.15}$$

Comparing the values of the stability factors of fixed bias and collector-to-base bias, we find that the stability factors in the second case are less compared to those in the first.

Advantages and disadvantages

The collector to base bias circuit arrangement has the following *advantages.*

(1) No resistance is required across the base-emitter junction.

(2) As the arrangement requires only one resistance R_B in the collector to base circuit, it is a simple method of biasing.

(3) The circuit provides stabilization of the operating point.

The collector to base bias arrangement suffers from the following *disadvantages.*

(1) The resistance R_B provides not only d.c feedback but also causes a.c feedback which reduces the gain of the amplifier.

(2) The circuit does not provide very good stabilization.

6.6.3 Self-bias or emitter bias or voltage-divider bias

The circuit diagram of the self-bias or the emitter bias or the voltage-divider bias applied to an *n-p-n* transistor in CE mode is shown in Fig. 6.5. It is the most commonly used voltage divider to establish a stable operating point.

The supply voltage V_{CC} supplies the forward bias of the emitter junction and the reverse bias of the collector junction through the resistances R_1, R_2, R_e and R_L. C_1 and C_2 are called the *coupling capacitors* with very small reactances at the lowest signal frequency and while they block d.c, they allow an easy flow of a.c. These capacitors ensure that the inclusion of input signal or connection of the output of this circuit to the next stage cannot change the d.c biasing.

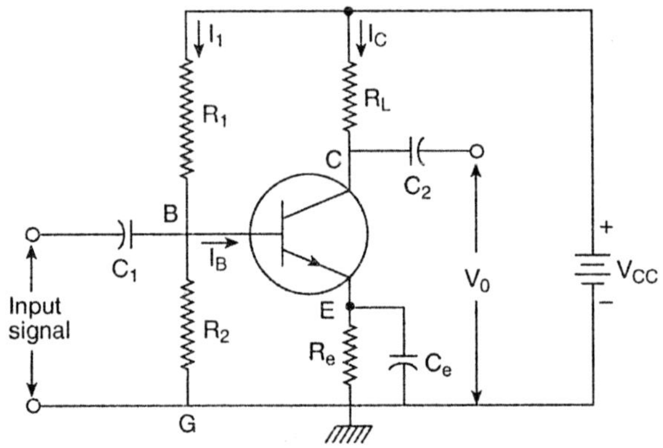

Fig. 6.5 Self-bias circuit for CE mode

The capacitor C_e is known as *bypass capacitor.* The small reactance of C_e prevents R_e from providing a.c negative feedback as the a.c signal is effectively short-circuited by C_e. R_1 and R_2 form a *potential divider* across V_{CC}. The p.d. across R_e tends to reverse bias the emitter junction but that across R_2 tends to bias it forward. When I_C increases due to increase in I_{CBO} caused by an increase in temperature, the p.d. across R_e increases. It makes the emitter junction less forward biased resulting in less base current. This, in turn, decreases the collector current I_C to bring it back to its initial value. Thus the Q-point is stabilized automatically and hence this circuit is called ***self-bias circuit.***

Analytical determination of Q-point

To determine the dc operating point, the effects of capacitors are neglected. If Thevenin's theorem is applied to the left of the terminals B and G of Fig. 6.5, we get the Thevenin voltage V_T and the Thevenin resistance R_T as

$$V_T = \frac{V_{CC} R_2}{R_1 + R_2} \tag{6.6.16}$$

and
$$R_T = \frac{R_1 R_2}{R_1 + R_2} \tag{6.6.17}$$

The resulting simplified circuit will be as shown in Fig. 6.6 below

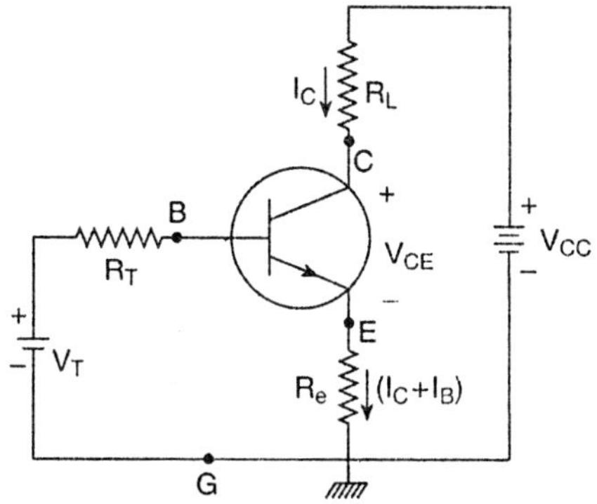

Fig. 6.6 Simplified version of self-bias circuit

Now, we apply KVL to the base circuit and get

$$V_T = I_B R_T + V_{BE} + I_E R_e \tag{6.6.18}$$

$$\text{or,} \quad V_T = I_B R_T + V_{BE} + (I_B + I_C) R_e, \quad (\because\ I_E = I_B + I_C)$$

$$\therefore \quad I_B = \frac{V_T - V_{BE} - I_C R_e}{R_T + R_e} \tag{6.6.19}$$

If we assume $I_{CBO} \ll I_B$, then from the relation $I_E = I_B + I_C$, we get $I_E \approx I_B + \beta I_B$ i.e.,

$$I_E = (1+\beta) I_B \tag{6.6.20}$$

Substituting this value of I_E in (6.6.18) and simplifying, we get

$$I_B = \frac{V_T - V_{BE}}{R_T + (1+\beta) R_e} \tag{6.6.21}$$

Once I_B is known, we can obtain I_C by the relation

$$I_C \approx \beta I_B \tag{6.6.22}$$

Application of KVL around the collector circuit gives,

$$\begin{aligned} V_{CC} &= I_C R_L + V_{CE} + I_E R_e \\ &= I_C R_L + V_{CE} + (I_C + I_B) R_e \\ &= I_C (R_L + R_e) + I_B R_e + V_{CE} \end{aligned} \tag{6.6.23}$$

Since $I_B \ll I_C$, $I_B R_e$ can be neglected compared to $I_C R_e$ and (6.6.23) becomes

$$V_{CC} \approx I_C (R_L + R_e) + V_{CE}$$

$$\text{or,} \quad V_{CE} = V_{CC} - I_C (R_L + R_e) \tag{6.6.24}$$

Stability factors

When Q-point is selected in the active region, the collector current is given by

$$I_C = \beta I_B + (1+\beta) I_{CBO} \tag{6.6.25}$$

Substituting for I_B from (6.6.19) into (6.6.25) we have,

$$I_C\left(1+\frac{\beta R_e}{R_T+R_e}\right)=\frac{\beta(V_T-V_{BE})}{R_T+R_e}+(1+\beta)I_{CBO}$$

Using the definitiions of stability factors, one can obtain from above relation,

$$S=\frac{\partial I_C}{\partial I_{CBO}}=\frac{1+\beta}{1+\dfrac{\beta R_e}{R_T+R_e}}=(1+\beta)\cdot\frac{1+\dfrac{R_T}{R_e}}{1+\beta+\dfrac{R_T}{R_e}} \tag{6.6.26}$$

$$S_V=\frac{\partial I_C}{\partial V_{BE}}=-\frac{\beta/R_e}{1+\beta+\dfrac{R_T}{R_e}} \tag{6.6.27}$$

and $$S_\beta=\frac{\partial I_C}{\partial\beta}=\frac{1}{\beta(1+\beta)}\left[I_C\frac{(R_T+R_e)(1+\beta)-\beta S\cdot R_e}{R_T+R_e}-I_{CBO}\cdot S\right] \tag{6.6.28}$$

● Since the resistances R_1 and R_2 in Fig. 6.5 serve as voltage divider, the circuit arrangement of self-bias is known as the *voltage-divider bias.*

Advantages : This self-bias circuit has a number of *advantages* and these may be highlighted as under.

1. The operating point is very stable, irrespective of the temperature-changes.
2. Several combinations of the potential divider resistors R_1 and R_2 can be easily arranged.
3. The gain of the transistor remains uninfluenced by the negative feedback.

6.7 Bias compensation technique

In the previous we have discussed different methods of transistor biasing for stabilization of Q-point. In those methods negative feedback is mainly used. There is another way to maintain the Q-point stable by using p-n junction diodes or temperature sensitive devices like thermistors, sensistor etc. That method is known as *bias compensation technique.* So, ***bias compensation technique** is the method of stabilization of the operating point by compensating the variations of current using p-n junction or any temperature sensitive device.*

Fig. 6.7(a) shows a bias compensating circuit using a p-n junction diode. The diode is of the same material as that of the transistor and it is reverse biased by V_{BE}. So, the reverse saturation current I_0 of the diode changes with temperature at the same rate as the transistor collector saturation current I_{CBO}.

From Fig. 6.7(a), we have

$$I=I_B+I_0$$

$$\text{or,}\quad I_B=I-I_0 \tag{6.7.1}$$

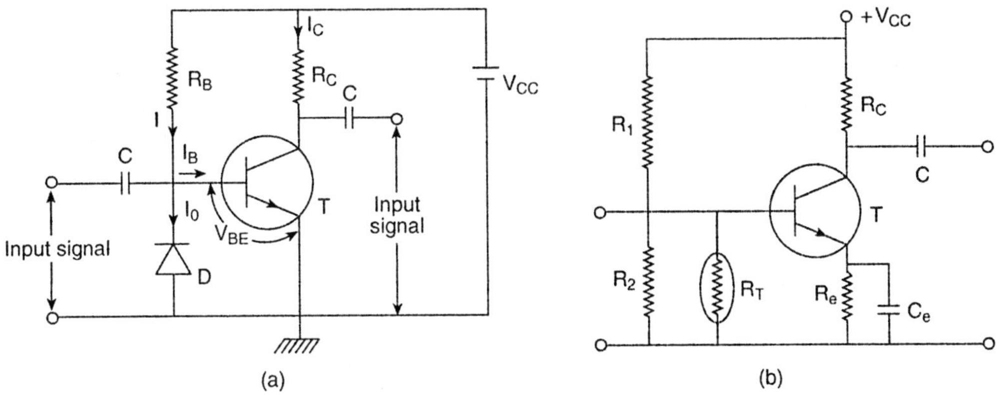

Fig. 6.7 Bias compensation (a) by using diode, (b) by using thermistor

The current I is given by

$$I = \frac{V_{CC} - V_{BE}}{R_B} \approx \frac{V_{CC}}{R_B} = \text{const.} \quad (\because V_{BE} \ll V_{CC}) \tag{6.7.2}$$

$$\begin{aligned} \text{Now, collector current } I_C &= \beta I_B + (1+\beta) I_{CBO} \\ &= \beta I - \beta I_0 + (1+\beta) I_{CBO}, \quad \text{using (6.7.1)}. \end{aligned} \tag{6.7.3}$$

If the temperature increases, the I_{CBO} increases in the transistor. The saturation current I_0 through the diode also increases. Since I $(= I_0 + I_B)$ is approximately constant by (6.7.2), an increase in I_0 causes the base current I_B to decrease so that I_C remains constant in accordance with (6.7.3) where $\beta \gg 1$.

Fig. 6.7(b) shows the bias compensation circuit using a thermistor R_T connected in parallel with R_2. Since a thermistor has a negative temperature coefficient of resistance, its resistance decreases with increase in temperature. So, forward bias of the base-emitter junction decreases. This decreases I_B and keeps I_C constant.

6.8 Illustrated Examples

▶ **Example 1.** *Calculate the collector current and the collector to emitter voltage for fixed bias circuit if* $V_{CC} = 9\,\text{V}$, $R_B = 300\,\text{k}\Omega$, $R_C = 2\,\text{k}\Omega$ *and* $\beta = 50$ *of a transistor.*

(Punjab Univ.)

Solution: Given, $V_{CC} = 9\text{V}$, $R_B = 300\,\text{k}\Omega$, $R_C = 2\,\text{k}\Omega$ and $\beta = 50$. Considering the base emitter loop of the fixed biased circuit,

$$I_B = \frac{V_{CC} - V_{BE}}{R_B}$$

Since $V_{CC} \gg V_{BE}$, the above equation becomes,

$$I_B = \frac{V_{CC}}{R_B} = \frac{9}{300 \times 10^3}\,\text{A} = 30\mu\text{A}$$

Again, $\beta = I_C/I_B \Rightarrow \boldsymbol{I_C} = \beta I_B = 50 \times 30\,\mu\text{A} = \mathbf{1.5\,mA}$

The saturation value of $I_{C,\text{sat}} = \dfrac{V_{CC}}{R_C} = \dfrac{9}{2 \times 10^3}\,\text{A} = 4.5\,\text{mA}$

Here, $I_C < I_{C,\text{sat}}$. So, the transistor is *not in saturation region.*

$$\therefore \quad \boldsymbol{V_{CE}} = V_{CC} - I_C R_C$$
$$= 9 - 1.5 \times 10^{-3} \times 2 \times 10^3\,\text{V} = \mathbf{6\,V}$$

▶ **Example 2.** *A silicon transistor of* $\beta = 100$ *is used in a collector feedback bias circuit* (Fig. 6.4) *with a source voltage of* 20 V. *If* $R_C = 1000\Omega$ *and* $R_B = 10^5\Omega$, *find the operating point. Given* $V_{BE} = 0.7$. **(Punjab Univ.)**

Solution: Given, $V_{CC} = 20\,\text{V}$, $R_C = 1000\Omega$, $R_B = 10^5\Omega$, $V_{BE} = 0.7$ and $\beta = 100$.

Applying the KVL to collector-emitter loop and considering $I_B \ll I_C$, we obtain

$$I_B = \frac{V_{CC} - V_{BE} - \beta I_B R_C}{R_B}$$
$$= \frac{20 - 0.7 - 100 I_B \times 10^3}{10^5}$$

or, $$10^5 I_B + 10^5 I_B = 19.3$$

or, $$I_B = 0.0965 \times 10^{-3}\,\text{A} = 0.0965\,\text{mA}$$

$\therefore$ The collector current $I_C = \beta I_B = 100 \times 0.0965\,\text{mA} = 9.65\,\text{mA}$

The collector-emitter voltage V_{CE} is given by

$$V_{CE} = V_{CC} - I_C R_C$$
$$= 20 - 9.65 \times 10^{-3} \times 10^3\,\text{V}$$
$$= 10.35\,\text{V}$$

$\therefore$ The ***operating point*** is given by $\boldsymbol{V_{CE}} = \mathbf{10.35\,V}$, $\boldsymbol{I_C} = \mathbf{9.65\,mA}$.

▶ **Example 3.** *It is required to set the operating point by biasing with feedback resistor at* $I_C = 1\,\text{mA}$, $V_{CE} = 10\,\text{V}$. *If* $\beta = 100$, $V_{CC} = 15\,\text{V}$, $V_{BE} = 0.3\,\text{V}$, *how will you do it?*

Solution: Given, $I_C = 1\,\text{mA}$, $V_{CE} = 10\,\text{V}$, $\beta = 100$, $V_{CC} = 15\,\text{V}$ and $V_{BE} = 0.3\,\text{V}$.

To set the operating point, the values of R_B and R_C shall have to be calculated. Considering the collector-emitter loop we can write,

$$V_{CC} = V_{CE} + I_C R_C$$

or, $$R_C = \frac{V_{CC} - V_{CE}}{I_C} = \frac{15 - 10}{1 \times 10^{-3}}\Omega = \mathbf{5\,k\Omega}$$

Again, $\beta = \dfrac{I_C}{I_B} \Rightarrow I_B = \dfrac{I_C}{\beta} = \dfrac{10^{-3}}{100} = 10^{-5}\,\text{A}$

Now,
$$V_{CC} = I_B R_B + V_{BE} + I_C R_C$$

$$\text{or,}\quad R_B = \frac{V_{CC} - V_{BE} - I_C R_C}{I_B}$$
$$= \frac{15 - 0.3 - 10^{-3} \times 5 \times 10^3}{10^{-5}}\,\Omega$$
$$= 970 \times 10^3\,\Omega = \mathbf{970\,k\Omega}$$

▶ **Example 4.** *In the circuit of* Fig. 6.5, $R_1 = 82\,\text{k}\Omega$, $R_2 = 16\,\text{k}\Omega$, $R_L = 2.2\,\text{k}\Omega$, $R_e = 750\Omega$, $V_{CC} = 22\,\text{V}$, $V_{BE} = 0.7\,\text{V}$ *and* $\beta = 100$. *Determine the operating point. Neglect* I_{CO}. **(U.P. Tech. Univ.)**

Solution: Given, $R_1 = 82\,\text{k}\Omega$, $R_2 = 16\,\text{k}\Omega$, $R_L = 2.2\,\text{k}\Omega$, $R_e = 750\Omega$, $V_{CC} = 22\,\text{V}$, $V_{BE} = 0.7\,\text{V}$ and $\beta = 100$.

The p.d. across R_2 is $V_2 = \dfrac{V_{CC}}{R_1 + R_2} R_2$, assuming I_B to be very small and current through R_1 and R_2 is equal.

$$\therefore\quad V_2 = \frac{22 \times 16 \times 10^3}{(82 + 16) \times 10^3}\,\text{V} = 3.59\,\text{V}$$

From the base circuit, we obtain

$$V_2 = V_{BE} + I_E R_e$$
$$= V_{BE} + I_C R_e, \quad (\because I_E \approx I_C)$$
$$\text{or,}\quad I_C = \frac{V_2 - V_{BE}}{R_e} = \frac{3.59 - 0.7}{750}\,\text{A} = 3.85\,\text{mA}$$

Applying the KVL to the output circuit, we have

$$V_{CC} = I_C(R_L + R_e) + V_{CE}$$
$$\text{or,}\quad V_{CE} = V_{CC} - I_C(R_L + R_e)$$
$$= 22 - 3.85 \times 10^{-3} \times (2.2 + 0.75) \times 10^3\,\text{V}$$
$$= 22 - 11.36\,\text{V} = 10.64\,\text{V}$$

$\therefore$ The ***operating point*** is specified by $\mathbf{I_C = 3.85\,mA}$ and $\mathbf{V_{CE} = 10.64\,V}$.

▶ **Example 5.** *A* Ge-*transistor is to be operated at zero signal* $I_C = 1\,\text{mA}$. *If the collector supply* $V_{CC} = 12\,\text{V}$, *determine* R_B *in the base resistor method of biasing,* $\beta = 100$, $V_{BE} = 0.3\,\text{V}$. *If another transistor of the same batch but with* $\beta = 50$ *is used, determine the new value of zero signal* I_C *for the same* R_B. *Make comments on the results obtained.*

Solution: Here, by the problem, $V_{CC} = 12\,\text{V}$, $\beta = 100$, $V_{BE} = 0.3$, zero signal collector current, $I_C = 1\,\text{mA} = \beta I_B$.

$$\therefore \text{ Zero signal base current, } I_B = \frac{I_C}{\beta} = \frac{1\,\text{mA}}{100} = 0.1\,\text{mA}$$

$$\text{Base resistance } \boldsymbol{R_B} = \frac{V_{CC} - V_{BE}}{I_B} = \frac{12 - 0.3}{0.01 \times 10^{-3}} = \mathbf{1170 \times 10^3\,\Omega}$$

$$\text{When } \beta = 50 \text{ and } R_B = 1170 \times 10^3\,\Omega,\ I_B = \frac{V_{CC} - V_{BE}}{R_B} = \frac{12 - 0.3}{1170 \times 10^3} = 0.01\,\text{mA}$$

$$\therefore\ \boldsymbol{I_C} = \beta I_B = 50 \times 0.01 = \mathbf{0.5\ mA}$$

● Above calculations show that with change in transistor parameter β, the zero signal collector current changes from 1 mA to 0.5 mA. Hence, the base resistor method *cannot provide stability.*

► **Example 6.** *For a n-p-n transistor circuit in CE mode using voltage divider bias, shown in* Fig. 6.5, $R_1 = 8\,\text{k}\Omega$, $R_2 = 4\,\text{k}\Omega$, $R_L = 1\,\text{k}\Omega$, $R_e = 2\,\text{k}\Omega$, $V_{CC} = 12\,\text{V}$.

Draw the dc load line. Calculate the value of collector circuit I_C, collector-emitter voltage V_{CE} and hence locate the operating point.

Solution: ***dc-loadline***: Collector voltage,

$$V_{CE} = V_{CC} - I_C(R_L + R_e) \tag{6.8.1}$$

When $I_C = 0$, $V_{CE} = V_{CC} = 12\,\text{V}$ (by the problem).

This locates the first point A, with $OA = 12\,\text{V}$, of the load line on V_{CE}-axis.

When $V_{CE} = 0$, we get from (6.8.1)

$$I_C = \frac{V_{CC}}{R_L + R_e} = \frac{12}{(1+2) \times 10^3}$$
$$= 4 \times 10^{-3}\,\text{A} = \mathbf{4\,mA}$$

This locates the second point B, with $OB = 4\,\text{mA}$, of the load line on the I_C-axis. So, the line joining the points A and B i.e., the line AB *gives the dc load line.*

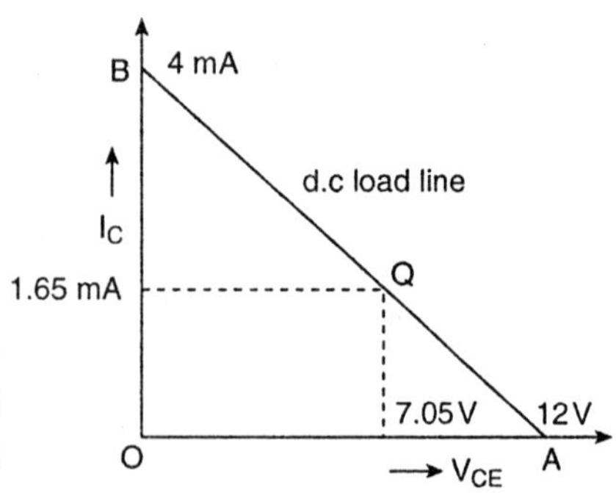

Fig. 6.8

Collector current: Collector current $I_C = \dfrac{V_{BO} - V_{BE}}{R_E}$

$$= \frac{V_2 - V_{BE}}{R_E} \quad (\because V_{BO} = V_2)$$

Now, voltage across resistance R_2 is: $V_2 = V_{BO} = \dfrac{V_{CC} R_2}{R_1 + R_2}$

$$\therefore \quad V_2 = \frac{12 \times 4 \times 10^3}{(8+4) \times 10^3} = 4\,\text{V}; \quad V_{BE} \text{ for silicon} = 0.7\,\text{V}$$

$$\therefore \quad I_C = \frac{4 - 0.7}{2 \times 10^3} = 1.65 \times 10^{-3}\,\text{A} = \mathbf{1.65\,mA}$$

Collector emitter voltage: $V_{CE} = V_{CC} - I_C(R_C + R_E)$

$$= 12 - 1.65 \times 10^{-3} \times (1+2) \times 10^3$$

$$= 12 - 4.95 = \mathbf{7.05\,V}$$

Operating point—The *operating point* Q on the load line AB is thus given by **(7.05 V, 1.65 mA)**.

▶ **Example 7.** *In a potential divider circuit, the zero signal operating point is fixed at* (1 mA, 3 V). *Given,* $V_{CC} = 12\,\text{V}$, $R_E = 1\,\text{k}\Omega$, $R_2 = 8\,\text{k}\Omega$, *find the value of* R_C *and* R_1.

Solution: By the problem, $I_C = 1\,\text{mA}$, $V_{CE} = 3\,\text{V}$, $V_{CC} = 12\,\text{V}$, $R_E = 1\,\text{k}\Omega$ and $R_2 = 8\,\text{k}\Omega$.

The circuit diagram for the probem is shown in Fig. 6.9

(i) Applying KVL for output side,

$$V_{CC} = I_C R_C + V_{CE} + I_E R_E$$

$$= I_C R_C + V_{CE} + I_C R_E \quad (\because I_E = I_C)$$

$$\therefore \quad 12\,\text{V} = (1\,\text{mA} \times R_C) + 3\,\text{V} + 1\,\text{V}$$

$$\text{or,} \quad \boldsymbol{R_C} = \frac{(12-4)\,\text{V}}{1\,\text{mA}} = \mathbf{8\,k\Omega}$$

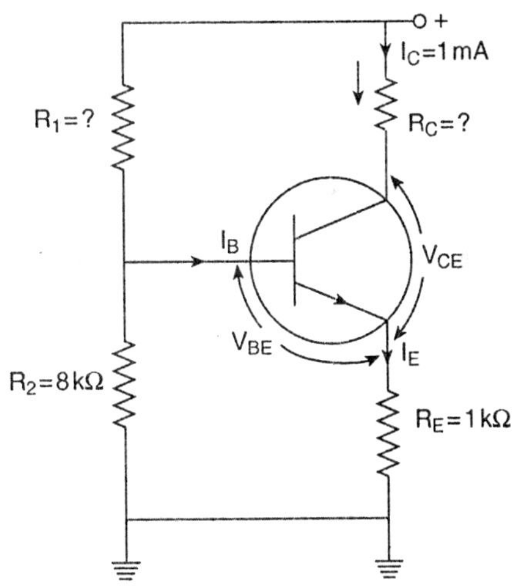

Fig. 6.9

(ii) Applying KVL for base-emitter circuit,

$$V_2 = V_{BE} + I_E R_E$$

$$= I_C R_E, \quad \text{neglecting } V_{BE} \text{ and}$$

replacing I_E by I_C.

$$\therefore \quad \boldsymbol{V_2} = 1\,\text{mA} \times 1\,\text{k}\Omega = 1\,\text{V}.$$

(iii) By solving the potential divider network,

$$V_2 = \frac{V_{CC}}{R_1 + R_2} R_2 \Rightarrow 3\,\text{V} = \frac{12\,\text{V}}{R_1 + 8\,\text{k}\Omega} \times 8\,\text{k}\Omega$$

$$\therefore \quad \boldsymbol{R_1} = \mathbf{88\,k\Omega}$$

▶ **Example 8.** *In* Fig. 6.10, $\alpha = 0.98$ *and* $V_{BE} = 0.7\,\text{V}$. *Find the resistance* R_1 *for an emitter current* $I_E = 2\,\text{mA}$. *Neglect* I_{CO}. **(Calcutta Hons.)**

Solution: Given, $\alpha = 0.98$.

$$\therefore \quad \beta = \frac{\alpha}{1-\alpha} = \frac{0.98}{1-0.98} = 49$$

$$\text{or,} \quad \frac{I_C}{I_B} = 49 \text{ or, } I_C = 49 I_B$$

$$\therefore \quad I_E = I_C + I_B = (49+1)I_B = 50 I_B$$

$$\text{or,} \quad I_B = \frac{I_E}{50} = \frac{2}{50}\,\text{mA} = 0.04\,\text{mA}$$

$$\text{and} \quad I_C = 49 I_B = 49 \times 0.04\,\text{mA} = 1.96\,\text{mA}$$

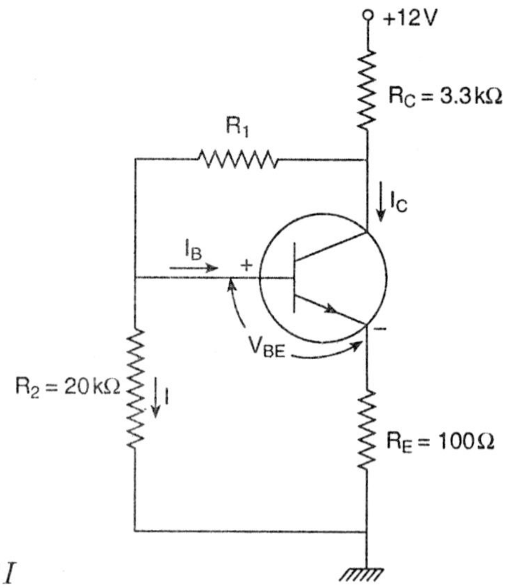

Fig. 6.10

Let I be the current through R_2.

$\therefore$ Applying the KVL to base-emitter circuit,

$$V_{BE} + 100 \times I_E = 20 \times 10^3 \times I$$

$$\text{or,} \quad 0.7 + 100 \times 2 \times 10^{-3} = 20 \times 10^3 I$$

$$\text{or,} \quad I = \frac{0.7+0.2}{20 \times 10^3} = 0.045 \times 10^{-3}\,\text{A} = 0.045\,\text{mA}$$

From the output circuit, we obtain

$$(I + I_B + I_C)R_C + (I + I_B)R_1 + IR_2 = 12$$

$$\text{or,} \quad (0.045 + 0.04 + 1.96) \times 10^{-3} \times 3.3 \times 10^3 + (0.045 + 0.04) \times 10^{-3} \times R_1$$
$$+ 0.045 \times 10^{-3} \times 20 \times 10^3 = 12$$

or, $\boldsymbol{R_1 = 51.2\,k\Omega}$, on simplification.

► **Example 9.** *In the circuit of* Fig. 6.11, *find the quiescent values of I_E and V_{CE}. Given $\beta = 90$ and $V_{BE} = 0.7\,V$. If β is incresed by 50%, what will be the effect on I_C and V_{CE}?* **(U.P. Tech. Univ.)**

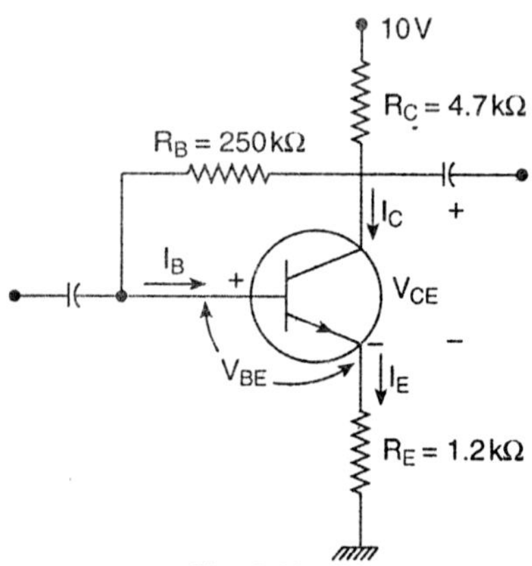

Fig. 6.11

Solution: Applying KVL to the base-emitter circuit and output circuit, we get

$$V_{CE} = I_B R_B + V_{BE} \tag{6.8.2}$$

$$\text{and} \quad (I_B + I_C)R_C + V_{CE} + I_E R_E = 10 \tag{6.8.3}$$

Putting, $I_C = \beta I_B$ and $I_E = (1+\beta)I_B$, we get from equation (6.7.2),

$$(1+90)I_B \times 4.7 \times 10^3 + (250 \times 10^3 I_B + 0.7) + (1+90)I_B \times 1.2 \times 10^3 = 10 \tag{6.8.4}$$

This gives, on simplification, $\quad I_B = 11.8 \times 10^{-6}\,\text{A} = 11.8\,\mu\text{A}$

$$\therefore \quad I_E = (1+\beta)I_B = 91 \times 11.8\,\mu\text{A} = 1.074 \times 10^{-3}\,\text{A} = \mathbf{1.074\,mA}$$

and
$$V_{CE} = 250 \times 10^3 I_B + V_{BE} = 250 \times 10^3 \times 11.8 \times 10^{-6}\,\text{V} + 0.7\,\text{V}$$
$$= \mathbf{3.65\,V}, \text{ on simplification.}$$

Collector current, $I_C = \beta I_B = 90 \times 11.8\mu\text{A} = 1.062\,\text{mA}$

When β is increased by 50%, the value of β becomes 135 which makes $I_C = 135 I_B$ and $I_E = 136 I_B$. Substituting these values of I_C and I_E in equation (6.7.2) and solving, we get, $I_B = 8.84\mu\text{A}$.

$\therefore$ $I_C = 135 \times 8.84\mu\text{A} = 1.193\,\text{mA}$ i.e., the value of I_C ***increases by***

$$\frac{1.193 - 1.062}{1.062} \times 100\% = \mathbf{12.3\%}.$$

Since V_{CE} is independent of β, the change in β has ***no effect on*** V_{CE}.

Review Questions

1. (a) What is d.c load line? How will you draw the d.c load line on the output characteristics of a transistor? What is its importance? **(Lucknow Univ.)**
   (b) Define Q-point of a transistor. How is it located?
   (c) What is meant by a.c load line? **(Bangalore Univ.)**
2. What do you understand by biasing of a transistor? Why is there a necessity of biasing? On what factors does the selection of an operating point depend?
3. (a) What do you mean by stabilisation as applied to a transistor amplifier? Why is stabilisation necessary?
   (b) What is meant by thermal runaway of a transistor? **(Calcutta Hons.)**
4. (a) What are the factors that affect the bias stability of a transistor? **(Calcutta Hons.)**
   (b) Define the stability factors with respect to changes in I_{CBO}, V_{BE} and β. Why is the stability with respect to change in V_{CE} not considered?
5. Draw of fixed base bias circuit and calculate the value of base resistance R_B and the collector current I_C. Derive an expression for its stability factor. What are the disadvantages and advantages of this circuit?
6. Explain the collector to base bias of a transistor. Calculate the stability factor and discuss the results. **(Punjab Univ.)**
7. What is self-bias? Draw the circuit diagram showing the self-bias of an *npn* transistor in the CE-configuration. Explain physically how the self-biasing resistor improves the stability. Explain the functions of the bypass and the coupling capacitors. **(Calcutta Hons.)**

8. State and explain the bias compensation technique to maintain the Q-point stable by using a diode or a thermistor.

Objective Questions

1. The smallest possible value of the stability factor, $S = \frac{\partial I_C}{\partial I_{CBO}}$ is

   (a) 0, (b) 1, (c) 0.5, (d) 2.

2. If biasing is not done in an amplifier circuit, it results in

   (a) decrease in base current, (b) unfaithful amplification,
   (c) excessive collector bias, (d) none of the above.

3. Thermal runway occurs when the

   (a) collector is reverse biased, (b) transistor is not biased,
   (c) emitter is forward biased, (d) junction capacitance is high.

4. The blocking capacitor is used in an amplifier circuit to

   (a) increase the gain, (b) block d.c and allow easy flow of a.c
   (c) increase the bandwidth, (d) none of the above.

5. The emitter resistance R_e is bypassed by a capacitor C_e so as to

   (a) avoid negative feedback gain loss, (b) reduce the noise,
   (c) improve the stability of Q-point, (d) reduce the distortion.

6. The collector-base bias in a BJT uses

   (a) positive feedback, (b) negative feedback,
   (c) bias compensation technique, (d) none of these.

7. In the designing of a biasing circuit, the value of collector load R_C is determined by considering

   (a) V_{CE}, (b) V_{BE}, (c) I_B, (d) none of these.

8. The circuit that provides the best stabilization of operating point is

   (a) fixed bias, (b) collector to base bias,
   (c) voltage divider bias, (d) none of these.

9. If the value of the collector current I_C increases, then the value of V_{CE}

(a) decreases, (b) increases,
(b) remains the same, (d) none of these.

10. For amplification with minimum distortion a BJT must be operated

(a) near the saturation region, (b) near the cut-off region,
(c) at the middle of the active region, (d) anywhere of the active region.

11. The d.c load line of a transistor circuit

(a) is a graph between I_C and I_B,
(b) is a graph between I_C and V_{CE},
(c) is a current line,
(d) none of the above.

12. Thermal runaway occurs if the quiescent point is such that

(a) $V_{CE} < V_{CC}$, (b) $V_{CE} > \frac{1}{2}V_{CC}$,
(c) $V_{CE} < 2V_{CC}$, (d) $V_{CE} < \frac{1}{2}V_{CC}$.

13. To achieve good stability in potential divider method, the current I_1 through R_1 and R_2 should be

(a) $I_1 > 2I_B$, (b) $I_1 > 5I_B$, (c) $I_1 > 10I_B$, (d) $I_1 < 5I_B$.

14. The stability factor of a BJT circuit is defined as

(a) $S_f = dI_E/dI_B$, (b) $S_f = dI_C/dI_E$,
(b) $S_f = dI_C/dI_{CO}$, (d) $S_f = dI_C/dI_B$.

15. Improper biasing of a transistor circuit leads to

(a) distortion in output signal,
(b) heavy loading of emitter terminal,
(c) excessive heat generation at collector terminal,
(d) none of the above.

16. The load line moves porallel to itself on the CE output characteristics of a transistor when

(a) V_{CC} changes, (b) R_L changes,
(b) both R_L and V_{CC} change, (d) V_i changes.

17. The self-bias arrangement gives a better Q-point stability when

(a) R_e is small, (b) β is small but R_e is large,
(c) both β and R_e are large, (d) both β and R_e are small.

18. In CE amplifier with resistive load, the phase difference between input and output is

(a) 0°, (b) 90°, (c) 180°, (d) 360°.

Problems

1. Find the Q-point of the fixed bias circuit of Fig. 6.3 with the following circuit parameters: $V_{CC} = 12\,\text{V}$, $R_C\ (\equiv R_L) = 1.2\,\text{k}\Omega$, $R_B = 240\,\text{k}\Omega$, $\beta = 100$. Assume $I_{CBO} \ll I_B$. **(GATE)** $[I_C = 4.7\,\text{mA},\ V_{CE} = 6.36\,\text{V}]$

2. Consider the circuit of Fig. 6.4 with $V_{CC} = 12\,\text{V}$, $R_C\ (\equiv R_L) = 1\,\text{k}\Omega$, $\beta = 100$ and $V_{BE} = 0.7\text{V}$. Find R_B which makes $V_{CE} = 6.32\text{V}$. Assume $I_{CBO} \ll I_B$.

(Calcutta Hons.) $[100\,\text{k}\Omega]$

3. Consider the self-bias arrangement of Fig. 6.5. Find the Q-point with the following specifications:

$$V_{CC} = 22\,\text{V},\quad R_C\ (\equiv R_L) = 10\,\text{k}\Omega,\quad R_e = 1.5\,\text{k}\Omega,$$

$$R_1 = 39\,\text{k}\Omega,\quad R_2 = 3.9\,\text{k}\Omega\ \text{ and }\ \beta = 140.$$

Assume $I_{CBO} \ll I_B$. **(GATE)** $[I_C = 0.85\,\text{mA},\ V_{CE} = 12.22\,\text{V}]$

4. Find the Q-point of a self-bias transistor circuit with the following specifications: $V_{CC} = 22.5\,\text{V}$, $R_L = 5.6\,\text{k}\Omega$, $R_e = 1\,\text{k}\Omega$, $R_1 = 90\,\text{k}\Omega$, $R_2 = 10\,\text{k}\Omega$, $V_{BE} = 0.7\,\text{V}$ and $\beta = 55$. Assume that $I_B \gg I_{CBO}$.

(Calcutta Hons.) $[V_{CE} = 13.8\,\text{V},\ I_C = 1.31\,\text{mA}]$

5. Calculate I_B, I_C, I_E and V_{CE} for the transistor circuit of Fig. 6.11 given beside. Take $\beta = 50$ and $V_{BE} = 0.7\,\text{V}$. **(Uttaranchal Univ.)**

$[0.243\,\text{mA},\ 12.2\,\text{mA},\ 12.4\,\text{mA},\ 7.8\,\text{V}]$

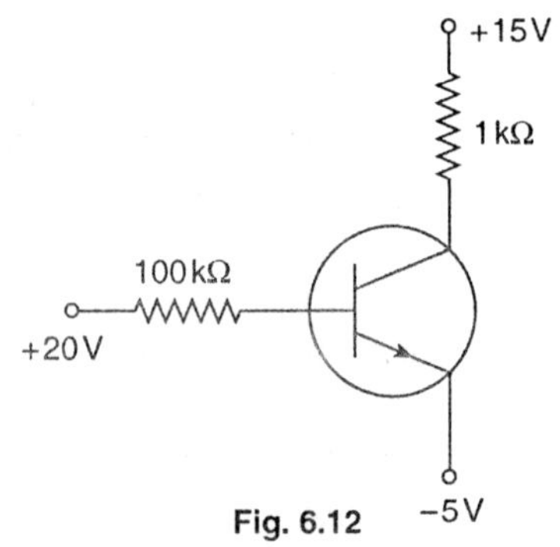

Fig. 6.12

6. For the silicon transistor circuit of Fig. 6.12, give the equation of the load line. What is I_C at the saturation point? Find V_{CE} at the cut-off point.

(U.P. Tech. Univ.)

$[V_{CE} + 10^3 I_C = 20;\ 19.8\,\text{mA},\ 20\,\text{V}]$

7. For the circuit as given in Fig. 6.13, find the values of β, V_{CC} and R_B. Draw the d.c load line and locate the operating point on it.

(U.P. Tech. Univ.) [153.3, 17.7 V, 745 kΩ]

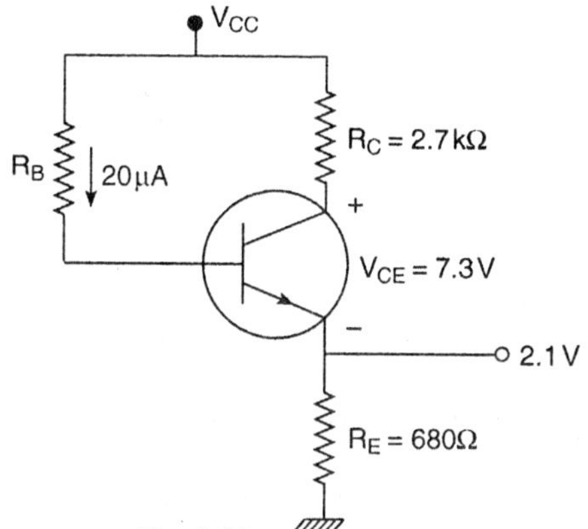

Fig. 6.13

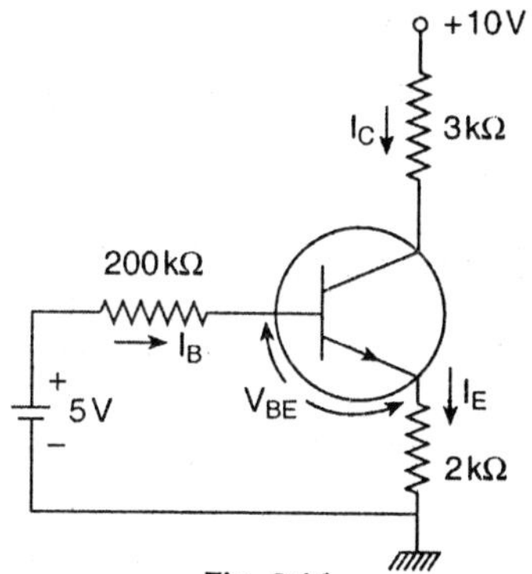

Fig. 6.14

8. Find the transistor currents in the circuit of Fig. 6.14. Assume $I_{CO} \ll I_C$, $\beta = 100$ and $V_{BE} = 0.7$ V.

(Calcutta Hons.)

$[I_B = 10.7\mu\text{A}, I_C = 1.07\,\text{mA}, I_E = 1.08\,\text{mA}]$

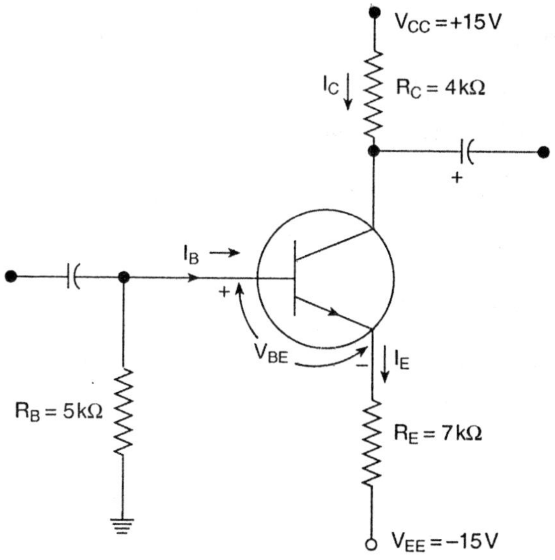

Fig. 6.15

9. In the circuit of Fig. 6.15, $\beta = 99$ and $V_{BE} = 0.7$ V. Calculate the quiescent values of I_B, I_C, I_E and V_{CE}. If β increases by 20%, what is the corresponding change in I_C? What is your comment on the result?

[20.3µA, 2.01 mA, 2.03 mA, 7.75 V, 0.5%; circuit is stable against changes in β]

10. In a CE transistor circuit, load resistance R_C is 5 kΩ and quiescent current I_C is 1.2 mA. Determine the operating point when the battery voltage $V_{CC} = 12\,V$.

(Punjab Univ.)

[**Hints.** Load resistance, $R_C = 5\,k\Omega$, quiescent current, $I_C = 1.2\,mA$ and battery voltage, $V_{cc} = 12$ volt.

$\therefore$ Collector emitter voltage, $V_{CE} = V_{CC} - I_C R_C = 12 - 1.2 \times 10^{-3} \times 5 \times 10^3 = 6$ V.

$\therefore$ Operating point is **6 V; 1.2 mA**]

11. For a voltage divider biasing circuit of a transistor, the data given are as under:

$$V_{CC} = 12\,V, \quad R_1 = 10\,k\Omega, \quad R_2 = 5\,k\Omega, \quad R_C = 1\,k\Omega, \quad R_E = 2\,k\Omega, \quad V_{BE} = 0.7\,V$$

$\beta = 100$. Draw d.c load line and determine the operating point.

[$OA = 12\,V$, $OB = 4\,mA$, (7.05 V, 1.65 mA)]

12. (a) A Ge-transistor is to be operated at zero signal collector current of 2 mA. If $V_{CC} = 10\,V$, what is the value of R_B? Take $\beta = 100$.

(b) If the transistor is replaced by a new one with $\beta = 60$, what will be the new value for zero signal collector current?

(c) Find the stability factor in the second case. [485 kΩ, 1.2 mA, 61]

[**Hints.** (a) Here $V_{CC} = 10\,V$, $\beta = 100$. Zero signal $I_C = 2\,mA$.

$I_B = I_C/\beta = 2/100 = 0.02\,mA$. Now, $V_{CC} = I_B R_B + V_{BE} \Rightarrow 10 = 0.02 \times R_B + 0.3 \Rightarrow$ $R_B = (10 - 0.3)\,V/0.02\,mA =$ **485 kΩ**, assuming V_{BE} for Ge-transistor.

(b) From $V_{CC} = V_{BE} + I_B R_B$, $10 = 0.3 + I_B \times 485 \times 10^3 \Rightarrow I_B = \dfrac{(10 - 0.3)\,V}{485 \times 10^3\,\Omega} = 0.02\,mA$.

So, zero signal, $I_C = \beta I_B = 60 \times 0.02 =$ **1.2 mA**

(c) Stability factor, $S = (1 + \beta) = 1 + 60 =$ **61**]

13. A silicon n-p-n transistor is used in the circuit shown in Fig. 6.16. From the output characteristics of the CE configuration of the transistor, draw the d.c load line and plot the Q-point. [Coordinates of Q-pt : (3.6 V, 2.5 mA)]

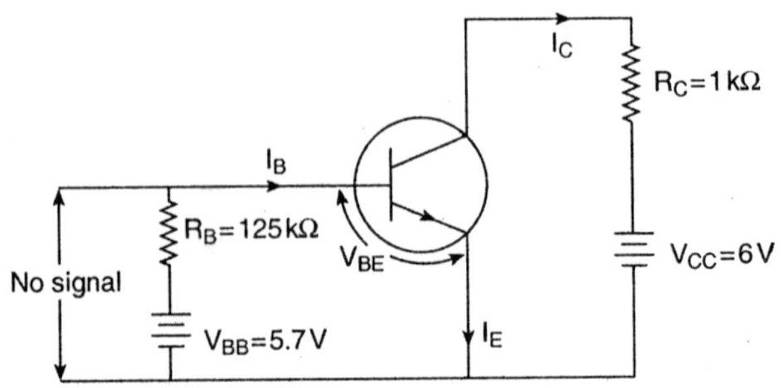

Fig. 6.16

14. A transistor is biased with feedback resistor $R_B = 200\,\text{k}\Omega$. If the collector load resistance is of $2\,\text{k}\Omega$ and the supply voltage is of 20 V, determine the collector current if β varies within the limits $50 < \beta < 200$. Comment on the result.

[$I_C = 3.33\,\text{mA}$, $I_C = 6.66\,\text{mA}$]

15. For a Ge-transistor, calculate the minimum and maximum value of emitter current and collector-to-emitter voltage for an emitter resistor biasing circuit having the following data:

$$R_B = 10\,\text{k}\Omega,\;\; R_C = 50\Omega,\;\; 100\Omega,\;\; V_C = 6\,\text{V} \text{ and } 50 \le \beta \le 200.$$

[$I_R = 19.25\,\text{mA}$, $V_{CE} = 3.1\,\text{V}$; $I_E = 38.2\,\text{mA}$, $V_{CE} = 0.3\,\text{V}$]

7

Transistor Hybrid Model

7.1 Introduction

The performance of a transistor amplifier circuit can be analysed by two different techniques depending on the nature of the input signal. The large-signal response of a transistor can be obtained graphically since the operation of a transistor in this case is not confined within the active region but extends to the saturation or cutoff region. For small signals and low frequencies, however, a tansistor can be operated in the active region with reasonable linearity and hence a non-linear transistor operated in active region can be replaced by a linear equivalent circuit that can be analysed by applying basic laws of electricity. Thus, a number of transistor linear models or equivalent circuit models have been developed.

In this chapter, however, we shall be interested in developing the most common ***hybrid parameter model*** only. The basic assumption in the process is that the variations of the parameters about the Q-point is very small and can essentially be taken to be constant. The quiescent values of I and V are determined by the methods of biasing, discussed in an earlier chapter.

7.2 Transistor as two-port device and h-parameters

For the linear performance of the amplifier, a transistor is connected in any one of the *three* modes (the CB, CE and CC) where the a.c. signals (the a.c. currents or voltages) are applied to a pair of input terminals (called the *input port*) and the amplified signal is obtained from a pair of output terminals (the *output port*) of the amplifier. The transistor amplifier can, therefore, be regarded as a '*blackbox*' of two-port (four terminals) network having a pair of input and a pair of output terminals. This is shown in Fig 7.1.

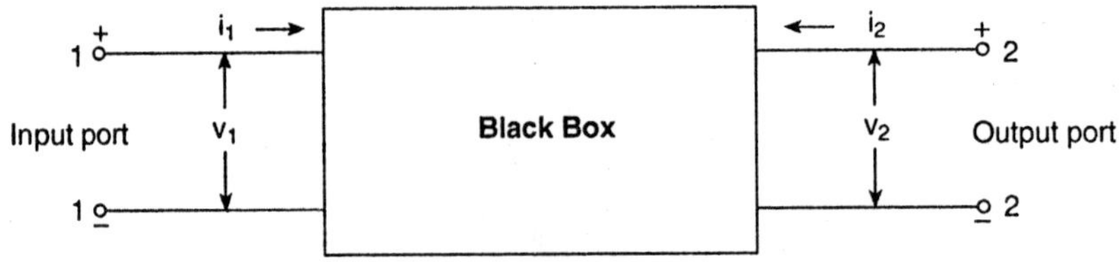

Fig. 7.1 Two-port representation of a transistor

The terminal behaviour of a two-port network is usually specified by two signal currents i_1 and i_2 and two signal voltages v_1 and v_2. To get the transistor hybrid model, the input current i_1 and the output voltage v_2 are chosen as *independent* variables and the input voltage v_1 and the output current i_2 as *dependent* variables. So, the four variables can be related by the following two equations.

$$v_1 = h_{11}i_1 + h_{12}v_2 \tag{7.2.1}$$

$$\text{and} \quad i_2 = h_{21}i_1 + h_{22}v_2 \tag{7.2.2}$$

Note that in the above equations h_{11} has the *dimensions of impedance* (measured in ohm) and h_{22} the *dimensions of admittance* (measured in mho) whereas h_{12} and h_{21} are *dimensionless* numbers (no unit). As these parameters have mixed or hybrid dimensions, they are called the ***hybrid parameters*** or simply ***h-parameters***. These four parameters can be defined by putting $v_2 = 0$ (i.e., the output port short-circuited to a.c) or $i_1 = 0$ (i.e., the input port open-circuited to a.c) in equations (7.2.1) and (7.2.2) respectively. Thus, we plainly have

$$h_{11} = \left.\frac{v_1}{i_1}\right|_{v_2=0} = \text{ } \textit{input impedance} \text{ with output terminals shorted to a.c}$$

$$h_{12} = \left.\frac{v_1}{v_2}\right|_{i_1=0} = \text{ } \textit{reverse voltage} \text{ gain with input terminals open-circuited to a.c}$$

$$h_{21} = \left.\frac{i_2}{i_1}\right|_{v_2=0} = \text{ } \textit{forward current gain} \text{ with output teminals shorted to a.c}$$

$$h_{22} = \left.\frac{i_2}{v_2}\right|_{i_1=0} = \text{ } \textit{output admittance} \text{ with the input terminals open-circuited to a.c}$$

There is however a general notations for h-parameters. These are $h_{11} = h_i$, $h_{12} = h_r$, $h_{21} = h_f$ and $h_{22} = h_o$ where the subscripts i, r, f and o stand for input, reverse, forward and output respectively. For transistor, a second subscript b, c or e is also added in order to designate the circuit configuration. For instance, b subscript would denote CB, c subscript would refer to CC and e subscript the CE configuration. So, for CE configuration, the parameters are denoted by h_{ie}, h_{re}, h_{fe} and h_{oe} respectively.

7.3 Hybrid equivalent circuit

For small signal operation of a transistor, the a.c voltages and currents are small compared to d.c voltages and currents. In this case, the hybrid parameters are real numbers and the equations (7.2.1) and (7.2.2) suggest to construct an equivalent circuit of the two-port network. This equivalent circuit is known as ***hybrid equivalent circuit*** or ***transistor hybrid model*** (Fig. 7.2).

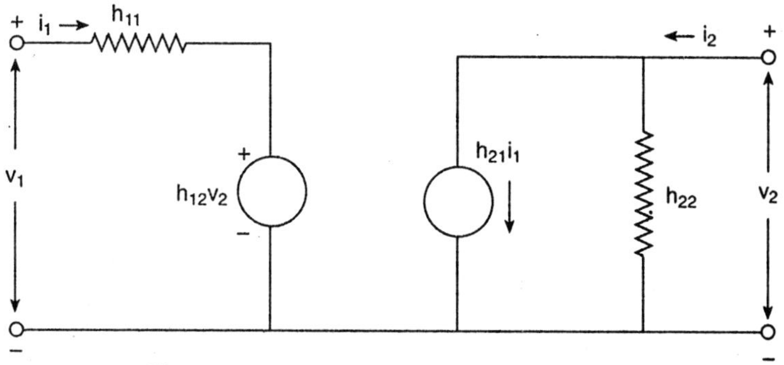

Fig. 7.2 *h*-equivalent circuit of a two-port network

Applications of KVL and KCL to the input and output circuits give the the relations (7.2.1) and (7.2.2) respectively. As the voltages and currents are a.c values of the signal, Fig. 7.2 is also the *a.c equivalent circuit.*

7.4 Advantages of *h*-parameters

The hybrid-parameter model has a number of *advantages* and some of the important ones are enumerated as under.

(i) The *h*-parameters can be easily determined from the static characteristic curves of a transistor.

(ii) At very low or audio frequencies, the *h*-parameters are all *real numbers.* It implies that reactive elements are absent in the two-port device.

(iii) The performance factors such as current, voltage and power gains etc. can be easily calculated from simple mathematical expressions.

(iv) The *h*-parameters for a transistor are usually specified by the manufacturers.

(v) The hybrid-parameter model is rather convenient to use in circuit analysis and in design.

7.5 *h*-parameters from transistor characteristics

The *h*-parameters can be easily determined from the input and output characteristics of a transistor. Since the CE mode is the most commonly used transistor configuration and the manufacturers generally specify the CE *h*-parameters, we are interested to discuss the determination of *h*-parameters in this mode.

In CE configuration, the instantaneous total base voltage v_B and the output collector current i_C are both functions of the instantaneous total base current i_B and the collector voltage v_C. Therefore, we get

$$v_B = \phi_1(i_B, v_C) \tag{7.5.1}$$

$$\text{and} \quad i_C = \phi_2(i_B, v_C) \tag{7.5.2}$$

By Taylor's series expansion of the above equations about the Q-point (I_B, V_C) and ignoring the higher order terms, we obtain

$$\Delta v_B = \left.\frac{\partial v_B}{\partial i_B}\right|_{V_C} \Delta i_B + \left.\frac{\partial v_B}{\partial v_C}\right|_{I_B} \Delta v_C \tag{7.5.3}$$

$$\text{and} \quad \Delta i_C = \left.\frac{\partial i_C}{\partial i_B}\right|_{V_C} \Delta i_B + \left.\frac{\partial i_C}{\partial v_C}\right|_{I_B} \Delta v_C \tag{7.5.4}$$

In the above equations Δi_B, Δi_C, Δv_B and Δv_C are the small-signal (a.c) base and collector currents and voltages. So, using the standard notations, equations (7.5.3) and (7.5.4) can be re-written in the following form

$$v_b = h_{ie} i_b + h_{re} v_c \tag{7.5.5}$$

$$\text{and} \quad i_c = h_{fe} i_b + h_{oe} v_c \tag{7.5.6}$$

$$\text{where} \quad h_{ie} = \left.\frac{\partial v_B}{\partial i_B}\right|_{V_C}, \qquad h_{re} = \left.\frac{\partial v_B}{\partial v_C}\right|_{I_B},$$

$$h_{fe} = \left.\frac{\partial i_C}{\partial i_B}\right|_{V_C} \quad \text{and} \quad h_{oe} = \left.\frac{\partial i_C}{\partial v_C}\right|_{I_B} \tag{7.5.7}$$

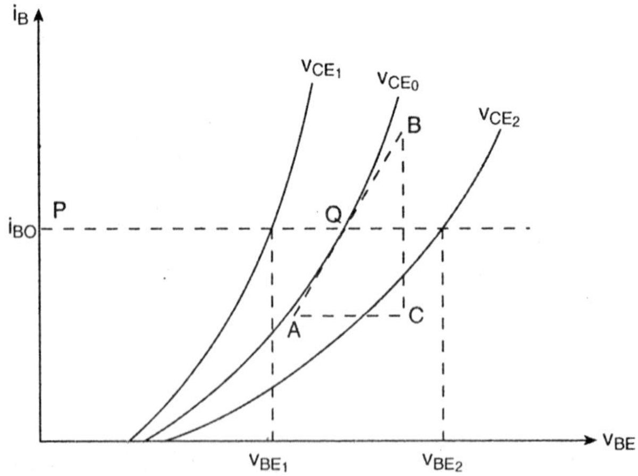

Fig. 7.3 Determination of h_{ie} and h_{re} from input characteristics

The parameters h_{ie} and h_{re} can be determined from the input characteristic curves. First, we are to locate the Q-point on the input characteristics as indicated in Fig. 7.3. To establish I_B = const. (i.e., i_{BO}), a horizontal line is drawn through the Q-point.

From the definition of h_{re}, we have

$$h_{re} = \left.\frac{\partial v_B}{\partial v_C}\right|_{I_B} = \left.\frac{v_{BE_2} - v_{BE_1}}{v_{CE_2} - v_{CE_1}}\right|_{i_{BO}} \tag{7.5.8}$$

The slope of the curve at the Q-point gives h_{ie}. Therefore, we have

$$h_{ie} = \left.\frac{\partial v_B}{\partial i_B}\right|_{V_C} = \frac{AC}{BC} \tag{7.5.9}$$

To find the h_{fe} and h_{oe} we first locate the Q-point on the output characteristic curves (Fig. 7.4). To establish V_C = constant (i.e., v_{CEO}), a vertical line is drawn through the Q-point.

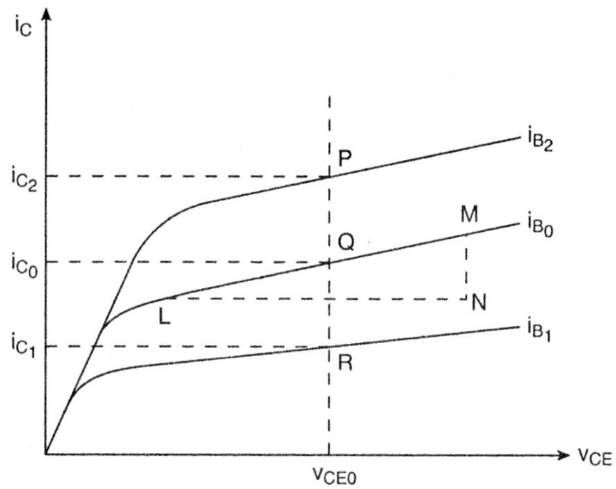

Fig. 7.4 Determination of h_{fe} and h_{oe} from output characteristics

From the definition of h_{fe}, we obtain

$$h_{fe} = \left.\frac{\partial i_C}{\partial i_B}\right|_{V_C} = \left.\frac{i_{C_2} - i_{C_1}}{i_{B_2} - i_{B_1}}\right|_{v_{CEO}} \tag{7.5.10}$$

If we now draw a tangent to the given curve at the point Q, we get the value of h_{oe} which is given by

$$h_{oe} = \left.\frac{\partial i_C}{\partial v_C}\right|_{I_B} = \frac{MN}{LN} \tag{7.5.11}$$

Similarly, the h-parameters for the other configurations (CB and CC) may be determined.

It is to be noted that the h-parameters are *not constants* but are dependent on the position of the Q-point and also on the temperature that affects the characteristic curves.

A set of typical h-parameters for a transistor of different configurations are given below in Table 7.1.

Table 7.1 : Values of typical h-parameters for a BJT

Parameters	CE	CB	CC
h_i	600 to 2000 Ω	30 to 80 Ω	600 to 2000 Ω
h_r	3×10^{-4}	5×10^{-6} to 200×10^{-6}	1
h_f	10 to 100	−0.90 to −0.989	−10 to −100
h_O	20×10^{-6} ℧	0.5×10^{-6} to 1.5×10^{6} ℧	20×10^{-6} ℧

7.6 Analysis of a transistor amplifier by h-parameters

The a.c equivalent circuit of a transistor amplifier using h-parameters is shown in Fig. 7.5. A signal source V_s having the output resistance R_s is connected across the input terminal $(1, 1')$ and a load resistance R_L is connected to the output terminal $(2, 2')$ of the amplifier. Since the Q-values do not appear in the a.c equivalent circuit, the biasing circuit is excluded. To understand the performance of the amplifier, we are interested to find the current gain, input impedance, voltage gain, output impedance and power gain. We shall consider here the general case by using h-parameters with their subscripts and assume the sinusoidal variations for the current and voltage to use the r.m.s. values of the a.c quantities for analysis.

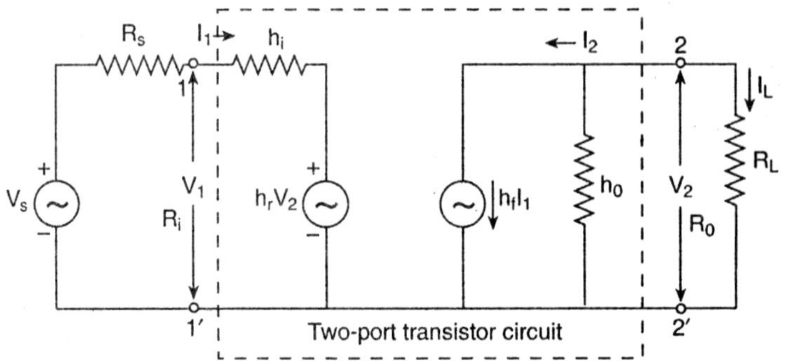

Fig. 7.5 h-parameter model of a transistor amplifier

(a) ***Current gain*:** The current gain or the current amplification A_I of a transistor amplifier is the ratio of the output current to the input current.

In reference to Fig. 7.5, therefore, we have:

$$\text{Current gain,} \quad A_I = \frac{I_L}{I_1} = -\frac{I_2}{I_1} \tag{7.6.1}$$

Applying the KCL to the output circuit, we get

$$I_2 = h_O V_2 + h_f I_1 = h_O(-I_2 R_L) + h_f I_1$$

$$(\because \; V_2 = I_L R_L = -I_2 R_L)$$

$$\therefore \quad (1 + h_O R_L) I_2 = h_f I_1$$

$$\Rightarrow \quad \frac{I_2}{I_1} = \frac{h_f}{1 + h_O R_L}$$

$$\therefore \quad \textit{Current gain, } \boldsymbol{A_I} = -\frac{I_2}{I_1} = -\frac{\boldsymbol{h_f}}{\boldsymbol{1 + h_O R_L}} \tag{7.6.2}$$

(b) ***Input impedance*****:** The input impedance or input resistance R_i is defined as the resistance obtained at the input terminal $(1, 1')$. It is the ratio of the input voltage V_1 at the input terminal $(1, 1')$ to the input current I_1.

$$\therefore \quad R_i = \frac{V_1}{I_1} \tag{7.6.3}$$

Applying the KVL to the input circuit, we immediately get

$$V_1 = h_i I_1 + h_r V_2$$

$$\text{or,} \quad \frac{V_1}{I_1} = h_i - h_r \frac{I_2 R_L}{I_1}, \quad \text{since } V_2 = -I_2 R_L$$

$$\text{or,} \quad R_i = h_i + h_r A_I R_L, \quad (\because \; A_I = -I_2/I_1)$$

Using (7.6.2) in the above equation, we finally get,

$$\boldsymbol{R_i = h_i - \frac{h_f h_r R_L}{1 + h_O R_L}} \tag{7.6.4}$$

(c) ***Voltage gain*****:** The voltage gain or voltage amplification factor A_V is defined as the ratio of the output voltage V_2 to the input voltage V_1. So,

$$A_V = \frac{V_2}{V_1} = -\frac{I_2 R_L}{I_1 R_i}$$

$$= A_I \frac{R_L}{R_i} \quad (\because \; A_I = -I_2/I_1)$$

Substituting for A_I from equation (7.6.2) and R_i from equation (7.6.4) in the above equation, we get

$$A_V = \frac{-h_f R_L}{(1 + h_O R_L)\left(h_i - \frac{h_f h_r R_L}{1 + h_O R_L}\right)}$$

$$= \frac{-h_f R_L}{h_i + (h_i h_O - h_f h_r) R_L}$$

$$\text{or,} \quad \boldsymbol{A_V = -\frac{h_f R_L}{h_i + \Delta h \cdot R_L}} \tag{7.6.5}$$

where Δh is given by $\Delta h = h_i h_O - h_f h_r$.

(d) ***Output resistance*****:** The output resistance R_O of the amplifier is the resistance we look into the output terminal $(2, 2')$ when the voltage source is replaced by its internal resistance R_s i.e., when $V_s = 0$ and $R_L = \infty$. In that case, the circuit in Fig. 7.5 reduces to that in Fig. 7.6 where the output terminal $(2, 2')$ is driven by a source V_2. If I_2 is the current delivered by V_2, then

Output impedance is: $$R_O = \frac{V_2}{I_2} \qquad (7.6.6)$$

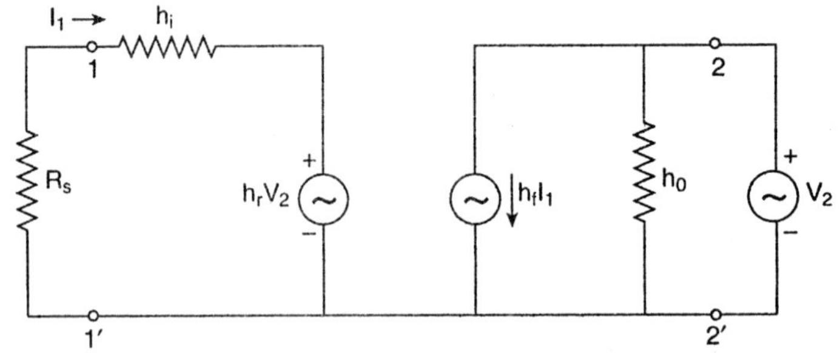

Fig. 7.6 Circuit to determine R_O

Applying the KVL to the input circuit of Fig. 7.6, we get

$$(R_s + h_i)I_1 + h_r V_2 = 0$$

$$\text{or,}\quad I_1 = -\frac{h_r V_2}{R_s + h_i} \qquad (7.6.7)$$

Again, applying the KCL for the output circuit, we obtain

$$I_2 = h_O V_2 + h_f I_1$$

$$= h_O V_2 - \frac{h_f h_r V_2}{R_s + h_i}, \quad \text{using (7.6.7) for } I_1$$

$$= V_2\left(h_O - \frac{h_f h_r}{R_s + h_i}\right)$$

$$\therefore\quad \boldsymbol{R_O} = \frac{V_2}{I_2} = \boldsymbol{\frac{R_s + h_i}{R_s h_O + h_i h_O - h_f h_r}} \qquad (7.6.8)$$

(e) ***Power gain:*** The power gain A_P is the ratio of the output a.c power to the input a.c power. Thus

$$A_P = \frac{V_2 I_L}{V_1 I_1} = -\frac{V_2}{V_1}\cdot\frac{I_2}{I_1} = A_V A_I$$

Substituting for A_V and A_I from (7.6.5) and equation (7.6.2), we get

$$\boldsymbol{A_P} = \left(-\frac{h_f R_L}{h_i + \Delta h R_L}\right)\cdot\left(-\frac{h_f}{1 + h_O R_L}\right)$$

$$= \boldsymbol{\frac{h_f^2 R_L}{(1 + h_O R_L)(h_i + \Delta h R_L)}} \qquad (7.6.9)$$

(f) *Overall current gain and overall voltage gain*: In order to calculate the *overall current gain* A_{I0} and the *overall voltage gain* A_{V0}, the effect of source resistance R_s is taken into account. To obtain the overall current gain we take the Norton's equivalent for the source (Fig. 7.7) at the input terminals $(1, 1')$ of the circuit of Fig. 7.5.

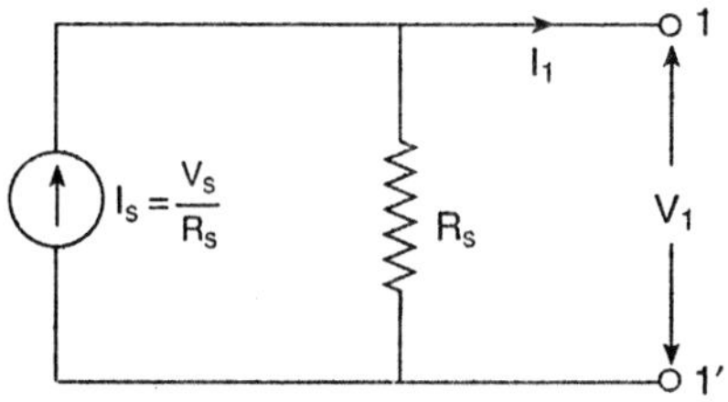

Fig. 7.7 Norton's equivalent of current source

Then, overall current gain is

$$A_{I0} = \frac{I_L}{I_s} = -\frac{I_0}{I_1}\cdot\frac{I_1}{I_s} = A_I\cdot\frac{I_1}{I_s} \tag{7.6.10}$$

Now, from Fig. 7.7 we get

$$(I_s - I_1)R_s = V_1$$

$$\text{or,}\quad \left(\frac{I_s}{I_1} - 1\right)R_s = \frac{V_1}{I_1} = R_i$$

$$\text{or,}\quad \frac{I_1}{I_s} = \frac{R_s}{R_i + R_s} \tag{7.6.11}$$

From (7.6.10) and (7.6.11), we have

$$A_{I0} = A_I\cdot\frac{R_s}{R_i + R_s} \tag{7.6.12}$$

Clearly, for an ideal current source ($R_s \to \infty$), $A_{IO} \to A_I$ i.e., the overall current gain reduces to current gain of the amplifier.

When the source resistance R_s is taken into account, the *overall voltage gain* is given by

$$A_{V0} = \frac{V_2}{V_s} = \frac{V_2}{V_1}\cdot\frac{V_1}{V_s} = A_V\cdot\frac{V_1}{V_s} \tag{7.6.13}$$

From Fig. (7.5) we have,

$$V_1 = I_1R_i = \frac{V_s}{R_i + R_s}R_i$$

$$\text{or,}\quad \frac{V_1}{V_s} = \frac{R_i}{R_i + R_s} \tag{7.6.14}$$

Using (7.6.14) in (7.6.13), we obtain

$$A_{V0} = A_V\cdot\frac{R_i}{R_i + R_s} = A_I\frac{R_L}{R_i + R_s} \tag{7.6.15}$$

since $A_V = A_IR_L/R_i$.

For an ideal voltage source $R_s = 0$. So, A_{V0} reduces to A_V. But, in practice $R_s \neq 0$ and hence $A_{V0} < A_V$.

The important results derived in this section are summarized below in Table 7.2 for ready reference.

Table 7.2 : Results of analysis of transistor amplifier

Current gain	$A_I = -\dfrac{h_f}{1 + h_O R_L}$
Input resistance	$R_i = h_i - \dfrac{h_f h_r R_L}{1 + h_O R_L}$
Voltage gain	$A_V = A_I \dfrac{R_L}{R_i} = -\dfrac{h_f R_L}{h_i + (h_i h_O - h_f h_r) R_L}$
Output resistance	$R_O = \dfrac{R_s + h_i}{R_s h_O + h_i h_O - h_f h_r}$
Power gain	$A_P = \dfrac{h_f^2 R_L}{(1 + h_O R_L)[h_i + (h_i h_O - h_f h_r) R_L]}$
Overall current gain	$A_{IO} = A_I \dfrac{R_s}{R_i + R_s}$
Overall voltage gain	$A_{VO} = A_V \dfrac{R_i}{R_i + R_s} = A_I \dfrac{R_L}{R_i + R_s}$

7.7 Conversion formulae for h-parameters

We have just considered the general case with h-parameters represented by their first subscript in obtaining the expression for A_I, R_i, A_V, R_O and A_P. These expressions, appropriate to a particular mode, can be achieved by adding the second subscript to the h-parameters for that mode. The transistor manufacturers generally supply the values of h-parameters for CE mode. So, it is necessary to convert the set of h-values of CE mode to the set of h-values for CB or CC mode.

The adjoining table gives the approximate relationships of CB and CC h-parameters in terms of CE h-parameters.

Table 7.3 : Conversion formulae

CB	CC
$h_{ib} = \dfrac{h_{ie}}{1 + h_{fe}}$	$h_{ic} = h_{ie}$
$h_{rb} = \dfrac{h_{ie} h_{oe}}{1 + h_{fe}} - h_{re}$	$h_{rc} = 1$
$h_{fb} = -\dfrac{h_{fe}}{1 + h_{fe}}$	$h_{fc} = -(1 + h_{fe})$
$h_{ob} = \dfrac{h_{oe}}{1 + h_{fe}}$	$h_{oc} = h_{oe}$

7.8 Emitter follower

A common collector amplifier using an n-p-n transistor is shown in Fig. 7.8. The output voltage is developed across the emitter resistance R_L. When the base voltage increases, it causes an increase in emitter current which results in an increase in the voltage drop across R_L. This makes the emitter potential more positive. The emitter potential thus follows the input base potential. Hence the circuit is called ***emitter follower***. It is to be noted that the circuit has approximately unit voltage gain with a very high input resistance and very low output resistance and is widely used as buffer stage between a high impedance source and a low impedance load. It increases the power level of the signal without causing any phase reversal. The h-parameter equivalent circuit of the emitter follower is shown in Fig. 7.9.

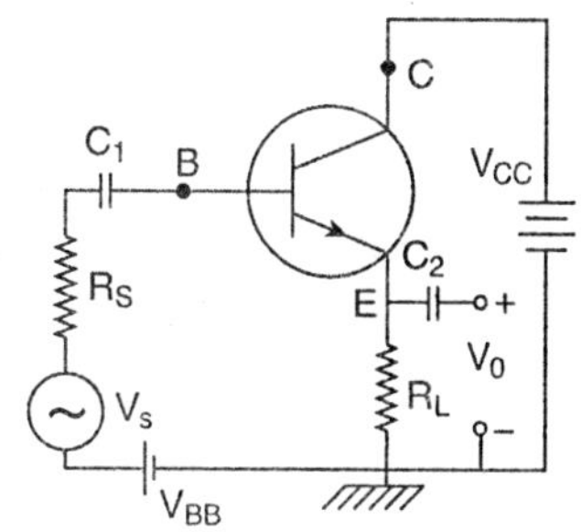

Fig. 7.8 Common collector amplifier: emitter follower

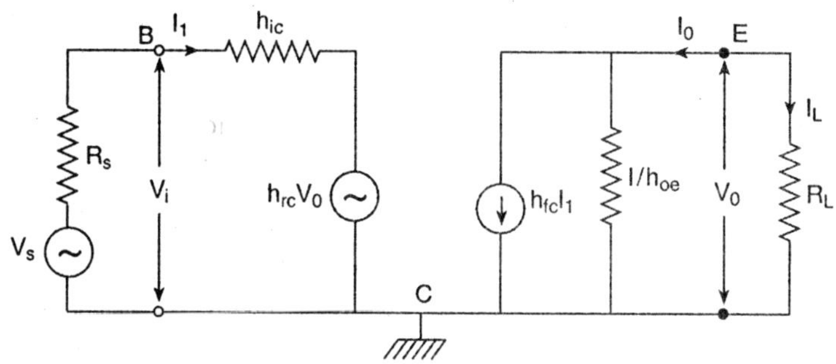

Fig. 7.9 h-parameter equivalent circuit of emitter follower

Current gain: The general expression for current gain is

$$A_I = -\frac{h_f}{1 + h_O R_L} \tag{7.8.1}$$

For CC amplifier, $h_f = h_{fc}$ and $h_O = h_{oc}$ and the conversion table gives

$$h_{fc} = -(1 + h_{fe}) \text{ and } h_{oc} = h_{oe}$$

$\therefore$ From (7.8.1) we have,

$$\textit{Current gain,} \quad \boxed{A_I = \frac{1 + h_{fe}}{1 + h_{oe} R_L}} \tag{7.8.2}$$

Input resistance: From the general relation, we may write the input resistance in terms of CC h-parameters as

$$R_i = h_{ic} + h_{rc} A_I R_L$$

From the conversion table, $h_{ic} = h_{ie}$ and $h_{rc} = 1$.

$$\therefore \quad R_i = h_{ie} + A_I R_L = h_{ie} + \frac{(1 + h_{fe}) R_L}{1 + h_{oe} R_L} \tag{7.8.3}$$

using the relation (7.8.2) for A_I.

Voltage gain: The voltage gain A_V is given by

$$A_V = A_I \cdot \frac{R_L}{R_i}. \tag{7.8.4}$$

For CC amplifier, we may assume $h_{fe} \gg 1$ and $h_{oe} \approx 0$. So, the equations (7.8.3) and (7.8.2) reduce respectively to

$$R_i = h_{ie} + h_{fe} R_L \tag{7.8.5}$$

$$\text{and } A_I = h_{fe}. \tag{7.8.6}$$

Using these values of R_i and A_I in equation (7.8.4), we finally get

$$\boxed{\boldsymbol{A_V = \frac{h_{fe} R_L}{h_{ie} + h_{fe} R_L}}} \tag{7.8.7}$$

Output resistance: Using the general relation, we get the output resistance for the CC amplifier as

$$R_0 = \frac{R_s + h_{ic}}{R_s h_{oc} + h_{ic} h_{oc} - h_{fc} h_{rc}}$$

From the conversion table, $h_{ic} = h_{ie},\ h_{oc} = h_{oe}$

$$h_{fc} = -(1 + h_{fe}),\ h_{rc} = 1$$

$$\therefore\ \boldsymbol{R_0 = \frac{R_s + h_{ie}}{R_s h_{oe} + h_{ie} h_{oe} + (1 + h_{fe})}} \tag{7.8.8}$$

7.9 Simplified hybrid model

Table 7.1 shows that the values of h-parameters vary widely for a given type of transistor. One can get exact values of A_I, A_V, R_i, R_0 etc. by detailed calculations, provided h-parameters are exactly known. Thus, it is necessary to have a simpler analysis so that an approximate solution may provide a better understanding of the physical process involved in the circuit. From the typical values of h-parameters of transistor in CE configuration, we observe that $h_{re} \approx 10^{-4}$ and $h_{oe} \approx 10^{-5}$ A/V. Since

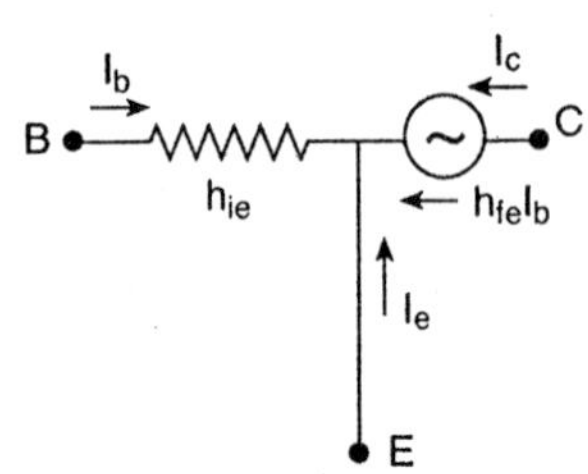

Fig. 7.10(a) Simplified hybrid model for any transistor configuration at low frequency

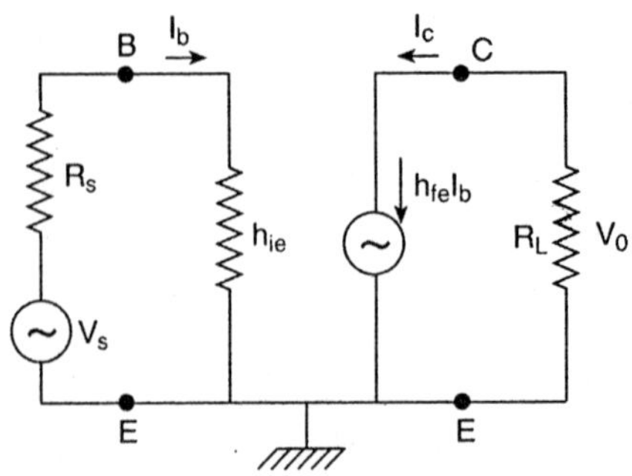

Fig. 7.10(b) Simplified hybrid model for CE configuration

h_{re} is usually very small, we can assume $h_{re}V_0 \approx 0$ and the resistance offered by $1/h_{oe}$ is often too large to be neglected in comparison with the load resistance R_L. So, h_{ie} and h_{fe} are the only h-parameters required for the approximate analysis of the amplifier at low frequency range. Moreover, it is found that the error in A_I, A_V, R_i and R_0 is less than 10% when it is assumed that $h_{fe} \gg 1$ and the load resistance $R_L < 0.1/h_{oe}$.

Considering all the conditions mentioned above, the simplified hybrid model for CE configuration at low frequency can be obtained as shown in Fig. 7.10(a). This model may also be used for any other configuration by connecting the appropriate node to ground. Thus the approximate CE mode hybrid model is obtained as shown in Fig. 7.10(b).

7.9.1 Analysis of CE amplifier using simplified hybrid model

The simplified h-parameter model of Fig. 7.10(b) can be used to obtain the approximate expression of the quantities given in Table 7.2.

Current gain—From the circuit of Fig. 7.10(b), we get $I_c = h_{fe}I_b$. Therefore, the current gain

$$A_I = -\frac{I_c}{I_b} = -h_{fe} \tag{7.9.1}$$

Input resistance—Input resistance, $R_i = \dfrac{\text{input voltage}}{\text{input current}}$

$$\therefore \quad R_i = \frac{V_b}{I_b} = \frac{h_{ie}I_b}{I_b} = h_{ie} \tag{7.9.2}$$

Voltage gain—Voltage gain, A_V is given by

$$A_V = \frac{V_0}{V_i} = A_I\frac{R_L}{R_i} = -\frac{h_{fe}R_L}{h_{ie}} \tag{7.9.3}$$

using (7.9.1) and (7.9.2).

Output resistance—To find the output resistance R_0, the source voltage V_s is replaced by a short circuit, R_L is removed and the output terminals are driven by an external source voltage V_0. If I_0 is the current delivered by V_0, then the output resistance $R_0 = V_0/I_0$. Thus we may write

$$R_0 = \left.\frac{V_0}{I_0}\right|_{\substack{V_S = 0 \\ R_L = \infty}} \tag{7.9.4}$$

Now, with $V_S = 0$ in Fig. 7.10(b), we have $I_b = 0$ and $I_0 = I_c = h_{fe}I_b = 0$. So, equation (7.9.4) gives $R_0 = \infty$.

Overall voltage gain and overall current gain—The overall voltage gain (A_{V0}), obtained by considering the source resistance R_s, is given by

$$A_{V0} = A_I \cdot \frac{R_L}{R_i + R_s}$$

Here, $A_I = -h_{fe}$. So, the overall voltage gain becomes

$$A_{VO} = -\frac{h_{fe}R_L}{R_i + R_s} \tag{7.9.5}$$

The overall current gain is given by

$$A_{IO} = A_I \cdot \frac{R_s}{R_i + R_s} = -\frac{h_{fe}R_s}{R_i + R_s} \tag{7.9.6}$$

7.9.2 Analysis of CB amplifier using simplified hybrid model

The simplified hybrid equivalent circuit of a CB amplifier is shown in Fig. 7.11. This circuit can be used to obtain the approximate expressions of current gain (A_I), input resistance (R_i), voltage gain (A_V) and output resistance (R_O). In this circuit, the input signal is applied to the emitter terminal and the load resistor is connected between the collector and the ground i.e., base terminal.

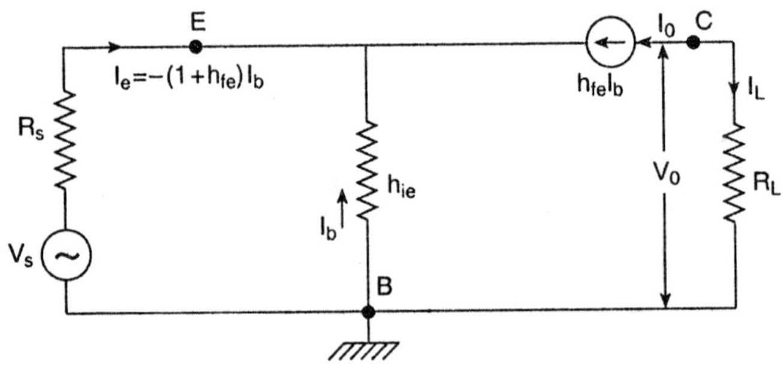

Fig. 7.11 Simplified hybrid model of CB amplifier

Current gain—From the circuit of Fig. 7.11, we have

$$\text{current gain, } A_I = \frac{I_L}{I_e} = \frac{-h_{fe}I_b}{-(1 + h_{fe})I_b}$$

$$= \frac{h_{fe}}{1 + h_{fe}} = -h_{fb} \tag{7.9.7}$$

Since $h_{fe} \gg 1$, the current gain $A_I \approx 1$.

Input resistance—Input resistance, R_i is given by

$$R_i = \frac{V_i}{I_e} = \frac{-h_{ie}I_b}{-(1 + h_{fe})I_b} = \frac{h_{ie}}{1 + h_{fe}} = h_{ib} \tag{7.9.8}$$

Voltage gain—The voltage gain, $A_V = \dfrac{V_O}{V_i} = \dfrac{I_L R_L}{I_e R_i} = A_I \dfrac{R_L}{R_i}$

Since $A_I = -h_{fb}$ and $R_i = h_{ib}$, we have from above

$$A_V = -\frac{h_{fb}R_L}{h_{ib}} \tag{7.9.9}$$

Output impedance—From the input circuit of Fig. 7.11, we get

$$V_s = I_e R_s - h_{ie} I_b = -(1 + h_{fe}) I_b \cdot R_s - h_{ie} I_b \qquad (7.9.10)$$

By definition, the output resistance R_0 is obtained by replacing V_s by a short circuit, removing the load resistance R_L and driving the output terminals with an external source voltage V_0. In that case, if I_0 be the current delivered by V_0 then $R_0 = V_0/I_0$.

Now, from (7.9.10), $V_s = 0$ implies that $I_b = 0$ and hence $R_0 = \infty$.

7.9.3 Analysis of CC amplifier using simplified hybrid model

The simplified hybrid equivalent circuit of a *CC* amplifier is shown in Fig. 7.12. In this circuit, the transistor is replaced by the approximate *CE* equivalent circuit. The supply voltage V_{CC} is replaced by a short circuit. The load resitance R_L is connected between the emitter and the collector. The collector is grounded for a.c.

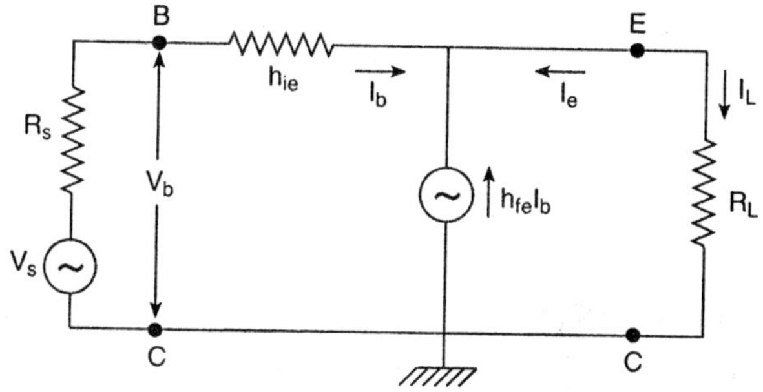

Fig. 7.12 Simplified hybrid model of CC configuration

Current gain—Applying KCL to the emitter junction E of Fig. 7.12, we get

$$I_e = -(I_b + h_{fe} I_b) = -(1 + h_{fe}) I_b$$

The load current $I_L = -I_e = (1 + h_{fe}) I_b \qquad (7.9.11)$

$\therefore$ The current gain, $A_I = \dfrac{I_L}{I_b} = 1 + h_{fe} \simeq h_{fe}, \quad (\because h_{fe} \gg 1). \qquad (7.9.12)$

Input resistance–Input voltage, $V_i = V_b = I_b h_{ie} + I_L R_L$

$$= [h_{ie} + (1 + h_{fe}) R_L] I_b, \text{ using (7.9.11).}$$

$\therefore$ Input resistance, $R_i = \dfrac{V_i}{I_b} = h_{ie} + (1 + h_{fe}) R_L \qquad (7.9.13)$

$$= h_{ie} + h_{fe} R_L, \quad \text{since } h_{fe} \gg 1. \qquad (7.9.14)$$

Voltage gain—Voltage gain, $A_V = A_I \dfrac{R_L}{R_i} = \dfrac{h_{fe} R_L}{h_{ie} + h_{fe} R_L} \qquad (7.9.15)$

using (7.9.12) and (7.9.13) for R_L and R_i respectively.

The *overall voltage gain*, A_{VO} including the source resistance R_s can be determined by using the relation (7.6.15) as

$$A_{VO} = A_I \frac{R_L}{R_i + R_s} = \frac{h_{fe}R_L}{R_s + h_{ie} + h_{fe}R_L} \tag{7.9.16}$$

Output resistance—To find the output resistance R_O, we replace V_s by a short circuit and consider an external driving voltage V_O at the output terminals after disconnecting the load resistance R_L (Fig. 7.13).

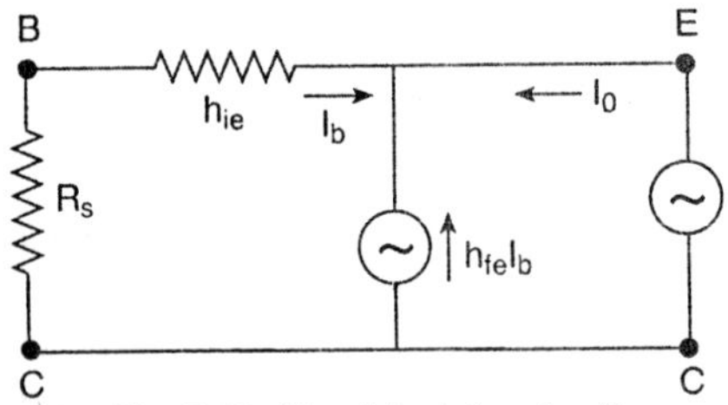

Fig. 7.13 Circuit to determine R_O

Thus $R_O = \left.\dfrac{V_O}{I_O}\right|_{\substack{V_s = 0 \\ R_L = \infty}}$

where I_O is the current delivered by V_O.

Now from Fig. 7.13, $\quad V_O = -I_bR_s - I_bh_{ie} = -(R_s + h_{ie})\,I_b \qquad (7.9.17)$

and $\quad I_O = -(1 + h_{fe})I_b \simeq -h_{fe}I_b, \quad$ since $h_{fe} \gg 1. \qquad (7.9.18)$

So, the output resistance, $\quad R_O = \dfrac{V_O}{I_O} = \dfrac{R_s + h_{ie}}{h_{fe}} \qquad (7.9.19)$

using (7.9.17) and (7.9.18).

7.10 Approximate expressions for performance quantities in CE, CB and CC modes

The approximate expressions for the performance quantities (A_I, R_i, A_V etc.) of an amplifier have been derived in the previous three sections for CE, CB and CC configurations. They are now given in Table 7.4 below for ready reference.

Table 7.4 Approximate expressions for A_I, R_i, A_v, A_{VO} etc.

Performance quantities	CE mode	CB mode	CC mode
A_I	$-h_{fe}$	$-h_{fb} \approx 1$	h_{fe}
R_i	h_{ie}	$h_{ib} = \dfrac{h_{ie}}{1 + h_{fe}}$	$h_{ie} + h_{fe}R_L$
A_V	$-\dfrac{h_{fe}R_L}{h_{ie}}$	$-\dfrac{h_{fb}R_L}{h_{ib}} = h_{fe}\dfrac{R_L}{R_i}$	$\dfrac{h_{fe}R_L}{h_{ie} + h_{fe}R_L}$
A_{VO}	$-\dfrac{h_{fe}R_L}{R_s + h_{ie}}$	$-\dfrac{h_{fe}R_L}{R_s + h_{fe} + h_{ie}}$	$\dfrac{h_{fe}R_L}{R_s + h_{ie} + h_{fe}R_L}$
R_O	∞	∞	$\dfrac{R_s + h_{ie}}{h_{fe}}$

7.11 Darlington pair

Darlington pair is a three-terminal device in which a pair of identical transistors are connected in such a way that the emitter of one transistor is connected to the base of the other so that it acts like a single unit with an extremely high current gain. Fig. 7.14(a) shows a *npn* Darlington pair and Fig. 7.14(b) a *pnp* Darlington pair.

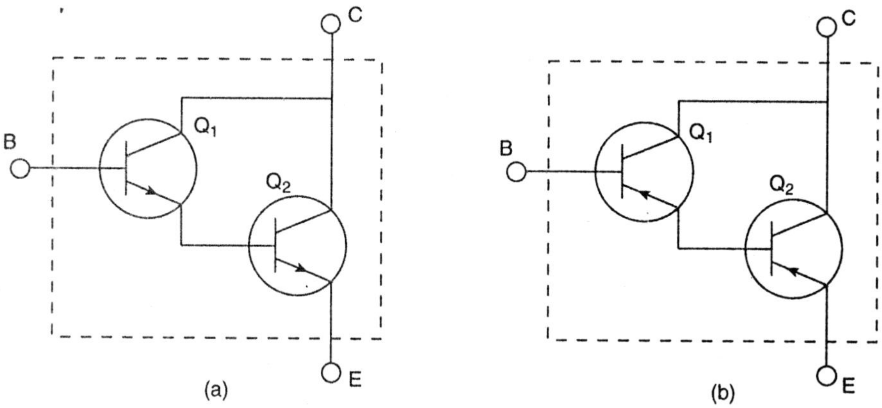

Fig. 7.14 Darlington pair: (a) *npn* (b) *pnp*

In both the above cases, the collectors of the component transistors are connected to a common point C. So, a Darlington pair has *three* terminals, namely, the base (B), the emitter (E) and the collector (C). In actual operation, a Darlington pair can treated as two cascaded emitter follower (Fig. 7.15).

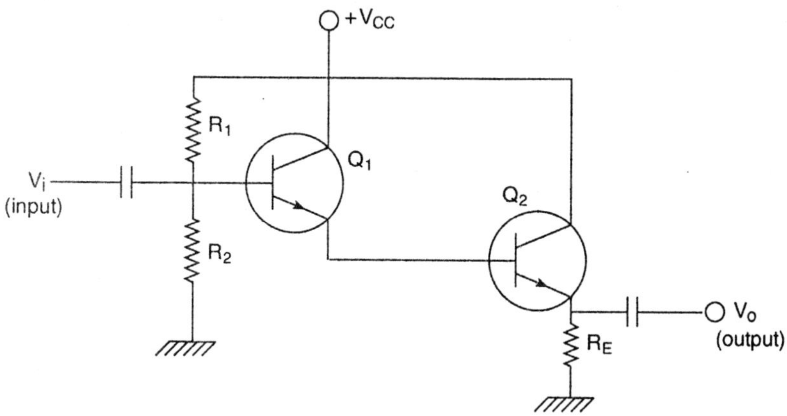

Fig. 7.15 Darlington connection

The emitter current of Q_1 is given by

$$I_{e_1} = I_{b_1} + I_{c_1} = (1 + \beta_1) I_{b_1} \tag{7.11.1}$$

Similarly, for Q_2, the emitter current is given by

$$I_{e_2} = (1 + \beta_2) I_{b_2} \tag{7.11.2}$$

Since, $I_{e_1} = I_{b_2}$, we obtain

$$I_{e_2} = (1+\beta_2)(1+\beta_1)I_{b_1}$$
$$\approx \beta_1\beta_2 I_{b_1}, \quad (\because\ \beta_1 \gg 1 \text{ and } \beta_2 \gg 1) \tag{7.11.3}$$

The overall current gain is, therefore, $$\frac{I_{e_2}}{I_{b_1}} = \beta_1\beta_2 \tag{7.11.4}$$

So, the *overall current gain is equal to the product of the individual current gains* of Q_1 and Q_2. Thus, a Darlington pair can be used to provide a very large current gain ($\sim 10^3$).

Since the Darlington pair has two emitter follower in cascade, the overall voltage gain is slightly less than unity. The input resistance ($\sim \beta_1\beta_2 R_E$) is higher and the output resistance is lower than that of the single-stage emitter follower. A Darlington pair may, therefore, be considered as an improvement over a single-stage emitter follower.

It is to be noted that the leakage current of the *first* transistor is amplified by the second one. This is a *disadvantage* of Darlington pair. However, Darlington pair is very *useful* in circuits of voltage regulators, current and power amplifiers.

7.12 Illustrated Examples

▶ **Example 1.** *The hybrid parameters of a transistor used as an amplifier in* CE *configuration are* $h_{ie} = 800\Omega$, $h_{fe} = 46$, $h_{oe} = 80 \times 10^{-6}\mho$ *and* $h_{re} = 5.4 \times 10^{-4}$. *If the load resistance is* $5\,\text{k}\Omega$ *and the effective source resistance is* 500Ω, *calculate* (i) *the current gain,* (ii) *the input resistance,* (iii) *voltage gain,* (iv) *output resistance and* (v) *the power gain.* **(Garhwal Univ.)**

Solution: Given, $h_{ie} = 800\Omega$, $h_{fe} = 46$, $h_{oe} = 80 \times 10^{-6}\mho$ and $h_{re} = 5.4 \times 10^{-4}$.

Load resistance, $R_L = 5\,\text{k}\Omega$ and effective source resistance, $R_s = 500\Omega$.

(i) The current gain, $$\boldsymbol{A_I} = -\frac{h_{fe}}{1+h_{oe}R_L} = -\frac{46}{1+80\times10^{-6}\times5\times10^3} = \mathbf{-32.9}$$

(ii) Input resistance, $$\boldsymbol{R_i} = h_{ie} + h_{re}A_I R_L = 800 - 5.4\times10^{-4}\times32.9\times5\times10^3 = \mathbf{711\Omega}$$

(iii) Voltage gain, $$A_V = A_I\frac{R_L}{R_i} = -32.9\times\frac{5\times10^3}{711} = -231.$$

The overall voltage gain, $$\boldsymbol{A_{V0}} = A_V\cdot\frac{R_i}{R_i+Rs} = -\frac{231\times711}{711+500} = \mathbf{-135.6}$$

(iv) Output resistance, $\boldsymbol{R_O} = \dfrac{R_s + h_{ie}}{R_s h_{oe} + h_{ie} h_{oe} - h_{fe} h_{re}}$

$$= \frac{500 + 800}{500 \times 80 \times 10^{-6} + 800 \times 80 \times 10^{-6} - 46 \times 5.4 \times 10^{-4}}\,\Omega$$

$= \mathbf{16.4\,k\Omega}$, on simplification.

(v) The power gain, $\boldsymbol{A_P} = A_I \cdot A_{VO} = -32.9 \times (-135.6) = \mathbf{4461}$

▶ **Example 2.** *A CE amplifier has a load resistance of 3 kΩ and an unbypassed 100Ω resistor in the emitter lead. The h-parameters are: h_{ie} = 4000Ω, $h_{re} = 7 \times 10^{-4}$, h_{fe} = 135 and h_{oe} = 25μA/V. Determine the current gain, the input resistance, the output resistance and the voltage gain.* **(Calcutta Hons.)**

Solution: Here, the load resistance, $R_L = 3\,\text{k}\Omega$; resistance connected to the emitter, $R_E = 100\Omega$; $h_{ie} = 4000\Omega$; $h_{re} = 7 \times 10^{-4}$; $h_{fe} = 135$ and $h_{oe} = 25\mu\text{A/V}$.

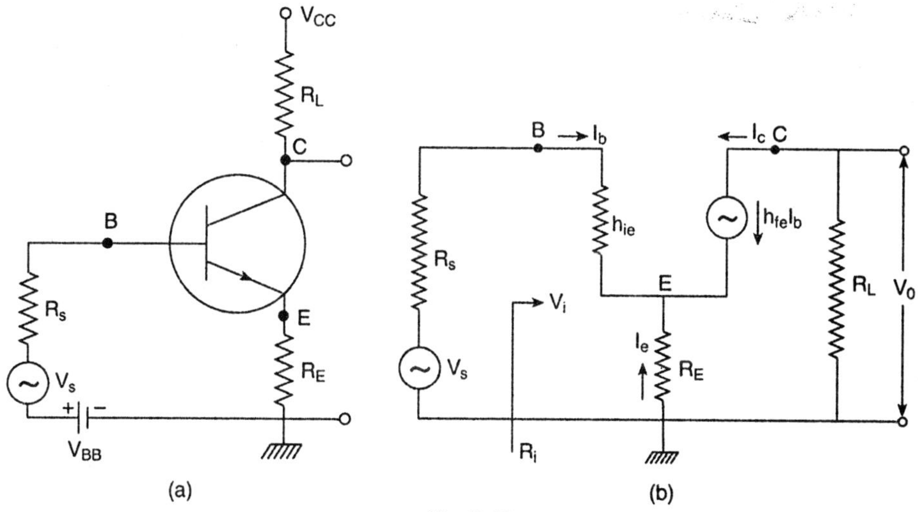

Fig. 7.16

Now, $h_{oe} \cdot R_L = 25 \times 10^{-6} \times 3 \times 10^3 = 0.075 < 0.1$ i.e., $R_L < \frac{1}{10} h_{oe}$

So, we can use the h-parameter equivalent circuit. The CE amplifier circuit and the equivalent circuit are shown in Fig. 7.16(a) and 7.16(b) respectively.

From Fig. 7.16, the ***current gain***, $\boldsymbol{A_I} = -\dfrac{I_c}{I_b} = -h_{fe} = \mathbf{-135}$.

The ***input resistance*** R_i is given by

$$R_i = \frac{V_i}{I_b} = \frac{I_b h_{ie} - I_e R_E}{I_b} = h_{ie} - \frac{I_e}{I_b} R_E$$

$$= h_{ie} + (1 + h_{fe}) R_E$$

$$\therefore\ \boldsymbol{R_i} = 4000 + (1 + 135) \times 100\Omega = \mathbf{17.6\,k\Omega}$$

$$\text{The } \textbf{voltage gain, } A_V = A_I \frac{R_L}{R_i} = -h_{fe} \cdot \frac{R_L}{h_{ie} + (1 + h_{fe}) R_E}$$

$$= -\frac{135 \times 3 \times 10^3}{17.6 \times 10^3} = \mathbf{-23}$$

When R_L is excluded, the ***output resistance*** becomes ***infinite*** due to the small value of h_{oe}. The output resistance becomes R_L i.e., $\mathbf{3\,k\Omega}$ when the load is included in the circuit.

► **Example 3.** *In the small signal amplifier as in* Fig. 7.17, $h_{fe} = 100$, $h_{ie} = 560\Omega$, $R_C = 2\,k\Omega$, $R_E = 1\,k\Omega$, $R_B = 600\,k\Omega$; *h_{re} and h_{oe} are negligible.*

Draw the h-parameter equivalent circuit for the amplifier. Calculate the input and the output impedances and voltage gain of the amplifier. Why are small signals used?

(Calcutta Hons.)

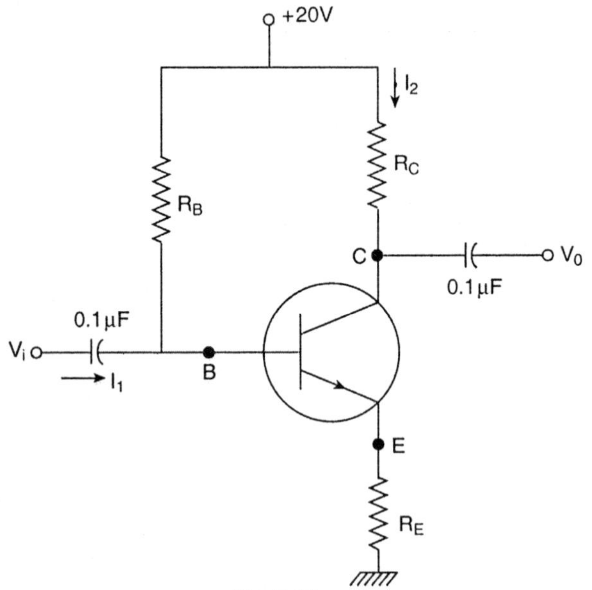

Fig. 7.17

Solution: Given, h_{re} and h_{oe} are both negligible. So, we can use the simplified equivalent circuit as illustrated in Fig. 7.18.

The input resistance (excluding R_B) is

$$R_i = \frac{V_i}{I_1} = \frac{I_1 h_{ie} - I_e R_E}{I_1} = h_{ie} + (1 + h_{fe}) R_E$$

$$= 560 + 101 \times 10^3 \Omega = 101.56\,k\Omega$$

When R_B is included, the ***input resistance*** becomes

$$\boldsymbol{R_i'} = R_i \| R_B = \frac{R_i R_B}{R_i + R_B} = \frac{101.56 \times 600}{101.56 + 600}\,k\Omega$$

$$= \mathbf{86.86\,k\Omega}$$

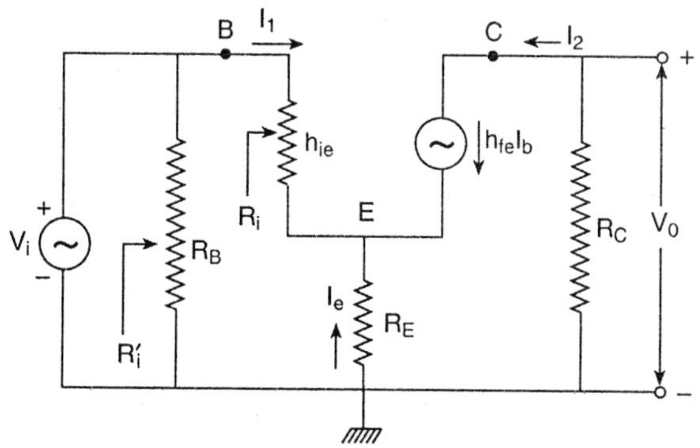

Fig. 7.18

Excluding R_C, the ***output resistance*** is ***infinity***. When the load R_C is included, the output resistance, $R_0 = R_C = \mathbf{2\,k\Omega}$.

The ***voltage gain,*** $\boldsymbol{A_V} = A_I \dfrac{R_C}{R_i} = -h_{fe} \dfrac{R_c}{h_{ie} + (1 + h_{fe})R_E}$

$$= -\frac{100 \times 2 \times 10^3}{101.56 \times 10^3} = \mathbf{-1.97}$$

The equivalent circuit holds good when small signals are used and the output waveform remains undistorted. So, small signals are used.

▶ **Example 4.** *A transistor amplifier in* CE *configuration couples a source of internal resistance* $1\,\text{k}\Omega$ *to a load of* $20\,\text{k}\Omega$. *Find the input and the output resistance if* $h_{ie} = 1\,\text{k}\Omega$, $h_{re} = 2.5 \times 10^{-4}$, $h_{fe} = 150$ *and* $1/h_{oe} = 40\,\text{k}\Omega$. **(Calcutta Hons.)**

Solution: Here, the current gain, $A_I = -\dfrac{h_{fe}}{1 + h_{oe}R_L} = -\dfrac{150}{1 + 20/40} = -100$

∴ Input resistance, $\boldsymbol{R_i} = h_{ie} + A_I h_{re} R_L$

$$= 1000 - 100 \times 2.5 \times 10^{-4} \times 20 \times 10^3$$

$$= 1000 - 500 = \mathbf{500\Omega}$$

Output resistance, $\boldsymbol{R_o} = \dfrac{R_s + h_{ie}}{R_s h_{oe} + h_{ie} h_{oe} - h_{fe} h_{re}}$

$$= \frac{1000 + 1000}{(1/40) + (1/40) - 150 \times 2.5 \times 10^{-4}} = \mathbf{160\,k\Omega}$$

▶ **Example 5.** *A* CE *amplifier uses a transistor with* $h_{ie} = 1\,\text{k}\Omega$, $h_{fe} = 100$, $h_{re} = 5 \times 10^{-4}$ *and* $h_{oe} = 25 \times 10^{-6}\Omega^{-1}$. *The load resistance is* $5\,\text{k}\Omega$. *Find the current amplification and the overall voltage and power gains for a source resistance of* $1\,\text{k}\Omega$.

Solution: Here, we have $h_{ie} = 1\,\text{k}\Omega$, $h_{fe} = 100$, $h_{re} = 5 \times 10^{-4}$, $h_{oe} = 25 \times 10^{-6}\Omega^{-1}$, $R_L = 5\,\text{k}\Omega$ and $R_s = 1\,\text{k}\Omega$.

$\therefore$ Current amplification, $\boldsymbol{A_I} = \dfrac{h_{fe}}{1 + h_{oe}R_L} = -\dfrac{100}{1 + 25 \times 10^{-6} \times 5 \times 10^3} = \mathbf{-88.89}$

Input resistance, $R_i = h_{ie} + A_I h_{re} R_L = 10^3 - 88.89 \times 5 \times 10^{-4} \times 5 \times 10^3 = 777.8\Omega.$

$\therefore$ Overall voltage gain (including source reistance)

$$\boldsymbol{A_{VO}} = A_I \frac{R_L}{R_s + R_i} = -88.89 \times \frac{5 \times 10^3}{10^3 + 777.8} = \mathbf{-250}$$

Overall power gain, $\boldsymbol{A_{PO}} = A_{VO} A_I = -250 \times (-88.89) = \mathbf{22.2 \times 10^3}$

► **Example 6.** *Calculate the voltage gain of a CE amplifier from the following data :*

$$h_{fe} = 100, \quad R_L = 15\,\text{k}\Omega$$

$$h_{ie} = 60\Omega \ \ \textit{and} \ \ \Delta h = h_{ie}h_{oc} - h_{re}h_{fe} = 12$$

Solution : The voltage gain A_v is given by

$$A_v = \frac{h_{fe}R_L}{h_{ie} + \Delta h R_L} = \frac{100 \times 15 \times 10^3\Omega}{60 \times 10^3\Omega + 12 \times 15 \times 10^3\Omega}$$

$$= \mathbf{6.3}$$

► **Example 7.** *A treansistor used in a common base (CB) amplier has the following values of h-parameter.*

$$h_{ib} = 28\Omega, \quad h_{fb} = -0.98, \quad h_{rb} = 5 \times 10^{-4} \ \ \textit{and} \ \ h_{ob} = 0.34 \times 10^{-6}\,\text{s}$$

Calculate the values of input impedance, output impedance, current gain and voltage gain if the load resistance R_L is 1.2 kΩ. Assume the source resistance R_s to be zero.

Solution : We have, $h_{ib} = 28\Omega$, $h_{fb} = -0.98$, $h_{rb} = 5 \times 10^{-4}$, $h_{ob} = 0.34 \times 10^{-6}\,\text{S}$, $R_L = 1.2\,\text{k}\Omega = 1200\Omega$ and $R_s = 0$.

$\therefore$ Current gain, $\boldsymbol{A_i} = \dfrac{h_{fb}}{1 + h_{ob}R_L} = \dfrac{-0.98}{1 + 0.34 \times 10^{-6} \times 1200} = \mathbf{0.98}$

Input impedance, $\boldsymbol{Z_{\text{in}}} = h_{ib} - \dfrac{h_{rb}h_{fb}R_L}{1 + h_{ob}R_L}$

$$= 28 - \frac{5 \times 10^{-4} \times (-0.98) \times 1200}{1 + (0.34 \times 10^{-6}) \times 1200} = \mathbf{28.6\Omega}$$

Output impedance, $\boldsymbol{Z_{\text{out}}} = \dfrac{R_s + h_{ib}}{(R_s + h_{ib})h_{ob} + h_{rb}h_{fb}}$

$$= \frac{0 + 28}{(0 + 28) \times 0.34 \times 10^{-6} + 5 \times 10^{-4} \times (-0.98)}$$

$$= 56{,}000 = \mathbf{56\ k\Omega}$$

Voltage gain, $\boldsymbol{A_v} = \dfrac{A_i R_L}{Z_{\text{in}}} = \dfrac{0.98 \times 1200}{28.6} = \mathbf{41}.$

▶ **Example 8.** *A junction transistor has the following h-parameters:*

$$h_{ie} = 2\,\text{k}\Omega,\ \ h_{re} = 1.6 \times 10^{-4},\ \ h_{fe} = 49\ \ \textit{and}\ \ h_{oe} = 50\mu\text{A/V}$$

Determine the current gain, voltage gain, input impedance and the output impedance of the CE amplifier if the load resistance is $30\,\text{k}\Omega$ *and source resistance is* 600Ω.

Solution: We have: $h_{ie} = 2\,\text{k}\Omega = 2000\Omega$, $h_{re} = 1.6 \times 10^{-4}$, $h_{fe} = 49$, $h_{oe} = 50\,\mu\text{A/V}$, $R_L = 30\,\text{k}\Omega = 30 \times 10^3\Omega$ and $R_s = 600\Omega$.

$$\therefore \text{ Current gain } \boldsymbol{A_i} = \frac{h_{fe}}{1 + h_{oe}R_L} = \frac{49}{1 + 50 \times 10^{-6} \times 30 \times 10^3} = \mathbf{19.6}$$

$$\text{Input impedance, } \boldsymbol{Z_{\text{in}}} = h_{ie} - \frac{h_{re}h_{fe}R_L}{1 + h_{oe}R_L} = 2000 - \frac{1.6 \times 10^{-4} \times 49 \times 30 \times 10^3}{1 + 50 \times 10^{-6} \times 30 \times 10^3}$$

$$= \mathbf{1906\Omega}$$

$$\text{Voltage gain, } \boldsymbol{A_V} = \frac{A_iR_L}{Z_{\text{in}}} = \frac{19.6 \times 30 \times 10^3}{1906} = \mathbf{308.5}$$

$$\text{Output impedance, } \boldsymbol{Z_{\text{out}}} = \frac{R_s + h_{ie}}{(R_s + h_{ie})h_{oe} - h_{re}h_{fe}}$$

$$= \frac{6000 + 2000}{(6000 + 2000) \times 50 \times 10^{-6} - 1.6 \times 10^{-4} \times 49}$$

$$= \mathbf{20.4\,k\Omega}$$

▶ **Example 9.** *A transistor in a common collector (or emitter follower) circuit as illustrated in Fig. 7.19 has the following set of h-parameters:*

$$h_{ic} = 2\,\text{k}\Omega,\ \ h_{fc} = -51,\ \ h_{rc} = 1\ \ \textit{and}\ \ h_{oc} = 25 \times 10^{-6}\mho.$$

Determine the values of input and output impedances, current and voltage gains of the amplifier stage.

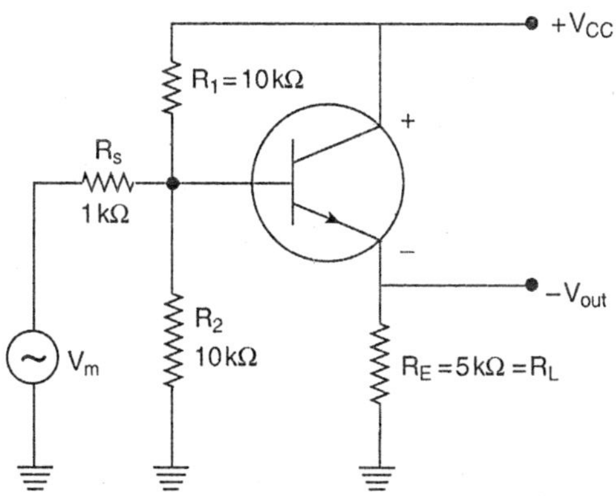

Fig. 7.19

Solution: Current gain, $A_i = \dfrac{h_{fc}}{1 + h_{oc}R_L} = \dfrac{-51}{1 + 25 \times 10^{-6} \times 5 \times 10^3} = 45.3$

$$\text{Input impedance, } Z_{\text{in}} = h_{ic} - \frac{h_{rc}h_{fc}R_L}{1 + h_{oc}R_L}$$

$$= 2 \times 10^3 - \frac{1 \times (-51) \times 5 \times 10^3}{1 + 25 \times 10^{-6} \times 5 \times 10^3}$$

$$= 228 \times 10^3 \Omega = 228\,\text{k}\Omega$$

$$\text{Input impedance of amplifier stage, } \boldsymbol{Z_{is}} = Z_{\text{in}} \| (R_1 \| R_2)$$

$$= 228 \| (10 \| 10) = \mathbf{4.9\,k\Omega}$$

$$\text{Output impedance, } Z_{\text{out}} = \frac{R_s + h_{ic}}{(R_s + h_{ic})h_{oc} - h_{rc}h_{fc}}$$

$$= \frac{1000 + 2000}{(1000 + 2000) \times 2.5 \times 10^{-6} - 1 \times (-51)}$$

$$= 58.8\Omega$$

$$\text{Output impedance of amplifier stage, } \boldsymbol{Z_s} = Z_{\text{out}} \| R_E$$

$$= 58.8 \| (5 \times 10^3) = \mathbf{58.1\Omega}$$

$$\text{Current gain of amplifier stage, } \boldsymbol{A_{is}} = \frac{A_i R_s}{R_s + Z_{is}} = \frac{45.3 \times 1000}{1000 + 4900} = \mathbf{7.7}$$

$$\text{Voltage gain, } \quad A_V = \frac{A_i R_L}{Z_{\text{in}}} = \frac{45.3 \times 5 \times 10^3}{228 \times 10^3} = 1$$

$$\text{Voltage gain of amplifier stage, } \boldsymbol{A_{VS}} = \frac{A_V Z_{is}}{R_s + Z_{is}} = \frac{1 \times 4900}{1000 + 4900} = \mathbf{0.83}$$

▶ **Example 10.** *Find the h-parameters for the two-port network of Fig. 7.20.*

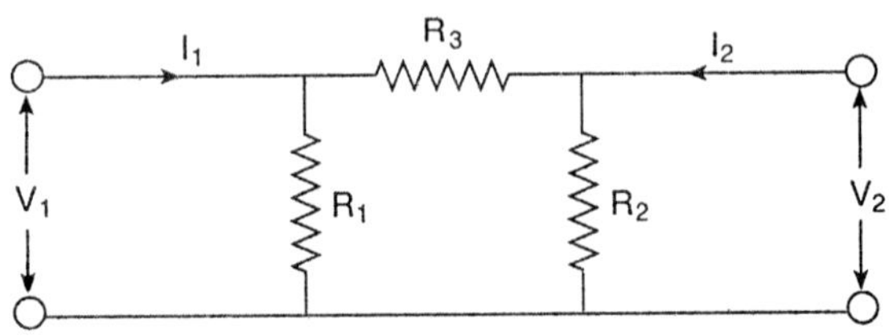

Fig. 7.20 **(Burdwan Hons.)**

Solution: The input voltage V_1 and the output current I_2 are given by

$$V_1 = h_{11}I_1 + h_{12}V_2$$

$$\text{and } I_2 = h_{21}I_1 + h_{22}V_2$$

Here, $h_{11} = \left.\frac{V_1}{I_1}\right|_{V_2=0}$ = input impedance with output short-circuited.

When V_2 is short-circuited, V_1 is given by

$$V_1 = I_1 \cdot \frac{R_1 R_3}{R + 1 + R_3}$$

$$\therefore \quad \boxed{h_{11} = \frac{R_1 R_3}{R_1 + R_3}}$$

Now, $h_{12} = \left.\frac{V_1}{V_2}\right|_{I_1=0}$ = reverse voltage gain with input open-circuited.

When input is open-circuited, we have

$$V_1 = \frac{R_1}{R_1 + R_3} V_2$$

$$\therefore \quad \boxed{h_{12} = \frac{R_1}{R_1 + R_3}}$$

The hybrid parameter, $h_{21} = \left.\frac{I_2}{I_1}\right|_{V_2=0}$ = forward current gain with the output short-circuited.

If $V_2 = 0$, then the current through R_3 is $\frac{R_1}{R_1 + R_3} I_1$ from left to right.

So, the current $I_2 = -\frac{R_1}{R_1 + R_3} I_1$

$$\therefore \quad \boxed{h_{21} = -\frac{R_1}{R_1 + R_3}}$$

Finally, $h_{22} = \left.\frac{I_2}{V_2}\right|_{I_1=0}$ = output admittance with input open-circuited.

When the input is open-circuited, we get

$$I_2 = \frac{V_2}{R_2 \| (R_1 + R_3)} = \frac{V_2(R_1 + R_2 + R_3)}{R_2(R_1 + R_3)}$$

$$\therefore \quad \boxed{h_{22} = \frac{R_1 + R_2 + R_3}{R_2(R_1 + R_3)}}$$

Review and Short Questions

1. Discuss the two-port representation of a transistor and hence define the h-parameters.
2. How can you determine CE h-parameters experimentally?
3. What are the advantages of using h-parameter model?
4. Why is this circuit not valid for high frequencies? **(Burdwan and Calcutta Hons.)**
5. What are h-parameters?

6. Obtain expressions for current gain, input impedance, voltage gain, output impedance, overall current gain, overall voltage gain and power gain of an amplifier in terms of h-parameters. **(Bangalore Univ.)**

7. Draw a neat diagram of a common emitter transistor amplifier as well as its equivalent circuit. Derive the expressions for the current gain, voltage gain, input resistance, output resistance and power gain. **(Guru Nanak Univ.)**

8. What is an emitter follower? What are its important characteristics? Draw the circuit diagram of an emitter follower and use simplified hybrid model to find expressions for voltage gain, input impedance and output impedance. **(Burdwan Hons.)**

9. Derive expressions for the voltage gain of a transistor amplifier including and excluding the output resistance of the signal source. Show that the power gain is the product of the voltage gain and the current gain. Explain the significance of the negative sign in the expressions for the current and the voltage gains.

10. Draw the approximate hybrid model for any transistor configuration at low frequencies. Show that only h_{ie} and h_{fe} are important in the model. Is the approxiation justified?

11. Derive the approximate CE hybrid model when $h_{oe}R_L \ll 0.1$ and show that the input resistance is then approximately h_{ie}. What are the approximate current and voltage gains, and the output resistance?

12. Draw the approximate hybrid equivalent circuit. Under what conditions is this approximate model valid? Justify the use of such simplified model.

13. Give the circuit diagram of an emitter follower and state one use of the circuit. **(Burdwan Hons.)**

14. What is a Darlington pair? What are its important characteristics?

15. What is the Darlington connection? Compare between an emitter follower and a Darlington pair.

Objective Questions

1. The h-parameter model is valid for
   (a) small signal and low frequency operation,
   (b) small as well as large signal operation,
   (c) small signal and high frequency operation,
   (d) none of these.

2. h-parameters of a transistor are

(a) four in number, (b) h_{ie}, h_{re}, h_{fb}, h_{oc},
(c) h_{ie}, h_{rb}, h_{fc}, h_{ob}, (d) h_{ib}, h_{rc}, h_{fc}, h_{oe}.

3. The dimensionless h-parameters are

(a) h_{11} and h_{12}. (b) h_{11} and h_{22},
(c) h_{12} and h_{21}, (d) h_{21} and h_{22}.

4. The h-parameter that has the largest value is

(a) h_{re}, (b) h_{oe}, (c) h_{ie}, (d) h_{fe}.

5. The parameter h_{ib} represents input impedance in

(a) CB-configuration with output shorted,
(b) CE-arrangement with output shorted,
(c) CB-arrangement with output open,
(d) CE-configuration with output open.

6. The typical value of h_{ic} is

(a) 1 kΩ, (b) 40 kΩ, (c) 100 kΩ, (d) None.

7. Emitter follower is the another name of transistor in

(a) CE mode, (b) CB mode, (c) CC mode, (d) all of these.

8. The values of h-parameters depend on

(a) configuration, (b) temperature, (c) operating point, (d) all of these.

9. Which of the following transistor configuration can act as a buffer stage?

(a) CE, (b) CB, (c) CC, (d) all of these.

10. If A_V and A_I and A_P are the voltage, current and power gains of a transistor amplifier, then

(a) $A_V = A_P A_I$, (b) $A_V = A_P/A_I$, (c) $A_P = A_V + A_I$, (d) $A_P = A_V - A_I$.

11. For an emitter follower, the voltage gain is

(a) greater than unity, (b) less than unity, (c) unity.

12. In a Darlington pair,

(a) the two transistors are connected in parallel,
(b) the emitter of the first transistor feeds the base of the second,

(c) the collector of the first transistor feeds the base of the second,
(d) the bases of the two transistors are connected together.

13. A Darlington pair consists of two stages:

(a) CE and CC, (b) both CE,
(c) both CC, (d) CE and CB.

14. The Darlington pair is mainly used for

(a) impedance matching, (b) wideband voltage amplification,
(c) power amplification, (d) reducing distortion.

15. If each transistor in a Darlington pair has a current gain 100, what would be the overall current gain?

(a) 10000, (b) 100,
(b) 200, (d) 1.

16. Darlington pair is used to get very high

(a) voltage gain, (b) current gain,
(c) power gain, (d) output voltage.

17. The h-parameters are called hybrid parameters because they

(a) are different from y- and z-parameters,
(b) are mixed with other parameters,
(c) apply to circuits contained in a box,
(d) are defined by using both open-circuit and short-circuit terminations.

18. Which of the following four h-parameters of a transistor has smallest value?

(a) h_i, (b) h_O, (c) h_r, (d) h_f.

19. Which of the following statements is *not* correct regarding the h-parameters of a transistor?

(a) The values of h-parameters can be obtained from transistor characteristics,
(b) Their values depend upon the transistor configuration,
(c) Their values depend on the operating point,
(d) They are four in numbers.

20. A transistor has $h_{fe} = 100$, $h_{ie} = 5.2\,\text{k}\Omega$ and $r_{bb} = 0$. At room temperature, $V_T = 26\,\text{mV}$. The collector current I_C will be

(a) 10 mA, (b) 5 mA, (c) 1 mA, (d) 0.5 mA.

Problems

1. A CE transistor amplifier is characterised by $h_{ie} = 2\,\text{k}\Omega$, $h_{re} = 2 \times 10^{-4}$, $h_{fe} = 50$ and $h_{oe} = 20 \times 10^{-6}$ A/V. If the load resistance is $4\,\text{k}\Omega$ and the source resistance is 200Ω, determine the input resistance, the output resistance and the voltage, current and power gains. **(Calcutta Hons.)**

   [$R_i = 1.963\,\text{k}\Omega$, $R_0 = 64.7\,\text{k}\Omega$, $A_V = -94.3$, $A_{V_0} = -85.6$, $A_I = -46.3$, $A_P = 4.37 \times 10^3$, $A_{P_0} = 3.96 \times 10^3$]

2. A transistor amplifier in CE configuration couples a source of internal resistance $1\,\text{k}\Omega$ to a load of $20\,\text{k}\Omega$. Find the input and the output resistances if $h_{ie} = 1\,\text{k}\Omega$, $h_{re} = 2.5 \times 10^{-4}$, $h_{fe} = 150$ and $1/h_{oe} = 40\,\text{k}\Omega$. **(Calcutta Hons.)**

   [$R_i = 500\Omega$, $R_0 = 160\,\text{k}\Omega$]

3. Consider a CE amplifier with a emitter resistance $R_E = 200\Omega$ and a load resistance $R_L = 1\,\text{k}\Omega$ as shown in Fig. 7.9. Calculate the current gain, voltage gain, input and output impedances. Assume $h_{ie} = 2000\Omega$, $h_{fe} = 100$, $h_{re} \approx 0$, $h_{oe} \approx 0$. **(Calcutta Hons.)**

   [$A_I = -100$, $A_V = -4.5$, $R_i = 22.2\,\text{k}\Omega$, $R_0 = \infty$]

4. Consider the emitter follower circuit of Fig. 7.8 with $R_L = 2.5\,\text{k}\Omega$ and $R_B = 200\,\text{k}\Omega$. Draw the h-parameter equivalent circuit and obtain voltage gain, input impedance, output impedance and current gain. Take $h_{ie} = 1.5\,\text{k}\Omega$, $h_{fe} = 100$ and neglect h_{re} and h_{oe}. **(Calcutta Hons.)** [0.994, $111.89\,\text{k}\Omega$, 14.8Ω, 101]

5. A transistor in CE class A amplifier has $h_{fe} = 100$ and $h_{oe} = 30 \times 10^6$s. If the load resistance is $3\,\text{k}\Omega$, find the current gain. **(Calcutta Hons.)** [−91.7]

6. For a transistor $h_{ib} = 50\Omega$, and $h_{fb} = -0.975$. Determine approximately the values of h_{ie} and h_{fe}. [$2\,\text{k}\Omega$, 39]

   Find h_{fb} and h_{fc}, if the value of h_{fe} of a transistor is 50. [−0.98, −51]

7. The hybrid parameters of a transistor used as an amplifier in CE configuration are $h_{ie} = 800\Omega$, $h_{fe} = 50$, $h_{oe} = 80 \times 10^{-6}\Omega^{-1}$ and $h_{re} = 5.5 \times 10^{-4}$. If the load resistance is $5\,\text{k}\Omega$ and the effective source resistance is 500Ω, determine the current gain, input resistance, voltage gain, power gain and output resistance.

   [−35.7; 702Ω; −254, −148; 5283; $17\,\text{k}\Omega$]

8. Design an emitter follower circuit having $R_i = 500\,\text{k}\Omega$, $R_0 = 20\Omega$ which uses a transistor with $h_{fe} = 50$, $h_{ie} = 1\,\text{k}\Omega$ and $h_{oe} = 25\,\mu$A/V. Find the values of A_i and A_v for this circuit. **(AMIE)** [$R_E = 10\,\text{k}\Omega$, $A_i = 51$, $A_v = 0.998$]

9. Establish the following approximate relations :

(a) $h_{fb} = -\dfrac{h_{fe}}{1+h_{fe}}$

(b) $h_{ib} = \dfrac{h_{ie}}{1+h_{fe}}$, and

(c) $h_{ob} = \dfrac{h_{oe}}{1+h_{fe}}$. **(North Bengal Hons.)**

10. Find the values of voltage gain, current gain, input impedance and power gain for a CE-amplifier with $R_L = 1600\Omega$ and $R_s = 1000\Omega$. The transistor has the following set of h-parameters : $h_{ie} = 1100\Omega$, $h_{re} = 2.5 \times 10^{-4}$, $h_{fe} = 25\,\mu\text{A/V}$.

[69.8, 48, 1084Ω, 3350]

11. A signal source with an open circuit voltage of 1 mV and internal impedance of 600 Ω is connected to the input of a transistor a.c 125 V in CE-configuration. The signal parameters measured at $V_{CB} = -5\,\text{V}$ and $I_E = 2\,\text{mA}$ are as follows : $h_{ie} = 1.7\,\text{k}\Omega$, $h_{re} = 6.5 \times 10^{-4}$, $h_{fe} = 125$, $h_{oe} = 80\,\text{mA/V}$.

Calculate the input and output impedances and signal amplification for a load of 5.6 kΩ. Also find the value of signal voltage at the output.

[$Z_{\text{in}} = 1.387\,\text{k}\Omega$, $Z_{\text{out}} = 22.3\,\text{k}\Omega$, 243.3 V]

12. A typical transistor has, as specified by manufacturer, the following h-parameters : $h_{ie} = 4$, $h_{fe} = 330$, $h_{re} = 2 \times 10^{-4}$ and $h_{oe} = 3 \times 10^{-5}\mho$. It is being used as a CE-amplifier with load resistance $R_L = 4\,\text{k}\Omega$ and internal resistance of signal source $R_s = 10\,\text{k}\Omega$. Calculate the following quantities : (a) current gain, (n) voltage gain, (c) power gain, (d) input impedance, (e) output impedance.

[(a) 287, (b) −341, (c) 97867, (d) 4.21 kΩ, (e) 39.3 kΩ]

8

Basic Voltage and Power Amplifiers

8.1 Introduction

An amplifier is an electronic circuit consisting of one or more active devices that can produce an enlarged and undistorted output signal (a.c voltage and current waveforms) of small input a.c signals by transferring power from the d.c supply. A properly biased transistor raises the strength of a weak signal and behaves like an amplifier. Almost all electronic equipments like radio receivers, tape recorders, microphones, loudspeakers, TVs, transducers used in scientific and medical investigations etc. require amplifiers. The radio receivers amplify the small signal voltage ($\sim 10^{-6}$ V) induced in the antena of the radio-set before it is fed to the loudspeaker, and transducers generate small signals ($\sim 10^{-6}$ V to 10^{-3} V) which are amplified million times to make them strong enough for being used in the indicating instruments. Modern electronic musical equipments such as electronic guitar, tanpura etc. also use the electronic amplifier circuits to produce sound waves of required frequencies and amplitudes.

8.2 Fundamentals of amplifiers

An ***amplifier*** is generally known as *an electronic circuit capable of producing an enlarged and undistorted output of the small input signal.*

So, the input is increased by an amplification factor of more than unity. The triangular symbol shown in Fig. 8.1 is used to represent an amplifier. It shows the direction of flow of signal through the circuit from the input to the output side.

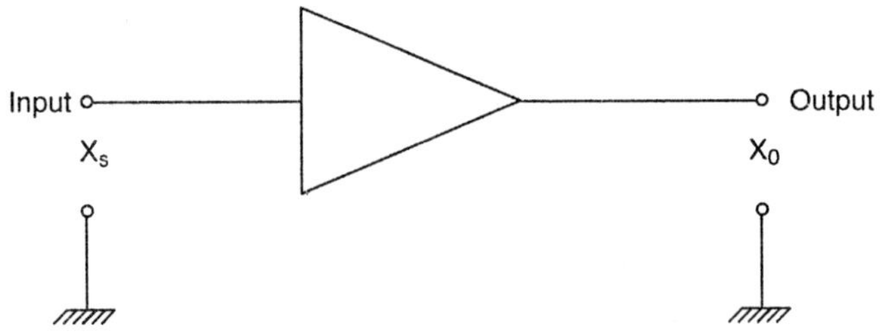

Fig. 8.1 Symbolic representation of an amplifier

Let X_s and X_0 represent the input (supply) and the output signals respectively. For a linear amplifier both X_s and X_0 have the same shape of waveforms with amplitude of X_0 being more or less than that of X_s. It is desirable that the change in the output signal ΔX_0 is greater than the change in the input signal ΔX_s for an amplifier. Moreover, the magnitude of the *amplifier gain*, $A = \Delta X_0/\Delta X_s$ should remain constant over a defined range of the signal frequency so that it can be applied in many fields. Since an amplifier changes the amplitude and phase of the input signal, the complex gain is plainly given by

$$A = A_0 e^{j\theta} \tag{8.2.1}$$

where A_0 is the amplitude of the gain and θ the phase shift.

Since both A and θ are functions of the frequency, the *frequency response* of an amplifier is very *important* for analysing the performance of an amplifier.

8.3 Classification of amplifiers

Linear amplifiers are classified in different ways depending on its transistor biasing conditions, interstage coupling, frequency range of operation, the type of load, the method of operation etc.

8.3.1 Classification of amplifiers: A, B, C and AB

Depending on the biasing conditions of the transistor i.e., the position of the Q-point and the amplitude of the input signal, the amplifiers can be classified into the following *four* different categories: A, B, C and AB.

(i) Class A amplifier—When the Q-point and the amplitude of the input signal are so chosen that the output current flows for the *full-cycle* of the input signal, then the amplifier is called a ***class A amplifier***. For the operation of a class A amplifier, the Q-point is confined to the linear portion of the dynamic transfer characteristic curve (Fig. 8.2).

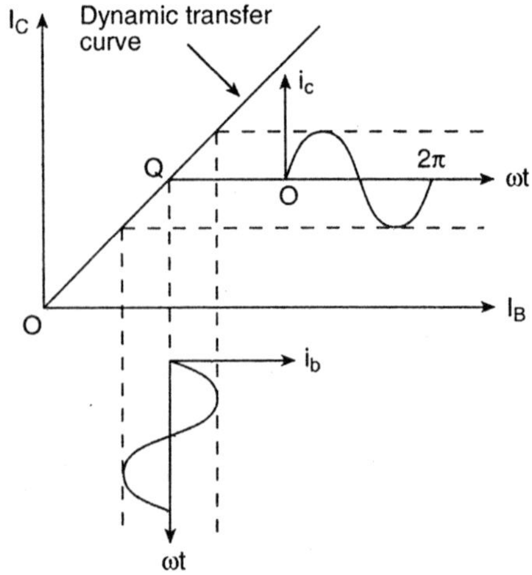

Fig. 8.2 CE operation of class A amplifier

(ii) Class B amplifier—When the transistor bias and the amplitude of the input signal are so chosen that the output current flows during *one half-cycle* of the input signal, the amplifier is called a ***class B amplifier***. For this operation, the Q-point is chosen near the extreme end of the load line (Fig. 8.3).

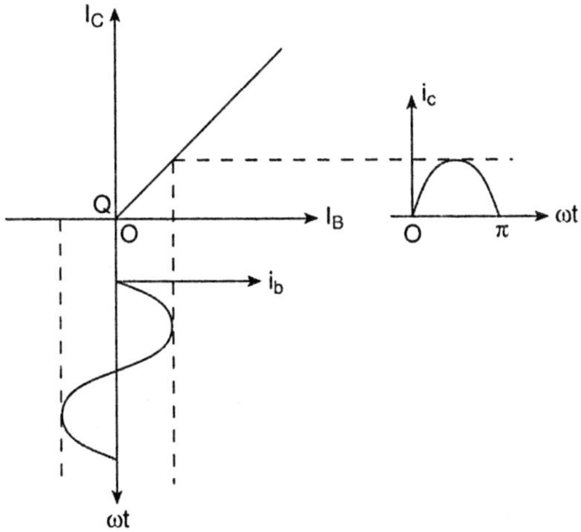

Fig. 8.3 CE operation of class B amplifier

(iii) Class C amplifier—If the bias and the amplitude of the input signal are such that the output current flows for appreciably *less than half-cycle* of the input signal, as illustrated in Fig. 8.4, the amplifier is termed ***class C amplifier***.

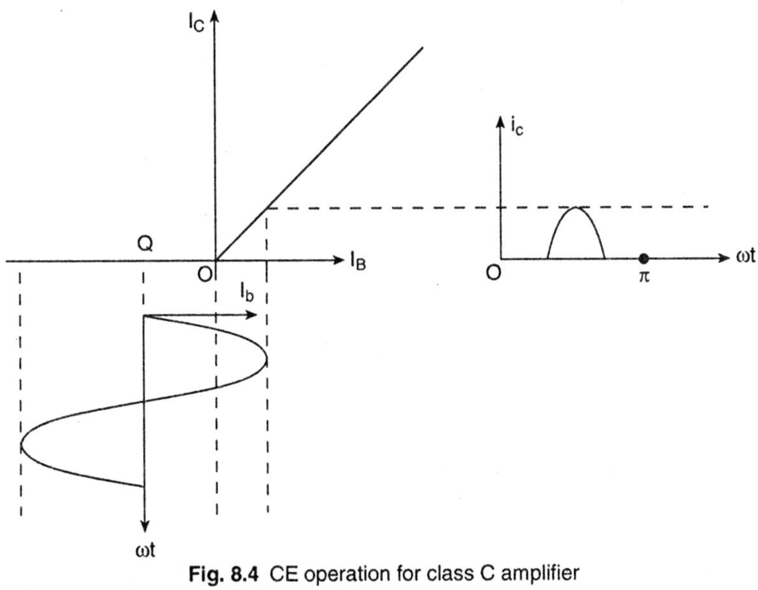

Fig. 8.4 CE operation for class C amplifier

(iv) Class AB amplifier—The operation of this type of amplifier lies in between class A and class B amplifiers i.e., the biasing conditions and amplitude of the input signal are such that the output current flows during appreciably *more than half but less than a complete cycle* as shown in Fig 8.5.

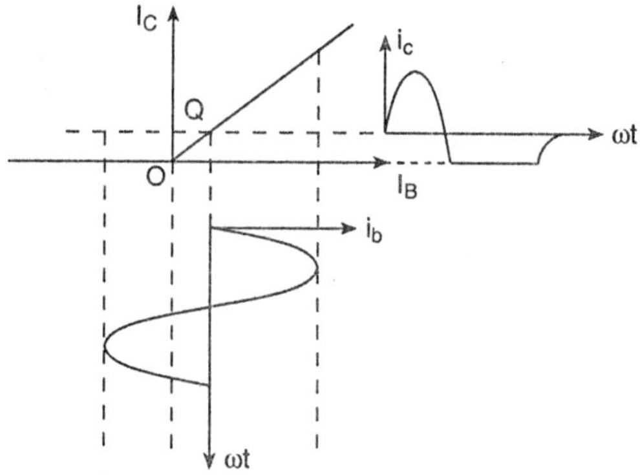

Fig. 8.5 CE operation for class AB amplifier

8.3.2 Audio, video and radio frequency amplifiers

The amplifiers are also classified, based on the frequency of operation, in the following *three* main categories.

(i) Audio frequency (AF) amplifiers—This type of amplifier are used to amplify audio frequency (~ 20 Hz to 20 kHz) signal for amplification in sound systems like phonograph, magnetic tape, microphones etc.

(ii) Video frequency (VF) amplifiers—This type of amplifiers are employed to amplify signals over a wide range of frequencies (~ a few Hz to a few MHz). It is therefore known as ***video*** or ***broadband amplifier***. These amplifiers are mainly used in TV to amplify the picture signals.

(iii) Radio frequency (RF) amplifiers—The amplifiers designed to amplify signals in the radio frequency range (~ 20 kHz to 30 MHz) are known as ***RF amplifiers***. These amplifiers are mainly used in radio receivers to amplify only a narrow band of frequencies about the carrier frequency.

8.3.3 Voltage, Current and Power amplifiers

There is another method of classifying the amplifiers.

When we are interested in the output of the amplifiers, we classify them as voltage, current and power amplifiers in order to produce more voltage, more current and more power respectively.

A ***voltage amplifier*** is used to amplify signal voltage when the power consumed is not of prime concern. In this amplifier, the input resistance is very high and the output impedance is very low.

An ideal voltage amplifier is characterised by a number of important properties which are enumerated below.

(i) The gain is independent of the signal amplitude and frequency.

(ii) It introduces no noise and distortion of the signal.

(iii) It has an infinite input resistance and zero output resistance.

Since the input resistance is infinite, the voltage gain is *independent* of the resistance of the input signal source. The zero output resistance implies that the voltage gain does not depend on the load resistance. All these requirements are not fulfilled in practical voltage amplifiers.

A ***current amplifier*** is designed to have maximum current amplification and it is the current gain which is of primary concern. This amplifier has very low input resistance and very high output impedance i.e., just the reverse of voltage amplifier.

The function of current amplifiers is similar to that of voltage amplifiers except that they primarily amplify current waveforms rather than voltage waveforms.

An ideal current amplifier has a number of important characteristics. And these are given as under.

(i) The current gain is independent of the signal amplitude and frequency.

(ii) It does no introduce any noise and distortion of the signal, and

(iii) It has zero input resistance and infinite output resistance.

The zero input resistance makes the current gain *independent* of the source resistance and the infinite output resistance provides the current gain *independent* of load resistance. The practical current amplifiers however do not fulfill all the above conditions.

When the power gain is of primary concern, the ***power amplifier*** is designed to get maximum output power. These amplifiers are generally used in radio receivers, loudspeakers etc. to supply more powers.

8.3.4 RC-, Transformer- and Direct-coupled amplifiers

Depending on the way how the interstage coupling is made, the amplifiers are further classified into *three* main categories as RC-coupled, transformer-coupled and direct-coupled amplifiers.

All these amplifiers are designed in such a way that the output of one stage is coupled to the input of the next stage through appropriate coupling devices. The output of the last stage of a multistage amplifying system supplies sufficiently high gain. The RC-coupled and the transformer-coupled amplifiers will be discussed in detail in the multistage or cascade amplifiers as they are too often used.

8.4 Principle of multistage amplifiers

When the gain of a single stage amplifier is found insufficient and the input and output impedances are not satisfactory for practical use, two or more amplifying stages are usually cascaded together to obtain the required level of the output signal. Such amplifiers are called ***multistage*** or ***cascade amplifiers***.

In these amplifiers the output of one stage is coupled to the input of the next stage through a coupling device. The schematic diagram of a two-stage cascade amplifier is represented in Fig. 8.6.

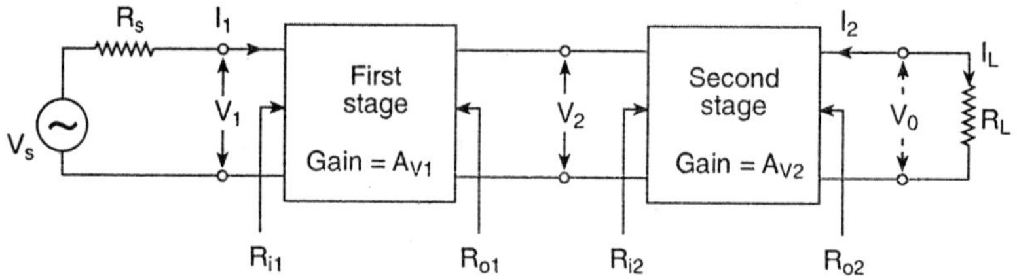

Fig. 8.6 Two-stage cascade amplifier

The overall voltage gain of the above amplifier is given by

$$A_V = \frac{\text{Output voltage of the second stage}}{\text{Input voltage of the first stage}}$$

$$= \frac{V_0}{V_1} = \frac{V_2}{V_1} \cdot \frac{V_0}{V_2} = A_{V_1} \times A_{V_2} \tag{8.4.1}$$

where A_{V_1} and A_{V_2} are respectively the voltage gains of the first and the second stage of the amplifier.

In a multi-stage amplifier, therefore, $A_V = A_{V_1} \times A_{V_2} \times A_{V_3} \times \cdots = \prod_i A_{V_i}$ where A_{V_i} is the voltage gain of the ith stage.

So, *the overall voltage gain is equal to the product of the voltage gains obtained in the individual stages.*

To obtain a *large voltage gain*, CC amplifiers and CB amplifiers are not suitable, since the voltage gain of CC configuration or CB configuration is less than unity. As the current gain of CE configuration is much larger than unity, the voltage gain can be increased by connecting a number of CE stages in cascade. Hence CE configuration must be used as intermediate stages in cascade amplifiers. The coupling elements RC network, transformer etc. are used to transmit a.c only. The d.c is blocked to avoid d.c-interference among the stages and the consequential shifting in Q-points.

8.4.1 Impedance matching

Impedance matching is an important part in designing amplifier circuits to produce *maximum* power transfer.

For maximum power transfer from a source to a load, the internal resistance of the source (R_s) must be *equal* to the load resistance (maximum power transfer theorem). So, in a single stage amplifier we must have $R_s = R_i$ and $R_L = R_O$ (Fig. 8.7(a)) and for a multistage amplifier, $R_s = R_{i1}$, $R_{01} = R_{i2}$, $R_{02} = R_L$ (Fig. 8.7(b)). Since the input and the output resistances differ widely in CB and CC modes, extra matching arrangements are *essential* for cascading the different stages. But in CE, the input and

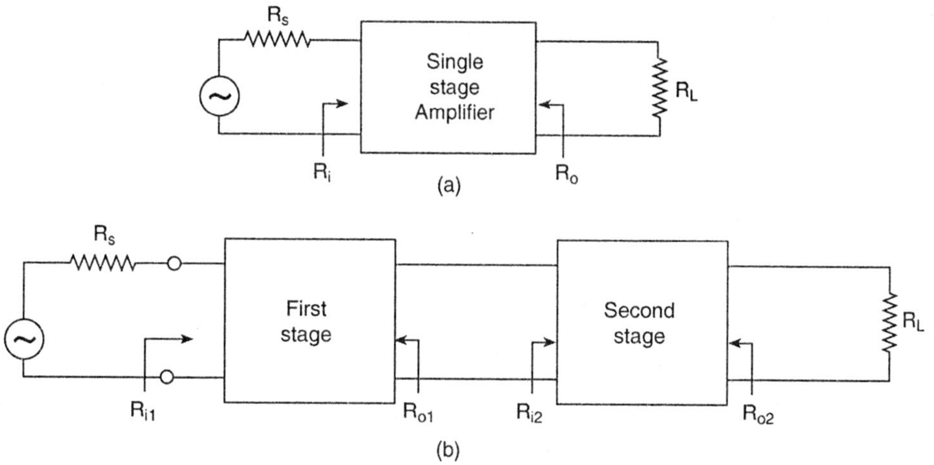

Fig. 8.7 Impedance matching (a) single stage amplifier (b) multistage amplifier

the output resistances are of the same order and hence they can be cascaded without any additional matching arrangement.

8.4.2 Decibel gain

The power gain A_p of an amplifier is *defined* as

$$A_p = \frac{\text{Power delivered to the load}}{\text{Power received from the source}} = \frac{P_O}{P_i}$$

Since the absolute power gains arise from cascading of amplifiers, they very often span a large range. If a *logarithmic scale* is used to express the gain, the variation becomes small and it allows the gains of individual stages to be added rather than multiplied to give the overall gain. The logarithm to the base 10 of the power gain expresses the value in a unit called *bel.* But, in practice, one-tenth of this unit, called the *decibel* (dB) is used. So, the power gain in dB or ***decibel power gain*** is given by

$$A_p(\text{dB}) = 10 \log_{10} A_p = 10 \log_{10}(P_O/P_i) \tag{8.4.2}$$

For a series of n number of amplifiers with power gains, $A_{p_1}, A_{p_2}, \cdots, A_{p_n}$. The overall gain is

$$\begin{aligned} A_p &= A_{p_1} \times A_{p_2} \times A_{p_3} \times \cdots \times A_{p_n} \\ \therefore\ A_p(\text{dB}) &= 10 \log_{10}(A_{p_1} \times A_{p_2} \times A_{p_3} \times \cdots \times A_{P_n}) \\ &= A_{p_1}(\text{dB}) + A_{p_2}(\text{dB}) + A_{p_3}(\text{dB}) + \cdots + A_{p_n}(\text{dB}) \end{aligned} \tag{8.4.3}$$

Since $P = I^2R = (V^2/R)$, the power gain for an impedance-matched system (i.e, with equal load resistances) can be expressed using (8.4.2) as

$$A_p(\text{dB}) = 20 \log_{10} \frac{V_O}{V_i} = 20 \log_{10} \frac{I_O}{I_i} \tag{8.4.4}$$

where V_O and V_i are the output and input voltages and I_O and I_i are the corresponding currents.

Obviously, the terms $20\log_{10}(V_O/V_i)$ and $20\log_{10}(I_O/I_i)$ are referred to as *decibel voltage gain* and *decibel current gain* respectively i.e.,

Decibel voltage gain, $A_V(\text{dB}) = 20\log_{10} A_V = 20\log_{10}(V_O/V_i)$ (8.4.5)

Decibel current gain, $A_I(\text{dB}) = 20\log_{10} A_I = 20\log_{10}(I_O/I_i)$ (8.4.6)

∴ For equal resistances we have from (8.4.4),

Power gain in dB = voltage gain in dB = current gain in dB

$$\text{or,}\quad A_p(\text{dB}) = A_V(\text{dB}) = A_I(\text{dB}) \tag{8.4.7}$$

When $A_p = \frac{1}{2}$ i.e., $P_O = \frac{1}{2}P_i$, then we get $V_O = \frac{1}{\sqrt{2}}V_i$ and

$$A_p(\text{dB}) = -10\log_{10} 2 = -3(\text{dB}) \tag{8.4.8}$$

So, in the frequency response curve of any amplifier, the *half power frequencies are* sometimes called the −3 dB *frequencies.*

8.5 RC-coupled transistor amplifier

When an *RC* network is used for connecting the output of one stage to the input of the next stage in a cascade amplifier, it is known as a resistance capacitance or ***RC-coupled amplifier.*** The circuit arrangement of a two-stage *RC*-coupled CE mode transistor amplifier is shown in Fig. 8.8. *RC*-coupled amplifiers are most widely used as AF-amplifier since they are *stable, less expensive with better frequency response.*

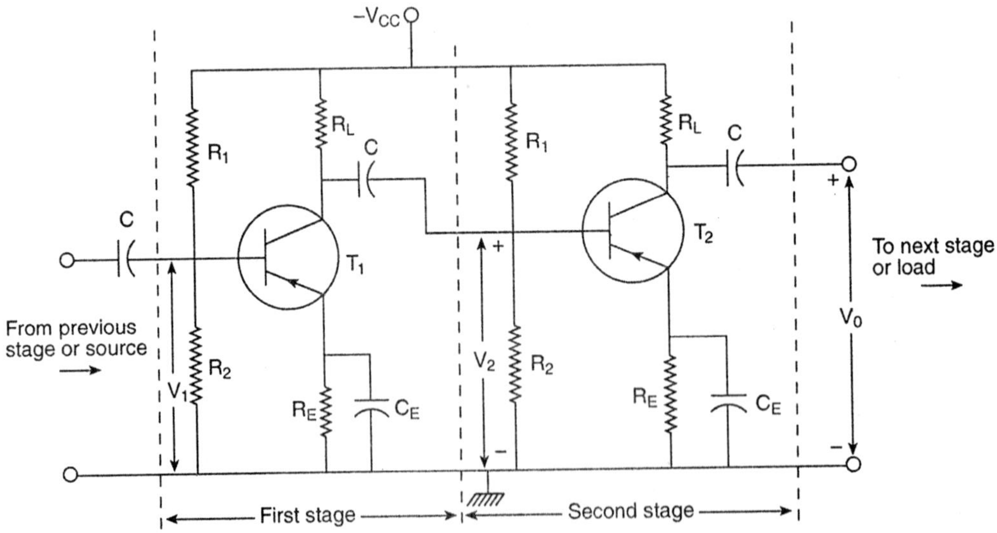

Fig. 8.8 RC-coupled transistor amplifier

The coupling capacitor C blocks the d.c currents or voltages but allows the a.c components of the output signal to pass through it to the input of the next stage. The supply V_{CC}, together with R_1, R_2, R_L and R_E, determine the proper biasing conditions i.e., the Q-point. The bypass capacitor C_E connected in parallel to R_E has a very high value of capacitance so that its reactance $1/\omega C_E$ at low frequency is very small and it bypasses the a.c components of the emitter current.

Operation—When an a.c signal is applied across the base-emitter circuit of the CE mode transistor in the first stage of a RC-coupled amplifier, the amplified output signal produced at the load resistance R_L is fed to the base-emitter circuit of the second stage through the coupling capacitor C. Since the capacitor C allows the passage of the amplified a.c signal and blocks the d.c components from stage I to stage II, the second stage produces further amplification of this signal. Thus the different amplifier stages, cascaded in this arrangement, amplifiy the input signal at the first stage to a proper value at the final stage in order to make the overall gain high enough for practical purposes.

8.5.1 Frequency response of RC-coupled transistor amplifier

The variation of the magnitude and the phase angle of the voltage gain of an amplifier with frequency of the input signal is known as the ***frequency response*** of the amplifer. A typical frequency response curve of a RC-coupled transistor amplifier is illustrated below in Fig. 8.9.

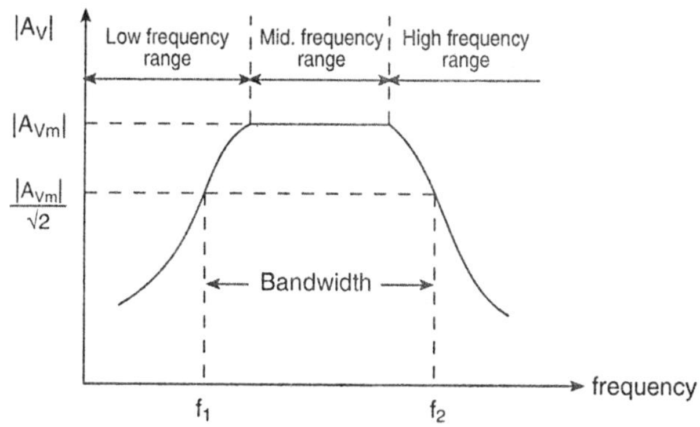

Fig. 8.9 Frequency response curve of a RC-coupled amplifier

The frequency response characteristic curve has *three* regions : low frequency range, mid-fequency range and high-frequency range. The voltage gain is almost constant in the mid-frequency range but falls off in both the low frequency and the high frequency ranges. The variation of the voltage gain is primarily determined by the coupling capacitors and the junction capacitances of the transistor at different frequencies in the above three frequency-ranges separately.

In order to derive approximate expressions for the voltage gain of one stage of an RC-coupled amplifier in the different frequency ranges, the following simplifying assumptions are made:

(i) The amplifier stages are all identical.

(ii) h_{ie} and h_{fe} of the transistors do not vary with frequency.

(iii) h_{re} and h_{oe} of the transistors are small enough to be neglected.

(iv) The emitter bypass capacitor has a negligible reactance in the range of frequencies of operation.

(v) The resistors R_1 and R_2 are sufficiently large and they have no effect on the ac behaviour of the circuit.

(i) Mid-frequency range—In the mid-frequency range (50 Hz to 20 kHz), the reactance of the coupling capacitor C and the bypass capacitor decreases with increase in frequency. This causes an increase in voltage gain and a simultaneous increase in the loading effect of the first stage due to increase in the reactance of the junction

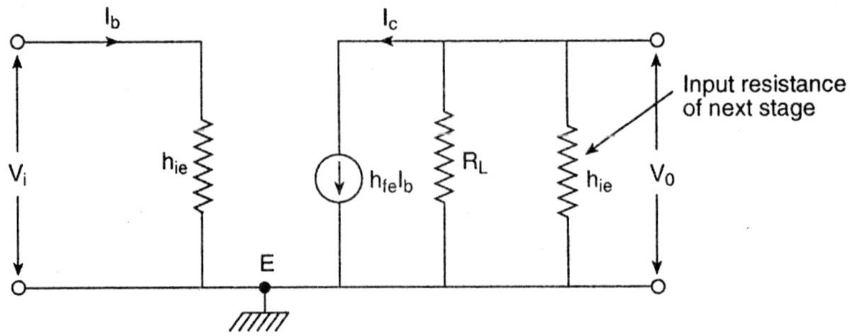

Fig. 8.10 Equivalent circuit at mid-frequency range

capacitance decreases the gain. As a result, the gain at the mid-frequency range becomes almost constant. In the mid-frequency range, the reactance of the coupling capacitor is very small. To obtain the expression for voltage gain, we draw the h-parameter *equivalent circuit* of the first stage of an RC-coupled amplifier which is illustrated in Fig. 8.10.

In this case, we assume that the transistors used in the two stages are identical and R_1 and R_2 are sufficiently large compared to h_{ie}.

The effective load resistance is given by

$$\frac{1}{Z_L} = \frac{1}{R_L} + \frac{1}{h_{ie}}$$

$$\therefore \quad Z_L = \frac{h_{ie} R_L}{h_{ie} + R_L} \tag{8.5.1}$$

The output voltage V_O is given by

$$V_O = -h_{fe} I_b Z_L = -h_{fe} I_b \frac{h_{ie} R_L}{h_{ie} + R_L} \tag{8.5.2}$$

on substituting the value of Z_L from the relation (8.5.1).

The input voltage V_i is given by

$$V_i = h_{ie} I_b. \tag{8.5.3}$$

∴ The voltage gain in the mid-frequency is

$$A_{V_M} = \frac{V_0}{V_i} = -\frac{h_{fe}R_L}{h_{ie} + R_L}, \tag{8.5.4}$$

using equations (8.5.2) and (8.5.3).

The above relation shows that *the voltage gain of the RC-coupled* CE *transistor amplifier is independent of the frequency in the mid-frequency range.*

(ii) Low frequency range—In the low frequency range (< 50 Hz), the reactance of the coupling capacitor increases with decreasing frequency resulting in the drop in voltage gain. Moreover, the reactance of the bypass capacitor and the effect of the various junction capacitances are negligible. So, there is a net drop in voltage gain. The h-parameter equivalent circuit of the RC-coupled CE transistor amplifier at low frequency range is shown in Fig. 8.11.

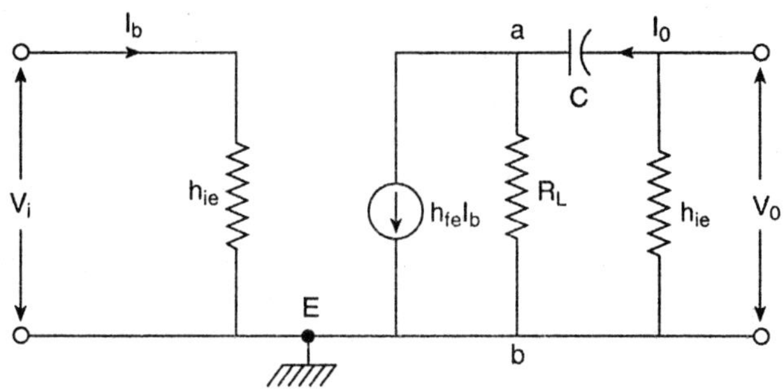

Fig. 8.11 Equivalent circuit of RC-coupled amplifier at low-frequency range

From Fig. 8.11, the effective load resistance is given by the relation

$$\frac{1}{Z_L} = \frac{1}{R_L} + \frac{1}{h_{ie} + \frac{1}{j\omega C}}$$

$$\therefore \quad Z_L = \frac{(h_{ie} - j/\omega C)}{R_L + h_{ie} - j/\omega C} \cdot R_L \tag{8.5.5}$$

The output current is: $$I_0 = -\frac{V'}{h_{ie} - j/\omega C} = \frac{h_{fe}I_bZ_L}{h_{ie} - j/\omega C}$$

$$= \frac{h_{fe}I_bR_L}{R_L + h_{ie} - j/\omega C}, \quad \text{using (8.5.5).} \tag{8.5.6}$$

where V' is the p.d across ab.

∴ The output voltage is $$V_0 = -I_0h_{ie}$$

$$= -\frac{R_Lh_{fe}I_bh_{ie}}{R_L + h_{ie} - j/\omega C}, \quad \text{using (8.5.6).} \tag{8.5.7}$$

And the input voltage V_i is given by

$$V_i = h_{ie}I_b. \tag{8.5.8}$$

So, the low frequency voltage gain is

$$A_{V_L} = \frac{V_0}{V_i} = -\frac{h_{fe}R_L}{R_L + h_{ie} - \frac{j}{\omega C}} \tag{8.5.9}$$

using (8.5.7) and (8.5.8).

From (8.5.9), using (8.5.4), $A_{V_L} = \dfrac{A_{V_M}}{1 - \frac{j}{\omega C(h_{ie}+R_L)}} = \dfrac{A_{V_M}}{1 - j(f_1/f)}$ (8.5.10)

where $f_1 = \dfrac{1}{2\pi C(h_{ie} + R_L)}$ and

$f = \dfrac{\omega}{2\pi}$ = frequency of the input signal.

Now, from equation (8.5.10), $|A_{V_L}| = \dfrac{|A_{V_M}|}{\sqrt{1+(f_1/f)^2}}$ (8.5.11)

The above relation thus shows that *the low frequency voltage gain decreases with the decrease in frequency.*

When $f = f_1$, we get $|A_{V_L}| = |A_{V_M}|/\sqrt{2}$ i.e., the corresponding power delivered to the load is reduced by a factor $\frac{1}{2}$. This f_1 is known as the ***lower half-power frequency***. So, the lower half-power frequency may be *defined as the frequency at which the magnitude of the voltage gain in the low frequency range falls to* $1/\sqrt{2}$ *times the mid-frequency voltage gain.*

- For a single stage, $\dfrac{A_{V_L}}{A_{V_M}} = \dfrac{1}{\sqrt{1+f_1^2/f^2}}$.

For n-indentical stages, $\dfrac{A_{V_L}}{A_{V_M}} = \left(\dfrac{1}{\sqrt{1+f_1^2/f^2}}\right)^n$,

since $A'_{V_L} = A_{V_{L_1}} \times A_{V_{L_2}} \times \cdots = (A_{V_L})^n$.

(iii) High frequency range—In the high frequency range (> 20 kHz), the reactance of the coupling capacitor is neglible but those of the capacitances that appear in the shunt across the output become appreciable. The shunt capacitance consists of the transistor collector capacitance and the stray wiring capacitance. The reactance of

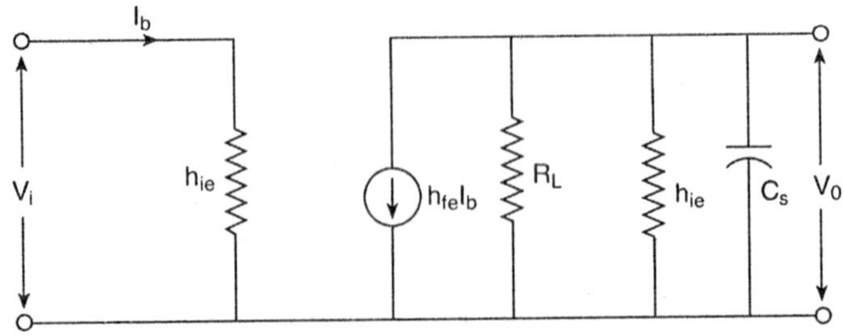

Fig. 8.12 *h*-parameter equivalent circuit of a one stage *RC*-coupled amplifier at high frequency range

these capacitances decreases with increase in frequency which causes to bypass a part of the output to the earth. So, the voltage gain falls. Moreover, the capacitive reactance of the base-emitter junction and the transistor β also decrease at high frequencies. So, there is a net loss in voltage gain. The high-frequency h-parameter equivalent circuit for one stage of an RC-coupled amplifier is shown in Fig. 8.12.

Let C_s be the equivalent shunt capacitance across the output terminals and Z_L be the effective load impedance.

$$\therefore \frac{1}{Z_L} = \frac{1}{R_L} + \frac{1}{h_{ie}} + j\omega C_s$$

$$\therefore Z_L = \frac{h_{ie}R_L}{h_{ie} + R_L + j\omega C_s h_{ie} R_L} \tag{8.5.12}$$

The input voltage is given by $$V_i = h_{ie} I_b \tag{8.5.13}$$

The output voltage is $$V_0 = -h_{fe} I_b Z_L \tag{8.5.14}$$

The high frequency voltage gain is therefore given by

$$A_{V_H} = \frac{V_0}{V_i} = -\frac{h_{fe}Z_L}{h_{ie}}$$

$$= -\frac{h_{fe}R_L}{h_{ie} + R_L + j\omega C_s h_{ie} R_L}, \text{ using (8.5.12)} \tag{8.5.15}$$

From (8.5.4), using (8.5.15), $$A_{V_H} = \frac{A_{V_M}}{1 + j\omega C_s \cdot \frac{h_{ie}R_L}{h_{ie}+R_L}} = \frac{A_{V_M}}{1 + j(f/f_2)} \tag{8.5.16}$$

where $$f_2 = \frac{1}{2\pi C_s} \cdot \frac{h_{ie} + R_L}{h_{ie}R_L} \quad (\because f = \omega/2\pi) \tag{8.5.17}$$

$$\therefore |A_{V_H}| = \frac{|A_{V_M}|}{\sqrt{1 + (f^2/f_2^2)}} \tag{8.5.18}$$

When $f = f_2$ we get, $|A_{V_H}| = \frac{1}{\sqrt{2}}|A_{V_M}|$ i.e., the corresponding power delivered to the load is reduced by a factor $\frac{1}{2}$. This f_2 is known as ***upper half-power frequency***. So, the upper half-power frequency may be defined as *the frequency at which the magnitude of the voltage gain in the high frequency range falls to* $\frac{1}{\sqrt{2}}$ *times the magnitude of the mid-frequency voltage gain.*

Bandwidth—The range of frequencies between the lower half-power frequency f_1 and the upper half-power frequency f_2 is called the ***bandwidth*** or *B. W* of the amplifier. Over this range, $|A_V|$ lies above 99.5% of the maximum value of A_{V_M}.

Gain-bandwidth product

The ***gain-bandwidth product*** of a RC-coupled amplifier is the *product* of the mid-frequency voltage gain and the bandwidth of the amplifier.

At the mid-frequencies, the voltage gain of the *first stage* is given by

$$A_{V_{M1}} = -\frac{h_{fe}R_L}{R_L + h_{ie}}$$

and that of the *second stage* is given by

$$A_{V_{M2}} = -\frac{h_{fe}R_L}{h_{ie}}$$

So, the overall voltage gain of two-stage amplifier at the mid-frequency range is

$$\begin{aligned} A &= A_{V_{M1}} \times A_{V_{M2}} \\ &= \left(\frac{h_{fe}}{h_{ie}}\right)^2 \cdot \frac{R_L^2 h_{ie}}{R_L + h_{ie}} \end{aligned} \tag{8.5.19}$$

Therefore, the *gain-bandwidth product* (GBP) is

$$\begin{aligned} A \cdot BW &= A(f_2 - f_1) \\ &= Af_2 \quad (\because f_2 \gg f_1 \text{ in normal operation}) \\ &= \left(\frac{h_{fe}}{h_{ie}}\right)^2 \times \frac{R_L^2 h_{ie}}{R_L + h_{ie}} \cdot \frac{1}{2\pi C_s} \cdot \frac{h_{ie} + R_L}{h_{ie} \cdot R_L} \\ &= \left(\frac{h_{fe}}{h_{ie}}\right)^2 \cdot \frac{R_L}{2\pi C_s} \end{aligned} \tag{8.5.20}$$

This is approximately constant. Equation (8.5.20) implies that an *RC*-coupled amplifier with *more gain* offers a *reduced bandwidth*. And conversely, *more bandwidth* results in *less gain*.

• The *half-power frequencies* are also referred to as the *cut off frequencies*: f_1 is the lower cut-off frequency and f_2 the upper cut-off frequency.

• Any connecting wire in a transistor circuit and the chassis ground on which the circuit is constructed act like the two plates of a capacitor. The capacitance that exists between this wire and the chassis ground is known as *stray-wiring capacitance.* This capacitance is one of the reasons for a loss of voltage gain because of its low capacitive reactance at a frequency that prevents the ac current from reacting the load resistor.

8.5.2 Miller effect

John M. Miller studied the effects of the plate impedance and voltage gain on the input admittance of a vacuum tube and hence the name *Miller effect.* The same effect is also observed in transistor amplifier. *The effect of amplification of capacitance at the input by the feedback from the output circuit of the amplifier through the internal base-collector junction capacitance (C_{bc}) is called the* ***Miller effect.***

At high frequencies, the reactances of the coupling capacitors and emitter bypass capacitors are negligible. But the base-collector junction capacitance (C_{bc}) and the base-emitter junction capacitance (C_{be}) become appreciable to shunt the load resistance. As a result, the effective ac load becomes smaller and hence the voltage gain decreases with increase in frequency. To understand the effect of C_{be} and C_{bc}, let us consider the equivalent circuit of a single stage amplifier in CE mode, as illustrated in Fig. 8.13 below.

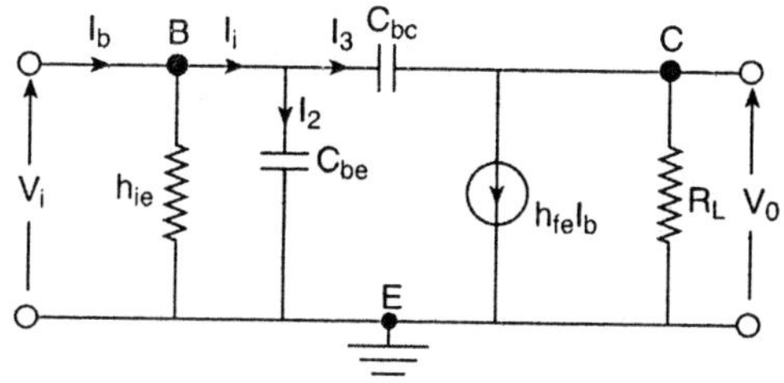

Fig. 8.13 Effect of junction capacitances

From Fig. 8.13,
$$I_i = I_2 + I_3 = \frac{V_i}{1/j\omega C_{be}} + \frac{V_i - V_0}{1/j\omega C_{bc}} \tag{8.5.21}$$

But
$$\frac{V_0}{V_i} \approx -\frac{h_{fe}}{h_{ie}} R_L \tag{8.5.22}$$

Using the value of V_0 in equation (8.5.21) we get,

$$\begin{aligned} I_i &= \frac{V_i}{1/j\omega C_{be}} + \frac{V_i + \frac{h_{fe}R_L}{h_{ie}} V_i}{1/j\omega C_{bc}} \\ &= jV_i\omega C_{be} + jV_i\omega C_{bc}\left(1 + \frac{h_{fe}}{h_{ie}}\right) \\ &= \frac{V_i}{1/j\omega C_s} \end{aligned}$$

where
$$C_s = C_{be} + \left(1 + \frac{h_{fe}}{h_{ie}}\right) C_{bc} \tag{8.5.23}$$

Equation (8.5.23) implies that C_{bc} provides feedback from output to the input and C_{be} and C_{bc} are in effect equivalent to a single capacitance C_s across the input resistance h_{ie}. Thus, an amplification of capacitance effect is observed at the input. This is known as the ***Miller effect.***

8.5.3 Phase characteristics of *RC*-coupled amplifier

The gain of an amplifier is plainly given by $A = A_0 e^{j\theta}$ where A_0 is the amplitude of the gain and θ the phase shift. Since both A and θ are functions of frequency, the frequency response of an amplifier is very important in analysing the performance of an amplifier. To preserve the form of the input signal at the output terminal, it is necessary to maintain a constant phase shift of the signal in going from the input to the output of the amplifier. So, the study of variation of phase shift with frequency i.e., phase characteristics is an important part in amplifier analysis.

The mid-frequency voltage gain of an RC-coupled amplifier is given by

$$A_{V_M} = -\frac{h_{fe}R_L}{h_{ie} + R_L} = |A_{V_M}|\, e^{j\pi} \tag{8.5.24}$$

where $|A_{V_M}| = \dfrac{h_{fe}R_L}{h_{ie} + R_L}$ = magnitude of the voltage gain. It is *independent* of the frequency and increases with R_L, approaching h_{fe} as $R_L \to \infty$.

Equation (8.5.24) implies that the output voltage leads over the input voltage by a constant phase angle $\theta_M = \pi$ i.e., 180° in the mid-frequency range.

From equation (8.5.10) we can write the low-frequency voltage gain as,

$$A_{V_L} = \frac{A_{V_M}}{1 - j(f_1/f)} = \frac{|A_{V_M}|}{\sqrt{1 + \frac{f_1^2}{f^2}}} e^{j\left(\pi + \tan^{-1}(f_1/f)\right)} \tag{8.5.25}$$

Obviously, the phase angle of A_{V_L} is given by

$$\theta_L = \pi + \tan^{-1}\frac{f_1}{f} \tag{8.5.26}$$

This equation implies that the phase shift θ_L increases with decreasing frequency f. At $f = f_1$ = lower half-power frequency, we get from (8.5.26)

$$\theta_L = \pi + \tan^{-1}\frac{f_1}{f_1} = \pi + \frac{\pi}{4} = \frac{5\pi}{4} = 225°$$

In the high-frequency range, the voltage gain is given by, using (8.5.16)

$$A_{V_H} = \frac{A_{V_M}}{1 + j(f/f_2)} = \frac{|A_{V_M}|}{\sqrt{1 + f^2/f_2^2}} e^{j\left(\pi - \tan^{-1}(f/f_2)\right)} \tag{8.5.27}$$

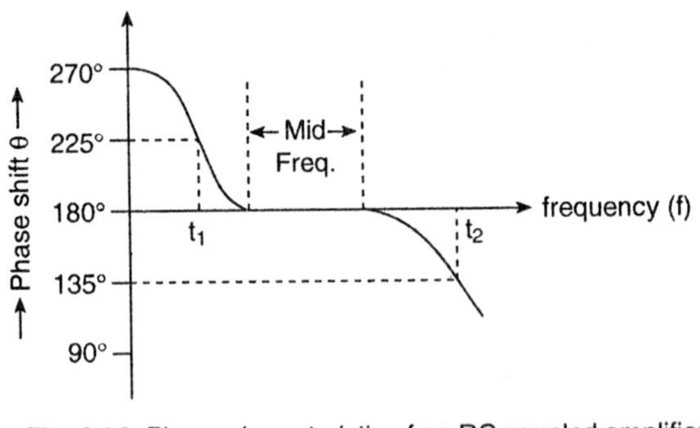

Fig. 8.14 Phase characteristic of an RC coupled amplifier

So, the phase angle of A_{V_H} is given by

$$\theta_H = \pi - \tan^{-1}\frac{f}{f_2} \tag{8.5.28}$$

Clearly, θ_H decreases with increase in frequency f. At $f = f_2$ = upper half-power frequency, we get from (8.5.28)

$$\theta_H = \pi - \tan^{-1}\frac{f_2}{f_2} = \pi - \frac{\pi}{4} = \frac{3\pi}{4} = 135°$$

The variation of the phase shift with frequency in different frequency ranges is shown in Fig. 8.14.

8.5.4 Bode plot

A ***Bode plot*** is a graph of decibel voltage gain versus frequency plotted on a semi-logarithmic graph paper to show the frequency response of an amplifier. Semilogarithmic graph paper is used to plot a quantity like voltage gain over a wide range of frequencies. The vertical axis uses a linear scale to plot voltage gain while the horizontal axis uses a logarithmic scale in order to expand the frequency range. When the gain of an amplifier is plotted in dB, straight line approximations can be employed to obtain the performance characteristics.

The low frequency voltage gain of an *RC*-coupled amplifier expressed in dB is obtained from equation (8.5.11) as

$$|A_{V_L}|_{\text{dB}} = |A_{V_M}|_{\text{dB}} - 20\log_{10}\sqrt{1+\frac{f_1^2}{f^2}} \tag{8.5.29}$$

The decibel voltage gain at high frequency range is given by

$$|A_{V_H}|_{\text{dB}} = |A_{V_M}|_{\text{dB}} - 20\log_{10}\sqrt{1+\frac{f^2}{f_2^2}} \tag{8.5.30}$$

using equation (8.5.18).

When equations (8.5.29) and (8.5.30) are plotted against frequency, we obtain the *Bode plot* of the *RC*-coupled amplifier, as shown by the solid line in Fig. 8.15.

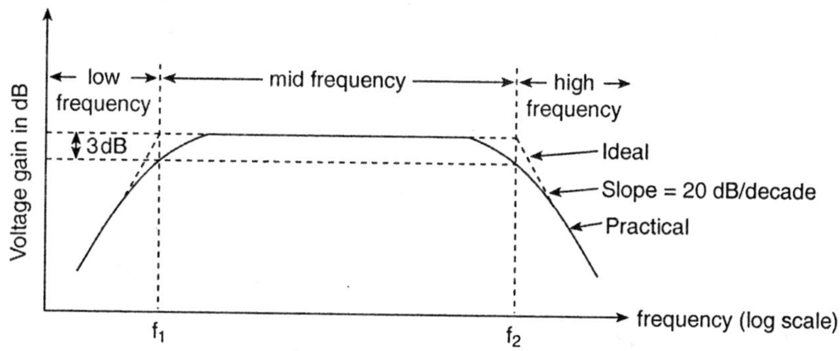

Fig. 8.15 Bode plot of an RC-coupled amplifier

From equation (8.5.29), we may write

$$|A_{V_L}|_{\text{dB}} = |A_{V_M}|_{\text{dB}} - 20\log_{10}(f_1/f), \quad \text{for } f \ll f_1 \tag{8.5.31}$$

and

$$|A_{V_L}|_{\text{dB}} = |A_{V_M}|_{\text{dB}}, \quad \text{for } f \gg f_1. \tag{8.5.32}$$

Clearly, equation (8.5.31) represents a *straight line* when decibel gain is plotted against frequency having a slope of 20 dB/decade. Equation (8.5.32) represents a straight line parallel to frequency axis. These two straight lines meet at a sharp corner at $f = f_1$.

From equation (8.5.30), we get

$$|A_{V_H}|_{\text{dB}} = |A_{V_M}|_{\text{dB}} - 20\log_{10}(f/f_2), \quad \text{for } f \gg f_2 \tag{8.5.33}$$

and $$|A_{V_H}|_{\text{dB}} = |A_{V_M}|_{\text{dB}}, \quad \text{for } f \ll f_2. \tag{8.5.34}$$

The above two relations represent two straight lines, one with a slope −20 dB/decade and the other parallel to frequency axis. These two straight lines produce a sharp corner at $f = f_2$. Note that (8.5.32) and (8.5.34) represent the same straight line.

The idealized Bode plot is shown by dashed line in Fig. 8.15. The frequencies f_1 and f_2 are termed the *corner frequencies* instead of cut-off frequencies because the ideal plot has a sharp corner at each cut-off frequency. The graph breaks at each cut-off frequency and then decreases at a rate of 20 dB/decade. So, the corner frequency is also termed as *break frequency*. Many technicians prefer to use ideal Bode plot because it is easy to draw and it gives approximately the same information as actual Bode plot does. Corrections can be made to obtain the actual curves near the corner frequencies. At the corner frequency f_1, we have from (8.5.29)

$$-|A_{V_L}|_{\text{dB}} + |A_{V_M}|_{\text{dB}} = 20\log_{10}\sqrt{2} = +3\,\text{dB}$$

This is the correction at f_1. Similarly, from (8.5.30), we can find a correction of −3 dB at f_2. So, the actual Bode plot may be easily obtained from the idealized plot.

8.6 Transformer-coupled transistor amplifier

The transistor amplifier in which transformers are employed to couple the input signal source to the first stage, the output of one stage to the input of the next stage and the output of the last stage to the load is called ***transformer-coupled amplifier***. Fig. 8.16 shows a two stage transformer-coupled amplifier circuit. T_1 is the coupling transformer which is used to transfer the output of the first stage to the input of the second. The

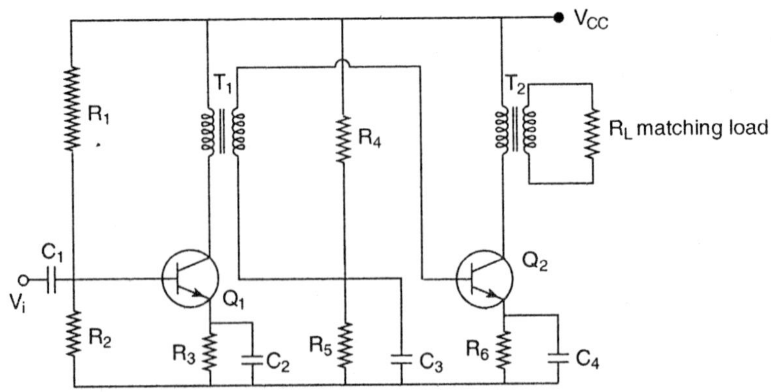

Fig. 8.16 Two-stage transformer-coupled amplifier

primary of this transformer serves as the collector load of the transistor Q_1 while the secondary gives the input to the base of the transistor Q_2. T_2, the output transformer, is used to transfer the output of the second stage to the load.

In Fig. 8.16, R_1, R_2, R_4 and R_5 are the biasing resistors for the transistors Q_1 and Q_2 forming the voltage divider circuit, where as R_3 and R_6 act as emitter stabilizing resistors. C_1 is the input coupling capacitor and C_2, C_3 and C_4 are the bypass capacitors.

Operation—The input signal which is applied to the base of the first transistor Q_1 appears in an amplified form across the primary of the coupling transformer T_1. It is then transferred to the base of Q_2 by the secondary of the transformer. The second transistor Q_2 further amplifies this signal and the more amplified voltage is obtained across the primary of the output transformer T_2. The secondary of T_2 then transfers this output to the load.

Frequency response—The *frequency response curve* for a transformer-coupled amplifier is illustrated in Fig. 8.17.

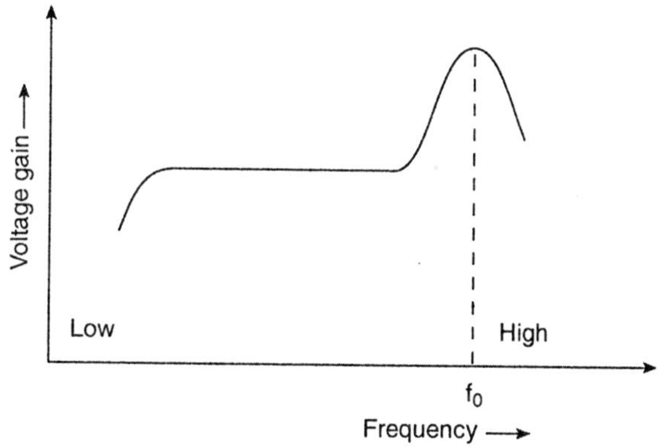

Fig. 8.17 Frequency response of TC-amplifier

The frequency response curve shows that

(a) the gain is constant only over a small range of frequency i.e., the frequency response is poor,

(b) there is a decrease in gain at both low and high frequencies, and

(c) there is a resonant rise in gain at a frequency f_0.

The output voltage of a *TC*-amplifier is the product of the collector current and the reactance of the primary of the coupling transformer. Since the reactance of the primary is small at low frequencies, it gives a lower gain. At high frequencies, the distributed capacitance between the turns acts as by-pass capacitor resulting in the decrease in output voltage and hence the gain. The resonant rise occurs at f_0 due to the resonance or tuning effect of inductance and distributed capacitors.

Advantages: (a) The low d.c resistance of the primary connected in the collector circuit makes a transformer-coupled amplifier more efficient than a *RC*-coupled system.

(b) A very good impedance matching can be provided by suitably adjusting the turns ratio of the transformer. Transformer coupling, therefore, provides a very good overall gain.

(c) There is no signal power loss in collector and base resistors.

Disadvantages: (a) The gain varies considerably with the frequency i.e., the frequency response is poor.

(b) The coupling transformers are bulky and costly.

(c) The windings in the coupling transformer tend to produce a '*hum*' in the output.

(d) Low frequency signals are less amplified compared to the high frequency ones. So, there is a large frequency distortion.

8.7 Power amplifiers

An amplifier, in general, is an electronic circuit capable of amplifying the strength of a weak signal. So far we discussed amplifiers either to obtain a large voltage gain or a large current gain but not both simultaneously. Most of electronic systems e.g., public address system (PAS) consisting of microphones, tape recorder, stereo system, radio or TV receivers, etc. are required to be provided with appreciable signal power. The voltage amplifiers, described in this chapter, can increase the amplitude of a signal many times but may not, on their own, be able to drive an output device such as a loudspeaker or a motor. Likewise, a current amplifier having a large current gain may amplify the current many times, but may be unable to supply sufficient current to the output device. So, in either case, the voltage or current amplifier alone does not have sufficient power (voltage × current). *The class of amplifiers which are capable of providing an appreciable amount of power and hence can drive low-impedance load* (e.g., loudspeaker) *are known as* ***power amplifiers***.

In most amplifier systems, the voltage level of a signal is first raised to a sufficiently high value by a multistage amplifier and that high voltage is then fed to a power amplifier. The resulting output voltage is then large enough to drive the final stage. So, a *power amplifier is a large signal amplifier* and *invariably a voltage amplifier precedes the power amplifier.*

It is to be noted that power amplifier *cannot generate* power. It takes extra power from the d.c sources connected to the output circuit and transforms it into useful a.c signal power to drive the load. Thus a *power amplifier* is a *d.c to a.c power converter* whose action is controlled by the input signal.

Since a power amplifier converts d.c power from the supply to a.c signal power at the load, a special type of bipolar transistor, called ***power transistor***, is employed as the active element in a power amplifier. The *special features* of a power transistor are

(i) The power ratings of power transistors are higher ($>$ 1 W) than those of transistors used in voltage amplifier.

(ii) The emitter and the base region are constructed with heavier doping.

(iii) The thickness of the base of a power transistor is greater than that of an ordinary transistor so that β is small (~ 50).

(iv) The area of contact between the emitter and the base is increased in a power transistor to reduce the ohmic resistance.

(v) The area of collector layer is large and it is connected to a metallic heat sink for quick removal of heat produced at the collector-base junction.

The *three* factors governing the limit of operation of a power transistor are the *voltage, current* and the *power dissipation ratings.* These are specified by the manufacturer. During operation, the power transistor must have working values of those parameters much less than the rated values. Otherwise, it may be destroyed due to overheating.

Since power amplifiers are large signal amplifiers, they use a large part of their a.c load line for thier normal operation and the excursion of the operating point (Q-point) may go beyond the *linear* region of the active device characteristics of the amplifier. So, the linear equivalent circuit method cannot be employed to analyse such circuits. Generally, the graphical method, using the concept of load line and the Q-point, is employed for the purpose.

Power amplifiers are classified into *four* types: *class A, class B, class C* and *class AB.* Of these, class A and class B push-pulll power amplifiers are most common and class C amplifiers have some specific uses.

8.7.1 Class A power amplifier with resistive load

The circuit diagram of a series-fed class A power amplifier using an *n-p-n* transistor is illustrated in Fig. 8.18. Since the CE mode gives the highest power gain among the three possible transistor configurations, the CE mode is generally employed in power amplifier circuits.

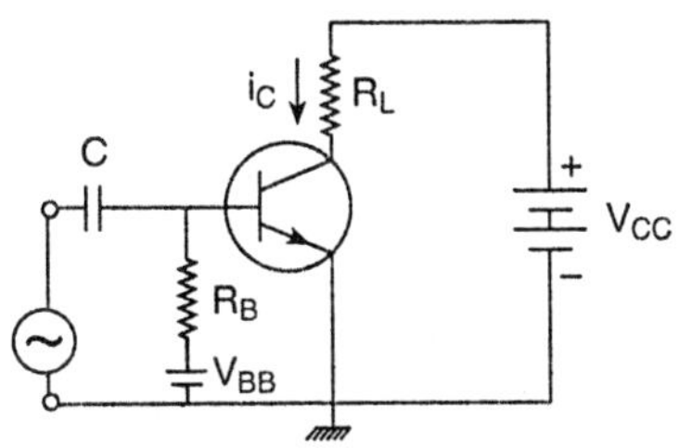

Fig. 8.18 Class A power amplifier with resistive load

Fig. 8.18 shows that the amplifier is directly coupled to the load resistance R_L with R_B as biasing resistance. The capacitor C is the blocking capacitor which prevents the interaction between the dc biasing voltage and the ac input signal voltage at the base circuit. For the class A operation, the collector current flows over the entire cycle of the input signal.

Analysis

To analyse the performance of the circuit, we use the output characteristics of the transistor (Fig. 8.19) and consider that the power dissipation by the transistor must lie within the safe limit specified by the manufacturers.

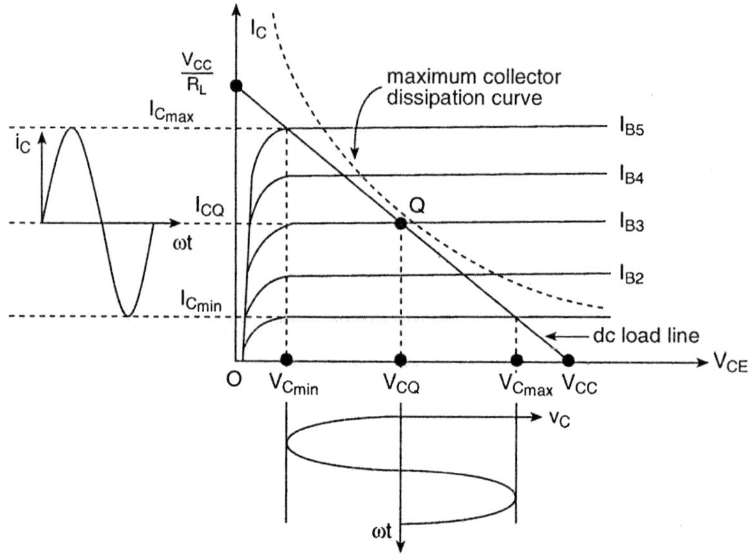

Fig. 8.19 Current and voltage waveforms for a class A power amplifier

The maximum power dissipation for a particular ambient temperature is given by

$$P_d = V_C' I_C' = \text{constant} \tag{8.7.1}$$

where V_C' and I_C' are the collector voltage and collector current respectively.

The dotted hyperbolic curve in the output characteristics of Fig. 8.19 represents the *maximum collector dissipation curve* in accordance with equation (8.7.1).

The dc load line is given by the equation

$$V_{CE} = V_{CC} - I_C R_L \tag{8.7.2}$$

The interaction of the load line with the output characteristic curve for the dc bias current determines the dc operating point or Q-point which is chosen near the center of the load line to allow maximum swing of collector current and collector voltage i.e., to achieve maximum ac power output.

If a sinusoidal input signal of proper amplitude is applied, then the Q-point moves along the load line between cut-off and saturation resulting in a sinusoidal output current and voltage as shown in Fig. 8.19. At cut-off, the collector current becomes minimum ($I_{C_{\text{min}}}$) and the collector voltage attains the maximum value ($V_{C_{\text{max}}}$). On the other hand, at the saturation, the collector current becomes maximum ($I_{C_{\text{max}}}$) and collector voltage minimum ($V_{C_{\text{min}}}$). If I_m and V_m correspond to the peak value of the output current and voltage respectively, then

$$I_m = \frac{I_{C_{\max}} - I_{C_{\min}}}{2} \tag{8.7.3}$$

and
$$V_m = \frac{V_{C_{\max}} - V_{C_{\min}}}{2} \tag{8.7.4}$$

The rms values of the output current and voltage are then given by

$$I_{\text{rms}} = \frac{I_m}{\sqrt{2}} = \frac{I_{C_{\max}} - I_{C_{\min}}}{2\sqrt{2}} \tag{8.7.5}$$

and
$$V_{\text{rms}} = \frac{V_m}{\sqrt{2}} = \frac{V_{C_{\max}} - V_{C_{\min}}}{2\sqrt{2}} \tag{8.7.6}$$

If I_{CQ} is the dc biasing current corresponding to the Q-point and $I = I_m \sin \omega t$ is the input signal current, then the instantaneous collector current is given by

$$i_C = I_{CQ} + I_m \sin \omega t \tag{8.7.7}$$

where ω is the angular frequency of the input signal.

∴ The average power dissipated in the load is

$$\begin{aligned} P_L &= \frac{1}{2\pi}\int_0^{2\pi} i_C^2 R_L d(\omega t) = \frac{1}{2\pi}\int_0^{2\pi} (I_{CQ} + I_m \sin \omega t)^2 R_L d(\omega t) \\ &= I_{CQ}^2 R_L + \tfrac{1}{2} I_m^2 R_L \end{aligned} \tag{8.7.8}$$

Clearly, $I_{CQ}^2 R_L$ is the dc power dissipated in R_L when the signal is absent. It is increased by $\frac{1}{2}I_m^2 R_L$ when the signal is applied. So, $\frac{1}{2}I_m^2 R_L$ represents the ac output power (P_{ac}) in the load. Thus, the ac output power in the load R_L is given by

$$P_{ac} = \frac{1}{2} I_m^2 R_L = \frac{1}{2} I_m \cdot V_m = \frac{I_m}{\sqrt{2}} \cdot \frac{V_m}{\sqrt{2}} = I_{rms} \cdot V_{rms} \tag{8.7.9}$$

$$= \frac{(I_{C_{\max}} - I_{C_{\min}})(V_{C_{\max}} - V_{C_{\min}})}{8}, \tag{8.7.10}$$

using equations (8.7.5) and (8.7.6).

The input dc power supplied by the source is given by

$$P_s = I_{CQ} \cdot V_{CC} \tag{8.7.11}$$

The *collector efficiency* or the *efficiency of power conversion* (η) is *defined* as the ratio of ac output power to the dc input power from the source. So, mathematically, the *collector efficiency* is given by

$$\eta = \frac{P_{ac}}{P_s} = \frac{(I_{C_{\max}} - I_{C_{\min}})(V_{C_{\max}} - V_{C_{\min}})}{8 I_{CQ} V_{CC}} \tag{8.7.12}$$

For ideal characteristics with maximum permissible signal strength, we may write $I_{C_{\min}} = 0$, $I_{C_{\max}} = 2I_{CQ}$, $V_{C_{\min}} = 0$ and $V_{C_{\max}} = V_{CC}$. So, equation (8.7.12) gives the maximum value of efficiency as

$$\eta_{\max} = \frac{2I_{CQ} \cdot V_{CC}}{8 I_{CQ} \cdot V_{CC}} = \frac{1}{4} = 25\% \tag{8.7.13}$$

Thus, the maximum conversion efficiency of a class A power amplifier directly coupled to a resistive load is only 25%. In practice, most series fed circuits provide efficiencies much less than 25%. The poor efficiency is mainly due to loss of energy in the form of heat at the load resistance R_L and the transistor.

- The instantaneous collector to emitter voltage drop is given by

$$v_C = V_{CC} - i_C R_L = V_{CC} - (I_{CQ} + I_m \sin \omega t)\, R_L \tag{8.7.14}$$

The average power dissipated in the transistor is

$$\begin{aligned} P_T &= \frac{1}{2\pi}\int_0^{2\pi} i_C v_C d(\omega t) \\ &= \frac{1}{2\pi}\int_0^{2\pi} (I_{CQ} + I_m \sin \omega t)\,(V_{CC} - I_{CQ}R_L - I_m R_L \sin \omega t)\, d(\omega t) \\ &= I_{CQ}\cdot V_{CC} - I_{CQ}^2 R_L - \tfrac{1}{2} I_m^2 R_L \end{aligned} \tag{8.7.15}$$

With no signal, $$P_T = I_{CQ}\cdot V_{CC} - I_{CQ}^2 R_L = I_{CQ}\cdot V_{CQ} \tag{8.7.16}$$

When the signal is applied, this P_T is decreased by $\frac{1}{2} I_m^2 R_L = P_{ac}$.

So, the *collector dissipation is maximum under quiescent conditions* i.e., *in the absence of the signal.*

Therefore, the total dc power supplied by the battery (source) is

$$P_s = P_L + P_T = I_{CQ}\cdot V_{CC},$$

using equations (8.7.8) and (8.7.15).

The above equation is the same as (8.7.11), obtained earlier.

8.7.2 Transformer-coupled class A power amplifier

In Art. 8.7.1, we discussed that the overall efficiency of a directly coupled class A amplifier does not exceed 25%. The reason is that the load resistance is directly

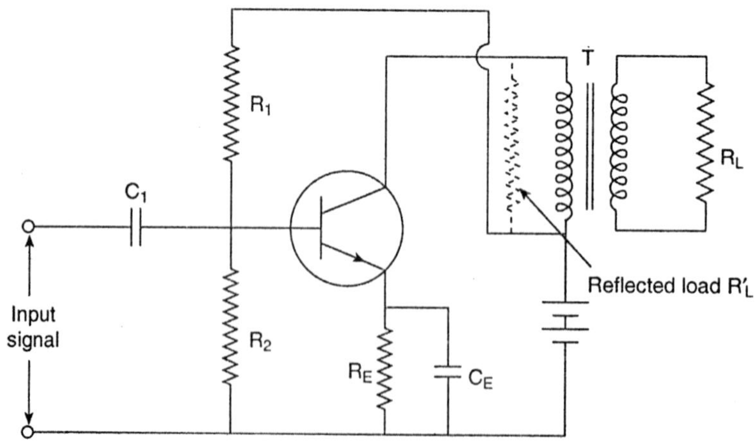

Fig. 8.20 Transformer-coupled class A power amplifier

connected to the output circuit and hence the quiescent current (I_{CQ}) passes through the resistance. A considerable amount of power is thus wasted as heat. The problem is solved by using a suitable transformer for coupling the load at the output stage of the amplifier (Fig. 8.20). This circuit is known as *transformer-coupled class A power amplifier.*

Since the primary winding of the transformer has a low resitance, the power absorbed in the winding is negligible compared to that in the resistive load. Moreover, the load resistance can be matched to the source for maximum power transfer by adjusting the turn ratio of the transformer. The resistors R_1, R_2, R_E and the d.c supply $+V_{CC}$ are adjusted for proper d.c biasing of the transistor so that it can be operated as a class A amplifier. C_1 and C_E are respectively the d.c blocking capacitor and the emitter bypass capacitor.

Analysis

To analyse the performance of the circuit, we use the output characteristics of the transistor (Fig. 8.21) and consider that the power dissipation by the transistor must lie within the safe limit specified by the manufacturer.

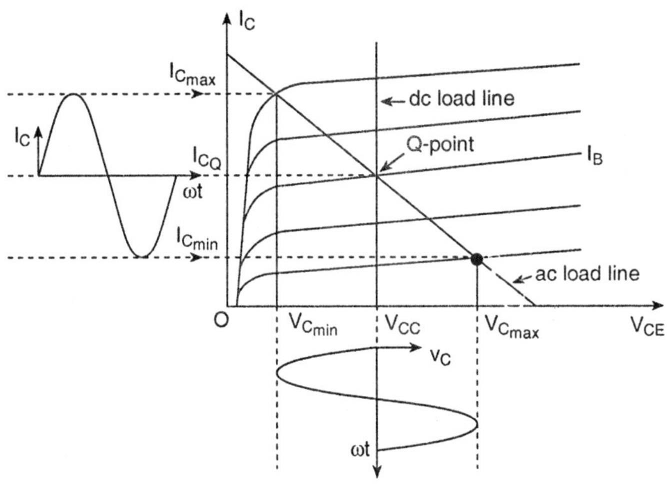

Fig. 8.21 Load lines and the current and voltage waveforms for a transformer-coupled class A power amplifier

As the primary winding of the transformer has a very low resistance (~ few Ω) and the emitter resistance is also small, we can neglect the d.c voltage drop acorss them.

∴ The d.c load line is given by

$$V_{CE} = V_{CC} \tag{8.7.17}$$

This equation represents a *straight line* passing though the point $(V_{CC}, 0)$ and parallel to the current axis, as shown in Fig. 8.21.

The interaction of the load line with the output characteristics for the d.c bias current determines the d.c operating point (Q-point). For a.c analysis, a.c load line is to be drawn. Here, the effective a.c load is $R'_L = n^2 R_L$ where n is the *turn ratio* of the transformer and R_L the load resistance (i.e., the resistance of the loudspeaker or other transducer). So, the a.c load line is a *straight line* with a slope $-1/R'_L$ and it passes through the d.c operating point Q.

If a sinusoidal input signal of proper amplitude is applied, then the Q-point moves along the load line between the cut-off and saturation resulting in a sinusoidal output current and voltage, as shown in Fig. 8.21. At *cut-off*, the collector current becomes minimum ($I_{C_{\min}}$) and the collector voltage attains the maximum value ($V_{C_{\max}}$). On the other hand, at saturation, the collector current becomes maximum ($I_{C_{\max}}$) and the collector voltage minimum ($V_{C_{\min}}$). If I_m and V_m the correspond to the peak value of the output current and voltage respectively, then

$$I_m = \frac{I_{C_{\max}} - I_{C_{\min}}}{2} \tag{8.7.18}$$

$$\text{and } V_m = \frac{V_{C_{\max}} - V_{C_{\min}}}{2} \tag{8.7.19}$$

The corresponding r.m.s values are then given by

$$I_{\text{rms}} = \frac{I_m}{\sqrt{2}} = \frac{I_{C_{\max}} - I_{C_{\min}}}{2\sqrt{2}} \tag{8.7.20}$$

$$\text{and } V_{\text{rms}} = \frac{V_m}{\sqrt{2}} = \frac{V_{C_{\max}} - V_{C_{\min}}}{2\sqrt{2}} \tag{8.7.21}$$

The a.c output power is thus

$$P_{\text{ac}} = V_{\text{rms}} \cdot I_{\text{rms}} = \frac{(V_{C_{\max}} - V_{C_{\min}})(I_{C_{\max}} - I_{C_{\min}})}{8} \tag{8.7.22}$$

The input d.c power supplied by the source is given by

$$P_s = I_{CQ} \cdot V_{CC} \tag{8.7.23}$$

∴ The *collector efficiency* or the *efficiency of power conversion* (η) is given by

$$\eta = \frac{P_{\text{ac}}}{P_s} = \frac{(V_{C_{\max}} - V_{C_{\min}})(I_{C_{\max}} - I_{C_{\min}})}{8 I_{CQ} \cdot V_{CC}} \tag{8.7.24}$$

For ideal characteristics with maximum permissible signal strength, we may write

$$I_{C_{\min}} = 0,\ I_{C_{\max}} = 2I_{CQ},\ V_{C_{\min}} = 0 \text{ and } V_{C_{\max}} = 2V_{CC}.$$

So, equation (8.7.24) gives the maximum value of efficiency as

$$\eta_{\max} = \frac{2V_{CC} \cdot 2I_{CQ}}{8 I_{CQ} V_{CC}} = \frac{1}{2} = 50\%. \tag{8.7.25}$$

Thus *the maximum conversion efficiency of a transformer-coupled class A power amplifier is* 50%. In practice, however, the efficiency is still smaller.

• The theoretical maximum efficiency of a transformer-coupled class A power amplifier is 50% which is *two times* (double) the *efficiency of a directly coupled class A power amplifier.*

• From equation (8.7.25), in the ideal situation, we get

$$P_{ac} = \tfrac{1}{2} P_s$$

When no signal is applied, the power dissipation in the collector junction is

$$P_T = I_{CQ} \cdot V_{CQ} \approx I_{CQ} \cdot V_{CC}, \quad (\because \; V_{CQ} \approx V_{CC})$$

The power given by d.c supply $+V_{CC}$ is

$$P_s = I_{CQ} \cdot V_{CC}$$

$$\therefore \; P_T = P_s = 2P_{ac} \tag{8.7.26}$$

So, the collector dissipation is nearly *double* the output a.c power in a transformer-coupled class A power amplifier.

8.7.3 Push-pull class B power amplifier

We already observed that the efficiency of a directly coupled class A power amplifier is 25%. When the load resistance at the output is replaced by a transformer, we get transformer-coupled power amplifier whose efficiency is 50%. So, there is still a loss of 50% of d.c power. The reason is that in both cases quiescent collector current I_{CQ} always flows through the transistor resulting in a d.c power loss of $I_{CQ} \cdot V_{CE}$ in the transistor itself. This is a considerable fraction of the total d.c power. So, to increase the efficiency further, the transistor should be operated in class B or class C mode where the current flows through the transistor only for one-half cycle or less of the input signal.

In a class B amplifier, the transistor conducts only for one-half cycle and remains off during the other half-cycle. This condition is achieved only by setting the Q-point at *cut-off* (i.e., at the lower end) of the load line. So, we obtain an output signal with either positive or negative half-cycles absent.

This gives a *larger distortion.* To avoid this, two identical transistors are connected such that one of the trasistors conducts during one half-cycle and the other conducts during the next half-cycle i.e., while the first transistor pushes the signal high during one half-cycle, the second one pulls it during the next half-cycle. Such an arrangement is known as ***push-pull arrangement.*** With this arrangement, greater conversion efficiency and less distortion of the signal can be achieved. Fig. 8.22 shows the circuit diagram of a transformer-coupled class B push-pull power amplifier.

The circuit consists of two identical transistors Q_1 and Q_2, and two centre-tapped transformers T_1 and T_2. The input transformer T_1 is called *phase splitter* which produces two signal voltages 180° out of phase with each other. These two signal voltages, with opposite polarity, are used to drive the transistors Q_1 and Q_2. The

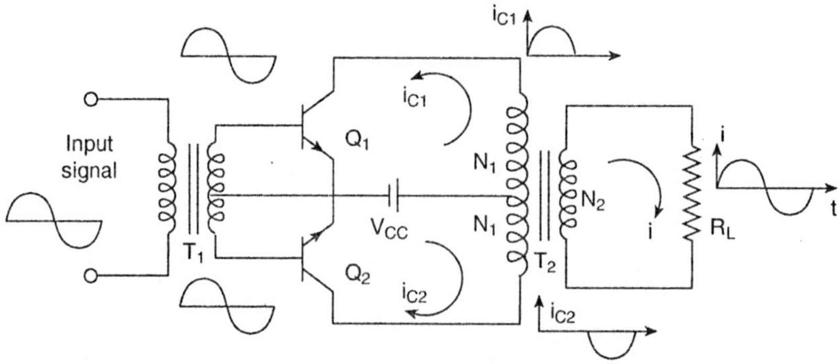

Fig. 8.22 Push-pull arrangement of class B power amplifier

output transformer T_2 is used to combine the a.c output signals from the collector to the load R_L. For normal operation of the circuit, Q_1 and Q_2 are biased at *cut-off.* This is achieved by connecting the emitters together to the centre-tap of the transformer T_1 and the power supply V_{CC}, to the centre-tap of the transformer T_2.

Operation

When no signal is applied, both the transistors Q_1 and Q_2 are *cut-off* and no current is drawn from V_{CC}. So, there is *no power loss* on stand by mode and the power dissipation in both the transistors is practically zero. During the positive half-cycle of the input signal, the base of Q_1 is positive and that of Q_2 is negative. So, Q_1 conducts and Q_2 remains *cut-off* for this half-cycle. When the input signal goes negative, Q_2 conducts and Q_1 turns off. The signal waveform at different points of the circuit are shown. It is seen that at any instant, only one transistor in this circuit conducts and each transistor handles one-half of the input signal. The output transformer combines the two halves to produce a *full sine-wave* in the load resistance R_L.

Analysis

To analyse the performance of the circuit, we consider the output characteristics of the transistors and load lines (both d.c and a.c) as shown in Fig. 8.23.

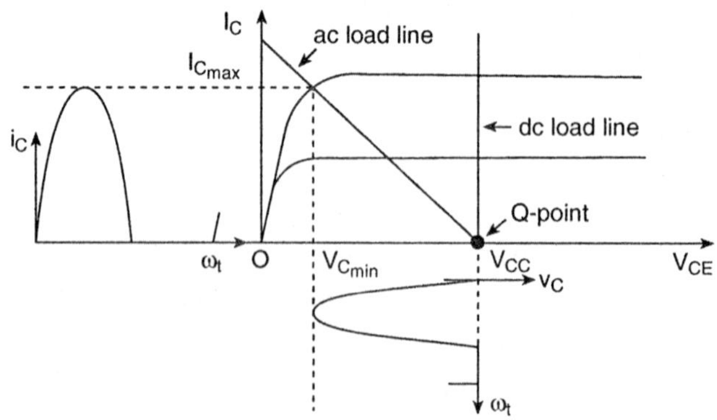

Fig. 8.23 Load lines and current and voltage waveforms for a class B push-pull amplifier

When no signal is applied, the transistor Q_1 (also Q_2) is at *cut-off*. So, $I_C = 0$ and $V_{CE} = V_{CC}$. So, the d.c load line is given by,

$$V_{CE} = V_{CC}. \tag{8.7.27}$$

The d.c load line is a *straight line* passing through the Q-point $(V_{CC}, 0)$ and parallel to I_C axis.

For the a.c analysis, it is necessary to determine a.c load line. The effective a.c load between the collector and the emitter of Q_1 is given by $R_L' = n^2 R_L$ where n is the *turn ratio* of the transformer T_2. So, the a.c load line is a *straight line* with slope $-1/R_L'$ and it passes through the Q-point.

When a sinusoidal input signal of proper amplitude is applied, the Q-point moves along the load line between *cut-off* and *saturation* resulting in a half-sinusoidal output current and voltage, as shown in Fig. 8.23.

The a.c output power at the transistor Q_1 is

$$\begin{aligned}(P_{\text{ac}})_1 &= V_{\text{rms}} \cdot I_{\text{rms}} = \frac{V_{CC} - V_{C_{\text{min}}}}{2} \cdot \frac{I_{C_{\text{max}}}}{2} \\ &= \tfrac{1}{4}(V_{CC} - V_{C_{\text{min}}}) \cdot I_{C_{\text{max}}}.\end{aligned} \tag{8.7.28}$$

∴ The total a.c output power due to Q_1 and Q_2 is

$$P_{\text{ac}} = 2(P_{\text{ac}})_1 = \tfrac{1}{2}(V_{CC} - V_{C_{\text{min}}}) \cdot I_{C_{\text{max}}} \tag{8.7.29}$$

Since half-sinusoidal current passes through each of Q_1 and Q_2, the average value of d.c current is $I_{C_{\text{max}}}/\pi$.

The d.c input power supplied by V_{CC} is

$$P_s = 2 \times V_{CC} \cdot \frac{I_{C_{\text{max}}}}{\pi} = \frac{2}{\pi} V_{CC} \cdot I_{C_{\text{max}}} \tag{8.7.30}$$

∴ The *efficiency of power* conversion is

$$\eta = \frac{P_{\text{ac}}}{P_s} = \frac{\pi}{4}\left(1 - \frac{V_{C_{\text{min}}}}{V_{CC}}\right) \tag{8.7.31}$$

When $V_{C_{\text{min}}} \ll V_{CC}$, we obtain from (8.7.31) the maximum efficiency,

$$\eta_{\text{max}} \simeq \frac{\pi}{4} = \frac{\pi}{4} \times 100\% = 78.5\% \tag{8.7.32}$$

So, the *theoretical maximum efficiency of a push-pull class B power amplifier is* 78.5%.

- In case of power amplifier, there is a large swing of input base current i_{b1}. For this case, the collector current i_{c1} of the transistor Q_1 can be expressed as

$$i_{c1} = A_0 + A_1 i_{b1} + A_2 i_{b1}^2 + A_3 i_{b1}^3 + \cdots \tag{8.7.33}$$

Since Q_1 and Q_2 are two *identical* transistors, the collector current i_{c2} of Q_2 can be written as

$$i_{c2} = A_0 + A_1 i_{b2} + A_2 i_{b2}^2 + A_3 i_{b2}^3 + \cdots \tag{8.7.34}$$

In push-pull arrangemnt $i_{b1} = -i_{b2}$ and i_{c1} and i_{c2} flow in opposite directions. So the *output curent* is given by

$$
\begin{aligned}
i_0 &= k(i_{c1} - i_{c2}), \quad \text{where } k \text{ is a constant} \\
&= k\left[A_1(i_{b1} - i_{b2}) + A_2(i_{b1}^2 - i_{b2}^2) + A_3(i_{b1}^3 - i_{b2}^3) + \cdots\right] \\
&= k\left[2A_1 i_{b1} + 2A_3 i_{b1}^3 + 2A_5 i_{b1}^5 + \cdots\right] \qquad (8.7.35)
\end{aligned}
$$

since $i_{b1} = -i_{b2}$.

Let $i_{b1} = I_m \cos\omega t$. Then the equation (8.7.35) gives,

$$
\begin{aligned}
i_0 &= 2A_1 k I_m \cos\omega t + 2A_3 I_m^3 \cos^3\omega t + \cdots \\
&= 2A_1 k I_m \cos\omega t + 2A_3 I_m^3 \cdot \tfrac{1}{4}(3\cos\omega t + \cos 3\omega t) + \cdots \\
&= B_1 \cos\omega t + B_2 \cos 3\omega t + \cdots \qquad (8.7.36)
\end{aligned}
$$

where $B_1, B_2, \cdots$ are constants.

Equation (8.7.36) indicates that *a push-pull circuit eliminates all the even harmonic distortions.* Here, the *third harmonic* is the *principal source of distortion.*

Advantages of class B push-pull power amplifier

The class B push-pull amplifier has the following important advantages over the class A power amplifier.

(i) The maximum efficiency of a class B push-pull amplifier is 78.5% which is much higher than that (50%) of class A amplifier.

(ii) The class B push-pull amplifier eliminates even order harmonics in the a.c output signal. So, there is less distortion for a given output power per transistor.

(iii) There is no d.c component in the output signal because the output currents of the two transistors flow in opposite directions and cancel out the d.c components. So, there is no possibility of core saturation of the output transformer and we can use small size cores in the transformer.

(vi) There is no collector current drain when there is no signal.

Disadvantages

(i) The use of transformer is the main disadvantage of class B push-pull power amplifier. The AF transformer which is used in this amplifier is rather bulky and expensive.

(ii) The power supply must also have a good regulation for the critical biasing condition.

8.7.4 Complementary symmetry push-pull amplifier

The main disadvantage of using transformers in class B push-pull power amplifier can be avoided by using transistors having complementary symmetry in the emitter follower configuration. Fig. 8.24 shows the circuit diagram of such a class B push-pull amplifier.

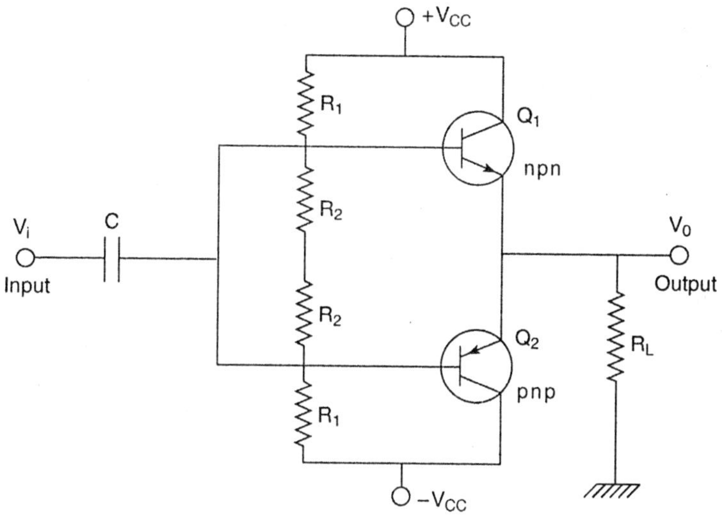

Fig. 8.24 Complementary symmetry push-pull amplifier

The term 'complementary' means that the circuit uses two transistors—one *n-p-n* and the other *p-n-p*, having identical input and output characteristics. The term 'symmetry' means that the biasing resistors are equal such that the emitter-base junction of each transistor is biased with equal voltage. In this circuit, while the input signal is coupled with capacitance, the output is directly coupled.

When no input signal is applied, none of Q_1 and Q_2 conducts. When the input signal is applied, the transistors being of opposite type conduct in alternate half-cycles of the input. During the positive half-cycle, the transistor Q_1 is *on* and Q_2 *off*. So, a half-cycle of the signal is produced across the load R_L. During the negative half-cycle of the input signal, Q_2 conducts and Q_1 remains *off*. Then another half-cycle is produced across R_L. Thus a complete cycle is obtained across the load.

The above circuit suffers from a *disadvantage* of obtaining a pair of transistors having exact completementary symmetry. So, the lack of balance in the characteristics of Q_1 and Q_2 produces considerable distortion and there is the presence of even harmonics in the output signal.

8.8 Distortion and noise in amplifiers

Distortion and noise are the two rather important factors that are to be addressed in studying all kinds of amplifiers.

8.8.1 Distortion

For an ideal amplifier it is expected that the output signal should be an enlarged but an identical reproduction of the input signal. But, in practice, it is seen that the output waveform does not ideally correspond to the input signal due to the inherent nonlinearity of the amplifying active element or the associated circuit elements.

The deviation of the output waveform from the input waveform is called ***distortion***. Different types of distortions occur in practical amplifiers. These are described and discussed below.

(a) *Amplitude distortion or nonlinear distortion*

The nonlinear input characteristics of a transistor amplifier indicates that the input resistance changes with the amplitude of the input signal. So, when a sinusoidal input signal voltage of large amplitude is applied, the base current waveform remains no longer sinusoidal. This type of distortion is called the ***input nonlinear distortion*** or ***amplitude distortion***. The output characteristics of the transistor are also not exactly parallel in the active region for equal changes in the base current. So, when a sinusoidal base current is applied, the collector current and also the collector voltage waveform do not remain exactly sinusoidal.

The output is associated with new harmonics which are not present in the original input signal. This type of distortion in amplifiers is called the ***output nonlinear distortion***.

(b) *Frequency distortion*

The internal capacitances of the active device or the reactances of the circuit components associated with the amplifier change with the frequency of the input signal. It makes the gain to be different for different signal components of different frequencies. This type of distortion is known as ***frequency distortion***.

(c) *Phase distortion*

The presence of the reactive and resistive circuit elements may cause different phase shifts for different signal frequencies. So, the output becomes distorted. This is known as ***phase*** or ***time delay distortion***.

(d) *Intermodulation distortion*

When the input signal of different frequencies are applied to an amplifier, the output is found the contain their harmonics, the sum and the difference of frequencies along with the original frequencies. This arises due to the nonlinearity of the active device. This type of distortion is called ***intermodulation distortion***.

8.8.2 Noise

Any kind of undesirable and randomly varying signals not related to the useful input signal in an electronic device is referred to as ***noise***. In almost all practical amplifiers, a small output signal, known as *noise*, is obtained even when there is no input signal. In radio receiver, it produces a *hiss* and in video amplifier *snow-like* appearance on the screen. Noise is superimposed on the signal. So, it may be difficult to distinguish the signal in the background of the noise if the input signal is very weak. Depending on the source, noise may be classified as *external* and *internal*. The external noise includes the disturbances caused by lightning, sparking, running of heavy electrical machines etc. The various sources of internal noise are as follows:

(a) *Thermal noise or Johnson noise*

The noise generated by the random thermal motion of current carriers in the resistive component of a circuit is called ***thermal*** or ***Johnson noise*** or ***resistance noise***. The r.m.s noise voltage V_n, generated across a resistance R at an absolute temperature T, is given by the expression

$$V_n = \sqrt{4kTBR}$$

where k is Boltzmann's constant and B the bandwidth of the noise.

(b) *Shot noise*

The noise introduced by the random fluctuations of the rate of injection of current carriers from the emitter into the base of a transistor is known as the ***shot noise***. The mean square shot noise current is given by

$$I_n^2 = 2eI_{dc}B$$

where e is the magnitude of the electronic charge, I_{dc} the d.c current and B the bandwidth of the noise.

(c) *Partition noise*

The noise generated by random fluctuations in the division of the emitter current into the collector and the base currents is called the ***partition noise***.

8.8.3 Signal-to-noise ratio and noise-figure

The *ratio of the signal power to the noise power* is called the ***signal-to-noise ratio*** (abbreviated as S/N). It is the measure of the noise contamination of the signal.

For a signal source of r.m.s voltage V_s and internal resistance R, the maximum possible S/N*-ratio* is

$$\frac{V_s^2}{V_n^2} = \frac{V_s^2}{4kTBR}$$

where the source of noise is the Johnson noise in R.

The *noisiness* of a device is measured by a quantity known as ***noise factor***. It is defined as *the ratio of input signal-to-noise power to the output signal-to-noise power*. It is thus given by

$$\textit{Noise factor},\ F = \frac{\textit{input } S/N}{\textit{output } S/N}$$

When the noise factor is expressed in decibel, then it is called ***noise figure*** (*NF*). The noise figure *NF* and the noise factor F is thus related as

$$\boxed{\mathbf{NF = 10\log_{10} F}}$$

If then *NF* is smaller, the device would be less noisy and conversely. For a noise-free ideal amplifier, $F = 1$ i.e., $NF = 0$.

8.9 Illustrated Examples

► **Example 1.** *In the three stages of a multistage amplifier the voltage gains are* 10, 20 *and* 30 *respectively. Calculate the overall voltage gain of the amplifier.*

(Kurukshetra Univ.)

Solution: The voltage gains at the first, second and the third stages are 10, 20 and 30 respectively. Thus, $A_{V_1} = 10$, $A_{V_2} = 20$ and $A_{V_3} = 30$.

$$\therefore \text{ The overall voltage gain, } \boldsymbol{A_V} = A_{V_1} \times A_{V_2} \times A_{V_3}$$
$$= 10 \times 20 \times 30 = \mathbf{6000}$$

► **Example 2.** *In a three-stage amplifier, the overall voltage gain is* 1728. *If each stage has equal voltage gain, find its value.*

Solution: Let n be the number of stages, A the voltage gain of each stage and A_n the overall voltage gain.

$$\therefore \quad A_n = A \times A \times A \times \cdots (n \text{ times}) = A^n$$

By the problem, $A^3 = 1728 \Rightarrow A = \sqrt[3]{1728} = \mathbf{12}.$

► **Example 3.** *The current gain of a single stage amplifier is* 60. *If the collector load resistance is* $50\,\Omega$ *and the input impedance is* $100\,\Omega$, *calculate the total voltage gain if two such stages are R-C coupled.*

Solution: By the problem, $A_i = 60$, $R_L = 500\Omega$ and $Z_i = 100\Omega$.

$$\therefore \quad \text{Voltage gain, } A_v = A_i \frac{R_L}{Z_i} = \frac{60 \times 500}{100} = 300$$

When two such stages are coupled, the total voltage gain A is given by

$$\boldsymbol{A} = 300 \times 300 = \mathbf{9 \times 10^4}$$

► **Example 4.** *A three-stage amplifier has voltage gains* 100, 200 *and* 400 *respectively in the first, second and the third stage. Express the total gain in* dB.

Solution: Voltage gain in stage 1 in dB $= 20 \log_{10} 100 = 40.$

Voltage gain in stage 2 in dB $= 20 \log_{10} 200 = 20 \times 2.3010$
$= 46.02$

Voltage gain in stage 3 in dB $= 20 \log_{10} 400 = 20 \times 2.602 = 52.04$

$$\therefore \quad \textit{\textbf{Tatal gain}} \text{ in dB} = 40 + 46.02 + 52.04 = \mathbf{138.06}$$

► **Example 5.** *In an amplifier, the impedance of a load to be connected in the output circuit should be* $500\,\Omega$ *for maximum transfer of power. If a loudspeaker of* $8\,\Omega$ *is to be connected, what should be the turn ratio of the transformer to be used?* **(Nagpur Univ.)**

Solution: For the maximum transfer of power, the impedance of the output stage of the amplifier, $R_0 =$ the load of $500\,\Omega$ i.e., $R_0 = 500\,\Omega$.

The impedance of the loudspeaker, $R_L = 8\,\Omega$. The turn ratio N_p/N_s is given by

$$R_0 = \left(\frac{N_p}{N_s}\right)^2 R_L$$

$$\therefore \quad \frac{N_p}{N_s} = \left(\frac{R_0}{R_L}\right)^{1/2} = \left(\frac{500}{8}\right)^{1/2} = 7.9 \approx 8$$

The turn ratio should, therefore, be **8 : 1**.

▶ **Example 6.** *A two stage RC-coupled amplifier uses the transistors having the h-parameters* $h_{ie} = 1100\Omega$ *and* $h_{fe} = 250$. *If the load resistance is* $10\,\text{k}\Omega$ *and the coupling capacitor is* $1\,\mu\text{F}$, *find the lower half-power frequency.*

Solution: Here,

$$h_{ie} = 1100\Omega, \quad h_{fe} = 250$$

$$R_L = 10\,\text{k}\Omega = 10 \times 10^3\Omega$$

$$C = 1\mu\text{F} = 1 \times 10^{-6}\,\text{F}$$

$\therefore$ The lower half-power frequency f_l is given by

$$f_l = \frac{1}{2\pi C(h_{ie} + R_L)}$$

$$= \frac{1}{2\pi \times 10^{-6}(1100 + 10 \times 10^3)}\ \text{Hz} = \mathbf{14.3\ Hz}$$

▶ **Example 7.** *The mid-band gain of an RC-coupled amplifier is* 120. *At frequencies of* 100 Hz *and* 100 kHz, *the gain falls to* 60. *Determine the lower and the upper half-power frequencies.*

Solution: The low-frequency gain is given by

$$|A_{V_L}| = \frac{|A_{V_M}|}{\sqrt{1 + (f_l/f)^2}}, \text{ where } f_l = \text{lower half-power frequency.}$$

Given, $|A_{V_L}| = 60$ for $f = 100\,\text{Hz}$ and $|A_{V_M}| = 120$.

$$\therefore\ 60 = \frac{120}{\sqrt{1 + (f_l/100)}} \Rightarrow \mathbf{f_l = 173.2\ Hz}$$

Also, high-frequency gain $|A_{V_H}| = \dfrac{|A_{V_M}|}{\sqrt{1 + (f/f_h)^2}}$, where $f_h =$ upper half-power frequency.

Putting $|A_{V_H}| = 60$, $|A_{V_M}| = 120$ and $f = 100$ kHz $= 100 \times 10^3$ Hz $= 10^5$ Hz.

$$60 = \frac{120}{\sqrt{1 + (10^5/f_h)^2}} \Rightarrow f_h = 57.7 \times 10^3\,\text{Hz} = \mathbf{57.7\ kHz}$$

▶ **Example 8.** *The h-parameters used in RC-coupled amplifier are* $h_{fe} = 600$ *and* $h_{ie} = 10\,\text{k}\Omega$. *If the shunt capacitance is* 610 pF, *coupling capacitance is* $0.5\,\mu$F *and load resistance is* $10\,\text{k}\Omega$, *calculate* (i) *the voltage gain at mid-frequencies and* (ii) *the upper cut-off frequency.*

Solution: Given, $h_{ie} = 10\,\text{k}\Omega = 10{,}000\,\Omega$, $h_{fe} = 600$, $R_L = 10\,\text{k}\Omega = 10{,}000\,\Omega$, $C_s = 610\,\text{pF} = 6.1 \times 10^{-10}\,\text{F}$ and $C_c = 0.5\,\mu\text{F} = 0.5 \times 10^{-6}\,\text{F}$.

(i) Voltage gain at mid-frequency, $$(A_\text{v})_\text{mid} = -\frac{h_{fe}R_L}{R_L + h_{ie}} = -\frac{600 \times 10{,}000}{10{,}000 + 10{,}000} = \mathbf{-300}$$

(ii) Upper cut-off frequency, $$f_u = \frac{1}{2\pi C_s}\left(\frac{1}{R_L} + \frac{1}{h_{ie}}\right)$$
$$= \frac{1}{2 \times 3.14 \times 6.1 \times 10^{-10}} \times \left(\frac{1}{10000} + \frac{1}{10000}\right)$$
$$= 52.2 \times 10^3\,\text{Hz} = \mathbf{52.2\ kHz}$$

▶ **Example 9.** *A single stage amplifier has a gain of* 50. *The collector load* $R_c = 300\,\Omega$ *and the input impedance is* $600\,\Omega$. *Find the overall gain when two such stages are cascaded by RC-coupling.*

Solution: Gain of second stage remains so as it is independent of loading of any stage. But due to the loading of input impedance of second stage, the gain of first stage is less than 50.

Effective load of first state $= R_c \| R_\text{in} = \dfrac{R_c R_\text{in}}{R_c + R_\text{in}} = \dfrac{300 \times 600}{300 + 600} = 200\,\Omega.$

$\therefore\ A_{V_1}$ = gain in first stage = $50 \times (200/300) = 100/3$

V_{V_2} = gain in second stage = 50

$\therefore$ Overall gain, $\boldsymbol{A_V} = A_{V_1} \times A_{V_2} = (100/3) \times 50 = 5000/3 = \mathbf{1667}$.

Review and Short Questions

1. What are the different classifications of amplifiers? Give a brief account of class A, B, AB and C amplifiers. **(Meerut Univ.)**
2. What are the fundamental differences among the class A, class B and class C amplifiers? **(Burdwan Hons.)**
3. Explain the principle of multistage amplifiers.
4. Define decibel power gain. What do you mean by 3 dB frequencies? What is meant by bandwidth?

5. What is the use of coupling in amplifiers? Draw a diagram of *RC*-coupled transistor amplifier. Explain its frequency response graph. What are its advantages? **(Kanpur Univ.)**

6. Draw a typical frequency response characteristic of an RC coupled amplifier indicating therein three freuqency ranges and bandwidth. **(Calcutta Hons.)**

7. Draw the circuit diagram and explain the working of a transformer-coupled transistor amplifier.

8. Draw the frequency response curve and explain why the amplifier gain falls off at low and high frequencies. **(Meerut Univ.)**

9. What are the advantages and disadvantages of a transformer-coupled amplifier?

10. Give the circuit of a *RC*-coupled amplifier. Obtain an expression for the voltage gain in mid-frequency range. **(Nagpur Univ.)**

11. Explain 'Miller effect' in the context of a high frequency transistor amplifier. **(Burdwan Hons.)**

12. In what range of frequencies is Miller effect active in an RC coupled amplifier? How does it control the upper cut-off frequency? **(Calcutta Hons.)**

13. Draw the circuit diagram of a two-stage RC-coupled CE transistor amplifier. Show how the magnitude and phase angle of its voltage gain vary with frequency. Qualitatively explain these variations. Define the half-power frequencies. **(Lucknow Univ.)**

14. What is a power transistor? Enumerate its special features.

15. What is a power amplifier? What do you mean by the conversion efficiency of a power amplifier? How does the efficiency change as the operation changes from class A to class C through class AB and class B?

16. Draw a neat circuit diagram of a series-fed class A power amplifier. Find expressions for its maximum output power and efficiency. **(Burdwan Hons.)**

17. Show that the maximum efficiency of a class A power amplifier directly coupled to the load resistance is 25%. **(Calcutta Hons.)**

18. Draw the circuit diagram of a class A direct coupled power amplifier. Show that the dissipation in the transistor is a maximum under quiescent condition. Also, obtain the expression for the maximum efficiency of the circuit.

19. Explain with a circuit diagram the operation of a transformer-coupled class A power amplifier. How do you get ac load line for this circuit? **(Osmania Univ.)**

20. (a) Show that the maximum efficiency of a class A transformer-coupled power amplifier is double that of a directly-coupled class A power amplifier.

(b) How is the collector dissipation related to the output ac power in a transformer-coupled class A power amplifier? What is the collector voltage rating for this amplifier?

21. (a) What is push-pull amplifier? What are the advantages of such amplifiers?

(b) Draw the circuit diagram of a class B push-pull amplifier using transistors. Find an expression for the efficiency of a class B push-pull amplifier.

(Calcutta Hons.)

22. (a) Show that a push-pull circuit eliminates all the even harmonic distortions.

(Calcutta Hons.)

(b) Explain why the class B amplifiers are preferred over class A ones for high power applications.

23. Draw the circuit diagram of a complementary symmetry push-pull class B power amplifier and explain its principle of operation. State its advantages and disadvantages.

24. Explain with a circuit diagram the operation of a single-tuned amplifier. Draw its ac equivalent circuit and find an expression for the voltage gain and bandwidth.

(Burdwan Hons.)

25. What is a tuned amplifier? In which range of frequencies are tuned amplifiers used? Draw the circuit diagram of a single-tuned amplifier.

26. What is a class C amplifier? Draw the circuit diagram of a class C power amplifier with a tuned load and explain its operation. What are the uses of such an amplifier ?

27. (a) 'A power amplifier is generally preceded by a voltage amplifier'—Explain the statement.

(b) Explain why class B and class C amplifiers are more efficient than class A amplifiers.

28. What is Bode plot? Draw the Bode plot of an RC-coupled amplifier and explain it. What are corner frequencies?

29. What do you mean by distortion in amplifiers? Name the different types of distortions.

30. Discuss the different types of distortions found in transistor amplifiers.

31. What is the desired position of the Q-point for a minimum distortion and why?

32. Discuss briefly the various sources of noise in transistor amplifiers.

(Calcutta Hons.)

33. What is noise in an amplifier? Define 'noise factor' and 'noise figure' and explain the significance of 'noise figure'.

Objective Questions

1. An audio amplifier works over the frequency range

   (a) 10 Hz to 10 MHz, (b) 20 Hz to 20 kHz, (c) 20 kHz to 30 MHz.

2. The low-frequency fall in gain in a RC-coupled amplifier is due to the increase in reactance of the

   (a) coupling capacitor, (b) bypass capacitor, (c) junction capacitance.

3. A video amplifier has a bandwidth of about

   (a) 10 MHz, (b) 4 MHz, (c) 4.5 kHz, (d) none.

4. The input non-linear distortion can be reduced by

   (a) decreasing the output resistance of the signal source,
   (b) increasing the output resistance of the signal source,
   (c) increasing the frequency of the input signal,
   (d) none of the above.

5. The output non-linear distortion occurs because

   (a) The source resistance is not negligible,
   (b) The input resistance of a transistor amplifier is not a constant,
   (c) The output characteristics of the transistor are not parallel lines,
   (d) Of none of the above.

6. For cascading, one should use

   (a) CE configuration, (b) CB configuration, (c) CC configuration.

7. If the collector current in a CE amplifier flows only during the one-half of the input cycle, then it is a

   (a) Class A amplifier, (b) Class B amplifier,
   (c) Class C amplifier, (d) Class AB amplifier.

8. If the voltage gains in a three-stage amplifier be 5, 10 and 15 respectively, then the overall voltage gain is

   (a) 30, (b) 350, (c) 750, (d) none of these.

9. For a class A power amplifier, supply dc voltage is ± 12 V, the quiescent collector current is 72 mA and the load resistance is 100Ω. If the output voltage across the load is 12 V peak-to-peak, the efficiency of the amplifier is (neglect the loss occurring in the bias resistors)

(a) 10.4%, (b) 20.8%, (c) 25%, (d) 33.3%. **(DRDO)**

10. Approximate equivalent noise temperature (deg.K) of an amplifier with a noise factor of 1.04 is

(a) 301.6, (b) 11.6, (c) 278.4, (d) 5.8. **(ISRO)**

11. Which of the following principles is applied while designing the output matching network for a high power class C amplifier?

(a) Maximum power transfer theorem, (b) Norton's theorem,
(c) Thevenin's theorem, (d) Power = $(\text{voltage})^2/R_L$. **(ISRO)**

12. For a class B amplifier providing a 20 V peak signal to 16 ohm load and a power supply of $V_{CC} = 30$ V, the efficiency will be

(a) 52.3%, (b) 25.65%, (c) 75%, (d) 78.6%. **(ISRO)**

13. A transistor amplifier with 85% of efficiency is likely to be

(a) class A, (b) class B, (c) class AB, (d) class C. **(ISRO)**

14. If the power level of an amplifier is halved then, in decibel scale, the change is

(a) 3 dB, (b) 0.5 dB, (c) 2 dB, (d) −3 dB.

15. The class of amplifiers which is most inefficient but has least distortion is

(a) class A, (b) class B, (c) class AB, (d) class C.

16. The maximum efficiency of a transformer-coupled class A power amplifier is

(a) 25%, (b) 50%, (c) 78.5%, (d) 81.2%.

17. Push-pull connection is mostly used with

(a) class A, (b) class B, (c) class AB, (d) class C.

18. The main purpose of using transformer coupling in class A amplifier is to make it more

(a) efficient, (b) distortion-free, (c) bulky, (d) costly.

19. Cross-over distortion behaviour is characteristic of

(a) class A output stage, (b) class B output stage,
(c) class AB output stage, (d) common-base output stage. **(GATE)**

20. In a "public address system" amplifier, the input stage is a

(a) current amplifier, (b) voltage amplifier,
(c) power amplifier, (d) combination of current and voltage amplifiers.

Problems

1. A RC-coupled amplifier has a mid-frequency gain of 100. The gain is 80 at 50 Hz and 100 kHz. Calculate the half-power frequencies. [37.5 Hz, 133.3 kHz]

2. The mid-frequency gain of a RC-coupled amplifier is 100. The lower half-power frequency is 50 Hz and the upper half-power frequency is 200 kHz. At what frequency is the gain 80? [66.6 Hz, 150 kHz]

3. A two stage RC-coupled amplifier uses transistors having h-parameters, $h_{ie} = 1\,\text{k}\Omega$, $h_{fe} = 100$. If the load resistance is 2.2 kΩ, calculate the overall mid-frequency gain. Neglect the effect of source resistance and biasing resistances. [15125]

4. The mid-band voltage gain of a RC-coupled amplifier is 150. The gain is 50 at a frequency of 40 Hz and 60 at a frequency of 500 kHz. Find the bandwidth of the amplifier. **(Calcutta Hons.)** [218.105 kHz]

5. In a transformer coupled amplifier, show that the effective a.c load R_L' is n^2R_L where n is the turn-ratio of the transformer and R_L is the load connected to the transformer secondary.

6. What turn ratio of transformer is needed to couple onto a load 8Ω so that it appears as a 5 kΩ effective load? **(Burdwan Hons.)** [25 : 1]

7. If f_1 and f_2 are the lower and upper half power frequencies of one stage, find the corresponding values for two stages in cascade. [$f_1' = 1.56f_1$, $f_2' = 0.64f_2$]

8. If the voltage gains of three stages of a three-stage amplifier are 50, 70 and 90 respectively, find the overall gain of the amplifier. [3.15×10^5]

9. With the given data of Illustrated Example 8, calculate the lower cut-off frequency. [15.9 Hz]

10. A RC-coupled transistor amplifier has the mid-frequency gain 100. The values of lower and upper cut-off frequencies are 50 Hz and 100 kHz respectively. Find the frequencies at which the gain would be reduced to 80. [37.5 Hz, 133.3 kHz]

11. A RC-coupled amplifier has $h_{ie} = 4.5\,\text{k}\Omega$, $h_{fe} = 330$. If the collector load resistance is 5.5 kΩ and the lower and upper cut-off frequencies are chosen as 30 Hz and 120 kHz respectively, calculate the values of (i) shunt capacitor and (ii) the coupling capacitor. [536 pF, 0.53 μF]

12. In a RC-coupled amplifier, the h-parameters of a transistor used are $h_{fe} = 100$ and $h_{ie} = 1.5\,\text{k}\Omega$., If the coupling capacitor is of $0.5\,\mu\text{F}$, collector load is $2.5\,\text{k}\Omega$ and shunt capacitance is 700 pF, calculate the

(a) voltage gain at mid-frequencies,

(b) lower half-power frequency, and

(c) upper half-power frequency. [−62.5, 79.6 Hz, 242.6 kHz]

13. An amplifier has an efficiency of 32% and a collector dissipation of 0.8 W. Calculate the dc power input and ac power output of the circuit.

[1.176 W, 0.376 W]

14. An amplifier has an efficiency of 31% and power output of 12 W. Is this likely to be a class A or class B amplifier? What is the dc input power to the circuit?

[class A, 38.7 W]

15. For a transistor amplifier, the voltage gain is 110. If the input impedance is $20\,\Omega$ and source impedance is $200\,\Omega$, what is the effective voltage gain of the amplifier?

[10]

16. A power transistor of a class A amplifier is supplied from a dc supply of 10 V. If the maximum collector current is 100 mA, find the power transferred to a 4Ω load, if load is (a) directly-coupled, and (b) transformer-coupled.

[20 mW, 500 mW]

17. A certain power transistor meant for class A operation has a zero signal power dissipation of 10 W. If the ac output power is 4 W, find (a) collector efficiency and (b) power rating of the transistor. [25%, 10 W]

18. A power transistor of a transformer-coupled class A amplifier has to deliver a maximum of 4 W of low frequency power to a 4Ω load with a power supply 24 V. Find the values of (a) dc input power, (b) quiescent voltage and current, (c) peak value of collector current, and (d) transformer turns ratio. Take $\mu = 0.4$.

[8 W; 24 V, 0.33 A; 0.4 A; 4.2]

19. Calculate the phase angle between input and output signal for lower half-power frequency. Also show that as the frequency $f \to 0$, the phase angle θ_L tends to $3\pi/2$. [225°]

20. Prove that for lower half-power frequency $|A_{V_M}| = \sqrt{2}|A_{V_L}|$.

9

Field Effect Transistor

9.1 Introduction

Field effect transistor (abbreviated as FET) is a *three-terminal* semiconductor device which depends for its operation on the control of current by an electric field across the conducting region. As the current in a FET is carried predominantly by one type of carriers (majority carriers), it is also called a ***unipolar transistor***, unlike bipolar one (BJT) where two types of carriers—electrons and holes—are involved.

It is commonly used for weak-signal amplification in many digitial and in some analog circuits. It can also be used in oscillator circuits and for switching d.c circuits. FETs are preferred in such circuits and systems that require high impedance like wireless communication and broadcast receivers.

There are *two* types of field effect transistors.

(i) Junction field-effect transistor, abbreviated as JFET, and

(ii) Insulated-gate field-effect transistor, IGFET or metal-oxide-semiconductor field-effect transistor, MOSFET.

In this chapter we shall mainly discuss the principles of operation and the basic characteristic features of these devices.

9.2 Junction field effect transistor (JFET)

Depending on the type of the semiconductor bar used for the construction, the JFET is again classified as : (i) n-channel JFET and (ii) p-channel JFET. In what follows, we shall essentially concern ourselves with their structures, their characteristics and the principles of operation.

9.2.1 Structure of JFET

The schematic diagram of an *n-channel* JFET is shown in Fig. 9.1(a). It consists of a uniformly doped n-type semiconductor bar of Si or GaAs, having ohmic contacts at the two ends and two p-n junctions formed by doping two sides heavily with p-type impurities. If the semiconductor bar is of p-type, the JFET is termed a *p-channel* JFET (Fig. 9.2(a)) and the p-n junctions are formed by doping heavily with n-type impurities.

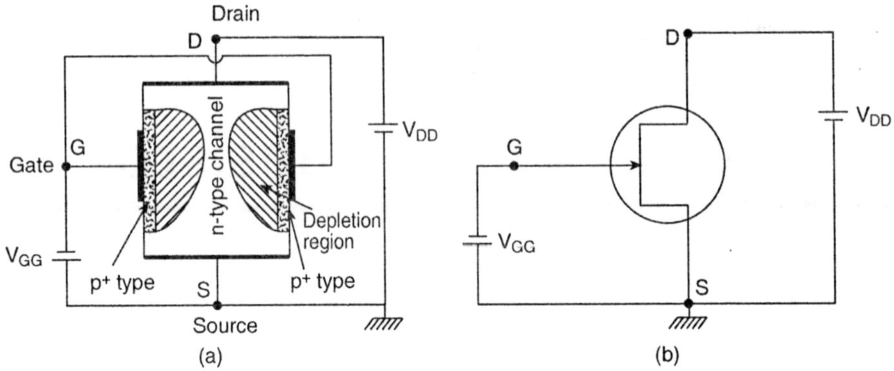

Fig. 9.1 (a) Basic structure (b) Symbol with supply voltages of *n*-channel JFET

When a voltage is applied between the end terminals, a current is caused to flow through the bar along its length due to the flow of majority carriers. The FET being a three-terminal device, the following notations are used for FET circuit connections.

(a) ***Source*** : The terminal through which the majority carriers enter the semiconductor bar is called the *source* S.

(b) ***Drain*** : The terminal through which the majority carriers leave the semiconductor bar is known as the *drain* D.

(c) ***Gate*** : The impurity regions on the two sides of the bar where p-n junctions are formed by doping heavily with impurities opposite to that of the bar are called the *gate* G.

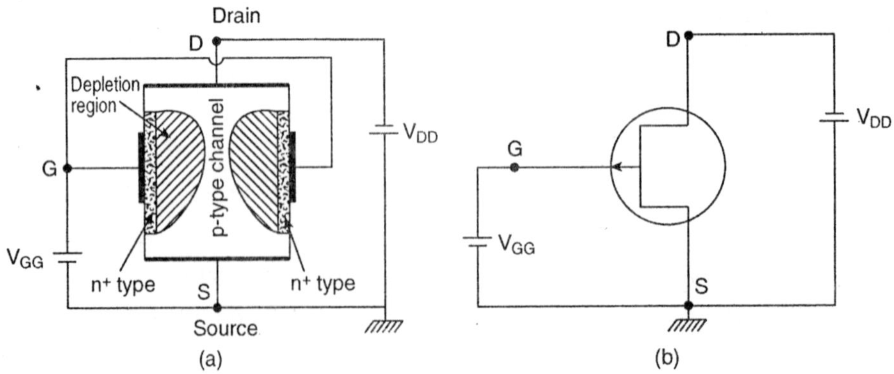

Fig. 9.2 (a) Basic structure (b) Symbol with supply voltages of *p*-channel JFET

The region of the bar through which the majority carriers move from source to drain during the current conduction is known as the ***channel***. The channel opening decreases with increasing depletion widths. The circuit symbols of n-channel JFET and p-channel JFET with supply voltages are shown in Fig. 9.1(b) and Fig. 9.2(b) respectively. The arrow head indicates the direction of current when the gate junction is forward biased.

9.2.2 Principle of operation

For the normal operation of a JFET, the gate junction is reverse-biased. To understand the principle of operation of a JFET, let us consider the n-channel JFET in common-source configuration (Fig. 9.1). On application of the reverse-bias voltage V_{GG} at the gate junctions, the depletion regions are formed that extend into the n-channel bar. During the current flow through the channel from source to drain, the gradual ohmic potential drop across the channel causes the gate junctions more reverse-biased near the drain-end than the source-end. So, the width of the depletion region is more near the drain terminal than the source-end of the gate i.e., the effective cross-sectional area through which the current flows in the bar decreases from source to drain terminal. If the gate reverse-bias voltage is increased, the effective channel cross-section for current flow and hence the drain current decreases. Thus the resistance of a JFET is controlled by the gate-voltage i.e., the JFET behaves as a ***voltage-controlled resistor***. Since the transverse electric field applied to the gate gives the effect of controlling the different parameters of the device, the name *field-effect* is used.

The principle of operation of a p-channel JFET can be understood by a similar explanation if the voltage polarities are reversed and the role of the electrons is replaced by that of the holes.

9.3 Static characteristics of JFET

The ***static characteristics*** of a JFET are *the graphical plots of the drain current I_D against the drain-to-source voltage V_{DS}, with gate-to-source voltage V_{GS} remaining constant.* It is also known as the *common-source drain characteristics* or the

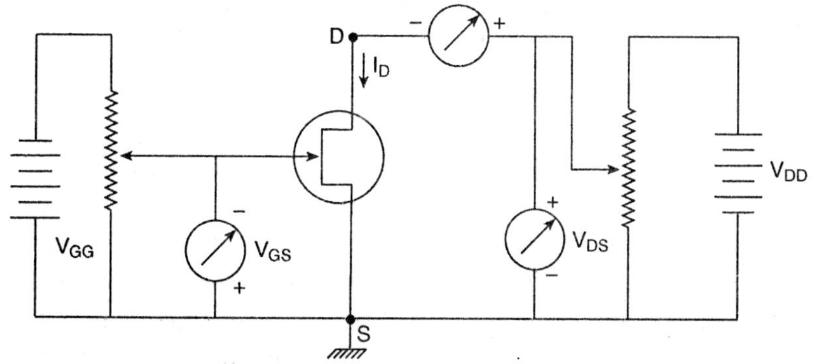

Fig. 9.3 Circuit diagram for drawing static characteristic curves of an n-channel JFET

output characteristics of a JFET. The circuit arrangement for obtaining the static characteristics of an n-channel JFET is illustrated in Fig. 9.3.

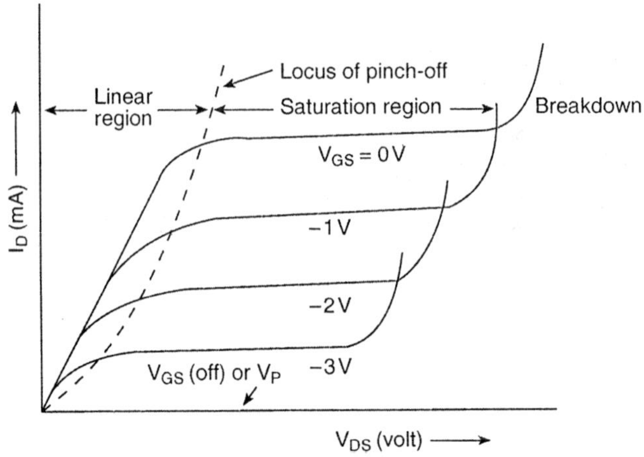

Fig. 9.4 Drain characteristics of *n*-channel JFET

A family of typical static characteristic curves obtained for an n-channel JFET is illustrated in Fig 9.4.

The statistic characteristics are divided into *three* distinct regions.

(i) Linear or ohmic region : In this region, the drain current I_D is nearly proportional to the drain-to-source voltage V_{DS}. Here V_{DS} is small and the channel behaves as a simple resistor.

(ii) Saturation or active region : In this region, the drain current I_D is approximately constant and is independent of V_{DS}. This constant current is known as the *saturation current* $I_{D_{sat}}$ and the minimum value of V_{DS} at which $I_{D_{sat}}$ is obtained is called the *saturation voltage* $V_{D_{sat}}$. The values of $I_{D_{sat}}$ and $V_{D_{sat}}$ decrease with the increase in reverse gate bias.

(iii) Breakdown region : In this region, I_D rises sharply with small increase in drain-to-source voltages i.e., in V_{DS}.

Explanation of the static characteristics—When $V_{GS} = 0$ and a +ve voltage V_{DS} of small value is applied to the drain with respect to the source, a current flows through the n-channel bar. As the voltage increases from zero at the source to the small value V_{DS} at the drain, the widths of the depletion regions remain very small and the channel is entirely open to act as a simple resistor. So, the current I_D increases linearly with V_{DS} in this region which is known as *linear* or *ohmic region.*

When the reverse bias across the gate-to-source is increased, the depletion regions formed at the gate junctions increase and the effective conducting channel cross-section decreases. The gradual potential drop along the length of the channel makes the gate more reverse-biased near the drain-end than the source-end. Therefore, the width of the depletion region increases when we move from source to drain, and the current flows through a wedge-shaped channel. As V_{DS} is increased slowly, the effective cross-sectional area for current flow in the channel decreases and the channel resistance increases. As a result, the drain current I_D increases non-linearly with V_{DS}. At a certain value of V_{DS}, the depletion regions meet each other near the drain and the channel is said to be *pinched off* [Fig. 9.5(a)]. The value of V_{DS} at which pinch off takes place is called the *pinch-off voltage*. Since the current beyond pinch-off [Fig. 9.5(b)] flows due to injection of carriers from channel to the depletion region through the point where the depletion regions meet, the current remains almost constant, $I_{D_{sat}}$ and the corresponding value of V_{DS} is the saturation voltage, $V_{D_{sat}}$. This explains the *saturation region* of static characteristics.

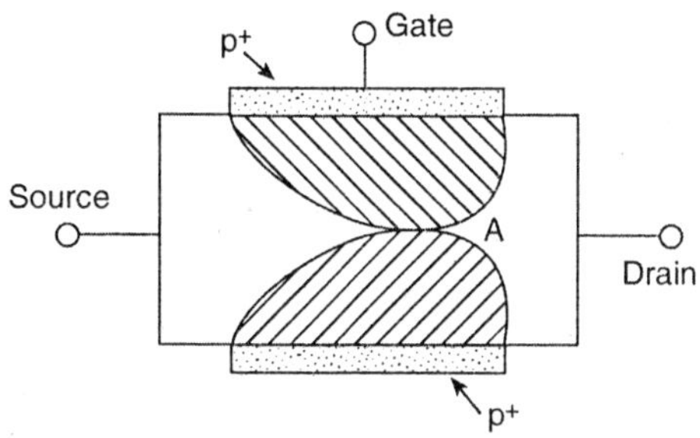

Fig. 9.5(a) JFET channel at pinch-off

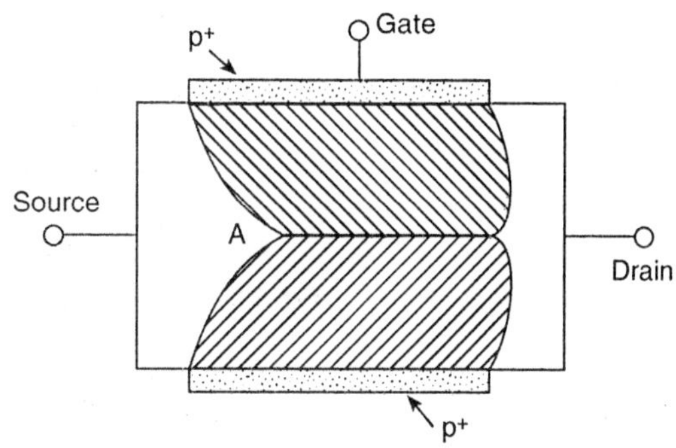

Fig. 9.5(b) JFET channel beyond pinch-off

On application of negative voltage (V_{GS}) between gate and source, additional reverse bias is provided between the channel and the gate. As a result, the maximum drain current ($I_{D_{sat}}$) reduces at saturation and pinch-off occurs earlier (i.e., at lower value of V_{DS}) than the previous one (i.e., with $V_{GS} = 0$). Also, the break down occurs at lower value of V_{DS} because both V_{GS} and V_{DS} contribute to the break down. In fact, the saturation level of I_D decreases as V_{GS} is made more and more negative as shown in Fig. 9.4. The dotted parabolic curve in Fig. 9.4 indicates how the pinch off voltage varies with the reverse bias V_{GS}.

At a certain value of V_{GS}, the gate voltage is made sufficiently negative such that the $I_{D_{sat}}$ is reduced to zero and the device is turned off for all practical purposes. This voltage is called *gate to source cut-off voltage*. It is labelled as V_{GS} (off) or V_P, as shown in Fig. 9.4.

At a large value of V_{DS}, the reverse bias at the gate junction becomes sufficient to cause an avalanche breakdown which results in a sharp rise in the drain current. Thus the *breakdown region* of the characteristics is explained.

- Since the drain current is controlled by the gate bias beyond the pinch-off, the amplification of an a.c. signal can be achieved by varying the gate bias in a JFET.

- If a forward bias is applied to the gate of a n-channel JFET, the holes are injected from the gate into the channel. As a result, the input impedance decreases and the field effect control is lost. So, a forward bias is not applied to the gate.

9.4 Transfer characteristic curve of JFET

The graphical plot indicating the variation of saturation drain current $I_{D_{\text{sat}}}$ with gate-to-source voltage V_{GS} for a constant drain-to-source voltage (greater than pinch-off voltage) is called the *transfer characteristic curve* of a JFET.

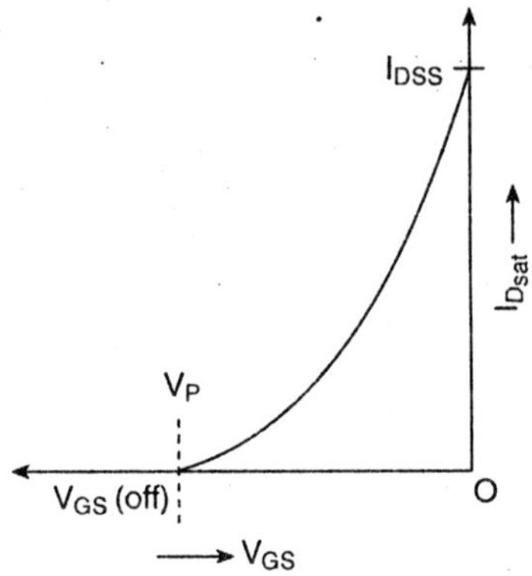

Fig. 9.6 Transfer characteristic of an *n*-channel JFET

Fig. 9.6 shows the transfer characteristic curve of an n-channel JFET. The shape of the curve is aproximately parabolic. The relation between I_D and V_{GS} is given by Shockley's equation

$$I_D = I_{DSS}\left(1 - \frac{V_{GS}}{V_P}\right)^2 \tag{9.4.1}$$

where I_{DSS} is the saturation drain current at $V_{GS} = 0$.

Equation (9.4.1) indicates that a JFET can be used as a *voltage-controlled current source* because the drain curent I_D can be controlled by varying the gate-to-source voltage V_{GS}. But, in case of a BJT, $I_C \simeq \beta I_B$ in the active region. So, the output current I_C is controlled by the input current I_B. Hence a BJT is a *current-controlled* electronic device.

9.5 JFET as a voltage variable resistor

The static characteristic curves of a JFET (Fig. 9.4) show that before pinch off, the I_D-V_{DS} curves are almost straight lines. The slopes of the curves are determined by the values of V_{GS}. So, before pinch off, a JFET can be used as a simple resistor the value of which is determined by the slope of I_D-V_{DS} curve that can be varied by V_{GS}. For $V_{GS} = 0$, the pinch off voltage is given by $V_P = V_{D_{\text{sat}}}$ and $I_D = I_{D_{\text{sat}}}$ and the ohmic d.c. resistance is approximately given by

$$R_{DSO} = \frac{|V_P|}{I_{D_{\text{sat}}}} \tag{9.5.1}$$

When a reverse bias voltage V_{GS} is applied to the gate junction, the depletion layers formed at the gate junctions decrease the channel cross-sectional area for current flow even at $V_{DS} = 0$. So, the resistance increases and saturation drain current decreases. In this situation, the pinch off and avalanche breakdown take place for smaller values of V_{DS}. The variation of the JFET resistance with V_{GS} is approximately given by

$$R_{DS} = \frac{R_{DSO}}{(1 - V_{GS}/V_P)^2} \tag{9.5.2}$$

9.6 The pinch-off voltage

To derive an approximate expression for the pinch-off voltage V_P, we consider the n-channel JFET of Fig. 9.1(a) where two sets of depletion regions, one on each side of the n-type rectangular bar, are formed. The two sets of depletion regions being similar, we consider only one set here. Since the gate is heavily doped with p-type impurity, the width of the depletion region is approximately equal to the width of depletion region in the n-type channel. For the calculation of pinch-off voltage, let us consider the approximate wedge-shape channel of Fig. 9.7 showing the variation of width of the depletion region with the distance from source to drain.

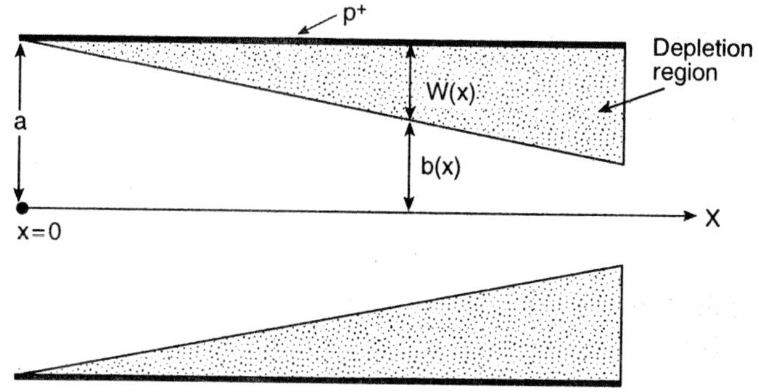

Fig. 9.7 Approximate wedge-shape channel

Suppose, a is the half-width of the bar and $b(x)$ the effective half-width of the conducting channel at a distance x from the source end. Then, the depletion-layer half-width at x can be written as

$$W(x) = a - b(x). \tag{9.6.1}$$

Since the junction is assumed to be abrupt and the p-n junction is heavily doped, the half-width of the depletion layer $W(x)$ is given by

$$W(x) = \sqrt{\frac{2\varepsilon}{eN_D} V(x)} \tag{9.6.2}$$

where, ε = permittivity of the material of the channel,

e = magnitude of electronic charge,

N_D = concentration of donor impurity,

and $V(x)$ = applied reverse bias across the space-charge region at x.

Now, we know that the pinch-off voltage V_P, also known as *gate cut-off voltage* V_{GS} (off), is the gate reverse voltage that removes all free charges from the channel and makes $I_D = 0$. Therefore, putting $b(x) = b = 0$ in equation (9.6.1) and $V(x) = V_P$ in (9.6.2) we get the expression for pinch-off voltage as

$$V_P = \frac{eN_D}{2\varepsilon} a^2 \tag{9.6.3}$$

For a *p*-channel JFET, N_D in equation (9.6.3) is to be replaced by N_A, the acceptor impurity concentration.

9.7 JFET parameters

The drain current I_D of a JFET is a function of the drain-to-source voltage V_{DS} and the gate-to-source voltage V_{GS}. For the analysis of JFET amplifier, the following *three parameters* are defined. These parameters can be determined by the slope of the static characteristic curves.

(i) Transconductance or mutual conductance (g_m) : It is *defined* as the ratio of small change in drain current to the corresponding small change in gate-to-source voltage, keeping the drain-to-source voltage constant.

$$\therefore \quad g_m = \left.\frac{\delta I_D}{\delta V_{GS}}\right|_{V_{DS}=\text{const.}} \tag{9.7.1}$$

(ii) Drain resistance (r_d) : It is *defined* as the ratio of small change in drain-to-source voltage to the corresponding change in drain current for a constant gate-to-source voltage i.e.,

$$r_d = \left.\frac{\delta V_{DS}}{\delta I_D}\right|_{V_{GS}=\text{const.}} \tag{9.7.2}$$

(iii) Amplification factor (μ) : It is *defined* as the ratio of the small change in the drain-to-source voltage to the corresponding change in gate-to-source voltage, with constant drain current i.e.,

$$\mu = -\left.\frac{\delta V_{DS}}{\delta V_{GS}}\right|_{I_D=\text{const.}} \tag{9.7.3}$$

The negative sign indicates that δV_{DS} and δV_{GS} have opposite sign at constant I_D, μ being a positive quantity.

Relation between μ, g_m and r_d : The drain current I_D is plainly a function of V_{GS} and V_{DS}. We, therefore, have

$$\therefore \quad I_D = f\,(V_{GS}, V_{DS})$$

$$\text{Thus,} \quad dI_D = \left.\frac{\partial I_D}{\partial V_{GS}}\right|_{V_{DS}} \cdot dV_{GS} + \left.\frac{\partial I_D}{\partial V_{DS}}\right|_{V_{GS}} \cdot dV_{DS} \tag{9.7.4}$$

If I_D remains constant for simultaneous change of V_{GS} and V_{DS}, then $dI_D = 0$. Then the equation (9.7.4) gives,

$$\left.\frac{\partial I_D}{\partial V_{GS}}\right|_{V_{DS}} + \left.\frac{\partial I_D}{\partial V_{DS}}\right|_{V_{GS}} \cdot \left.\frac{\partial V_{DS}}{\partial V_{GS}}\right|_{I_D} = 0$$

$$\text{or,}\quad g_m + \frac{1}{r_d}(-\mu) = 0$$

$$\text{or,}\quad \boxed{\mu = g_m \times r_d} \tag{9.7.5}$$

$\therefore$ ***Amplification factor = Transconductance × Drain resistance***

9.8 Metal-oxide-semiconductor field effect transistor (MOSFET)

Metal-oxide-semiconductor field-effect-transistor (abbreviated as MOSFET) is the field effect transistor which is commercially more important than JFET. MOSFET is extensively used in digital circuits because it occupies less space in IC and is suitable for very large scale integration. The MOSFET is also used as power device. Since metal, oxide and semiconductor layers are used in constructing its special structure, the name MOS comes to specify this type of FET.

The MOSFETs are usually classified into *two* types: (i) enhancement MOSFET and (ii) depletion MOSFET.

9.8.1 Enhancement MOSFET

Structure : The basic structure of a p-channel ***enhancement MOSFET*** is shown in Fig. 9.8. It consists of a lightly doped n-type semiconductor material, called the ***substrate***, into which two highly doped p^+ regions are formed. One of these regions acts as the ***source*** and the other as the ***drain***. Then, a thin layer of SiO_2 is grown over the surface. The insulated metal contact made on the SiO_2 layer is known as ***gate***. Two

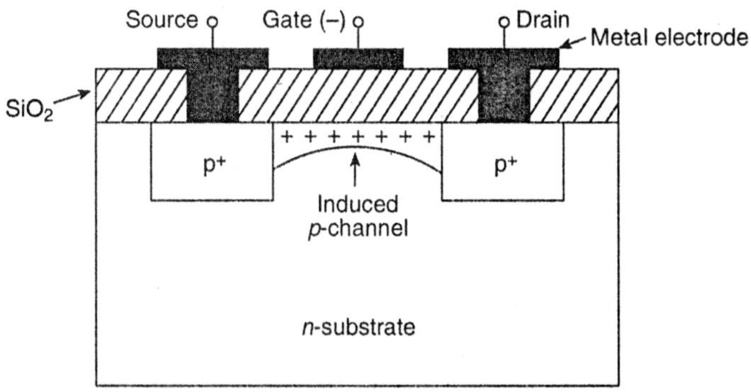

Fig. 9.8 Schematic diagram of p-channel enhancement MOSFET

other metal contacts are made to the source and the drain through holes cut into the SiO_2 layer. This metal oxide-semiconductor structure helps to control the operation by applying proper electric field to the gate, the source and the drain. Hence this device is called *metal-oxide-semiconductor field-effect transistor* or simply MOSFET. Again, the gate is insulated by SiO_2 layer. So, it is also known as *insulated gate field effect transistor* or IGFET. Since the insulating SiO_2 layer offers a very high input resistance ($\sim 10^6\Omega$ to $10^{15}\Omega$), the gate current of a MOSFET is very small compared to JFET.

The n-channel enhancement MOSFET has a similar structure in which n^+ regions are formed on p-type semiconductor substrate.

Circuit symbols : The circuit symbols of n-channel and p-channel enhancement type MOSFETs are illustrated below in Fig. 9.9.

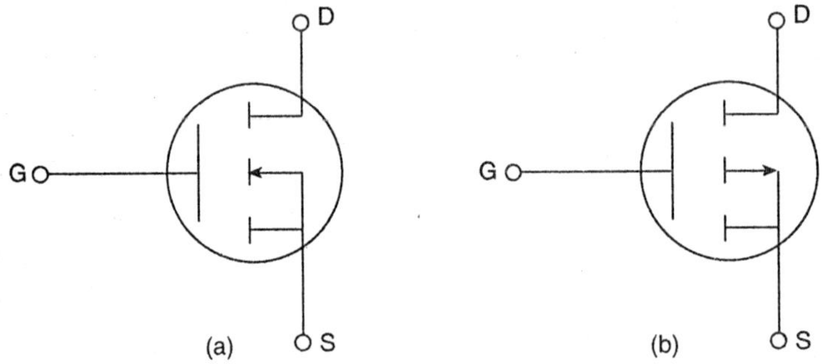

Fig. 9.9 Enhancement MOSFET (a) n-channel (b) p-channel

The arrow indicates the formation of inversion layer and the direction of drain current. There is no conducting link between the gate and the rest of the device.

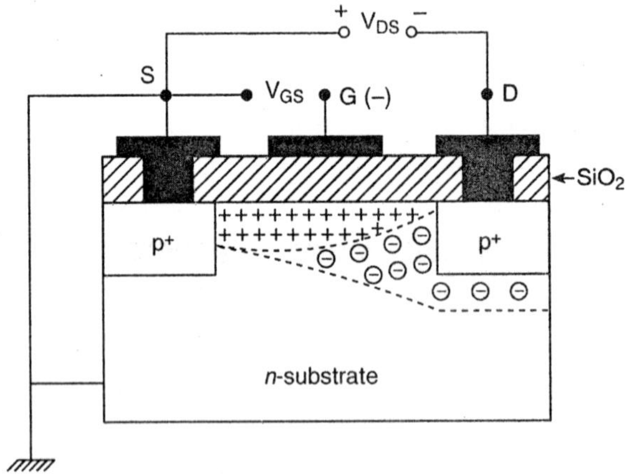

Fig. 9.10 Operation of enhancement MOSFET

Operation : To understand the operation of an enhancement MOSFET, suppose the source and the substrate of a n-channel MOSFET be grounded and a negative voltage is applied to the gate with drain-to-source voltage set to zero. This negative gate voltage repels the electrons from the substrate below the gate oxide layer, leaving only the uncovered donor ions of the n-type substrate. If V_{GS} is made more negative, the minority holes from the n-substrate are attracted to the gate and are accumulated below the gate SiO_2 layer. When the gate-to-source voltage exceeds a critical value V_T, known as *threshold voltage*, these holes form an *inversion layer* from source to the drain below the gate oxide layer. Current can now flow through this induced channel from source to drain if a negative voltage is applied to the drain w.r.t. the source, due to the drift of holes towards the drain. As V_{GS} is made more negative, more carriers are drawn to the channel causing thereby more drain current i.e., the negative gate voltage enhances the drain current as well as the conductance of the channel. Hence the device is called the enhancement MOSFET or E-MOSFET.

9.8.2 Depletion MOSFET

Depletion MOSFET is a type of MOSFET which is normally 'on' state. It came as a part of the evolution towards the enhancement MOSFET.

Structure : The basic structure of a n-channel depletion MOSFET is shown in Fig. 9.11. The process of construction of a depletion MOSFET is similar to that of enhancement MOSFET, the only difference is that a narrow region of the semiconductor substrate between the source and the drain is lightly diffused with the same type of impurity as that of the source and the drain before the SiO_2 layer is grown over the surface of the structure.

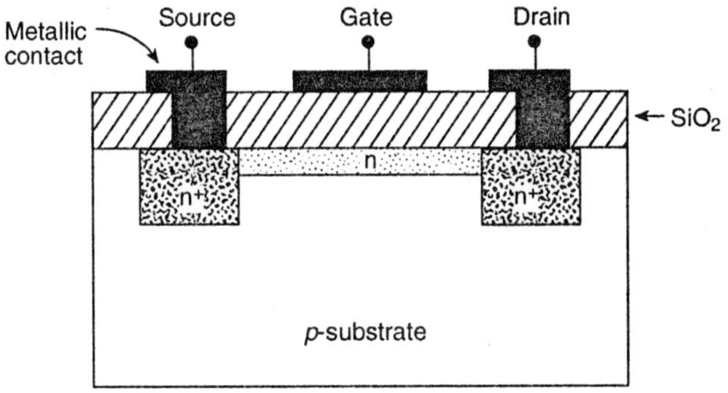

Fig. 9.11 Structure of n-channel depletion MOSFET

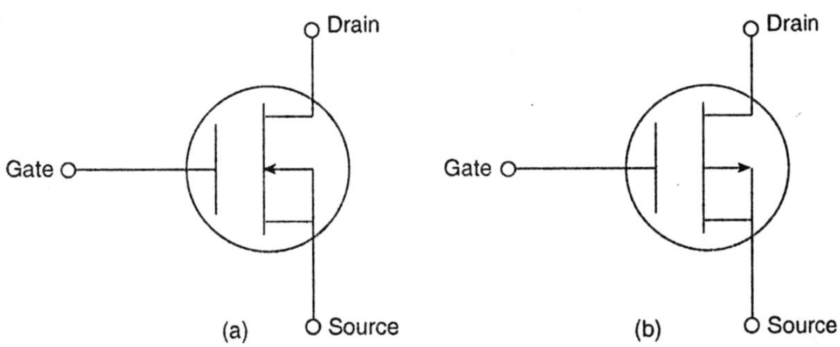

Fig. 9.12 Depletion MOSFET (a) n-channel (b) p-channel

Circuit symbols : Fig. 9.12 shows the circuit symbols of n-channel and p-channel depletion type MOSFET.

The space between the gate and the other terminals indicates that the gate is insulated. The arrow head points to the n-type material of the device.

Operation : For the operation of a depletion MOSFET, the source and the substrate are grounded. When the gate-to-source voltage is zero, i.e, $V_{GS} = 0$ and a positive voltage is applied to the drain w.r.t. the source, an appreciable drain current is obtained due to the existence of diffused channel between the source and the drain. When a negative voltage is applied to the gate, the electrons from the channel region are repelled towards the p-substrate and the holes are attracted from the substrate

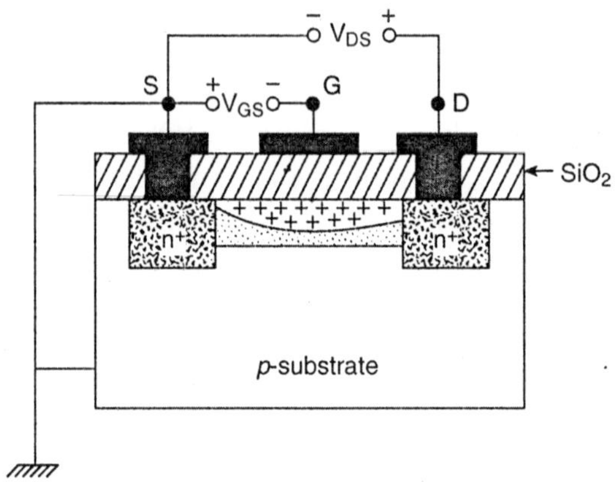

Fig. 9.13 Operation of depletion MOSFET

towards the channel. As a result of the recombination of electrons and holes in the channel region, a depletion of majority carriers takes place (Fig. 9.13). This is why the device is called the *depletion* MOSFET or simply D-MOSFET. When the gate voltage is sufficiently negative, the depletion region extends completely across the channel and the current between the source and the drain is cut off. The negative voltage at which the channel becomes non-conducting is known as the *gate-source cut-off voltage* or the *threshold voltage*, V_T.

When a positive gate voltage is applied and V_{DS} is gradually increased, the MOSFET is said to be operated in enhancement mode.

9.9 JFET as amplifier

9.9.1 Common source amplifier

A common source amplifier using an n-channel JFET is shown in Fig. 9.14. This FET configuration is similar to common emitter configuration of bipolar junction transistor. Here, the a.c signal is connected in series with the battery V_{GG}, the gate bias. This causes a variation of v_{gs} to the total of gate-to-source voltage. Let the corresponding

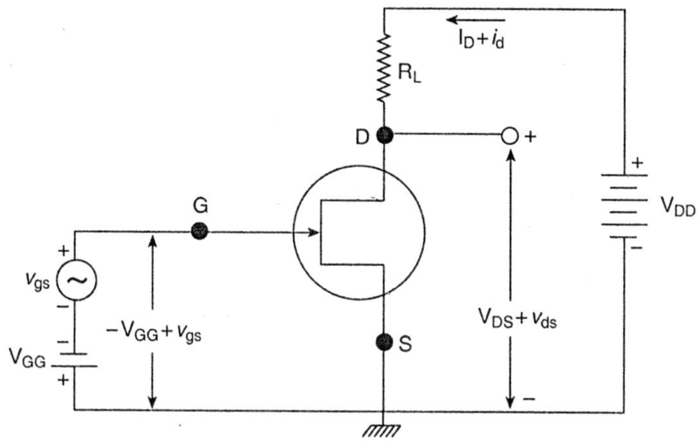

Fig. 9.14 Common source amplifier

output voltage change be v_{ds} which appears across the load resistance R_L connected across the drain-source terminal. When no input signal is applied, we get from the output circuit

$$V_{DD} = V_{DS} + I_D R_L \tag{9.9.1}$$

But, using KVL for a.c and d.c components, we have

$$V_{DD} = V_{DS} + v_{ds} + (I_D + i_d)R_L \tag{9.9.2}$$

Subtracting equation (9.9.1) from (9.9.2), we obtain

$$i_d R_L + v_{ds} = 0 \tag{9.9.3}$$

For small signal voltages, equation (9.7.4) provides

$$i_d = g_m v_{gs} + \frac{v_{ds}}{r_d} \tag{9.9.4}$$

Substituting (9.9.4) in (9.9.3), we get

$$\left(g_m v_{gs} + \frac{v_{ds}}{r_d}\right) R_L + v_{ds} = 0$$

$$\text{or,}\quad \frac{v_{ds}}{v_{gs}} = -\frac{g_m R_L}{(1 + R_L/r_d)} = -\frac{g_m r_d R_L}{R_L + r_d}$$

$$= -\frac{\mu R_L}{R_L + r_d}, \quad (\because \mu = g_m r_d)$$

$$\therefore \text{ Voltage gain,} \quad \boldsymbol{A_V} = \frac{v_{ds}}{v_{gs}} = -\mu \frac{\boldsymbol{R_L}}{\boldsymbol{R_L + r_d}} \tag{9.9.5}$$

The negative sign indicates that input and output voltages *differ in phase* by π.

The output resistance is a parallel combination of a.c drain resistance r_d and the load resistance R_L.

$$\therefore \text{ Output resistance,} \quad \boldsymbol{R_0} = \frac{\boldsymbol{R_L r_d}}{\boldsymbol{R_L + r_d}} \tag{9.9.6}$$

9.9.2 Common drain amplifier

A ***common drain amplifier***, using an n-channel JFET, is shown in Fig. 9.15. This FET configuration is similar to common collector configuration of bipolar junction

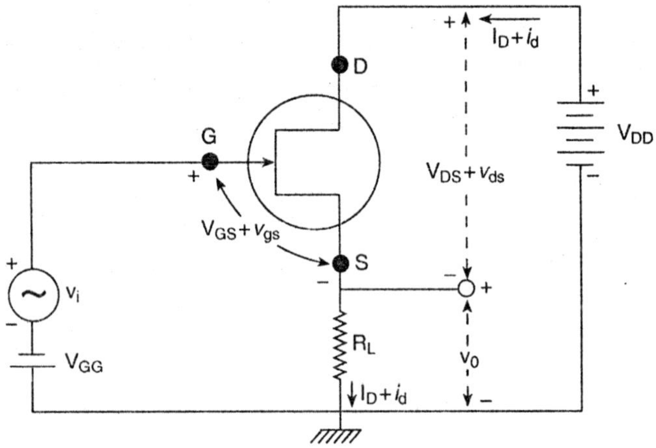

Fig. 9.15 Common drain amplifier

transistor. Here, v_i is the a.c input gate voltage and v_0 the corresponding output voltage appearing across the load R_L. The d.c and a.c components of voltages and currents are shown by usual notations.

Applying KVL to the output circuit for a.c component, we get

$$i_d R_L + v_{ds} = 0 \tag{9.9.7}$$

For small signal, we get $$i_d = g_m v_{gs} + \frac{v_{ds}}{r_d}$$

$\therefore$ From (9.9.7), $$\left(g_m v_{gs} + \frac{v_{ds}}{r_d}\right) R_L + v_{ds} = 0 \tag{9.9.8}$$

Applying KVL to the input circuit a.c, we obtain

$$v_{gs} = v_i - i_d R_L = v_i + v_{ds}$$

$\therefore$ From (9.9.8), $$R_L \left[g_m(v_i + v_{ds}) + \frac{v_{ds}}{r_d}\right] + v_{ds} = 0$$

or, $$v_{ds}\left[1 + \frac{R_L}{r_d} + g_m R_L\right] = -g_m v_i R_L$$

or, $$v_{ds} = -\frac{g_m r_d v_i R_L}{r_d + R_L + g_m r_d R_L}$$

$$= -\frac{\mu v_i R_L}{r_d + (1+\mu) R_L} \qquad (\because\ \mu = g_m r_d)$$

Again, $v_0 = -v_{ds}$. $\therefore \quad -\frac{v_{ds}}{v_i} = \frac{v_0}{v_i} = \frac{\mu R_L}{r_d + (1+\mu)R_L}$

$\therefore$ Voltage gain is:

$$\boxed{A_V = \frac{\mu R_L}{r_d + (1+\mu)R_L}}$$

$$\text{If} \quad R_L \gg \frac{r_d}{1+\mu}, \text{ then } A_V = \frac{\mu}{1+\mu}.$$

In this case, if $\mu \gg 1$ then $A_V = 1$. The voltage gain unity implies that the output voltage (at the source) follows the input voltage (at the gate). This is why a common drain amplifier is also called a ***source follower***.

9.10 Complementary metal-oxide-semiconductor (CMOS)

Complementary metal-oxide-semiconductor, abbreviated CMOS, is a semiconductor technology used in manufacturing most of the microchips in modern computer. In CMOS, two types of enhancement MOSFETs (PMOS and NMOS) are interconnected in a complementary symmetry arrangement on the same substrate such that an increase in gate bias of the device enhances the conductivity between the drain and the source. Since the transistors are arranged in complementary symmetry, only one of the two MOSFETs is turned on at a time. So, power dissipation is controlled by the leakage currrent of the '*off*' transistor and it is of extremely low value. Hence, CMOS technique is preferred to conventional MOSFET circuits in critical power saving applications and fabrication of ICs.

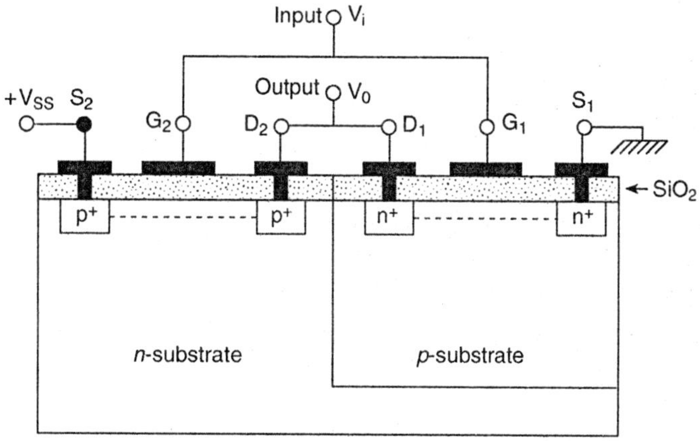

Fig. 9.16 Cross-sectional view of CMOS inverter

Fig. 9.16 shows a cross-sectional view of a simple CMOS inverter circuit consisting of a combination of NMOS and a PMOS on the same substrate. Fig. 9.17 shows the symbolic version of the same circuit.

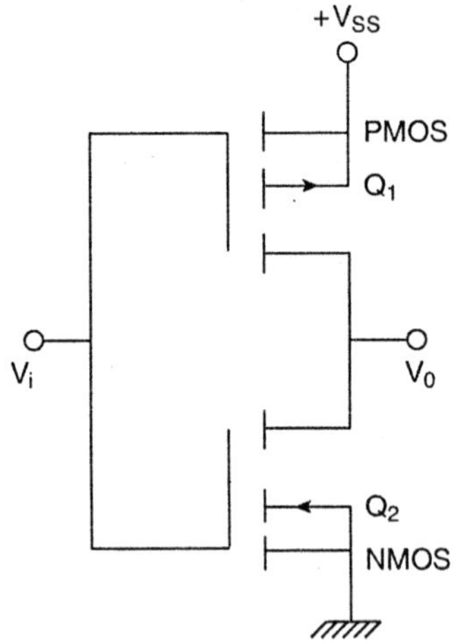

Fig. 9.17 CMOS inverter circuit

Here, Q_1 is a PMOS where the source is connected to the applied voltage V_{SS} and Q_2 is an NMOS with source grounded. The gate terminals of both the MOSFETs are connected together to the input V_i while the drain terminals are connected together to form the output terminal V_0.

When the input V_i is high ($> V_T$), the transistor Q_2 (i.e., NMOS) is *on* and Q_1 (i.e., PMOS) is *off*. So, the output V_0 is low. On the other hand, when V_i is low, the transistor Q_1 is switched *on* but Q_2 becomes cut *off*. As a result, the output V_0 is high. Thus, in complementary symmetry configuration while one transistor is *on*, the other remains *off*.

Since the two transistors are in series, the drain current that flows is limited by the off transistor to the leakage value. So, the quiescent current practically does not flow and the power dissipation is caused only by the leakage current of the *off*-transistor. The power dissipation is thus virtually zero ($\sim$ a few nanowatt). The extremely low value of power dissipation or consumption of CMOS has led it to fabricate many other logic circuits — digital and analog.

Advantages and disadvantages

The CMOS has a number of *advantages* over other FETs. These are the following.

1. Extremely low power consumption,
2. Fast switching speed,
3. Very high input impedance.
4. Ability to operate in a large temperature range (−40°C to +85°C)
5. Ability to operate under varying operating conditions.
6. Ability to function without being disturbed by external noise.
7. High packing density i.e., large number of CMOS on a small Si-wafer.

The *only disadvantge* of CMOS is its susception to electrostatic discharging. Even the small static charges deposited on the user's hand may even permanently damage the device.

9.11 Illustrated Examples

▶ **Example 1.** *It is found that as V_{GS} changes from $-1\,$V to $-1.5\,$V, keeping V_{DS} constant, the I_D of a FET drops from 7 to 5 mA. Find the transconductance of the FET. If the a.c drain resistance is 200 kΩ, find the amplificatioon factor of the FET.*

Solution : Transconductance, $g_m = \left.\frac{\partial I_D}{\partial V_{GS}}\right|_{V_{DS}} = \frac{7-5}{-1-(1.5)} = \mathbf{4\,mA/V}$

Amplification factor, $\mu = r_d \times g_m = 200 \times 10^3 \times 4 \times 10^{-3} = \mathbf{800}$

▶ **Example 2.** *The following readings were obtained in an experiment with FET.*

V_{GS} (Volt)	0	0	0.3
V_{DS} (Volt)	6	16	16
I_D (mA)	12	12.3	12.0

Calculate (i) *a.c drain resistance* (ii) *transconductance and* (iii) *amplification factor.*

(Himachal Univ.)

Solution : (i) The a.c drain resistance is defined as $r_d = \left.\frac{\partial V_{DS}}{\partial I_D}\right|_{V_{GS}}$

As V_{GS} is constant in the first two readings, we get

$$\partial V_{DS} = 16 - 6 = 10\,\text{V} \quad \text{and} \quad \partial I_D = 12.3 - 12 = 0.3\,\text{mA}$$

$$\therefore \quad r_d = \frac{10}{0.3 \times 10^{-3}}\Omega = \mathbf{33.3\,k\Omega}$$

(ii) Transconductance, $$g_m = \left.\frac{\partial I_D}{\partial V_{GS}}\right|_{V_{DS}}$$

Here, V_{DS} is constant in the second and the third readings. So, we have

$$g_m = \frac{(12.3 - 12.0) \times 10^{-3}}{0.3 - 0}\mho = \mathbf{10^{-3}\mho}$$

(iii) Amplification factor, $\mu = r_d \times g_m = 33.3 \times 10^3 \times 10^{-3} = \mathbf{33.3}$

▶ **Example 3.** *If the load resistance, the a.c drain resistance and the transconductance of a common source* JFET *amplifier are* 500 kΩ, 100 kΩ *and* 0.2 mA/V, *obtain the voltage gain of the amplifier.* **(Calcutta Hons.)**

Solution : Given, a.c drain resistance $r_d = 100\,\text{k}\Omega = 10^5\Omega$

load resistance $R_L = 500\,\text{k}\Omega = 5 \times 10^5\Omega$

transconductacne $g_m = 0.2\,\text{mA/V} = 0.2 \times 10^{-3}\,\text{A/V}$

$\therefore$ Amplification factor, $\mu = r_d \times g_m = 10^5 \times 0.2 \times 10^{-3} = 20$

$\therefore$ Voltage gain, $\boldsymbol{A_V} = -\frac{\mu R_L}{R_L + r_d} = -\frac{20 \times 5 \times 10^5}{5 \times 10^5 + 10^5} = -\frac{100}{6} = \mathbf{-16.67}$

The negative sign indicates that input and output voltages *differ in phase* by π.

► **Example 4.** *An n-channel, JFET has $I_{DSS} = 12\,\text{mA}$ and pinch off voltage $V_p = -4\,\text{V}$. Find the drain current for $V_{GS} = -2\,\text{V}$.*

Solution: By the problem, $I_{DSS} = 12\,\text{mA} = 12 \times 10^{-3}\,\text{A}$, $V_p = -4\,\text{V}$ and $V_{GS} = -2\,\text{V}$.

$\therefore$ Drain current, $I_{DS} = I_{DSS}(1 - V_{GS}/V_P)^2$

$$= 12 \times 10^{-3}\left[1 - \left(\frac{-2\,\text{V}}{-4\,\text{V}}\right)\right]^2$$

$$= 12 \times 10^{-3} \times \tfrac{1}{4}$$

$$= 3 \times 10^{-3}\,\text{mA} = \mathbf{3\,mA}$$

► **Example 5.** *An n-channel JFET has pinch off voltage of $-4.5\,\text{V}$, $I_{DSS} = 9\,\text{mA}$. At what value of V_{GS} will I_{DS} equal to 3 mA and what will be its g_m at that I_{DS}?*

Solution: By the problem, $V_P = -4.5\,\text{V}$, $I_{DSS} = 9\,\text{mA}$ and $I_{DS} = 3\,\text{mA}$.

Now, the drain-source current is

$$I_{DS} = I_{DSS}\left(1 - \frac{V_{GS}}{V_P}\right)^2 \Rightarrow V_{GS} = V_P\left(1 - \sqrt{\frac{I_{DS}}{I_{DSS}}}\right)$$

$$\therefore \quad V_{GS} = -4.5\left(1 - \sqrt{\frac{3}{9}}\right) = \mathbf{-1.902\,V}$$

$$\text{Transconductance, } g_m = -\frac{2I_{DSS}}{V_P}\left(1 - \frac{V_{GS}}{V_P}\right)$$

$$= \frac{-2 \times 9 \times 10^{-3}}{-4.5}\left(1 - \frac{-1.902}{-4.5}\right)$$

$$= \mathbf{2.31\,mA/V}, \text{ on simplification}$$

► **Example 6.** *An n-channel JFET has $I_{DSS} = 10\,\text{mA}$ and $V_P = -4.0\,\text{V}$. Determine the minimum value of V_{DS} for pinch-off region and drain current I_{DS} for $V_{GS} = -2\,\text{V}$ in pinch-off region*

Solution: Given,, $I_{DSS} = 10\,\text{mA} = 10 \times 10^{-3}\,\text{A}$, $V_{GS} = -2\,\text{V}$ and $V_P = -4.0\,\text{V}$. So, the drain current

$$I_{DS} = I_{DSS}\left(1 - \frac{V_{GS}}{V_P}\right)^2 = 10 \times 10^{-3} \times \left(1 - \frac{-2}{-4}\right)^2 = \mathbf{2.5\,mA}$$

The minimum value of V_{DS} for pinch-off region is equal to V_P. So, the value of

$$[V_{DS}]_{\text{min}} = V_P = \mathbf{-4.0\,V}$$

▶ **Example 7.** *A common source* FET *amplifier uses a load resistance* $R_L = 200\,\text{k}\Omega$. *If the a.c drain resistance and transconductance of the* FET *are* $100\,\text{k}\Omega$ *and* $0.15\,\text{mA/V}$ *respectively, calculate the magnitude of the voltage gain of the amplifier.* **(Punjab Univ.)**

Solution: Given, load resistance $R_L = 200\,\text{k}\Omega = 200 \times 10^3\Omega$

a.c drain resistance $r_d = 100\,\text{k}\Omega = 100 \times 10^3\Omega$

transconductance $g_m = 0.15\,\text{mA/V} = 0.15 \times 10^{-3}\,\text{A/V}$

∴ Magnitude of voltage gain is

$$|\boldsymbol{A_V}| = \frac{\mu R_L}{R_L + r_d} = \frac{g_m r_d R_L}{R_L + r_d}, \quad (\because \mu = g_m r_d)$$

$$= \frac{0.15 \times 10^{-3} \times 100 \times 10^3 \times 200 \times 10^3}{200 \times 10^3 + 100 \times 10^3} = \mathbf{10}$$

▶ **Example 8.** *In a bar type* JFET, *the voltage across the space charge region at a distance x from the source along the channel is given by*

$$V(x) = C[a - b(x)]^2$$

where $a =$ zero-bias channel half-width, $b(x) =$ effective channel half-width at x and C is a constant. Find the pinch-off voltage V_p. Also find the the channel half-width when $V = V_p/2$. **(Calcutta Hons.)**

Solution: At pinch-off, the effective channel half-width becomes zero.

∴ Putting $b(x) = b = 0$ in the given expression, we get

$$V(x) = \boldsymbol{V_p = Ca^2} \Rightarrow C = V_p/a^2,$$

∴ The given relation reduces to $V(x) = \dfrac{V_p}{a^2}[a - b(x)]^2$

$$= V_p\left[1 - \frac{1}{a}b(x)\right]^2$$

Now, putting $V(x) = \frac{1}{2}V_p$ in the above equation, we at once have

$$\frac{V_p}{2} = V_p\left[1 - \frac{b(x)}{a}\right]^2$$

$$\therefore \; b(x) = \left(1 - \frac{1}{\sqrt{2}}\right)a = 0.29a$$

Therefore, the required ***channel half-width*** is $\mathbf{0.29}\boldsymbol{a}$.

▶ **Example 9.** *An n-channel silicon (dielectric constant* $= 12$*) FET with a channel half-width* $a = 3 \times 10^{-6}\,\text{m}$ *is doped with a concentration* $N_D = 10^{21}\,\text{m}^{-3}$. *Find the pinch-off voltage.* **(Calcutta Hons.)**

Solution : By the problem, channel half-width, $a = 3 \times 10^{-6}$ m

donor concentration, $N_D = 10^{21}\ \text{m}^{-3}$

dielectric constant of Si, $k = 12$.

$$\therefore \text{ The pinch-off voltage, } V_P = \frac{eN_D a^2}{2\varepsilon} = \frac{1.6 \times 10^{-19} \times 10^{21} \times (3 \times 10^{-6})^2}{2 \times 12 \times 8.854 \times 10^{-12}}\ \text{V}$$

$$= \mathbf{6.8\ V}$$

▶ **Example 10.** *The pinch-off voltage of a p-channel JFET is $V_P = 5\,\text{V}$ and $I_{DSS} = -40\,\text{mA}$. The drain-source voltage V_{DS} is such that a saturated drain current $I_{DS} = -15\,\text{mA}$ is maintained. Find the gate-source voltage V_{GS}.* **(GATE)**

Solution : Here, $V_P = 5\,\text{V}$, $I_{DSS} = -40\,\text{mA}$, $I_{DS} = -15\,\text{mA}$.

$$\text{Now,} \quad I_{DS} = I_{DSS}\left(1 - \frac{V_{GS}}{V_P}\right)^2$$

Substituting the values in the above relation, we obtain

$$-15 = -40\left(1 - \frac{V_{GS}}{5}\right)^2$$

$$\text{or, } V_{GS} = 5\left(1 - \sqrt{\frac{15}{40}}\right) = 1.94$$

The gate-source voltage is therefore **1.94 *V***.

▶ **Example 11.** *Prove the following two relations:* (a) $V_{GS(off)} = 2I_{DSS}/g_m$ *and* (b) $g_m = g_{mo}(1 - V_{GS}/V_{GS(off)})$

Solution : **(a)** Taking the derivative of the relation

$$I_D = I_{DSS}\left(1 - \frac{V_{GS}}{V_{GS(off)}}\right)^2,$$

$$\frac{dI_D}{dV_{GS}} = g_m = 2I_{DSS}\left(1 - \frac{V_{GS}}{V_{GS(off)}}\right)\left(-\frac{1}{V_{GS(off)}}\right).$$

$$\therefore\ g_m = -\frac{2I_{DSS}}{V_{GS(off)}}\left(1 - \frac{V_{GS}}{V_{GS(off)}}\right) \tag{9.11.1}$$

When $V_{GS} = 0$, we get: $\quad g_{mo} = -2I_{DSS}/V_{GS(off)}$ (9.11.2)

Rearranging, $\quad \boldsymbol{V_{GS(off)} = -2I_{DSS}/g_m}$

(b) Substituting (9.11.2) in (9.11.1), we obtain

$$\boldsymbol{g_m = g_{mo}\left(1 - \frac{V_{GS}}{V_{GS(off)}}\right)}$$

Review and Short Questions

1. What is field effect transistor? Why is it so named?

2. Describe the construction and working of a junction field effect transistor (JFET) clearly explaining the terms source, drain, gate and channel. On what factors does the drain current depend? **(Punjabi Univ.)**

3. With a neat sketch, describe the construction of an n-channel JFET. **(Burdwan Hons.)**

4. Describe the principle of operation of an n-channel JFET.

5. Draw a family of common source drain characteristics of a typical n-channel JFET. Explain the nature of the curves. **(Burdwan Hons.)**

6. How can you obtain the static characteristics of a JFET? Give the necessary circuit diagram. **(Calcutta Hons.)**

7. "A JFET can be used as a voltage variable resistor."—Explain.

8. Define the terms (a) transconductance g_m, (b) a.c drain resistance r_d, (c) amplification factor μ of a FET. Deduce the relation $\mu = g_m r_d$. **(Lucknow Univ.)**

9. When is the channel of a JFET said to be pinched off? Define the pinch-off voltage.

10. What is a MOSFET? Why is it so called? Why is it commercially more important than JFET? What is the main physical difference between them?

11. What is the difference between an enhancement-type and a depletion-type MOSFET?

12. Give a sketch of the basic structure of an n-channel enhancement-type MOSFET. Explain the operation and sketch the drain characteristics of the MOSFET. **(Himachal Univ.)**

13. What is a depletion MOSFET? Sketch the basic structure and operation of an n-channel depletion MOSFET. What is threshold voltage? **(Lucknow Univ.)**

14. Discuss FET as an amplifier having common source configuration. Obtain an expression for its voltage gain and output resistance. **(Calcutta Hons.)**

15. Find an expression for the voltage gain and output resistance of a common drain FET amplifier. Why is it known as source follower? **(Himachal Univ.)**

16. "FET is a voltage-controlled device whereas BJT is a current-controlled device"—Explain. **(Calcutta Hons.)**

17. (a) What is CMOS? What are its advantages over FET? Explain the operation of any CMOS application.

    (b) Draw the cross-sectional view of a CMOS. **(Calcutta Hons.)**

Objective Questions

1. FET is a

(a) *p-n* junction diode, (b) *n*- or *p*-channel unipolar transistor,
(c) bipolar transistor, (d) none of these. **(HImachal Univ.)**

2. A FET consists of

(a) source, (b) drain, (c) gate, (d) all of these.
(Purvanchal Univ.)

3. A JFET is a

(a) unipolar device, (b) voltage control device,
(c) current control device, (d) both (a) and (b).

4. The channel of JFET is between

(a) gate and drin, (b) drain and source,
(c) gate and source, (d) input and output.

5. Enhancement MOSFET biasing is done by

(a) voltage divider bias, (b) source bias,
(c) self-bias, (d) collector feedback bias.

6. The input impedance of MOSFET as compared to FET is

(a) equal, (b) less, (c) more, (d) none.
(Himachal Univ.)

7. MOSFET differs from JFET because

(a) of power rating,
(b) the MOSFET has two gates,
(c) the JFET has a *PN*-junction,
(d) MOSFET has a physical channel.

8. Which of the following are true for a field effect transistor (FET)?

(a) A FET has a very high power gain,
(b) A FET has a negative temperature coefficient cut-off resistance,
(c) A FET has a very high input impedance,
(d) A FET is a current controlled device. **(GATE)**

9. A field effect transistor is a

(a) unipolar device,
(b) special type of bipolar junction transistor,
(c) unijunction device,
(d) device with low input impedance. **(GATE)**

10. A junction field effect transistor behaves as a

(a) voltage controlled current source,
(b) voltage controlled voltage source,
(c) current controlled voltage source,
(d) current controlled current source. **(GATE)**

11. The high input impedance of field effect transistor (FET) amplifier is due to

(a) the pinch-off voltage,
(b) its very low gate current,
(c) the source and drain being far apart,
(d) the geometry of the FET. **(GATE)**

12. A JFET always operates with

(a) gate to source PN-junction reverse bias,
(b) gate to source PN-junction forward bias,
(c) drain connected to ground,
(d) gate connected to source.

13. In a CMOS,

(a) when one transistor is OFF, the other remains ON,
(b) both the transistors remain OFF at the same time,
(c) both the transistors remain ON at the same time.

14. The zero bias pinch-off voltage has the same magnitude as the

(a) gate-source voltage,
(b) gate-source cut-off voltage,
(c) drain-source voltage,
(d) gate voltage.

15. In enhanced n-channel MOSFET, an induced n-type channel can be produced between the source and the drain if

(a) $V_{GS} = 0$,
(b) $V_{GS} < 0$,
(c) $V_{GS} > 0$.

16. The pinch-off voltage is the voltage

(a) at which gate-source junction breaks down,
(b) which causes the depletion region to meet,

(c) applied between the drain and the source,
(d) none of the above.

17. For operation of enhancement n-channel MOSFET

(a) V_{GS} should be positive and V_{DS} negative,
(b) both V_{GS} and V_{DS} should be positive,
(c) both V_{GS} and V_{DS} should be negative,
(d) V_{GS} should be negative but V_{DS} positive.

18. The main factor which differentiates D-MOSFET from E-MOSFET is the absence of

(a) channel, (b) gate, (c) pn-junction, (d) insulated gate.

Problems

1. Calculate the dynamic resistance of a JFET having amplification factor 80 and transconductance $200\mu℧$. **(Punjab Univ.)** [$400\,\text{k}\Omega$]

2. For a constant gate-to-source voltage, a change in drain-to-source voltage of 3 V produces a change in drain current of $30\,\mu$A. Calculate the a.c drain resistance of FET. [$100\,\text{k}\Omega$]

3. A common source FET amplifier has a load resistance $R_L = 500\,\text{k}\Omega$. If the a.c drain resistance r_d and amplification factor μ of the FET are $100\,\text{k}\Omega$ and 24 respectively, calculate the voltage gain of amplifier. **(Guru Nanak Univ.)** [20]

4. In an experiment for the determination of the drain resistance r_d and the transconductance g_m of a JFET following data are obtained

V_{GS} (volt)	0	0	−0.2
V_{DS} (volt)	7	15	15
I_D (mA)	10	10.25	9.65

Calculate the values of r_d and g_m. **(Calcutta Hons.)** [$32\,\text{k}\Omega$, $3\,\text{mA/V}$]

[**Hints.** $$r_d = \left.\frac{\Delta V_{DS}}{\Delta I_D}\right|_{V_{GS}=\text{const.}} = \frac{(15-7)\,\text{V}}{(10.25-10)\,\text{mA}} = \frac{8}{0.25}\,\text{k}\Omega = \mathbf{32\,k\Omega};$$

$$g_m = \left.\frac{\Delta I_D}{\Delta V_{GS}}\right|_{V_{DS}=\text{const.}} = \frac{(9.65-10.25)\,\text{mA}}{-0.2\,\text{V}} = \frac{-0.6\,\text{mA}}{-0.2\,\text{V}} = \mathbf{3\,mA/V}$$

Note that μ can also be determined from the above data, for $\mu = r_d \times g_m = (32\,\text{k}\Omega) \times (3\,\text{mA/V}) = 96$]

5. A FET amplifier in the common-source configuration uses a load resistance of 250 kΩ. The a.c drain resistance of the device is 100 kΩ and the transconductance is 0.5 mA/V. Calculate the voltage gain and all possible output resistances of the amplifier. **(Calcutta Hons.)** [−35.71; 100 kΩ, 71.4 kΩ]

6. A common source FET amplifier uses a load resistance $R_L = 250\,\text{k}\Omega$. If the a.c drain resistance and transconductance of the FET are 200 kΩ and 0.1 mA/V respectively, calculate the magnitude of the voltage gain of the amplifier. **(Guru Nanak Univ.)** [11.1]

7. A common-drain amplifier uses a JFET having $r_d = 100\,\text{k}\Omega$ and $g_m = 3500\mu\text{s}$. If the load resistance is 5 kΩ, calculate the voltage-gain and the output resistance including load resistance. [0.943, 269Ω]

8. An n-channel JFET gives a saturation voltage $V_{D\,\text{sat}} = 4\,\text{V}$ for $V_{GS} = -1\,\text{V}$. Another n-channel JFET gives $V_{D\text{sat}} = 6.5$ for the same V_{GS}. Compare the pinch-off voltages of the two JFETs. [2 : 3]

9. A JFET has $V_p = -4.5\,\text{V}$, $I_{DSS} = 10\,\text{mA}$ and $I_{DS} = 2.5\,\text{mA}$. Determine the transconductance. [2.22 mA/V]

10. Determine the value of transconductance of a FET, when I_D changes from 1 mA to 1.8 mA with a change in gate-source voltage from −2.2 V to 2 V. [4 mA/V]

11. A JFET has $I_{DSS} = 16\,\text{mA}$ and $V_p = 3\,\text{V}$. What are the maximum drain current, the gate-source cut-off voltage and the value of r_{ps}? [16 mA, −3 V, 187.5 Ω]

12. In JFET with $V_{GS} = 0$, when V_{DS} is changed from 1 to 2 V, the drain current I_D changes from 5 to 10 mA. If V_{DS} is changed from 15 to 22 V, I_D changes from 20 to 20.2 mA. Calculate the dynamic drain resitance for small V_{DS} (ohmic resitance) and large V_{DS} (pinch-off region). [200 Ω, 35 kΩ]

13. An EFT is characterised by $V_P = 8\,\text{V}$, $V_{GS} = 4.0\,\text{V}$ and $I_D = 6.0\,\text{mA}$. Calculate I_{DSS}. [24 mA]

14. (a) When a reverse gate voltage of 12 V is applied to JFET, the gate current is 1 mA. What is the resistance between the gate and the source?

    (b) When the reverse gate voltage of JFET changes from 4.0 to 3.9 V, the drain current changes from 1.3 to 1.6 mA. What is the transconductance?

    [(a) 12 kΩ, (b) 3 mS]

15. An n-channel silicon JFET has channel width 6 μm and doping concentration $N_D = 10^{21}\,\text{m}^{-3}$. Find the pinch-off voltage (dielectric constant of silicon $\varepsilon = 12$). [6.78 V]

16. A common source JFET amplifier uses a load resistor of $R_L = 500\,\text{k}\Omega$. Assuming $g_m = 0.2\,\text{m A/V}$ and a.c drain resistance $r_d = 100\,\text{k}\Omega$, calcualte the voltage gain of the amplifier. **(Calcutta Hons.)** [−16.6]

17. An n-channel silicon JFET has a donor concentration of $2 \times 10^{21}/\text{m}^3$ and a channel width of $4\,\mu\text{m}$. If the dielectric constant of silicon is 12, find the pinch-off voltage. If the FET operates with a gate-source voltage $-2\,\text{V}$, what is the saturation voltage $V_{D_{\text{sat}}}$? (Calcutta Hons.) [6.02 V, 4.02 V]

18. An n-channel Si JFET has a donor concentrationn of $10^{21}\,\text{m}^{-3}$ and a channel width of $4\,\mu\text{m}$. If the dielectric constant of Si is 11.8, calculate the saturation drain current for $V_{GS} = -2\,\text{V}$ and $I_{DSS} = 12\,\text{mA}$. [1.44 mA]

19. For a JFET with a pinch-off voltage $V_P = 4\,\text{V}$ and a saturation current $I_{DSS} = 10\,\text{mA}$, find (a) the value of the curent if $V_{GS} = -2V$, assume $V_{DS} > V_P$; and (b) the value of source-to-gate voltage for the JFET, if $I_D = 7\,\text{mA}$ and the device operates with $V_{DS} > V_P$. [2.5 mA, −0.64 V]

20. Prove, by using Shockley equation, that the transconductance of JFET is given by

$$\begin{aligned} g_m &= g_{mo}\,(1 - V_{GS}/V_P) \\ &= -\frac{2}{V_P}\,(I_{DSS} I_{DS})^{\frac{1}{2}}\,. \end{aligned}$$

10

Feedback Amplifiers

10.1 Introduction

The word '*feedback*' means 'the return of a fraction of the output of an amplifier, a microphone or such other device to the input, causing distortion or a whistling sound'. Feedback plays an important role in almost all the electronic circuits and makes an amplifier highly precise. An *amplifier* in which a part or the whole of the output signal is fed back and superimposed on the input signal for significant modifications in the performance of the amplifier is called a ***feedback amplifier***.

There are *two* types of feedback in amplifiers:

(a) the *positive* or *direct* or *regenerative feedback,* and

(b) the *negative* or *inverse* or *degenerative feedback.*

When the feedback signal is *in phase* with the input signal, it *enhances* the magnitude of the input signal. This type of feedback is known as a ***positive feedback***. When however the feedback signal is *out of phase* with the input signal, it leads to a *diminution* in the magnitude of the input signal and such a feedback is called a ***negative feedback***.

While the positive feedback increases the gain of the amplifier, the negative feedback reduces it. In spite of the reduction in overall gain, the ***negative feedback*** offers a number of *advantages*. At the cost of the gain, it *reduces distortion, improves stability of gain, changes the frequency response and modifies the input and the output impedances.* So, the negative feedback is extensively used in electronic circuits. ***Positive feedback***, on the other hand, introduces an *instability in the gain* of the amplifier and *distorts the* *input signal.* So, it is used in oscillators which can convert d.c energy into a periodically varying electrical output. Since the positive feedback increases the signal power, it is sometimes exploited in narrowband amplifiers and radio receivers.

10.2 Principle of feedback amplifier

The block diagram of a feedback amplifier is shown in Fig. 10.1. It essentially consists of *two* parts: a basic or *internal amplifier A* and a *feedback network* β for the sampling and the reverse transmission of a part (or whole) of the output signal to the input of the system.

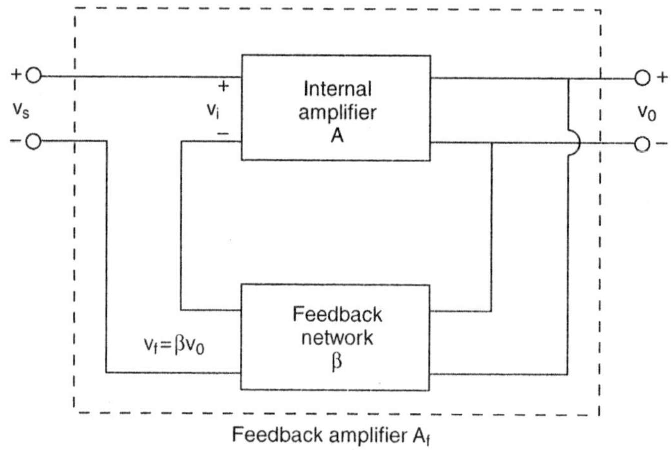

Fig. 10.1 Block diagram of a feedback amplifier

If v_i be the input signal to the internal amplifier and v_0 the output voltage, then the *gain* is plainly given by

$$A = \frac{v_0}{v_i} \tag{10.2.1}$$

The feedback network containing *passive elements* such as resistors, inductors or capacitors and *active elements* such as transistors can extract a portion $v_f = \beta v_0$ of the output voltage and return it to the externally applied input signal. The fraction $\beta = v_f/v_0$ is variously termed as the ***feedback ratio***, the *feedback fraction*, the *reverse transfer ratio* or the *reverse transmission factors*.

Both A and β can be real (positive or negative) quantities or complex functions of the frequency and may introduce phase changes. For *positive feedback*, v_f is in phase with the input signal v_s and for *negative feedback* it is out of phase with the input signal. So, the input voltage of the basic amplifier becomes

$$v_i = v_s \pm v_f \tag{10.2.2}$$

The negative sign corresponds to the negative feedback for which we have

$$v_i = v_s - v_f \tag{10.2.3}$$

Now, from (10.2.1), for *negative* feedback,

$$v_0 = Av_i = A(v_s - v_f) = A(v_s - \beta v_0)$$

$$\text{or,} \quad (1 + A\beta)v_0 = Av_s$$

$$\text{or,} \quad \boxed{\frac{v_0}{v_s} = \frac{A}{1 + A\beta}} \tag{10.2.4}$$

The *transfer gain* A_f of the feedback amplifier (shown within the dashed box in Fig. 10.1) is the ratio of the output voltage v_0 to the externally applied input signal voltage v_s.

$$\therefore \quad A_f = \frac{v_0}{v_s} = \frac{A}{1 + A\beta} \tag{10.2.5}$$

Similarly, considering only the magnitudes of A_f, A and β, we can write for *positive* feedback

$$\boxed{A_f = \frac{A}{1 - A\beta}} \tag{10.2.6}$$

The quantity $A\beta$ is called the ***loop gain.*** It is also known as the *feedback factor* or the *return ratio* or the *loop transmission.* $(1 \pm A\beta)$ is referred to as *return difference.*

The gain A_f is sometimes called the ***closed-loop gain.*** When $\beta = 0$, eq. (10.2.5) or (10.2.6) gives $A_f = A$. Thus A represents the *transfer gain without feedback* and A is also known as ***open-loop gain.***

For *positive feedback* $|1 - A\beta| < 1$ so that $|A_f| > |A|$. For *negative feedback* $|1 + A\beta| > 1$ and $|A_f| < |A|$. In case of positive feedback when $|1 - A\beta| = 0$ i.e., $|A\beta| = 1$, then $|A_f| = \infty$. This implies that *the amplifier gives an output signal even without an input signal.* The amplifier then becomes an ***oscillator.***

The feedback in an amplifier is generally expressed in *decibel* (dB) by the relation

$$F = \text{dB of feedback} = 20\log_{10}\left|\frac{A_f}{A}\right| = 20\log\left|\frac{1}{1 + A\beta}\right| \tag{10.2.7}$$

• For *negative feedback,* $|1 + A\beta| > 1$ and F is *negative.* For *positive feedback,* F is *positive,* since $|1 + A\beta| < 1$.

10.3 Feedback topologies

The feedback signal in a feedback amplifier can be taken out from the output voltage across the load or from the output current in the load. When the feedback network is connected in shunt with the output and the feedback signal is proportional to the output voltage, it is termed ***voltage feedback*** or *voltage sampling.*

If however the feedback network is connected in series with the output and the sampled signal is proportional to the output current, then it is called ***current feedback*** or *current sampling.*

For each of these two types of sampling, the feedback signal can be transmitted to the input either in series with the externally applied signal, known as ***series mixing*** or in *shunt* with the external signal, known as ***shunt mixing***. Thus we have the following *four* feedback amplifier topologies.

1. Voltage-series feedback or voltage sampling-series mixing.
2. Voltage-shunt feedback or voltage sampling-shunt mixing.
3. Current-series feedback or current sampling-series mixing.
4. Current-shunt feedback or current sampling-shunt mixing

We shall now discuss these amplifier topologies sequentially in what follows.

10.3.1 Voltage-series feedback

The schematic diagram of a *voltage-series feedback* configuration is shown in Fig. 10.2(a). This is basically a voltage amplifier. Here, the feedback signal is obtained by connecting the feedback network in parallel with the load and is proportional to the output-voltage. Since the feedback network forms a shunt circuit with the output and a series circuit with the input, this arrangement is also known as *shunt-series* feedback.

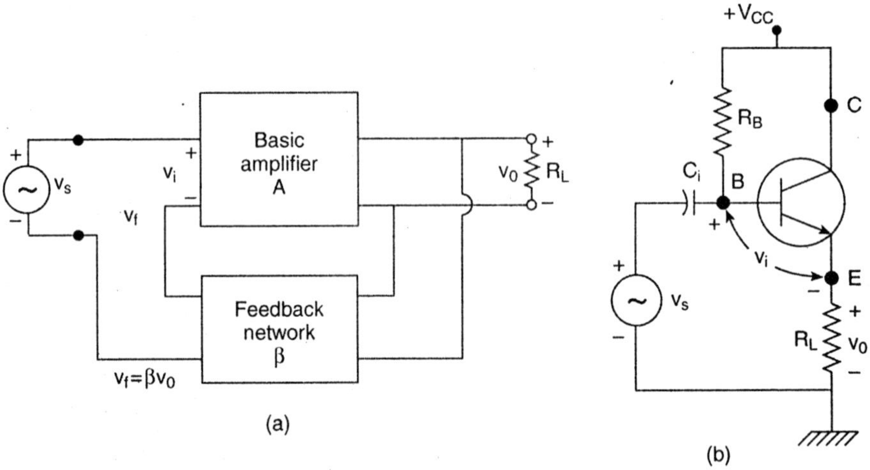

Fig. 10.2 (a) Block diagram of voltage-series feedback
(b) Example of voltage-series feedback(emitter follower)

The *emitter follower* circuit of Fig. 10.2(b) is an example of this type of feedback. Here, the output voltage v_O across the load resistance R_L is totally returned to the input. So, the feedback voltage $v_f = v_O$ and the feedback ratio β is unity. The resistance R_B is used to keep the emitter-base junction forward-biased and the capacitance C_i to decouple the signal source from the supply V_{CC}. C_i acts as a short-circuit for a.c. Since the sampled signal is a voltage and is returned in series with the external input signal voltage, the emitter follower circuit is ***voltage-series*** or *voltage-voltage* or *shunt-series* feedback.

Here, the input voltage to the transistor is given by the relation

$$v_i = v_s - v_O$$

and the overall gain is given by the expression

$$A_f = \frac{A}{1 + A\beta} \tag{10.3.1}$$

If the load resistance R_L is moved from the emitter to the collector, the gain becomes A, the gain of the amplifier without the feedback.

Then, we have $$A \approx \frac{h_{fe}R_L}{h_{ie}}, \quad \text{(from Art. 7.8)}$$

Since $\beta = 1$ here, the equation (10.3.1) becomes, using the above relation,

$$A_f \approx \frac{h_{fe}R_L}{h_{ie} + h_{fe}R_L} \tag{10.3.2}$$

which appears to be in agreement with the equation (7.8.7).

10.3.2 Voltage-shunt feedback

Fig. 10.3(a) schematically illustrates, in block diagram, the voltage-shunt feedback configuration.

Here, the feedback signal is the current i_f which is proportional to the output voltage v_O. Since i_f is fed in parallel with the external input signal current, this is a *voltage-shunt feedback.* Again, the feedback network is in parallel with both the output and the input circuits. So, it is a *shunt-shunt* feedback.

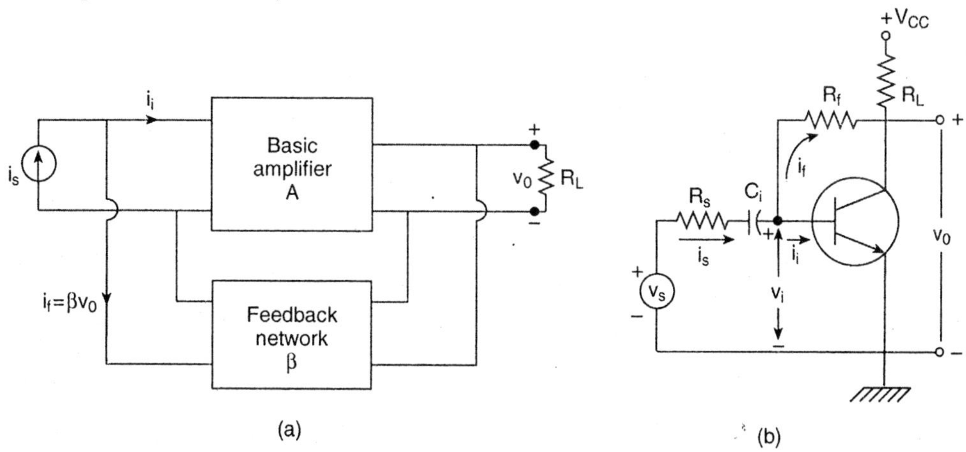

Fig. 10.3 (a) Block diagram of voltage-shunt feedback
(b) Circuit showing voltage-shunt feedback

Fig. 10.3(b) illustrates the circuit of such a type of feedback. The resistor R_f provides the feedback from the collector to the base of the transistor which is in CE mode. The capacitor C_i decouples the signal source from the supply voltage V_{CC} and behaves as a short-circuit for a.c. Here, the output voltage v_O is much greater than the input voltage v_i and is 180° out of phase with v_i. An increase in signal voltage increases the signal current i_s. Since $i_s = i_i + i_f$, the increase in i_i is much less than that in i_s showing that the circuit provides a *negative* feedback.

The feedback current is given by

$$i_f = \frac{v_s - v_O}{R_f} \approx -\frac{v_O}{R_f}, \quad \text{since } |v_O| \gg |v_s|. \tag{10.3.3}$$

Thus the feedback current i_f is proportional to the output voltage v_O. So, the above circuit is a ***voltage-shunt feedback*** amplifier.

10.3.3 Current-series feedback

The block diagram of a current-series feedback amplifier is shown in Fig. 10.4(a). Here, the feedback network is connected in series with the load and a feedback signal voltage v_f which is proportional to the output current i_O is produced. This v_f is applied in series with the input signal voltage v_s. So, this type of feedback is known as *current-series* feedback or *series-series* feedback.

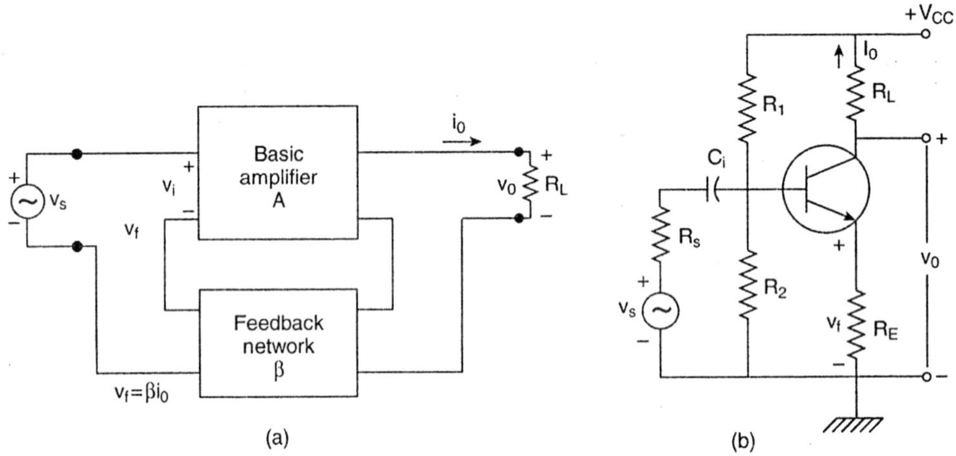

Fig. 10.4 (a) Block diagram of current-series feedback
(b) Circuit showing current-series feedback

Fig. 10.4(b) illustrates the circuit of this type of feedback. Here, the voltage appearing across the non-bypassed emitter resistor R_E acts as the feedback voltage v_f. Since the base current is very small, we may write $v_f = -i_O \cdot R_E$ where i_O is the output collector current. Since the output current on passing through R_E results in a feedback voltage v_f that acts in series with the input signal, the above circuit is referred to as the ***current-series feedback*** or *series-series feedback*.

10.3.4 Current-shunt feedback

Fig. 10.5(a) shows schematically the *current-shunt feedback* configuration. Here, the feedback current i_f is proportional to the output current i_O and it is fed in parallel to the input signal current source. So, this arrangement is called *current-shunt* or *series-shunt* feedback. Fig. 10.5(b) provides the circuit diagram of such a feedback amplifier. This is a *two-stage* amplifier. The resistance R_f in the circuit diagram connected between the emitter of T_2 and the base of T_1 gives the path of the feedback.

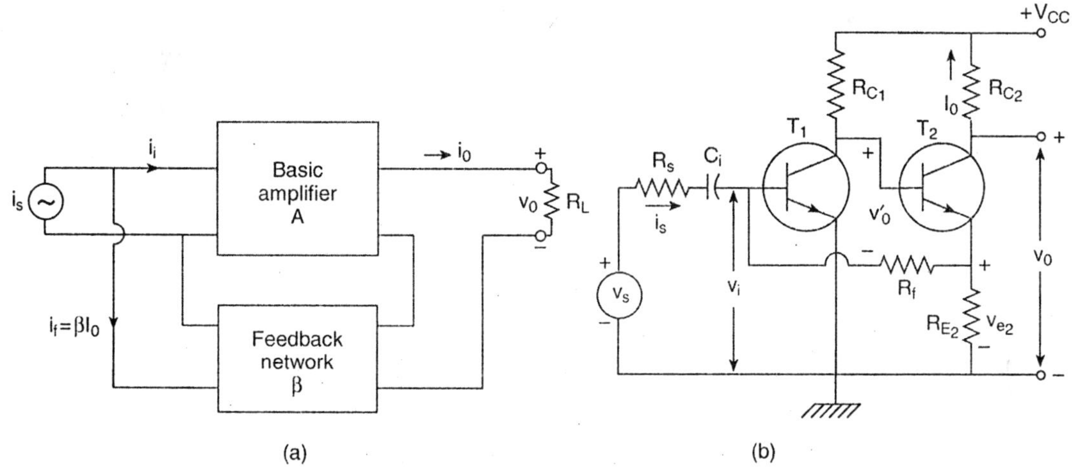

Fig. 10.5 (a) Block diagram of current-shunt feedback
(b) Circuit showing current-shunt feedback

Here, the feedback current is given by

$$i_f = \frac{v_i - v_{e2}}{R_f} \approx -\frac{v_{e2}}{R_f}, \quad \text{since } v_{e_2} \gg v_i. \tag{10.3.4}$$

Again, since the base current is very small compared to the collector current i_0 of T_2, we may plainly write

$$v_{e2} = (i_f - i_0)\,R_{E_2}$$

$$\text{or,} \quad i_f = \frac{i_0 - i_f}{R_f} \cdot R_{E_2}$$

$$= \frac{R_{E_2}}{R_f + R_{E_2}} \cdot i_0 \tag{10.3.5}$$

Thus, the feedback current i_f is proportional to the output current. So, this is a ***current-shunt feedback*** or *series-shunt feedback* amplifier.

10.4 Effects of negative feedback

On completion of the discussion on feedback topologies, we are now in a position to study the *effects of negative feedback*. The negative feedback plainly reduces the gain of an amplifier. In spite of the reduction in the overall gain, it introduces the following modifications in the amplifier characteristics, improving the working of an amplifier.

(a) ***Stability of gain***: With the reduction in gain, the negative feedback stabilizes the gain w.r.t the changes in transistor parameters like h_{fe} and voltage supply.

(b) ***Reduction in nonlinear distortion***: The nonlinear distortion can be reduced to a great extent by using negative feedback and this increases the signal handling capacity of the amplifier.

(c) ***Reduction in phase and frequency distortions***: The negative feedback also decreases the phase distortion and the frequency distortion.

(d) ***Reduction in output noise***: The noise level is much lowered in negative feedback amplifiers.

(e) ***Increase in bandwidth***: The negative feedback changes both the lower and the upper half-power frequencies in such a way that the bandwidth increases.

(f) ***Changes in input and output impedances***: The input and output impedances which affect the efficiency of power transfer in an amplifier may also be changed by means of negative feedback.

10.4.1 Stability of gain

The overall gain A_f of the negative feedback amplifier is given by

$$A_f = \frac{A}{1 + A\beta} \tag{10.4.1}$$

where, we have, $(1 + A\beta) > 1$.

If $A\beta \gg 1$, the above equation immediately reduces to

$$A_f = \frac{1}{\beta} \tag{10.4.2}$$

Thus the overall gain of the feedback amplifier depends on β, the value of which is controlled by the feedback network and does *not* depend upon other circuit parameters such as A of the internal amplifier. Since the feedback circuit usually consists of stable passive elements, the stability of gain is very high. The supply voltage variations, transistor parameters, the working temperature and ageing of the components have marked effects on the open-loop gain A, but their effects on A_f are *not significant.*

Differentiating (10.4.1) we get,

$$dA_f = \frac{dA}{(1 + A\beta)^2} \tag{10.4.3}$$

Now, dividing (10.4.3) by (10.4.1), at once we have,

$$\frac{dA_f}{A_f} = \frac{dA}{A} \cdot \frac{1}{(1 + A\beta)}$$

$$\text{or,} \quad \left(\frac{dA_f}{A_f}\right) \Big/ \left(\frac{dA}{A}\right) = \frac{1}{1 + A\beta} \tag{10.4.4}$$

The quantity dA/A measures the *fractional change in voltage gain of the internal amplifier* and dA_f/A_f measures that of the *feedback amplifier.*

The ratio $\left(\frac{dA_f}{A_f}\right)\Big/\left(\frac{dA}{A}\right)$ is called the *sensitivity* S of the transfer gain to changes in the internal amplifier gain. Thus, the ***gain sensitivity*** is given by

$$\boxed{S = \frac{1}{1 + A\beta}} \tag{10.4.5}$$

In negative feedback amplifier, $(1 + A\beta) > 1$, which implies that $S < 1$. Thus, the percentage change in gain with feedback, A_f is much smaller than that without feedback, A. The *gain stability thus greatly improves with negative feedback.*

10.4.2 Nonlinear distortion

When the input signal of an amplifier has a very large amplitude, the output signal operation extends into the nonlinear region of the transfer characteristic causing thereby a distortion in the output signal. This is known as ***nonlinear distortion.*** Negative feedback can be used to reduce this distortion.

Let the input signal be sinusoidal and the distortion voltage consists only of the second harmonic component. If v_0 and v_d be the signal components of the output voltage and the distortion voltage respectively then, in absence of feedback, the output voltage is given by

$$v_0 = Av_s + v_d \tag{10.4.6}$$

In presence of negative feedback, the output becomes

$$\begin{aligned} v_{of} &= A(v_s - \beta v_{of}) + v_d \\ &= \frac{Av_s}{1 + A\beta} + \frac{v_d}{1 + A\beta} \end{aligned} \tag{10.4.7}$$

To compare the performance of the amplifier with and without feedback under the same output signal components, we have to increase the input signal from v_s to $v_s(1 + A\beta)$ in the case of feedback. Then (10.4.7) becomes

$$v_{of} = Av_s + \frac{v_d}{1 + A\beta} \tag{10.4.8}$$

Comparing (10.4.6) and (10.4.8), we may conclude that the use of negative feedback reduces the nonlinear distortion by a factor $1/(1 + A\beta)$.

10.4.3 Phase distortion

When the coupling impedance or the load of an amplifier has a reactive part, the gain is a complex quantity.

Let $A(\omega)$ be the gain of the amplifier without the feedback. Then, we have

$$A(\omega) = |A|e^{j\theta} \tag{10.4.9}$$

where θ is the phase shift of the component ω in transferring from the input to the output.

For an amplifier with negative feedback, we may write

$$A_f(\omega) = |A_f|e^{j\theta_f} = \frac{|A|e^{j\theta}}{1+\beta|A|e^{j\theta}}$$

where θ_f represents the phase shift for the component ω with feedback.

Now,
$$A_f(\omega) = \frac{|A|e^{j\theta}}{(1+\beta|A|\cos\theta) + j\beta|A|\sin\theta}$$
$$= \frac{|A|e^{j\theta}}{|B|e^{j\phi}} \tag{10.4.10}$$

where
$$|B| = \sqrt{(1+\beta|A|\cos\theta)^2 + (\beta|A|\sin\theta)^2}$$

and
$$\phi = \tan^{-1}\frac{\beta|A|\sin\theta}{1+\beta|A|\cos\theta} \tag{10.4.11}$$

Comparing (10.4.10) with the relation $A_f(\omega) = |A_f|e^{j\theta_f}$, we get

$$|A_f| = \frac{|A|}{|B|} \tag{10.4.12}$$

$$\text{and } \theta_f = \theta - \phi \tag{10.4.13}$$

Clearly, the negative feedback *reduces the phase angle* of the gain by an amount ϕ. When $|A|\beta \gg 1$, the equation (10.4.11) gives $\phi = \tan^{-1}(\tan\theta) = \theta$ and hence $\theta_f = 0$, i.e., *no phase distortion* would occur.

10.4.4 Frequency distortion

The frequency distortion takes place when the gain of an amplifier varies with the frequency of the input signal. Negative feedback reduces this type of distortion as well.

The gain of an amplifier with negative feedback is given by

$$A_f = \frac{A}{1+A\beta}$$

where $(1+A\beta) > 1$.

Differentiating both sides with respect to ω, we at once get

$$\frac{dA_f}{d\omega} = \frac{1}{(1+A\beta)^2}\cdot\frac{dA}{d\omega} \tag{10.4.14}$$

Since $|1+A\beta| > 1$, we have $\quad \dfrac{dA_f}{d\omega} < \dfrac{dA}{d\omega}.$

So, this shows that the *negative feedback reduces the frequency distortion.*

10.4.5 Reduction of output noise

Let v_n be the noise voltage at the input of the amplifier with a voltage gain A when no feedback is applied. Therefore, the noise voltage at the output is given by

$$v_{on} = Av_n \tag{10.4.15}$$

In presence of the feedback, the output noise voltage however becomes

$$(v_{on})_f = A_f v_n = \frac{Av_n}{1+A\beta} = \frac{v_{on}}{1+A\beta} \tag{10.4.16}$$

where β is the feedback fraction and A_f the voltage gain with feedback.

Since $|1 + A\beta| > 1$, it follows from equation (10.4.16) that

$$(v_{on})_f < v_{on},$$

i.e., the *output noise-level gets reduced.*

10.4.6 Increase of bandwidth

The bandwidth of an amplifier can also be increased by using negative feedback. In Art. 8.5, we have seen that the high frequency voltage gain of an RC-coupled amplifier is given by equation (8.5.11), i.e.,

$$A_{V_H} = \frac{A_{V_M}}{1+j\dfrac{f}{f_2}} \tag{10.4.17}$$

where A_{V_M} and f_2 represent respectively the *mid-frequency voltage gain* and the *upper half-power frequency.*

For a negative feedback amplifier, the voltage gain can be given in the form

$$\begin{aligned} A_{V_H f} &= \frac{A_{V_H}}{1+A_{V_H}\cdot\beta} \\ &= \frac{A_{V_M}\Big/\left(1+j\dfrac{f}{f_2}\right)}{1+\beta\cdot\dfrac{A_{V_M}}{1+j\frac{f}{f_2}}} \\ &= \frac{A_{V_M}}{1+A_{V_M}\beta+j\dfrac{f}{f_2}} = \frac{A_{V_M f}}{1+j\dfrac{f}{f_2'}} \end{aligned} \tag{10.4.18}$$

where $\quad A_{V_M f} = \dfrac{A_{V_M}}{1+\beta A_{V_M}}$ is the *mid-frequency gain* with the feedback and f_2' is given by

$$f_2' = f_2\left(1 + A_{V_M}\beta\right), \tag{10.4.19}$$

the *upper half-power frequency* in presence of the feedback.

In a similar manner, considering the low frequency gain, as given by (8.5.6), we obtain the low frequency gain with the feedback as

$$A_{V_L}f = \frac{A_{V_M}f}{1 - j\dfrac{f_1'}{f}}$$

where $$f_1' = \frac{f_1}{1 + A_{V_M}\beta} \qquad (10.4.20)$$

is the *lower half-power frequency* with the feedback.

So, the equations (10.4.19) and (10.4.20) show that the negative feedback *increases* the *upper half-power frequency* and *decreases* the *lower half-power frequency*. Thus, the bandwidth $(f_2' - f_1')$ with the feedback is greater than the bandwidth $(f_2 - f_1)$ without the feedback. Fig. 10.6 shows the *frequency response* of a typical RC-coupled amplifier with and without the feedback.

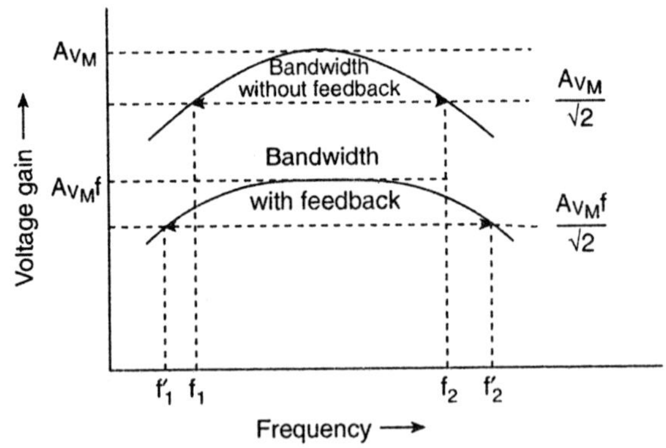

Fig. 10.6 Frequency response of a typical RC-coupled amplifier with and without the feedback

10.4.7 Effect of feedback on input impedance

To find the change in input impedance in negative feedback, let us first consider the series-mixing feedback circuit as illustrated in Fig. 10.7.

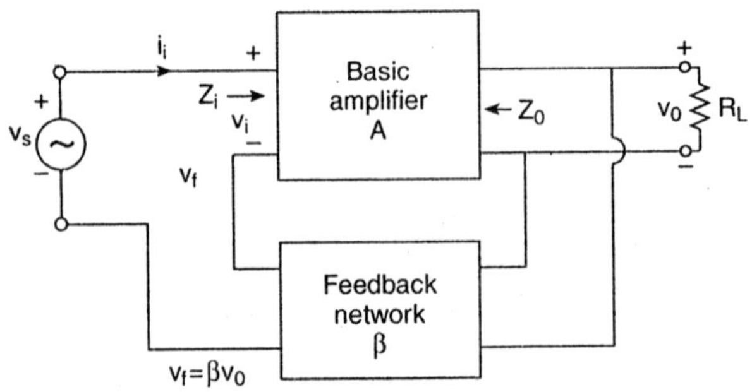

Fig. 10.7 Determination of input impedance for series-mixing feedback

Here the input impedance of the basic amplifier is given by

$$Z_i = \frac{v_i}{i_i} \tag{10.4.21}$$

and the input impedance of the feedback amplifier is given by

$$Z_{if} = \frac{v_s}{i_i} \tag{10.4.22}$$

Now, the external input signal voltage

$$\begin{aligned} v_s &= v_i + v_f = v_i + \beta v_O \\ &= v_i + A\beta v_i \end{aligned}$$

where v_f is the feedback voltage, $\beta = v_f/v_O$ = feedback ratio and $A = v_O/v_i$ = voltage gain of the internal amplifier.

$$\therefore \quad v_s = v_i(1 + A\beta) \tag{10.4.23}$$

Equation (10.4.22) gives, on using equation (10.4.23)

$$Z_{if} = \frac{v_i(1 + A\beta)}{i_i} = Z_i(1 + A\beta), \tag{10.4.24}$$

using also the equation (10.4.21).

For the negative feedback $|1 + A\beta| > 1$. So, the equation (10.4.24) implies that $Z_{if} > Z_i$ i.e., the *series-mixing of negative feedback signal increases the input impedance for the feedback.*

Consider now the shunt-mixing feedback as illustrated in Fig. 10.8 below.

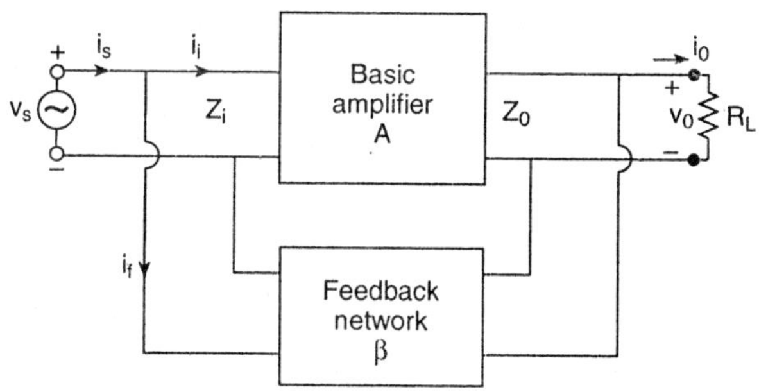

Fig. 10.8 Determination of input impedance for shunt-mixing feedback

In this case, the input impedance Z_i is given by

$$Z_i = \frac{v_s}{i_i} \tag{10.4.25}$$

and the input impedance for the feedback is given by

$$Z_{if} = \frac{v_s}{i_s} \tag{10.4.26}$$

Now, the input signal current

$$i_s = i_i + i_f = i_i + \beta i_0$$
$$= i_i + A\beta i_i = (1 + A\beta)i_i, \quad (10.4.27)$$

since $\beta = i_f/i_0$ = reverse transmission factor and $A = i_0/i_i$ = current gain.

∴ Equation (10.4.26) gives,

$$Z_{if} = \frac{v_s}{i_i(1 + A\beta)} = \frac{Z_i}{1 + A\beta}, \quad \text{using (10.4.25).} \quad (10.4.28)$$

Since $|1 + A\beta| > 1$ for negative feedback, the equation (10.4.28) implies that *shunt-mixing feedback signal decreases the input impedance.*

10.4.8 Effect of feedback on output impedance

To find the output impedance of an amplifier we are to make the input signal voltage $v_s = 0$ and remove the load resistance R_L. We then connect a hypothetical voltage source v across the load terminals.

To determine the output impedance in negative feedback, let us consider the voltage-series feedback circuit as shown in Fig. 10.9.

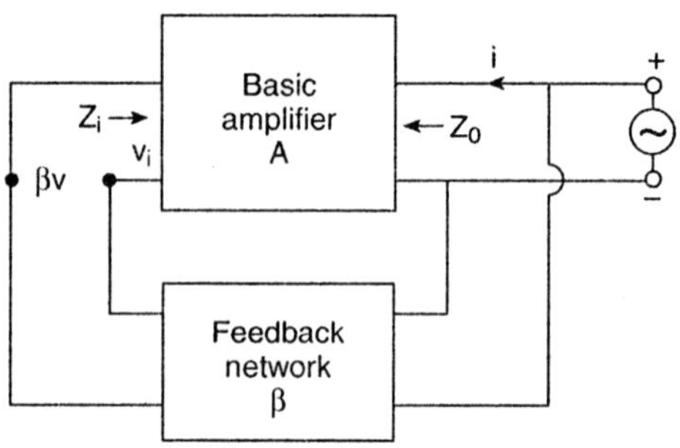

Fig. 10.9 Determination of output impedance for voltage-series feedback

Here, the output impedance of the basic amplifier is given by

$$Z_0 = \frac{v}{i'} \quad (10.4.29)$$

where i' is the current delivered by v with *no* feedback.

The input voltage of the basic amplifier with feedback is

$$v_i = v_s - v_f = -v_f = -\beta v, \quad (\because\ v_s = 0)$$

Therefore, the voltage appearing at the output, in opposition to the hypothetical voltage v, due to this input voltage is

$$Av_i = -A\beta v$$

∴ The total voltage at the output is given by

$$v'_0 = v - Av_i = v + A\beta v = (1 + A\beta)v. \tag{10.4.30}$$

The total current with the feedback is therefore

$$i = \frac{v(1 + A\beta)}{Z_0} \tag{10.4.31}$$

So, the output impedance with the feedback is given by

$$Z_{of} = \frac{v}{i} = \frac{Z_0}{1 + A\beta} \tag{10.4.32}$$

Since $(1 + A\beta) > 1$ for the negative feedback, the equation (10.4.32) implies that the output impedance is decreased by a factor $1/(1 + A\beta)$ in voltage-series feedback i.e., *voltage sampling decreases the output impedance.*

Consider now the current feedback amplifier as shown in block diagram in Fig. 10.10.

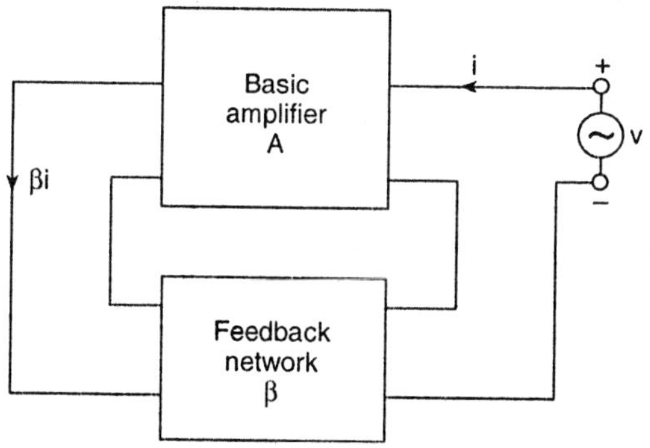

Fig. 10.10 Determination of output impedance for current-mixing feedback

Let i be the current drawn from the source v by the amplifier with feedback. Then the fraction $-\beta i$ of this current is the *feedback* to the input. This current gets amplified by the gain A and appears at the output as $-A\beta i$. Therefore,

$$i = \frac{v}{Z_0} - A\beta i$$

$$\text{or,} \quad (1 + A\beta)i = \frac{v}{Z_0} \tag{10.4.33}$$

The output impedance with feedback is thus given by

$$Z_{of} = \frac{v}{i} = Z_0(1 + A\beta). \tag{10.4.34}$$

For a *negative* feedback $(1+A\beta) > 1$. This implies that the *negative current feedback increases the output impedance.*

10.5 Regulated power supplies

It is a well known fact that almost all the electronic devices and circuits require a constant d.c. source of voltage for their operation. Primarily, this requirement is met by dry cells or batteries for ordinary portable low power systems. But these are expensive. So, it is necessary to construct a device or circuit that converts domestic a.c supply to d.c in order to supply constant d.c voltage to most of the electronic devices. *The electronic device or circuit that provides a sustaintially constant voltage, even if the a.c supply voltage or d.c load current changes, is called a* ***regulated*** *or* ***stabilized power supply.***

A regulated power supply consists of the following *three* parts:

1. a rectifier, 2. a filter and 3. a voltage regulator.

The system is illustrated in the block diagram of Fig. 10.11.

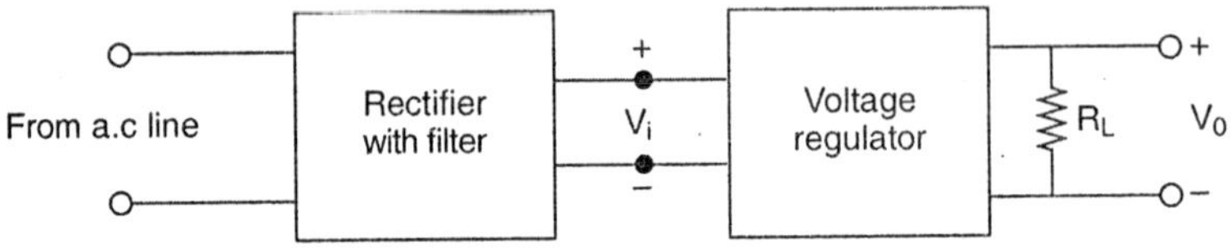

Fig. 10.11 Block diagram of regulated power supply

The rectifiers and filters have already been discussed in Chapter 4. We shall now study the different type of regulated power supplies in the following sections.

10.5.1 Transistor series regulator

In Chapter 3, we observed that a Zener diode can be used as a voltage regulator. But that was the simplest type voltage regulator. To increase the current handling capacity of the regulator we have to modify that circuit by using transistor. Fig. 10.12 shows the modified simple series voltage regulator circuit. In this circuit, the transistor is connected in series with the load. So, this circuit is called ***transistor series regulator.***

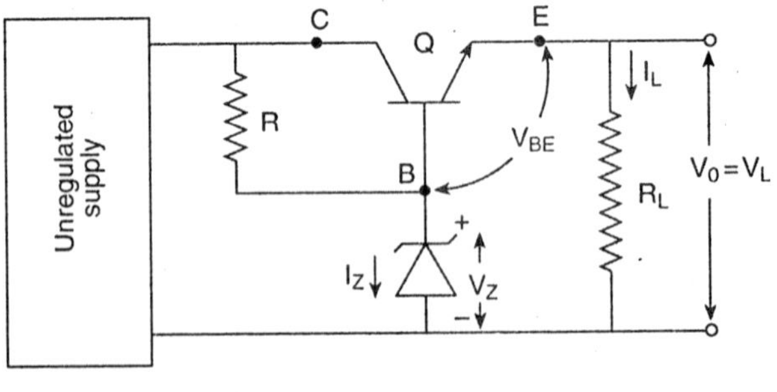

Fig. 10.12 Circuit diagram of transistor series regulator

The transistor in the circuit behaves as a variable resistance whose value is determined by the amount of base current.

The output voltage is given by

$$V_O = V_L = V_Z - V_{BE}$$

$$\text{or,} \quad V_{BE} = V_Z - V_L \tag{10.5.1}$$

Since the collector terminal and the emitter terminal are in series with the load, the load current must pass through the transistor Q. So, Q is sometimes called a ***pass transistor***.

Principle of operation—To understand the working of the series regulator, suppose the load resistance is increased. As a result, the load current decreases and the load voltage (V_L) tends to increase. But (10.5.1) shows that the increase in V_L will decrease V_{BE} since the Zener voltage V_Z acts as reference voltage and remains fixed. Now the decrease in V_{BE} causes the forward bias of the transistor to reduce. This reduces the level of conduction of the transistor Q. Thus the collector to emitter voltage (V_{CE}) increases. This in turn will slightly reduce the input current to compensate for the increase in load resistance so that the load voltage may remain at a constant value. Thus the output voltage is regulated or stabilized at constant voltage even when the load resistance is varied. The action of the circuit may be summerized in the form as given below.

$$V_L \uparrow \longrightarrow V_{BE} \downarrow \longrightarrow V_{CE} \uparrow \longrightarrow V_L \downarrow$$

• The output voltage of a transistor series regulator is approximately equal to Zener voltage (V_Z). This regulator may also be used for larger load current.

10.5.2 Controlled transistor series regulator

The transistor series regulator circuit of Fig. 10.12 can be improved further to obtain the ***controlled transistor series regulator*** as shown in Fig. 10.13.

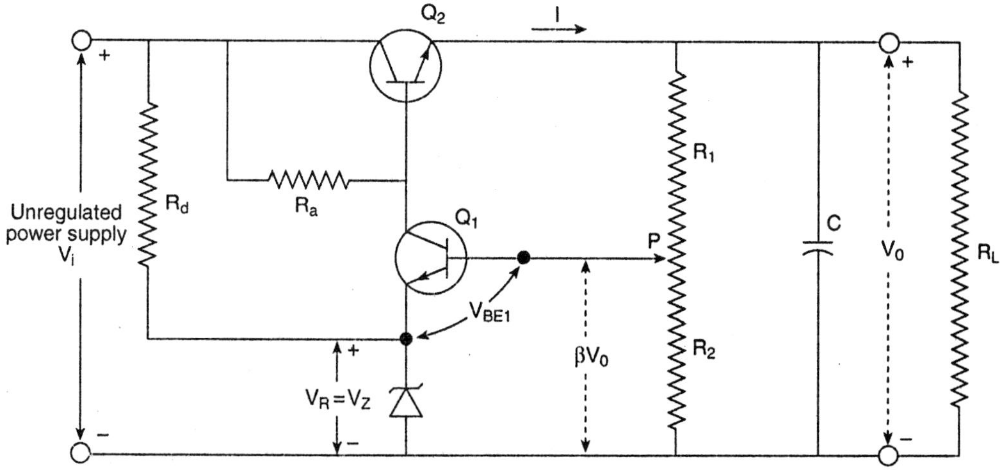

Fig. 10.13 Controlled transistor series regulator circuit

In this regulator circuit two transistors Q_1 and Q_2 are employed. The main section of the circuit consists of *three* parts.

1. Control transistor : The transistor Q_2, known as *pass transistor*, acts as the control transistor since it tries to maintain the output voltage constant.

2. Error amplifier : The transistor Q_1 along with the resistance R_a functions as voltage comparator and dc amplifier. So, Q_1 is known as *amplifier transistor.*

3. Reference voltage : The voltage (V_Z) taken across the Zener diode is called *reference voltage* (V_R). The resistance R_d acts as a Zener series limiting resistance.

The resistors R_1 and R_2 function as sampling network to provide a sample signal βV_0 to the transistor Q_1 where $\beta = R_2/(R_1+R_2)$. This part is arranged by a potentiometer.

Principle of stabilization : In the circuit of Fig. 10.13, the regulation of dc output voltage V_0 occurs due to variation of the current I i.e., the load current. Since an increase in the base voltage increases the forward bias of base-emitter junction of the transistor Q_2, this current I is a function of base voltage.

Suppose, there is an increase in load voltage due either to input voltage variation or change of load of Q_1. The increase in V_0 increases the feedback voltage βV_0. This sample signal voltage βV_0 is then compared with the reference voltage V_R $(= V_Z)$ across the Zener diode by the error amplifier. By adjusting the sampling network (i.e., potentiometer P or variable resistor), the voltage βV_0 is made nearly equal to V_R, and the difference $(\beta V_0 - V_R)$ is then amplified by the amplifier and is applied to the base of Q_2. The polarity of the difference is maintained in such a condition that $\beta V_0 > V_R$ and the base of the transistor Q_1 becomes more positive. This causes the collector current of Q_1 to increase and the base-emitter voltage V_{BE1} of Q_1 to go up. As a result, Q_1 conducts more current and less current passes through the base of the pass transistor Q_2. So, the load current reduces and brings down the output voltage back to the original value. Thus any change in output voltage is acted upon by a change in the current I which forces V_0 to remain stabilized i.e., constant.

From Fig 10.13, the output voltage is given by

$$V_0 = V_Z + V_{BE1} + \frac{R_1}{R_1 + R_2} V_0$$

$$\text{or,} \quad \frac{R_2}{R_1 + R_2} V_0 = V_Z + V_{BE1}$$

$$\text{or,} \quad \beta V_0 = V_Z + V_{BE1}$$

$$\text{or,} \quad V_0 = \frac{1}{\beta}(V_Z + V_{BE1}) \tag{10.5.2}$$

$$\therefore \quad V_0 = (V_Z + V_{BE1})\left(1 + \frac{R_1}{R_2}\right) \tag{10.5.3}$$

Thus the output voltage can be controlled by adjusting the ratio of R_1/R_2 i.e., by changing the position of the slider of the potentiometer P.

10.5.3 Transistor shunt regulator

The voltage regulator in which the control element is connected in shunt with the load resistance is called a ***shunt regulator***. Fig. 10.14 shows the circuit diagram of a transistor shunt regulator.

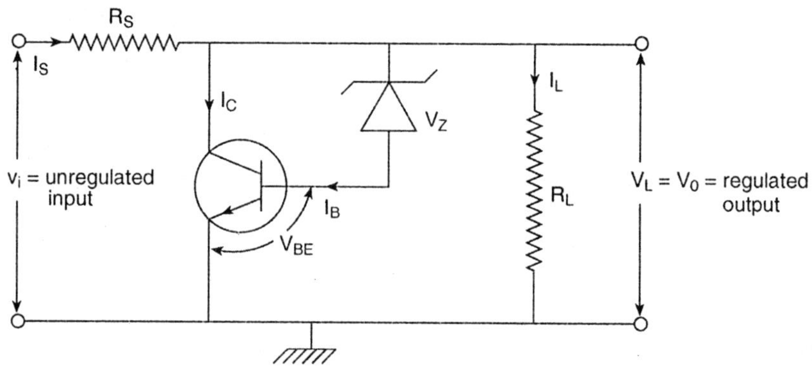

Fig. 10.14 Transistor shunt regulator

From the above circuit, we observe that the output voltage i.e., the load voltage V_L is equal to the sum of the Zener voltage V_Z and the base-to-emitter voltage V_{BE} of the transistor. So, we have

$$V_0 = V_L = V_Z + V_{BE} \tag{10.5.4}$$

$$\text{or,} \quad V_{BE} = V_L - V_Z \tag{10.5.5}$$

Since V_Z is fixed for a given Zener diode, any increase or decrease in load voltage has a corresponding effect on V_{BE}.

Principle of regulation—To study the *principle of regulation* of transistor shunt voltage regulator, suppose, the unregulated input voltage V_i somehow increases. This causes an increase in V_L. According to equation (10.5.5), we find that V_{BE} also increases. As a result of this, the base current I_B of the transistor increases which causes the collector current to increase. So, there is a net increase in the input current I_S which in turn increases the potential drop V_{R_S} across the series resistance R_S. This larger voltage drop in R_S tries to reduce the V_L back to the original value and hence the output voltage V_0 or V_L is stabilized.

The action of the circuit may be summarized in the form as given below.

$$V_i \uparrow \longrightarrow V_L \uparrow \longrightarrow V_{BE} \uparrow \longrightarrow I_C \uparrow \longrightarrow V_{R_S} \uparrow \longrightarrow V_L \downarrow$$

10.5.4 Transistor current regulator

A ***current regulator*** is a circuit that gives a constant dc output current irrespective of the variation in load current. In some applications, such circuits are required. Fig. 10.15 illustrates the circuit diagram of a current regulator using a *pnp* transistor and a Zener diode.

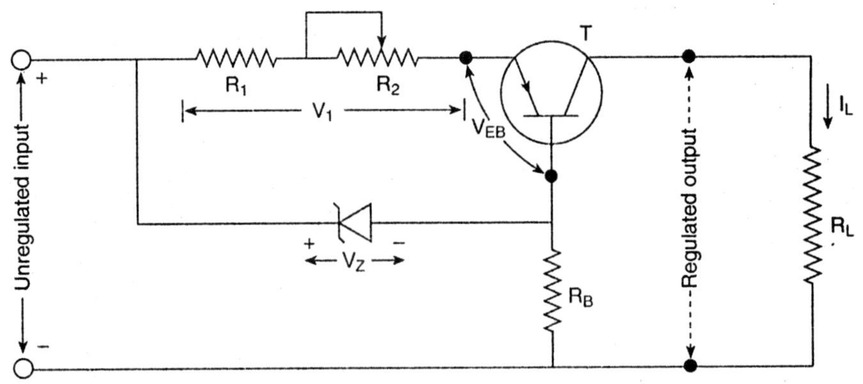

Fig. 10.15 Transistor current regulator circuit

The load current I_L is carried by the transistor T. So the circuit is a *series current regulator.* The change in load current I_L is counteracted by the change in the internal resistance of the transistor T due to the negative feedback of the circuit. Applying KVL to the circuit, we have,

$$V_1 + V_{EB} = V_Z$$

$$\text{or,} \quad V_{EB} = V_Z - V_1 \tag{10.5.6}$$

where V_{EB} is the emitter to base voltage of the transistor T, V_Z is the voltage across the Zener diode and V_1 is the voltage drop across the resistance R_1 and R_2.

Principle of regulation—To study the *principle of regulation,* suppose the load current I_L is increased due to a change in the load resistance R_L. Since $I_L = I_C$ for the transistor T, an increase in I_L results in an increase in the emitter current I_E of T, causing thereby enhancement in the voltage drop V_1. But V_Z is constant. So (10.5.6) indicates that V_{EB} decreases when V_1 increases. Consequently, it brings the collector current or the load current back to its original value. Thus current regulation is achieved.

The output current is fixed at a desired level by varying the resistance R_2. The circuit thus maintains the current at that level, irrespective of the variation of the load resistance R_L.

The current for the Zener diode operation is determined by the resistance R_B connected to the base of the transistor. The transistor base current supplies a small part of the current through R_B such that a small current passes through the Zener diode. The R_B should be of proper value such that a sufficient current passes through it in order to keep the Zener diode active in the circuit when R_L changes. When the load resistance R_L is sufficiently large, the voltage drop across it is equal to the voltage drop across R_B and the current I_L remains approximately constant.

● It is to be noted that the internal impedance of an ideal current source is infinite but that of an ideal voltage source is zero. While a current regulator acts as an ideal current source, a voltage regulator behaves like an ideal voltage source.

10.6 Illustrated Examples

▶ **Example 1.** *An amplifier has a gain of* 100 *without feedback. This gain is reduced to* 50 *with negative feedback. Find the feedback fraction of the amplifier.* **(Himachal Univ.)**

Solution: Given: $A = 100$, $A_f = 50$. Therefore, we have

$$50 = \frac{100}{1+100\beta} \Rightarrow 1+100\beta = 2$$

$$\therefore \quad 100\beta = 1 \Rightarrow \boldsymbol{\beta = 0.01}$$

▶ **Example 2.** *The overall gain of a multistage amplifier is* 100. *When negative feedback is applied, the gain is reduced to* 10. *Find the fraction of the output that is fed back to the input.* **(Punjab Univ.)**

Solution: Given, the voltage gain without feedback, $A = 100$, and voltage gain with feedback, $A_f = 10$.

Now, we have $$A_f = \frac{A}{1+A\beta},$$ where β is the feedback fraction.

$$\therefore \quad 10 = \frac{100}{1+100\beta} \quad \text{or,} \quad \beta = 0.09$$

$\therefore$ The ***fraction of the output*** that is fed back to the input is **0.09**.

▶ **Example 3.** *An amplifier with an open loop gain of* 25 *is subjected to negative feedback of* 10%. *If the open loop gain thereby increases by* 5%, *find the percentage change in gain with feedback.*

Solution: The open loop gain, $A = 25$; feedback factor, $\beta = 10\% = 0.1$.

$$\therefore \text{ Overall gain with feedback, } A_f = \frac{A}{1+A\beta} = \frac{25}{1+25\times 0.1} = 7.14$$

The open loop gain, by the problem, is increased by $5\% = 1/20$.

$$\therefore \text{ The increased open loop gain, } A' = 25 + \frac{1}{20}\times 25 = 26.25$$

$$\therefore \text{ The increased overall gain, } A'_f = \frac{A'}{1+A'\beta} = \frac{26.25}{1+26.25\times 0.1} = 7.24$$

$$\therefore \textit{ Percentage change} \text{ in gain with feedback} = \frac{7.24-7.14}{7.14}\times 100\% = \mathbf{1.4\%}.$$

▶ **Example 4.** *An amplifier has a voltage gain of* -500. *The gain is reduced to* -100 *when negative feedback is applied. Determine the reverse transmission factor* β *and express the amount of feedback in dB.* **(Punjab Univ.)**

Solution: Voltage gain without the feedback, $A = -500$ and the voltage gain with the feedback, $A_f = -100$.

Now, we have
$$A_f = \frac{A}{1 + A\beta}$$
$$\Rightarrow \ 1 + A\beta = \frac{A}{A_f} = \frac{-500}{-100} = 5 \Rightarrow A\beta = 4$$
$$\therefore \ \beta = \frac{4}{A} = \frac{4}{-500} = \mathbf{-0.008}$$

$$\therefore \ \textit{dB of feedback} = 20\log_{10}\left|\frac{A_f}{A}\right| = 20\log_{10}\left|\frac{1}{1+A\beta}\right| = 20\log_{10}\frac{1}{5}$$
$$= 20 \times (-0.699) = \mathbf{-13.98}$$

▸ **Example 5.** *An amplifier gain equal to* 10^3 *produces* 15% *harmonic distortion at full output. What value of* β *is needed to reduce the distortion to a tolerance level of* 5%*?*

Solution: Given, the voltage gain without the feedback, $A = 10^3$.

Harmonic distortion, $dA/A = 15\%$.

Tolerance level of distortion with feedback, $dA_f/A_f = 5\%$.

Now, we have
$$\frac{dA_f/A_f}{dA/A} = \frac{1}{1 + A\beta}$$
$$\Rightarrow \ \frac{5}{15} = \frac{1}{1 + 1000\beta}$$
or, $\beta = \mathbf{0.002}$, on simplification.

▸ **Example 6.** *A transistor amplifier has a voltage gain of* 50*. The input resistance of the amplifier is* 1 kΩ*. The amplifier is now provided with* 10% *negative feedback in series with the input. Calculate the voltage gain and the input impedance with feedback.*

(Nagpur Univ.)

Solution: Given, the voltage gain without feedback, $A = 50$. So, the feedback fraction, $\beta = 10\% = 0.1$.

$\therefore$ The ***voltage gain*** with feedback A_f is given by
$$A_f = \frac{A}{1 + A\beta} = \frac{50}{1 + 50 \times 0.1} = \mathbf{8.33}$$

Input resistance without feedback, $Z_i = 1\,\text{k}\Omega = 1000\,\Omega$.

$\therefore$ ***Input impedance*** with feedback is given by
$$Z_{if} = Z_i(1 + A\beta) = 1000 \times (1 + 50 \times 0.1)\Omega = \mathbf{6000\Omega}$$

► **Example 7.** *The overall gain of a negative feedback amplifier is* 80. *Variations of this gain of only* ±1 *per cent can be tolerated for some specific use. If the open loop gain variations of* ±10% *are expected owing to production spreads in device characteristics, determine the minimum value of feedback fraction* β *and also the gain of the internal amplifier to satisfy the above condition.* **(Punjab Univ.)**

Solution : If $S' = dA_f/A_f$ measures the fractional change in voltage gain with feedback and $= dA/A$ measures the same quantity without feedback (or internal amplifier), then

$$\frac{dA_f}{A_f} = \frac{dA}{A} \cdot \frac{1}{1 + A\beta}$$

Now, by the problem, $dA_f/A_f = \pm 1\%$, $dA/A = \pm 10\%$, $A_f = 80$.

$$\therefore \quad 1 + A\beta = \left(\frac{dA}{A}\right) \Big/ \left(\frac{dA_f}{A_f}\right) = 10 \qquad (10.6.1)$$

$$\text{Also,} \qquad A_f = \frac{A}{1 + A\beta} \qquad (10.6.2)$$

∴ Gain of the internal amplfier, $\boldsymbol{A} = A_f(1 + A\beta) = 80 \times 10 = \mathbf{800}$

Since $1 + A\beta = 10$ and $A = 800$, the minimum value of feedback fraction, $\boldsymbol{\beta} = (10 - 1)/A = 9/800 = \mathbf{0.01125}$.

► **Example 8.** *The open-loop gain of an amplifier changes by* 20% *due to changes in the parameters of the active amplifying device. If a change of gain by* 20% *is allowable, what type of feedback has to be applied? If the amplifier gain with feedback is* 10, *calculate the minimum value of the feedback ratio and the open loop gain.*

Solution : For gain stability, negative feedback has to be applied. Here, by the problem,

$$dA/A = 20\%, \quad dA_f/A_f = 2\%$$

Using the realtion, $\quad \dfrac{dA_f}{A_f} = \dfrac{1}{1 + A\beta} \cdot \dfrac{dA}{A}$, we obtain

$$1 + A\beta = (dA/A)/(dA_f/A_f) = 20/2 = 10$$

∴ Gain with feedback, $A_f = \dfrac{A}{1 + A\beta} \Rightarrow 10 = A/10$. ∴ $\mathbf{A = 100}$

which is the *open-loop gain.*

Since $1 + A\beta = 10$, $1 + 100\beta = 10$. This implies that feedback ratio, $\boldsymbol{\beta} = \mathbf{0.09}$.

► **Example 9.** *The voltage gain of an amplifier is* 200 *and its band width is* 200 kHz. *If a negative feedback with feedback fraction* 0.02 *is applied, find* (a) *the gain of amplifier with feedback and* (b) *the band with of the amplifier with feedback.* **(Madurai Univ.)**

Solution : Given, voltage gain without feedback, $A = 200$, feedback fraction, $\beta = 0.02$

$\therefore$ Voltage gain with feedback, $\boldsymbol{A_f} = \dfrac{A}{1 + A\beta} = \dfrac{200}{1 + 200 \times 0.02} = 200/5 = \mathbf{40}.$

If f_l and f_h be the lower and the upper frequency limits respectively without feedback, then we have

$$\text{Bandwidth } \textit{without feedback}, \; BW = f_h - f_l = 200 \text{ kHz} \;\; \text{(Given)}$$

If f_{lf} and f_{hf} be the lower and upper frequency limits respectively *with feedback*,

$$\begin{aligned} f_{lf} &= f_l(1 + A\beta) \\ \text{and } f_{hf} &= f_h(1 + A\beta) \end{aligned}$$

$$\begin{aligned} \therefore \text{ Bandwidth with feedback, } \boldsymbol{(BW)_f} &= f_{hf} - f_{lf} = (f_h - f_l)(1 + A\beta) \\ &= BW(1 + A\beta) \\ &= 200 \times (1 + 200 \times 0.02) \\ &= 200 \times (1 + 4) = \mathbf{1000 \text{ kHz}} \end{aligned}$$

Review and Short Questions

1. What do you mean by feedback in an amplifier? Explain the difference between the positive and the negative feedbacks. **(Calcutta Hons. and Meerut Univ.)**

2. Explain with the help of a block diagram the working principle of a feedback amplifier. Find an expression for the voltage gain with feedback. **(Calcutta Hons.)**

3. State the effects of negative feedback on an amplifier.

4. How is the bandwidth of a *RC*-coupled amplifier modified when the negative feedback is used? **(Calcutta Hons.)**

5. Define the following terms: Regenerative feedback, degenerative feedback, feedback ratio, dB of feedback.

6. What are the different ways of sampling the output signal in a feedback amplifier? Name the four feedback topologies.

7. What are the advantages of negative feedback in amplifiers? **(Burdwan Hons.)**

8. (a) Find the relation between the open loop gain and the closed loop gain of a negative feedback amplifier.

   (b) Hence show that the negative feedback can improve the stability of gain of an amplifier.

   (c) When does a feedback amplifier become oscillatory? **(Burdwan Hons.)**

9. How does the negative feedback reduce (a) nonlinear distortion (b) noise level (c) phase distortion and (d) frequency distortion? **(Lucknow Univ.)**

10. Could there be an amplifier giving an output signal even without an input signal? Explain.

11. (a) What is a regulated power supply?

(b) Draw the circuit diagram of a series voltage regulator and explain its principle of operation. **(Calcutta Hons.)**

12. What is a current regulator? Explain with a circuit diagram, the operation of a transistor current regulator.

13. Explain with a circuit diagram, the operation of a controlled transistor series regulator.

14. Draw the circuit diagrams of series and shunt voltage regulators, and compare between them.

Objective Questions

1. Emitter follower is an example of

(a) voltage-shunt feedback, (b) voltage-series feedback,
(c) current-shunt feedback, (d) current-series feedback.

2. Positive feedback

(a) reduces distortion, (b) increases bandwidth,
(c) reduces noise, (d) makes the gain unstable.

3. Series-mixing of negative feedback in an amplifier

(a) decreases the input resistance, (b) increases the input resistance,
(c) keeps the input resistance unchanged, (d) none of the above.

4. Negative feedback

(a) makes the gain unstable, (b) decreases the gain,
(c) decreases the bandwidth, (d) increases phase-distortion.

5. A voltage regulator normally uses

(a) negative feedback, (b) positive feedback,
(b) no feedback, (d) clipping network.

6. If the power and the voltage gains of an amplifier are 25,000 and 250 respectively, then the current gain will be

(a) 250, (b) 100, (c) 62.5×10^{-4}, (d) None.

7. The output voltage of a voltage regulator changes from 10 V to 9.8 V when load current varies from zero to its maximum value. Then the load regulation is

(a) 98%, (b) 2%, (c) 19.8%, (d) 0.2%.

8. Which of the following is *not* an essential element of a d.c power supply?

(a) Filter, (b) Rectifier,
(c) Voltage regulator, (d) Voltage amplifier.

9. An ideal voltage regulator has a voltage regulation of

(a) 100, (b) 1, (c) 50, (d) 0.

10. A negative feedback is applied to an amplifier with the feedback voltage proportional to the output current. This feedback increases the

(a) Input impedance of the amplifier, (b) distortion of the amplifier,
(c) Output impedance of the amplifier, (d) gain of the amplifier. **(DRDO)**

11. For a practical feedback circuit to have sustained oscillation, the most appropriate value of the loop gain is

(a) 1, (b) −1, (c) −1.02, (d) 1.02. **(DRDO)**

12. Negative feedback in an amplifier

(a) reduces the gain, (b) increases frequency and phase distortions,
(b) reduces the bandwidth, (d) increases the noise. **(GATE)**

Problems

1. An amplifier has a voltage gain 1000. If the negative feedback is applied to the input circuit, the voltage gain reduces to $\frac{1}{10}$th of its value without feedback. Find the fraction of the output that is fed back to the input. **(Madurai Univ.)** [0.009]

2. An amplifier has a gain of 100 without feedback. This gain is reduced to 50 with negative feedback. Find the feedback fraction of the amplifier.

(Himachal Univ.) [0.01]

3. An amplifier has a gain of 300. When the feedback is applied, it reduces to 200. Calculate the feedback factor. **(Madurai Univ.)** [1/600]

4. An amplifier with negative feedback has a gain of 100. It is found that without feedback an input signal of 50 mV is required to produce a given output, whereas with feedback the input signal must be 600 mV for the same output . Find the value of A (or A_V) and β. **(Osmania Univ.)** [1200, 11/12%]

5. With negative feedback, an amplifier gives an output of 10.5 V with an input of 1.12 V. When the feedback is removed, it requires 0.15 V input for the same output. Find the feedback ratio β and express the amount of feedback in decibel (dB). **(Punjabi Univ.)** $[-0.0924, -17.462]$

6. An amplifier has a gain with feedback of -30. The gain of the amplifier without feedback is -50. Calculate the feedback factor. **(Guru Nanak Univ.)** $[-0.0133]$

7. What is closed loop gain for an amplifier with open circuit gain 100, if 10% of the output voltage is fed back in such a manner that it is 180° out of phase with input? **(Punjab Univ.)** [9.1]

8. Calculate the gain with negative feedback of an amplifier with initial gain, $A = 100$ and feedback factor, $\beta = 0.1$ **(Punjab Univ.)** [9.1]

9. The total harmonic distortion in an amplifier is reduced from 8% to 2% with 5% negative feedback. What was the initial voltage gain? What will be the gain with feedback? **(Nagpur Univ.)** [60, 15]

10. The voltage gain of a transistor amplifier is 50. Its input and output resistances are 1 kΩ and 40 kΩ respectively. If the amplifier is provided with 10% negative voltage feedback in series with the input, calculate the voltage gain and the input and the output resistances. **(Kurukshetra Univ.)** [8.33, 6 kΩ, 6.67 kΩ]

11. The gain of an amplifier without feedback is -500. It is used in a negative feedback loop with feedback ratio 0.01. If the gain without feedback changes by 20%, what would be the corresponding change in gain with feedback? Also, determine the gain with feedback. [3.3%, −83.3]

12. An amplifier has a voltage gain of 200 without feedback. If 1/600th of the output voltage is applied to the input, calculate the voltage gain if the feedback is negative. **(Meerut Univ.)** [150]

13. The voltage gain of an amplifier is 100, and can have a variation of 20%. How can the variation of gain be reduced to 5%? **(Madurai Univ.)** [by 3% feedback]

[**Hints.** Variation in voltage gain implies harmonic distortion

$$\frac{S'}{S} = \frac{5}{20} = \frac{1}{1 + A\beta} = \frac{1}{1 + 100\beta}$$

$$\Rightarrow \quad 1 + 100\beta = 4 \quad \therefore \ \beta = 3/100 = 3\%]$$

14. With negative voltage feedback, an amplifier gives an output of 10 V with an input of 0.5 V. When feedback is removed, it required 0.25 V input for the same output. Calculate the gain without feedback and the feedback fraction β.

(Punjab Univ.) [40, 0.025]

[**Hints.** Voltage gain with feedback, $A_f = 10/0.5 = 20$

Voltage gain without feedback, $A = 10/0.25 = \mathbf{40}$

Since $A_f = A/(1 + A\beta)$, $1 + A\beta = A/A_f = 40/20 = 2$

$$\therefore\ A\beta = 2 - 1 = 1 \Rightarrow \beta = 1/A = 1/40 = \mathbf{0.025}]$$

15. The open-loop gain of an amplifier is -80 and it gives an output distortion voltage of 0.1 V. If the tolerable output distortion voltage with negtive feedback is 0.05,V, find the reverse transmission factor. [-0.0125]

16. An amplifier has a voltage gain of -100. The feedback ratio is -0.04. Find (i) the voltage gain with feedback, (ii) the amount of feedback in dB, (iii) the output voltage of the feedback amplifier for an input voltage of 40 mV and (iv) the feedback voltage. [-20, -13.98 dB, -0.8 V, -4.32 mV]

17. An amplifier with a mid-frequency gain of -50 has a feedback ratio -0.01. The lower and the upper half power frequencies without feedback are 200 Hz and 50 kHz respectively. Calculate these frequencies in the presence of feedback. What is the per cent change in the bandwith due to feedback.

[133.3 Hz, 75 kHz, 50.4%]

18. Consider the series regulator circuit of Fig. 10.12 with unregulated input $V_i =$ 15 V, Zener voltage $V_Z = 12$ V, $R = 100\,\Omega$ and $R_L = 1\,\text{k}\Omega$. Find the output voltage V_0 and the current through the Zener diode. Assume $h_{FE} = 100$.

[11.3 V, 29.9 mA]

19. A CE amplifier has a load resistance R_L and a source resistance R_g. What are the input and the outpout resistances of the amplifier? Hence determine these resistances of an emitter follower from the concept of feedback.

$$[h_{ie}, R_L, h_{ie} + h_{fe}R_L, \tfrac{(R_g + h_{ie})R_L}{R_g + h_{ie} + h_{fe}R_L}]$$

11

Oscillators

11.1 Introduction

An *oscillator* is a mechanical or electronic device in which a periodic fluctuation of a physical quantity occurs about an equilibrium value. An electronic device, consisting of active and passive circuit elements and producing a sinusoidal or other repeatitive waveforms of known frequency at the output without the application of any external input signal is called an *electronic oscillator* or simply ***oscillator***. Since an oscillator does not require any external input signal to produce the output signal, it is *a regenerative circuit of infinite gain.* It simply converts the d.c power from the supply source into the a.c power in the load. So, the function of an oscillator is just the *reverse* of a rectifier. Oscillators are widely used in electronics having applications in radio and television transmitters and receivers, quartz clocks, video games, electronic sound systems, computers etc.

In a computer, a special type of oscillator known as ***clock*** serves as a pacemaker for the microprocessor. Its frequency is usually specified in MHz and is an important factor in determining the rate at which a computer can perform the instructions or the commands it receives.

11.2 Classification of oscillators

Depending on the type of the output waveform generated, the oscillators are usually classified in *two* main categories. These are

(a) *the sinusoidal* or *harmonic oscillators*, and

(b) *the relaxation oscillators.*

An oscillator which generates a sinusoidal, or nearly so, waveform with a desired frequency is known as a ***sinusoidal*** or ***harmonic oscillator***. On the other hand, an oscillator which produces non-sinusoidal waveforms (e.g., square wave, sawtooth wave etc.) is termed a ***relaxation oscillator***.

According to the fundamental mechanism involved in producing the output waveform, the sinusoidal and the relaxation oscillators are further classified as

(i) *negative resistance oscillators*, and

(ii) *feedback oscillators*.

The ***negative resistance oscillators*** use an active device whose current-voltage characteristic has a *negative slope* within some range of its operation. A tunnel diode oscillator is an example.

When an oscillator contains a regenerative positive feedback amplifier whose loop gain is adjusted for an infinite gain, the oscillator is called a ***feedback oscillator***.

Depending upon the type of the associated circuits and the components, the feedback type sinusoidal oscillators can be further classified as

(i) *LC* (Inductor-capacitor) *oscillators*

(ii) *RC* (resistor-capacitor) *oscillators*, and

(iii) *Crystal oscillators*.

● The oscillators are also grouped in accordance with the frequency of the generated signals by them and is shown in the following table.

Table 11.1

Type of oscillator	Range of frequency
AF (audio frequency)	20Hz – 20 kHz
RF (radio frequency)	20 kHz – 30 MHz
VHF (very high frequency)	30 MHz – 300 MHz
UHF (ultra high frequency)	300 MHz – 3 GHz
Microwave oscillator	3 GHz – several GHz

The *tuned collector, Hartley* and *Colpitt* oscillators are examples of RF oscillators. They are also LC feedback oscillators. *Phase-shift* and *Wien-bridge* oscillators belong to AF oscillators. Both of them are RC feedback oscillators.

11.3 Oscillatory circuit (Tank circuit)

The frequency of an oscillator is determined by the amplifier circuit parameters. But it is desired that the frequency of the oscillators should be predictable as well as adjustable. A *frequency determining circuit* is thus essential in all oscillators. The LC and RC networks perform this particular function.

A circuit which produces electrical oscillations of any desired frequency is called an ***oscillatory circuit*** or ***tank circuit***. Since an oscillatory circuit determines and stores the frequency of oscillation, it is aptly named *tank circuit.*

A simple oscillatory circuit consisting of a capacitor C and an inductor L in parallel combination connected to a battery is shown in Fig. 11.1. Here the frequency of the electrical oscillation is determined by the values of L and C.

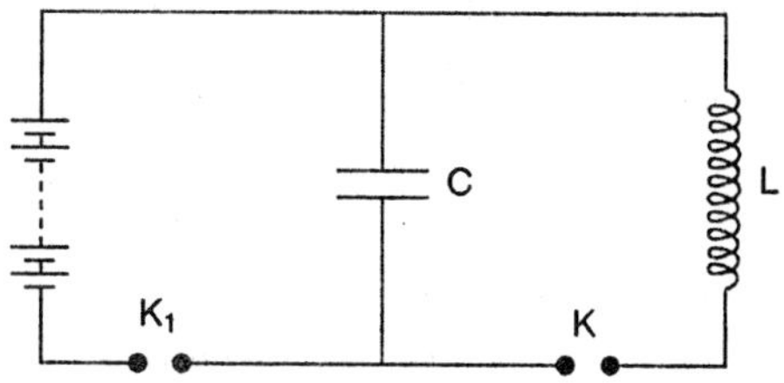

Fig. 11.1 Simple oscillatory circuit

Let the capacitor be initially charged from the battery by connecting K_1. The polarities are indicated in Fig. 11.2(a). Further operations step-wise are as follows.

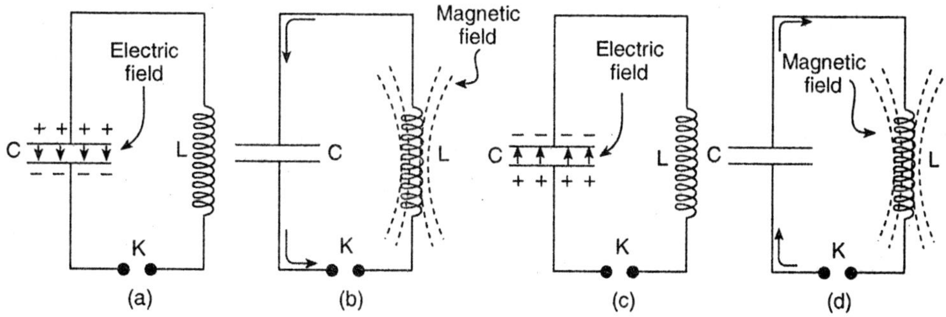

Fig. 11.2 Step-wise operations of an LC oscillatory circuit

Step 1. Fig. 11.2(a) shows that the upper plate is deficient in electrons and the lower plate has an excess of electrons. This gives rise to an electric field across the capacitor and it stores the electrostatic energy given by

$$U_E = \frac{1}{2}\frac{Q_0^2}{C} \tag{11.3.1}$$

where C is the capacitance and Q_0 the maximum charge on the capacitor.

2. When the switch K is closed and K_1 is switched off, the capacitor discharges through the inductor L, the direction of electron flow being shown by the arrowhead. This gives rise to a current flow which sets up a magnetic field around L, and the current grows up slowly to a maximum value I_0 when the capacitor is fully discharged. At this instant, the electrostatic energy becomes zero and the magnetic field energy becomes maximum given by

$$U_M = \tfrac{1}{2}LI_0^2 \tag{11.3.2}$$

where L is the self-inductance of the coil.

3. As the capacitor is fully discharged, the magnetic field begins to collapse, producing a counter emf which, in accordance with Lenz's law, makes the current flow in the same direction. As a result, the capacitor is now charged with opposite polarity as shown in Fig. 11.2(c). Finally, the magnetic field is fully collapsed. The magnetic field energy is thus zero and the electrostatic energy is again a maximum at this particular stage.

4. After the collapsing of the magnetic field, the capacitor begins to discharge and current flows in the opposite direction and a magnetic field is again set up around the inductor as shown in Fig. 11.2(d).

Thus the sequence of charging and discharging results in an alternating motion of electrons or an alternating (oscillating) current. The interchange of energy between L and C is repeated over and over again. This gives rise to an electrical oscillation, the frequency of which we shall now determine.

Frequency—Let Q be the charge on the capacitor and I the current through the inductor at any time t during oscillations. Then, the total energy is

$$U = \frac{1}{2}\frac{Q^2}{C} + \frac{1}{2}LI^2 \tag{11.3.3}$$

Assuming that the circuit has zero resistance so that there is no loss of energy due to Joule heating or any other cause, we may write

$$U = \frac{1}{2}\frac{Q^2}{C} + \frac{1}{2}LI^2 = \text{constant}$$

$$\therefore \quad \frac{dU}{dt} = \frac{Q}{C}\frac{dQ}{dt} + LI\frac{dI}{dt} = 0$$

$$\text{or,} \quad \frac{Q}{C} + L\frac{d^2Q}{dt^2} = 0, \text{ since } \frac{dQ}{dt} = I$$

$$\text{or,} \quad \frac{d^2Q}{dt^2} + \frac{1}{LC}Q = 0 \tag{11.3.4}$$

The solution of the second order differential equation (11.3.4) is given by

$$Q = Q_0 \cos \omega t \tag{11.3.5}$$

where ω is given by $\omega = 1/\sqrt{LC}$.

$$\therefore \quad \frac{dQ}{dt} = -Q_0\omega \sin \omega t$$

$$\text{or,} \quad \boxed{I = I_0 \sin \omega t} \tag{11.3.6}$$

where we have, $I_0 = -Q_0\omega$.

The circuit is thus ***oscillatory*** and the ***frequency*** *of oscillation* is given by

$$f = \frac{\omega}{2\pi} = \frac{1}{2\pi\sqrt{LC}} \tag{11.3.7}$$

If there are no losses of energy during the process of interchange of energy between L and C, the oscillation in the tank circuit would continue indefinitely. However, in practice, there is radiation loss in addition to the resistive and other losses in the coil and the dielectric losses in the capacitor. During each cycle, a part of the original energy is thus used up in overcoming these losses. So, the amplitude of the oscillating current goes on decreasing and finally goes down to zero when all the energy is completely used up. Thus, the oscillations produced in such a circuit are ***damped oscillations*** as has been illustrated in Fig. 11.3.

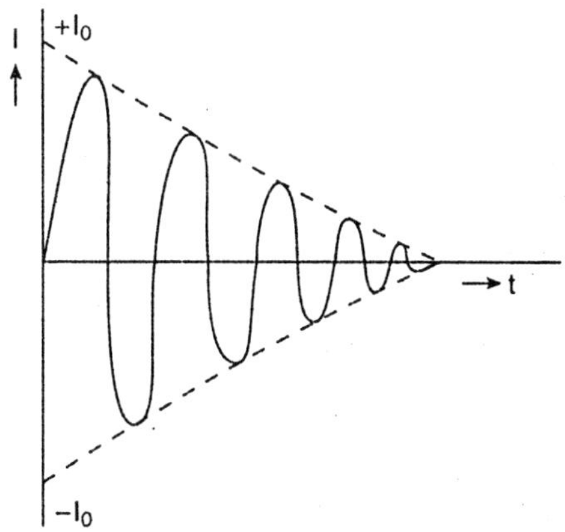

Fig. 11.3 Damped oscillations

We have just seen that the oscillations produced in a tank circuit are damped oscillations. But, in practice, we need continuous undamped oscillations to operate electronic equipments. So, to maintain the continuous undamped oscillations, the appropriate amount of energy must be delivered to the tank circuit at proper time intervals and in correct phase to meet the losses. In order to ensure this, the following conditions must be fulfilled.

(i) The amount of energy supplied to the tank circuit should meet the losses in the circuit and the a.c energy loss due the load or electromagnetic radiation.

(ii) The supplied energy must have the *same frequency* as that of the oscillations in the tank circuit.

(iii) The supplied energy must also be *in phase* with the oscillations set up in the tank circuit so that it might assist the oscillations.

11.4 Feedback oscillator concept and Barkhausen criterion

From the discussions of the previous section, it is clear that an oscillator is basically an amplifier provided with a feedback circuit. It amplifies the oscillations produced by the tank circuit and renders them undamped by feeding the proper amount of energy in correct phase to the tank circuit through the feedback network. So, an oscillator has the following *essential* components.

1. *Tank circuit*: This circuit consists of an inductor L in parallel with a capacitor C. It produces oscillations of frequency, $f = 1/2\pi\sqrt{LC}$.

2. *Transistor amplifier*: This part amplifies the oscillations produced by the tank circuit by drawing power from a battery. This amplified output is supplied to the tank circuit through the feedback network to meet the losses.

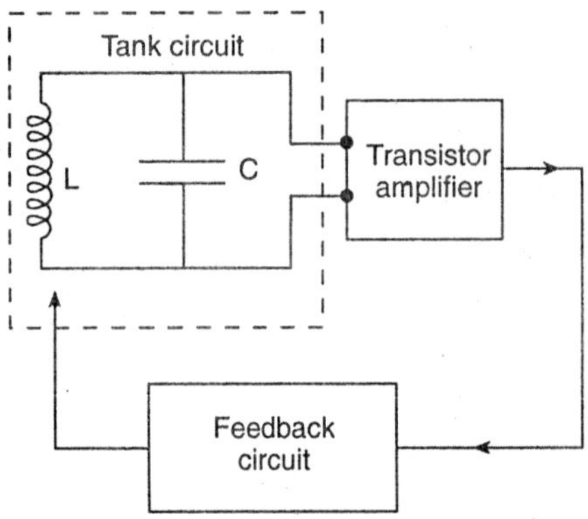

Fig. 11.4 Oscillator components

3. ***Feedback circuit***: To make up the losses in the tank circuit, a part of the output energy is required to be transferred to the input, in appropriate amount at proper intervals and in correct phase. For this purpose *a positive feedback network is necessary in an oscillator.* This positive feedback, being in phase with the input signal, increases the gain of the amplifier and this voltage gain A_f of a positive feedback amplifier is given by

$$\boxed{A_f = \frac{A}{1 - A\beta}} \tag{11.4.1}$$

where A is the voltage gain without the feedback and β the feedback fraction. Thus $A_f > A$ for a positive feedback i.e., the gain of the amplifier is increased.

Now, if the loop gain $A\beta = 1$, equation (11.4.1) gives $A_f = \infty$ i.e., the amplifier then gives an output voltage without requiring any externally applied input voltage. This means that the *amplifier becomes an oscillator.*

To understand the situation, let us assume that a transient disturbance (it may be the noise voltage) gives an output voltage V_0 when no external input voltage is applied. If a part βV_0 of the output signal is fed back to the input, it appears at the output as an enhanced signal $A\beta V_0$. If this signal equals V_0, we get

$$A\beta V_0 = V_0$$

$$\text{or,}\ A\beta = 1 \tag{11.4.2}$$

This implies that the spurious output voltage degenerates itself and the amplifier oscillates.

In complex notation, we can write the above condition as $A\beta = 1 + j0$ which indicates that $|A\beta| = 1$ and the phase of $A\beta = 2n\pi$ where $n = 0, 1, 2, 3, \cdots$.

This condition of ***unity loop gain*** i.e., $A\beta = 1$ is known as the ***Barkhausen criterion*** for sustained oscillation. The condition means that $|A\beta| = 1$ and the phase angle of $A\beta$ is zero or an integral multiple of 360°. So, the ***two conditions*** for a feedback amplifier to work as an oscillator are

(i) The *feedback must be positive* or *regenerative.*

(ii) The *loop gain must be unity.*

11.5 Tuned-collector oscillator

The circuit diagram of a tuned-collector oscillator using a *npn*-transistor in CE configuration is shown in Fig. 11.5.

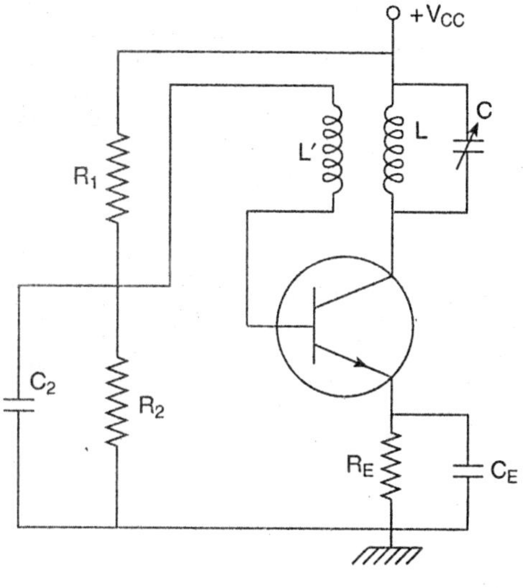

Fig. 11.5 Tuned-collector oscillator

Here, the necessary d.c bias of the transistor is determined by the resistances R_1, R_2, R_E and the d.c voltage source V_{CC}. The variable capacitor C, connected with the transformer primary inductance L constitutes the *tank circuit* i.e., the frequency determining circuit. The capacitors C_2 and C_E bypass R_2 and R_E respectively so that they have no effect on the a.c operation of the circuit. Moreover, C_2 provides a.c ground for the transformer secondary L'. Since the LC tuned circuit is connected to the collector, this justifies for the name 'tuned-collector oscillator'. The feedback between the output (collector) and the input (base) is provided by the secondary L' of the transformer which is mutually coupled to L. As the transistor is connected in CE-configuration, there is a phase shift of 180° between its input and the output circuits. The transformer introduces another 180° phase shift. Thus a total phase shift of 360° between the output and the input is produced, resulting in a positive feedback. Now, if the amplifier provides sufficient gain to make the loop gain $|A\beta| = 1$, then oscillations can be obtained.

Principle of operation: When the d.c supply voltage V_{CC} is switched on, a transient current is developed in the tank circuit which initializes natural oscillations. These natural oscillations induce a small emf in L' by mutual induction. This causes variations in the base current I_B which are amplified β times and appear in the collector circuit. A portion of this amplified energy is used to meet the losses in the tuned circuit and the rest is radiated in the form of electromagnetic waves. The frequency of oscillation is practically equal to the natural resonant frequency of LC tuned circuit and is given by

$$f = \frac{1}{2\pi\sqrt{LC}}$$

Analysis: To analyse the performance of a tuned-collector oscillator, we consider the approximate h-parameter equivalent circuit as shown in Fig. 11.6.

The output voltage V_0 developed across the tuned circuit is given by

$$V_0 = -(R + j\omega L)I_L$$

The feedback voltage is the induced voltage in the coil L' and is

$$V_f = j\omega M I_L$$

∴ Feedback fraction,

$$\beta = \frac{V_f}{V_0} = -\frac{j\omega M}{R + j\omega L} \tag{11.5.1}$$

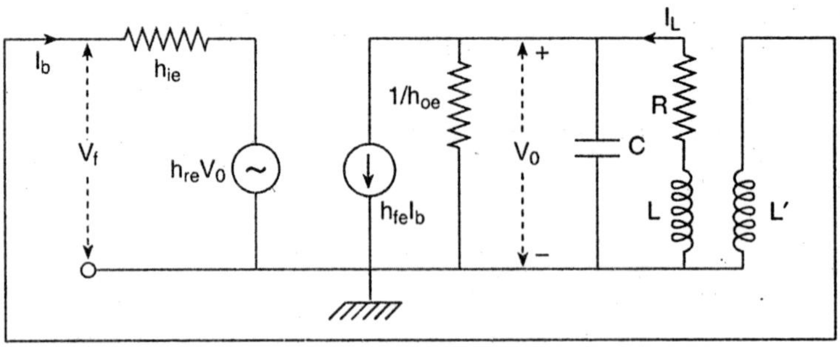

Fig. 11.6 *h*-parameter equivalent circuit of a tuned collector oscillator

The voltage gain of the amplifier without feedback is given by

$$A = -\frac{h_{fe}Z_L}{h_{ie} + Z_L\Delta h} \tag{11.5.2}$$

where $\Delta h = h_{ie}h_{oe} - h_{re}h_{fe}$ and Z_L is the *effective load* given by

$$\frac{1}{Z_L} = \frac{1}{R + j\omega L} + \frac{1}{1/j\omega C}$$

$$\text{or,}\quad Z_L = \frac{R + j\omega L}{1 - \omega^2 LC + j\omega RC} \tag{11.5.3}$$

To satisfy the Barkhausen criterion for oscillation, we must have $A\beta = 1$. Therefore, using (11.5.1) and (11.5.2), we get

$$\frac{h_{fe}Z_L}{h_{ie} + Z_L\Delta h} \times \frac{j\omega M}{R + j\omega L} = 1$$

$$\text{or,}\quad \frac{j\omega M h_{fe}}{R + j\omega L} = \frac{h_{ie}}{Z_L} + \Delta h$$

Substituting Z_L from (11.5.3) and simplyfying, we get

$$j\omega M h_{fe} = h_{ie}(1 - \omega^2 LC + j\omega RC) + \Delta h(R + j\omega L) \tag{11.5.4}$$

Equating the real parts from both sides of (11.5.4), we get

$$h_{ie}(1 - \omega^2 LC) + R\cdot \Delta h = 0$$

$$\text{or,}\quad \omega^2 = \frac{h_{ie} + R\cdot \Delta h}{h_{ie}\cdot LC}$$

$\therefore$ The *frequency of oscillation is* given by

$$f = \frac{\omega}{2\pi} = \frac{1}{2\pi\sqrt{LC}}\sqrt{1 + \frac{R\Delta h}{h_{ie}}} \tag{11.5.5}$$

Since $R\cdot \Delta h \ll h_{ie}$ for a practical circuit, the equation (11.5.5) reduces to

$$f \approx \frac{1}{2\pi\sqrt{LC}} \tag{11.5.6}$$

Equating the imaginary parts from both sides of (11.5.4), we get

$$\omega M h_{fe} = \omega RC h_{ie} + \omega \Delta h L$$

$$\therefore \quad \boxed{M = \frac{RCh_{ie} + L\Delta h}{h_{fe}}}$$

This gives the ***condition for sustained oscillations of the oscillator.***

11.6 Hartley oscillator

The circuit diagram of a Hartley oscillator using a *npn*-transistor in CE mode is illustrated beside in Fig. 11.7.

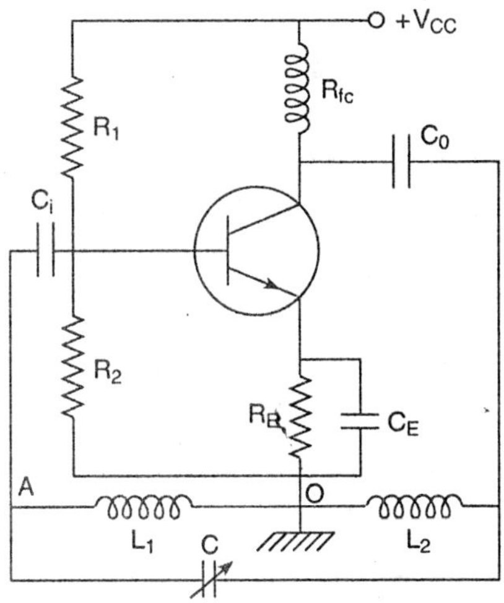

Fig. 11.7 Hartley oscillator circuit

The stabilized d.c-bias of the transistor is provided by the resistors R_1, R_2, R_E and the d.c supply voltage by V_{CC}. The capacitors C_i and C_o are the coupling capacitors and C_E is the bypass capacitor. Instead of usual R_C, a radio frequency choke (RFC) is connected in the collector circuit to present a high impedance at the frequency of oscillation and to prevent the radio frequency current from reaching the d.c supply. The R_{fc} also prevents the d.c supply from short-circuiting the a.c output developed across L_2. Here, the tank circuit consists of a parallel resonant circuit made of a variable capacitor C and a trapped inductor coil AB of inductance $L = L_1 + L_2$. The junction O of L_1 and L_2 is grounded, as shown.

Principle of operation : When the d.c power supply is switched on, the transient current generated in the tank circuit initializes oscillations. The voltage developed across L_2 is the output and that across L_1 is the feedback voltage V_f which is transformed to the input of CE amplifier through the capacitor C_i for amplification and to meet the losses in the tank circuit. Since the point O is earthed, the phase difference between the signals at A and B is 180°. As the transistor is connected in CE mode, it introduces an additional 180° phase shift. Thus the phase shift around the closed loop is 360° which satisfies the condition of phase shift for oscillation. Now, if the amplifier provides the sufficient condition for loop gain $|A\beta| = 1$, then sustained oscillations can take place.

In Hartley oscillator, the frequency of the oscillation is given by

$$f = \frac{1}{2\pi\sqrt{(L_1 + L_2)C}} \tag{11.6.1}$$

Analysis: To analyse the performance of a Hartley oscillator, we consider the approximate h-parameter equivalent circuit as shown in Fig. 11.8.

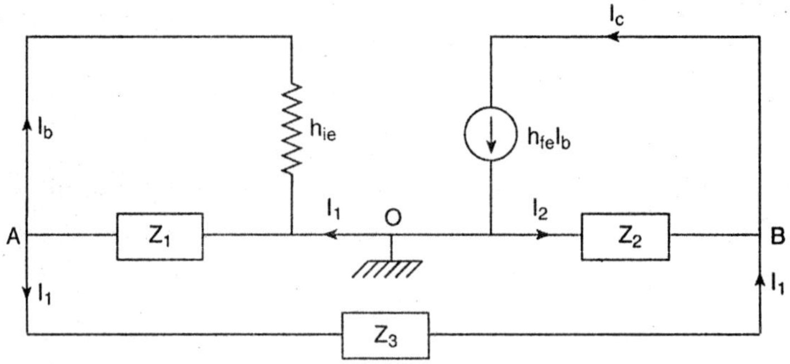

Fig. 11.8 *h*-parameter equivalent circuit of Hartley oscillator

Let Z_1, Z_2 and Z_3 represent the inductive or capacitive reactances of the oscillator such that $Z_1 = jX_1 = j\omega(L_1 + M)$, $Z_2 = jX_2 = j\omega(L_2 + M)$ and $Z_3 = jX_3 = -j/\omega C$, ω being the angular frequency of oscillation. Since h_{re} of a transistor is negligible, the feedback source $h_{re}V_0$ has been omitted. The output resistance $1/h_{oe}$ is very large and is in parallel to Z_2. So, it has also been neglected.

Let Z' be the equivalent impedance of the parallel combination of h_{ie}, the input resistance of the transistor and the impedance Z_1. Therefore,

$$\frac{1}{Z'} = \frac{1}{h_{ie}} + \frac{1}{Z_1}$$

$$\text{or } Z' = \frac{Z_1 h_{ie}}{Z_1 + h_{ie}} \tag{11.6.2}$$

Now Z' is in series with Z_3 and this series combination is in parallel with Z_2. Therefore, the load impedance Z_L between the output terminals B and O is given by

$$\frac{1}{Z_L} = \frac{1}{Z_2} + \frac{1}{Z' + Z_3}$$

$$\text{or, } Z_L = \frac{Z_2(Z' + Z_3)}{Z_2 + Z' + Z_3}$$

Substituting for Z' from (11.6.2) and simplifying, we obtain

$$Z_L = \frac{Z_2[Z_1 Z_3 + h_{ie}(Z_1 + Z_3)]}{h_{ie}(Z_1 + Z_2 + Z_3) + Z_1 Z_2 + Z_1 Z_3} \tag{11.6.3}$$

Now, the output voltage between B and O is

$$V_0 = -I_1(Z' + Z_3)$$

and the feedback voltage between O and A is given by

$$V_{fb} = -I_1 Z'$$

The feedback ratio is thus $\beta = \dfrac{V_{fb}}{V_0} = \dfrac{Z'}{Z_3 + Z'}$

Now from Barkhausen criterion for sustained oscillation, we must have $A\beta = 1$.

$$\therefore \left(-\frac{h_{fe}Z_L}{h_{ie}}\right) \times \left(\frac{Z'}{Z' + Z_3}\right) = 1,$$

$$(\because A = -h_{fe}Z_L/h_{ie} = \text{voltage gain without feedback})$$

Using (11.6.2) and (11.6.3) in the above relation and simplifying, we get

$$h_{ie}(Z_1 + Z_2 + Z_3) + Z_1Z_2(1 + h_{fe}) + Z_1Z_3 = 0 \qquad (11.6.4)$$

This is the ***general equation*** of Hartley oscillator.

Now, putting $Z_1 = j\omega L_1 + j\omega M$, $Z_2 = j\omega L_2 + j\omega M$ and $Z_3 = 1/j\omega C = -j/\omega C$ in equation (11.6.4), we get

$$j\omega h_{ie}\left[L_1 + L_2 + 2M - \frac{1}{\omega^2 C}\right] - \omega^2(L_1 + M)(L_2 + M)(1 + h_{fe}) + \frac{L_1 + M}{C} = 0 \quad (11.6.5)$$

Equating the imaginary parts from both sides of (11.6.5), we have

$$L_1 + L_2 + 2M - 1/\omega^2 C = 0 \qquad (11.6.6)$$

The ***frequency*** *of the oscillation* is thus given by

$$f = \frac{\omega}{2\pi} = \frac{1}{2\pi\sqrt{(L_1 + L_2 + 2M)C}} \qquad (11.6.7)$$

Equating the real parts from both sides of (11.6.5) to zero, we obtain the ***condition for sustained oscillation*** that runs as

$$(L_2 + M)(1 + h_{fe}) - 1/\omega^2 C = 0$$

$$\text{or, } 1 + h_{fe} = \frac{1}{\omega^2 C(L_2 + M)} \qquad (11.6.8)$$

$$= \frac{L_1 + L_2 + 2M}{L_2 + M}, \quad \text{using (11.6.6)}$$

$$= 1 + \frac{L_1 + M}{L_2 + M}$$

$$\Rightarrow \boxed{h_{fe} = \frac{L_1 + M}{L_2 + M}} \qquad (11.6.9)$$

If we neglect the mutual inductive effect M between the coils L_1 and L_2, then both the frequency of oscillation and the condition for sustained oscillations get modified to the expressions as are given below:

$$f = \frac{1}{2\pi\sqrt{(L_1 + L_2)C}} \qquad (11.6.10)$$

$$\text{and} \qquad h_{fe} = \frac{L_1}{L_2} \qquad (11.6.11)$$

11.7 Colpitt oscillator

The circuit diagram of a Colpitt oscillator using a *npn*-transistor in CE configuration is shown in Fig. 11.9. The resistors R_1, R_2, R_E and d.c supply V_{CC} provide the necessary self-bias. C_i and C_o are the coupling capacitors and C_E the bypass capacitor. The radio frequency choke R_{fc} does not allow the RF current to reach the d.c supply and also prevents the d.c supply from short-circuiting the a.c output across L. Here the tank circuit consists of an inductor L and two capacitors C_1 and C_2. The junction O of C_1 and C_2 is grounded.

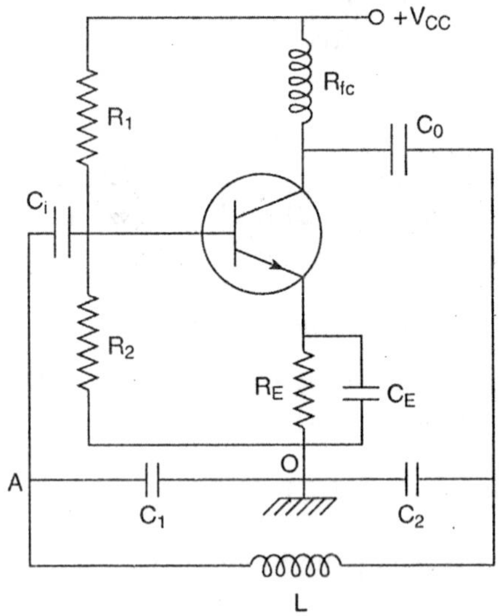

Fig. 11.9 Colpitt oscillator

Working principle : When the d.c power supply is switched on, the capacitors C_1 and C_2 get charged. Thereafter, they discharge through the coil L producing a damped harmonic oscillation in the tank circuit. The voltage across C_1 is fed back to the input of the CE amplifier. Since the point O is earthed, the oscillatory voltage at A will have 180° phase shift with respect to A at resonance. Again, the CE amplifier introduces an additional 180° phase shift. As a result, a total phase shift of 360° is produced. Thus the Barkhausen phase shift requirement is fulfilled. Now, if the amplifier provides a unity loop gain i.e., $|A\beta| = 1$, *sustained oscillations* will take place.

The frequency of oscillation of Colpitt oscillator is given by

$$f = \frac{1}{2\pi\sqrt{LC}} \tag{11.7.1}$$

$$\text{where} \quad C = \frac{C_1 C_2}{C_1 + C_2}$$

Analysis : To analyse the performance of a Colpitt oscillator we may again consider the h-parameter equivalent circuit shown in Fig. 11.8 and arrive at the general equation of oscillation, i.e. equation (11.6.4), given by

$$h_{ie}(Z_1 + Z_2 + Z_3) + Z_1 Z_2 (1 + h_{fe}) + Z_1 Z_3 = 0$$

But, here, $\quad Z_1 = -j/\omega C_1, \quad Z_2 = -j/\omega C_2$ and $Z_3 = j\omega L$.

Substitution of these values in the general equation readily gives

$$-jh_{ie}\left(\frac{1}{\omega C_1} + \frac{1}{\omega C_2} - \omega L\right) - \frac{1 + h_{fe}}{\omega^2 C_1 C_2} + \frac{L}{C_1} = 0 \tag{11.7.2}$$

Equating the imaginary parts from both sides of (11.7.2), we obtain,

$$\frac{1}{\omega C_1} + \frac{1}{\omega C_2} - \omega L = 0$$

$$\text{or,} \quad \omega^2 = \frac{1}{L}\left(\frac{1}{C_1} + \frac{1}{C_2}\right) \tag{11.7.3}$$

∴ The ***frequency*** *of oscillation* is given by

$$f = \frac{\omega}{2\pi} = \frac{1}{2\pi}\sqrt{\frac{C_1 + C_2}{LC_1C_2}} \tag{11.7.4}$$

$$= \frac{1}{2\pi\sqrt{LC}} \tag{11.7.5}$$

where $C = C_1C_2/(C_1 + C_2)$.

Now, equating the real parts from both sides of the equation (11.7.2) to zero, the ***condition for sustained oscillation*** is obtained as

$$\frac{1 + h_{fe}}{\omega^2 C_1 C_2} = \frac{L}{C_1}$$

$$\text{or,} \quad \omega^2 LC_2 = 1 + h_{fe}$$

$$\text{or,} \quad \left(\frac{1}{C_1} + \frac{1}{C_2}\right) C_2 = 1 + h_{fe}, \quad \text{using (11.7.3)}$$

$$\text{or,} \quad \boxed{h_{fe} = \frac{C_2}{C_1}}$$

11.8 Phase-shift oscillator

11.8.1 Principle and circuit diagram

Let a sinusoidal voltage $v_i = v_0 \sin \omega t$ of frequency $f = \omega/2\pi$ be applied in series to an RC-network as shown in Fig. 11.10. The a.c output voltage v_R across R leads the input voltage by an angle ϕ given by

$$\tan\phi = \frac{1}{\omega CR} = \frac{1}{2\pi fCR} \tag{11.8.1}$$

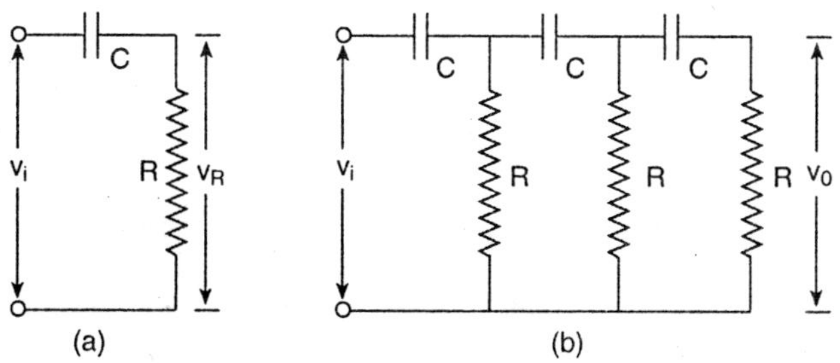

Fig. 11.10 (a) Simple RC network (b) RC ladder network

By adjusting the values of R and C, one can make ϕ equal to 60° for a given frequency. In a phase shift oscillator, a total phase shift of 180° is produced between the input and the output voltages by using a ladder network comprising of three identical RC-sections, each producing a phase shift of 60°.

Circuit diagram: The circuit diagram of a phase shift oscillator using a *npn*-transistor in CE configuration is shown in Fig. 11.11.

The resistors R_1, R_2, R_E and R_L and the d.c power supply V_{CC} determine the necessary bias conditions. C_E is the emitter bypass capacitor. The ladder network

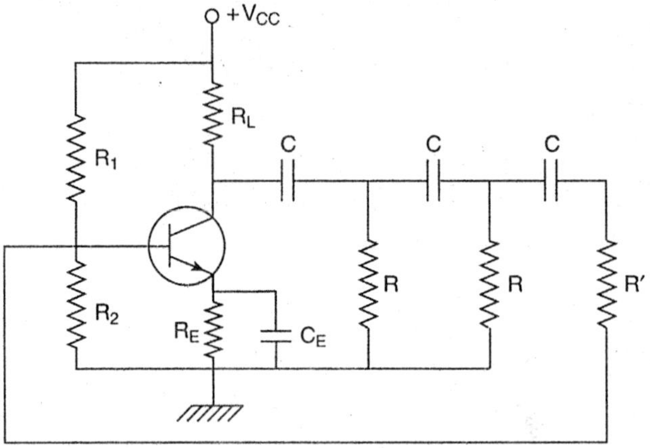

Fig. 11.11 Phase-shift oscillator circuit

consisting of three identical RC-sections is used for feedback of energy from the output (collector) to the input (base) of the amplifier. In the last section, the resistance R' is so chosen that $R' + h_{ie} = R$ where h_{ie} is the input resistance of the transistor.

11.8.2 Principle of operation and analysis

When the d.c power supply V_{CC} is switched on, oscillations set in the circuit by any random variation caused in the base current or by the noise inherent in the transistor. This variation is amplified in the collector circuit and fed back to the RC network. The ladder network produces a phase shift of 180° and the CE amplifier produces additional 180° phase shift. So, the total phase shift around the loop is 360°. When the transistor provides sufficient condition for unity loop again i.e. $|A\beta| = 1$, then the circuit acts as an oscillator and sustained oscillations take place.

The *frequency of oscillation* is given by

$$f = \frac{1}{2\pi C\sqrt{4RR_L + 6R^2}} \tag{11.8.2}$$

Analysis: To analyse the performance of a phase-shift oscillator, we consider the a.c equivalent circuit of phase shift oscillator as shown in Fig. 11.12(a). Since $1/h_{oe}$ is much greater than R_L and the value of h_{re} is very small ($\ll 1$), the circuit of

Fig. 11.12(a) can be modified by neglecting $h_{re}V_2$ and replacing the current source $h_{fe}I_3$ by its equivalent Thevenin voltage source $h_{fe}I_3R_L$ as shown in Fig. 11.12(b).

Applying KVL to the *three* meshes (1), (2) and (3) respectively (Fig. 11.12(b)), we obtain

$$(R + R_L - jX_C)I_1 - RI_2 + h_{fe}R_LI_3 = 0 \quad (11.8.3)$$

$$-RI_1 + (2R - jX_C)I_2 - RI_3 = 0 \quad (11.8.4)$$

$$\text{and} \quad -RI_2 + (2R - jX_C)I_3 = 0 \quad (11.8.5)$$

where $X_C = 1/\omega C$ = reactance of the capacitor, and I_1, I_2, and I_3 are the currents in mesh (1), (2) and (3) respectively (Fig. 11.12(b)).

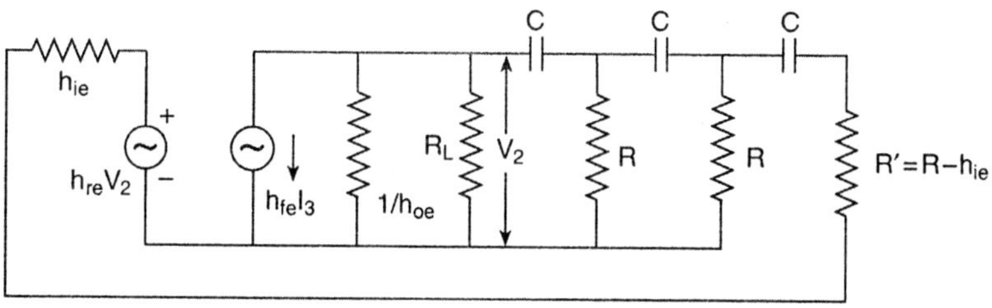

(a)

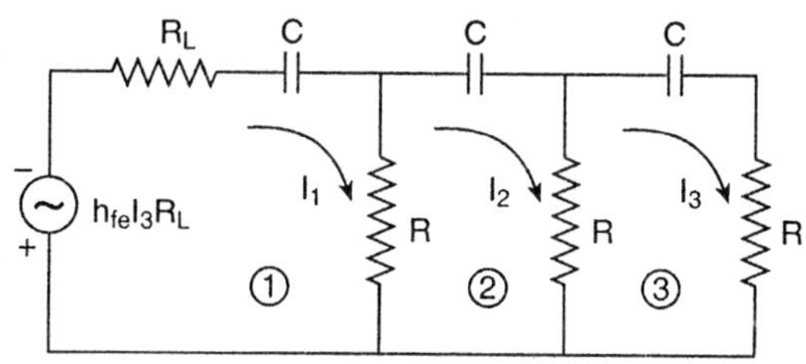

(b)

Fig. 11.12 (a) AC equivalent circuit of phase-shift oscillator (b) Modified circuit diagram

Since I_1, I_2 and I_3 are non-zero, the determinant of their coefficients must be zero.

$$\therefore \quad \begin{vmatrix} R + R_L - jX_C & -R & h_{fe}R_L \\ -R & 2R - jX_C & -R \\ 0 & -R & 2R - jX_C \end{vmatrix} = 0$$

$$\text{or,} \quad (R + R_L - jX_C)(3R^2 - 4RjX_C - X_C^2) - R^2(2R - jX_C) + h_{fe}R_LR^2 = 0 \quad (11.8.6)$$

Equating the imaginary parts from both sides of (11.8.6), we have

$$-4RX_C(R + R_L) - X_C(3R^2 - X_C^2) + R^2X_C = 0$$

$$\text{or,} \quad X_C^2 = 4RR_L + 6R^2 \quad (11.8.7)$$

$$\text{or,} \quad \frac{1}{\omega^2 C^2} = 4RR_L + 6R^2$$

$$\text{or,} \quad \omega^2 = \frac{1}{(4RR_L + 6R^2)C^2}$$

The ***frequency of oscillation*** is therefore given by

$$f = \frac{\omega}{2\pi} = \frac{1}{2\pi C\sqrt{4RR_L + 6R^2}} \tag{11.8.8}$$

The condition for sustained oscillation can be obtained by equating the real parts from both sides of (11.8.6) to zero, i.e., from the relation

$$(R + R_L)(3R^2 - X_C^2) - 4RX_C^2 - 2R^3 + h_{fe}R_L R^2 = 0$$

$$\text{or,} \quad 3R^2 R_L + R^3 - X_C^2(5R + R_L) + h_{fe}R_L R^2 = 0 \tag{11.8.9}$$

Substituting for X_C^2 from (11.8.7) in (11.8.9) and simplifying,

$$h_{fe} = 23 + 29\frac{R}{R_L} + 4\frac{R_L}{R} \tag{11.8.10}$$

Usually, $R_L = R$. So, from the equation (11.8.10), we have the ***condition for sustained oscillation*** simply as

$$h_{fe} = 56$$

11.9 Wien-bridge oscillator

The Wien-bridge oscillator is an RC oscillator which uses a nearly balanced bridge as the feedback network. Fig. 11.13 illustrates a Wien-bridge oscillator.

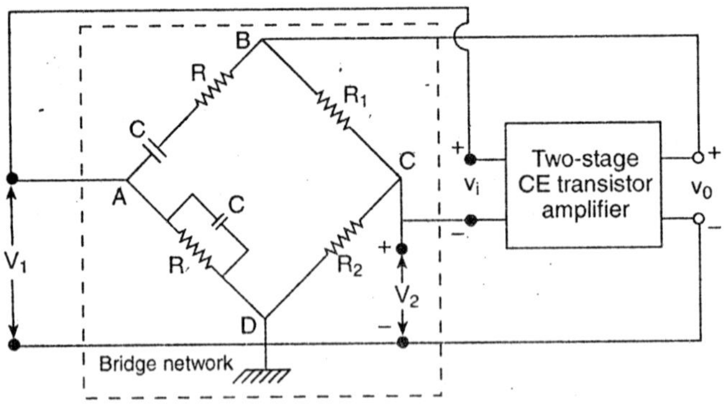

Fig. 11.13 Wien-bridge oscillator

The output voltage V_0 of the two stage CE transistor amplifier is the input voltage of the bridge. The output voltage V_i of the bridge is the input voltage for the amplifier. When the bridge is exactly balanced, V_i becomes zero and hence $\beta = 0$. But for sustained oscillation, β cannot be zero. So, the bridge is made slightly unbalanced by adjusting the ratio between the resistances R_1 and R_2. Since the bridge itself does not introduce any phase shift between its input and output voltages, the phase shift around the loop is zero. The two stage CE transistor amplifier serves the purpose of phase shift by introducing a phase shift of 180° at each stage.

Analysis : The total impedance of the arm AB of the bridge containing R and C in series is given by

$$Z_1 = R + \frac{1}{j\omega C} \tag{11.9.1}$$

The impedance Z_2 of the arm AD of the bridge containing R and C in parallel is given by the relation

$$\frac{1}{Z_2} = \frac{1}{R} + \frac{1}{1/j\omega C}$$

$$\text{or,}\ Z_2 = \frac{R/j\omega C}{R + 1/j\omega C} \tag{11.9.2}$$

When the bridge is in a *balanced condition*, we get

$$\frac{R_1}{R_2} = \frac{Z_1}{Z_2} = \frac{(R + 1/j\omega C)^2}{R/j\omega C} = 2 + j\frac{R^2 - 1/\omega^2 C^2}{R/\omega C} \tag{11.9.3}$$

Equating the real and imaginary parts from both sides of (11.9.3), we obtain

$$\frac{R_1}{R_2} = 2 \tag{11.9.4}$$

$$\text{and} \qquad \frac{R^2 - 1/\omega^2 C^2}{R/\omega C} = 0$$

$$\Rightarrow \quad R^2 = \frac{1}{\omega^2 C^2} \quad \Rightarrow \quad R = \frac{1}{\omega C} \tag{11.9.5}$$

$$\therefore \quad \omega = \frac{1}{RC}$$

Therefore, the ***frequency*** *of oscillation* is given by

$$f = \frac{\omega}{2\pi} = \frac{1}{2\pi RC} \tag{11.9.6}$$

When the bridge is *balanced*, the voltage across the arms AD and CD must be equal and we shall have

$$V_1 = V_2 = \frac{V_0}{R_1 + R_2} \cdot R_2 = \frac{1}{3} V_0, \quad (\because\ R_1 = 2R_2) \tag{11.9.7}$$

$$\therefore \text{ Voltage gain,} \qquad A = \frac{V_0}{V_1} = \frac{V_0}{V_2} = \frac{R_1 + R_2}{R_2} = 3 \text{ at } \textit{balance.}$$

In general, $$\boxed{A = 1 + \frac{R_1}{R_2}} \tag{11.9.8}$$

At *balance*, $V_i = V_1 - V_2 = 0$. But the output voltage of the bridge i.e., V_i must not be zero for oscillations to occur. So, the ratio $R_2/(R_1 + R_2)$ is taken to be less than 1/3. Let us assume that

$$\frac{R_2}{R_1 + R_2} = \frac{1}{3} - \frac{1}{\delta}, \quad \text{where } \delta > 3. \tag{11.9.9}$$

The bridge is then slightly *unbalanced* and gives a feedback voltage V_i to the input of the amplifier.

The bridge is led to unbalanced condition by adjusting the ratio R_1/R_2 such that V_1/V_0 remains 1/3 but V_2/V_0 becomes smaller than 1/3

$$\therefore \quad \frac{V_2}{V_0} = \frac{R_2}{R_1 + R_2} = \frac{1}{3} - \frac{1}{\delta} \tag{11.9.10}$$

Feedback voltage, $$V_i = V_1 - V_2 = \left(\frac{1}{3} - \frac{1}{3} + \frac{1}{\delta}\right) V_0 = \frac{V_0}{\delta} \tag{11.9.11}$$

Clearly, V_i and V_0 are *in phase*. Now, the feedback ratio is given by

$$\beta = \frac{V_i}{V_0} = \frac{1}{\delta} \tag{11.9.12}$$

Obviously, the condition $A\beta = 1$ is satisfied by making the voltage gain of the amplifier greater than 3 i.e.,

$$A = \frac{V_0}{V_i} = \delta > 3 \tag{11.9.13}$$

So, *the amplifier used in Wien bridge oscillator must have a gain greater than* 3.

The frequency of oscillation can be changed continuously by varying simultaneously the capacitances of the two capacitors. Using gauged variable air capacitors, the frequency can be changed in the ratio 10 : 1. For wider change in frequency, different values of the two identical resistors R are used.

Advantages of Wien-bridge oscillator: The Wien-bridge oscillator has the following advantages:

(i) *Good frequency stability*: The frequency of oscillation in Wien-bridge oscillator is primarily determined by passive elements like R and C and the values of these elements can be kept fixed against temperature. So frequency remains markedly constant over a fairly long time interval. The frequency stability is thus good.

(ii) *High overall gain*: Since a two-stage CE transistor amplifier is used in Wien-bridge oscillator, the overall gain is high.

(iii) *Good sine-wave output*: A very good sine wave is obtained at the output of the Wien-bridge oscillator.

(iv) *Easy frequency variation*: The frequency of oscillation can be varied easily by changing the capacitances of the two capacitors.

(v) *Good amplitude stabilization*: If the resistance R_1 is replaced by a *thermistor* which has negative temperature coefficient of resistance and R_2 is replaced by a *sensistor* which has a positive temperature coefficient of resistance, the amplitude of oscillation is stabilized against variations caused by ageing of the transistor and other circuit components.

Disadvantages: The advantages are the following:

(i) The circuit cannot produce very high frequency.

(ii) A large number of components are required for the construction of the two-stage amplifier.

11.10 Crystal oscillator

Principle: When a suitably cut non-centro-symmetric crystal like quartz is subjected to a mechanical stress, an electric potential difference appears across its faces. The polarity of the potential reverses when the stress changes from compression to tension.

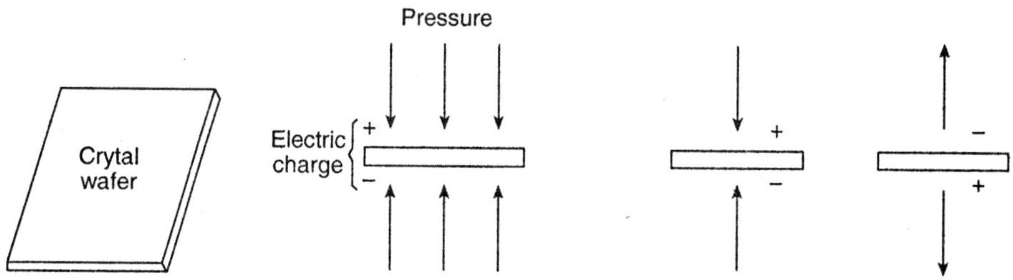

Fig. 11.14 Piezoelectric effect

The converse of the phenomenon is also true. When an appropriate alternating voltage is applied to a wafer of quartz, it undergoes mechanical vibrations at the frequency of the applied voltage. This property interconnecting the mechanical and the electrical behaviour of a crystal is known as ***piezo-electric effect*** (Fig. 11.14).

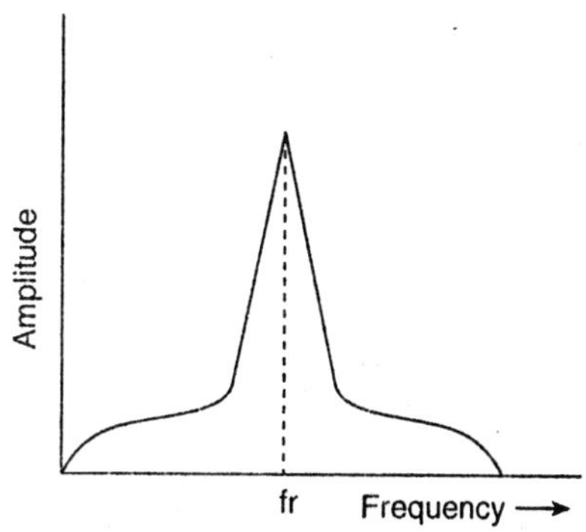

Fig. 11.15 Frequency response of a crystal

The amplitude of the mechanical oscillations attains a maximum value when the frequency of the applied a.c voltage is equal to the natural frequency of the crystal, called the *resonant frequency* as indicated in Fig. 11.15.

The symbol of a piezoelectric crystal and the electrical equivalent circuit of a vibrating piezoelectric crystal are shown in Fig. 11.16(a) and 11.16(b) respectively.

The equivalent circuit represents a series resonant LCR circuit shunted by a capacitor C'. The inductor L, the capacitor C and the resistor R are the analog of the mass, the compliance and the viscous damping factor of the mechanical system. The capacitance C' is the electrostatic capacitance between the mounting electrodes with the crystal as the dielectric.

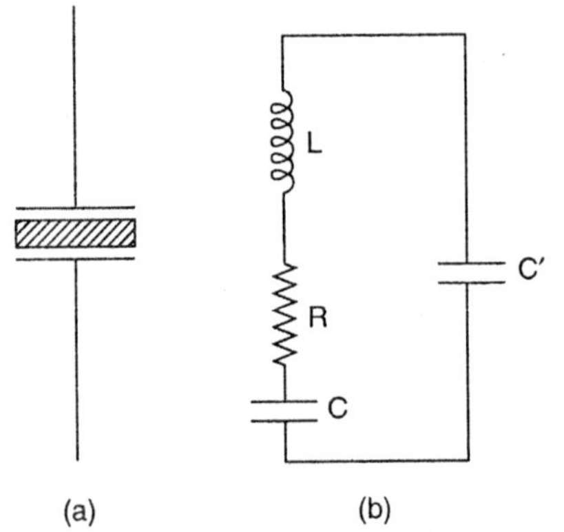

Fig. 11.16 (a) Symbol of piezoelectric crystal (b) Equivalent circuit

The equivalent circuit of Fig. 11.16(b) is characterized by two resonant frequencies. At series resonant frequency, the reactance of the series arm is zero i.e.,

$$\omega_s L - \frac{1}{\omega_s C} = 0$$

$$\text{or,} \quad f_s = \frac{\omega_s}{2\pi} = \frac{1}{2\pi\sqrt{LC}} \tag{11.10.1}$$

At this frequency, the impedance of the series arm becomes a pure resistance having a low R-value.

When $\omega > \omega_s$, the reactance of the series arm becomes inductive and the resonance, called *parallel resonance*, occurs when the inductive reactance is equal to the capacitive reactance of C'.

$$\therefore \quad \omega_p L - \frac{1}{\omega_p C} = \frac{1}{\omega_p C'}$$

$$\text{or,} \quad \omega_p = \sqrt{\frac{1}{L}\left(\frac{1}{C} + \frac{1}{C'}\right)}$$

$$\text{or,} \quad f_p = \frac{1}{2\pi}\sqrt{\frac{C + C'}{LCC'}} \tag{11.10.2}$$

At parallel resonance frequency, the impedance is very high. Since $C' \gg C$, the equation (11.10.2) gives

$$f_p \approx \frac{1}{2\pi\sqrt{LC}}$$

$$\therefore \quad f_p \approx f_s$$

The resonant frequencies of crystals are inversely proportional to the thickness. Since a very thin crystal is fragile and crystals of very large thickness are not available, only commercially available crytals can generate signals in the frequency range from 15 kHz to 10 MHz.

Circuit diagram : The circuit diagram of a crystal oscillator using a *npn*-transistor in CE mode is shown in Fig. 11.17. The d.c bias is determined by the resistors R_1, R_2, R_E, C_E and the d.c supply voltage by V_{CC}. The tank circuit consists of the capacitor C_1 and the primary inductance L_1 of the transformer. The secondary L_2 is connected to the base of the transistor through the crystal and the other end is grounded. The resonance frequency of L_1C_1 circuit is made approximately equal to the natural frequency of the crystal.

Working: When the d.c power supply V_{CC} is switched 'on', a transient current flows through the tank circuit producing oscillations in it. By mutual induction, a voltage is induced in L_2. There is a 180° phase shift between the input and the output. An additional 180° phase shift is introduced due to CE transistor amplifier. So, a total 360° phase shift is produced. This causes positive feedback. Now, if the transistor amplifier provides a sufficient gain to make the loop gain $A\beta = 1$, then sustained oscillations take place. Since the crystal is connected in the base circuit, the frequency of oscillation is controlled by the crystal and the tank circuit oscillates at the natural frequency of the crystal.

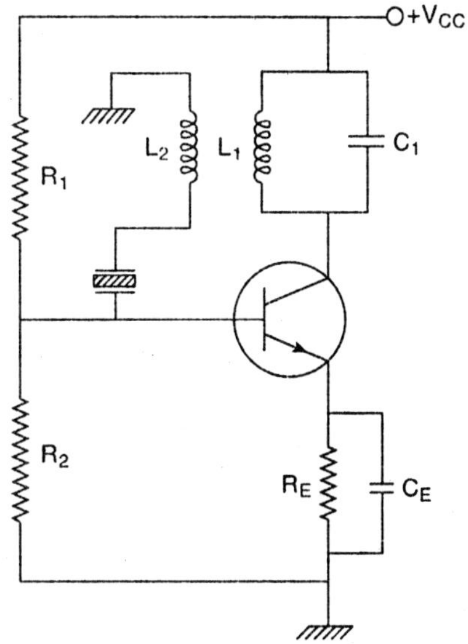

Fig. 11.17 Crystal oscillator circuit

11.11 Multivibrators

A ***multivibrator*** is an electric oscillator consisting of two active devices, usually transistors, interconnected in an electrical network to generate a continuous non-sinusoidal waveforms.

Basically, the multivibrator is a two-stage amplifier operating in two mode or states in such a way that each amplifier stage feeds back the other so that the active element of one stage is driven to saturation and the other to cut-off. Then a set of actions producing the opposite effects follows. So, it changes the saturation stage to cut-off and conversely. The output of both the stages are complementary. Thus a multivibrator forms an important type of switching circuit which has *two* definable stable or quasi-stable states between which the switching occurs.

There are *three* types of multivibrators depending on the stability of the states: *astable, monostable* and *bistable* multivibrators.

(a) Astable multivibrators—In this type of multivibrator, the states are ***not*** *stable.* The circuit continually switches from one state to the other at a rate determined by the circuit components. It does not require any external trigger pulse to change the states. It is also known as ***free-running multivibrator***.

(b) Monostable multivibrator—In this type of multivibrator, the circuit has one stable state and a transient state. A trigger pulse causes the circuit to go over to the transient state from the stable or the quiescent state. After entering the unstable state, the circuit returns to its stable one automatically after a time interval determined by the circuit components. This multivibrator is also known as ***single-shot*** or ***one-shot multivibrator***.

(c) Bistable multivibrator—In this type of multivibrator, the circuit has both the states stable. It requires an external trigger pulse to change the states. It continues in the existing state till the trigger pulse is applied. This multivibrator circuit is also known as ***Eccles-Jordan*** or ***flip-flop multivibrator***.

We shall now discuss these multivibrators in somewhat details.

11.11.1 Astable multivibrator

The multivibrator that has no stable but only two quasi-stable states between which it keeps on oscillating automatically at a rate determined by the circuit components, is known as an ***astable*** or ***free-running multivibrator***.

Fig. 11.18 shows the circuit diagram of an astable multivibrator using two similar *npn*-transistors Q_1 and Q_2. This is basically a symmetrical two-stage RC-coupled amplifier, the output of each stage driving the input of the other stage. R_{B_1} and R_{B_2} are the biasing resistors and R_{C_1} and R_{C_2} are the collector load resistors. The capacitor C_1 couples the output of Q_1 to the input of Q_2 whereas C_2 couples the output of Q_2 to the input of Q_1. In practical circuits, we use $R_{C_1} = R_{C_2}$, $R_{B_1} = R_{B_2}$ and $C_1 = C_2$. Each CE amplifier stage produces a phase shift of 180°. So, the total phase shift between the input and the output is 360°. Hence a positive feedback takes place and the circuit oscillates. The feedback signal is so strong that the transistors are driven either to the cut-off or to the saturation.

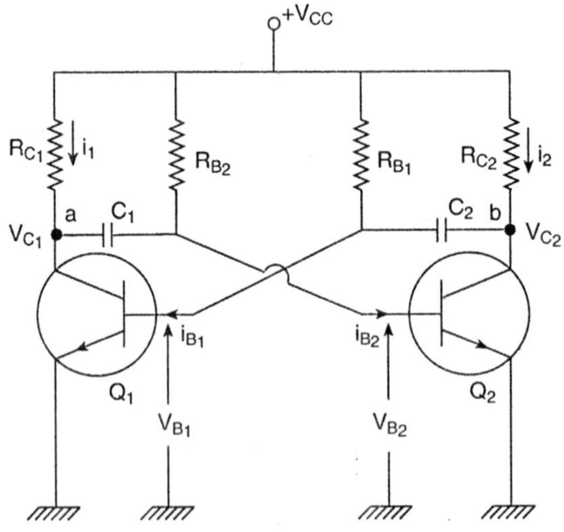

Fig. 11.18 Astable multivibrator

Operation: The operation of a multivibrator is based on the fact that no two active elements have exactly identical characteristics. So, when the power supply

V_{CC} is switched on, one of the transistors, say Q_1, starts conducting before Q_2 and carries a larger collector current. The currents i_1 and i_2 flowing through R_{C_1} and R_{C_2} respectively produce a change in voltage at the points a and b. With $i_1 > i_2$, an increase in i_1 increases the potential difference $i_1 R_{C_1}$ and thus causes lowering of the collector voltage V_{C_1} of Q_1. Due to the coupling capacitor C_1, this decrease in V_{C_1} decreases the base current i_{B_2} of Q_2 and consequently i_2 decreases. resulting in an increase in collector voltage V_{C_2} of Q_2. Due to the existence of coupling through C_2, this increase in V_{C_2} increases the base current i_{B_1} of Q_1. This results in a further increase of i_1 . These cumulative actions drive Q_1 quickly to saturation and Q_2 to cut-off. Thus, V_{C_1} is nearly equal to $0\,V$ (actually $V_{CE,\text{sat}}$) and $V_{C_2} \approx V_{CC}$. The duration of existence of this state depends on the *time constant* RC of the coupling circuits. Now, as $V_{C_1} = 0$, the capacitor C_1 starts charging through R_{B_2} towards V_{CC} with a time constant $R_{B_2}C_1$. When the potential drop across C_1 increases to cut-in voltage, the base-emitter junction of Q_2 becomes forward biased. So, Q_2 begins conducting and quickly goes to saturation.

At this stage, V_{C_2} decreases to $0\,V$ (actually $V_{CE,\text{sat}}$) and the transistor Q_1 is driven to cut-off from the saturation due to coupling. The voltage V_{C_1} rises to V_{CC}. Now, since $V_{C_2} = 0$, the capacitor C_2 starts charging through R_{B_1} towards V_{CC} with a time constant $R_{B_1}C_2$. When the voltage across C_2 rises to cut-in voltage, it makes Q_1 conducting. Thus the whole cycle is repeated. So, the above circuit oscillates between the two states. In one state, Q_1 is 'on' and Q_2 is 'off' and in the other state, Q_1 is 'off' and Q_2 is 'on'. The waveforms at the collector and at the base voltages are shown in Fig. 11.19 (a) and Fig. 11.19(b) respectively.

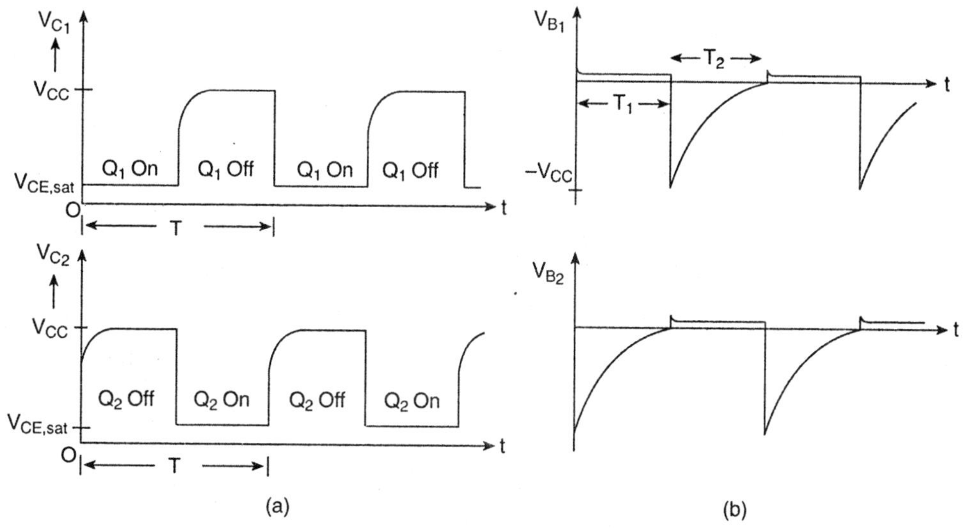

Fig. 11.19 (a) Waveform at the collectors (b) Waveform at the base

Switching times: The multivibrator circuit oscillates between a state in which Q_1 is 'on' and Q_2 is 'off' and a state in which Q_1 is 'off' and Q_2 is 'on'.

The 'on' time of Q_1 is given by

$$T_1 = R_{B_2} C_1 \ln 2 \tag{11.11.1}$$

The 'on' time for Q_2 is similarly given by

$$T_2 = R_{B_1} C_2 \ln 2 \tag{11.11.2}$$

The total time period T of the wave is therefore given by their sum.

$$\therefore \quad T = T_1 + T_2 = (R_{B_2} C_1 + R_{B_1} C_2) \ln 2 \tag{11.11.3}$$

$$= 0.69(R_{B_2} C_1 + R_{B_1} C_2) \tag{11.11.4}$$

In practice, however, $R_{B_1} = R_{B_2} = R$ (say) and $C_1 = C_2 = C$, say i.e., the two stages are fully symmetrical.

$$\therefore \quad T = 0.69(RC + RC) = 1.38RC \tag{11.11.5}$$

Therefore, the frequency of the oscillation is plainly given by

$$f = \frac{1}{T} = \frac{1}{1.38RC} = \frac{0.7}{RC} \tag{11.11.6}$$

Applications: The astable multivibrators are widely used in pulse circuits. Usually, these vibrators find applications as generators of square waves. The other merit is that since no triggerning signal is required, they serve as sources of fast waveforms.

11.11.2 Monostable multivibrator

The circuit diagram of a monostable multivibrator using two similar transistors Q_1 and Q_2 is shown in Fig. 11.20. This is basically a two-stage RC-coupled amplifier the output of each stage driving the input of the other. R_{B_1} and R_{B_2} are the biasing

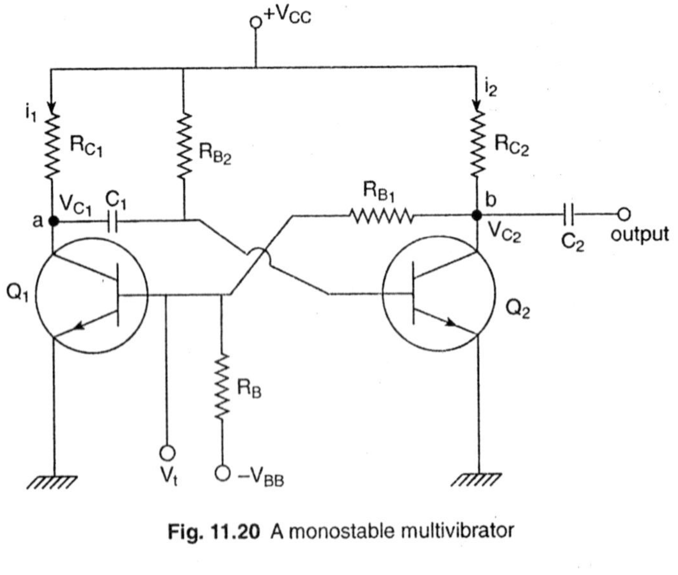

Fig. 11.20 A monostable multivibrator

resistors and R_{C_1} and R_{C_2} are the collector load resistors. The capacitor C_1 couples the output of Q_1 to the input of Q_2 while the output of Q_2 is directly coupled to the input of Q_1 through the base resistor R_{B_1}.

Since a monostable multivibrator has one stable state and one quasi-stable state, the transistor Q_2 is forward biased by V_{CC} and R_{B_2} to drive it saturated in the stable state. The stable state is ensured when the biasing resistor R_{B_2} is so selected that it permits sufficient base current in Q_2 for operation in the saturation region.

The transistor Q_1 remains cut-off in the stable state when it is reverse biased by $-V_{BB}$ and R_B. The base of Q_1 is provided with V_t to apply a suitable positive trigger pulse so that the emitter-base junction of the transistor Q_1 becomes forward biased and Q_1 starts conducting for a certain interval of time and returns automatically to its original stable state.

Operation

When the power supply V_{CC} is switched on but no trigger pulse is applied, the transistor Q_1 is in cut-off condition by $-V_{BB}$ and R_B; and the transistor Q_2 conducts at saturation since it is forward biased by V_{CC} and R_{B_2}. Thus the *initial stable state* of the circuit is given with Q_2 *conducting at saturation and* Q_1 *at cut-off.*

When the trigger pulse is applied at the base of Q_1, the base-emitter junction is forward biased and Q_1 starts conducting. When Q_1 conducts, the collector current i_1 through R_{C_1} produces a change in voltage at the point 'a'. An increase in i_1 increases the potential drop $i_1 R_{C_1}$ and thus lowers the collector voltage V_{C_1} of Q_1. Due to the coupling capacitor C_1, this decrease in V_{C_1} decreases the base current i_{B_2} of Q_2 and consequently i_2 decreases, resulting in an increase in collector voltage V_{C_2} of Q_2. Due to the existence of connection through R_{B_1}, this increase in V_{C_2} increases the base current i_{B_1} of Q_1. This results in further increase of i_1. This cumulative actions drive Q_1 quickly to saturation and Q_2 to cut-off. Thus, V_{C_1} is nearly equal to 0 V (actually $V_{CE,\text{sat}}$) and $V_{C_2} \approx V_{CC}$. So, the result of the series of actions after the *application of trigger pulse* is that the transistor Q_1 *becomes conducting* at saturation and the transistor Q_2 is *cut-off.*

Now, as $V_{C_1} = 0$, the capacitor C_1 starts charging through R_{B_2} towards V_{CC} with a time constant $R_{B_2}C_1$. When the potential difference across C_1 increases to cut-in voltage (~ 0.7 V), the base-emitter junction of Q_2 becomes forward biased. So, Q_2 begins conducting and quickly goes to saturation. At this stage, V_{C_2} decreases to 0 V (actually $V_{CE,\text{sat}}$) and the transistor Q_1 is driven to cut-off from the saturation due to connection through R_{B_1} and the potential V_{C_1} increases to V_{CC}. Thus the circuit returns to the original stable state with Q_2 saturated and Q_1 cut-off after a time interval equal to the time constant $R_{B_2}C_1$. The circuit remains in this state until another trigger pulse causes the circuit to repeat itself the same cycle. The waveform of the output voltage is shown in Fig. 11.21. The multivibrator produces one output pulse for every input trigger pulse. Hence monostable multivibrator circuit is sometimes called *one-shot multivibrator.*

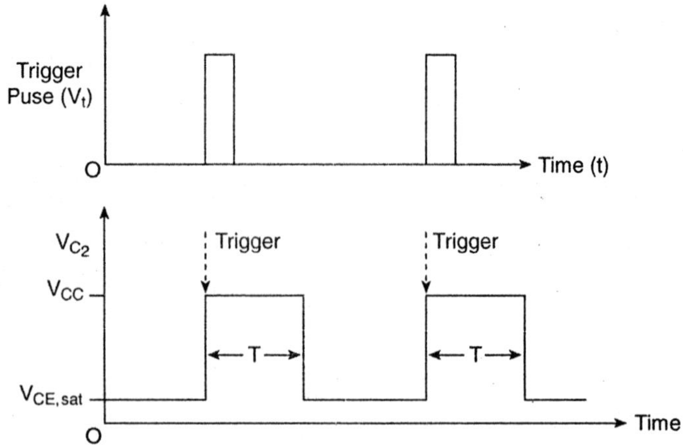

Fig. 11.21 Waveform of output voltage of a monostable multivibrator with trigger pulse

The width of thc duration of output pulse is given by

$$T = R_{B_2} C_1 \ln 2 = 0.693\, R_{B_2} C_1 \tag{11.11.7}$$

Applications: A monostable multivibrator has a number of applications. For example:

(i) The falling part of the output of a monostable multivibrator is often used for triggering another pulse generator. A pulse delayed by a time T with respect to the input pulse may be produced.

(ii) Different types of pulses used in telecommunication systems and computers become somewhat distorted during their use. These pulses can be used to generate new fresh and sharp pulses by employing monostable multivibrator.

11.11.3 Bistable multivibrator

The circuit diagram of a bistable multivibrator using two similar *npn* transistor Q_1 and Q_2 is shown in Fig. 11.22. This is basically a symmetric two-stage *RC*-coupled amplifier, the output of each stage driving the input of the other. R_{C_1} and R_{C_2}, usually equal, are the collector load resistors. The resistor R_{B_2} couples the output of Q_2 to the input of Q_1 whereas R_{B_1} couples the output of Q_1 to the input of Q_2. Both the transistors are provided with a voltage $-V_{BB}$ for maintaining stable state, and V_{t_1} and V_{t_2} to apply trigger pulse to the base of the transistor Q_1 and Q_2 respectively. A bistable multivibrator, as its very name signifies, has two stable states. The circuit can stay in one of the stable states indefinitely (as long as d.c power is on) until a proper external trigger pulse makes it to change to the other stable state. If another suitable trigger pulse is applied at the other input, it can be brought back to the original stable state. Thus one trigger pulse causes to *flip* the multivibrator from one stable state to another and the next pulse causes it to *flop* back to the initial stable state. So, a bistable multivibrator is called a *flip-flop*.

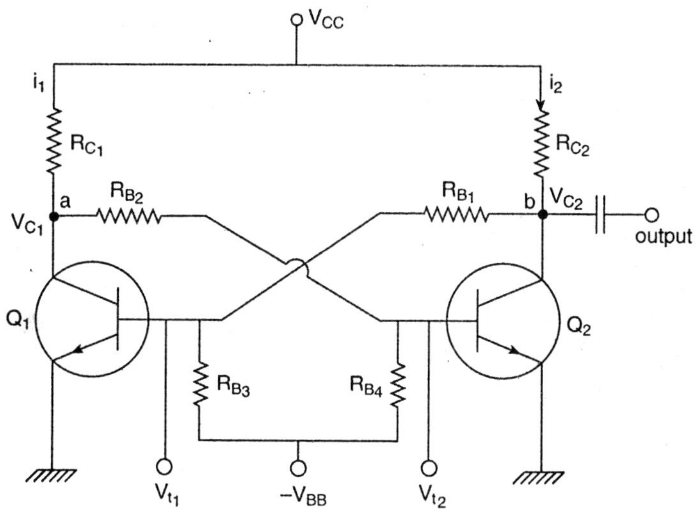

Fig. 11.22 Circuit diagram of bistable multivibrator

Operation

When the power supply V_{CC} is switched on, one of the transistors (say Q_1) starts conducting slightly more than the other (Q_2) due to small difference in the characteristics of the transistors. The current i_1 and i_2 flowing through R_{C_1} and R_{C_2} respectively produce a change in voltage at the points 'a' and 'b'. With $i_1 > i_2$ an increase in i_1 increases the potential difference $i_1 R_{C_1}$ and thus causes lowering of the collector voltage V_{C_1} of Q_1. Due to the resistor R_{B_2}, this decrease in V_{C_1} decreases the base current i_{B_2} of Q_2 and consequently i_2 decreases resulting in an increase in collector voltage V_{C_2} of Q_2. Due to the existence of coupling through R_{B_1}, the increase in V_{C_2} increases the base current i_{B_1} of Q_1. This results in a further increase of i_1. These cumulative actions drive Q_1 to *saturation* and Q_2 to *cut-off*. Thus V_{C_1} is nearly equal to 0 V (actually $V_{CE,\text{sat}}$) and $V_{C_2} \approx V_{CC}$. This is the *first stable state* of bistable multivibrator.

The multivibrator can be driven from the first stable state (i.e., Q_1 *on* and Q_2 *off*) to the other stable state (i.e., Q_1 *off* and Q_2 *on*) by applying either a suitable negative trigger pulse to the base of Q_1 or a suitable positive trigger pulse to the base of Q_2.

Now, suppose a negative trigger pulse of sufficient magnitude is applied to the base of Q_1. It decreases the forward bias of base-emitter junction of Q_1 and causes a reduction of collector current and thereby an increase in V_{C_1}. The increase in V_{C_1} appears across the emitter-base junction of Q_2 as it is connected to V_{C_1} by R_{B_2}. As a result, the collector current i_2 increases and V_{C_2} falls. The decreasing V_{C_2} appears across the emitter-base junction of Q_1 where it further increases the reverse bias. These cumulative actions drive Q_2 into saturation and Q_1 to cut-off. This is the *second stable state* of the multivibrator. The circuit will stay in this state (i.e., Q_2 *on* and Q_1 *off*) until a positive trigger pulse is applied to the base of Q_1 or a negative pulse to the base of the transistor Q_2.

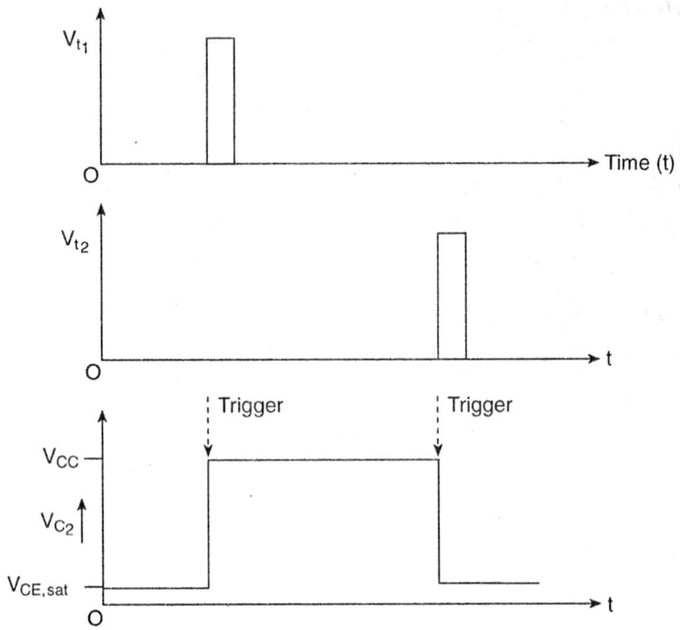

Fig. 11.23 Timing diagram of a bistable multivibrator

The timing diagram of a bistatble multivibrator is shown in Fig. 11.23. It is to be noted that the spacing of the trigger pulse determines the output waveform.

Applications : The bistable multivibrator, also known as *flip-flop multivibrator*, is employed in many digital operations such as counting and storing binary information, generating pulses etc. It finds wide applications in digital circuits like frequency divider, counter and memory element.

11.12 Illustrated Examples

► **Example 1.** *The voltage gain of a transistor is* 100. *The transistor is to be used as an oscillator. What should be the type of feedback provided to input circuit? Calculate the feedback factor. What will be the frequency of oscillations if* $L = 0.1\,\text{mH}$ *and* $C = 400\,\text{pF}$*?* **(Madurai Univ.)**

Solution : To use a transistor as an oscillator, the type of feedback provided to the input circuit should be positive. When a transistor behaves as an oscillator, the closed loop gain becomes unity i.e., $A\beta = 1$.

Here, inductance, $L = 0.1\,\text{mH} = 10^{-4}\,\text{H}$, voltage gain, $A = 100$ and capacitance, $C = 400\,\text{pF} = 4 \times 10^{-10}\,\text{F}$.

Therefore the feedback factor, $$\beta = \frac{1}{A} = \frac{1}{100} = \mathbf{0.01}.$$

The frequency of oscillation, $$f = \frac{1}{2\pi\sqrt{LC}} = \frac{1}{2\pi\sqrt{10^{-4} \times 4 \times 10^{-10}}}\ \text{Hz}$$

$$= \mathbf{795\ kHz}.$$

▶ **Example 2.** *The tuned-collector oscillator circuit used in the local oscillator of a radio receiver makes use of an LC tuned circuit with* $L = 58.6\mu\text{H}$ *and* $C = 300\,\text{pF}$. *Calculate the frequency of oscillation.* **(Punjab Univ.)**

Solution: Given, the inductance, $L = 58.6\,\mu\text{H} = 58.6 \times 10^{-6}\,\text{H}$

Capacitance, $C = 300\,\text{pF} = 3 \times 10^{-10}\,\text{F}$

$\therefore$ The frequency of oscillation,

$$f = \frac{1}{2\pi\sqrt{LC}}$$

$$= \frac{1}{2\pi\sqrt{58.6 \times 10^{-6} \times 3 \times 10^{-10}}}\,\text{Hz}$$

$$= 1200 \times 10^3\,\text{Hz} = \mathbf{1.2\ MHz}$$

▶ **Example 3.** *Calculate the operatiing frequency for the Hartley oscillator when* $L_1 = 1000\,\mu\text{H}$, $L_2 = 100\,\mu\text{H}$, $C = 20\,\mu\text{F}$ *and* $M = 20\,\mu\text{H}$. **(Punjab Univ.)**

Solution: Taking mutual induction under consideration, frequency of the oscillator is

$$f = \frac{1}{2\pi\sqrt{(L_1 + L_2 + 2M)C}}$$

By the problem,

$$L_1 = 1000\,\mu\text{H} = 1000 \times 10^{-6}\,\text{H}$$

$$L_2 = 100\,\mu\text{H} = 100 \times 10^{-6}\,\text{H}$$

$$M = 20\,\mu\text{H} = 20 \times 10^{-6}\,\text{H}, \quad C = 20\,\mu\text{F} = 20 \times 10^{-6}\,\text{F}$$

$$\therefore \quad f = \frac{1}{2\pi\sqrt{(1000 + 100 + 40) \times 10^{-6} \times 20 \times 10^{-6})}}$$

$$= 1.053 \times 10^6\,\text{Hz}, \quad \text{on simplification}$$

$$= 1053 \times 10^3\,\text{Hz} = \mathbf{1053\ kHz}$$

▶ **Example 4.** *What is the frequency of Hartley's oscillator if the total inductance is* $80\,\mu\text{H}$ *and the capacitance* $180\,\mu\text{F}$? **(Nagpur Univ.)**

Solution: Given, the total inductance of the Hartley's oscillator

$$L_1 + L_2 = 80\,\mu\text{H} = 80 \times 10^{-6}\,\text{H}$$

Capacitance, $C = 180\,\mu\text{F} = 180 \times 10^{-6}\,\text{F}$

$\therefore$ The frequency of oscillation,

$$f = \frac{1}{2\pi\sqrt{(L_1 + L_2)C}}$$

$$= \frac{1}{2\pi\sqrt{80 \times 10^{-6} \times 180 \times 10^{-6}}}\,\text{Hz}$$

$$= \mathbf{1.3\ kHz}$$

▶ **Example 5.** *A coil of* 10 μH *used in a Colpitt oscillator generates a wave of frequency* 16 MHz. *The transistor has* $h_{fe} = 50$ *and* $\Delta h_e = 0.5$. *Find the values of the capacitances.* **(Himachal Univ.)**

Solution: Given, the frequency of oscillation, $f = 16\,\text{MHz} = 16 \times 10^6\,\text{Hz}$,
and the inductance, $L = 10\mu\text{H} = 10 \times 10^{-6}\,\text{H} = 10^{-5}\,\text{H}$.

Let C_1 and C_2 be the capacitances of the tank circuit. Now, the frequency

$$f = \frac{1}{2\pi}\sqrt{\frac{1}{L}\left(\frac{1}{C_1} + \frac{1}{C_2}\right)}$$

$$\text{or,} \quad (2\pi f)^2 = \frac{1}{L}\left(\frac{1}{C_1} + \frac{1}{C_2}\right)$$

$$\text{or,} \quad \frac{1}{C_1} + \frac{1}{C_2} = 4\pi^2 f^2 \times L$$

$$= 4\pi^2 \times (256 \times 10^{12}) \times 10^{-5}$$

$$= 1024\pi^2 \times 10^7 \qquad (11.12.1)$$

The condition for sustained oscillations is thus given by

$$\frac{C_1}{C_2} = \frac{h_{fe}}{\Delta h_e} = \frac{50}{0.5} = 100$$

$$\text{or,} \quad C_1 = 100C_2 \qquad (11.12.2)$$

Solving (11.12.1) and (11.12.2), we finally have

$$\mathbf{C_1 = 1000\,pF} \text{ and } \mathbf{C_2 = 10\,pF}$$

▶ **Example 6.** *In a phase-shift oscillator, the load resistance* $R_L = 3.3\,\text{k}\Omega$, *the resistance of* RC *network,* $R = 5.6\,\text{k}\Omega$ *and the capacitance of* RC *network,* $C = 0.01\,\mu\text{F}$. *Calculate the frequency of the oscillator. What is the* h_{fe} *of the transistor if the oscillations are sustained in the circuit?* **(Himachal Univ.)**

Solution: Here, the load resistance, $R_L = 3.3\,\text{k}\Omega = 3.3 \times 10^3\Omega$

Resistance of the RC network, $R = 5.6\,\text{k}\Omega = 5.6 \times 10^3\Omega$

Capacitance of the RC network, $C = 0.01\,\mu\text{F} = 0.01 \times 10^{-6}\,\text{F}$.

∴ The frequency of the oscillator,

$$f = \frac{1}{2\pi C\sqrt{4RR_L + 6R^2}}$$

$$= \frac{1}{2\pi \times 0.01 \times 10^{-6}\sqrt{[4 \times 5.6 \times 10^3 \times 3.3 \times 10^3 + 6 \times (5.6 \times 10^3)^2]}}$$

$= \mathbf{982\ Hz}$, on simplification.

$$\text{The value of } h_{fe} = 23 + 29\frac{R}{R_L} + 4\frac{R_L}{R}$$

$$= 23 + 29 \times \frac{5.6 \times 10^3}{3.3 \times 10^3} + 4 \times \frac{3.3 \times 10^3}{5.6 \times 10^3}$$

$$= \mathbf{74.6}, \text{ on simplification.}$$

▶ **Example 7.** *A Wien-bridge oscillator is used for a frequency range* 30 Hz *to* 3 kHz. *The variable capacitor has a range of* 50 pF *to* 500 pF. *Find the resistance values required. If the resistances in the arms are in the ratio* 5 : 1, *find out the gain of the amplifier.*

(Calcutta Hons.)

Solution : In a Wien-bridge oscillator, the frequency of oscillation is given by

$$f = \frac{1}{2\pi RC}$$

The maximum capacitance corresponds to the minimum frequency, i.e., $f = 30\,\text{Hz}$ when $C = 500\,\text{pF} = 500 \times 10^{-12}\,\text{F}$.

$$\therefore\ 30 = \frac{1}{2\pi R \times 500 \times 10^{-12}}$$

$$\text{or, } R = 10.6 \times 10^6\,\Omega = 10.6\,\text{M}\Omega$$

Since the capacitance varies in the ratio 50 pF : 500 pF = 1 : 10, the oscillator spans a frequency range of 30 Hz to 300 Hz with $R = 10.6\,\text{M}\Omega$. For the frequency range of 300 Hz to 3 kHz, R is to be $\frac{1}{10}$th of 10.6 MΩ i.e. the new value of R is 1.06 MΩ. Hence the two values of R are **10.6 MΩ** and **1.06 MΩ**.

The resistances in the arms are in the ratio 5 : 1. $\therefore\ R_1/R_2 = 5:1$

$$\text{Gain of the amplifier is thus: } \boldsymbol{A} = \frac{R_1 + R_2}{R_2} = 1 + \frac{R_1}{R_2} = \mathbf{6}.$$

▶ **Example 8.** *Calculate the frequency of oscillations when a phase shift oscillator has* $R = 10\,\text{k}\Omega$, $C = 0.01\,\mu\text{F}$ *(RC-network) and* R_L *(load resistance)* $= 2.2\,\text{k}\Omega$. *What will be the minimum current gain needed for the purpose?*

Solution : Frequency of oscillation is given by

$$f = \frac{1}{2\pi C\sqrt{4RR_L + 6R^2}}$$

By the problem, $C = 0.01 \times 10^{-6}\,\text{F}$, $R = 10 \times 10^3\,\Omega$, $R_L = 2.2 \times 10^3\,\Omega$.

$$\therefore\ f = \frac{1}{2\pi \times 0.01 \times 10^{-6}\sqrt{(4 \times 10 \times 10^3 \times 2.2 \times 10^3 + 6 \times 10^2 \times 10^6)}}$$

$$= \frac{10^4}{2\pi \times \sqrt{6.88}} = \mathbf{606.7\,Hz}, \text{ on simplification.}$$

Current gain, $A_I = \dfrac{I_C}{I_B} = -\dfrac{h_{fe}}{1 + h_{oe}R_L}$. Since h_{oe} is very small compared to 1,

$$\text{Current gain} = |h_{fe}| = 4\frac{R_L}{R} + 23 + 29\frac{R}{R_L}$$

$$= 4 \times (2.2/10) + 23 + 29 \times (10/2.2)$$

$$= \mathbf{155.7.}$$

▶ **Example 9.** *In an astable multivibrator, the value of $R_1 = R_2 = 15\,\text{k}\Omega$ and $C_1 = C_2 = 0.005\,\mu\text{F}$. Calculate the frequency of the oscillation.* **(Andhra Univ.)**

Solution: Here, $R_1 = R_2 = 15\,\text{k}\Omega = 15 \times 10^3\Omega = R$ (say).

$C_1 = C_2 = 0.005\,\mu\text{F} = 0.005 \times 10^{-6}\,\text{F} = C$ (say).

∴ The frequency of the oscillation is given by

$$f = \frac{1}{1.38\,RC} = \frac{1}{1.38 \times (15 \times 10^3) \times (0.005 \times 10^{-6})}\,\text{Hz}$$

$$= \mathbf{9.66 \times 10^3\ Hz}$$

▶ **Example 10.** *A piezoelectric crystal used in a crystal oscillator is resonant at 450 kHz. For this frequency, its equivalent inductance and series resistance are 3.5 H and 9050 Ω, respectively. What is the quality factor Q of the crystal?*

Solution: The quality factor Q of the crystal in a crystal oscillator is given by

$$Q = \frac{\omega L}{R}$$

where ω is the angular frequency, L the equivalent inductance and R series resistance.

By the problem, $\omega = 2\pi \times 450 \times 10^3$ rad/s, $L = 3.5\,\text{H}$ and $R = 9050\,\Omega$.

$$\therefore\ \boldsymbol{Q} = \frac{2\pi \times 450 \times 10^3 \times 3.5}{9050} = \mathbf{1093}$$

▶ **Example 11.** *In a transistorized astable multivibrator, the biasing resistors are $R_{B_1} = R_{B_2} = 10\,\text{k}\Omega$, collector circuit resistors are $R_{C_1} = R_{C_2} = 1\,\text{k}\Omega$ and coupling copacitors are $C_1 = C_2 = 0.01\,\mu\text{F}$. Find the frequency of oscillation of the circuit. Also, find the minimum value of β for the transistor.*

Solution: Here, $R_{B_1} = R_{B_2} = 10\,\text{k}\Omega$, $R_{C_1} = R_{C_2} = 1\,\text{k}\Omega$, and $C_1 = C_2 = 0.01\,\mu\text{F} = 0.01 \times 10^{-6}\,\text{F}$

∴ The frequency of oscillation is

$$f = \frac{0.7}{R_B C} = \frac{0.7}{10 \times 10^3 \times 0.01 \times 10^{-6}}\,\text{Hz} = \mathbf{7\ kHz}.$$

The minimum value of β for the transistor is given by

$$\beta_{\text{min}} = \frac{R_B}{R_C} = \frac{10 \times 10^3}{1 \times 10^3} = \mathbf{10}$$

▶ **Example 12.** *The electrical equivalent circuit of a crystal consists of* $L = 137\,\text{H}$, $C = 0.0235\,\text{pF}$, $R = 15\,\text{k}\Omega$ *connected in parallel with a capacitor* $C' = 3.5\,\text{pF}$. *Find the resonant frequencies.*

Solution: By the problem, $L = 137\,\text{H}$, $C = 0.0235\,\text{pF}$, $R = 15\,\text{k}\Omega$ and $C' = 3.5\,\text{pF}$.
∴ The series resonant frequency is

$$f_s = \frac{1}{2\pi\sqrt{LC}} = \frac{1}{2\pi\sqrt{1.37 \times 0.0235 \times 10^{-12}}}\,\text{Hz}$$

$$= 88.7 \times 10^3\,\text{Hz} = \mathbf{88.7\ kHz}$$

and, the parallel resonant frequency is

$$f_p = \frac{1}{2\pi}\sqrt{\frac{C + C'}{L(CC')}} = \frac{1}{2\pi}\sqrt{\frac{3.5235 \times 10^{-12}}{137 \times 0.0235 \times 3.5 \times 10^{-24}}}$$

$$= 89 \times 10^3\,\text{Hz} = \mathbf{89\ kHz.}$$

Review Questions

1. What is an oscillator? What are its various types? Is any input necessary to get an output from an oscillator? **(Utkal and Lucknow Univ.)**

2. Why and which type of feedback is necessary in oscillator circuits? Explain the condition under which a feedback amplifier works as an oscillator. **(Meerut and Punjabi Univ.)**

3. What is Barkhausen criterion? State the basic conditions for oscillations in a feedback amplifier. What are the primary requirements to obtain steady oscillations at a fixed frequency? **(Calcutta Hons.)**

4. Why is a frequency determining circuit necessary in an oscillator? Discuss a tank circuit and oscillations in it. How can it be used to generate undamped electromagnetic waves? **(Guru Nanak Univ.)**

5. (a) What are the essentials of an oscillator?

   (b) Why is an amplifier necessary in an oscillator?

   (c) What do you mean by positive feedback? What happens to the overall gain when $A\beta = 1$? **(Punjab Univ.)**

6. (a) Give the basic principle of operation of a tuned collector oscillator.
   (b) Write an expression for the frequency of oscillation of a tuned collector oscillator and obtain the condition for sustained oscillations in the circuit. **(Calcutta Hons.)**

7. Describe a Hartley oscillator with a circuit diagram. Find an approximate expression for the frequency of oscillation. Also, obtain the condition for sustained oscillations. **(Calcutta Hons.)**

8. (a) Give the circuit diagram of a Colpitt oscillator and explain its operation.
   (b) Derive the condition for sustained oscillations and the frequency of oscillation in a Colpitt oscillator.
   (c) What is the difference between a Colpitt and a Hartley oscillator? **(Calcutta Hons.)**

9. With the help of a circuit diagram, explain the operation of a phase-shift oscillator and find its frequency of oscillation. What is the minimum gain required for sustained oscillations? **(Burdwan Hons.)**

10. Give a neat circuit diagram of Wien bridge oscillator and explain how it works.

    Find an expression for the frequency of oscillation and the gain of the amplifier used in Wien bridge oscillator. **(Calcutta Hons, Osmania Univ.)**

11. Draw a circuit of an astable multivibrator and explain its operation. What is the frequency of the circuit? **(Madurai Univ.)**

12. (a) Discuss the principle of operation of a crystal oscillator. Mention its two advantages.
    (b) What is the range of frequencies of normal operation of crystal oscillators? Why?
    (c) What is meant by the frequency stability of an oscillator? Why is the frequency stability of an oscillator high when a piezoelectric crystal is used? **(Calcutta Hons.)**

13. Assuming $R = 0$ in the equivalent circuit of a vibrating piezoelectric crystal, obtain an expression for the crystal impedance and show its variation with frequency. **(Calcutta Hons.)**

14. What is a multivibrator? Name the different classes of multivibrators and briefly distinguish among them.

15. What is a bistable multivibrator? Explain its operation with the help of a circuit diagram.

16. Give the circuit diagram of a monostable multivibrator and explain its operation.

17. (a) Prove that the gain of the amplifier used in a Wien-bridge oscillator must be greater than 3 for sustained oscillations.

(b) Mention one advantage of the Wien-bridge oscillator over the phase-shift oscillator.

(c) State the advantages and the disadvantages of the Wien-bridge oscillator.

18. Draw the switching waveforms for the astable multivibrator.

Objective Questions

1. For generating an AF signal, the most suitable circuit is

   (a) Colpitt oscillator, (b) Hartley oscillator,
   (c) Wien-bridge oscillator, (d) tuned collector oscillator.

2. Crystal oscillators are used for

   (a) low-frequency generation, (b) high frequency stability,
   (c) high output voltage, (d) low output voltage.

3. A bistable multivibrator has

   (a) one stable and one unstable state, (b) two stable states,
   (c) two unstable states, (d) none of these.

4. An astable multivibrator produces

   (a) a sine wave, (b) a square wave,
   (c) a triangular wave, (d) a saw-tooth wave.

5. For the generation of a 1 MHz signal, one must choose

   (a) Wien-bridge oscillator, (b) Colpitt oscillator,
   (c) phase-shift oscillator, (d) Hartley oscillator.

6. In LC circuit, there is an exchange of

   (a) magnetic and sound energy, (b) magnetic and light energy,
   (c) electrostatic and magnetic energy, (d) light and mechanical energy.

   (Himachal Univ.)

7. Positive-feedback is used in

   (a) low gain amplifiers, (b) high gain amplifiers, (c) oscillators.

   (Himachal Univ.)

8. The negative feedback in an amplifier increases the

   (a) gain, (b) noise, (c) band width, (d) frequency distortion.

9. A feedback amplifier will work as an oscillator if the loop gain $A\beta$ is

(a) zero,
(b) 1 and feedback is positive,
(c) < 1,
(d) 1 and feedback is negative.

10. Barkhausen criterion for sustained oscillations in an oscillator is

(a) $A\beta = 0$,
(b) $A\beta = 1$,
(c) $A\beta < 1$,
(d) $A\beta > 1$.

11. For sustained oscillations in a Hartley oscillator

(a) $h_{fe} = (L_1 + M)/(L_2 + M)$.
(b) $h_{fe} = L_1 + M$,
(c) $h_{fe} = L_2 + M$,
(d) $h_{fe} = L_1 + L_2 + 2M$.

12. Wien-bridge oscillator is most often used whenever

(a) wide range of high purity sine waves are to be generated,
(b) high feedback ratio is needed,
(c) square output waves are required,
(d) extremely high resonant frequencies are required.

13. Crystal oscillators are superior to tuned LC oscillators mainly because of their

(a) exceptionally high degree of frequency stability,
(b) high value of Q-factor,
(c) size of the crystal,
(d) both (a) and (b) above.

14. The primary advantage of a crystal oscillator is that

(a) it can oscillate at any frequency,
(b) it gives a high output voltage,
(c) it operates on a very low dc supply voltage,
(d) its frequency of oscillation remains almost constant.

15. Which of the following factors do no affect the frequency stability of an oscillator?

(a) inter-element capacitances and stray capacitances,
(b) output load,
(c) temperature variation,
(d) coil size.

16. An astable multivibrator is also called free-running relaxation oscillator since

(a) it oscillates continuously between two unstable states,
(b) it runs and relaxes alternately,
(c) it requires no dc power for its operations,
(d) none of the above.

17. A relaxation oscillator is one which

(a) has two stable states, (b) relaxes indefinitely,
(c) produces non-sinusoidal output, (d) oscillates continuously.

18. The RC network, as illustrated in Fig. 11.24 below, can provide maximum theoretical phase-shift of

(a) 90°, (b) 180°, (c) 270°, (d) 360°.

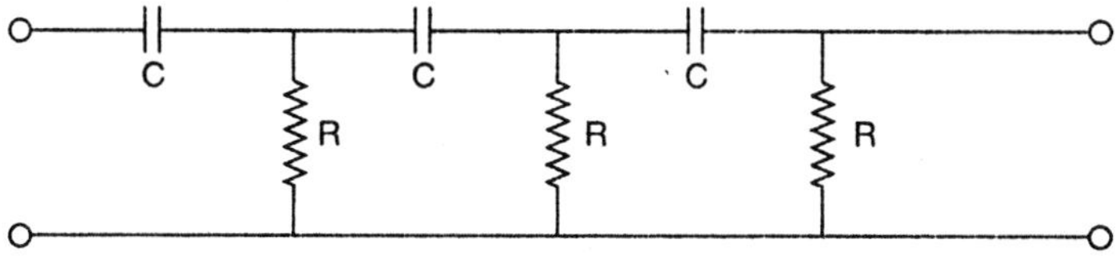

Fig. 11.24

19. A monostable multivibrator is frequently used

(a) in memory and timing circuits,
(b) in counting circuits,
(c) for producing triangular waves,
(d) for regeneration of distorted waves.

20. In an astable multivibrator

(a) $\beta = 1$, (b) $\beta A = 1$, (c) $\beta > 1$, (d) $\beta < 1$.

Problems

1. In a Hartley oscillator, the values for $L_1 = 0.3\,\text{mH}$ and $L_2 = 0.2\,\text{mH}$. If the capacitor C is set at $0.03\mu\text{F}$, calculate the frequency produced by the oscillator. **(Punjabi Univ.)** [130.3 kHz]

2. Calculate the frequency of oscillations of a Hartley oscillator having $L_1 = 0.5\,\text{mH}$, $L_2 = 1\,\text{mH}$ and $C = 0.2\mu\text{F}$. **(Bangalore Univ.)** [0.29 MHz]

3. A Hartley oscillator uses a transistor in CE mode with the following h-parameters:

$$h_{fe} = 50 \quad \text{and} \quad \Delta h_e = 0.5$$

The variable capacitor has a range 50 pF to 450 pF. The oscillations are to vary from a frequency 50 kHz to a frequency 150 kHz. Determine the required inductance of the coils. Neglect the mutual inductance between the coils.

[0.223 mH, 22.3 mH]

4. In a phase-shift oscillator, $R = R_L = 1\,\text{M}\Omega$ and $C = 75\,\text{pF}$. At what frequency does the circuit oscillate? **(Himachal Univ.)** [670 Hz]

5. A Wien-bridge oscillator has a frequency of 1000 Hz and a capacitance of 100 pF. Find the resistance. If the amplifier gain is 10, obtain the ratio of the resistances in the other arms. **(Calcutta Hons.)** [1.59 MΩ, 7 : 23]

6. A Wien-bridge oscillator is to a span a range of frequencies from 30 Hz to 30 kHz. The variable capacitance can be changed from 50 pF to 500 pF. Find the resistances needed to span the frequency range. If the gain of the amplifier is 6, what must be the ratio of the resistances in the other arms of the bridge? **(Calcutta Hons.)** [10.6 MΩ, 1.06 MΩ, 106 kΩ, 5 : 1]

7. A Colpitt oscillator has a coil with inductance of 40 μH and is tuned by a capacitor of 750 pF across the amplifier input and 2500 pF across the output. Find the frequency of oscillation and the minimum gain which the amplifier must provide to maintain oscillations. **(Burdwan Hons.)** [1.05 MHz, 3.3]

8. Calculate the frequency of oscillation of a Colpitt oscillator with $C_1 = C_2 = 500\,\text{pF}$ and $L = 1\,\text{mH}$. [318.4 kHz]

9. A Hartley oscillator has tank circuit inductances $L_1 = L_2 = 100\mu\text{H}$. It is required to design the oscillator to produce oscillations at 50 kHz. Obtain the exact value of the tank circuit capacitance for the above requirement. Draw the tank circuit with capacitance and inductance values. **(Bangalore Univ.)** [0.05 μF]

10. In a Hartley oscillator, the vahues of inductance coils used in the tank circuit are $L_1 = 1000\,\mu\text{H}$ and $L_2 = 100\,\mu\text{H}$. Calculate the value of h_{fe} if the mutual inductance between the coils be 20 μH. [8.5]

11. In a Hartley oscillator $L_1 = 100\,\mu\text{H}$, $L_2 = 1\,\text{mH}$ and the mutual inductance between the coils is $M = 10\,\mu\text{H}$. If the value of the capacitor is $C = 20\,\text{pF}$, calculate the frequency of oscillations. Also, calculate the value of h_{fe} for sustained oscillations. [1.06 MHz, 0.117]

12. The electrical equivalent circuit of a crystal consists of a series $L = 1\,\text{H}$, $C = 0.01\,\mu\text{F}$, $R = 1\,\text{k}\Omega$ connected in parallel with $C' = 10\,\text{pF}$. Find the series and parallel resonant frequencies of the crystal. Find Q at each frequency. [1.591 MHz, 1.592 MHz, 9.996, 10.003]

13. Consider a transistorized astable multivibrator circuit with biasing resistances $R_{B1} = R_{B2} = 1\,\text{k}\Omega$. Determine the value of the coupling capacitors required to produce a train of pulses 10 μs wide at a repeatition rate of 10 kHz. [0.014 μF, 0.126 μF]

12

Operational Amplifiers

12.1 Introduction

The operational amplifier, abbreviated OP AMP, is a direct-coupled, high gain, differential amplifier. OP AMPs were so named as they were originally designed to perform the basic mathematical operations like addition, subtraction, integration, differentiation etc. in electronic analog computers. Today, OP AMPs are among the most widely applied electronic devices and are used in signal amplification, waveforming, servo-controls, impedance transformation, active filters, oscillators, analog to digital and digital to analog converters etc. Large variety of OP AMPs in IC forms are available. They are less expensive and have thermal stabilization. An IC OP AMP acts as a complete functional building block with some external pins for connecting power supply voltages and external circuit components to use it for a specific purpose. So, an user needs to be acquainted with its terminal properties but not to deal with the detailed internal circuit configuration of it. An OP AMP uses negative feedback and its performance is solely controlled by the feedback elements, independent of the transistor characteristics, resistors, capacitors which constitute the internal OP AMP circuit. Since the circuit elements of the feedback network are usually passive, the circuit operation is highly stable and predictable.

12.1.1 OP AMP Circuit

An OP AMP is used to perform various mathematical operations. It requires a large number of transistors connected in various modes to execute the operations. So, the internal circuit design of an IC OP AMP is very complicated. Fig. 12.1 shows the circuit diagram of a general purpose IC OP AMP.

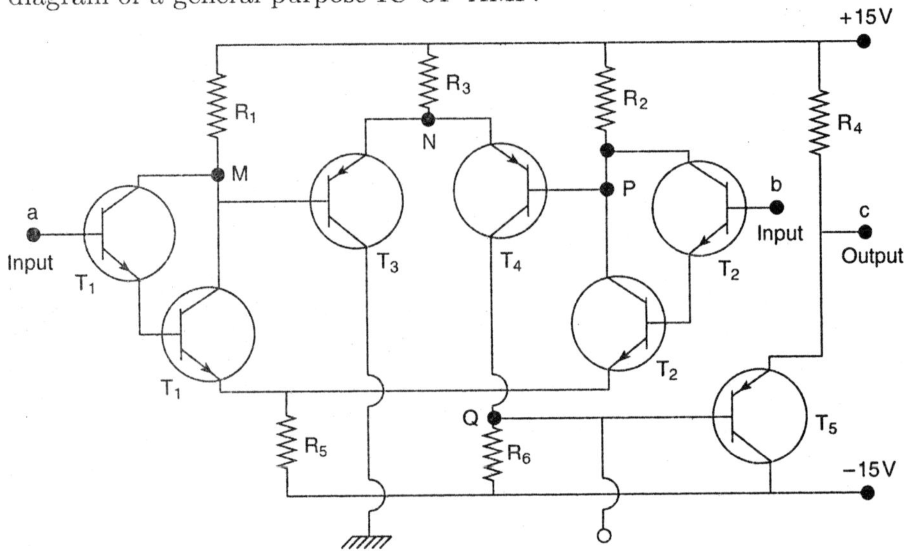

Fig. 12.1 Simplified circuit of an IC OP AMP

The terminals labelled 'a' and 'b' are the two inputs of the circuit. The Darlington connection of the two *npn* transistors T_1 or T_2 provides the high input impedance. The terminal labelled 'c' is the output terminal which is taken from a single-ended emitter follower. Thus, the circuit consists of two differential gain stages depending on the two inputs. The transistors T_1 and T_2 are identical. T_3 and T_4 are also the same. For better performance, R_1 must be equal to R_2. Since the corresponding components of each pair are affected approximately equally due to their close spacing in the IC chip, the errors from the thermal drift are reduced.

To understand the operation of the OP AMP circuit (Fig. 12.1), let us consider the case when the input terminal 'a' is grounded and a small positive voltage v_1 is applied to the input terminal 'b'. This will reduce the potential of the point P and make T_4 to conduct more. As a result, the potential of the point Q increases causing T_5 to conduct less and the output voltage at the terminal 'c' to be more positive. Thus the signal applied at the terminal 'b', with the terminal 'a' grounded, appears in an amplified form at the output terminal 'c' with the same polarity. The output voltage is then given by

$$v_{01} = A_1 v_1 \tag{12.1.1}$$

where A_1 is the voltage gain.

Clearly, A_1 is positive since v_{01} and v_1 both have the same polarity.

When the terminal 'b' is grounded and a small positive voltage v_2 is applied to the terminal 'a', the potential of the point M becomes less positive that makes T_3 to conduct more. Consequently, the potential of the point N reduces and T_4 conducts less. So, the potential of Q becomes more negative causing T_5 to conduct more and the potential of the output terminal 'c' reduces. Thus the signal applied at the terminal 'a', with the terminal 'b' grounded, appears in an amplified form at the output terminal 'c' with reverse polarity. The output is given by

$$v_{02} = A_2 v_2 \tag{12.1.2}$$

where A_2 is the voltage gain which is negative for opposite polarities of v_{02} and v_2.

When v_1 and v_2 are applied simultaneously to the input terminals 'b' and 'a' respectively, the output voltage becomes

$$v_0 = v_{01} + v_{02} = A_1 v_1 + A_2 v_2 \tag{12.1.3}$$

For an ideal OP AMP, $A_1 = -A_2$. So, we get $v_0 = 0$ for $v_1 = v_2$.

12.1.2 Circuit symbols

The standard circuit symbol of the OP AMP is shown in Fig. 12.2. It has two inputs (labelled 'a' and 'b') and one output (labelled 'c'). The input terminal 'a' marked '$-$' is known as *inverting input terminal* whereas the input terminal 'b' marked '+' is known as the *non-inverting input terminal.* When a signal voltage is applied at the inverting input terminal, keeping the non-inverting input terminal at ground, the output appearing at c is inverted (i.e., 180° out of phase) with respect to the input. On the other hand, when the signal is applied at the non-inverting input terminal, keeping the inverting input terminal at ground, the output appearing at c is in phase with respect to the input.

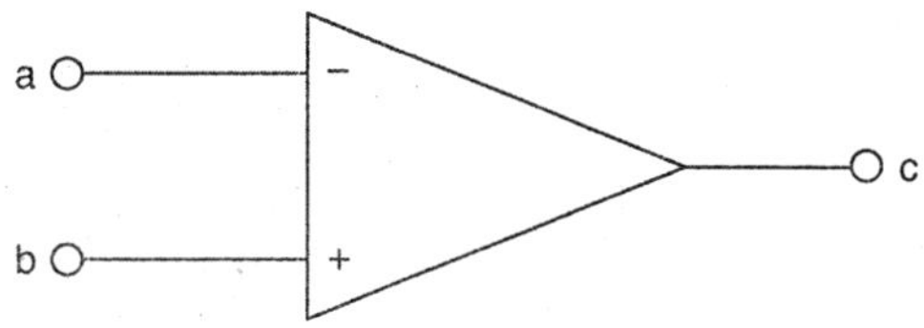

Fig. 12.2 Circuit symbol of an OP AMP

When two input voltages are applied simultaneously at the two input terminals, the output voltage at c is proportional to the difference of the two input voltages. The proportionality constant is called the *open-loop voltage gain A* of the OP AMP. This is a real constant approaching infinity for all frequencies in an ideal OP AMP.

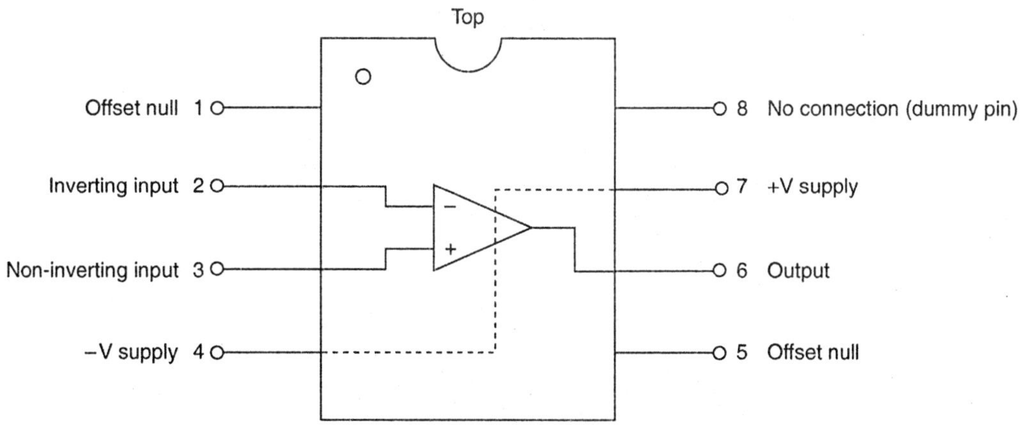

Fig. 12.3 Pin diagram of an OP AMP IC (A741)

Fig. 12.3 shows the pin diagram of an IC OP AMP (A 741) available as 8-lead dual-in-line package (DIP) with two parallel rows of pins.

Pin number 2 and pin number 3 are the inverting and non-inverting inputs respectively. Pin number 6 is the output terminal. Pin number 7 and 4 are connected to the positive (+V) and negative (−V) supply voltages respectively. Pins 1 and 5 are used to provide offset null voltage adjustments. Pin number 8 is a dummy pin which therefore requires no external connection.

12.2 Characteristics of an ideal OP AMP

An *ideal* OP AMP has a number of important characteristics. These are enumerated as under.

1. An infinite voltage gain,
2. infinite bandwidth,
3. An infinite input impedance,
4. Zero output impedance
5. Perfect balance i.e., the output voltage is zero when the input voltages are equal.
6. Temperature stability i.e., the characteristics do not change with temperature.

However, in practice, an OP AMP deviates from these *ideal* characteristics. An IC OP AMP A 741 has the d.c or the low frequency voltage gain of the order of 10^5 and a finite bandwidth. The gain is very high but not infinite. The gain remains constant up to a few hundred kHz and then decreases with increase in frequency. The input impedance is $\sim 2\,\text{M}\Omega$ and the output impedance $\sim 75\,\Omega$. A perfect balance *cannot* be achieved with this OP AMP.

12.3 AC equivalent circuit of an OP AMP

The a.c. equivalent circuit of a practical OP AMP is shown in Fig. 12.4(a). R_i is the input resistance between the inverting input terminal and the ground, R_n is that

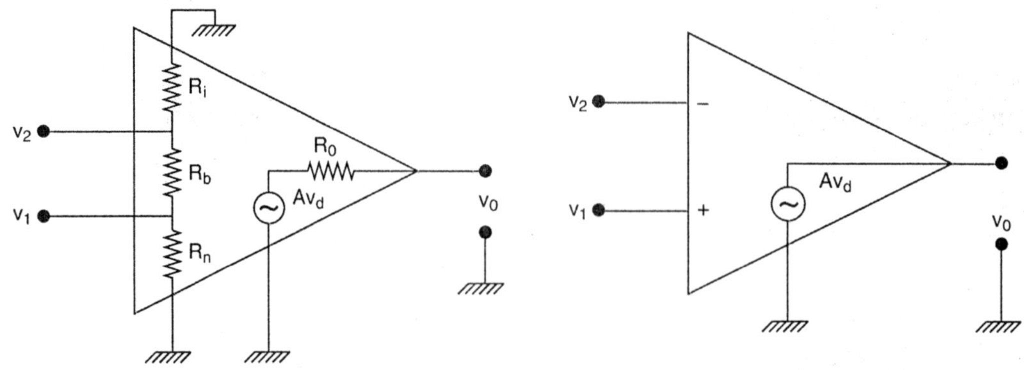

Fig. 12.4 AC equivalent circuit of an OP AMP (a) practical, (b) ideal

between the non-inverting input terminal and the ground. The resistance between the two input terminals is R_b. All the three resistances R_i, R_n and R_b are typically of very high values.

The output voltage is taken through the output resistance R_0 and it is given by

$$v_0 = Av_d \tag{12.3.1}$$

where A is the amplifier gain and v_d $(= v_1 - v_2)$ is the difference input signal.

Here the amplifier gain A is the voltage gain A_d for the difference signal and is typically very high.

For an *ideal* OP AMP, the input resistances R_i, R_n and R_b are infinite, the output resistance R_0 is zero and the voltage gain A_d is infinite. Fig. 12.4(b) shows the a.c. equivalent circuit of an *ideal* OP AMP.

12.4 Architecture of IC OP AMP

Since an IC OP AMP uses a large number of transistors to perform various mathematical operations, the design of its internal circuit is very complicated. The circuit consists mainly of *four* units—(a) differential amplifier, (b) more gain stages, (c) level shifter and (d) emitter follower. The architecture or block diagram of a typical IC OP AMP is shown below in Fig. 12.5.

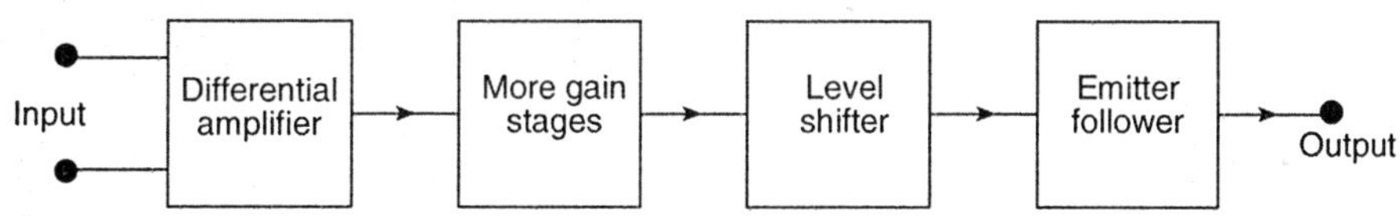

Fig. 12.5 Architecture of an IC OP AMP

Here the input stage is a *differential amplifier* which provides the inverting and non-inverting inputs with high input resistance, high voltage gain and large CMRR (see Section 12.8.7). The differential amplifier is followed by *more gain stages* which is a cascaded high gain voltage amplifier that increases the voltage gain to a higher value. On an IC chip, large capacitors cannot be fabricated. So, the gain stages are directly coupled by adjusting the dc level of one stage before applying its output to the next stage. The *level shifter* performs this operation. Finally, the output is taken from a single-ended *emitter follower* that provides the low output resistance.

12.5 Offset voltages and currents in OP AMP

In an ideal OP AMP, a perfect symmetry is assumed between the two halves of the differential amplifier used as the input stage. So, an *ideal* OP AMP is *perfectly balanced* i.e., the output voltage $v_0 = 0$ when the two inputs v_1 and v_2 are equal. However, in practice, an OP AMP shows an imbalance due to mismatch of the built-in transistors following the inverting and non-inverting input terminals. This mismatch causes unequal bias currents flowing through the input terminals and results in a non-zero output voltage without any input signal. For the convenience of discussion, we assume that the output is processed by an *input offset voltage* (v_{io}). The input offset voltage (v_{io}) of an OP AMP is the voltage to be applied between the input terminals to balance the amplifier [Fig. 12.6(a)]. The *output offset voltage* (v_{oo}) of an OP AMP is defined as the voltage at the output terminal when the two inputs are grounded [Fig. 12.6(b)].

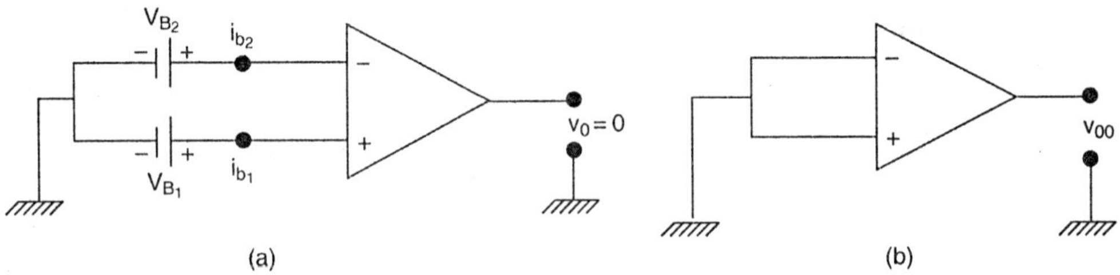

Fig. 12.6 (a) Input offset voltage (b) Output offset voltage

The input offset voltage is equal to the output voltage for zero input voltage divided by the open-loop voltage gain of the amplifier. For OP AMP consisting of transistors at the input terminals, the input offset voltage is in the range of 10 μV to 7 mV and for OP AMP with FETs at the input terminals, it is 0.5 mV to 15 mV. The small value of the offset voltage is amplified by the circuit and may not be neglected at the output. So, most of OP AMP are provided with some means of *nulling out* the effect of input offset voltage.

Since the input circuit of an OP AMP is always a differential amplifier and in most cases BJT are employed, the transistors of the differential amplifier must be provided with base bias current. The two base currents must be equal to keep the circuit balanced. But there is always some difference between the two base bias currents due to manufacturing imperfections. So the term *input bias current* is introduced.

The *input bias current* is the *average* of the bias currents entering the two input terminals of a balanced amplifier [Fig. 12.6(a)]. Thus the input bias current is given by

$$i_B = \left.\frac{i_{b_1} + i_{b_2}}{2}\right]_{v_O=0} \tag{12.5.1}$$

Typically, i_B is in the range of 10 nA to 50 nA for BJT input OP AMP and 10 pA to 100 pA for FET input OP AMP.

The *input offset current* (i_{io}) is defined as the *difference* between the individual bias currents entering into the input terminals of a balanced amplifier. Thus,

$$i_{io} = i_{b_1} - i_{b_2}\Big]_{v_O=0} \tag{12.5.2}$$

Typically, i_{io} is in the range of 3 – 20 nA for BJT input OP AMP and a few pA for FET input OP AMP.

12.6 Open-loop characteristic of OP AMP

The *open-loop characteristic* of an OP AMP is the graphical plot of the output voltage v_O against the difference input signal v_d $(= v_1 - v_O)$. Fig. 12.7 shows a typical open-loop characteristic of an OP AMP.

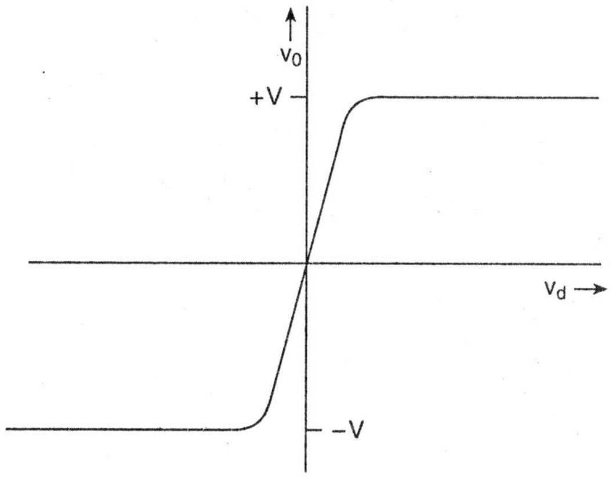

Fig. 12.7 Open-loop characteristic of an OP AMP

Since the open-loop gain is very high, the output becomes approximately equal to the positive dc supply voltage +V for v_d slightly positive. If v_d increases further, the output voltage v_O saturates to +V. On the other hand, when v_d is slightly negative, the output voltage v_O attains nearly the negative dc supply voltage −V. With more negative value of v_d, v_O remains saturated at −V. The +V and −V voltages are known as the positive saturated voltage $+V_{sat}$ and the negative saturated voltage $-V_{sat}$ respectively.

Slew rate

The *slew rate* (SR), sometimes called *slewing*, is a *measure of the maximum rate of change of the output voltage* v_O *with respect to time.* So,

$$\text{SR} = \left(\frac{dv_O}{dt}\right)_{\text{max}} \tag{12.6.1}$$

When a practical OP AMP is connected as an amplifier with a closed-loop gain of A_c, the expected output voltage is A_c for the input dc voltage of 1 V. In practice, the circuit takes some time to reach that value of output voltage due to the inherent *time constant* of the OP AMP. For a monolithic OP AMP, the typical value of SR is 1 V/μs. For an ideal OP AMP, it is infinitely large.

12.7 Concept of virtual ground

The concept of *virtual ground* makes the analysis and understanding of the operation of many OP AMP circuits very simple, interesting and instructive. In order to explain this, let us consider the basic circuit of an inverting OP AMP as has been illustrated in Fig. 12.8.

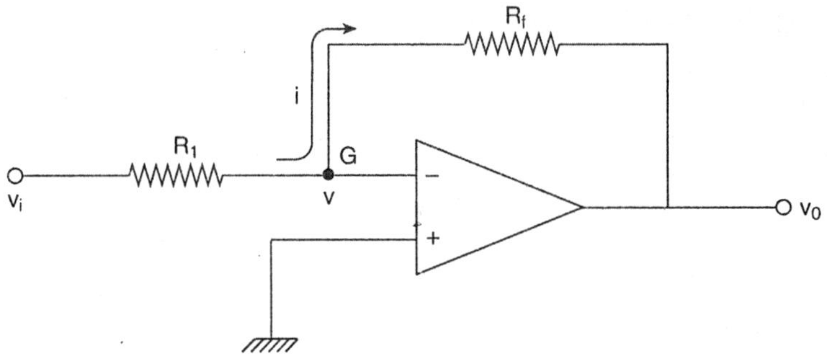

Fig. 12.8 Inverting amplifier

The input voltage v_i is connected to the inverting input terminal through R_1 and the feedback resistance R_f connects the output terminal to the input terminal providing a negative feedback. The non-inverting terminal is grounded. The negative feedback tends to reduce the voltage v at the inverting input terminal. Since $v_0 = Av$ and $A \to \infty$ for an OP AMP, for a finite v_0 we must have $v \to 0$. So, the inverting input terminal G is *practically* at the ground potential. Since G is *not actually* connected to the ground, it is called *virtually at ground* whatever be the values of v_i and v_0.

When a terminal is actually grounded, it behaves as a sink for infinite current. But, in the case of an OP AMP, no current can flow into the OP AMP IC through G due to the infinite input impedance. So, the terminal G behaves as *virtual ground* and cannot serve as a sink for any current.

12.8 Applications of OP AMP

An OP AMP finds a large variety of applications in consumer, industrial and scientific devices. The applications are rather too many to exhaust. Some of the important applications are only being considered hereunder.

12.8.1 Inverting amplifier

The amplifier whose output voltage is inverted with respect to the input voltage is called ***inverting amplifier***.

Fig. 12.8 shows the basic circuit diagram of an inverting amplifier using OP AMP with a very small input signal v_i applied to the inverting input terminal. R_1 and R_f are the input resistance and the feedback resistance respectively. The non-inverting terminal is grounded. Since the open-loop gain of an OP AMP is infinite and the input resistance of the OP AMP is very high (ideally infinite), no current enters into the OP AMP and the point G behaves as 'virtual ground' with a potential $v \approx 0$. Therefore, the current through R_1 is equal to the current through R_f i.e.,

$$i = \frac{v_i - v}{R_1} = \frac{v - v_0}{R_f} \tag{12.8.1}$$

$$\text{or,} \quad \frac{v_i}{R_1} = -\frac{v_0}{R_f} \quad (\because\ v \approx 0)$$

$$\text{or,} \quad \frac{v_0}{v_i} = -\frac{R_f}{R_1} \tag{12.8.2}$$

Thus the closed-loop voltage gain of an inverting amplifier is given by the ratio of the feedback resistance R_f to the input resistance R_1 and it is *independent* of OP AMP parameters. The negative sign indicates that with respect to the input, the output is 180° out of phase i.e., the output is inverted with respect to the input. Thus, the above circuit represents an *inverting amplifier.*

Effect of finite value of open-loop gain

The effect of finite loop gain can be realised if we simplify the equation (12.8.1). This equation gives

$$R_f(v_i - v) = R_1(v - v_0)$$

$$\text{or,} \quad v_0 R_1 = (R_1 + R_f)v - v_i R_f$$

$$\text{or,} \quad v_0 R_1 - v(R_1 + R_f) = -v_i R_f$$

$$\text{or,} \quad v_0\left[R_1 + \frac{R_1 + R_f}{A}\right] = -v_i R_f, \quad [\because\ v_0 = -Av]$$

So, the voltage gain with feedback is given by

$$A_f = \frac{v_0}{v_i} = -\frac{R_f}{\left(R_1 + \frac{R_1+R_f}{A}\right)}$$

$$= -\frac{R_f}{R_1} \cdot \frac{1}{1 + \frac{1}{A\beta}} \tag{12.8.3}$$

where $\beta = \dfrac{R_1}{R_1 + R_f}$ = feedback factor.

If $A \to \infty$, then $A_f = -R_f/R_1$ which is true for closed loop voltage gain of an ideal OP AMP as given in equation (12.8.2).

12.8.2 Phase shifter

If the resistance R_1 and R_f of the circuit of Fig. 12.8 are replaced respectively by impedances Z_1 and Z_f having equal magnitudes but different phase angles, ϕ_1 and ϕ_f, then the voltage gain is given by

$$\frac{v_0}{v_i} = -\frac{Z_f}{Z_1} = -\frac{|Z_f|e^{j\phi_f}}{|Z_1|e^{j\phi_1}}$$

$$= -e^{j(\phi_f - \phi_1)} \quad (\because\ |Z_f| = |Z_1|)$$

$$= e^{j(\pi + \phi_f - \phi_1)} \quad (\because\ e^{j\pi} = -1) \tag{12.8.4}$$

Equation (12.8.4) shows that the output v_0 leads the input v_i by a phase angle $(\pi + \phi_f - \phi_1)$ with $|v_0| = |v_i|$. This implies that the circuit shifts the phase of the input signal voltage, leaving its magnitude unchanged. The phase-shift can have any value between 0 to 2π. Thus an OP AMP can be used as a ***phase shifter***.

12.8.3 Scale changer

In Fig. 12.8, if the ratio $R_f/R_1 = k$ (a real constant), then we have

$$\frac{v_0}{v_i} = -\frac{R_f}{R_1} = -k$$

$$\text{or,}\quad v_0 = -kv_i \tag{12.8.5}$$

This shows that the output voltage scale is obtained by multiplying the input voltage by a factor $-k$, known as *scale factor*. The accurate value of k can be determined by using precision resistors and thus the inverting amplifier can work as a ***scale changer***. A low voltage can be measured by applying it to a scale changer and then dividing the amplified voltage by the scale factor.

12.8.4 Non-inverting amplifier

The amplifier whose output is *in phase* with the input is called ***a non-inverting amplifier***.

Fig 12.9 shows the circuit of a basic non-inverting amplifier using OP AMP with a small input signal voltage v_i applied at the non-inverting terminal. The inverting terminal is connected to the ground through the resistor R_1 and the output is fed back

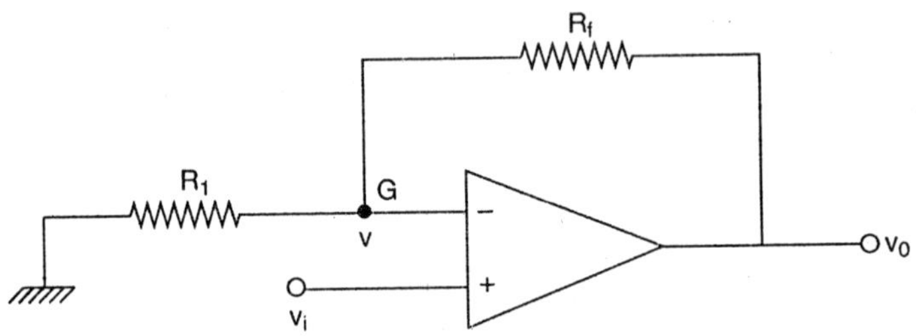

Fig. 12.9 Non-inverting amplifier

through the resistor R_f. Let v be the potential at the point G. The output voltage is then given by

$$v_0 = A(v_i - v) \tag{12.8.6}$$

Since the voltage gain $A \to \infty$ for an OP AMP, we must have $(v_i - v) \to 0$ i.e., $v_i \approx v$. Hence the potential at G may be taken to be v_i i.e., there exists a *virtual short circuit* at the input terminals. Again, the current flowing through the OP AMP is negligible due to the infinite input impedance. So, almost the same current i flows through R_1 and R_f. Hence, applying KCL at G, we have

$$i = \frac{v_i - 0}{R_1} = \frac{v_O - v_i}{R_f}$$

$$\text{or,} \quad \frac{v_O - v_i}{v_i} = \frac{R_f}{R_1}$$

$$\text{or,} \quad \frac{v_O}{v_i} = 1 + \frac{R_f}{R_1}$$

$$\text{i.e., Voltage gain, } A_V = \frac{v_O}{v_i} = 1 + \frac{R_f}{R_1} \tag{12.8.7}$$

The positive sign of the voltage gain indicates that the output is *in phase* with the input. Thus the circuit represents a non-inverting amplifier.

Effect of finite value of open-loop gain

If A be the open-loop gain, then the output $v_O = A\,(v_i - v)$ where $(v_i - v)$ is the differential input to the amplifier.

Now, from Fig. 12.9, we have

$$v = \frac{v_O \cdot R_1}{R_1 + R_f} = \beta v_O \quad \text{where } \beta = \frac{R_1}{R_1 + R_f}$$

Therefore, we have

$$v_O = A(v_i - \beta v_O)$$

$$\text{or,} \quad \frac{v_O}{v_i} = \frac{A}{1 + A\beta}$$

So, the voltage gain with feedback is

$$A_f = \frac{v_O}{v_i} = \frac{A}{1 + A\beta} \tag{12.8.8}$$

If $A \to \infty$, then the equation (12.8.8) reduces to

$$A_f = \frac{1}{\beta} = \frac{R_1 + R_f}{R_1} = 1 + \frac{R_f}{R_1}$$

which is true for the closed-loop gain of an ideal OP AMP.

12.8.5 Voltage follower

The voltage gain of a non-inverting amplifier has just been seen to be given by

$$A_V = 1 + \frac{R_f}{R_1}$$

If we choose $R_f = 0$ and $R_1 = \infty$, the gain becomes unity and the circuit of Fig. 12.9 reduces to that of Fig. 12.10.

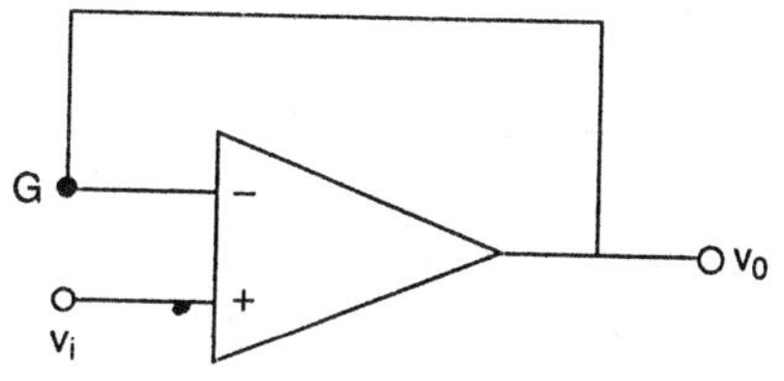

Fig. 12.10 Voltage follower

The unity gain implies that $v_O = v_i$ i.e., the output voltage is equal to the input voltage. Also, the positive sign indicates that the output voltage is in phase with the input. So, the output voltage faithfully follows the input voltage at all instants and hence the circuit of Fig. 12.10 represents a ***unity-gain buffer*** or a ***voltage-follower***. This allows the input voltage to be transferred to the output without any change and avoids the loading of the source. Since this circuit offers a high input impedance and low output impedance, it can be used as an *impedance matching device* between a high impedance source and a low impedance load.

12.8.6 Adder or summing amplifier

An ***adder*** is an inverting amplifier whose output is the inverted algebraic sum of the constituent inputs.

Fig. 12.11 shows an adder or summing amplifier circuit. The input signals $v_1, v_2, v_3, \cdots, v_n$ are applied to the inverting input terminal of the OP AMP through the resistors $R_1, R_2, R_3, \cdots, R_n$ respectively. The non-inverting terminal is grounded. Since the point G is the virtual ground and no current flows through the OP AMP due to the high input impedance, the current through R_f is equal to the algebraic sum of the currents $i_1, i_2, i_3, \cdots, i_n$ through $R_1, R_2, R_3, \cdots, R_n$ respectively. So,

$$i_1 + i_2 + i_3 + \cdots + i_n = i$$

$$\text{or,} \quad \frac{v_1}{R_1} + \frac{v_2}{R_2} + \frac{v_3}{R_3} + \cdots + \frac{v_n}{R_n} = -\frac{v_0}{R_f}$$

$$\text{or,} \quad v_0 = -R_f\left(\frac{v_1}{R_1} + \frac{v_2}{R_2} + \frac{v_3}{R_3} + \cdots + \frac{v_n}{R_n}\right) \tag{12.8.9}$$

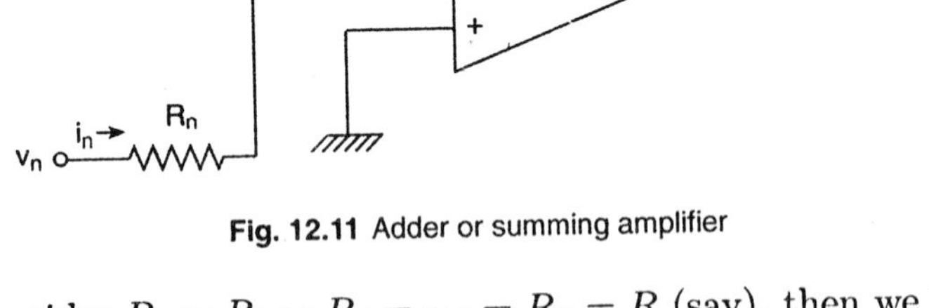

Fig. 12.11 Adder or summing amplifier

If we now consider $R_1 = R_2 = R_3 = \cdots = R_n = R$ (say), then we have

$$v_0 = -\frac{R_f}{R}(v_1 + v_2 + v_3 + \cdots + v_n) \tag{12.8.10}$$

Further, if we make $R_f = R$, then the above equation (12.8.10) reduces to

$$v_0 = -(v_1 + v_2 + v_3 + \cdots + v_n) \tag{12.8.11}$$

The equation (12.8.11) shows that *the output voltage is exactly equal to the algebraic sum of the input voltages with a change in sign.* So, the above circuit works as an ***inverting adder.***

12.8.7 Differential amplifier

A *differential amplifier* is one whose function is to amplify the difference between the two input voltages.

Fig. 12.12 shows the circuit diagram of a basic differential amplifier using OP AMP. Two input signals v_1 and v_2 are applied to the inverting and non-inverting input terminals respectively through resistances R_1, R_2 as shown in Fig. 12.12.

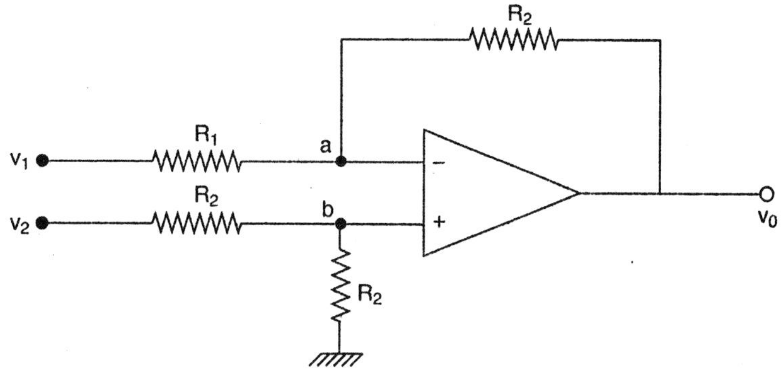

Fig. 12.12 Differential amplifier

Since there exists a virtual short circuit between the inverting and the non-inverting input terminals of the OP AMP due to infinite open-loop gain and infinite input impedance, the points a and b will have the same potential, say v. Also, the currents through R_1 and R_2 are equal. So, we may write

$$\frac{v_1 - v}{R_1} = \frac{v - v_0}{R_2} \tag{12.8.12}$$

and $$\frac{v_2 - v}{R_1} = \frac{v - 0}{R_2} \tag{12.8.13}$$

Subtracting equation (12.8.12) from equation (12.8.13) we have,

$$\frac{v_2 - v_1}{R_1} = \frac{v_0}{R_2}$$

or, $$\boxed{v_0 = \frac{R_2}{R_1}(v_2 - v_1)} \tag{12.8.14}$$

Thus the difference of the two input signals is amplified and the circuit behaves as a ***differential*** (or difference) ***amplifier.***

From (12.8.14), the output of an ideal OP AMP may be rewritten as

$$v_0 = A_d(v_2 - v_1)$$

where $A_d(= R_2/R_1)$ is the voltage gain for the difference signal $v_d = v_2 - v_1$. However, if $v_1 = v_2$, then one should have $v_0 = 0$. But, in practice, this does not happen due to lack of perfect circuit symmetry. The output depends not only on the difference signal $v_d = v_2 - v_1$, but also on the average signal $v_c \left(= \frac{1}{2}(v_1 + v_2)\right)$, known as *common-mode signal.* Therefore, the output may be written as

$$v_0 = A_d v_d + A_c v_c$$

where A_c is the voltage gain for common-mode signal.

For an ideal OP AMP, $A_d = \infty$ and $A_c = 0$. In practice, we expect A_d to be very large and A_c very small. So, a figure of merit known as *common-mode rejection ratio* (CMRR) of the OP AMP is introduced. It determines the success of a differential amplifier in providing the voltage gain for the difference mode signal and rejecting the common-mode signal. The CMRR is defined as

$$\text{CMRR} = \left|\frac{A_d}{A_c}\right| \tag{12.8.15}$$

For better performance, the value of CMRR must be larger than unity. Again, the value of CMRR being very large, we may express it in decibel (dB) as

$$\text{CMRR (dB)} = 20 \log_{10}\left(\frac{A_d}{A_c}\right)$$

Another important performance parameter of an OP AMP is the ***supply voltage rejection ratio (SVRR)*** or ***power supply rejection ratio (PSRR)***. It is *defined* as the ratio of the change in input offset voltage to the corresponding change in one power supply voltage with all other supply voltages remaining constant. Mathematically,

$$\boldsymbol{SVRR} \text{ or } \boldsymbol{PSRR} = \frac{\Delta V_{io}}{\Delta V_{\text{supply}}} = \frac{\text{Change in input offset voltage}}{\text{Change in supply voltage}}.$$

So, PSRR is a measure of the ability of an OP AMP to disregard changes in power supply voltage. PSRR is expressed in μV/V or in decibel (dB).

12.8.8 Subtractor

The ***subtractor*** is a special type of differential amplifier which produces an output equal to the difference of the two input signals.

If we choose $R_1 = R_2$ in Fig. 12.12, then (12.8.14) reduces to $v_0 = v_2 - v_1$ i.e., the output voltage v_0 gives the difference of the input voltages v_2 and v_1. Thus the circuit 12.12 would then serve as a simple *subtractor.*

12.8.9 Differentiator

If the output voltage of a device is proportional to the time derivative of the input voltage, the device is called a ***differentiator***.

Fig. 12.13 shows the circuit diagram of a differentiator using OP AMP with a input voltage v_i applied to the inverting terminal through the capacitor C. The non-inverting terminal is connected to the ground.

The point G is the virtual ground due to the infinite voltage gain of the OP AMP. If q be the charge on the capacitor C, then

$$v_i = \frac{q}{C} \tag{12.8.16}$$

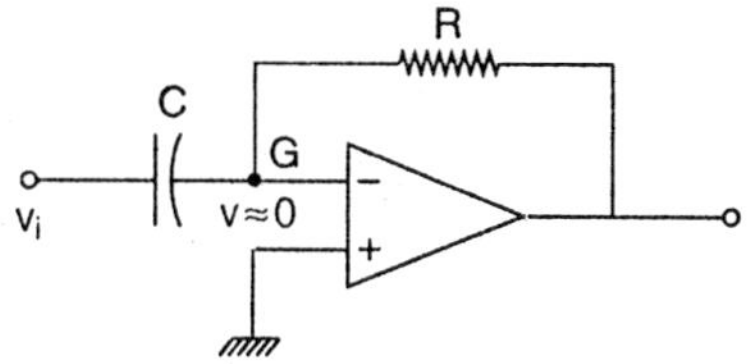

Fig. 12.13 Differentiator

Differentiating (12.8.16) with respect to time, we get

$$\frac{dv_i}{dt} = \frac{1}{C}\frac{dq}{dt} = \frac{i}{C} \tag{12.8.17}$$

where $i = dq/dt$ is the current through the capacitor. Since G is the virtual ground, the current i flows through R, the feedback resistance. Therefore,

$$i = \frac{0 - v_0}{R} = -\frac{v_0}{R} \tag{12.8.18}$$

From equations (12.8.17) and (12.8.18), we obtain

$$\boxed{v_0 = -CR\frac{dv_i}{dt}} \tag{12.8.19}$$

This relation shows that the output voltage is proportional to the time derivative of the input voltage. Hence the circuit is referred to as a ***differentiator***.

12.8.10 Integrator

An ***integrator*** is a device whose output is the integral of the input.

Fig. 12.14 shows the circuit diagram of an integrator using OP AMP with an input voltage v_i applied at the inverting terminal through a resistance R. The capacitor C acts as the feedback element. The infinite voltage gain of the OP AMP makes G a virtual ground. The current i flowing through the resistance R is given by

$$i = \frac{v_i}{R} \tag{12.8.20}$$

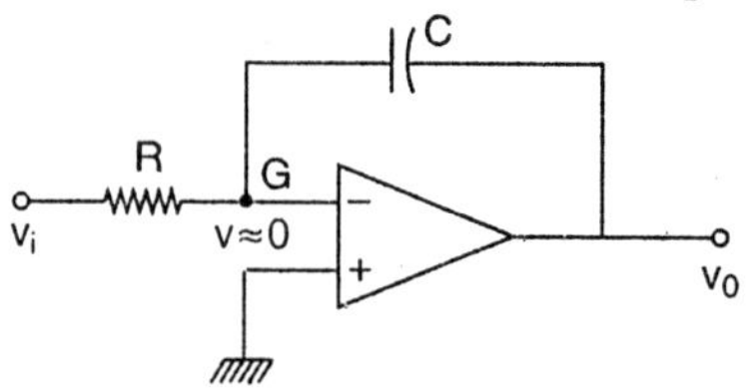

Fig. 12.14 Integrator

Since the input impedance of the OP AMP is very high (ideally infinite), the current i produces the output voltage v_0 across the feedback capacitor C.

$$\therefore \quad C(0 - v_0) = Q, \quad Q \text{ is the charge on capacitor}$$

$$\text{or,} \qquad v_0 = -\frac{1}{C}\int_0^t i\,dt, \quad \left[\because\ Q = \int_0^t i\,dt\right]$$

$$= -\frac{1}{CR}\int_0^t v_i\,dt \tag{12.8.21}$$

Thus the output voltage is proportional to the *time integral* of the input voltage. Hence the circuit is referred to as an ***integrator*** circuit.

12.8.11 Voltage to current converter

The ***voltage to current converter*** is an electronic device whose output current is proportional to its input signal voltage.

Fig. 12.15 shows the circuit diagram of a voltage to current converter by using OP AMP. The circuit is basically the non-inverting amplifier circuit. The output current i_L is the current through the load resistance R_L. Since the infinite voltage gain

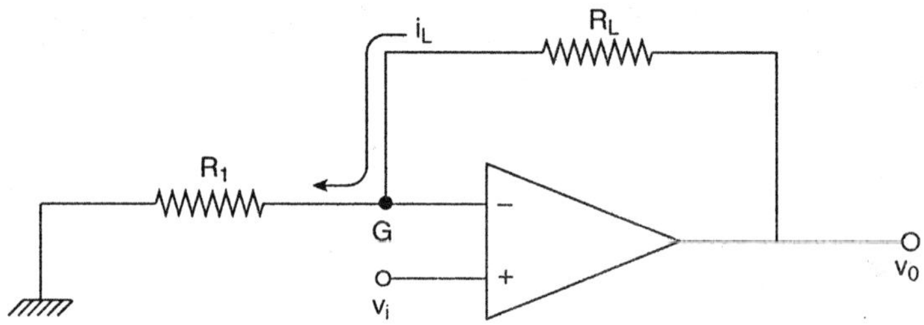

Fig. 12.15 Voltage to current converter

makes the input terminals of a non-inverting amplifier 'virtually shorted', the potential at G is also v_i. So, the load current i.e., the output current is given by

$$i_L = \frac{v_i}{R_1} \tag{12.8.22}$$

Thus the output current is proportional to the input voltage v_i and does not depend on the load resistance R_L. Hence this circuit is aptly referred to as a ***voltage to current converter.***

12.8.12 Current to voltage converter

*The **current to voltage converter*** is a device whose output voltage is proportional to its input signal current.

Fig. 12.16 shows the circuit diagram of a current to voltage converter using OP AMP. The input signal current i_s is supplied by the output current of a photocell or photo-multiplier tube. Since the point G behaves as a *virtual* ground, almost the total current passes through the feedback resistance R_f to produce the output voltage v_0. So, we shall have

$$v_0 = -i_s \cdot R_f \tag{12.8.23}$$

Thus the output voltage v_0 is proportional to the input signal current i_s and hence the circuit is called a ***current to voltage converter.***

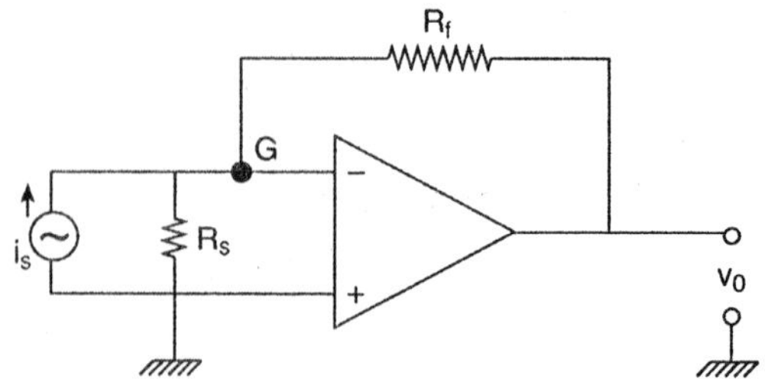

Fig. 12.16 Current to voltage converter

12.8.13 Logarithmic amplifier

The ***logarithmic amplifier*** is a device whose output voltage is proportional to the logarithm of the input voltage.

If the feed back resistor R_f in Fig. 12.8 is replaced by a diode, the output changes as the logarithm of the input voltage v_i and the resulting circuit behaves as a *logarithmic amplifier.* Fig. 12.17 below shows the circuit diagram of a logarithmic amplifier using OP AMP.

The diode current i for the forward voltage v_f is given by

$$i = I_s \left[\exp\left(\frac{ev_f}{\eta kT}\right) - 1\right] \tag{12.8.24}$$

where I_s is reverse saturation current. $\eta = 1$ for Ge and approximately 2 for Si.

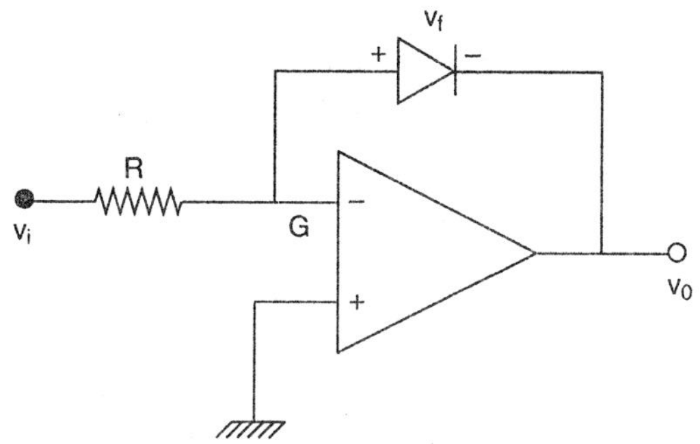

Fig. 12.17 Logarithmic amplifier

If $\dfrac{ev_f}{\eta kT} \gg 1$, we get from equation (12.8.24)

$$i = I_s \exp\left(\frac{ev_f}{\eta kT}\right)$$

$$\text{or,} \quad \frac{ev_f}{\eta kT} = \ln\left(\frac{i}{I_s}\right)$$

$$\text{or,} \quad v_f = \frac{\eta kT}{e} \ln\left(\frac{i}{I_s}\right) \tag{12.8.25}$$

Since the point G acts as a *virtual* ground in Fig. 12.17, we have

$$i = \frac{v_i - 0}{R} = \frac{v_i}{R}$$

and the output-voltage becomes

$$v_O = -v_f = -\frac{\eta kT}{e} \ln\left(\frac{v_i}{I_s R}\right) \tag{12.8.26}$$

Thus the output voltage is proportional to the logarithm of the input voltage.

• If the position of R and the diode are interchanged, we get the circuit of an amplifier known as ***antilogarithmic amplifier.***

12.8.14 Active diode circuit or precision rectifier

An ordinary diode rectifier can only rectify a sinusoidal voltage having peak value greater than the cut-in voltage V_γ (0.7 V for Si) of the diode. In power amplifier, the input voltages being much larger than the cut-in voltages, this limitation does not impose a serious problem. But in some applications, particularly in instrumentation, the 0.7 V cut-in voltage is large enough to affect the performance of the circuit. This cut-in voltage problem can be virtually eliminated by placing the diode in the feedback loop of an OP AMP (Fig. 12.18). The combination of a diode and an OP AMP behaves like an ideal diode and a superior rectification is obtained. So, this rectifier is called ***precision rectifier***. It is also known as ***super diode***. Fig. 12.18 shows a simple precision rectifier circuit.

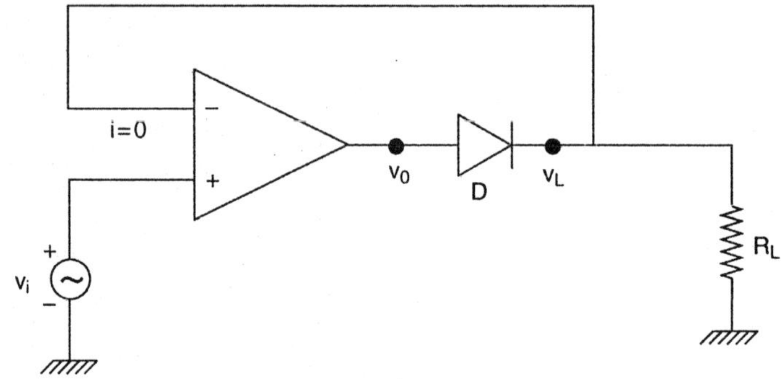

Fig. 12.18 A precision rectifier

For $V_\gamma = 0.7\,\text{V}$ and open-loop gain $A = 10^5$, we get $V_i = V_\gamma/A = 7\mu\text{V}$. So, when the input voltage v_i is slightly positive ($v_i > 7\mu\text{V}$), the output voltage v_O becomes much greater than cut-in voltage V_γ of the diode due to the large open-loop gain of the OP AMP and the diode becomes forward biased to conduct. As the feedback loop of the OP AMP is closed and the circuit behaves as a voltage follower i.e., $v_L = v_i$, v_L follows the positive variations of v_i.

When the input voltage v_i is negative, the output voltage v_O is also negative. As a result, the diode is reverse biased and the feedback loop opens. So, v_O no longer follows v_i and the current i at the inverting input terminal of the OP AMP is zero.

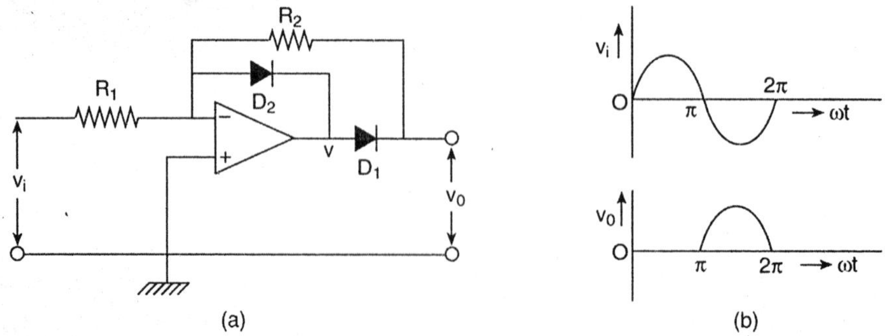

Fig. 12.19 (a) Precision half-wave rectifier (b) Input and output waveforms

The load current i_L and the load voltage v_L are also zero. Thus the output voltage v_L across the load R_L is almost a perfect half-wave. The basic circuit configuration shown in Fig. 12.18 has a problem of poor frequency response. So this configuration is not commonly used. An improved and alternative version of the precision half-wave rectifier is shown in Fig. 12.19(a).

During the positive half-cycle of the input voltage v_i, the OP AMP output v remains negative because v_i is fed to the inverting terminal of the OP AMP. So D_1 is reverse biased and D_2 is forward biased. As a result, diode D_2 is *on* and D_1 is *off*, providing a negative feedback for the input current. The output v_0 is zero because one terminal of R_2 is connected to the virtual ground G and no current flows through it.

During the negative half-cycle of the input voltage v_i, the OP AMP output v is positive. In this condition, D_1 is forward biased and D_2 is reverse biased. So, D_1 is *on* and D_2 is *off*. If we neglect the diode resistance, then the gain of the amplifier will be $-R_2/R_1$ and hence the output $v_0 = -(R_2/R_1)v_i$ is positive as illustrated in Fig. 12.19(b).

The circuit diagram of a precision full wave rectifier is shown in Fig. 12.20. In this circuit, two half-wave rectifier circuits are combined. The input connections for the second OP AMP are reversed with respect to the first OP AMP.

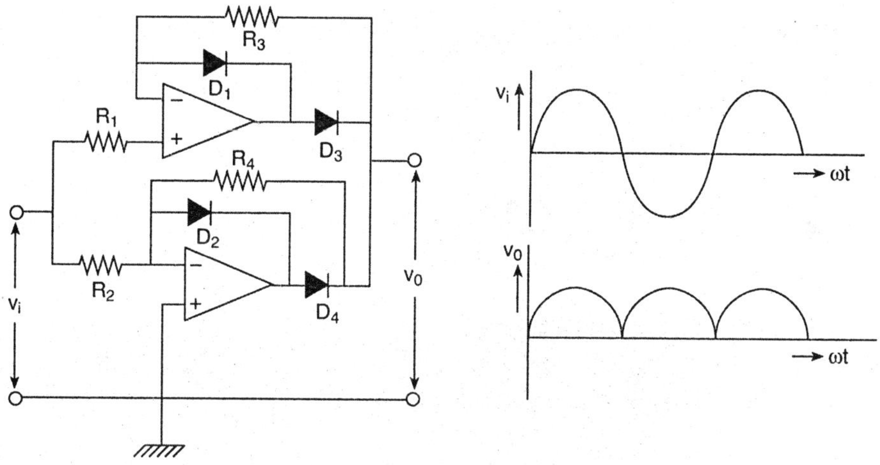

(a) (b)

Fig. 12.20 (a) Precision full wave rectifier (b) Input and output voltage waveforms

12.8.15 Peak detector

The circuit diagram of an ***active peak detector*** using OP AMP is shown in Fig. 12.21.

In this circuit, the capacitor connected at the output terminal is charged by the amplifier output current through the diode D. When the input voltage v_i is greater than the output voltage v_0, the OP AMP voltage v becomes positive and the diode D is forward biased and hence it conducts. As the circuit behaves as a voltage follower,

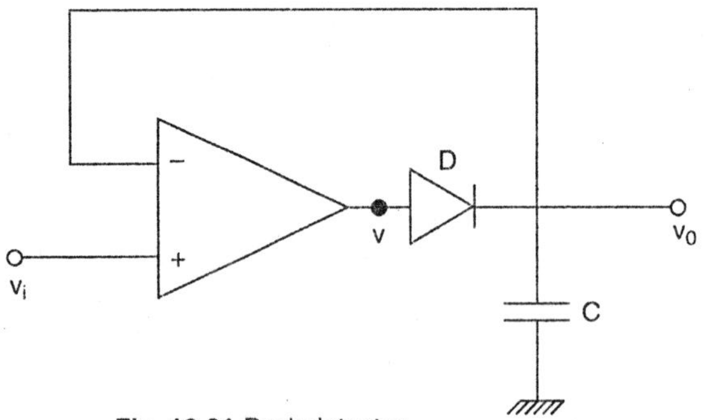

Fig. 12.21 Peak detector

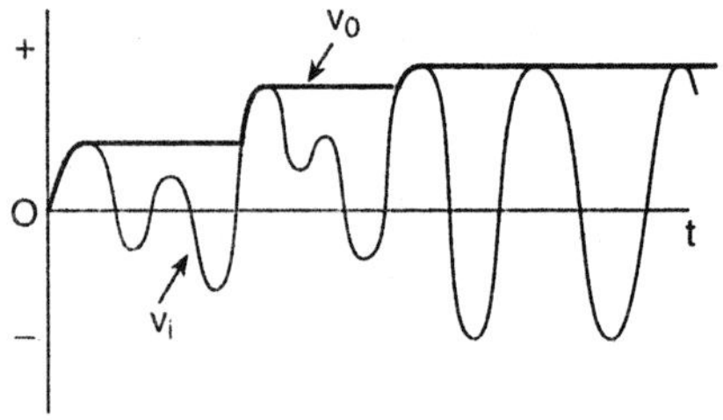

Fig. 12.22 Input and output voltage waveforms

the output of the amplifier $v = v_i$ and the capacitor is charged by the amplifier output current through D. When the input voltage v_i drops below the capacitor voltage v_0, the OP AMP output v becomes negative and the diode D is reverse biased and it turns *off*. In this condition, the feedback path becomes open and the capacitor cannot discharge. It then holds the most positive peak value (Fig.12.22) even when the input signal is removed. After the operation, the circuit is reset by placing a low-leakage switch such as a MOSFET across the capacitor to discharge it.

12.8.16 Voltage comparator

A ***voltage comparator*** is a device which can compare two voltage levels. It is a switching device which gives an output voltage when one input voltages is larger, and another output voltage when the other input voltage is larger. Thus the output of the comparator indicates which of the two input voltages is greater. An OP AMP operating in the open-loop condition can be used as a comparator. The two voltages to be compared are then applied to the inverting and non-inverting inputs. When the voltage to the non-inverting input terminal (v_1) is slightly greater than the voltage to the inverting input terminal (v_2), the OP AMP output quickly switches to maximum positive output voltage +V. When v_2 is slightly greater than v_1, the OP AMP switches to the maximum value of negative output voltage −V. The output voltage switches to zero when $v_1 = v_2$, i.e., $v_d = v_1 - v_2 = 0$.

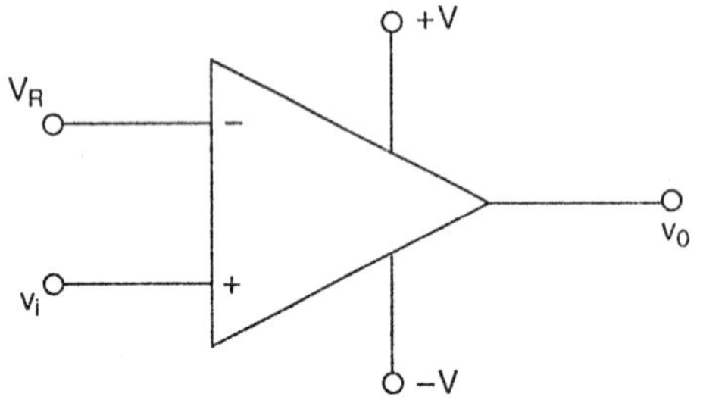

Fig. 12.23 Voltage comparator

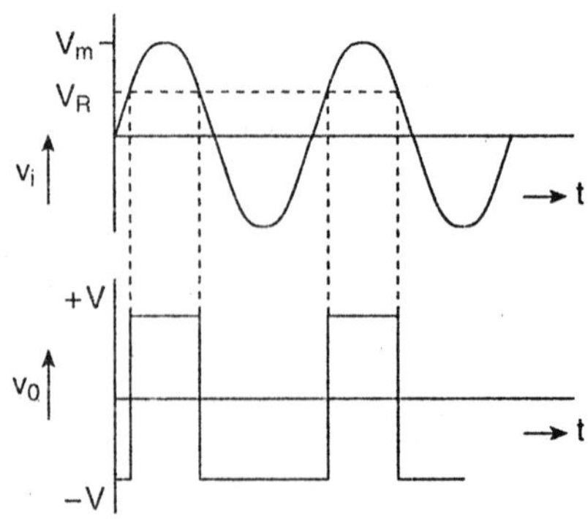

Fig. 12.24 Input and output waveforms of voltage comparator

Fig. 12.23 shows an OP AMP comparator which is in open-loop condition with voltages +V and −V. This comparator circuit compares the input signal $v_i = V_m \sin \omega t$ applied to the non-inverting input terminal with a reference voltage V_R applied to the inverting terminal. It is to be noted that $V > V_m > V_R$. When v_i exceeds V_R, the output voltage v_O switches to +V. As long as $v_i > V_R$, the output voltage stays at +V. When v_i drops below V_R, the output voltage of the comparator becomes −V. Fig. 12.24 shows the input and output voltages of the comparator.

A comparator may be used as a *zero-detector* if the inverting or non-inverting input terminal is grounded. When the inverting input terminal is grounded (i.e., $V_R = 0$), the output v_O switches to +V when v_i is slightly positive i.e., v_i passes through zero towards positive values. On the other hand, when v_i crosses zero towards negative values, the output v_O is driven to −V. Thus a zero-crossing detector can be used to generate a symmetric square wave from sine waves (Fig. 12.25).

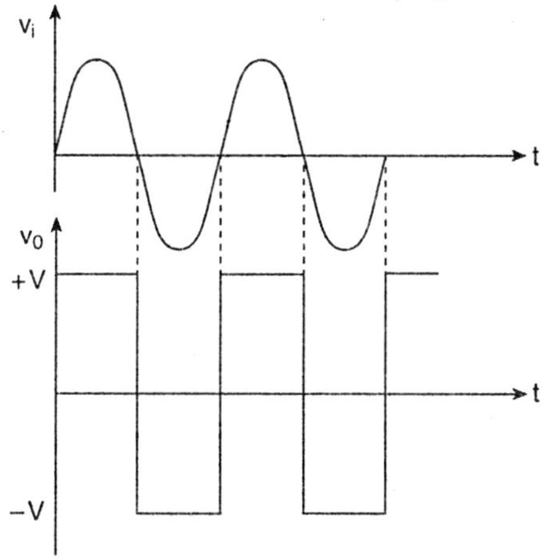

Fig. 12.25 Generation of symmetric square wave from sinusoidal wave by zero-crossing detector

12.9 Schmitt trigger or regenerative comparator

A comparator circuit which can switch from one voltage level to another by using positive or regenerative feedback and can exhibit the phenomenon of *hysteresis* or *backlash* is called a ***regenerative comparator*** or more commonly, a ***Schmitt trigger***, after the original developer of vacuum tube version of circuit.

The circuit of an OP AMP Schmitt trigger is shown in Fig. 12.26. The input voltage v_i is applied to the inverting terminal and the feedback voltage v_f to the non-inverting terminal.

Using the superposition principle, we have the feedback voltage

$$v_f = \frac{R_2 V_R}{R_1 + R_2} + \frac{R_1 v_O}{R_1 + R_2} \qquad (12.9.1)$$

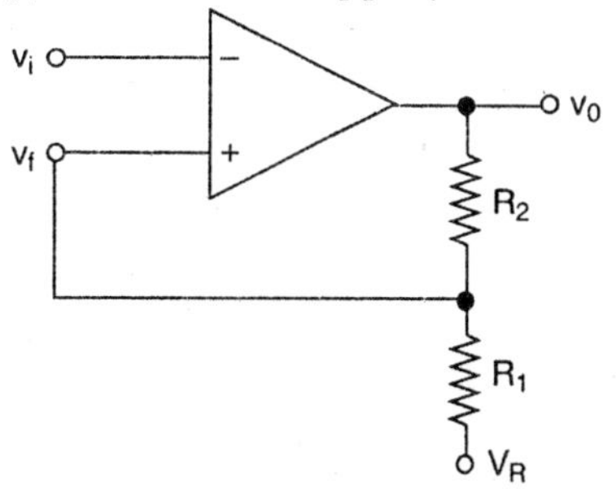

Fig. 12.26 OP AMP Schmitt trigger

If $v_i \ll v_f$, the output v_O will saturate to the positive saturation voltage $+V_{sat}$ giving $v_O = V_{sat}$. Then the equation (12.9.1) gives

$$v_f = \frac{R_2 V_R}{R_1 + R_2} + \frac{R_1 V_{sat}}{R_1 + R_2} = V_1 \text{ (say)}. \qquad (12.9.2)$$

If now v_i is increased in the positive direction, the output v_O will remain unchanged at V_{sat} until $v_i = V_1$, the *threshold* or *critical* or *trigger voltage*. At the trigger voltage V_1, the output of the amplifier regeneratively switches to $v_O = -V_{sat}$, the negative saturation level. The output remains at this value as long as $v_i > V_1$. Fig. 12.27(a) corresponds to this transfer characteristic.

The feedback voltage v_f at the non-inverting terminal for $v_i > V_1$ is given by

$$v_f = \frac{R_2 V_2}{R_1 + R_2} - \frac{R_1 V_{sat}}{R_1 + R_2} = V_2 \text{ (say)}. \qquad (12.9.3)$$

If v_i is now decreased, the output will remain unchanged at $-V_{sat}$ until $v_i = V_2$.

At $v_i = V_2$, the output regeneratively switches to $v_O = +V_{sat}$. Fig. 12.27(b) shows this transfer characteristic. The overall transfer characteristic of the Schmitt trigger is shown in Fig. 12.27(c). This curve has the form of magnetic hysteresis loop and hence, called a *hysteresis curve*. This curve indicates that the circuit triggers at a higher voltage for increasing input signals than for the decreasing input signals. The two input voltages at which the transitions of output voltage occur are called *trip points*. V_1 and V_2 are known as the *upper trip point* (UTP) and the *lower trip point* (LTP) respectively. The difference between these two points is the width of the hysteresis loop. The width is thus given by

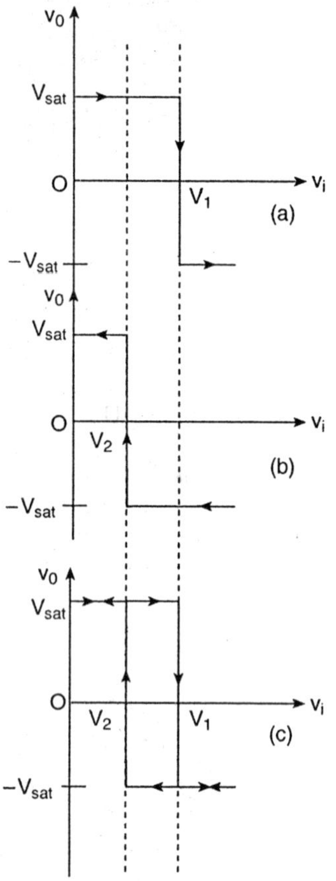

Fig. 12.27 Transfer characteristics of Schmitt trigger

$$V_H = V_1 - V_2 = \frac{2R_1 V_{sat}}{R_1 + R_2} \qquad (12.9.4)$$

- The Schmitt trigger may remain in one of its two stable states indefinitely. When a triggering signal is applied, it then changes state. So, a Schmitt trigger circuit behaves as a *bistable multivibrator*.

- In a Schmitt trigger, hyteresis is desirable to prevent the noise associated with the input signal. For $V_H = 0.3\,\text{V}$, no false triggering in the Schmitt trigger occurs as long as the peak-to-peak noise voltage is smaller than $0.3\,\text{V}$

● Since the circuit of a Schmitt trigger compares the input voltage with the trigger level V_1 and V_2 for the transition or swing of the output voltage between $-V_{sat}$ to $+V_{sat}$, it is a *comparator*. It is a *regenerative comparator* because it uses positive or regenerative feedback.

An *important application* of the Schmitt trigger is to convert a slowly varying input voltage into a square wave output voltage. For this, consider an arbitrary (symmetric or asymetric) input signal v_i plotted against t as shown in Fig. 12.28(a). The signal has an amplitude large enough to exceed beyond V_1 and V_2. When the input voltage v_i exceeds V_1, the output voltage v_0 switches abruptly to $-V_{sat}$ and when v_i drops below V_2, the output swings to $+V_{sat}$. Thus the output voltage is a square wave (Fig. 12.28b) with peak value $+V_{sat}$ to $-V_{sat}$ and is independent of the shape of the applied input signal (Fig. 12.28). Used in this way, Schmitt trigger acts as a symmetric or a non-symmetric *squaring circuit* or ***squarer***.

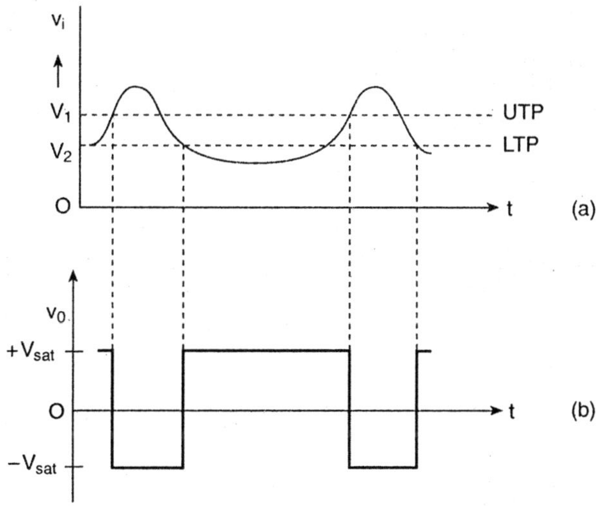

Fig. 12.28 Conversion of a slowly varying input voltage into a square wave by Schmitt trigger

12.10 OP AMP as analog computer

An ***analog computer*** is a computer which is used to process analog data. Analog computers store data in a continuous form of physical quantities and perform calculations with the help of measures. OP AMPs are the basic building blocks of an electronic analog computer. Although digital computers are mostly used in different fields of science, there are few areas where analog computers are suitably used. An analog computer does not require any programming language. So, this is an *advantage* of the analog computer over the digital one. To illustrate the usefulness of OP AMP in analog computer, we shall consider the following two cases: the solutions of (i) second order differential equation and (ii) simultaneous linear algebraic equations.

12.10.1 Solution of differential equation

Consider the following second order differential equation

$$\frac{d^2v}{dt^2} + P\frac{dv}{dt} + Qv = \omega \qquad (12.10.1)$$

where P and Q are real constants and ω is a given function of time.

For convenience, the above equation may be rewritten as

$$\frac{d^2v}{dt^2} = -P\frac{dv}{dt} - Qv + \omega \qquad (12.10.2)$$

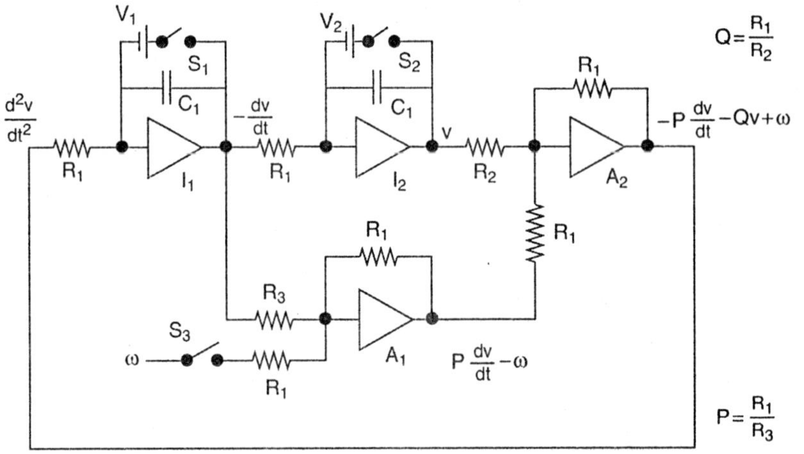

Fig. 12.29 Solution of differential equation by analog computer

To start with, let us assume that d^2v/dt^2 is available in voltage form at the input of the first integrator I_1 of the circuit of Fig. 12.29. If we choose the time constant R_1C_1 of the integrator as 1 second, then its output will be $-dv/dt$. The output of integrator I_1 i.e., $-dv/dt$ is now applied to the input of a second identical integrator I_2 which gives an output v.

The output of the first integrator (i.e., $-dv/dt$) is also applied to the input of an adder A_1 where the given function ω in the form of voltage is also applied as second input. The output of A_1 is thus $P\frac{dv}{dt} - \omega$ where $P = R_1/R_3$. Now the output of the second integrator (i.e., v) and that of the first adder A_1 $\left(\text{i.e., } P\frac{dv}{dt} - \omega\right)$ are applied to the input of a second adder A_2. The output of A_2 will be $-P\frac{dv}{dt} - Qv + \omega$ where $Q = R_1/R_2$. Equation (12.10.2) indicates that this output is equal to d^2v/dt^2 which is the input of the integrator I_1. So, in order to supply the assumed voltage d^2v/dt^2 at the input of the integrator I_1, the output of A_2 is connected to the input of the first integrator.

For the incorporation of initial conditions i.e., setting the values of dv/dt and v at $t = 0$, proper dc voltages are applied across the capacitors of both the integrators I_1

and I_2. The complete solution of equation (12.10.1) is then obtained at the output of the second integrator I_2 with switches S_1 and S_2 opened and S_3 closed simultaneously by means of relay at $t = 0$.

The solution of the given equation may be observed by employing a low frequency CRO. The variation of dv/dt with time can also be visualised by connecting the output of the integrator I_2 to the CRO.

● Integrators are always preferred to differentiators to construct an electronic analog computer to solve differential equations. There is a good number of reasons for this and they are enumerated below.

(a) The gain of an integrator decreases with increasing frequency whereas the gain of a differentiator increases with increasing frequency. So, the stability of an integrator is better than that of a differentiator with respect to high frequency spurious signals.

(b) An integrator is less sensitive to noise voltage than a differentiator.

(c) It is easier to incorporate initial conditions to an integrator.

(d) For a rapidly changing input signal, a differentiator is more likely to be over loaded.

12.10.2 Solution of simultaneous linear algebraic equations

Let us consider the following two linear simultaneous equations:

$$a_1x + b_1y = c_1 \tag{12.10.3}$$

$$\text{and} \quad a_2x + b_2y = c_2 \tag{12.10.4}$$

where a_1, b_1, c_1, a_2, b_2 and c_2 are all constants.

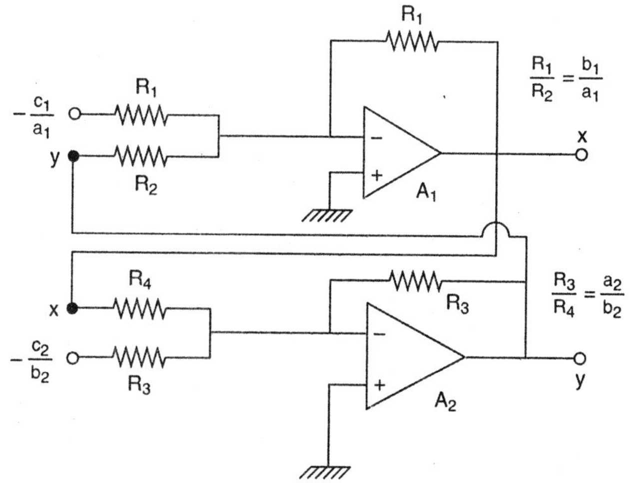

Fig. 12.30 Solution of simultaneous linear algebraic equations by analog computer

To solve for x and y, we rewrite the above two equation as

$$x = \frac{c_1}{a_1} - \frac{b_1}{a_1}y \tag{12.10.5}$$

$$\text{and } y = \frac{c_2}{b_2} - \frac{a_2}{b_2}x \tag{12.10.6}$$

These two equations can be implemented by using OP AMP circuit of Fig. 12.30.

To start with, let us assume that $-c_1/a_1$ and y are available in the form of voltages which are the two inputs of the adder A_1.

The output of A_1 will appear as $x = c_1/a_1 - (b_1/a_1)y$. if we set $R_1/R_2 = b_1/a_1$. According to equation (12.10.6) y is obtained as the output of the second adder A_2 if $-c_2/b_2$ and x are available as the two input voltages and $R_3/R_4 = a_2/b_2$.

Thus the proper assembling of resistances and the inputs of two adders will give the solutions in terms of voltages at the outputs of the two adders. For the accurate resistance ratios and input voltages, high resolution potentiometers are employed in practical circuits.

12.11 Illustrated Examples

► **Example 1.** *An inverting amplifier has $R_1 = 20\,\text{k}\Omega$ and $R_f = 100\,\text{k}\Omega$. Find the output voltage and the input current for an input voltage of $1\,\text{V}$.* **(Punjab Univ.)**

Solution: Here, the input resistance, $R_1 = 20\,\text{k}\Omega$

Feedback resistance, $R_f = 100\,\text{k}\Omega$

and the input voltage, $v_i = 1\,\text{V}$.

The closed-loop voltage gain of an inverting amplifier gives,

$$\frac{v_0}{v_i} = -\frac{R_f}{R_1}$$

$$\therefore \textbf{\textit{Output voltage,}} \quad v_0 = -v_i\frac{R_f}{R_1} = -\left(1 \times \frac{100 \times 10^3}{20 \times 10^3}\right) = \mathbf{-5V}$$

$$\text{The } \textbf{\textit{input current}} = \frac{v_i}{R_1} = \frac{1}{20 \times 10^3}\,\text{A} = 0.05 \times 10^{-3}\,\text{A} = \mathbf{0.05\,mA}$$

► **Example 2.** *Calculate the actual output voltage of an integrator after $2\,\text{s}$ for the input voltage of $1\,\text{V}$ d.c. Given that the input resistance $= 100\,\text{k}\Omega$ and the feedback capacitance $= 1\,\mu\text{F}$.* **(Calcutta Hons.)**

Solution: Given, the input resistance, $R = 100\,\text{k}\Omega = 100 \times 10^3\Omega$

Feedback capacitance, $C = 1\,\mu\text{F} = 1 \times 10^{-6}\,\text{F}$

Input voltage, $v_i = 1\,\text{V}$ and time, $t = 2\,\text{s}$

$$\therefore \text{ The } \textbf{\textit{output voltage}}, \ v_0 = -\frac{1}{CR}\int_0^t v_i dt$$
$$= -\frac{1}{10^{-6} \times 100 \times 10^3}\int_0^2 1\, dt$$
$$= -\frac{1}{10^{-1}} \times 2 \text{ volt} = \mathbf{-20 \text{ volt}}$$

▶ **Example 3.** *A ramp voltage of* 1.5 V per millisecond (Fig. 12.31) *is applied to an OP AMP differentiator for having $R = 2\,\text{k}\Omega$ and $C = 0.01\,\mu\text{F}$. Find the output voltage and its waveform.* **(Calcutta Hons.)**

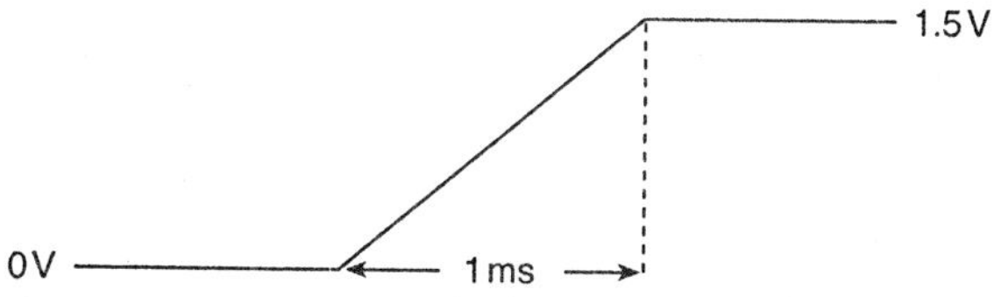

Fig. 12.31

Solution: From Fig. 12.31, $\frac{dv_i}{dt} = \frac{1.5\,\text{V}}{1\,\text{ms}} = 1.5\,\text{V/ms}$, for $0 < t < 1\,\text{ms}$

and $\frac{dv_i}{dt} = 0$, for outside this interval.

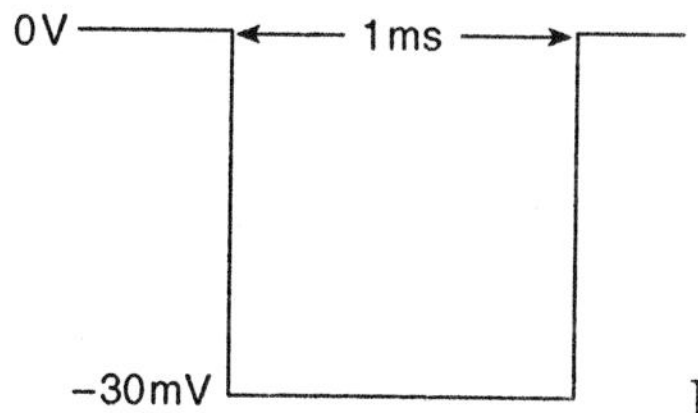

Fig. 12.32

Given, $R = 2\,\text{k}\Omega = 2 \times 10^3 \Omega$

and $C = 0.01\,\mu\text{F} = 0.01 \times 10^{-6}\,\text{F}$.

$\therefore \quad RC = 2 \times 10^3 \times 0.01 \times 10^{-6}\,\text{s} = 0.02\,\text{ms}$

Hence the ***output voltage*** for $0 < t < 1$ ms is given by

$$v_0 = -RC\frac{dv_i}{dt} = -(0.02 \times 1.5)\,\text{V} = -0.03\,\text{V} = \mathbf{-30\,mV}$$

Otherwise, however, $\boldsymbol{v_0 = 0}$.

The ***waveform*** of the output voltage is illustrated in Fig. 12.32 above.

▶ **Example 4.** *Calculate the voltage V_0 in the circuit of* Fig. 12.33. **(Calcutta Hons.)**

Solution: Since the gain of OP AMP is infinite, the potential of the point X in the circuit is 4 V. Applying KCL at X,

$$\frac{V_0 - 4}{10\,\text{k}\Omega} = \frac{4-5}{5\,\text{k}\Omega},$$

since the input impedance of OP AMP is infinite.

$$\therefore \quad \mathbf{V_0 = 2V}$$

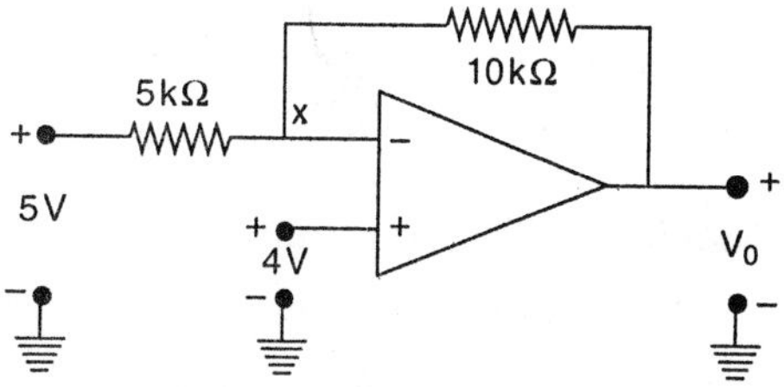

Fig. 12.33

► **Example 5.** *Draw a circuit using one or more OP AMPs whose output v_0 is given as $v_0 = 4v_1 + 6v_2$, where v_1 and v_2 are the two input signals.* **(Burdwan Hons.)**

Solution: To design the circuit for the given output voltage, we have to first construct an inverting adder by a proper choice of feedback resistance and input resistances to produce the factors 4 and 6. Then this inverting adder must be followed by an inverter.

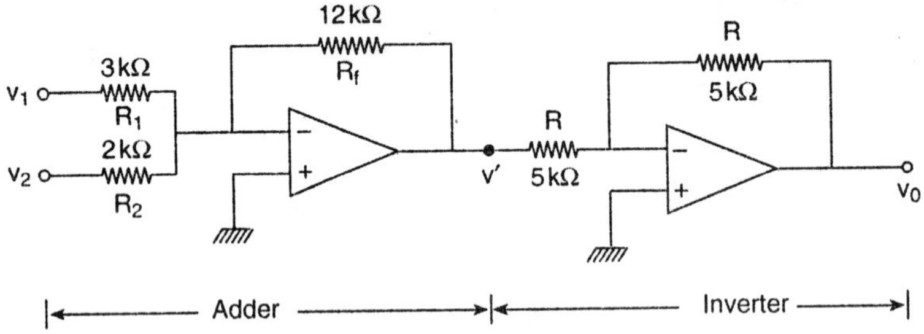

Fig. 12.34

Fig. 12.34 shows that the output of the adder is

$$v' = -\left(\frac{R_f}{R_1}v_1 + \frac{R_f}{R_2}v_2\right) = -(4v_1 + 6v_2)$$

and the inverter produces the output voltage

$$v_0 = -v' = 4v_1 + 6v_2$$

► **Example 6.** *Show that the output voltage of the circuit in* Fig. 12.35 *is given by*

$$v_0 = v_2\,\frac{R_1 + R_2}{R_3 + R_4}\cdot\frac{R_4}{R_1} - v_1\,\frac{R_2}{R_1}$$

(Burdwan Hons.)

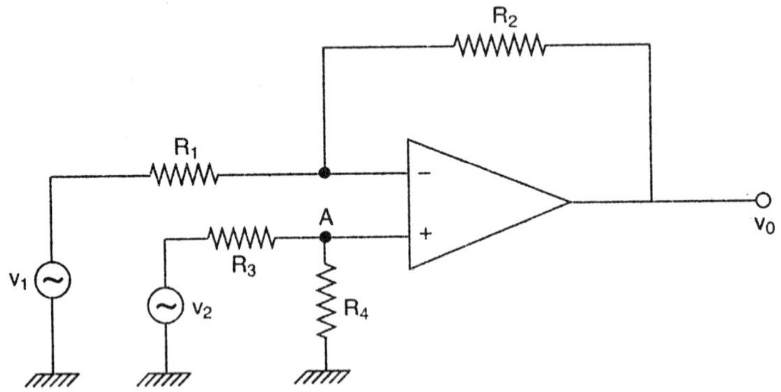

Fig. 12.35

Solution: The resultant output voltage is obtained by the superposition principle.

The input voltage v_1 alone produces an output given by

$$v' = -\frac{R_2}{R_1}v_1.$$

The voltage at the non-inverting terminal is given by

$$v_A = \frac{v_2}{R_3 + R_4} R_4.$$

But v_A alone produces an output given by

$$v'' = \left(1 + \frac{R_2}{R_1}\right) v_A = \frac{v_2 R_4}{R_3 + R_4} \cdot \frac{R_1 + R_2}{R_1}$$

$$= v_2 \cdot \frac{R_1 + R_2}{R_3 + R_4} \cdot \frac{R_4}{R_1}$$

The resultant output v_0 is, therefore, given by, $v_0 = v'' + v'$.

$$\therefore \quad \boldsymbol{v_0 = v_2 \frac{R_1 + R_2}{R_3 + R_4} \cdot \frac{R_4}{R_1} - v_1 \frac{R_2}{R_1}}$$

▶ **Example 7.** *In reference to the circuit given in Fig.* 12.35, *if* $R_1 = 1\,\text{k}\Omega$, $R_2 = 5\,\text{k}\Omega$, $R_3 = 1\,\text{k}\Omega$, $R_4 = 1\,\text{k}\Omega$, $v_1 = 1\,\text{V}$ *and* $v_2 = 4\,\text{V}$, *compute separately the output voltage produced by input voltage* v_1 *alone and that produced by the voltage at the non-inverting terminal alone. Hence find the total output.*

Solution: Output voltage due to v_1 alone is $v' = -\frac{R_2}{R_1} v_1 = -\frac{5}{1} \times 1 = \mathbf{-5\,V}$.

Voltage v_A at the non-invewrting terminal A is given by

$$v_A = \frac{v_2}{R_3 + R_4} R_4$$

So, the output produced by v_A alone is

$$v'' = \left(1 + \frac{R_2}{R_1}\right) v_A = \frac{R_1 + R_2}{R_1} \times \frac{v_2 R_4}{R_3 + R_4}$$

$$= \frac{(R_1 + R_2) R_4}{(R_3 + R_4) R_1} v_2 = \frac{1 + 5}{1 + 1} \times \frac{1}{1} \times 4$$

$$= \mathbf{12\,V}.$$

Hence, the total output, $v_0 = v' + v'' = (-5 + 12)\text{V} = \mathbf{7\,V}$

▶ **Example 8.** *In a differential amplifier with two inputs, the output is* 2.01 mV *when the inputs are* $110\,\mu\text{V}$ *and* $90\,\mu\text{V}$ *but the output is* 2 mV *when the inputs are* $10\,\mu\text{V}$ *and* $-10\,\mu V$. *Find the CMRR of the amplifier.* **(Burdwan Hons.)**

Solution: In the first case, $v_1 = 110\,\mu\text{V}$, $v_2 = 90\,\mu\text{V}$ and $v_0 = 2.01\,\text{mV}$.

$$\therefore \quad v_d = 20\,\mu\text{V} \;\text{ and }\; v_c = \frac{110 + 90}{2}\,\mu\text{V} = 100\,\mu\text{V}$$

In the second case, $v_1 = 10\,\mu\text{V}$, $v_2 = -10\,\mu\text{V}$, $v_0 = 2\,\text{mV}$. So, $v_d = 20\,\mu\text{V}$ and $v_c = 0$.

Applying the relation $v_0 = A_d v_d + A_c v_c$ to the second case,

$$2 \times 10^{-3} = A_d \times 20 \times 10^{-6} + A_c \cdot 0$$
$$\text{or,}\quad A_d = 100$$

Applying $v_0 = A_d v_d + A_c v_c$, to the first case, we obtain

$$2.01 \times 10^{-3} = 100 \times 20 \times 10^{-6} + A_c \times 100 \times 10^{-6}$$
$$\Rightarrow A_c = \frac{1}{10}$$

$\therefore$ The value of **CMRR** is $\left|\frac{A_d}{A_c}\right| = \left|\frac{100}{1/10}\right| = \mathbf{1000}$.

▶ **Example 9.** *Consider the differential input amplifier of* Fig. 12.35. *Assuming infinite input resistance, zero output resistance and finite differential gain* $A_V = V_0/(V_1 - V_2)$, *show that closed loop gain is given by* $A_{V_f} = A_V\big/\left(1 + \frac{A_V}{1+n}\right)$.

Explain the case when $A_V \to \infty$.

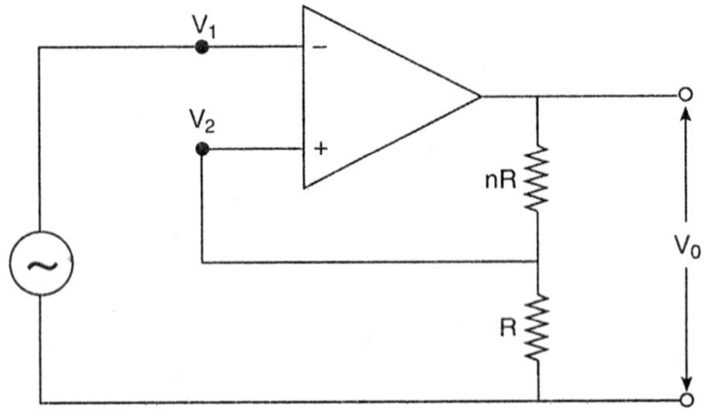

Fig. 12.36

Solution : Here, the input voltage, $V_1 = V_s$ and

$$\text{the input voltage, } V_2 = \frac{R}{nR + R} V_0 = \frac{V_0}{1+n}$$

$$\therefore \text{ The output voltage, } V_0 = A_V(V_1 - V_2) = A_V\left(V_s - \frac{V_0}{1+n}\right), \text{ and}$$

$$\text{the closed loop gain, } \boldsymbol{A_{V_f}} = \frac{V_0}{V_s} = \frac{\boldsymbol{A_V}}{1 + \dfrac{\boldsymbol{A_V}}{\mathbf{1+n}}}$$

Now, $A_{V_f} = \dfrac{1}{\left(\frac{1}{A_V} + \frac{1}{1+n}\right)}$. So, when $A_V \to \infty$ we have $\boldsymbol{A_{V_f} \to (1+n)}$.

Review Questions

1. What is an OP AMP? What are the characteristics of an ideal OP AMP? Give its circuit symbols and state some of its applications. **(Punjab Univ., Calcutta Hons.)**

2. (a) Describe inverting and non-inverting operational amplifiers. **(Punjabi Univ.)**
   (b) Discuss OP AMP as scale changer and voltage follower. **(Osmania Univ.)**

3. (a) Explain the 'virtual ground' concept in OP AMP with diagram.
   (b) Show how an OP AMP can be used as (a) an adder (or summing amplifier) and (b) a differential amplifier. **(Calcutta Hons.)**

4. Show how an OP AMP may be used as (a) a phase shifter, (b) an integrator and (c) a differentiator. **(Nagpur Univ., Calcutta Hons.)**

5. Explain the use of OP AMP as (a) voltage to current converter and (b) current to voltage converter. **(Nagpur Univ.)**

6. (a) What is an OP AMP? Why is it so-called?
   (b) How do the characteristics of an ideal OP AMP differ from those of a practical OP AMP?

7. (a) What type of feedback is used in an OP AMP adder? Justify your answer. **(Calcutta Hons.)**
   (b) Draw the pin diagram of an OP AMP IC and mention the lead connections. What are the advantages of using OP AMP ICs?

8. Explain clearly the concept of 'virtual ground' at the input terminals of a non-inverting amplifier.

9. (a) Define the following terms in connection with an operational amplifier: Difference signal, common-mode signal, common-mode rejection ratio, input bias current, input offset current, input offset voltage, output offset voltage, power supply rejection ratio and slew rate.
   (b) How is a potentiometer used with a general purpose OP AMP for offset null adjustment?

10. Derive expressions for the voltage gain and the input resistance of an inverting amplifier using an OP AMP. How is the effect of the input bias current minimized?

11. Draw the circuit diagram of a difference amplifier using an OP AMP and find an expression for the output voltage. **(Calcutta Hons.)**

12. Show how a logarithmic amplifier can be built with an OP AMP. What happens if the diode and the resistor R in Fig. 12.17 are interchanged?

13. (a) How can an OP AMP be used as (i) a comparator? (ii) a precision rectifier and (iii) a square wave generator?

(b) Explain how you can construct using an OP AMP and other components (i) a peak detector, (ii) a half-wave rectifier and (iii) a full-wave rectifier.

14. What is Schmitt trigger? Draw schematically an OP AMP Schmitt trigger and explain its operation. Name an application of the Schmitt trigger.

15. Construct an OP AMP circuit to solve the following differential equation:

$$\frac{d^2v}{dt^2} + P\frac{dv}{dt} + Qv = v_1$$

where P and Q are real positive constants and v_1 is a given function of t.

16. Construct an analog computer by which you can solve the following simultaneous equations:

$$a_1x + b_1y = c_1$$

$$\text{and} \quad a_2x + b_2y = c_2$$

17. What is slew rate? How does it limit the high frequency operation of an OP AMP?

18. Sketch the input-output characteristic of an OP AMP. Hence explain why an open-loop OP AMP is unstable for linear applications.

19. Explain how an OP AMP may be used as a unity gain buffer or a voltage follower.

20. (a) Sketch the transfer characteristic of a Schmitt trigger and indicate the hysteresis.

(b) What are the parameters that determine the hysteresis?

(c) Indicate one method of obtaining a non-symmetrical square-wave.

(Calcutta Hons.)

21. Design a circuit using one or more OP AMPs whose output is given as $v_0 = 2v_1 + 5v_2$ where v_1 and v_2 are the two inputs.

Objective Type Questions

1. The output of the circuit given below will be

(a) 1 V, (b) 11 V, (c) −10 V, (d) 0 V.

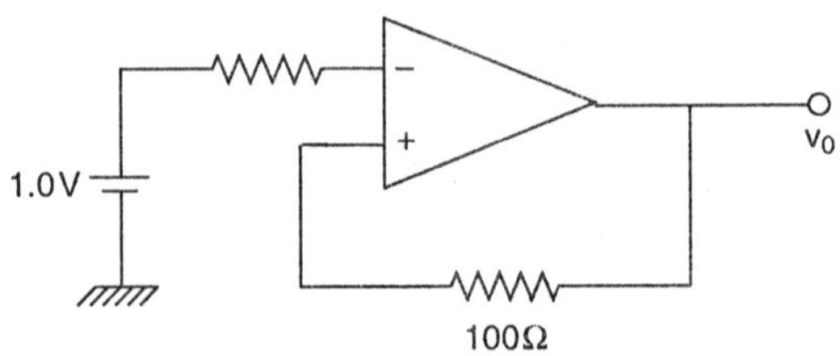

Fig. 12.37

(GATE)

2. The inverting input terminal of an operational amplifier (OP AMP) is shorted with the output terminal, apart from being grounded. A voltage v_i is applied to the non-inverting input terminal of the OP AMP. Under this configuration, the OP AMP functions as

(a) an open loop inverter, (b) a voltage to current converter

(c) a voltage follower, (d) an oscillator. **(GATE)**

3. The output V_0 of the ideal OP AMP circuit shown in Fig. 12.38 is given by

(a) -7 V, (b) -5 V, (c) 5 V, (d) 7 V.

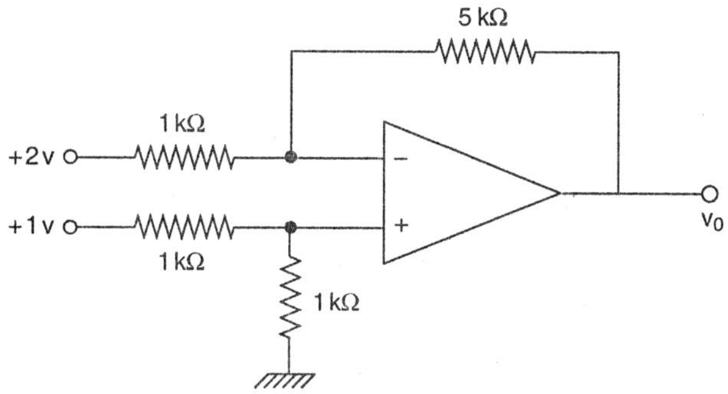

Fig. 12.38 **(GATE)**

4. The voltage gain of an OP AMP non-inverting amplifier is

(a) less than unity, (b) equal to unity, (c) greater than unity.

5. An OP AMP can amplify

(a) only a.c, (b) only d.c, (c) both a.c and d.c.

6. The open-loop gain of a practical OP AM is of the order of

(a) 10^5, (b) 10^3, (c) 10^9, (d) 1.

7. Inside a 741 OP AMP, the functional block is a

(a) differential amplifier, (b) level shifter,

(b) class-A power amplifier, (d) class-AB power amplifier. **(DRDO)**

8. The feedback scheme used in a simple OP AMP based non-inverting amplifier is

(a) series-series, (b) series-shunt,

(c) shunt-series, (d) shunt-shunt. **(DRDO)**

9. A differential amplifier is used in the input state of OP AMP to achieve very high

(a) Open-loop gain, (b) Band width,
(c) Slew rate, (d) CMRR. **(ISRO)**

10. The input stage of a typical OP AMP is usually

(a) a CE amplifier, (b) a differential amplifier,
(c) an emitter follower, (d) a class-B push-pull amplifier.

11. If a square wave is applied to the input of an OP AMP integrator circuit, the output would be a

(a) triangular wave, (b) sine wave,
(c) pulse, (d) ramp.

12. CMRR of a good OP AMP is

(a) less than 1, (b) equal to 1,
(c) very large, (d) equal to common-mode gain.

13. The slew rate of an ideal OP AMP is

(a) infinite, (b) zero, (c) 1 V/s, (d) 1 V/μs

14. An OP AMP comparator circuit employs

(a) positive feedback, (b) no feedback, (c) negative feed back.

Problems

1. An OP AMP inverting amplifier has an input resistor of $10\,\text{k}\Omega$ and a feedback resistor of $50\,\text{k}\Omega$. If the input voltage is 0.5 V, find the output voltage and the input current. **(Calcutta Hons.)** [-2.5 V, 50μA]

2. Compute the voltage gain for the amplifier shown in Fig. 12.39. Find the output voltage v_{out} if the input voltage is $v_{\text{in}} = 0.5\sin 100\pi t$ volt.

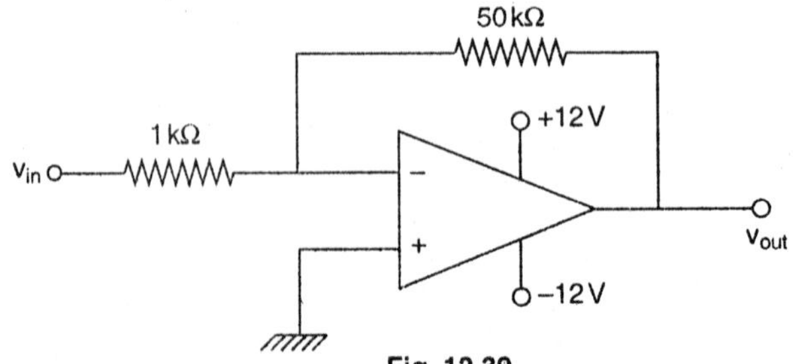

Fig. 12.39 **(Calcutta Hons.)**

[$-25\sin 100\pi t$, for linear operation]

3. Calculate the voltage v_0 in the following circuit of Fig. 12.40.

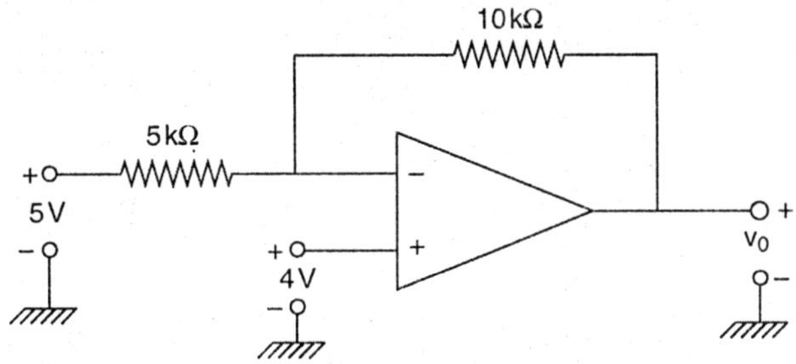

Fig. 12.40 (Calcutta Hons.) [2 V]

4. In the circuit of Fig. 12.14, $R = 100\,\text{k}\Omega$ and $C = 1\,\mu\text{F}$. If the input voltage v_i is a $\pm 10\,\text{V}$, 250 Hz square wave, determine the output voltage v_0. (Calcutta Hons.)
[A triangular wave of amplitude 0.2 V and frequency 250 Hz]

5. Design an OP AMP circuit which will give an output voltage $v_0 = 0.5v_1 - 2v_2$ for two input voltages v_1 and v_2. (Calcutta Hons.)

6. In the following circuit of Fig. 12.41, determine the value of V_0 in volt.

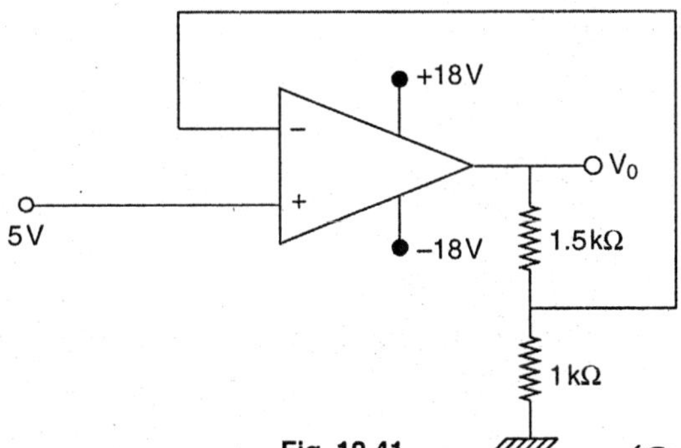

Fig. 12.41 (Calcutta H.) [12.5 V]

7. If in the circuit of Fig. 12.42, the resistance R changes from $1\,\text{k}\Omega$ to $2\,\text{k}\Omega$, explain whether the current through the 500Ω resistance will change. Find the voltage V_0 in each case.

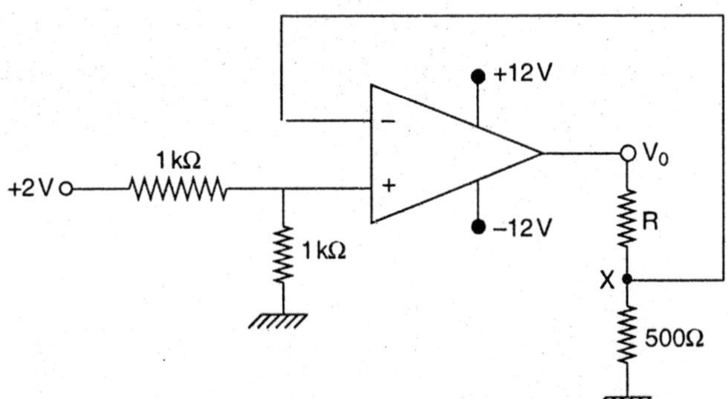

Fig. 12.42 (Calcutta Hons.) [No; 3V, 5V]

8. A voltage $5\sin 2000\pi t$ mV is applied to the input of an OP AMP integrator (Fig. 12.14) for which $R = 100\,\text{k}\Omega$ and $C = 1\,\mu\text{F}$. Calculate the output voltage.

(Agra Univ.) $\left[\frac{1}{40\pi}(\cos 2000\pi t - 1)\,\text{mV}\right]$

9. What is the output voltage in the circuit of Fig. 12.43? Explain the operation of the circuit.

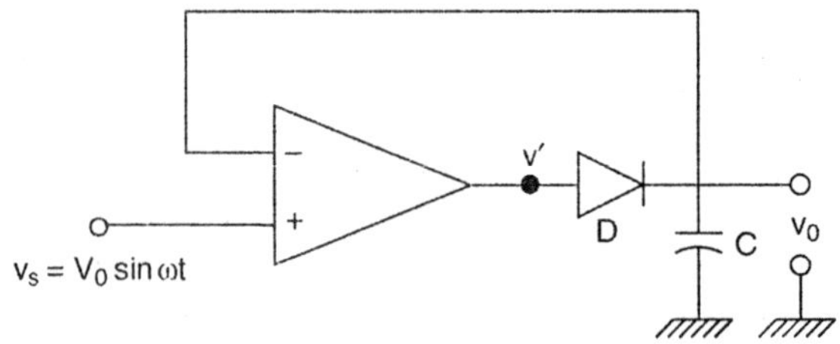

Fig. 12.43

(Calcutta Hons.)

10. An inverting OP AMP has an input resistor of $10\,\text{k}\Omega$, a feedback resistor of $100\,\text{k}\Omega$ and a power supply voltage ± 12V. If the input voltage is 0.5 V, find the output voltage and also the input current. Also find the output corresponding to an input of 3 V. **(Calcutta Hons.)** $[-5\text{V}, 50\mu\text{A}, -12\text{V}]$

11. Design a circuit using one or more OP AMPs whose output is given as $v_0 = 2v_1 + 5v_2$ where v_1 and v_2 are the two inputs. **(Burdwan Hons.)**

12. For the circuit of Fig. 12.44, show that the output voltage is given by

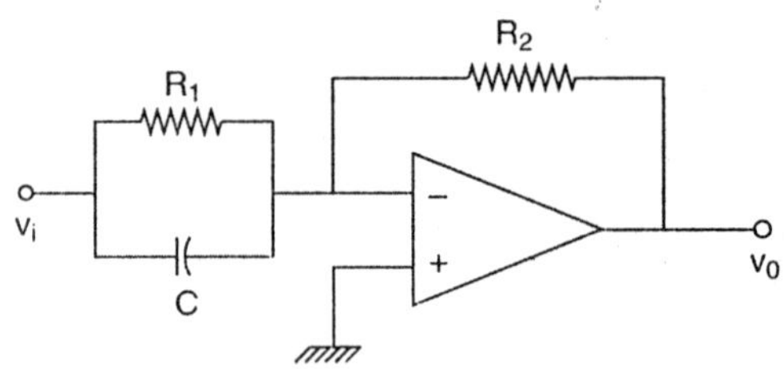

Fig. 12.44

$$v_0 = -\frac{R_2}{R_1}v_i - CR_2\frac{dv_i}{dt}.$$

If the input is a ramp of the form $v_i = \alpha t$ where α is a positive constant, find the output voltage v_0. **(GATE)**

13. Design an OP AMP based circuit whose output would be

$$v_0(t) = 2v_1(t) - 3\int v_2(t)dt$$

where $v_1(t)$ and $v_2(t)$ are the two input signals to the circuit. **(Burdwan Hons.)**

14. If a sinusoidal voltage $v_i = 10\sin 2000\pi t$ mV is applied to the input of the OP AMP differentiator of Fig. 12.13 with $C = 1\,\mu\text{F}$ and $R = 100\,\text{k}\Omega$, what would be the output voltage? $[-2\pi\cos 200\pi t \text{ volt}]$

15. Draw an analog computer circuit using OP AMP to solve the following simultaneous equations:

$$x + 10y = 3$$
$$2x + 5y = -5$$

(Calcutta Hons.)

13

Communication Principles

13.1 Introduction

The history of the growth and development of radio communication is both remarkable and long, too long to narrate here in full. We shall therefore cut the long story short to suit our purpose.

The first demonstration of the production and reception of radio waves was made in Germany in 1887 by Hertz who was interested only in studying the properties of the waves and not in their applications. Oliver Lodge developed a system of wireless communication in England in 1894 following which many new types of crystal detectors were invented by J. C. Bose, Marconi, De Forest, and others. The first thermionic electron tube—a vacuum diode detector tube—for radio use was patented by Fleming in England in 1904. The three-element vacuum tube (triode) was invented by De Forest in 1906 and was shown subsequently that it could be used as a radio-frequency (RF) or audio-frequency (AF) *amplifier* or even as a *modulator*. One of the most remarkable applications of the device was to the *feedback* or *regenerative circuit*. The greatest single impetus in radio history was the introduction of radio-broadcasting in 1921. The first broadcasting signals were exclusively *amplitude modulated* (AM) and many types of receiving sets—regenerative, super regenerative, super heterodyne etc. came into being. To eliminate certain limitations in AM-operation, Armstrong began working on *frequency modulation* (FM) in 1914 and could perfect his system only in 1933. It is impractical to transmit audible signals by wireless. To transmit a 'message' in the form of articulated speech or music, a high frequency wave was used as a vehicle or *carrier* for the relatively low-frequency message. The exact mechanism consists in modifying or modulating the carrier signal in some manner by the audio-signal and the resulting radiated signal is known as a *modulated wave.* For the detection or restoration of the signal at the receiving end, the modulated wave naturally must undergo *demodulation* or *detection.*

The development of transistor finally overthrew the thermionic vacuum tubes and an unheard of revolution was brought about not only in the field of science but in the social field as well.

13.2 Modulation

Audio signal in the form of sound waves always require a material medium for their transmission and their velocity in air medium ($\sim$ 332 m/s) is much less than that of the electromagnetic waves ($\sim 3 \times 10^8$ m/s). Moreover, these audio signals cannot be transmitted through air over an appreciable distance as they get attenuated very quickly. Even if the audio signals like speech, music, songs etc. are converted into the corresponding electrical signal (20 Hz–20 kHz) with the help of microphone, it cannot be directly and efficiently radiated out into space in the form of electromagnetic radio waves by using a transmitting antenna. Several difficulties are encountered in the propagation of electromagnetic waves at the frequencies corresponding to the audio spectrum. Some of these are:

1. For efficient radiation and reception, the transmitting and receiving antennas would have to have heights comparable with the wavelength of the signal being transmitted. For an audio signal of frequency 20 kHz, the value of the wavelength $\lambda = (3 \times 10^8)/(20 \times 10^3)\,\text{m} = 15\,\text{km}$. This corresponds to a half-wave antenna of length $= \lambda/2 = 7.5\,\text{km}$. To have a vertical antenna of such an unusual height is just not practical.
2. For efficient radiations of electrical energy, frequencies should be very high ($> 20\,\text{kHz}$). In this context, an audio frequency radiation has very low efficiency and is thus not practicable.
3. Direct transmission of audio signals over long distances is not a feasible proposition, for an enormous power would be needed for the purpose.
4. In direct transmission, the audio signals at different frequencies get mixed up and a proper reception of these signals at the receiving end would be extremely difficult, if not impossible.

To overcome these problems for the transmission of messages (low frequencies) to long distances, it is essential that the audio frequency signal must first be used to *modulate* some characteristics of a high frequency wave, called the *carrier wave* and the resulting wave, called the *modulated wave*, be transmitted efficiently into space by using conveniently smaller size antennas. At the receiving end, the messages are then recovered, that is, *demodulated.* Thus, ***modulation*** *is the process in which some characteristics of a high frequency* ***carrier wave*** *are modified in accordance with the instantaneous value of a low frequency audio signal, termed* ***modulating wave.***

Let a sinusoidal carrier wave be represented by

$$v_c = V_c \cos(\omega_c t + \phi) \tag{13.2.1}$$

where v_c is the instantaneous value of the sine wave, V_c its amplitude or peak value, ω_c the angular frequency and ϕ the phase relation of the wave with respect to some standard or reference.

In the above equation, there are *three* characteristic parameters—***amplitude*** $\boldsymbol{V_c}$, ***frequency*** $\boldsymbol{\omega_c}$ or $\boldsymbol{f_c}$ $(= \omega_c/2\pi)$ and ***phase*** $\boldsymbol{\phi}$. Any one of these three may be varied by the modulating signal resulting in *three* basic types or methods of modulation. These are the following

1. ***Amplitude modulation*** (AM),
2. ***Frequency modulation*** (FM) and
3. ***Phase modulation*** (PM).

We shall now define, describe and discuss the different types of modulation, one after the other, in what follows.

13.3 Amplitude modulation (AM)

In ***amplitude modulation***, *the amplitude of the high frequency carrier wave is modified in accordance with the instantaneous amplitude of the low frequency modulating wave* or audio frequency signal. *The change in amplitude* from the unmodulated value *is proportional to the instantaneous value of the modulating signal and is independent of its frequency.*

Let the carrier wave be given by

$$v_c = V_c \cos(\omega_c t + \phi) \tag{13.3.1}$$

Assuming, for simplicity, the modulating signal to be sinusoidal, let the modulating voltage be represented by

$$v_m = V_m \cos \omega_m t \tag{13.3.2}$$

where V_m is the amplitude and ω_m the angular frequency of the modulating signal. It is assumed that the modulating frequency $\omega_m \ll \omega_c$, the carrier frequency.

Since in the case of amplitude modulation, the amplitude of the carrier wave is varied according to the instantaneous amplitude of the modulating wave, the amplitude modulated carrier wave can be represented by

$$\begin{aligned} v(t) &= (V_c + K_a V_m \cos \omega_m t) \cos(\omega_c t + \phi) \\ &= V_c(1 + m_a \cos \omega_m t) \cos(\omega_c t + \phi) \end{aligned} \tag{13.3.3}$$

where K_a is the proportionality constant and $m_a = (K_a V_m)/V_c$ is another constant, variously known as ***modulating index***, *modulating factor, depth or degree of modulation* or *modulation coefficient.*

To avoid *distortion* due to amplitude modulation, V_m should be less than V_c and the degree of modulation is reflected by the ***modulation index***. *It is defined as the ratio of the maximum deviation of the amplitude from the unmodulated value to the amplitude of the unmodulated carrier wave.*

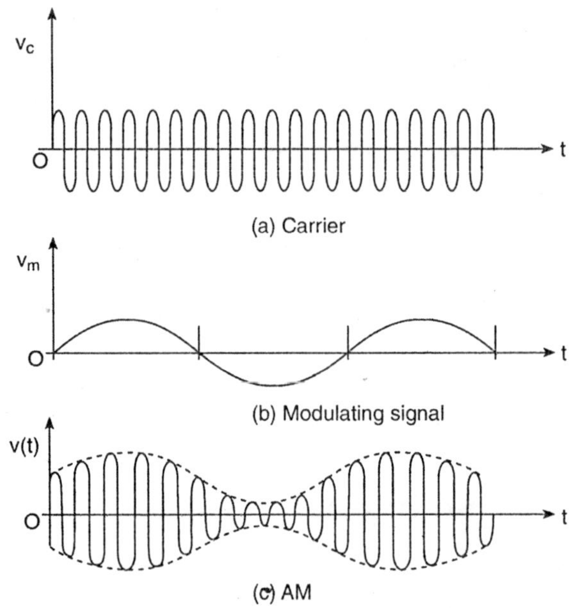

Fig. 13.1 AM and FM modulations

Typical waveforms of the unmodulated carrier and the modulaing signal are shown in Fig. 13.1(a) and 13.1(b) respectively. Fig. 13.1(c) represents the resultant amplitude of the modulated carrier. The outline of the modulated carrier has the same form as the modulating wave.

From (13.3.3), the maximum and the minimum amplitudes in AM wave are obtained and we can write

$$V_{\max} = V_c(1 + m_a) \tag{13.3.4}$$

$$\text{and} \quad V_{\min} = V_c(1 - m_a) \tag{13.3.5}$$

These two equations give the modulation index, m_a as

$$m_a = \frac{V_{\max} - V_{\min}}{V_{\max} + V_{\min}} = \frac{V_{\max} - V_{\min}}{2V_c} = \frac{2\Delta V}{2V_c} = \frac{\Delta V}{V_c} \tag{13.3.6}$$

where ΔV is the maximum deviation of the modulated carrier amplitude with respect to the unmodulated one.

By measuring the quantities $V_{\max}, V_{\min}$, V_c and ΔV from the waveform of the modulated carrier displayed on a cathode ray oscilloscope, one can determine the value of m_a from (13.3.6).

Usually, the value of m_a lies between 0 and 1. When $m_a = 1$, the wave is said to be *fully* (100%) *modulated* and $m_a > 1$ corresponds to *overmodulation* which produces undesirable *distortion* in the modulated signal.

13.3.1 Frequency spectrum of AM wave

Equation (13.3.3) gives,

$$v(t) = V_c \cos(\omega_c t + \phi) + \frac{m_a V_c}{2} \cos[(\omega_c + \omega_m)t + \phi]$$
$$+ \frac{m_a V_c}{2} \cos[(\omega_c - \omega_m)t + \phi] \tag{13.3.7}$$

The above equation shows that the amplitude modulated wave contains the original carrier frequency ω_c and two more frequencies $(\omega_c + \omega_m)$ and $(\omega_c - \omega_m)$.

The sum of the carrier frequency and the modulating frequency i.e., $(\omega_c + \omega_m)$ is called the ***upper side-band*** (USB) ***frequency*** and the difference of frequencies i.e., $(\omega_c - \omega_m)$ is called the ***lower side-band*** (LSB) ***frequency***. The sidebands contain the full information or intelligence while the carrier contains no information but helps only in the transmission process.

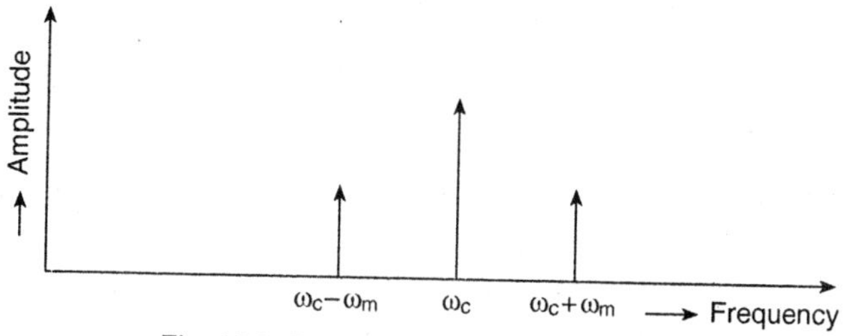

Fig. 13.2 Frequency spectrum of AM wave

The frequency spectrum as shown in Fig. 13.2 consists of *three* discrete frequencies. The central frequency i.e., the carrier, has the highest amplitude. The other two having equal amplitudes and spaced symmetrically about it can never exceed half the carrier amplitude. These two represent the side-band frequencies. In an AM wave, the *bandwidth* is $(\omega_c + \omega_m) - (\omega_c - \omega_m) = 2\omega_m$. So, for full transmission of the AM wave, *twice the modulating signal frequency* is needed.

13.3.2 Power in AM wave

Equation (13.3.7) shows that the AM wave consists of the carrier wave of amplitude V_c and two more side-band components of equal amplitudes, namely $\frac{1}{2}m_a V_c$.

The power associated with the carrier alone is

$$P_c = \left(\frac{V_c}{\sqrt{2}}\right)^2 \Big/ R = \frac{V_c^2}{2R} \tag{13.3.8}$$

where $V_c/\sqrt{2}$ is the r.m.s. voltage and R is the resistance in which power is dissipated.

If P_1 and P_2 are the powers for the lower and the uper side-band frequency components respectively, then we have

$$P_1 = P_2 = \left(\frac{m_a V_c}{2\sqrt{2}}\right)^2 \Big/ R = \frac{m_a^2 V_c^2}{8R} = \frac{1}{4} m_a^2 P_c \tag{13.3.9}$$

using the equation (13.3.8).

∴ The total power in the two side-band frequencies is given by

$$P_{SB} = P_1 + P_2 = 2 \times \frac{m_a^2 P_c}{4} = \frac{1}{2} m_a^2 P_c \tag{13.3.10}$$

$$\Rightarrow \quad P_{SB} \propto m_a^2$$

This shows that the total side-band power is proportional to m_a^2.

∴ The total power of the AM wave is

$$P = P_c + P_{SB} = \left(1 + \frac{m_a^2}{2}\right) P_c \tag{13.3.11}$$

Therefore, $P > P_c$ i.e., the *total power of the AM wave is greater than that of the unmodulated carrier wave.*

For 100% modulation, $m_a = 1$ and equations (13.3.10) and (13.3.11) reduce to

$$P_{SB} = \tfrac{1}{2} P_c \tag{13.3.12}$$

$$\text{and} \quad P = \tfrac{3}{2} P_c \tag{13.3.13}$$

Thus, we observe that for 100% modulation, the *total side-band power is half of the unmodulated carrier power* and the *total power of* AM *wave is* 1.5 *times the unmodulated carrier power.* So, *one-third* of the total power is the *useful power* contained in the side-band components and the rest *two-third* lies in the carrier component. A single side-band (SSB) contains only 1/6th of the total power. SSB transmission is thus more power-efficient.

13.3.3 AM wave generation

The block diagram of an AM wave amplifier or modulator is shown in Fig. 13.3. The carrier signal and the modulating signal are fed to the modulating amplifier whose gain is varied by the modulating signal so that the output becomes an AM wave.

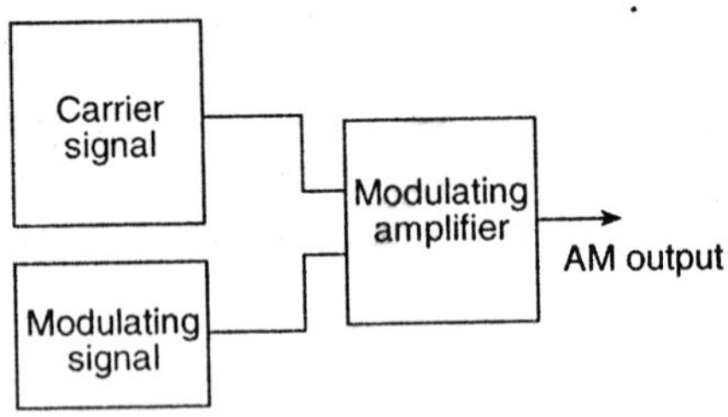

Fig. 13.3 Block diagram of AM wave modulator

The AM waves can be generated by various methods, broadly classified into (i) *linear modulators* and (ii) *square law modulators.* While normal broadcasting uses linear modulators, square law modulators are used for low voltage application, e.g., mobile communication. Fig. 13.4 shows a BJT collector modulator circuit using a *npn*-transistor in CE mode. The carrier voltage is fed to the base through a tuned circuit and the modulating signal is coupled to the collector to control the gain of the amplifier so that the output gives an AM waveform.

The resistors R_1, R_2 and the supply voltage V_{CC} provide the required d.c operating point of the amplifier. The capacitor C_E is so chosen as to bypass the carrier frequency signal and the capacitors C_B and C_C act as short circuit for a.c but block the d.c.

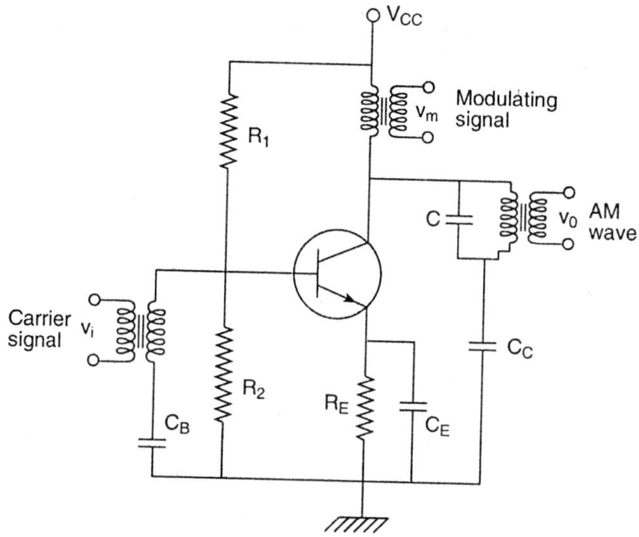

Fig. 13.4 BJT collector modulator

The modulating signal applied to the collector causes the collector current to vary about the d.c value such that

$$i_c = I_c + K_1 V_m \cos\omega_m t \tag{13.3.14}$$

where K_1 is a proportionality constant and $V_m \cos\omega_m t$ is the modulating signal.

Since i_c controls the voltage gain A_V of the amplifier, we can write

$$A_V = K_2 i_c = K_2(I_c + K_1 V_m \cos\omega_m t) \tag{13.3.15}$$

where K_2 is another proportionality constant.

Let the input carrier voltage be $v_i = V_c \cos(\omega_c t + \phi)$. Then the output of the amplifier is given by

$$\begin{aligned} v_0 = A_V v_i &= K_2 V_c (I_c + K_1 V_m \cos\omega_m t)\cos(\omega_c t + \phi) \\ &= A(1 + m_a \cos\omega_m t)\cos(\omega_c t + \phi) \end{aligned} \tag{13.3.16}$$

where $A = K_2 I_c V_c$ and $m_a = K_1 V_m / I_c$.

This shows that the amplifier output is an amplitude modulated wave.

13.4 Frequency modulation (FM)

Frequency modulation *is the process in which the frequency of the carrier wave is varied in accordance with the instantenous value of the modulating signal.* Here the amplitude and the phase remain constant.

Let the carrier wave be given by

$$v_c = V_c \cos(\omega_c t + \phi) = V_c \cos\theta(t) \tag{13.4.1}$$

where $\theta(t) = \omega_c + \phi$ and the modulating signal be represented by

$$v_m = V_m \cos \omega_m t \tag{13.4.2}$$

In the case of FM, the frequency of the carrier is varied according to the instantaneous value of the modulating signal. So, the angular frequency of the frequency-modulated wave can be represented as

$$\omega(t) = \omega_c + KV_m \cos \omega_m t \tag{13.4.3}$$

where K is a proportionality constant.

The frequency of the carrier wave is $f_c = \omega_c/2\pi$ and the instantaneous frequency of the FM wave is $f = \omega/2\pi$.

The maximum and the minimum frequencies of the FM wave are given by

$$f_{\max} = f_c + \frac{KV_m}{2\pi} \tag{13.4.4}$$

$$\text{and } f_{\min} = f_c - \frac{KV_m}{2\pi} \tag{13.4.5}$$

$\therefore$ The maximum deviation of the frequency from the unmodulated carrier frequency f_c is given by

$$\Delta f = f_{\max} - f_c = \frac{KV_m}{2\pi} \tag{13.4.6}$$

This relation shows that Δf is *proportional to* the amplitude V_m of the modulated signal but is *independent of* f_m $(= \omega_m/2\pi)$.

The phase angle at any instant t of the FM wave is given by

$$\begin{aligned} \theta(t) &= \int_0^t \omega(t)dt = \int_0^t (\omega_c + KV_m \cos \omega_m t)dt \\ &= \omega_c t + \frac{KV_m}{\omega_m} \sin \omega_m t \\ &= \omega_c t + m_f \sin \omega_m t \end{aligned} \tag{13.4.7}$$

where $m_f = KV_m/\omega_m$ is a constant known as the ***modulation index*** in FM.

Now, using (13.4.6),
$$m_f = \frac{KV_m}{\omega_m} = \frac{KV_m}{2\pi f_m} = \frac{\Delta f}{f_m} \tag{13.4.8}$$

Thus, the ***modulation index*** is *defined as the ratio of the maximum frequency deviation to the modulating frequency.*

Therefore, the FM wave may be written as

$$v(t) = V_c \cos \theta(t) = V_c \cos(\omega_c t + m_f \sin \omega_m t) \tag{13.4.9}$$

Fig. 13.5 shows the waveforms of carrier wave, modulating wave and the frequency modulated or FM wave.

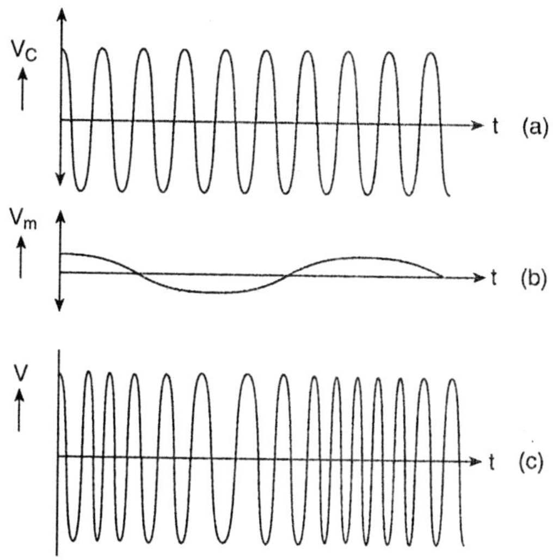

Fig. 13.5 Waveforms (a) carrier wave, (b) modulating wave, (c) FM wave

Energy spectrum of FM wave—It can be shown that the frequency modulated wave contains an infinite number of side-frequencies on either side of the carrier wave. These are $(f_c \pm f_m)$, $(f_c \pm 2f_m)$, $(f_c \pm 3f_m)$ and so on. The amplitudes of these side-frequency components $(f_c \pm nf_m)$ are proportional to $J_n(f_m)$ where J_n is the *Bessel function of the first kind of order n.*

In practice, however, one need not consider all the side-frequencies, since beyond a certain frequency range from the carrier, the amplitudes are negligibly small. As in AM wave, the information is contained in side-frequencies. So, if f_m is the modulating frequency, the bandwidth required for transmission or reception of FM signal is $2nf_m$, n being the number of significant side-frequency pairs.

If the modulating wave does not have a single frequency, the frequency-spectrum of FM wave becomes complicated. The bandwidth is now given by $2nf_{\max}$ where $f_{\max}$ is the highest modulating frequency. ***Carlson's rule** of thumb* states that *the bandwidth needed to pass an FM signal is twice the sum of the maximum frequency deviation and the highest modulation frequency,* i.e., a bandwidth

$$2(\Delta f + f_{\max}) = 2(m_f + 1)f_{\max}$$

Energy in FM wave—The frequency modulation does not alter the amplitude V_c of the carrier, the average energy-values in modulated and unmodulated waves are equal. The energy of the side-frequency components is obtained by reduction of energy of the carrier.

FM-wave generation—In an LC-oscillator tank circuit, if the capacitance or inductance is varied, the frequency of oscillation also varies. If this variation can be made proportional to the modulating voltage, one can generate a true FM-wave.

There are many devices, e.g., BJT, FET, varactor diode whose reactance can be varied by the applied voltage. Fig. 13.6 illustrates a BJT reactance modulator circuit connected to the tank circuit of an LC-oscillator to generate FM wave.

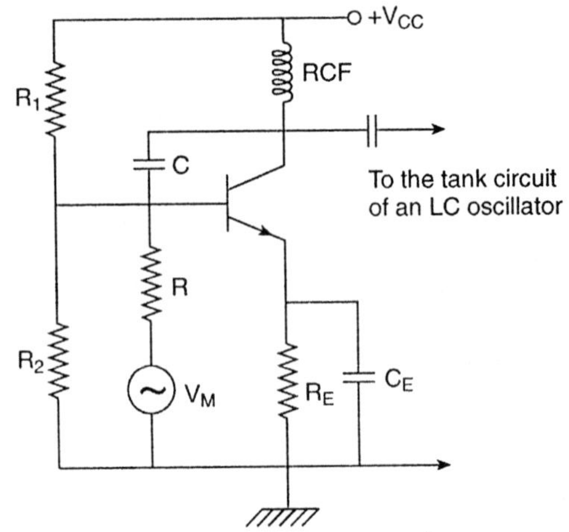

Fig. 13.6 BJT reactance modulator for FM generation

• A ***comparative study*** of AM and FM may be instructive. The bandwidth of FM is more than that of AM. Since the AM signal picks up both the atmospheric and the man-made noises, AM receivers are far more noisy compared to the FM ones. Further, the information carrying efficiency through AM is comparatively poorer. FM systems however are relatively complex and costlier and the range of service is also restricted to short distances.

13.5 Phase modulation (PM)

Phase modulation *is the process in which the phase angle of the carrier wave is varied in accordance with the instantaneous value of the modulating signal.* Here the amplitude and the frequency of the carrier remain constant. The instantaneous change in phase angle from the unmodulated value is proportional to the modulating signal and is independent of its frequency.

Let the carrier wave and the modulating signal be represented respectively by

$$v_c = V_c \cos \omega_c t \tag{13.5.1}$$

$$\text{and } v_m = V_m \cos \omega_m t \tag{13.5.2}$$

Therefore, the instantaneous phase angle of the phase-modulated (PM) wave can be written as

$$\begin{aligned} \theta(t) &= \omega_c t + K V_m \cos \omega_m t \\ &= \omega_c t + m_p \cos \omega_m t \end{aligned} \tag{13.5.3}$$

where K is a proportionality constant and $m_p = KV_m$ is another constant known as the ***modulation index*** in phase modulation.

So, the phase modulated wave may be written as

$$v = V_c \cos(\omega_c t + m_p \cos \omega_m t) \tag{13.5.4}$$

Comparing the equations (13.4.9) and (13.5.4), it is found that the frequency modulated (FM) and the phase modulated (PM) waves are very similar in form, the modulation component representing a deviation of the phase angle superimposed on the phase angle $\omega_c t$ of the unmodulated carrier.

Now, the instantaneous angular frequency $\omega(t)$ of the phase modulated wave is

$$\omega(t) = \frac{d\theta}{dt} = \omega_c - m_p\omega_m \sin\omega_m t \tag{13.5.5}$$

The instantaneous frequency is given by

$$f(t) = \frac{\omega(t)}{2\pi} = f_c - m_p f_m \sin\omega_m t \tag{13.5.6}$$

The maximum deviation of frequency from the unmodulated carrier frequency is Δf and is given by

$$\Delta f = [f(t) - f_c]_{\max} = m_p f_m = KV_m f_m \tag{13.5.7}$$

A comparison of (13.4.6) and (13.5.7) shows that the maximum frequency-deviation in FM is *proportional to the amplitude* V_m only and is *independent of the frequency* of modulating signal, but in the case of PM, Δf is proportional to both the *amplitude and the frequency of the modulating signal.*

13.6 Demodulation or Detection

The process of restoring or recovering the modulating signal from the modulated carrier wave is called the ***demodulation*** or the ***detection.*** It is the demodulation that extracts the intelligence from the modulated wave. Thus demodulation is just the *reverse process of modulation* and is naturally performed at the receiving end. The circuit that is utilized for the purpose of detection or demodulation is known as the ***detector*** or the ***demodulator.***

13.6.1 Demodulation of AM wave

For the detection or demodulation of the AM wave, *two* operations are to be performed and these are:

(i) *the rectification of modulated wave,* and

(ii) *the elimination of carrier component* of the modulated wave.

For the detection of AM wave, we normally use a ***diode detector circuit.*** This circuit is also known as the ***envelope detector*** or the ***linear detector circuit.***

The most widely used diode envelope detector circuit is shown in Fig. 13.7. The circuit is basically a half-wave rectifier with an RC-filter. It takes a high frequency AM signal as input across the primary of the transformer T and produces an output voltage which is the envelope of the AM wave that contains the intelligence in AF range. Hence

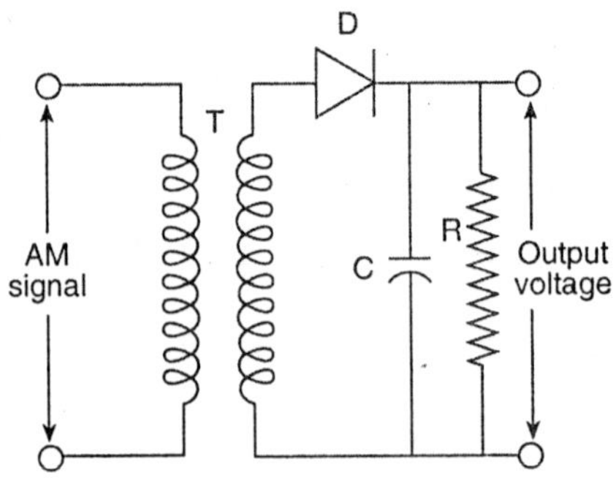

Fig. 13.7 An AM voltage detector

it is called *envelope detector*. Again, the output voltage is proportional to the input voltage. So, this is why such a detector is called a ***linear detector***.

The capacitor in the circuit stores up charges close to the peak value during each positive half-cycle of the carrier wave and releases it slowly through the resistor R when the signal falls during the negative half-cycles. The value of C is so chosen that the *time constant* RC of the circuit is long enough compared to the carrier cycle but shorted to the cycle of modulating signal. As a result, the output across the capacitor tends to follow the envelope of the modulated wave but involves some jaggedness as is illustrated in Fig. 13.8.

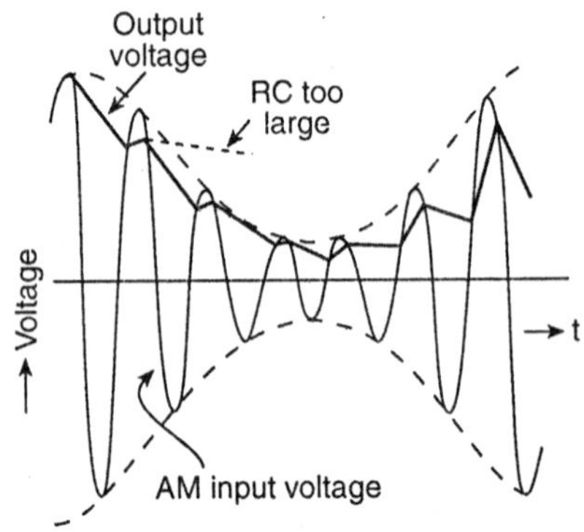

Fig. 13.8 Waveforms showing the operation of a diode detector

In practice, however, the carrier frequency being much higher than the modulating frequency, the detector output reproduces only the modulation envelope of the input, without marking the jaggedness. In this way, the modulating signal is recovered or restored at the output of the detector.

Selection of RC time constant

The selection of RC time constant is rather important. If it is *too low*, the fluctuations of output voltage of the detector are too large due to *quick decay* of the capacitor charge. If it is *too high*, the declining part of the envelope is not faithfully followed due to *slow decay* of the capacitor charge.

We calculate, in what follows, the *maximum permissible value* of the RC time constant.

Calculations—Let the envelope of the modulated wave be represented by

$$v = V_c(1 + m_a \cos \omega_m t) \tag{13.6.1}$$

At an instant $t = t_0$, the slope of the envelope is

$$\left.\frac{dv}{dt}\right|_{t=t_0} = -V_c m_a \omega_m \sin \omega_m t_0 \tag{13.6.2}$$

Let the capacitor be charged to envelope-voltage at $t = t_0$ and then start discharging. At any time $t > t_0$, the voltage across C is given by

$$v_c = v_0 e^{-(t-t_0)/RC} \tag{13.6.3}$$

The rate of change of voltage across C at $t = t_0$, is given by

$$\left.\frac{dv_c}{d(t-t_0)}\right|_{t_0} = -\frac{v_0}{RC} = \frac{V_c}{RC}(1 + m_a \cos \omega_m t_0) \tag{13.6.4}$$

If the envelope function is to be faithfully followed, the slope of the curve of capacitor voltage at $t = t_0$ must be equal to or less than the slope of the envelope.

$$\therefore \quad -\frac{V_c}{RC}(1 + m_a \cos \omega_m t_0) \leq -V_c m_a \omega_m \sin \omega_m t_0$$

$$\text{or,} \quad RC \leq \frac{1}{\omega_m}\left[\frac{m_a \sin \omega_m t_0}{1 + m_a \cos \omega_m t_0}\right]^{-1} \tag{13.6.5}$$

The worst situation occurs when the term in the bracket of equation (13.6.5) is a maximum, i.e., when

$$\frac{d}{dt_0}\left[\frac{m_a \sin \omega_m t_0}{1 + m_a \cos \omega_m t_0}\right] = 0$$

$$\text{or,} \quad \cos \omega_m t_0 = -m_a \;\text{ and }\; \sin \omega_m t_0 = \sqrt{1 - m_a^2}$$

Substituting these values in (13.6.5), we have

$$RC \leq \frac{1}{\omega_m}\frac{\sqrt{1 - m_a^2}}{m_a} \tag{13.6.6}$$

This is an *impractical* result since for 100% modulation, i.e., for $m_a = 1$, $RC \leq 0$, i.e., $C = 0$. At this condition, the output would contain the modulating signal and the carrier. When the modulation index m_a is small, $\sqrt{1 - m_a^2} \simeq 1$ and (13.6.6) gives the value of RC for detection with tolerable distortion as

$$RC \leq \frac{1}{\omega_m m_a} \tag{13.6.7}$$

Thus, the maximum or the highest angular modulating frequency detectable with tolerable distortion is

$$(\omega_m)_h = \frac{1}{m_a RC} \tag{13.6.8}$$

• The diode detector is widely used in AM radio broadcast receivers for a number of *advantages.* These are: (i) it can *handle* comparatively *large input signal,* (ii) it can *operate as a linear* or *power detector,* (iii) the *distortion* is *tolerable* and (iv) it can be readily adopted for *use in automatic gain control circuits.*

• The *disadvantage* a diode detector suffers from is that the detector circuit *consumes appreciable power* when the diode is conducting.

13.6.2 Demodulation of FM wave

The circuit that is used to extract the original intelligence from the frequency modulated wave is called ***FM detector*** or ***discriminator.*** For the detection of FM wave, *two* operations are to be performed:

1. The *frequency variation of FM is converted into amplitude variation,* and
2. *Recovery of the modulating signal from this voltage with amplitude variation.*

There are different types of FM detectors. The simplest type of FM detector is the *slope detector* or *single tuned discriminator* shown in Fig. 13.9.

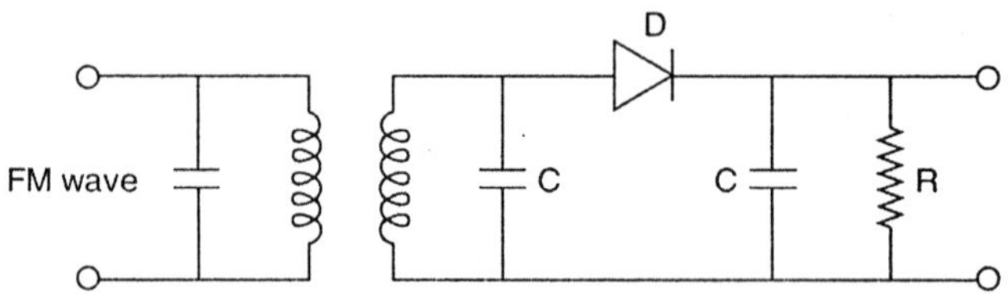

Fig. 13.9 Single tuned discriminator

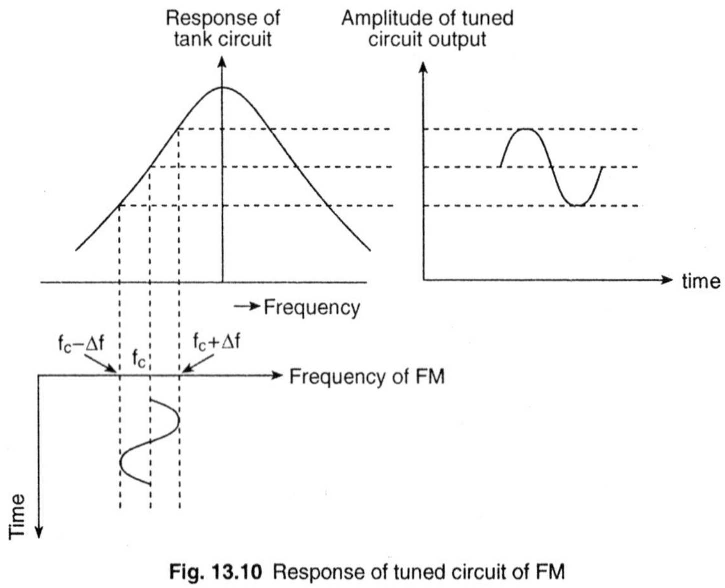

Fig. 13.10 Response of tuned circuit of FM

The FM wave is applied to the parallel resonant circuit (LC tank circuit) for tuning to a frequency slightly different from the carrier frequency f_c such that f_c lies in the middle of the most linear region of the response curve (Fig. 13.10). Due to the modulating frequency f_m, the frequency of the FM wave varies between $f_c - \Delta f$ and $f_c + \Delta f$. This corresponds to an amplitude variation and the tuned circuit output becomes an AM wave with frequency f_m. Thus the slope detector transforms FM to AM. The intelligence contained in the envelope of this AM wave is then recovered by a diode detector.

A slope detector suffers from the *drawback* that it is not suitable for high frequency application. This is because the response curve of the tank circuit deviates from the linearity at very high values of frequency. Because of this non-linearity in response curve, the output voltage becomes distorted with large frequency deviation. The circuit cannot also reject the amplitude variation of input FM.

13.7 Radio communication : Basic elements

Radio communication is the process of *communication by means of radio waves* which are electromagnetic waves of high frequencies ranging from 10 kHz to 300 GHz. The radio waves propagate in space from the transmitting to the receiving antenna to convey information.

Basic elements—The *basic elements* of a radio communication system consist of

1. A ***radio transmitter*** for the generation of radio frequency (RF) waves, called the *carrier waves.*

2. A ***microphone*** that converts the audio signals into electric pulses to modulate the carrier waves in accordance with the information to be transmitted.

3. A ***transmitting antenna*** to radiate the modulated carrier waves in some specified or in all directions.

The above *three* elements essentially constitute the *transmitting station* of a communication system.

In the *receiving station* of the system, the *basic elements* are the following.

4. A ***receiving antenna*** that intercepts and receives a part of the transmitted wave.

5. A ***radio receiver*** that receives, selects, amplifies and detects (i.e., demodulates) the information cloaked in the modulate radio waves from the transmitting station.

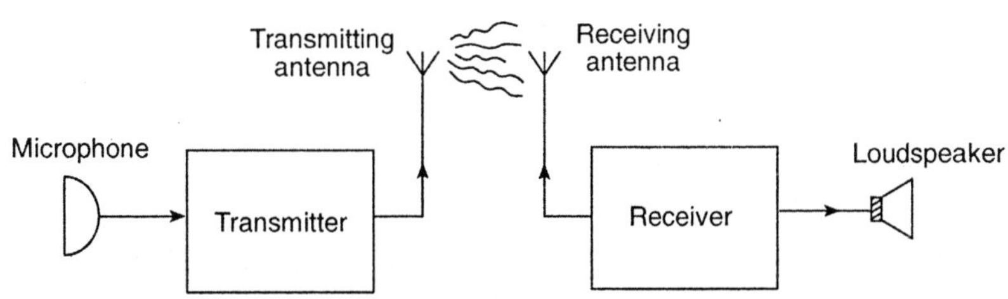

Fig. 13.11 Basic elements of a radio communication system

6. A ***loudspeaker*** for converting the demodulated or the detected electric signal back into sound and thereby reproduce the original information.

Fig. 13.11 illustrates the basic elements of a radio communication system in the form of a block diagram.

• *Antenna* is a system of elevated conductors that couple the transmitter or the receiver to space.

13.8 Modes of propagation of radio waves

Radio waves launched from a transmitting antenna propagate outward without being markedly affected by the atmosphere, rain, etc. Their movement, however, is best in free space and they require no material medium for their propagation. They readily pass through non-metallic substances. But in passing through metals, they are markedly attenuated.

The radio waves may travel from the transmitter to the receiver in a number of different ways. Of these the important ones are:

(i) *Ground or surface waves,*

(ii) *Space or tropospheric waves* and

(iii) *Sky waves.*

The medium of propagation is the atmosphere of the earth.

I. Ground or surface waves—In ground wave propagation, the radio waves are guided by the earth and travel along the curved surface of the ground from the transmitting to the receiving antenna. During propagation, the wave induces current in the ground over which it travels and loses thereby some energy owing to absorption by the ground. The ground wave propagation is thus *useful* only at *low frequencies,* since higher the frequency, higher is the absorption. Ground waves below 500 kHz can be used in communication up to distances of about 1500 km. AM radio broadcast in *medium frequency band* can thus be made primarily with the ground waves by covering local areas.

Irrespective of the atmospheric conditions, the radio transmission by ground waves is *quite reliable.*

II. Space waves—Space waves propagate in straight lines, unless however the earth and the atmosphere introduce any change in their path. Radio waves move as space waves within about 15 km over the earth's surface. As the propagation is through the troposphere, space waves are also called *tropospheric waves.* They have *two* components—(a) *direct* or *line-of-slight wave* and (b) the *ground reflected wave.* Fig. 13.12 shows these two components.

The *direct wave* travels from the transmitting to the receiving antenna almost in a right line. Deviations from perfect straight path occur due to the varying density of atmosphere and the diffraction effect.

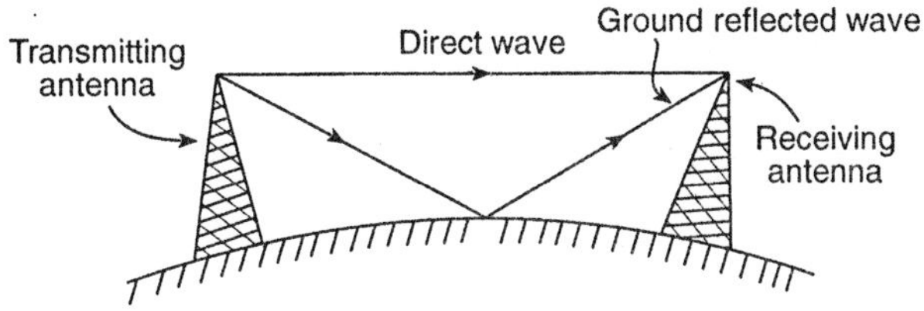

Fig. 13.12 Space wave propagation

The *ground-reflected wave* travels from the transmitting antenna to the earth and on being reflected from the earth's surface reaches the receiving antenna. At the point of reflection, a phase change of π (= 180°) occurs in it. If the path difference between the direct and the reflected waves be an even-multiple of $\lambda/2$ (= half wavelength), the two waves arrive at the receiving antenna in opposite phase and destroy each other's effect. In case, the path difference is an odd-multiple of $\lambda/2$, the phase difference is zero or 2π and the two in-phase waves reinforce each other.

By varying the height of the receiving antenna, the pathlengths of the two waves can be changed and alternate enhancement and reduction of the signal strength can be made. This phenomenon is called ***selective fading.***

• Space wave is *not* continuously absorbed by the earth's surface. It can thus cover a greater range than the ground wave. Transmission in UHF (ultra high frequency) band is made entirely by direct waves. But the FM and TV-programmes in VHF (very high frequency) are transmitted by the direct waves and ground-reflected waves.

• Direct waves cannot make transmission round the globe, since the earth's curvature obstructs the direct view between the transmitting and the receiving antennas. But due to refraction or scattering of the wave by the earth's atmosphere and diffraction by rough earth's surface, the signal can be received over a distance slightly beyond the direct line-of-sight. This extra distance is known as ***diffraction zone*** or ***shadow zone.***

Duct propagation—An interesting type of propagation of ground waves is the *duct propagation.* The density of air and its refractive index decreases rapidly with height above the ground. This result in a curvature of the rays passing through the atmosphere, more than that of the earth. A complete bending of the waves thereby occurs (Fig. 13.13) and they are again reflected by the earth. The waves are thus continuously refracted and reflected so that the rays initially moving parallel to the earth's surface get trapped and propagate around the curved surface of the earth by

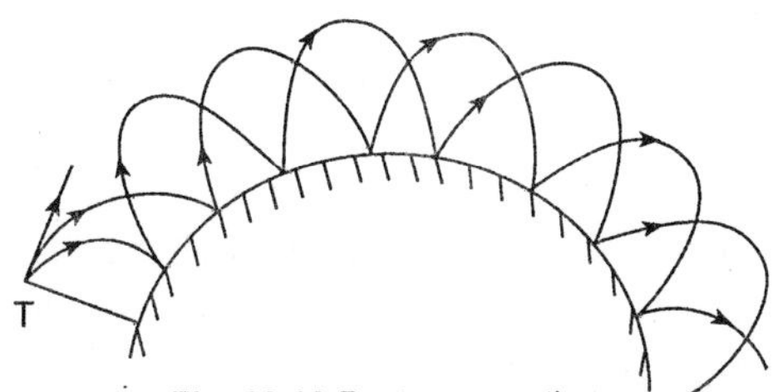

Fig. 13.13 Duct propagation

successive hoppings of reflections. This phenomenon is called *super refraction* or the ***duct propagation***.

● The duct propagation is possible only if the height of the transmitted antenna is less than the height of the top of atmospheric duct.

III. Sky waves—In sky wave propagation, the radio waves reach the receiving antenna by reflection from ionized stratified layers of the upper atmosphere—the *ionosphere*. Short wave transmission girdling the globe is made with sky waves via successive reflections (multi-hop transmission) at the ionosphere and the earth's surface (Fig. 13.14).

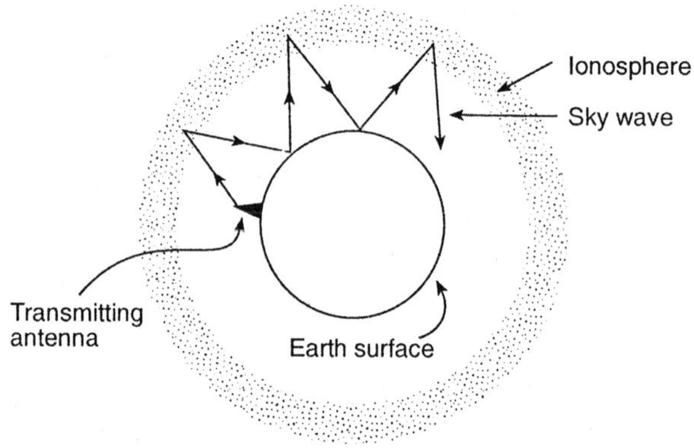

Fig. 13.14 Girdling the globe by reflection of sky waves

Sky wave propagation is not possible in space devoid of air. Being subject to fading and changes in the properties of ionosphere, the sky wave transmission is more erratic, compared to the surface wave transmission. The ionospheric properties vary widely with time of the day and night, seasons of the year and the atmospheric changes. Nevertheless, irrespective of all these, the sky wave is the primary means of short wave radio communication.

13.9 Ionosphere: Different layers

Ionosphere is the ionized region of the upper atmosphere of the earth. It extends from about 40 km to a height exceeding 400 km above the earth's surface. It consists of electrons, positive and negative ions and neutral particles. It is rather a *weakly ionized medium* since the concentration of neutral particles far exceeds that of charged particles. The origin of the ionosphere lies in simultaneous production of electrons and positive ions by direct collision of oxygen and nitrogen of the atmosphere with X-rays and UV-radiations from the sun. Some of these electrons adhere to neutral particles to create negative ions, some again recombine with positive ions to re-produce the neutral particles. Thus, the production and loss of electrons occur simultaneously in ionosphere. In *equilibrium*, the net concentration of electrons and negative ions together is equal to that of the positive ions. The ionosphere is thus *electrically neutral*.

Different layers—The electrons are not uniformly distributed throughout the ionosphere and the variation in electron concentration with height leads to the formation of different strata. A *stratum* of the ionosphere where the electron concentration attains a maximum at a particular height and falls off on either side, is called a *layer*. The maximum concentration in a layer is called the *peak concentration* and the corresponding height the *peak height* of the layer.

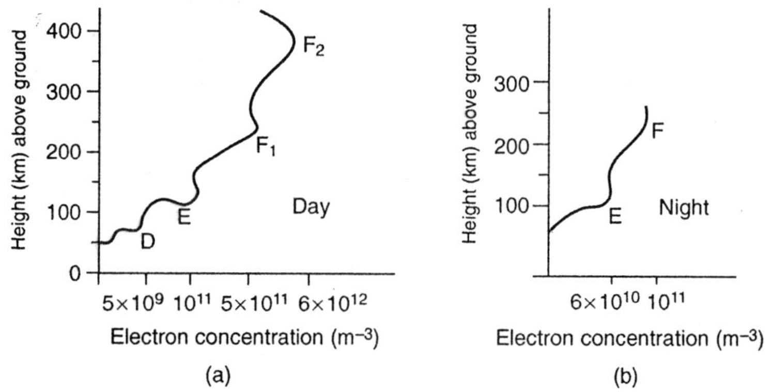

Fig. 13.15 Different ionosopheric layers during (a) day and (b) night and the height distribution of electron concentration

During day time, there occurs mainly *four* layers: $\boldsymbol{D, E, F_1}$ and $\boldsymbol{F_2}$ from the lowest to the highest, E and F_1 are the intermediate layers lying respectively above D and below F_2. Above F_2, the electron concentration rapidly decreases with height. *At night*, the D-layer is practically non-existent and F_1, F_2 coalesce into a single layer, called F *layer*. Fig. 13.15 illustrates the height distribution of electron concentration of different layers during day (Fig. 13.15a) and during night (Fig. 13.15b). The electron concentration at all heights in E and F layers markedly diminishes at night, compared to their diurnal values.

The peak electron concentrations and the peak heights of E, F_1 and F_2 show diurnal and seasonal variations and also the 11-year sun-spot cycle variations. These variations being regular and predictable, the above layers are termed *regular layers*. The variation of F_2 is more than that of E and F_1. In sky wave propagation, the E, F_1 and F_2 layers are used to reflect the waves back to the earth.

The D-layer attenuates the radio waves propagating through it. This is why at night, when the D-layer is practically absent, the strength of radio waves reflected from the ionosphere is greater. *The reception of sky waves is thus better at night than at day.*

13.10 Reflection of radio waves from ionosphere

Radio waves, in short wave region, can propagate over large distances by way of reflections from the ionosphere. We shall now analyse the mechanism of the reflection by considering the motion of electrons in the ionosphere.

Let $E = E_0 \sin \omega t$ be the electric field associated with the passing radio waves of angular frequency ω. Ignoring the collisions of electrons and neutral particles, the equation of motion of an electron is given by

$$m\frac{dv}{dt} = -eE_0 \sin \omega t \tag{13.10.1}$$

where v is the instanteneous velocity of electron of charge $-e$ and mass m.

Integrating (13.10.1) with respect to t,

$$v = \frac{eE_0}{m\omega} \cos \omega t + c_1 \text{ (const.)} \tag{13.10.2}$$

At $E = 0$, $v = 0$ and hence, from (13.10.2), $c_1 = 0$. And we have

$$v = \frac{eE_0}{m\omega} \cos \omega t \tag{13.10.3}$$

If N be the concentration of free electrons at any height in the atmosphere, the conduction current density is

$$J_c = -Nev = -\frac{Ne^2 E_0}{m\omega} \cos \omega t \tag{13.10.4}$$

As the ionosphere is a weakly ionized medium, its permittivity $\varepsilon \simeq \varepsilon_0$, the free-space permittivity. So, the displacement current density

$$J_d \simeq \varepsilon_0 \frac{\partial E}{\partial t} = \varepsilon_0 \omega E_0 \cos \omega t \tag{13.10.5}$$

$\therefore$ Total current density due to the electrons in the ionosphere is

$$J = J_c + J_d = \omega E_0 \left(\varepsilon_0 - \frac{Ne^2}{m\omega^2} \right) \cos \omega t \tag{13.10.6}$$

$\therefore$ *Effective permittivity* of the ionized region is

$$\varepsilon = \varepsilon_0 - \frac{Ne^2}{m\omega^2} = \varepsilon_0 \left(1 - \frac{Ne^2}{m\omega^2 \varepsilon_0} \right) \tag{13.10.7}$$

which shows that $\varepsilon < \varepsilon_0$.

Substituting $e = 1.6 \times 10^{-9}$C, $m = 9.1 \times 10^{-31}$ kg, $\varepsilon_0 = 8.854 \times 10^{-12}$ F/m and $\omega = 2\pi f$, we get

$$\varepsilon = \varepsilon_0 \left(1 - \frac{81N}{f^2} \right) \tag{13.10.8}$$

where f is the frequency of the radio waves.

$\therefore$ The refractive index of ionized medium, using Snell's law and e.m. theory, is

$$n = \frac{\sin i}{\sin r} = \sqrt{\frac{\varepsilon}{\varepsilon_0}} = \sqrt{1 - \frac{81N}{f^2}} \tag{13.10.9}$$

where i = angle of incidence and r = angle of refraction.

The ionosphere thus behaves as a *rarer* medium. So, at the interface of the media, the incident ray bends and gradually deviates further and further away from the normal to the interface to eventually get reflected from the earth.

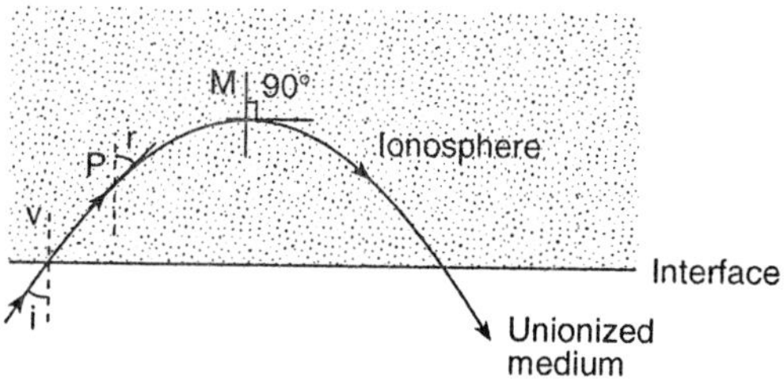

Fig. 13.16 Ray bending and reflection in ionosphere

At the top point M (Fig. 13.16) of the trajectory, $r = 90°$, the ray is parallel to the earth's surface and thereafter it moves downward to the earth. M is thus called the *point of reflection*, where

$$n = \sin i = \sqrt{1 - \frac{81N_r}{f^2}}, \quad \text{using (13.10.9)} \qquad (13.10.10)$$

N_r being the electron concentration at the point of reflection.

Equation (13.10.10) shows that greater the i-value, higher is the index n and lower is N_r. So, for more oblique incident rays, the point of reflection is lower, the electron concentration decreasing with decreasing height.

For a sufficiently small i, equation (13.10.10) is not satisfied even with largest N_r. The *ray then penetrates the layer and docs not return to the earth.* This is shown in (Fig. 13.18) for the ray 1..

For normal or vertical incidence of radio waves, $i = 0$ and hence $n = 0$ at the point of reflection. The maximum frequency of the wave that can be reflected to the earth from ionosphere for normal incidence is called the ***critical frequency*** f_c and is related to the peak concentration N_p of the layer by

$$f_c = 9\sqrt{N_p} \qquad (13.10.11)$$

In terms of f_c, equation (13.10.10) assumes the form

$$n = \sin i = \sqrt{1 - (f_c/f)^2} \qquad (13.10.12)$$

Radio waves with frequencies $f < f_c$ are reflected whatever be the angle of incidence. With $f > f_c$, the reflection can occur for oblique incidence. From (13.10.12) then

$$\left(\frac{f_c}{f}\right)^2 = 1 - \sin^2 i = \cos^2 i$$

$$\text{or,} \quad \boxed{f = f_c \sec i} \qquad (13.10.13)$$

The above relation is called the ***secant law*** that relates the frequency f of the reflected wave for an angle of incidence i to the critical frequency f_c. If the ray

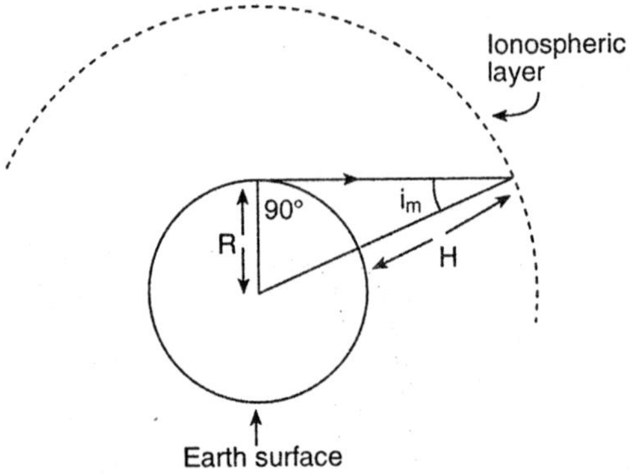

Fig. 13.17 Ray leaving tangential to the earth

is incident tangential to the earth (Fig. 13.17), the angle of incidence i attains its maximum value i_m and is given by

$$\sin i_m = \frac{R}{R+H} \tag{13.10.14}$$

where R = earth's radius and H the height of the ionized layer.

13.11 Some terminologies in sky wave communication

We shall discuss in this section some terms relating to sky-wave communication.

1. Skip distance—With relatively large angle of incidence, it is observed that as the angle of incidence increases, the distance between the transmitting and the receiving points increases (rays 2, 3, 4). There is however a particular angle of incidence for which the distance from the transmitter at which the reflected ray of a given frequency returns to the ground from ionosphere is *minimum*. This *minimum distance* is termed ***skip distance*** (Fig. 13.18) for that frequency.

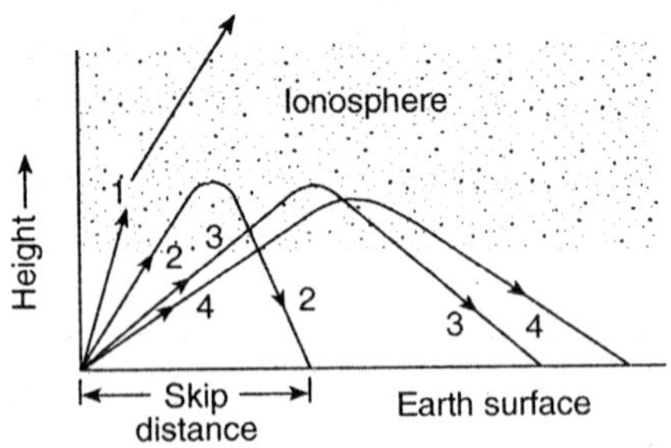

Fig. 13.18

The skip distance increases with frequency and for a given frequency, it depends on the height of the reflecting layer, increasing with increasing height of the layer.

2. Maximum usable frequency (MUF)—The frequency of the radio wave which makes the distance of a given receiving point equal to the skip distance is termed the

maximum usable frequency. If the frequency of the transmitted wave is more than MUF, the sky wave will fail to reach the receiver. MUF is thus the maximum frequency with which ionospheric radio communication is possible. For sky wave communication between two points, the optimum frequency is taken to be about 85% of MUF.

3. Single and multi-hop transmissions—The transmission is called a *single-hop transmission* if the radio wave launched by a transmitter reaches the receiver after the first reflection from the atmosphere and the path is referred to as a *single-hop path.* If the ray leaves the earth's surface tangentially at the transmitting point, the distance between the transmitting and the receiving pints is a maximum for single-hop.

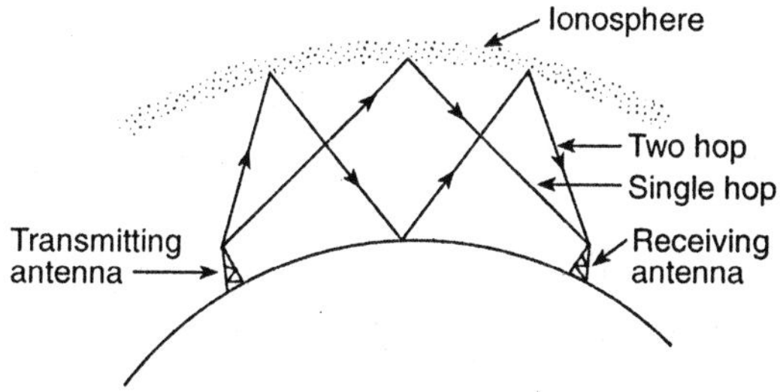

Fig. 13.19 Single and two-hop transmission

Transmission that uses multiple reflections at the ionosphere is known as *multiple-hop transmission.* Fig. 13.19 shows a single-hop and a two-hop paths involving one and two reflections at the atmosphere. Since in multi-hop transmission, the path length is very long the attenuation of the siganl is large, the transmitter power should be enhanced to avoid the feebleness of the signal at the receiving point.

4. Fading—One of the problems in radio communication using ionospheric reflection is *fading* of a signal. The sky wave signal strength is found to vary with time because of fluctuations in the properties of ionosphere e.g., irregular variations in ionization concentrations, movement of reflecting layers, interference between waves arriving by different number of hops or by reflection from different layers. The fluctuation of the received sky wave signal existing over short periods of the order of a minute or less is called the ***fading*** which is an undesirable phenomenon in the radio communication system.

13.12 Radio transmitters and receivers

In a radio transmitter, a radio frequency (RF) wave, called *carrier wave,* is generated. The electrical form of intelligence (e.g. speech, music) to be broadcast is superposed on the carrier to modulate its amplitude or the frequency. The modulated carrier wave is radiated from the transmitting antenna.

A radio receiver recovers the intelligence from the modulated carrier arriving at the receiving antenna.

AM transmiter—The block diagram of a typical AM transmitter is illustrated in Fig. 13.20. A stable *crystal oscillator* generates the desired RF waves. A chain of

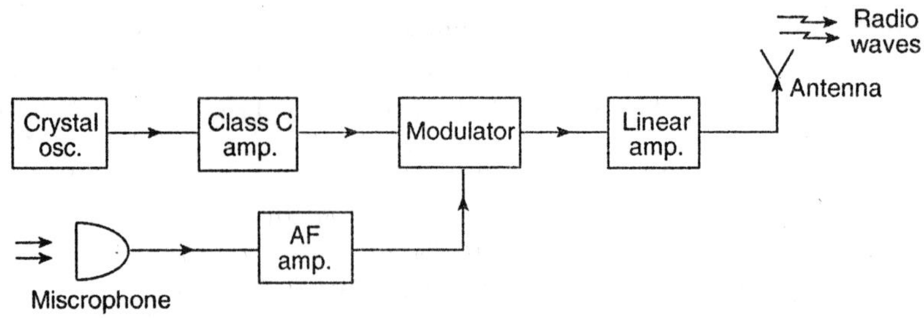

Fig. 13.20 Block diagram of AM transmitter

Class C amplifiers increases the power level. The intelligence (e.g. speech, music) to be transmitted are converted into electric signals by using *microphones.* This audio frequency (AF) signal is amplified by *AF amplifier* and combined with RF carrier wave in the *modulator.* The generated modulated carrier is further amplified by *linear amplifiers* and finally fed to the transmitting *antenna.*

AM receiver—The block diagram of a widely used superheterodyne type AM radio receiver is shown in Fig. 13.21. The radio waves arrive from different stations and induce their own voltage in the antenna. The receiver is tuned to a particular station using resonant circuits.

The RF *selection* is a single tuned RF amplifier that selects the desired frequency and amplifies the RF signal. The amplified RF signal is next mixed with the output of RF oscillator in a *mixer.* The output of the mixer is a signal of frequency equal

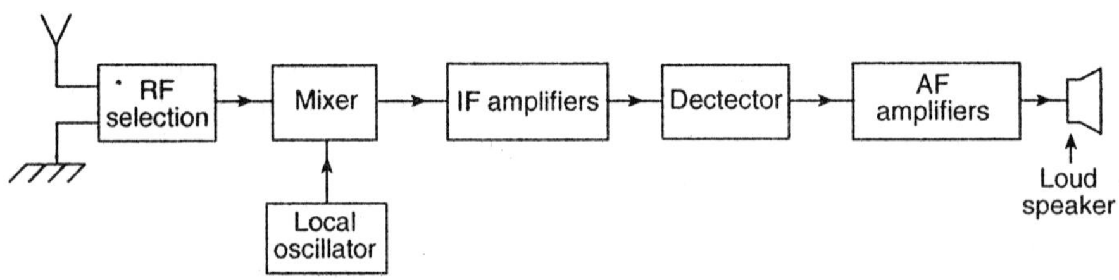

Fig. 13.21 Block diagram of a superterodyne type AM receiver

to the difference of the two frequencies. This is termed *intermediate frequency* (IF). The IF signal is then amplified by IF *amplifiers* and the output of the last amplifier is fed to a *detector.* The detected signal is next amplified by AF *amplifiers* and fed to a *loudspeaker* that converts the electric signal back into the original sound signal.

● The process of mixing two frequencies to produce the sum or difference frequency is called *heterodyning.* Hence the name of the circuit.

FM transmitter—The block diagram of a typical FM transmitter using a reactance modulator and an automatic frequency control system (AFC) is shown in Fig. 13.22. The audio signal from a microphone is amplified by the *audio amplifier.* The *reactance modulator* varies the frequency of the LC *oscillator* according to the variation of the audio signal and thereby produces FM signal. The *frequency multiplier* increases the frequency to the desired value. The power level of FM signal is increased by the *Class C power amplifier* and fed into the *antenna.* A sample of the output FM

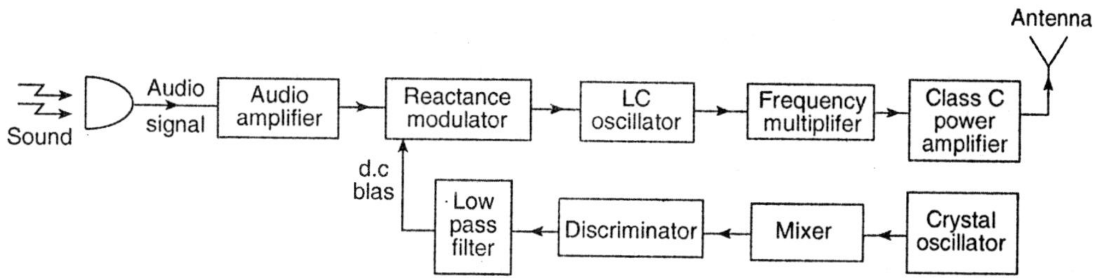

Fig. 13.22 Block diagram of a typical FM transmitter

signal is compared with that of a *crystal oscillator* in a mixer. The difference frequency signal generated by the *mixer* is passed through a phase *discriminator* which gives zero d.c output when the transmitter has the correct carrier frequency. In case, the average value of the transmitter frequency gets drifted, the discriminator develops a d.c voltage of such a polarity that when applied to the reactance modulator, the drift in frequency is rectified.

FM receiver—The block diagram of a typical FM receiver is shown in Fig. 13.23. It is essentially a superheterodyne receiver.

The RF *amplifier* is a tuned amplifier that selects and amplifies the desired FM signal of a particular radio station. The amplified RF signal is then mixed with the output of a local RF oscillator in a *mixer* to give an output signal of frequency equal

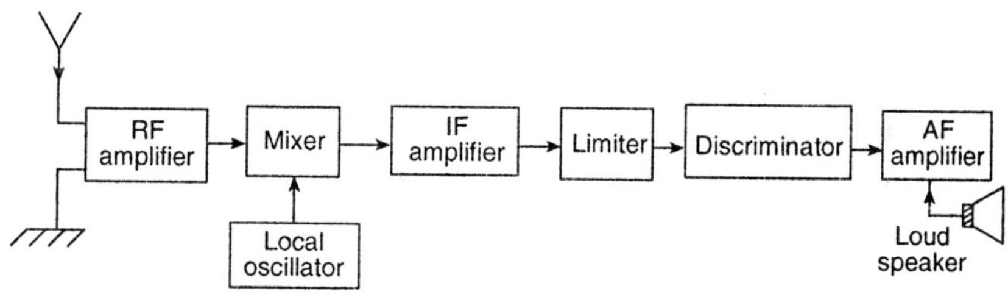

Fig. 13.23 Block diagram of a typical FM receiver

to the difference of the two frequencies, i.e., an IF signal. The *limiter* removes all amplitude variations due to noise or interference from the FM wave. The output of the limiter is passed to the *discriminator* whose output is the original signal. It is further *amplified* and passed to a *loudspeaker.*

13.13 Satellite communication : An outline

Reliable communication by sky waves is affected by fading, atmospheric disturbances etc. Artificial satellites offer reliable communication links over long distances.

An artificial satellite is launched in space in a circular orbit in the equatorial plane at a height of 36000 km above the earth. Its period of revolution is 24 hours, being equal to that of the earth about its axis. To an observer on the earth's surface, therefore, the satellite appears stationary and hence is called ***geostationary satellite***. The wave containing the information is transmitted to the satellite from a transmitter situated on the earth's surface. The signal is then processed and amplified in the satellite and re-transmitted towards the receiving point on the earth. As the satellite is geostationary, the transmitting and the receiving antennas on the earth can be oriented in fixed directions.

A satellite communication system, illustrated schematically in Fig. 13.24, essentially consists of *three* main sections. These are:

(i) an *uplink*

(ii) a *satellite transponder*, and

(iii) a *downlink*.

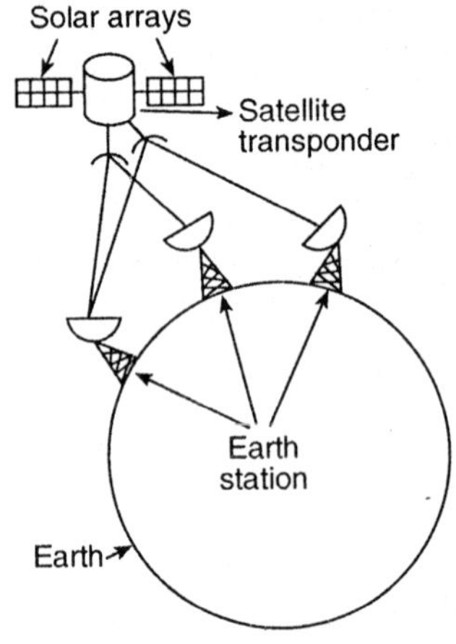

Fig. 13.24 A satellite communications system

The *uplink section* is the earth station transmitter that radiates the wave to the satellite; usually, the carrier frequency for the wave transmitted—the *uplink frequency*—is 6 GHz. The *satellite transponder* has a *frequency translator* that converts the uplink frequency to the *downlink frequency*—the carrier frequency of the wave from the satellite to the earth station—to 4 GHz. Different uplink and downlink frequencies are used so as to avoid an unwanted feed back. The signal is amplified and transmitted to the earth station receivers. The *downlink section* contains the earth station receivers that recovers the signal from the modulated carrier.

The circuits on board the satellite are powered by means of solar cells mounted on panels in the satellite.

Communication between satellites can be made via *cross links* or *intersatellite links*, as illustrated in Fig. 13.25. Conventional speech, telephone signals, data and TV pictures and sound are all conveyed through the satellite system.

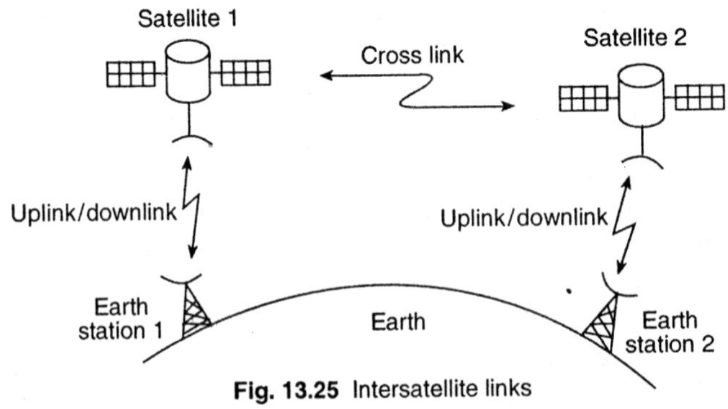

Fig. 13.25 Intersatellite links

Satellite communication has a number of *advantages* such as (i) good quality reception, (ii) large bandwidth, (iii) large area coverage, (iv) handling of huge number of signals and data, etc. One of the *disadvantages* is the high installation cost.

• A geostationary satellite can cover approximately one-third of the earth's surface so that three equally spaced geostaionary satellites can cover almost the entire earth.

13.14 Illustrated Examples

▶ **Example 1.** *An audio signal of* 1 kHz *is used to modulate a carrier of* 500 kHz. *Determine the side-band frequencies and bandwidth required.* **(Guru Nanak Univ.)**

Solution: Frequency of the carrier wave, $f_c = 500\,\text{kHz}$

Frequency of the modulating signal, $f_m = 1\,\text{kHz}$

$\therefore$ The ***upper side-band*** frequency $= f_c + f_m =$ **501 kHz**

and the ***lower side-band*** frequency $= f_c - f_m =$ **499 kHz.**

$$\therefore\ \textit{Bandwidth} = (f_c + f_m) - (f_c - f_m) = 2f_m = \mathbf{2\,kHz}$$

▶ **Example 2.** *A carrier of* 200 V, 1.5 MHz *is modulated by a* 50 V, 1 MHz *sine wave signal. Find the modulation factor.* **(Punjab Univ.)**

Solution: Given, the voltage amplitude of the carrier wave, $V_c = 200\,\text{V}$

and the voltage amplitude of the modulating signal, $V_m = 50\,\text{V}$.

Taking $K_a = 1$, the ***modulation factor***, $m_a = K_a \dfrac{V_m}{V_c} = \dfrac{50}{200} = \dfrac{1}{4} = \mathbf{25\%}$

▶ **Example 3.** *A* 5 kW *carrier wave is amplitude modulated to a modulation index* 60%. *Find the total power in the modulated wave.* **(Punjab Univ.)**

Solution: Power of the unmodulated wave, $P_c = 5\,\text{kW}$

Also, by the problem, modulation index, $m_a = 60\% = 0.6$

The total power of the modulated wave is therefore given by

$$P = \left(1 + \frac{m_a^2}{2}\right) P_c$$

$$= \left(1 + \frac{0.6 \times 0.6}{2}\right) \times 5\ \text{kW}$$

$$= \mathbf{5.90\,kW}$$

▶ **Example 4.** *Calculate the power developed by an AM wave in a load of* 100Ω *when the peak voltage of the carrier wave is* 100 V *and the modulation factor or index is* 0.4*?* **(Gauhati and Calcutta Hons.)**

Solution: Here, the load resistance $R = 100\Omega$.

Peak voltage of the carrier $V_c = 100\,\text{V}$.

Modulation factor $m_a = 0.4$.

$\therefore$ The power developed by the AM wave in the load is given by

$$\boldsymbol{P} = \left(1 + \frac{m_a^2}{2}\right) P_c$$

$$= \left(1 + \frac{m_a^2}{2}\right) \times \frac{V_c^2}{2R}, \quad \left(\because P_c = \frac{V_c^2}{2R}\right)$$

$$= \left(1 + \frac{0.4 \times 0.4}{2}\right) \times \frac{(100)^2}{2 \times 100} \text{ W}$$

$$= \mathbf{54 \, W}$$

► **Example 5.** *An amplitude modulated wave is represented by the expression*

$$\rho_m = 5(1 + 0.8 \cos 10^4 t) \sin 220 \times 10^4 t.$$

What are the minimum and maximum amplitudes of the AM wave? What are frequency components contained in the modulated wave and what is the amplitude of each component? **(Himachal Univ.)**

Solution: The AM wave is given by the expression

$$\rho_m = 5(1 + 0.8 \cos 10^4 t) \sin 220 \times 10^4 t$$

The instantaneous voltage of the AM wave is represented by

$$\rho_m = V_c(1 + m_a \cos \omega_m t) \cos(\omega_c t + \phi)$$

Comparing these two equations, we get $V_c = 5$ V and $m_a = 0.8$.

$\therefore$ The ***maximum amplitude*** of AM wave is $\boldsymbol{V_{\text{max}}} = V_c(1 + m_a) = 5(1 + 0.8) = \mathbf{9 \, V}$

and the ***minimum amplitude***, $\boldsymbol{V_{\text{min}}} = V_c(1 - m_a) = 5(1 - 0.8) \text{ V} = \mathbf{1 \, V}$

Frequency of the carrier wave, $\boldsymbol{f_c} = \dfrac{\omega_c}{2\pi} = \dfrac{220 \times 10^4}{2\pi} \text{ Hz} = \mathbf{350 \ kHz}$

Frequency of the modulating signal, $\boldsymbol{f_m} = \dfrac{\omega_s}{2\pi} = \dfrac{10^4}{2\pi} \text{ Hz} = \mathbf{1.59 \, kHz}$

Lower side-band ***frequency***, $\boldsymbol{f_c - f_m} = (350 - 1.59) \text{ kHz} = \mathbf{348.41 \, kHz}$

Upper side-band ***frequency***, $\boldsymbol{f_c + f_m} = \mathbf{351.59 \, kHz}$

Amplitudes of the components of AM wave are:

(i) ***carrier wave***, $V_c = \mathbf{5 \, V}$

(ii) ***Lower side-band component***, $\dfrac{m_a V_c}{2} = \dfrac{0.8 \times 5}{2} = \mathbf{2 \, V}$

(iii) ***Upper side-band component***, $\dfrac{m_a V_c}{2} = \mathbf{2 \, V}$

▶ **Example 6.** *Find the carrier and modulation frequencies, the modulation index and maximum deviation of the FM wave represented by the voltage equation*

$$\rho_m = 12\sin(6\times10^8 t + 5\sin 1250t)$$ **(Punjabi Univ.)**

Solution : The given equation for FM wave is

$$\rho_m = 12\sin(6\times10^8 t + 5\sin 1250t)$$

Comparing this equation with $\rho_m = V_c\sin(\omega_c t + m_f\sin\omega_m t)$, we get

$$\textit{Carrier frequency, } f_c = \frac{\omega_c}{2\pi} = \frac{6\times10^8}{2\pi}\,\text{Hz} = \mathbf{95.5\,MHz}$$

$$\textit{Modulating frequency, } f_m = \frac{\omega_m}{2\pi} = \frac{1250}{2\pi} = \mathbf{199\,Hz}$$

$$\textit{Modulation index, } m_f = \mathbf{5}.$$

Maximum deviation in frequency, $f = m_f\cdot f_m = 5\times199 = \mathbf{995\,Hz}$

▶ **Example 7.** *The peak-to-peak voltage of an AM wave is* 224 volt *and the minimum dip-to-dip voltage is* 96 volt. *Calculate* (i) *the modulation index,* (ii) *the percentage modulation,* (iii) *amplitude of the carrier and* (iv) *amplitude of sidebands.*

Solution : (i) ***Modulation index***, $m_a = \dfrac{V_{\max}-V_{\min}}{V_{\max}+V_{\min}} = \dfrac{(224-96)/2}{(224+96)/2} = \mathbf{0.4}$

(ii) ***Percentage modulation*** $= m_a\times100\% = 0.4\times100\% = \mathbf{40\%}$

(iii) ***Amplitude of carrier***, $V_c = \dfrac{V_{\max}-V_{\min}}{2m_a} = \dfrac{(224-96)/2}{2\times0.4} = \mathbf{80\,V}$

(iv) ***Amplitude of sidebands*** $= m_aV_c/2 = 0.4\times80/2 = \mathbf{16\,V}$

▶ **Example 8.** *The total power of an AM wave is* 1.5 kW *at a depth of modulation of* 80%. *Determine the power of the carrier and each sideband.*

Solution : We have, $P_{av} = P_c(1+m_a^2/2)$, with usual notations.

$$\therefore\ P_c = \frac{P_{av}}{1+m_a^2/2} = \frac{1.5}{1+0.8^2/2}\,\text{kW} = \mathbf{1.14\ kW}$$

Power of each side-band is

$$P_{SB}/2 = P_c\cdot\frac{m_a^2}{4} = 1.14\times\frac{0.8^2}{4} = 0.18\,\text{kW} = \mathbf{180\,W}$$

▶ **Example 9.** *A diode envelope detector uses a parallel RC network with* $R = 220\,\text{k}\Omega$ *and* $C = 200\,\text{pF}$. *If an AM wave with* 40% *modulation is fed to the detector, find the highest modulation frequency that can be detected with tolerable distortion.*

Solution: For tolerable distortion, the RC time constant is to be limited as

$$RC \leq \frac{1}{m_a \omega_m} \Rightarrow \omega_m \leq \frac{1}{m_a RC}$$

$$\therefore \ f_m \leq \frac{1}{2\pi m_a RC}$$

$$\therefore \ (f_m)_{\max} = \frac{1}{2\pi \times 0.4 \times 220 \times 10^3 \times 200 \times 10^{-12}}, \text{ with given data.}$$

$$= \mathbf{9.04\,kHz}$$

► **Example 10.** *An FM transmitter sends out a* 100 MHz *carrier, modulated by a* 15 kHz *sinunoidal audio signal. The maximum frequency deviation is* 30 kHz.

Find (i) *the modulation index,* (ii) *three significant-pairs of side frequencies and* (iii) *the channel width required for these three side-frequencies.* **(Calculatta Hons.)**

Solution: (i) Frequeny ***modulation index***, $m_f = \frac{\Delta f}{f_m} = \frac{30}{15} = \mathbf{2}.$

(ii) Three significant pairs of **side-frequencies** are: $f_c \pm f_m$, $f_c \pm 2f_m$ and $f_c \pm 3f_m$. They are, therefore, given by

$$\mathbf{100\,MHz \pm 15\,kHz, \ 100\,MHz \pm 30\,kHz} \text{ and } \mathbf{100\,MHz \pm 45\,kHz}.$$

(iii) The required ***channel width*** is: $2 \times 3 \times 15 \text{ kHz} = \mathbf{90 \ kHz}$

► **Example 11.** *An amplitude modulated broadcast radio transmitter radiates at* 20 kW *with modulated index* 75%. *How much is the carrier power?* **(Madurai Univ.)**

Solution: Total power of modulated broadcast wave is

$$P_T = 20\,\text{kW}$$

Modulation index, $m_a = 75\% = 0.75$. Let P_c be the carrier power.

From the relation, $P_T = P_c(1 + m_a^2/2)$, we have

$$20 = P_c(1 + 0.75^2/2) = 1.281 P_c$$

$$\therefore \ \textit{Carrier power}, \ P_c = 20/1.281 = \mathbf{15.16\,kW}$$

► **Example 12.** *A frequency modulated signal has a resting frequency of* 102 MHz *and the highest frequency of* 102.02 MHz *when modulated by a modulating signal of frequency* 4 kHz. *Determine* (i) *frequency deviation,* (ii) *modulation index,* (iii) *carrier swing and* (iv) *the lowest frequency of FM wave.*

Solution: By the problem, $\nu_0 = 102\,\text{MHz}$, $\nu_{\max} = 102.02\,\text{MHz}$ and $\nu_m = 4\,\text{kHz}$.

$\therefore$ (i) ***Frequency deviation***, $\Delta\nu = \nu_{\max} - \nu_0 = 102.02 - 102 = 0.02\,\text{MHz} = \mathbf{20 \ kHz}$.

(ii) ***Modulation index***, $m_v = \Delta\nu/\nu_m = 20\,\text{kHz}/4\,\text{kHz} = \mathbf{5}$.

(iii) ***Carrier swing***, $CS = 2 \times \Delta\nu = 2 \times 20 = \mathbf{40\,kHz}$.

(iv) ***Lowest frequency*** **of FM wave**, $\nu_{\text{min}} = \nu_0 - \Delta\nu = 102 - 0.02 = \mathbf{101.98\,MHz}$.

▶ **Example 13.** *A sinusoidal carrier wave of frequency* 1 MHz *and amplitude* 150 V *is amplitude modulated by a modulating signal of* 5 kHz. *If the modulation is* 60%, *find the frequency of upper and lower sidebands and also the amplitude of sidebands.*

Solution: By the problem, we have: $\nu_c = 1\,\text{MHz} = 1000\,\text{kHz}$, $\nu_m = 5\,\text{kHz}$. $m_a = 60\% = 0.6$ and $E_c = 150\,\text{V}$.

$$\therefore \textit{Frequency of upper sideband} = \nu_c + \nu_m = 1000 + 5 = \mathbf{1005\,KHz}$$

$$\textit{Frequency of lower sideband} = \nu_c - \nu_m = 1000 - 5 = \mathbf{995\,kHz}$$

$$\textit{Amplitude of each sideband} = \tfrac{1}{2} m_a E_c = \tfrac{1}{2} \times 0.6 \times 150\,\text{V} = \mathbf{45\,V}$$

▶ **Example 14.** *An envelope detector is used for detecting amplitude modulated waves with* 80% *modulation. The maximum frequency of audio signal is* 1 kHz. *If the load resistance is* 2.5 kΩ, *find the capacitance of the capacitor for minimum distortion.*

Solution: For minimum distortion, we have the relation

$$\omega_s = \frac{1}{m_a RC}.$$

Here, $R = 2.5\,\text{k}\Omega = 2500\Omega$, $m_a = 80\% = 0.8$ and maximum frequency of audio signal, $f_s = 1000\,\text{Hz}$. So, $\omega_s = 2\pi f_s = 2\pi \times 1000$.

We thus have, $2\pi \times 1000 = 1/(0.8 \times 2500 \times C)$

$$\text{or,}\ \ C = \frac{1}{2\pi \times 1000 \times 0.8 \times 2500} = \mathbf{0.08\mu F}$$

▶ **Example 15.** *The maximum electron concentration of an ionospheric layer is* $10^{12}/m^3$. *Calculate the critical frequency for this layer.*

Solution: The critical frequency f_c is related to the maximum electron concentration N of the layer as

$$f_c = 9/\sqrt{N} = 9\sqrt{10^{12}}\,\text{Hz} = 9 \times 10^6\,\text{Hz} = \mathbf{9\,MHz}$$

▶ **Example 16.** *There is an ionized layer at a height of* 100 km *from the earth's surface. The critical frequency for the layer is* 3 MHz. *Its electron concenration* N (*in* m^{-3}) *increases linearly with height* h (*in* km) *measured from the bottom of the layer and bears the relationship:* $N = 5 \times 10^{10} + 10^9 h$. *If a radio wave is sent vertically upward from the earth's surface, determine the height of the point above the ground whence the wave would be reflected back.*

Solution: If N_p be the electron concentration at the reflecting point P, the critical frequency f_c is given by

$$f_c = 9\sqrt{N_p} \Rightarrow N_p = f_c^2/81$$

Here, we have, $f_c = 3\,\text{MHz} = 3 \times 10^6\,\text{Hz}$. Therefore,

$$N_p = \frac{(3 \times 10^6)^2}{81} = \frac{9 \times 10^{12}}{81} = 1.11 \times 10^{11}\,\text{m}^{-3}$$

If h be the height of the reflecting point from the bottom of the ionized layer, then

$$N_p = 5 \times 10^{10} + 10^9 h \quad \text{(given)}$$

$\therefore$ We have:

$$1.11 \times 10^{11} = 5 \times 10^{10} + 10^9 h$$

$$\text{or,} \quad h = \frac{6.1 \times 10^{10}}{10^9} = 61\,\text{km}$$

$\therefore$ Required ***height of the reflecting point*** above the ground is plainly: $100\,\text{km} + 61\,\text{km} =$ **161 km**.

Review and Short Questions

1. What is modulation? Why is modulation necessary in radio transmission? State the different methods of modulation. **(Lucknow and Burdwan Hons.)**

2. What do you mean by (i) carrier wave, (ii) modulating wave and (iii) modulated wave? When and why are modulation and demodulation required?

3. What is amplitude modulation? Derive an expression for an amplitude modulated carrier wave. What is modulation index? **(Punjabi Univ.)**

4. Show that the total power for a fully amplitude modulated wave is 1.5 times the unmodulated carrier power. **(Calcutta Hons.)**

5. Explain with a circuit diagram how the amplitude modulated signals can be produced. **(Burdwan Hons.)**

6. Show that an amplitude-modulated wave can be represented by a carrier and two side-frequencies for each modulation frequency. **(Calcutta Hons.)**

7. Find the expressions for the carrier power and the sideband power. **(Burdwan Hons.)**

8. Show that for a 100% modulated AM wave only one-third of the total power is contained in the sidebands. **(North Bengal Hons.)**

9. Explain with a neat circuit diagram the working of an AM wave generator.

10. Explain why it is not possible to transmit audio signals in free space.

Why cannot electrical signals having low-frequency be transmitted directly?

(Himachal Univ.)

11. What do you mean by demodulation or detection? Explain with diagram the working of a diode envelope detector for AM, indicating the sources of distortion of the detected output.

12. What is frequency modulation? Derive an expression for an FM wave with sinusoidal modulation. Define modulation index. Show that the frequency modulation index is directly proportional to its amplitude and inversely proportional to the frequency of modulating signal. **(Osmania and Guru Nanak Univ.)**

13. What is frequency modulation? Show that the amount of frequency deviation is independent of the modulating frequency. Draw the waveform of an FM wave.

14. Make a comparative study of AM and FM systems.

15. Point out the differences between AM and FM signals. **(Burdwan Hons.)**

16. What is the advantage of FM over AM. **(Calcutta Hons.)**

17. What is phase modulation? Obtain an expression for a phase-modulated wave when the modulating signal is sinusoidal. Compute Δf and show that in PM it is proportional to both the amplitude and the frequency of the modulating signal.

18. A diode, with a load resistance R in shunt with a small capacitance C, is used to detect an amplitude modulated input carrier signal. If the maximum modulation factor is m, find the highest modulation frequency that can be detected without excessive distortion. **(Calcutta Hons.)**

19. Define (i) upper side-frequency and (ii) lower side-frequency in connection with frequency spectrum of an AM wave modulated by a single frequency signal.

What are sidebands of AM waves? How can you find the band width from the sidebands?

20. Name the basic elements of a radio communication system, stating in brief the function of each element.

21. Mention the different ways in which a radio-wave can travel from a transmitting antenna to a receiving antenna.

What are ground waves? Where are they used? What are their advantages?

22. What is meant by space waves? Where are they used? Explain the phenomenon of selective fading in communication through tropospheric waves.

What is duct propagation?

23. What are sky waves? What frequency band is used in sky wave propagation?

24. What is ionosphere? How is it formed? 'Sky wave reception is better at night than at day'—Why?

25. Explain briefly how a radio wave is reflected from ionosphere? **(Burdwan Hons.)**

26. What is meant by the term 'maximum usable frequency' in ionspheric propagation of a radio wave? **(Burdwan Hons.)**

27. Define the critical frequency of an ionospheric layer. Show that the critical frequency f_c is related to peak electron concentration N_p of the reflection layer by $f_c = 9\sqrt{N_p}$.

28. Discuss the mechanism of reflection of e.m. waves in ionospheric layers. What do you mean by skip distance and how does it vary with the hour of the day and the frequency of the wave? **(Calcutta Hons.)**

29. How are the shortcomings of a straight forward AM radio receiver eliminated in a superheterodyne receiver? Show the basic blocks of the syperheterodyne receiver and explain the function of each block. **(Burdwan Hons.)**

Objective Questions

1. Modulation is necessary since

(a) audio signals cannot traverse long distances,
(b) audio waves require large antennas for transmissions,
(c) simultaneous transmission of several audio signals mix up,
(d) all the above are true.

2. In amplitude modulation, the characteristic of carrier wave that changes in accordance with the modulating signal is

(a) phase, (b) frequency, (c) amplitude, (d) none.

3. With modulation factor $m_a = 1$ in amplitude modulation, the ratio of power in sidebands to the total power in modulated wave is

(a) 2/3, (b) 1/3, (c) 3/2, (d) 3.

4. For maximum power transmission by a given AM wave, the depth of modulation must be

(a) 0.75, (b) 1.0, (c) 0.50, (d) 0.25.

5. The modulation index in FM

(a) is always < 1, (b) is always > 1,
(c) can be > 1, (d) is always 0.

6. The depth of modulation in AM is given by

(a) $\frac{V_{\max} - V_{\min}}{V_{\max} + V_{\min}}$, (b) $\frac{V_{\max} - V_c}{V_{\min} + V_c}$,

(c) $\frac{V_{\max} + V_{\min}}{V_{\max} - V_{\min}}$, (d) $\frac{V_{\max} - V_m}{V_{\max} + V_m}$.

7. Demodulation is required at

(a) the receiving end, (b) the transmitting end,

(c) both receiving and transmitting ends, (d) anywhere in space.

8. If f_c and f_m be the frequency of the carrier and the modulating signal, then

(a) $f_c = f_m$, (b) $f_c > f_m$, (c) $f_c < f_m$.

9. In AM wave, the amplitude variation is proportional to

(a) the amplitude of the modulating signal,

(b) the frequency of the modulating signal,

(c) the instantaneous value of the modulating signal,

(d) none of the above.

10. In AM, the full information can be conveyed by transmitting only the

(a) upper sideband, (b) lower sideband,

(c) any of the sidebands, (d) the carrier wave.

11. For 100% modulation (AM), the fraction of total energy carried by each sideband is

(a) 1/3, (b) 2/3, (c) 1/2, (d) 1/6.

12. In FM, if $\Delta\nu$ be the frequency deviation and ν_m the modulating frequency, then the modulation index is

(a) $\nu_m \times \Delta\nu$, (b) $\Delta\nu/\nu_m$, (c) $\nu_m/\Delta\nu$, (d) $\nu_m - \Delta\nu$.

13. A detector circuit has two main parts, namely

(a) rectifier and amplifier, (b) amplifier and filter circuit,

(c) rectifier and filter circuit, (d) amplifier and oscillator.

14. The upper and lower side frequencies in amplitude modulation are

(a) ω_c, $(\omega_c + \omega_m)$, (b) ω_c, $(\omega_c - \omega_m)$,

(c) $(\omega_c + \omega_m)$, $(\omega_c - \omega_m)$, (d) $(\omega_c - \omega_m)$, $(\omega_c + \omega_m)$.

15. The critical frequency f_c of an ionospheric layer is related to its peak (maximum) electron concentration N_p as

(a) $f_c \propto N_p$, (b) $f_c \propto \sqrt{N_p}$, (c) $f_c \propto 1/N_p$, (d) $f_c \propto 1/\sqrt{N_p}$.

16. Number of sideband components in an FM wave is

(a) two, (b) three, (c) ten, (d) infinite,

17. Medium frequency radio waves correspond to the frequency range

(a) 3–30 kHz, (b) 30–300 kHz, (c) 3–30 MHz, (d) None.

18. Indicate if the following statements are true (T) or false (F).

(a) Higher the frequency of the wave, longer is the length of antenna.
(b) A transmitting antenna can be used as a receiving antenna.
(c) The maximum usuable frequency of communication is fixed and does not vary from time to time.
(d) Radio waves have frequencies in the range 50 Hz to 1 kHz.
(e) For carrier frequencies above 50 MHz, communication via sky wave is not possible.

19. Percentage modulation of an AM wave having a power content of 8 kW at carrier frequency and 2 kW in each of its side bands is

(a) 60%, (b) 70%, (c) 100%, (d) 80%. **(ISRO)**

20. If the total power of an amplitude modulated signal is 600 W and the carrier power is 400 W, the modulation index is

(a) 0.75, (b) 0.5, (c) 0.25, (d) 1. **(ISRO)**

21. The modulation index of an amplitude modulated wave is changed from 0 to 1. The transmitted power is

(a) doubled, (b) halved, (c) increased by 50%, (d) unchanged. **(ISRO)**

22. In which of the following, an increase in the modulation index leads to an increase in bandwidth?

(a) PM, (b) FM, (c) AM, (d) both (a) and (c). **(ISRO)**

23. In phase modulated signal, the frequency deviation is proportional to

(a) frequency only, (b) amplitude only,
(c) both (a) and (b), (d) none of these. **(ISRO)**

Problems

1. How much audio power is necessary to fully modulate a 50 kW carrier if the modulating system efficiency is 72%? **(Punjab Univ.)** [34.7 kW]

2. A 500 W carrier is modulated to a depth of 60%. Calculate the total power in the modulated wave. **(Punjab Univ.)** [590 W]

3. An AM broadcast radio transmitter radiates at 20 kW at modulation index 75%. What is the carrier power? **(Himachal Univ.)** [15.61 kW]

4. What is the power developed in an AM wave in a load of 200 Ω when the peak voltage of the carrier is 120 V and modulation factor is 0.5 ? [40.5 W]

5. An amplitude modulated wave is represented by

$$\rho_m = 10(1 + 0.6 \cos 6280t) \sin 211 \times 10^4 \text{ volt}$$

Calculate (a) the minimum and maximum amplitudes of the AM wave, (b) the frequencies and the amplitudes of the components contained in the modulated wave. [(a) 16 V, 4 V; (b) 33.43×10^4 Hz, 10^3 Hz; 10 V, 6 V]

6. A 1000 kHz carrier wave is amplitude modulated by audio signals between 400 to 1600 Hz. Find (a) the frequency span of each sideband, (b) the maximum upper side-frequency, (c) the minimum lower side-frequency and (d) the channel width. **(Calcultta Hons.)** [1200 Hz; 1.16 MHz, 998.4 kHz, 3.2 kHz]

7. A carrier wave of 400 watt is subjected to 80% amplitude modulation. Determine (a) power in sidebands and (b) total power of modulated wave. [128 W, 528 W]

8. Find the value of modulation index of a frequency modulated carrier having a carrier swing of 120 kHz and a modulated signal of 4 kHz. [15]

9. The frequency deviation in a frequency modulated wave transmission is 15 kHz. If it broadcasts in the 88–108 MHz frequency band, calculate the percentage modulation. [20%]

10. A linear diode detector is used to detect an amplitude modulated wave of 75% modulation by using a capacitor of 225 pF capacity with parallel combination of a load resistance 150 kΩ. Calculate the maximum modulation frequency which it can detect. [6.29 kHz]

11. The peak-to-peak value of an AM voltage has a maximum value of 8 V and a minimum value of 2 V. Calculate the percentage modulation and the amplitude of the unmodulated wave. [60%, 2.5 V]

12. Find the audio power required to amplitude modulate a 20 kW carrier to a depth of 0.5, if the efficiency of the modulation system is 70%. [3.57 kW]

13. An AM broadcast transmitter radiates 10 kW of unmodulated carrier power and 12.5 kW of total power when the carrier is modulated by a sinusoidal signal. Determine the depth of modulation. If another sinusoidal signal modulate the carrier simultaneously, to a depth of 60%, what will be the total radiated power?. [0.707, 14.3 kW]

14. A frequency modulated wave is given by $v = V_0 \cos(\omega_c t + m_f \sin \omega_m t)$, the symbols having their usual meaning.

Show that if the modulation index $m_f \ll 1$, the bandwidth of FM is about $2\omega_m$.

[**Hints.** $v = V_0 \cos \omega_c t \cos(m_f \sin \omega t) - V_0 \sin \omega_c t \sin(m_f \sin \omega_m t)$

$= V_0 \cos \omega_c t - V_0 \sin \omega_c t \cdot m_f \sin \omega_m t$, using $\cos\theta \simeq 1$ and $\sin\theta \simeq \theta$ for small θ.

$$= V_0 \cos \omega_c t - \frac{V_0 m_f}{2}[\cos(\omega_c - \omega_m)t - \cos(\omega_c + \omega_m)t]$$

Plainly, the bandwidth $= (\omega_c + \omega_m) - (\omega_c - \omega_m) = \mathbf{2\omega_m}$]

15. Let an AM wave have 80% modulation. Calculate the percent of power saved when a single sideband is transmitted instead of the total AM wave. [87.8%]

[**Hints.** $P_{av} = P_c(1 + m_a^2/2) = P_c(1 + 0.8^2/2) = 1.32P_c$. Power associated with a single sideband, $P_{SSB} = P_{SB}/2 = P_c \times m_a^2/4 = \frac{0.8^2}{4} P_c = 0.16P_c$.

$$\therefore \text{ Percentage of power saved} = \frac{1.32P_c - 0.16P_c}{1.32P_c} \times 100 = \mathbf{87.8\%}]$$

16. Audio signals in the range 300 Hz to 3.4 kHz are used to amplitude modulate a carrier of frequency 200 kHz. Determine (i) the bandwidth of the modulated signals, (ii) frequency range of the lower sidebands and (iii) the frequency range of upper sidebands. **(Calcutta Hons.)** [6.8 kHz, 3.1 kHz, 3.1 kHz]

17. A 1 kW carrier is amplitude modulated by a 5 kHz modulating signals to a depth of 50%. Determine the total power of the AM wave. **(Burdwan Hons.)** [1.25 kW]

18. An AM wave with 50% modulation is applied to a diode detector with a load resistance of 220 kΩ and a shunt capacitor. If the frequency of the modulating voltage lies in the range 1 kHz of 15 kHz, find the value of the shunt capacitor for detection with tolerable distortion. **(Burdwan Hons.)** [9.6 pF]

19. An ionosphere radio link is set up between two points 1000 km apart. The height and critical frequency of the reflecting layer are 350 km and 10 MHz respectively. Determine the maximum usable frequency. [17.43 MHz]

20. A radio link is set up between two points on the earth's surface, 1200 km apart. Single-hop transmission of radio wave reflected from E-layer is used. The height and electron density of the reflecting layer are 110 km and $1.5 \times 10^5\ \text{cm}^{-3}$ respectively. Find the angle with the horizontal at which the ray is launched at the transmitting point. Given, the earth's radius = 6400 km. [10.39°].

14

Digital Electronics

14.1 Decimal number system

Since our boyhood days, we are familiar with the decimal system which has ten numerals : $0, 1, 2, 3, 4, 5, 6, 7, 8, 9$. The *base* (or *radix*) of the system is 10. A number in this system is expressed in terms of the position values of the numerals.

For instance, the number 50142 is represented as

$$50142 = 5 \times 10^4 + 0 \times 10^3 + 1 \times 10^2 + 4 \times 10^1 + 2 \times 10^0.$$

While the digit 2 has the position value 10^0, the digit 5 has the position value 10^4. So, 2 is the least significant digit (LSD) and 5 the most significant digit (MSD).

Let us now express the number 6245.937 in terms of the positional values.

$$6245.937 = 6 \times 10^3 + 2 \times 10^2 + 4 \times 10^1 + 5 \times 10^0 + 9 \times 10^{-1} + 3 \times 10^{-2} + 7 \times 10^{-3}.$$

• It is to be particularly noted that the powers to the base 10 are numbered to the left of the decimal point starting from 0 and those to the right of the decimal point starting from -1.

14.2 Binary number system

In binary system of representation of numbers, the *base* (or *radix*) is 2 and it uses only two numerals : 0 and 1. In a digital system, there are only two possible states or conditions, e.g., a situation True or False, a switch closed or open, a voltage signal high or low. These states or conditions are designated as 1 and 0 respectively. The binary digits, 0 and 1, are called *bits*, and like the decimal system, the binary system also has positional-value representation.

For instance, the number 31 of decimal system is written in binary system as 11111, for, we have

$$11111 = 1 \times 2^4 + 1 \times 2^3 + 1 \times 2^2 + 1 \times 2^1 + 1 \times 2^0 = 31$$

This is also written in the form $11111_2 = 31_{10}$, the subscript points to the number system. Plainly, the bit 1 at the extreme left is MSB, the most significant bit and the bit 1 at the extreme right is LSB, the least significant bit.

Similarly, 1101.011 in the binary system is written as 13.375 in decimal system, for

$$\begin{aligned} 1101.011 &= 1 \times 2^3 + 1 \times 2^2 + 0 \times 2^1 + 1 \times 2^0 + 0 \times 2^{-1} + 1 \times 2^{-2} + 1 \times 2^{-3} \\ &= 8 + 4 + 0 + 1 + 0 + \frac{1}{4} + \frac{1}{8} = 13.375 \end{aligned}$$

$$\therefore \quad 1101.011_2 = 13.375_{10}$$

Note that the positional values of bits in a binary number are given by ascending powers of 2 to the left of the 'binary point' starting from 0 and to the right of binary point in descending powers of 2 starting from −1. The word 'decimal point' in decimal numbers has been replaced by 'binary point' in binary numbers.

14.3 Conversion of decimal number into binary

1. Integral decimal number—The given integral decimal number is *progressively divided* by 2 *till* 0 *is reached.* The *remainders* in each division, taken in *the reverse order*, will give the corresponding binary number. The following examples would enlight.

Let us convert the decimal number 21 and 39 into the equivalent binary numbers.

$$\begin{array}{r|l} 2 & 21 \\ 2 & 10 \;-\; 1 \uparrow \\ 2 & 5 \;-\; 0 \\ 2 & 2 \;-\; 1 \\ 2 & 1 \;-\; 0 \\ & 0 \;-\; 1 \end{array} \qquad \begin{array}{r|l} 2 & 39 \\ 2 & 19 \;-\; 1 \uparrow \\ 2 & 9 \;-\; 1 \\ 2 & 4 \;-\; 1 \\ 2 & 2 \;-\; 0 \\ 2 & 1 \;-\; 0 \\ & 0 \;-\; 1 \end{array}$$

Taking the remainders in the reverse order, we have

$$21_{10} = 10101_2; \qquad 39_{10} = 100111_2$$

2. Fractional decimal number—The given number is *progressively multiplied* by 2. For each step that results in an 1 in units place, *tansfer the* 1 *to the binary record* and repeat the mutiplication process with the fractional number. If however the result of multiplication does not give an 1 units place, i.e., the product is less than unity, *transfer* 0 *to the binary record* and repeat the process. The last step is attained if the fractional part is zero or it is terminated if the desired accuracy is reached. The transfers in the forward direction, i.e., from top to bottom will give the equivalent binary. For clarification, we give an example converting 0.5219 into its binary equivalent.

$$\begin{array}{rcllll}
0.5219 \times 2 & = & 1.0438 = 0.0438 & \text{with} & \text{a carry} & 1 \\
0.0438 \times 2 & = & 0.0876 = 0.0876 & \cdots & \cdots & 0 \\
0.0876 \times 2 & = & 0.1752 = 0.1752 & \cdots & \cdots & 0 \\
0.1752 \times 2 & = & 0.3504 = 0.3504 & \cdots & \cdots & 0 \\
0.3504 \times 2 & = & 0.7008 = 0.7008 & \cdots & \cdots & 0 \\
0.7008 \times 2 & = & 1.4016 = 0.4016 & \cdots & \cdots & 1 \downarrow
\end{array}$$

We terminate the process for an *approximate* result, representing 0.5219 by six binary digits. We thus obtain

$$0.5219_{10} = 0.100001_2$$

3. Mixed number—To find the binary equivalent of mixed number—partly integral, partly fractional—e.g., 33.565, the number is split into an integer and a fraction and the binary equivalent of the given number is obtained separately using the methods 1 and 2 as given above. An example follows with 33.565

We first obtain the binary equivalent of 33, by method 1.

$$\begin{array}{r|ll}
2 & 33 & \\
2 & 16 & -1 \uparrow \\
2 & 8 & -0 \\
2 & 4 & -0 \\
2 & 2 & -0 \\
2 & 1 & -0 \\
 & 0 & -1
\end{array} \qquad \therefore\ 33_{10} = 100001_2$$

Next, we find the binary equivalent of 0.565

$$\begin{array}{l}
0.565 \times 2 = 1.130 = 0.130 \text{ with a carry } 1 \\
0.130 \times 2 = 0.260 = 0.260 \text{ with a carry } 0 \\
0.260 \times 2 = 0.520 = 0.520 \text{ with a carry } 0 \\
0.520 \times 2 = 1.040 = 0.040 \text{ with a carry } 1 \downarrow
\end{array}$$

We terminate the process for an *approximate* result by 4 binary digits.

$$\therefore\ 0.565_{10} = 1001_2$$

$$\therefore\ 33.565_{10} = 100001.1001_2$$

14.4 Binary addition

The rules of addition of two binary numbers are the same as in decimal *except* however that in binary the digits of the sum in any significant position could be either 0 or 1 only. If there is any 'carry' obtained in a significant position, it is used by the pair of digits one significant position higher.

The following four rules dictate the binary addition.

1. $0 + 0 = 0$
2. $0 + 1 = 1$
3. $1 + 0 = 1$
4. $1 + 1 = 10$ (not ten, but one zero)
   $= 0$, with carry of 1

Example. Add the binary numbers 10110 and 10011.

$$\begin{array}{rr} & 10110 \\ & +10011 \\ \hline \text{Sum :} & 101001 \end{array}$$

14.5 Binary subtraction

The binary subtraction is guided by the following four rules :

1. $0 - 0 = 0$
2. $1 - 0 = 1$
3. $1 - 1 = 0$
4. $10 - 1 = 1$

The 4th rule shows that $10_2 = 2_{10}$ and if 1 is subtracted from 2, we get 1. Note the following example.

Subtract the binary number 100111 from the number 101101.

$$\begin{array}{rrl} & 101101 & \text{(minuend)} \\ & -100111 & \text{(subtrahend)} \\ \hline \text{Difference :} & 000110 & \end{array}$$

14.6 Complement method of subtraction

1. 1's complement—The 1's complement of a binary number is the number that is obtained by changing each 0 into an 1 and each 1 into a 0 of the given binary number. For instance, the 1's complement of the binary number 10110 is 01001.

An *alternative method* of subtraction is by 1's complement. The sequential procedure is as follows.

1. Compute the 1's complement of the subtrahend.
2. Add the 1's complement to the minuend.

3. When there is no end-around carry, then the answer is recomplemented and a negative sign is attached.
4. If, however, there is an end-around carry of 1, remove the carry 1 from the addition and add it to the remainder. No recomplementing is needed.

We shall now subtract 1101_2 from 1010_2 by the 1's complement method.

$$\begin{array}{r l} 1010 & \\ +0010 & \text{(1's complement of 1101)} \\ \hline 1100 & \leftarrow \text{No end-around carry} \end{array}$$

∴ 1's complement of 1100 is 0011. ∴ Final answer = −0011.

● Students should check their answer by using the conventional method.

Next, we shall subtract 1001 from 1010 by complement method. 1's complement of the subtrahend is 0110.

$$\therefore \begin{array}{r l} 1010 & \\ +0110 & \\ \hline 10000 & \leftarrow \text{end-around carry} \end{array} \qquad \therefore \begin{array}{r l} 0000 & \text{removing the carry 1} \\ +\ 1 & \\ \hline 0001 & \leftarrow \text{Final answer} \end{array}$$

2. 2's complement—The 2's complement of a binary number is the number that is obtained by adding 1 to its 1's complement. For example, the 2's complement of the binary number 10110 is 01001 + 1 = 01010.

Binary subtraction can be carried out by using 2's complement method. The procedure is as follows.

1. Compute the 2's complement of the subtrahend.
2. Add the 2's complement to the minuend.
3. When there is no end-around carry, the 2's complement is computed and a negative sign is attached. So, the answer of the subtraction is negative.
4. If, however, there is an end-around carry of 1, the result is positive and it is obtained by removing the end-around carry. No recomplementing is needed.

Let us now subtract 1101_2 from 1010_2 by the 2's complement method.

$$\begin{array}{r l} 1010 & \\ +0011 & \text{(2's complement of 1101)} \\ \hline \square 1101 & \leftarrow \text{No end-around carry} \end{array}$$

The 2's complement of 1101 is 0010 + 1 = 0011

∴ The final answer = −0011.

● This answer agrees with the result we had obtained by 1's complement method discussed earlier.

We shall now subtract 1001_2 from 1010_2 by 2's complement method. The 2's complement of the subtrahend is $0110 + 1 = 0111$.

$$
\begin{array}{r}
\therefore \quad 1010 \\
+0111 \\
\hline
\boxed{1}0001
\end{array}
\quad \leftarrow \text{end-around carry of 1}
$$

Removing the carrying we get 0001.

$\therefore$ The final answer = 0001.

14.7 Binary multiplication

The binary multiplication is carried out as in decimal system and is guided by the following rules.

1. $0 \times 0 = 0$
2. $0 \times 1 = 0$
3. $1 \times 0 = 0$
4. $1 \times 1 = 1$

The following example is rather illustrative. Multiply 11001×110.

$$
\begin{array}{r}
11001 \\
110 \\
\hline
00000 \\
11001 \\
11001 \\
\hline
10010110
\end{array}
\qquad
\begin{array}{l}
\text{Verification} \\
25 \times 6 \\
= 150
\end{array}
$$

$\therefore$ The required result of multiplication is : 10010110.

14.8 Binary division

The binary division is carried out as in the decimal system. The example would be illustrative. Let us divide 1110 by 110.

$$
\begin{array}{r|l}
 & 10 \\
\cline{2-2}
110 & 1110 \\
 & 110 \\
\cline{2-2}
 & 10
\end{array}
\qquad
\begin{array}{r|l}
\multicolumn{2}{c}{\text{Verification}} \\
\multicolumn{2}{c}{14 \div 6} \\
 & 2 \\
\cline{2-2}
6 & 14 \\
 & 12 \\
\cline{2-2}
 & 2
\end{array}
$$

$\therefore$ Quotient : 10
Remainder : 10

Quotient : 2
Remainder : 2

14.9 Octal number system

In this system, the base or the radix is 8 and hence the name. Plainly, the eight numerals are $0, 1, 2, 3, 4, 5, 6$ and 7. To count beyond 7, 2-digit combinations are formed by taking the second digit followed by the first, then the second digit followed by the second and so on and so forth.

So, after 7, the next number in the octal system is, as stated, 10 (second digit followed by the first), the next number is 11 (second digit followed by the second). The position value of each digit is in ascending powers of 8 for integers and in descending powers of 8 for fractions.

$$\leftarrow \; 8^3 \; 8^2 \; 8^1 \; \bullet \; 8^{-1} \; 8^{-2} \; 8^{-3} \; \rightarrow$$

$$\uparrow$$

octal point

The following example will illustrate how to convert octal number to the decimal. Let the number be 345.450.

$$\begin{aligned} 345.450 &= 3 \times 8^2 + 4 \times 8^1 + 5 \times 8^0 + 4 \times 8^{-1} + 5 \times 8^{-2} + 0 \times 8^{-3} \\ &= 3 \times 64 + 4 \times 8 + 5 \times 1 + 4 \times \tfrac{1}{8} + 5 \times \tfrac{1}{64} \\ &= 192 + 32 + 5 + 0.5 + 0.078 = 229.578_{10} \end{aligned}$$

14.10 Hexadecimal number system

In the hexadecimal number system, the base or the radix is 16 and the sixteen numerals are $0, 1, 2, 3, 4, 5, 6, 7, 8, 9, A, B, C, D, E, F$, where A represents 10, B respresents 11, $\cdots$, E represents 14 and F 15. After F, the next numbers are formed by two digit combinations taking the second digit followed by the first, then the second followed by the second and so on. So, in hexadecimal system 10 (the second followed by the first) represents 16 in the decimal system.

The position value for each digit is in ascending powers of 16 for integers and in descending powers of 16 for fractions.

Examples of conversion of hexadecimal to decimal :

$$\begin{aligned} 2B8_{16} &= 2 \times 16^2 + 11 \times 16^1 + 8 \times 16^0 = 512 + 176 + 8 = 696_{10} \\ 12{\cdot}CF_{16} &= 1 \times 16^1 + 2 \times 16^0 + 12 \times 16^{-1} + 15 \times 16^{-2} = 16 + 2 + \tfrac{12}{16} + \tfrac{15}{256} \\ &= 18 + 0.75 + 0.058 = 18.808_{10} \end{aligned}$$

14.11 Boolean algebra

The algebra of logic used in the operation of computer device was developed by George Boole and is therefore known as ***Boolean algebra***. It is a two-valued (binary) logic permitting only two values or states for its variables. The two states are '*true*' and '*false*' in logic and can be represented respectively by the 'on' and the 'off' states of electronic circuits.

The two variables in Boolean algebra are represented by 0 and 1. So, every variable is either a 0 or an 1.

Negative or fractional numbers do not occur. Logically, therefore,

(i) if $X = 0$, then $X \neq 1$; (ii) if $X = 1$, then $X \neq 0$.

Boolean algebra utilizes only the following *three* operations on its variables.

1. OR addition, represented by $+$ sign.
2. AND multiplication, represented by either $\times$ sign or $\cdot$ sign. Usually, the adjacent letters are written without the dot.
3. NOT operation, represented by a bar over the variable.

14.12 Boolean addition and multiplication

In Boolean addition, the variables have values of either a binary 1 or a binary 0 and the basic rules run as

$$0 + 0 = 0$$
$$0 + 1 = 1$$
$$1 + 0 = 1$$
$$1 + 1 = 1$$

When applied to logic circuits, the *Boolean addition is identical with* OR.

The basic rules for Boolean multiplication are given by

$$0 \cdot 0 = 0$$
$$0 \cdot 1 = 0$$
$$1 \cdot 0 = 0$$
$$1 \cdot 1 = 1$$

Boolean multiplication is the same as AND, *when applied to logic circuits.*

14.13 Logic expressions

NOT : The operation of a NOT circuit can be symbolically expressed as follows. Let the input variable be A and the output variable be X. Then $X = \bar{A}$, implying that the output is the complement of input (Fig. 14.1).

$$\therefore \quad \text{If } A = 0, \quad \text{then } X = 1$$
$$\text{If } A = 1, \quad \text{then } X = 0.$$
$$\text{So (i) } \bar{0} = 1, \text{ (ii) } \bar{1} = 0.$$

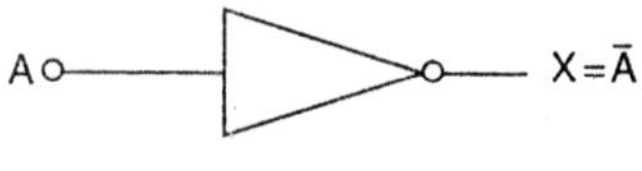

Fig. 14.1 Symbol of NOT gate

AND : The operation of AND gate involving two inputs can be expressed in equational form.

Let one input variable be A and the other be B, and the output variable be X. Then, the Boolean expression for this basic gate function (Fig. 14.2) is

Fig. 14.2 Symbol of a two-input AND gate

$$X = AB$$

OR : The operation of OR gate involving two inputs can also be expressed is equational form. Let one input be A and the other be B, and the output be X. Then, the Boolean expression for OR function is (Fig. 14.3) given by

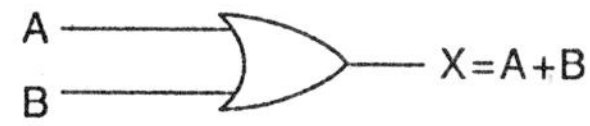

Fig. 14.3 Symbol of a two-input OR gate

$$X = A + B$$

14.14 Commutative laws

1. The commutative law of addition of two variables A and B is

$$A + B = B + A$$

This implies that the order in which the variables are OR-ed is immaterial, that is, it makes no difference.

2. The commutative law of multiplication of two variables A and B is givem by

$$AB = BA$$

which implies that the order in which the variables are AND-ed is immaterial for this makes no difference.

14.15 Associative laws

1. The associative law of addition for *three* variables is stated as follows :

$$A + (B + C) = (A + B) + C$$

It implies that in OR-ing of several variables, the result is the same regardless of their grouping.

2. The associative law of multiplication, involving *three* variables, is given by

$$A(BC) = (AB)C$$

The law implies that when AND-ing several variables, the result obtained is the same regardless of the order of grouping.

14.16 Distributive law

The distributive law involving three variables runs as

$$A(B + C) = AB + AC$$

which implies that OR-ing several variables and AND-ing the result with a single variable is equivalent to AND-ing the single variable with each of the several variables and then OR-ing the products.

14.17 Rules for Boolean algebra

The basic rules for the Boolean algebra are the following

1. $A+0=A$	7. $A\cdot A=A$
2. $A+1=1$	8. $A\cdot \bar{A}=0$
3. $A\cdot 0=0$	9. $\bar{\bar{A}}=A$
4. $A\cdot 1=A$	10. $A+AB=A$
5. $A+A=A$	11. $A+\bar{A}B=A+B$
6. $A+\bar{A}=1$	12. $(A+B)(A+C)=A+BC$

1. A variable A OR-ed with a 0 equals the value of the variable $A+0=A$ (Fig. 14.4) Similarly, the other rules can also be proved readily.

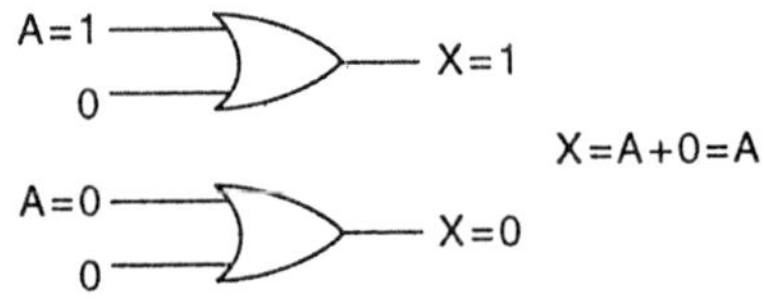

Fig. 14.4 Realisation of addition rule

14.18 De Morgan's theorems

There are *two* De Morgan's theorems. These are :

$$1.\quad \overline{AB}=\bar{A}+\bar{B}$$
$$2.\quad \overline{A+B}=\bar{A}\bar{B}$$

The *first theorem states* that the complement of the product of two variables is equal to the sum of the complements of the variables, i.e., the complement of two variables AND-ed is the same as the OR of the complements of each variable (Fig. 14.5)

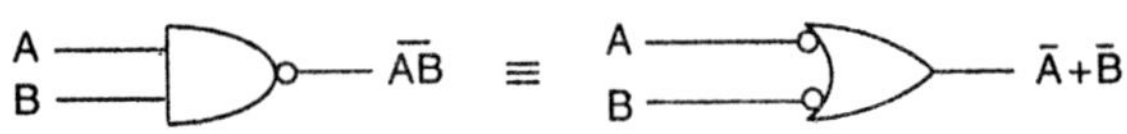

Fig. 14.5 Logic representation of De Morgan's first theorem

The *second theorem states* that the complement of the sum of two variables is equal to the product of the complements of the variables, i.e., the complement of two variables OR-ed is the same as the AND of the complements of each variable (Fig. 14.6).

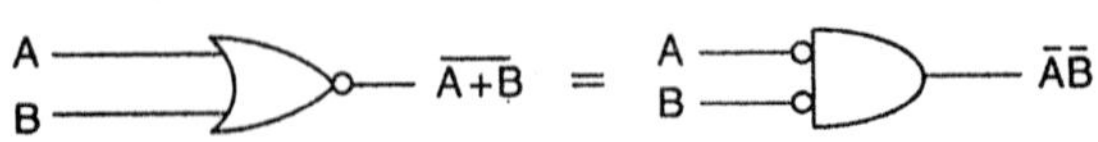

Fig. 14.6 Logic representation of De Morgan's second theorem

Proofs. The variable A or B takes a value 0 or 1. So, four combinations of the two variables are possibvle. These combinations are applied to prove the theorems.

First theorem

(a) When $A = 0$ and $B = 0$, we have $\overline{A\cdot B} = \overline{0\cdot 0} = \bar{0} = 1$, and $\bar{A} + \bar{B} = \bar{0} + \bar{0} = 1 + 1 = 1$. Therefore, $\overline{AB} = \bar{A} + \bar{B}$.

(b) When $A = 0$ and $B = 1$, then $\overline{AB} = \overline{0\cdot 1} = \bar{0} = 1$, and $\bar{A}+\bar{B} = \bar{0}+\bar{1} = 1+0 = 1$. So, $\overline{AB} = \bar{A} + \bar{B}$.

(c) If $A = 1$ and $B = 0$, then $\overline{AB} = \overline{1\cdot 0} = \bar{0} = 1$, and $\bar{A} + \bar{B} = \bar{1} + \bar{0} = 0 + 1 = 1$. Therefore, $\overline{AB} = \bar{A} + \bar{B}$.

(d) When $A = 1$ and $B = 1$, we get $\overline{AB} = \overline{1\cdot 1} = \bar{1} = 0$, and $\bar{A}+\bar{B} = \bar{1}+\bar{1} = 0+0 = 0$. Therefore, $\overline{AB} = \bar{A} + \bar{B}$.

The equality of both sides for all the four cases above proves De Morgan's first theorem i.e., $\overline{AB} = \bar{A} + \bar{B}$.

Second theorem

(a) When $A{=}0$ and $B{=}0$ we have $\overline{A+B} = \overline{0+0} = \bar{0} = 1$ and $\bar{A}\cdot\bar{B} = \bar{0}\cdot\bar{0} = 1\cdot 1 = 1$. Therefore, $\overline{A+B} = \bar{A}\cdot\bar{B}$.

(b) When $A = 0$ and $B = 1$, then $\overline{A+B} = \overline{0+1} = \bar{1} = 0$ and $\bar{A}\cdot\bar{B} = \bar{0}\cdot\bar{1} = 1\cdot 0 = 0$. Therefore, $\overline{A+B} = \bar{A}\cdot\bar{B}$.

(c) When $A = 1$ and $B = 0$, then $\overline{A+B} = \overline{1+0} = \bar{1} = 0$ and $\bar{A}+\bar{B} = \bar{1}\cdot\bar{0} = 0$. So, $\overline{A+B} = \bar{A}\cdot\bar{B}$.

(d) If $A = 1$ and $B = 1$, then $\overline{A+B} = \overline{1+1} = \bar{1} = 0$ and $\bar{A}\cdot\bar{B} = \bar{1}\cdot\bar{1} = 0\cdot 0 = 0$. Therefore, $\overline{A+B} = \bar{A}\cdot\bar{B}$.

The equality of both sides for all the four cases proves De Morgan's second theorem i.e., $\overline{A+B} = \bar{A}\cdot\bar{B}$.

14.19 Digital logic gates

A logic gate is an electronic circuit that performs logic functions or takes a logic decision. It has one output and one or more inputs. The output signal is produced when there is a specified combination of input signals. The logic gates, the building block of digital systems, work on the logical Boolean algebra. The Boolean operations, namely OR-operation, AND- operation and NOT operation are implemented by *three* logic gates called OR-gate, AND-gate and NOT-gate.

14.19.1 OR gate

The ***OR-gate*** implements Boolean OR operation and has two or more inputs and one output. OR-*gate is a logic gate whose output is* 1 *state if any or all of the inputs are in* 1 *state.* The principle of its operation is illustrated in Fig. 14.7.

Fig. 14.7(a) shows a two-input OR-gate with two ideal diodes D_1, D_2. While A, B represent two inputs, Y the output and R the output load resistor. Fig. 14.7(b) gives the the symbolic representation of OR-gate.

Let either 0 or 1 V be the two input voltages.

1. When both $A = 0$ and $B = 0$, the diodes D_1 and D_2 do not conduct. So, the current through R is zero. So, the output voltage $Y = 0$.

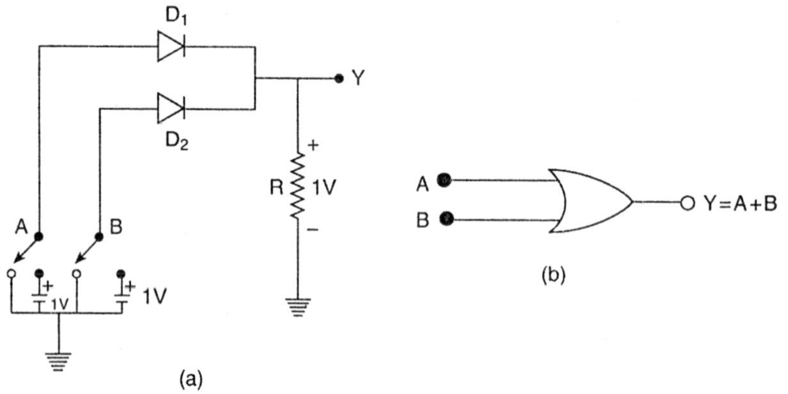

Fig. 14.7 (a) A two-input diode OR gate, (b) Logic symbol of OR gate

2. When $A = 1$, but $B = 0$, the diode D_1 is forward biased and hence conducts. D_2 is non-conducting. Current through R flows due to D_1. So, output voltage, $Y = 1$.

3. When $A = 0$, but $B = 1$, D_2 is forward biased and conducts, but D_1 is non-conducting. Current flows through R due to D_2. The output voltage $Y = 1$.

4. When both $A = 1$ and $B = 1$, each of D_1, D_2 are forward biased and hence they conduct, so the current through R flows and the output voltage $Y = 1$.

The above operations can all be summarised in a tabular form called the ***Truth table***, defined as a table that provides the output states for all the possible input combinations.

Truth table

A	B	$Y = A + B$
0	0	0
1	0	1
0	1	1
1	1	1

14.19.2 AND gate

AND gate implements the Boolean AND-operation and has two or more inputs and one output. AND-gate is a logic gate whose output is 1, *if and only if* all the inputs are in 1-state. The gate is illustrated in Fig. 14.8(a) where a two-input AND-gate has been used. Fig. 14.8(b) stands for its symbolic representation. A and B are the two inputs, Y the output and R the output load resistor.

Let either 0 or 1 V be the two input voltages.

1. When both $A = 0$ and $B = 0$, the inputs are short-circuited to ground. The battery in output side biases D_1 and D_2 forward. Hence both D_1, D_2 conduct. But through the diodes, the output is also shorted to ground so that the output $Y = 0$.

2. When $A = 0$, but $B = 1$, the diode D_1 conducts and the output is shorted through the diode to ground. So, the output $Y = 0$.

3. When $A = 1$, but $B = 0$, the diode D_2 conducts and the output is shorted to ground through this diode. So, the output $Y = 0$.

4. When both $A = 1$ and $B = 1$, neither diode conducts. No current flows through R and the output, $Y = 1$.

The truth table, as given below, summarises the above results.

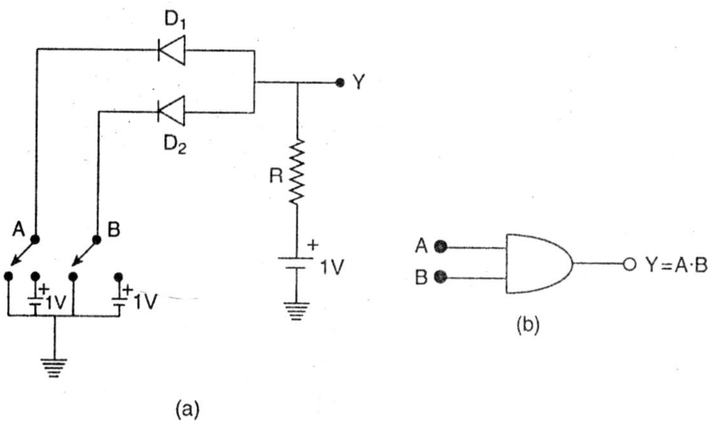

Truth table

A	B	$Y = A \cdot B$
0	0	0
0	1	0
1	0	0
1	1	1

Fig. 14.8 (a) A two-input diode AND gate, (b) Logic symbol of AND gate

14.19.3 NOT gate

The ***NOT gate***-circuit has only one input and one output and inverts the polarity of a pulse. If the input is 1, the output is 0 and conversely. The truth table below summarises the results. Fig. 14.9(a) shows a NOT gate using transistor and its symbol is shown in Fig. 14.9(b). See overleaf.

1. When no signal is applied to the input, i.e., $A = 0\,\text{V}$, the transistor is cut-off and the collector current is zero. So, the potential drop across R is zero. The supplied voltage V_{cc} appears at the output terminal. So, the output $Y = V_{cc}$.

2. When a positive pulse is applied to A, i.e., $A = 1$, the transistor conducts drawing the maximum collector current. So, the entire V_{cc} drops across R and the output $Y = 0$.

Truth table

Input	Output
A	$Y = \bar{A}$
0	1
1	0

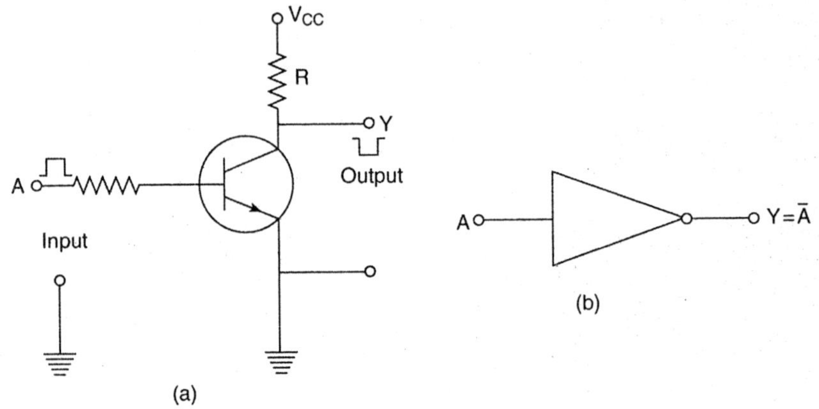

Fig. 14.9 (a) A NOT gate using BJT, (b) Logic symbol of NOT gate

14.19.4 NAND gate

When an AND gate and a NOT gate are combined in such a way that NOT gate follows the AND gate, the combination is referred to as a NOT-AND or ***NAND gate***. The output of a NAND gate is '0' when all the inputs are 1. In all other cases, the output is 1. The split-up logic circuit is illustrated in Fig. 14.10(a) where a two-input AND gate is followed by a NOT gate. Fig. 14.10(b) stands for its symbolic representation. A and B are the two inputs, Y the output.

The Boolean expression for the NAND gate is $Y = \overline{A \cdot B}$ and the adjacent truth table summarises the results of NAND operation.

Truth table

A	B	$Y = \overline{A \cdot B}$
0	0	1
0	1	1
1	0	1
1	1	0

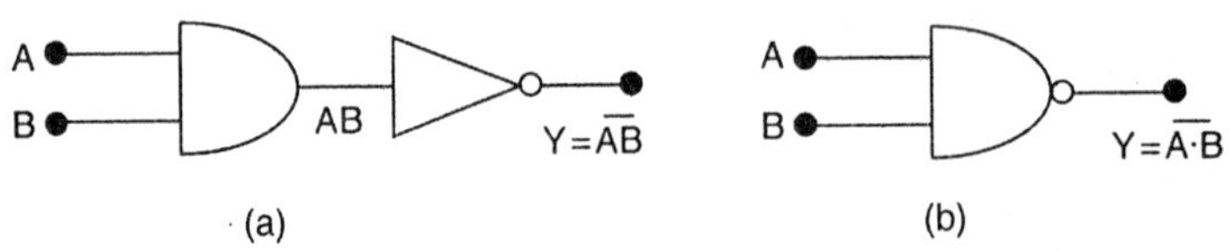

Fig. 14.10 (a) Split-up logic circuit of NAND gate, (b) Logic symbol of NAND gate

14.19.5 NOR gate

When an OR gate and a NOT gate are combined in such a way that the NOT gate follows the OR gate, the combination is referred to as a NOT-OR or ***NOR gate***. Here the output is 1 when all the inputs are 0, and in all other cases the output is 0. The split-up logic circuit is shown in Fig. 14.11(a) where a two-input OR gate is followed by a NOT gate and Fig. 14.11(b) stands for its symbolic representation. A and B are the two inputs, Y the output.

Truth table

A	B	$Y = \overline{A + B}$
0	0	1
0	1	0
1	0	0
1	1	0

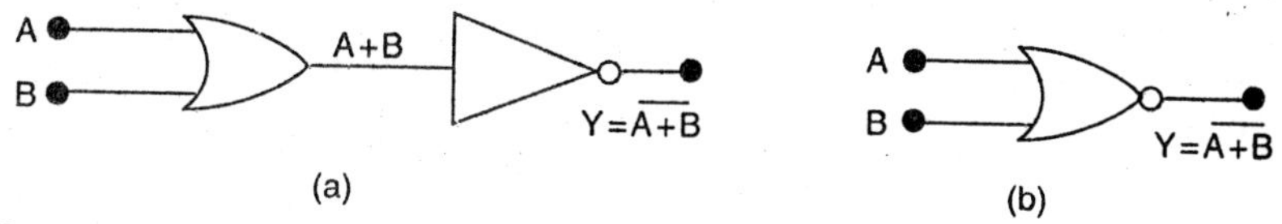

Fig. 14.11 (a) Split-up logic circuit of NOR gate, (b) Logic symbol of a two-input NOR gate

The Boolean expression for the NOR gate is $Y = \overline{A+B}$. The truth table as given summarises the results of NOR operation.

14.19.6 Exclusive-OR gate

Exclusive OR gate, abbreviated as XOR or EOR, is a two-input logic gate whose output takes the state 1 only when the two inputs are not equal. Thus XOR gate acts as an *inequality detector.* The logic symbol of the XOR gate is depicted in Fig. 14.12. A and B are the two inputs and Y the output. The Boolean expression for the output of this gate is $Y = A \oplus B$. The adjacent truth table summarizes the results of XOR operation.

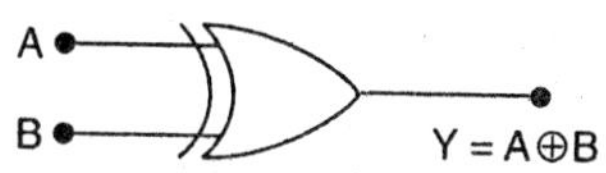

Fig. 14.12 Logic symbol of XOR gate

Truth table

A	B	$Y = A \oplus B$
0	0	0
0	1	1
1	0	1
1	1	0

The function $Y = A\bar{B} + \bar{A}B$ satisfies the truth table of XOR gate and hence the XOR operation can be realised by combining AND, OR and NOT gates. Fig. 14.13 shows the circuit implementation of this function using basic gates.

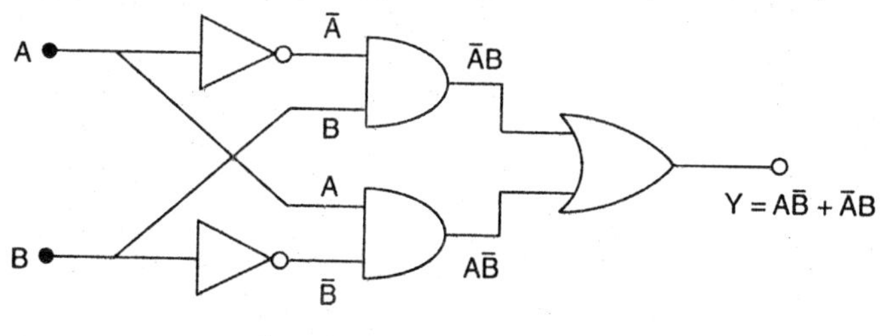

Fig. 14.13 Logic circuit diagram of XOR gate

14.19.7 Exclusive-NOR gate

The ***exclusive-NOR gate***, abbreviated as XNOR or ENOR, is a two-input logic gate whose output is just the complement to the XOR gate i.e., the output is in the state 1 when both the inputs are equal (either 0 or1). Thus the XNOR gate acts as an *equality detector* or a *coincidence checker.* Fig. 14.14 stands for the symbolic representation of XNOR gate where A and B are the two inputs and Y the output.

The Boolean expression for the output of the XNOR gate is $Y = \overline{A \oplus B} = A \odot B$. The adjacent truth table summarises the results of XNOR operation.

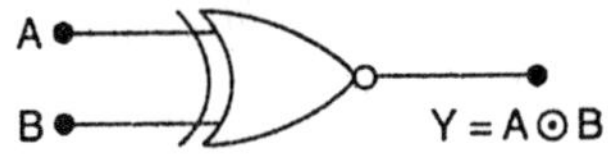

Fig. 14.14 Logic symbol of XNOR gate

Truth table

A	B	$Y = A \odot B$
0	0	1
0	1	0
1	0	0
1	1	1

The Boolean function $Y = AB + \bar{A}\bar{B}$ satisfies the truth table of XNOR and hence the circuit implementation of this function can be realised by basic gates as shown in Fig. 14.15.

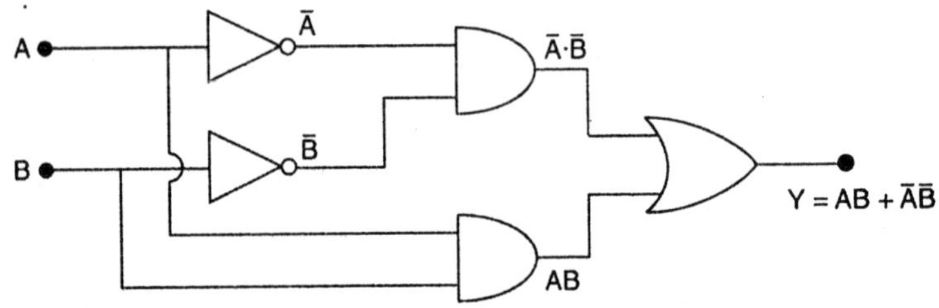

Fig. 14.15 XNOR gate realisation

14.20 Digital comparator

A *digital comparator* is a combinational circuit system that compares two binary numbers. If A and B are two binary numbers, then a digital comparator determines whether A is greater than, equal to or less than B.

To construct a logic circuit diagram of a digital comparator, let us consider two variables P and Q defined as $P = A\bar{B}$ and $Q = \bar{A}B$.

Clearly, $P = 1$ for $A = 1$ and $B = 0$ i.e., $A > B$

and $Q = 1$ for $A = 0$ and $B = 1$ i.e., $A < B$

Again, the output of XNOR is given by

$$R = \overline{\bar{A}B + A\bar{B}}$$

Now, $R = 1$ for $A = B$ (either 0 or 1)

$= 0$ for $A \neq B$

Thus, XNOR determines the condition of $A = B$ i.e., the equality condition of binary numbers A and B.

So, we are at the position to construct the logic block diagram of a digital comparator. Fig. 14.16 shows the required block diagram.

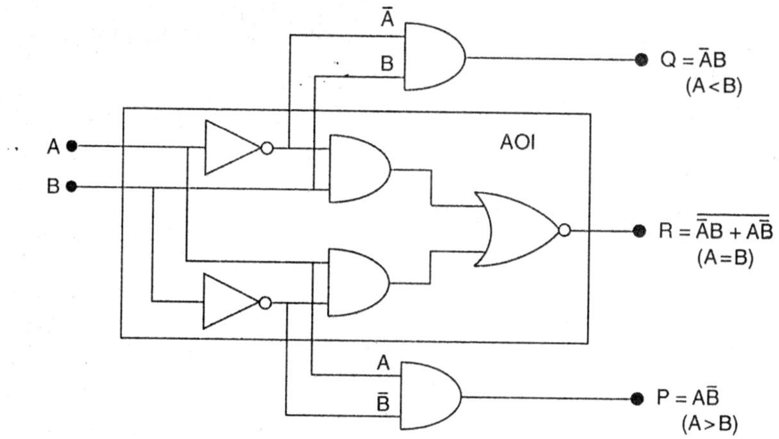

Fig. 14.16 Logic block diagram of digital comparator

The block diagram consists of two inverters, two AND gates and an AND-OR-INVERTER (AOI) block. The AOI block comprises of an XNOR and two AND gates. All the three outputs P, Q and R are also shown. Digital comparators are usually used in computer system.

14.21 Boolean function representation

A Boolean function corresponding to a given truth table may be obtained by two different methods.

1. Sum-of-product (SOP) method, and
2. Product-of-sum (POS) method.

SOP method—To find the Boolean expression in SOP method, the following steps are to be followed.

(i) Consider those rows of the truth table for which $Y = 1$.

(ii) Determine the corresponding *logical product* or *minterm* of the input variables obtained by keeping a variable unchanged if it is 1 and complementing the variable if it is 0.

(iii) The sum or the OR combination of the minterms gives the required Boolean expression.

To explain this method let us consider the truth table of OR gate as given below.

Input		Output
A	B	Y
0	0	0
0	1	1
1	0	1
1	1	1

The second, third and fourth rows correspond to the output $Y = 1$ and the corresponding minterms are respectively $\bar{A}B$, $A\bar{B}$ and AB. Therefore, the logic function i.e., the Boolean expression is SOP form is

$$Y = \bar{A}B + A\bar{B} + AB \tag{14.21.1}$$

This expression is not in simple form. Before implementing the logic circuit, it needs to be simplified.

$$\begin{aligned} \text{Now, } Y &= \bar{A}B + A(B + \bar{B}) \\ &= \bar{A}B + A \quad (\because\ B + \bar{B} = 1) \\ &= \bar{A}B + A(1 + B) \\ &= \bar{A}B + A + AB \\ &= A + (\bar{A} + A)B \\ &= A + B \end{aligned} \tag{14.21.2}$$

Thus the logic expression for OR gate given by (14.21.1) and (14.21.2) are equivalent.

POS method—Sometimes SOP forms are not convenient and we have to follow the alternative POS method. The procedure is as follows.

(i) Consider those rows of the truth table for which $Y = 0$.

(ii) Find the corresponding *logical sum* or *maxterm* obtained by keeping a variable unchanged if it is 0 and complementing the variable if it is 1.

(iii) The product or the AND combination of the maxterms gives the required Boolean expression.

To explain this method, we consider the truth table of XOR gate [Art. 14.19.6]. The first and the fourth rows correspond to $Y = 0$. The maxterms are $(A + B)$ and $(\bar{A} + \bar{B})$ respectively. Therefore, the Boolean expression is given by

$$Y = (A + B)(\bar{A} + \bar{B}) \tag{14.21.3}$$

On simplification, the expression becomes,

$$\begin{aligned} Y &= A\bar{A} + A\bar{B} + \bar{A}B + B\bar{B} \\ &= A\bar{B} + \bar{A}B \quad (\because\ A\bar{A} = B\bar{B} = 0) \\ &= A \oplus B \end{aligned} \tag{14.21.4}$$

So, (14.21.3) and (14.21.4) represent equivalent Boolean expressions for XOR.

14.22 Karnaugh map

Boolean expressions are generally simplified by using the laws and the theorems of Boolean algebra. This method requires some experience and intuitions. When the number of variables is not large (less than six), a simple graphical method, known as ***Veitch***

diagram or the ***Karnaugh map*** or ***K-map*** can be used to simplify a Boolean expression. This K-map is also used to find a logic expression from a given truth table. K-map is actually a two dimensional representation of a truth table. For n-variables, the K-map consists of 2^n cells or locations. To apply the Karnaugh map to simplify a given Boolean expression of two or more variables, we have to follow sequentially the procedures as mentioned below.

(i) The Boolean function or expression is first specified by a truth table.

(ii) The variables are then represented along the horizontal and vertical axes in such a way that the state of only one variable changes in going from one cell to an adjacent one. For three or more variable K-map, the variables are suitably grouped together for horizontal and vertical arrangement.

(iii) The groups of variables for which $Y = 1$ are then identified from the truth table prepared and the corresponding cells are entered with 1 s. The remaining cells are put 0 s.

(iv) The horizontal or the vertical adjacent cells having 1 entries are encircled and the variables on which Y depends are identified. Then the logic product or the minterms are formed with these variables.

(v) The OR combination of the minterms obtained in each encircled group gives the simplified expression in SOP form.

14.22.1 Two-variable Karnaugh map

To apply the K-map for simplification of Boolean expression, let us consider the Boolean function $Y = AB + \bar{A}B$. The truth table for this function is shown below. Fig. 14.17 shows the K-map for the two variables A and B. A and B are represented along the horizontal and the vertical axes respectively and are arranged in such a way that only one variable changes from one cell to an adjacent one.

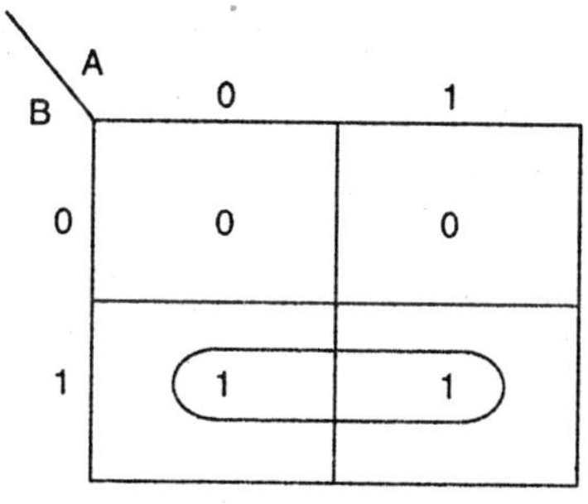

Fig. 14.17 K-map entries from truth table and encircling

Truth table

Inputs		Output
A	B	Y
0	0	0
0	1	1
1	0	0
1	1	1

Identifying from the truth table the rows for which $Y = 1$, we put 1s in the corresponding cells in the K-map (Fig. 14.17). The remaining K-map cells are then filled with 0 s.

To get the simplified Boolean expression, we encircle the adjacent cells known as *adjacencies*, as shown in Fig. 14.17. The encircled pair of 1*s* show that the variable B remains in the state 1 but A assumes both the states 0 and 1. Thus the output Y does not depend on A and it remains in the state 1 i.e., the state of B. Therefore, the function is simplified to $Y = B$.

Using Boolean identities we can verify the result as follows.

$$Y = AB + \bar{A}B = B(A + \bar{A}) = B \quad (\because\ A + \bar{A} = 1)$$

14.22.2 Three-variable Karnaugh map

To apply the Karnaugh map for simplification of Boolean expression let us consider the Boolean function

$$Y = \bar{A}\bar{B}C + \bar{A}B\bar{C} + A\bar{B}C + AB\bar{C}$$

At first we construct the truth table for the given expression as shown.

From the truth table, the combinations of variables giving $Y = 1$ are noted and 1s are entered in the corresponding cells of the K-map (Fig. 14.18). The remaining cells are filled with 0s.

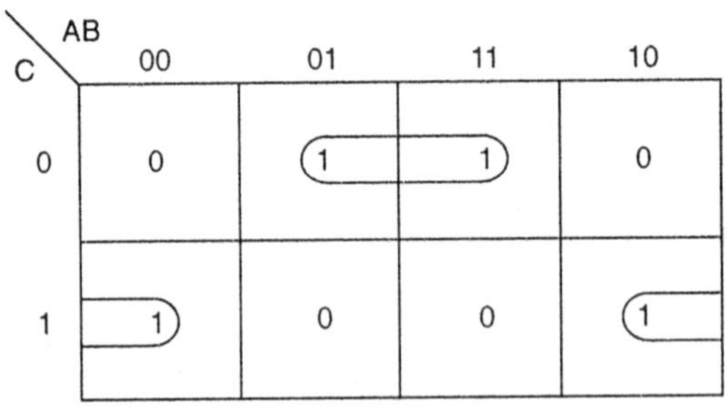

Fig. 14.18 K-map entries from truth table and encircling adjacencies

Truth table

Inputs			Output
A	B	C	Y
0	0	0	0
0	0	1	1
0	1	0	1
0	1	1	0
1	0	0	0
1	0	1	1
1	1	0	1
1	1	1	0

To get the terms of the simplified expression, we encircle the adjacencies with 1 entries. Here we obtained two blobs i.e., encircled groups. The cells at the edges of the K-map are also considered as adjacent. The blobs in the first row shows that $Y = 1$ for $B = 1$ and $C = 0$ whatever be the value of A. This gives the term $B\bar{C}$. The blobs of the edge cells show that $Y = 1$ for $B = 0$ and $C = 1$. This gives the term $\bar{B}C$. So the simplified expression is $Y = B\bar{C} + \bar{B}C$.

14.23 Combinational logic circuits

The combinational logic circuits are the digital circuits whose instantaneous outputs depend on the inputs appearing at that instant of time, the previous history of the inputs to the gate being immaterial. These logic circuits thus have no memory. In the following sections, we shall discuss some of the combinational logic circuits.

14.23.1 Binary adders

Addition and subtraction are the basic arithmetic operations that are performed by a digital computer. Multiplication and division are in fact repeated additions and subtractions respectively. So, a digital computer must have circuits which can perform addition and subtraction. Half adder and full adder are the logic circuits used for binary addition.

(A) Half adder— A half adder is a two-input and two-output logic circuit that can add two binary digits at a time, producing a sum and a carry as outputs. To complete a binary addition, two of these type of circuits are required. Hence this circuit is called ***half adder***.

The symbol of a half-adder is shown in Fig. 14.19. A and B represent the two bits to be added and S and C denote the sum and the carry outputs. The truth table of half adder is shown below.

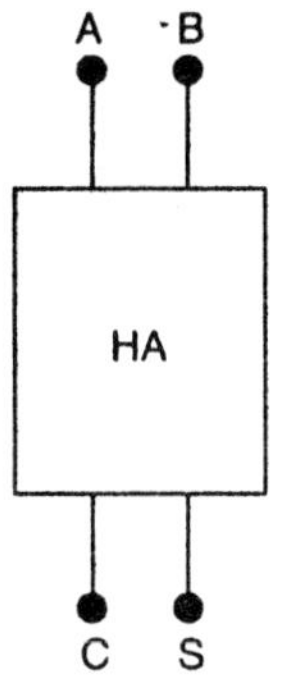

Fig. 14.19 Symbol of half adder

Truth table

Inputs		Outputs	
A	B	S	C
0	0	0	0
0	1	1	0
1	0	1	0
1	1	0	1

In the first three rows of the truth table, the sum is given by a single digit S. But in the last row the sum is $1 + 1 = 10$. So, $S = 0$ and $C = 1$ taken together represent the sum of A and B.

By inspection of the truth table it is seen that S can be realised by a XOR gate and C by an AND gate. So, the logical expression for S and C are

$$S = \bar{A}B + A\bar{B}$$
$$= A \oplus B$$
$$\text{and } C = AB$$

The logic circuit of Fig. 14.20 represents a half adder realised by logic gates.

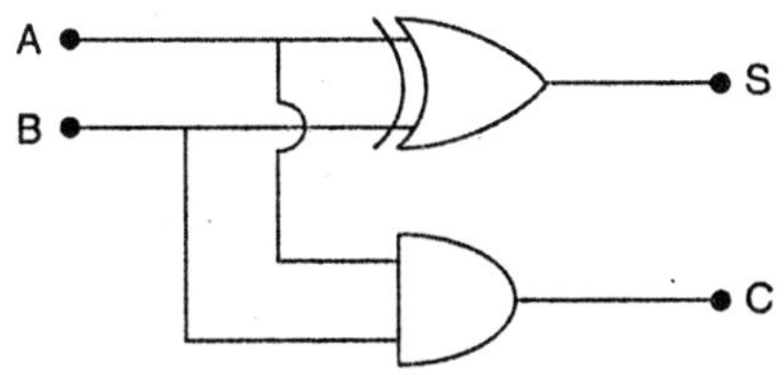

Fig. 14.20 Half adder realisation by logic gates

(B) Full adder—A half adder adds only two bits. It accomplishes the first step in addition of bits and does not handle a carry coming from the previous addition. But to add the binary numbers $A = A_n A_{n-1} \cdots A_2 A_1 A_0$ and $B = \cdots B_n B_{n-1} \cdots B_2 B_1 B_0$ at nth order bits, A_n and B_n, one must consider the carry bit C_{n-1} coming from previous addition. A full adder performs this operation. So, a ***full adder*** is defined as a logic circuit that can add three bits—two bits to be added and a carry bit coming from the previous addition, resulting in a sum and a carry.

The symbol of a full adder is shown in Fig. 14.21. A_n and B_n are the bits to be added and C_{n-1} is the carry-in from the previous addition. S_n and C_n denote the sum and carry-out. A_n, B_n, C_{n-1} represent the inputs and S_n, C_n the outputs. The truth table of full adder is shown in Table below.

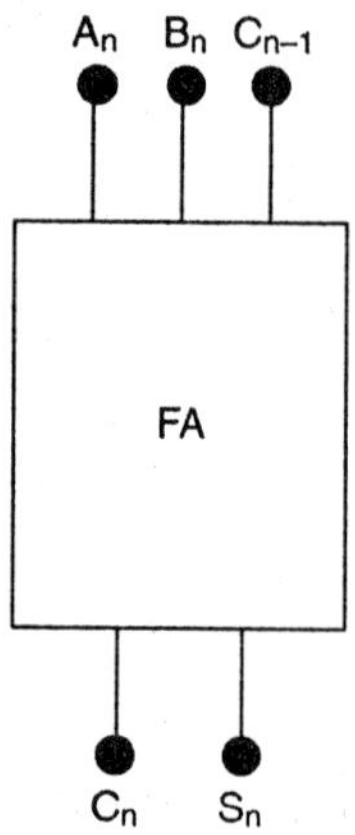

Fig. 14.21 Symbol of full adder

Truth table

Inputs			Output	
A_n	B_n	C_{n-1}	S_n	C_n
0	0	0	0	0
0	0	1	1	0
0	1	0	1	0
0	1	1	0	1
1	0	0	1	0
1	0	1	0	1
1	1	0	0	1
1	1	1	1	1

From the truth table the sum S_n and the carry C_n can be expressed in SOP form as

$$\begin{aligned} S_n &= \bar{A}_n \bar{B}_n C_{n-1} + \bar{A}_n B_n \bar{C}_{n-1} + A_n \bar{B}_n \bar{C}_{n-1} + A_n B_n C_{n-1} \\ &= \left(\bar{A}_n \bar{B}_n + A_n B_n\right) C_{n-1} + \left(\bar{A}_n B_n + A_n \bar{B}_n\right) \bar{C}_{n-1} \\ &= \bar{X} C_{n-1} + X \bar{C}_{n-1}, \quad \text{where } X = \bar{A}_n B_n + A_n \bar{B}_n = A_n \oplus B_n \\ &= X \oplus C_{n-1} = A_n \oplus B_n \oplus C_{n-1} \end{aligned} \tag{14.23.1}$$

$$\begin{aligned} \text{and } C_n &= \bar{A}_n B_n C_{n-1} + A_n \bar{B}_n C_{n-1} + A_n B_n \bar{C}_{n-1} + A_n B_n C_{n-1} \\ &= X C_{n-1} + A_n B_n = \left(A_n \oplus B_n\right) C_{n-1} + A_n B_n \end{aligned} \tag{14.23.2}$$

The relations (14.23.1) and (14.23.2) indicate that a full adder can be realized by using two half-adders and an OR gate as shown in Fig. 14.22.

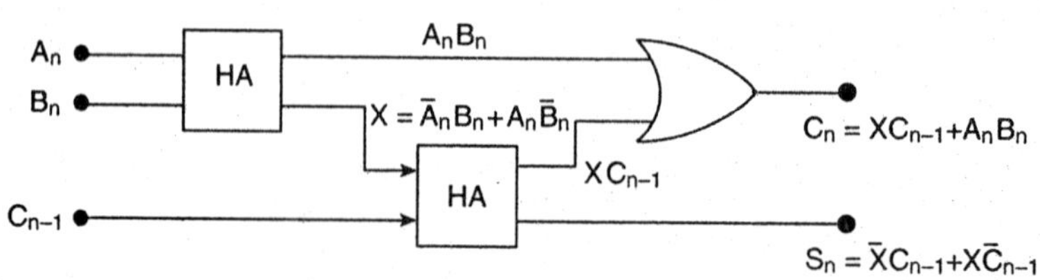

Fig. 14.22 Realisation of a full adder

14.23.2 Multiplexer

The ***multiplexer***, abbreviated as MUX, is a combinational logic circuit which selects one out of many input signals and transmits the selected one to the output with the help of control signals. Since a multiplexer selects one out of many inputs and directs it to the output, it is also named as *data selector.* Fig. 14.23 show the block diagram of a 4-to-1 multiplexer with two select inputs. Fig. 14.24 is the realisation of the multiplexer circuit using AND-OR logic gates.

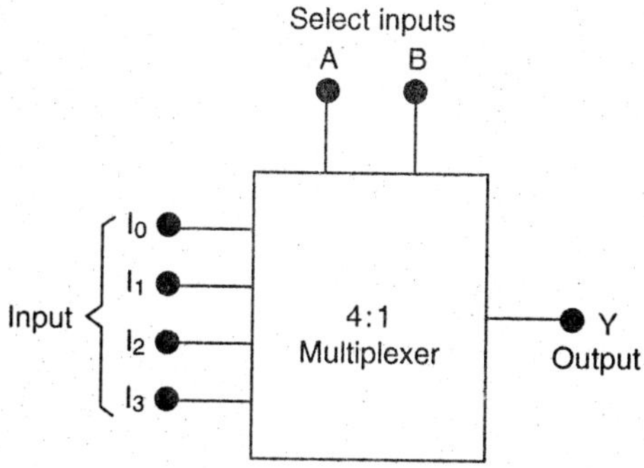

Fig. 14.23 Block diagram of a 4:1 multiplexer

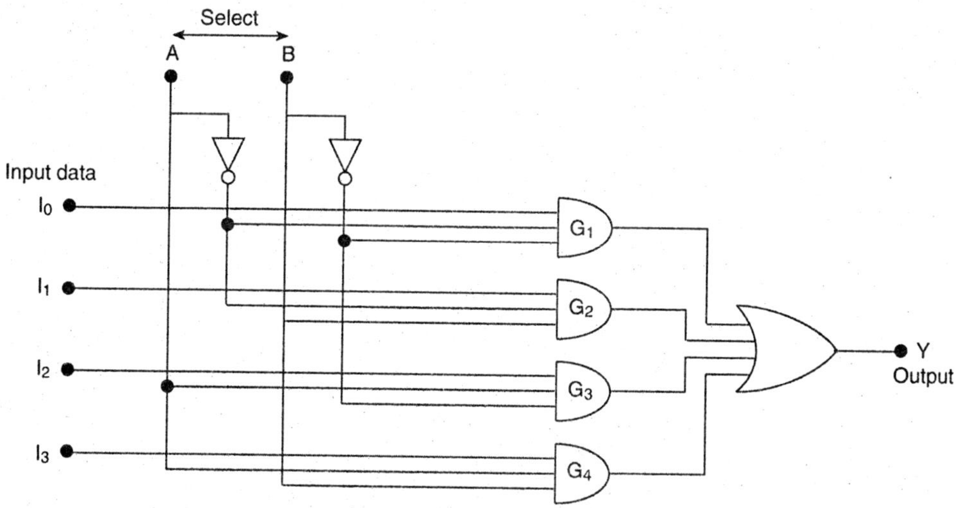

Fig. 14.24 Multiplexer (4:1) circuit using logic gates

The logical expression for the output may be written as

$$Y = I_0\bar{A}\bar{B} + I_1\bar{A}B + I_2A\bar{B} + I_3AB$$

To understand the performance of the mutiplexer, suppose $AB = 00$ i.e., $A = 0$ and $B = 0$. For this combination of A and B, the AND gate G_1 is enabled and all other gates are diabled resulting in $Y = I_0$ i.e., the data bit I_0 is transmitted to the output.

For other combinations of AB, we get only one data bit at the output as shown in the truth table below.

Truth table

Select Inputs		Output
A	B	Y
0	0	I_0
0	1	I_1
1	0	I_2
1	1	I_3

Multiplexers are usually available in IC form having 2, 4, 8 and 16 inputs. N select lines and 2^N AND gates are employed for 2^N data inputs. Multiplexers are widely used for the generation of combinational logic functions.

14.23.3 Encoder

In a digital system the circuits are operated in a binary manner. So, for the processing of the information like decimal numbers, alphabets or special characters, it is essential to convert them all into proper binary form before they are used in the digital circuits. A combinational logic circuit that executes this conversion of decimal numbers, alphabets or special characters into unique binary coded information is called an ***encoder***.

When one of the several input lines is in the state 1, a binary code is produced at the output lines of the encoder. An encoder generally has 2^n or less number of input terminals for n output terminals. The encoder in a computer key board consists of 84 inputs (26 capital and 26 small letters, 10 numerals and 22 special characters) and 7 bits at the output (since $2^6 < 84 < 2^7$).

To illustrate the design of an encoder, let us consider the encoder that transforms the decimal numbers 0 to 9 into 4 bit binary code. Fig. 14.25 shows the block diagram of the encoder. Let $I_1, I_2, \cdots, I_9$ denote the decimal inputs corresponding to the decimal numbers $1, 2, \cdots, 9$. To represent a particular decimal digit, only the corresponding input is at higher level (state 1) and all other terminals are at lower level (state 0). Exceptionally, the 0 state of all the inputs correspond to decimal 0.

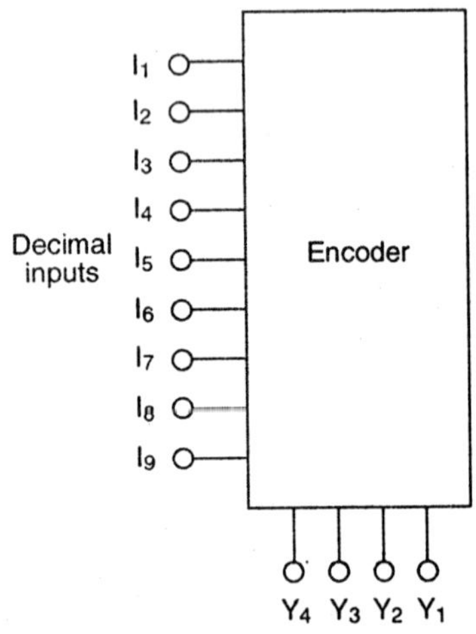

Fig. 14.25 Block diagram of a decimal-BCD encoder

$Y_4Y_3Y_2Y_1$ represents the BCD (binary coded decimal number). The truth table for the decimal digit 0 to 9 is given in the next page.

Truth table

Inputs									Output			
I_9	I_8	I_7	I_6	I_5	I_4	I_3	I_2	I_1	Y_4	Y_3	Y_2	Y_1
0	0	0	0	0	0	0	0	0	0	0	0	0
0	0	0	0	0	0	0	0	1	0	0	0	1
0	0	0	0	0	0	0	1	0	0	0	1	0
0	0	0	0	0	0	1	0	0	0	0	1	1
0	0	0	0	0	1	0	0	0	0	1	0	0
0	0	0	0	1	0	0	0	0	0	1	0	1
0	0	0	1	0	0	0	0	0	0	1	1	0
0	0	1	0	0	0	0	0	0	0	1	1	1
0	1	0	0	0	0	0	0	0	1	0	0	0
1	0	0	0	0	0	0	0	0	1	0	0	1

The truth table indicates that $Y_1 = 1$ when $I_1 = 1$ or $I_3 = 1$ or $I_5 = 1$ or $I_7 = 1$ or $I_9 = 1$. So, we can readily write

$$
\begin{aligned}
Y_1 &= I_1 + I_3 + I_5 + I_7 + I_9 \\
Y_2 &= I_2 + I_3 + I_6 + I_7 \\
Y_3 &= I_4 + I_5 + I_6 + I_7 \\
\text{and } Y_4 &= I_8 + I_9
\end{aligned}
$$

The implementation of this encoder using OR gates is shown in Fig. 14.26.

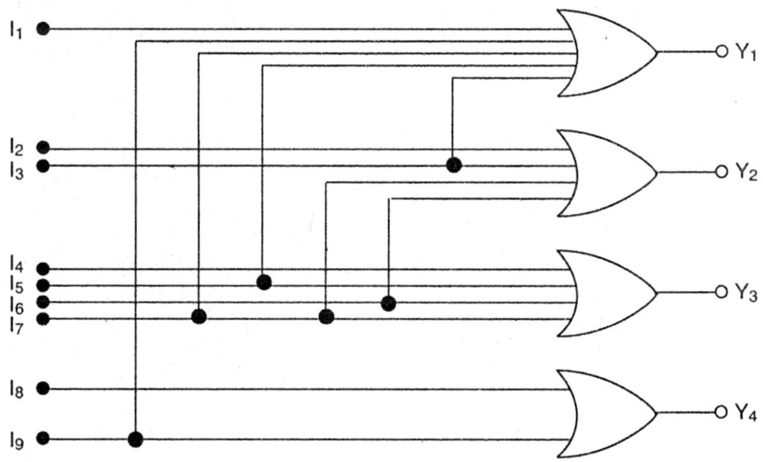

Fig. 14.26 Circuit realisation of decimal-BCD encoder

Clearly, when I_8 is 'on' i.e. at the state 1, we get $Y_1 = 0$, $Y_2 = 0$, $Y_3 = 0$ and $Y_4 = 1$. I_8 corresponds to the decimal number 8 and the output represents the binary code 1000.

Thus, we have: $(8)_{10} = (1000)_2$.

14.23.4 Decoder

The digital circuit that can perform the inverse operation of an encoder is known as a ***decoder***. So, the function of a decoder is to energize a particular output(s) depending on the binary coded input(s), i.e., a decoder identifies or decodes the particular binary code used at the input. In general form, a decoder with n input terminals to handle n-bit word can produce 2^n distinct form of coded information. Thus, *a decoder may be defined as a combinational logic circuit that converts binary coded information from n inputs to a maximum of* 2^n *unique outputs.*

The block diagram of a BCD to decimal decoder is shown in Fig. 14.27. A, B, C and D are the inputs through which binary coded information is given and the corresponding output is produced at one of the output terminals. For example, if $ABCD$ represents 1001, then one finds that the output Y_9 is high. Y_9 corresponds to the decimal number 9. Therefore $(1001)_2 = (9)_{10}$. Thus $ABCD$ i.e., 4-bit BCD may produce $2^4 = 16$ distinct coded informations representing decimal equivalent numbers. The block diagram indicates 4-to-16 decoder.

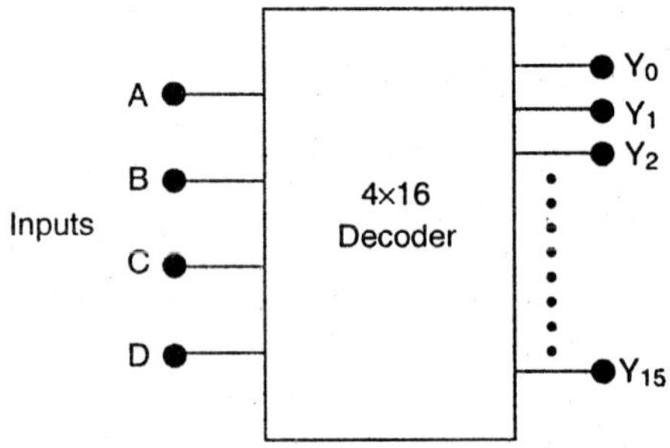

Fig. 14.27 Block diagram of a decoder

The realisation of 4 to 10 line decoder using AND gates is shown in Fig. 14.28. It implies that a 4-bit input code selects 1 out of 10 outputs.

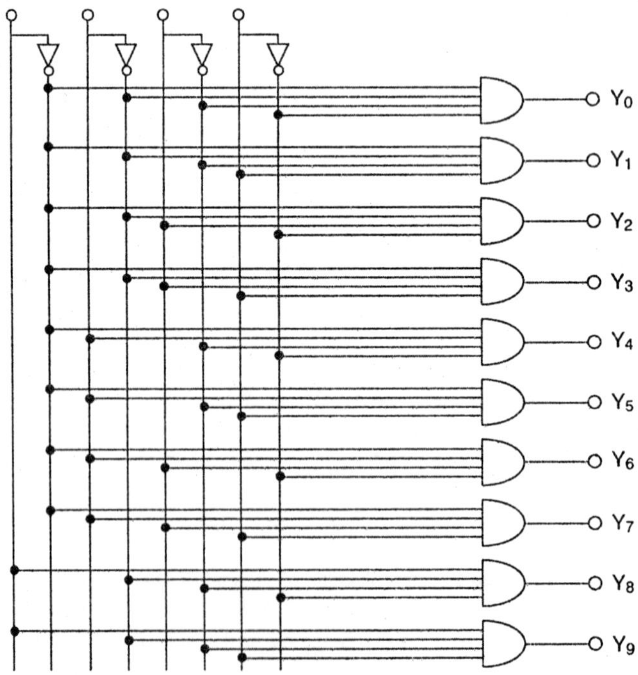

Fig. 14.28 BCD-to-decimal decoder

14.24 Sequential logic circuits

So far we have discussed different types of combinational logic circuits. But for the implementation of many digital systems we require some circuits whose instantaneous outputs depend on the present inputs as well as the past inputs and (or) outputs. Since the outputs of such circuits depend on the sequence of the inputs applied, they are called ***sequential logic circuits***. Thus sequential logic circuits possess a memory that can store past information. Fig. 14.29, (a) and (b), show the block diagrams of a combinational circuit and a sequential circuit respectively.

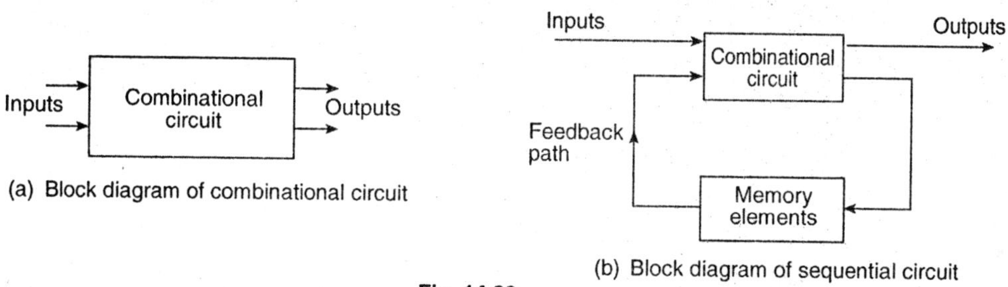

Fig. 14.29

The circuit of Fig. 14.29(a) gives a particular output with a *combined* set of inputs applied simultaneously to it. So, this type of circuit is called a ***combinational circuit.*** On the other hand, the type of circuit shown in Fig. 14.29(b) possesses a memory in addition to the present inputs and works in sequence by taking into account the previous inputs as well as the present inputs. So, this type of circuit is called a ***sequential circuit***. The differences between the combinational circuit and the sequential circuit are given in the following table.

Combinational circuit vs. sequential circuit

Combinational circuit	Sequential circuit
1. In this logic circuit, the output at any instant depends solely on the combined set of inputs applied simultaneously to it at that instant.	**1.** In this logic circuit, the output depends on the present inputs as well as the previous sequence of inputs.
2. No memory element is used in it.	**2.** At least one memory element is used.
3. Since no memory is used, it is easy to design.	**3.** Since memory is used, it is difficult to design.
4. It requires more hardware for its realisation.	**4.** It requires less hardware for its realisation.
5. Since the smultaneously applied inputs are primary inputs, it is faster in speed.	**5.** Since the secondary inputs are applied after a delay, it is comparatively slower.
6. It is rather costly.	**6.** It is cheaper in price.

14.24.1 Types of sequential circuits

A sequential circuit is generally specified by a time sequence of inputs, outputs and the internal states of the circuit. So, depending on the timing of the signals, the sequential circuits in digital system are classified into two categories, namely, (a) *clocked* or *synchronuous*, and (b) *unlocked* or *asynchronous.*

A. Clocked or synchronous sequential circuit

When a sequential circuit operates in synchronism with a pulse train, it is called a ***clocked*** or ***synchronous sequential circuit.*** The train of pulses known as *clock pulses* (abbreviated Ck or CLk) are generated by a master oscillator to provide the regular timing pulses. During the timing pulses the events are allowed to occur, otherwise the circuit remains inert and has only propagation delays. Typical clock pulses are shown in Fig. 14.30.

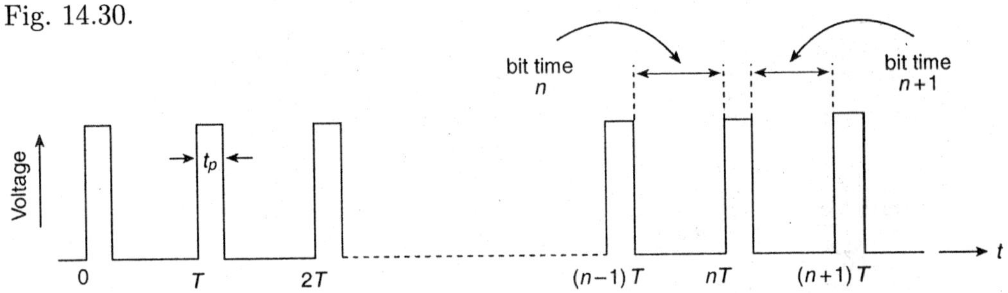

Fig. 14.30 Clock pulse train

The time for which the pulse remains in higher state is called the *pulse width* t_p. The time interval between the arrival of two successive clock pulses is called a *period* T. The pulse width t_p is very small compared to the time period T. Obviously, $1/T$ gives the *clock frequency* i.e., the number of clock pulses per second. During the interval between two pulses, the binary state 0 or 1 at each node in the digital system does not change. This time interval is referred to as *bit time.* The transition of one binary state to another occurs only at the arrival of the clock pulse. An example of the clocked sequential circuit is the magnetic tape reader where the reading occurs only when the part of the tape to be read comes under the read head.

The block diagram of a synchronous or clocked sequential circuit is shown in Fig. 14.31.

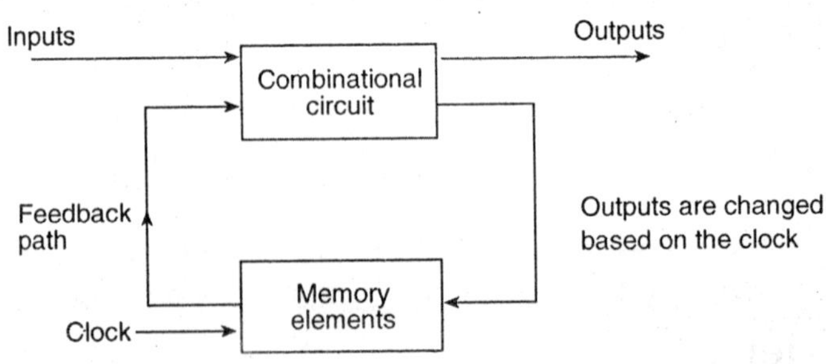

Fig. 14.31 Synchronous or clocked sequential circuit

B. Unclocked or asynchronous sequential circuit

The behaviour of this type of circuit depends upon the order in which the input changes and the state of the circuit can be affected at any instant of time i.e., an ***unclocked*** or ***asynchronous circuit*** does not wait for the timing pulses and an event takes place after the completion of a previous event. As an example of this type of circuit, we may consider the telephone dial system in which the user dials the next number as soon as the dialing of the previous number is completed. Thus an unclocked circuit possesses an interval logic circuit delay.

14.25 Flip-flops and latches

The sequential logic circuits require a memory that can store past information. A bistable multivibrator (discussed in Chapter 11), also known as *flip-flop*, can serve as a memory element in digital systems. So, flip-flops are the basic building blocks of sequential logic circuits. The output of a flip-flop can remain either in 0 (low voltage) state or in 1 (high voltage) state indefinitely unless an external trigger is applied to change the state. So, a flip-flop is also referred to as a *binary multivibrator.*

Moreover, a flip-flop retains or remembers a state indefinitely after the removal of the trigger. So, it possesses a 1 *bit memory.* The most basic type of circuits operated with signal levels in which informations are locked or latched are referred to as ***latches***. The latches are the basic circuits from which all flip-flops are constructed. In digital system, the term latch is generally used to mean a type of bistable device which is different from flip-flop. In latches or in flip-flops the outputs are connected back to the opposite inputs. The basic difference between latches and flip-flops is the method of changing their state. A latch stores a 0 or a 1 at its output and no clock is used in its design. When a latch is operated in synchronism with a clock pulse, it is called a ***flip-flop***. So, latches are used in asynchronous sequential logic circuits while the flip-flops are used in synchronous sequential logic circuits.

14.26 RS latch

The simplest type of latch is the ***Reset-Set latch***, abbreviated as *RS latch.* It has two inputs namely reset R and set S inputs and two outputs Q and $\bar{Q}$ which are complements to each other The RS latch can be constructed by using (i) two cross-coupled transistor switches as shown in Fig. 14.32 or (ii) two cross-coupled NOR gates (Fig. 14.33) or (iii) two cross-coupled NAND gates (Fig. 14.39).

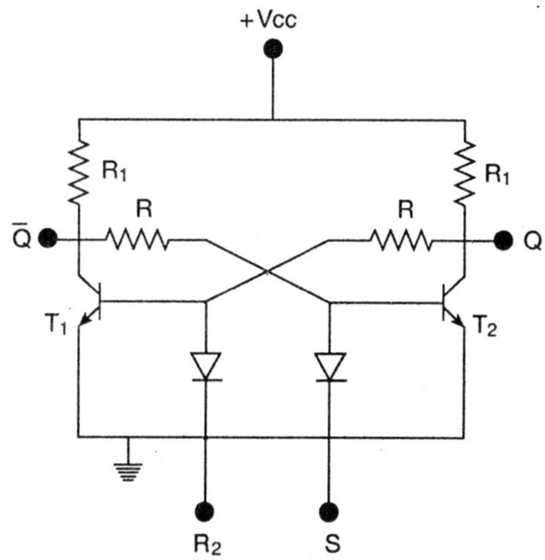

Fig. 14.32 Two cross-coupled transistors connected as RS latch

Operation of RS latch

To explain the operation of a RS latch, let us consider the RS latch constructed by two cross coupled NOR gates as shown in Fig. 14.33. Here the output of NOR gate I is connected to the input of NOR gate II and vice-versa.

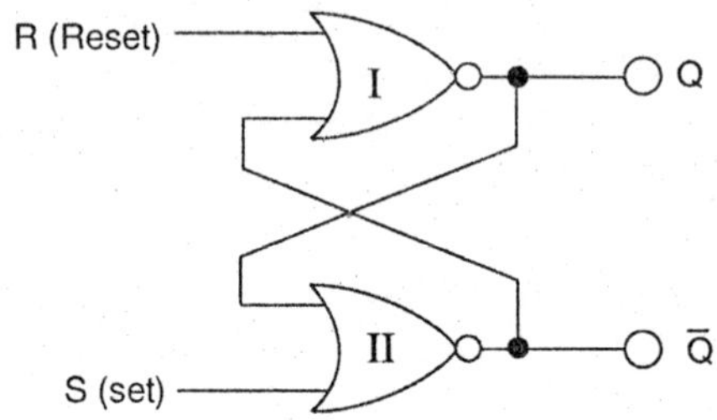

Fig. 14.33 RS latch using NOR gates

As the logic 1 at any input of a NOR gate forces its output to a logic 0, we may find the outputs Q and $\bar{Q}$ for the following four sets of inputs.

Case 1: *When* $R = 0$ *and* $S = 1$.

In this case, input S of NOR gate II is 1 which makes the output $\bar{Q} = 0$. So, both inputs of NOR gate I are 0 so that its output $Q = 1$. In this case, the latch is said to be *set* as 1 is stored in the output Q. Fig. 14.34 shows this step of operation.

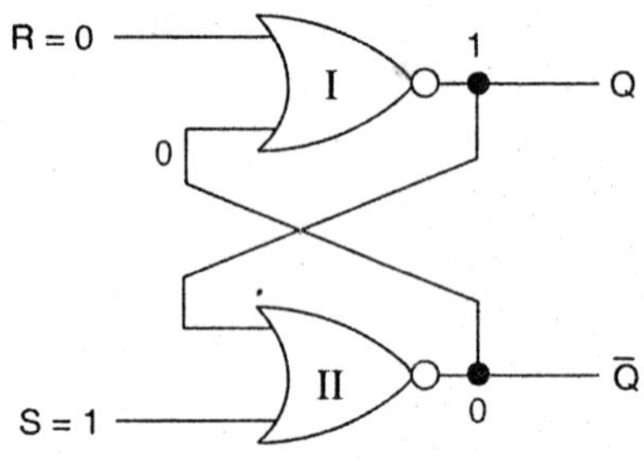

Fig. 14.34 Set 1 of inputs

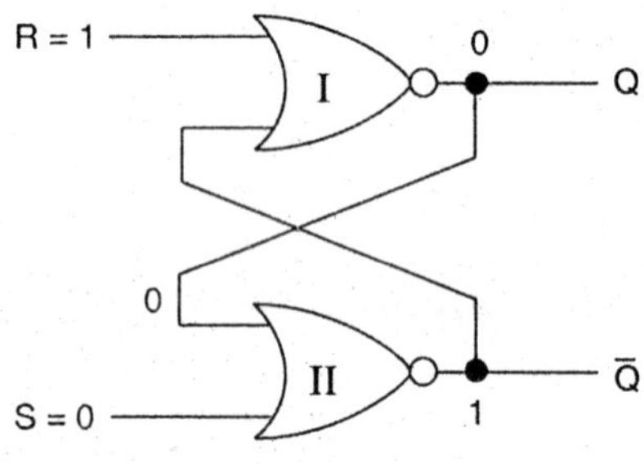

Fig. 14.35 Set 2 of inputs

Case 2: *When* $R = 1$ *and* $S = 0$

In this case, the input $R = 1$ makes the output of NOR gate I i.e., $Q = 0$. This indicates that both inputs of NOR gate II are now 0 so that its output $\bar{Q} = 1$ as shown in Fig. 14.35. In this case, the latch is said to be *cleared* or *reset* as 0 is stored at the output Q.

Case 3: *When* $R = 0$ *and* $S = 0$

For this case, we assume $Q = 1$ and $\bar{Q} = 0$ are the initial conditions of the outputs. With $\bar{Q} = 0$ both inputs of NOR gate I are 0 so that its output $Q = 1$. Again, with $Q = 1$, the output of NOR gate II is $\bar{Q} = 0$. Thus the outputs remain the same as initial

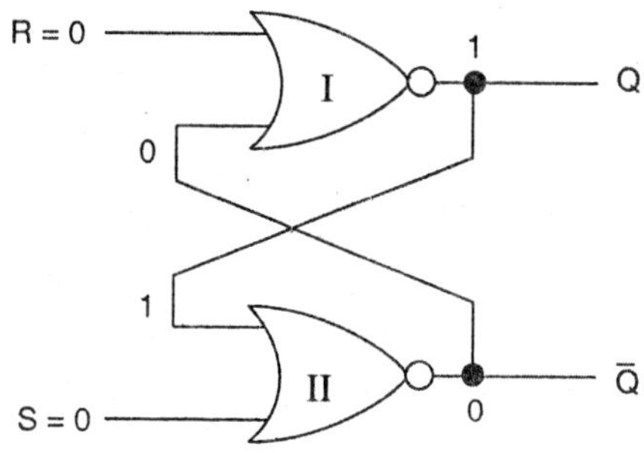

Fig. 14.36 Set 3 of inputs

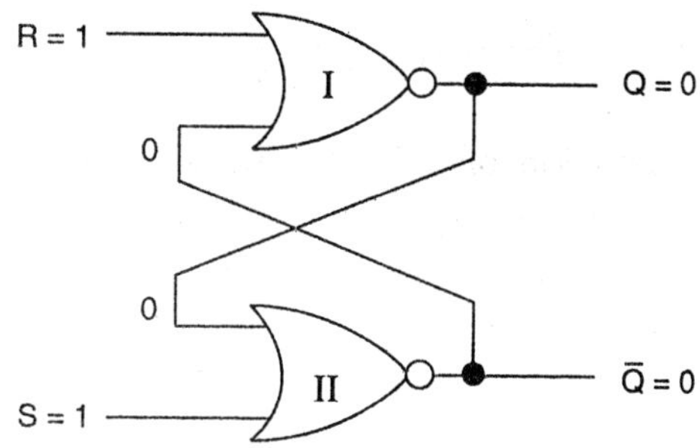

Fig. 14.37 Set 4 of inputs

values. So, with $R = 0$ and $S = 0$, the output remain unchanged. The same result is obtained with initially $Q = 0$ and $\bar{Q} = 1$. Fig. 14.36 shows this step of operation.

Case 4 : *When* $R = 1$ *and* $S = 1$

In this case, both inputs are 1. So, they force both outputs to be 0 i.e., $Q = 0$ and $\bar{Q} = 0$ simultaneously. This is logically inconsistent and hence it is prohibited. For normal operation, this condition must be avoided. Fig. 14.37 shows this operation.

The symbol of a *RS* latch is shown in Fig. 14.38 and the adjacent truth table summarises the results.

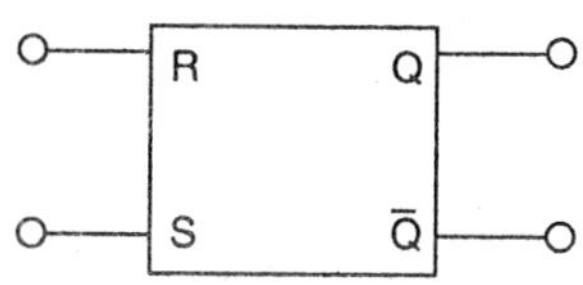

Fig. 14.38 Symbol of RS latch

Truth Table

Input		Output		Comments
R	S	Q	$\bar{Q}$	
0	0	NC	NC	No change
0	1	1	0	Set
1	0	0	1	Reset
1	1	×	×	Indeterminate

RS latch can also be realised by using NAND gates as shown in Fig. 14.39. Its truth table is the same as that of NOR latch.

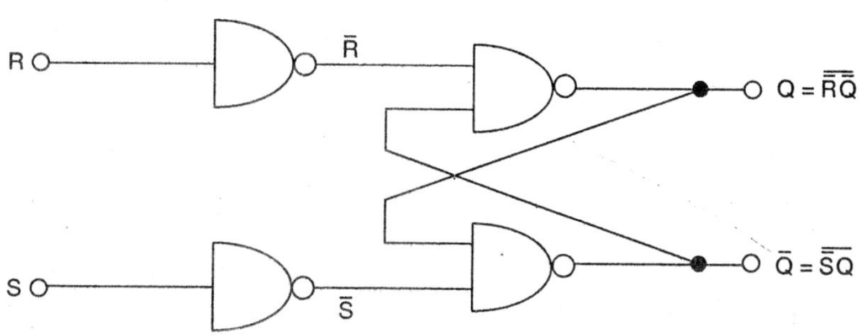

Fig. 14.39 RS latch using NAND gates

14.27 Types of flip-flops

The operation of a latch can be modified by providing an additional control input that determines the point of time when the state of the circuit is to be changed. This additional control input is called the *clock pulse* or simply *clock*. A latch with a clock is known as *flip-flop* which can work in synchronous mode. Depending on the number of inputs and the manner in which the inputs affect the outputs, the flip-flops are classified in many categories. Some of the important flip-flops are mentioned below.

1. Clocked RS flip-flop,
2. JK flip-flop,
3. Master-slave JK flip-flop,
4. D-type flip-flop,
5. T-type flip-flop,
6. Edge-triggered flip-flop

We shall now discuss about them in somewhat details.

14.27.1 Clocked RS flip-flop

When a clock input is added to an ordinary RS flip-flop, a *clocked RS flop-flop* with synchronous mode of operation is achieved. The logic diagram of Fig. 14.40(a) shows a clocked RS flip-flop using NAND gates. Fig. 14.40(b) represents its symbol.

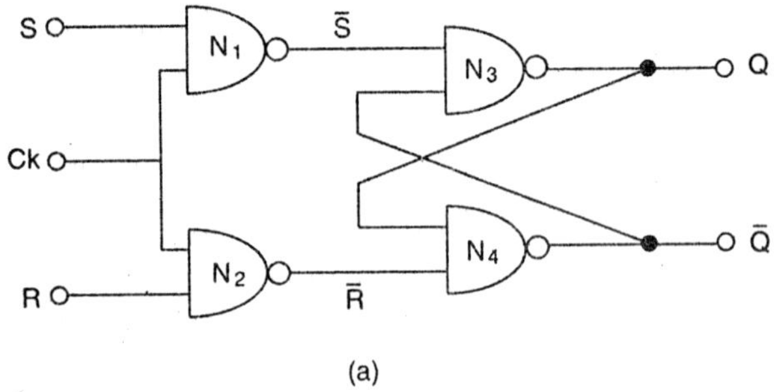

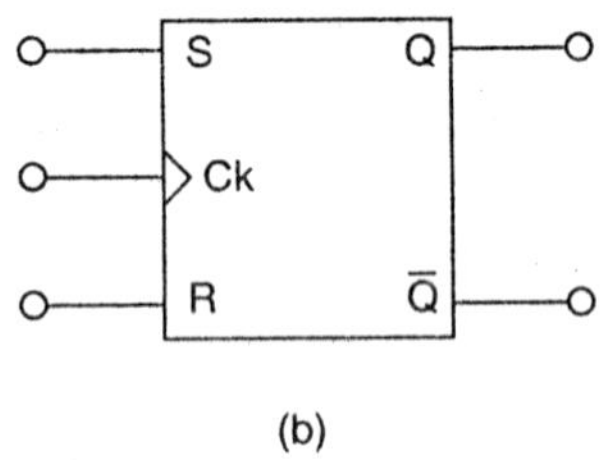

Fig. 14.40 (a) Clocked RS flip-flop, (b) its symbol

The circuit consists of two parts. The NAND gates N_3 and N_4 form a level sensitive flip-flop called latch and the gates N_1 and N_2, called the *steering gates*, to monitor the state of the flip-flop with the application of the clock pulse. During a bit time, the flip-flop does not change the state.

Operation

When clock pulse is absent i.e., $Ck = 0$, the output of N_1 and N_2 are 1 irrespective of the values of R and S i.e., the circuit retains the same state when clock is absent. On the other hand, when the clock pulse is present i.e., $Ck = 1$, the circuit responds to R and S in the same way as in the case of RS latch. The truth table of a clocked RS flip-flop is given below.

Truth Table of ***RS*** flip-flop

Inputs			Outputs		Comments
Ck	R	S	Q_{n+1}	$\bar{Q}_{n+1}$	
0	0	0	Q_n	$\bar{Q}_n$	NC (No change)
0	0	1	Q_n	$\bar{Q}_n$	NC
0	1	0	Q_n	$\bar{Q}_n$	NC
0	1	1	Q_n	$\bar{Q}_n$	NC
1	0	0	Q_n	$\bar{Q}_n$	NC
1	0	1	1	0	Set
1	1	0	0	1	Reset
1	1	1	1	1	Invalid (Race)

For stable operation of the circuit, it is necessary that R and S should never be high simultaneously when the clock is present.

A clocked RS flip-flop circuit can be realised by using a basic NOR latch and two AND gates at the input as shown in Fig. 14.41.

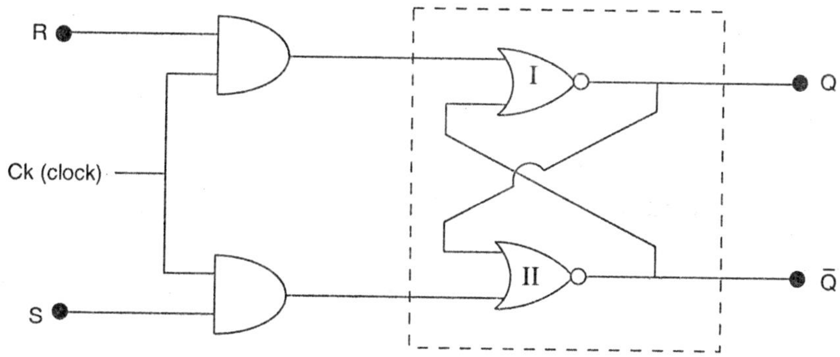

Fig. 14.41 Clocked RS flip-flop using NOR and AND gates

14.27.2 JK flip-flop

The modified or the improved version of RS flip-flop where the ambiguous result of the output for the two high input of S and R has been eliminated, is known as ***JK flip-flop***. Fig. 14.42 shows the JK flip-flop configuration.

Here, two AND gates A_1 and A_2 are used at the input terminals of the RS flip-flop. The data input J, used for *set*, is applied together with $\bar{Q}$ to the input of A_1 so that $S = J\bar{Q}$. The data input K, used for *reset*, is fed together with Q to the input of A_2 so that $R = KQ$. With these modifications, JK flip-flop becomes the most versatile and

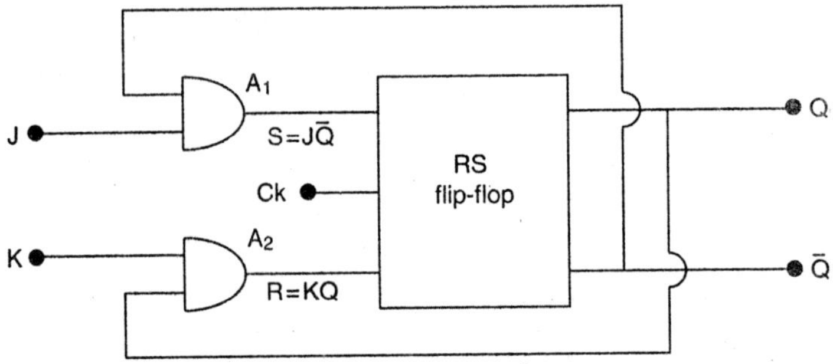

Fig. 14.42 A JK flip-flop

widely used flip-flop and an essential building block of sequential circuits like counters and registers. The detailed truth table of the system is given below.

Detailed truth table of JK flip-flop

J_n	K_n	Q_n	$\bar{Q}_n$	S_n	R_n	Q_{n+1}	Last column
0	0	0	1	0	0	Q_n	Q_n
0	0	1	0	0	0	Q_n	
1	0	0	1	1	0	1	1
1	0	1	0	0	0	Q_n	
0	1	0	1	0	0	Q_n	0
0	1	1	0	0	1	0	
1	1	0	1	1	0	1	$\bar{Q}_n$
1	1	1	0	0	1	0	

The reduced truth table concerning J_n, K_n and Q_{n+1} is obtained from the above table and is given below. The circuit symbol of a JK flip-flop is shown in Fig. 14.43.

Reduced Truth table of JK flip-flop

Inputs		Outputs	Comments
J_n	K_n	Q_{n+1}	
0	0	Q_n	No change
0	1	0	Reset
1	0	1	Set
1	1	$\bar{Q}_n$	Toggle

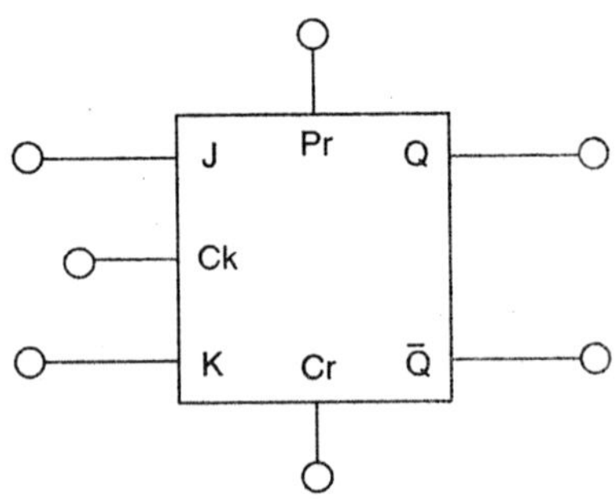

Fig. 14.43 Logic symbol of a JK flip-flop

The truth table of the JK flip-flop gives the output conditions as a function of the inputs J and K when the clock pulse is present. However, the output is uncertain when the clock pulse is absent.

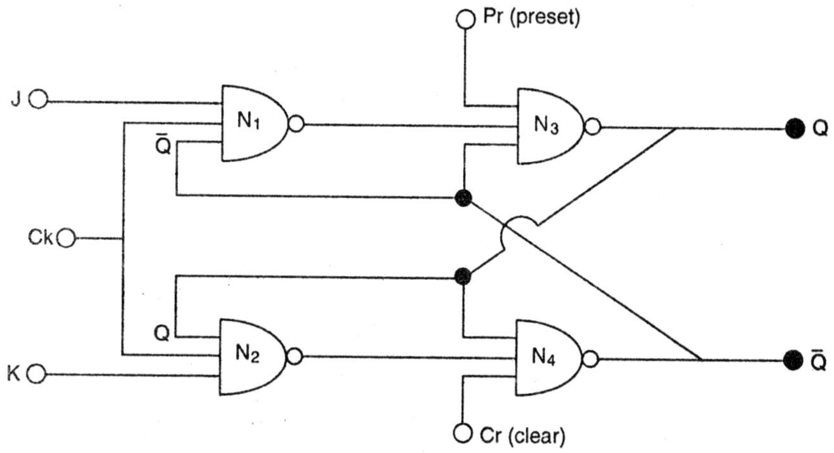

Fig. 14.44 Realisation of JK flip-flop by using NAND gates only

The JK flip-flop can also be realised by using NAND gates only, as shown in Fig. 14.44. Its truth-table is the same as that of the previous one.

Preset and clear

The output state of a flip-flop is uncertain before the clock pulse is applied i.e., the initial state of the output Q may be 1 (set) or 0 (reset) state in the absence of clock pulse. However, in many applications it is necessary to assign the initial state of the flip-flop. This is accomplished by the direct or asynchronous inputs, known as *preset* (Pr) and *clear* (Cr) inputs. Fig. 14.44 shows a JK flip-flop with preset and clear inputs. When $Cr = 0$ and $Pr = 1$, we get $Q = 0$ i.e., flip-flop is *reset*. On the other hand, when $Cr = 1$ and $Pr = 0$, the output $Q = 1$ and the flip-flop is *set*. When $Cr = 1$ and $Pr = 1$, the circuit operates in normal fashion. The condition $Pr = Cr = 0$ must be avoided since it leads to uncertain output Q.

14.27.3 Master-slave JK flip-flop

Race around condition

When a pulse is applied at the input of a logic gate, the output does not change immediately because of the propagation delay time. The *propagation delay time* Δt is the approximate time interval between the application of an input pulse and its arrival at the output terminal. This delay arises due to the presence of many active devices like diodes, transistors, MOSFETs etc. between the input and the output of a gate. However, in IC form of the flip-flop, Δt is negligibly small ($\sim 10^{-9}$ s). But, in the case of JK flip-flop, it creates a real problem when $J = K = Ck = 1$ i.e., JK flip-flop is in *toggle mode.* Since the propagation delay time is much smaller than the width of the clock pulse t_p i.e., $\Delta t \ll t_p$ (as shown in Fig. 14.45) and there is feedback connection from output to input in JK flip-flop, the input changes and the output toggles more than once during a pulse time t_p. Thus at the end of the pulse the output becomes ambiguous i.e., uncertain. This situation is known as ***racing*** or ***race around condition***.

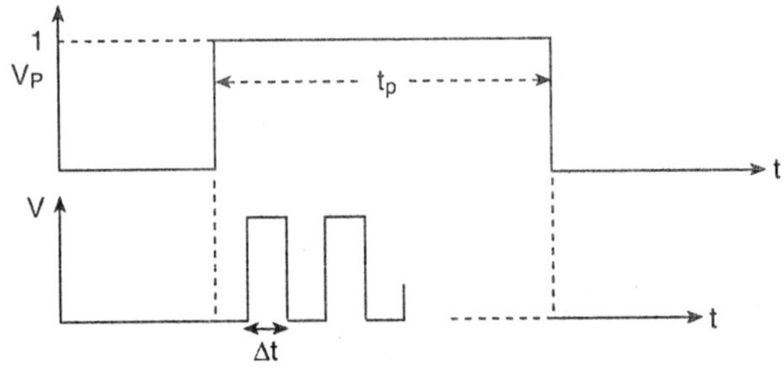

Fig. 14.45 $\Delta t << t_p$

The race around problem can be avoided if $t_p < \Delta t < T$. However, there are electronic methods to overcome this difficulty. One practical solution is offered by the ***master-slave JK flip-flop*** which is described below.

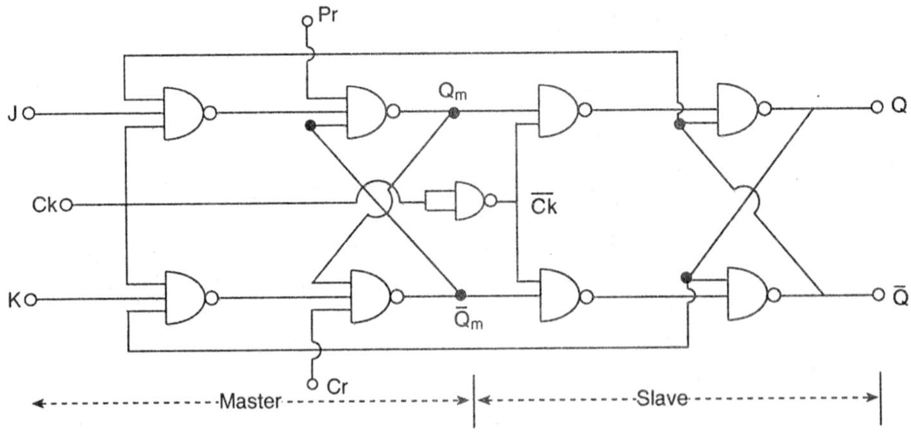

Fig. 14.46 A master-slave JK flip-flop

The configuration of a master-slave flip-flop is shown in Fig. 14.46. It is a cascade of two RS-flip-flops. The first one is called the *master* and the second one the *slave*. The output of the slave is fedback to the input of the master. The positive clock pulses are applied directly to the master and those pulses after inversion are applied to the slave. With $Pr = 1$, $Cr = 1$ and $Ck = 1$, the master is *enabled* and it follows the truth table of JK flip-flop.

Also, since $\overline{Ck} = 0$, the slave remains *disabled* and hence cannot change state. The information of the master's output Q_m is stored in the input of the slave during the time t_p. As a result, the race around condition does not take place. When $Ck = 0$, we have $\overline{Ck} = 1$. At this condition, the master is *disabled* and the slave is *enabled* to transmit the information Q_m to its output Q. The slave now operates in accordance with the truth table of clocked RS flip-flop. Thus the race around condition does not arise in master-slave JK flip-flop.

14.27.4 *D*-type flip-flop

In a RS flip-flop, the data must be supplied on two input lines R and S. Moreover, the two data inputs must not be high simultaneously. In some practical applications such as serial data transfer, flip-flop with a single input is required. Also, there are some synchronous digital systems in which a signal is to be delayed by exactly one cycle of the clock. The flip-flop which removes the two disadvantages of RS flip-flop and fulfils the needs of the above mentioned practical purposes, is the ***D-type flip-flop***. Fig. 14.47(a) shows a D-type flip-flop with logic symbol shown in Fig. 14.47(b).

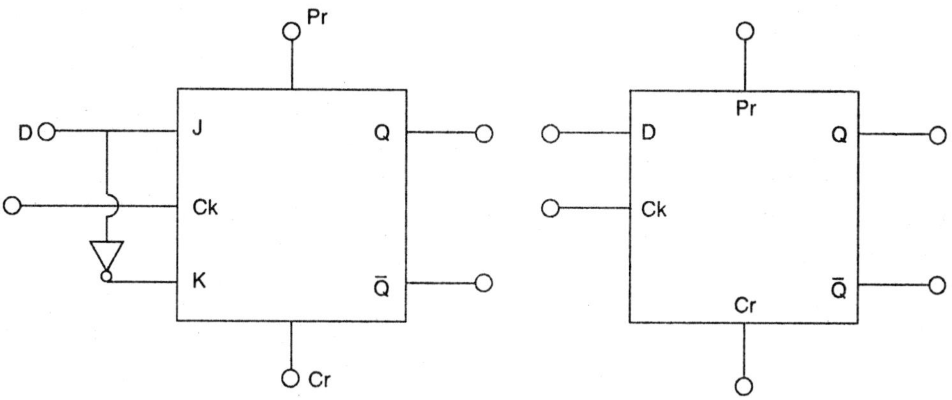

Fig. 14.47 (a) A D-type flip-flop, (b) logic symbol

Fig. 14.47(a) indicates that D-type flip-flop possesses only one data input D with a NOT gate connected between the inputs J and K, making J and K complementary to each other. With this arrangement, J and K never assume 1 (or 0) at the same time.

To understand the action of the D-type flip-flop, let initially $Q = 0$.

Now, if $D = 0$, it makes $J = 1$ and $K = 0$. As the clock pulse goes high i.e., $Ck = 1$, we have the output $Q = 1$ according to the truth table of JK flip-flop. Thus, the data input $D = 1$ is transferred to the output $Q(= 1)$. Similarly, if $D = 0$, then $J = 0$ and $K = 1$ resulting in $Q = 0$. In this case also, the data input $D = 0$ is transferred to the output Q. So, we get the truth table of D-type flip-flop as given below.

The truth table of D-type flip-flop indicates that the output Q_{n+1} at the end of a clock pulse is equal to the data input D_n before the application of the clock pulse. So, the data input is transferred to the output on application of a clock pulse. When the clock pulse goes low, Q retains or stores the last value of D. D-type flip-flop is often called a D-*latch* (D for data).

Truth table for *D*-type flip-flop

Input	Output
D_n	Q_{n+1}
0	0
1	1

Also, the data is not transferred to the output, unless a clock pulse arrives. The action of this flip-flop is thus delayed by one clock pulse. So, it is also called a *delay* (D for delay) *type* flip-flop.

14.27.5 T-type flip-flop

In many applications of sequential circuits, it is necessary to have a toggle function other than *set* and *reset* functions.

When J and K inputs of a JK flip-flop are connected together, we get a T (toggle)-***type flip-flop***. The action of a T-type flip-flop is a toggle switch i.e., it can toggle its state with each clock pulse input only when $T = 1$. Fig. 14.48 shows a T-type flip-flop and its symbol. The adjacent table is the truth table.

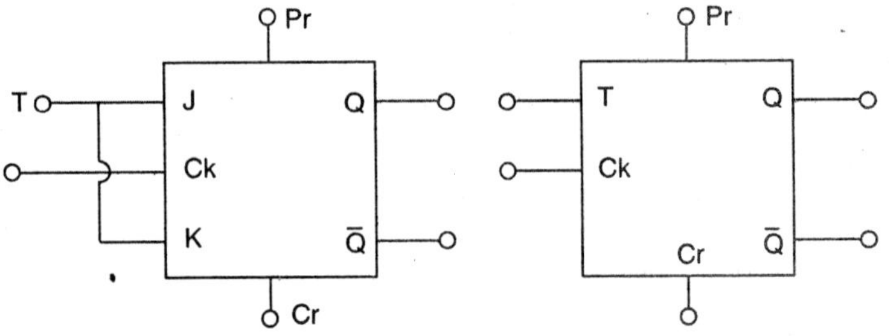

Truth table for T-type flip-flop

Input	Output
T_n	Q_{n+1}
0	Q_n
1	$\bar{Q}_n$

Fig. 14.48 (a) T-type flip-flop, (b) logic symbol

- T-type flip-flop is also called a *binary counter, divide-by-two, scale of two* or *trigger* flip-flop.

14.27.6 Edge triggered flip-flop

The ***edge-triggered flip-flops*** are defined as the flip-flops which respond either at positive edge (i.e., rising edge) or at the negative edge (i.e., falling edge) of the clock pulse. When clock pulses are applied to *RC differentiator circuit* with time constant $\tau \ll t_p$ (the pulse width), the edge trigger spikes are generated. Fig. 14.49 shows the generation of spikes from a clock pulse.

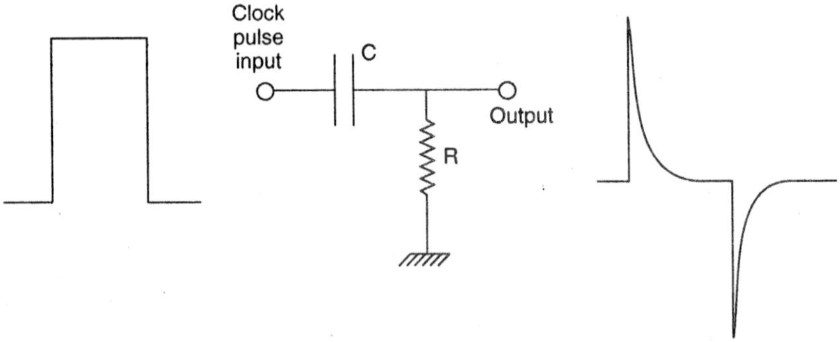

Fig. 14.49 RC differentiator circuit generating spikes from clock pulse

When the clock signal changes states from 0 to 1 or from 1 to 0, the output of a flip-flop changes during the short interval of time with the edge trigger spikes. Since the edge triggering reduces the width of clock pulse, it helps to avoid race around condition.

• A small triangle is used at the clock input inside the logic symbol of a flip-flop to indicate positive edge triggering. An additional bubble is used with the triangle to indicate negative triggering of a flip-flop (Fig. 14.50).

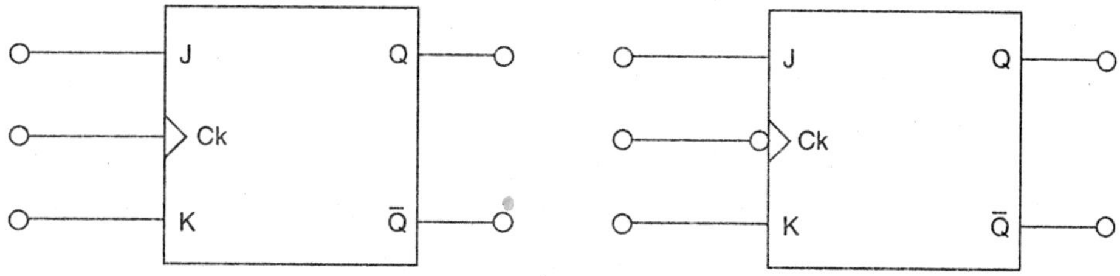

Fig. 14.50 Logic symbol of JK flip-flop (a) positive edge-triggered (b) negative edge-triggered

14.28 Illustrated Examples

▶ **Example 1.** *Determine the binary equivalent of* 576. **(Calcutta Hons.)**

Solution: Successive divison by 2 Remainders

Divisor	Number		Remainder
2	576		
2	288	—	0 ↑
2	144	—	0
2	72	—	0
2	36	—	0
2	18	—	0
2	9	—	0
2	4	—	1
2	2	—	0
2	1	—	0
	0	—	1

Taking the remainders in the reverse order, we have: $(576)_{10} = (1001000000)_2$.

▶ **Example 2.** *Convert from binary system to the decimal system* 11.1101.

Solution: The decimal equivalent of the integer part 11 is $1 \times 2^1 + 1 \times 2^0 = 2 + 1 = 3$ and the decimal equivalent of the fraction .1101 is $1 \times 2^{-1} + 1 \times 2^{-2} + 0 \times 2^{-3} + 1 \times 2^{-4} = 0.5 + 0.25 + 0 + 0.0625 = 0.8125$.

$$\therefore \quad 11.1101_2 = 3.8125_{10}$$

▶ **Example 3.** *Convert the decimal number* 2478 *to a hexadecimal number. Convert the hexadecimal number* $C5E2$ *to binary number.* **(Calcutta Hons.)**

Solution: To convert a decimal number to its equivalent hexa-decimal number we have to divide successively the given decimal number by 16 till the remainder becomes less than 16. Then the remainders taken in the reverse order give the hexadecimal equivalent.

Now,

Divisor	Number		Remainder	Hex notation
16	2478			
16	154	—	14	E ↑
16	9	—	10	A
	0	—	9	9

$$\therefore \ 2478_{10} = 9AE_{16}$$

To convert a hexadecimal number to a binary number, we replace each hex digit by its 4-bit binary equivalent. So, we get

Hex:	C	5	E	2
Binary:	1100	0101	1110	0010

$$\therefore \ C5E2_{16} = 1100010111100010_2$$

▶ **Example 4.** *Convert the octal number* 472 *to its equivalent binary number.*

Solution: To convert an octal number to its equivalent binary, we replace each octal digit by its 3-bit binary equivalent. So, by the process, we have

Octal:	4	7	2
Binary:	100	111	010

$$\therefore \ 472_8 = 100111010_2$$

▶ **Example 5.** *Convert the decimal number* $(12.625)_{10}$ *into its binary equivalent.*

Solution: Integral part

Divisor	Number	Remainder
2	12	
2	6	0
2	3	0
2	1	1
	0	1

$$\therefore \ (12)_{10} = (1100)_2$$

Fractional part

		carry
0.625×2	$= 1.25$	1
0.25×2	$= 0.50$	0
0.50×2	$= 1.00$	1
0.00×2	$= 0.00$	0

$$\therefore \ (0.625)_{10} = (0.1010)_2$$

So, we have, $$\mathbf{(12.625)_{10} = (1100.1010)_2}$$

▶ **Example 6.** *Covert the binary number* $(1001.111)_2$ *into its decimal equivalent.*

Solution : Integral part :

$$\begin{aligned} & 1\times 2^3+0\times 2^2+0\times 2^1+1\times 2^0 \\ &= 1\times 8+0\times 4+0\times 2+1\times 1 \\ &= 8+0+0+1 \\ &= 9 \end{aligned}$$

$$\therefore \quad (1001)_2=(9)_{10}$$

Fractional part :

$$\begin{aligned} 1\times 2^{-1}+1\times 2^{-2}+1\times 2^{-3} &= 1\times \tfrac{1}{2}+1\times \tfrac{1}{4}+1\times \tfrac{1}{8} \\ &= 1\times 0.5+1\times 0.25+1\times 0.125 \\ &= 0.875 \end{aligned}$$

$$\therefore \quad (0.111)_2=(0.875)_{10}$$

So, we have, $\mathbf{(1001.111)_2=(9.875)_{10}}$

▶ **Example 7.** *Covert the octal number* $(1573)_8$ *into its decimal equivalent.*

Solution :

$$\begin{aligned} & 1\times 8^3+5\times 8^2+7\times 8^1+3\times 8^0 \\ &= 1\times 512+5\times 64+7\times 8+3\times 1 \\ &= 512+320+56+3=891 \end{aligned}$$

So, $\mathbf{(1573)_8=(891)_{10}}$

▶ **Example 8.** *Perform the binary subtraction* 1101 − 111.

Solution :

$$\begin{array}{rl} 1101 & \text{(minuend)} \\ -\ 111 & \text{(subtrahend)} \\ \hline \mathbf{110} & \end{array}$$

▶ **Example 9.** *Perform the following binary addition*

$$11111+1011+101+10+1.$$

Also show the corresponding decimal addition. (Calcutta Hons.)

Solution : Here,

$$\begin{aligned} 11111_2 &= 1\times 2^4+1\times 2^3+1\times 2^2+1\times 2^1+1\times 2^{\circ}=31_{10} \\ 1011_2 &= 1\times 2^3+0\times 2^2+1\times 2^1+1\times 2^{\circ}=11_{10} \\ 101_2 &= 1\times 2^2+0\times 2^1+1\times 2^{\circ}=5_{10} \\ 10_2 &= 2_{10} \end{aligned}$$

For binary addition we take the numbers in pairs. The corresponding decimal addition is also shown parallely.

Binary addition	Decimal addition
$\begin{array}{r} 11111 \\ +\ 1011 \\ \hline 101010 \end{array}$	$\begin{array}{r} 31 \\ +\ 11 \\ \hline 42 \end{array}$
$\begin{array}{r} 101010 \\ +\ 101 \\ \hline 101111 \end{array}$	$\begin{array}{r} 42 \\ +\ 5 \\ \hline 47 \end{array}$
$\begin{array}{r} 101111 \\ +\ 10 \\ \hline 110001 \end{array}$	$\begin{array}{r} 47 \\ +\ 2 \\ \hline 49 \end{array}$
$\begin{array}{r} 110001 \\ +\ 1 \\ \hline 110010 \end{array}$	$\begin{array}{r} 49 \\ +\ 1 \\ \hline 50 \end{array}$

$\therefore$ The required binary sum is 110010 and the decimal addition gives 50.

► **Example 10.** *Prove the following Boolean identities.*

(i) $\overline{\overline{AB} + \bar{A} + AB} = 0$

(ii) $ABC + A\bar{B}C + AB\bar{C} = AC + AB\bar{C} = A(B + C)$. **(Calcutta Hons.)**

Solution: (i)

$$\begin{aligned} \overline{\overline{AB} + \bar{A} + AB} &= \overline{(\overline{AB} + AB) + \bar{A}} \\ &= \overline{1 + \bar{A}} \quad (\because \overline{AB} + AB = 1) \\ &= \bar{1} \quad (\because 1 + \bar{A} = 1) \\ &= \mathbf{0} \end{aligned}$$

(ii)

$$\begin{aligned} ABC + A\bar{B}C + AB\bar{C} &= AC(B + \bar{B}) + AB\bar{C} \\ &= AC \cdot 1 + AB\bar{C} \quad (\because B + \bar{B} = 1) \\ &= AC + AB\bar{C} = A(C + B\bar{C}) \\ &= A(C + B) = A(B + C) \quad (\because C + B\bar{C} = C + B) \end{aligned}$$

► **Example 11.** *A three variable truth table has high output for these input conditions:* $111, 010, 100$ *and* 110. *Find the Boolean expression and the corresponding logic circuit.* **(Calcutta Hons.)**

Solution: Let A, B and C are the three input variables. The sum of the fundamental product rule gives the output as

$$Y = ABC + \bar{A}B\bar{C} + A\bar{B}\bar{C} + AB\bar{C}$$

and the corresponding logic circuit is shown below in Fig. 14.51

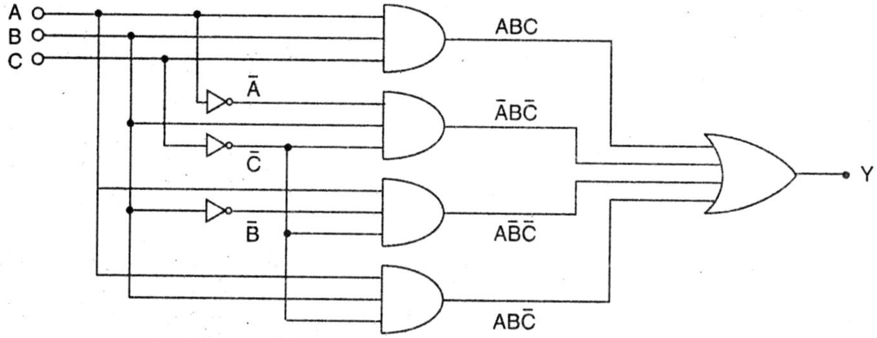

Fig. 14.51

▶ **Example 12.** *Implement* (i) AND *with* OR *and* NOT *gates.*
(ii) OR *with* AND *and* NOT *gates.*

Solution: (i) Construction of AND gate by using OR and NOT gate:

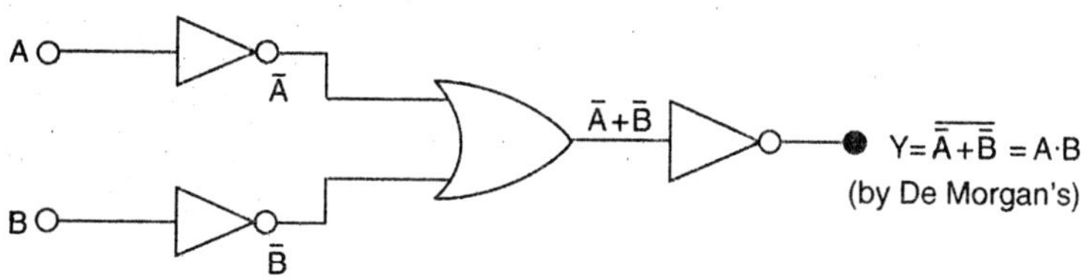

Fig. 14.52

(ii) Construction of OR by using AND and NOT gates:

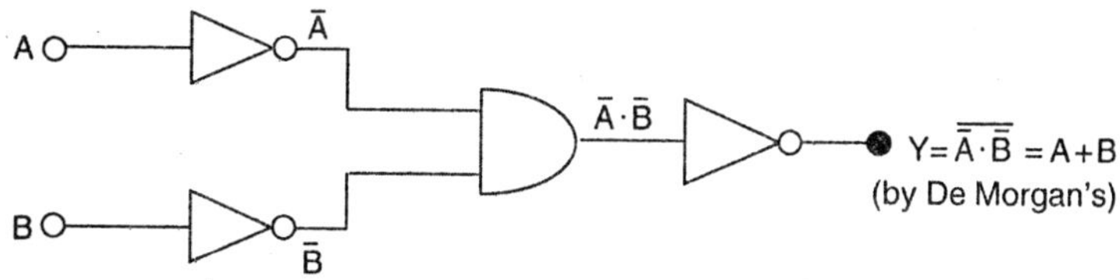

Fig. 14.53

▶ **Example 13.** *An* AND *gate is followed by a* NOT *gate. With two inputs A and B, obtain the Boolean expression of the output C.* **(Calcutta Hons.)**

Solution: A and B are the two inputs of the AND gate. So the output of AND gate is AB. This is the input of the NOT gate which gives the final output $C = \overline{A\cdot B}$. The logic circuit is shown in Fig. 14.54 below.

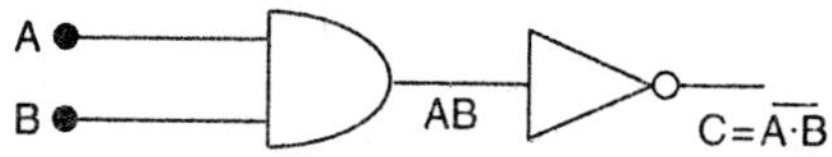

Fig. 14.54

▶ **Example 14.** (a) *Show how you can form* (i) *an* AND *gate and* (ii) *an* OR *gate from* NAND *gates.* **(Calcutta Hons.)**

(b) *Show how you can form* (i) *an* AND *gate and* (ii) *an* OR *gate from* NOR *gates.*

Solution : (a) (i) AND gate by using NAND gates only.

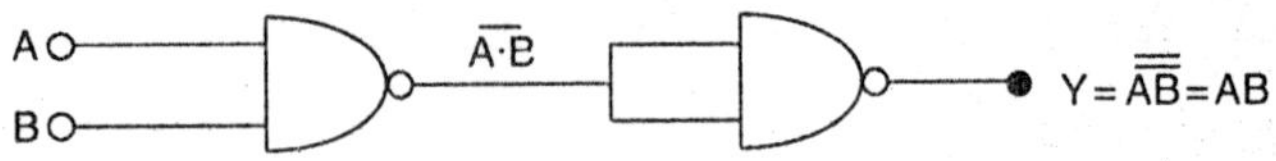

Fig. 14.55

(ii) OR gate using NAND gates only.

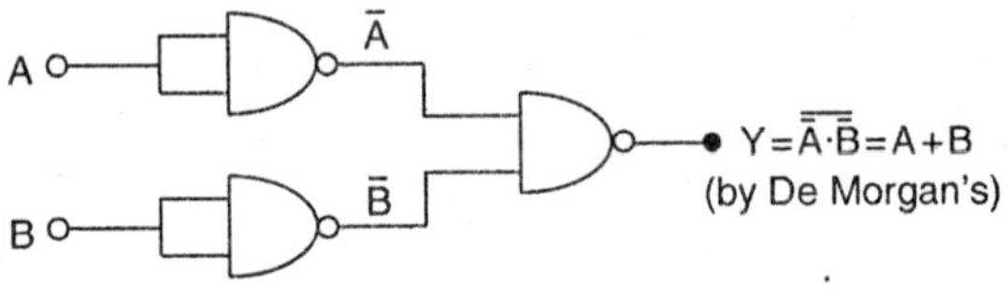

Fig. 14.56

(b) (i) AND gate using NOR gates only.

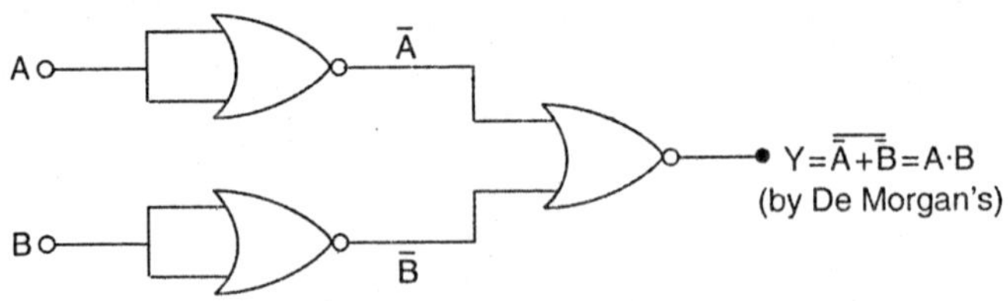

Fig. 14.57

(ii) OR gate using NOR gates only.

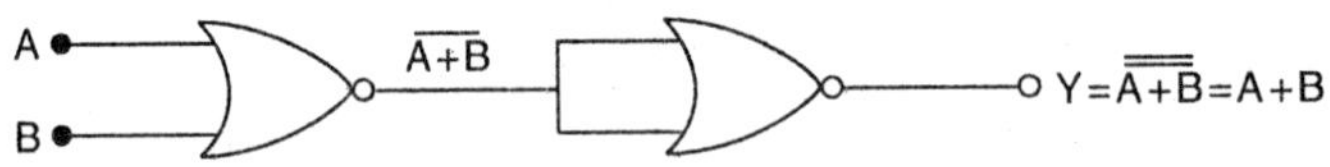

Fig. 14.58

▶ **Example 15.** *Show that* $(A \oplus B) \oplus C = A \oplus (B \oplus C)$. **(Calcutta Hons.)**

Solution :

$$\begin{aligned}\text{L.H.S.} &= (A \oplus B) \oplus C \\ &= (A\bar{B} + \bar{A}B) \oplus C \quad (\because\ A \oplus B = A\bar{B} + \bar{A}B) \\ &= (A\bar{B} + \bar{A}B)\bar{C} + \overline{(A\bar{B} + \bar{A}B)}C \\ &= A\bar{B}\bar{C} + \bar{A}B\bar{C} + \overline{A\bar{B}} \cdot \overline{\bar{A}B} \cdot C, \text{ using De Morgan's}\end{aligned}$$

$$= A\bar{B}\bar{C} + \bar{A}B\bar{C} + (\bar{A} + \bar{\bar{B}})\cdot(\bar{\bar{A}} + \bar{B})\cdot C$$
$$= A\bar{B}\bar{C} + \bar{A}B\bar{C} + (\bar{A} + B)(A + \bar{B})\cdot C$$
$$= A\bar{B}\bar{C} + \bar{A}B\bar{C} + (\bar{A}A + \bar{A}\bar{B} + AB + B\bar{B})\cdot C$$
$$= A\bar{B}\bar{C} + \bar{A}B\bar{C} + \bar{A}\bar{B}C + ABC \quad (\because\ A\bar{A} = B\bar{B} = 0)$$

$$\text{R.H.S.} = A \oplus (B \oplus C)$$
$$= A \oplus (B\bar{C} + \bar{B}C)$$
$$= A(\overline{B\bar{C} + \bar{B}C}) + \bar{A}\cdot(B\bar{C} + \bar{B}C)$$
$$= A\cdot(\overline{B\bar{C}}\cdot\overline{\bar{B}C}) + \bar{A}B\bar{C} + \bar{A}\bar{B}C$$
$$= A\cdot(\bar{B} + \bar{\bar{C}})\cdot(\bar{\bar{B}} + \bar{C}) + \bar{A}B\bar{C} + \bar{A}\bar{B}C$$
$$= A\cdot(\bar{B} + C)\cdot(B + \bar{C}) + \bar{A}B\bar{C} + \bar{A}\bar{B}C$$
$$= A(\bar{B}B + \bar{B}\bar{C} + BC + C\bar{C}) + \bar{A}B\bar{C} + \bar{A}\bar{B}C$$
$$= A\bar{B}\bar{C} + ABC + \bar{A}B\bar{C} + \bar{A}\bar{B}C \quad (\because\ \bar{B}B = \bar{C}C = 0)$$
$$= A\bar{B}\bar{C} + \bar{A}B\bar{C} + \bar{A}\bar{B}C + ABC$$
$$\therefore\ \boldsymbol{(A \oplus B) \oplus C = A \oplus (B \oplus C)}.$$

▶ **Example 16.** *Design a two-input* XOR *gate exclusively with the help of* NAND *gates.*

(Calcutta Hons.)

Solution: If A and B are two inputs of XOR gate, then the output is given by $Y = A\bar{B} + \bar{A}B$.

Now, $$Y = A\bar{B} + \bar{A}B$$
$$= \overline{(\overline{A\bar{B}})\cdot(\overline{\bar{A}B})}, \text{ using De Morgan's law.}$$

This expression indicates that XOR can be constructed using NAND gates only.

The logic circuit is illustrated in Fig. 14.50.

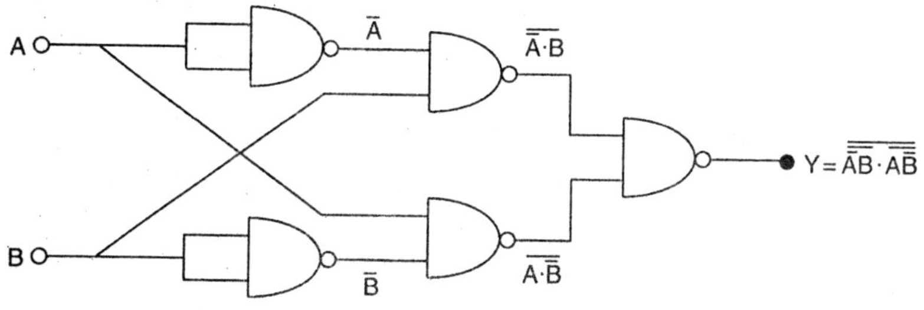

Fig. 14.59

▶ **Example 17.** *A binary signal* $(1100112)_2$ *is given to the input of a NOT gate. Find the output.*

Solution: For NOT gate, if input is 1, then the output is zero and if the input is 0, then the output is 1.

$$\therefore \text{ If } \quad A = (1100111)_2, \text{ then the output is}$$

$$\boldsymbol{Y = (0011000)_2}.$$

▶ **Example 18.** *Two binay signals* $A = (100101)_2$ *and* $B = (110110)_2$ *are applied as inputs of a NOR gate. Determine the output.*

Solution: For NOR gate, the output $Y = \overline{A + B}$.

Since, here, we have $A + B = 110111$,the output is given by

$$\overline{A + B} = 001000$$

$$\text{or, output } \boldsymbol{Y = (001000)_2}$$

Review Questions

1. (a) What are binary numbers?
   (b) What are the bases of decimal, binary, octal and hexadecimal number systems?
   (c) Explain the terms bit, MSB and LSB in binary number system?
   (d) What are the advantages of using $(0, 1)$ notation in the binary system?
   (Calcutta Hons.)

2. (a) What is a logic gate?
   (b) Define (i) AND, (ii) OR and (iii) NOT gate. Give their symbols and the truth tables.

3. Draw a positive-logic OR gate with two diodes and explain its operation. Give its logic symbol and the truth table. **(Calcutta Hons.)**

4. Draw a two-input AND circuit and explain its operation.

5. (a) What is Boolean algebra?
   (b) Show the rules for OR, AND and NOT operations in a tabular form.
   (c) Give the Boolean expression for OR, AND and NOT functions.
   (Calcutta Hons.)

6. State and prove De Morgan's theorems. **(Burdwan Hons.)**

7. Explain the operation of a NOT circuit. Give its truth table and logic symbol.
   (Bangalore Univ.)

8. What is Boolean algebra? Mention its unique feature. Discuss the fundamental laws of Boolean algebra. **(Karnatak Univ.)**

9. (a) What are NOR and NAND gates? Give their truth tables.
   (b) Is it possible to get AND gate by using a few NOR gates? How? **(Calcutta Hons.)**

10. (a) Give the logic symbol, the Boolean expression and the truth table of an XOR gate and an XNOR gate.
    (b) Design a XNOR gate using minimum number of NAND gates.

11. Explain the operation of a digital comparator.

12. What is a half adder? Give its symbol and truth table. How can it be implemented with logic gates?

13. What is a full adder? What is the difference between a half adder and a full adder? Give the truth table of a full adder and hence show that a full adder can be constructed using two half adders and an OR gate.

14. (a) What is a multiplexer? Draw a logic block diagram of a 4 : 1 multiplexer.
    (b) Design a 4-to-1 multiplexer using basic gates. **(Burdwan Hons.)**

15. What is an encoder? Explain how the decimal digits 0 to 9 can be encoded as BCD using logic gates.

16. What is decoder? Explain how BCD digits can be decoded as decimal digits in a decoder.

17. What is a Karnaugh map? Explain with a suitable example how it can be used to simplify logic expressions.

18. What do you mean by SOP and POS representation of Boolean function? Explain with suitable examples.

19. Define sequential logic system. How does it differ from combinational logic system? **(Calcutta Hons.)**

20. What is a flip-flop? Is there any difference between latch and flip-flop? What is the importance of flip-flop in a digital system?

21. What is RS latch? Give its logic symbol, truth table and circuit realisation using NOR/NAND gates. Explain the operation.

22. (a) What is RS flip-flop? Give its logic symbol, truth table and circuit realisation using NOR gates.
    (b) Show how an RS flip-flop can be converted into a JK flip-flop. **(Burdwan Hons.)**

23. What is meant by race-around condition in a flip-flop? How can it be avoided? What do you mean by edge-triggering in a flip-flop? **(Calcutta Hons.)**

24. With the help of a truth table explain the operation of a JK flip-flop having preset and clear input facilities. **(Calcutta Hons.)**

25. (a) Draw a master-slave JK flip-flop system using universal gate. Explain its operation. What are the functions of 'preset' and 'clear' inputs?

(b) Explain how to convert a JK flip-flop into D-type and T-type flip-flops.

26. (a) Sketch the logic system for a clocked SR flip-flop.

(b) Verify that the state of the system does not change in between clock pulse.

(c) Give the truth table.

(d) Explain the entries in the truth table.

27. (a) Explain clearly the difference between combinational and sequential circuits.

(b) Explain the difference between racing and toggling.

(c) Explain the difference between synchronous and asynchronous flip-flops.

Objective Questions

Select the correct alternative in the following questions from among those supplied.

1. 10101 binary number corresponds to decimal number

(a) 31, (b) 21, (c) 11, (d) 3. **(GATE)**

2. For binary number system, the base is

(a) 8, (b) 10, (c) 16, (d) 2.

3. The number $(23)_{10}$ in binary system is

(a) $(10111)_2$, (b) $(11011)_2$, (c) $(11101)_2$, (d) $(11110)_2$.

4. In a logic circuit the output Y in terms of inputs A, B, C is $Y = ABC + A\bar{B}C + AB\bar{C}$. This circuit can be replaced by another logic circuit having

(a) $Y = ABC$, (b) $Y = A(B + C)$,

(c) $Y = B(A + C)$, (d) $Y = C(A + B)$. **(GATE)**

5. The Boolean expression $Y = \bar{A}\bar{B}\bar{C}D + \bar{A}\bar{B}CD + \bar{A}B\bar{C}D + \bar{A}BCD$ reduces to

(a) $\bar{A}B$, (b) D, (c) $\bar{A}$, (d) $\bar{A}D$. **(GATE)**

6. The circuit shown in the figure (Fig. 14.60) functions as

(a) an OR gate, (b) an AND gate
(c) a NOR gate, (d) a NAND gate

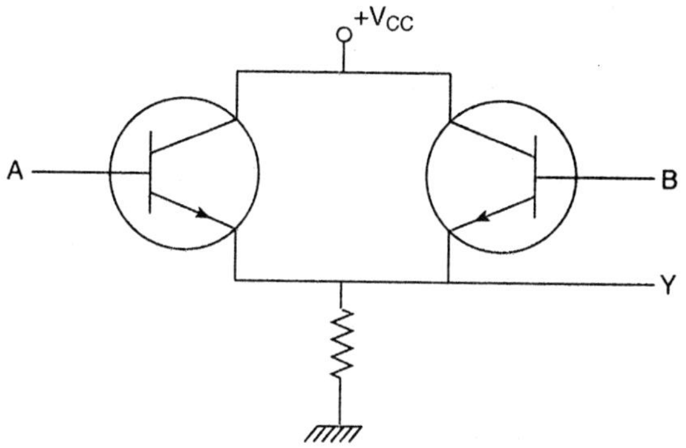

Fig. 14.60 **(GATE)**

7. The circuit shown below in Fig. 14.61 can be used as

(a) NOR gate, (b) OR gate, (c) NAND gate, (d) AND gate.

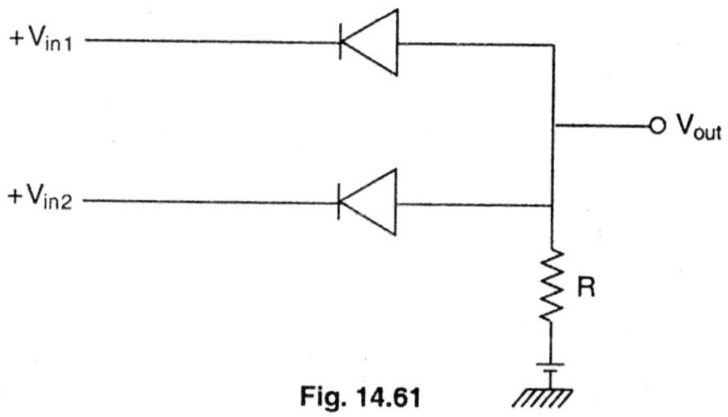

Fig. 14.61

8. The Boolean expression : $B(A+B)+A\cdot(\bar{B}+A)$ can be realized using minimum number of

(a) One AND gate, (b) Two AND gates,
(c) One OR gate, (d) Two OR gates. **(GATE)**

9. If the output of the logic circuit shown in Fig. 14.62 is 1, the input could be

(a) $A=1,\ B=1,\ C=1,\ D=0$ (b) $A=1,\ B=1,\ C=0,\ D=0$
(c) $A=1,\ B=0,\ C=1,\ D=1$ (d) $A=0,\ B=1,\ C=1,\ D=1$

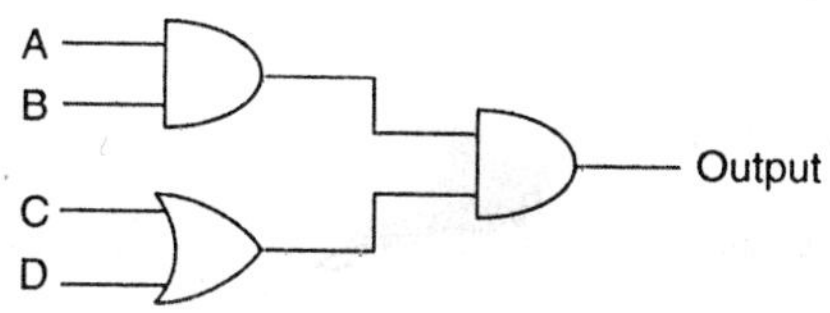

Fig. 14.62

10. To get an output 1 from the circuit as shown in Fig. 14.63, the input ABC must be

(a) 010, (b) 100, (c) 101, (d) 110.

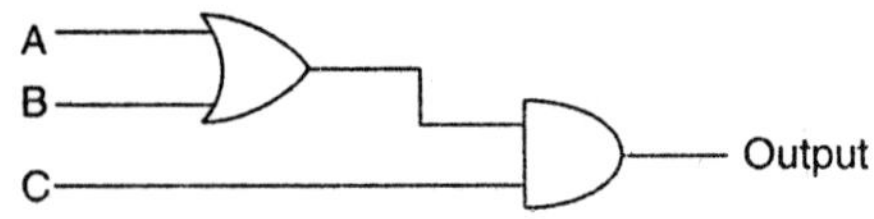

Fig. 14.63 (GATE)

11. The circuit as shown in Fig. 14.64 acts as a

(a) NOR gate, (b) NAND gate, (c) AND gate, (d) XOR gate.

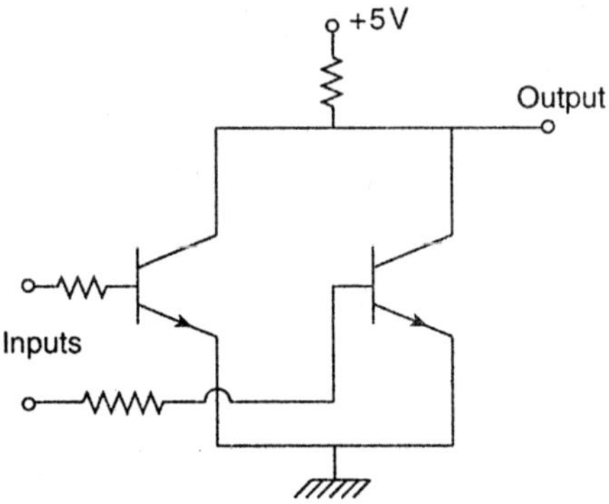

Fig. 14.64 (GATE)

12. An inverter gate can be developed by using

(a) Two diodes, (b) A transistor,
(c) A resistance and a capacitance, (d) An inductance and a capacitance.

13. Boolean algebra is based on

(a) numbers, (b) symbols, (c) logic, (d) either of these.

14. NOR gate is a combination of

(a) OR and OR gates, (b) OR and NOT gates,
(c) AND and OR gates, (d) NOT and NOT gates.

15. NAND gate is a combination of

(a) AND and NOT gates, (b) AND and OR gates,
(c) OR and NOT gates, (d) NOT and NOT gates.

16. De Morgan's theorem is

(a) $\overline{A+B} = \bar{A} \cdot \bar{B}$, (b) $\bar{A} + \bar{B} = \bar{A} \cdot \bar{B}$,
(c) $\bar{A} - \bar{B} = \overline{A \cdot B}$, (d) $\bar{A} - \bar{B} = \bar{A} \cdot \bar{B}$

17. In a three variable K-map, the number of cells will be

(a) 3, (b) 5, (c) 8, (d) 16.

18. The complement of the Boolean expression $F = (X + \bar{Y} + Z)(\bar{X} + \bar{Z})(X + Y)$ is

(a) $XYZ + X\bar{Z} + \bar{Y}Z$, (b) $\bar{X}Y\bar{Z} + XZ + \bar{X}\bar{Y}$,
(c) $\bar{X}Y\bar{Z} + XZ + \bar{Y}\bar{Z}$, (d) $XYZ + \bar{X}\bar{Y}$. **(DRDO)**

19. The Boolean function $F = \bar{A}\bar{D} + \bar{B}D$ can be realised by

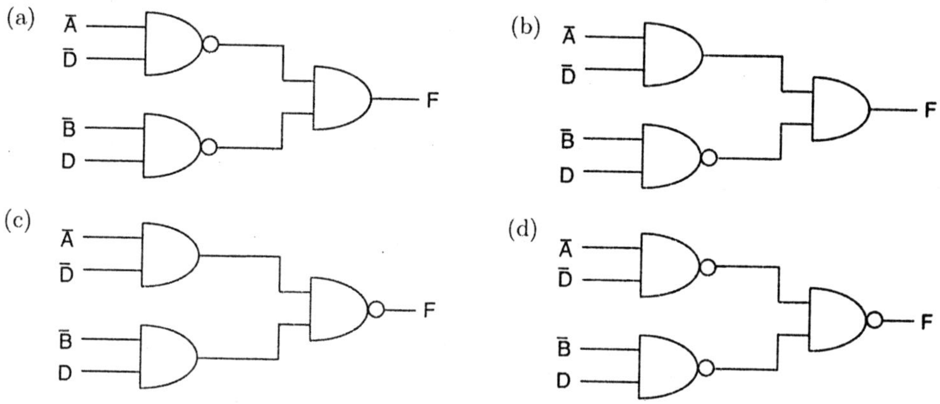

Fig. 14.65

(DRDO)

20. Which of the following is TRUE?

(a) Both latch and flip-flop are edge triggered,
(b) A latch is level triggered and a flip-flop is edge triggered,
(c) A latch is edge triggered and a flip-flop is level triggered,
(d) Both latch and flip-flop are level triggered. **(DRDO)**

21. In a JK flip-flop, we have $J = Q'$ and $K = 1$. Assuming that the flip-flop was initially cleared and clocked frequency 6 pulses, the sequence at the Q output will be

(a) 010000, (b) 011001, (c) 010010, (d) 010101. **(ISRO)**

22. In the logic equation $A(A + \bar{B}\bar{C} + C) + \bar{B}(\bar{C} + \bar{A} + BC)(A + \bar{B}C + A\bar{C}) = 1$, if $C = \bar{A}$, then

(a) $A + B = 1$, (b) $\bar{A} + B = 1$, (c) $A + \bar{B} = 1$, (d) $A = 1$.

(DRDO)

23. In sequential circuits, the output states depend upon

(a) past input states, (b) present input states,
(c) present as well as past input, (d) none of these.

24. For the multiplexer, as shown in Fig. 14.66, the Boolean expression for the output Y will be

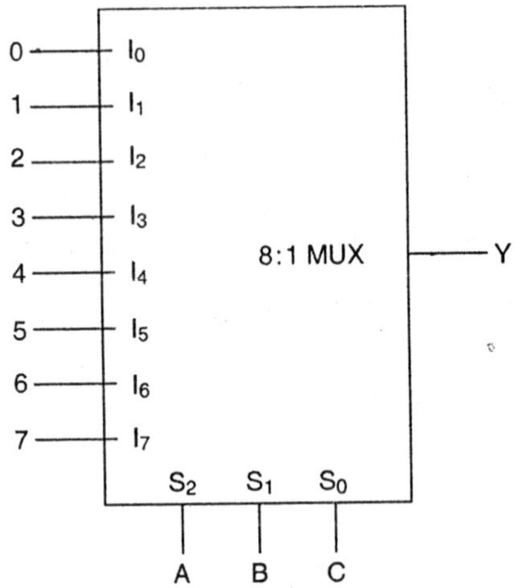

Fig. 14.66

(a) $\bar{A}\bar{B} + \bar{B}\bar{C} + AC$, (d) $A\bar{B} + \bar{B}\bar{C} + A\bar{C}$,

(c) $A\bar{B} + \bar{B}C + AC$, (d) $\bar{A}\bar{B} + \bar{B}C + \bar{A}C$. **(DRDO)**

25. A gate having two inputs (A, B) and one output (Y) is implemented using a 4-to-1 multiplexer as shown in Fig. 14.67. A_1 (MSB) and A_0 are the control bits and I_0, I_3 are the inputs to the multiplexer.

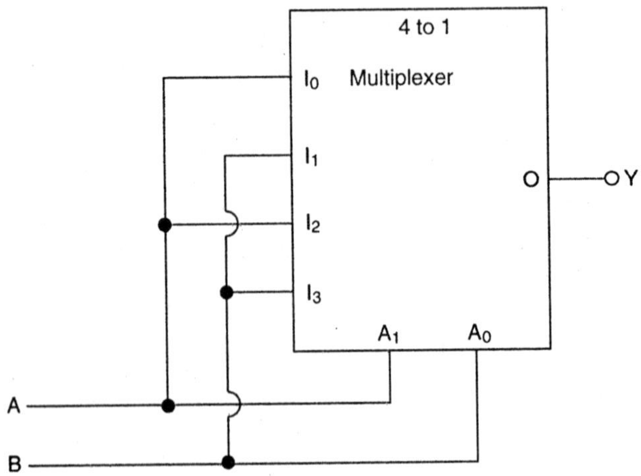

Fig. 14.67

The gate is

(a) NAND, (b) NOR, (c) XOR, (d) OR.

26. Sequential circuits contain

(a) no memory element, (b) all inputs applied simultaneously,
(c) at least one memory element, (d) none of these.

27. D flip-flop is used as

(a) divider circuit, (b) differentiator,
(c) delay switch, (d) all of these.

28. T flip-flop is used as

(a) toggle switch, (b) transfer data circuit,
(c) time delay switch, (d) none of these.

29. In T-flip-flop, the output frequency is

(a) the same as input frequency, (b) one-half of its input frequency,
(b) double of its input frequency, (d) none of these.

30. Race condition occurs in

(a) asynchronous circuits, (b) synchronous circuits,
(c) combinational circuits, (d) all the digital circuits.

31. Following flip-flop is used to eliminate race around problem:

(a) RS flip-flop, (b) Master-slave JK flip-flop,
(b) JK flip-flop, (d) None of these.

32. Karnaugh map is used

(a) to draw the digital circuit layout,
(b) to locate different gates in a digital circuit,
(c) to simplify logical functions,
(d) none of the above.

Problems

1. Determine the binary equivalents of

(a) 25 (b) 23.8125 (c) 129 (d) 15.005 (e) −15
(f) 17.35 (g) −23 (h) 923 (i) 4.9 (j) 125

(Calcutta and Burdwan Hons.)

[(a) 11001, (b) 10111.1101, (c) 10000001, (d) 1111.00001, (e) −1111, (f) 10001.01011, (g) −1011, (h) 1110011011, (i) 100.111001, (j) 1111101]

2. Obtain the decimal equivalent of (a) 1111, (b) 111.1101, (c) 11001011, (d) 11011010, (e) 10101.101, (f) 11010.1. **(Calcutta and Burdwan Hons.)**

[(a) 15, (b) 7.8125, (c) 203, (d) 218, (e) 21.625, (f) 26.5]

3. Perform the following binary operations:

(a) $101010 + 1111$ (b) 101101×101 (c) $111 + 111 + 111$
(d) $11010 - 1101$ (e) $11001 + 01111$ (f) $11001 - 10011$
(g) 1010×111 (h) $111.11 + 101.10$ (i) $11001 - 10101$
(j) $10001 - 1110$

[(a) 111001, (b) 11100001, (c) 10101, (d) 1101, (e) 101000, (f) 110, (g) 1000110, (h) 1101.01, (i) 100, (j) 11]

4. Convert $(110111010.101101)_2$ to its hexadecimal equivalent.

(Calcutta Hons.) [$1BA.B4$]

5. Convert the following hexadecimal numbers into their equivalent binary numbers.

(a) $E5$ (b) $B4D$ (c) $7AF4$ **(Bangalore Univ.)**

[11100101, 101101001101, 0111101011110100]

6. (a) Convert the following octal numbers to their equivalent decimal numbers and binary numbers: (a) 327 and (b) 567. **(Bangalore Univ.)**

[(a) $215_{10}, 011010111_2$; (b) $375_{10}, 101110111_2$]

(b) Convert $(362.3)_{10}$ into octal numbers. [$(552.2314)_8$)]

7. Perform the binary addition corresponding to the decimal addition $2 + 2 = 4$.

(Calcutta Hons.)

8. Perform the binary addition $100101 + 101 + 1101 + 100$. Express the answer in both binary and decimal codes. [111011, 59]

9. Solve the Boolean expression $C = A + B$ for the following inputs: $A = 0$, $B = 0$; $A = 1$, $B = 0$; $A = 1$, $B = 1$. **(Calcutta Hons.)** $[0, 1, 1]$

10. Establish the following identities of Boolean algebra:

(a) $A(A + B) = A$
(b) $\overline{AB} + \bar{A} + AB = 1$ **(Calcutta Hons.)**
(c) $A + AB = A$
(d) $\overline{\bar{A} + B} + \overline{\bar{A} + \bar{B}} = A$ **(Kurukshetra Univ.)**
(e) $(A + B)(\bar{A} + C) = AC + \bar{A}B$ **(N. Bengal Hons.)**
(f) $(A + B)(A + \bar{B})(\bar{A} + C) = AC$ **(Calcutta Hons.)**
(g) $AB + \bar{A}\bar{B} = \overline{A\bar{B} + \bar{A}B}$ **(Burdwan Hons.)**

(h) $A + \bar{A}B = A + B$

(i) $AB + AC + B\bar{C} = AC + B\bar{C}$ **(Burdwan Hons.)**

(j) $(A + B)(B + C)(C + A) = AB + BC + CA$ **(Burdwan Hons.)**

(k) $A + AB + \bar{A}B = \overline{\bar{A}\bar{B}}$

(l) $AC + ABC = AC$

(m) $(AB + C)(AB + D) = AB + CD$

(n) $(A + B)(A + C) = A + BC$

11. Reduce the expression $\overline{\overline{AB} + \bar{A} + AB}$, using Boolean algebraic theorems.

(GATE) [1]

12. Simplify the Boolean expressions:

(a) $Y = (A + B)(A + \bar{B})(\bar{A} + B)$ **(Calcutta Hons.)**

(b) $Y = (\overline{A + B}(\bar{A} + \bar{C})(\bar{B} + \bar{C})$ **(Mangalore Univ.)**

[(a) $Y = AB$, (b) $Y = \bar{A}\bar{B}$]

13. Explain how an OR gate can be implemented with AND and NOT gates.

(Kalyani Hons.)

14. Two digital signals $A = 101101$ and $B = 110101$ are applied to a two-input AND gate. Sketch the input and output signals and give the equivalent binary numbers of the output. **(Calcutta Hons.)**

15. Use both 1's and 2's complements separately to perform the following binary subtractions:

(a) 11011 – 01101 (b) 1010 – 1101

(c) 11001 – 1010 (d) 110101 – 10101

(e) 1100 – 1001 **(Calcutta Hons.)**

[(a) 01110, (b) –0011, (c) 1111, (d) 100000, (e) 11]

16. Given the logic equation $Y = A\bar{B} + \bar{A}B$, draw its circuit using AND, OR and NOT gates. **(GATE)**

17. In the logic circuit shown in Fig. 14.68, write down the Boolean equation expressing the output X in terms of the input A, B and C. Simplify the equation and draw the simplified logic circuit using appropriate gates. **(GATE)**

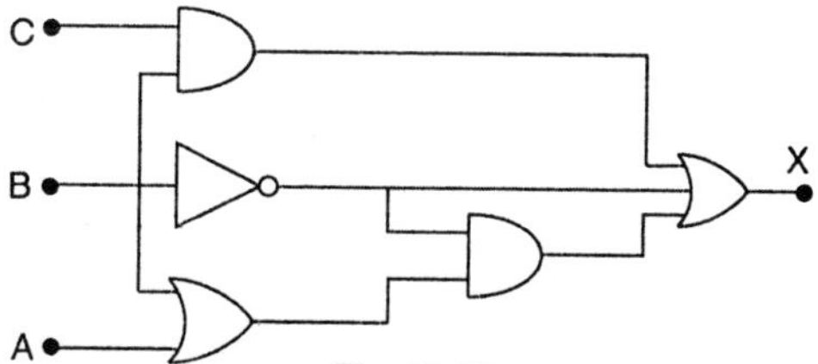

Fig. 14.68

18. Simplify of function $Y = \bar{A}BC + A\bar{B}C + AB\bar{C} + ABC$ by using a Karnaugh map and draw the simplified logic circuit.

19. Design a two-input XOR gate exclusively with the help of NOR gates.

(Calcutta Hons.)

20. Noting that $A \oplus B = \bar{A}B + A\bar{B}$ and $\overline{A \oplus B} = AB + \bar{A} \cdot \bar{B}$, prove that

$$A \oplus B = \overline{\overline{A + B} + \overline{\bar{A} + \bar{B}}}$$

21. Simplify the following Boolean expression: $Y = A\bar{B}D + A\bar{B}\bar{D}$. [$Y = A\bar{B}$]

22. Reduce the expression

$$AB + \overline{AC} + \overline{AB}C\,(AB + C) \qquad [AB + \bar{A} + \bar{C}]$$

23. Draw logic gates combination to express the Boolean expression $A + \bar{B}C$.

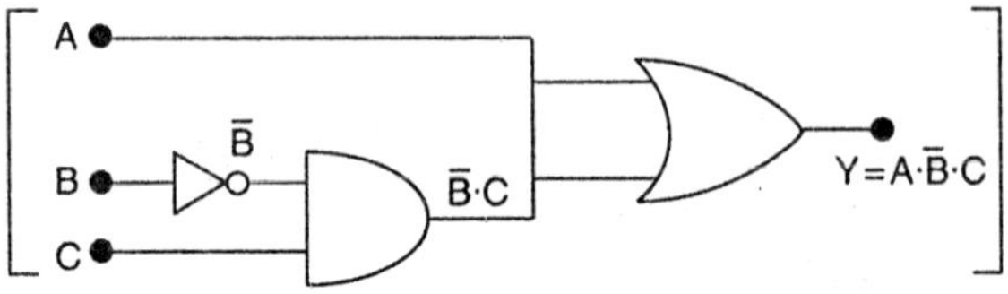

Fig. 14.69

15

Some Electronic Measuring Instruments

In this chapter we shall give a brief account of some important electronic measuring instruments such as *electronic multimeter, digital voltmeter* and the *cathode ray oscilloscope*. An *electronic measuring instrument* is a device, based on electronic and electric principle, for measuring physical quantities rapidly, highly reliably and with precision so much so that they are fast replacing the purely electrical ones of the past. The old order changes yielding place to the new is probably more true in science than in any thing else.

15.1 Electronic multimeter

We are familiar with *ordinary voltmeter*—a permanent magnet moving-coil instrument. This electrical instrument suffers from low sensitivity and is not appropriate for highly accurate measurements. Here comes the ***electronic voltmeter*** having in-built amplifying device. Its high impedance and good frequency-response make it suitable for measuring even very low voltages. With associated additional circuits, an electronic voltmeter can measure current and resistance as well. It is then called an ***electronic multimeter*** or ***volt-ohm-multimeter*** (VOM).

The *basic circuitry* of a solid state d.c. voltmeter using a balanced bridge amplifier is illustrated in Fig. 15.1. The upper arms of the bridge are two identical FETs, F_1 and F_2. The resistors R_1, R_2 and R_3 form the lower arms of the bridge, R_3 acting as *zero-adjustment resistor*. The meter movement is connected between the source terminals of FETs. The circuit is balanced with *no input signal* condition when the

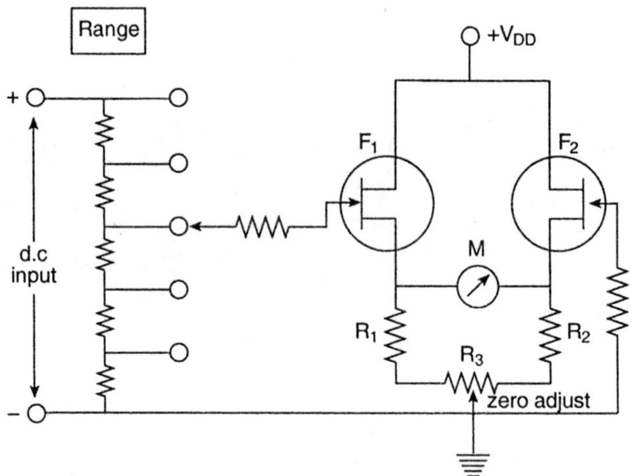

Fig. 15.1 Electronic voltmeter (d.c)

meter deflection reads *zero*. When positive voltage is applied to the gate of F_1, its drain current increases with consequential rise in the voltage at its source terminal. The resulting *unbalanced voltage* between the source terminals of the two FETs causes some current to flow through M, the scale of which is calibrated to agree with the input voltage.

To measure *a.c. voltages*, the input is first *rectified* to a proportional d.c. voltage which is then measured by d.c. voltmeter. The block diagram, as given in Fig. 15.2, is self-explanatory. The *function switch* selects the nature of input voltage, the *rectifier* then converts the input a.c. to a proportional d.c.; the *range switch* selects the range

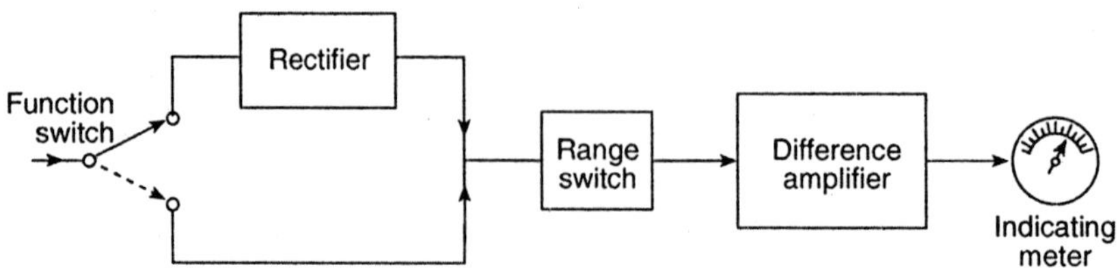

Fig. 15.2 Basic functional block of elctronic voltmeter

of input and the *difference amplifier* gives the necessary gain to increase the sensitivity of the voltmeter. The *indicating meter* shows deflection directly proportional to the input voltage.

To measure *current and resistance*, additional circuitry is needed. For instance, the current I to be measured is passed through a known resistance R and the voltage drop IR is determined by the above voltmeter. Since R is known, I can be easily evaluated.

For measuring *unknown resistance* R_x, a voltage drop V_x is produced in R_x using a battery of known voltage V_b and a known resistance (R_1, R_2 or R_3). As V_x is

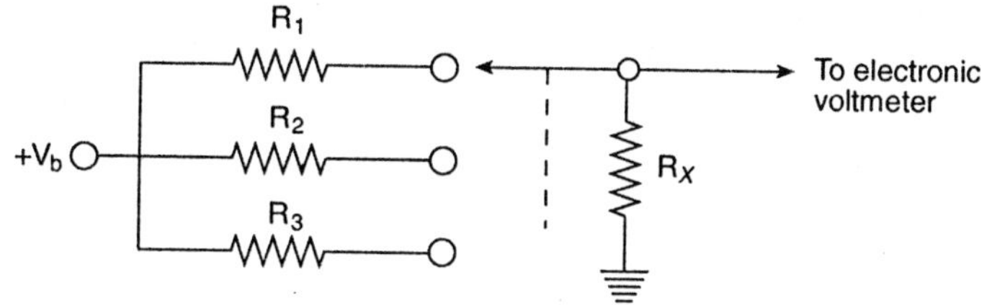

Fig. 15.3 Resistance measurement with electronic voltmeter

proportional to R_x, the meter scale can be calibrated in terms of the resistance. Fig. 15.3 shows a typical circuit for measuring resistance with electronic voltmeter.

15.2 Digital voltmeter

The ***digital voltmeter*** is one that displays the measured values of voltages directly as *discrete numbers*, unlike the analog type that uses the deflection of a moveable pointer and a scale. Digital type instrument is extremely convenient as the numerical readouts reduce the errors of human reading and interpolation. It also eliminates errors due to parallax and increases the speed of reading.

A ***digital voltmeter*** **(DVM)** is a basic device in a digital multimeter which can measure voltages, currents and also resistances. Additional circuitry with a DVM converts it into a *multimeter* which, apart from voltage, current and resistance, can also measure the inductance and the capacitance and even some non-electric quantities with suitable transducers.

A digital voltmeter may be of different categories of which the ***dual slope*** **DVM** is most widely used and we describe here only this type of DVM.

Dual slope DVM—The *principle* of dual slope DVM consists in first converting the input voltage to a positive charge by integration over a set of period. The quantity of charge so obtained is determined by counting the number of pulses of negative charge needed to discharge the integrator capacitor C back to zero and this number is *a measure of the input voltage.*

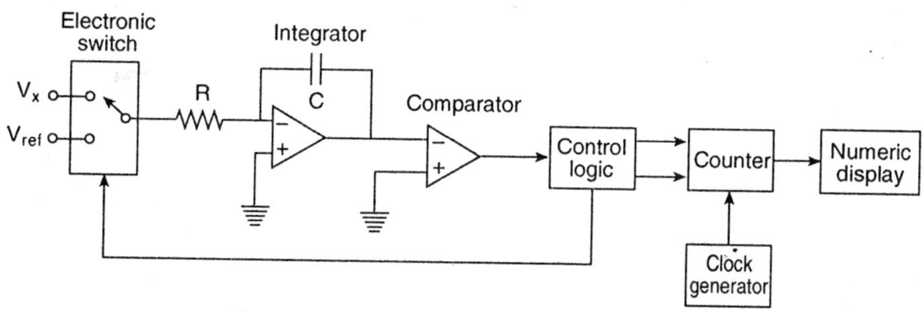

Fig. 15.4 Block diagram of a dual slope DVM

A dual slope DVM is illustrated in Fig. 15.4 in block diagram. The *electronic switch* (usually FET type) first switches the *integrator* to the unknown voltage V_x for a known period T_1 during which the output voltage of the capacitor C changes from 0 to V_p, say where

$$V_p = \frac{1}{RC} V_x T_1 \qquad (15.2.1)$$

If V_x is constant, the output is a *ramp voltage* having a slope proportional to V_x. The switch now connects the integrator to V_{ref}, the *reference voltage of opposite polarity.* The capacitor now discharges and the integrator output becomes a *ramp voltage* proportional to V_{ref}. If T_2 be the period of ramp from output V_p to zero volt, then we have

$$V_p = -\frac{1}{RC} V_{\text{ref}} T_2 \qquad (15.2.2)$$

From equations (15.2.1) and (15.2.2), we get

$$V_x = V_{\text{ref}}(T_2/T_1)$$

$$\therefore \quad \boxed{\boldsymbol{V_x \propto T_2}}$$

The time interval T_2 is determined with an electronic interval counter and hence V_x is measured.

Resolution and sensitivity of DVM—The ***resolution*** of a DVM is *defined* as the *number of digits used in the meter.*

For instance, a three-digit meter for 0 – 1 V range can display 000 to 999 in mV with a resolution of 1 mV.

The ***sensitivity*** of a DVM is *defined* as the *smallest change in input which the meter can detect.* It is expressed as

$$\boxed{S = (\nu_s)_{\text{min}} R}$$

where $(\nu_s)_{\text{min}}$ = lowest full scale value of DVM and R = resolution in decimal.

• By means of *additional circuitry*, the DVM can be utilized to measure both the a.c and d.c currents, voltages, d.c resistances when it is aptly termed as a ***digital multimeter***.

15.3 Cathode ray oscilloscope (CRO)

The ***cathode ray oscilloscope***, abbreviated as ***CRO***, is a versatile laboratory instrument which is employed for the measurement and analysis of different types of waveforms. Using appropriate transducers, many physical quantities such as pressure, strain, temperature, acceleration etc. can be converted to voltages which then can be displayed in a CRO. It is basically a fast X-Y plotter that shows an input signal vs. another signal or time.

A *CRO* has a number of components. The main component of a *CRO* is the ***cathode ray tube*** (*CRT*) which we shall first describe in the following section.

15.3.1 Cathode ray tube (CRT)

It consists essentially of a *highly evacuated* funnel-shaped *glass tube* (Fig. 15.5). It has an ***electron gun*** that generates a sharply focussed beam of electrons of high velocities. The electrons are emitted from a *cathode K*, heated indirectly by a *filament FF*. The electrons thus emitted pass through a *small aperture O*, drilled on the axis of a tubular *control grid G*. The electron-flow can be controlled by changing the control grid bias potentiometrically.

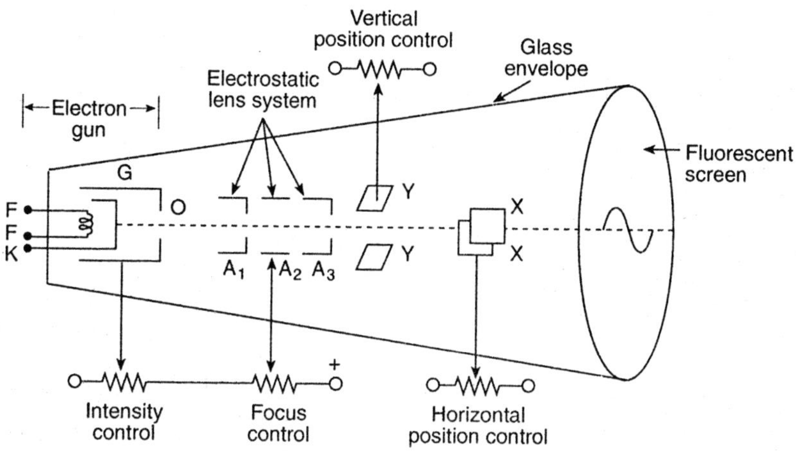

Fig. 15.5 Cathode ray tube (CRT)

On emergence from G (i.e., on emergence from the electron-gun), the electron beam is accelerated and focussed by the *electrostatic lens system* A_1, A_2, A_3 and passes through two pairs of ***deflecting plates***—*vertical* or *Y-plates* (Y, Y) and *horizontal* or *X-plates* (X, X). Usually, the voltage under study is applied to Y-plates that deflects the electron beam vertically and another voltage, proportional to time, called *time base*, is applied to X-plates. Eventually, the electron beam impinges onto the ***fluorescent screen*** to generate a *visible pattern* of the signal.

The screen usually consists of a *flat glass plate*—optically polished outside and coated inside with a *phosphor*, a fluorescent material. With no voltage to X or Y plates, the fast electrons strike the screen and produces a visible spot or dot on it.

• A CRT should be operated with a low brightness of the spot for if the intense light spot is left on the screen for a long time, the overheating may damage the screen—the *burning of the screen.*

15.3.2 Basic CRO

While the main part of a CRO is CRT which we have just described, the other major components of it are (i) ***attenuator***, (ii) ***vertical amplifier***, (iii) ***delay line***, (iv) ***triggering circuit***, (v) ***time-base or sweep generator***, (vi) ***horizontal amplifier***

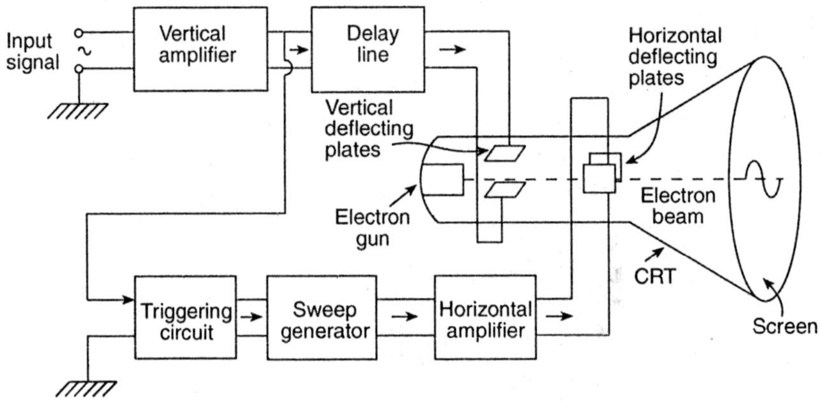

Fig. 15.6 Block diagram of a CRO

and (vii) ***power supply***. The components, except the *attenuator*, are shown in the block diagram of Fig. 15.6.

The signal is applied to the Y-plates of CRT through the *attenuator*, the *vertical amplifier* and the *delay line*. The gain of amplifier is set by the attenuator, calibrated in volts/div. The delay line allows the input signal to reach the Y-plates after the beam starts moving horizontally.

The linear time-base voltage feeds the X-plates through the horizontal amplifier of adequate gain, causing the electron move in a horizontal line across the screen from left to right. During fly back, the return trace is blanked out by a high negative voltage to grid. For a stationary pattern, the horizontal deflection should start at the same point of input signal each time it sweeps. This is ensured by the *triggering circuit* that synchronizes the frequency of sweep signal to that of input signal. A part of input signal is used to the trigger circuit that gives rise to a trigger pulse at a selected point on input wave and starts the *time-base generator* each time. The trigger circuit is the link between the vertical input and the horizontal time-base.

The process involves a time-interval ($\sim$ 100 ns) and a part of the signal (which is used for triggering) is not displayed. For this, the *delay line* is used to retard the arrival of the signal of Y-plates until the trigger and the sweep circuits start the sweep of the beam.

- Besides the above, there are two *power supply* sections—a high voltage supply to operate CRT and a low voltage d.c. supply to all circuits.

- There is also a ***CRO probe*** which is a shielded co-axial cable used to connect the signal source to CRO without loading or perturbing the test circuit.

15.4 Deflection and focussing systems in CRO

Usually, the electrostatic method of deflection of an electron beam is employed in a general purpose CRO. The beam is deflected on the screen by applying voltages on the vertical or the horizontal deflecting plates. The d.c. or peak-to-peak a.c. voltage applied to the deflecting plates to displace the spot by 1 mm on the screen is called the ***deflection factor***. Its reciprocal is termed the *deflection sensitivity*. Hence, ***the deflection sensitivity S is defined as the deflection of the spot on the screen corresponding to*** 1 volt ***applied to deflecting plates.***

The deflection sensitivity in CRO is usually expressed in mm/V and its reciprocal the deflection factor in V/mm.

We shall now find expressions for the *electrostatic deflection* and the *deflection sensitivity* in CRO.

Electrostatic deflection : Expression—Let l be the length of each deflecting plate, s the separation between them (Fig. 15.7) and L' their distance from the screen

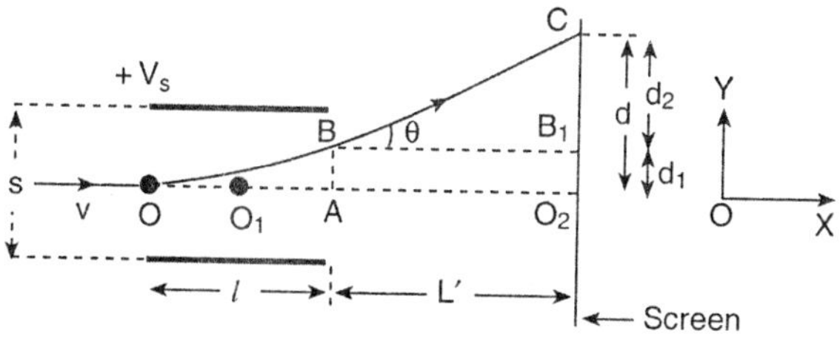

Fig. 15.7 Electrostatic deflection of electron beam

(from near ends). Also, let V_s be the deflecting voltage applied between the plates and V_a the final accelerating anode voltage. Then, the velocity v with which an electron enters the region between the plates along x-axis is

$$\tfrac{1}{2}mv^2 = eV_a \Rightarrow v^2 = \frac{2eV_a}{m} \tag{15.4.1}$$

where e and m are respectively the charge and mass of the electron.

∴ Force on the electron the towards positive deflection plate is

$$F = \frac{eV_s}{s} \tag{15.4.2}$$

∴ Acceleration of the electron is: $$f = \frac{eV_s}{ms} \tag{15.4.3}$$

Now, the time t taken by the electron to pass through deflecting plates is l/v. So, the upward velocity of electron on emergence from the deflecting system is

$$v_y = ft = f\frac{l}{v} = \frac{eV_s}{ms}\cdot\frac{l}{v} \tag{15.4.4}$$

From the entry point O to the exit point B, the electron is deflected to the positive plate in a *parabolic trajectory.* The vertical displacement of electron during this time-interval is

$$O_2B_1 = d_1 = \frac{1}{2}ft^2 = \frac{1}{2}f\left(\frac{l}{v}\right)^2 = \frac{eV_s}{2ms}\left(\frac{l}{v}\right)^2 \tag{15.4.5}$$

No deflecting force exists on the electron after its exit from the region of deflecting plates and it then moves in a straight line inclined at θ with the initial direction.

$$\therefore \quad \tan\theta = \frac{v_y}{v} = \frac{d_2}{L'}, \tag{15.4.6}$$

d_2 being the additional deflection on the screen.

$$\therefore \quad d_2 = \frac{L'v_y}{v} = \frac{eV_s l}{msv^2}L', \quad \text{using (15.4.6) and (15.4.4)} \tag{15.4.7}$$

$\therefore$ *Total deflection* on screen is: $\quad d = d_1 + d_2 = \dfrac{eV_s}{ms} \cdot \dfrac{l}{v^2}\left(\dfrac{l}{2} + L'\right) \qquad (15.4.8)$

using (15.4.5) and (15.4.7).

If L be the distance of the screen from the centre O_1 of the plates, then $L = l/2 + L'$ and (15.4.8) reduces to

$$\boxed{d = \frac{\boldsymbol{lLV_s}}{\boldsymbol{2sV_a}}} \tag{15.4.9}$$

The expression shows that for a given anode voltage V_a and for a given CRT, the *deflection d is proportional to the deflecting voltage* V_s. The CRT can thus be used as a *linear voltage indicator.*

Now, by definition, the ***deflection sensitivity*** of CRT is

$$S = \frac{s}{V_s} = \frac{lL}{2sV_a}\ \text{m/V} = \frac{500lL}{sV_a}\ \text{mm/V} \tag{15.4.10}$$

and its reciprocal $1/S$ is the *deflection factor* G of CRT.

From (15.4.10) it is observed that *deflection sensitivity is independent of the deflecting voltage, but is proportional inversely to the accelerating anode voltage* V_a. The sensitivity S can thus be increased by reducing V_a. But this reduces the brightness of the spot. The problem however is removed in modern CRTs, by an extra-electrode called *post-deflection accelerating electrode.*

- Typical values of deflection sensitivity is about 1 mm/V to 0.1 mm/V.

Magnetic deflection: Deflection of electron beams in CRT may as well be made by magnetic field, as in TV-picture tubes. Here the electrons are deflected by a perpendicular magnetic field produced by a current-carrying coil and they experience a force in a direction at right angles to the direction of their motion and the field. On emerging from the field, the electrons move in a straight path at an angle with their original direction.

The deflection on a screen corresponding to unit magnetic flux is known as the ***magnetic deflection sensitivity.*** We shall now derive expressions for both the deflection and the deflection sensitivity.

Expression for deflection—Let a uniform magnetic field B act on an electron emitted from the electron gun over a length l of its path and $V\,(\equiv V_a)$ the accelerating voltage (Fig. 15.8). Then, we have

$$\frac{mv^2}{r} = Bev \Rightarrow r = \frac{mv}{Be} \tag{15.4.11}$$

$$\text{and } \tfrac{1}{2}mv^2 = eV \quad \Rightarrow v = \sqrt{2eV/m} \tag{15.4.12}$$

where r = radius of the circular path of electron in the magnetic field, v = velocity of electron of mass m and charge e.

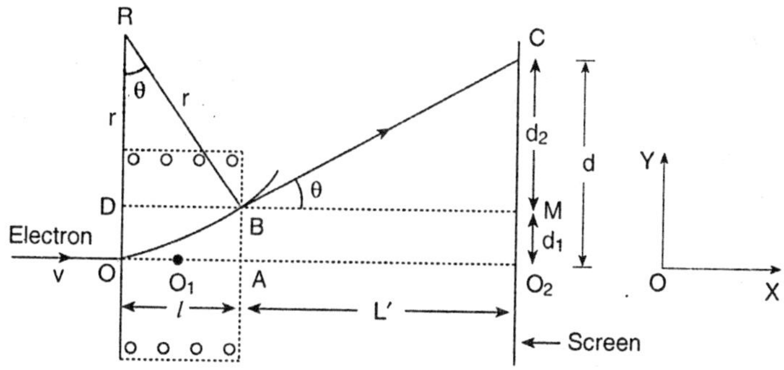

Fig. 15.8 Magnetic deflection of electron beam

Using (15.4.11) and (15.4.12), $$r = \frac{1}{B}\sqrt{\frac{2mV}{e}} \tag{15.4.13}$$

On emergence from the field at B, the electron travels along the line BC, hitting the screen at C, BC being tangential to the arc OB.

At the point of exit B from the field, the deflection of electron on the screen is:

$$d_1 = O_2M = AB = OD = OR - DR = r(1 - \cos\theta) \tag{15.4.14}$$

Outside the magnetic field, the deflection of the electron is

$$d_2 = BM\tan\theta = L'\tan\theta \tag{15.4.15}$$

In practice, the field exists over a short distance and θ is small. So, we have

$$\tan\theta \simeq \theta = l/r; \quad \cos\theta \simeq 1 - \theta^2/2 \simeq 1 - l^2/2r^2.$$

∴ Total deflection of electron on the screen is

$$d = d_1 + d_2 = r(1-\cos\theta) + L'\tan\theta$$

$$= r(l^2/2r^2) + L'(l/r) = \frac{l}{r}\left(L' + \frac{l}{2}\right) = \frac{lL}{r} \tag{15.4.16}$$

where $L = (L' + l/2) = O_1O_2$ = the distance of the screen from the centre of the field region.

Finally, we get

$$\boxed{d = BlL\sqrt{\frac{e}{2mV_a}}} \tag{15.4.17}$$

by substituting for r from (15.4.13) into (15.4.16).

By definition, the ***magnetic deflection sensitivity***

$$S_m = \frac{d}{B} = lL\sqrt{\frac{e}{2mV_a}} \tag{15.4.18}$$

which shows that *the magnetic deflection sensitivity is proportional inversely to the square root of the accelerating voltage V_a.*

● If l and L are in meter, e in Coulomb, V in volt and m in kg, S_m is in m/tesla and can be converted to mm/gauss, using the relation 1 tesla = 10^4 gauss.

Electrostatic focussing of electron beam—Electrons emerging from the control grid tend to diverge from each other due to their negative charge and variation in energy. Such an electron beam would give rise to a broad, ill-defined spot on the screen. For a sharp image on the screen, it is thus imperative to *focus* the beam. This is done by a system of three cylindrical electrodes A_1, A_2, A_3 (Fig. 15.9), called ***electrostatic lens.***

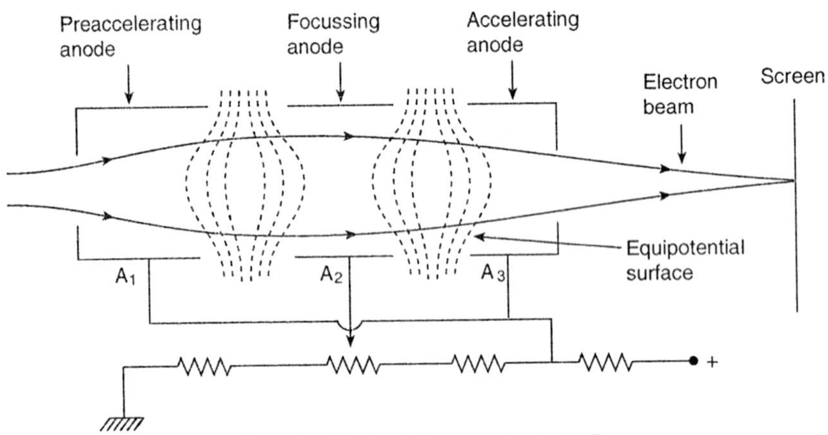

Fig. 15.9 Electrostatic focussing in CRT

Lens A_2, *termed focussing anode*, is given a potential slightly lower than the accelerating anodes A_1, A_3. This difference in potentials generate a *non-uniform* electric field between them. The *equipotential surfaces* between the electrodes take up the shape of a *convex lens* (Fig. 15.9). As the electron beam passes through these regions, the electrons are under a force in a direction normal to the equipotential surfaces. And the beam gets focussed just as a light beam is focussed by a convex lens. The focal length of the electrostatic lens can be adjusted by varying the voltage of the focussing anode A_2 and keeping A_1, A_3 at a fixed high potential.

It would be worthwhile to make a comparison of the electrostatic and the magnetic defections and this is done in what follows.

Comparison of electrostatic and magnetic deflections:

(1) The electrostatic deflection requires little power for deflection, whereas a relatively large power is consumed in the electromagnets of the magnetic deflection system.

(2) The electrostatic deflection can be used at higher frequencies than the magnetic deflection.

(3) The electrostatic deflection depends on the square of electron velocities, whereas the magnetic deflection depends on the first power. This makes the defocussing effect due to unequal velocities more serious in electrostatic deflection.

(4) In electrostatic deflection, the deflection sensitivity falls more rapidly with increase in anode voltage compared to that in magnetic deflection.

(5) Along with electrons, there may be negative ions inside. In magnetic deflection, the ions are deflected to a small degree and leads to a stationary '*ion spot*' at the centre of the scanned area. No such undesirable 'ion spot' occurs in electrostatic defection as ions and electrons are equally deflected.

(6) Electrostatic deflection is inversely proportional to the accelerating voltage V_a, whereas the magnetic deflection is inversely proportional to the square root of the accelerating voltage V_a. So, for operation with high accelerating voltage, magnetic deflection is preferred. And this is why high voltage TV picture tubes commonly use the magnetic deflection while for the general purpose CRT, the electrostatic deflection is preferred.

15.5 Time-base generator

The ***time base generator***, also called ***sweep generator***, is needed for visual display of a signal waveform. It develops a *saw-tooth waveform* that is used as a *time base* in CRO. Each cycle of this periodic waveform (Fig. 15.10) consists of two parts—in the

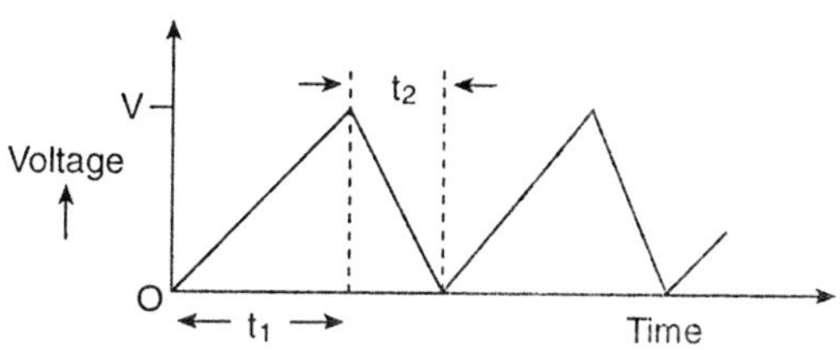

Fig. 15.10 Saw-tooth wave

first part the *voltage increases linearly* from 0 to V in time t_1, called the *sweep time* and in the second part, the *voltage decreases linearly* but quickly from V to 0 in time t_2, called the *return time* or the *fly back time*. The return time is a small fraction of the sweep time, i.e., $t_2 \ll t_1$.

There are several circuits for producing saw-tooth waveform. We shall discuss here how the *unijunction transistor* (UJT) can be used for this.

UJT saw-tooth generator—A simple circuit using UJT is shown in Fig. 15.11(a). The capacitor C is charged by the d.c. supply V_s through R. The voltage across the capacitor increases with time along the path OA (Fig. 15.11b) according to

$$V = V_s\left(1 - e^{-t/CR}\right)$$

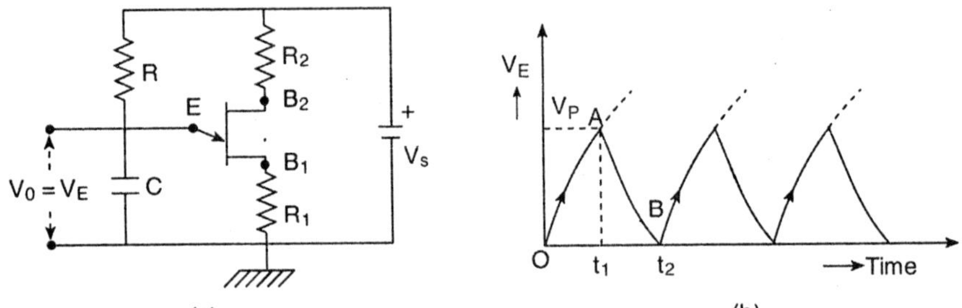

(a) (b)

Fig. 15.11 (a) UJT sweep generator (b) output waveform across C

When the voltage reaches a definite value V_p, termed the *firing potential*, the UJT is turned on (emitter junction *forward* biased) and C discharges through R_1. R_1 is of a small value. So, the voltage across C falls sharply along the path AB (Fig. 15.11b) to a low value at which the emitter-base-one junction is *reverse* biased. The UJT then switches to off-state and the voltage across C builds up again to repeat the cycle to generate the saw-tooth waveform. By adjusting C and R, the period T of the waveform can be adjusted and one must select $CR \gg T$ to make the capacitor voltage rise linearly.

On neglecting R_1, R_2, the fraction of V_s appearing across emitter-base-one (B_1) is

$$\frac{R_{B_1}}{R_{B_1} + R_{B_2}} \cdot V_s = \eta V_s$$

where R_{B_1} and R_{B_2} are respectively the resistance between $E - B_1$ and $E - B_2$ and $\eta = R_{B_1}/(R_{B_1} + R_{B_2})$, called *intrinsic stand-off ratio.* The emitter junction is forward biased when V_E rises above ηV_s $(= V_p)$. Assuming discharge time to be zero, T is given by the time taken by V_E to rise to ηV_s $(= V_P)$.

$$\therefore \quad V_P = \eta V_s = V_s\left[1 - e^{-T/CR}\right]$$

$$\Rightarrow \quad T = CR \ln \frac{1}{1-\eta} \qquad (15.5.1)$$

- The saw-tooth waveform, as described, is not sufficiently linear for precisional work. There is however other circuits by which linearity can be improved. We describe below one such method.

The method consists in charging the capacitor by a *constant current source* (with infinite internal impedance). The voltage across C under constant charging current I is given by

$$V = \frac{1}{C}\int I dt = It/C \qquad (15.5.2)$$

$$\therefore \quad V \propto t.$$

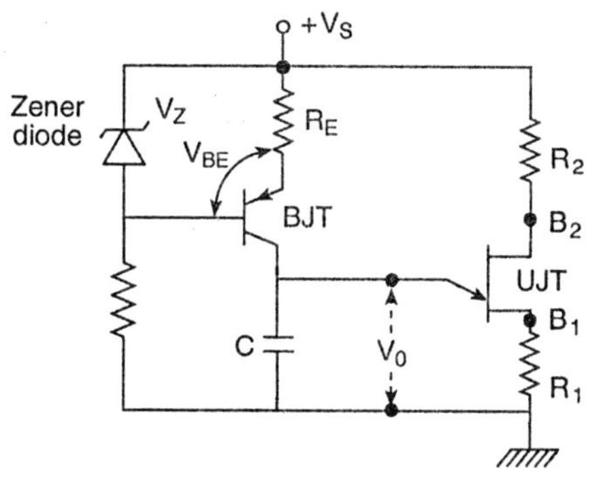

Fig. 15.12 A practical linear time base generator

Fig. 15.12 illustrates a practical *sweep generator* with BJT as a constant current source and UJT as a voltage controlled switch. The constant charging is

$$I = I_C \simeq I_E = (V_z - V_{BE,\,\mathrm{sat}})/R_E$$

And, the capacitor voltage is given by equation (15.5.2).

If V_2 be the *firing potential* of UJT, then the period T of the saw-tooth is approximately given by

$$T = CV_2/I \tag{15.5.3}$$

15.6 Applications of CRO

Some of the important applications of cathode ray oscillocope are given below.

1. Display and study of waveforms—The signal, whose waveform is under study, is applied to the vertical input terminals of CRO. The time-variation of the signal is observed by means of the sweep generator displacing the spot in proportion to time along the horizontal direction. The nature of the signal can be studied from the trace on the screen. For instance, one can see if the signal is sinusoidal or rich in harmonics.

2. Measurement of frequency: (a) Using time-base—The sweep rate is adjusted such that the signal frequency is N times the repeatition frequency of saw-tooth. Then, N complete cycles of signal will be displayed on the screen. If the time base is set to t sec/division and N cycles of signal occupy d divisions on the horizontal scale, then, the time period $T = (t/N)d$ sec.

$$\therefore \ \textit{Frequency}, \ f = 1/T = \frac{N}{t.d} \ \text{Hz}.$$

(b) Using Lissajous figure—*Lissajous figures* are patterns on CRO-screen on simultaneous applications of sine waves to horizontal and vertical deflection plates. This signal under study is applied to Y-plates and an accurately calibrated standard variable frequency oscillator is employed to X-plates, with the internal sweep generator off. The frequency of the standard oscillator is next continuously varied till a simple Lissajous figure (e.g., a straight line, ellipse, circle etc.) is obtained on the screen. Then, the unknown frequency is equal to that of the standard source.

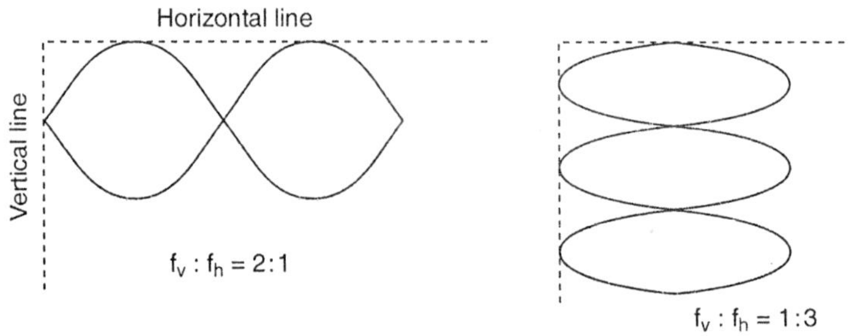

Fig. 15.13 Lissajous patterns with different frequency ratios

It may not be possible, due to the frequency range of the standard, to get such simple Lissajous figures. However, other *closed-loop figures* may be observed. Then

$$\frac{f_{\text{vertical}}}{f_{\text{horizontal}}} = \frac{\text{No. of loops touching a horizontal tangent line}}{\text{No. of loops touching a vertical line}}$$

If the frequency f_h of the standard is known, the unknown frequency f_v can be determined.

Two Lissajous patterns with different frequency ratios are shown in Fig. 15.13.

3. Measurement of phase : (a) Using dual-trace CRO—The *phase difference* ϕ between two a.c signals can be measured by displaying them simultaneously as two separate traces.

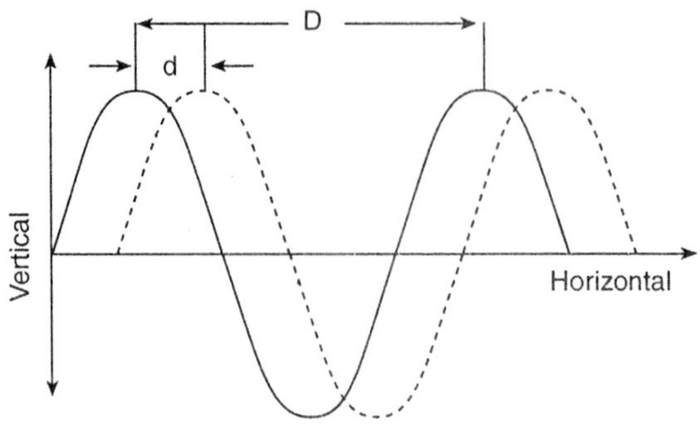

Fig. 15.14 Two sinusoidal signals of same frequency with a phase difference

If D be the distance between two successive peaks of a single wave on the horizontal scale, it corresponds to 2π phase difference. If then the distance between two neighbouring peaks (Fig. 15.14) of the two waveforms be d, then the phase difference of the two signals is

$$\phi = \frac{2\pi}{D} d$$

(b) Using Lissajous figures—Lissajous figures can be used to measure the *phase difference* between two sinusoidal voltages of the same amplitude and frequency. The signals are simultaneously applied to the horizontal and the vertical plates. Let the signals be represented by

$$v_x = V \sin \omega t \tag{15.6.1}$$

$$v_y = V \sin(\omega t + \phi) \tag{15.6.2}$$

where ϕ is the phase angle by which v_y leads v_x.

Eliminating t from (15.6.1) and (15.6.2), we have

$$\frac{v_x^2}{V^2} + \frac{v_y^2}{V^2} - \frac{2v_x v_y}{V^2} \cos\phi = \sin^2\phi \tag{15.6.3}$$

The Lissajous figure is thus an *ellipse* given by (15.6.3).

In reference to Fig. 15.15, if A be the point of intersection of the ellipse with y-axis and if B be the maximum y-displacement, then $B = V$ and $A^2 = B^2 \sin^2 \phi$.

$$\therefore \ \sin\phi = \pm\frac{A}{B} \Rightarrow \phi = \sin^{-1}(\pm A/B) \tag{15.6.4}$$

By measuring the distances A and B from the screen, ϕ can be evaluated.

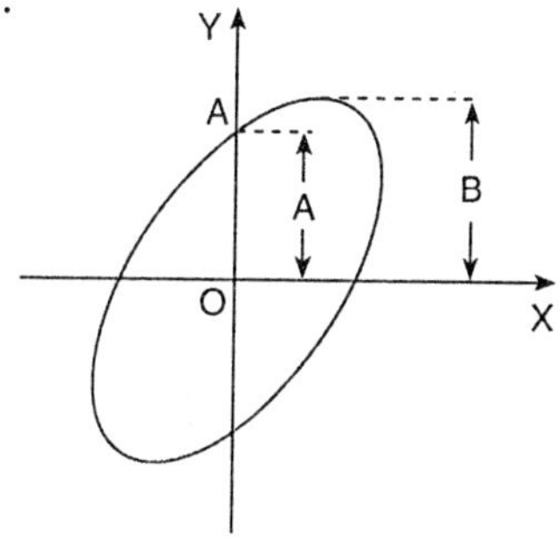

Fig. 15.15 Determination of phase difference

- Note that $0 < \phi < 90°$, if the major axis of ellipse lies in the first and third quadrants and $90° < \phi < 180°$, if it lies in the second and fourth quadrants.
- If $\phi = 0$, the ellipse reduces to a *straight line* of slope angle 45°. If $\phi = 180$, the ellipse reduces also to a *straight line* of slope -1. If $\phi = 90°$, then (15.6.3) reduces to $v_x^2 + v_y^2 = V^2$, which is a *circle* of radius V.

15.7 Illustrated Examples

▶ **Example 1.** *The beam of electron in a cathode ray tube enters a magnetic deflection system after being accelerated through a p.d.* 1 kV. *The magnetic field of the deflection system is* 1.5×10^{-2} T *and acts over an axial length of* 1 cm. *Compute:* (i) *the deflection of the spot on the fluorescent screen at a distance of* 20 cm *from the centre of the deflection system and* (ii) *the magnetic deflection sensitivity.*

Solution: Here, $B = 1.5 \times 10^{-2}$ tesla, $V = 1\,\text{kV} = 10^3\,\text{V}$, $l = 1\,\text{cm} = 0.01\,\text{m}$ and $L = 20\,\text{cm} = 0.2\,\text{m}$. Also, $e = 1.6 \times 10^{-19}\,\text{C}$, $m = 9.1 \times 10^{-31}\,\text{kg}$. Therefore,

(i) Deflection of spot, $\boldsymbol{d} = BlL\sqrt{\dfrac{e}{2V_a m}}$

$$= 1.5 \times 10^{-2} \times 0.01 \times 0.2 \times \sqrt{\frac{1.6 \times 10^{-19}}{2 \times 10^3 \times 9.1 \times 10^{-31}}}$$

$$= 0.281\,\text{m} = \mathbf{28.1\,cm}$$

(ii) Magnetic deflection sensitivity

$$S_m = \frac{d}{B} = \frac{0.281\,\text{m}}{1.5\times10^{-2}\,\text{T}} = \frac{281\,\text{mm}}{150\,\text{gauss}} = \mathbf{1.87\ mm/gauss}$$

► **Example 2.** *The deflection sensitivity of a* CRT *is* 0.02 mm/volt. *When an unknown voltage is applied to the horizontal plates, the spot shifts horizontally by* 2 mm. *Find the value of the unknown voltage.*

Solution: Here, $S = 0.02$ mm/volt, $d = 2$ mm, $V = ?$

Now, the deflection sensitivity, $S = \frac{d}{V} \Rightarrow V = d/S.$

$\therefore$ Unknown voltage, $\mathbf{V} = d/S = \frac{2\,\text{mm}}{0.02\,\text{mm/V}} = \mathbf{100\ V}$

► **Example 3.** *An electron beam in a CRT, accelerated by a p.d. of* 1000 V, *passes through a pair of deflecting plates of axial length* 2 cm *and separated by* 5 mm. *The distance of the fluorescent screen from the centre of the deflecting plates is* 25 cm. *If the p.d. between the plates be* 20 V, *determine* (i) *the transit time of the beam through the plates,* (ii) *the transverse acceleration imparted to the electrons by the deflecting voltage and* (iii) *the deflection of the spot on the CRT-screen. Compute also the deflection sensitivity.*

Solution: By the problem, $V_a = 1000$ V, $l = 2$ cm $= 2\times10^{-2}$ m, $l = 25$ cm $= 25\times10^{-2}$ m, $V_d = 20$ V, $s = 5$ mm $= 5\times10^{-3}$ m.

Also, $e = 1.6\times10^{-19}$ C and $m = 9.1\times10^{-31}$ kg.

$\therefore$ Axial velocity of electron, $v = \sqrt{\frac{2eV_a}{m}} = \sqrt{\frac{2\times1.6\times10^{-19}\times1000}{9.1\times10^{-31}}}$

$$= 1.87\times10^{7}\ \text{m/s}$$

(i) ***Transit time,*** $\mathbf{t} = \frac{l}{v} = \frac{2\times10^{-2}}{1.87\times10^{7}} = \mathbf{1.07\times10^{-9}s}$

(ii) ***Transverse acceleration,*** $\mathbf{f} = \frac{eV_d}{sm} = \frac{1.6\times10^{-19}\times20}{5\times10^{-3}\times9.1\times10^{-31}}$

$$= \mathbf{0.702\times10^{15}\ m/s^2}$$

(iii) ***Deflection of spot,*** $\mathbf{d} = \frac{lL}{2s}\cdot\frac{V_d}{V_a} = \frac{2\times10^{-2}\times25\times10^{-2}\times20}{2\times5\times10^{-3}\times1000}$

$$= \frac{1}{100}\,\text{m} = \mathbf{1\ cm.}$$

$\therefore$ ***Deflection sensitivity,*** $\mathbf{S} = \frac{d}{V_d} = \frac{1}{20}\frac{\text{cm}}{\text{V}} = \mathbf{0.5\ \frac{mm}{V}}$

▶ **Example 4.** *A Lissajous figure with three tangencies with the horizontal and five tangencies with the vertical is obtained on a CRO screen by applying two sinusoidal signals simultaneously to the horizontal and the vertical plates. If the frequency of the signal applied to the horizontal plates is known to be* 12 kHz, *find the freuquency of the signal applied to the vertical plates.* **(Calcutta Hons.)**

Solution: Let f_v and f_h be the freuquencies in Hz of the vertical and horizontal signals respectively. Then, by the problem,

$$\frac{f_v}{f_h} = \frac{\text{Number of horizontal tangencies}}{\text{Number of vertical tangencies}} = \frac{3}{5}$$

$$\therefore \ \boldsymbol{f_v} = f_h \times \frac{3}{5} = \frac{12 \times 3}{5} = \mathbf{7.2\ kHz} \quad (\because\ f_h = 12\ \text{kHz})$$

▶ **Example 5.** *The deflection system in a CRT employs a magnetic field of* 10^{-4} T *acting over an axial length of* 4 cm *and is placed* 24 cm *from the screen. If the accelerating voltage is* 600 V, *find the deflection of the spot on the screen.*

Solution: By the problem, $B = 10^{-4}$ tesla, $V = 600$ volt, $l = 4\,\text{cm} = 0.04\,\text{m}$ $L = 24\,\text{cm} = 0.24\,\text{m}$. Also, $e = 1.6 \times 10^{-19}\,\text{C}$, $m = 9.1 \times 10^{-31}\,\text{kg}$.

$$\therefore \ \text{Deflection of spot,}\ \boldsymbol{d} = BlL\sqrt{\frac{e}{2Vm}}$$

$$= 10^{-4} \times 0.04 \times 0.24\sqrt{\frac{1.6 \times 10^{-19}}{2 \times 600 \times 9.1 \times 10^{-31}}}\ \text{m}$$

$$= 0.0116\,\text{m}\mathbf{1.16\ cm}$$

▶ **Example 6.** *The length of deflecting plates in a CRO is* 5 cm. *They are separated by* 4 mm. *The distance of the fluorescent screen from the nearest edge of the deflecting plates in* 15 cm. *A d.c voltage of* 25 V *is applied to the deflecting plates. If the acelerating potentyial difference is* 1000 V, *find the displacement of the spot on the screen.* **(Bangalore Univ.)**

Solution: By the problem, $l = 5\,\text{cm} = 0.05\,\text{m}$, $s = 4\,\text{mm} = 0.004\,\text{m}$, $L' = 15\,\text{cm}$.

$$\therefore \quad L = \left(\frac{l}{2} + L'\right) = (2.5 + 15)\,\text{cm} = 17.5\,\text{cm} = 0.175\,\text{m},$$

And, $V_s = 25\,\text{V}$ and $V_a = 1000\,\text{V}$

$\therefore$ The displacement of the spot on the screen

$$d = \text{total deflection}$$

$$= \frac{lLV_s}{2sV_a} = \frac{0.05 \times 0.175 \times 25}{2 \times 0.004 \times 1000}\ \text{m}$$

$$= 27.34 \times 10^{-3}\,\text{m} = \mathbf{27.34\ mm}$$

► **Example 7.** *Determine the deflection sensitivity of a signal for a CRT in which* $l = 2\,\text{cm}$, $L = 30\,\text{cm}$, $s = 0.5\,\text{cm}$ *and* $V_a = 2000\,\text{V}$. **(Punjabi Univ.)**

Solution: Given, $l = 2\text{ cm} = 0.2\,\text{m}$, $L = 30\text{ cm} = 0.3\,\text{m}$

$$s = 0.5\text{ cm} = 0.005\,\text{m} \text{ and } V_a = 2000\,\text{V}.$$

∴ The electrostatic deflection sensitivity for the CRT is given by

$$S = \frac{lL}{2sV_a}$$

$$= \frac{0.02 \times 0.3}{2 \times 0.005 \times 2000}\,\text{m/V}$$

$$= 3 \times 10^{-4}\,\text{m/V}$$

$$= \mathbf{0.3\,mm/V}$$

Review Questions

1. What is a digital multimeter? Give the advantages of a digital meter over an analog one.

   State the principle of operation of a dual slope digital voltmeter.

2. Explain, with the basic circuit diagram, the action of a solid-state electronic d.c. voltmeter. How can you modify it to measure a.c. voltages?

3. What is a cathode ray tube? Draw a neat sketch of a CRT with electrostatic focussing and deflection system. Explain the function of each part. **(Calcutta Hons.)**

4. How is the electron beam in a CRT deflected horizontally and vertically? Distinguish between electrostatic and magnetic deflections? **(North Bengal Hons.)**

5. What are 'deflection sensitivity' and 'deflection factor'? Derive an expression for the deflection sensitivity for electrostatic deflection of an electron beam in a CRT. **(Calcutta Hons.)**

6. Explain briefly how a waveform is displayed on the screen of a CRO.

   How can you measure the frequency of a signal voltage by means of a CRO? **(Burdwan Hons.)**

7. How can the phase difference between two a.c. voltages be measured by a CRO? **(Calcutta Hons.)**

8. What is a time base? Draw a simple circuit diagram of a RC time base generator. Explain its action. Why is it unsuitable for use in a laboratory type CRO?
(Calcutta Hons.)

9. What is the time base in a CRO? Sketch a graph showing the variation of time base voltage with time. (Calcutta Hons.)

10. Obtain an expression for the magnetic deflection sensitivity in a CRT.

What is post-acceleration and why is it used?

Why is CRO operated with a low spot brightness?

11. In connection with a CRO, explain the following terms: (i) sweep voltage, (ii) synchronization, (iii) time base and (iv) blanking.

Short and Objective Questions

1. Draw the basic circuit in a solid state electronic d.c voltmeter. Explain its action.
2. Explain, in brief, the working principle of a digital voltmeter.
3. What do you mean by deflection sensitivity of a CRT, using electrostatic deflection? (Calcutta Hons.)
4. Discuss the relative merits of electrostatic and magnetic deflection systems.
5. Distinguish between a CRT and a CRO.
6. Draw the block diagram of a general purpose CRO and indicate its basic components. (Calcutta Hons.)
7. What are the limitations of the basic saw-tooth generating circuit for use in a laboratory CRO. Calcutta Hons.)
8. Explain the necessity of time base circuit. (Nagpur Univ.)
9. The waveform of time base signal is

   (a) saw-tooth, (b) sinusoidal, (c) square wave, (d) none.

10. The deflection sensitivity S of a CRT using electrostatic deflection depends on the accelerating voltage V_a as

    (a) $S \propto V_a$, (b) $S \propto 1/V_a$, (c) $S \propto 1/\sqrt{V_a}$, (d) $S \propto \sqrt{V_a}$.

11. A multimeter can measure

    (a) voltage only, (b) voltage and current,
    (c) voltage and resistance, (d) voltage, current and resistance.

12. The purpose of synchronizing circuit in CRO is to

(a) get a focussed spot on the screen,
(b) get a stationary pattern on the screen,
(c) change the number of cycles displayed,
(d) change the amplitude of the signal.

13. The control of intensity in a CRO is made by

(a) changing the trigger level,
(b) changing the control grid bias,
(c) applying voltage to the deflecting plates,
(d) none of the above.

14. Which one of the following is *not* true for an electronic voltmeter?

(a) It has low input impedance,
(b) It does not load test circuit,
(c) It can measure low voltage,
(d) It has good frequency response.

15. Position control in a CRO is obtained by using

(a) a small a.c voltage to the deflecting plates,
(b) a small d.c voltage to the deflecting plates,
(c) the time base generator,
(d) the synchronizing control.

16. The sensitivity of a voltmeter is usually expressed in units of

(a) Ω/A, (b) Ω/V, (c) A/V, (d) A.

17. Fill in the blanks:

(a) In electron gun, emission occurs ________
(b) ________ is the main component of a CRO.
(c) Usually, in television ________ deflection is used.
(d) The CRO is operated with a low spot brightness to avoid ________ problems.

Problems

1. A sinusoidal wave is displayed on the screen of a cathode ray oscilloscope, with the calibrated time base set at 0.1 ms/div. If one cycle of the displayed wave form is spread over 10 divisions along the horizontal axis, calculate the frequency of the waveform. [1 kHz]

2. A Lissajous pattern is obtained on a CRO screen when sinusoidal voltages are applied to the two sets of deflecting plates. The figure makes 3 tangencies with the horizontal and 5 tangencies with the vertical. The frequency of the horizontal signal is 2 kHz. Find the frequency of the vertical signal. [1.2 kHz]

3. A cathode ray tube is designed for a deflection sensitivity 0.6 mm/V. The deflecting plates are 3 cm long and 5 mm apart. The distance of the screen, as measured from the centre of the plates, is 25 cm. Determine the voltage to be applied to the final accelerating anode. [1250 V]

4. Calculate the deflection sensitivity of a CRT from the following data.

   Length of the deflecting plates = 1.6 cm

   Spacing of the plates = 5 mm

   Distance of screen from the centre of plates = 20 cm

   Final anode voltage = 1200 V. [0.26 mm/V]

5. Two sinusoidal voltages of the same amplitude and frequency are simultaneously applied to the vertical and horizontal deflection systems of a CRO. This resulted in the trace of an ellipse on the fluorescent screen, with a negative slope of the major axis. The trace has a maximum vertical value of 2.6 divisions and it crosses the vertical axis at 1.1 divisions above the origin. Determine the phase difference between the voltages. [155°]

6. Two a.c signals of the same frequency but having a phase difference are displayed simultaneously on a dual-trace CRO. If the horizontal separation between two neighbouring peaks of the displayed waveform corresponds to 2 divisions and the distance between two consecutive peaks of a single wave corresponds to 12 divisions of the horizontal scale, calculate the phase difference between the signals.

   [60°]

7. The accelerating voltage of a CRT is 1 kV.

   A sinusoidal voltage is applied to a set of deflecting plates of axial length 2 cm. Determine the frequency of the deflecting voltage if the electrons remain between the plates for one-half cycle. [467.5 MHz]

8. The deflecting plates in a CRT have an axial length of 1 cm and a plate separation of 0.6 cm. If the screen is distant 30 cm from the centre of the plates and the electron beam enters the deflecting system with an energy 1 keV, determine (i) the deflection factor and (ii) the maximum spot displacement on the screen.

   [4 V/mm, 18 cm]

9. The deflection system of a CRT employs a magnetic field of 10^{-4} T acting over an axial length of 3 cm and is placed 20 cm from the screen. If the accelerating voltage is 800 V, find the deflection of spot on the fluorescent screen. Also,

calculate the deflection for an ion of electronic charge but of mass 1600 times the electronic mass. [0.67 cm, 0.016 cm]

[**Hints.** Here $B = 10^{-4}$ T, $l = 3 \times 10^{-2}$ m, $L = (20 + 1.5) \times 10^{-2}$ m, $V_a = 800$ volt.

So, deflection of spot $d = BlL\sqrt{e/2mV_a} = 10^{-4} \times 3 \times 10^{-2} \times 21.5 \times 10^{-2} \times$ $\sqrt{1.6 \times 10^{-19}/2 \times 9.1 \times 10^{-31} \times 800}$ m $=$ **0.67 cm.**

For the ion of mass $1600\,m$, deflection $= BlL\sqrt{e/2 \times 1600\,m \times V_a} = 0.67/\sqrt{1600}$ cm $=$ **0.016 cm**]

10. A certain Lissajous pattern is produced by applying sinusoidal voltage to the vertical and horizontal input terminals of a CRO. The pattern makes five tangencies with the vertical and three with the horizontal. Calculate the frequency of the signal applied to the vertical amplifier if the frequency of the horizontal input voltage is 3 kHz. **(Calcutta Hons.)** [1.8 kHz]

11. A cathode ray tube has the following dimensions: length of plates 4.0 cm, separation of plates 0.5 cm, distance between screen and the plates 16 cm. Find

    (a) the displacement produced by a deflecting voltage of 20 volts.

    (b) the angle of the beam with the tube axis on energing from the field, if the final anode voltage is 2 kV, and

    (c) the electrostatic deflection sensitivity. [0.72 cm, 2.545°, 3.6×10^{-6} m/V]

Bibliography

1. ALLISON, J., *Electronic Semiconductors and Devices,* Mc-Graw Hill, 1990.
2. ASHCRAFT, N.W. AND MERMIN, N.D., *Solid State Physics*, CBS Publishing Asia Ltd., 1988.
3. BARNA, A., *Operational Amplifiers*, John Wiley and Sons, Inc., 1971.
4. BLACMORE, J.S., *Solid State Physics*, Cambridge University Press, 1985.
5. BOYLESTAD, R.L. AND NASHELSKY, L., *Electronic Devices and Circuit theory*, Prentice Hall of India, 2001.
6. CHAIKIN, P.M. AND LUBENSKY, T.C., *Principles of Condensed Matter Physics*, Cambridge University Press, 1999.
7. DEKKER A.J., *Solid State Physics*, Macmillan India Ltd., 1981.
8. GOSLING, W., TOWNSEND, W.G. AND WATSON, J., *Field-Effect Electronics*, John Wiley and Sons, Inc., 1971.
9. GRAY, P.E., *Introduction to Electronics,* John Wiley and Sons, Inc., 1967.
10. GUPTA, H.C., *Solid State Physics*, Vikas Publishing House, New Delhi, 1997.
11. INNES A.C.-ROSE AND RHODERICK, E.H., *Introduction to Superconductivity,* Pergamon Press, 1969.
12. JACOBOWITZ, H., *Electronics Made Simple*, Doubleday & Co., New York, 1984.
13. KACHHAVA, C.M., *Solid State Physics*, Tata Mc-Graw Hill Publishers Co. Limited, 1990.
14. KALSI, H.S., *Electronic Instumentation*, Mc-Graw Hill Education (India) Private Limited, 2013.
15. KEER, H.V., *Principle of the Solid State*, Wiley Eastern Limited, 1993.
16. KITTEL C., *Introduction to Solid State Physics*, John Wiley and Sons, 1995.
17. MALVINO, A.P., *Electronic Principles*, Tata Mc-Graw Hill, 1984.
18. McKELVEY, J.P., *Solid State and Semiconductor Physics*, Harper and Row, 1966.

19. Millman, J. and Halkias, C.C., *Integrated Electronics*, Tata Mc-Graw Hill, 1991.

20. Omar, M.A., *Elementary Solid State Physics*, Pearson Education Inc., 2006.

21. Patterson, J.D., *Introduction to the Theory of Solid State Physics*, Addison-Wesley, 1971.

22. Peatman, J.B., *The Design of Digital System*, Mc-Graw Hill Book Co., 1972.

23. Roy Chaudhuri, D., *Digital Circuits,* Platinum Publishers, 2009.

24. Roy, D.K., *Physics of Semconductor Devices,* Universities Press, 2011.

25. Rudden, M.N. and Wilson, J., *Elements of Solid State Physics*, John Wiley and Sons Limited, 1980.

26. Ryder, J.D., *Introduction to Circuit Analysis*, Prentice-Hall, Inc., 1973.

27. Ryder, J.D., *Electronic Fundamentals and Applications,* PHL Learning (P) Ltd., 2009.

28. Srivastava, J.P., *Elementary Solid State Physics*, Prentice-Hall of India, 2001.

29. Streetman, B.G., *Solid State Electronic Devices*, Prentice-Hall, 1990.

30. van der Ziel A., *Solid State Physical Electronics,* Prentice-Hall, 1968.

31. Wahab, M.A., *Solid State Physics*, Narosa Publishing House, 1999.

32. Wert, C.A. and Thomson, R.M., *Physics of Solids*, Mc-Graw Hill Book Co., 1970.

Index : Solid State Physics

D

E

F

G

H

I

J, K

L

M

N

O

P

Q

R

S

T

U

V

W

X, Y, Z

Index : Electronics

A

B

C

D

E

F, G

H

I

J, K

L

M

N

O

P, Q

R

S

T

U, V

W, X, Y, Z